ISOLATED NEUTRON STARS: FROM THE SURFACE TO THE INTERIOR

Edited by:

Silvia Zane
University College London, United Kingdom

Roberto Turolla
University of Padova, Italy

Dany Page
Universidad Nacional Autonoma de Mexico, Ciudad Universitaria, Mexico

Reprinted from *Astrophysics and Space Science*
Volume 308, Nos. 1–4, 2007

Library of Congress Cataloging-in-Publication Data is available

ISBN 978-1-4020-5997-1 (HB)
ISBN 978-1-4020-5998-8 (eBook)

Published by Springer,
P.O. Box 17, 3300 AA Dordrecht, The Netherlands.

Cover image: University College London (UCL)

Printed on acid-free paper

Printed in the Netherlands

TABLE OF CONTENTS

Astrophys Space Sci (2007) 308: 1–11
DOI 10.1007/s10509-007-9309-y

ORIGINAL ARTICLE

Recent progress on anomalous X-ray pulsars

Victoria M. Kaspi

Received: 20 July 2006 / Accepted: 25 September 2006 / Published online: 16 March 2007

Abstract I review recent observational progress on Anomalous X-ray Pulsars, with an emphasis on timing, variability, and spectra. Highlighted results include the recent timing and flux stabilization of the notoriously unstable AXP 1E 1048.1–5937, the remarkable glitches seen in two AXPs, and the newly recognized variety of AXP variability types, including outbursts, bursts, flares, and pulse profile changes. I also discuss recent discoveries regarding AXP spectra, including their surprising hard X-ray and far-infrared emission, as well as the pulsed radio emission seen in one source. Much has been learned about these enigmatic objects over the past few years, with the pace of discoveries remaining steady. However additional work on both observational and theoretical fronts is needed before we have a comprehensive understanding of AXPs and their place in the zoo of manifestations of young neutron stars.

Keywords Pulsars · Magnetars · Variability · X-ray spectra · Timing

PACS 97.60.Jd · 97.60.Gb · 98.70.Qy

1 Introduction

Although very few in number, the eight, and possibly nine, known so-called “Anomalous X-ray Pulsars” (AXPs; see Table 1) are potentially very powerful for making progress on the physics of neutron stars. AXPs embody properties that are highly reminiscent of two other, very different classes of neutron star: the spectacular Soft Gamma Repeaters (SGRs; see contributions by Mereghetti and others in this volume), and conventional radio pulsars. The great similarity of AXPs to SGRs is what makes the case for AXPs being magnetars so compelling and also offers the hope of constraining magnetar physics. The intriguing similarities with radio pulsars offer the promise of solving long-standing problems in our theoretical understanding of the latter.

In this review, I describe the most recent observational progress on AXPs. The review will be divided into sections on timing (Sect. 2), variability (Sect. 3), and spectra (Sect. 4), choices that, unfortunately, may exhibit some personal bias, necessary given the limited space available. I hope to show that recent AXP progress has been significant, however ultimately, much observational and theoretical work remains to be done before a complete picture of AXPs and their place in the neutron-star zoo becomes clear.

Note that the most comprehensive and recent review of AXPs and magnetars in general is that by Woods and Thompson (2004). In this review, I make use of the detailed, online summary of magnetar properties and references maintained at McGill University by C. Tam (http://www.physics.mcgill.ca/~pulsar/magnetar/main.html).

2 Timing

Timing observations of AXPs hold considerable information about both their surroundings via the external torques they feel, as well as potentially about their internal structure, via the “glitches” they experience. Here we summarize what is known regarding AXP timing properties, highlighting the most interesting issues.

V.M. Kaspi (✉)
Physics Department, McGill University, Montreal, Canada, H3A 2T8
e-mail: vkaspi@physics.mcgill.ca

Table 1 Properties of known and candidate AXPs

Name	P (s)	$\dot{P}$[a] ($\times 10^{-11}$)	B ($\times 10^{14}$ G)	Timing[b] properties	X-ray Variability[c] properties	Waveband[d] detections	Association
CXOU J010043.1–721134	8.02	1.9	3.9	S?	S	X O?	SMC
4U 0142+61	8.69	0.2	1.3	S G?	M P	H X O I	...
1E 1048.1–5937	6.45	2.7	4.2	N	S F F B	H X I?	...
CXOU J164710.2–455216	10.61	0.16	1.3	G?	B P	X	Westerlund 1
1RXS J170849.0–400910	11.00	1.9	4.7	S G G	S	H X I	...
XTE J1810–197	5.54	0.5	1.7	N	O B	X I R	...
1E 1841–045	11.78	4.2	7.1	N	S	H X I?	SNR Kes 73
AX J1845–0258	6.97	...	...	...	O	X	SNR G29.6+0.1
1E 2259+586	6.98	0.048	0.59	S G	S O B P	H X I	SNR CTB 109

[a]Long-term average value

[b]S = stable (i.e. can be phase-connected over many-month intervals, generally); N = noisy (i.e. generally difficult to phase-connect over many-month intervals); G = one glitch

[c]S = stable (i.e. no variability, generally); M = modest variability; F = one flare; O = one outburst; B = bursts; P = pulse profile changes

[d]H = hard X-ray; X = X-ray; O = optical; I = infrared; R = radio

2.1 Stability and phase-coherent timing

Studies of the timing properties of AXPs can reasonably be categorized as pre- and post-Rossi X-ray Timing Explorer (RXTE), launched in late 1995. Prior to RXTE, timing studies were limited to occasional observations spread over many years. These characterized the overall spin-down behavior of several AXPs, and also suggested some deviations from simple spin-down (e.g. Baykal and Swank 1996; Corbet and Mihara 1997; Baykal et al. 2000; Paul et al. 2000). However the nature of these deviations could not be determined, because of the paucity of observations. Careful searches for Doppler shifts of the observed periodicities had been done (e.g. Iwasawa et al. 1992; Baykal and Swank 1996); typical upper limits on $a \sin i$ were $\sim$0.1 lt-s for a variety of orbital periods.

RXTE and its Proportional Counter Array (PCA) revolutionized the timing of AXPs. The first PCA studies of AXPs reduced the limits on $a \sin i$ to 0.03 lt-s for 1E 2259+586 and 0.06 lt-s for 1E 1048.1–5937, effectively ruling out any main-sequence star companion and rendering binary accretion models highly problematic (Mereghetti et al. 1998).

Subsequently, a program of regular monthly monitoring of the AXPs with the PCA on RXTE was approved and showed that fully phase-coherent timing of AXPs could be done over years, assuming spin-down models consisting of very few parameters (Kaspi et al. 1999). For example, Gavriil and Kaspi (2002) showed that in 4.5 yr of RXTE monitoring, pulse times of arrival for 1E 2259+586 could be predicted to within 1% of the pulse period using ν, $\dot{\nu}$ and $\ddot{\nu}$ only. Such stability is comparable to that of conventional young radio pulsars and very much unlike the large amplitude torque noise commonly seen in accreting neutron stars. Long-term, regular monthly (or even bi-monthly) observations of AXPs continue today, with four of the five persistent confirmed Galactic sources (4U 014+61, 1RXS J170849.0–400910, 1E 1841–045 and 1E 2259+586) generally exhibiting stability that allows phase coherence with few parameters over years (Kaspi et al. 2000; Gavriil and Kaspi 2002; Gotthelf et al. 2002; Kaspi and Gavriil 2003; Dib et al. 2006). A summary of the timing properties of the known AXPs is given in Table 1.

One of the established persistent Galactic AXPs, 1E 1048.1–5937, has been much less stable than the others, such that phase-coherent timing with unambiguous pulse counting over more than a few weeks or months has been difficult to achieve (Kaspi et al. 2001; Gavriil and Kaspi 2004). This poor stability is apparent in the source's frequency evolution (Fig. 1). Recently, however, during an extended period of pulsed flux stability following two long-lived X-ray flares (see Sect. 3.3), the timing has also become quite stable, such that unambiguous phase coherence could be maintained over a nearly 2-yr interval from MJD 53158 to 53858 using 4 spin parameters, although significant residuals remain (Fig. 2). Details of these results will be described elsewhere. It remains to be seen if an end to this stability will be accompanied by additional radiative events, a result that would hopefully be useful for strongly constraining models for the torque noise and flares.

2.2 Glitches

The impressive timing stability seen in most AXPs, in which pulse periods could easily be determined to better than a

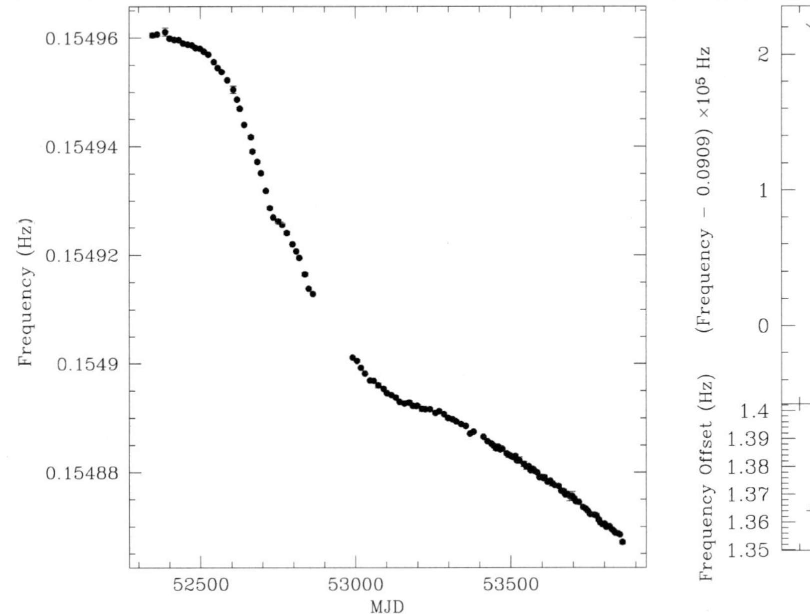

Fig. 1 Spin-frequency history of 1E 1048.1–5937 from RXTE monitoring (work in preparation). Note the simple spin down in the last 1.9 yr

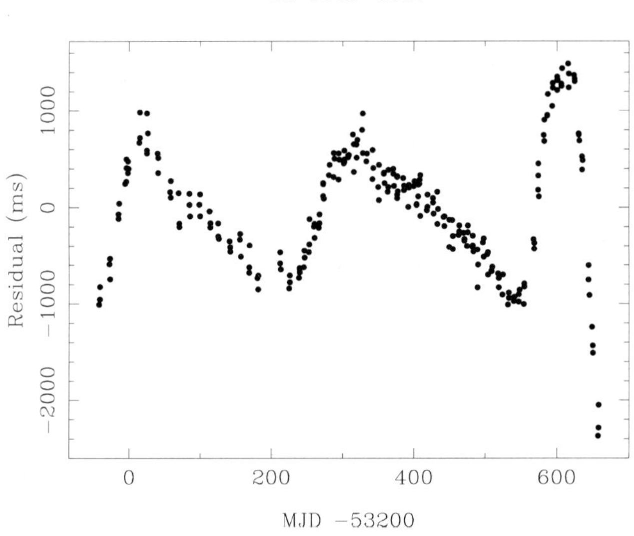

Fig. 2 Phase residuals following removal of four frequency derivatives for the last 1.9 yr of data for AXP 1E 1048.1–5937. The features are roughly 300 d apart and are presently consistent with being due to random timing noise (work in preparation)

part in a billion, permitted for the first time the unambiguous detection of sudden spin-up glitches in AXPs. 1RXS J170849.0–400910 was the first pulsar seen to glitch (Kaspi et al. 2000). This glitch had a fractional frequency increase of 6×10^{-7}, similar to what is seen for the Vela pulsar and other comparably young objects. Continued monitoring of the same AXP revealed a second, larger glitch 1.5 yr later (Kaspi and Gavriil 2003; Dall'Osso et al. 2003), with a

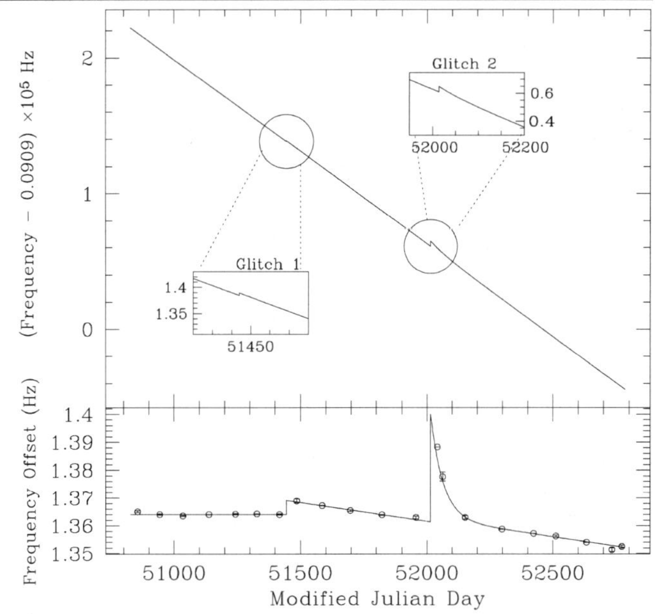

Fig. 3 Long-term RXTE timing data for 1RXS J170849.0–400910 showing the two glitches observed thus far (from Kaspi and Gavriil 2003)

nearly complete recovery of the frequency jump, unusual by pulsar standards (see Fig. 3).

In June 2002, at the time of a major radiative outburst (see Sect. 3), the otherwise stable AXP 1E 2259+586 exhibited a large glitch (fractional frequency increase of 4×10^{-6}) which was accompanied by significant and truly remarkable recovery in which roughly a quarter of the frequency jump relaxed on a ~40-day scale. As part of that recovery, the measured spin-down rate of the pulsar was temporarily larger than the long-term pre-burst average by a factor of 2! This recovery was similar to, though better sampled than, the second glitch seen in 1RXS J170849.0–400910. The 1E 2259+586 event was the first (and still the only unambiguous, but see Sect 3.3) time a spin-up glitch in a neutron star was accompanied by any form of radiative change. This event suggests that large glitches in AXPs are generally associated with radiative events; perhaps such an event occurred at the time of the second glitch in 1RXS J170849.0–400910 but went unnoticed due to sparse monitoring. The very interesting recoveries of the 1E 2259+586 and second 1RXS J170849.1–400910 glitches seem likely to be telling us something interesting about the interior of a magnetar and how it is different from that of a conventional, low-field neutron star. This seems worth more attention than it has received in the literature thus far.

Recently a timing anomaly was reported in AXP 4U 0142+61 (Dib et al. 2006). This is work in progress and will be reported on elsewhere. However, Morii et al. (2005) and Dib et al. (2006) suggest that this pulsar may have glitched during a large observing gap in 1998–1999, a possibility supported by apparent changes in the pulse profile that seem to have occurred in the same interval (Dib et al. 2006).

3 AXP variability

Arguably one of the most interesting recent discoveries in the study of AXPs is the range and diversity of their X-ray variability properties. Pre-RXTE, there was variability reported (e.g. Baykal and Swank 1996; Corbet and Mihara 1997; Oosterbroek et al. 1998; Paul et al. 2000), however those relatively sparse observations were made using different instruments aboard different observatories, some imaging, some not, and were often presented without uncertainties, making their interpretation difficult. Moreover, given the sparseness of the observations, identifying a time scale for the variations, or being certain the full dynamic range was being sampled, was not possible.

Post-RXTE, and, additionally, with contemporaneous observations made using Chandra and XMM, the picture has become much clearer. Presently there appear to be at least four types of X-ray variability in AXP pulsed and persistent emission: outbursts (sudden large increases in the pulsed and persistent flux, accompanied by bursts and other radiative and timing anomalies, which decay on time scales of weeks or months), bursts (sudden events, lasting milliseconds to seconds), long-term flux changes (time scales of years), and pulse profile changes (on time scales of days to years). We examine each of these phenomena in turn.

3.1 Outbursts and transients

The current best example of an AXP outburst is that seen in June 2002 from 1E 2259+586 (Kaspi et al. 2003; Woods et al. 2004). This outburst, which lasted only a few hours, fortuitously occurred during a few-hour monthly RXTE monitoring observation. During the outburst, the pulsed and persistent fluxes increased by a factor of ~20 (see Fig. 4), there were over 80 short SGR-like bursts in a few-hour period (see Sect. 3.2), there were substantial pulse profile changes (see Sect. 3.4), there was a short-term decrease in the pulsed fraction, the spectrum hardened dramatically, there was a large glitch (see Sect. 2.2 above), and there was an infrared enhancement (see Sect. 3.5). All this came after over 4 yr of otherwise uneventful behavior (Gavriil and Kaspi 2002). Note that had RXTE not been observing the source during the outburst, the entire event would have appeared, from the monitoring data, to consist principally of a glitch. With only monthly monitoring, all but the longest-term radiative changes would have been missed. Interestingly, this outburst was notably different from SGR outbursts; for the AXP, the energy in the outburst "afterglow" ($\sim 10^{41}$ erg; see Fig. 4) greatly exceeded that in the bursts ($\sim 10^{37}$ erg). This is in direct contrast to the giant flares of SGRs. The reason for this difference is unknown.

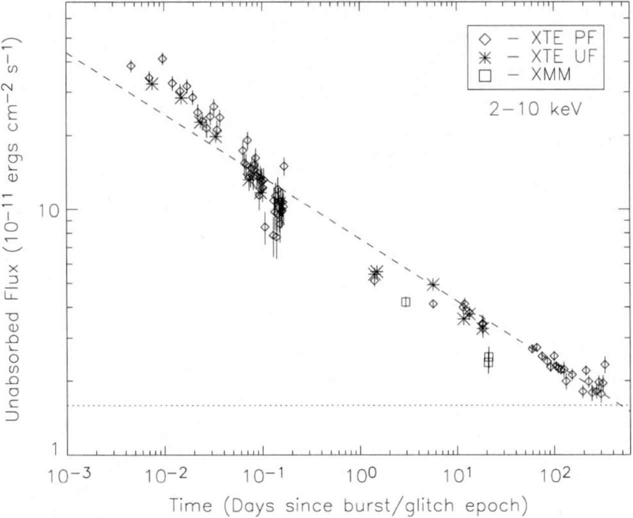

Fig. 4 Flux time series plotted logarithmically following the 2002 1E 2259+586 outburst (Woods et al. 2004). *Diamonds* denote inferred unabsorbed flux values calculated from RXTE PCA pulsed-flux measurements. *Asterisks* and *squares* mark independent phase-averaged unabsorbed flux values from RXTE and XMM, respectively. The *dotted line* denotes the flux level measured using XMM 1 fortuitously week prior to the glitch. The *dashed line* is a power-law fit to the PCA flux measurements during the observations containing the burst activity (<1 day)

The outburst of 1E 2259+586 seems likely to be a good model for the behavior of the "transient" AXP XTE J1810–197. This source, unknown prior to 2003, was discovered as a bright 5.5 s X-ray pulsar, upon emerging from behind the Sun in that year (Ibrahim et al. 2004), and has been fading ever since. The source's spin down and spectrum are consistent with it being an AXP. Gotthelf et al. (2004) showed from archival ROSAT observations that in the past, the source, in quiescence, was nearly two orders of magnitude fainter than in outburst. See the contribution by Gotthelf et al. in this volume for more details. An important question raised by the discovery of XTE J1810–917 is "how many more such objects exist in our Galaxy"? This has important implications for AXP birthrates.

Such an outburst may also explain the behavior of the unconfirmed transient AXP AX J1845–0258 (see Table 1). Pulsations with a period of 7 s were observed in an archival 1993 ASCA observation (Torii et al. 1998; Gotthelf and Vasisht 1998), however subsequent observations of the target revealed a large drop in flux, and pulsations have not been redetected. Although no $\dot{\nu}$ has been measured for this source, it seems likely to be an AXP given its period and location at the center of a supernova remnant (Gaensler et al. 1999). It thus is plausible that the 1993 outburst was similar to that of 1E 2259+586 or XTE J1810–197, although the likely dynamic range for this source is unprecedented for an

AXP. See the contribution to these proceedings by Tam et al. and Tam et al. (2006) for more details.

One of the most puzzling aspects of transient AXPs is why the quiescent source is so faint. In the standard magnetar model, the requisite magnetic field decay energy input is persistent, as is the magnetospheric twist for a fixed magnetar-strength magnetic field (Thompson and Duncan 1996; Thompson et al. 2002). Although much attention has been paid to why a magnetar's flux might skyrocket suddenly—a sudden reconfiguration of the surface following a crustal yield—relatively little attention has been paid to how to stop or hide the bright X-ray emission from a neutron star presumably harboring a magnetar-strength field.

3.2 Bursts

During the 2002 1E 2259+586 outburst, over 80 short, SGR-like X-ray bursts were seen superimposed on the overall flux trend over the course of the few-hour RXTE observation (Kaspi et al. 2003). Some were super-Eddington, though only on very short (few ms) time scales. In a detailed analysis of these bursts, Gavriil et al. (2004) found that in most respects, they are identical to SGR bursts. Specifically, the durations, differential fluence distribution, the burst morphologies, the wait-time distribution, and the correlation between fluency and duration, are all SGR-like. However a few of the burst properties were different from those of SGR bursts: for example, the AXP bursts had a wider range of durations, the AXP bursts were on average less energetic than in SGRs, and the more energetic AXP bursts have the hardest spectra—the opposite of what is seen for SGRs. Unlike in SGRs, the AXP bursts were correlated with pulsed intensity.

Bursts in AXPs were first reported by Gavriil et al. (2002) who discovered two such events in archival RXTE data from the direction of 1E 1048.1–5937. A third burst, found nearly 3 yr later (Gavriil et al. 2006), unambiguously identified the AXP as the origin thanks to a simultaneous pulsed flux increase. The bursting behavior in this source is not obviously correlated with any other property, although we note tentatively that no bursting has been seen during the most recent 2 yr, when the pulsar has been exhibiting much improved timing (see Sect. 2, Fig. 1) and also pulsed flux stability. Perhaps the bursting was symptomatic of whatever activity also caused the timing instability and pulsed flux flares (see Sect. 3.3).

Four bursts have also been seen from the transient AXP XTE J1810–197 (Woods et al. 2005). These events consisted of a $\sim$1 s spike followed by an extended tail in which the pulsed flux was enhanced, similar to the third burst of 1E 1048.1–5937 and a handful from 1E 2259+586. The XTE J1810–197 and 1E 1048.1–5937 bursts also showed a correlation with pulsed intensity, as did some of the 1E 2259+586 bursts.

These observations led Woods et al. (2005) to hypothesize that there are in fact two distinct classes of bursts, which they named Type A and Type B. Type A bursts are similar to SGR bursts, in that they are uncorrelated with pulse phase and have no extended tails. Type B bursts, on the other hand, thus far observed exclusively in AXPs, are correlated with pulsed intensity, generally have long extended tails, and those tails tend to contain more energy than the burst itself. As both types of burst were seen in 1E 2259+586, clearly they are not mutually exclusive, even during the same event. Woods et al. (2005) speculate that Type A bursts have a magnetospheric trigger, whereas Type B bursts originate from crust fractures.

Most recently, five small, sub-Eddington X-ray bursts have been seen between April and June 2006 from 4U 0142+61 (Kaspi et al. 2006; Dib et al. 2006, work in preparation). The latter four, all within a single RXTE orbit, were clearly accompanied by a pulsed flux increase (by a factor of $\sim$4 relative to the long-term average) which decayed on a time scale of minutes. This suggests, as does the accompanying pulse profile change and timing anomaly, that this source may have entered an extended active phase.

With at least half of the known AXPs now established as capable of bursting, it is clear that the production of occasional though clustered short SGR-like bursts is a generic AXP phenomenon.

3.3 Long-term flux variations

Variability in AXP 1E 1048.1–5937 had been reported for years (e.g. Corbet and Mihara 1997; Oosterbroek et al. 1998; Baykal et al. 2000; Mereghetti et al. 2004). RXTE monitoring determined the time scale of the changes and the morphology of the pulsed light curve with far superior time sampling than in previous studies (Gavriil and Kaspi 2004). Specifically, the RXTE monitoring showed that over $\sim$7 yr, the source exhibited two extended pulsed flux "flares," the first lasting $\sim$100 days and the second lasting over a year, each with rise times of several weeks. Assuming a distance to the source of 5 kpc (which is not especially well established), Gavriil and Kaspi (2004) estimated the total energy released in the pulsed components of the first and second flares to be 3 and 30 $\times 10^{40}$ erg, respectively. Subsequently, Tiengo et al. (2005), using XMM, which (unlike RXTE) is sensitive to pulsed fraction, showed that in fact the pulsed fraction is anti-correlated with the phase-averaged flux, suggesting the total energy released was at least twice that in the pulsed component.

During these flares, the spin-down rate fluctuated by at least a factor of 10 (Gavriil and Kaspi 2004). However, there was no obvious correlation between the detailed evolution of the spin-down rate and flux. Recently, while its flux has been stable, 1E 1048.1–5937 has shown much greater timing stability (see Sect. 2). This suggests that the large torque noise and flux flaring were causally related; we must await another such event to confirm this.

The flaring observed in 1E 1048.1–5937, a new phenomenon not yet observed in any other AXP, is well understood in the context of the twisted magnetosphere model (Thompson et al. 2002). The flux enhancements can be seen as being due to increased twisting of the magnetosphere by currents originating from the stressed crust. If so, a harder spectrum is expected when the pulsar is brighter. Unfortunately the existing data cannot confirm this important prediction for this source. Decoupling of the torque from the pulsed flux can occur in this model depending on the location of the enhanced magnetospheric current; a current near the pole will have a disproportionate impact on the spin-down. Gavriil and Kaspi (2004) argued that the absence of predicted torque–luminosity variations in this source are problematic for models in which the X-ray emission originates from accretion from a fossil disk (e.g. Chatterjee et al. 2000; Alpar 2001, but see Ertan et al. contribution, this issue).

The twisted magnetosphere model prediction that flux should be correlated with hardness, though unconfirmed in 1E 1048.1–5937, does seem to be borne out in observations of 1RXS J170849.0–400910 (Rea et al. 2005, see contribution by Rea et al., this volume). Moreover, those authors suggest that the epochs of greatest hardness occur near those in which glitches were detected in this source (see Sect. 2.2), with subsequent softening post-glitch. At least one additional glitch needs to occur, with better observational coverage, before this conclusion is firm.

Recently, much longer-term radiative variations have been identified in AXP 4U 0142+61, in which the pulsed flux appears to be increasing since 2000, such that there was a $\sim$20% change by early 2006, just prior to its exhibiting a sudden pulse profile change and bursts. This behavior is described in more detail in the contribution by Dib et al. to this volume as well as in Dib et al. (2006). One particularly interesting implication of an increase in total X-ray flux from this source is that the phenomenon provides a simple test of the irradiated fall-back disk model for the near-IR emission (see Sect. 4.3). If the source of irradiation, the AXP, increases in brightness, the disk ought to as well.

3.4 Pulse profile changes

The first observed pulse profile change in an AXP was reported by Iwasawa et al. (1992) using *GINGA* data obtained in 1989. They witnessed a large change in the ratio of the amplitudes of the two peaks in the pulse profile of 1E 2259+586, namely from near unity to over two. They also reported a contemporaneous timing anomaly which, in hindsight, is consistent with a spin-up glitch.

A very similar pulse profile change was witnessed during and immediately following the 2002 outburst of 1E 2259+586 (Kaspi et al. 2003; Woods et al. 2004). Here, the ratio of the amplitudes of the two pulses in the profile went from unity pre-outburst to roughly two mid-outburst, relaxing back to normal on a time scale of a few weeks (Fig. 5). Curiously, the temporarily larger peak in the 2002 outburst appeared to be the temporarily smaller peak in 1989, suggesting that even if the physical origin of the events is the same, the details of the geometry were different. Given the nature of this event, a likely explanation for the pulse profile change is a magnetospheric reconfiguration following a crustal fracture that simultaneously affected the inside and outside of the star. This very strongly suggests that the Iwasawa et al. (1992) pulse profile change was observed not long after a similar event; this is consistent with their reported timing anomaly, and suggests such events occur roughly every 1–2 decades in this source.

Most recently, long-term (i.e. time scale of several years) pulse profile variations have been reported for AXP 4U 0142+61 (Dib et al. 2006, see contribution by Dib et al., these proceedings). These may accompany a long-term pulsed flux increase (see Sect. 3.3). The profile changes are consistent with a significant event having occurred some time between mid-1998 and 2000, in which the second and higher harmonics of the profile increased dramatically, and have been returning to their pre-event level ever since. This gradual evolution was, however, interrupted by an apparent sudden activity episode, in which an SGR-like burst was detected along with a timing anomaly in April 2006 (see Sect. 3.2; Kaspi et al. 2006). The analysis of the latter data are in progress.

3.5 Near-IR variability

The large number of recent near-IR detections of AXPs has revealed an interesting new variability phenomenon in these sources. Following the 2002 outburst of 1E 2259+586, there was an infrared (K_s) enhancement by a factor of $\sim$3, 3 days post-outburst. This then decayed together with the X-ray pulsed flux, with identical power-law indices (see Fig. 4, Sect. 3.1; Tam et al. 2004). Those authors concluded that the origins of both flux increases were magnetospheric (but see contribution by Ertan et al., these proceedings for the fossil-disk viewpoint). A similar correlation was also reported for AXP XTE J1810–197 (Rea et al. 2004).

Meanwhile, significant near-IR variability has also been reported for AXP 1E 1048.1–5937 (Israel et al. 2002; Wang and Chakrabarty 2002; Durant and van Kerkwijk 2005). However, if anything, the near-IR is *anti-correlated* with the X-ray flux. This is puzzling given the 1E 2259+586 and XTE J1810–197 results. It is worth keeping in mind, however, that there is now some evidence that 1E 1048.1–5937 has been in an active phase from which it may have recently emerged (see Sect. 2); it will be interesting to see how its

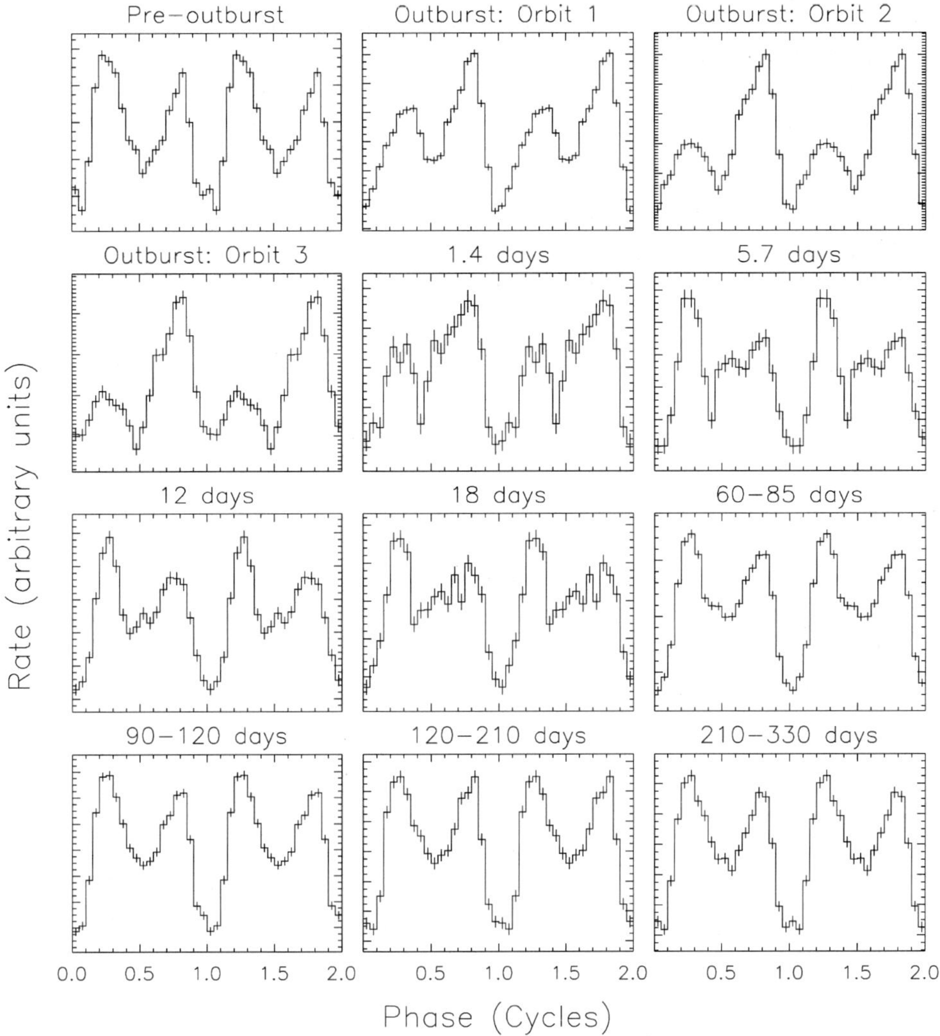

Fig. 5 Pulse profile changes in 1E 2259+586 following its 2002 outburst (Woods et al. 2004)

near-IR flux varies now that both the X-ray flux and timing have stabilized (see Fig. 1).

In addition, Durant and van Kerkwijk (2006a) report on near-IR observations of 4U 0142+61 and report no apparent correlation with the X-ray pulsed flux. They argue that the situation for this source is unclear and warrants additional, more frequent observations. Simultaneous far-IR observations would also be of interest to establish conclusively that they originate from a separate mechanism, namely radiation from a passive, irradiated fall-back disk (Wang et al. 2006).

4 AXP spectra

The description of spectra of AXPs has, until very recently, been limited to the soft (0.5–10 keV) X-ray band, as that is where AXPs were discovered and have been traditionally studied. However the recent discoveries of optical and near-IR emission have cast them firmly into the multi-wavelength regime, and the even more recent detections in the hard X-ray band as well as in the far-IR and radio bands broadens the situation even further.

4.1 X-ray spectra

In the traditional 0.5–10 keV X-ray band, AXPs have long been known to show two-component spectra, which are well described by a blackbody plus a power law (e.g. White et al. 1996; Israel et al. 2001; Morii et al. 2003). It is currently thought that the blackbody arises from internal heating due to the decaying intense magnetic field, while the non-thermal component is a result of resonant scattering of the thermal seed photons off magnetospheric currents in the twisted magnetosphere (Thompson et al. 2002). Detailed spectral modelling in this framework appears to describe the spectrum of 1E 1048.1–5937 (Lyutikov and Gavriil 2006), but see the contribution by N. Rea in these proceedings. Although some attempts were made to model the entire spectrum using a single blackbody distorted by the effects of the intense magnetic field on the atmosphere (e.g. Özel 2001), these could not reproduce the non-thermal component adequately (e.g. Perna et al. 2001; Thompson et al. 2002; Lai and Ho 2003).

There is evidence supporting a physical connection between the two components, such as the very slow evolution of the pulse profile with energy (e.g. Israel et al. 2001; Gavriil and Kaspi 2002). For some AXPs (for example 1E 1048.1–5937 and 1E 1841–045), there is little if any variation in the pulse profile, with no obvious difference between profiles in energy bands that are thermally and non-thermally dominated. Even in other AXPs for which the profile is energy-dependent, the profiles in bands that are thermally and non-thermally dominated are still very similar. This is in stark contrast to the situation for rotation-powered pulsars, for which the thermal and non-thermal components generally have radically different X-ray pulse profiles (e.g. Harding et al. 2002).

Recently, Halpern and Gotthelf (2005) have suggested on purely theoretical grounds that the spectrum of XTE J1810–197 is more appropriately described by a two-component model consisting of two blackbodies (see also contribution by Gotthelf et al. in this volume). Their main argument for this interpretation is that an extrapolation of the power-law component to low energies greatly exceeds that expected if the seeds are thermal photons from the surface, as the blackbody eventually runs out of photons to supply. Moreover, they argue, the expected blackbody cut-off would then result in a substantial underestimate of the infrared flux, assuming the latter is part of the non-thermal spectrum. Very recently, Durant and van Kerkwijk (2006c) have shown using independently measured interstellar column densities that the intrinsic spectra really are cut off, i.e. the power-law component does not in fact extend far below the observable band. If correct, the rationale for preferring the double blackbody over the blackbody plus power law would not apply.

This section would not be complete without some discussion of the interesting recent results of Durant and van Kerkwijk (2006b, 2006c), some of which are reported on by Durant et al. in this volume. Specifically, using high-resolution X-ray spectroscopy, they identified absorption edges whose amplitudes determine N_{H} under reasonable assumptions, independent of the overall continuum spectral modelling. Using these newly measured values of N_{H} and a novel distance estimation technique using reddening runs with distance for red clump stars, they were able to first improve the spectral fits for several AXPs significantly, and second determine much improved distance estimates for them. Amazingly, among other things, they find that the soft X-ray luminosities of practically *all* AXPs are consistent with the value $\sim 1.3 \times 10^{35}$ erg s^{-1}. Durant and van Kerkwijk (2006b) argue that this is consistent with the magnetar model's prediction that there should be a saturation luminosity above which internal neutrino cooling is at work.

4.2 Hard X-ray spectra

A particularly interesting recent AXP discovery is that they are copious hard X-ray emitters (Molkov et al. 2004; Kuiper et al. 2004). Though their spectra in νF_ν appear to fall off in the softer X-ray band, they turn up again above $\sim$10 keV. The luminosities above 10 keV independently greatly exceed the available spin-down power by factors of over 100. This requires a new mechanism for accelerating particles in the magnetosphere, in addition to an energy source, presumably the decaying magnetic field. SGR 1806–20 has also been detected in this energy range, though interestingly it is somewhat softer than the AXPs (Mereghetti et al. 2005). Kuiper et al. (2006) have further shown that the hard X-ray emission is a generic property of AXPs, and for at least three sources, extends without a break up to 150 keV. They also show from *COMPTEL* upper limits that the break must lie under $\sim$750 keV. Determining the location of the break could greatly constrain emission models, possibly even providing independent evidence for the magnetar-strength field. See the contribution by P. den Hartog et al. in this volume for details regarding hard-X-ray emission from 4U 0142+61.

This hard X-ray emission, apart for being interesting for constraining the physics of magnetars, offers a unique way of detecting AXPs throughout the Galaxy, since the soft X-ray emission suffers from high absorption, especially in the inner Galaxy. A focussing hard X-ray instrument (like the NASA mission concept *NuSTAR*) would have the capability of detecting every magnetar in the Galaxy, provided their hard X-ray emission is generic, even in quiescence.

4.3 Optical and infrared spectra

Currently, five of the known AXPs have been conclusively detected in the near-IR, with only 4U 0142+61 (the closest, least absorbed AXP) having been detected optically. None of the SGRs has been detected optically, and only one has been seen in the near-IR (SGR 1806–20; Kosugi et al. 2005), and only during a particularly active phase. For a summary of these detections, see www.physics.mcgill.ca/~pulsar/magnetar/main.html and references therein.

The near-IR spectra of AXPs are an interesting puzzle. First, given the variability in the near-IR (Sect. 3.5), piecing together an accurate instantaneous spectrum using photometry requires contemporaneous observations, not always possible. Even more of a problem has been the generally unknown reddening toward the sources, which have an enormous impact on the inferred intrinsic spectrum. Nevertheless, some information regarding the optical and near-IR spectra of AXPs is known. Overall, the major mystery is how the optical and near-IR emission connects with the X-ray spectrum. The blackbody emission seen in X-rays grossly underpredicts that in optical/near-IR, while a simple extrapolation of the non-thermal component (when the spectrum is described in this way—see Sect. 4.1) generally overpredicts it. Expecting at least the optical emission to connect spectrally with the X-rays is reasonable given that the latter is pulsed in 4U 0142+61 (Kern and Martin 2002; Dhillon et al. 2005) hence seems likely to originate in the magnetosphere, as does, presumably, the non-thermal X-ray emission.

Very recently, Wang et al. (2006) have shown using *Spitzer* data of 4U 0142+61 that in the far-IR, there appears to be a component that is spectrally distinct from the near-IR emission. They interpret this component as resulting from a passive debris disk irradiated by the central X-ray source. They suggest that such disks, which originate from matter that falls back following the supernova explosion, may be ubiquitous around neutron stars. They further suggest that the correlated near-IR/X-ray flux decay observed by Tam et al. (2004) following the 2002 outburst of 1E 2259+586 and for XTE J1810–197 (see Sect. 3.5) could be a result of a disk around that AXP as well, a possibility also discussed by Ertan et al. (2006) and Ertan et al. in these proceedings. If the fallback-disk interpretation is correct for 4U 0142+61, the pulsed X-ray flux change recently detected in this source (Sect. 3.3) may provide an opportunity for a test of the disk hypothesis (Dib et al. 2006, and Dib et al. in this volume), as there should be a correlated increase in the near-IR flux. This idea requires that the overall X-ray flux, not just the pulsed component, also be increasing, which requires observations with an imaging X-ray telescope to verify.

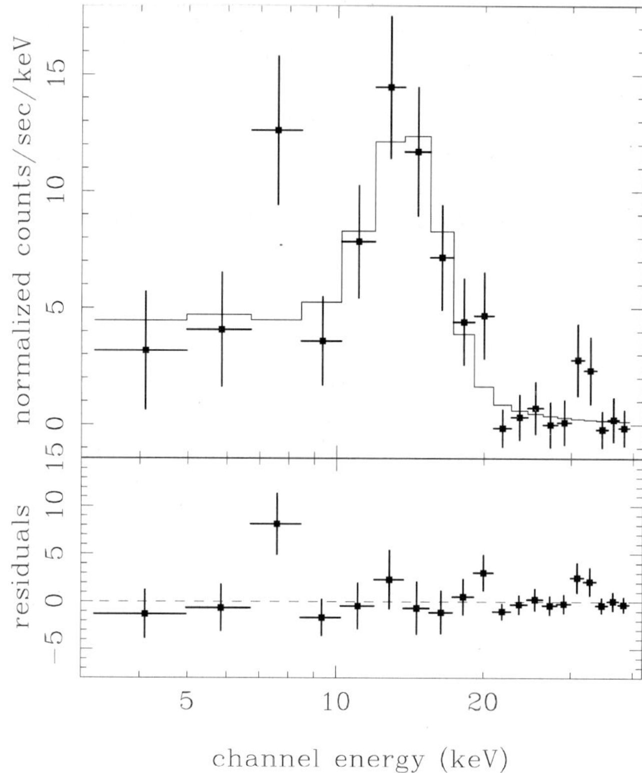

Fig. 6 Spectrum of the 1 s after the peak of the first burst observed from 1E 1048.1–5937 (Gavriil et al. 2002)

4.4 Radio spectrum

Very recently (in fact *after* this meeting took place!), Camilo et al. (2006) reported the detection of radio pulsations from XTE J1810–197. This was a magnetar first and in many ways a welcome discovery, having provided a nice radiative link between magnetars and radio pulsars, in addition to the similarity already established from timing observations (see Sect. 2). It also demonstrated that pulsed radio emission can definitely be produced in magnetar-strength fields in contrast to some predictions (e.g. Baring and Harding 1998). Previous searches of other non-transient AXPs had come up empty (e.g. Burgay et al. 2006; Crawford et at. 2006, and see contribution by Burgay et al., this volume), suggesting the radio emission here might somehow be related to this source's transient nature. Given the small radio beaming fraction reasonably expected for such slow pulsars, the absence of radio pulsations from other sources could also be due to small-number statistics.

Also interesting is the unusual spectrum of the radio emission seen from XTE J1810–197. It has an unusually flat spectrum, with spectral index > -0.5, whereas radio pulsars have very steep spectra, with most indices between -1 and -3. XTE J1810–197 is the brightest radio pulsar known at frequencies >20 GHz. Why this should be the case is an interesting new puzzle for magnetar physics, one which has the potential to shed crucial new light on the long-standing problem of the origin of radio emission in rotation-powered pulsars.

4.5 Spectral features

Finally, there have been reports of features in AXP spectra. The first such report was for 1E 1048.1–5937 during the first of its two observed 2001 bursts (Gavriil et al. 2002). The feature, which was most prominent in the first 1-s of the burst, was extremely broad and seen apparently in emission at a central energy of 14 keV (see Fig. 6). In terms of flux, it was comparable to the burst continuum emission. It was very significant; Monte Carlo simulations showed that such a feature at any energy would be seen only 0.01% of the time. Gavriil et al. (2006) saw a similar feature in the third observed 1E 1048.1–5937 burst. In that case, the central energy measured was also ~13 keV. In addition, Woods et al. (2005) observed a similar feature in one burst from XTE J1810–197, this time at energy 12.6 keV, with probability of chance occurrence 4×10^{-6}. If interpreted as proton cyclotron lines, the energies of these features imply a magnetic field above $\sim 10^{14}$ G, consistent with the magnetar hypothesis. However if the features are interpreted as electron cyclotron lines, the implied field is correspondingly $\sim$2000 times lower. The latter would not necessarily be inconsistent with the magnetar interpretation, as it is unclear what altitude above the neutron-star surface these lines originate.

Note that similar spectral features during SGR bursts have also been reported in the tails of a few SGR bursts (Ibrahim et al. 2002, 2003, see contribution by Ibrahim et al., these proceedings). However, recently their statistical significance has been questioned (see contribution by Woods et al.).

Rea et al. (2005) reported spectral features at certain rotational phases from 1RXS 170849.0–400910 from *BeppoSax* data. These were claimed at the time to be significant at the $\sim 4\sigma$ level. Observations with XMM of the same source saw no such features, implying either that the *BeppoSax* features were spurious or that the effect is time variable. See the contribution by Rea et al. in these proceedings for more information.

5 Conclusions

My hope in writing this review is to demonstrate that there remain many important unsolved problems in the study of AXPs. Overall, the basic issue of what differentiates AXPs from SGRs remains, as does the origin of the intense magnetic fields inferred. Other important issues for which there simply was not enough space for discussion here include the possible association of AXPs (and SGRs) with massive star progenitors (e.g. Figer et al. 2006; Muno et al. 2006, see contribution by Gaensler et al.), the puzzling lack of "anomalously" bright X-ray emission from high-magnetic field radio pulsars (see contribution by Gonzalez et al.), and the proposed connection between magnetars and the so-called "RRATS" (McLaughlin et al. 2006, and see contribution by Lyne in this volume).

As for how some of these problems will be solved, continued monitoring observations with RXTE, as well as targeted studies with Chandra, XMM and INTEGRAL, will obviously be of use. Greater concerted efforts in the optical and near- and far-IR are warranted, as are careful and repeated radio searches for transient pulsations or other phenomena. Finally, target-of-opportunity observations at the times of the rare moments of AXP activity are definitely crucial for unravelling the overall physical picture of these very interesting sources.

Acknowledgements I thank C. Tam for maintaining the online McGill Magnetar Catalog as well as R. Dib, F. Gavriil, and P. Woods for useful conversations.

References

Alpar, M.A.: Astrophys. J. **554**, 1245 (2001)
Baring, M.G., Harding, A.K.: Astrophys. J. **507**, L55 (1998)
Baykal, A., Strohmayer, T., Swank, J., et al.: Mon. Not. Roy. Astron. Soc. **319**, 205 (2000)
Baykal, A., Swank, J.: Astrophys. J. **460**, 470 (1996)
Burgay, M., Rea, N., Israel, G., et al.: Mon. Not. Roy. Astron. Soc. **372**, 410 (2006)
Camilo, F., Ransom, S., Halpern, J., et al.: Nature **442**, 892 (2006)
Chatterjee, P., Hernquist, L., Narayan, R.: Astrophys. J. **534**, 373 (2000)
Corbet, R.H.D., Mihara, T.: Astrophys. J. **475**, L127 (1997)
Crawford, F., Hessels, J., Kaspi, V.M.: Astrophys. J. (submitted)
Dall'Osso, S., Israel, G., Stella, L., et al.: Astrophys. J. **599**, 485 (2003)
Dhillon, V.S., Marsh, T., Hulleman, F., et al.: Mon. Not. Roy. Astron. Soc. **363**, 609 (2005)
Dib, R., Kaspi, V.M., Gavriil, F.P.: Astrophys. J. (2006a, submitted)
Dib, R., Kaspi, V.M., Gavriil, F.P., Woods, P.M.: Astron. Telegr. 845 (2006b)
Durant, M., van Kerkwijk, M.H.: Astrophys. J. **627**, 376 (2005)
Durant, M., van Kerkwijk, M.H.: Astrophys. J. **652**, 576 (2006a)
Durant, M., van Kerkwijk, M.H.: Astrophys. J. **650**, 1070 (2006b) (astro-ph/0606027)
Durant, M., van Kerkwijk, M.H.: Astrophys. J. **650**, 1082 (2006c) (astro-ph/0606604)
Ertan, Ü., Göğüş, E., Alpar, M.A.: Astrophys. J. **640**, 435 (2006)
Figer, D.F., MacKenty, J., Robberto, M., et al.: Astrophys. J. **643**, 1166 (2006)
Gaensler, B.M., Gotthelf, E.V., Vasisht, G.: Astrophys. J. **526**, L37 (1999)
Gavriil, F.P., Kaspi, V.M.: Astrophys. J. **567**, 1067 (2002)
Gavriil, F.P., Kaspi, V.M.: Astrophys. J. **609**, L67 (2004)
Gavriil, F.P., Kaspi, V.M., Woods, P.M.: Nature **419**, 142 (2002)
Gavriil, F.P., Kaspi, V.M., Woods, P.M.: Astrophys. J. **607**, 959 (2004)
Gavriil, F.P., Kaspi, V.M., Woods, P.M.: Astrophys. J. **641**, 418 (2006)
Gotthelf, E.V., Gavriil, F., Kaspi, V.M., et al.: Astrophys. J. **564**, L31 (2002)
Gotthelf, E.V., Halpern, J.P., Buxton, M., Bailyn, C.: Astrophys. J. **605**, 368 (2004)
Gotthelf, E.V., Vasisht, G.: New Astron. **3**, 293 (1998)
Halpern, J.P., Gotthelf, E.V.: Astrophys. J. **618**, 874 (2005)
Harding, A.K., Strickman, M.S., Gwinn, C., et al.: Astrophys. J. **576**, 376 (2002)
Ibrahim, A.I., Markwardt, C.B., Swank, J.H., et al.: Astrophys. J. **609**, L21 (2004)
Ibrahim, A.I., Safi-Harb, S., Swank, J.H., et al.: Astrophys. J. **574**, L51 (2002)
Ibrahim, A.I., Swank, J.H., Parke, W.: Astrophys. J. **584**, L17 (2003)
Israel, G., Oosterbroek, T., Stella, L., et al.: Astrophys. J. **560**, L65 (2001)
Israel, G.L., Covino, S., Stella, L., et al.: Astrophys. J. **580**, L143 (2002)
Iwasawa, K., Koyama, K., Halpern, J.P.: Publ. Astron. Soc. Jpn. **44**, 9 (1992)
Kaspi, V.M., Gavriil, F.P.: Astrophys. J. **596**, L71 (2003)
Kaspi, V.M., Chakrabarty, D., Steinberger, J.: Astrophys. J. **525**, L33 (1999)
Kaspi, V.M., Lackey, J.R., Chakrabarty, D.: Astrophys. J. **537**, L31 (2000)
Kaspi, V.M., Gavriil, F.P., Chakrabarty, D., et al.: Astrophys. J. **558**, 253 (2001)
Kaspi, V.M., Gavriil, F.P., Woods, P.M., et al.: Astrophys. J. **588**, L93 (2003)
Kaspi, V.M., Dib, R., Gavriil, F.P.: Astron. Telegr. 794 (2006)
Kern, B., Martin, C.: Nature **415**, 527 (2002)
Kosugi, G., Ogasawara, R., Terada, H.: Astrophys. J. **623**, L125 (2005)
Kuiper, L., Hermsen, W., Mendez, M.: Astrophys. J. **613**, 1173 (2004)
Kuiper, L., Hermsen, W., den Hartog, P., Collmar, W.: Astrophys. J. **645**, 556 (2006)
Lai, D., Ho, W.C.G.: Astrophys. J. **588**, 962 (2003)

Lyutikov, M., Gavriil, F.P.: Mon. Not. Roy. Astron. Soc. **368**, 690 (2006)
McLaughlin, M.A., Lyne, A.G., Lorimer, D.R., et al.: Nature **439**, 817 (2006)
Mereghetti, S., Israel, G.L., Stella, L.: Mon. Not. Roy. Astron. Soc. **296**, 689 (1998)
Mereghetti, S., Tiengo, A., Stella, L., et al.: Astrophys. J. **608**, 427 (2004)
Mereghetti, S., Götz, D., Mirabel, I.F., Hurley, K.: Astron. Astrophys. **433**, L9 (2005)
Molkov, S.V., Cherepaschchuk, A.M., Lutovinov, A.A., et al.: Sov. Astron. Lett. **30**, 534 (2004)
Morii, M., Sato, R., Kataoka, J., Kawai, N.: Publ. Astron. Soc. Jpn. **55**, L45 (2003)
Morii, M., Kawai, N., Shibazaki, N.: Astrophys. J. **622**, 544 (2005)
Muno, M.P., Clark, J.S., Crowther, P.A., et al.: Astrophys. J. **636**, L41 (2006)
Oosterbroek, T., Parmar, A.N., Mereghetti, S., Israel, G.L., Astron. Astrophys. **334**, 925 (1998)
Özel, F.: Astrophys. J. **563**, 276 (2001)
Paul, B., Kawasaki, M., Dotani, T., Nagase, F.: Astrophys. J. **537**, 319 (2000)
Perna, R., Heyl, J., Hernquist, L., et al.: Astrophys. J. **557**, 18 (2001)
Rea, N., Testa, V., Israel, G., et al.: Astron. Astrophys. **425**, L5 (2004)
Rea, N., Oosterbroek, T., Zane, S., et al.: Mon. Not. Roy. Astron. Soc. **361**, 710 (2005)
Tam, C.R., Kaspi, V.M., van Kerkwijk, M.H., Durant, M.: Astrophys. J. **617**, L53 (2004)
Tam, C.R., Kaspi, V.M., Gaensler, B.M., Gotthelf, E.V., Astrophys. J. (2006, in press) (astro-ph/0602522)
Thompson, C., Duncan, R.C.: Astrophys. J. **473**, 322 (1996)
Thompson, C., Lyutikov, M., Kulkarni, S.R.: Astrophys. J. **574**, 332 (2002)
Tiengo, A., Mereghetti, S., Turolla, R., et al.: Astron. Astrophys. **437**, 997 (2005)
Torii, K., Kinugasa, K., Katayama, K., et al.: Astrophys. J. **503**, 843 (1998)
Wang, Z., Chakrabarty, D.: Astrophys. J. **579**, L33 (2002)
Wang, Z., Chakrabarty, D., Kaplan, D.L.: Nature **440**, 772 (2006)
White, N.E., Angelini, L., Ebisawa, K., et al.: Astrophys. J. **463**, L83 (1996)
Woods, P.M., Thompson, C.: In: Lewin, W.H.G., van der Klis, M. (eds.) Compact Stellar X-ray Sources. Cambridge University Press, Cambridge (2004)
Woods, P.M., Thompson, C., Kaspi, V.M., et al.: Astrophys. J. **605**, 378 (2004)
Woods, P.M., Kouveliotou, C., Gavriil, F.P., et al.: Astrophys. J. **629**, 985 (2005)

Astrophys Space Sci (2007) 308: 13–23
DOI 10.1007/s10509-007-9384-0

ORIGINAL ARTICLE

XMM–Newton observations of soft gamma-ray repeaters

Sandro Mereghetti · Paolo Esposito · Andrea Tiengo

Received: 20 July 2006 / Accepted: 24 August 2006 / Published online: 29 March 2007

Abstract All the confirmed Soft Gamma-ray Repeaters have been observed with the EPIC instrument on the XMM–Newton satellite. We review the results obtained in these observations, providing the most accurate spectra on the persistent X-ray emission in the 1–10 keV range for these objects, and discuss them in the context of the magnetar interpretation.

Keywords Gamma-rays: observations · Pulsars: individual SGR 1806-20, SGR 1900+14, SGR 1627-41 · Pulsars: general

PACS 97.60.Jd · 98.70.Qy

1 Introduction

Soft Gamma-ray Repeaters (SGRs) were discovered as sources of short intense bursts of gamma-rays, and for a long time were considered as a puzzling category of Gamma-ray bursts. Their neutron star nature was immediately suggested by the 8 s periodicity seen in the famous event of 5 March 1979, but it was only with the discovery of their pulsating counterparts in the few keV region that this was finally proved. Although the main motivations for the magnetar model (Duncan and Thompson 1992; Thompson and Duncan 1995) were driven by the high energy properties of the SGRs bursts and giant flares, X-ray observations of the "quiescent" emission have provided fundamental information to understand the nature of these objects (Woods and Thompson 2004).

Extensive observational programs have been carried out with the RossiXTE satellite, focusing mainly on the SGRs timing properties. Long term studies based on phase-connected timing analysis revealed significant deviations from a steady spin-down (Woods et al. 1999a, 2000, 2002), larger than the timing noise seen in radio pulsars and not linked in a simple way with the bursting activity. RossiXTE has also been used to investigate the variations of the pulse profiles as a function of energy and time (Göğüş et al. 2002), and to study the statistical properties of the bursts (Göğüş et al. 2001). However, the RossiXTE observations are not ideal to accurately measure the flux and spectrum of these relatively faint sources located in crowded fields of the Galactic plane, since its non-imaging instruments suffer from source confusion and large uncertainties in the background estimate. Imaging satellites like BeppoSAX, ASCA and Chandra have yielded useful spectral information, but it is only with the advent of the large collecting area XMM–Newton satellite that high quality spectra of SGRs have been obtained, in particular with the EPIC instrument (Strüder et al. 2001; Turner et al. 2001).

Here we review the results obtained with the XMM–Newton satellite for the three confirmed SGRs in our Galaxy. There are also some XMM–Newton data on SGR 0526-66 in the Large Magellanic Cloud, but they are of limited use due to the contamination from diffuse emission from the surrounding supernova remnant and will not be discussed here.

S. Mereghetti (✉) · A. Tiengo
INAF—Istituto di Astrofisica Spaziale e Fisica Cosmica, Milano, Italy
e-mail: sandro@iasf-milano.inaf.it

P. Esposito
Dipartimento di Fisica Nucleare e Teorica and INFN-Pavia, Università di Pavia, Pavia, Italy

2 SGR 1806-20

SGR 1806-20 is probably the most prolific and the best studied of the known SGRs. It showed several periods of bursting activity since the time of its discovery in 1979 (Laros et al. 1986) and recently attracted much interest since it emitted the most powerful giant flare ever observed from an SGR (Hurley et al. 2005; Palmer et al. 2005; Mereghetti et al. 2005a).

The low energy X-ray counterpart of SGR 1806-20 was identified with the ASCA satellite (Murakami et al. 1994), thanks to the detection and precise localization of a burst simultaneously seen at higher energy with BATSE. Subsequent observations with RossiXTE led to the discovery of pulsations with $P = 7.5$ s and $\dot{P} = 8 \times 10^{-11}$ s s^{-1} (Kouveliotou et al. 1998).

Possible associations of SGR 1806-20 with the variable non-thermal core of a putative radio supernova remnant (Frail et al. 1997) and with a luminous blue variable star (van Kerkwijk et al. 1995) were disproved when a more precise localization of the SGR could be obtained with the Interplanetary Network (Hurley et al. 1999b) and later improved with Chandra (Kaplan et al. 2002). The transient radio source observed with the VLA after the December 2004 giant flare (Cameron et al. 2005) led to an even smaller error region and, thanks to the superb angular resolution (FWHM $\sim 0.1''$) available with adaptive optics at the ESO Very Large Telescope, a variable near IR counterpart (K_s=19.3–20), could be identified (Israel et al. 2005), the first one for an SGR.

The distance of SGR 1806-20 is subject of some debate (Cameron et al. 2005), and is particularly relevant for its implications on the total energetics of the 2004 giant flare. A firm lower limit of 6 kpc can be derived from the HI absorption spectrum (McClure-Griffiths and Gaensler 2005), but the likely associations with a massive molecular cloud and with a cluster of massive stars indicate a distance of $\sim$15 kpc (Corbel and Eikenberry 2004; Figer et al. 2005). In the following we will adopt this value.

Before the XMM–Newton observations, the most accurate spectral measurements for SGR 1806-20 in the soft X-ray range were obtained with BeppoSAX in 1998–1999 (Mereghetti et al. 2000). They showed that a power law with photon index $\Gamma = 1.95$ or a thermal bremsstrahlung with temperature $kT_{tb} = 11$ keV were equally acceptable fits. All the observations indicated a fairly constant flux, corresponding to a 2–10 keV luminosity of $\sim 3 \times 10^{35}$ erg s^{-1}.

Being located at only $\sim 10°$ from the Galactic center direction, SGR 1806-20 has been extensively observed with the INTEGRAL satellite since 2003. A few hundreds bursts have been detected with the IBIS instrument in the 15–200 keV range, leading to the discovery of a hardness intensity anti-correlation and allowing to extend the number-flux relation of bursts down to fluences smaller than 10^{-8} erg cm^{-2} (Götz et al. 2004, 2006b). In addition, it was discovered with INTEGRAL that the persistent emission from SGR 1806-20 extends up to 150 keV (Mereghetti et al. 2005b; Molkov et al. 2005). The hard X-ray emission, well fit by a power law with photon index $\Gamma \sim 1.5$–1.9, seems to correlate in hardness and intensity with the rate of burst emission, that reached a maximum in Fall 2004.

The bursting activity of SGR 1806-20 culminated with the giant flare of 2004 December 27 (Borkowski et al. 2004; Hurley et al. 2005; Palmer et al. 2005), that produced the strongest flash of gamma-rays at the Earth ever observed. The emission was so intense to cause saturation of most in-flight detectors, significant ionization of the upper atmosphere (Campbell et al. 2005), and a detectable flux of radiation backscattered from the Moon (Mazets et al. 2005; Mereghetti et al. 2005a). Other observations of this exceptional event, and their implications for the physics of neutron stars, are discussed during the High Energy Density Laboratory Astrophysics proceedings (Stella 2006; Israel 2006). Comparing this giant flare with those seen from SGR 0526-66 and SGR 1900+14, it is found that the energy in the pulsating tails of the three events was roughly of the same order ($\sim 10^{44}$ ergs), while the energy in the initial spike of SGR 1806-20 (a few 10^{46} ergs) was at least two orders of magnitude higher than that of the other giant flares. This indicates that the magnetic field in the three sources is similar. In fact the pulsating tail emission is thought to originate from the fraction of the energy released during the initial hard pulse that remains magnetically trapped in the neutron star magnetosphere, forming an optically thick photon-pair plasma (Thompson and Duncan 1995). The amount of energy that can be confined in this way is determined by the magnetic field strength, which is thus inferred to be of several 10^{14} G in these three magnetars.

SGR 1806-20 is the target of an ongoing campaign of XMM–Newton observations aimed at studying in detail the long term variations in the properties of its persistent emission. These observations, coupled with similar programs carried out with ESO telescopes in the infrared band (Israel 2006) and at hard X-ray energy with INTEGRAL (Götz et al. 2006a) and Suzaku, can be used to study the connection between the persistent emission and the source activity level, as manifested by the emission of bursts and flares.

2.1 XMM–Newton results

Seven XMM–Newton observations of SGR 1806-20 have been carried out to date (see Table 1). Four were obtained from April 2003 to October 2004, before the giant flare (Mereghetti et al. 2005c). At the time of the giant flare the source was not visible by XMM–Newton (and most other satellites) due to its proximity to the Sun, thus the next observation could be done only in March 2005 (Tiengo et al.

Table 1 XMM–Newton observations and timing results for SGR 1806-20

Obs.	Date	Duration (ks)	Mode[a] and exp. time PN camera	Mode[a] and exp. time MOS1/2 cameras	Pulse period (s)
A	2003 Apr 3	32	FF (5.4 ks)	LW (6 ks)	7.5311 ± 0.0003
B	2003 Oct 7	21	FF (13.4 ks)	LW (17 ks)	7.5400 ± 0.0003
C	2004 Sep 6	51	SW (36.0 ks)	LW (51 ks)	7.55592 ± 0.00005
D	2004 Oct 6	18	SW (12.9 ks)	Ti (18 ks)	7.5570 ± 0.0003
E	2005 Mar 7	24	SW (14.7 ks)	Ti/FF (24 ks)	7.5604 ± 0.0008
F	2005 Oct 4	33	SW (22.8 ks)	Ti/FF (33 ks)	7.56687 ± 0.00003
G	2006 Apr 4	29	SW (20.5 ks)	Ti/FF (29 ks)	7.5809 ± 0.0002

[a]FF = Full Frame (time resolution 73 ms); LW = Large Window (time resolution 0.9 s); SW = Small Window (time resolution 6 ms); Ti = Timing (time resolution 1.5 ms)

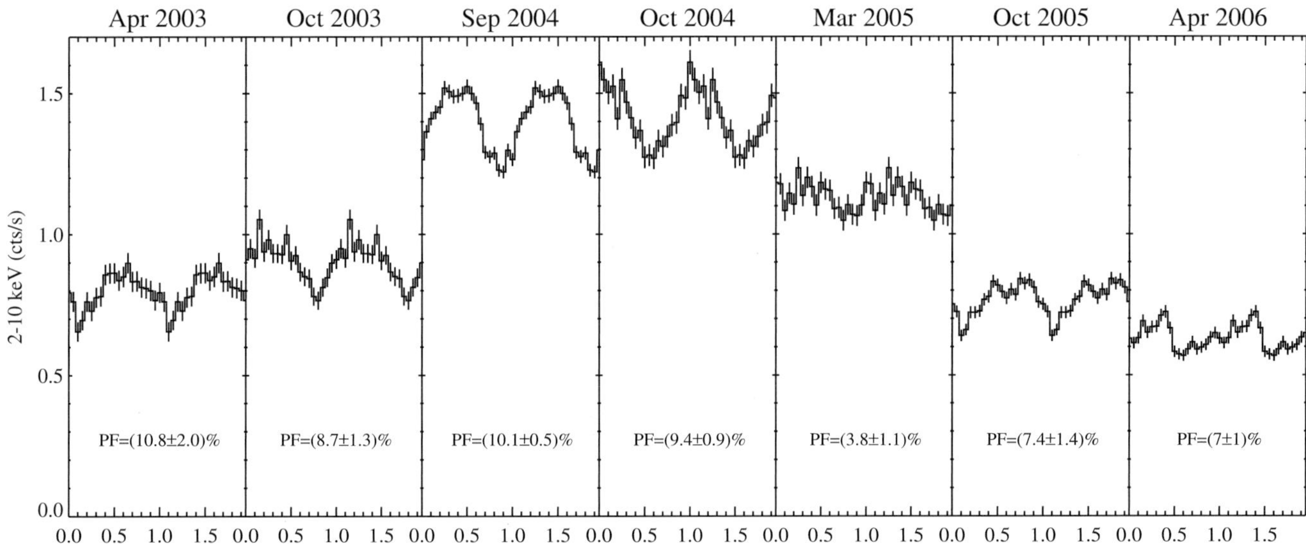

Fig. 1 *Folded light curves* of SGR 1806-20 obtained with the EPIC pn instrument in the seven XMM–Newton observations. Note the flux increase in the two observations before the December 2004 giant flare and the small pulsed fraction in the first 2005 observation

2005). This was followed by another observation in October (Rea et al. 2005) and a most recent one in April 2006, the results of which are presented here for the first time.

The bursts detected in some of these observations (mostly in September-October 2004) were excluded by appropriate time selections to derive the spectral results reported below. After screening out the bursts, the source pulsations were clearly detected in all the observations. The corresponding folded light curves are shown in Fig. 1, where all the panels have the same scale in count rate to facilitate a comparison of the flux variations between the observations. The main spectral results are summarized in Table 2, where we have reported only the best fit parameters for the power law plus blackbody model (see Mereghetti et al. 2005c; Tiengo et al. 2005 for more details).

Indeed the strong requirement for a blackbody component is one of the main results of the high quality XMM–Newton spectra. In this respect the most compelling evidence comes from the September 2004 observation (obs. C), which, thanks to the high source count rate and long observing time, provided the spectra with the best statistics. A fit with an absorbed power law yields a relatively high χ^2 value ($\chi^2_{\rm red} = 1.37$) and structured residuals, while a much better fit ($\chi^2_{\rm red} = 0.93$) can be obtained by adding a blackbody component. The best fit parameters are photon index $\Gamma = 1.2$, blackbody temperature $kT_{BB} = 0.8$ keV and absorption $N_H \sim 6.5 \times 10^{22}$ cm^{-2}. Although some of the observations with lower statistics give acceptable fits also with a single power law, the results reported in Table 2 indicate that all the observations are consistent with the presence of an additional blackbody component with similar parameters. As an example we show in Fig. 2 (panels A and B) the power law plus blackbody fit of the April 2006 EPIC pn spectrum. The residuals (panel B) show a deviation at $\sim 3\sigma$

Table 2 Summary of the spectral results[a] for SGR 1806-20

Obs.	Absorption 10^{22} cm^{-2}	Power law photon index	kT_{BB} (keV)	R_{BB} (km)[b]	Flux[c] 10^{-11} erg cm^{-2} s^{-1}	χ^2_{red} (d.o.f.)
A	6.6 (5.6–8.4)	1.4 (1.0–1.7)	0.6 (0.4–0.9)	2.6 (0.7–13.9)	1.23	1.01 (56)
B	6.0 (4.9–6.6)	1.2 (0.5–1.4)	0.7 (0.6–1.0)	1.8 (1.2–2.9)	1.39	0.97 (68)
C	6.5 (6.2–6.9)	1.21 (1.09–1.35)	0.8 (0.7–0.9)	1.9 (1.6–2.6)	2.66	0.93 (70)
D	6.5 (5.9–7.1)	1.2 (0.9–1.4)	0.8 (0.6–0.9)	2.2 (1.6–3.5)	2.68	0.90 (69)
E	6.0 (5.8–6.2)	0.8 (0.5–1.0)	0.91 (0.86–1.05)	1.9 (1.6–2.1)	1.92	1.02 (70)
F	6.4 (6.0–6.8)	1.4 (1.1–1.7)	0.7 (0.6–0.8)	2.2 (1.7–3.3)	1.34	1.11 (69)
G	6.2 (5.6–6.6)	1.2 (0.9–1.4)	0.7 (0.6–0.8)	2.0 (1.6–2.7)	1.07	1.09 (68)

[a]Errors are at the 90% c.l. for a single interesting parameter

[b]Radius at infinity assuming a distance of 15 kpc

[c]Absorbed flux in the 2–10 keV energy range

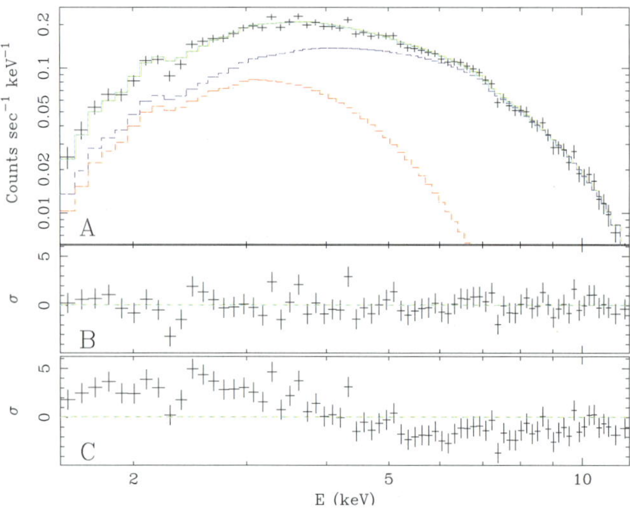

Fig. 2 **A**: EPIC pn spectrum of obs. G fitted with a power law (*blue line*) plus blackbody (*red line*) model. **B**: residuals of the fit in units of standard deviations. **C**: residuals of the spectrum of obs. G fitted with the best fit model of obs. C (rescaled in normalization) in order to illustrate the softening after the giant flare

near 2.3 keV. However, this possible feature is not confirmed by the MOS data, and is not present in the spectra of the other XMM–Newton observations (see Sect. 6).

The second new result derived from these observations is the long term flux variability. All the observations of SGR 1806-20 in the 1–10 keV range obtained in the previous years with ROSAT, ASCA and BeppoSAX were consistent with a flux of $\sim 10^{-11}$ erg cm^{-2} s^{-1}. On the other hand, the XMM–Newton data showed a doubling of the flux in September-October 2004 followed by a gradual recovery to the "historical" level during the observations performed after the giant flare. Interestingly, the same trend was seen above 20 keV with INTEGRAL (Mereghetti et al. 2005b; Götz et al. 2006a), as well as in the flux of the NIR counterpart (Israel et al. 2005; Israel 2006).

The observations performed before and after the giant flare show significant differences also in the source pulsed fraction and spectral shape. The pulsed fraction in the first observation after the flare was the smallest seen with XMM–Newton, while it increased again in the following observations. The spectral hardness followed a similar trend: the four pre-flare observations give marginal evidence for a gradual hardening, while the spectrum was definitely softer in the post-flare observations. This is illustrated in the bottom panel of Fig. 2, which shows the residuals of the April 2006 spectrum fitted with the pre-flare model (obs. C): the trend in the residuals clearly indicate the spectral softening.

Finally, we can compare the XMM–Newton spectra with those obtained in the previous years with other satellites. We base this comparison on the fits with a single power law, since this model was successfully used to describe the ASCA and BeppoSAX data. The average photon index in the four XMM–Newton observations of 2003–2004 ($\Gamma = 1.5 \pm 0.1$) was significantly smaller than that observed in 1993 with ASCA ($\Gamma = 2.25 \pm 0.15$, Mereghetti et al. 2002) and in 1998–2001 with BeppoSAX ($\Gamma = 1.97 \pm 0.09$, Mereghetti et al. 2002). This indicates that a spectral hardening occurred between September 2001 and April 2003.

A long term variation occurred also in the average spin-down rate: while the early sparse period measurements with ASCA and BeppoSAX (Mereghetti et al. 2002), as well as a phase-connected RossiXTE timing solution spanning February-August 1999 (Woods et al. 2000), were consistent with an average $\dot{P} \sim 8.5 \times 10^{-11}$ s s^{-1}, subsequent RossiXTE data indicate a spin-down larger by a factor $\sim$4 (Woods et al. 2002) and the four XMM–Newton period measurements before the giant flare show a further increase to an average $\dot{P} = 5.5 \times 10^{-10}$ s s^{-1} (Mereghetti et al. 2005c).

As shown in Fig. 3, the changes in spectral hardness and spin-down rate of SGR 1806-20 follow the correlation between these quantities discovered in the sample of AXPs

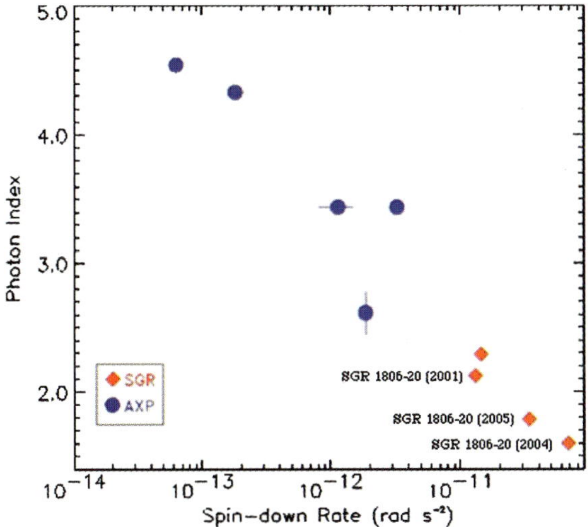

Fig. 3 Correlation between power law photon index and spin-down rate in Anomalous X-ray Pulsars and SGRs (adapted from Marsden and White 2001). Each point refers to a different source, except the three points for SGR 1806-20 in different time periods

and SGRs by comparing different objects: the sources with the harder spectrum have a larger long term spin-down rate (Marsden and White 2001). These results indicate, for the first time, that such a correlation also holds within different states of a single source.

3 SGR 1900+14

Bursts from this SGR were discovered with the Venera satellites in 1979 (Mazets et al. 1979). No other bursts were detected until thirteen years later, when four more events were seen with BATSE in 1992 (Kouveliotou et al. 1993). In the meantime the X-ray counterpart had been discovered with ROSAT (Vasisht et al. 1994), and later found to pulsate at 5.2 s with ASCA (Hurley et al. 1999c). Subsequent observations with the RossiXTE satellite confirmed the pulsations and established that the source was spinning down rapidly, with a period derivative of $\sim 10^{-11}$ s s^{-1} (Kouveliotou et al. 1999).

The peak of the activity from SGR 1900+14 was reached on 1998 August 27, with the emission of a giant flare (Hurley et al. 1999a; Feroci et al. 1999), resembling the only similar event known at that time, the exceptional burst of 5 March 1979. The 1998 giant flare from SGR 1900+14 could be studied much better than that of SGR 0526-66. The flare started with a short ($\sim$0.07 s) soft spike, followed by a much brighter short and hard pulse that reached a peak luminosity of $\sim 10^{45}$ erg s^{-1}. The initial spike was followed by a softer gamma-ray tail modulated at 5.2 s (Hurley et al. 1999a; Mazets et al. 1999), which decayed in a quasi exponential manner over the next $\sim$6 minutes (Feroci et al. 2001). Integrating over the entire flare assuming isotropic emission, at least 10^{44} erg were released in hard X-rays above 15 keV (Mazets et al. 1999). SGR 1900+14 also emitted another less intense flare on 18 April 2001 (Feroci et al. 2003; Guidorzi et al. 2004), which based on its energetics was classified as an "intermediate" flare.

Despite being the less absorbed of the galactic SGRs ($N_H \sim 2 \times 10^{22}$ cm^{-2}) no optical/IR counterpart has been yet identified for SGR 1900+14. Its possible association with a young cluster of massive stars (Vrba et al. 2000), where the SGR could have been born, gives a distance of $\sim$15 kpc, that we will adopt in the following.

Recently, persistent emission from SGR 1900+14 has been detected also in the 20–100 keV range thanks to observations with the INTEGRAL satellite (Götz et al. 2006b).

3.1 XMM–Newton results

SGR 1900+14 lies in a sky region that, until recently, was not observable by XMM–Newton due to technical constraints in the satellite pointing. Thus the first observation of SGR 1900+14 could be obtained only in September 2005 (Mereghetti et al. 2006b). This observation occurred during a long period of inactivity (the last bursts before the observations were reported in November 2002, Hurley et al. 2002).

The spectrum could not be fit satisfactorily with single component models, while a good fit was obtained with the sum of a power law and a blackbody, with photon index $\Gamma = 1.9 \pm 0.1$, temperature kT $= 0.47 \pm 0.02$ keV, absorption $N_H = (2.12 \pm 0.08) \times 10^{22}$ cm^{-2}, and unabsorbed flux $\sim 4.8 \times 10^{-12}$ erg cm^{-2} s^{-1} (2–10 keV). An acceptable fit could also be obtained with the sum of two blackbodies with temperatures of 0.53 and 1.9 keV.

The XMM–Newton power law plus blackbody parameters are in agreement with previous observations of this source carried out with ASCA (Hurley et al. 1999c), BeppoSAX (Woods et al. 1999b; Esposito et al. 2007) and Chandra (Kouveliotou et al. 2001), but the flux measured in September 2005 is the lowest ever seen from SGR 1900+14. A $\sim$30% decrease of the persistent emission, compared to the "historical" level of $\sim 10^{-11}$ erg cm^{-2} s^{-1}, had already been noticed in the last BeppoSAX observation (Esposito et al. 2007), that was carried out in April 2002, six month earlier than the last bursts reported before the recent reactivation. The long term fading experienced by SGR 1900+14 in 2002–2005 might be related to the apparent decrease in the bursting activity in this period.

A second XMM–Newton observation was carried out on 1 April 2006, as a target of opportunity following the source reactivation indicated by a few bursts detected by Swift (Palmer et al. 2006) and Konus-Wind (Golenetskii et al. 2006). The spectral shape was consistent with that measured in the first observation, but the flux was $\sim 5.5 \times 10^{-12}$ erg cm^{-2} s^{-1} (Mereghetti et al. 2006b).

The spin period was 5.198346 ± 0.000003 s in September 2005 and 5.19987 ± 0.00007 s in April 2006. In both observations the pulsed fraction was $\sim$16% and no significant changes in the pulse profile shape were seen after the burst reactivation.

For both observations we performed phase-resolved spectroscopy extracting the spectra for different selections of phase intervals. No significant variations with phase were detected, all the spectra being consistent with the model and parameters of the phase-averaged spectrum, simply rescaled in normalization.

4 SGR 1627-41

From the point of view of the bursts and timing properties, SGR 1627-41 is one of the less well studied SGRs. This source was discovered during a period of bursting activity that lasted only six weeks in 1998 (Woods et al. 1999c; Hurley et al. 1999d; Mazets et al. 1999). Since then no other bursts were observed.

With a column density of $N_H \sim 10^{23}$ cm^{-2}, corresponding to $A_V \sim 40$–50, SGR 1627-41 is the most absorbed of the known SGRs. Thus it is not surprising that little is known on its possible counterparts. Near IR observations (Wachter et al. 2004) revealed a few objects positionally consistent with the small Chandra error region, but they are likely foreground objects unrelated to the SGR.

A distance of 11 kpc is generally assumed for SGR 1627-41, based on its possible association with the radio complex CTB 33, comprising the supernova remnant SNR G337.0-0.1 and a few HII regions (Corbel et al. 1999).

During the active period a soft X-ray counterpart with flux $F_x \sim 7 \times 10^{-12}$ erg cm^{-2} s^{-1} was identified with BeppoSAX (Woods et al. 1999c). However it was not possible to reliably measure a periodicity (a marginal detection at 6.4 s was not confirmed by better data).

Observations carried out in the following years with BeppoSAX, ASCA and Chandra, showed a monotonic decrease in the luminosity, from the value of $\sim 10^{35}$ erg s^{-1} (for $d = 11$ kpc) seen in 1998 down to $\sim 4 \times 10^{33}$ erg s^{-1}.

4.1 XMM–Newton results

A 52 ks long XMM–Newton pointing on SGR 1627-41 was done in September 2004 (Mereghetti et al. 2006a), while some other information could be extracted from two observations in which SGR 1627-41 was serendipitously detected at an off-axis angle of $\sim 10'$. These $\sim$30 ks long observations, whose main target was IGR J16358-4726, were carried out in February and September 2004.

All the XMM–Newton data showed a rather faint source ($\sim 9 \times 10^{-14}$ erg cm^{-2} s^{-1}) with a soft spectrum. Unfortunately the source faintness did not allow a sensitive search for periodic pulsations. Both a steep power law (photon index $\Gamma = 3.7 \pm 0.5$) and a blackbody with temperature $kT_{BB} = 0.8^{+0.2}_{-0.1}$ keV gave acceptable fits. The absorption was consistent with that measured in all the previous observations, $N_H = 9 \times 10^{22}$ cm^{-2}. There is evidence that the spectrum softened between the two Chandra observations carried out in September 2001 and August 2002 (Kouveliotou et al. 2003). The photon index measured with XMM–Newton is consistent with that of the last Chandra observation but, due to the large uncertainties, also a further softening cannot be excluded.

The XMM–Newton flux measurements are compared with those obtained in previous observations in Fig. 4. The two panels refer to the observed (top) and emitted (bottom) fluxes in the 2–10 keV range and for a common value of the absorption in all the observations ($N_H = 9 \times 10^{22}$ cm^{-2}, see Mereghetti et al. 2006a for details). The long term decrease in luminosity is clear, but, owing to the source spectral variations, the details of the decay light curve are different for the observed and unabsorbed flux.

If one considers the observed fluxes, the Chandra and XMM–Newton data suggest that SGR 1627-41 continued to fade also after September 2001, while the unabsorbed values indicate a possible plateau level at $\sim 2.5 \times 10^{-13}$ erg cm^{-2} s^{-1}. It is important to realize that the quantity most relevant for theoretical modeling, i.e. the emitted

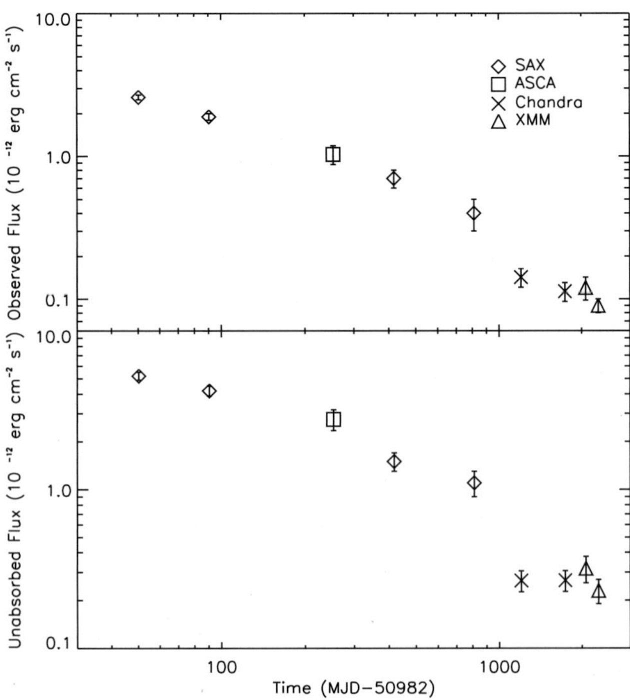

Fig. 4 Long term *light curve* of SGR 1627-41 based on data from different satellites (Mereghetti et al. 2006a). *Top panel*: absorbed flux in the 2–10 keV range. *Bottom panel*: unabsorbed flux in the 2–10 keV range. For clarity, the XMM–Newton points of 2004 September 4, which are consistent with the last measurement, are not plotted

Table 3 Main properties of the four confirmed SGRs

	SGR 1627-41	SGR 1806-20	SGR 1900+14	SGR 0526-66
Coordinates	$16^h\ 35^m\ 51.83^s$	$18^h\ 08^m\ 39.337^s$	$19^h\ 07^m\ 14.33^s$	$05^h\ 26^m\ 00.89^s$
	$-47°\ 35'\ 23.3''$	$-20°\ 24'\ 39.85''$	$+09°\ 1'9\ 20.1''$	$-66°\ 04'\ 36.3''$
Error	$0.2''$ [a]	$0.06''$ [b]	$0.15''$ [*d*]	$0.6''$ [g]
Distance	11 kpc	15 kpc	15 kpc	50 kpc
Period	–	7.6 s	5.2 s	8 s
Period derivative ($\mathrm{s\,s^{-1}}$)	–	$(8.3–81) \times 10^{-11}$ [c]	$(6.1–20) \times 10^{-11}$ [e]	6.6×10^{-11} [g]
Magnetic field[a]	–	$(8–25) \times 10^{14}$ G	$(6–10) \times 10^{14}$ G	7×10^{14} G
Flux range[b] ($\mathrm{erg\,cm^{-2}\,s^{-1}}$)	$(0.025–0.6) \times 10^{-11}$	$(1.3–3.8) \times 10^{-11}$	$(0.5–2.7) \times 10^{-11}$	0.07×10^{-11}
Typical flux[b] ($\mathrm{erg\,cm^{-2}\,s^{-1}}$)	$\sim 3 \times 10^{-13}$	$\sim 1.5 \times 10^{-11}$	$\sim 10^{-11}$	$\sim 10^{-12}$
20–60 keV flux ($\mathrm{erg\,cm^{-2}\,s^{-1}}$)	–	$(3–5) \times 10^{-11}$	$\sim 1.5 \times 10^{-11}$	–
Optical/IR	J > 21.5, H > 19.5,	$K_s = 20–19.3$ [b]	J > 22.8, $K_s > 20.8$ [f]	V > 27.1, I > 25 [h]
	$K_s > 20.0$ [a]	J > 21.2, H > 19.5		
Luminosity[c] ($\mathrm{erg\,s^{-1}}$)	$\sim 4 \times 10^{33}$	$\sim 4 \times 10^{35}$	$\sim 3 \times 10^{35}$	$\sim 2 \times 10^{35}$
Photon index	3	1.2	2	3.1
Blackbody kT	–	0.8 keV	0.45 keV	0.53 keV
N_H	$9 \times 10^{22}\ \mathrm{cm^{-2}}$	$6.5 \times 10^{22}\ \mathrm{cm^{-2}}$	$2.2 \times 10^{22}\ \mathrm{cm^{-2}}$	$0.55 \times 10^{22}\ \mathrm{cm^{-2}}$
Giant Flare	–	December 27, 2004	August 27, 1998	March 5, 1979
Initial spike energy (erg)	–	$(1.6–5) \times 10^{46}$	$> 6.8 \times 10^{43}$	1.6×10^{44}
Pulsating tail energy (erg)	–	1.3×10^{44}	5.2×10^{43}	3.6×10^{44}
Most active periods	1998 Jun–Jul	1983–1985, 1996–1999,	1979 Mar, 1992, 1998–1999,	1979 Mar–Apr,
		2003–2004	2001–2002, 2006	1981 Dec–1983 Apr

References: [a] Wachter et al. (2004); [b] Israel et al. (2005); [c] Woods et al. (2007); [d] Frail et al. (1999); [e] Woods et al. (2002); [f] Kaplan et al. (2002); [g] Kulkarni et al. (2003); [h] Kaplan et al. (2001)

[a] Assuming spin-down due to dipole radiation: $B = 3.2 \times 10^{19} (P\dot{P})^{1/2}$ G

[b] Unabsorbed flux in the 2–10 keV energy range

[c] Luminosity in the 2–10 keV energy range assuming the distances reported above

flux, is subject to the uncertainties in the spectral parameters. This is particularly important for high N_H values and small fluxes, as in the case of SGR 1627-41.

The long term luminosity decrease of SGR 1627-41 was interpreted as evidence for cooling of the neutron star surface after the deep crustal heating that occurred during the period of SGR activity in 1998. The decay light curve was fitted with a model of deep crustal heating requiring a massive neutron star ($M > 1.5 M_{\odot}$) which could well explain a plateau seen between days 400 and 800 (Kouveliotou et al. 2003). However, the evidence for such a plateau is not so compelling, according to our reanalysis of the BeppoSAX data. In fact all the BeppoSAX and ASCA points in the top panel of Fig. 4, before the rapid decline seen with Chandra in September 2001, are well fit by a power law decay, $F(t) \propto (t - t_0)^{-0.6}$, where t_0 is the time of the discovery outburst.

5 XMM–Newton results on the SGRs bursts

Up to now limited spectral information has been obtained for SGR bursts below 20 keV. In particular, before our XMM–Newton observations, spectra with good energy resolution and sensitivity at a few keV were lacking. However, some studies have provided evidence that the optically thin thermal bremsstrahlung model, which gives a good phenomenological description of the burst spectra in the hard X-ray range, is inconsistent with the data below 15 keV (Fenimore et al. 1994; Olive et al. 2004; Feroci et al. 2004). We therefore tried to address this issue using the XMM–Newton data.

Several tens of bursts were detected during some of the XMM–Newton observations of SGR 1806-20 (while none was seen in SGR 1900+14 and SGR 1627-41). These bursts had durations typical of the short bursts more commonly observed at higher energy. Since the individual bursts had too few counts for a meaningful spectral analysis, we extracted a cumulative spectrum by summing all the bursts detected during the 2004 observations. The resulting spectrum corresponds to a total exposure of 12.7 s and contains about 2000 net counts in the 2–10 keV range. We checked that pile-up effects were not important (see Mereghetti et al. 2005c for details). The spectrum of the remaining observing time was used as background.

All the fits with simple models (power law, blackbody, thermal bremsstrahlung) gave formally acceptable χ^2 values, but the power law and the bremsstrahlung required a large absorption ($N_H = 10^{23}$ cm^{-2}), inconsistent with the value seen in the persistent emission. We therefore favor the blackbody model, which yields $kT_{BB} = 2.3 \pm 0.2$ keV and $N_H = 6 \times 10^{22}$ cm^{-2}, in agreement with the value determined from the spectrum of the persistent emission.

The residuals from this best fit showed a deviation at 4.2 keV. Although the deviation is formally at 3.3σ, it could not be reproduced in spectra obtained with different data selections and binning criteria. Therefore we consider it as only a marginal evidence for an absorption line (Mereghetti et al. 2005c).

6 (Absence of) spectral lines

In models involving ultra-magnetized neutron stars, proton cyclotron features are expected to lie in the X-ray range, for surface magnetic fields strengths of $\sim 10^{14}$–10^{15} G. Detailed calculations of the spectrum emerging from the atmospheres of magnetars in quiescence have confirmed this basic expectation (Zane et al. 2001; Ho and Lai 2001). Model spectra exhibit a strong absorption line at the proton cyclotron resonance, $E_{c,p} \sim 0.63 z_G (B/10^{14}\ \mathrm{G})$ keV, where z_G, typically in the 0.70–0.85 range, is the gravitational red-shift at the neutron star surface. No evidence for persistent cyclotron features has been reported to date in SGRs, despite some features have been possibly detected during bursts (see e.g. Strohmayer and Ibrahim 2000; Ibrahim et al. 2003).

A sensitive search for spectral lines was among the main objectives of the XMM–Newton observations of SGRs. However no evidence for emission or absorption lines, was found by looking at the residuals from the best fit models. In the case of SGR 1806-20 and SGR 1900+14 the upper limits are the most constraining ever obtained for these sources in the ~1–10 keV energy range. They are shown in Fig. 5 for the most sensitive observation of each source, i.e. those of September 2004 for SGR 1806-20 and of September 2005 for SGR 1900+14. The plotted curves represent the upper limits on the equivalent widths as a function of the assumed line energy and width. They were derived by adding Gaussian components to the best fit models and computing the allowed range in their normalization.

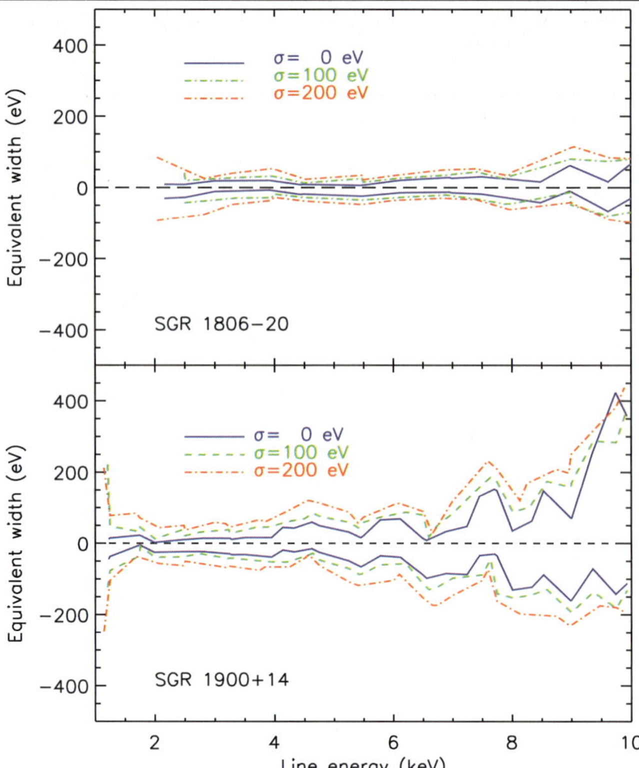

Fig. 5 Upper limits (at 3σ) on spectral features in the persistent emission of SGR 1806-20 (*top*) and SGR 1900+14 (*bottom*)

Some reasons have been proposed to explain the absence of cyclotron lines in magnetars, besides the obvious possibility that they lie outside the sampled energy range. Magnetars might differ from ordinary radio pulsars because their magnetospheres are highly twisted and therefore can support current flows (Thompson et al. 2002). The presence of charged particles (e^- and ions) produces a large resonant scattering depth and since (i) the electron distribution is spatially extended and (ii) the resonant frequency depends on the local value of the magnetic field, repeated scatterings could lead to the formation of a hard tail instead of a narrow line. A different explanation involves vacuum polarization effects. It has been calculated that in strongly magnetized atmospheres this effect can significantly reduce the equivalent width of cyclotron lines, thus making difficult their detection (Ho and Lai 2003).

7 Conclusions

Many of the results presented above fit reasonably well with the magnetar model interpretation. However, there are also a few aspects that require more theoretical and observational efforts to be interpreted in this framework, in particular when one considers the variety of different behaviors shown by these sources and their close relatives like the Anomalous X-ray pulsars (Kaspi 2006).

The long term variations seen in SGR 1806-20, the source observed more often with XMM–Newton, are qualitatively consistent with the predictions of the magnetar model involving a twisted magnetosphere (Thompson et al. 2002). As mentioned above, according to this model, resonant scattering from magnetospheric currents leads to the formation of a high-energy tail. A gradually increasing twist results in a larger optical depth that causes a hardening of the X-ray spectrum. At the same time, the spin down rate increases because, for a fixed dipole field, the fraction of field lines that open out across the speed of light cylinder grows. In addition, the stresses building up in the neutron star crust and the magnetic footprints movements lead to crustal fractures causing an increase in the bursting activity. Since spectral hardening, spin-down rate, and bursting rate increase with the twist angle, it is not surprising that these quantities varied in a correlated way in SGR 1806-20 (see Fig. 6 and Götz et al. 2006a). However, as visible in Fig. 6, while the spectral hardening took place gradually over several years, the spin-down variation occurred more rapidly in 2000. A recent analysis of RossiXTE data around the time of the giant flare (Woods et al. 2007) shows that the correlation between spectral and variability parameters is indeed rather complex.

The long term flux evolution of SGR 1900+14 is shown in Fig. 7. It can be seen that, excluding the enhancements seen in correspondence of the flares, the luminosity remained always at the same level in the years 1997–2001, and then decreased slightly until the lowest value seen with XMM–Newton in September 2005. The following observation of April 2006 showed that the decreasing luminosity trend has been interrupted by the recent onset of bursts emission. However, the moderate flux increase was not as-

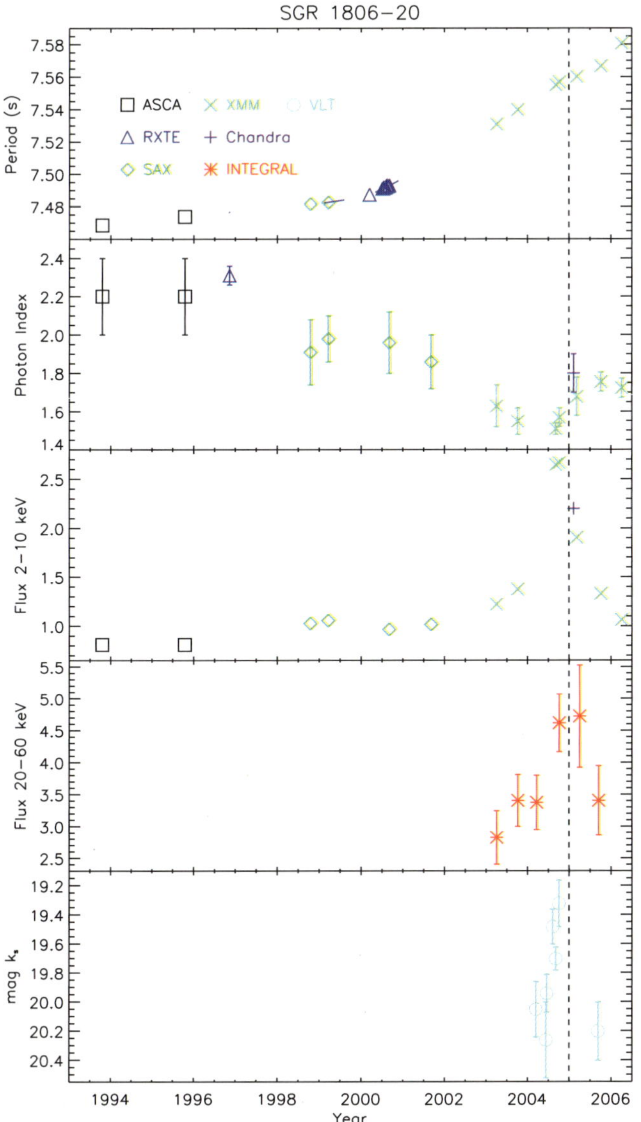

Fig. 6 From *top* to *bottom*: long term evolution of the pulse period, photon index, X-ray flux (2–10 keV), hard X-ray flux (20–60 keV), and infrared magnitude of SGR 1806-20. Fluxes are in units of 10^{-11} erg cm^{-2} s^{-1}. The *vertical dashed line* indicates the December 2004 giant flare

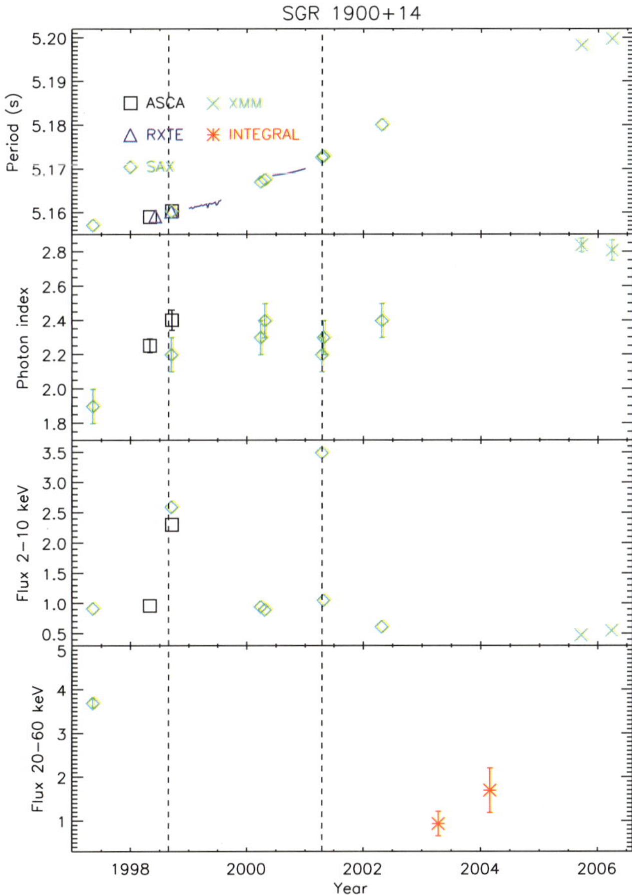

Fig. 7 From *top* to *bottom*: long term evolution of the pulse period, photon index, X-ray flux (2–10 keV), and hard X-ray flux (20–60 keV) of SGR 1900+14. Fluxes are in units of 10^{-11} erg cm^{-2} s^{-1}. The *vertical dashed lines* indicate the 27 August 1998 giant flare and the 18 April 2001 intermediate flare

sociated with significant changes in the X-ray spectral and timing properties, probably because the source is, up to now, only moderately active.

The luminosity now reached by SGR 1627-41, $\sim 3.5 \times 10^{33}$ erg s^{-1} is the smallest ever observed from a SGR, and is similar to that of the low state of the transient anomalous X-ray pulsar XTE J1810-197 (Ibrahim et al. 2004; Gotthelf et al. 2004). This low luminosity might be related to the long period ($\sim$6 years) during which SGR 1627-41 has not emitted bursts. However, the behavior of this source differs from that of the other SGRs that during periods of apparent lack of bursts changed only moderately their luminosity. In fact no bursts were detected from SGR 1900+14 in the three years preceding the XMM–Newton observations: had its luminosity decreased with the same trend exhibited by SGR 1627-41 it would have been much fainter than observed by XMM–Newton. Even more striking is the case of SGR 0526-66, which has a high luminosity ($\sim 2 \times 10^{35}$ erg s^{-1}), despite no signs of bursting activity have been observed in the last 15 years (unless weak bursts from this source have passed undetected due to its larger distance and location in a poorly monitored sky region).

It thus seems that, similarly to the case of Anomalous X-ray Pulsars, SGRs comprise both "persistent" sources (SGR 1806-20, SGR 1900+14 and SGR 0526-66) and "transients" (SGR 1627-41). The reasons behind this difference are currently unclear and not necessarily the same that differentiate between SGRs and AXPs.

Acknowledgements This work has been supported by the Italian Space Agency through the contract ASI-INAF I/023/05/0. XMM-Newton is an ESA science mission with instruments and contributions directly funded by ESA Member States and NASA.

References

Borkowski, J., Götz, D., Mereghetti, S., et al.: GRB Circular Network 2920 (2004)
Cameron, P.B., Chandra, P., Ray, A., et al.: Nature **434**, 1112 (2005)
Campbell, P., Hill, M., Howe, R., et al.: GRB Circular Network 2932 (2005)
Corbel, S., Eikenberry, S.S.: Astron. Astrophys. **419**, 191 (2004)
Corbel, S., Chapuis, C., Dame, T.M., et al.: Astrophys. J. **526**, L29 (1999)
Duncan, R.C., Thompson, C.: Astrophys. J. **392**, L9 (1992)
Esposito, P., Mereghetti, S., Tiengo, A., et al.: Astron. Astrophys. **461**, 605 (2007)
Fenimore, E.E., Laros, J.G. Ulmer, A.: Astrophys. J. **432**, 742 (1994)
Feroci, M., Frontera, F., Costa, E., et al.: Astrophys. J. **515**, L9 (1999)
Feroci, M., in't Zand, J.J.M., Soffitta, P., et al.: GRB Circular Network 1060 (2001)
Feroci, M., Mereghetti, S., Woods, P., et al.: Astrophys. J. **596**, 470 (2003)
Feroci, M., Caliandro, G.A., Massaro, E., et al.: Astrophys. J. **612**, 408 (2004)
Figer, D.F., Najarro, F., Geballe, T.R., et al.: Astrophys. J. **622**, L49 (2005)
Frail, D.A., Vasisht, G., Kulkarni, S.R.: Astrophys. J. **480**, L129 (1997)
Frail, D.A., Kulkarni, S.R., Bloom, J.S.: Nature **398**, 127 (1999)
Golenetskii, S., Aptekar, R., Mazets, E., et al.: GRB Circular Network 4936 (2006)
Gotthelf, E.V., Halpern, J.P., Buxton, M., et al.: Astrophys. J. **605**, 368 (2004)
Götz, D., Mereghetti, S., Mirabel, I.F., et al.: Astron. Astrophys. **417**, L45 (2004)
Götz, D., Mereghetti, S., Hurley, K.: These proceedings (2006a)
Götz, D., Mereghetti, S., Tiengo, A., et al.: Astron. Astrophys. **449**, L31 (2006b)
Göğüş, E., Kouveliotou, C., Woods, P.M., et al.: Astrophys. J. **558**, 228 (2001)
Göğüş, E., Kouveliotou, C., Woods, P.M., et al.: Astrophys. J. **577**, 929 (2002)
Guidorzi, C., Frontera, F., Montanari, E., et al.: Astron. Astrophys. **416**, 297 (2004)
Ho, W.C.G., Lai, D.: Mon. Not. Roy. Astron. Soc. **327**, 1081 (2001)
Ho, W.C.G., Lai, D.: Mon. Not. Roy. Astron. Soc. **338**, 233 (2003)
Hurley, K., Cline, T., Mazets, E., et al.: Nature **397**, 41 (1999a)
Hurley, K., Kouveliotou, C., Cline, T., et al.: Astrophys. J. **523**, L37 (1999b)
Hurley, K., Kouveliotou, C., Woods, P., et al.: Astrophys. J. **519**, L143 (1999c)
Hurley, K., Li, P., Kouveliotou, C., et al.: Astrophys. J. **510**, L111 (1999d)
Hurley, K., Mazets, E., Golenetskii, S., et al.: GRB Coordinates Network 1715 (2002)
Hurley, K., Boggs, S.E., Smith, D.M., et al.: Nature **434**, 1098 (2005)
Ibrahim, A.I., Swank, J.H., Parke, W.: Astrophys. J. **584**, L17 (2003)
Ibrahim, A.I., Markwardt, C.B., Swank, J.H., et al.: Astrophys. J. **609**, L21 (2004)
Israel, G.L.: Astrophys. Space Sci. **308**. DOI 10.1007/s10509-007-9321-2 (2007)
Israel, G.L., Covino, S., Mereghetti, S., et al.: The Astronomer's Telegram 378 (2005)
Kaplan, D.L., Kulkarni, S.R., van Kerkwijk, M.H., et al.: Astrophys. J. **556**, 399 (2001)
Kaplan, D.L., Fox, D.W., Kulkarni, S.R., et al.: Astrophys. J. **564**, 935 (2002)
Kaspi, V.: Astrophys. Space Sci. **308**. DOI 10.1007/s10509-007-9309-y (2007)
Kouveliotou, C., Fishman, G.J., Meegan, C.A., et al.: Nature **362**, 728 (1993)
Kouveliotou, C., Dieters, S., Strohmayer, T., et al.: Nature **393**, 235 (1998)
Kouveliotou, C., Strohmayer, T., Hurley, K., et al.: Astrophys. J. **510**, L115 (1999)
Kouveliotou, C., Tennant, A., Woods, P.M., et al.: Astrophys. J. **558**, L47 (2001)
Kouveliotou, C., Eichler, D., Woods, P.M., et al.: Astrophys. J. **596**, L79 (2003)
Kulkarni, S.R., Kaplan, D.L., Marshall, H.L., et al.: Astrophys. J. **585**, 948 (2003)
Laros, J.G., Fenimore, E.E., Fikani, M.M., et al.: Nature **322**, 152 (1986)
Marsden, D., White, N.E.: Astrophys. J. **551**, L155 (2001)
Mazets, E.P., Golenetskii, S.V., Guryan, Y.A. Sov. Astron. Lett. **5**, 343 (1979)
Mazets, E.P., Aptekar, R.L., Butterworth, P.S., et al.: Astrophys. J. **519**, L151 (1999)
Mazets, E.P., Cline, T.L., Aptekar, R.L., et al.: Astrophysics e-prints, astro-ph/0502541 (2005)
McClure-Griffiths, N.M., Gaensler, B.M.: Astrophys. J. **630**, L161 (2005)
Mereghetti, S., Cremonesi, D., Feroci, M., et al.: Astron. Astrophys. **361**, 240 (2000)

Mereghetti, S., Feroci, M., Tavani, M., et al.: Mem. Soc. Astron. Ital. **73**, 572 (2002)
Mereghetti, S., Götz, D., von Kienlin, A., et al.: Astrophys. J. **624**, L105 (2005a)
Mereghetti, S., Götz, D., Mirabel, I.F., et al.: Astron. Astrophys. **433**, L9 (2005b)
Mereghetti, S., Tiengo, A., Esposito, P., et al.: Astrophys. J. **628**, 938 (2005c)
Mereghetti, S., Esposito, P., Tiengo, A., et al.: Astron. Astrophys. **450**, 759 (2006a)
Mereghetti, S., Esposito, P., Tiengo, A., et al.: Astrophys. J. **653**, 1423 (2006b)
Molkov, S., Hurley, K., Sunyaev, R., et al.: Astron. Astrophys. **433**, L13 (2005)
Murakami, T., Tanaka, Y., Kulkarni, S.R., et al.: Nature **368**, 127 (1994)
Olive, J.-F., Hurley, K., Sakamoto, T., et al.: Astrophys. J. **616**, 1148 (2004)
Palmer, D., Barthelmy, S., Barbier, L., et al.: GRB Circular Network 2945 (2005)
Palmer, D., Sakamoto, T., Barthelmy, S., et al.: The Astronomer's Telegram 789 (2006)
Rea, N., Israel, G., Covino, S., et al.: The Astronomer's Telegram 645 (2005)
Stella, L.: In: Proc. High Energy Density Laboratory Astrophysics (2006)
Strohmayer, T.E., Ibrahim, A.I.: Astrophys. J. **537**, L111 (2000)
Strüder, L., Briel, U., Dennerl, K., et al.: Astron. Astrophys. **365**, L18 (2001)
Thompson, C., Duncan, R.C.: Mon. Not. Roy. Astron. Soc. **275**, 255 (1995)
Thompson, C., Lyutikov, M., Kulkarni, S.R.: Astrophys. J. **574**, 332 (2002)
Tiengo, A., Esposito, P., Mereghetti, S., et al.: Astron. Astrophys. **440**, L63 (2005)
Turner, M.J.L., Abbey, A., Arnaud, M., et al.: Astron. Astrophys. **365**, L27 (2001)
van Kerkwijk, M.H., Kulkarni, S.R., Matthews, K., et al.: Astrophys. J. **444**, L33 (1995)
Vasisht, G., Kulkarni, S.R., Frail, D.A., et al.: Astrophys. J. **431**, L35 (1994)
Vrba, F.J., Henden, A.A., Luginbuhl, C.B., et al.: Astrophys. J. **533**, L17 (2000)
Wachter, S., Patel, S.K., Kouveliotou, C., et al.: Astrophys. J. **615**, 887 (2004)
Woods, P.M., Thompson, C.: In: Lewin, W.H.G., van der Klis, M. (eds.) Compact Stellar X-ray Sources (2004), astro-ph/0406133
Woods, P.M., Kouveliotou, C., van Paradijs, J., et al.: Astrophys. J. **524**, L55 (1999a)
Woods, P.M., Kouveliotou, C., van Paradijs, J., et al.: Astrophys. J. **518**, L103 (1999b)
Woods, P.M., Kouveliotou, C., van Paradijs, J., et al.: Astrophys. J. **519**, L139 (1999c)
Woods, P.M., Kouveliotou, C., Finger, M.H., et al.: Astrophys. J. **535**, L55 (2000)
Woods, P.M., Kouveliotou, C., Göğüş, E., et al.: Astrophys. J. **576**, 381 (2002)
Woods, P.M., Kouveliotou, C., Finger, M.H., et al.: Astrophys. J. **654**, 470 (2007)
Zane, S., Turolla, R., Stella, L., et al.: Astrophys. J. **560**, 384 (2001)

Astrophys Space Sci (2007) 308: 25–31
DOI 10.1007/s10509-007-9321-2

ORIGINAL ARTICLE

MMIV: de SGR 1806–20 Anno Mirabili

Unveiling the AXP/SGR connection

GianLuca Israel

Received: 27 July 2006 / Accepted: 23 August 2006 / Published online: 21 March 2007

Abstract On 27th December 2004 SGR 1806–20, one of the most active Soft γ-ray Repeaters (SGRs), displayed an extremely rare event, also known as giant flare, during which up to 10^{47} ergs were released in the ~1–1000 keV range in less than 1 s. Before and after the giant flare we carried out IR observations by using adaptive optics (NAOS-CONICA) mounted on VLT which provided images of unprecedented quality (FWHM better than $0.1''$). We discovered the likely IR counterpart to SGR 1806–20 based on positional coincidence with the VLA uncertainty region and flux variability of a factor of about 2 correlated with that at higher energies.

Moreover, by analysing the Rossi-XTE/PCA data we have discovered rapid Quasi-Periodic Oscillations (QPOs) in the pulsating tail of the 27th December 2004 giant flare of SGR 1806–20. QPOs at ~92.5 Hz are detected in a 50 s interval starting 170 s after the onset of the giant flare. These QPOs appear to be associated with increased emission by a relatively hard unpulsed component and are seen only over phases of the 7.56 s spin period pulsations away from the main peak. QPOs at ~18 and ~30 Hz are also detected ~200–300 s after the onset of the giant flare. This is the first time that QPOs are unambiguously detected in the flux of a Soft Gamma-ray Repeater, or any other isolated neutron star. We interpret the highest QPOs in terms of the coupling of toroidal seismic modes with Alfvén waves propagating along magnetospheric field lines. The lowest frequency QPO might instead provide indirect evidence on the strength of the internal magnetic field of the neutron star.

G. Israel (✉)
I.N.A.F., Osservatorio Astronomico di Roma, Via Frascati 33, I-00040 Monteporzio Catone (RM), Italy
e-mail: gianluca@mporzio.astro.it

Keywords Stars: neutron · Stars: oscillations · Pulsars: individual: SGR 1806–20 · Infrared: stars · X-rays: bursts

PACS 97.10.Sj · 97.60.Gb · 97.60.Jd · 98.38.Jw · 98.70.Qy

1 Introduction

Soft Gamma-ray Repeaters (SGRs) are characterized by short and recurrent bursts (<1 s) of soft γ rays. Only four confirmed SGRs are known, three in the Galaxy and one in the Large Magellanic Cloud (see Woods and Thompson 2006 for a recent review). The nature of SGRs has remained a mystery for many years. The ~8 s periodicity clearly seen in the tail of the 1979 March 5th giant flare of SGR 0526–66 suggested an association of SGRs with neutron stars. Several observational properties of SGRs are successfully modelled in terms of "magnetars," isolated neutron stars in which the dominant source of free energy is their intense magnetic field (B $\sim 10^{14}$–10^{15} G; Duncan and Thompson 1992; Thompson and Duncan 1995). The "magnetar" model is founded on two observational facts: firstly, the rotational energy loss inferred from the SGR and AXP spin-down is insufficient to power their persistent X-ray luminosity of $\sim 10^{34}$–10^{36} erg s^{-1}; secondly, there is no evidence for a companion stars which could provide the mass to power the X-ray emission through accretion.

Bursting activity from SGR 1806–20 resumed at the end of 2003 displaying an increase in both the γ-ray burst rate and the hard X-ray persistent emission (Mereghetti et al. 2005b) throughout 2004, and culminating with the giant flare of 27th December 2004 (Borkowski et al. 2004, during which $\sim 10^{47}$ erg were released for a distance of about 10 kpc; Cameron et al. 2005; McClure-Griffiths and Gaensler 2005). The event was so intense that caused a strong perturbation in the Earth ionosphere and saturated the detectors on every high-energy satellite. Few days after this event, SGR 1806–20 was detected in the radio band for the first time, providing a very accurate position (VLA; Cameron et al. 2005; Gaensler et al. 2005a). The radio polarization and flux decay were consistent with synchrotron radiation from an expanding nebula (Gaensler et al. 2005b).

Thanks to a Target of Opportunity (ToO) observational campaign on SGR 1806–20 carried out during 2004 with the ESO VLT we likely discovery of the IR counterpart to SGR 1806–20 based on positional coincidence with the radio position and flux variability. Moreover, based on serendipitous high-time resolution data obtained with the Rossi X-ray Timing Explorer (RXTE) Proportional Counter Array (PCA), we carried out the first detailed X-ray timing analysis of the 2004 December 27th hyperflare of SGR 1806–20, and we discovered rapid quasi periodic oscillations (QPOs) in its X-ray flux few minutes after the onset of the giant flare.

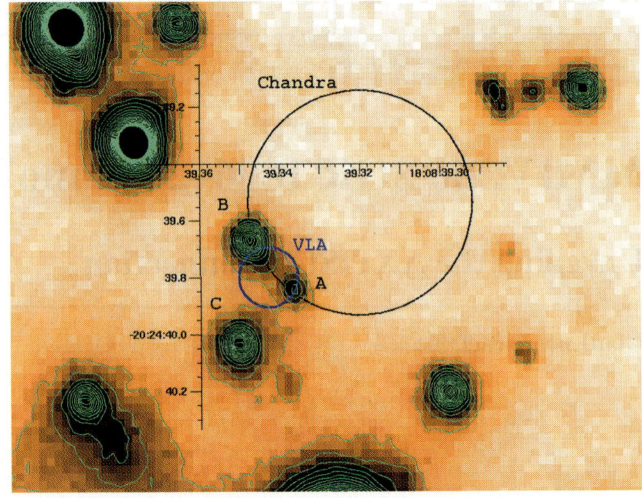

Fig. 1 NAOS-CONICA Ks band image of the 1.5″ × 1.5″ portion of the sky around the 1σ Chandra and VLA uncertainty circles (radius of 0.3″, 0.04″ and 0.1″, respectively) with the proposed counterpart marked with A

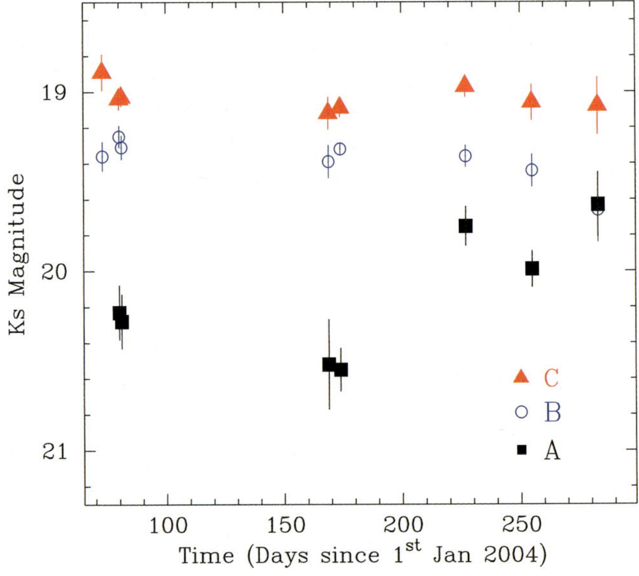

Fig. 2 NAOS-CONICA Ks light curves (*right panel*) of the proposed counterpart to SGR 1806–20 (*A*) with two nearby objects

2 VLT NAOS-CONICA IR observations

The data were acquired at VLT with the Nasmyth Adaptive Optics System and the High Resolution Near IR Camera (NAOS-CONICA). Data were reduced following standard procedures for photometry and astrometry (see Israel et al. 2005a for details of the observations and data reduction). Absolute photometry was derived by analysis of the best seeing frames, and cross-checked by means of archival ISAAC data of the same region and about 100 isolated stars taken from the 2MASS catalog and within the instrument FOV: the results were in agreement to within 0.05 Ks magnitudes. In particular, the final astrometric image accuracy is of ∼0.1″ (2MASS absolute accuracy included).

Source *A*, a relatively faint (Ks ∼ 20) object, at the sky position R.A. = $18^{h}\,08^{min}\,39.337''$, Dec. = $-20°\,24'\,39.85''$ (equinox 2000, 90% uncertainty of 0.06″), is found to be consistent with the Chandra and VLA positional uncertainty circles superimposed on our IR astrometry–corrected frame (see Fig. 1). Objects *B* and *C* (∼0.23″ and 0.27″ away from *A*, respectively) are only marginally consistent with the X-ray and radio positions, though statistically plausible. Light curves of the *A*, *B* and *C* objects are shown in Fig. 2. Candidate *A* is the only one showing a clear brightening (a factor of ∼2) in the IR flux between June and October 2004 (see Israel et al. 2005a for more details).

Both the XMM-Newton (Mereghetti et al. 2005a) and INTEGRAL (Mereghetti et al. 2005b) persistent fluxes of SGR 1806–20 showed an increase across the two semesters of 2004 by a factor of $1.94^{+0.01}_{-0.02}$ and $1.7^{+0.4}_{-0.3}$ in the 2–10 keV and 20–100 keV bands, respectively.[1] During the same time interval the NAOS-CONICA Ks flux increased by a factor of $2.4^{+0.9}_{-0.5}$, consistent with high energy flux variations, supporting the identification of object A as the correct IR counterpart of SGR 1806–20.

Independent from our work, the object *A* has been proposed as the IR counterpart to SGR 1806–20 (Kosugi et al. 2005; their object *B*3). A comparison of their photometry

[1] For the 2–10 keV 2004 first semester flux we assumed that of October 2003, based on the unvaried INTEGRAL flux between October 2003 and February–April 2004.

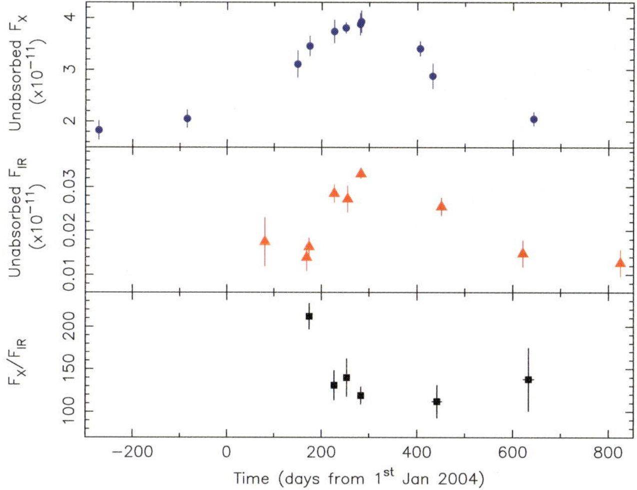

Fig. 3 The 2–10 keV XMM-Newton and Chandra (*upper panel*) and the NAOS-CONICA Ks (*central panel*) unabsorbed flux light curves of SGR 1806–20, together with the ratios between the two bands (*lower panel*; Israel et al. 2007)

with our shows that nearly all the Ks magnitudes have an offset of about 0.2, with the important exception of objects A and C which are 1.6 and 0.6 Ks magnitudes brighter than the corresponding objects $B3$ and $B1$ in Kosugi et al. (2005), respectively. Even though we do not have a clear explanation for the observed differences, we note that in our images we did not see any evidence for (i) a brightening of objects B and C, and (ii) an increase of the local background around object A (based on our best datasets with FWHM $\leq 0.1''$), in contrast to Kosugi et al. (2005).

An unusually high background level (regardless of its origin) may of course result in a flux underestimation of a source that lies in the same area.

During 2005 and 2006 we further monitor SGR 1806–20 both in the IR and X-ray bands by means of VLT (IR band), XMM-Newton and Chandra (X-rays): an updated light curve inferred from these instruments is shown in Fig. 3, together with the ratio between the Ks band and the 2–10 keV unabsorbed fluxes.

3 RXTE observation of the 27th December 2004 Giant Flare

The giant flare of SGR 1806–20 was serendipitously recorded at 21:31:30.7 UT on 2004 December 27th, during a RXTE observation of the cluster of galaxies A2163. The PCA was observing in its GoodXenon mode, resulting in full timing ($\sim$1 μs) covering the nominal energy range $\sim$2–120 keV. We restricted our timing analysis to the gap-free interval, starting 12.8 s after the onset of the initial spike. We accumulated a light curve from all PCA channels with a resolution of 1/256 s (a 0.5 s binned light curve is shown in Fig. 4).

In addition to complex signals below $\sim$30 Hz (see below) in the dynamical power spectrum density, a peak in the power spectra around 90 Hz was detected around 170–220 s from T_0 (see Fig. 5 and Israel et al. 2005b for the analysis details). Visual inspection of the 2 s spectrogram showed that this signal was significantly present only during a $\sim$50 s increase in the DC component underlying the $\sim$7.56 s pulsations in tail of the flare (*gray area* in Fig. 5b). The peak shown in Fig. 5d has a centroid frequency of 92.5 ± 0.2 Hz, a FWHM of $1.7^{+0.7}_{-0.4}$ Hz (uncertainties are at 1σ confidence level), and a signal coherence of $Q \sim 50$. The integrated fractional rms of the peak was $7.3 \pm 0.7\%$, and a chance probability, calculated by using the prescription of Israel and Stella (1996), of $\sim 1.5 \times 10^{-5}$–1.3×10^{-4} after normalization for the number of trial periods (128–1024) over which the search was carried out. We thus consider rather robust the 92.5 Hz QPO detection.

Selecting 3 phase intervals centered around the main peak at phase $\sim$0.4, the second peak at phase $\sim$0.7, and the minimum around phase 0, it is evident that the 92.5 Hz oscillations are absent in the main peak (3σ upper limit of 4.1% rms) clearly detected in the minima (rms amplitude of $10.7 \pm 1.2\%$) and still present, though at a lower significance, in the second peak (rms amplitude of $8.0 \pm 0.9\%$; see panels e, f and g of Fig. 5). The results of this pulse-phase timing analysis are summarised in panels e, f and g which show the power spectra for the main peak, the second peak and the minima, respectively. It is evident that the 92.5 Hz oscillations are absent in the main peak (3σ upper limit of 4.1% rms) clearly detected in the minima (rms amplitude of $10.7 \pm 1.2\%$) and still present, though at a lower significance, in the second peak (rms amplitude of $8.0 \pm 0.9\%$).

In order to check for correlations which might provide clues on the origin of the oscillations we study the 92.5 Hz QPOs with respect to the other timing parameters of the hyperflare. Figure 4 shows the result of this analysis where the PCA light curve is overimposed to some of these parameters: the DC level (gray filled circles), the count rates of the main peak (gray filled squares), of the second peak (light dark filled squares) and of all components together (filled stars). From the comparison between Figs. 5 and 4 it is evident that the time interval over which the 92.5 Hz QPOs are significantly detected coincides with a bell-shape enhancement in the DC level about 200 s after the beginning of the flare. Moreover, it is apparent that the intensity of the DC level and that of the main peak of the pulse are anti-correlated (the first peak component shows a marked decrease corresponding to the rise of the bell-shaped intensity bump of the DC), likely implying two distinct emission components. It is also worth noting that the hardness ratio (HR) between the 1–9 keV and 9–25 keV nominal energy band light curves shows that the pulse minima are al-

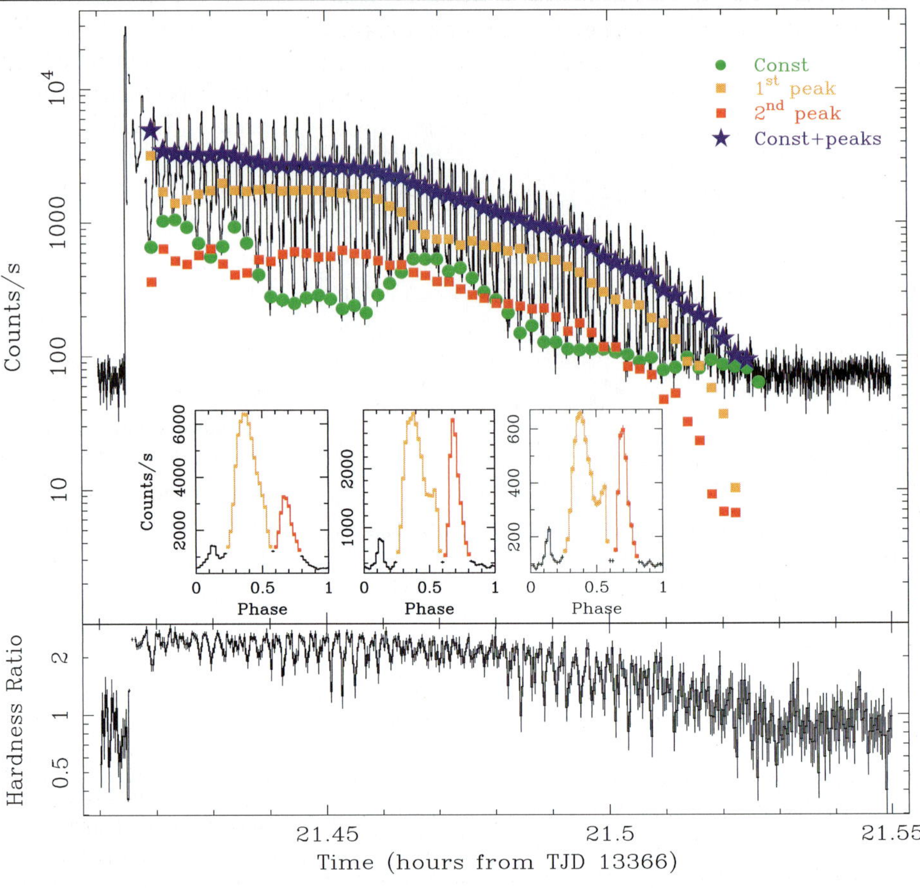

Fig. 4 The RXTE PCA 0.5 s binned light curve of the SGR 1806–20 hyperflare (*upper panel*). Overimposed are four parameters resulting from the fitting of each 7.56 s pulse with a constant plus five Gaussian model: the DC level (*gray filled circles*), the intensity of the main peak (*gray filled squares*), of the second peak (*light dark filled squares*), and of all components (*filled stars*). The *three insets* show the average pulse shapes as a function of time with the two main peak marked in *gray* and *light dark lines*. Time intervals over which the *light curve* have been folded are approximatively those where insets lie (before, during and after the gray region of Fig. 5b). *Lower panel* represents the PCA 9–25 keV over 1–9 keV energy band light curve ratio (see text for details)

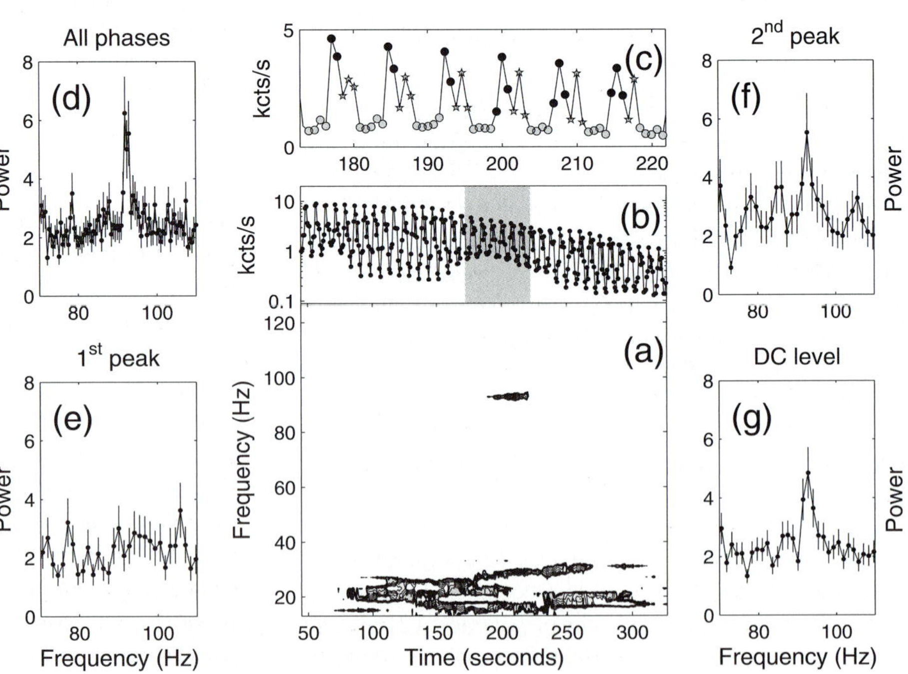

Fig. 5 The ~92 Hz oscillation as seen by the PCA. **a** Spectrogram with 2 s time step and resolution. The contours represent Leahy powers from 3.2 to 3.7; **b** *Light curve* corresponding to the same time axis as panel (**a**). The time resolution is 0.75 s. The *gray-shaded area* indicates the time interval shown in panel (**c**); **c** Close-up of the gray area in panel (**b**). The different symbols mark the first peak (*black circles*), the second peak (*stars*) and the DC level (*gray circles*); **d** average power spectrum (at 0.5 Hz resolution) of the gray area in panel (**b**); **e** average power spectrum (at 1.33 Hz resolution) of the phase interval including the first peak as seen in panel (**c**); **f** same as (**e**) for the second peak; **g** same as (**e**) for the DC level

ways softer than peaks, except during the bell-shaped bump where minima are as hard as the maxima (see lowest panel of Fig. 4). These findings clearly suggest that the bell-shaped bump in the decaying pulsating tail of the SGR 1806–20 hyperflare represents an additional unpulsed component underlying the main pulse component.

From Fig. 5a, one can see also a significant amount of power in the 20–30 Hz interval. The structure of this signal is complex and the accumulation of all power spectra leads only to a broad excess. Therefore, we subdivided the light curve into smaller intervals and checked for significant signal below 40 Hz. A fit to the continuum (constant plus a power-law) yields two QPOs at 18.1 ± 0.3 Hz and 30.4 ± 0.3 Hz with a single trial significance of 3.6 and 4.7σ, respectively. A weak excess is also visible at $\sim$95 Hz, indicating a possible time evolution of the $\sim$92.5 Hz QPO frequency. The relatively low statistics in the time intervals $\sim$200–300 s prevented us from checking whether these additional QPOs show a 7.56 s pulse phase dependence like that displayed by the 92.5 Hz QPO.

4 Discussion

4.1 The IR flux of SGR 1806–20 and the comparison with AXPs and radio pulsars

The high spatial resolution of the NAOS-CONICA images allowed us to identify the likely IR counterpart of SGR 1806–20, and to monitor its IR flux for seven months in 2004, during which an increase by a factor of $\sim$2 was detected, correlated with the flux in the high energy bands.

IR variability has been detected in nearly all AXPs with known IR counterpart. In particular, for 1E 1048.1–5937, XTE J1810–197 and 2E 2259+586, IR variability has been found, or suspected, to be correlated with the persistent X-ray emission (Israel et al. 2002; Rea et al. 2004; Tam et al. 2004). Based on the NACO results we can conclude that the IR/X-ray correlation observed in AXPs also holds for SGR 1806–20. The total fluency of the IR enhancement between June and October 2004 is about 10^{41} erg (we assumed $A_V = 29 \pm 2$; see Eikenberry et al. 2004), a factor of about 100 smaller than that in the 2–10 keV band.

Based on the above reported findings we note that the SGR 1806–20 emission varies in a similar fashion (in terms of timescale and amplitude of variation) over more than five orders of magnitude in photon energy. The similar flux variation in the IR and X-ray bands suggests that the emission in the two bands has a similar, if not the same, origin. Moreover, it has become evident that X-ray flux enhancement of the persistent emission of SGRs is correlated with their burst rate, making it difficult to compare the fluxes among different SGRs without knowing their burst history (see Woods and Thompson 2006). Tam et al. (2004) argued that IR thermal surface emission (within the magnetar model) is ruled out during the correlated X-ray/IR flux decay phases of 1E 2259+586 (implausibly high implied brightness temperature), suggesting the magnetospheric origin for the IR enhancement. Alternatively, the IR flux can be due to re irradiation by material in the vicinity of the pulsar. This model naturally predicts a correlation between the IR and the X-ray flux (Perna et al. 2000; Rea et al. 2004).

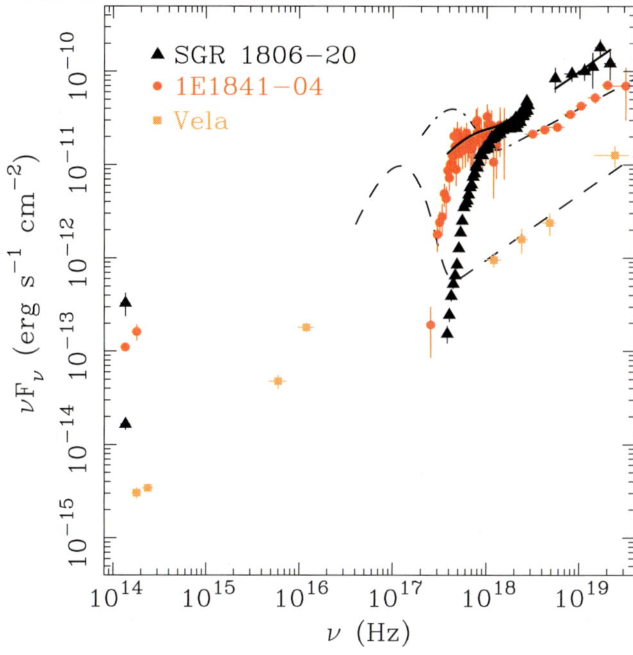

Fig. 6 Broad band energy spectrum of SGR 1806–20 (*triangles*) and, as a comparison, the AXP 1E 1841−045 (*circles*) and the radio pulsar Vela (*squares*). In the case of SGR 1806–20 and 1E 1841−045 high energy data are taken from XMM-Newton (Mereghetti et al. 2005b; referring to 6th October 2004 for SGR 1806–20), Chandra (for 1E 1841−045) and INTEGRAL (Mereghetti et al. 2005b; 21st September–14th October 2004 for SGR 1806–20). Absorbed and unabsorbed IR fluxes ($A_V = 29 \pm 2$, 5th October 2004 NACO observation) are shown in the case of SGR 1806–20, unabsorbed ($A_V = 13 \pm 1$) IR fluxes are instead reported for the likely candidate of 1E 1841−045 (*circles*; Israel et al. 2007). All the data for Vela are taken form literature (Kaspi et al. 2006 and references therein). *Solid curves* (*continuous*, *stepped* and *dot-stepped*) are the unabsorbed fluxes for the black body plus power law model used to fit the high energy part of spectra

This is the first time that the broad band energy properties of an SGR can be compared, over a similar energy band, with those of other classes of isolated neutron stars, such as AXPs and radio pulsars. In Fig. 6 we show the "nearly simultaneous" broad band energy spectrum of SGR 1806–20 from the IR to γ rays (high energy data are taken form Mereghetti et al. 2005b; see caption for details). The high energy part of the spectrum is clearly consistent with being non-thermal emission (a power-law model is generally used) from the source. We also plot the spectrum from the AXP 1E 1841−045, for which 20–200 keV band data are available (Kuiper et al. 2004); a similar non-thermal component is displayed by the source. Non-thermal components are also seen in radio pulsars and modelled with power-law components (see Kaspi et al. 2006 for a recent review). In some cases there is a smooth connection between optical, X-rays and γ-ray emission (Crab), while in other cases the extrapolation is plausible (Vela; see Fig. 6). It is worth noting the similar flux ratios in the IR and hard X-ray bands for the three objects, and the significant difference of the char-

acteristic temperature of thermal soft X-ray components between radio pulsars ($\leq$0.1 keV) and SGRs/AXPs (0.4–0.8 keV for a BB fit and 0.2–0.5 keV for a magnetic atmosphere fit, Perna et al. 2000), suggesting a significantly larger energy injection on the neutron star surface in "magnetar" candidates than in radio pulsars. Looking at Fig. 3 it also evident that both the rising and decaying IR and soft X-ray flux phases (before and after the giant flare) are correlated (so far only the decaying phases of AXPs were monitored). Moreover, it is evident that SGR 1806–20 seems to be intrinsically more IR bright than the AXPs where the F_{IR}/F_X is of the order of 10^3 (Israel et al. 2007).

Future detailed multi-wavelength observations campaigns of AXPs and SGRs will likely help clarifying the link between IR and high energy bands. Furthermore, the detection of the quiescent IR flux level of SGR 1806–20 will allow to compare the net energy released by the source in the IR and X-ray/γ-ray bands during its bursting active phase.

4.2 The X-ray fast oscillations and their nature

We discovered rapid quasi periodic X-ray oscillations in the evolving X-ray flux of the 2004 December 27th hyperflare of SGR 1806–20, the first ever for a magnetar candidate. The higher frequency QPOs at $\sim$92.5 Hz were detected in association with an emission bump that occurred in the DC component. These QPOs were detected only in the spin phase intervals away from the main peak and reached maximum amplitude corresponding to the DC component phase intervals. Evidence for $\sim$18 and $\sim$30 Hz QPOs was found between 200 and 300 s from the onset of the hyperflare.

In the context of the magnetar scenario, the main spike of the giant flare arises from a fireball of pair-dominated plasma that expands at relativistic speeds, while the energy deposited in the magnetosphere can give rise to a "trapped fireball" that remains confined to the star's closed magnetic field lines. The long pulsating tail of giant flares probably arises from the cooling of plasma that remains confined in such a trapped fireball. The bump in the DC component of the decay some 200 s after the main spike during which the $\sim$92.5 Hz QPOs were seen, might be due to a temporary enhancement of the Alfvén wave emission due to dissipation of the seismic energy. This may lead to the formation of a hot pair-dominated corona which partially enshrouds the trapped fireball, thereby reducing the amplitude of main peak of the 7.56 s modulation. A similar interpretation was originally proposed by Feroci et al. (2001) (see also Thompson and Duncan 2001) to explain the smooth flux decay in the first tens of seconds of the 1998 giant flare from SGR 1900+14. Indeed, a variety of seismic modes are expected to be excited as consequence of the magnetically-induced large scale fracturing of the crust which gives rise to giant SGR flares.

Out of the variety of non-radial neutron star modes studied by McDermott et al. (1998), there are several classes that have characteristic frequencies in the $\sim$10–100 Hz range. Toroidal modes appear to be especially promising because they should be easily excited by the large crustal fracturing. Moreover, these torsional modes couple more easily with the external magnetic field lines than modes originating deeper in the stellar interior (Blaes et al. 1989; Duncan 1998). The fundamental toroidal mode of a rigid neutron star's crust corresponds to a period of $\sim$33.6 ms, somewhat dependent on the mass, radius and crustal magnetic field (McDermott et al. 1998; Duncan 1998). Therefore, the 30.4 Hz ($\sim$32.8 ms) oscillation could be easily identified with the ${}_2t_0$ mode, and the 92.5 Hz ($\sim$10.8 ms) QPO would thus correspond to a higher harmonic: indeed, it matches well the expected frequency of the $l = 7$ mode, suggesting a relatively small-scale structure in the seismic wave pattern (and thus the magnetic multipole structure). This could be due, for example, to the principal mode inducing further fractures at various sites in the crust. The shortest duration of the higher frequency QPO is qualitatively in accord with the expectation that the damping rate of the oscillations strongly increases with frequency (Duncan 1998). A large ($\sim$5 km) crustal fracturing on the surface of SGR 1806–20 was inferred from a $\sim$5 ms rise timescale observed during the onset of the hyperflare (Schwartz et al. 2005). We note that such fracturing can easily excite the toroidal modes with characteristic frequencies at which QPOs have been detected.

The 18 Hz oscillation, on the other hand, might be associated with a different mode which must couple to the magnetosphere as well. A poloidal component of the core magnetic field supports a torsional mode with a frequency $\nu_{core} \simeq 2.5 B_{z,15}$ Hz (Thompson and Duncan 2001) with $R \sim 10$ km and a core density $\simeq 10^{15}$ g cm^{-3}, $B_{z,15}$ being the core poloidal field in units of 10^{15} G: a strong $B_{z,15} \simeq 7$ would be required to match the observed 18 Hz. Although extremely strong, such a field is fully plausible given that a (mainly toroidal) field in excess of 10^{16} G is required to power the SGR 1806–20 flux during its whole life if one speculate that repeated giant flares of this magnitude over the $\sim 10^4$ yr lifetime occurred in the past.

Motivated by our discovery of QPOs (or GSOs) in the X-ray flux of SGR 1806–20, Strohmayer and Watts (2005) looked for such kind of QPOs in the X-ray flux of the giant flare of SGR 1900+14, serendipitously recorded again by RXTE on 28th August 1998. They detected transient $\sim$ 84 Hz oscillations during a 1 s interval beginning approximatively 60 s after the initial spike. Moreover, additional QPOs at 28, 53.3 and 155.1 Hz were detected in the average power spectrum in time intervals centered on the rotational phase at which the 85 Hz signal was detected (Strohmayer and Watts 2005). Indeed, an association of these frequencies with $l = 2$, 4, 7 and 13 toroidal modes appear plausible and suggest a common origin with those detected in SGR 1806–20.

Moreover, Watts and Strohmayer (2006) using timing data from RHESSI, have confirmed our detection of the 92.5 Hz QPOs in the tail of the giant flare of SGR 1806–20. Intriguingly, they also found another, stronger, QPO at higher energies, at 626.5 Hz. Both QPOs are visible at particular (but different) rotational phases, implying an association with a specific area of the neutron star surface or magnetosphere (Watts and Strohmayer 2006). The most likely crust mode candidate for the latter QPOs is the $n = 1$ toroidal modes, the frequency of which agrees extremely well with the most recent models (Piro 2005). A re-analysis of the RXTE data of the giant flare of SGR 1806–20 allowed to detect the 626.5 Hz oscillations though at a slightly different frequency (see Watts, this book proceedings, for more details).

In conclusion, the detection of QPOs from these sources has opened a new observational windows for the study of neutron stars. Much theoretical effort is clearly needed, though, in the simplest interpretation, the toroidal modes seems to already fit well the theoretical expectations.

Acknowledgements We thank the ESO Director's Discretionary Time Committee for accepting the observation of SGR 1806–20 few hours after the 5th October 2004 intense X-ray burst. We are also indebted with VLT personnel for their continuous help in optimising and performing the NACO observations. We thanks D. Dobrzycka and W. Hummel for their help in analysing NAOS-CONICA images. This work is partially supported through ASI and Ministero dell'Istruzione, Università e Ricerca Scientifica e Tecnologica (MIUR–COFIN), and Istituto Nazionale di Astrofisica (INAF) grants. N.R. is supported by a Marie Curie Training Grant (HPMT-CT-2001-00245). The findings here reported have been obtained thanks to the work of a large number of people. Below is the complete list of the team: T. Belloni, S. Covino and S. Campana (INAF, Osservatorio Astronomico di Brera), L. Stella, S. Dall'Osso and V. Testa (INAF, Osservatorio Astronomico di Roma), P. Casella (University of Amsterdam), Elisa Nichelli (INAF, Osservatorio Astronomico di Roma and University of Roma "Tor Vergata"), Y. Rephaeli (School of Physics and Astronomy, Tel Aviv University and Center for Astrophysics and Space Sciences, University of California), D.E. Gruber (Eureka Scientific Corporation), N. Rea (SRON, National Institute for Space Research), M. Persic (INAF, Osservatorio Astronomico di Trieste), R.E. Rothschild (Center for Astrophysics and Space Sciences, University of California), R. Mignani (Mullard Space Science Laboratory), G. Marconi and G. Lo Curto (European Southern Observatory), S. Mereghetti and D. Götz (INAF, Istituto di Astrofisica Spaziale e Fisica Cosmica "G. Occhialini"), and Rosalba Perna (Department of Astrophysical and Planetary Sciences and JILA, University of Colorado).

References

Blaes, O., Blandford, R.D., Goldreich, P., et al.: Astrophys. J. **343**, 839 (1989)

Borkowski, J., Götz, D., Mereghetti, S., et al.: GCN **2920** (2004)

Cameron, P.B., Chandra, P., Ray, A., et al.: Nature **434**, 1112 (2005)

Duncan, R.C.: Astrophys. J. **498**, L45 (1998)

Duncan, R.C., Thompson, C.: Astrophys. J.: **392**, L9 (1992)

Eikenberry, S.S., Matthews, K., LaVine, J.L., et al.: Astrophys. J. **616**, 506 (2004)

Feroci, M., Hurley, K., Duncan, R.C., et al.: Astrophys. J. **549**, 1021 (2001)

Gaensler, B.M., Kouveliotou, C., Garrett, M., et al.: GCN **#2929** (2005a)

Gaensler, B.M., Kouveliotou, C., Gelfand, J.D., et al.: Nature **434**, 1104 (2005b)

Israel, G.L., Stella, L.: Astrophys. J. **468**, 369 (1996)

Israel, G.L., Covino, S., Stella, L., et al.: Astrophys. J. **580**, L143 (2002)

Israel, G.L., Belloni, T., Stella, L., et al.: Astrophys. J. **628**, L53 (2005a)

Israel, G.L., Covino, S., Mignani, R., et al.: Astron. Astrophys. **438**, L1 (2005b)

Israel, G.L., et al.: In preparation (2007)

Kaspi, M.V., Roberts, M.S.E., Harding, A.: In: Compact Stellar X-ray Sources, p. 279. Cambridge University Press, Cambridge (2006) (see also astro-ph/0406133)

Kosugi, G., Ogasawara, R., Terada, H.: Astrophys. J. **623**, L125 (2005)

Kuiper, L., Hermsen, W., Mendez, M.: Astrophys. J. **613**, 1173 (2004)

McClure-Griffiths, N.M., Gaensler, B.M.: Astrophys. J. **630**, L161 (2005), astro-ph/0503171

McDermott, P.N., Van Horn, H.M., Hansen, C.J.: Astrophys. J. **325**, 725 (1988)

Mereghetti, S., Götz, D., Andersen, M.I., et al.: Astron. Astrophys. **433**, L9 (2005a)

Mereghetti, S., Tiengo, A., Esposito, P., et al.: Astrophys. J. **628**, 938 (2005b)

Perna, R., Hernquist, L., Narayan, R.: Astrophys. J. **541**, 344 (2000)

Piro, A.L.: Astrophys. J. **634**, L153 (2005)

Rea, N., Testa, V., Israel, G.L., et al.: Astron. Astrophys. **425**, L5 (2004)

Schwartz, S.J., Zane, S., Wilson, R.J., et al.: Astrophys. J. **627**, L129 (2005)

Tam, C.R., Kaspi, V.M., van Kerkwijk, M.H., et al.: Astrophys. J. **617**, L53 (2004)

Thompson, C., Duncan, R.C.: Mon. Not. Roy. Astron. Soc. **275**, 255 (1995)

Thompson, C., Duncan, R.C.: Astrophys. J. **561**, 980 (2001)

Strohmayer, T.E., Watts, A.L.: Astrophys. J. **632**, L111 (2005)

Watts, A.L., Strohmayer, T.E.: Astrophys. J. **637**, L117 (2006)

Woods, P., Thompson, C.: In: Compact Stellar X-ray Sources, p. 547. Cambridge University Press, Cambridge (2006) (see also astro-ph/0406133)

Astrophys Space Sci (2007) 308: 33–37
DOI 10.1007/s10509-007-9300-7

ORIGINAL ARTICLE

Long term spectral variability in the soft gamma-ray repeater SGR 1900+14

Andrea Tiengo · Paolo Esposito · Sandro Mereghetti · Lara Sidoli · Diego Götz · Marco Feroci · Roberto Turolla · Silvia Zane · Gian Luca Israel · Luigi Stella · Peter Woods

Received: 7 July 2006 / Accepted: 29 August 2006 / Published online: 23 March 2007

Abstract We present a systematic analysis of all the BeppoSAX data of SGR1900+14. The observations spanning five years show that the source was brighter than usual on two occasions: ~20 days after the August 1998 giant flare and during the 10^5 s long X-ray afterglow following the April 2001 intermediate flare. In the latter case, we explore the possibility of describing the observed short term spectral evolution only with a change of the temperature of the blackbody component. In the only BeppoSAX observation performed before the giant flare, the spectrum of the SGR1900+14 persistent emission was significantly harder and detected also above 10 keV with the PDS instrument. In the last BeppoSAX observation (April 2002) the flux was at least a factor 1.2 below the historical level, suggesting that the source was entering a quiescent period.

Keywords Stars: individual (SGR 1900+14) · Stars: neutron · X-rays: stars

PACS 97.60.Jd · 98.70.Qy

A. Tiengo (✉) · P. Esposito · S. Mereghetti · L. Sidoli
INAF-IASF Milano, via Bassini 15, I-20133 Milan, Italy
e-mail: tiengo@iasf-milano.inaf.it

P. Esposito
Dipartimento di Fisica Nucleare e Teorica, Università di Pavia, and INFN-Pavia, via Bassi 6, I-27100 Pavia, Italy

D. Götz
CEA Saclay, DSM/DAPNIA/Service d'Astrophysique, F-91191 Gif-sur-Yvette, France

M. Feroci
INAF-IASF Roma, via Fosso del Cavaliere 100, I-00133 Roma, Italy

R. Turolla
Dipartimento di Fisica, Università di Padova, via Marzolo 8, I-35131 Padova, Italy

S. Zane
Mullard Space Science Laboratory, University College London, Holmbury St. Mary, Dorking Surrey RH5 6NT, United Kingdom

G.L. Israel · L. Stella
INAF, Osservatorio Astronomico di Roma, via Frascati 33, I-00040 Monteporzio Catone, Italy

P. Woods
Dynetics, Inc., 1000 Explorer Boulevard, Huntsville, AL 35806, USA

1 Introduction

The soft gamma-ray repeater (SGR) SGR 1900+14 was discovered in 1979 through a series of short and soft gamma-ray bursts (Mazets et al. 1979). Many years later, its persistent pulsating X-ray counterpart was discovered in the 2–10 keV energy band (Hurley et al. 1999b). More recently, it was also detected in the hard X-ray range (20–100 keV) with the INTEGRAL satellite, becoming the second SGR with a measurable hard X-ray tail in the spectrum of its persistent emission (Götz et al. 2006).

The bursting activity of SGR 1900+14 is rather discontinuous (see Fig. 1, bottom panel) and culminated on 1998 August 27 with the emission of a Giant Flare, when more than 10^{44} ergs of γ-rays were emitted in less than one second (Hurley et al. 1999a; Mazets et al. 1999; Feroci et al. 2001). This was one of the three Giant Flares detected up to now from three different SGRs and was interpreted as a new evidence in favor of the *magnetar* model. In this model (Thompson and Duncan 1995, 1996), the SGRs and the Anomalous X-ray Pulsars (AXPs, another class of X-ray

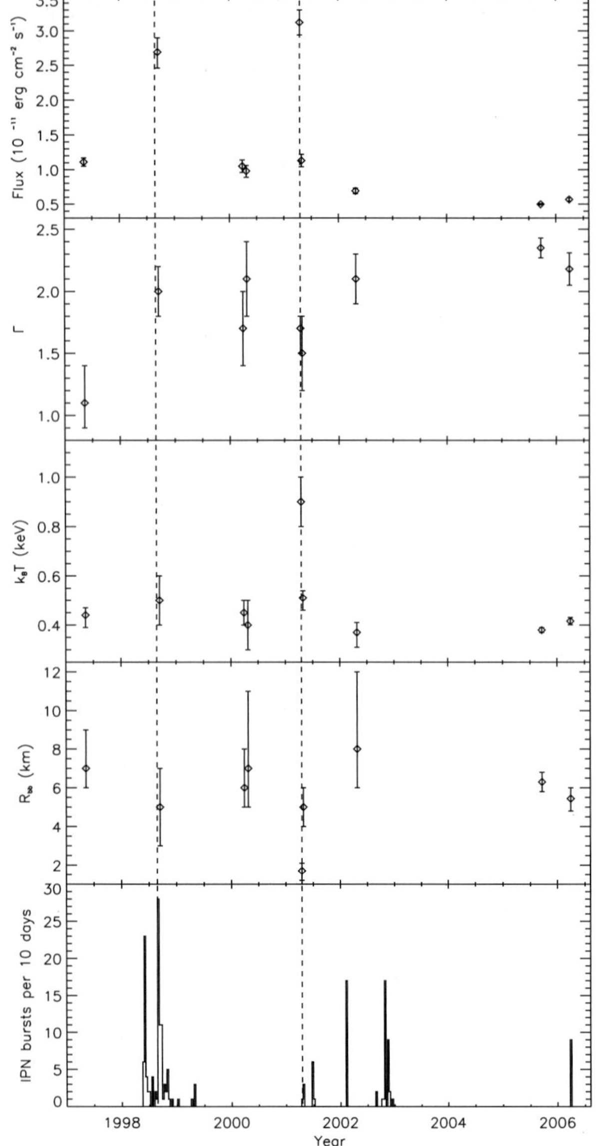

Fig. 1 Long term evolution of the 2–10 keV unabsorbed flux, the spectral parameters (for an absorbed power-law plus blackbody model, assuming $N_{\mathrm{H}} = 2.55 \times 10^{22}$ cm^{-2}) and the burst activity (as observed by interplanetary network) of SGR 1900+14. The *vertical dashed lines* indicate the times of the Giant and Intermediate Flare (1998 August 27 and 2001 April 18, respectively). We also plotted the 2005–2006 points obtained with XMM-Newton observations (Mereghetti et al. 2006)

Table 1 Log of the BeppoSAX observations of SGR 1900+14

Obs	Date	Instrument	Exp. time
A	1997-05-12	SAX/MECS	46 ks
		SAX/PDS	20 ks
B	1998-09-15	SAX/MECS	33 ks
		SAX/PDS	16 ks
C	2000-03-30	SAX/MECS	40 ks
		SAX/PDS	18 ks
D	2000-04-25	SAX/MECS	40 ks
		SAX/PDS	19 ks
E	2001-04-18	SAX/MECS	46 ks
		SAX/PDS	17 ks
F	2001-04-29	SAX/MECS	58 ks
		SAX/PDS	26 ks
G	2002-03-09	SAX/PDS	48 ks
H	2002-04-27	SAX/MECS	83 ks

sources with similar properties, Mereghetti et al. 2002) are believed to be neutron stars powered by the decay of their extremely intense magnetic field ($B \sim 10^{14}$–10^{15} G).

Here we present the analysis of the persistent emission of SGR 1900+14 both in the soft and hard X-ray range and its evolution across the Giant Flare and in relation to its bursting activity.

2 Soft X-ray emission

2.1 Observations and data analysis

We have analyzed the X-ray observations of SGR 1900+14 performed with the BeppoSAX satellite (see Table 1).

The spectra were extracted from the MECS (Boella et al. 1997) and LECS (Parmar et al. 1997) instruments using circles with radii 4′ and 8′, respectively. The background spectra were extracted in all cases from nearby regions. Time filters were applied to both the source and background spectra to exclude the SGR bursts detected during observation A, B and F. The standard response matrices were used for the MECS and LECS spectra.

2.2 Spectral results

We have first tried to fit the spectra with an absorbed power-law model, but three observations give unacceptable values of the χ^2 and structured residuals. For these observations, a good fit is obtained with the addition of a blackbody component. Since such a two-components model is typical of the magnetar candidates (Woods and Thompson 2004), we have used this model to fit all the available spectra, obtaining the results reported in Fig. 1. As can be seen in the upper panel, the flux varies by a factor >5, with the highest values observed during observations B and E. These two observations were taken shortly after extreme bursting events. The former was performed 20 days after the Giant Flare, that was followed by a ∼2 months period of enhanced X-ray flux (Woods et al. 1999). The latter started only 7.5 hours after the 2001 April 18 Intermediate Flare which had a fluence ∼20 times lower than that of the Giant Flare (Feroci et al. 2003). The afterglow following this bright burst is clearly visible during the BeppoSAX observation as a decrease in the X-ray flux, accompanied by a significant softening of the spectrum (Feroci et al. 2003). In addition to the afterglow analysis already reported by Feroci et al. (2003), we have performed a time resolved spectroscopy of the afterglow by dividing observation E into 5 time intervals. Although the 5 spectra can be fitted by a variety of models, the spectral evolution of the afterglow is well represented by an additional blackbody component with fixed emitting area (∼1.5 km, for a source distance of 15 kpc) and progressively decreasing temperature (k_BT∼1.3–0.9 keV), that can be interpreted as due to a portion of the neutron star surface heated during the flare.

Excluding the two observations taken after the exceptional explosive events (B and E), the flux of SGR 1900+14 had a rather constant value of $\sim 10^{-11}$ erg cm^{-2} s^{-1} from 1997 to 2001. On the other hand a significantly lower flux level was seen in the following observations. The flux decrease actually started when the source was still moderately active (the flux in observation H is at least 1.2 times lower than in all the previous quiescent observations) and has been interrupted by a slight rise in coincidence with the March 2006 burst reactivation, as shown by recent XMM-Newton observations (Mereghetti et al. 2006).

Although the flux of the only pre-Giant Flare observation is compatible with that of the quiescent post-flare observations taken before 2002, its spectrum is significantly harder, as indicated by the comparison of the photon indexes plotted in the second panel of Fig. 1. The spectral change between observation A and the following quiescent observations is well illustrated by the clear trend of the residuals from the simultaneous fit of the spectra of observations A, C and D (Fig. 2).

The blackbody parameters are compatible in all the available observations, except for that taken during the afterglow

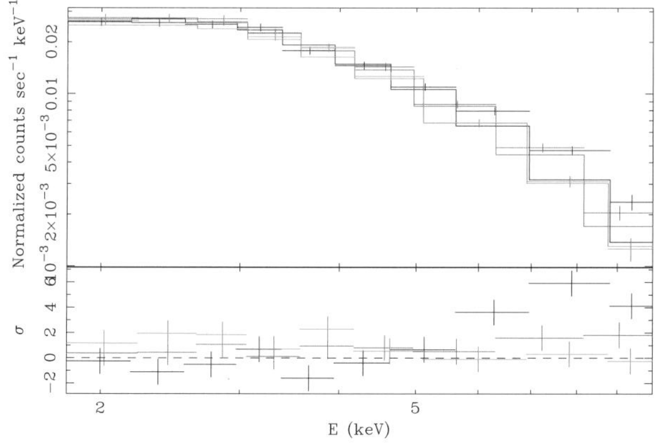

Fig. 2 BeppoSAX/MECS spectra of observations A (*black*), C (*red*) and D (*green*) simultaneously fit, with the same parameters, to an absorbed power-law plus blackbody model. The data have been rebinned graphically to emphasize the trend in the spectral residuals of observation A

of the Intermediate Flare (observation E). This indicates that a constant blackbody component with $k_BT \sim 0.4$ keV and emitting area with $R \sim 6$–7 km might be a permanent feature of the X-ray spectrum of SGR 1900+14.

3 Hard X-ray emission

3.1 Detection with the PDS instrument

To study the high energy emission from SGR 1900+14 we used the BeppoSAX PDS instrument, which operated in the 15–300 keV range. The PDS instrument was more sensitive than INTEGRAL in this energy band, but it had no imaging capabilities and therefore the possible contamination from nearby sources must be taken into account. The field of view (FoV) of the PDS instrument was 1.3° (FWHM) and the background subtraction was performed through a rocking system that pointed to two 3.5° offset positions every 96 s. In the case of SGR 1900+14, the background pointings were free of contaminating sources, as confirmed by the identical count rates observed in the two offset positions during each observation. The field of SGR 1900+14 is instead rather crowded, with three transient sources, the X-ray pulsars 4U 1907+97 (Giacconi et al. 1971; Liu et al. 2000) and XTE J1906+09 (Marsden et al. 1998), and the black hole candidate XTE J1908+94 (in't Zand et al. 2002), located at angular distances of 47′, 33′ and 24′ from the SGR, respectively. The pulsations of the two pulsars are clearly visible in the PDS data below 50 keV when they are active, while XTE J1908+94, if in outburst, is clearly visible in the simultaneous MECS and LECS images and, being very bright, also in the lightcurve collected by the All Sky Monitor (ASM) on board the RossiXTE satellite. We have found that at least one of these contaminating sources was on in all the BeppoSAX observations except for the first one (see Table 2). Thus, only the 1997 observation (obs. A), during which a significant signal was detected in the background subtracted PDS data, can be used to study SGR 1900+14 without the problem of known contaminating sources.

We searched for the SGR pulsation period (5.15719 s, as measured in the simultaneous MECS data) in the PDS data, but the result was not conclusive, giving only a 3σ upper limit of 50% to the pulsed fraction of a sinusoidal periodicity, to be compared to the ∼20% pulsed fraction observed below 10 keV.

3.2 Spectral analysis

The background subtracted PDS spectrum of observation A can be well fit by a power-law with photon index $\Gamma = 1.6 \pm 0.3$, significantly flatter than that measured by INTEGRAL ($\Gamma = 3.1 \pm 0.5$, see Fig. 3) in 2003/2004. The

Table 2 Status of the three transient sources within the PDS field of view during the BeppoSAX observations. The presence of the two X-ray pulsars (4U 1907+97 and XTE J1906+09) is confirmed by the presence of their pulsations in the PDS data, while that of the black hole candidate (XTE J1908+94) by its detection in the MECS and LECS images and in the RossiXTE ASM lightcurve

Obs	4U 1907+97	XTE J1906+09	XTE J1908+94
A	OFF	OFF	OFF
B	ON	ON	OFF
C	OFF	ON	OFF
D	ON	ON	OFF
E	OFF	ON	OFF
F	ON	ON	OFF
G	OFF	ON	ON

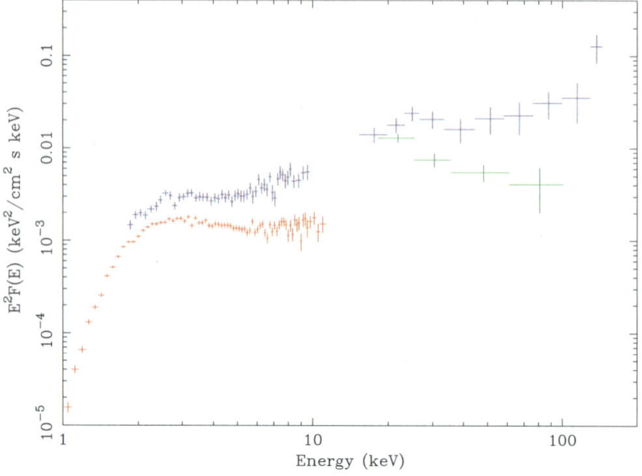

Fig. 3 Broad band spectra of SGR 1900+14 taken on 1997 May 12 (observation A) with BeppoSAX (*blue*, both MECS and PDS data), on 2005 September 20–22 by XMM-Newton (*red*) and from March 2003 to June 2004 with INTEGRAL

corresponding 20–100 keV flux is 6×10^{-11} erg cm^{-2} s^{-1}, a factor $\sim$4 higher than during the INTEGRAL observations, which confirms that before the Giant Flare the hard X-ray tail of SGR 1900+14 was brighter.

The INTEGRAL spectrum was collected during $\sim$2.5 Ms of different observations performed between March 2003 and June 2004, and thus it represents the hard X-ray emission of SGR 1900+14 averaged over that long time period. Therefore, its relation to the soft X-ray spectrum can be studied only comparing the spectra taken by other instruments in a similar time period, as shown for example in Fig. 3. The PDS instrument, instead, being a high sensitivity hard X-ray detector coupled to the MECS and LECS soft X-ray cameras, gives us the chance to study the broad band spectrum of SGR 1900+14 during a single observation. Fitting the 1–150 keV BeppoSAX spectrum of observation A, we obtain a good result ($\chi^2 = 1.17$ for 136 degrees of freedom) simply extrapolating to higher energies the best-fit model found in the soft X-ray range. In fact, a fit with an absorbed power-law plus blackbody model gives the following parameters: photon index $\Gamma = 1.04 \pm 0.08$, blackbody temperature $k_B T = 0.50 \pm 0.06$, radius $R_{bb} = 5 \pm 2$ km, and absorption $N_{\rm H} = (1.8 \pm 0.5) \times 10^{22}$ cm^{-2}.

4 Conclusions

We have studied the variability of SGR 1900+14, both in the hard and in the soft X-ray range, finding the following results:

- Except for the observations immediately following exceptional flares, the flux level in soft X-rays was stable while the source was moderately active and progressively decreased when it entered a 3 years long quiescent period.
- The Intermediate Flare of 2001 April 18 was followed by an X-ray afterglow that can be successfully interpreted as due to the heating of a significant fraction of the neutron star surface, that then cools down in $\sim$1 day. This is consistent with the interpretation of similar events in other magnetar candidates (Thompson et al. 2002).
- The soft X-ray spectrum during the only available pre-flare observation was harder than in the following quiescent observations. This is similar to what observed in SGR 1806–20, the only other SGR that could be monitored before and after a Giant Flare (Mereghetti et al. 2005; Rea et al. 2005; Tiengo et al. 2005).
- Comparing the hard X-ray spectrum of SGR 1900+14 recently observed with INTEGRAL to that observed with the PDS instrument in 1997, we find evidence for variations in flux and spectral slope.
- The reduction of the X-ray tail in coincidence with the Giant Flare is supported by the count rates detected in the PDS instrument above 50 keV during the different BeppoSAX observations, that indicate how they significantly decreased already in the first post-flare observation. Since the hard X-ray tail in the spectrum of SGR 1900+14 might contain most of its total emitted energy, its variability in relation to the bursting activity is a key point to try to understand the SGR emission processes.

References

Boella, G., Chiappetti, L., Conti, G., et al.: Astron. Astrophys. Suppl. Ser. **122**, 327 (1997)

Feroci, M., Hurley, K., Duncan, R.C., et al.: Astrophys. J. **549**, 1021 (2001)

Feroci, M., Mereghetti, S., Woods, P., et al.: Astrophys. J. **596**, 470 (2003)

Giacconi, R., Kellogg, E., Gorenstein, P., et al.: Astrophys. J. **165**, L27 (1971)

Götz, D., Mereghetti, S., Tiengo, A., et al.: Astron. Astrophys. **449**, L31 (2006)
Hurley, K., Cline, T., Mazets, E., et al.: Nature **397**, 41 (1999a)
Hurley, K., Kouveliotou, C., Woods, P., et al.: Astrophys. J. **510**, L107 (1999b)
in't Zand, J.J.M., Miller, J.M., Oosterbroek, T., et al.: Astron. Astrophys. **394**, 553 (2002)
Liu, Q.Z., van Paradijs, J., van den Heuvel, E.P.J.: Astron. Astrophys. Suppl. Ser. **147**, 25 (2000)
Marsden, D., Gruber, D.E., Heindl, W.A., et al.: Astrophys. J. **502**, L129 (1998)
Mazets, E.P., Cline, T.L., Aptekar, R.L., et al.: Astron. Lett. **25**, 628 (1999)
Mazets, E.P., Golenetskii, S.V., Guryan, Y.A.: Sov. Astron. Lett. **5**, 343 (1979)
Mereghetti, S., Chiarlone, L., Israel, G.L., et al. In: Becker, W., Lesch, H., Trümper, J. (eds.) Neutron Stars, Pulsars, and Supernova Remnants (2002), ArXiv: astro-ph/0205122
Mereghetti, S., Tiengo, A., Esposito, P., et al.: Astrophys. J. **628**, 938 (2005)
Mereghetti, S., Esposito, P., Tiengo, A., et al.: Astrophys. J. **653**, 1423 (2006)
Parmar, A.N., Martin, D.D.E., Bavdaz, M., et al.: Astron. Astrophys. Suppl. Ser. **122**, 309 (1997)
Rea, N., Israel, G., Covino, S., et al.: Astron. Telegr. **645** (2005)
Thompson, C., Duncan, R.C.: Mon. Not. Roy. Astron. Soc. **275**, 255 (1995)
Thompson, C., Duncan, R.C.: Astrophys. J. **473**, 322 (1996)
Thompson, C., Lyutikov, M., Kulkarni, S.R.: Astrophys. J. **574**, 332 (2002)
Tiengo, A., Esposito, P., Mereghetti, S., et al.: Astron. Astrophys. **440**, L63 (2005)
Woods, P.M., Kouveliotou, C., van Paradijs, J., et al.: Astrophys. J. **518**, L103 (1999)
Woods, P.M., Thompson, C. In: Lewin, W.H.G., van der Klis, M. (eds.) Compact Stellar X-ray Sources (2004), ArXiv: astro-ph/0406133

Astrophys Space Sci (2007) 308: 39–42
DOI 10.1007/s10509-007-9317-y

ORIGINAL ARTICLE

The radio nebula produced by the 27 December 2004 giant flare from SGR 1806-20

Joseph D. Gelfand

Received: 10 July 2006 / Accepted: 21 August 2006 / Published online: 21 March 2007

Abstract On 27 December 2004, just the third giant flare was observed from a magnetar, in this case SGR 1806-20. This giant flare was the most energetic of the three, and analysis of a Very Large Array observation of SGR 1806-20 after the giant flare revealed the existence of a new, bright, transient radio source at its position. Follow-up radio observations of this source determined that initially, this source underwent a mildly relativistic one-sided expansion which ceased at the same time as a temporary rebrightening of the radio source. These observational results imply that the radio emission is powered by $\sim 10^{24}$ g of baryonic material which was ejected off the surface on the neutron star during the giant flare.

Keywords Pulsars: individual SGR 1806-20 · Pulsars: general

PACS 97.60.Jd

1 Introduction

Soft γ-ray repeaters are believed to be magnetars, neutron stars with very high magnetic field strengths, $B \sim 10^{14}$ G as opposed to $B \sim 10^{12}$ G observed from canonical radio pulsars (Thompson and Duncan 1995). Many observational properties of magnetars are significantly different from that of "normal" radio pulsars, however the biggest difference between these two classes of neutron stars are the giant γ-ray ($L_\gamma \gtrsim 10^{44}$ ergs) flares emitted by the former class of objects which, according to the magnetar model, is due to fracturing of the neutron star caused by magnetic stresses between the crustal and external magnetic field (Feroci et al. 2001; Thompson and Duncan 2001). The brightest of these giant flares was observed on 27 December 2004 from SGR 1806-20 (Hurley et al. 2005; Palmer et al. 2005), and assuming a distance of ~ 15 kpc (Corbel and Eikenberry 2004), the γ-ray energy of this flare was $\sim 2 \times 10^{46}$ ergs (Palmer et al. 2005)—roughly two orders of magnitude larger than the energy emitted in the previous two giant flares observed in the population of Soft γ-ray Repeaters. Analysis of a Very Large Array (VLA) observation of SGR 1806-20 ~ 7 days after the giant flare led to the discovery of a new, bright (~ 200 mJy at 1.4 GHz; Gaensler et al. 2005) radio source (Cameron et al. 2005; Gaensler et al. 2005), which follow-up radio observations determined was evolving on short time scales. In this paper, we report on an ongoing radio monitoring campaign of this source (Sect. 2) as well as a model for the observed radio emission (Sect. 3).

J.D. Gelfand (✉)
Harvard-Smithsonian Center for Astrophysics, 60 Garden Street, Cambridge, MA 02138, USA
e-mail: jgelfand@cfa.harvard.edu

2 Radio observations

As mentioned above, at the time of the first detection the radio source produced by the 27 December 2004 Giant Flare was extremely bright, and as a result we have been able to track the evolution of the radio flux, size, and position of this source for the past 18 months using observations conducted by a suite of radio telescopes. The 4.8 and 8.5 GHz light curve for this source is shown in Fig. 1. As one can see, the behavior of the light curve is the same at all of the observed wavelength, and the flux of this source goes through multiple phases:

– Before Day 9 after the giant flare, the flux decreases as $t^{-1.5}$ to t^{-2}, with a possible dependence between the

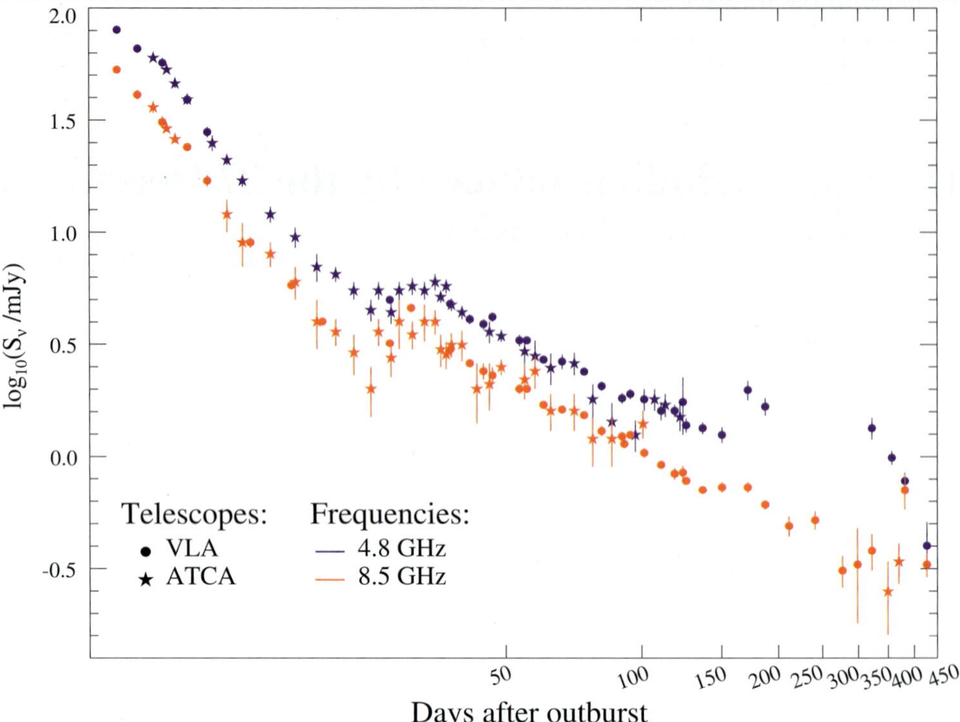

Fig. 1 4.8 and 8.5 GHz *light curve* of the radio source produced by SGR 1806-20 during the 27 December 2004 giant flare, spanning from 7 to 450 days after the giant flare

power law index of the decay and the observation frequency (Gaensler et al. 2005).

- Between ~ 9 and ~ 25 days after the giant flare, the flux decreases as t^{-3} across the observed band (Gaensler et al. 2005; Gelfand et al. 2005).
- Between ~ 25 and ~ 30 days after the giant flare, the flux increases by a factor of ~ 2 at all of the observed frequencies (Gelfand et al. 2005).
- After Day ~ 30, the flux decreases roughly as t^{-1}.

Using the VLA, as well as the Very Long Baseline Array (VLBA) and Multi-Element Radio Linked Interferometer Network (MERLIN), we have also been able to track the size, orientation, and position of this radio source. As a result of these observations, we have learned that:

- Before Day ~ 9 after the giant flare, the position and size of the radio source were roughly constant.
- Between ~ 9 and ~ 30 days after the giant flare a steady change in the position and size of the source was observed (Taylor et al. 2005), and the source is significantly elongated with a constant position axis during this time (Fender et al. 2006). The observed proper motion corresponds to a velocity of $\sim 0.3c$, and its alignment with the source's major axis implies that it was undergoing a one-sided expansion with velocity $v \sim 0.8c$ (Taylor et al. 2005; Fender et al. 2006).
- After Day ~ 30, there has not been a significant change detected in the position or size of the radio source.

To verify that the radio source underwent a one-sided expansion, we recently observed SGR 1806-20 with the VLA in its A-array plus Pie-Town configuration in order to make a spatially resolved image of this source. From the resultant image, it is apparent that the radio source has moved and grown since the initial VLA observation and that the radio emission is predominantly one-sided—confirming previously published results.

3 Neutron star ejecta model for radio emission

A natural explanation for the one-sided nature of the radio emission is that it is the result of a one-sided expulsion of material from the neutron star during the giant flare (Gaensler et al. 2005; Granot et al. 2006). A general schematic of this model is shown in Fig. 2. Initially, this material was freely-expanding inside a low-density cavity surrounding SGR 1806-20 (panel (a) in Fig. 2; Gaensler et al. 2005; Granot et al. 2006). However, some time before the first radio observation, this ejecta collided with a pre-existing shell in the interstellar medium (ISM) (panel (b) in Fig. 2). This collision shocked both the neutron star ejecta and the material inside surrounding shell, and it is emission from material recently shocked as a result of this collision that dominates the radio emission before Day 9. As a result of this collision, the neutron star ejecta is compressed into a shell. The ejecta resumed its expansion into the surrounding ISM on Day 9 (panel (c) in Fig. 2), which accounts for

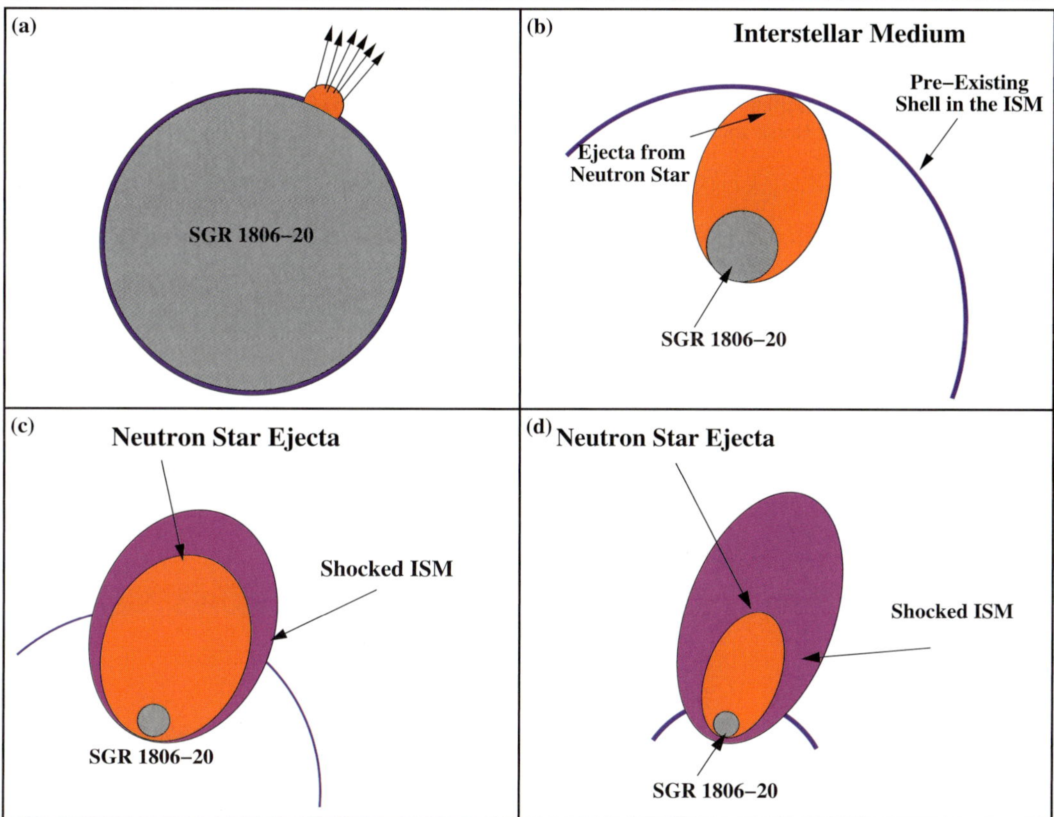

Fig. 2 Schematic for the neutron star ejecta model for the radio emission detected in the aftermath of the 27 December 2004 giant flare from SGR 1806-20. Initially, the ejecta from the neutron star surface was freely expanding (**a**), but collided with a pre-existing shell in the interstellar medium (**b**). After this collision, the neutron star ejecta continued to expand, sweeping up and shocking interstellar material (**c**). Eventually, the swept-up material decelerates the neutron star ejecta, and dominates the observed radio emission (**d**)

the proper motion and increase in size observed to begin at this time (Taylor et al. 2005; Fender et al. 2006), as well the break in the light curve (Gaensler et al. 2005). The steep decline in flux during this expansion period is the result of the previous compression of the ejecta into a shell. As the ejecta expands into the ISM, it sweeps up and shocks the surrounding material. As a result, it is surrounded by a thin shell of shocked ISM, whose mass increases as the ejecta expands. This swept-up material decelerates the ejecta, and the rebrightening observed ~ 25 days after the giant flare, as well as the observed deceleration of the growth and proper motion at this time, are the result of the swept-up ISM beginning to dominate both the dynamics and emission from the radio source (panel (d) in Fig. 2).

The constant motion and growth observed between Day ~ 9 and the rebrightening observed on Day ~ 25 imply that the outflow from the neutron star is baryon-dominated (Gelfand et al. 2005; Taylor et al. 2005; Granot et al. 2006). If so, we are able to estimate the mass of the material ejecta from the neutron star from the observed rebrightening (Gelfand et al. 2005) and the observed deceleration (Taylor et al. 2005). Both methods give similar estimates for the ejected mass, $> 10^{24}$ g (Gelfand et al. 2005; Taylor et al. 2005). If correct, this implies that at early times the mass outflow would have been opaque to γ-rays and, as a result, the giant flare would not have been observed. However, if the material and radiation was generated from different locations on the neutron star (Gelfand et al. 2005; Granot et al. 2006), this is no longer an issue. Another alternative is that the mass outflow was not baryon-dominated, but powered by the expulsion of a magnetic flux loop from the surface of the neutron star that triggered the giant flare (Lyutikov 2006). In this scenario, the radio source should expand and move ultra-relativistically ($\Gamma \gg 10$), as opposed to the mildly-relativistic ($\Gamma \sim 2$) expansion and motion observed, though this discrepancy can be explained if the radio source is moving away from us. In this case, Doppler beaming is a major consideration and it is hard to explain the sharp decrease seen in emission as the source is decelerated by the swept-up ISM. In either case, understanding the nature of the observed radio emission is important in understanding the mechanism behind the giant flares observed from magnetars.

Acknowledgements This work would not be possible if it was not for the effort of a large number of people including Bryan Gaensler,

Chryssa Kouveliotou, Greg Taylor, David Eichler, Yuri Lyubarsky, Yoni Granot, Enrico Ramirez-Ruiz, Katherine Newton-McGee, Dick Hunstead, Duncan Campbell-Wilson, Rob Fender, Naomi McClure-Griffiths, Maura McLaughlin, Michael Garrett, Dave Palmer, Neil Gehrels, Peter Woods, Alexander van der Horst, and Ralph Wijers.

References

Cameron, P.B., Chandra, P., Ray, A., et al.: Nature **434**, 1112 (2005)
Corbel, S., Eikenberry, S.S.: Astron. Astrophys. **419**, 191 (2004)
Fender, R.P., Muxlow, T.W.B., Garrett, M.A., et al.: Mon. Not. Roy. Astron. Soc. **367**, L6 (2006)
Feroci, M., Hurley, K., Duncan, R.C., et al.: Astrophys. J. **549**, 1021 (2001)
Gaensler, B.M., Kouveliotou, C., Gelfand, J.D., et al.: Nature **434**, 1104 (2005)
Gelfand, J.D., Lyubarsky, Y.E., Eichler, D., et al.: Astrophys. J. Lett. **634**, L89 (2005)
Granot, J., Ramirez-Ruiz, E., Taylor, G.B., et al.: Astrophys. J. **638**, 391 (2006)
Hurley, K., Boggs, S.E., Smith, D.M., et al.: Nature **434**, 1098 (2005)
Lyutikov, M.: Mon. Not. Roy. Astron. Soc. **367**, 1594 (2006)
Palmer, D.M., Barthelmy, S., Gehrels, N., et al.: Nature **434**, 1107 (2005)
Taylor, G.B., Gelfand, J.D., Gaensler, B.M., et al.: Astrohys. J. Lett. **634**, L93 (2005)
Thompson, C., Duncan, R.C.: Mon. Not. Roy. Astron. Soc. **275**, 255 (1995)
Thompson, C., Duncan, R.C.: Astrophys. J. **561**, 980 (2001)

Astrophys Space Sci (2007) 308: 43–50
DOI 10.1007/s10509-007-9311-4

ORIGINAL ARTICLE

The continuum and line spectra of SGR 1806-20 bursts

Alaa I. Ibrahim · William C. Parke · Jean H. Swank · Hisham Anwer · Roberto Turolla · Silvia Zane · M.T. Hussein · T. El-Sherbini

Received: 27 July 2006 / Accepted: 12 October 2006 / Published online: 20 March 2007

Abstract The defining property of Soft Gamma Repeaters is the emission of short, bright bursts of X-rays and soft γ-rays. Here we present the continuum and line spectral properties of a large sample of bursts from SGR 1806-20, observed with the Proportional Counter Array (PCA) onboard the Rossi X-ray Timing Explorer (RXTE). Using 10 trail spectral models (5 single and 5 two component models), we find that the burst continua are best fitted by the single component models: cutoff power-law, optically thin bremsstrahlung, and simple power-law. Time resolved spectroscopy show that there are two absorption lines at $\sim$5 keV and 20 keV in some bursts. The lines are relatively narrow with 90% upper limit on the line widths of 0.5–1.5 keV for the 5 keV feature and 1–3 keV for the 20 keV feature. Both lines have considerable equivalent width of 330–850 eV for the 5 keV feature and 780–2590 eV for the 20 keV feature. We examined whether theses spectral lines are dependent upon the choice of a particular continuum model and find no such dependence. Besides, we find that the 5 keV feature is pronounced with high confidence in the cumulative joint spectrum of the entire burst sample, both in the individual detectors of the PCA and in the co-added detectors spectrum. We confront the features against possible instrumental effects and find that none can account for the observed line properties. The two features do not seem to be connected to the same physical mechanism because (1) they do not always occur simultaneously, (2) while the 5 keV feature occurs at about the same energy, the 20 keV line centroid varies significantly from burst to burst over the range 18–22 keV, and (3) the centroid of the lines shows anti-correlated red/blue shifts. The transient appearance of the features in the individual bursts and in portions of the same burst, together with the spectral evolution seen in some bursts point to a complex emission mechanism that requires further investigation.

Keywords Pulsar: individual (SGR 1806-20) · Pulsars: general · Stars: magnetic fields · Stars: magnetar · Gamma-rays: bursts · X-rays: bursts

PACS 95.85.Pw · 95.85.Nv · 96.12.Hg · 97.60.Gb

A.I. Ibrahim (✉) · H. Anwer · M.T. Hussein · T. El-Sherbini
Department of Physics, Faculty of Science, Cairo University, Cairo, Egypt
e-mail: alaa@gwu.edu

A.I. Ibrahim · H. Anwer · M.T. Hussein · T. El-Sherbini
Center of Advanced Interdisciplinary Science, Astrophysics Division, Faculty of Science, Cairo University, Cairo, Egypt

A.I. Ibrahim · W.C. Parke
Department of Physics, George Washington University, Washington, DC 20052, USA

J.H. Swank
Astrophysics Science Division, NASA Goddard Space Flight Center, Greenbelt, MD 20771, USA

R. Turolla
Department of Physics, University of Padova, Padova, Italy

S. Zane
Mullard Space Science Lab, University College of London, Holmbury St Mary, Surrey, UK

1 Introduction

SGR 1806-20 is one of four confirmed soft gamma repeaters (SGRs) that, together with the anomalous X-ray pulsars (AXPs), are considered to host a "magnetar", i.e. an ultra-magnetized neutron star. SGRs and some AXPs are characterized by the emission of short ($<$0.1 s), bright ($L \sim 10^{39}$–10^{42} ergs s^{-1}) bursts of X-rays and soft γ-rays (see Woods

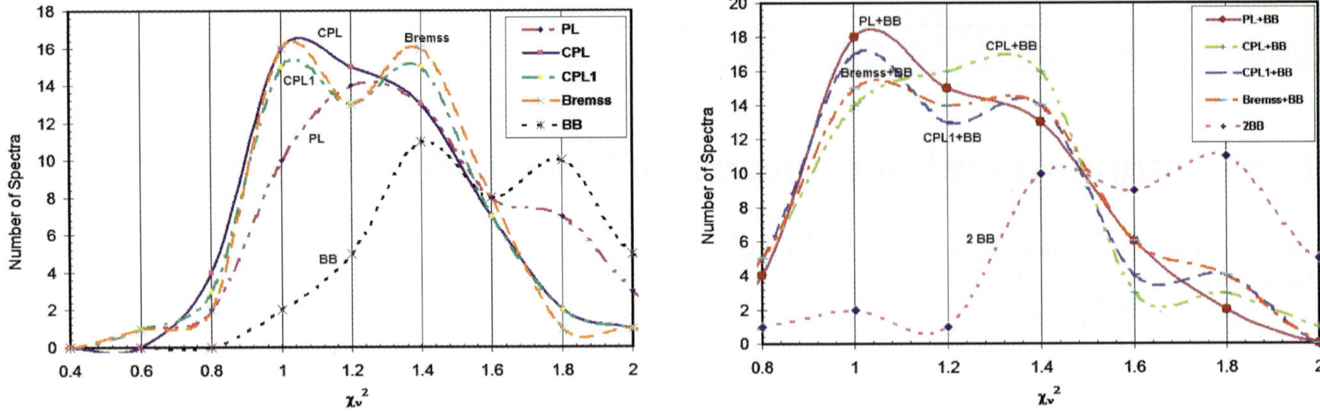

Fig. 1 Reduced χ^2 distributions for the 10 trial models used to fit the SGR 1806-20 burst spectra. The *left panel* shows the single component models and the *right panel* shows the two component models (*Bremss* ≡ Optically Thin Thermal Bremsstrahlung, *CPL* ≡ Cutoff Power-law, *CPL1* ≡ CPL with the photon index frozen to 1, *PL* ≡ Power-law, *BB* ≡ Blackbody). The single component models CPL and Bremss provide the best fits with average χ_ν^2 of 1.15 and 1.17, respectively, whereas PL comes next with a χ_ν^2 of 1.25. CPL1 and the two component models are therefore less possible

and Thompson 2005 for a review). In the magnetar model (Thompson and Duncan 1995), the burst emission results when magnetic stresses build up sufficiently to crack a patch of the neutron star crust, inducing a *starquake* that ejects hot plasma particles into the magnetosphere. The burst temporal profiles vary from single peak to multi-peak to multi-burst events. The burst continua have been fitted to optically thin thermal bremsstrahlung (Strohmayer and Ibrahim 1999, 2000; Ibrahim et al. 2001), cutoff-power law (Mazets et al. 1999), simple power-law (Strohmayer and Ibrahim 1999; Ibrahim et al. 2002, 2003; Gavriil et al. 2004), and the sum of two blackbody models (Oliver et al. 2004; Feroci et al. 2004). Here we study the burst continuum with all such models using RXTE PCA observations of SGR 1806-20 bursts, report on the presence of some spectral lines, and investigate the veracity of such lines.

2 Observation and data analysis

2.1 Burst continuum

We obtained the RXTE PCA observations of the 1996 burst activity of SGR 1806-20 and reduced the data with HEASoft/Ftools 6.06 as explained in Ibrahim et al. (2002, 2003). All 5 detectors of the PCA were operating normally and their photons were co-added. The data were recorded in the event-mode and good Xenon configurations, with 125 μs time resolution and 64 energy channels covering 2–60 keV. We excluded faint bursts that do not have enough counts to accumulate an individual spectrum, and very bright bursts with count rates exceeding 90 000 counts s^{-1} where dead-time and pileup effects become significant. We classified the bursts according to their light curve morphology into single-peak and multi-peak bursts and fitted each emission peak separately. This gave rise to a total of 56 spectra. We started by fitting the data with each of the following single component model (that includes photoelectric absorption): optically thin thermal bremsstrahlung (Bremss), cutoff power-law (CPL), cutoff power-law with the photon index frozen to unity (CPL1), simple power-law (PL), and blackbody radiation (BB). Figure 1 (left panel) shows the reduced χ^2 distributions of such 5 models. CPL, Bremss, and CPL1 provided the best fits with average χ_ν^2 of 1.15, 1.17, and 1.18, respectively. PL still provided acceptable fit with $\chi_\nu^2 = 1.24$. BB provided the worst fit with $\chi_\nu^2 = 1.80$. We next investigated the effect of adding a BB component to each of the above 5 models. The new χ_ν^2 distribution is shown in the right panel of Fig. 1. The best fit models: PL+BB, Bremss+BB, CPL+BB, CPL1+BB provided comparable average χ_ν^2 of 1.12, 1.15, 1.15, and 1.16, respectively. The 2 BB model provided the worst fit with $\chi_\nu^2 = 1.86$.

2.2 Line spectra

In some individual burst spectra, we found an absorption line at ∼5 keV that was significant at ∼4σ in such bursts (see Ibrahim et al. 2002, 2003 for details). To investigate any dependence of the line on a particular continuum model we fitted the spectra of these bursts to the best fit models we introduced above. Figure 2 shows the fit residuals with the single-component models Bremss, PL, CPL, and CPL1, as well as with the two-component model CPL1+BB. The line is minimally affected, indicating no real dependence on a specific continuum.

Detailed examination of the spectral fits in the full PCA bandpass (2–60 keV) showed a number of spectra with absorption structure in the vicinity of 20 keV. Most of these bursts also showed the 5 keV feature (see Fig. 3). As done

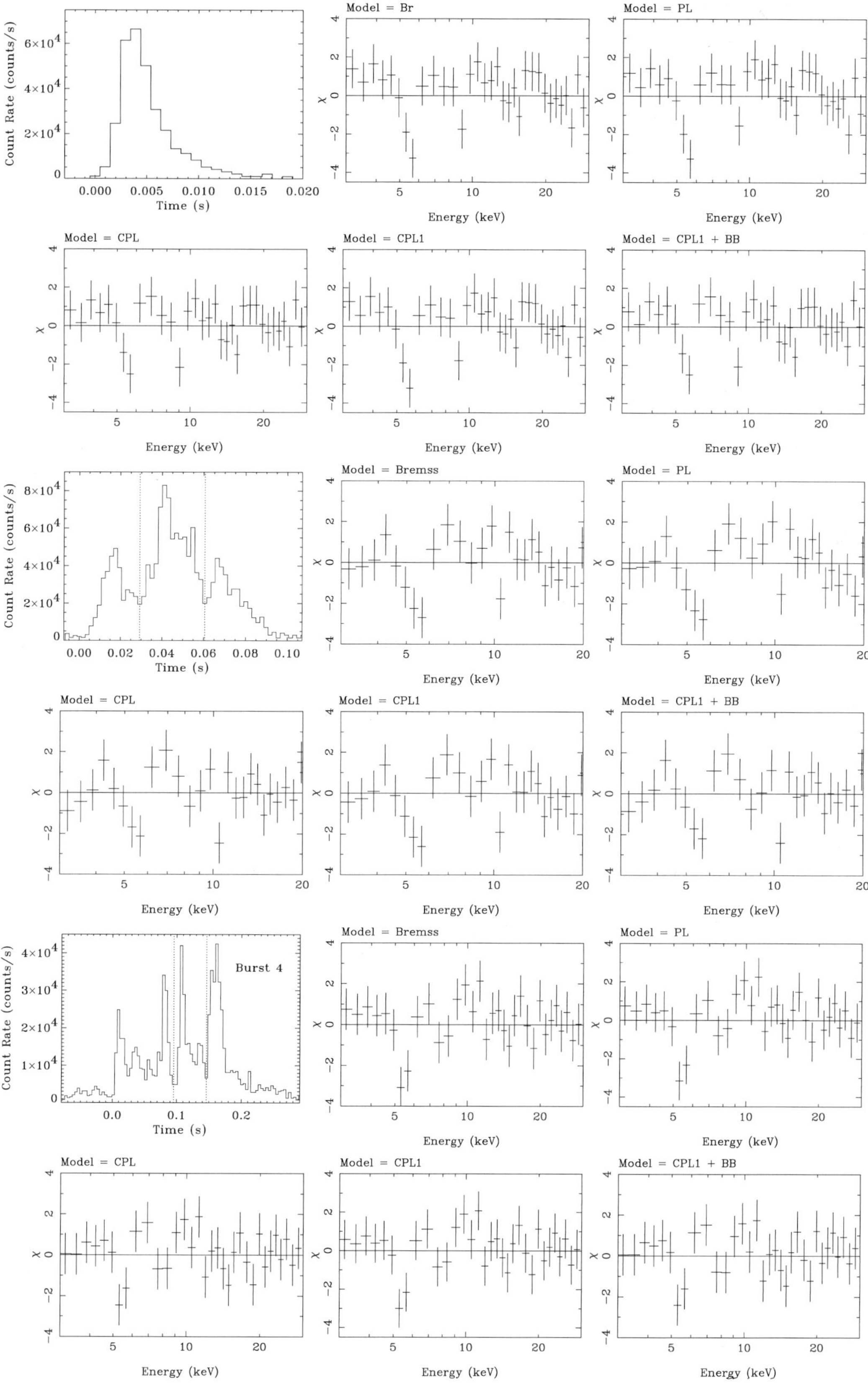

Fig. 2 SGR 1806-20 bursts from Ibrahim et al. 2003 that showed evidence for the 5 keV feature, fitted by different continuum models. Next to the *burst light curve* we show the fit residuals of the model indicated on the *top left corner*. The feature show no significant correlation with a particular model

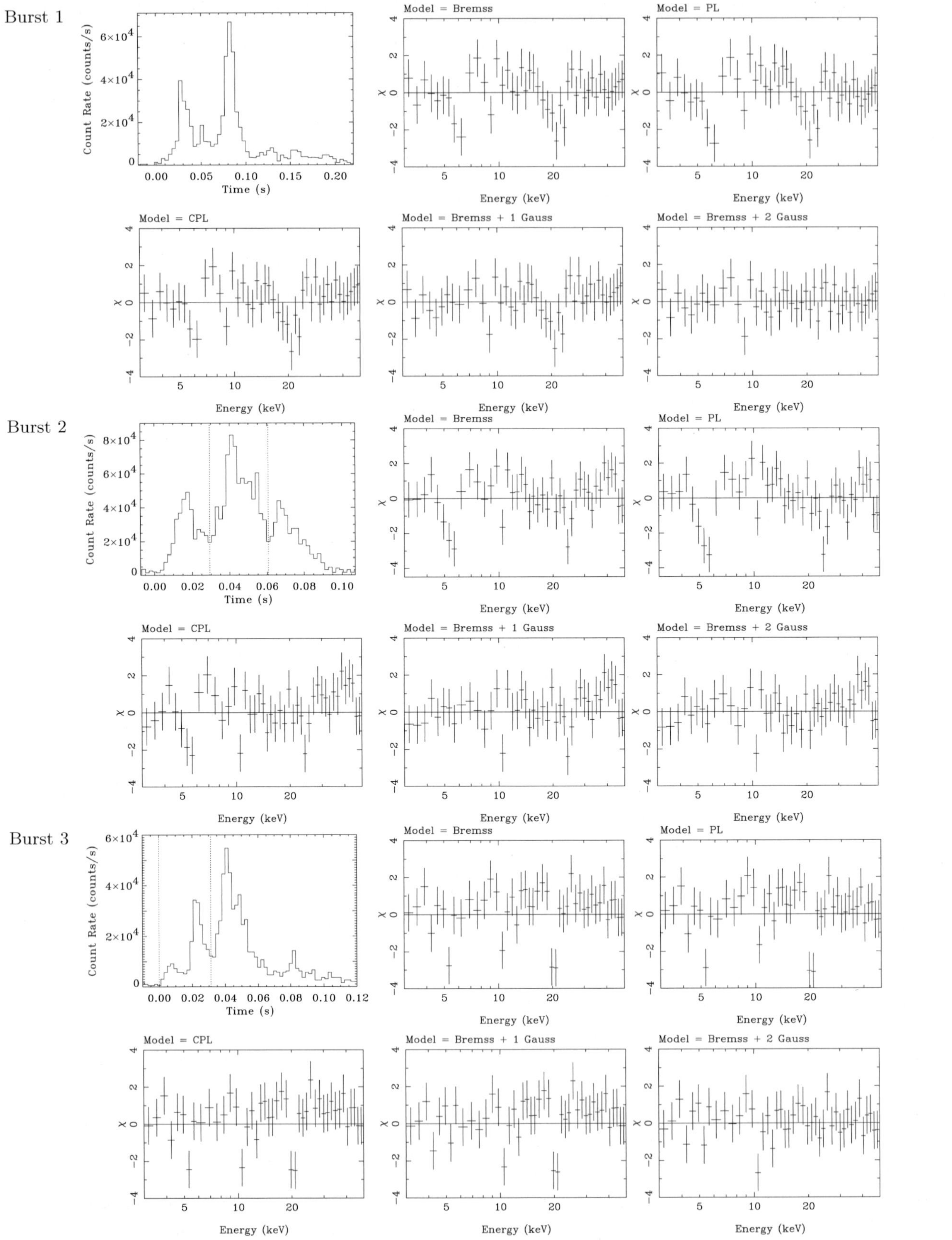

Fig. 3 Bursts from SGR 1806-20 that showed evidence for the ~20 keV feature. The *first two bursts* also showed evidence for the 5 keV feature. The *third and fourth bursts* showed the 20 keV feature with weak and no evidence for the 5 keV line, respectively. The features are present in the whole burst in the *first and fourth bursts* and in the indicated interval in the *second and third bursts*

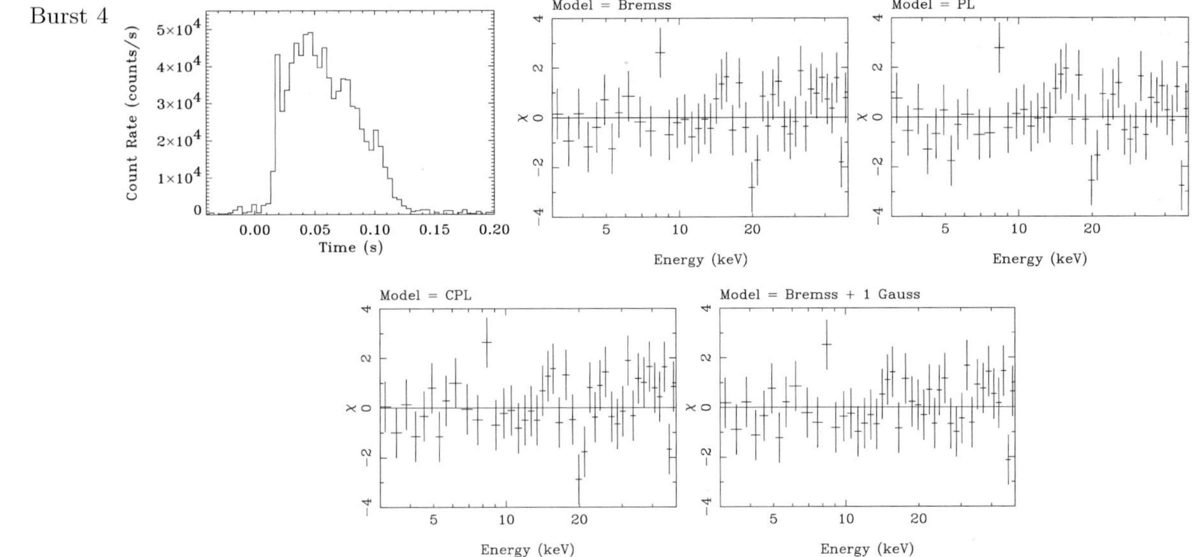

Fig. 3 Continued

Table 1 Spectral properties of SGR 1806-20 bursts with evidence for the 20 keV absorption feature

Burst	kT (keV)	N_H (10^{22} cm^{-2})	χ_1^2/ν_1	Feature	E (keV)	EW[a] (keV)	Line width[b] (keV)	χ_2^2/ν_2	$P_{\text{F-test}}$
1	47.80 ± 9.67	16.63 ± 1.67	49.2/43	5 keV	5.67 ± 0.18	0.33	(<0.91)		
				+20 keV	19.0 ± 0.60	2.02	1.61 ± 0.81 (<3.12)	23.6/38	2.4×10^{-5}
2	19.0 ± 2.53	16.7 ± 1.86	63.6/43	5 keV	5.23 ± 0.01	0.44	(<0.45)		
				+20 keV	22.1 ± 0.61	1.65	0.81 ± 0.73 (<2.29)	34.3/38	1.9×10^{-4}
3	17.9 ± 5.16	15.9 ± 3.73	61.5/43	5 keV	5.25 ± 0.24	0.49	(<1.54)		
				+20 keV	18.7 ± 0.57	2.59	(<1.24)	31.2/38	6.4×10^{-5}
4	38.7 ± 4.83	12.7 ± 1.02	62.1/46	20 keV	18.4 ± 0.16	0.78	(<1.10)	49.8/44	7.8×10^{-3}

Errors quoted are the 1σ error

[a]Equivalent width

[b]The 90% confidence upper limit on the line width is given between parentheses

with the 5 keV feature, we examined the 20 keV residuals with the different continuum models as shown in Fig. 3. The residuals are almost unaffected by the choice of the continuum among the three best fit single models. To weigh the statistical significance of the new feature we added a Gaussian absorption line near 20 keV to the fit model and examined the change in χ^2 using the F-test. As shown in Table 1, the fits were improved considerably, revealing an absorption line whose energy varied from burst to burst between ∼18 and 22 keV. In contrast, the 5 keV line energy varied only minimally. The ∼20 keV line was also broader and showed a larger equivalent width. Given that we studied a total of 56 spectra and found the feature at the chance probabilities listed in Table 1, we can use the same approach at the end of Sect. 2 in Ibrahim et al. (2003) to estimate the overall significance of the feature. This gives a chance probability of 2×10^{-8}.

2.3 The grand spectrum and individual PCA detectors

To further assess the significance of the absorption features reported above, we co-added the 56 burst spectra to form a cumulative grand spectrum. The first panel of Fig. 4 shows the spectral continuum and fit residuals with the Bremss model when all 5 PCU detectors of the PCA are added. The 5 keV absorption line is pronounced at a better than 10σ significance. Since the ∼20 keV feature occurred at different energies in the different bursts, we do not expect it to persist in this grand spectrum.

The grand spectrum provided sufficiently high signal to noise ratio that allowed us to accumulate a spectrum from

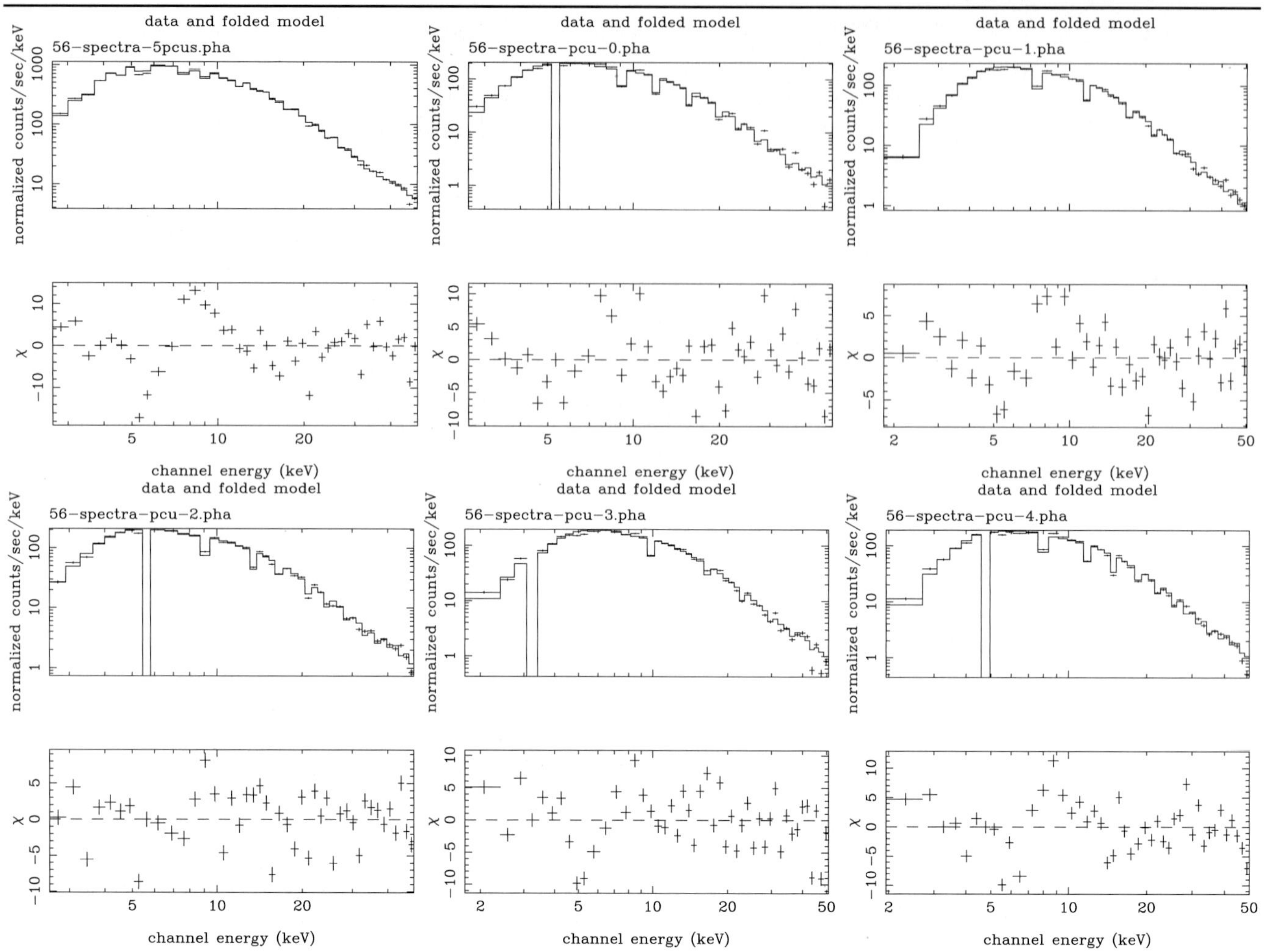

Fig. 4 The spectral continuum and fit residuals (w.r.t. Bremss model) of the grand spectrum of the co-added 56 bursts from the co-added 5 PCUs of the PCA (*first panel*) and the individual PCU detectors (*next 5 panels*)

each of the 5 individual PCU detectors. In the next panels of Fig. 4 we show the grand spectrum and fit residuals from each of the 5 PCUs. PCU 0, 2, 3, and 4 show data gaps in the spectral continua due to shifts in the pulse-height bin alignment that took place onboard the spacecraft. This known effect gives rise to no data (or a gap) at a particular energy bin in the aforementioned PCUs. Its net effect is therefore to mask the source counts at such energy bin. This is evident at 3 keV in PCU 3 and at 5 keV in PCU 0. The detection of the 5 keV feature should therefore be improved if we exclude PCU 0. This is indeed the case as shown in Fig. 5, where the spectrum is only accumulated from PCUs 1-4. The top panel of Fig. 5 shows the spectrum and fit residuals for the 56 co-added bursts and the bottom panel shows the same plots for an individual burst (burst 2 in Fig. 3) (middle panel). The line profile is clearly improved (compare with Fig. 3) and its significance is also enhanced. In PCUs 1 and 3 in which this effect does not occur or overlap with the line energy, the 5 keV feature is clearly evident in the residuals of their grand spectrum.

2.4 Testing for instrumental and systematic effects

Our last test for instrumental and systematic effects involves an RXTE Crab Nebula observation that took place a day before those of SGR 1806-20, i.e. on 1998 Nov. 17. The top panel of Fig. 6 shows the light curve, which shows a count rate close to those of the bursts. The spectrum is well represented by an absorbed power law, showing no evidence for systematic residuals near 5 keV or 20 keV, see bottom panels of Fig. 6. The fit is not improved by adding an absorption line near 5 keV (systematic errors of 1% were included). The 90% upper limit on the equivalent width is 40 eV. The spectrum shown is for the entire observation ($\sim$850 s). The observation was taken at 16 s time resolution which did not allow time resolved spectroscopy on a time scale comparable with that of the bursts.

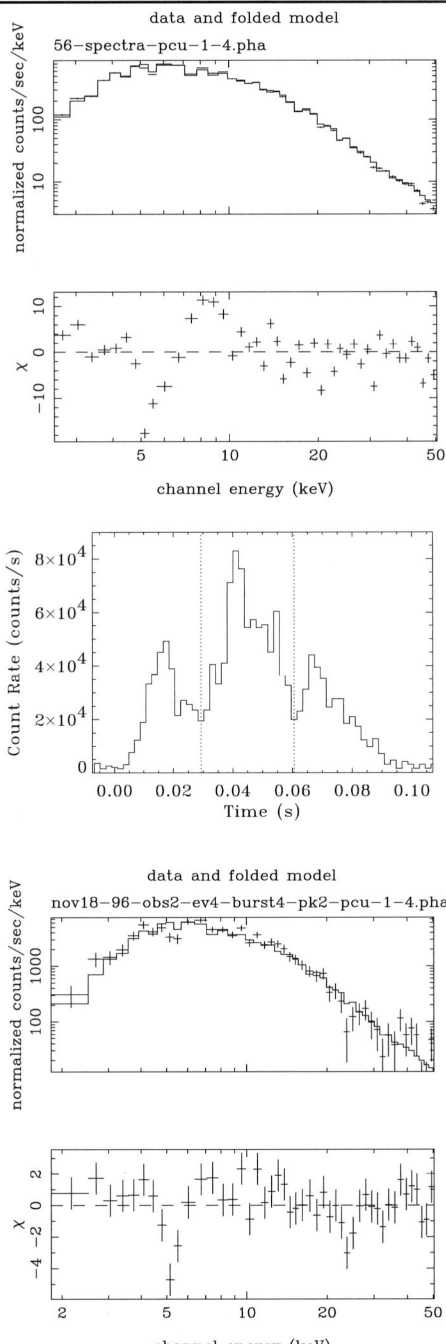

Fig. 5 *Top*: The spectral continuum and fit residuals (w.r.t. Bremss model) of the grand spectrum of the co-added 56 bursts from PCUs 1-4. *Bottom*: The spectral continuum and fit residuals (w.r.t. Bremss model) of burst 2 in Fig. 3 (middle panel) from PCUs 1-4. The line profile is improved due to the removal of PCU 0 that masks the source counts at 5 keV as explained in the text

3 Conclusion

We have studied the continuum spectrum of a large sample of bursts from SGR 1806-20 and here we outline the main conclusions. Further discussion on the origin of the 20 keV feature and the details of the spectral continuum fits will be presented elsewhere.

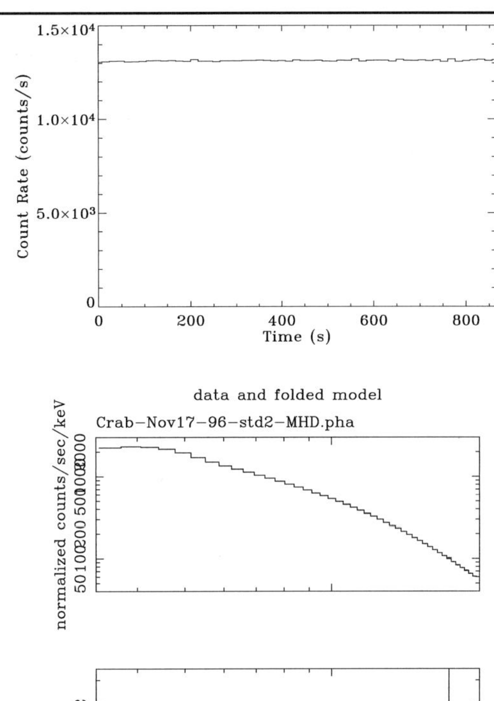

Fig. 6 *Top*: The light curve of the 1996 Nov. 17 observation of the Crab nebula with RXTE PCA. *Bottom*: The spectral continuum and fit residuals (w.r.t. PL model), showing no evidence for systematic residuals at 5 keV or 20 keV.

(1) The burst Continuum is found to be well fitted by the single component models Bremss, CPL and PL. The two component models that include a BB component do not provide a significant improvement and may therefore not be justified. In the magnetar model, the burst emission comes from an electron positron pair plasma, trapped in the magnetosphere. The emission from the neutron star surface during the bursts is quite insignificant. This picture is well represented by the single component models (Bremss, CPL, and PL) and is consistent with our results.

(2) The 5 keV absorption line is detected with high confidence both in the individual bursts and in the grand burst spectrum. The feature is pronounced when the PCA detectors are co-added and is equally significant in the individual detectors. The feature is found to be independent on the choice of the continuum model and none of the presently understood PCA instrumental effects can account for its observed properties.

(3) With its varying line centroid, broader width, larger equivalent width, and appearance with and without the

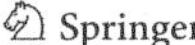

5 keV feature, the ∼20 keV absorption line is quite puzzling. The feature also shows opposite red/blue shift correlation with respect to the 5 keV feature in the same pairs of bursts (e.g. bursts 1 and 2). Like the 5 keV line, we found that the 20 keV feature is independent upon a particular continuum model.

(4) The 5 keV feature has been interpreted as proton cyclotron resonance in an ultra-string magnetic field of $\approx 10^{15}$ G (Ibrahim et al. 2002; Ibrahim et al. 2003). However, while atomic transitions are not categorically ruled out for the 5 keV feature, the large variance in line centroid of the ∼20 keV feature strongly argues against such an interpretation. In the proton/ion cyclotron resonance interpretation of the 5 keV feature, the 20 keV feature cannot be another ion cyclotron resonance because in bursts 1 and 2 where both features appear and would therefore be subjected to the same respective magnetic field and gravitational red shift we find that: 1) the variation in the line centroid of the 20 keV feature is very large compared to that of the 5 keV, and 2) the features show anti-correlated red/blue shift (while the 5 keV is red shifted from burst 1 to burst 2, the 20 keV is blue shifted).

The only other remaining interpretation that is currently available in the literature is vacuum polarization resonance (Pavlov and Shibanov 1979; Ventura et al. 1979), which in this case comes with a set special circumstances that will be discussed elsewhere. However, we do not rule out the possibility that the 5 keV feature is an atomic transition line whereas the ∼20 keV feature is a proton cyclotron resonance, implying a stellar magnetic field of $\sim 4 \times 10^{15}$ G. The lack of atomic calculation models for the magnetar environment does not allow us to further constrain this possibility.

References

Feroci, M., Caliandro, G.A., Massaro, E., et al.: Astrophys. J. **612**, 408–413 (2004)

Gavriil, F., Kaspi, V., Woods, P.: Astrophys. J. **607**, 959–969 (2004)

Ibrahim, A.I., Strohmayer, T.E., Woods, P.M., et al.: Astrophys. J. **558**, 237–252 (2001)

Ibrahim, A.I., Safi-Harb, S., Swank, J.H., et al.: Astrophys. J. **574**, L51–L55 (2002)

Ibrahim, A.I., Swank, J.H., Parke, W.C.: Astrophys. J. Lett. **584**, L17–L21 (2003)

Mazets, E.P., Cline, T.L., Aptekar, R.L., et al.: Astron. Lett. **25**, 628–634 (1999)

Olive, J.-F., Hurley, K., Atteia, J.-L., et al.: Astrophys. J. **616**, 1148–1158 (2004)

Pavlov, G.G., Shibanov, Yu.A.: Sov. Phys. JETP **49**, 741 (1979)

Strohmayer, T.E., Ibrahim, A.I. In: Meegan, C.A., Preece, R.D., Koshut, T.M. (eds.) Fourth Huntsville Symp. on Gamma-Ray Bursts, Huntsville, AL. AIP Conf. Proc., vol. 428, p. 947. AIP, New York (1999)

Strohmayer, T.E., Ibrahim, A.I.: Astrophys. J. Lett. **537**, L111–L114 (2000)

Thompson, C., Duncan, R.: Mon. Not. Roy. Astron. Soc. **725**, 255–300 (1995)

Ventura, J., Nagel, W., Meszaros, P.: Astrophys. J. **233**, L125–L128 (1979)

Woods, P.M., Thompson, C.: In: Lewin, W.H.G., van der Klis, M. (eds.) Compact Stellar X-Ray Sources. Cambridge Univ. Press, Cambridge (2005, in press), astro-ph/0406133

Astrophys Space Sci (2007) 308: 51–59
DOI 10.1007/s10509-007-9379-x

ORIGINAL ARTICLE

Unveiling soft gamma-ray repeaters with INTEGRAL

Diego Götz · Sandro Mereghetti · Kevin Hurley

Received: 30 June 2006 / Accepted: 14 September 2006 / Published online: 20 March 2007

Abstract Thanks to INTEGRAL's long exposures of the Galactic Plane, the two brightest Soft Gamma-Ray Repeaters, SGR 1806-20 and SGR 1900+14, have been monitored and studied in detail for the first time at hard-X/soft gamma rays.

This has produced a wealth of new scientific results, which we will review here. Since SGR 1806-20 was particularly active during the last two years, more than 300 short bursts have been observed with INTEGRAL and their characteristics have been studied with unprecedented sensitivity in the 15–200 keV range. A hardness-intensity anticorrelation within the bursts has been discovered and the overall Number-Intensity distribution of the bursts has been determined. In addition, a particularly active state, during which 100 bursts were emitted in 10 minutes, has been observed on October 5 2004, indicating that the source activity was rapidly increasing. This eventually led to the Giant Flare of December 27th 2004, for which a possible soft gamma-ray (>80 keV) early afterglow has been detected.

The deep observations allowed us to discover the persistent emission in hard X-rays (20–150 keV) from 1806-20 and 1900+14, the latter being in a quiescent state, and to directly compare the spectral characteristics of all Magnetars (two SGRs and three Anomalous X-ray Pulsars) detected with INTEGRAL.

Keywords Gamma-rays: observations · Pulsars: individual SGR 1806-20, SGR 1900+14 · Pulsars: general

PACS 95.85.Pw · 95.85.Nv · 96.12.Hg · 97.60.Gb

D.G. acknowledges the French Space Agency (CNES) for financial support. Based on observations with INTEGRAL, an ESA project with instruments and the science data centre funded by ESA member states (especially the PI countries: Denmark, France, Germany, Italy, Switzerland, Spain), Czech Republic and Poland, and with the participation of Russia and the USA. ISGRI has been realized and maintained in flight by CEA-Saclay/DAPNIA with the support of CNES. K.H. is grateful for support under NASA's INTEGRAL U.S. Guest Investigator program, Grants NAG5-13738 and NNG05GG35G.

D. Götz (✉)
CEA-Service d'Astrophysique, Orme des Merisiers, Bat. 709, 91191 Gif-sur-Yvette, France
e-mail: diego.gotz@cea.fr

S. Mereghetti
INAF—Istituto di Astrofisica Spaziale e Fisica Cosmica, Milano, Italy

K. Hurley
Space Sciences Laboratory, University of California at Berkeley, Berkeley, CA, USA

1 Introduction

Soft gamma-ray repeaters (SGRs, for a recent review see Woods and Thompson (2003)) are a small group (4–7) of peculiar high-energy sources generally interpreted as "magnetars", i.e. strongly magnetised ($B \sim 10^{15}$ G), slowly rotating ($P \sim 5-8$ s) neutron stars powered by the decay of the magnetic field energy, rather than by rotation (Duncan and Thompson 1992; Paczynski 1992; Thompson and Duncan 1995). They were discovered through the detection of recurrent short ($\sim$0.1 s) bursts of high-energy radiation in the tens to $\sim$hundred keV energy range, with peak luminosity up to $10^{39}-10^{42}$ erg s^{-1}, above the Eddington limit for neutron stars. The rate of burst emission in SGRs is highly variable. Bursts are generally emitted during sporadic periods of activity, lasting days to months, followed by long "quiescent" time intervals (up to years or decades) during which no bursts are emitted. Occasionally SGRs emit "giant flares", that last up to a few hundred seconds and have peak luminosity up to $10^{46}-10^{47}$ erg s^{-1}. Only three giant flares have been observed to date, each one from a different source (see, e.g., Mazets et al. 1979a for 0526-66, Hurley et al. 1999 for 1900+14, Palmer et al. 2005; Mereghetti et al. 2005a; Hurley et al. 2005 for 1806-20).

Persistent (i.e. non-bursting) emission is also observed from SGRs in the soft X–ray range (<10 keV), with a typical luminosity of $\sim 10^{35}$ erg s^{-1}, and, in three cases, periodic pulsations with periods of 5–8 seconds have been detected. Such pulsations proved the neutron star nature of SGRs and allowed the derivation of spin-down at rates of $\sim 10^{-10}$ s s^{-1}, consistent with dipole radiation losses for magnetic fields of the order of $B \sim 10^{14}-10^{15}$ G. The X-ray spectra are generally described with absorbed power laws, but in some cases strong evidence has been found for the presence of an additional blackbody-like component with a typical temperature of $\sim$0.5 keV (Mereghetti et al. 2005b).

Over the last few years the INTEGRAL satellite (Winkler et al. 2003), launched in 2002 and operating in the 15 keV-10 MeV energy range, has provided a wealth of new results concerning the two brightest SGRs, 1806-20 and 1900+14. Most aspects concerning the SGRs, short bursts, giant flares, and persistent emission, have been investigated, and new results have been found for each of them. We will review them here.

2 SGR 1806-20

SGR 1806-20 was discovered by the Interplanetary Network (IPN) in 1979 (Laros et al. 1986). It lies in a crowded region close to the galactic centre. Kouveliotou et al. (1998) discovered a quiescent X-ray pulsating ($P = 7.48$ s) counterpart, which was spinning down rapidly ($\dot{P} = 2.8 \times 10^{-11}$ s s^{-1}). If this spindown is interpreted as braking by a magnetic dipole field, its strength is $B \sim 10^{15}$ G. The source activity is variable, alternating between quiet periods and very active ones.

After a period of quiescence, SGR 1806-20 became active in the Summer of 2003 (Hurley et al. 2003). Its activity then increased in 2004 (see e.g. Mereghetti et al. 2004; Golenetskii et al. 2004). A strong outburst during which about one hundred short bursts were emitted in a few minutes occurred on October 5 2004 (Götz et al. 2006a). Finally a giant flare, whose energy (a few 10^{46} erg) was two orders of magnitude larger than those of the previously recorded flares from SGR 0526-66 and SGR 1900+14, was emitted on December 27th 2004 (see e.g. Palmer et al. 2005; Mereghetti et al. 2005a; Hurley et al. 2005).

2.1 Short bursts

The results presented in this Section are based on observations obtained with the IBIS coded mask telescope (Ubertini et al. 2003), and in particular with its low-energy (15 keV–1 MeV) detector ISGRI (Lebrun et al. 2003). More than 400 short bursts have been detected with IBIS/ISGRI. They have been identified either using the triggers provided by the INTEGRAL Burst Alert System (IBAS, Mereghetti et al. 2003), or by computing light curves with 10 ms time resolution and looking for significant excesses corresponding to the direction of SGR 1806-20.

2.1.1 Spectral evolution

All the bursts detected by IBIS are typical in terms of duration and spectra. The new result provided by the analysis of the INTEGRAL sample is the spectral evolution within the bursts. By computing time resolved hardness ratios, Götz et al. (2004, 2006a) showed that some bursts evolve significantly with time, especially the ones with a Fast Rise Exponential Decay (FRED) profile. The hardness ratios have been computed using the background subtracted light curves in two energy bands (20–40 (S) and 40–100 (H) keV) and were defined as $\mathrm{HR} = (H - S)/(H + S)$. It turns out that the bursts' peaks tend to be spectrally softer than the bursts' tails. This behaviour had been reported earlier only for two peculiar bursts originating from SGR 1900+14. These two bursts were quite different from usual bursts, lasting about 1 s and having a very hard spectrum ($kT \sim 100$ keV, Woods et al. 1999). One example of this kind of evolution detected in regular bursts for SGR 1806-20 is shown in Fig. 1.

The spectral behaviour described above gives rise to a global hardness-intensity anti-correlation. In fact, by considering all the individual time bins of all the bursts this anti-correlation within the bursts has been discovered (see Fig. 2). To investigate the statistical robustness of the correlation found, the Spearman rank-order correlation coefficient of the 217 data points, R_s, has been computed, which is –0.49. This corresponds to a chance probability of 4×10^{-15} (7.4σ) that the distribution is due to uncorrelated data. According to an F-test the data are significantly (8σ) better described by a linear fit ($\mathrm{HR} = 0.47 - 0.22 \times \log I$) than by a constant. This correlation still lacks of a solid theoretical interpretation since the current Magnetar scenario does not provide a clear prediction of the burst spectral evolution with time.

2.1.2 Number-intensity distribution

We derived the fluences of 224 bursts using the vignetting and dead-time corrected light curves. We applied a conversion factor between counts and physical units derived by

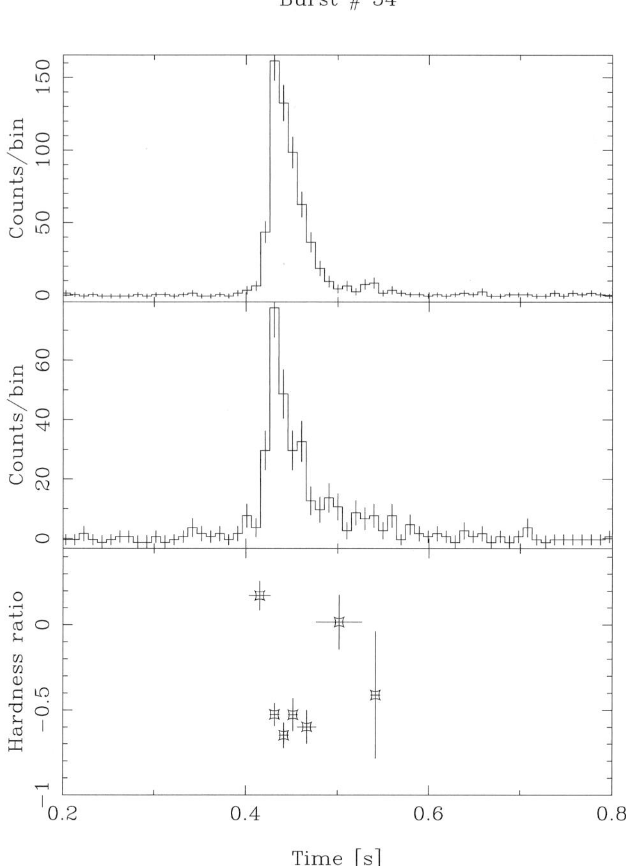

Fig. 1 IBIS/ISGRI light curves in the soft (20–40 keV, *upper panel*) and hard (40–100 keV, *middle panel*) energy range and hardness ratio (*lower panel*) for a short burst from SGR 1806-20

the spectral analysis of the brightest bursts and assuming that the averaged burst spectra do not change much between bright and faint bursts; for details see Götz et al. (2006a). These fluences have been used to compute the number-intensity distribution (Log N − Log S) of the bursts. The experimental distribution deviates significantly from a single power-law (Fig. 3). This is first of all due to the fact that the source has been observed at different off-axis angles. The faintest bursts are missed when the source is observed at large off-axis angles. In order to correct for this effect we have computed the effective exposure of the source, taking into account the variation of sensitivity at various off-axis angles. This yields the exposure-corrected cumulative distribution shown by the dashed line in Fig. 3.

Since the numbers at each flux level are not statistically independent, one cannot use a simple χ^2 minimisation approach to fit the cumulative number-intensity distribution. So we have used the unbinned detections and applied the Maximum Likelihood method (Crawford et al. 1970), assuming a single power-law distribution for the number-flux relation ($N(>S) \propto S^{-\alpha}$). We have used only the part of the distribution where completeness was achieved (i.e.

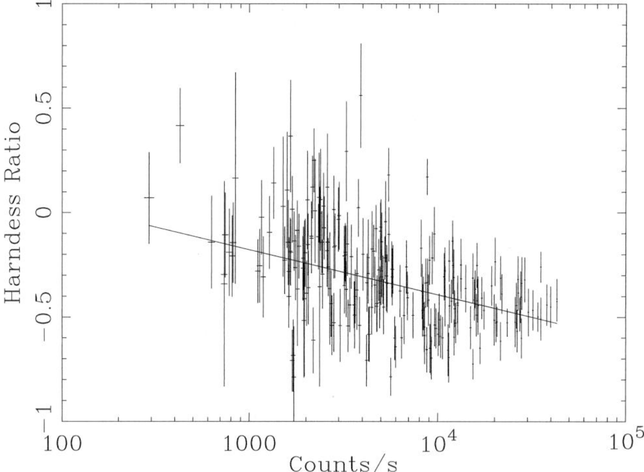

Fig. 2 Hardness ratio ($(H - S)/(H + S)$) versus total count rate (20–100 keV, corrected for vignetting). The points are derived from the time resolved hardness ratios of the bursts with the best statistics. The *line* indicates the best fit with a linear function given in the text. From Götz et al. (2006a)

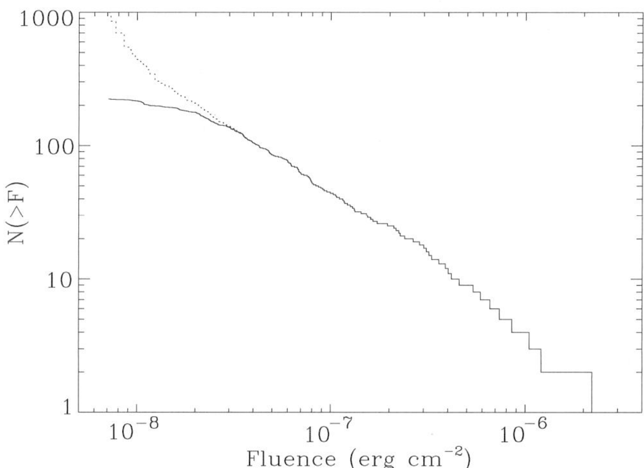

Fig. 3 Number-intensity distribution of all the bursts detected by INTEGRAL in 2003 and 2004. The *continuous line* represents the experimental data, while the *dashed line* represents the data corrected for the exposure. From Götz et al. (2006a)

$S \geq 3 \times 10^{-8}$ erg cm^{-2}). In this case the expression to be maximised is

$$\mathcal{L} = T \ln \alpha - \alpha \sum_i \ln S_i - T \ln(1 - b^{-\alpha}) \quad (1)$$

where S_i are the unbinned fluxes, b is the ratio between the maximum and minimum values of the fluxes, and T is the total number of bursts. This method yields $\alpha = 0.91 \pm 0.09$. If a single power-law model is an adequate representation of data, the distribution of the quantities

$$y_i = \frac{1 - S_i^{-\alpha}}{1 - b^{-\alpha}} \quad (2)$$

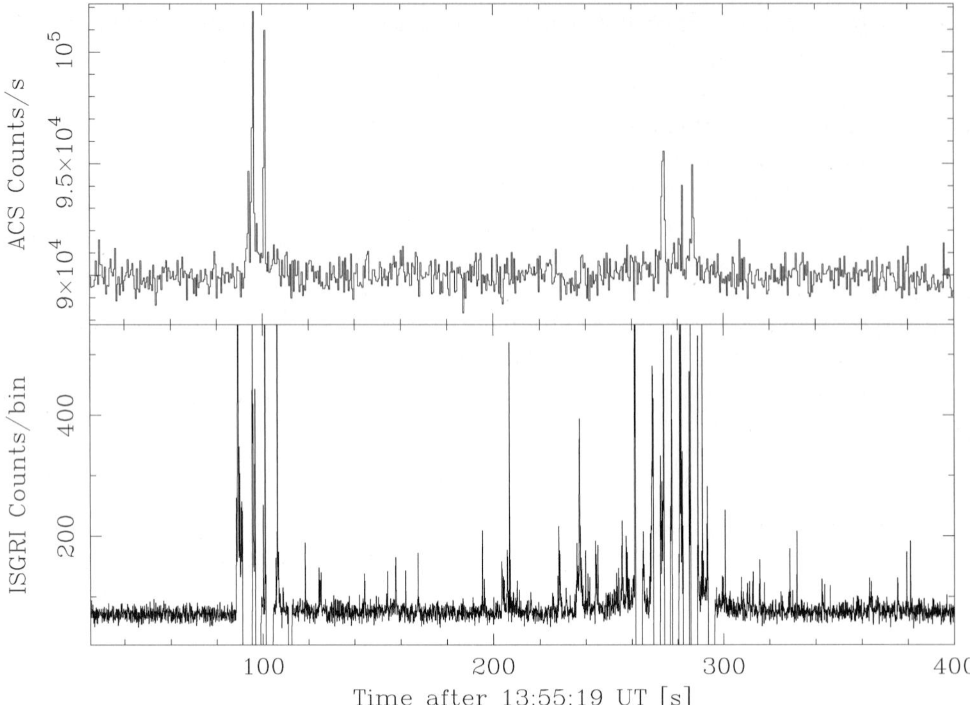

Fig. 4 Light curves of the initial part of the October 5, 2004 outburst of SGR 1806-20. *Upper panel*: light curve at energy greater than ~80 keV obtained with the SPI Anti-Coincidence System in bins of 0.5 s. *Bottom panel*: light curve in the 15–200 keV energy range obtained with the IBIS/ISGRI instrument (bin size 0.1 s). The gaps in the IBIS/ISGRI light curve are due to saturation of the satellite telemetry. From Götz et al. (2006a)

should be uniform over the range $(0, 1)$. In our case, a Kolmogorov-Smirnov (K-S) test shows that a power law is an appropriate model, yielding a probability of 98.8% that the data are well described by our model.

We then divided the bursts into two samples comprising 51 and 173 bursts respectively. The division is based on the periods of different activity of the source: the 51 bursts were detected in 1 year and the 173 in 2.5 months. The two slopes derived with the Maximum Likelihood method are $\alpha = 0.9 \pm 0.2$ for the low level activity period and $\alpha = 0.88 \pm 0.11$ for the high level one. The two slopes are statistically consistent with each other and a K-S test shows that the probability that the two distributions are drawn from the same parent distribution is 93%. Thus we conclude that the relative fraction of bright and faint bursts is not influenced by the level of activity of the source.

2.1.3 *The large outburst of October 5 2004*

On October 5 2004 IBAS triggered at 13:56:49 UT on a series of bursts originating from SGR 1806-20. Detailed analysis of this event showed the presence of more than 100 bursts; the activity ended at 14:08:03 UT. Some bursts were so bright that they saturated the available telemetry share for IBIS, generating some data gaps lasting up to 10–20 s. The initial part of the outburst is shown in Fig. 4.

The fluence of the entire outburst as measured by ISGRI is 1.5×10^{-5} erg cm^{-2}, with a spectrum which is considerably harder than that of the usual short bursts: $kT = 58 \pm 2$ keV, using a thermal bremsstrahlung model. This fluence value is however heavily affected by the saturation of the brightest bursts and represents only a lower limit to the real fluence. In order to recover the complete fluence of the event we used the data from the Anticoincidence Shield (ACS) of the INTEGRAL spectrometer SPI (Vedrenne et al. 2003). As can be seen in Fig. 4 (upper panel), only the brightest bursts are visible in these data and hence they represent complementary information to the ISGRI data.

We used the Monte Carlo package MGGPOD (Weidenspointner et al. 2005) and a detailed mass modelling of SPI and the whole satellite (see Weidenspointner et al. 2003 and references therein) to derive the effective area of the ACS for the direction of SGR 1806-20. We computed the ACS light curve with a binsize of 0.5 s and estimated the background by fitting a constant value to all the data of the same pointing excluding the bursts. We used the background subtracted light curve to compute the fluence of each burst cluster in counts. The ACS data do not provide any spectral information, so we computed the conversion factor to physical units based on the spectral shapes derived from ISGRI data and on the effective area computed through our simulations. The resulting fluences above 80 keV are 1.2×10^{-5} and 9.4×10^{-6} erg cm^{-2} for the first and second clusters respectively. Converting these fluences to the 15–100 keV band one obtains 7.4×10^{-5} and 3.2×10^{-5} erg cm^{-2} respectively. By adding these results to the ones obtained for the ISGRI total spectrum, one can derive the total energy output during the whole event, which is 1.2×10^{-4} erg cm^{-2}. This corresponds to 3.25×10^{42} erg for an assumed distance of 15 kpc (McClure-Griffiths and Gaensler 2005).

These results can be explained in the framework of a recent evolution of the magnetar model, where Lyutikov (2003) explains SGR bursts as generated by loss of magnetic equilibrium in the magnetosphere, in close analogy to solar flares: new current-carrying magnetic flux tubes rise continuously into the magnetosphere, driven by the deformations of the neutron star crust. This in turn generates an increasingly complicated magnetic field structure, which at some point becomes unstable to resistive reconnection. During these reconnection events, some of the magnetic energy carried by the currents associated with the magnetic flux tubes is dissipated. The large event described here can be explained by the simultaneous presence of different active regions (where the flux emergence is especially active) in the magnetosphere of the neutron star. In fact, a long outburst with multiple components is explained as the result of numerous avalanche-type reconnection events, as reconnection at one point may trigger reconnection at other points. This explains the fact that the outburst seems to be composed by the sum of several short bursts. This kind of event may be correlated with a particularly complicated magnetic field structure. A large part of the energy stored in the manetosphere has then been released during the giant flare on December 27, when a global restructuring may have taken place. This mode also suggests that short events are due to reconnection, while longer events have in addition a large contribution from the surface, heated by the precipitating particles, and are harder. This may explain the generally harder spectra observed. However more "classical" scenarios involving only crust fracturing with a large-scale shear deformation of the crust involving the collective motion of many small units, without an internal contribution, cannot be ruled out, see e.g. Thompson and Duncan (2001).

The October 5th event fits the trend of increasing source bursting activity shown by SGR 1806-20 in 2003 and 2004. In the same time span also the luminosity and spectral hardness of the persistent emission at high (20–150 keV, see below) and low (2–10 keV, Mereghetti et al. 2005b) energies increased. On the other hand, this peculiar event did not mark a peak or a turnover in the SGR activity. In fact the two XMM observations of SGR 1806-20 performed just before (September 6 2004) and the day after this large outburst (as a ToO in response to it) yielded similar spectral parameters, fluxes and pulse profiles, and bursts were seen in both observations (Mereghetti et al. 2005b).

Thus events like these release a small (compared to giant flares) fraction of the energy stored in the twisted magnetic field of the neutron star, not allowing the magnetic field to decay significantly. They are rather related to phases of high activity due to large crustal deformations (indicating that a large quantity of energy is still stored in magnetic form) and can be looked at as precursors of a major reconfiguration of the magnetic field.

2.2 The giant flare of December 27 2004

A giant flare from SGR 1806-20 was discovered with the INTEGRAL gamma-ray observatory on 2004 December 27 (Borkowski et al. 2004), and detected with many other satellites (e.g. Palmer et al. 2005; Hurley et al. 2005). The analysis of the SPI-ACS data (>80 keV) of the flare, presented in Mereghetti et al. (2005a), show that the giant flare is composed by an initial spike lasting 0.2 s followed by a $\sim$400 s long pulsating tail, modulated at the neutron star period of 7.56 s. The initial spike was so bright that it saturated the ACS, so we could derive only a lower limit on its fluence, which turned out to be two orders of magnitude brighter (10^{46} ergs, see e.g. Terasawa et al. 2005) than the previously observed giant flares from SGR 1900+14 (Hurley et al. 1999), and SGR 0526–66 (Mazets et al. 1979a). The energy contained in the tail (1.6×10^{44} ergs), on the other hand, was of the same order as the one in the pulsating tails of the previously observed giant flares.

A $\sim$0.2 s long small burst was detected in the ACS data 2.8 s after the initial spike. It is superposed on the pulsating tail and has no clear association with the pulse phase. This burst has been interpreted by Mereghetti et al. (2005a) as the reflection by the Moon of the initial spike of the giant flare. In fact this delay corresponds to the light travel time between INTEGRAL, the Moon, and back. A similar detection was reported with the Helicon-Coronas-F satellite (Mazets et al. 2005).

The most striking feature provided by the INTEGRAL data is the detection of a possible early high-energy afterglow emission associated with the giant flare. At the end of the pulsating tail the count rate increased again, forming a long bump which peaked around $t \sim 700$ s and returned to the pre-flare background level at $t \sim 3000-4000$. This component decays as $\sim t^{-0.85}$, and is shown in blue in Fig. 5, while the overall long term background trend is shown in yellow, and the giant flare itself in red. The association of this emission with SGR 1806-20 is discussed in Mereghetti et al. (2005a). The fluence contained in the 400–4000 s time interval is approximately the same as that in the pulsating tail. With simple gamma-ray burst afterglow models based on synchrotron emission one can derive the bulk Lorentz factor Γ from the time t_0 of the afterglow onset: $\Gamma \sim 15(E/5 \times 10^{43}\ \text{ergs})^{1/8}\ (n/0.1\ \text{cm}^{-3})^{-1/8}\ (t_0/100)^{-3/8}$, where n is the ambient density. This is consistent with the mildly relativistic outflow inferred from the radio data (Granot et al. 2006).

2.3 Discovery of the persistent emission

In 2005 two groups reported independently the discovery of persistent hard X-ray emission originating from SGR 1806-

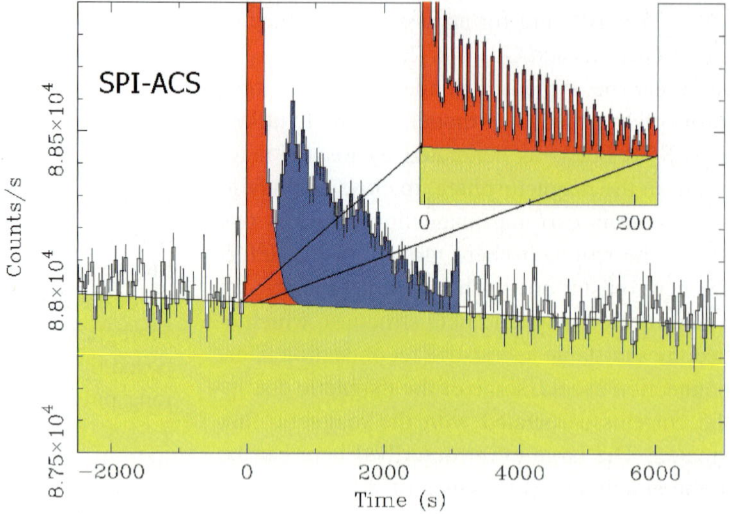

Fig. 5 Light curve of the Giant Flare of December 27 2004 as measured with SPI-ACS above 80 keV. The light curve is binned at 50 s, and hence the pulsating tail is not visible (it is visible in the inset where the light curve is binned at 2.5 s). (*yellow*: instrumental background, *red*: Flare tail, *blue*: high-energy afterglow, see text)

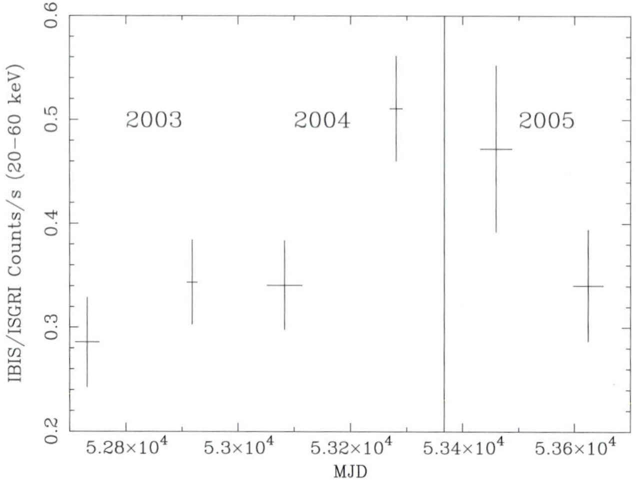

Fig. 6 Long term light curve of SGR 1806-20, as measured with IBIS. The *vertical line* represents the time of the giant flare of December 27 2004

20 (Mereghetti et al. 2005c; Molkov et al. 2005). Up to then, spectral information on the persistent emission of SGRs was known only below 10 keV. The low energy spectrum is usually well described by the sum of a power law component and a black body.

The spectrum above 20 keV is rather hard with a photon index between 1.5 and 2.0 and extends up to 150 keV without an apparent cutoff. It connects rather well with the low energy ($<$10 keV) spectrum (Mereghetti et al. 2005b), and the intensity and spectral hardness are correlated with the degree of bursting activity of the source (Mereghetti et al. 2005c; Götz et al. 2006a) and with the IR flux (Israel et al. 2005). Our group is continuously monitoring the hard X-ray flux of SGR 1806-20, and the long term light curve of the source is shown in Fig. 6. As can be seen, the persistent flux increased in 2003 and 2004 up to the giant flare (which is marked with a vertical line in the plot), and then decreased in 2005.

This behaviour can be interpreted as an increase of the twist angle in the magnetar magnetic field, which in turn increases the burst emission rate, and produces harder spectra, as predicted by Thompson et al. (2002).

3 SGR 1900+14

SGR 1900+14 was discovered in 1979 by Mazets et al. (1979b) when it emitted 3 bursts in 2 days. Since then short bursts were observed from this source with BATSE, RXTE and Interplanetary Network satellites in the years 1979-2002. SGR 1900+14 emitted a giant flare on August 27 1998 (e.g. Hurley et al. 1999), followed by less intense "intermediate" flares on August 29 1998 (Ibrahim et al. 2001) and in April 2001 (Lenters et al. 2003). The last bursts reported from SGR 1900+14 were observed with the Third Interplanetay Network (IPN) in November 2002 (Hurley et al. 2002). No bursts from this source were revealed in all the INTEGRAL observations from 2003 to 2005, but Swift has detected renewed activity in 2006 (Palmer et al. 2006a).

3.1 Discovery of the persistent emission

Using 2.5 Ms of INTEGRAL data, Götz et al. (2006b) reported the discovery of persistent hard X-ray emission, this time from a quiescent SGR 1900+14. This emission extended up to $\sim$100 keV, but with a softer spectrum compared to SGR 1806-20, having a photon index of 3.1$\pm$0.5. Also the luminosity is dimmer in this case, being $\sim$4 $\times$ 10^{35} erg s^{-1}, a factor of three lower than SGR 1806-20. The INTEGRAL observations spanned March 2003 to June 2004, and did not include the recent reactivation of the source in March 2006 (Palmer et al. 2006a), when the source emitted a few tens of regular bursts plus an intense burst series,

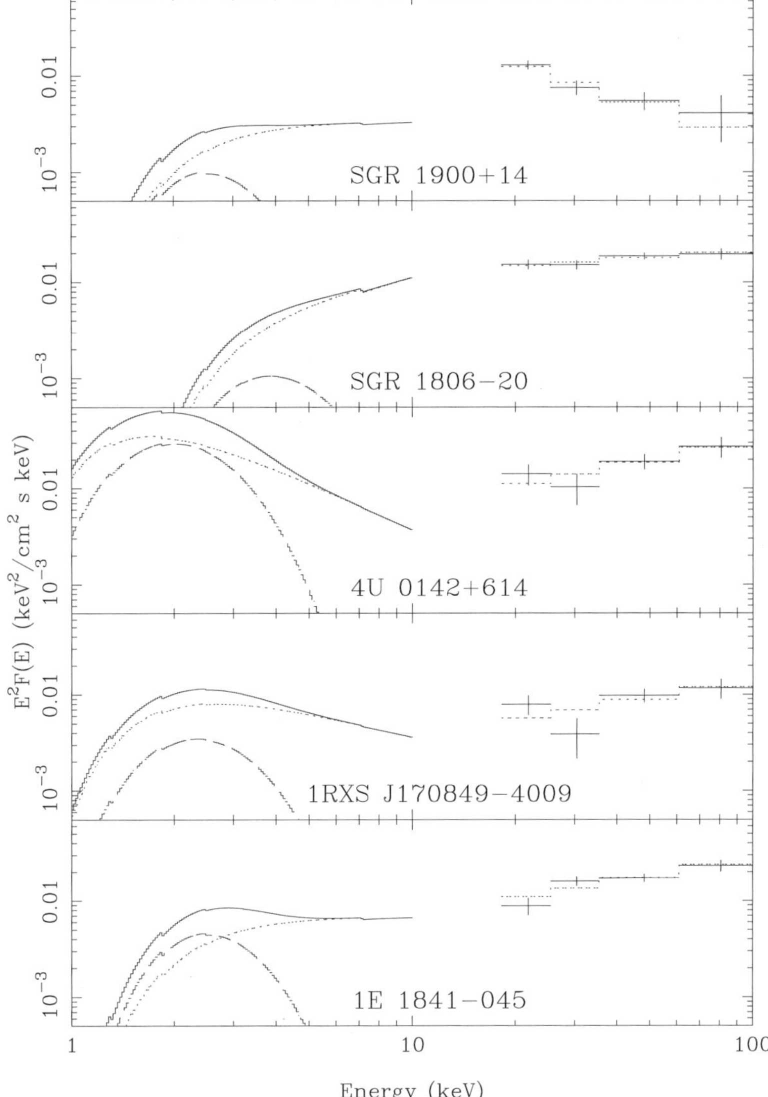

Fig. 7 Broad band X-ray spectra of the five magnetars detected by INTEGRAL. The data points above 18 keV are the INTEGRAL spectra with their best fit power-law models (*dotted lines*). The *solid lines* below 10 keV represent the absorbed power-law (*dotted lines*) plus blackbody (*dashed lines*) models taken from Woods et al. (2001) (SGR 1900+14, during a quiescent state in spring 2000), Mereghetti et al. (2005b) (SGR 1806-20, observation B, when the bursting activity was low), Göhler et al. (2005) (4U 0142+614), Rea et al. (2005) (1RXS J170849-4009), and Morii et al. (2003) (1E 1841-045). From Götz et al. (2006b)

lasting ~30 s (Palmer et al. 2006b), reminiscent of the October 5 2004 event from SGR 1806-20. We recently analysed the INTEGRAL data spanning from August 2004 to March 2006, and found that the hard X-ray flux of the source flux did not increase up to a few weeks before its reactivation. This indicates that the reactivation was not triggered by a flux increase, at least on the time scale of a few months sampled by INTEGRAL.

The soft and constant spectrum of SGR 1900+14 is possibly related to the fact that this source is still in a rather quiescent state.

4 Comparison with the anomalous X-ray pulsars

Hard X-ray persistent emission (>20 keV) has recently been detected from another group of sources, the Anomalous X-ray Pulsars (AXPs, Mereghetti and Stella 1995), which share several characteristics with the SGRs and are also believed to be magnetars (see Woods and Thompson 2003). Hard X-ray emission has been detected from three AXPs with INTEGRAL: 1E 1841-045 (Molkov et al. 2004), 4U 0142+61 (den Hartog et al. 2006) and 1RXS J170849-400910 (Revnivtsev et al. 2004). The presence of pulsations seen with RXTE up to ~200 keV in 1E 1841-045 (Kuiper et al. 2006) proves that the hard X-ray emission originates from the AXP and not from the associated supernova remnant Kes 73. The discovery of (pulsed) persistent hard X-ray tails in these three sources was quite unexpected, since below 10 keV the AXP have soft spectra, consisting of a blackbody-like component ($kT \sim 0.5$ keV) and a steep power law (photon index ~3–4).

In order to coherently compare the broad band spectral properties of all the SGRs and AXPs detected at high energy, we analysed all the public INTEGRAL data using the same procedures. Our results are shown in Fig. 7, where the INTEGRAL spectra are plotted together with the results of observations at lower energy taken from the literature (see figure caption for details).

As can be seen, AXPs generally present harder spectra than SGRs in hard X-rays. In particular, for the three AXPs, a spectral break is expected to occur between 10 and 20 keV in order to reconcile the soft and the hard parts of the spectrum. On the other hand SGRs, present a softer spectrum at higher energies also implying a break around 15 keV (especially for SGR 1900+14) but in the opposite sense with respect to the AXPs. The fact that the spectral break is more evident in SGR 1900+14 could be due to the fact that its level of activity was much lower during our observations, compared to SGR 1806-20. All three AXPs, on the other hand, can be considered to have been in a quiescent state since no bursts were been reported from them during the INTEGRAL observation.

The magnetar model, in its different flavours, explains this hard X-ray emission as powered by bremsstrahlung photons produced either close to the neutron star surface, or at a high altitude ($\sim$100 km) in the magnetosphere (Thompson et al. 2002; Thompson and Belobodorov 2005). The two models can be distinguished by the presence of a cutoff at $\sim$100 keV or $\sim$1 MeV. Unfortunately current experiments like INTEGRAL are not sensitive enough to firmly assess the presence of the cutoffs and hence to distinguish between the two models.

5 Conclusions

Thanks to INTEGRAL, and in particular to its imager IBIS, we have been able to study most of the magnetars' phenomenology with unprecedented sensitivity at high energies. One of the most striking results is the discovery, which was particularly unexpected for AXPs, of the persistent hard X-ray emission. This discovery, which can be considered one of the most important INTEGRAL results at all, represents a new important input for theoreticians who started to include it in the magnetar model (see e.g. Belobodorov 2007).

Also, the fact that short bursts evolve with time is a new feature that has to be considered with care within the magnetar model: up to now no clear explanation has been provided for this.

The large number of detected short bursts from SGR 1806-20 allowed us a good determination of the shape and slope of their Number-Intensity distribution, showing that a single power law holds over 2.5 orders of magnitude.

In addition, the fact that SGR 1806-20 has been particularly active in these last years, also emitting a once-in-a-lifetime event such as the giant flare (and its possible high energy afterglow), has allowed observations of relatively rapid changes of the bursting and persistent emission of a Magnetar and to interpret then with the evolution of a very strong and complicated magnetic field, confirming the magnetic field as the dominant source of energy in Soft Gamma-Ray Repeaters and Anomalous X-ray Pulsars.

References

Belobodorov, A.M.: Astrophys. Space Sci., DOI 10.1007/s10509-007-9318-x (2007)
Borkowski, J., Götz, D., Mereghetti, S., et al.: GCN 2920 (2004)
Crawford, D.F., Jauncey, D.L., Murdoch, H.S.: Astrophys. J. **162**, 405 (1970)
Duncan, R.C., Thompson, C.: Astrophys. J. **392**, L9 (1992)
Göhler, E., Wilms, J., Staubert, R.: Astron. Astrophys. **433**, 1079 (2005)
Golenetskii, S., Aptekar, R., Mazets, E., et al.: GCN 2665 (2004)
Götz, D., Mereghetti, M.F.I., et al.: Astron. Astrophys. **417**, L45 (2004)
Götz, D., Mereghetti, S., Molkov, S., et al.: Astron. Astrophys. **445**, 313 (2006a)
Götz, D., Mereghetti, S., Tiengo, A., et al.: Astron. Astrophys. **449**, L31 (2006b)
Granot, J., Ramirez-Ruiz, E., Taylor, G.B., et al.: Astrophys. J. **638**, 391 (2006)
den Hartog, P.R., Hermsen, W., Kuiper, L., et al.: Astron. Astrophys. **451**, 587 (2006)
Hurley, K., Cline, T., Mazets, E., et al.: Nature **397**, 41 (1999)
Hurley, K., Mazets, E., Golenetskii, S., et al.: GCN Circ. 1715 (2002)
Hurley, K., Atteia, J.-L., Kawai, N., et al.: GCN 2308 (2003)
Hurley, K., Boggs, S.E., Smith, D.M., et al.: Nature **434**, 1098 (2005)
Ibrahim, A.I., Strohmayer, T.E., Woods, P.M., et al.: Astrophys. J. **558**, 237 (2001)
Israel, G., Covino, S., Mignani, R., et al.: Astron. Astrophys. **438**, L1 (2005)
Kouveliotou, C., Dieters, S., Strohmayer, T., et al.: Nature **393**, 235 (1998)
Kuiper, L., Hermsen, W., den Hartog, P.R., et al.: Astrophys. J. **645**, 556 (2006)
Laros, J., Fenimore, E.E., Fikani, M.M., et al.: Nature **322**, 152 (1986)
Lebrun, F., Leray, J.P., Lavocat, P., et al.: Astron. Astrophys. **411**, L141 (2003)
Lenters, G.T., Woods, P.M., Goupell, J.E., et al.: Astrophys. J. **587**, 761 (2003)
Lyutikov, M.: Mon. Not. Roy. Astron. Soc. **346**, 540 (2003)
Mazets, E.P., Golentskii, S.V., Ilinskii, V.N., et al.: Nature **282**, 587 (1979a)
Mazets, E.P., Golenetskij, S.V., Guryan, Y.A., Sov. Astron. Lett. **5**, 343 (1979b)
Mazets, E.P., Cline, T.L., Aptekar, R.L., et al.: astro-ph/0502541 (2005)
McClure-Griffiths, N.M., Gaensler, B.M.: Astrophys. J. **630**, L161 (2005)
Mereghetti, S., Stella, L.: Astrophys. J. **442**, L17 (1995)
Mereghetti, S., Götz, D., Borkowski, et al.: Astron. Astrophys. **411**, L291 (2003)
Mereghetti, S., Götz, D., Mowlavi, N., et al.: GCN 2647 (2004)
Mereghetti, S., Götz, D., von Kienlin, A., et al.: Astrophys. J. **624**, L105 (2005a)
Mereghetti, S., Tiengo, A., Esposito, P., et al.: Astrophys. J. **628**, 938 (2005b)
Mereghetti, S. , Götz, D., Mirabel, I.F., et al..: Astron. Astrophys. **433**, L9 (2005c)

Molkov, S.V., Cherepashchuk, A.M., Lutovinov, A.A., et al.: Astron. Lett. **30**, 534 (2004)
Molkov, S., Hurley, K., Sunyaev, R., et al.: Astron. Astrophys. **433**, L13 (2005)
Paczynski, B., Astron. Astrophys. **42**, 145 (1992)
Palmer, D.M., Barthelmy, S., Gehrels, N., et al.: Nature **434**, 1107 (2005)
Morii, M., Sato, R., Kataoka, J., Kawai, N., Publ. Astron. Soc. Pac. **55**, L45 (2003)
Palmer, D.M., Sakamoto, T., Barthelmy, S., et al.: ATel 789 (2006a)
Palmer, D.M., Sakamoto, T., Barthelmy, S., et al.: GCN Circ. 4949 (2006b)
Rea, N., Oosterbroek, T., Zane, S., et al.: Mon. Not. Roy. Astron. Soc. **361**, 710 (2005)
Revnivtsev, M.G., Sunyaev, R.A., Varshalovich, D.A., et al.: Astron. Lett. **30**, 382 (2004)
Terasawa, T., Tanaka, Y.T., Takei, Y., et al.: Nature **434**, 1110 (2005)
Thompson, C., Beloborodov, A.M.: Astrophys. J. **634**, 565 (2005)
Thompson, C., Duncan, R.C.: Mon. Not. Roy. Astron. Soc. **255** 275 (1995)
Thompson, C., Duncan, R.C.: Astrophys. J. **561**, 980 (2001)
Thompson, C., Lyutikov, M., Kulkarni, S.R.: Astrophys. J. **574**, 332 (2002)
Ubertini, P., Lebrun, F., Di Cocco, G., et al.: Astron. Astrophys. **411**, L131 (2003)
Vedrenne, G., Roques, J.-P., Schönfelder, V., et al.: Astron. Astrophys. **411**, L63 (2003)
Weidenspointner, G., Kiener, J., Gros, M., et al.: Astron. Astrophys. **411**, L113 (2003)
Weidenspointner, G., Harris, M.J., Sturner, S., et al.: Astrophys. J. Suppl. Ser. **156**, 69 (2005)
Winkler, C., Courvoisier, T.J.-L., Di Cocco, G., et al.: Astron. Astrophys. **411**, L1 (2003)
Woods, P.M., Thompson, C.: In: Lewin, W.H.G., van der Klis, M. (eds.) Compact Stellar X-ray Sources, astro-ph/0406133 (2004)
Woods, P.M., Kouveliotou, C., van Paradijs, J., et al.: Astrophys. J. **527**, L47 (1999)
Woods, P.M., Kouveliotou, C., Göğüş, E., et al.: Astrophys. J. **552**, 748 (2001)

Astrophys Space Sci (2007) 308: 61–65
DOI 10.1007/s10509-007-9306-1

ORIGINAL ARTICLE

Our distorted view of magnetars: application of the resonant cyclotron scattering model

Nanda Rea · Silvia Zane · Maxim Lyutikov · Roberto Turolla

Received: 20 July 2006 / Accepted: 9 September 2006 / Published online: 15 March 2007

Abstract The X-ray spectra of the magnetar candidates are customarily fitted with an empirical, two component model: an absorbed blackbody and a power-law. However, the physical interpretation of these two spectral components is rarely discussed. It has been recently proposed that the presence of a hot plasma in the magnetosphere of highly magnetized neutron stars might distort, through efficient resonant cyclotron scattering, the thermal emission from the neutron star surface, resulting in production of non-thermal spectra. Here we discuss the Resonant Cyclotron Scattering (RCS) model, and present its XSPEC implementation, as well as preliminary results of its application to Anomalous X-ray Pulsars and Soft Gamma-ray Repeaters.

Keywords Neutron stars · Pulsars · Magnetars · X-ray · Resonant cyclotron scattering

N.R. is supported by an NWO Post-Doctoral Fellowship. S.Z. thanks the Particle Physics and Astronomy Research Council, PPARC, for support through an Advanced Fellowship.

N. Rea (✉)
SRON Netherlands Institute for Space Research, Sorbonnelaan, 2, 3584CA, Utrecht, The Netherlands
e-mail: N.Rea@sron.nl

S. Zane
Mullard Space Science Laboratory, University College of London, Holbury St. Mary, Dorking Surrey, RH5 6NT, UK

M. Lyutikov
University of British Columbia, 6224 Agricultural Road, Vancouver, BC, V6T 1Z1, Canada

R. Turolla
Physics Department, University of Padua, via Marzolo 8, 35131, Padova, Italy

PACS 97.60.Jd · 97.60.Gb

1 Introduction

Anomalous X-ray Pulsars (AXPs) and Soft Gamma-ray Repeaters (SGRs) are a small class of slowly rotating (5–12 s) neutron stars with emission properties much at variance with those of ordinary X-ray pulsars, both the young radio pulsars and the X-ray binary pulsars. They are called "anomalous" because their high X-ray luminosity (10^{34}–10^{36} erg/s) cannot be easily explained in terms of the conventional processes which apply to other classes of pulsars, i.e. accretion from a binary companion or injection of rotational energy in the pulsar wind/magnetosphere. On the other hand, measurements of spin periods and period derivatives, when the latter are interpreted as due to electromagnetic dipolar losses, suggest that these objects may host "magnetars," i.e. neutron stars endowed with an ultra-strong magnetic field (10^{14}–10^{15} G, see Duncan and Thompson 1992; Thompson and Duncan 1993, 1996). The magnetar scenario appears so far very promising in explaining both the main energy source of these objects (the decay of the superstrong field) and the emission of the short, energetic bursts. Moreover, the magnetar scenario can account for the properties of Giant Flares, extremely energetic transient events ($L \approx 10^{44}$–10^{47} erg/s) detected from SGRs. However, alternative scenarios to explain the enigmatic properties of these sources have been invoked. Among these, models involving accretion from a fossil disk, left by the supernova event which gave birth to the neutron star, are still largely plausible (van Paradijs et al. 1995; Chatterjee et al. 2000). Very recently such a (maybe passive) disk has been indeed observed in the IR around AXP 4U 0142 + 61 (Wang et al. 2006). Magnetars candidates are strong X-ray sources, and their

X-ray spectra in the 0.5–10 keV band are usually fit with a two component model consisting of (besides interstellar absorption) a thermal blackbody with a typical temperature $kT \sim 0.3$–0.4 keV and a power-law with a relatively steep photon index, $\Gamma \sim 2$–4 (see Kaspi 2006 in this volume, and Woods and Thompson 2004 for recent reviews). In some cases, SGR spectra have been fit with a single power-law component, but recent longer observations showed that, also for these sources, a blackbody component is often required (Mereghetti et al. 2005a, 2005b).

Despite the fact that the blackbody plus power-law spectral model has been largely applied to magnetar spectra for many years, a reliable physical interpretation of these two components is still missing. Quite recently, an attempt to interpret the observed spectrum and long term variations of 1E 1048.1–5937 in terms of a (non-resonant) Comptonization model has been presented by Tiengo et al. (2005). Furthermore, the recent discovery of magnetar counterparts in the radio to the γ-ray bands (Camilo et al. 2006; Hulleman et al. 2000; Kuiper et al. 2004) enforced the idea that their multiwavelength spectral energy distribution is by far more complicated than a simple blackbody plus power-law distribution (see also Den Hartog et al. contribution in this volume).

2 The Resonant Cyclotron Scattering (RCS) model applied to magnetars spectra

Following the original suggestion by Thompson et al. (2002), it has been proposed by Lyutikov and Gavriil (2006) that the presence of a warm non-relativistic plasma in the magnetosphere of magnetars, might be responsible, through efficient resonant electron cyclotron scattering, of distorting the thermal X-ray emission from the star surface. In particular, if a large volume of the neutron star magnetosphere of the is partially filled by $e^{\pm}$-currents, the thermal (or quasi-thermal) cooling radiation emerging from the star surface will experience repeated scatterings at the cyclotron resonance.

The efficiency of the process is quantified by the scattering optical depth, $\tau_{\rm res}$. Following again Lyutikov and Gavriil (2006), and assuming a dipolar configuration for the magnetic field, this can be written as

$$\tau_{\rm res} = \int \sigma_{\rm res} n \, dl = \tau_0 (1 + \cos^2\alpha)$$

where $\sigma_{\rm res}$ is the cross section for electron scattering in the magnetized regime, n is the electrons number density, α is the angle between the photon propagation direction and the local magnetic field, and

$$\tau_0 = \frac{\pi^2 e^2 n r}{3 m_e c \omega_B}.$$

Here r is the radial coordinate, $\omega_B = eB/m_e c$ is the electron cyclotron frequency, and B is the local value of the magnetic field. At variance with the case of unmagnetized Thomson scattering, in presence of a magnetic field $\sigma_{\rm res}$ is resonant and is given by

$$\sigma_{\rm res} = \frac{\sigma_T}{4} \frac{(1+\cos^2\alpha)\omega^2}{(\omega - \omega_B)^2 + \Gamma^2/4}$$

where $\Gamma = 4e^2\omega_B^2/3m_e c^3$ is the natural width of the first cyclotron harmonic and σ_T is the cross-section for unmagnetized Thomson scattering. The crucial point is that, in this model, the plasma distribution is spatially extended. Since the value of the resonant energy depends on the local value of the magnetic field, repeated scatterings of photons could lead to the formation of a high energy tail instead of a narrow line. At relatively large distances from the surface, the magnetic field drops to typical values such that scattering is resonant for soft X-ray photons, below $\sim$10 keV. On the other hand, at these energies the resonant scattering optical depth greatly exceeds that for Thomson scattering, $\tau_T \sim rn\sigma_T$,

$$\frac{\tau_{\rm res}}{\tau_T} \sim \frac{\pi}{8} \frac{c}{r_e \omega_B} \sim 10^5 \left(\frac{1\ \text{keV}}{\hbar\omega_B}\right)$$

where $r_e = e^2/mc^2$ is the classical electron radius. This implies that even a relatively small amount of plasma suspended in the magnetosphere of the neutron star may considerably modify the emergent spectra. There is now increasing support to the idea that the magnetar magnetospheres might be twisted (Thompson et al. 2002), in which case they may be partially filled by e^- currents with densities well in excess of the Goldreich–Julian density (which is expected in the case of a simple dipolar configuration, see Goldreich and Julian 1969).

A further effect is related to the fact that the spatial region, over which such currents are distributed, is permeated by a highly inhomogeneous magnetic field. Although for soft X-ray photons the scattering conserves energy in the electron rest frame, thermal/bulk electron motion produces an energy transfer in the observer rest frame. Multiple resonant cyclotron scatterings of thermal photons will, on average, up-scatter the transmitted radiation, resulting in the formation of a non-thermal spectrum. The emerging spectrum will then appear distorted and, as it is the case for thermal non-magnetic comptonization, its shape may be well represented by the superposition of thermal and non-thermal components (see Fig. 1).

Here we present an application of these models to magnetar spectra. Theoretical models are computed as in Lyutikov and Gavriil (2006), under the assumption that scattering occurs in a non-relativistic, static, warm medium and neglecting electron recoil, i.e. in the limit $\beta_{th} < 1$ (where β_{th}, a model parameter, is the thermal velocity of electrons

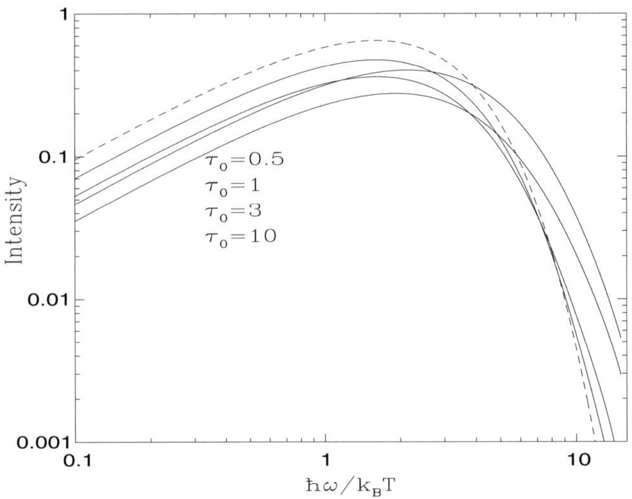

Fig. 1 Modification of an initially Plankian spectrum (*dashed line*) by multiple cyclotron scattering for different values of τ_0 and $\beta_{th} = 0.3$ (Lyutikov and Gavriil 2006)

in units of the speed of light) and for low energy photons, $\epsilon \ll mc^2$. Within the range of parameter values allowed by the numerical code, the non-thermal effects are most prominent for models in which the plasma is mildly relativistic, $\beta_{th} \lesssim 1$, and the optical depth is reaching its boundary value $\tau_{\rm res} \sim 10$.

3 The RCS XSPEC model

In order to perform a quantitative comparison between fits to magnetar spectra performed with the model described above and with the canonical blackbody plus power-law model, we developed a code to upload the RCS model into XSPEC. We created a grid of intensity tables (through a Monte-Carlo simulation) for a set of values of the three model parameters, i.e. β_{th}, τ_0 and T, where T is the temperature of the seed thermal surface emission (assumed to be a blackbody). The parameters ranges used were $0.1 < \beta_{th} < 0.5$ (step 0.1), $1 < \tau < 10$ (step 1) and $0.1\ \mathrm{keV} < T < 3\ \mathrm{keV}$ (step 0.05 keV). For each model, the spectrum has been computed in the energy range 0.1–12 keV (step 0.05 keV). The final XSPEC *atable* spectral model has therefore four parameters, the three listed above plus the last one being the normalization constant, which are simultaneously varied during the spectral fitting following the standard χ^2 minimization technique.

Note that a previous attempt to apply the Lyutikov and Gavriil (2006) model to spectral data was presented by the authors themselves. However, in that case β_{th} and τ_0 were fixed a priori during each fit and eventually varied manually. The fit itself was done only by varying T and the normalization factor, while in this case we are able to fit simultaneously all the four parameters, resulting in a more precise determination of their best fitting values and in a more reliable χ^2 determination.

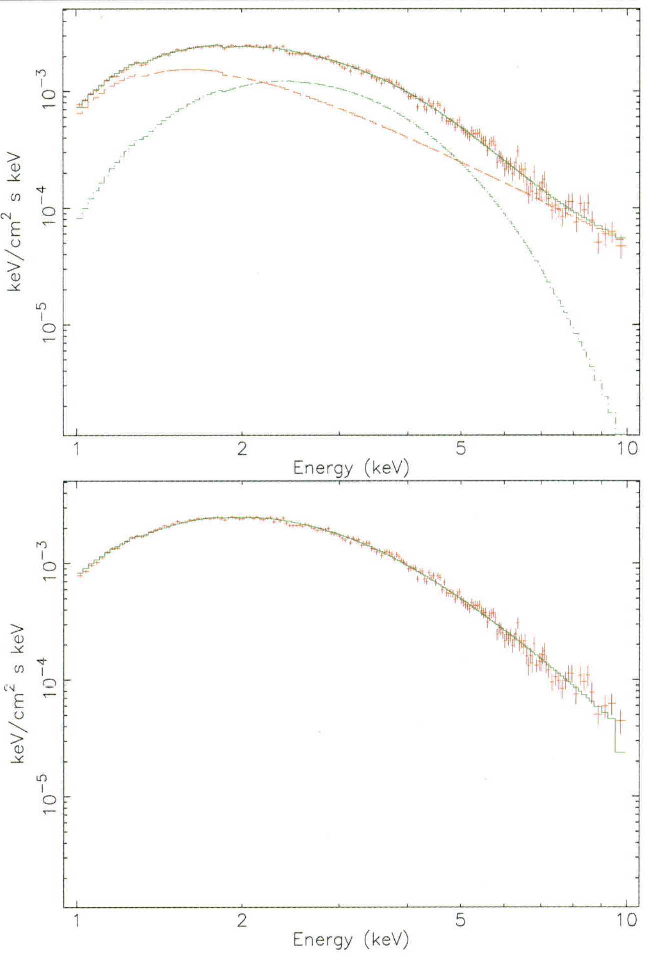

Fig. 2 The XMM–Newton spectrum of 1E 1048–5937 fitted with an absorbed power-law plus a blackbody (*top panel*) and with the RCS model (*bottom panel*)

4 Preliminary results

Here we present the preliminary results of the spectral modelling of magnetars with the Resonant Cyclotron Scattering model (RCS). We concentrate on the soft X-ray spectra (1–10 keV) of three sources, 1E 1048–56, 1RXS J1708–4009 and SGR 1806–20, and we use data obtained with the XMM-Newton satellite. Detailed information about the observations, data reduction and analysis can be found in Mereghetti et al. (2004) and Rea et al. (2005a, 2005b), for the three sources, respectively.

The fit of the X-ray spectrum of 1E 1048–56 is presented in Fig. 2 and Table 1. In this case we find that the RCS model is successful in reproducing the data, as well as the canonical blackbody plus power-law model (BB+PL). The two fits have the same number of degrees of freedom. The

Table 1 Best fit values of the spectral parameters obtained by fitting the XMM–Newton spectrum of 1E 1048–5937 with the BB+PL and an RCS model (for more details about this observation see Mereghetti et al. (2004)). Errors are at 1σ confidence level, and the reported flux is absorbed and in the 2–10 keV energy range

	BB + PL	RCS model
N_H (10^{22} cm^{-2})	$1.12^{+0.04}_{-0.02}$	$0.49^{+0.01}_{-0.02}$
kT (keV)	$0.64^{+0.01}_{-0.01}$	$0.53^{+0.01}_{-0.01}$
BB norm	$1.03^{+0.02}_{-0.01} \times 10^{-4}$	–
Γ	$3.3^{+0.2}_{-0.1}$	–
PL norm	$1.10^{+0.13}_{-0.04} \times 10^{-2}$	–
β_{th}	–	$0.41^{+0.01}_{-0.02}$
τ_0	–	$1.9^{+0.1}_{-0.1}$
RCS norm	–	$0.22^{+0.01}_{-0.03}$
χ^2_ν	0.99	1.08
Flux (erg s^{-1} cm^{-2})	$7.9^{+0.1}_{-0.1} \times 10^{-12}$	$7.9^{+1.5}_{-1.7} \times 10^{-12}$

value of the column density found with the RCS model is lower (although both consistent with Durant and van Kerkwijk 2006), while the temperature of the thermal component T and the total flux are consistent with that found with the BB+PL modelling. These results are in agreement with the previous findings by Lyutikov and Gavriil (2006), who presented a similar attempt by using Chandra data and varying manually some of the model parameters to reproduce the spectrum.

As mentioned before, the technique used here is more reliable in providing a determination of the best fitting parameters. In this case, the source spectrum can be reproduced by invoking scattering by e^- with thermal velocity $\beta_{th} = 0.41$ and optical depth $\tau_0 = 1.9$.

On the other hand, the cases of 1RXS J1708–4009 and SGR 1806–20 are more problematic. These sources have much harder X-ray spectra (in the 1–10 keV band), that cannot be accounted for by the RCS model alone (see Fig. 3) at least when the spectral deformation resulting from the resonant scattering process is computed in the simple way proposed by Lyutikov and Gavriil (2006).

However, it is worth noting that, at variance with 1E 1048–56, these two sources exhibit strong high-energy power-law tails which extend up to the γ-rays (Kuiper et al. 2006; Molkov et al. 2005; Mereghetti et al. 2005a). Motivated by that, we fitted the whole 1–50 keV spectra with the RCS model plus a power-law. For both sources, we obtained a good fit with a reduced χ^2 of 1.08 and 1.1, respectively. Although extremely promising, these analysis are very preliminary and they still require a careful assessment of intercalibration issues which arise when different detectors are used. Therefore, we do not discuss the results in more detail nor we present here best fitting parameter values. These results will be presented in more details in a forthcoming paper (Rea et al. 2006, in preparation).

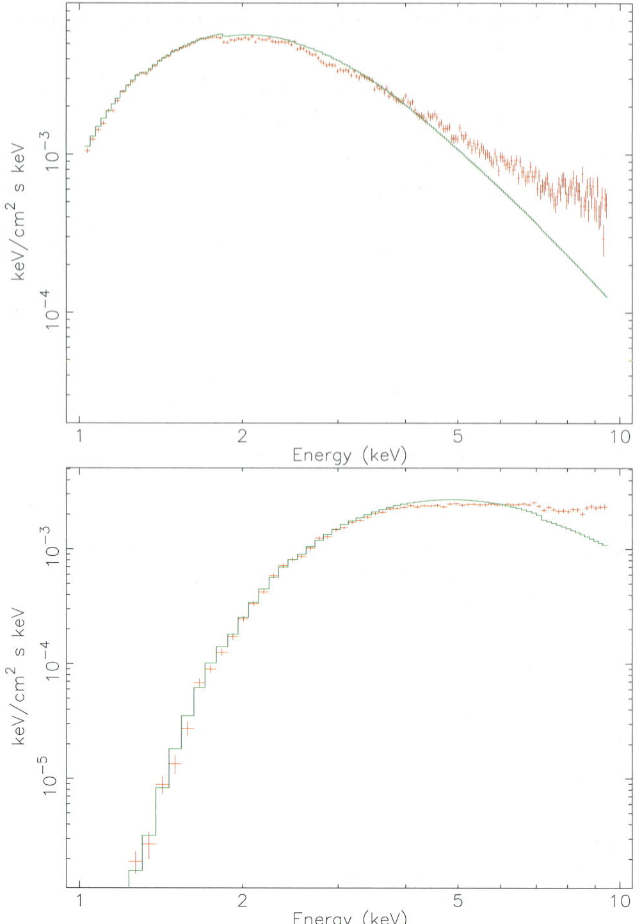

Fig. 3 XMM–Newton spectra of 1RXS J1708–4009 (*top panel*) and SGR 1806–20 (*bottom panel*) fitted with he RCS model. The black-body plus power-law fits might be found in Rea et al. (2005a) and Mereghetti et al. (2005a, 2005b)

5 Conclusion

The preliminary results reported here show that the X-ray spectra of the AXP 1E 1048–56 can be successfully reproduced by the RCS model. The quality of the fit is comparable to that obtained by using the canonical blackbody plus power-law model, with the advantage that the RCS model spectra provide a better physical understanding of the presence of the thermal/non-thermal components in the observed spectrum. On the other hand, this is not totally unexpected since this source has a non-thermal tail which is relatively soft with respect to other AXPs, and a similar result has already been found by Lyutikov and Gavriil, although their fitting technique was less accurate. For the other two sources considered here, 1RXS J1708–4009 and SGR 1806–20, we find that the RCS model alone cannot account for the relatively hard non-thermal component observed below ~10 keV.

On the other hand, in both cases a composite model consisting of a RCS model plus a power-law can fit the X-ray spectra in the 1–100 keV band. This may suggest that, in the 6–10 keV energy range, the X-ray emission of these sources is contaminated by the hard non-thermal component which has been recently observed with INTEGRAL at much higher energies (Kuiper et al. 2006). A similar hard tail has not been observed, at least so far, in the case of 1E 1048–56. A detailed paper with a systematic application of the RCS model to the whole class of magnetars and a study of their spectral and flux variability in this scenario is in preparation, accompanied by a more complete discussion (Rea et al. 2006, in preparation).

Acknowledgements We thank Valentina Bianchin and Gavin Ramsay for their kind help in building the XSPEC RCS model. We also thank Fotis Gavriil for having kindly allowed us to look into his preliminary model, Andrea Tiengo, Gianluca Israel and the anonymous referee for useful comments. NR thanks the Mullard Space Science Laboratory, where this work was partially done, for the warm hospitality.

References

Camilo, F., Ransom, S., Halpern, J., et al.: Nature **442**(7105), 892–895 (2006) Preprint (astro-ph/0605429)
Chatterjee, P., Hernquist, L., Narayan, R.: Astrophys. J. **534**, 373 (2000)
Duncan, R.C., Thompson, C.: Astrophys. J. **392**, L9 (1992)
Durant, M., van Kerkwijk, M.H.: Astrophys. J. **650**, 1070 (2006) (astro-ph/0606604)
Goldreich, P., Julian, W.H.: Astrophys. J. **157**, 869 (1969)
Hulleman, F., van Kerkwijk, M.H., Kulkarni, S.R.: Nature **408**, 689 (2000)
Kuiper, L., Hermsen, W., Mendez, M.: Astrophys. J. **613**, 1173 (2004)
Kuiper, L., Hermsen, W., den Hartog, P.R., Collmar, W.: Astrophys. J. **645**, 556 (2006)
Lyutikov, M., Gavriil, F.: Mon. Not. Roy. Astron. Soc. **368**, 690 (2006)
Mereghetti, S., Tiengo, A., Stella, L., et al.: Astrophys. J. **608**, 427 (2004)
Mereghetti, S., Tiengo, A., Esposito, P., et al.: Astrophys. J. **628**, 938 (2005a)
Mereghetti, S., Gotz, D., Mirabel, I.F., Hurley, K.: Astron. Astrophys. **433**, L9 (2005b)
Molkov, S., Hurley, K., Sunyaev, R., et al.: Astron. Astrophys. **433**, L13 (2005)
Rea, N., Oosterbroek, T., Zane, S., et al.: Mon. Not. Roy. Astron. Soc. **361**, 710 (2005a)
Rea, N., Israel, G.L., Covino, S., et al.: Astron. Telegr., #645 (2005b)
Thompson, C., Duncan, R.C.: Astrophys. J. **408**, 194 (1993)
Thompson, C., Duncan, R.C.: Astrophys. J. **473**, 322 (1996)
Thompson, C., Lyutikov, M., Kulkarni, S.R.: Astrophys. J. **574**, 332 (2002)
Tiengo, A., Mereghetti, S., Turolla, R., et al.: Astron. Astrophys. **437**, 997 (2005)
van Paradijs, J., Taam, R.E., van den Heuvel, E.: Astron. Astrophys. **299**, L41 (1995)
Wang, Z., Chakrabarty, D., Kaplan, D.L.: Nature **440**(7085), 772–775 (2006)
Woods, P.M., Thompson, C.: Preprint (2004) (astro-ph/0406133)

Astrophys Space Sci (2007) 308: 67–71
DOI 10.1007/s10509-007-9304-3

ORIGINAL ARTICLE

Spitzer space telescope observations of SGR and AXP environments

Stefanie Wachter · Chryssa Kouveliotou · Sandeep Patel · Don Figer · Peter Woods

Received: 14 July 2006 / Accepted: 21 August 2006 / Published online: 20 March 2007

Abstract Both Anomalous X-ray Pulsars (AXPs) and Soft Gamma Repeaters (SGRs) are thought to be manifestations of magnetars. However, the specific physical characteristics that differentiate the two classes of objects remain unclear. There is some evidence that the progenitors of these sources and/or the environment in which they form might influence the type of phenomena the resulting magnetar displays. Several of the AXPs appear to be associated with supernova remnants, while embedded clusters of massive stars have been found in the immediate vicinity of some SGRs. Since both AXPs and SGRs are distributed close to the Galactic plane, high extinction makes studies in the optical difficult. We present early results from our Spitzer program aimed at probing the environmental factors that might contribute to the difference in the observed characteristics between AXPs and SGRs.

Keywords SGR · AXP · Infrared · Environment

PACS 97.60.Jd

S. Wachter (✉)
Spitzer Science Center/California Institute of Technology, Pasadena, CA, USA
e-mail: wachter@ipac.caltech.edu

C. Kouveliotou
MSFC/NASA HQ, Huntsville, AL, USA

S.K. Patel
NSSTC/USRA/MSFC, Huntsville, AL, USA

D.F. Figer
Rochester Institute of Technology, Rochester, NY, USA

P.M. Woods
Dynetics, Huntsville, AL, USA

1 Introduction

Both Anomalous X-ray Pulsars (AXPs) and Soft Gamma Repeaters (SGRs) are thought to be manifestations of magnetars. However, the specific physical characteristics that differentiate the two classes of objects remain unclear. It has been suggested that the two are evolutionary links (e.g. Gotthelf et al. 1999), with AXPs evolving into SGRs as they age (since at most one SGR is convincingly associated with a SNR). However, in this scenario, systematically longer spin periods for SGRs than AXPs and a difference in the inferred magnetic field strength would be expected, contrary to observations (Gaensler et al. 2001). We now also know of a small sample of radio pulsars with magnetic field strengths approaching or overlapping those of the SGRs and AXPs, but they do not display the same type of high energy properties (e.g. Kaspi and McLaughlin 2005). This raises the question why some neutron stars are "normal" radio pulsars, while others are X-ray and γ-ray emitting magnetars and emphasizes that dipole magnetic field strength alone cannot be the sole factor in determining whether a neutron star exhibits magnetar characteristics. Possible "second parameters" that have been suggested include an intrinsic difference in magnetic field geometry (e.g. Kulkarni et al. 2003), and that magnetars are formed from more massive progenitors than radio pulsars (Eikenberry et al. 2004; Gaensler et al. 2005b).

The potential connection to the progenitor mass is particularly intriguing, since there is ample evidence that SGRs and AXPs are associated with sites of massive star formation. Contrary to the standard picture that most massive stars will collapse into black holes, Heger et al. (2003) present evolutionary models that show that even very massive stars can end their lives as neutron stars due to strong stellar winds that reduce the final core mass before the supernova occurs. This scenario is supported by the recent detection of a large radio shell around the AXP 1E 1048.1-5937, which Gaensler et al. (2005b) interpret as the wind blown bubble from a 30–40 $M_{\odot}$ progenitor to the AXP. They also suggest that the absence of a SNR association in this case is due to the expansion of ejecta into a low density bubble, failing to produce observable emission. Thus, the environments of magnetars provide unique insight and constraints not only on their origin and evolution but also on those of massive stars in general.

To explore the role of the environment further, we are undertaking a uniform study of the large scale neighborhood for several members of the SGRs and AXPs with the Spitzer Space Telescope (Werner et al. 2004).

2 Spitzer space telescope observations

The data presented in the remainder of this article represent a combination of observations from our own Spitzer cycle 2 program and the GLIMPSE (Benjamin et al. 2003 and http://www.astro.wisc.edu/sirtf/) and MIPSGAL (http://ssc.spitzer.caltech.edu/legacy/abs/carey.html) legacy projects. We have acquired imaging data with all three instruments aboard Spitzer as part of our program, but the preliminary results discussed here only utilize 3.6–8.0 μm observations obtained with the InfraRed Array Camera (IRAC, Fazio et al. 2004) and the 24 μm array of the Multiband Imaging Photometer for Spitzer (MIPS, Rieke et al. 2004).

The recent discovery of a debris disk around AXP 0142+61 with Spitzer (Wang et al. 2006) underscores the exciting prospect of directly detecting the magnetar counterparts. However, the moderate spatial resolution ($1.2''$/pixel) of IRAC (and to an even greater extent of MIPS with $2.55''$/pixel) and the low flux levels predicted for these sources in combination with bright neighbor contamination in the typically crowded fields of SGRs and AXPs severely limits our ability to image such counterparts.

In the following sections we will present two specific examples from our Spitzer survey, SGR 1806-20 and AXP 1841-045.

2.1 SGR 1806-20

SGR 1806-20 lies within a cluster of massive stars (Fuchs et al. 1999) containing a luminous blue variable (LBV) and several Wolf–Rayet and OB supergiant stars (Figer et al. 2005). The distance to SGR 1806-20 is currently a matter of debate. Published estimates include 9.8 kpc (Cameron et al. 2005), 11.8 kpc (Figer et al. 2004), and ~ 15 kpc (Corbel et al. 1997; McClure-Griffiths and Gaensler 2005). Increased bursting activity from SGR 1806-20 was detected in 2003/2004, culminating in a giant flare on 27 December 2004. Radio (Cameron et al. 2005; Gaensler et al. 2005b) and near-IR observations following this flare established the first likely near-IR candidate for an SGR (Israel et al. 2005; Kosugi et al. 2005).

Figure 1 shows a three-color image composed of the IRAC 4.5 μm (blue), 8.0 μm (green), and MIPS 24 μm (red) data sets. The image covers $\sim 8' \times 11'$, north is up and east to the left. At IRAC wavelengths, only the stellar cluster members are visible and the cluster does not stand out strongly against the general background of stars. The 24 μm image, however, reveals prominent diffuse emission enveloping the cluster. The emission appears approximately centered on the stellar density peak of the cluster (see also the contours in Fig. 2a) and extends $\sim 1.5'$ from north to south and $\sim 0.9'$ from east to west, corresponding to 3.9×6.6 pc at a distance of 15 kpc. Figure 2 contrasts the difference between the 24 μm and radio 20 cm contours (taken from the MAGPIS survey, Helfand et al. 2005), overlaid on the 24 μm image. The radio position of the SGR is indicated by a white circle. The central portion of the 24 μm emission strongly resembles the distribution of the 12–18 μm emission seen with ISO, which Fuchs et al. (1999) interpret as the nascent dust cloud of the cluster. Our 24 μm observations reveal a much more extended emission, particularly along the north/south axis. The radio nebula (note that this is *not* the radio emission detected after the giant flare) is not centered on the position of the SGR and is thought to be generated by the LBV in the cluster, the bright source visible to the east of the SGR in Fig. 1. We will have to combine the 24 μm data with our IRS 16 μm and MIPS 70 μm observations before we can determine whether we can detect multiple components (e.g. in temperature) within this extended diffuse emission.

2.2 AXP 1841-045

AXP 1841-045 lies at the center of the shell-type SNR Kes 73. Deep near-IR searches for a counterpart to the AXP by Wachter et al. (2004) and subsequently Durant (2005) failed to identify promising IR counterpart candidates. Kes 73 exhibits prominent emission at both X-ray and radio wavelengths. Gotthelf and Vasisht (1997) suggest that the SNR is young (< 2000 yr) and the result of a type II/Ib SN, indicating a massive progenitor. Sanbonmatsu and Helfand (1992) determined a distance of 7.0 kpc to the remnant based on HI measurements.

Kes 73 is invisible in all 4 channels of IRAC, but appears prominently in the MIPS 24 μm image. Figure 3 shows a 3-color composite of the IRAC 8.0 μm (blue) and MIPS 24 μm (green) Spitzer images and the MAGPIS radio 20 cm map (red). For a more detailed comparison of just the MIPS 24 μm and radio image see also Fig. 4. Both the MIPS and radio image show enhanced emission at the western edge

Fig. 1 Three-color image of a $8' \times 11'$ area around SGR1806-20, composed of IRAC 4.5 μm (*blue*), 8.0 μm (*green*), and MIPS 24 μm (*red*) data sets. North is *up*, east to the *left*. The position of the SGR is indicated by a *green circle*

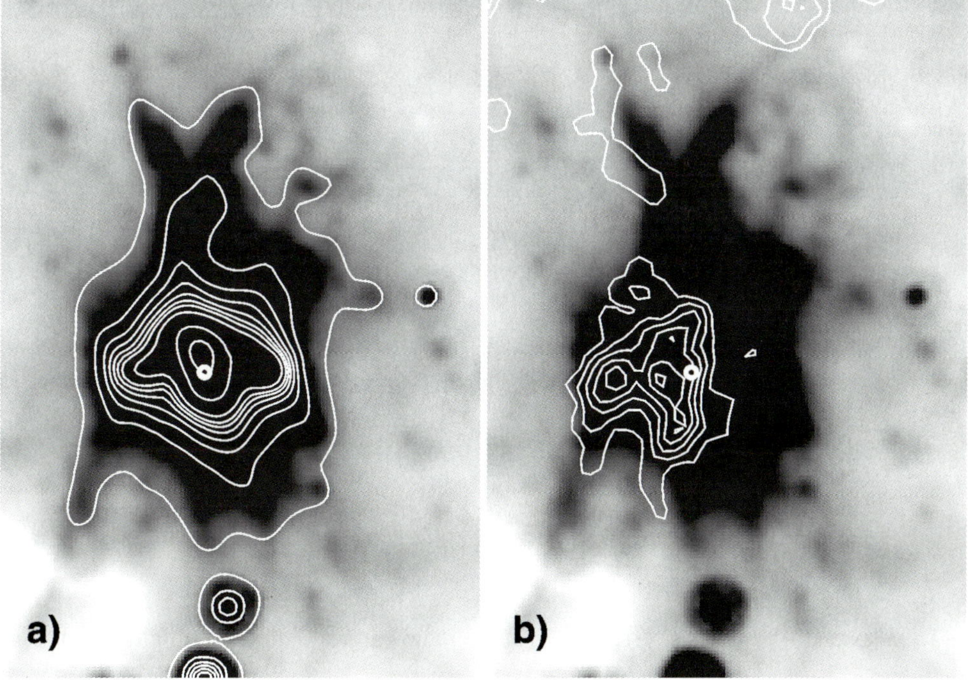

Fig. 2 The 24 μm (**a**) and 20 cm radio contours (**b**) overlaid on the 24 μm image. The radio position of the SGR is indicated by a *white circle*

of the remnant. In addition, the MIPS image reveals several embedded sources immediately adjacent to this western edge indicating that the expansion of the SNR might be stopped by interaction with denser material. These embedded sources could represent star formation triggered by compression of the molecular material by the expanding SNR, or indicate a site of active star formation that also gave birth to the AXP. In general, the 24 μm and radio emission highlight similar structures, especially several arc like features can be seen in both images (Fig. 4). On the other hand, there are also distinct differences, in particular, enhanced 24 μm emission is visible in the central area

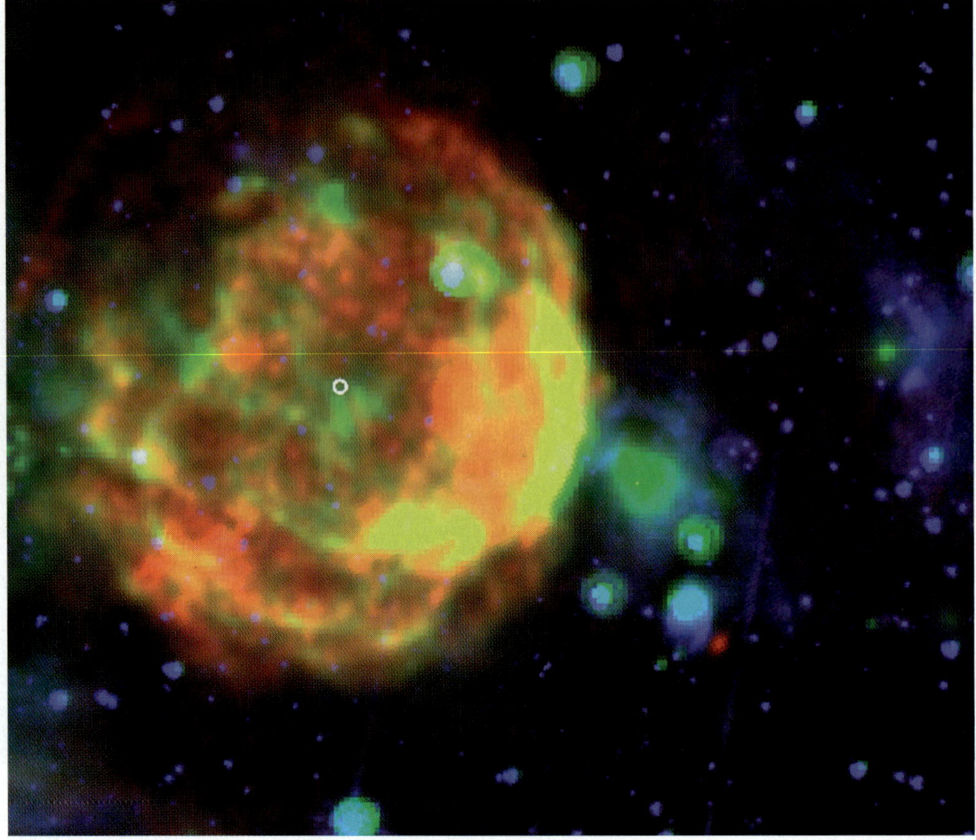

Fig. 3 A 3-color image of the area surrounding AXP 1841-045 and its associated SNR Kes 73, composed of the IRAC 8.0 μm (*blue*), MIPS 24 μm (*green*) Spitzer images and the MAGPIS radio 20 cm map (*red*). The Chandra position of the AXP is marked with a *white circle*. North is *up* and east to the *left*

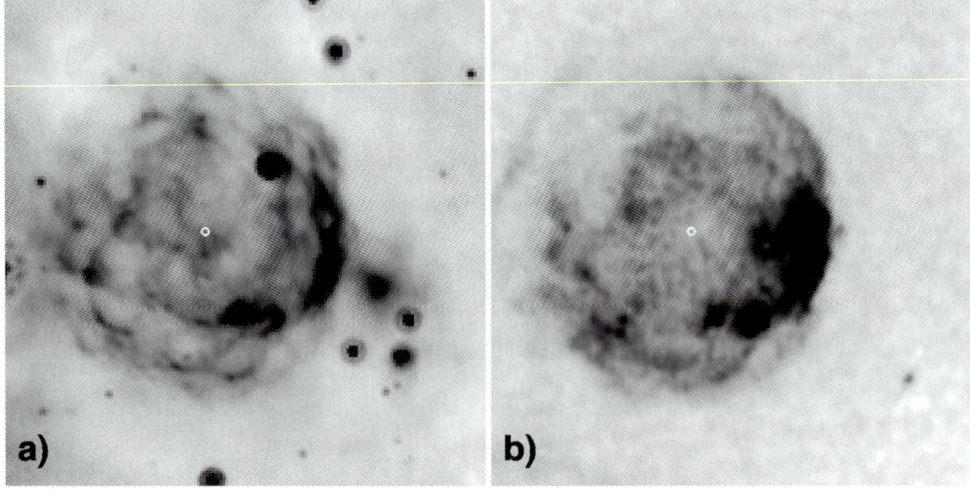

Fig. 4 SNR Kes 73, **a** Spitzer MIPS 24 μm image, **b** MAGPIS 20 cm radio image

and also extending past the western edge of the radio emission.

3 Discussion

Spitzer observations are ideally suited to study the immediate environment in which magnetars form. Since SGRs and AXPs are concentrated towards the plane of the Galaxy, observations of these sources in the optical are strongly hindered by high interstellar obscuration. At levels of $A_V =$ 30–60 observed for some SGRs, even the near-IR is still heavily affected, whereas extinction effects are minimized at Spitzer wavelengths. In addition, the radiation observed at Spitzer wavelengths originates largely from continuum and/or line emission of dust with different grain sizes and temperatures and traces the conditions in the material out of which the magnetar might have formed.

For each AXP and SGR in our program we have searched our Spitzer data for

- evidence for embedded clusters of young stars,
- evidence for supernova remnants (SNRs), and
- evidence for wind blown bubbles from the magnetar progenitor.

Embedded clusters of massive stars have been found in the vicinity of some SGRs, notably SGR 1806-20 and SGR 1900+14. If magnetars are formed in a cluster environment, the masses and ages of the stars in the cluster can be used to constrain these quantities for the magnetar progenitor. Hence, finding additional clusters or evidence for sites of ongoing star formation would be very useful. As the example of SGR 1806-20 illustrates, our IRAC observations largely trace the stellar population in the cluster while the 24 μm observations highlight diffuse extended emission enveloping the cluster. For massive stars, we expect a large portion of the mid-IR flux to come from free-free emission in a stellar wind.

Little is known about the appearance of SNRs in the mid-IR. Surveys conducted with IRAS (e.g. Saken et al. 1992) and IRAC/Spitzer (Reach et al. 2006) highlight the large variety of mechanisms that can contribute to SNR emission at Spitzer wavelengths. So far, MIPS 24 μm observations have proven particularly effective in revealing SNRs. For example, Morris et al. (2006) discovered a SNR that is only detected through strong line emission in the 24 μm band and is not visible at IRAC wavelengths nor in archival radio and X-ray data. In our data set, all of the known SNRs associated with our targets are prominently detected at 24 μm, while the IRAC observations typically show no evidence of emission associated with the SNRs. Kes73 (Fig. 3) is a very typical example in this regard.

The discovery of a possible wind blown bubble in the radio around AXP 1E 1048.1-5937 by Gaensler et al. (2005b) provides for a new avenue to characterize the properties of the magnetar progenitors. The mid-IR is ideally suited to search for such structures, since the radiatively heated dust swept up by the winds is a strong emitter at these wavelengths. Large scale bubbles around hot stars were first seen with IRAS (van Buren and McCray 1988). We expect these bubbles around SGRs and AXPs to be large, with a radii of ~ 3–$20'$ (depending on the distance), if the structure detected around AXP 1E 1048.1-5937 indicates typical size scales. Our Spitzer data reveal faint, shell or bubble-like structures around several of our targets, including AXP 1E 1048.1-5937. However, it is difficult to disen-tangle such morphologies from the surrounding ISM emission. A confirmed association of these bubbles with any particular AXP/SGR will have to await a careful analysis and comparison of the data from all three Spitzer instruments.

Acknowledgements This work is based on observations made with the Spitzer Space Telescope, which is operated by the Jet Propulsion Laboratory, California Institute of Technology under a contract with NASA. Support for this work was provided by NASA through an award issued by JPL/Caltech.

References

Benjamin, R.A., Churchwell, E., Babler, B.L., et al.: Publ. Astron. Soc. Pac. **115**, 953 (2003)

Cameron, P.B., Chandra, P., Ray, A., et al.: Nature **434**, 1112 (2005)

Corbel, S., Wallyn, P., Dame, T.M., et al.: Astrophys. J. **478**, 624 (1997)

Durant, M.: Astrophys. J. **632**, 563 (2005)

Eikenberry, S.S., Matthews, K., LaVine, J.L., et al.: Astrophys. J. **616**, 506 (2004)

Fazio, G.G., Hora, J.L., Allen, L.E., et al.: Astrophys. J. Suppl. Ser. **154**, 10 (2004)

Figer, D.F., Najarro, F., Kudritzki, R.P.: Astrophys. J. **610**, L109 (2004)

Figer, D.F., Najarro, F., Geballe, T.R., et al.: Astrophys. J. **622**, L49 (2005)

Fuchs, Y., Mirabel, F., Chaty, S., Claret, A., Cesarsky, C.J., Cesarsky, D.A.: Astron. Astrophys. **350**, 891 (1999)

Gaensler, B.M., Slane, P.O., Gotthelf, E.V., Vasisht, G.: Astrophys. J. **559**, 963 (2001)

Gaensler, B.M., Kouveliotou, C., Gelfand, J.D., et al.: Nature **434**, 1104 (2005)

Gaensler, B.M., McClure-Griffiths, N.M., Oey, M.S., Haverkorn, M., Dickey, J.M., Green, A.J.: Astrophys. J. Lett. **620**, L95 (2005)

Gotthelf, E.V., Vasisht, G.: Astrophys. J. **486**, L133 (1997)

Gotthelf, E.V., Vasisht, G., Dotani, T.: Astrophys. J. **522**, L49 (1999)

Heger, A., Fryer, C.L., Woosley, S.E., Langer, N., Hartmann, D.H.: Astrophys. J. **591**, 288 (2003)

Helfand, D.J., Becker, R.H., White, R.L., et al.: astro-ph/0510468 (2005)

Israel, G., Covino, S., Mignani, R., et al.: Astron. Astrophys. **438**, L1 (2005)

Kaspi, V.M., McLaughlin, M.A.: Astrophys. J. Lett. **618**, L41 (2005)

Kosugi, G., Ogasawara, R., Terada, H.: Astrophys. J. **623**, L125 (2005)

Kulkarni, S.R., Kaplan, D.L., Marshall, H.L., Frail, D.A., Murakami, T., Yonetoku, D.: Astrophys. J. **585**, 948 (2003)

McClure-Griffiths, N.M., Gaensler, B.M.: Astrophys. J. **630**, L161 (2005)

Morris, P.W., Stolovy, S., Wachter, S., Noriega-Crespo, A., Pannuti, T.G., Hoard, D.W.: Astrophys. J. **640**, L179 (2006)

Reach, W.T., Rho, J., Tappe, A., et al.: Astron. J. **131**, 1479 (2006)

Rieke, G.H., Young, E.T., Engelbracht, C.W., et al.: Astrophys. J. Suppl. Ser. **154**, 25 (2004)

Saken, J.M., Fesen, R.A., Shull, J.M.: Astrophys. J. Suppl. Ser. **81**, 715 (1992)

Sanbonmatsu, K.Y., Helfand, D.J.: Astron. J. **104**, 2189 (1992)

van Buren, D., McCray, R.: Astrophys. J. **329**, L93 (1988)

Wachter, S., Patel, S.K., Kouveliotou, C., et al.: Astrophys. J. **615**, 887 (2004)

Wang, Z., Chakrabarty, D., Kaplan, D.L.: Nature **440**, 772 (2006)

Werner, M.W., Roellig, T.L., Low, F.J., et al.: Astrophys. J. Suppl. Ser. **154**, 1 (2004)

Astrophys Space Sci (2007) 308: 73–77
DOI 10.1007/s10509-007-9328-8

ORIGINAL ARTICLE

Anomalous X-ray pulsars: persistent states with fallback disks

Ü. Ertan · M.A. Alpar · M.H. Erkut · K.Y. Ekşi · Ş. Çalışkan

Received: 21 August 2006 / Accepted: 3 November 2006 / Published online: 16 March 2007

Abstract The anomalous X-ray pulsar 4U 0142+61 was recently detected in the mid infrared bands with the SPITZER Observatory (Wang, Chakrabarty and Kaplan: Nature 440, 772 (2006)). This observation is the first instance for a disk around an AXP. From a reanalysis of optical and infrared data, we show that the observations indicate that the disk is likely to be an active disk rather than a passive dust disk beyond the light cylinder, as proposed in the discovering paper. Furthermore, we show that the irradiated accretion disk model can also account for all the optical and infrared observations of the anomalous X-ray pulsars in the persistent state.

Keywords Neutron stars · Pulsars · Accretion and accretion disks

PACS 97.60.Jd · 97.60.Gd · 97.10.Gz

We acknowledge support from the Astrophysics and Space Forum at Sabancı University. S.Ç. acknowledges support from the FP6 Marie Curie Reintegration Grant, INDAM.

Ü. Ertan (✉) · M.A. Alpar · M.H. Erkut
Sabancı University, 34956, Orhanlı Tuzla, İstanbul, Turkey
e-mail: unal@sabanciuniv.edu

M.A. Alpar
e-mail: alpar@sabanciuniv.edu

M.H. Erkut
e-mail: hakane@sabanciuniv.edu

K.Y. Ekşi
İstanbul Technical University (İTÜ), 34469, İstanbul, Turkey
e-mail: eksi@itu.edu.tr

Ş. Çalışkan
Boğaziçi University, Bebek, 34342, İstanbul, Turkey
e-mail: scaliskan@sabanciuniv.edu

1 Introduction

Anomalous X-ray Pulsars (AXPs) and Soft Gamma-ray Repeaters (SGRs) (Mereghetti et al. 2002; Hurley 2000; Woods and Thompson 2006) are characterized mainly by their X-ray luminosities ($L_X \sim 10^{34}$–10^{36} erg s^{-1}), which are orders of magnitude higher than their rotational powers $\dot{E}_{\rm rot} = I\Omega\dot{\Omega}$. Their spin periods are clustered to a very narrow range (5–12 s). Most of SGRs and AXPs show repetitive, short ($\lesssim 1$ s) super-Eddington bursts with luminosities up to 10^{42} erg s^{-1}. Three giant flares with peak luminosities $L_{\rm p} > 10^{44}$ erg s^{-1} and durations of a few minutes were observed from three different SGRs (Mazets et al. 1979, 1999; Hurley et al. 1999; Palmer et al. 2005).

Energetics of the bursts strongly indicate a magnetar mechanism for these bursts. Magnetar models (Thompson and Duncan 1995, 1996) adopt strong magnetic fields with magnitude $B_* > 10^{14}$ G on the stellar surface to explain the burst energetics. These bursts are likely to be local events near the neutron star and their energies could be originating either from the dipole component or from the higher multipoles of the stellar magnetic field. In the current magnetar models, it is the dipole component of the magnetic field which must be of magnetar strength to account for the spin down properties of the AXPs and SGRs.

In the alternative fallback disk model (Chatterjee et al. 2000; Alpar 2001), the source of the X-rays is accretion onto the neutron star, while the optical/IR light originates from the accretion disk. Rotational evolution of the neutron star is determined by the interaction between the disk and the dipole component of the neutron star's magnetic field ($B_* \sim 10^{12}$–10^{13} G). Fallback disk models can account for the period clustering of AXPs and SGRs as the natural outcome of disk-magnetosphere interaction during their lifetimes (Alpar 2001; Ekşi and Alpar 2003). As higher multipole fields rapidly decrease with increasing radial distance (as r^{-5} for quadrupole component), the dipole component of the magnetic field determines the interaction and the angular momentum transfer between the disk and the neutron star. The strength of magnetic dipole field in order to explain the period clustering of the AXPs and SGRs over their $\dot{M}$ history is $B_* \sim 10^{12}$–10^{13} G (Alpar 2001; Ekşi and Alpar 2003). Therefore, disk models are consistent with the magnetar fields in the quadrupole or higher components, while they are not compatible with magnetar fields in the dipole component.

Fallback disk models can also explain the enhancements observed in the persistent luminosities of SGRs and AXPs. The X-ray enhancement of the SGR 1900+14 following its giant flare can be explained by the relaxation of a disk which has been pushed back by a preceding burst (Ertan and Alpar 2003). The same model with similar disk parameters can also reproduce the correlated X-ray and IR enhancement of AXP 2259+58, which lasted for ~1.5 years, if this is triggered by a burst, with a burst energy estimated to have remained under the detection limits (Ertan et al. 2006). The suggestion of fallback disks has motivated observational searches for disk emission in the optical and IR bands, and resulted in various constraints on the models. Some of the AXPs were observed in more than one IR band (Hulleman et al. 2001, 2004; Israel et al. 2002, 2003, 2004; Wang and Chakrabarty 2002; Kaspi et al. 2003; Tam et al. 2004; Morii et al. 2005; Durant and van Kerkwijk 2006). AXP 4U 0142+61 is the source with most extended observations, as it was also observed in the optical R and V bands (Hulleman et al. 2000; Hulleman et al. 2004; Dhillon et al. 2005; Wang et al. 2006). Discovery of modulation in the R band luminosity of 4U 0142+61 at the neutron star's rotation period P = 8.7 s, with a pulsed fraction 27% (Kern and Martin 2002; Dhillon et al. 2005), is particularly significant. This fraction is much higher than the pulsed fraction of the X-ray luminosity of this source, indicating that the origin of the pulsed optical emission cannot be the reprocessed X-rays by the disk. Magnetospheric models for these pulsations can be built with either a dipole magnetar field or within disk-star dynamo model (Cheng and Ruderman 1991), in which a magnetospheric pulsar activity is sustained by a stellar dipole field of $\sim 10^{12}$ G and a disk protruding within the magnetosphere. Ertan and Cheng (2004) showed that this pulsed optical component of the AXP 4U 0142+61 can be explained by both types of magnetospheric models. Thus the presence of strong optical pulsations from the magnetosphere does not rule out the possibility of a fallback disk together with 10^{12}–10^{13} G surface dipole magnetic field.

In the present work, we concentrate on the unpulsed optical/IR emission from the AXPs and SGRs in their persistent states, and test the expectations of the irradiated accretion disk model through the observations in different optical/IR energy bands (V, R, I, J, H, K, and K_s). The optical/IR emission expected from the irradiated fallback disks was first computed and discussed by Perna et al. (2000) and Hulleman et al. (2000). Using similar irradiation strengths, Perna et al. (2000) and Hulleman et al. (2000) found similar optical fluxes that remain well beyond those indicated by the observations of AXP 4U 0142+61 and AXP 1E 2259+586. To explain this result, Perna et al. (2000) suggested that the inner disk regions could be cut by an advection dominated flow, while Hulleman et al. (2000) concluded that the then existing optical data of the AXP 4U 0142+61 (in I, R, V bands) can only be accounted for by an extremely small outer disk radius, around a few $\times 10^9$ cm (see Ertan et al. 2007 for detailed discussion). In present work, we show that the optical/IR data of the AXPs can be explained by the irradiated accretion disk model without any implausible constraints on the outer and inner disk radii. The main reason for the difference between our results and those of earlier works is that both Hulleman et al. (2000) and Perna et al. (2000) assumed a particular irradiation strength, while we keep it as a free parameter considering the observations of the low mass X-ray binaries (LMXBs) which indicate varying irradiation strengths (see Ertan and Çalışkan (2006) for a detailed discussion) We give the details of the disk model in Sect. 2. We discuss our results in Sect. 3, and summarize the conclusions in Sect. 4.

2 Optical/IR emission from the irradiated disk

Model fits to the contemporaneous X-ray and IR data of the AXP 1E 2259+586 favor the irradiated disk models, though they do not exclude the nonirradiated thin disk model (Ertan et al. 2006). We start by assuming that the AXP disks are irradiated and include the irradiation strength as a free parameter through our calculations.

We calculate the disk blackbody emission taking both the intrinsic dissipation and the irradiation flux into account. A steady disk model is a good approximation for the present evolution of the AXP and SGR disks in their persistent states. For a steady thin disk, the intrinsic dissipation can be written as

$$D = \frac{3}{8\pi} \frac{GM\dot{M}}{R^3} \tag{1}$$

(see, e.g. Frank et al. 2002) where $\dot{M}$ is the disk mass flow rate, M is the mass of the neutron star and R is the radial distance from the neutron star. In the absence of irradiation, the effective temperature $T_{\rm eff}$ of the disk is proportional to $R^{-3/4}$ for a given $\dot{M}$. For an irradiated disk, the irradiation flux can be written as

$$F_{\rm irr} = \sigma T_{\rm irr}^4 = C \frac{\dot{M} c^2}{4\pi R^2} \tag{2}$$

with

$$C = \eta(1-\epsilon)\frac{H}{R}\left(\frac{d\ln H}{d\ln R} - 1\right) \quad (3)$$

(Shakura and Sunyaev 1973), where η is the conversion efficiency of the rest mass energy into X-rays, ϵ is the X-ray albedo of the disk face, c is the speed of light and H is the pressure scale height of the disk. Since the ratio H/R is roughly constant along the disk, the irradiation strength is expected to be constant along the disk. Depending on the geometry and the temperature of the scattering source, efficiency of the irradiation could vary from source to source. Estimates for the parameter C are usually in the range 10^{-4}–10^{-3} for the low mass X-ray binaries (de Jong et al. 1996; Dubus et al. 1999; Ertan and Alpar 2002).

For a constant C along the disk, the irradiation temperature $T_{\rm irr} = (F_{\rm irr}/\sigma)^{1/4}$ is proportional to $R^{-1/2}$ (2). For small radii, dissipation is the dominant source of the disk emission. At a critical radius $R_{\rm c}$, the irradiation flux becomes equal to the dissipation rate, and beyond $R_{\rm c}$, the dominant heating mechanism is X-ray irradiation. Equating $F_{\rm irr}$ to D (1) and (2), the critical radius is found to be

$$R_{\rm c} = \frac{3}{2}\frac{GM_*}{Cc^2} \simeq \left(\frac{10^{-4}}{C}\right) 3\times 10^9 \text{ cm}. \quad (4)$$

The effective temperature profile of the disk can be written as

$$T_{\rm eff} = \left(\frac{D}{\sigma} + \frac{F_{\rm irr}}{\sigma}\right)^{1/4} \quad (5)$$

where σ is the Stefan–Boltzmann constant.

To find the model disk flux in a given observational band, we integrate the calculated blackbody emissions of all radial grids radiating in this band. For comparison with data, we calculate the model disk fluxes along the optical/IR bands V, R, H, I, J, K and $\rm K_s$. For all sources we set $\cos i = 1$ where i is the angle between the disk normal and the line of sight of the observer. We equate the disk mass flow rate $\dot{M}$ to the accretion rate onto the neutron star, thus assuming the mass loss due to the propeller effect is negligible. We first adjust $\dot{M}$ to obtain the observed X-ray flux. Next, using this value of $\dot{M}$ and taking the strength of the magnetic dipole field $B_* = 10^{12}$ G on the surface of the neutron star we calculate the Alfvén radius $R_{\rm A}$ which we take to be the inner radius of the disk. Then, we look for a good fit to the overall available optical/IR data by adjusting the irradiation strength C within the uncertainties discussed in Sect. 3. For the sources with more than one detections in a particular IR/optical band we take the observation nearest to the date of X-ray observation.

3 Results and discussion

We summarize our results in Table 1. For each source, the first column gives the unabsorbed flux data obtained from the observed magnitudes and the estimated $A_{\rm V}$ values (for references see Table 1), and the second column gives the model fluxes. For the AXP 4U 0142+61, reasonable range of reddening is $2.6 < A_{\rm V} < 5.1$ (Hulleman et al. 2004). We obtain a good fit with $A_{\rm V} = 3.5$ (Fig. 1). Table 1 shows that the irradiated steady disk model is also in agreement with all the other AXPs observed in the optical and IR bands. The parameters of the model for each source are given in Table 2.

At present, AXP 4U 0142+61, which has been observed in nine different optical/IR bands from 8 μm to V in the same X-ray luminosity regime, seems to be the best source to study the properties of AXPs in the persistent state. The irradiation parameter C obtained from our model fits remains in the range ($10^{-4} < C < 10^{-3}$) estimated from the observations of LMXBs and the disk stability analyses of the soft X-ray transients (de Jong et al. 1996; Dubus et al. 1999; Ertan and Alpar 2002). Within the critical radius $R_{\rm c}$ given by (4), dissipation is the dominant heating mechanism. For

Table 1 The data flux values were calculated by using the magnitudes and $A_{\rm V}$ values given in the references below. References: (J1708-40) Durant and van Kerkwijk 2006, Rea et al. 2003; (1E 2259+586) Hulleman et al. 2001, Woods et al. 2004; (1E 1841-45) Wachter et al. 2004, Morii et al. 2003; (1E 1048-59) Wang and Chakrabarty 2002, Mereghetti et al. 2004

Band	J1708-40		1E 2259+58		1E 1841-045		1E 1048-59	
	Flux (10^{-15} erg s^{-1} cm^{-2})		Flux (10^{-15} erg s^{-1} cm^{-2})		Flux (10^{-15} erg s^{-1} cm^{-2})		Flux (10^{-15} erg s^{-1} cm^{-2})	
	Data ($A_{\rm V} = 7.8$)	Model	Data ($A_{\rm V} = 6.1$)	Model	Data ($A_{\rm V} = 8.4$)	Model	Data ($A_{\rm V} = 5.6$)	Model
$\rm K_s$	49	44	3.7	3.6	68	68	29	22
K		54		4.5		84		27
H	51	57		4.8		89	22	28
J	50	53		4.4		83	33	26
I		56	<15	4.4		88		24
R		65	<42	4.5	$< 3.8 \times 10^5$	100		26
V		48		3.0		76		18

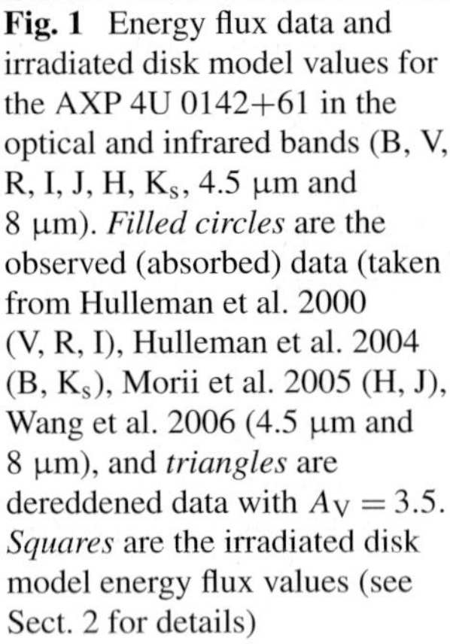

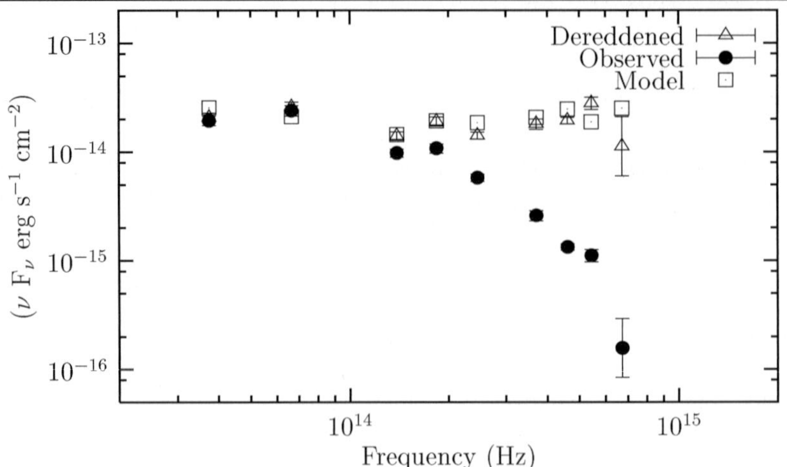

Fig. 1 Energy flux data and irradiated disk model values for the AXP 4U 0142+61 in the optical and infrared bands (B, V, R, I, J, H, K_s, 4.5 μm and 8 μm). *Filled circles* are the observed (absorbed) data (taken from Hulleman et al. 2000 (V, R, I), Hulleman et al. 2004 (B, K_s), Morii et al. 2005 (H, J), Wang et al. 2006 (4.5 μm and 8 μm), and *triangles* are dereddened data with $A_V = 3.5$. *Squares* are the irradiated disk model energy flux values (see Sect. 2 for details)

Table 2 The parameters of the irradiated disk model which gives the optical/IR flux values seen in Table 1. For all the sources, we set $\cos i = 1$ where i is the inclination angle between the disk normal and the line of sight of the observer, and we take the outer disk radius $R_{\mathrm{out}} = 2 \times 10^{12}$ cm. See Sect. 3 for details

	1RXS J1708-40	1E 2259+58	4U 0142+61	1E 1841-045	1E 1048-59
R_{in} (cm)	1.2×10^9	2.3×10^9	1.0×10^9	1.3×10^9	3.3×10^9
C	5.0×10^{-4}	1.6×10^{-4}	1.0×10^{-4}	7.2×10^{-4}	7.0×10^{-4}
d (kpc)	5	3	3	7	3
$\dot{M}$ (g s^{-1})	1.0×10^{15}	9.1×10^{13}	4.8×10^{14}	2.2×10^{15}	1.3×10^{14}

the disk model of the AXP 4U 0142+61, $R_c \simeq 3 \times 10^9$ cm and $R_{\mathrm{in}} = 1 \times 10^9$ cm. The innermost disk emitting mostly in the UV bands also contributes to the optical emission. The radial distance at which the disk blackbody temperatures peak at the optical bands (R, V) is about 10^{10} cm. Peak temperatures of the IR bands from I to SPITZER mid IR bands (4.5 μm and 8 μm) lie between $R \sim 2 \times 10^{10}$ cm and $R \sim 10^{12}$ cm. There are several observations of this source in some of the IR bands; we adopt the data nearest to the X-ray observation epoch used in our analysis here.

For AXP J1708-40, Durant and van Kerkwijk (2006) recently found that the previously reported IR data in K_s, H, and J bands are likely to be a background star. They found another object within the positional error cycle and argued that this second object is more likely to be the IR counterpart to the AXP J1708-40. For this source, we adopt the IR (K_s, H, J) data set reported by Durant and van Kerkwijk (2006).

For the AXP 1E 2259+586, we use the data taken before the X-ray enhancement phase of this source (Hulleman et al. 2001). This source was detected in K_s band and there are upper limits for I and R bands. Our model flux values are three and ten times below the upper limits reported for I and R bands respectively.

AXP 1E 1048-59 was detected in K_s, H and I bands (Wang and Chakrabarty 2002). Observed X-ray flux from this source between December 2000 to January 2003 show a variation within a factor of 5 (Mereghetti et al. 2004). We use the X-ray flux obtained from the nearest X-ray observation to the date of the IR observations when the source was in the persistent state.

AXP 1E 1841 was detected only in the K_s band, and there is a high upper limit in the R band (Wachter et al. 2004). Model estimates in other optical/IR bands for this source (and the other AXPs) can be tested by future optical and IR observations.

We note that our model does not address the uncertainties in the innermost disk, the contributions from the magnetospheric pulsed emission, and the shielding effects which might decrease the irradiation strength at some regions of the disk. All the model flux values remain within 30% of the data points, which is a reasonable fit considering the uncertainties mentioned above.

Since there is no detections in short wavelength optical bands for the AXPs (except for AXP 4U 0142+61), model fits are not sensitive to the chosen inner disk radii. For the model fits, we equate the inner disk radii to the Alfvén radii (Table 2) corresponding to a magnetic field with magnitude $B_* = 10^{12}$ G on the stellar surface and the accretion rates derived from the estimated X-ray luminosities (see Table 1 for references). On the other hand, optical data of the AXP 4U 0142+61 in R and V bands provide a constraint for the inner disk radius, and thereby for the strength of the magnetic dipole field of this source (see Sect. 4).

4 Conclusion

We have shown that the optical, infrared and X-ray observations of the AXPs in their persistent states can be explained with irradiated disk models. Among the AXPs, 4U 0142+61 is currently the only source which provides an upper limit for the inner disk radius through its optical (R, V) data. For the best model fit for this source, which we have obtained with $A_V = 3.5$, the model inner disk radius ($\sim 10^9$ cm) is around the Alfvén radius for the accretion rate, estimated from the X-ray luminosity, together with a dipole magnetic field strength $B_* \simeq 10^{12}$ G on the neutron star surface. Nevertheless, it is possible to obtain reasonable fits by increasing the inner disk radius and decreasing the reddening accordingly. For $A_V = 2.6$, the minimum value of the reddening in the reasonable range $2.6 < A_V < 5.1$ (Hulleman et al. 2004), we obtain the best fit with $R_{in} \simeq 8 \times 10^9$ cm which corresponds to the maximum reasonable dipole field strength $B_* \simeq 4 \times 10^{13}$ G on the pole and half of this on the equator of the neutron star. Nevertheless, we note that these limits could be increased depending on the amount of possible mass loss due to propeller effect and/or on how much the inner disk radius penetrates inside the Alfvén radius (see Ertan et al. 2007 for a detailed discussion for 4U 0142+61). On the other hand, even including these possibilities, very recent analysis concluding $A_V = 3.5 \pm 0.4$ (Durant and van Kerkwijk 2007) implies surface dipole magnetic field strengths less than about 10^{13} G. The magnetar fields ($B_* > 10^{14}$ G) in multipoles, that could be responsible for the burst energies, are compatible with this picture, while the optical data excludes a hybrid model involving a disk surrounding a dipole field of magnetar strength.

On the other hand, existing IR data of the AXPs, including recent observations of 4U 0142+61 by SPITZER in 4.5 μm and 8 μm bands (Wang et al. 2006), do not put an upper limit for the extension of the outer disk radius R_{out}. The lower limit for R_{out} provided by the longest wavelength IR data of the AXP 4U 0142+61 is around 10^{12} G. Further observations in the longer wavelength infrared bands by SPITZER space telescope will provide valuable information about the structure and possibly the extension of the fallback disks around these systems.

References

Alpar, M.A.: Astrophys. J. **554**, 1245 (2001)
Chatterjee, P., Hernquist, L., Narayan, R.: Astrophys. J. **534**, 373 (2000)
Cheng, K.S., Ruderman, M.: Astrophys. J. **373**, 187 (1991)
de Jong, J.A., van Paradijs, J., Augusteijn, T.: Astron. Astrophys. **314**, 484 (1996)
Dhillon, V.S., Marsh, T.R., Hulleman, et al.: Mon. Not. Roy. Astron. Soc. **363**, 609 (2005)
Dubus, G., Lasota, J.-P., Hameury, J.-M., Charles, P.: Mon. Not. Roy. Astron. Soc. **303**, 139 (1999)
Durant, M., van Kerkwijk, M.H.: Astrophys. J. **648**, 534 (2006)
Durant, M., van Kerkwijk, M.H.: Astrophys. J. **650**, 1082 (2007) (astro-ph/0606604)
Ekşi, Y.K., Alpar, M.A.: Astrophys. J. **559**, 450 (2003)
Ertan, Ü., Alpar, M.A.: Astron. Astrophys. **393**, 205 (2002)
Ertan, Ü., Alpar, M.A.: Astrophys. J. **593**, L93 (2003)
Ertan, Ü., Cheng, K.S.: Astrophys. J. **605**, 840 (2004)
Ertan, Ü., Çalışkan, Ş.: Astrophys. J. **649**, L87 (2006)
Ertan, Ü, Erkut, M.H., Ekşi, Y.K., Alpar, M.A.: Astrophys. J. (2007, in press). Preprint astro-ph/0606259
Ertan, Ü., Göğüş, E., Alpar, M.A.: Astrophys. J. **640**, 435 (2006)
Frank, J., King, A.R., Raine, D.: Accretion Power in Astrophysics. Cambridge University Press, Cambridge (2002)
Hulleman, F., van Kerkwijk, M.H., Kulkarni, S.R.: Nature **408**, 689 (2000)
Hulleman, F., Tennant, A.F., van Kerkwijk, M.H., et al.: Astrophys. J. **563**, L49 (2001)
Hulleman, F., van Kerkwijk, M.H., Kulkarni, S.R.: Astron. Astrophys. **416**, 1037 (2004)
Hurley, K., Cline, T., Mazets, E., et al.: Nature **397**, 41 (1999)
Hurley, K.: In: Kippen, R.M., Mallozzi, R.S., Fishman, G.J. (eds.) Gamma-Ray Bursts: Fifth Huntsville Symposium, AIP Conference Proceedings, vol. 526, p. 763. AIP, New York (2000)
Israel, G.L., Covino, S., Stella, L., et al.: Astrophys. J. **580**, L143 (2002)
Israel, G.L., Covino, S., Perna, R., et al.: Astrophys. J. **589**, L93 (2003)
Israel, G.L., Rea, N., Mangano, V., et al.: Astrophys. J. **603**, L97 (2004)
Kaspi, V.M., et al.: Astrophys. J. **588**, L93 (2003)
Kern, B., Martin, C.: Nature **417**, 527 (2002)
Mazets, E.P., Golenetskii, S.V., Il'inskii, V.N., Aptekar, R.L., Guryan, Y.A.: Nature **282**, 587 (1979)
Mazets, E.P., Cline, T., Aptekar, R.L., et al.: Astron. Lett. **25**, 635 (1999)
Mereghetti, S., Tiengo, A., Stella, L., et al.: Astrophys. J. **608**, 427 (2004)
Mereghetti, S., Chiarlone, L., Israel, G.L., Stella, L.: In: Becker, W., Lecsch, H., Trumper, J. (eds.) Proc. of the 270th WE-Heraus Seminar on Neutron Stars, Pulsars and Supernova Remnants, p. 29. MPE, Garching (2002) (MPE Rep. 278)
Morii, M., Sato, R., Kataoka, J., Kawai, N.: Publ. Astron. Soc. Jpn. **55**, L45 (2003)
Morii, M., Kawai, N., Kataoka, J., et al.: Adv. Space Res. **35**, 1177 (2005)
Palmer, D.M., et al.: Nature **434**, 1107 (2005)
Perna, R., Hernquist, L., Narayan, R.: Astrophys. J. **541**, 344 (2000)
Rea, N., et al.: Astrophys. J. **586**, L65 (2003)
Shakura, N.I., Sunyaev, R.A.: Astron. Astrophys. **24**, 337 (1973)
Tam, C.R., Kaspi, V.M., van Kerkwijk, M.H., Durant, M.: Astrophys. J. **617**, L53 (2004)
Thompson, C., Duncan, R.C.: Mon. Not. Roy. Astron. Soc. **275**, 255 (1995)
Thompson, C., Duncan, R.C.: Astrophys. J. **473**, 322 (1996)
Wachter, S., Patel, S.K., Kouveliotou, C., et al.: Astrophys. J. **615**, 887 (2004)
Wang, Z., Chakrabarty, D.: Astrophys. J. **579**, L33 (2002)
Wang, Z., Chakrabarty, D., Kaplan, D.: Nature **440**, 772 (2006)
Woods, P.M., Thompson, C.: In: Lewin, W.H.G., van der Klis, M. (eds.) Compact Stellar X-ray Sources (2006)
Woods, P.M., et al.: Astrophys. J. **605**, 378 (2004)

Astrophys Space Sci (2007) 308: 79–87
DOI 10.1007/s10509-007-9327-9

ORIGINAL ARTICLE

The anatomy of a magnetar: XMM monitoring of the transient anomalous X-ray pulsar XTE J1810–197

E.V. Gotthelf · J.P. Halpern

Received: 21 July 2006 / Accepted: 23 August 2006 / Published online: 15 March 2007

Abstract We present the latest results from a multi-epoch timing and spectral study of the Transient Anomalous X-ray Pulsar XTE J1810–197. We have acquired seven observations of this pulsar with the Newton X-ray Multi-mirror Mission (XMM-Newton) over the course of two and a half years, to follow the spectral evolution as the source fades from outburst. The spectrum is arguably best characterized by a two-temperature blackbody whose luminosities are decreasing exponentially with $\tau_1 = 870$ d and $\tau_2 = 280$ d, respectively. The temperatures of these components are currently cooling at a rate of 22% per year from a nearly constant value recorded at earlier epochs of $kT_1 = 0.25$ keV and $kT_2 = 0.67$ keV, respectively. The new data show that the temperature T_1 and luminosity of that component have nearly returned to their historic quiescent levels and that its pulsed fraction, which has steadily decreased with time, is now consistent with the previous lack of detected pulsations in quiescence. We also summarize the detections of radio emission from XTE J1810–197, the first confirmed for any AXP. We consider possible models for the emission geometry and mechanisms of XTE J1810–197.

Keywords Pulsars: general · Stars: individual (XTE J1810–197) · Stars: neutron · X-rays: stars

PACS 95.85.-e · 95.85.Nv · 97.60.Jd · 97.60.Gb · 95.75

XMM-Newton is an ESA science mission with instruments and contributions directly funded by ESA Member States and NASA. This research is supported by XMM-Newton grant NNG05GJ61G and NASA ADP grant ADP04-0059-0024.

E.V. Gotthelf (✉) · J.P. Halpern
Columbia Astrophysics Laboratory, Columbia University,
550 West 120th Street, New York, NY 10027-6601, USA
e-mail: eric@astro.columbia.edu

1 Introduction

Neutron star (NS) astronomy has been recently invigorated by the identification of a new class of magnetically dominated emitters. Known as anomalous X-ray pulsars (AXPs) and soft gamma-ray repeaters (SGRs), these objects are apparently young, isolated neutron stars (NSs), whose properties differ markedly from those of the Crab pulsar, previously considered prototypical of the young NSs (for a review see Mereghetti et al. 2002). All AXPs and at least 3 of the 4 SGRs are identified as relatively slow (5–12 s) pulsars. Many are located at the centers of recognized, young supernova remnants (SNRs), directly associating them with their supernova explosions. These objects emit predominantly at X-ray energies and are distinguished by their characteristic spectral signature. This radiation cannot be accounted for by rotational energy losses alone, as for the radio pulsars, but is most likely powered by the decay of an enormous magnetic field characterized by a dipole with $B_p \gtrsim 4.4 \times 10^{13}$ G, at the pole. Collectively, these isolated NSs are understood within the context of the magnetar theory (Duncan and Thompson 1992; Thompson and Duncan 1996).

A unique pulsar has been discovered whose remarkable properties offer great promise for deciphering the emission mechanism(s) of magnetars. XTE J1810–197 is a 5.54 s X-ray pulsar whose measured and derived physical parameters are fully characteristic of an AXP, but expresses behavior not previously associated with any such object (Gotthelf et al. 2004, hereafter paper I). XTE J1810–197 is a transient AXP (TAXP)—it was discovered during a bright impulsive outburst (Ibrahim et al. 2004) that is still fading steadily. Even more surprising is the discovery of highly variable radio emission, providing the first confirmed example of radio flux from an AXP (Becker 2005b), and the subsequent detection of radio pulsations at the X-ray period (Camilo et al. 2006).

The outburst that resulted in the detection of the transient AXP occurred sometime between 2002 November and 2003 January (Ibrahim et al. 2004). Since then, over the course of a year, regular scans of the region with RXTE recorded an exponential flux decay with a time-constant of 269 ± 25 days from a maximum of $F(2\text{–}10\ \text{keV}) \approx 8 \times 10^{-11}$ erg cm^{-2} s^{-1}. In comparison, the previous average quiescent flux, with its softer spectrum, gives $F(0.5\text{–}10\ \text{keV}) \approx 5.5 \times 10^{-13}$ erg cm^{-2} s^{-1} (paper I). This contrast is unprecedented for an AXP. Further RXTE observations yielded SGR-like episodes of bursts (Woods et al. 2005), similar to those seen from 1E 2259+586 (Kaspi et al. 2003). A search for an optical/IR counterpart detected a fading IR source within the Chandra error circle, similar to ones associated with other AXPs, confirming its identification with XTE J1810–197 (Rea et al. 2004a, 2004b).

In this paper we present the latest results from TAXP XTE J1810–197 including new XMM-Newton observations extending to 2006 March. These observations allow us to characterize the spectral evolution of a TAXP in outburst as it returns to quiescence. We show that the spectrum has now begun a marked transition back to the nominal quiescent state, with a distinct temperature and flux evolution. This long-term evolution provides the unique possibility of decomposing its fading spectral components in a manner unavailable for other magnetars (Halpern and Gotthelf 2005, hereafter paper II). In the following, we used a revised distance to the NS, $d_{3.3}$, quoted in units of 3.3 kpc and based on the radio pulsar DM measurement of Camilo et al. (2006).

2 Long-term spectral and temporal evolution

Herein we present a total of seven XMM-Newton observations of XTE J180–197 of which four have been previously described in papers I, II, and (Gotthelf and Halpern 2005, hereafter paper III). The three new data sets were obtained using similar observing modes as for previous observations and reduced and analyzed in an identical manner. A log of these observations is recorded in Table 1. A full report on the reduction and analysis of these data sets will be presented in Gotthelf et al. (in preparation).

Figure 1 presents an up-to-date light curve of TAXP XTE J180–197 derived by adding the XMM-Newton flux measurements to the RXTE results of Ibrahim et al. (2004). For comparison, the XMM-Newton fluxes were extracted assuming a simple power-law spectral model fitted in the 2–10 keV energy band. The combined data points are well fitted by an exponential decay model with overall time-constant of 233.5 d, somewhat shorter but consistent with the initial RXTE trend, given the extended monitoring interval. To allow for a systematic offset in the RXTE

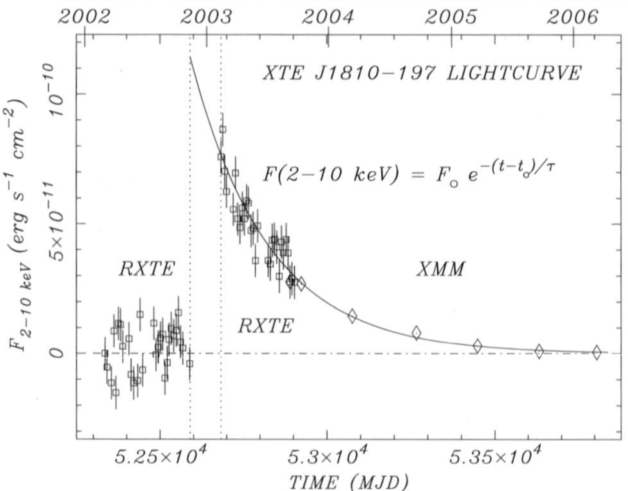

Fig. 1 Long-term light-curve of XTE J1810–197 using measurements obtained with RXTE (*squares*) and XMM-Newton (*diamond*). The RXTE data are from Ibrahim et al. (2004), renormalized to match the XMM-Newton results. The XMM-Newton data points are for the observations of Table 1, but fitted with a power-law model in the 2–10 keV energy band, for direct comparison with the RXTE results. The *solid line* is a best combined fit to an exponential decay model (see text for parameters)

flux measurements,[1] these data were rescaled by a factor of 1.42 to match the XMM-Newton results. The implied flux range for the initial outburst is $F(2\text{–}10\ \text{keV}) = (11\text{–}8) \times 10^{-11}$ erg s^{-1} cm^{-2}, over the interval during which the target was inaccessible to RXTE. Clearly the flux in this band has all but returned to its pre-outburst value.

Spectra of AXPs are nominally modeled assuming a two-component power-law plus blackbody model. The spectral and temporal trends for XTE J1810–197 allow us to strongly reject this model (see paper II for a comprehensive discussion). Instead, we find that a two-temperature blackbody model gives equally acceptable spectral fits but is better motivated physically (see Sect. 5). A summary of spectral results is presented in Table 1 and the spectra, fitted with the double blackbody model, are displayed in Fig. 2 (the 2003 October observation is excluded here for clarity). Although the hot blackbody component (BB2; Fig. 3) initially dominated the emission at low energies, it fades more rapidly than the cooler "warm" emission component (BB1; Fig. 3). Evidently the spectral fits after 2004 March deviate significantly at the lower end of the XMM-Newton energy band, below 0.7 keV. It is not yet clear if this is an instrumental artifact, however, taken at face value there seems to be an absorption feature at ≈350 eV with a Gaussian width of $\sigma \approx 160$ eV. The mean equivalent width of this feature is 600 eV, but its strength is somewhat time dependent. Perhaps as the hotter

[1] Ibrahim et al. (2004) assumed a nominal 2–10 keV conversion factor of 2.27 counts s^{-1} per CPU = 2.4×10^{-11} erg s^{-1} cm^{-2}.

Table 1 XMM-Newton spectral results for XTE J1810–197

Parameter	2003 September 8	2003 October 12	2004 March 11	2004 September 18	2005 March 18	2005 September 20	2006 March 12
Expo (ks)[a]	11.5/8.1	6.9/6.2	17.0/15.8	26.5/24.4	39.8/37.2	39.5/37.8	41.7/38.8
N_H (cm^{-2})[b]	0.65 ± 0.04	0.65 ± 0.04	0.65 ± 0.04	0.65 (fixed)	0.65 (fixed)	0.65 (fixed)	0.65 (fixed)
kT_1 (keV)	0.25 ± 0.02	0.29 ± 0.04	0.27 ± 0.02	0.25 ± 0.01	0.22 ± 0.01	0.20 ± 0.01	0.19 ± 0.01
kT_2 (keV)	0.68 ± 0.02	0.71 ± 0.03	0.70 ± 0.01	0.67 ± 0.01	0.60 ± 0.01	0.52 ± 0.01	0.46 ± 0.02
A_1 (cm^2)	5.6×10^{12}	2.9×10^{12}	3.3×10^{12}	4.0×10^{12}	4.9×10^{12}	6.6×10^{12}	7.2×10^{12}
A_2 (cm^2)	2.8×10^{11}	2.2×10^{11}	1.3×10^{11}	8.7×10^{10}	6.0×10^{10}	3.7×10^{10}	3.6×10^{10}
BB1 flux[c]	4.2×10^{-12}	5.4×10^{-12}	3.5×10^{-12}	2.6×10^{-12}	1.6×10^{-12}	1.0×10^{-12}	7.5×10^{-13}
BB2 flux[c]	3.5×10^{-11}	3.0×10^{-11}	1.8×10^{-11}	1.0×10^{-11}	4.0×10^{-12}	1.3×10^{-12}	6.8×10^{-13}
Total flux[c]	3.93×10^{-11}	3.84×10^{-11}	2.13×10^{-11}	1.29×10^{-11}	5.67×10^{-12}	2.35×10^{-12}	1.44×10^{-12}
L_{BB1}(bol)[d]	2.4×10^{34}	2.3×10^{34}	1.7×10^{34}	1.6×10^{34}	1.2×10^{34}	1.0×10^{34}	8.6×10^{33}
L_{BB2}(bol)[d]	6.3×10^{34}	5.7×10^{34}	3.1×10^{34}	1.8×10^{34}	7.9×10^{33}	2.8×10^{33}	1.7×10^{33}
χ^2_ν(dof)	1.1 (187)	1.1 (84)	1.1 (194)	1.2 (188)	1.6 (152)	1.6 (80)	1.6 (117)

Note. Uncertainties in spectral parameters are 90% confidence for two interesting parameters

[a] EPIC-pn exposure/livetime in units of ks

[b] Interstellar hydrogen absorbing column density in units of 10^{22} cm^{-2}

[c] Absorbed 0.5–10 keV flux in units of erg cm^{-2} s^{-1}

[d] Unabsorbed bolometric luminosity in units of erg s^{-1} assuming a distance of $d = 3.3$ kpc

component contributes less, this feature becomes more evident, suggesting an association exclusively with the cooler component. Further research is needed to quantify this feature and consider its reality, or physical implications.

The spectrum of XTE J1810–197 can be thought of as the combined flux from two concentric emitters, "hot spots," whose temperature and size are evolving at different rates, effectively changing the overall shape of the spectrum with time (paper II). With a set of spectral measurements spanning two and a half years, a clear trend has emerged. The bolometric luminosities of the two components are shown in Fig. 3. This reveals a cooler component decreasing exponentially in time with $\tau_1 = 870$ d, while the hotter temperature flux declines with $\tau_2 = 280$ d. As might be expected, the shorter time constant is very close to the one described by the long-term flux above 2 keV (Fig. 1), where the hotter blackbody component dominates. However, the latest two data points show that this component has now fallen faster than the original exponential. Although the *bolometric luminosities* of the two components are quite different at the latest epoch, their measured *fluxes* are now nearly equal in the 0.5–8 keV range. We also note that, while it is possible to fit an alternative power-law decay model to the hot blackbody flux, such a model would require a decay index that steepens with time.

Based on the decay rates and approximate outburst time we estimate an initial bolometric luminosity for the two spectral components of $L_1 \approx 3 \times 10^{34} d^2_{3.3}$ erg s^{-1} and $L_2 \approx 2 \times 10^{35} d^2_{3.3}$ erg s^{-1}, respectively. The total luminosity at the peak of the outburst is comparable to that of a persistent AXP, suggesting that this TAXP is anomalously faint. We can now compute the implied energies of $f_1 \approx 2 \times 10^{42} d^2_{3.3}$ erg and $f_2 \approx 4 \times 10^{42} d^2_{3.3}$ erg, for the two components, respectively. The energy of the outburst event is many orders-of-magnitude lower than that for a typical SGR flare.

The temperatures and inferred blackbody emitting areas have also evolved measurably with time. Figure 4 displays the time histories of these parameters for each spectral component, obtained from the model fits. Based on the three latest data points, the temperatures now show a definite cooling trend for each spectral component, as suggested by the broken line in Fig. 4. Prior to mid 2004, the temperatures likely remained nearly constant, after which they both fell at a rate of $\approx$22% per year ($\Delta kT_1 = -0.051$ keV yr^{-1}; $\Delta kT_2 = -0.15$ keV yr^{-1}). The size of the effective hot spots (derived from the blackbody emission areas) also followed distinct evolutions. The hot component has been shrinking exponentially since the initial XMM-Newton observation, while the warm component has steadily increased in size. A notable exception is the initial data point which deviates from the trend for both the areas and temperatures. This may be associated with a glitch or rotation instability, as suggested by the pulse timing results around this epoch (Sect. 3).

The area of the warm component (A_1; Table 1) may be reaching a maximum, corresponding to the whole NS surface. This would then provide a lower-limit on the NS radius of $\gtrsim 7.5\, d_{3.3}$ km, ignoring relativistic effects (redshift

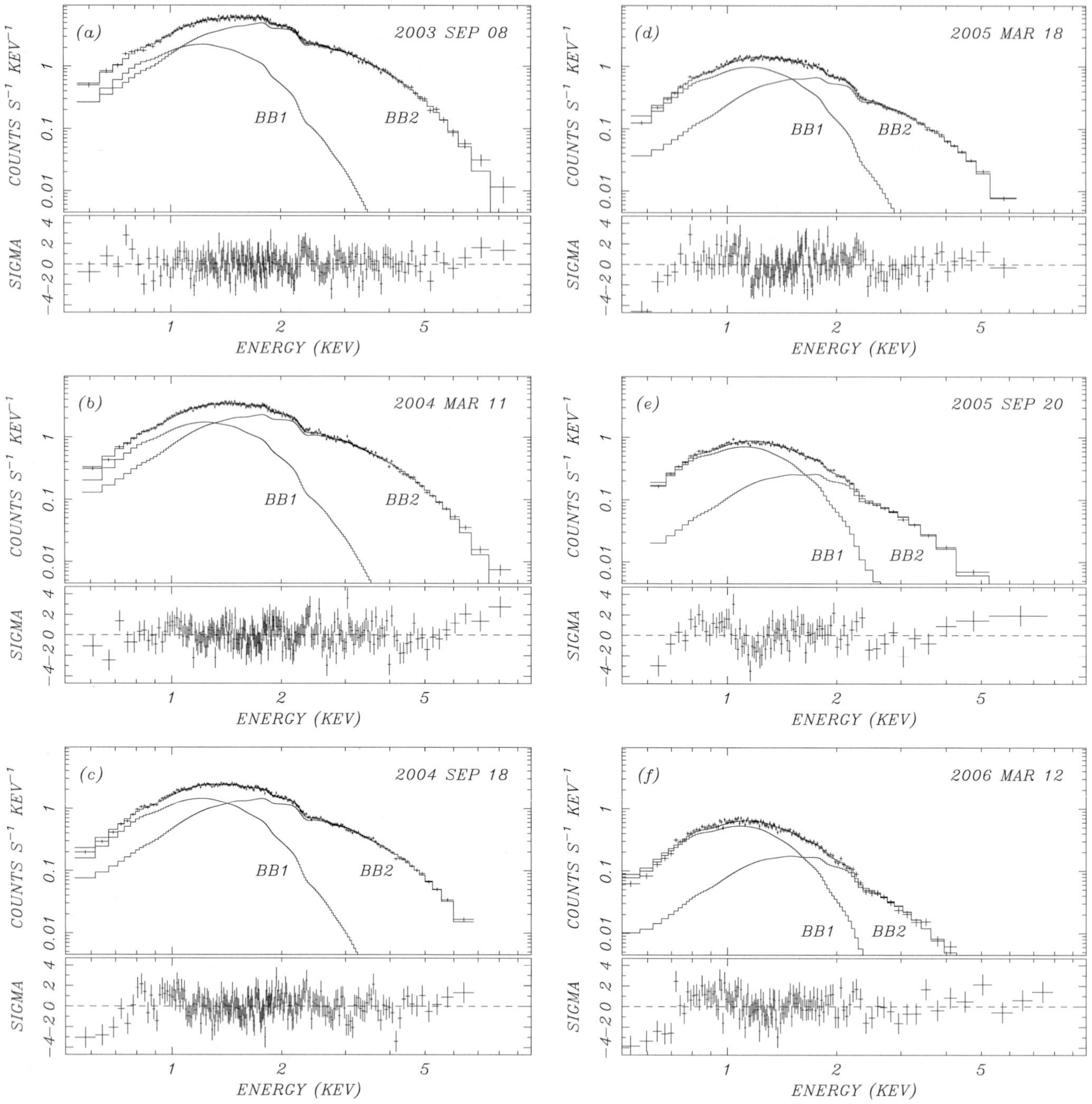

Fig. 2 XMM-Newton EPIC pn spectra of XTE J1810–197 from the earliest to the latest epochs, in six months intervals (the 2003 October spectrum is excluded, for clarity). These spectra are shown with the best-fit two-temperature blackbody model specified in Table 1. Although the temperatures of the blackbody components have not changed greatly between epochs, the flux of the hot component (BB2) has decayed rapidly relative to that of the cooler one (BB1). Also shown are the residuals from the best-fit models. The nature of the deviations to the model below 0.7 keV has yet to be determined (see text)

and light bending). In contrast, the area of the hotter component continues to shrink and its contribution to the flux is severely diminished.

As shown in Fig. 3, the X-ray luminosity of TAXP XTE J1810–197 has nearly returned to its historic quiescent level. This is also true of the temperature and inferred blackbody area. The pre-burst ROSAT spectrum of 1992 March 7 is reasonably well fitted with a single blackbody of $kT = 0.18 \pm 0.02$ keV covering $5.2 \times 10^{12}\, d_{3.3}^2$ cm^2 and $L_{\rm BB}(\rm bol) = 5.6 \times 10^{33}\, d_{3.3}^2$ erg s^{-1} (paper I). The latest XMM-Newton spectrum that overlaps the ROSAT band is again dominated by the cooler blackbody component. Fur-

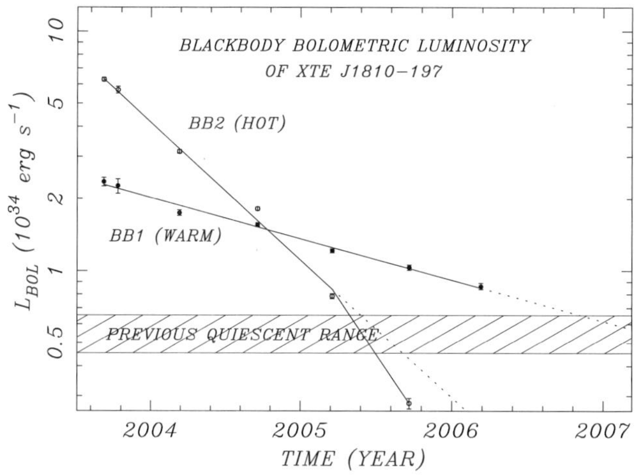

Fig. 3 A semi-log plot of the bolometric luminosity of XTE J1810–197 as a function of time (*crosses*) for each temperature component of the double blackbody model given in Table 1. The light-curves are fit assuming an exponential decays model (*solid lines*); the corresponding e-folding times is $\tau_1 = 280$ d and $\tau_2 = 870$ d for the warm and hot temperature components, respectively. The recent data for the hot component now deviates from this model. The fitted quantities have been extrapolated to the (1σ) quiescent range measured in paper I (*cross-hatched area*)

thermore, the current pulsed fraction is consistent with the quiescent upper-limits. Thus the double blackbody model is sufficient to describe the observed properties, and may not require a third "quiescent" component to explain the pre-outburst observations. XMM-Newton data from the next observing window (2006 September) is likely to be consistent with the ROSAT result in the soft energy band, effectively defining a return to the quiescence state.

3 Spin-down evolution of XTE J1810–197

The barycentered pulse period of XTE J1810–197 measured at each XMM-Newton observing epoch is shown in Fig. 5. These data points were derived following the method outlined in paper I. Prior to epoch 2004, the spin-down rate derived from the RXTE and initial XMM-Newton measurements were highly erratic, ranging from $\dot{P} = (0.8 - 2.2) \times 10^{-11}$ s s^{-1} (Ibrahim et al. 2004; paper III). This temporal behavior is likely associated with the outburst event. The latest period measurements suggest that the spin-down rate has settled down somewhat. A linear model fit to the last 5 period measurements yields a period derivative of $\dot{P} = 1.26 \pm 0.04 \times 10^{-11}$ s s^{-1} with a $\chi^2_\nu = 1.2$ for 3 DoF, corresponding to a null hypothesis probability of 0.27. Compared to the radio $\dot{P} = 1.016 \pm 0.001 \times 10^{-11}$ s s^{-1} of 17 March–7 May (Camilo et al. 2006), there is evidently still timing noise in the pulsar's spin-down. The revised X-ray rate implies a nominal characteristic age $\tau_c = 7.0$ kyr, surface magnetic field $B_s = 2.7 \times 10^{14}$ G, and spin-down power $\dot{E} = 2.9 \times 10^{33}$ erg s^{-1}, values typical for a magnetar.

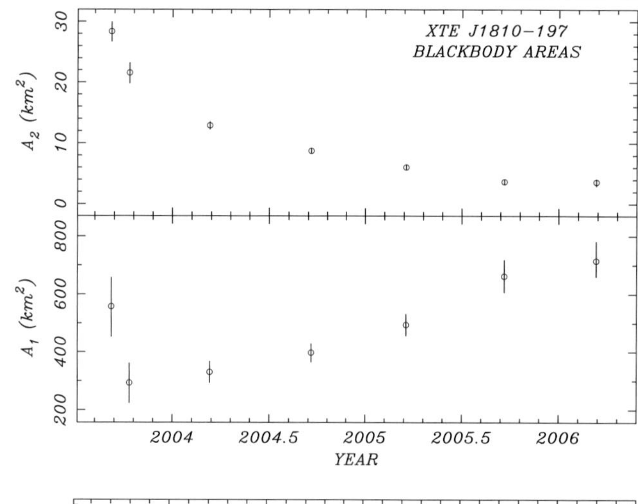

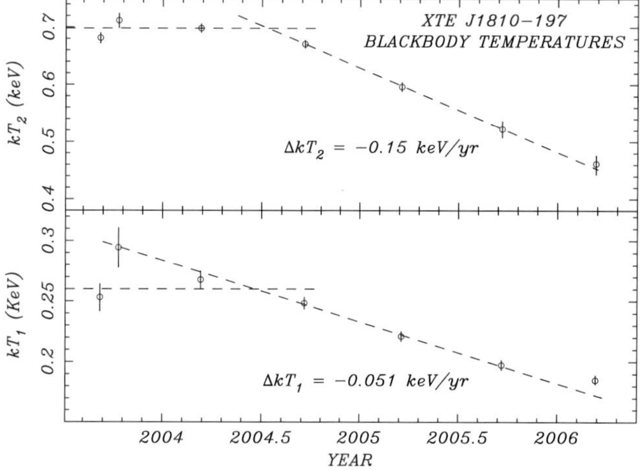

Fig. 4 The time history of the best-fit XMM-Newton component temperatures and surface emitting areas for XTE J1810–197 based on the double blackbody spectral model (Table 1). The evolving areas (*top*) indicates that the warm component is expanding to cover the NS surface, while the hot component is shrinking rapidly. The corresponding temperatures (*bottom*) are declining at a steady rate of 22% per year since mid 2004. Prior to that time the temperature of the hotter component was essentially constant. This trend for the warm component is somewhat ambiguous

The energy-dependent modulation of the pulsar has evolved noticeably over time with the declining flux. Figure 6 displays the pulse profiles at each XMM-Newton epoch, folded at the best determined periods. The data has been broken up into six energy bands, with the corresponding background in each band subtracted. The pulsed fractions, defined as the pulsed emission divided by the total flux, are indicated on the corner of each panel of the figure. These profiles are aligned so that phase zero is the same at each epoch, for ease of comparison. In any case, within an observation, the phase zero does not change with energy. This suggests a geometric interpretation for the modulation, such as from localized emission on the rotating NS star sur-

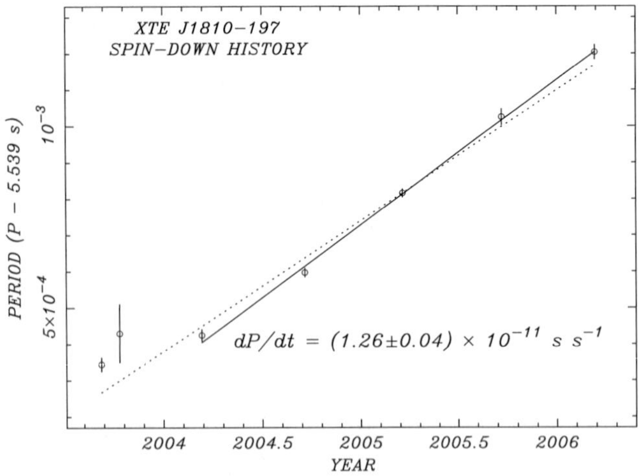

Fig. 5 The spin-down history of XTE J1810–197 pulse period as measured using the XMM-Newton data of Table 1. The initially erratic spin-down appears to be settling down – the last 5 data points are well-modeled by a linear spin-down model (*solid line*); in contrast, a poor fit is obtained using all points (*dotted line*)

face. The pulsed fraction has generally decreased with time, most notably at X-ray energies below $E < 2$ keV; at higher energies the trend is less clear due to the large uncertainties derived for those pulsed fractions. Furthermore, at all epochs, the pulsed fraction clearly increases with energy, an equally interesting result discussed below.

The evolving spectral shape and pulse profiles of TAXP XTE J1810–197 provide an important and unique (so far) diagnostic of the emission geometry, and ultimately, the emission mechanism(s) of NSs. Because the spectrum is well modeled by two blackbody components, and the phase alignment of the pulse profiles are energy independent, it is most natural to consider emission from two concentric regions on the NS surface. The hot component is associated with a smaller hot-spot, while the blackbody model predicts a larger annulus for the warm component. The complete collection of observed pulsed fractions is consistent with this model. The smaller, hotter spot always dominates the spectrum above 2 keV and offers a natural explanation for the higher modulation at these energies. However, at lower energies, this contribution gradually fades relative to cooler emission (which is less modulated), contributing less and less to the pulsed emission over time.

We note in passing that a fitted power-law spectral component would have to dominate the soft (<2 keV) X-rays at *all* epochs, while varying in its contribution to the hard (>2 keV) X-rays. Such a power law would drive an evolution of the pulse shapes that is opposite of what is observed. Thus, we find that the detailed evolution of the X-ray emission from XTE J1810–197 further supports the assumption of a purely thermal spectral model, and leaves no evidence of a steep power law as is commonly fitted to individual observations of AXPs.

Since the first observation of XTE J1810–197 it was apparent that the broad-band pulse shape is not a simple sine function; the pulse peak is relatively sharp with a broader inter-pulse trough (paper I). This effect is more pronounced at higher energy, where the profile is nearly triangular in shape. This suggests that the pulse profiles can be decomposed into a triangle function and a sinusoidal function, a model that was explored for the earlier data sets and detailed in papers II and III. This model continues to be appropriate for the new data and is used to extract unbiased pulsed fraction measurements for the profiles shown in Fig. 6.

Given the success in modeling the pulse profiles with these functions it is natural to associate the two pulsed components uniquely with the two spectral components, i.e., the triangle shape for the hotter blackbody component and the sinusoidal shape for the cooler one. However, we find that we can not model all the profiles in a consistent manner with just a simple superposition of these temporal components, based on the implied flux ratio in each energy band (Gotthelf and Halpern 2005). Instead we conclude that either the two spectral components contain an admixture of the two shapes or there is a third, unmodeled spectral component present. This is a direction of active research, to consider the correct characterization of the phase dependent spectral contribution.

4 Radio observations of XTE J1810–197

Considering that the absence of radio emission is a defining characteristic of AXPs, the serendipitous radio detection from XTE J1810–197 came as a great surprise. A chance search of VLA data taken about a year after outburst reveals an unresolved point source with a flux of 4.5 mJy at 1.43 GHz, located at the precise coordinates of the TAXP (Becker 2005b). Other archival VLA data obtained at various frequencies provide upper-limits before and after outburst. A follow-up VLA observation in 2006 March yielded a flux of 12.9 mJy at 1.43 GHz. Together, these results indicate highly erratic radio emission. A second surprise came with the detection around that time of pulsed radio emission at the X-ray period (Camilo et al. 2006). This search, at $\nu = 1.4$ GHz using the Parkes radio observatory, revealed a narrow, bright pulse with a high degree of linearly polarization (Camilo et al. 2006). A series of multi-frequency measurements in 2006 April–May confirms the erratic flux behavior and provides an unusual spectral slope of $\alpha \gtrsim -0.5$, where $S_\nu \propto \nu^\alpha$ (cf. $\alpha \sim -1.6$ for a typical radio pulsar). A revised distance to the pulsar of 3.3 kpc is inferred from the pulsar's dispersion. A search for pulsations in archival data acquired in 1997/1998 produced a null result and argues against significant emission prior to the outburst event. A summary of these radio observations is shown in Fig. 7.

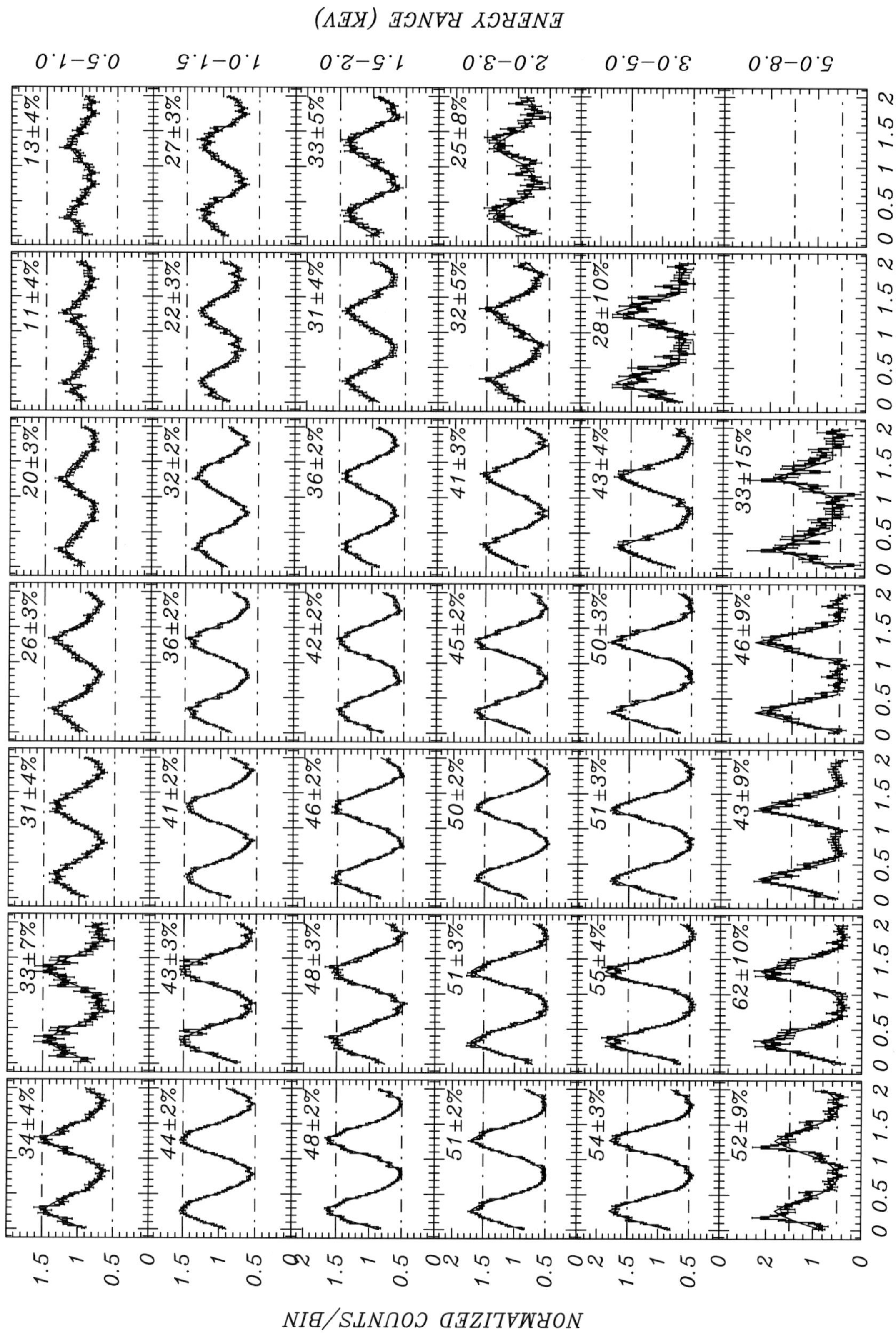

Fig. 6 Energy-dependent pulse profiles of XTE J1810–197 obtained with the XMM-Newton EPIC pn detector for all seven epochs of Table 1. The background for each profile has been subtracted and phase zero is aligned for each epoch, for comparison. The phase of the peak is seen to be energy-independent. The pulsed fraction at low X-ray energies has decreased with time, while remaining essentially unchanged at high energy. Also shown is the best fit to the two-component model for the pulse profile (*solid line*) described in the text (see Sect. 2)

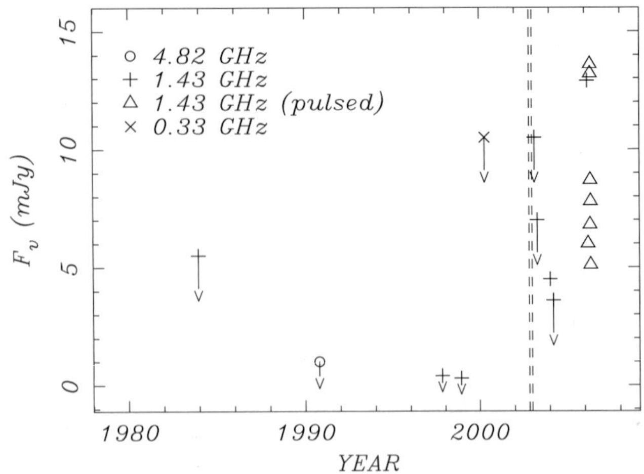

Fig. 7 A set of archival and dedicated radio observations of XTE J1810–197 measured at three frequencies. Prior to outburst (denoted by *vertical lines*) no radio emission is found corresponding to the plotted upper-limits. Following the X-ray outburst, large variations in flux are inferred from several measurements and upper-limits. Pulsed radio emission is also recorded at the X-ray period in the post-outburst era following the 2006 discovery. A search of archival data prior to the outburst places an upper-limit of ~0.4 mJy at 1.4 GHz. For clarity, multi-frequency measurements coinciding with the 1.4 GHz pulsed emission data points are not shown (from Becker 2005b; Camilo et al. 2006)

5 Emission geometry: models and theory

Although transient AXPs such as XTE J1810–197 are relatively rare, their short active duty cycle suggests the existence of a larger population of unrecognized young NSs. XTE J1810–197 provides a unique window into this population, with prior measurements in the quiescent state and detailed observations during its active, pulsed phase. Ultimately, we hope to use this TAXP to give insight into the emission mechanisms of magnetars, in general. With the discovery of pulsed radio flux, this may carry over to interpreting emission mechanism of radio pulsars, as well. Below we now summarize our initial attempts to interpret the XTE J1810–197 results in the context of a physical model. First we outline the basic arguments for the two-temperature blackbody spectral model as a plausible alternative to the nominal power-law plus blackbody model.

While both models give reasonable fits to the spectrum, it's important to note that the power-law component is used to model the *low energy residuals and not any high-energy tail*. Without a low energy cut-off of this component, it is not possible to connect it with the observed IR emission without an energy catastrophe. As detailed in paper II, invoking synchrotron self-absorption is excluded by radius/magnetic-field inconsistently. In contrast, the extrapolated spectrum of the warm blackbody component does not exceed the measured IR flux. Furthermore, currently there is no acceptable physical model to anchor a power-law component. As discussed earlier, the two components of the double blackbody model can be associated with a pair of hot spots on the surface of the star. The warm temperature component, covering a large fraction of the NS surface, is consistent with the decreased pulsed fraction at lower energies. The hotter blackbody component is consistent with a smaller emission area and greater modulation at higher energies. A comprehensive discussion of this issue can be found in paper II.

Determining the emission geometry on the NS is of great interest. The natural conclusion of applying the double blackbody model to the time resolved spectra and pulse profiles is that concentric hot regions give rise to the observed modulation. We consider a model of the phase-resolved spectrum taking into account the viewing geometry and offset of the hot spots from the rotation axis (Perna and Gotthelf 2006, in preparation). This is based on the NS emission model given in (Perna et al. 2001) that includes general relativistic effects (redshift and light bending). Our goal is to match spectrum and energy dependent pulse shape in order to determine viewing geometry, distance, and NS size. Preliminary work is able to reproduce the pulsed fraction to a reasonable degree but the pulse shape remains elusive.

A framework for a theoretical interpretation of the emission from XTE J1810–197 is suggested by the magnetar coronal model of Beloborodov and Thompson (2006). According to this model, the large outburst was generated by a starquake that resulted in a transition to an active coronal state that caused energy to be stored in the twisted B-field of the coronal loop. Particles (mostly e^+e^-) are accelerated within this loop and impact the NS surface with GeV energy. This heats up the loop footprint resulting in the observed hot spot emission. The decay timescale of the coronal loop, and thus the hotter component of the double blackbody luminosity, is of order of a few years. The decay rate is determined by ohmic dissipation of current in the excited loop. A cooler component likely arises from deep crustal heating associated with the initial outburst and possibly earlier ones. The features of this model are in general accordance with the observational properties and inferred model for XTE J1810–197.

Acknowledgements We thank Fred Jansen and Norbert Schartel for providing the four XMM-Newton Targets of Opportunity observations of XTE J1810–197.

References

Becker, R.H.: Astrophys. J. **632**, L29 (2005)
Beloborodov, A.M., Thompson, C.: Astrophys. J. (2006, in press); arXiv:astro-ph/0602417
Camilo, F., Ransom, S., Halpern, J., et al.: Nature **442**, 892 (2006) (astro-ph/0605429)
Duncan, R.C., Thompson, C.: Astrophys. J. Lett. **392**, L9 (1992)
Gotthelf, E.V., Halpern, J.P.: Paper III. Astrophys. J. **632**, 1075 (2005)
Gotthelf, E.V., Halpern, J.P., Buxton, M., et al.: Paper I. Astrophys. J. **605**, 368 (2004)

Halpern, J.P., Gotthelf, E.V.: Paper II. Astrophys. J. **618**, 874 (2005)
Ibrahim, I.A., et al.: Astrophys. J. Lett. **609**, L21 (2004)
Kaspi, V.M., Gavriil, F.P., Woods, P.M., et al.: Astrophys. J. **588**, 93 (2003)
Mereghetti, S., Chiarlone, L., Israel, G.L., et al.: Proc. of the 270. WE-Heraeus Seminar, vol. 29 (2002)
Perna, R., Heyl, J.S., Hernquist, L.E., et al.: Astrophys. J. **557**, 18 (2001)
Rea, N., et al.: Astron. Teleg. **284**, 1 (2004a)
Rea, N., et al.: Astron. Astrophys. **425**, L5 (2004b)
Thompson, C., Duncan, R.C.: Astrophys. J. **473**, 322 (1996)
Woods, P.M., Kouveliotou, C., Gavriil, F.P., et al.: Astrophys. J. **629**, 985 (2005)

Astrophys Space Sci (2007) 308: 89–94
DOI 10.1007/s10509-007-9305-2

ORIGINAL ARTICLE

PSR J1119–6127 and the X-ray emission from high magnetic field radio pulsars

M.E. Gonzalez · V.M. Kaspi · F. Camilo · B.M. Gaensler · M.J. Pivovaroff

Received: 17 July 2006 / Accepted: 9 October 2006 / Published online: 22 March 2007

Abstract The existence of radio pulsars having inferred magnetic fields in the magnetar regime suggests that possible transition objects could be found in the radio pulsar population. The discovery of such an object would contribute greatly to our understanding of neutron star physics. Here we report on unusual X-ray emission detected from the radio pulsar PSR J1119–6127 using XMM–Newton. The pulsar has a characteristic age of 1,700 yrs and inferred surface dipole magnetic field strength of 4.1×10^{13} G. In the 0.5–2.0 keV range, the emission shows a single, narrow pulse with an unusually high pulsed fraction of $\sim$70%. No pulsations are detected in the 2.0–10.0 keV range, where we derive an upper limit at the 99% level for the pulsed fraction of 28%. The pulsed emission is well described by a thermal blackbody model with a high temperature of $\sim2.4\times10^6$ K. While no unambiguous signature of magnetar-like emission has been found in high-magnetic-field radio pulsars, the X-ray characteristics of PSR J1119–6127 require alternate models from those of conventional thermal emission from neutron stars. In addition, PSR J1119–6127 is now the radio pulsar with the smallest characteristic age from which thermal X-ray emission has been detected.

M.E. Gonzalez (✉) · V.M. Kaspi
Department of Physics, McGill University, Rutherford Physics Building, Montreal, QC H3A 2T8, Canada
e-mail: gonzalez@physics.mcgill.ca

F. Camilo
Columbia Astrophysics Laboratory, Columbia University, 550 West 120th Street, New York, NY 10027, USA

B.M. Gaensler
Harvard–Smithsonian Centre for Astrophysics, 60 Garden Street, Cambridge, MA 02138, USA

Present Address:
B.M. Gaensler
School of Physics, The University of Sydney, Sydney, NSW 2006, Australia

M.J. Pivovaroff
Lawrence Livermore National Laboratory, P.O. Box 808, L-258, Livermore, CA 94550, USA

Keywords ISM: individual (G292.2–0.5) · Pulsars: individual (PSR J1119–6127) · X-rays: ISM

PACS 97.60.Jd · 95.85.Nv · 97.60.Gb

1 Introduction

The emission from the $\sim$1,500 radio pulsars (PSRs) discovered to date is generally thought to be powered by the loss of rotational kinetic energy due to magnetic braking. Radio pulsars with implied magnetic fields in the range $\sim10^{13-14}$ G have now been discovered, showing that radio emission can be produced in neutron stars with fields above the quantum critical field $B_c = 4.4\times10^{13}$ G. More exotic neutron stars with magnetic fields in the range $\sim10^{14-15}$ G have been observed at high energies and are believed to be powered by the decay of their large magnetic fields. Anomalous X-ray pulsars (AXPs) and Soft-Gamma Repeaters (SGRs) make up this class of so-called "magnetars" (Woods and Thompson 2006). As the inferred magnetic field strengths of radio pulsars and magnetars are now found to overlap, the underlying physical reasons for the differences in their emission properties remains a puzzle. To date, no radio pulsar has been observed to exhibit magnetar-like emission at high energies, posing an interesting challenge to current emission theories.

Here we summarize the analysis and results obtained from an XMM–Newton observations of PSR J1119–6127, originally published by Gonzalez et al. (2005). PSR J1119–6127 is one of the youngest radio pulsars known and also has one the highest inferred magnetic fields in the radio pulsar population. Our observation reveals unusual thermal emission from this object which may represent the first evidence for high-magnetic-field effects in the emission from a "normal" radio pulsar. We discuss these results in light of recent theoretical work on emission from highly magnetized neutron stars and observations of similar sources.

PSR J1119–6127, with spin period $P = 0.408$ s, is among the youngest radio pulsars known (Camilo et al. 2000). The measured braking index for the pulsar of $n = 2.91 \pm 0.05$ ($\dot{\nu} \propto -\nu^n$) implies an upper limit for the age of 1,700 yr. The pulsar has spin-down luminosity $\dot{E} \equiv 4\pi^2 I \dot{P}/P^3 = 2.3 \times 10^{36}$ erg s^{-1} (for a moment of inertia $I = 10^{45}$ g cm^2) and an inferred surface dipole magnetic field strength at the equator of $B \equiv 3.2 \times 10^{19} (P\dot{P})^{1/2}$ G $= 4.1 \times 10^{13}$ G. This value of B is among the highest known in the radio pulsar population. PSR J1119–6127 powers a small ($3'' \times 6''$) X-ray pulsar wind nebula (PWN, Gonzalez and Safi-Harb 2003), which results from the confinement of the pulsar's relativistic wind of particles and electromagnetic fields by the ambient medium. The pulsar lies close to the center of a $15'$-diameter "shell" supernova remnant (SNR), G292.2–0.5 (Crawford et al. 2001; Pivovaroff et al. 2001), located at a distance of 8.4±0.4 kpc, as determined from neutral hydrogen absorption measurements (Caswell et al. 2004).

2 Observation

We observed PSR J1119–6127 using the European Photon Imaging Camera (EPIC) aboard the XMM–Newton satellite on June 26, 2003. The MOS and PN instruments were operated in full-window and large-window mode, respectively. The temporal resolution was 2.6 s for MOS and 48 ms for PN. The data were analyzed using the Science Analysis System software (SAS v6.2.0) and standard reduction techniques. The effective exposure time was 48 ks for MOS1/MOS2 and 43 ks for PN.

Figure 1 shows the combined MOS image of the system in the 0.3–1.5 keV (*red*), 1.5–3.0 keV (*green*) and 3.0–10.0 keV (*blue*) bands.[1] The bright source at the center has coordinates $\alpha_{2000} = 11^{\mathrm{h}}19^{\mathrm{m}}14.65^{\mathrm{s}}$ and $\delta_{2000} = -61°27'50.2''$ ($4''$ error). This position coincides with the Chandra and radio coordinates of PSR J1119–6127. Although the spatial resolution of XMM–Newton (half power diameter of $15''$) permits the PWN to be neither resolved nor separated from the pulsar emission itself, XMM–Newton's sensitivity and high time resolution allow us to separate the pulsar's emission spectrally and temporally. The image reveals for the first time the detailed morphology of the SNR at X-ray energies as XMM–Newton's sensitivity was needed due to the remnant's low surface brightness. The large east-west asymmetry at low energies has been attributed to the presence of a molecular cloud on the east side of the field (Pivovaroff et al. 2001).

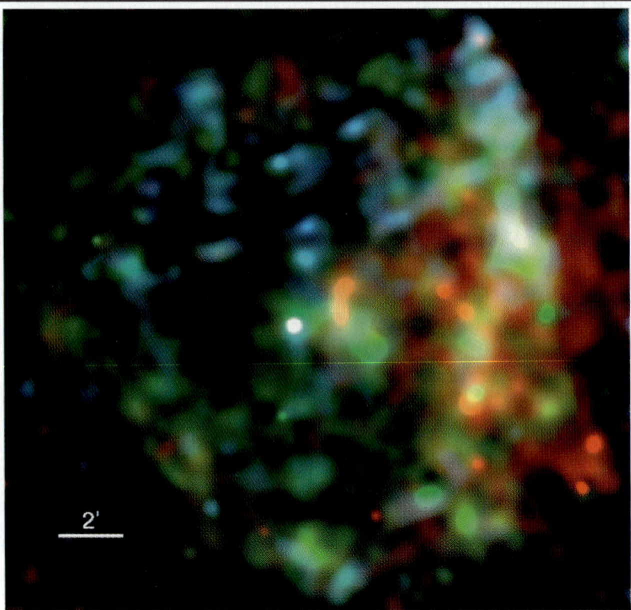

Fig. 1 Combined MOS image of G292.2–0.5 and PSR J1119–6127 in the 0.3–1.5 keV (*red*), 1.5–3.0 keV (*green*) and 3.0–10.0 keV (*blue*) bands. Individual MOS1/MOS2 images were first binned into pixels of $2.5'' \times 2.5''$ and unrelated point sources in the field were excluded. These images were then added and adaptively smoothed with a Gaussian with $\sigma = 5'' - 15''$ to obtain signal-to-noise ratios higher than 3. Background images and exposure maps at each energy band were similarly obtained and used to correct the final images

3 Results

3.1 Timing analysis

The PN data were used to search for pulsations from PSR J1119–6127. A circular region of $25''$-radius centered at the X-ray coordinates of the pulsar was used to extract the source photons. The data were divided into different energy ranges at 0.5–10.0 keV (620±30 photons), 0.5–2.0 keV (340±25 photons) and 2.0–10.0 keV (275±22 photons). The most significant signal was detected in the 0.5–2.0 keV range with $Z_2^2 = 52.8$ (6.6σ significance) at a frequency of 2.449726(6) Hz (1σ errors). This frequency is in agreement

[1] The detailed spatial distribution in the 3.0–10.0 keV (*blue*) image should be examined with caution as this energy band suffered from a high degree of stray-light contamination from a nearby high-energy source.

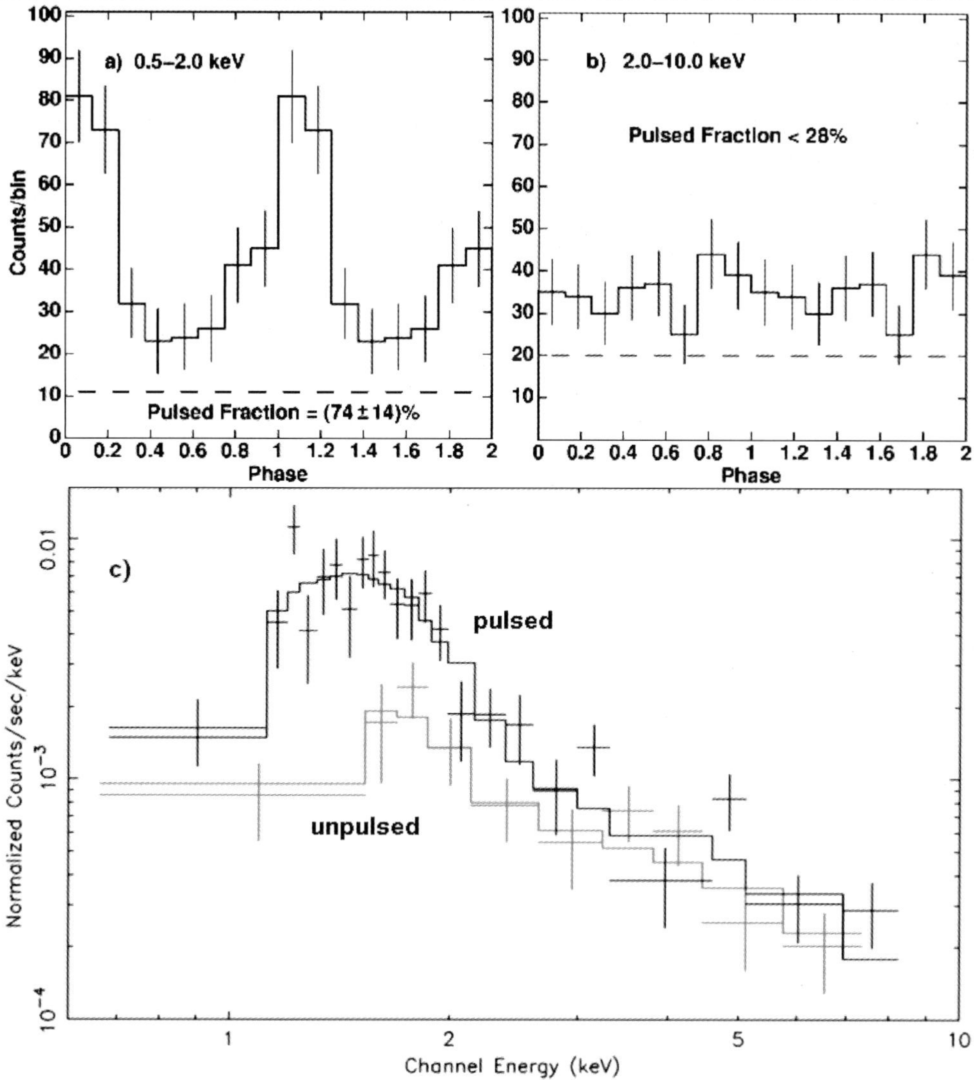

Fig. 2 *Top*: X-ray pulse profiles of PSR PSR J1119–6127 in the 0.5–2.0 keV (*left*) and 2.0–10.0 keV (*right*) ranges. Errors bars are 1σ and two cycles are shown. The *dashed lines* represent our estimates for the contribution from the pulsar's surroundings (see Sect. 3.1). *Bottom*: EPIC-PN spectra obtained for the pulsed (*black*) and unpulsed (*grey*) regions of the pulse profile with their respective best-fit blackbody plus power-law model (*solid curves*)

with the radio prediction for PSR J1119–6127 obtained using regular monitoring with the Parkes telescope. In the 2.0–10.0 keV range, no signal was found with a significance $>2.8\sigma$. In the 0.5–10.0 keV range, the above signal was detected with a 5.2σ significance.

Figure 2 shows the resulting pulse profiles at 0.5–2.0 keV (top, left) and 2.0–10.0 keV (top, right). The background was estimated from a nearby region away from bright SNR knots. The horizontal dashed lines represent our estimates for the contribution from the pulsar's surroundings obtained using the Chandra observation of the pulsar (Gonzalez and Safi-Harb 2003). The Chandra count rate excluding the pulsar was used to estimate the XMM–Newton counts using *WebPIMMS*. The resulting pulsed fraction [$\mathrm{PF} \equiv (F_{\max} - F_{\min})/(F_{\max} + F_{\min})$] is labeled in Fig. 2 ($1\sigma$ statistical errors). In the 2.0–10.0 keV range, we derive an upper limit for the pulsed fraction of 28% (at the 99% confidence level; see, e.g., Vaughan et al. 1994; Ransom et al. 2002). The radio emission from PSR J1119–6127 consists of a single peak of duty cycle 5% and luminosity at 1.4 GHz of 28 mJy kpc^2 (Camilo et al. 2000). Phase zero in the X-ray profiles corresponds to the radio peak (determined with 3 ms uncertainty). Taking into account the low temporal resolution of the XMM–Newton observation (phase bin width of 51 ms), this result suggests that the radio peak is in phase with the X-ray peak within our uncertainties, or possibly just slightly ahead. Additional observations at higher temporal resolution will help to confirm and constrain these results.

3.2 Spectral analysis

The EPIC data were used to perform a spectral analysis of PSR J1119–6127. Circular regions with radii of $20''$ and $25''$ were used for MOS and PN, encompassing ~75% and ~78% of the source photons, respectively. The derived fluxes have been corrected accordingly. Background regions were chosen from nearby areas away from bright SNR knots. The spectra were fit in the 0.5–10.0 keV range us-

Table 1 Fits to the XMM–Newton phase-averaged spectrum of PSR J1119–6127

Parameter	PL+PL ($\pm 1\sigma$)	BB+PL ($\pm 1\sigma$)	Atm[a]+PL ($\pm 1\sigma$)
N_H (10^{22} cm^{-2})	$2.3^{+0.4}_{-0.3}$	$1.6^{+0.4}_{-0.3}$	$1.9^{+0.5}_{-0.3}$
χ^2(dof)	79(66)	78(66)	78(66)
	Soft component characteristics		
Γ or T^{∞}	6.5±0.9	$2.4^{+0.3}_{-0.2}$ MK	0.9±0.2 MK
R^{∞} (km)	...	$3.4^{+1.8}_{-0.3}$	12 (fixed)
d (kpc)	8.4 (fixed)	8.4 (fixed)	$1.6^{+0.2}_{-0.9}$
f_{abs}[b] (10^{-14})	$2.1^{+2.3}_{-0.9}$	$1.5^{+1.8}_{-0.2}$	$1.7^{+7.0}_{-0.4}$
f_{unabs}[b] (10^{-13})	63^{+57}_{-32}	$2.4^{+3.0}_{-0.5}$	$7.2^{+31}_{-1.6}$
L_X[b] (10^{33})	53^{+50}_{-27}	$2.0^{+2.5}_{-0.4}$	$0.22^{+0.88}_{-0.05}$
	Hard component characteristics		
Γ	$1.3^{+0.5}_{-0.2}$	$1.5^{+0.3}_{-0.2}$	$1.5^{+0.2}_{-0.3}$
f_{abs}[b] (10^{-14})	$7.1^{+10}_{-1.5}$	$7.4^{+3.6}_{-1.0}$	$7.3^{+4.7}_{-2.7}$
f_{unabs}[b] (10^{-13})	$1.0^{+1.6}_{-0.2}$	$1.1^{+0.6}_{-0.2}$	$1.1^{+0.8}_{-0.3}$
L_X[b] (10^{33})	$0.8^{+1.3}_{-0.2}$	$0.9^{+0.5}_{-0.1}$	0.04±0.02

[a]The atmospheric model was computed with $B = 10^{13}$ G and pure hydrogen composition. The local values for the temperature T and radius $R = 10$ km of the star have been redshifted to infinity according to the formulae $T^{\infty} = T(1 - 2GM/Rc^2)^{1/2}$ and $R^{\infty} = R(1 - 2GM/Rc^2)^{-1/2}$, with $M = 1.4\ M_{\odot}$

[b]The 0.5–10.0 keV absorbed and unabsorbed fluxes f_{abs} and f_{unabs} have units of ergs s^{-1} cm^{-2}. The 0.5–10.0 keV X-ray luminosity L_X, at the distance d, is in units of ergs s^{-1}

ing XSPEC (v.11.3.0) with a minimum of 20 counts per bin from a total of 240±19, 210±18, and 620±30 background-subtracted counts in MOS1, MOS2 and PN, respectively.

Two-component models were needed in order to describe the low and high energy portions of the spectra. A non-thermal power-law component with photon index $\Gamma \sim 1.5$ described the high-energy emission in the spectra well. In turn, various models were used to describe the low-energy emission. The derived fits are summarized in Table 1.

We also extracted PN spectra from the "pulsed" and "unpulsed" regions of the pulse profile, at phases 0.7–1.3 (430±28 counts) and 0.3–0.7 (200±20 counts), respectively. These spectra are shown in Fig. 2 (*bottom*) and were well fit by two-component models that agree with those derived for the phase-averaged spectra. For example, a blackbody plus power-law model yielded $T^{\infty} = 2.8\pm0.4$ MK and $\Gamma = 1.4^{+0.5}_{-0.2}$ (1σ errors). The main difference between the pulsed and unpulsed spectra was found to be the relative contributions of the model components. The pulsed spectrum is dominated by the soft component below ~2 keV, while the unpulsed spectrum is dominated by the hard, power-law component at all energies.

4 Discussion

4.1 Observed emission characteristics

The pulse profile in the 0.5–2.0 keV range shows a single, narrow pulse with a high pulsed fraction. Modeled as a Gaussian, the full-width at half maximum is $0.26^{+0.08}_{-0.06}P$ (1σ errors) with $\chi^2(\text{dof}) = 2.1(4)$ and a probability of 0.72. A sinusoidal fit resulted in $\chi^2(\text{dof}) = 8.9(5)$ with a probability of 0.11. Due to the limited statistics available, the Gaussian fit is preferred only at the 2.4σ level (according to the F-test). A two-component profile cannot be ruled out with the present data (e.g., sine curve plus a narrow peak, or two narrow peaks at phases ~0.8 and ~1.1). Additional X-ray observations at higher temporal resolution are needed to further constrain the pulse shape.

The observed spectrum from PSR J1119–6127 requires two-component models. The hard, non-thermal component is consistent with arising from the pulsar's surroundings. Using the high-resolution Chandra data we estimate that the PWN plus SNR emission within a 25″ radius, excluding the pulsar, is well described by a power-law model with $\Gamma = 1.8^{+0.8}_{-0.6}$ and unabsorbed flux in the 0.5–10.0 keV range of $0.9^{+0.6}_{-0.5} \times 10^{-13}$ erg s^{-1} cm^{-2} (1σ errors). These values are in good agreement with the hard power-law component shown in Table 1. Although a small contribution from the pulsar to this hard emission cannot be ruled out, it would not affect our results on the pulsar's soft emission.

The soft spectral component must then arise from the pulsar. Non-thermal X-ray spectra from radio pulsars have photon indices in the range $0.5 < \Gamma < 2.7$ (Becker and Aschenbach 2002; Pavlov et al. 2002). Models for synchrotron emission in the pulsar magnetosphere (Cheng and Zhang 1999; Rudak and Dyks 1999) predict $\Gamma \lesssim 2$. The pulsed emission we detect from PSR J1119–6127, if interpreted as non-thermal in origin, has a steeper spectrum ($\Gamma = 6.5 \pm 0.9$, Table 1) than those observed or predicted. Therefore, a thermal origin for the observed emission is strongly favored with the present data. In addition, atmospheric models require small distances (or conversely, implausibly large emitting radii at 8.4 kpc) to account for the observed emission. We therefore favor a blackbody model to account for the observed emission.

The blackbody temperature of the X-ray emission from PSR J1119–6127 is $T^{\infty}_{\text{bb}} = (2.4^{+0.3}_{-0.2}) \times 10^6$ K. This temperature is among the highest seen in radio pulsars; while it is naively similar to those found in much older pulsars, none of them exhibits a higher temperature at statistically significant levels (e.g., PSR J0218+4232 has a characteristic age of $\tau_c = P/2\dot{P} = 5.1 \times 10^8$ yr and a blackbody temperature of $T^{\infty}_{\text{bb}} = 2.9 \pm 1.1 \times 10^6$ K, 3σ errors (Webb et al. 2004). PSR J1119–6127 is now the youngest radio pulsar from which thermal emission has been detected, the next youngest being Vela (τ_c = 11 kyr) with a blackbody temperature $T_{\mathrm{bb}}^{\infty} = (1.47 \pm 0.18) \times 10^6$ K (3σ range (Pavlov et al. 2001)). Moreover, the pulsed fraction of the thermal emission from PSR J1119–6127 is significantly higher than is seen in other radio pulsars; thermal sources where the emission arises from the entire surface or for localized regions show pulsed fractions at low energies of $\lesssim$40% (Pavlov et al. 2001, 2002; Becker and Aschenbach 2002). The narrow peak in the pulse profile points to yet another difference from what is normally seen in thermal emission from radio pulsars at low energies (Pavlov et al. 2002; Becker and Aschenbach 2002), namely broad pulsations.

4.2 Thermal emission mechanisms

Conventional models for thermal emission from neutron stars cannot account for the observed characteristics in PSR J1119–6127. Thermal emission from polar-cap reheating has been well studied and, whether the required return currents arrive from the outer gap region (Cheng and Zhang 1999) or from close to the polar cap (Harding and Muslimov 2001), the X-ray luminosity is constrained to be $\lesssim 10^{-5}\dot{E}$ for sources as young as PSR J1119–6127. This is at least 2 orders of magnitude below what we observe.

Thermal emission may also arise from the surface due to initial cooling. The observed luminosity is consistent with predictions from standard models of cooling neutron stars (Yakovlev et al. 2004; Page et al. 2006). However, the effective blackbody temperature is higher than predicted and the observed blackbody radius is smaller than allowed from neutron star equations of state (Lattimer and Prakash 2000). The very high observed pulsed fraction is also consistent with emission arising from a small fraction of the neutron star surface.

On the other hand, recent work on surface emission from highly magnetized neutron stars has explored the effects of a high magnetic field, as heat conductivity is expected to be suppressed perpendicular to the field lines and will be channeled along the lines instead (Geppert et al. 2004; Pérez-Azorín et al. 2006a). This will produce a highly anisotropic temperature distribution on the surface of the star, with small, hot regions at the magnetic poles. Highly modulated thermal emission with high temperatures will then be produced. These results have been applied to model the observed emission from RXJ 0720.4–3125, a highly magnetized neutron star, with apparent success (Perez-Azorin et al. 2006). Pulsed fractions as high as ~30% in the case of isotropic blackbody emission have been reported (Geppert et al. 2005). Therefore, it remains to be shown whether the same models can be applied to reproduce the observed emission characteristics in PSR J1119–6127. In this case, PSR J1119–6127 would be the first radio pulsar to show the effects of a high magnetic field through its X-ray emission.

We also point out the thermal emission with high temperature and high pulsed fraction that was found for PSR J1852+0040 (Gotthelf et al. 2005). Although a detailed timing solution has not been reported, initial estimates suggest a characteristic age of $\tau_c > 24$ kyr and low magnetic field of $B < 3 \times 10^{12}$ G. If these estimates are correct, existing theories for thermal emission from neutron stars cannot readily account for the observed characteristics, including those involving high-magnetic-field effects as mentioned above.

4.3 Other high-magnetic field pulsars

Many radio pulsars having inferred magnetic fields in the range 10^{13-14} G have now been discovered. A sample of these pulsars with associated X-ray observations is shown in Table 2. Most of these sources have proved to be very faint in X-rays. Only two pulsars, PSRs J1846–0258 and B1509–58, are bright non-thermal sources and power bright PWNe. As expected, and in agreement with normal radio pulsars, they are young and very energetic ($\tau_c < 2{,}000$ yr and $\dot{E} > 10^{36}$ erg s^{-1}).

On the other hand, the older and less energetic pulsars in Table 2 have not been detected in X-rays ($\tau_c > 10{,}000$ yr and $\dot{E} < 10^{33}$ erg s^{-1}). This includes PSR J1847–0130, the radio pulsar with highest inferred magnetic field discovered

Table 2 High-magnetic field radio pulsars

PSR	J1847–0130	J1718–3718	J1814–1744	J1846–0258	B0154–61	B1509–58
P (sec)	6.7	3.4	4	0.32	2.35	0.15
B (10^{13} G)	9.4	7.4	5.5	4.8	2.1	1.5
τ_c (kyr)	83	34	85	0.72	197	1.7
$\dot{E}$ (ergs s^{-1})	1.7×10^{32}	1.5×10^{33}	4.7×10^{32}	8×10^{36}	5.7×10^{32}	1.8×10^{37}
D (kpc)	~8	4–5	~10	~19	~1.7	~5
L_X (ergs s^{-1})	$<5\times10^{33}$	$\sim10^{30}$	$<6\times10^{35}$	6.4×10^{34}	$<1.4\times10^{32}$	2.4×10^{34}
T or Γ	–	$T\sim1.6$ MK	–	$\Gamma\sim1.4$	–	$\Gamma\sim1.4$
Ref.	(McLaughlin et al. 2003)	(Kaspi and McLaughlin 2005)	(Pivovaroff et al. 2000)	(Helfand et al. 2003)	(Gonzalez et al. 2004)	(Gaensler et al. 2002)

to date (0.9×10^{14} G). These pulsars then show no enhancement of high-energy emission despite having inferred magnetic fields in the magnetar range. One intermediate case is that of PSR J1718–3718, which does have a faint X-ray counterpart seen with Chandra. However, the detailed characteristics of this emission (e.g., thermal vs. non-thermal) could not be constrained with the data.

PSR J1119–6127 is then an interesting and puzzling source. Despite being young and energetic, it does not power a bright PWN and we have found its emission to be dominated by a thermal component. This is in direct contrast to the sources mentioned above, particularly PSR J1846–0258 with which PSR J1119–6127 shares almost identical spin characteristics and even similar surroundings in their respective SNRs (Helfand et al. 2003; Gonzalez and Safi-Harb 2005). It is therefore unclear what the physical reasons are behind their vastly different X-ray emission. In addition, while it is possible that the characteristics observed in PSR J1119–6127 may be due to heat conductivity effects on a highly magnetized atmosphere, the emission is not magnetar-like.

5 Conclusion

The X-ray emission from the young, high magnetic field radio pulsar PSR J1119–6127 shows a thermal spectrum with high temperature and small emitting radius, making it the radio pulsar with smallest characteristic age from which thermal X-ray emission has been detected. The pulse profile of this emission is consistent with a single, narrow pulse with a high pulsed fraction. Hot spots heated by back-flowing particles from the magnetosphere are not expected in such a young source, while the X-ray characteristics are not consistent with cooling emission from the whole surface. However, a highly anisotropic temperature distribution on the surface due to a high magnetic field may be able to account for the observed characteristics. This would make PSR J1119–6127 the first radio pulsar to exhibit high-magnetic-field effects on its X-ray emission. Additional X-ray observations, particularly with improved temporal resolution, will help to confirm and constrain the observed characteristics.

Many high magnetic field radio pulsars have now been observed in X-rays, some with very similar spin characteristics to PSR J1119–6127 and others with higher inferred fields, but it remains unclear why PSR J1119–6127 is to date the only one to show such effects. Recently, the discovery of radio emission from a magnetar (Camilo et al. 2006) has shown that such emission is possible in these sources, contributing to our understanding of the mechanisms at work. However, we still lack an understanding for the absence of magnetar-like emission from radio pulsars with high magnetic fields and additional observations of these objects are then needed.

References

Becker, W., Aschenbach, B.: In: Becker, W., Lesch, H., Trümper, J. (eds.) Proc. 270th WE-Heraeus Seminar on Neutron Stars, Pulsars, and Supernova Remnants. MPE, Garching, p. 64 (2002) (astro-ph/0208466, MPE Rep. 278)

Camilo, F., Kaspi, V.M., Lyne, A.G., et al.: Astrophys. J. **541**, 367 (2000)

Camilo, F., Ransom, S., Halpern, J., et al.: Nature **442**, 892 (2006) (astro-ph/0605429)

Caswell, J.L., McClure-Griffiths, N.M., Cheung, M.C.M.: Mon. Not. Roy. Astron. Soc. **352**, 1405 (2004)

Cheng, K.S., Zhang, L.: Astrophys. J. **515**, 337 (1999)

Crawford, F., Gaensler, B.M., Kaspi, V.M., et al.: Astrophys. J. **554**, 152 (2001)

Gaensler, B.M., Arons, J., Kaspi, V.M., et al.: Astrophys. J. **569**, 878 (2002)

Geppert, U., Küker, M., Page, D.: Astron. Astrophys. **426**, 267 (2004)

Geppert, U., Kueker, M., Page, D., ArXiv Astrophys. e-prints (2005), astro-ph/0512530

Gonzalez, M., Safi-Harb, S.: Astrophys. J. **591**, L143 (2003)

Gonzalez, M., Safi-Harb, S.: Astrophys. J. **619**, 856 (2005)

Gonzalez, M.E., Kaspi, V.M., Camilo, F., et al.: Astrophys. J. **630**, 489 (2005)

Gonzalez, M.E., Kaspi, V.M., Lyne, A.G., et al.: Astrophys. J. **610**, L37 (2004)

Gotthelf, E.V., Halpern, J., Seward, F.D.: Astrophys. J. **627**, 390 (2005)

Harding, A.K., Muslimov, A.G.: Astrophys. J. **556**, 987 (2001)

Helfand, D.J., Collins, B.F., Gotthelf, E.V.: Astrophys. J. **582**, 783 (2003)

Kaspi, V.M., McLaughlin, M.A.: Astrophys. J. **518**, 41 (2005)

Lattimer, J.M., Prakash, M.: Phys. Rev. **333**, 121 (2000)

McLaughlin, M.A., Stairs, I.H., Kaspi, V.M., et al.: Astrophys. J. **591**, L135 (2003)

Page, D., Geppert, U., Weber, F.: Nucl. Phys. A **777**, 492 (2006) (astro-ph/0508056)

Pavlov, G.G., Zavlin, V.E., Sanwal, D.: In: Becker, W., Lesch, H., Trümper, J. (eds.) Proc. 270th WE-Heraeus Seminar on Neutron Stars, Pulsars, and Supernova Remnants. MPE, Garching, p. 273 (2002) (astro-ph/0206024, MPE Rep. 278)

Pavlov, G.G., Zavlin, V.E., Sanwal, D., et al.: Astrophys. J. **552**, L129 (2001)

Pérez-Azorín, J.F., Miralles, J.A., Pons, J.A.: Astron. Astrophys. **451**, 1009 (2006a)

Perez-Azorin, J.F., Pons, J.A., Miralles, J.A., et al.: Astron. Astrophys. **459**, 175 (2006) (astro-ph/0603752)

Pivovaroff, M., Kaspi, V.M., Camilo, F.: Astrophys. J. **535**, 379 (2000)

Pivovaroff, M.J., Kaspi, V.M., Camilo, F., et al.: Astrophys. J. **554**, 161 (2001)

Ransom, S.M., Eikenberry, S.S., Middleditch, J.: Astron. J. **124**, 1788 (2002)

Rudak, B., Dyks, J.: Mon. Not. Roy. Astron. Soc. **303**, 477 (1999)

Vaughan, B.A., van der Klis, M., Wood, K.S., et al.: Astrophys. J. **435**, 362 (1994)

Webb, N.A., Olive, J.-F., Barret, D.: Astron. Astrophys. **417**, 181 (2004)

Woods, P.M., Thompson, C.: In: Lewin, W.H.G., van der Klis, M. (eds.) Compact Stellar X-ray Sources. Cambridge University Press, Cambridge, pp. 547–586 (2006) (astro-ph/0406133)

Yakovlev, D.G., Gnedin, O.Y., Kaminker, A.D., et al.: Adv. Space Res. **33**, 523 (2004)

Astrophys Space Sci (2007) 308: 95–99
DOI 10.1007/s10509-007-9352-8

ORIGINAL ARTICLE

Chandra smells a RRAT

X-ray detection of a rotating radio transient

Bryan M. Gaensler · Maura McLaughlin · Stephen Reynolds · Kazik Borkowski · Nanda Rea · Andrea Possenti · Gianluca Israel · Marta Burgay · Fernando Camilo · Shami Chatterjee · Michael Kramer · Andrew Lyne · Ingrid Stairs

Received: 5 July 2006 / Accepted: 21 August 2006 / Published online: 15 March 2007

Abstract "Rotating RAdio Transients" (RRATs) are a newly discovered astronomical phenomenon, characterised by occasional brief radio bursts, with average intervals between bursts ranging from minutes to hours. The burst spacings allow identification of periodicities, which fall in the range 0.4 to 7 seconds. The RRATs thus seem to be rotating neutron stars, albeit with properties very different from the rest of the population. We here present the serendipitous detection with the *Chandra X-ray Observatory* of a bright point-like X-ray source coincident with one of the RRATs. We discuss the temporal and spectral properties of this X-ray emission, consider counterparts in other wavebands, and interpret these results in the context of possible explanations for the RRAT population.

Keywords Pulsars: individual (J1819-1458) · Stars: flare, neutron · X-rays: stars

PACS 97.60.Gb · 97.60.Jd · 98.70.Qy

B.M.G. acknowledges the support of NASA through LTSA grant NAG5-13023 and of an Alfred P. Sloan Fellowship.

B.M. Gaensler (✉) · S. Chatterjee
Harvard-Smithsonian Center for Astrophysics,
Cambridge, MA, USA
e-mail: bgaensler@cfa.harvard.edu

Present Address:
B.M. Gaensler · S. Chatterjee
The University of Sydney, Sydney, NSW, Australia

M. McLaughlin
University of West Virginia, Morgantown, WV, USA

S. Reynolds · K. Borkowski
North Carolina State University, Raleigh, NC, USA

N. Rea
SRON Netherlands Institute for Space Research, Utrecht, Netherlands

A. Possenti · M. Burgay
Osservatorio Astronomico di Cagliari, Cagliari, Italy

G. Israel
Osservatorio Astronomico di Roma, Roma, Italy

F. Camilo
Columbia University, New York, NY, USA

M. Kramer · A. Lyne
Jodrell Bank Observatory, Macclesfield, UK

I. Stairs
University of British Columbia, Vancouver, BC, Canada

1 Introduction

Astronomers are still coming to terms with the fact that the zoo of isolated neutron stars harbours an increasingly diverse population. Objects of considerable interest include radio pulsars, anomalous X-ray pulsars, soft gamma repeaters, central compact objects in supernova remnants (SNRs), and dim isolated neutron stars. A startling new discovery has now forced us to further expand this group: McLaughlin et al. (2006) have recently reported the detection of eleven "Rotation RAdio Transients", or "RRATs", characterised by repeated, irregular radio bursts, with burst durations of 2–30 ms, and intervals between bursts of $\sim$4 min to $\sim$3 h. The RRATs are concentrated at low Galactic latitudes, with distances implied by their dispersion measures of $\sim$2$-$7 kpc.

For ten of the eleven RRATs discovered by McLaughlin et al. (2006), an analysis of the spacings between repeat

Table 1 Properties of the three RRATs with measured period derivatives

Source	P (s)	τ_c (Myr)	B (10^{12} G)
RRAT J1317-5759	2.6	3.3	5.83
RRAT J1819-1458	4.3	0.12	50
RRAT J1913+3333	0.92	1.9	2.7

bursts reveals an underlying spin period, P, and also in three cases, a spin period derivative, $\dot{P}$. The observed periods fall in the range 0.4 s $< P < 7$ s, which generally overlap with those seen for the radio pulsar population.

For the three RRATs with values measured for both P and $\dot{P}$, a characteristic age, τ_c, and a dipole surface magnetic field, B, can both be inferred, as listed in Table 1. These sources can also be placed on the standard "$P-\dot{P}$ diagram", and can thus be compared to other populations of rotating neutron star:

- RRAT J1317-5759 has properties typical of radio pulsars, except for a relatively long spin period.
- RRAT J1819-1458 is very young, with a high magnetic field. On the $P-\dot{P}$ diagram, it is located in the upper-right region, in the same area occupied by the magnetars and by the high-field radio pulsars.
- RRAT J1913+3333 has spin properties indistinguishable from the bulk of radio pulsars.

We here present the X-ray detection of RRAT J1819-1458, which provides a further point of comparison with the various neutron star population. The details of this detection and its interpretation are discussed by Reynolds et al. (2006).

2 X-ray emission from RRAT J1819-1458

2.1 Detection

As mentioned above, RRAT J1819-1458 is very young, and has a high surface magnetic field. This source bursts every ~3 min, making it the most active of the known RRATs. Its dispersion measure of 196 ± 3 pc cm^{-3} corresponds to an estimated distance of 3.6 kpc, with considerable uncertainty.

RRAT J1819-1458 sits only $\sim 11'$ from the young SNR G15.9+0.2. This source was the target of a 30 ks observation with *Chandra* ACIS in May 2005, and RRAT J1819-1458 fortuitously falls within the field of view (Reynolds et al. 2006). As shown in Fig. 1, there is a clear detection of a bright, unresolved X-ray source within the error ellipse for RRAT J1819-1458. Using the X-ray differential source count distribution for the Galactic plane of Sugizaki et al. (2001), we find that the probability of finding a field source of this count rate at this position is $<10^{-4}$, and conclude that

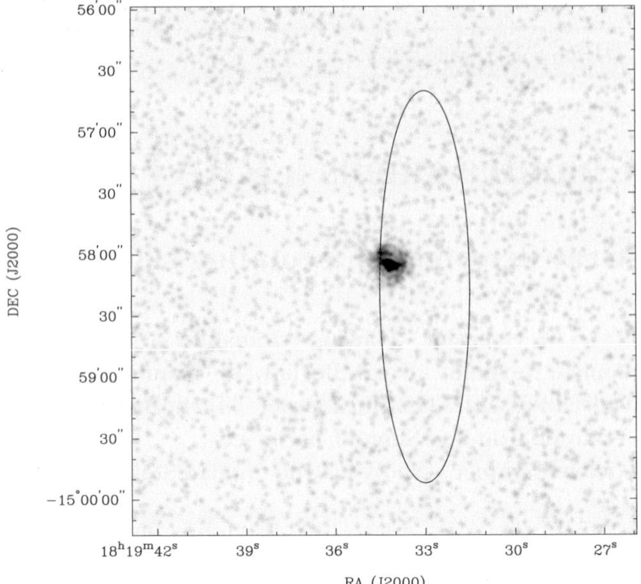

Fig. 1 A *Chandra* ACIS image of RRAT J1819-1458, observed $11'$ off-axis, and smoothed with a Gaussian of FWHM $2''$. The ellipse shows the 3-σ error ellipse for the position of the RRAT, as derived by McLaughlin et al. (2006) from radio timing

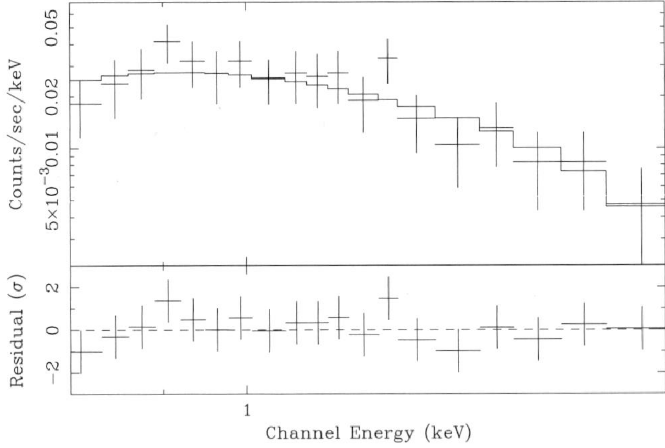

Fig. 2 The X-ray spectrum of RRAT J1819-1458, as observed with *Chandra* ACIS (Reynolds et al. 2006). The data points show the measurements, while the solid line shows the best-fit absorbed blackbody described in the text

we have both identified and localised the X-ray counterpart of RRAT J1819-1458.

2.2 Spectrum and variability

We have extracted 524 ± 24 counts from the X-ray counterpart to RRAT J1819-1458, the spectrum from which is shown in Fig. 2. While this is insufficient for detailed spectral modelling, it still yields crucial information.

Fitting simple absorbed power-law and blackbody models, we find that the spectrum is a poor fit to the former, but a good fit to the latter (Reynolds et al. 2006). At a distance

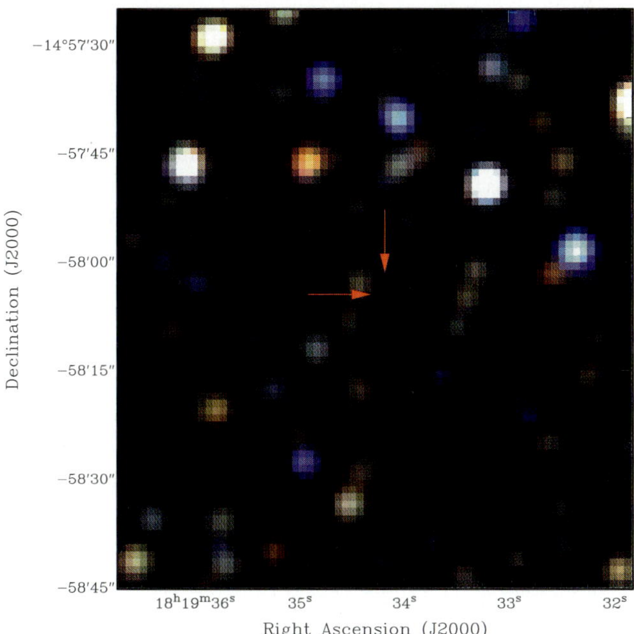

Fig. 3 A 2MASS image of the field surrounding RRAT J1819-1458, with J, H and K bands shown in *blue*, *green* and *red*, respectively. The arrows mark the position of the RRAT, as derived from *Chandra* observations. The upper limits on the infrared emission from RRAT J1819-1458 are $J > 15.6$, $H > 15.0$ and $K > 14.0$

of $3.6d_{3.6}$ kpc, the inferred blackbody radius (as viewed at infinity) is $20d_{3.6}$ km, which is consistent with standard neutron star equations of state, given the uncertainty in the distance estimate. For this blackbody fit, the foreground absorbing column is $N_H = 7^{+7}_{-4} \times 10^{21}$ cm^{-2}, and the surface temperature (at infinity) is $kT_\infty = 120 \pm 40$ eV. In the energy range 0.5–8.0 keV, the unabsorbed flux is 2×10^{-12} ergs cm^{-2} s^{-1} and the isotropic luminosity is $3.6d_{3.6}^2 \times 10^{33}$ ergs s^{-1}.

Given the extreme variability of the source seen at radio wavelengths, it is of interest to quantify the level of X-ray variability seen in this observation. Examination of individual CCD frames (at a time-resolution of 3.2-seconds) shows no evidence for individual brief X-ray bursts that might be a counterpart to the RRAT phenomenon, with a 3-σ upper limit on the observed fluence of any burst of 3×10^{-11} ergs cm^{-2} in the 0.5–8 keV energy range. The data also show no evidence of variability on any timescale ranging from 3.2 s to 5 days (the time-span covered by the observations). Although the time resolution is only slightly shorter than the spin period, $P = 4.3$ s, we do have limited sensitivity to an aliased pulsed signal. However, we find no pulsations down to a 3-σ pulsed fraction limit of 70% for a sinusoidal pulse profile (Reynolds et al. 2006).

2.3 Data at other wavelengths

With an accurate position for RRAT J1819-1458 in hand from our *Chandra* position, we have searched for a counterpart to this source at other wavelengths, using archival VLA, 2MASS and GLIMPSE data. No detection is made in any of these data.

The non-detection in 2MASS data, shown in Fig. 3, allows us to put a lower limit on the X-ray to infrared flux ratio of $f_X/f_{\rm IR} > 0.7$, which rules out most stellar counterparts (see Kaplan et al. 2004). Furthermore, the foreground column density inferred from the X-ray spectral fit is substantially smaller than the integrated column through the Galaxy at this position, arguing against the X-ray emission being from a background galaxy or cluster. These arguments, combined with the good blackbody fit to the X-ray spectrum, imply that the most reasonable interpretation for the X-ray emission is thermal emission from a neutron star surface.

3 Comparison with other sources

It is unclear whether the RRATs are a completely new group of neutron stars, or are a new manifestation of one of the previously known classes of object. Since all classes of young neutron star are X-ray emitters, the detection reported here provides vital new information which allows us to begin to discriminate between various interpretations for the RRATs.

We first note that while many neutron stars are pulsed in X-rays, those which produce thermal emission inevitably exhibit low pulsed fractions, due to gravitational bending of the light emanating from their surfaces (e.g., Psaltis et al. 2000). Thus our relatively poor limit on pulsed fraction is unsurprising and unconstraining.

Despite its location near the magnetars on the $P-\dot{P}$ diagram, the X-ray properties of RRAT J1819-1458 are distinct from those seen for magnetars: the RRAT is much colder and less luminous than the magnetars, and apparently lacks the hard X-ray tail seen for these sources, at our current level of sensitivity. Furthermore, the magnetar birth rate is well below that estimated for the RRAT population (Popov et al. 2006). The only possible link to magnetars is with the transient magnetar XTE J1810-197 which, while in quiescence, had a surface temperature $kT_\infty \approx 150-180$ eV (Ibrahim et al. 2004; Gotthelf et al. 2004), comparable to that seen here.

The X-ray temperature of RRAT J1819-1458 is also well below those of the "central compact objects", while its temperature and luminosity are above those of most of the dim isolated neutron stars.

The one population whose X-ray properties provide a reasonable match to that of the RRAT is that of radio pulsars, for which sources with ages around 100 kyr show spectra very similar to what we have found here. For example,

PSR J0538+2817 is 30 kyr old and has $kT_\infty = 160$ eV, while PSR B0656+14 is 110 kyr old and has $kT_\infty = 70$ eV (see for details Reynolds et al. 2006). The X-ray emission from RRAT J1819-1458 thus suggests that this source is a normal radio pulsar, albeit one that produces unusual radio bursts. Additional evidence to support this possibility is the subsequent discovery from PSR B0656+14 of RRAT-like behaviour (Weltevrede et al. 2006). If PSR B0656+14 was placed at the distance of RRAT J1819-1458, its occasional bright radio bursts would still be seen, but not the underlying regular train of pulsations.

One point to note is that the inferred surface magnetic field strength of RRAT J1819-1458 is more than an order of magnitude greater than those of PSRs J0538+2817 and B0656+14 discussed above. Two radio pulsars with comparable magnetic fields that have been detected in X-rays are PSRs J1718-3718 (Kaspi and McLaughlin 2005) and J1119-6127 (Gonzalez et al. 2005). These sources show temperatures ($kT \sim 150-200$ eV) and luminosities ($\sim 10^{32}-10^{33}$ ergs s^{-1}) comparable to that of RRAT J1819-1458, although both sources are probably much younger (35 and 1.7 kyr, respectively) and, in contrast to RRAT J1819-1456, have X-ray luminosities less than their spin-down luminosities.

4 Possible interpretations

If the RRATs are normal radio pulsars as proposed above, we then need to explain their transient behaviour. One possibility, discussed by Zhang et al. (2007), is that RRATs are pulsars that are no longer active, but for which a temporary "star spot" with multipole field components emerges above the surface. This magnetic field component could temporarily reactivate the radio beaming mechanism, producing the observed bursts.

The difficulty with this possibility is that none of the RRATs with known values of $\dot{P}$ appear to be near the "death line" in the $P-\dot{P}$ diagram, beyond which the pulsar mechanism is expected to turn off. To account for this, Zhang et al. (2007) have proposed that RRATs have dipole fields offset from their centres, causing their magnetic fields to be over-estimated. However, the X-ray temperature seen for RRAT J1819-1458 is consistent with it being a $\sim$100 kyr old neutron star (see Sect. 3 above), in reasonable agreement with its characteristic age as listed in Table 1, and confirming that if this is a radio pulsar, it is as yet nowhere near death.

Zhang et al. (2007) also consider the possibility that RRATs are caused by a brief reversal of the direction in which radio beams are emitted. They draw upon the previous work of Gil et al. (1994), who showed that whenever PSR B1822-09 produces "nulls" in its main pulse, it produces a much weaker interpulse, which has been interpreted by Dyks et al. (2005) to correspond to radiation temporarily emitted in the opposite direction. If one invokes an emitting geometry in which the main pulse is never seen, then the interpulse alone, appearing only when the unseen main pulse nulls, might correspond to RRAT-like emission. The difficulty with this interpretation is that in PSR B1822-09 this reversal lasts for several minutes, in contrast to the RRATs, for which multiple bursts in succession are yet to be observed.

Finally, the RRAT mechanism might be produced by interaction of the neutron star with an equatorial fallback disk or with orbiting circumpolar debris. Accretion from a disk should usually quench the radio emission mechanism, but sporadic drops in the accretion rate could allow the radio beam to turn on for a fraction of a second, producing the RRAT phenomenon (Li 2006). However, this possibility is at odds with the behaviour seen by Weltevrede et al. (2006) for PSR B0656+14, in which the RRAT-like bursts are superimposed on an underlying persistent series of faint radio pulsations. Alternatively, episodic injection of material from a circumpolar asteroid belt could temporarily activate a quiescent region of the magnetosphere, producing a RRAT burst, in some cases from a radio pulsar that is normally beaming away from us (Cordes and Shannon 2007).

5 Future observations and conclusions

A variety of forthcoming observations should be able to cast more light on the possibilities discussed above, and on the spatial and spin distributions of these sources. Several groups have already undertaken new radio searches for RRAT-like emission from other classes of neutron stars. Near-infrared observations with the VLT have also been obtained for some RRATs, to provide stronger constraints on the X-ray to infrared flux ratio of these sources, and to look for hints of a fossil/fallback disk (see Wang et al. 2006). Meanwhile, in a project called "Astro Pulse", the data from the Arecibo SETI@Home project are being reanalysed for short transient signals. With the sensitivity of Arecibo, many RRATs may be found in this analysis.

Finally, following on from our X-ray detection discussed here, *XMM-Newton* observations of two RRATs have been approved for Cycle 5: of RRAT J1819-1458 (to obtain a better spectrum and to properly search for pulsations) and of RRAT J1317-5759 (to see if it too emits detectable X-rays).

A final point to make is that it is already clear from the 11 RRATs known that the birth rate of these objects is $\sim 3-4$ times that previously estimated for all radio pulsars (McLaughlin et al. 2006; Popov et al. 2006). We are thus only seeing the upper tip of a large transient distribution. When one combines this with the populations of nullers (Backer 1970), "bursters" (e.g., PSR J1752+2359; Lewandowski et al. 2004), "winkers" (PSR B1931+24; Kramer et al. 2006) and "burpers" (GCRT J1745-3009; Hyman et al. 2005), it becomes clear that a full study of the transient radio sky is needed (Cordes et al. 2004). The next generation of radio telescopes, with the capability of monitoring very wide fields of view (e.g., LOFAR, xNTD and the SKA) should make many further discoveries of such phenomena.

References

Backer, D.C.: Nature **228**, 42 (1970)
Cordes, J.M., Shannon, R.M.: Astrophys. J. (2007, submitted), astro-ph/0605145
Cordes, J.M., Lazio, T.J.W., McLaughlin, M.A.: New Astron. Rev. **48**, 1459 (2004)
Dyks, J., Zhang, B., Gil, J.: Astrophys. J. **626**, L45 (2005)
Gil, J.A., Jessner, A., Kijak, J., et al.: Astron. Astrophys. **282**, 45 (1994)
Gonzalez, M.E., Kaspi, V.M., Camilo, F., et al.: Astrophys. J. **630**, 489 (2005)
Gotthelf, E.V., Halpern, J.P., Buxton, M., Bailyn, C.: Astrophys. J. **605**, 368 (2004)
Hyman, S.D., Lazio, T.J.W., Kassim, N.E., et al.: Nature **434**, 50 (2005)
Ibrahim, A.I., Markwardt, C.B., Swank, J.H., et al.: Astrophys. J. **609**, L21 (2004)
Kaplan, D.L., Frail, D.A., Gaensler, B.M., et al.: Astrophys. J. Suppl. **153**, 269 (2004)
Kaspi, V.M., McLaughlin, M.A.: Astrophys. J. **618**, L41 (2005)
Kramer, M., Lyne, A.G., O'Brien, J.T., et al.: Science **312**, 549 (2006)
Lewandowski, W., Wolszczan, A., Feiler, G., et al.: Astrophys. J. **600**, 905 (2004)
Li, X.-D.: Astrophys. J. **646**, L139 (2006)
McLaughlin, M.A., Lyne, A.G., Lorimer, D.R., et al.: Nature **439**, 817 (2006)
Popov, S.B., Turolla, R., Possenti, A.: Mon. Not. Roy. Astron. Soc. **369**, L23 (2006)
Psaltis, D., Özel, F., De Deo, S.: Astrophys. J. **544**, 390 (2000)
Reynolds, S., Borkowski, K., Gaensler, B., et al.: Astrophys. J. **639**, L71 (2006)
Sugizaki, M., Mitsuda, K., Kaneda, H., et al.: Astrophys. J. Suppl. **134**, 77 (2001)
Wang, Z., Chakrabarty, D., Kaplan, D.L.: Nature **440**, 772 (2006)
Weltevrede, P., Stappers, B.W., Rankin, J.M., Wright, G.A.E.: Astrophys. J. **645**, L149 (2006)
Zhang, B., Gil, J., Dyks, J.: Mon. Not. Roy. Astron. Soc. **374**, 1103 (2007)

Astrophys Space Sci (2007) 308: 101–107
DOI 10.1007/s10509-007-9325-y

ORIGINAL ARTICLE

QED can explain the non-thermal emission from SGRs and AXPs: variability

Jeremy S. Heyl

Received: 25 July 2006 / Accepted: 30 October 2006 / Published online: 21 March 2007

Abstract Owing to effects arising from quantum electrodynamics (QED), magnetohydrodynamical fast modes of sufficient strength will break down to form electron-positron pairs while traversing the magnetospheres of strongly magnetised neutron stars. The bulk of the energy of the fast mode fuels the development of an electron-positron fireball. However, a small, but potentially observable, fraction of the energy ($\sim 10^{33}$ erg) can generate a non-thermal distribution of electrons and positrons far from the star. This paper examines the cooling and radiative output of these particles. Small-scale waves may produce only the non-thermal emission. The properties of this non-thermal emission in the absence of a fireball match those of the quiescent, non-thermal radiation recently observed non-thermal emission from several anomalous X-ray pulsars and soft-gamma repeaters. Initial estimates of the emission as a function of angle indicate that the non-thermal emission should be beamed and therefore one would expect this emission to be pulsed as well. According to this model the pulsation of the non-thermal emission should be between 90 and 180 degrees out of phase from the thermal emission from the stellar surface.

Keywords Gamma-rays: observations · Pulsars: individual SGR 1806-20, AXP 4U 0142+61, AXP 1E 1841-045 · Radiation mechanisms: non-thermal

The Natural Sciences and Engineering Research Council of Canada, Canadian Foundation for Innovation and the British Columbia Knowledge Development Fund supported this work. J.S.H. is a Canada Research Chair.

J.S. Heyl (✉)
University of British Columbia, 6224 Agricultural Road, Vancouver, BC, Canada V6T 1Z1
e-mail: heyl@phas.ubc.ca

PACS 97.60.Jd · 98.70.Rz · 12.20.Ds · 52.35.Tc

1 Introduction

Simply put magnetars are neutron stars whose magnetic fields dominate their emission, evolution and manifestations. In the late 1970s and early 1980s, a fleet of sensitive detectors of high-energy radiation uncovered two new phenomena, the soft-gamma repeater and the anomalous X-ray pulsar. Strongly magnetized neutron stars provide the most compelling model for both types of object, and observations over the past few years indicate that these phenomena are two manifestations of the same type of object. Soft-gamma repeaters exhibit quiescent emission similar to that of anomalous X-ray pulsars (e.g. Rothschild et al. 1994; Murakami et al. 1994; Hurley et al. 1996, 1999), and anomalous X-ray pulsars sometimes burst (Gavriil et al. 2002; Kaspi et al. 2003). What makes magnetars a hot topic of research is the rich variety of physical phenomena that strong magnetic fields exhibit.

This article will focus on the quiescent emission from these interesting objects rather than the bursts themselves (the reader may wish to refer to the seminal work of Thompson and Duncan (1995) for details of the burst but may also want to look at Heyl and Hernquist (2005a) for an alternative). Furthermore, the article will concentrate on a possible model for the recently detected non-thermal emission from these objects.

In earlier work, Lars Hernquist and I analysed wave propagation through fields exceeding the quantum critical value $B_{\rm QED} \equiv m^2c^3/e\hbar \approx 4.4 \times 10^{13}$ G, and demonstrated circumstances under which electromagnetic (Heyl and Hernquist 1998b) and some MHD waves, particularly fast modes (Heyl and Hernquist 1999) evolve in a non-linear manner and eventually exhibit discontinuities similar to hydrodynamic shocks, owing to vacuum polarisation from quantum electrodynamics (QED). In paper I (Heyl and Hernquist 2005a), we developed a theory to account for bursts from SGRs and AXPs based on "fast-mode breakdown," in which the wave energy is dissipated into electron-positron pairs when the scale of these discontinuities becomes comparable to an electron Compton wavelength. We showed that, under appropriate conditions, an extended, optically thick pair-plasma fireball would result, radiating primarily in hard X-rays and soft γ-rays. In paper II (Heyl and Hernquist 2005b) we developed the theory of the non-thermal emission from the fast-mode cascade under a series of assumptions and approximations. In particular, only if the amplitude of the MHD fast mode is large and it passes through a large volume will the decay of the fast mode generate enough pairs to make the region near the neutron star optically thick and produce a pair fireball. On the other hand, smaller waves will still dissipate through pair-production but the emission will escape without appreciable absorption and appear non-thermal. Heyl and Hernquist (2005a) estimates the Thomson optical depth through the pair plasma created by a passing fast mode (their Fig. 3). If this optical depth exceeds unity, one would expect a fireball to form.

In what follows, I extend our previous investigation of fast-mode breakdown to estimate the spectrum of non-thermal emission expected outside the region containing an optically thick fireball with particular emphasis on the assumptions made in paper II and on the angular dependence of this emission.

2 The observed non-thermal emission

What would be the typical flux of these small-scale fast modes? Thompson and Duncan (1996) and Heyl and Kulkarni (1998) have argued that the quiescent emission of SGR and AXP neutron stars may be powered by the decay of the magnetic field. The quiescent thermal emission may only be a small fraction of the total energy released by the decay of the magnetic field. Recent observations of SGRs and AXPs indicate that the thermal radiation may indeed be just the tip of the iceberg (Molkov et al. 2005; Mereghetti et al. 2005; Kuiper et al. 2004); therefore, following the discussion of paper II, the amplitude of the thermal and non-thermal radiation are taken to be comparable, and the pair cascade operates beyond a certain radius from the star or equivalently below a certain magnetic field strength ($B_{\max}$). This model has two parameters, $B_{\max}$ determines the position of the two breaks in the spectrum and the total normalisation. The process of fast-mode breakdown according to paper II predicts a particular relationship between the location of the two breaks in the spectrum and particular slopes between and beyond the breaks.

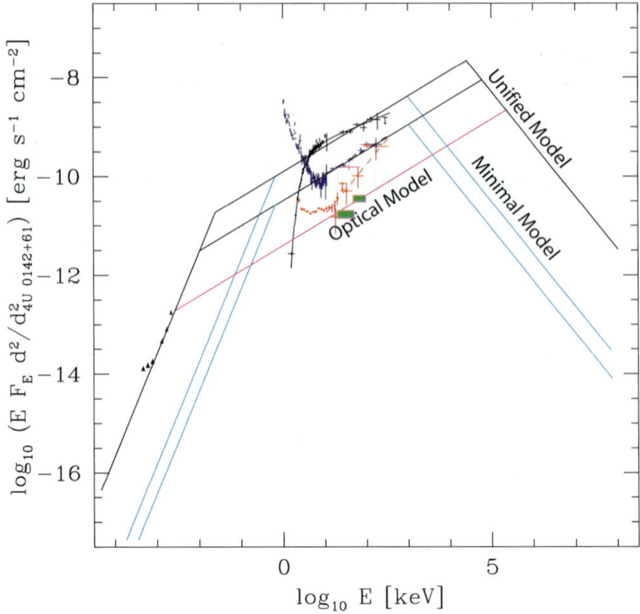

Fig. 1 The spectrum produced by fast-mode breakdown is superimposed over the observed thermal and non-thermal emission from several AXPs and SGRs for models that fit either the optical or INTEGRAL data solely and one that fits both sets of data. The unabsorbed optical data are from Hulleman et al. (2000) via Özel (2004) for AXP 4U 0142+61. The *uppermost black symbols* are the hard X-ray band (from Molkov et al. 2005) for SGR 1806-20. Mereghetti et al. (2005) obtained similar results for the SGR. The *middle sets* of points in the hard X-ray data (*blue* is total flux and *red* is pulsed flux) are from Kuiper et al. (2004) for AXP 1E 1841-045. The *green squares* plot the INTEGRAL data reported by den Hartog et al. (2004) for AXP 4U 0142+61. The den Hartog et al. (2004) results are normalised using the observations of the Crab by Jung (1989). We scaled the emission from the three sources by assuming that they all lie at the distance of AXP 4U 0142+61. The assumed distances are 3 kpc for AXP 4U 0142+61 (Hulleman et al. 2000), 7.5 kpc for AXP 1E 1841-045 (Sanbonmatsu and Helfand 1992) and 15 kpc for SGR 1806-20 (Molkov et al. 2005)

Figure 1 shows the observed broad-band spectrum of several AXPs and SGRs. To compare the spectra the more distant objects (AXP 1E 1841-045 and SGR 1806-20) whose hard X-ray emission was discovered with INTEGRAL have been placed at the distance of 4U 0142+61 whose optical emission (Hulleman et al. 2000) is very likely to be non-thermal (Özel 2004). The observed spectra depart from a power law for $E < 1$ keV because in this region the thermal radiation from the neutron star begins to dominate the non-thermal component.

Because the location of the two breaks in the spectrum both depend on the strength of the magnetic field at the inner edge of the breakdown region, the presence of extensive non-thermal optical emission indicates that the non-thermal hard X-ray emission should peak at about 30 MeV, a factor of two hundred beyond the observed spectrum. The best limits in this energy range are provided by Comptel (Kuiper et al. 2006) and appear to exclude both the "optical model" and the "unified model" for 4U 0142+61.

This conclusion assumes that the sources SGR 1806-20 and AXP 1E 1841-045 have a similar optical excess to 4U 0142+61. A more conservative assumption would be that the hard X-ray emission does not extend far beyond the observations from INTEGRAL with spectral breaks at about 1 MeV and 650 eV. This situation is somewhat natural. The fast-mode cascade is limited to pairs with sufficient energy to produce photons with $E > 1$ MeV that can subsequently pair produce. Lower energy electrons simply cool, giving the observed cooling spectrum in the hard X-rays. The total energy in the non-thermal emission is also reduced by a factor of a few. In the context of the fast-mode cascade it is difficult to have $E_{\text{break}} < 2mc^2$. This model is denoted as the "Minimal Model" in Fig. 1 because E_{break} takes on the minimal value that makes sense physically; i.e. $\approx$1 MeV; this is equivalent to assuming that magnetic field is much weaker than the quantum-critical limit in the pair-production region.

The analysis reviewed here from paper I assumed that the fast-mode was travelling perpendicular to the field and that the amplitude of the wave and the magnetic field where the shock forms are independent. The more detailed analysis that follows shows that the location of shock formation depends on the product of initial amplitude of the wave and its wavenumber, its direction of propagation and the strength of the dipole component of the star's magnetic field. The latter two parameters could potentially be determined independently of the emission spectrum, leaving a single free parameter to describe the emission, the product of initial amplitude of the wave and its wavenumber.

3 Spectrum

The previous calculation of the spectrum assumed that the energy dissipated by the wave over a given range of magnetic field strengths is constant and that the dissipation only occurs below a particular magnetic field strength. A direct calculation the evolution of the wave as it travels away from the star relaxes both of these assumptions. This determines that amount of energy dumped into pairs at various magnetic field strengths; however, it does not determine how this energy is finally dissipated. It can be emitted locally and promptly as classical synchrotron radiation or Landau transition radiation or elsewhere as curvature emission.

3.1 Weak-field regime

If the magnetic field is much weaker than the quantum-critical limit, the spectrum of the cooling pairs at a particular value of the magnetic field is given by

$$\frac{dE}{dE_\gamma} = \frac{E_\perp}{2(E_{\text{break}}^{1/2} - E_0^{1/2})} E_\gamma^{-1/2} \tag{1}$$

where $E_0 = m_e c^2 \xi$, $E_{\text{break}} \approx m_e c^2/(40\xi)$ and $\xi = B/B_{\text{QED}}$ for $B \ll B_{\text{QED}}$. This assumes that the final generation of pairs emits classical synchrotron radiation as described in paper II and that only the momentum perpendicular to the magnetic field is dissipated as synchrotron emission.

3.2 Strong-field regime

For $\xi \gtrsim 0.1$ the analysis of the preceding section breaks down. This is signaled by the fact that $E_0 \gtrsim E_{\text{break}}$, so (1), for example, does not make sense. In this regime both the processes of pair production and synchrotron emission are strong affected. Additionally photon splitting may play a role. Restricting the results to $\xi \ll 1$ yields the results of (Heyl and Hernquist 2005b).

Specifically in the strong-field regime, neither the primary pairs nor the secondaries can be assumed to have relativistic motion perpendicular to the magnetic field. The shock travels at a velocity (Heyl and Hernquist 1997b),

$$v = \frac{c}{n_\perp} \approx c\left[1 - \frac{\alpha_{\text{QED}}}{4\pi} X_1\left(\frac{1}{\xi}\right) \sin^2\theta\right]^{-1} \tag{2}$$

where the function X_1 is defined in Heyl and Hernquist (1997a) and well approximated by the fitting formula

$$X_1\left(\frac{1}{\xi}\right) \approx -\frac{14}{45}\xi^2 \frac{1 + \frac{6}{5}\xi}{1 + \frac{4}{3}\xi + \frac{14}{25}\xi^2} \tag{3}$$

from Potekhin et al. (2004). The value of X_1 quantifies by how much the index of refraction of photons polarized parallel to the external field differs from unity.

The primary pairs typically have a Lorentz factor of

$$\gamma = \left(1 - \frac{v^2}{c^2}\right)^{-\frac{1}{2}} \approx \frac{1}{|\xi \sin\theta|}\left(\frac{45\pi}{7\alpha_{\text{QED}}} \frac{1 + \frac{4}{3}\xi + \frac{14}{25}\xi^2}{1 + \frac{6}{5}\xi}\right)^{\frac{1}{2}}. \tag{4}$$

The typical Landau level occupied by these primary pairs is

$$n \approx \frac{45}{14} \frac{\pi}{\alpha_{\text{QED}}} \frac{1}{\xi^3} \frac{1 + \frac{4}{3}\xi + \frac{14}{25}\xi^2}{1 + \frac{6}{5}\xi} \tag{5}$$

independent of angle, so for $\xi > 27$ only the ground Landau level is occupied by the primary pairs. Usov and Melrose (1995) found a limit of $\xi > 0.1$ for pairs to be created of pairs in the ground Landau level. The limit found here is quite a bit different simply because the pair cascade discussed here is due to photons that are essentially created above the pair-production threshold in the shock. The direction of the photons is constant relative to the field direction; it is their energy that increases above the pair-production threshold; they carry much momentum perpendicular to the field. On the other hand, Usov and Melrose (1995) examine the pair production by photons travelling nearly parallel to the magnetic field. The momentum perpendicular to the field is small so the ground Landau level is preferred at much lower fields.

The Landau level of the secondaries is a factor of sixteen smaller, so for $\xi > 8$ only the ground level is occupied by the secondary pairs. In these cases there is no prompt emission at all from the pairs. The energy of the initial photon goes into the rest mass of the pairs and their motion along the magnetic field. If n is not large, the emission is better described by cyclotron emission than by synchrotron. The emission is not cutoff by considerations of the total pair-production optical depth as in the weak-field case Heyl and Hernquist (2005b) but by the Landau levels of the pairs produced in the cascade. In the weak-field limit, Heyl and Hernquist (2005b) found that the energy of the final pairs in the cascade was a factor of 1,700 lower than the primaries, so for $\xi > 0.08$ these final pairs occupy the ground Landau level and the synchrotron analysis of the cascade emission must be modified; the limiting field for the final pairs in a cascade to be in ground Landau level is similar to that found by Usov and Melrose (1995) for the pulsar cascade. Comparing the values of E_0 and $E_{\rm break}$ yields the same limiting field.

3.3 The total spectrum

The analysis of the pair cascade in the strong-field limit is beyond the scope of this small contribution. Instead I shall focus on the regime where the entire cascade is in the weak-field limit. This may occur for a very strong wave propagating near a weakly magnetised star or for a relatively weak wave travelling near a strongly magnetised star.

In an arbitrary magnetic field, the evolution of a fast mode can be parametrized by the integral of the "opacity to shocking"

$$\kappa = \frac{k\sin\theta}{2}\frac{b}{B}\frac{e^2}{hc}\left[\sin^2\theta\kappa_1 + (4-3\sin^2\theta)\kappa_2\right] \tag{6}$$

along the path of the wave. In this expression, b is the amplitude of the magnetic field of the fast mode, B is the amplitude of the local magnetic field, θ is the angle between the direction of propagation and the magnetic field and k is the wavenumber of the fast mode.

Although this quantity is not strictly an opacity in the sense of radiative transfer, it does characterize the evolution of the wave in an analogous way. In particular, the shape of the wave depends on the optical depth for shocking

$$\tau = \int \kappa\, dl. \tag{7}$$

The shock forms at $\tau = 1$, and the power carried by the wave decreases as τ increases above unity.

The quantity in brackets is the generalization of the function $\xi F(\xi)$ where $F(\xi)$ was defined by Heyl and Hernquist (1998b). The two terms are

$$\kappa_1 = -\xi^{-3} X_0^{(3)}\left(\frac{1}{\xi}\right), \tag{8}$$

$$\kappa_2 = X_0^{(2)}\left(\frac{1}{\xi}\right)\xi^{-2} - X_0^{(1)}\left(\frac{1}{\xi}\right)\xi^{-1} \tag{9}$$

where $X_0^{(n)}(x)$ denotes $d^n X_0(x)/dx^n$.

Figures 2 and 3 depict the opacity as a function of angle and magnetic field strength. For weak fields the opacity is proportional to $\sin\theta$; while in strong fields, it is proportional to $\sin\theta(2-\sin^2\theta)$.

The definition of $X_0(x)$ from Heyl and Hernquist (1997a) yields

$$\kappa_1 = \frac{2}{3} + \frac{1}{\xi} + \frac{1}{\xi^2} - \frac{1}{2\xi^3}\Psi^{(1)}\left(\frac{1}{2\xi}\right) \tag{10}$$

$$\approx \frac{8}{15}\xi^2 - \frac{32}{21}\xi^4 + \frac{128}{15}\xi^6 - \frac{2560}{33}\xi^8 + \mathcal{O}(\xi^{10}) \tag{11}$$

$$\approx \frac{2}{3} - \frac{1}{\xi} + \frac{1}{\xi^2} - \frac{\pi^2}{12}\frac{1}{\xi^3} + \frac{\zeta(3)}{2}\frac{1}{\xi^4} + \mathcal{O}\left(\frac{1}{\xi^5}\right) \tag{12}$$

$$\kappa_2 = \frac{2}{3} + \frac{\ln(4\pi\xi) - 2\ln\Gamma\left(\frac{1}{2\xi}\right) + 1}{\xi} + \frac{\Psi\left(\frac{1}{2\xi}\right) - 1}{\xi^2} \tag{13}$$

$$\approx \frac{8}{45}\xi^2 - \frac{32}{105}\xi^4 + \frac{128}{105}\xi^6 - \frac{2560}{297}\xi^8 + \mathcal{O}(\xi^{10}) \tag{14}$$

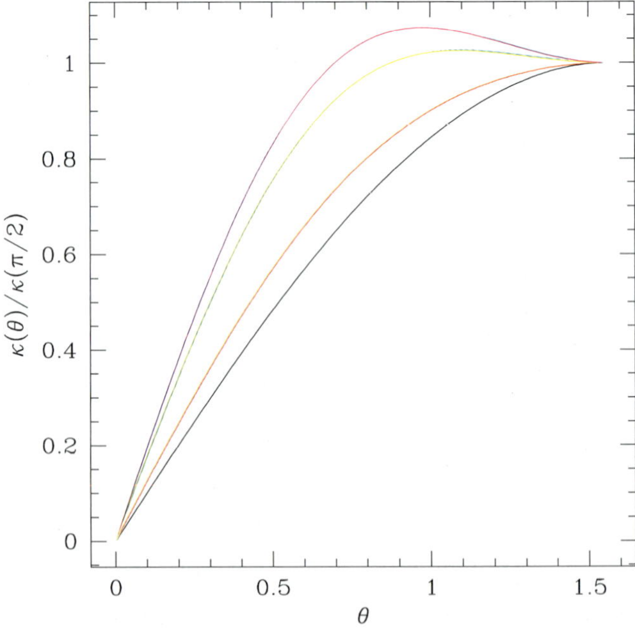

Fig. 2 The opacity as a function of angle at various magnetic field strengths. The curves *from bottom to top* are for $\xi = 0.1, 1, 10$ and 100. For weak fields, the opacity peaks at $\theta = \pi/2$. For strong fields the opacity peaks at $\theta = \cos^{-1}(1/\sqrt{3}) \approx 55°$

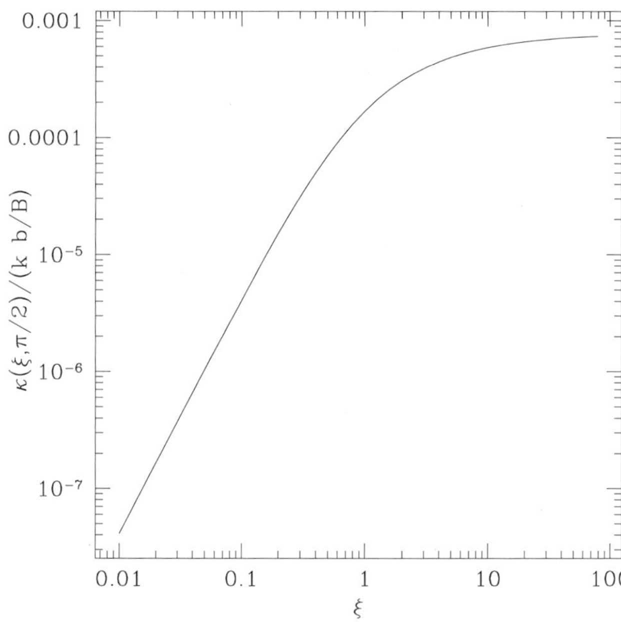

Fig. 3 The opacity as a function of the strength of the field at $\theta = \pi/2$ in units of k with $b = B$. The reciprocal of the ordinate is the number of wavelengths that a wave of amplitude equal to the background field will travel before forming a shock

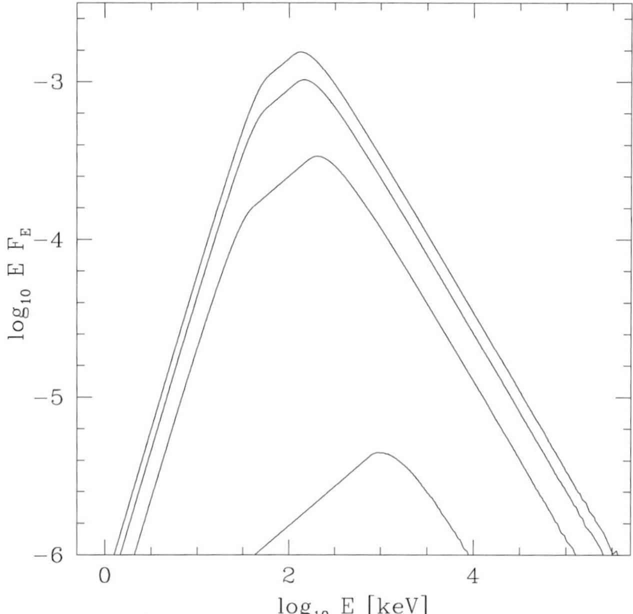

Fig. 4 Total spectrum for strong-field neutron star ($B_{\rm NS} = 30 B_{\rm QED}$) as a function of angle. The initial wave has $\lambda \sim 100$ m and $b \sim 2 B_{\rm QED}$. The curves from shortest to tallest are for $\theta = 60°$, $70°$, $80°$ and $90°$

$$\approx \frac{2}{3} + \frac{\ln(\frac{\pi}{\xi})}{\xi} - \frac{1}{\xi^2} + \frac{\pi^2}{24}\frac{1}{\xi^3} - \frac{\zeta(3)}{6}\frac{1}{\xi^4} + \mathcal{O}\left(\frac{1}{\xi^5}\right) \quad (15)$$

where $\Psi(x) \equiv d \ln \Gamma/dx$, $\Psi^{(1)}(x) = d\Psi/dx$ and $\zeta(3) \approx 1.202$. Both the opacity and the energy available for prompt synchrotron emission drop as the angle of propagation departs from perpendicular.

The total optical depth is simply the integral of the opacity over the path of the wave. If the wave is spherical, the value of b decreases as $1/r$ as the wave travels away from the star. For simplicity the angle of the wave propagation with respect to the field is held constant; this is appropriate for waves travelling radially and holds approximately for waves that have travelled several stellar radii from the surface. The dissipation of the energy of the wave is given by (72) in Heyl and Hernquist (1998b) and Fig. 2 in Heyl and Hernquist (2005a).

Summing over all of the magnetic fields where the dissipation occurs yields

$$\frac{dE_\perp}{dB} = -(1 - \cos\theta)\frac{dP}{dB} \quad (16)$$

where P is the power remaining in the wave and θ is the angle that the direction of propagation of the wave makes with the magnetic field. The angular dependence here comes from the assumption that only the component of the momentum perpendicular to the field yields prompt and local emission. Furthermore, this equation also assumes that the emitting pairs are produced in a weak-field region; they occupy high Landau levels, so the classical treatment of synchrotron emission holds.

For large distances from the neutron star, dP/dB is a constant and the fields are weak; this in combination with the results of Sect. 3.1 yields the spectra given in Heyl and Hernquist (2005b) and shown in Fig. 4 where the opacity depends on angle as in (6). The vertical axis gives the fraction of the energy of the initial wave that is dissipated over a factor of $e \approx 2.7$ in energy. For the weak-field treatment to be accurate, both the frequency and the amplitude of the wave must be sufficiently small to delay the formation of the shock until the weak-field regime. This results in a wave that only dissipates a small fraction of its energy into pairs.

In this case the total optical depth to shocking is only slightly greater than unity ($\tau_{90} = 1.08$) so the spectrum depends strongly on the angle of propagation as depicted in Fig. 4. The radiation is emitted only if $\tau > 1$, so the radiation is emitted only for $\sin\theta\tau_{90} > 1$ if the bulk of the opacity was in weak magnetic fields, and $\sin\theta(2 - \sin^2\theta)\tau_{90} > 1$ if it was in weak magnetic fields. Because the star in this case has $\xi = 30$ but the shock does not form until $\xi < 0.1$, one would expect the cutoff to lie between these extremes. These two estimates yield cutoff angles of 67° and 33°. The detailed calculations find that the emission cuts off within about 59° of the field direction.

If the field at the surface of the star is less than about one-tenth of $B_{\rm QED}$, the discussion of Sect. 3.1 applies, regardless of where the shock forms, yielding the results of Fig. 5.

In this case the shock may form relatively close to the star and a large fraction of the initial energy in the wave is dissi-

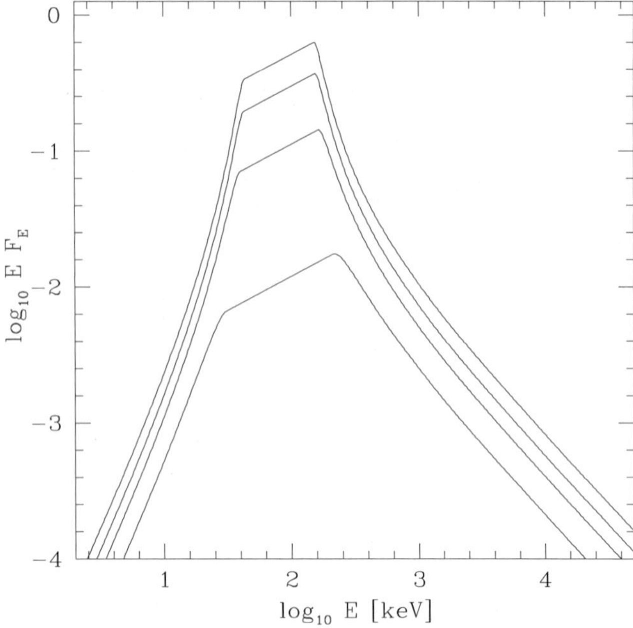

Fig. 5 Total spectrum for weak-field neutron star ($B_{\rm NS} = 0.1 B_{\rm QED}$) as a function of angle. The initial wave has $\lambda \sim 20\mu$ and $b \sim 0.01 B_{\rm QED}$. The curves from shortest to tallest are for $\theta = 22.5°$, 45°, 67.5° and 90°

pated over a narrow range of magnetic field strengths where the shock forms. The bulk of the energy may be released in the regime where dP/dB is not constant so the characteristic spectrum discussed in Heyl and Hernquist (2005b) does not emerge. For large angles with respect to the magnetic field, the shock forms further from the star and the bulk of the energy dissipates in the constant-dP/dB regime. Again the emission cuts off beyond a critical angle. Here $\tau_{90} = 6.82522$ yielding a cutoff angle of about 8°. Because the optical depth in the perpendicular direction is so large, the bulk of the angular dependence in the emission comes not from the opacity but from the $(1 - \cos\theta)$ factor in (16).

4 Comparison with observations

The model outlined in the preceding sections yields at least one important prediction. The hard X-ray emission if it is produced by fast-mode breakdown will be largest where the line of sight to the pulsar is perpendicular to the local magnetic field. This happens when the equatorial regions of the pulsar face the observer. On the other hand, the thermal emission from the pulsar is expected to be strongest when the polar regions face the observer (e.g. Heyl and Hernquist 1998a); consequently the hard and soft X-ray emission should be between 90 and 180 degrees out of phase relative to each other.

If the magnetic field of the neutron star consists of a dipole whose dipole moment makes an angle β with the spin axis, it is straightforward to determine the variability of the non-thermal emission. Furthermore, the line of sight makes an angle α with the spin axis. If $\alpha + \beta \geq 90°$, the hard emission peaks at a phase

$$\phi = \arccos(-\cot\alpha \cot\beta), \tag{17}$$

relative to the main pulse of soft emission (for $\alpha + \beta > 90°$ there can be a second pulse of soft emission from the second polar region). Because α and β lie between zero and ninety degrees, their cotangents are positive, so ϕ must lie between ninety and one hundred eighty degrees. On the other hand, if $\alpha + \beta < 90°$, the hard emission peaks 180° out of phase from the soft emission.

Kuiper et al. (2006) observed the X-ray emission from several anomalous X-ray pulsars from 0.5 to 300 keV. The AXP 1RXS J1708-4009 exhibits variability similar to that described here. In the 1.3–3.9 keV-band the emission peaks around phase 0.25; while above 8 keV the emission peaks around phase 0.55 (90 degrees away). The peak remains at phase 0.55 to the highest energies measured about 300 keV.

The AXP 4U 0142+61 shows a similar effect albeit less dramatically. At the lower energies that Kuiper et al. (2006) examined (0.5–1.7 keV), the pulse profile is double peaked with two peaks about 160° out of phase. As the energy of the radiation increases to 50 keV, the second peak increases in amplitude and the first peak practically vanishes. The AXP 1E 2259+586 shows similar trends but the variability is only detectable up to about 25 keV. On the other hand, the pulse profile of AXP 1E 1841-045 does not change much from 2.1 keV to 100 keV. This might mean that this object is an exception or that the soft thermal emission only begins to dominate at lower energies.

5 Discussion

This paper has examined the angular dependence of the non-thermal emission from AXPs and SGRs in the weak-field regime. It is a natural extension of the model for bursts and non-thermal emission presented in papers I and II. When a dislocation of the surface of the neutron star is sufficiently large, the resulting fast modes will produce sufficient pairs to make the inner magnetosphere of the neutron star opaque to X-rays, generating a fireball. Some small but observable fraction of the energy initially in the fast modes is dissipated outside the opaque region yielding a characteristic fast-mode breakdown spectrum. A second possibility is that the crust of the neutron star is constantly shifting over small scales, generating fast modes whose breakdown is insufficient to produce a fireball. In this case we would associate the non-thermal radiation with the quiescent thermal radiation from the surface of the star. Both are produced by the quasi-continuous decay of the star's magnetic field.

Observations of the AXPs and SGRs continue to surprise, as do theoretical investigations of ultramagnetised neutron stars. This and the preceding papers have presented a unified model for the thermal burst emission and non-thermal emission from ultramagnetised neutron stars. The model has few underlying assumptions: magnetars produce fast modes sufficient to power the non-thermal emission and, more rarely, the bursts, the magnetic field far from the star is approximately dipolar and quantum electrodynamics can account for the dynamics of pairs and photons in strong magnetic fields. The model for the non-thermal emission from a particular fast mode depends only on the product of the strength of the wave and its wavenumber, the angle the wave propagates relative to the magnetic field direction and the strength of the dipole component of the field. The intensity of the emission depends strongly on the direction of the initial wave. Waves that travel sufficiently close to the direction of magnetic field do not produce non-thermal emission by this mechanism at all. The emission is therefore beamed perpendicular to the magnetic field lines, so the non-thermal emission should be between ninety and one hundred eighty degrees (depending on the viewing geometry) out of phase relative to the thermal emission which is expected to be strongest along the magnetic field lines.

Further calculations of the fate of the pairs produced in the strong field regions ($\xi \gtrsim 0.1$) are needed to understand the spectra produced by fast-mode breakdown in general; however, the beaming of the radiation does appear robust because it results from the variation of opacity and available energy with angle that is well constrained even for strong fields. Further observations can easily verify or falsify this model and potentially provide direct evidence for the ultramagnetised neutron stars that power AXPs and SGRs and the macroscopic manifestations of QED processes that account for their unique attributes.

Acknowledgements This research has made use of NASA's Astrophysics Data System Bibliographic Services.

References

den Hartog, P.R., Kuiper, L., Hermsen, W., et al.: Astron. Telegr. **293**, 1 (2004)
Gavriil, F.P., Kaspi, V.M., Woods, P.M.: Nature **419**, 142 (2002)
Heyl, J.S., Hernquist, L.: Phys. Rev. D **55**, 2449 (1997a)
Heyl, J.S., Hernquist, L.: J. Phys. A **30**, 6485 (1997b)
Heyl, J.S., Hernquist, L.: Mon. Not. Roy. Astron. Soc. **300**, 599 (1998a)
Heyl, J.S., Hernquist, L.: Phys. Rev. D **58**, 043005 (1998b)
Heyl, J.S., Hernquist, L.: Phys. Rev. D **59**, 045005 (1999)
Heyl, J.S., Hernquist, L.: Astrophys. J. **618**, 463 (2005a) (paper I)
Heyl, J.S., Hernquist, L.: Mon. Not. Roy. Astron. Soc. **362**, 777 (2005b) (paper II)
Heyl, J.S., Kulkarni, S.R.: Astrophys. J. Lett. **506**, 61 (1998)
Hulleman, F., van Kerkwijk, M.H., Kulkarni, S.R.: Nature **408**, 689 (2000)
Hurley, K., Li, P., Vrba, F., et al.: Astrophys. J. Lett. **463**, L13+ (1996)
Hurley, K., Li, P., Kouveliotou, C., et al.: Astrophys. J. Lett. **510**, L111 (1999)
Jung, G.V.: Astrophys. J. **338**, 972 (1989)
Kaspi, V.M., Gavriil, F.P., Woods, P.M., et al.: Astrophys. J. Lett. **588**, L93 (2003)
Kuiper, L., Hermsen, W., Mendez, M.: Astrophys. J. **613**, 1173 (2004)
Kuiper, L., Hermsen, W., den Hartog, P.R., et al..: Astrophys. J. **645**, 556 (2006)
Mereghetti, S., Gotz, D., Mirabel, I.F., et al.: Astron. Astrophys. **433**, L9 (2005)
Molkov, S., Hurley, K., Sunyaev, R., et al.: Astron. Astrophys. **433**, L13 (2005)
Murakami, T., Tanaka, Y., Kulkarni, S.R., et al.: Nature **368**, 127 (1994)
Özel, F.: ArXiv Astrophysics e-prints, astro-ph/0404144 (2004)
Potekhin, A.Y., Lai, D., Chabrier, G., et al..: Astrophys. J. **612**, 1034 (2004)
Rothschild, R.E., Kulkarni, S.R., Lingenfelter, R.E.: Nature **368**, 432 (1994)
Sanbonmatsu, K.Y., Helfand, D.J.: Astron. J. **104**, 2189 (1992)
Thompson, C., Duncan, R.C.: Mon. Not. Roy. Astron. Soc. **275**, 255 (1995)
Thompson, C., Duncan, R.C.: Astrophys. J. **473**, 322 (1996)
Usov, V.V., Melrose, D.B., Aust. J. Phys. **48**, 571 (1995)

Astrophys Space Sci (2007) 308: 109–118
DOI 10.1007/s10509-007-9326-x

ORIGINAL ARTICLE

Resonant Compton upscattering in anomalous X-ray pulsars

Matthew G. Baring · Alice K. Harding

Received: 13 August 2006 / Accepted: 12 October 2006 / Published online: 21 March 2007

Abstract A significant new development in the study of Anomalous X-ray Pulsars (AXPs) has been the recent discovery by INTEGRAL and RXTE of flat, hard X-ray components in three AXPs. These non-thermal spectral components differ dramatically from the steeper quasi-power-law tails seen in the classic X-ray band in these sources. A prime candidate mechanism for generating this new component is resonant, magnetic Compton upscattering. This process is very efficient in the strong magnetic fields present in AXPs. Here an introductory exploration of an inner magnetospheric model for upscattering of surface thermal X-rays in AXPs is offered, preparing the way for an investigation of whether such resonant upscattering can explain the 20–150 keV spectra seen by INTEGRAL. Characteristically flat emission spectra produced by non-thermal electrons injected in the emission region are computed using collision integrals. A relativistic QED scattering cross section is employed so that Klein–Nishina reductions are influential in determining the photon spectra and fluxes. Spectral results depend strongly on the magnetospheric locale of the scattering and the observer's orientation, which couple directly to the angular distributions of photons sampled.

This work was supported in part by the NASA INTEGRAL Theory program, and the NSF Stellar Astronomy and Astrophysics program through grant AST 0607651.

M.G. Baring (✉)
Department of Physics and Astronomy, Rice University, MS 108, 6100 Main St., Houston, TX 77005, USA
e-mail: baring@rice.edu

A.K. Harding
Gravitational Astrophysics Laboratory, Exploration of the Universe Division, NASA Goddard Space Flight Center, Code 663, Greenbelt, MD 20771, USA
e-mail: harding@twinkie.gsfc.nasa.gov

Keywords Non-thermal radiation mechanisms · Magnetic fields · Neutron stars · Pulsars · X-rays

PACS 95.30.Cq · 95.30.Gv · 95.30.Sf · 95.85.Nv · 97.60.Gb · 97.60.Jd

1 Introduction

Over the last decade, there has been a profound growth in evidence for a new class of isolated neutron stars with ultra-strong magnetic fields, so-called *magnetars* that include Soft-Gamma Repeaters (SGRs) and Anomalous X-ray Pulsars (AXPs). Such a class was first postulated as a model for SGRs by Duncan and Thompson (1992), and later for AXPs (Thompson and Duncan 1996). The AXPs, are a group of six or seven pulsating X-ray sources with periods around 6–12 seconds. They are bright, possessing luminosities $L_X \sim 10^{35}$ erg s^{-1}, show no sign of any companion, are steadily spinning down and have ages $\tau \lesssim 10^5$ years (e.g. Vasisht and Gotthelf 1997). The steady X-ray emission has been clearly observed in a number of AXPs (e.g., see Tiengo et al. 2002, for XMM observations of 1E 1048.1-5937; Juett et al. 2002 and Patel et al. 2003, for the *Chandra* spectrum of 4U 0142+61), and also SGRs (see Kulkarni et al. 2003 for *Chandra* observations of the LMC repeater, SGR 0526-66). A nice summary of spectral fitting of ASCA X-ray data from both varieties of magnetars is given in Perna et al. (2001). This emission displays both thermal contributions, which have $kT \sim 0.5 - 1$ keV and so are generally hotter than those in isolated pulsars, and also non-thermal components with steep spectra that can be fit by power-laws $dn/dE \propto E^{-s}$ of index in the range $s = 2$–3.5.

Flux variability in AXPs is generally small, suggesting that even the non-thermal components experience a moderating influence of the stellar surface, rather than some more dynamic dissipation in the larger magnetosphere. Yet the recent observation (Gavriil and Kaspi 2004) of long-lived pulsed flux flares on the timescale of several months in AXP 1E 1048.1-5937 resembles earlier reports (Baykal and Swank 1996; Oosterbroek et al. 1998) of modest flux instability. There are also correlated long term variations in X-ray flux and non-thermal spectral index in the source 1RXS J170849.0-400910, as identified by Rea et al. (2005). Moreover, Kaspi et al. (2003) and Gavriil et al. (2002, 2004) reported bursting activity in the AXPs 1E 2259+586 and 1E 1048.1-5937, suggesting that anomalous X-ray pulsars are indeed very similar to SGRs, a "unification paradigm" that is currently gathering support, but remains to be established.

The recent detection by INTEGRAL and RXTE of hard, non-thermal pulsed tails in three AXPs has provided an exciting new twist to the AXP phenomenon. In all of these, the differential spectra above 20 keV are extremely flat: 1E 1841-045 (Kuiper et al. 2004) has a power-law energy index of $s = 0.94$ between around 20 keV and 150 keV, 4U 0142+61 displays an index of $s = 0.2$ in the 20–50 keV band, with a steepening at higher energies implied by the total DC+pulsed spectrum (Kuiper et al. 2006), and RXS J1708-4009 has $s = 0.88$ between 20 keV and 150 keV (Kuiper et al. 2006); these spectra are all much flatter than the non-thermal spectra in the <10 keV band. Also, no clear tail has been seen in 1E 2259+586, yet there is a suggestion of a turn-up in its spectrum in the interval 10–20 keV (Kuiper et al. 2006). The identification of these hard tails was enabled by the IBIS imager on INTEGRAL and secured by a review of archival RXTE PCA and HEXTE data. These tails do not continue much beyond the IBIS energy window, since there are strongly constraining upper bounds from Comptel observations of these sources that necessitate a break and steepening somewhere in the 150–750 keV band (see Figs. 4, 7 and 10 of (Kuiper et al. 2006)). Interestingly, Molkov et al. (2005) and Mereghetti et al. (2005) also reported evidence for hard tails in SGR 1806-20, so that the considerations here are germane also to SGRs in quiescence.

Explaining the generation of these hard tails forms the motivation for this paper, which presents an initial exploration of the production of non-thermal X-rays by inverse Compton heating of soft, atmospheric thermal photons by relativistic electrons. The electrons are presumed to be accelerated either along open or closed field lines, perhaps by electrodynamic potentials, or large scale currents associated with twists in the magnetic field structure (e.g., see Thompson and Beloborodov 2005). In order to power the AXP emission, they must be produced with highly super-Goldreich–Julian densities. In the strong fields of the inner magnetospheres (i.e. within 10 stellar radii) of AXPs, the inverse Compton scattering is predominantly resonant at the cyclotron frequency, with an effective cross section above the classical Thomson value. Hence, proximate to the neutron star surface, in regions bathed intensely by the surface soft X-rays, this process can be extremely efficient for an array of magnetic colatitudes. Here, an investigation of the general character of emission spectra is presented, using collision integral analyses that will set the scene for future explorations using Monte Carlo simulations. This scenario forms an alternative to recent proposals (Thompson and Beloborodov 2005; Heyl and Hernquist 2005) that the new components are of synchrotron or bremsstrahlung origin, and at higher altitudes than considered here. The efficiency of the resonant Compton process suggests it will dominate these other mechanisms if the site of electron acceleration is sufficiently near the stellar surface. This prospect motivates the investigation of resonant inverse Compton models.

2 The Compton resonasphere

2.1 Energetics

The scattering scenario for AXP hard X-ray tail formation investigated here assumes that the seed energization of electrons arises within a few stellar radii of the magnetar surface. This can in principal occur on either open or closed field lines, so both possibilities will be entertained. The key requirement is the presence of ultra-relativistic electrons moving along **B**, with an abundance satisfying the energetics of AXPs implied by their intense X-ray luminosities, $L_{\mathrm{X}} \gtrsim 10^{35}$ erg/s above 10 keV (Kuiper et al. 2006). The hard X-ray tails have luminosities that are 2–3 orders of magnitude greater than the classical spin-down luminosity $\dot{E}_{\mathrm{SD}} \sim 8\pi^4 B_p^2 R^6/(3P^4c^3)$ due to magnetic dipole radiation torques. Here B_p is the surface polar field strength. This signature indicates that other dissipation mechanisms, such as structural rearrangements of crustal magnetic fields, power the AXP emission (e.g. Thompson and Duncan 1995, 1996). Let n_e be the number density of such electrons, $\langle\gamma_e\rangle$ be their mean Lorentz factor, and ϵ_{rad} be the efficiency of them radiating during their traversal of the magnetosphere (either along open or closed field lines). Then one requires that $L_{\mathrm{X}} \sim \epsilon_{\mathrm{rad}}\langle\gamma_e\rangle m_e c^2(4\pi n_e R_c^2 c)$ if the emission column has a base that is a spherical cap of radius R_c. This yields number densities $n_e \sim 3\times 10^{17} L_{\mathrm{X},35}/\epsilon_{\mathrm{rad}}\langle\gamma_e\rangle$ cm^{-3} for scaled luminosities $L_{\mathrm{X},35} \equiv L_{\mathrm{X}}/10^{35}$ erg/s, if $R_c \sim 10^6$ cm. Therefore large densities are needed, though not impossible ones, since optically thin conditions for the surface thermal X-rays prevail provided that $\langle\gamma_e\rangle \gg 1$, and ϵ_{rad} is not minuscule.

Comparing en_e to the classic Goldreich–Julian (1969) density $\rho_{\mathrm{GJ}} = \nabla.\mathbf{E}/4\pi = -\mathbf{\Omega}.\mathbf{B}/(2\pi c)$ for force-free, magnetohydrodynamic rotators, one arrives at the ratio

$$\frac{en_e}{|\rho_{\mathrm{GJ}}|} \approx \frac{4{,}670}{\epsilon_{\mathrm{rad}}\langle\gamma_e\rangle}\,\frac{L_{\mathrm{X},35}P}{B_{15}R_6^2}, \tag{1}$$

for AXP pulse periods P in units of seconds, polar magnetic fields B_{15} in units of 10^{15} Gauss, and cap radii R_6 in units of 10^6 cm. Large electron Lorentz factors of $\gamma_e \gg 10^2$–10^3 and their efficient resonant Compton cooling (i.e. $\epsilon_{\rm rad} \sim 0.01$–$1$) are readily attained in isolated pulsars with $B \sim 0.1$ (e.g., see Sturner 1995; Harding and Muslimov 1998; Dyks and Rudak 2000), and such conditions are expected to persist into the magnetar field regime. For $\epsilon_{\rm rad}\langle\gamma_e\rangle \gtrsim 10^3$ and $R_6 \sim 1$, the requisite density n_e is super-Goldreich–Julian, but not dramatically so. This situation corresponds, however, to acceleration zone cap radii R_c that are considerably larger than standard polar cap radii $R\theta_{\rm cap}$ for AXPs. Here $\theta_{\rm cap} = \arcsin\{(2\pi R/Pc)^{1/2}\}$ for a pulsar of radius R and period P. For $P = 10$ s, this yields $R\theta_{\rm cap} \sim 4.6 \times 10^3$ cm. Concentrating the relativistic electrons in such a narrow column yields charge densities far exceeding the Goldreich–Julian benchmark implied by (1). However, since AXPs possess luminosities $L_X \gg \dot{E}_{\rm SD}$, profound collatitudinal confinement may not prove necessary. Energetically, non-dipolar structure at the surface is easily envisaged, since the dissipation mechanism that powers AXP emission can restructure the fields; there is suggestive evidence for these reconfigurations from variations seen in the pulse profiles, particularly after flaring activity (e.g., see Kaspi et al. 2003). The electron energization zone may then cover a much larger range of colatitudes than is assigned to a standard polar cap, and may extend to closed field lines in equatorial regions. Of course, deviations from dipole structure will modify the contours for the resonasphere considerably from those illustrated just below, though mostly in the inner magnetosphere.

2.2 Resonant Compton scattering

In strong neutron star fields the cross section for Compton scattering is resonant at the cyclotron energy and a series of higher harmonics (e.g., see Daugherty and Harding 1986), effectively increasing the magnitude of the process over the Thomson cross section σ_T by as much as the order of $1/(\alpha_f B)$, where α_f is the fine structure constant. Here, as throughout the paper, magnetic fields are written in units of $B_{\rm cr} = m_e^2c^3/(e\hbar) = 4.413 \times 10^{13}$ Gauss, the quantum critical field strength. Klein–Nishina-like declines operate in supercritical fields, reducing the effective cross section at the resonance (e.g. Gonthier et al. 2000). In the non-relativistic, Thomson regime (e.g., see Herold 1979), only the fundamental resonance is retained. For the specific case of ultra-relativistic electrons colliding with thermal X-rays, in the electron rest frame (ERF), the photons initially move mostly almost along $\mathbf{B}$, and the cyclotron fundamental is again the only resonance that contributes (Gonthier et al. 2000).

The dominance of this resonance in forming upscattering spectra leads to an effective kinematic coupling between the energies $\varepsilon_\gamma m_e c^2$ and $\gamma_e m_e c^2$ of colliding photons and electrons, respectively, and the angle of the initial photon θ_γ to the magnetic field lines: the cyclotron fundamental is sampled when

$$\gamma_e\varepsilon_\gamma(1 - \cos\theta_\gamma) \approx B, \quad \text{for } \gamma_e \gg 1. \tag{2}$$

The simplicity of this coupling automatically implies that integration over an angular distribution of incoming photons results in a flat-topped emission spectrum for Compton upscattering of isotropic photons in strong magnetic fields. This characteristic is well-documented in the literature in the magnetic Thomson limit (e.g., see Dermer 1990; Baring 1994; for old gamma-ray burst scenarios, and Daugherty and Harding 1989; Sturner et al. 1995 for pulsar contexts), specifically for collisions between $\gamma_e \gg 1$ electrons and thermal X-rays emanating from a neutron star surface.

It is instructive to compute the zones of influence of the resonant Compton process for magnetic dipole field geometry. For X-ray photons with momentum vector $\mathbf{k}$, emanating from a single point at position vector $\mathbf{R}_e$ on the stellar surface with colatitude θ_e (i.e., $\theta_e = 0$ corresponds to the magnetic pole), the kinematic criterion in (2) selects out a single photon angle θ_γ for a given local B, both of which are dependent on the altitude and colatitude of the point of interaction. This assumes that the photon propagates with no azimuthal component to its momentum, a specialization that will be remarked upon shortly. In the absence of rotation, the resonance criterion defines a surface that is azimuthally symmetric about the magnetic field axis. The locus of the projection of this surface onto a plane intersecting the magnetic axis can be found through elementary geometry, assuming a flat space-time. Observe that light bending due to the stellar gravitational potential will modify these loci significantly in inner equatorial regions. At an altitude r and colatitude θ, denoted by position vector $\mathbf{r}$, the magnetic dipole polar coordinate components are

$$B_r = \frac{B_p R^3}{r^3}\cos\theta, \qquad B_\theta = \frac{B_p R^3}{2r^3}\sin\theta \tag{3}$$

where R is the neutron star radius and B_p is the surface field strength at the magnetic pole. At this location $\mathbf{r}$, the corresponding polar coordinate components k_r and k_θ of the photon momentum are given by

$$\begin{aligned} \frac{k_r}{|\mathbf{k}|} &= \frac{\chi - \cos(\theta - \theta_e)}{\sqrt{1 - 2\chi\cos(\theta - \theta_e) + \chi^2}}, \\ \frac{k_\theta}{|\mathbf{k}|} &= \frac{\sin(\theta - \theta_e)}{\sqrt{1 - 2\chi\cos(\theta - \theta_e) + \chi^2}} \end{aligned} \tag{4}$$

where $\chi = r/R$ is the scaled altitude. The geometry of the magnetic dipole then uniquely determines the angle $\theta_\gamma =$

$\theta_\gamma(r/R, \theta; \theta_e)$ of the photon to the field via the relation $\cos\theta_\gamma = \mathbf{k}.\mathbf{B}/|\mathbf{k}|.|\mathbf{B}| = (k_r B_r + k_\theta B_\theta)/|\mathbf{k}|.|\mathbf{B}|$:

$$\cos\theta_\gamma = \frac{2\cos\theta[\chi - \cos(\theta-\theta_e)] + \sin\theta\sin(\theta-\theta_e)}{\sqrt{1+3\cos^2\theta}\sqrt{1-2\chi\cos(\theta-\theta_e)+\chi^2}}. \quad (5)$$

Inserting this into (2), for the specific case of $\theta_e = 0$, yields the equation for the locus defining the surface of resonant scattering for outward-going electrons:

$$\chi^3 = \Psi\frac{\sqrt{1+3\cos^2\theta}}{1-\cos\theta_\gamma}, \quad \Psi = \frac{B_p}{2\gamma_e\varepsilon_\gamma}. \quad (6)$$

Here Ψ is the key parameter that scales the altitude of the resonance locale, and typically might be in the range $1 - 10^3$ for magnetars of $\gamma_e \sim 10^2$–10^4. Equation (6) can be rearranged into polynomial form, but must be solved numerically. For the special case of soft photons emitted from the surface pole ($\theta_e = 0$), the surfaces of resonant scattering for different Ψ are illustrated in Fig. 1 as the heavy blue contours. The shadows of the emission point are also indicated to demarcate the propagation exclusion zone for the chosen emission colatitude.

The altitude of resonance is clearly much lower in the equatorial regions, since the photons tend to travel more across field lines in the observer's frame, and so access the resonance in regions of higher field strength. In contrast, at small colatitudes above the magnetic pole, θ_γ is necessarily small, pushing the resonant surface to very high altitudes where the field is much lower. In this case, the loci asymptotically approach the polar axis at infinity, satisfying $\chi^3\sin^2\theta \approx 16\Psi$ for the depicted case of $\theta_e = 0$. For $\theta_e \neq 0$ cases (not depicted), the contours are morphologically similar, though they incur significant deviations from those in Fig. 1, both in equatorial and polar regions; for example, when $\theta \to 0$, the loci do not extend to infinity for $\theta_e > 0$. Clearly, by sampling different emission colatitudes θ_e these surfaces are smeared out into annular volumes. Furthermore, since most photons not emitted at the poles possess azimuthal components to their momenta, propagation out of the plane of the diagram must also be considered, modifying (5) and (6). In the interests of compactness, such algebra is not offered here. It suffices to observe that introducing an azimuthal component to the photon momentum generally tends to increase propagation across the field in the observer's frame, i.e. increasing θ_γ, so that the resonance criterion in (2) is realized at lower altitudes and higher field locales. Hence, taking into account azimuthal contributions to photon propagation in the magnetosphere, loci like those depicted in Fig. 1 actually represent the outermost extent of resonant interaction, and so are *surfaces of last resonant scattering*, i.e. the outer boundaries to the *Compton resonasphere*. It is evident that, for the majority of closed field lines for long period AXPs, this resonasphere is confined to

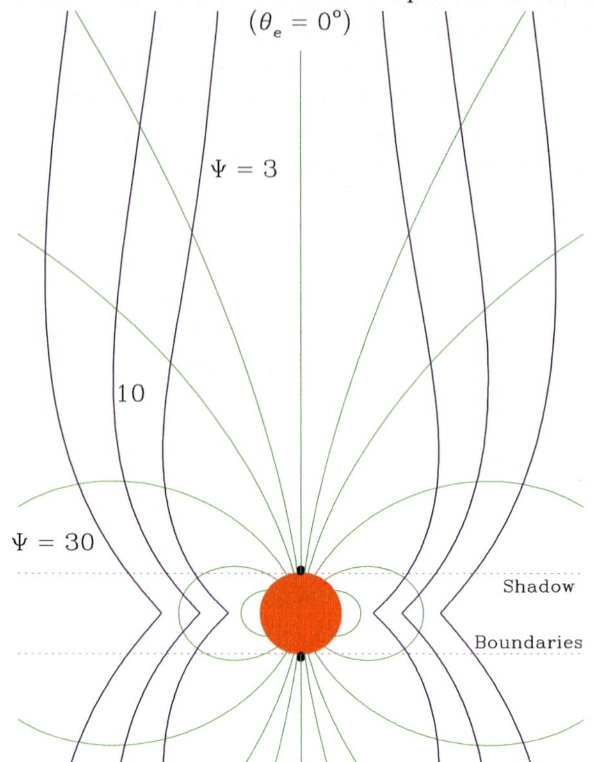

Fig. 1 Contours in a section of a pulsar magnetosphere that depict cross sections of the surfaces of last resonant scattering, i.e. the maximal extent of the Compton resonasphere. The heavyweight contours (in *dark blue*) are computed for different values of the resonance parameter Ψ defined in (6), and at extremely high altitudes asymptotically approach the magnetic axis (*vertical line*). The *filled red circle* denotes the neutron star, whose radius R establishes the spatial scale for the figure. The case illustrated is for photons emanating from the polar axis (i.e. $\theta_e = 0°$), denoted by *black dots*, for which the neutron star shadow regions are demarcated by the *dotted boundaries*, and only the surfaces (azimuthally-symmetric about the magnetic axis) are accessible to resonant Compton interactions. In computing the contours, the upscattering electrons were assumed ultra-relativistic, and the depicted spatial scales are linear

within a few stellar radii of the surface. Clearly, introducing more complicated, non-dipolar field topologies will also tend to lower the altitudes of the resonasphere.

3 Compton upscattering spectra in AXPs

To gain an initial idea of what emission spectra might be produced in the resonasphere of AXPs, collision integral calculations of upscattering spectra are performed. Here results are presented for monoenergetic electrons of Lorentz factor γ_e, from which spectral forms for various electron distributions can easily be inferred. This approach forgoes considerations of electron cooling, which naturally generates quasi-power-law distributions, at least in the magnetic Thomson limit (e.g., see Dermer 1990; Baring 1994); such issues will be addressed for supercritical fields in future presentations. To simplify the formalism for the spectra, monoenergetic incident photons of dimensionless energy $\varepsilon_\gamma = \varepsilon_s$ will be assumed, with the implicit understanding that $\varepsilon_s \sim 3kT/m_e c^2$, so that values of $\varepsilon_s \sim 0.003$ are commensurate with thermal photon temperatures $kT \sim 0.5$–1 keV observed in AXPs (see Perna et al. 2001). Distributing ε_s via a Planck spectrum provides only small changes to the spectra illustrated here, serving only to smear out modest spectral structure at the uppermost emergent photon energies.

Central to the characteristics of the spectral shape for resonant upscattering problems are the kinematics associated with both the Lorentz transformation from the observer's or laboratory frame (OF) to the electron rest frame (ERF), and the scattering kinematics in the ERF. Since choices of photon angles in the two reference frames are not unique, a statement of the conventions adopted here is now made to remove any ambiguities. Let the electron velocity vector in the OF be $\boldsymbol{\beta}_e$. This will be parallel to $\mathbf{B}$ due to rampant cyclo-synchrotron cooling perpendicular to the field. The dimensionless pre- and post-scattering photon energies (i.e. scaled by $m_e c^2$) in the OF are ε_i and ε_f, respectively, and the corresponding angles of these photons with respect to $-\boldsymbol{\beta}_e$ (i.e. field direction) are Θ_i and Θ_f, respectively. Observe that $\Theta_i \to \theta_\gamma$ establishes a connection to the notation used in Sect. 2. With this definition,

$$\cos\Theta_{i,f} = -\frac{\boldsymbol{\beta}_e.\mathbf{k}_{i,f}}{|\boldsymbol{\beta}_e|.|\mathbf{k}_{i,f}|}, \tag{7}$$

and the zero angles are chosen anti-parallel to the electron velocity. Here, $\mathbf{k}_i$ and $\mathbf{k}_f$ are the initial and final photon three-momenta in the OF. Boosting by $\boldsymbol{\beta}_e$ to the ERF then yields pre- and post-scattering photon energies in the ERF of ω_i and ω_f, respectively, with corresponding angles with respect to $-\boldsymbol{\beta}_e$ of θ_i and θ_f. The relations governing this Lorentz transformation are

$$\begin{aligned} \omega_{i,f} &= \gamma_e \varepsilon_{i,f}(1+\beta_e\cos\Theta_{i,f}), \\ \cos\theta_{i,f} &= \frac{\cos\Theta_{i,f}+\beta_e}{1+\beta_e\cos\Theta_{i,f}}. \end{aligned} \tag{8}$$

The inverse transformation relations are obtained from these by the interchange $\theta_{i,f} \leftrightarrow \Theta_{i,f}$ and the substitutions $\omega_{i,f} \to \varepsilon_{i,f}$ and $\beta_e \to -\beta_e$. The form of (8) guarantees that for most Θ_i, the initial scattering angle θ_i in the ERF is close to zero when $\gamma_e \gg 1$, exceptions being cases when $\cos\Theta_i \approx -\beta_e$. These exceptional cases form a small minority of the upscattering phase space, and indeed a small contribution to the emergent spectra, and so are safely neglected in the ensuing computations. This $\theta_i \approx 0$ approximation yields dramatic simplification of the differential cross section for resonant Compton scattering, and motivates the particular laboratory frame angle convention adopted in (7).

The scattering kinematics in the ERF differ from that described by the familiar Compton formula in the absence of magnetic fields (e.g., see Herold 1979; Daugherty and Harding 1986). In the special case $\theta_i \approx 0$ that is generally operable for the scattering scenario here, the kinematic formula for the final photon energy ω_f in the ERF can be approximated by

$$\omega_f = \omega'(\omega_i,\theta_f) \equiv \frac{2\omega_i r}{1+\sqrt{1-2\omega_i r^2\sin^2\theta_f}} \tag{9}$$

where

$$r = \frac{1}{1+\omega(1-\cos\theta_f)} \tag{10}$$

is the ratio ω_f/ω_i that would correspond to the non-magnetic Compton formula, which in fact does result if $\omega_i r^2 \sin^2\theta_f \ll 1$. Equation (9) can be found in (15) of Gonthier et al. (2000), and is realized for the particular case where electrons remain in the ground state (zeroth Landau level) after scattering. Such a situation occurs for the resonant problem addressed in this paper, a feature that is discussed briefly below.

Let n_γ be the number density of photons resulting from the resonant scattering process. For inverse Compton scattering, an expression for the spectrum of photon production $dn_\gamma/(dt\,d\varepsilon_f\,d\mu_f)$, differential in the photon's post-scattering laboratory frame quantities ε_f and $\mu_f = \cos\Theta_f$, was presented in (A7–A9) of Ho and Epstein (1989), valid for general scattering scenarios, including Klein–Nishina regimes. This was used by Dermer (1990) and Baring (1994) in the magnetic Thomson domain, i.e. when $B \ll 1$ and the photon energy in the electron rest frame (ERF) is far inferior to $m_e c^2$. Such specialization is readily extended to the magnetar regime by incorporating into the Ho and Epstein formalism the magnetic kinematics and the QED cross section for fully relativistic cases. The result can be integrated over μ_f and then written as

$$\begin{aligned} \frac{dn_\gamma}{dt\,d\varepsilon_f} &= \frac{n_e n_s c}{\mu_+ - \mu_-}\int_{\mu_l}^{\mu_u} d\mu_f \int_{\mu_-}^{\mu_+} d\mu_i \delta[\omega_f - \omega'(\omega_i,\theta_f)] \\ &\quad\times \frac{1+\beta_e\mu_i}{\gamma_e(1+\beta_e\mu_f)}\frac{d\sigma}{d(\cos\theta_f)}, \end{aligned} \tag{11}$$

noting that the angle convention specified in (7) requires the substitution $\beta_e \to -\beta_e$ in (A7–A9) of Ho and Epstein (1989). Here, the notation $\mu_i = \cos\Theta_i$ and $\mu_f = \cos\Theta_f$ is used for compactness, n_e is the number density of relativistic electrons, and n_s is that for the soft photons. These incident monoenergetic photons are assumed to possess a uniform distribution of angle cosines μ_i in some range $\mu_- \le \mu_i \le \mu_+$, which is generally broad enough to encompass the resonance, i.e. the value $\mu_i = [B/(\gamma_e\varepsilon_s) - 1]/\beta_e$. The bounds on the μ_f integration, defining the observable range of angle cosines with respect to the field direction, will be specialized to $\mu_l = -1$ and $\mu_u = 1$ in the illustration here, though dependence of the emergent spectra Θ_f values will be discussed below. The function $\omega'(\omega_i, \theta_f)$, which appears in the delta function in (11) that encapsulates the scattering kinematics, is that defined in (9).

Fully relativistic, quantum cross section formalism for the Compton interaction in magnetic fields can be found in Herold (1979), Daugherty and Harding (1986), and Bussard et al. (1986). These extend earlier non-relativistic quantum mechanical formulations such as in Canuto et al. (1971), and Blandford and Scharlemann (1976). The differential cross section, $d\sigma/d\cos\theta_f$, appearing in (11) is taken from (23) of Gonthier et al. (2000), and incorporates the relativistic QED physics. Yet, it is specialized to the case of scatterings that leave the electron in the ground state, the zeroth Landau level that it originates from. This expedient choice is entirely appropriate, since it yields the dominant contribution to the cross section at and below the cyclotron resonance (e.g. Daugherty and Harding 1986; Gonthier et al. 2000). For considerations below, where information on the final polarization state of the photon is retained, (22) of Gonthier et al. (2000) is used; this simplifies to the forms

$$\frac{d\sigma_{\parallel,\perp}}{d\cos\theta_f} = \frac{3\sigma_{\rm T}}{32}\frac{(\omega_f)^3 T_{\parallel,\perp}\exp\{-\omega_f^2\sin^2\theta_f/[2B]\}}{\omega_i[1+\omega_i(1-\cos\theta_f)-\omega_f\sin^2\theta_f]} \times\left\{\frac{1}{(\omega_i-B)^2+(\Gamma_{\rm cyc}/2)^2}+\frac{1}{(\omega_i+B-\zeta)^2}\right\} \tag{12}$$

for the differential cross sections, where the cyclotron decay width $\Gamma_{\rm cyc}$ (discussed below) has been introduced to render the resonance finite in the form of a Lorentz profile. The exponential factor in (12) is a relic of the Laguerre functions that signal the discretization of momentum/energy states perpendicular to the field. Furthermore, the factor in square brackets in the denominator is always positive, being proportional to $1-\omega_f r\sin^2\theta_f$, which can be shown to be precisely the square root appearing in (9). Also, $\zeta = \omega_i\omega_f(1-\cos\theta_f)$, and

$$\begin{aligned} T_\parallel &= 2\cos^2\theta_f + \omega_i(1-\cos\theta_f)^2 - \omega_f\sin^2\theta_f, \\ T_\perp &= 2 + \omega_i(1-\cos\theta_f)^2 - \omega_f\sin^2\theta_f. \end{aligned} \tag{13}$$

Here, the standard convention for the labelling of the photon linear polarizations is adopted: $\parallel$ refers to the state with the photon's *electric* field vector parallel to the plane containing the magnetic field and the photon's momentum vector, while $\perp$ denotes the photon's electric field vector being normal to this plane.

For magnetic Compton scattering, in the particular case of photons propagating along **B** prior to scattering (i.e. $\theta_i = 0$), the differential cross sections are independent of the initial polarization of the photon (Gonthier et al. 2000); this property is a consequence of circular polarizations forming the natural basis states for $\theta_i = 0$. Therefore, transitions $\perp\to\parallel$ and $\parallel\to\parallel$ yield identical forms for the cross sections, and separately so do the transitions $\perp\to\perp$ and $\parallel\to\perp$. Accordingly, the cross sections in (12) are labelled only by the post-scattering linear polarization state of the photon; they are summed when polarization-independent results are desired, i.e. $d\sigma/d\cos\theta_f = (d\sigma_\parallel/d\cos\theta_f + d\sigma_\perp/d\cos\theta_f)$. Therefore, clearly the upscattered photon spectra presented here are insensitive to the initial polarization level (zero or otherwise) of the soft photons.

The integrals in (11) can be manipulated using the Jacobian identity $d\mu_i\, d\mu_f = d\omega_i\, d\omega_f/(\gamma_e^2\beta_e^2\varepsilon_i\varepsilon_f)$ to change variables to ω_i and ω_f. Observe that the values of $\mu_\pm$ do not impact these integrations provided that the resonance condition in (2) is sampled. The ω_f integration is then trivial. The ω_i integration is more involved, but can be developed by suitable approximation, as follows. The relativistic Compton cross section is strongly peaked at the cyclotron fundamental (see Fig. 2 of Gonthier et al. 2000) due to the appearance of the resonant denominator $1/[(\omega_i - B)^2 + (\Gamma_{\rm cyc}/2)^2]$, where $\Gamma_{\rm cyc} \ll B$ is the dimensionless cyclotron decay rate from the first Landau level. Therefore, this Lorentz profile can be approximated by a delta function in ω_i space of identical normalization in an integration over ω_i:

$$\frac{1}{(\omega_i-B)^2+(\Gamma_{\rm cyc}/2)^2} \to \frac{2\pi}{\Gamma_{\rm cyc}}\delta(\omega_i - B). \tag{14}$$

This mapping, which was adopted in the magnetic Thomson limit by Dermer (1990), renders the ω_i trivial, with the non-resonant term in the cross section in (12) being neglected, and the evaluation of the integrals in (11) complete. The spectra scale as the inverse of the decay rate $\Gamma_{\rm cyc}$, whose form can be found, for example, in (13) or (23) of Baring et al. (2005; see also Latal 1986; Harding and Lai 2006). For $B \ll 1$, $\Gamma_{\rm cyc} \approx 4\alpha_f B^2/3$, which traces classical cyclotron cooling, while for $B \gg 1$, quantum effects and recoil reductions generate $\Gamma_{\rm cyc} \approx (\alpha_f/e)\sqrt{B/2}$.

Representative spectral forms are depicted in Fig. 2, for the situation where the emergent polarization is not observed. Because of the approximation to the resonance in (14), non-resonant scattering contributions were omitted when generating the curves; these contributions produce steep wings to the spectra at the uppermost and lowermost energies (not shown), and a slight bolstering of the resonant portion. This resonant restriction suffices for the purposes of this paper, and kinematically limits the range of emergent photon energies ε_f to

$$\gamma_e(1-\beta_e)B \le \varepsilon_f \le \frac{\gamma_e(1+\beta_e)B}{1+2B}. \tag{15}$$

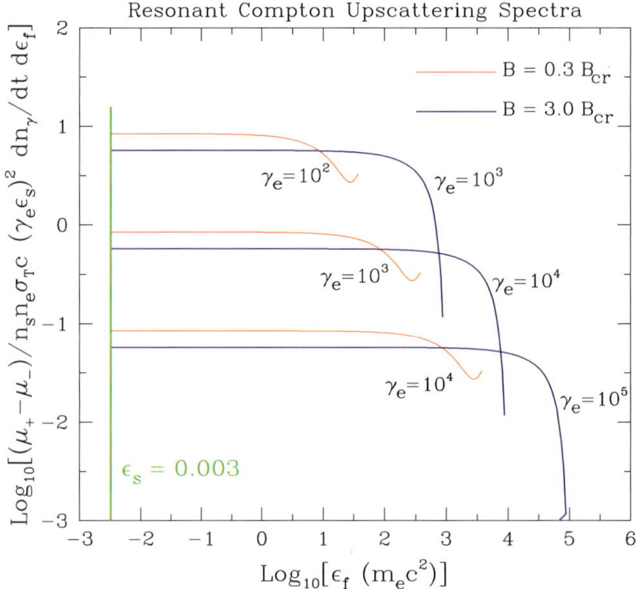

Fig. 2 Resonant Compton upscattering spectra (scaled) such as might be sampled in the magnetosphere of an AXP, for different relativistic electron Lorentz factors γ_e, as labelled. The emergent photon energy ε_f is scaled in terms of $m_e c^2$. The chosen magnetic field strengths of $B = 3B_{\rm cr}$ (heavyweight, *blue*) and $B = 0.3B_{\rm cr}$ (lighter weight, *red*) correspond to different altitudes and perhaps colatitudes. Results are depicted for seed photons of energy $\varepsilon_s = 0.003$ (marked by the *green vertical line*), typical of thermal X-rays emanating from AXP surfaces; downscattering resonant emission at $\varepsilon_f < \varepsilon_s$ was not exhibited

This range generally extends below the thermal photon seed energy ε_s. Such downscatterings correspond to forward scatterings $\cos\theta_f \approx 1$ in the ERF such that $\theta_f \lesssim \theta_i \sim 1/\gamma_e$. Since they constitute a minuscule portion of the angular phase space (and energy budget), only upscattering spectra are exhibited in Fig. 2.

In the $B = 0.3$ case, a quasi-Thomson regime, the spectra show a characteristic flat distribution that is indicative of the kinematic sampling of the resonance in the integrations (e.g., see Dermer 1990; Baring 1994). For much of this spectral range, forward scattering in the observer's frame is operating: $\mu_f = \cos\Theta_f \approx 1$. This establishes almost Thomson kinematics, with the scattered photon energy in the ERF satisfying $\omega_f \approx \omega_i$. Substituting these approximations into (12) and (13), and then summing over final polarizations yields a total approximate form for the flat portions of the spectral production rate in (11) for cases $\gamma_e \gg 1$:

$$\frac{dn_\gamma}{dt\,d\varepsilon_f} \approx \frac{n_e n_s \sigma_{\rm T} c}{\mu_+ - \mu_-}\,\frac{3\pi B^2}{4\Gamma_{\rm cyc}\gamma_e^3\varepsilon_s^2}. \tag{16}$$

The magnetic field and γ_e dependences are evident in Fig. 2 (remembering that the curves are multiplied by γ_e^2 for the purposes of illustration), and since $\Gamma_{\rm cyc} \propto B^2$ when $B \ll 1$, the normalization of the flat spectrum is independent of the field strength in the magnetic Thomson regime. Only at the highest energies does the spectrum begin to deviate from flat (i.e. horizontal) behavior, and this domain corresponds to significant scattering angles in the ERF, i.e. cosines $1 - \cos\theta_f$ not much less than unity. Then the mathematical form of the differential cross section becomes influential in determining the spectral shape. Specifically, for $\theta_f \sim \pi/2$, $T_\parallel$ drops far below $T_\perp$, as is evident in (13), causing the observed dip in the spectra. At slightly higher energies ε_f, there is a recovery when $T_\parallel$ rises as $\theta_f \to \pi$. Observe that $T_\perp$ is far less sensitive to scattering angles θ_f in the ERF, except for supercritical fields when recoil becomes significant. The sum of the two contributions yields a slight cusp at the maximum energy $\varepsilon_f \approx \gamma_e(1+\beta_e)B/(1+2B)$ when $B \lesssim 0.5$, which disappears for higher fields when the recoil reductions of $d\sigma/d\cos\theta_f$ become dominant.

In AXPs, the $B = 0.3$ case best represents higher altitude locales for the resonasphere, such as at smaller colatitudes near the polar axis. For $B = 3$, more typical of equatorial resonance locales, the flat spectrum still appears at energies $\varepsilon_s \le \varepsilon_f \ll \gamma_e(1+\beta_e)B/(1+2B)$, when again $\cos\theta_f \approx 1$. Yet the curves in Fig. 2 display more prominent reductions at the uppermost energies $\varepsilon_f \sim \gamma_e(1+\beta_e)B/(1+2B)$ due to the sampling of $1/2 \le \omega_f \ll \omega_i$ values in the ERF that correspond to strong electron recoil effects. Photons emitted in this regime have $1 - \cos\theta_f \sim 1$ in the ERF and are highly beamed along the field in the observer's frame, as will become apparent shortly. At these maximum energies, the approximations $\cos\theta_f \approx 1$ and $\omega_f \approx B/(1+2B)$ yield an analytic result for the emission rate, precisely (16) multiplied by the factor $1/(1+2B)^2$ that controls the severity of the reduction at the uppermost resonant energies. Note that Klein–Nishina reductions in the cross section regulate the overall normalizations of the $B = 3$ curves, as does the fact that the cyclotron width $\Gamma_{\rm cyc}$ no longer scales with field strength as B^2.

The dimensionless electron energy loss rate $d\gamma_e/dt$ can be obtained by multiplying the differential spectrum $dn_\gamma/dt\,d\varepsilon_f$ in (11) by ε_f/n_e and integrating over ε_f. Because of the flat nature of the spectrum, this receives a dominant contribution from near the maximum upscattered energy $\gamma_e(1+\beta_e)B/(1+2B)$. Then, using (16), in the magnetic Thomson limit $B \ll 1$, since energies $\varepsilon_f \sim 2\gamma_e B$ contribute most, the result $d\gamma_e/dt \propto B^4/\Gamma_{\rm cyc} \propto B^2$ is obviously realized for $\Gamma_{\rm cyc} \approx 4\alpha_f B^2/3$. The full integration of (11) over ε_f (or equivalently $\cos\theta_f$) is analytically tractable, leading to

$$\frac{d\gamma_e}{dt} \approx \frac{n_s \sigma_{\rm T} c}{\mu_+ - \mu_-}\,\frac{3\pi B^2}{4\gamma_e\varepsilon_s^2}, \qquad B \ll 1. \tag{17}$$

This result is commensurate with the form derived in (24) of Dermer (1990). In contrast, when $B \gg 1$, the overall normalization of the spectrum at energies $\varepsilon_f \ll \gamma_e(1+$

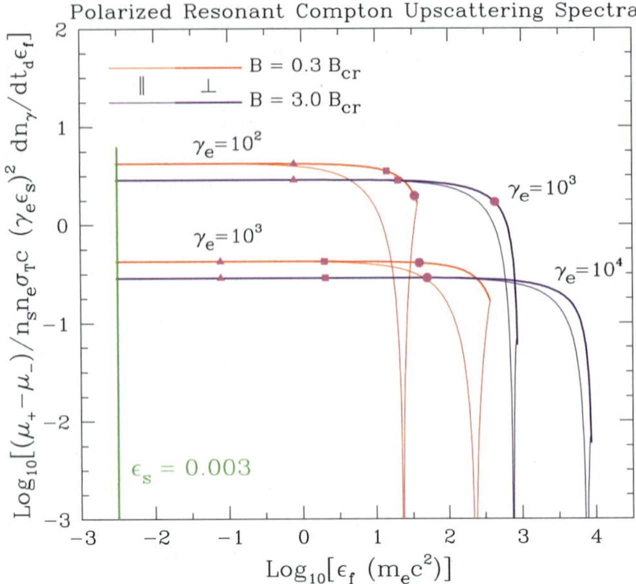

Fig. 3 Resonant Compton upscattering spectra appropriate for an AXP magnetosphere, scaled as in Fig. 2, but this time displaying the two polarizations $\perp$ (heavyweight) and $\|$ (lightweight) for the produced photons; the dominance of the $\perp$ polarization near the uppermost energies is evident. Again, the emergent photon energy ε_f is scaled in terms of $m_e c^2$, and results are presented for different relativistic electron Lorentz factors γ_e, as labelled. Specific emergent angles of the emission in the observer's frame, with respect to the magnetic field direction, are indicated by the filled magenta symbols, with triangles denoting $\Theta_f = 5°$, *squares* corresponding to $\Theta_f = 1°$, and *circles* representing $\Theta_f = 0.2°$. The magnetic field strengths of $B = 3B_{\rm cr}$ (*blue*) and $B = 0.3B_{\rm cr}$ (*red*) correspond perhaps to lower and higher altitudes, respectively. Again, the *green vertical line* marks the soft photon energy $\varepsilon_s = 0.003$, with downscattering resonant emission at $\varepsilon_f < \varepsilon_s$ not being exhibited

$\beta_e)B/(1 + 2B)$ is still controlled by (16), but now the cyclotron decay rate dependence $\Gamma_{\rm cyc} \propto B^{1/2}$ and recoil reductions at the highest ε_f come into play, so that the cooling rate $d\gamma_e/dt$ possesses a dependence on the field strength that is much weaker than $B^{3/2}$.

The resonant upscattering spectra are potentially polarized, perhaps strongly. Isolating the specific polarization forms in (13), the polarization-dependent resonant Compton spectra are readily computed and are illustrated in Fig. 3. It is clear from (13) that the emissivities for photons of final polarization state $\perp$ should always be superior to those for the $\|$ state. In a classical description, this is a consequence of the physical ease with which an oscillating electron can resonantly drive emission with electric field vectors perpendicular to **B**. Yet for much of the range of emergent photon energies above $\varepsilon_s = 0.003$, there is no material difference between fluxes for the two final polarizations. This case corresponds to $\cos\theta_f \approx 1$, i.e. photon emission along the field in the ERF, and hence induces zero linear polarization by symmetry, but permits emergent circular polarization. At the highest produced energies, significant differences between $\perp$ and $\|$ emission appear, with $1 - \cos\theta_f$ no longer very small (see (13) to help identify this characteristic). This domain motivates the development of medium energy gamma-ray polarimeters as a tool for geometry diagnostics.

Another feature of the upscattering process that is highlighted in the Fig. 3 is the intense beaming of radiation along the field in the observer's frame of reference, and the profound correlation of the angle of emission Θ_f with the emergent photon energy ε_f. This tight correspondence is mathematically guaranteed by the appearance of the kinematic delta function in (11) together with the intrinsically narrow nature of the scattering resonance, which permits the delta function approximation in (14). The extremely narrow range of Θ_f for each observed energy ε_f is broadened when non-resonant scattering is introduced. In the resonant case here, the $\gamma_e \gg 1$ regime dictates that most of the emission is collimated to within 5° of the field direction, and rapidly becomes beamed to within 0.2° as the final photon energy increases towards its maximum. This kinematic characteristic guarantees that spectral formation in Compton upscattering models is extremely sensitive to the observer's viewing perspective in relation to the magnetospheric geometry, offering useful probes if pulse-phase spectroscopy is achievable.

It should be noted that the spectra in Figs. 2 and 3 are in principal subject to attenuation by magnetic pair production, $\gamma \to e^+e^-$, reprocessing the highest energy photons to lower energies. This process is sensitive to the angle Θ_f of the scattered photons to the field, which, from the results depicted here, is strongly coupled to their energy ε_f. Gonthier et al. (2000) demonstrated (see their Fig. 7) that generally, for an extended range of values for B, pair creation attenuation would only operate for $\omega_i \gtrsim 10$–30 in the ERF, since the resonant Compton process couples the values of ω_f and θ_f. For the resonant scattering considerations here, this criterion translates to $\gamma \to e^+e^-$ being rife for local fields $B \gtrsim 10$ and being marginal, or more probably ineffective, at lower field strengths. For $B \gtrsim 10$ scattering circumstances, any pair creation that ensues does so by generating pairs in the ground state (e.g. Usov and Melrose 1995; Baring and Harding 2001), with Lorentz factors less than γ_e since $\varepsilon_f < \gamma_e$. Accordingly the cascading would consist at first of generations of upscattering and subsequent pair creation, until high enough altitudes are encountered for pair production to access excited Landau levels for the pairs, and then synchrotron/cyclotron radiation can ensue and complicate the cascade.

4 Discussion

It is clear that the spectra exhibited in Fig. 2 are considerably flatter than the hard X-ray tails ($\sim\varepsilon_f^{-1}$) seen in the AXPs, and extend to energies much higher than can be permitted by the Comptel upper bounds to these sources. However, they represent a preliminary indication of how flat the resonant scattering process can render the emergent spectrum, and what an observer detects will depend critically on his/her orientation and the magnetospheric locale of the scattering. The one-to-one kinematic correspondence between ε_f and μ_f (illustrated via the filled symbols in Fig. 3), imposed by $\omega_f \cong \gamma_e \varepsilon_f (1 - \mu_f) = \omega'(\omega_i)$ with $\omega_i = B$, implies that the highest energy photons are beamed strongly along the local field direction. This may or may not be sampled by an instantaneous observation, which varies with the rotational phase. Realistically, depending on the pulse phase, angles corresponding to $\mu_f < 1$ will be predominant, lowering the value of ε_f. Yet how low is presently unclear, and remains to be explored via a model with full magnetospheric geometry, an essential step. One can also expect substantial spectral differences between scattering locales attached to open and closed field lines, and also between dipolar and more complicated field morphologies with smaller radii of curvature.

Distributing the electron γ_e such as through resonant cooling will generate a convolution of the spectra depicted in Fig. 2; observe that the γ_e^2 scaling of the y-axis implies that the normalization of the curves is a strongly-declining function of γ_e. This can clearly steepen the continuum for a particular range of μ_f. For example, since $\varepsilon_f \propto \gamma_e$ near the maximum photon energy for a fixed electron Lorentz factor, integration over a truncated γ_e^{-p} distribution, where $\gamma_e \geq \gamma_{e,\min}$, naturally yields a photon spectrum $\varepsilon_f^{-(p+2)}$ at energies above the critical value $\varepsilon_f \sim 2\gamma_{e,\min} B/(1 + 2B)$ where the resonant flat top turns over. In particular, cooling the electrons as they propagate in the magnetosphere can lead to significant and possibly dominant contributions from Lorentz factors $\gamma_e \lesssim 10$ at higher altitudes and lower B that can evade the Comptel constraints on the AXPs (see Kuiper et al. 2006). In the Thomson regime, the cooling tends to steepen the continuum (e.g. Baring 1994) in the X-ray band due to a pile-up of electrons at low γ_e. If electrons propagate to high altitudes in magnetars, a similar steepening should be expected. As the polarized signal appears at the highest energies for each γ_e, a somewhat broad range of energies will exhibit polarization above around 50–100 keV when integrating over an entire cooled electron distribution. Note also that the cooling may persist down to mildly-relativistic energies, i.e., $\gamma_e \sim 1$, in which case it can seed a multiple scattering Comptonization of thermal X-rays that may generate the steep non-thermal continuum observed in AXPs below 10 keV. Such resonant, magnetic Comptonization has been explored by Lyutikov and Gavriil (2006), and provides very good fits to both Chandra (Lyutikov and Gavriil 2006) and XMM (see Rea et al. 2006) spectral data for the AXP 1E 1048.1-5937.

Finally, the introduction of non-resonant contributions can have a significant impact on the spectral shape. The resonasphere is spatially confined, almost to a surface for a given photon trajectory, so that for most of an X-ray photon's passage from the stellar surface, it scatters only out of the cyclotron resonance with outward-going electrons, particularly if rapid cooling is operating (a common circumstance). Moreover, inward-moving electrons traversing closed field lines participate in head-on collisions with surface X-rays, and so have great difficulty accessing the resonance. These circumstances provide ample opportunity for the emergent spectrum to develop a significant non-resonant component that does not acquire the characteristically flat spectral profiles exhibited here. Since (15) establishes $\varepsilon_f \lesssim 2\gamma_e B$ for $B \ll 1$, then this upper bound becomes inferior to the classical, non-magnetic inverse Compton result $\varepsilon_f \sim 4\gamma_e^2 \varepsilon_s/3$ when $B \lesssim \gamma_e \varepsilon_s$. This then defines a global criterion for when non-resonant Compton cooling dominates the resonant process. Assessing the relative weight of the resonant and non-resonant contributions requires a detailed model of magnetospheric photon and electron propagation and Compton scattering; this will form the focal point of our upcoming modeling of the high energy emission tails from Anomalous X-ray Pulsars.

Acknowledgements We thank Lucien Kuiper and Wim Hermsen for discussions concerning the INTEGRAL/RXTE data on the hard X-ray tails in Anomalous X-ray Pulsars, and the anonymous referee for suggestions helpful to the polishing of the manuscript.

References

Baring, M.G.: In: Fishman, G., Hurley, K., Brainerd, J.J. (eds.) Gamma-Ray Bursts, AIP Conference Proceedings, vol. 307, p. 572. Amer. Inst. Phys., New York (1994)
Baring, M.G., Harding, A.K.: Astrophys. J. **547**, 929 (2001)
Baring, M.G., Gonthier, P.L., Harding, A.K.: Astrophys. J. **630**, 430 (2005)
Baykal, A., Swank, J.: Astrophys. J. **460**, 470 (1996)
Blandford, R.D., Scharlemann, E.T.: Mon. Not. Roy. Astron. Soc. **174**, 59 (1976)
Bussard, R.W., Alexander, S.B., Mészáros, P.: Phys. Rev. D **34**, 440 (1986)
Canuto, V., Lodenquai, J., Ruderman, M.: Phys. Rev. D **3**, 2303 (1971)
Daugherty, J.K., Harding, A.K.: Astrophys. J. **309**, 362 (1986)
Daugherty, J.K., Harding, A.K.: Astrophys. J. **336**, 861 (1989)
Dermer, C.D.: Astrophys. J. **360**, 197 (1990)
Duncan, R.C., Thompson, C.: Astrophys. J. **392**, L9 (1992)
Dyks, J., Rudak, B.: Astron. Astrophys. **360**, 263 (2000)
Gavriil, F.P., Kaspi, V.M.: Astrophys. J. **609**, L67 (2004)
Gavriil, F.P., Kaspi, V.M., Woods, P.M.: Nature **419**, 142 (2002)
Gavriil, F.P., Kaspi, V.M., Woods, P.M.: Astrophys. J. **607**, 959 (2004)
Goldreich, P., Julian, W.H.: Astrophys. J. **157**, 869 (1969)
Gonthier, P.L., Harding A.K., Baring, M.G., et al.: Astrophys. J. **540**, 907 (2000)
Harding, A.K., Lai, D.: Rep. Prog. Phys. **69**, 2631 (2006)
Harding, A.K., Muslimov, A.G.: Astrophys. J. **508**, 328 (1998)
Herold, H.: Phys. Rev. D **19**, 2868 (1979)
Heyl, J., Hernquist, L.E.: Mon. Not. Roy. Astron. Soc. **362**, 777 (2005)
Ho, C., Epstein, R.I.: Astrophys. J. **343**, 227 (1989)

Juett, A.M., Marshall, H.L., Chakrabarty, D., et al.: Astrophys. J. **568**, L31 (2002)

Kaspi, V.M., Gavriil, F.P., Woods, P.M., et al.: Astrophys. J. **588**, L93 (2003)

Kuiper, L., Hermsen, W., Mendeź, M.: Astrophys. J. **613**, 1173 (2004)

Kuiper, L., Hermsen, W., den Hartog, P.R., et al.: Astrophys. J. **645**, 556 (2006)

Kulkarni, S.R., Kaplan, D.L., Marshall, H.L., et al.: Astrophys. J. **585**, 948 (2003)

Latal, H.G.: Astrophys. J. **309**, 372 (1986)

Lyutikov, M., Gavriil, F.P.: Mon. Not. Roy. Astron. Soc. **368**, 690 (2006)

Mereghetti, S., Götz, D., Mirabel, I.F., et al.: Astron. Astrophys. Lett. **433**, L9 (2005)

Molkov, S., Hurley, K., Sunyaev, R., et al.: Astron. Astrophys. Lett. **433**, L13 (2005)

Oosterbroek, T., Parmar, A.N., Mereghetti, S., et al.: Astron. Astrophys. **334**, 925 (1998)

Patel, S.K., Kouveliotou, C., Woods, P.M., et al.: Astrophys. J. **587**, 367 (2003)

Perna, R., Heyl, J.S., Hernquist, L.E., et al.: Astrophys. J. **557**, 18 (2001)

Rea, N., Oosterbroek, T., Zane, S., et al.: Mon. Not. Roy. Astron. Soc. **361**, 710 (2005)

Rea, N., Zane, S., Lyutikov, M., et al.: Astrophys. Space Sci. (2006, in press), astro-ph/0608650

Sturner, S.J.: Astrophys. J. **446**, 292 (1995)

Sturner, S.J., Dermer, C.D., Michel, F.C.: Astrophys. J. **445**, 736 (1995)

Thompson, C., Beloborodov, A.M.: Astrophys. J. **634**, 565 (2005)

Thompson, C., Duncan, R.C.: Mon. Not. Roy. Astron. Soc. **275**, 255 (1995)

Thompson, C., Duncan, R.C.: Astrophys. J. **473**, 332 (1996)

Tiengo, A., et al.: Astron. Astrophys. **383**, 182 (2002)

Usov, V.V., Melrose, D.B.: Aust. J. Phys. **48**, 571 (1995)

Vasisht, G., Gotthelf, E.V.: Astrophys. J. **486**, L129 (1997)

Astrophys Space Sci (2007) 308: 119–124
DOI 10.1007/s10509-007-9323-0

ORIGINAL ARTICLE

Newborn magnetars as sources of gravitational radiation: constraints from high energy observations of magnetar candidates

S. Dall'Osso · L. Stella

Received: 27 July 2006 / Accepted: 9 November 2006 / Published online: 21 March 2007

Abstract Two classes of high-energy sources, the Soft Gamma Repeaters and the Anomalous X-ray Pulsars are believed to contain slowly spinning "magnetars," i.e. neutron stars the emission of which derives from the release of energy from their extremely strong magnetic fields ($>10^{15}$ G). The enormous energy liberated in the 2004 December 27 giant flare from SGR 1806-20 ($\sim 5 \times 10^{46}$ erg), together with the likely recurrence time of such events, points to an internal magnetic field strength of $\geq 10^{16}$ G. Such strong fields are expected to be generated by a coherent $\alpha - \Omega$ dynamo in the early seconds after the Neutron Star (NS) formation, if its spin period is of a few milliseconds at most. A substantial deformation of the NS is caused by such fields and, provided the deformation axis is offset from the spin axis, a newborn millisecond-spinning magnetar would thus radiate for a few days a strong gravitational wave signal the frequency of which ($\sim$0.5–2 kHz range) decreases in time. This signal could be detected with Advanced LIGO-class detectors up to the distance of the Virgo cluster, where ≥ 1 yr^{-1} magnetars are expected to form. Recent X-ray observations revealed that SNRs around magnetar candidates do not appear to have received a larger energy input than in standard SNRs (see Vink and Kuiper, Mon. Not. Roy. Astron. Soc. 319, L14 (2006)). This is at variance with what would be expected if the spin energy of the young, millisecond NS were radiated away as electromagnetic radiation and/or relativistic particle winds. In fact, such energy would be transferred quickly and efficiently to the expanding gas shell. This may thus suggest that magnetars did not form with the expected very fast initial spin. We show here that these findings can be reconciled with the idea of magnetars being formed with fast spins, if most of their initial spin energy is radiated through GWs. In particular, we find that this occurs for essentially the same parameter range that would make such objects detectable by Advanced LIGO-class detectors up to the Virgo Cluster. If our argument holds for at least a fraction of newly formed magnetars, then these objects constitute a promising new class of gravitational wave emitters.

S. Dall'Osso (✉) · L. Stella
INAF, Osservatorio Astronomico di Roma, via di Frascati 33, 00040 Monteporzio Catone (Roma), Italy
e-mail: dallosso@mporzio.astro.it

L. Stella
e-mail: stella@mporzio.astro.it

Keywords Gravitational waves · Stars: magnetic fields · Stars: neutron · Stars: individual · SGR 1806-20

PACS 97.60.Jd · 97.60.Bw · 04.30.Db · 95.85.Sz

1 Introduction

The Soft Gamma Repeaters, SGRs, and the Anomalous X-ray Pulsars, AXPs, have a number of properties in common (Mereghetti and Stella 1995; Kouveliotou et al. 1998; Woods and Thompson 2004). They have spin periods of $\sim 5 \div 10$ s, spin-down secularly with $\sim 10^4 \div 10^5$ yr timescale, are isolated and in some cases associated to supernova remnants with $\sim 10^3 \div 10^4$ yr ages. Rotational energy losses are $10 \div 100$ times too low to explain the $\sim 10^{34} \div 10^{35}$ erg/s persistent emission of these sources. Both AXPs and SGRs have periods of intense activity during which recurrent, subsecond-long bursts are emitted (peak luminosities of $\sim 10^{38} \div 10^{41}$ erg/s). The initial spikes of giant flares have comparable duration but 3 to 6 orders of magnitude larger luminosity. Giant flares are rare, only three have been observed in about 30 yr of monitoring. Given the highly super-Eddington luminosities of recurrent bursts and, especially, giant flares, accretion models are not viable.

In the magnetar model, SGRs and AXPs derive their emission from the release of the energy stored in their extremely high magnetic fields (Duncan and Thompson 1992; Thompson and Duncan 1993, 1995, 1996, 2001). This is the leading model for interpreting the unique features of these sources. According to it, a wound-up, mainly toroidal magnetic field characterizes the neutron star interior ($B > 10^{15}$ G).

The emerged (mainly poloidal) field makes up the neutron star magnetosphere; dipole strenghts ($B_d \sim$ few $\times 10^{14}$ G) are required to generate the observed spin-down (Thompson and Duncan 1993; Thompson and Murray 2001). Impulsive energy is fed to the neutron star magnetosphere through Alfvén waves driven by local "crustquakes" and producing recurrent bursts with a large range of amplitudes. Giant flares likely originate in large-scale rearrangements of the toroidal inner field or catastrophic instabilities in the magnetosphere (Thompson and Duncan 2001; Lyutikov 2003). Most of this energy breaks out of the magnetosphere in a fireball of plasma expanding at relativistic speeds which produces the initial spike of giant flares. The oscillating tail that follows this spike, displaying many tens of cycles at the neutron star spin, is interpreted as due to a "trapped fireball," which remains anchored inside the magnetosphere (the total energy released in this tail is $\sim 10^{44}$ erg in all three events detected so far, comparable to the energy of a $\sim 10^{14}$ G trapping magnetospheric field).

2 The 2004 December 27 event and the internal magnetic field of magnetars

The 2004 December 27 giant flare from SGR1806-20 provides a new estimate of the internal field of magnetars. About 5×10^{46} erg were released during the ~ 0.6 s long initial spike of this event (Terasawa et al. 2005; Hurley et al. 2005). This is more than two decades higher than the energy of the other giant flares observed so far, the 1979 March 5 event from SGR 0526-66 (Mazets et al. 1979) and the 1998 August 27 event from SGR 1900+14 (Hurley et al. 1999; Feroci et al. 1999). Only one such powerful flare has been recorded in about 30 yr of monitoring of the ~ 5 known magnetars in SGRs. The recurrence time in a single magnetar implied by this event is thus about ~ 150 yr. The realisation that powerful giant flares could be observed from distances of tens of Mpc (and thus might represent a sizable fraction of the short Gamma Ray Burst population) motivated searches for 2004 December 27-like events in the BATSE GRB database (Lazzati et al. 2005; Popov and Stern 2006). The upper limits on the recurrence time of powerful giant flares obtained in these studies range from $\tau \sim 130$ to 600 yr per galaxy, i.e. ~ 4 to 20 times longer than inferred above. Therefore these authors conclude that on 2004 December 27 we have witnessed a "statistically unlikely" event. On the other hand, Tanvir et al. (2005) find that the location of 10–25% of the short GRBs in the BATSE catalogue correlates with the position of galaxies in the local universe (<110 Mpc), suggesting that a fraction of the short GRBs may originate from a population of powerful Giant Flare-like events.

A 2004 December 27-like event in the Galaxy could not be missed, whereas several systematic effects can reduce the chances of detection from large distances (see, e.g. the discussion in Lazzati et al. (2005); Nakar et al. (2005)). Rather than regarding the 2004 December 27 event as statistically unlikely, one can thus evaluate the chances of a recurrence time of hundreds of years, *given* the occurrence of the 2004 December 27 hyperflare. We estimate that, having observed a powerful giant flare in our galaxy in ~ 30 yr of observations, the Bayesian probability that the galactic recurrence time is $\tau > 600$ yr is $\sim 10^{-3}$, whereas the 90% confidence upper limit is $\tau \sim 60$ yr. We thus favor smaller values and assume in the following $\tau \sim 30$ yr.

In $\sim 10^4$ yr (that we adopt for the SGR lifetime), about 70 very powerful giant flares should be emitted by an SGR, releasing a total energy of $\sim 4 \times 10^{48}$ erg. We note that if the giant flares' emission were beamed in a fraction b of the sky (and thus the energy released in individual flares a factor of b lower), the recurrence time would be a factor of b shorter. Therefore the total release of energy would remain the same. If this energy originates from the magnetar's internal magnetic field, this must be $\geq 10^{15.7}$ G (Stella et al. 2005; Terasawa et al. 2005). This value should be regarded as a *lower limit*. Firstly, the magnetar model predicts a conspicuous neutrino luminosity from ambipolar diffusion-driven field decay, an energy component that is not available to flares. Including this, we estimate that the limit above increases by $\sim 60\%$ and becomes $B \geq 10^{15.9}$ G. Secondly, ambipolar diffusion and magnetic dissipation should take place at a faster rate for higher values of the field (Thompson and Duncan 1996). Therefore estimates of the internal B-field based on present day properties of SGRs likely underestimate the value of their initial magnetic field.

Very strong toroidal B fields are expected to be generated inside a differentially rotating fast spinning neutron star, subject to vigorous neutrino-driven convection instants after its formation (Duncan and Thompson 1992). A field of several $\times 10^{16}$ G can be generated in magnetars that are born with spin periods of a few milliseconds (Thompson and Duncan 1993). As discussed by Duncan (1998), values up to $\sim 10^{17}$ G cannot be ruled out.

In the following we explore the consequences of these fields for the generation of gravitational waves from newborn magnetars. We parametrize their (internal) toroidal field with $B_{t,16.3} = B_t/2 \times 10^{16}$ G, (external) dipole field with $B_{d,14} = B_d/10^{14}$ G and initial spin period with $P_{i,2} = P_i/(2\ \text{ms})$.

3 Magnetically-induced distortion and gravitational wave emission

The possibility that fast-rotating, magnetically-distorted neutron stars are conspicuous sources of gravitational radiation has been discussed by several authors (Bonazzola and Marck 1994; Bonazzola and Gourgoulhon 1996). More recently, work has been carried out in the context of the magnetar model, for internal magnetic fields strengths of $\sim 10^{14}$–10^{16} G (Konno et al. 2000; Palomba 2001; Cutler 2002). In the following we show that for the range of magnetic fields discussed in Sect. 2, newly born, millisecond spinning magnetars are conspicuous sources of gravitational radiation that will be detectable up to Virgo cluster distances (Stella et al. 2005).

The anisotropic pressure from the toroidal B-field deforms a magnetar into a prolate shape, with ellipticity $\epsilon_B \sim -6.4 \times 10^{-4}(\langle B^2_{t,16.3}\rangle)$, where the brackets indicate a volume-average over the entire core (Cutler 2002). As long as the axis of the magnetic distortion is not aligned with the spin axis, the star's rotation will cause a periodic variation of the mass quadrupole moment, in turn resulting in the emission of gravitational waves, GWs, at twice the spin frequency of the star.

Free precession of the ellipsoidal NS is also excited and, as shown by Mestel and Takhar (1972) and Cutler (2002), its viscous damping drives the symmetry axis of the magnetic distortion orthogonal to the spins axis, if the ellipsoid is prolate, i.e. if the magnetic field is toroidal. Therefore, viscous damping of free precession in newly born magnetars leads to a geometry that maximizes the time-varying mass quadrupole moment, and GW emission accordingly.

However, the power emitted in GWs scales as $\propto P^{-6}$. Therefore the GW signal, for a given toroidal B-field, depends critically on the initial value and early evolution of the spin period. The spin evolution of a newborn magnetar is determined by angular momentum losses from GWs, electromagnetic dipole radiation and relativistic winds. According to Thompson et al. (2004), the latter mechanism is negligible except for external dipole fields $<(6 \div 7) \times 10^{14}$ G and we will neglect it here.

The spin evolution of a newborn magnetar under the combined effects of GW and electromagnetic dipole radiation is given by

$$\dot{\omega} = -K_d\omega^3 - K_{\text{gw}}\omega^5 \tag{1}$$

where $\omega = 2\pi/P$ is the angular velocity, $K_d = (B_d^2 R^6)/(6\,Ic^3)$ and $K_{\text{gw}} = (32/5)(G/c^5)I\epsilon_B^2$, with R the neutron star radius, G the gravitational constant and c the speed of light. This gives a spin-down timescale of

$$\tau_{\text{sd}} \equiv \frac{\omega}{2\dot{\omega}} \simeq 10 \cdot P_{i,2}^2\left(B_{d,14}^2 + 1.15 B_{t,16.3}^4 P_{i,2}^{-2}\right)^{-1} \text{d}. \tag{2}$$

The condition for the newly formed magnetar to become an orthogonal rotator before loosing a significant fraction of its initial spin energy is (Stella et al. 2005):

$$\frac{\tau_{\text{sd}}}{\tau_{\text{ort}}} \simeq 26\,\frac{B^2_{t,16.3}}{B^2_{d,14}P^{-1}_{i,2} + 1.15\,B^4_{t,16.3}P^{-3}_{i,2}} > 1. \tag{3}$$

If condition (3) is met, the magnetar quickly becomes a maximally efficient GW emitter, while its spin period is still close to the initial one. In this case, the instantaneous signal strain can be expressed as:

$$h \sim 3 \times 10^{-26} d_{20}^{-1} P_2^{-2} B^2_{t,16.3} \tag{4}$$

where the distance $d_{20} = d/(20\ \text{Mpc})$ is in units of the Virgo Cluster distance and the angle-averaged strain is that given by Ushomirsky et al. (2000). We estimate the characteristic amplitude, $h_c = hN^{1/2}$, where $N \simeq \tau_{\text{sd}}/P_i$ is the number of cycles over which the signal is observed. Using Eq. (4) we obtain:

$$h_c \simeq 6 \times 10^{-22}\frac{B^2_{t,16.3}}{d_{20}P^{3/2}_{i,2}(B^2_{d,14} + 1.15 B^4_{t,16.3}P^{-2}_{i,2})^{1/2}}. \tag{5}$$

Under the conditions discussed above, strong GW losses are not quenched immediately after the magnetar birth but rather extend in time, typically from days to a few weeks, before fading away as a result of the star spin-down. The characteristic amplitude in (5) is within reach of GW interferometers of the Advanced LIGO class. In order to assess the detectability of these GW signal we compute the optimal (matched-filter) signal-to-noise ratio for a signal sweeping the 500 Hz–2 kHz band by using the current baseline performance of Advanced LIGO (details are given in Stella et al. (2005)). Figure 1 shows lines of constant S/N for a source at $d_{20} = 1$ and selected values of the initial rotation period ($P_i = 1.2$, 2 and 2.5 ms) in the (B_t, B_d) plane: GWs from newborn magnetars can produce $S/N > 8$ for $B_t \geq 10^{16.5}$ G and $B_d \leq 10^{14.5}$ G. This is the region of parameter space that offers the best prospects for detection as we now discuss. Matched-filtering represents the optimal detection strategy for long-lived, periodic signals, also in those cases where their frequency is slowly evolving over time. This process involves the correlation of the data stream with a discrete set of template signals that probe the relevant space of unknown parameters: the sky position (2 *extrinsic* parameters) and the 2 *intrinsic* parameters that control the evolution of the GW phase (see (1)). Template spacing in the parameter space is chosen appropriately, so as to reduce to a value

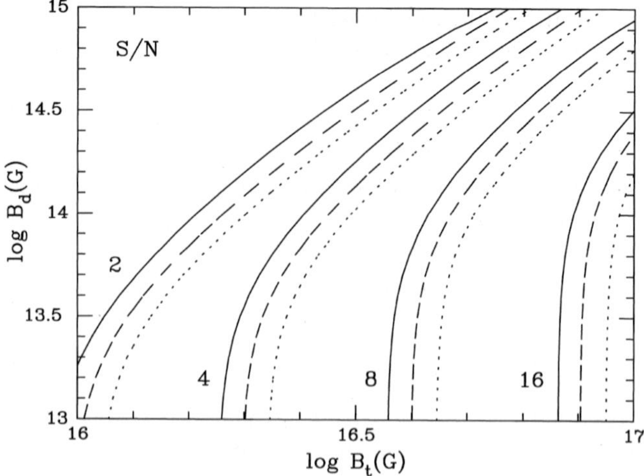

Fig. 1 Lines of constant S/N for selected values of the initial spin period in the internal toroidal magnetic field B_t vs. external dipole field B_d plane for a source at the distance of the Virgo Cluster ($d_{20} = 1$). *Solid*, *dashed* and *dotted curves* correspond to an initial spin period of P_i =1.2, 2 and 2.5 ms, respectively. The calculations take into account the time required for the toroidal magnetic field axis to become orthogonal to the spin axis. Note that according to Thompson et al. (2004), strong angular momentum losses by relativistic winds set in and dominate the spin down for $B_d > 6 \div 7 \times 10^{14}$ G; the curves for such values of B_d should thus be treated with caution

less than, say, 10% (depending on the sensitivity one wants to achieve) the fraction of the intrinsic signal-to-noise ratio that is lost in the cross-correlation. Dall'Osso and Re (2007) have recently investigated in detail a matched-filtering (coherent) search strategy for the expected signal from newly formed magnetars in the Virgo cluster. Given the two (uncorrelated) parameters involved in the signal frequency evolution, this approach implies an unaffordable computational cost. A huge number of templates $\sim 10^{18}$ would be required to cover the relevant parameter space (the plane B_d vs. B_t) with a sufficiently fine grid that the intrinsic signal-to-noise ratio would not be too degraded. Although this calculation is rather idealized, the resulting number of templates is so large that no realistic calculation could decrease it by the several orders of magnitude needed to make the search feasible. The search for this new class of signals represents however a challenge that must be investigated in greater depth. We are currently studying a hierarchical approach to the problem, a process where coherent and incoherent stages of the search are alternated as to reduce the computational requirements by a large factor, at the price of a relatively modest loss in sensitivity to the signal.

4 Observational constraints on GW emission from newly formed magnetars

A NS spinning at $\sim$ms period has a spin energy $E_{\mathrm{spin}} \approx 2.8 \times 10^{52} (P_i/1\ \mathrm{ms})^{-2}$ erg and the spin-down timescale

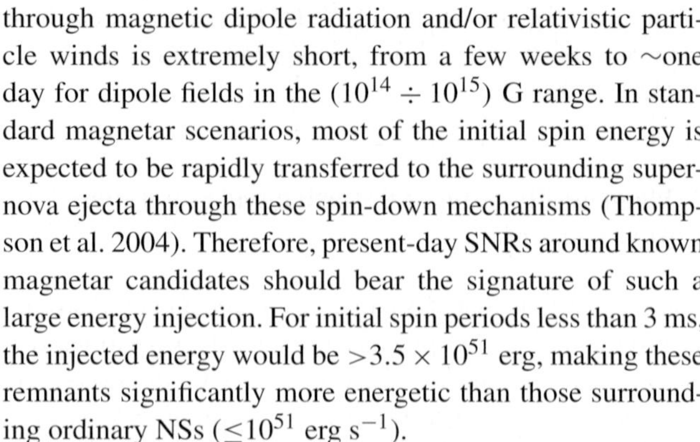

through magnetic dipole radiation and/or relativistic particle winds is extremely short, from a few weeks to $\sim$one day for dipole fields in the $(10^{14} \div 10^{15})$ G range. In standard magnetar scenarios, most of the initial spin energy is expected to be rapidly transferred to the surrounding supernova ejecta through these spin-down mechanisms (Thompson et al. 2004). Therefore, present-day SNRs around known magnetar candidates should bear the signature of such a large energy injection. For initial spin periods less than 3 ms, the injected energy would be $>3.5 \times 10^{51}$ erg, making these remnants significantly more energetic than those surrounding ordinary NSs ($\leq 10^{51}$ erg s^{-1}).

The X-ray spectra of the SNRs surrounding known magnetar candidates (two APXs and two SGRs) studied by Vink and Kuiper (2006), do not show any evidence that their total energy content differs from that in remnants surrounding common NSs ($\approx 10^{51}$ erg): this result constrains magnetar parameters at birth. Vink and Kuiper (2006) deduced from their measurements an initial spin $P_i \geq (5 \div 6)$ ms for the above mentioned sources, assuming that all the spin-down energy is emitted through electromagnetic radiation and/or particle winds, and thus absorbed by the surrounding ejecta in the early days of spin-down. Their limit period is long enough to rise a serious question as to the viability of the $\alpha - \Omega$ dynamo scenario for generating the large-scale magnetic fields of magnetars. Models that do not rely upon very short spin periods at birth, such as the flux-freezing scenario suggested by Ferrario and Wickramasinghe (2006), would be favored by these results.

However, given the possibility that newly formed magnetars be strong GW emitters in their early days, we show that the results by Vink and Kuiper (2006) can be accounted for within this framework. Most of a magnetar's initial spin energy could indeed be released through GWs, without being absorbed by the expanding remnant shell. Therefore, the results of X-ray studies can be used within our model to constrain the initial combination of P_i, B_d, B_t.

4.1 Strong GW emission at birth?

The general expression for the total energy emitted via GWs is given by

$$E_{\mathrm{gw}}^{\mathrm{TOT}} = -\int_{t_i}^{\infty} \dot{E}_{\mathrm{gw}}\, dt = -\int_{\omega_i}^{0} \frac{\dot{E}_{\mathrm{gw}}}{\dot{\omega}}\, d\omega. \quad (6)$$

We insert (1) and the expression for the GW luminosity of an elliptically distorted, spinning object into (6) to obtain the following analytical solution:

$$E_{\mathrm{gw}}^{\mathrm{TOT}} = I \int_{0}^{\omega_i} \frac{\omega^3}{\omega^2 + A}\, d\omega = I \left[\frac{\omega^2}{2} - \frac{A}{2} \ln(\omega^2 + A) \right]_0^{\omega_i} \quad (7)$$

where $A = K_d/K_{gw}$. Since E_{gw}^{TOT} amounts to a fraction α of the initial spin energy of the NS, we can write:

$$(1-\alpha)\omega_i^2 = A \ln\left(\frac{\omega_i^2 + A}{A}\right). \tag{8}$$

Finally, defining the ratio of the GW over magnetodipole torque as $x \equiv (\omega_i^2/A)$:

$$x \approx 2.25 \frac{B_{t,16.3}^4}{B_{d,14}^2\, P_{i,2}^2}, \qquad 1-\alpha = \frac{\ln(1+x)}{x}. \tag{9}$$

The above expressions provide us with two relations between the five parameters $(\omega_i, B_d, B_t, x, \alpha)$. A third one derives from the fact that the remaining fraction $(1-\alpha)$ of the initial spin energy is available for being transferred to the ejecta through magnetodipole radiation. By assuming that all this energy is effectively transferred to the SNR, we get

$$\omega_i^2 = \frac{2\, E_{SNR}^{inj}}{(1-\alpha)I} = \frac{2x\, E_{SNR}^{inj}}{\ln(1+x)I} \tag{10}$$

where E_{SNR}^{inj} is the spin-down energy injected in the expanding shell, constrained by the results of Vink and Kuiper (2006) to be $\simeq 10^{51}$ erg. We use the results of Lattimer and Prakash (2001), according to which a 1.4 $M_\odot$ NS has a radius $R \simeq 12$ km and a moment of inertia $I \approx 0.35MR^2 \simeq 1.4 \times 10^{45}$ g cm^2. Note that, by these numbers, one obtains the following relation between the measured P and $\dot{P}$ of NSs and the corresponding value of the dipolar magnetic field at the magnetic pole (our B_d):[1]

$$B_d \simeq 4.4 \times 10^{19}\,(P\dot{P})^{1/2}\ \text{G}. \tag{11}$$

As can be seen from Fig. 1, dipole magnetic fields stronger than $(5 \div 6) \times 10^{14}$ G rapidly quench the expected S/N ratio of GW signals from newly formed magnetars. Therefore, we restrict our investigation to polar dipole fields $B_d < 6$. From (10) we see that, once x is given, both α and ω_i are determined (one can indeed use any of the three as the free parameter). The first expression of (9) thus provides a relation between B_d and B_t.

We have repeated this procedure for five values of the ratio of the GW over the magnetodipole torque (x) and, for each of them, obtained the implied values of ω_i and α and a curve in the B_d vs. B_t plane.

Figure 2 summarizes our results. Loci in the B_d vs. B_t plane for which fast spinning, ultramagnetized magnetars are consistent with the energetic constraints derived by Vink and Kuiper (2006) are drawn for the five chosen values of x or, equivalently, the initial spin period (since E_{SN}^{inj} is fixed). Details are given in the caption.

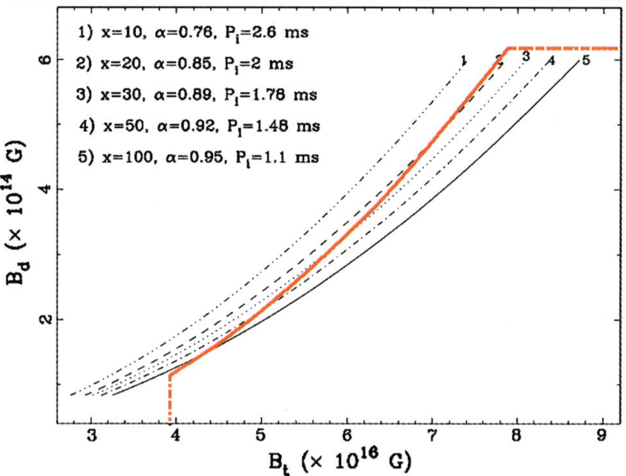

Fig. 2 Loci in the B_d vs. B_t plane for which $E_{SN}^{inj} \simeq 10^{51}$ ergs. Curves are labelled for different values of the ratio (x) of the GW to magnetodipole torques. Through (10), it is seen that fixing this ratio is equivalent to defining the corresponding initial spin. The corresponding locus is thus uniquely determined through the first expression in (9). Note that, given x, the corresponding curve shifts to the right for decreasing values of E_{SNR}^{inj}. The *red curve* represents the $S/N = 8$ curve of Fig. 1. In the framework of our model, thus, fast spinning newly formed magnetars that do emit $\leq 10^{51}$ ergs of spin-down energy through electromagnetic radiation can be expected to be detectable with LIGO II up to the distance of the Virgo cluster. Stated the other way, objects whose GW signal could be detectable by LIGO II up to the distance of the Virgo Cluster would inject—as found in the case of magnetar candidates in the Galaxy—$\leq 10^{51}$ erg in the expanding shells of the surrounding SNR

In summary, we have first identified a range of initial conditions (spin period, internal and external magnetic field), within which newly formed magnetars can be interesting targets for next generation GW detectors. Then we have calculated that, within most of that same region, magnetars should emit less than 10^{51} ergs through magnetodipole radiation (cf. Dall'Osso et al. (2007) and Fig. 2). This is compatible with the limits inferred through recent X-ray observations of SNRs around present-day magnetar candidates.

5 Discussion

The energy liberated in the 2004 December 27 flare from SGR 1806-20, together with the likely recurrence rate of these events, points to a magnetar internal field strength of $\sim 10^{16}$ G or greater. Such a field likely results from differential rotation in a millisecond spinning proto-magnetar and deforms the star into a prolate shape. Magnetars with these characteristics are expected to be very powerful sources of gravitational radiation in the first days to weeks of their life.

[1] This is a factor 1.5 less than usually assumed with $I = 10^{45}$ g cm^2 and $R = 10$ km. Use of the most up to date parameters is required, given the strong dependence of the two competing torques on the exact value of the magnetic fields.

An evolving periodic GW signal at $\sim$1 kHz, whose frequency halves over weeks, would unambiguously reveal the early days of a fast spinning magnetar.

Prospects for revealing their GW signal depend on the birth rate of these objects. The three associations between an AXP and a supernova remnant (ages in the $10^3 \div 10^4$ yr range) imply a magnetar birth rate of $\geq 0.5 \times 10^{-3}$ yr^{-1} in the Galaxy (Gaensler et al. 1999). Therefore the chances of witnessing the formation of a magnetar in our Galaxy are slim. A rich cluster like Virgo, containing $\sim$2000 galaxies, is expected to give birth to magnetars at a rate of ≥ 1 yr^{-1}. A fraction of these might have sufficiently high toroidal fields that a detectable GW signal is produced.

It has been recently found that SNR shells around some magnetar candidates have comparable expansion energies to standard SNR shells. This implies that either the NS was not initially spinning as fast as required for an $\alpha - \Omega$ dynamo to amplify its field to magnetar strengths, or that most of its initial spin energy was emitted in a way that did not interact with the ejecta. GWs have indeed such property. If the internal magnetic field of newly formed magnetars is $> 10^{16}$ G, comparable to the lower limit estimated through the energetics of the December 27 Giant Flare, then their GW emission can be strong enough to radiate away most of their initial spin energy. The required amount of energy emitted through GWs is indeed such that, had these magnetar candidates been at the distance of the Virgo cluster, they would have been revealed by a LIGO II-class GW detector.

Therefore, GWs from newly formed magnetars can account naturally for the recent X-ray observations of SNRs around galactic magnetars. The main conclusion that can be drawn at present is that newborn, fast spinning magnetars represent a potential class of GW emitters over Virgo scale distances that might well be within reach for the forthcoming generation of GW detectors.

References

Bonazzola, S., Marck, J.A.: Ann. Rev. Nucl. Part. Sci. **45**, 655 (1994)
Bonazzola, S., Gourgoulhon, E.: Astron. Astrophys. **312**, 675 (1996)
Cutler, C.: Phys. Rev. D **66**, 084025 (2002)
Dall'Osso, S., Re, V.: Phys. Rev. D (2007, submitted)
Dall'Osso, S., et al.: in preparation
Duncan, R.C.: Astrophys. J. Lett. **498**, L45 (1998)
Duncan, R.C., Thompson, C.: Astrophys. J. Lett. **392**, L9 (1992)
Feroci, M., et al.: Astrophys. J. Lett. **515**, L9 (1999)
Ferrario, L., Wickramasinghe, D.: Mon. Not. Roy. Astron. Soc. **367**, 1323 (2006)
Gaensler, B.M., Gotthelf, E.V., Vasisht, G.: Astrophys. J. Lett. **526**, L37 (1999)
Hurley, K., et al.: Nature **397**, 41 (1999)
Hurley, K., et al.: Nature **434**, 1098 (2005)
Konno, K., Obata, T., Kojima, Y.: Astron. Astrophys. **356**, 234 (2002)
Kouveliotou, C., et al.: Nature **393**, 235 (1998)
Lattimer, J.M., Prakash, M.: Astrophys. J. **500**, 426 (2001)
Lazzati, D., Ghirlanda, G., Ghisellini, G.: Mon. Not. Roy. Astron. Soc. **362**, L8 (2005)
Lyutikov, M.: Mon. Not. Roy. Astron. Soc. **346**, 540 (2003)
Mazets, E.P., Golentskii, S.V., Ilinskii, V.N., Aptekar, R.L., Guryan, Iu.A.: Nature **282**, 587 (1979)
Mereghetti, S., Stella, L.: Astrophys. J. Lett. **442**, L17 (1995)
Mestel, L., Takhar, H.S.: Mon. Not. Roy. Astron. Soc. **156**, 419 (1972)
Nakar, E., Gal-Yam, A., Piran, T., Fox, D.B.: astro-ph/0502148v1 (2005)
Palomba, C.: Astron. Astrophys. **367**, 525 (2001)
Popov, S.B., Stern, B.E.: Mon. Not. Roy. Astron. Soc. **365**, 885 (2006)
Stella, L., Dall'Osso, S., Israel, G.L., Vecchio, A.: Astrophys. J. Lett. **634**, L165 (2005)
Tanvir, N.R., Chapman, R., Levan, A.J., Priddey, R.S.: Nature **438**, 991 (2005)
Terasawa, T., et al.: Nature **434**, 1110 (2005) (astro-ph/0502315)
Thompson, C., Duncan, R.C.: Astrophys. J. **408**, 194 (1993)
Thompson, C., Duncan, R.C.: Mon. Not. Roy. Astron. Soc. **275**, 255 (1995)
Thompson, C., Duncan, R.C.: Astrophys. J. **473**, 322 (1996)
Thompson, C., Duncan, R.C.: Astrophys. J. **561**, 980 (2001)
Thompson, C., Murray, N.C.: Astrophys. J. **560**, 339 (2001)
Thompson, T.A., Chang, P., Quataert, E.: Astrophys. J. **611**, 380 (2004)
Ushomirsky, G., Cutler, C., Bildsten, L.: Mon. Not. Roy. Astron. Soc. **319**, 902 (2000)
Vink, J., Kuiper, L.: Mon. Not. Roy. Astron. Soc. **370**, L14 (2006)
Woods, P.M., Thompson, C.: astro-ph/0406133 v3 (2004)

Astrophys Space Sci (2007) 308: 125–132
DOI 10.1007/s10509-007-9324-z

ORIGINAL ARTICLE

Astrophysical input for gravitational wave searches

D.I. Jones

Received: 13 July 2006 / Accepted: 19 August 2006 / Published online: 15 March 2007

Abstract We describe several areas where the newly emerging field of gravitational wave astronomy would benefit from exploiting the expertise of the broader astrophysics community. We deal specifically with searches for long-lived gravitational wave signals from neutron stars, paying particular attention to the known radio pulsar population and supernova remnants.

Keywords Gravitational waves · Neutron stars

PACS 95.85.Sz · 95.30.Sf

1 Introduction

The main purpose of this contribution is to increase awareness of the newly emerging branch of gravitational wave astronomy, and to try and spell one various ways in which the vast astrophysical expertise of the electromagnetic astronomy community might be brought to bear on gravitational wave detection. This may lie in helping gravitational wave astronomers to make an initial detection, or by helping to extract useful astronomical information form such a detection. Either way, the impressive array of expertise of the delegates that assembled in London last April, reaching all of way from the radio band to gamma rays, will surely be of use in gravitational wave astronomy, and the two fields of observation–gravitational and electromagnetic–need to forge close links to maximise the science return of the experimental efforts. In this article I will talk about a few concrete areas where one field could benefit from close links with the other.

D.I. Jones (✉)
School of Mathematics, University of Southampton, Highfield, Southampton SO17 1BJ, UK
e-mail: D.I.Jones@soton.ac.uk

2 Gravitational wave emission

We will deal here exclusively with long-lived approximately monochromatic sources of gravitational waves, commonly known as *continuous sources*. Transient or burst like sources such as might be produced by binary inspiral or stellar core collapse will not be considered.

The extreme compactness and rapid rotation rates of neutron stars makes them ideal candidates for detectable gravitational wave emission, providing there exists some mechanism perturbing the star away from a state of axisymmetry. The gravitational radiation itself is typically emitted at (or close to) the stellar rotation frequency or one of its harmonics, or at a characteristic oscillation frequency of the star (of order kHz or higher), and so we will be concerned only with gravitational wave detection by the ground-based instruments such as LIGO, GEO600 and VIRGO. The space-based LISA instrument will operate mainly in the frequency interval 10^{-4}–10^{-1} Hz; at such low frequencies signals from spinning neutron stars are likely to be too weak to be detectable.

There exist three principle such mechanisms which deform compact objects away from axisymmetry, allowing gravitational wave emission. We will describe each briefly; see Jones (2002a, 2002b) for more detail.

Firstly, and most simply, strains in the solid crust or deformations induced by the internal magnetic field may cause the star's moment of inertia tensor to be triaxial. Steady rotation about a principal axis then produces gravitational radiation with an amplitude h proportional to the difference in the moments of inertia ΔI about two axis perpendicular to the rotation axis:

$$h = \left(\frac{2}{15}\right)^{1/2} \frac{G}{c^4} \frac{8\Omega^2}{d} \Delta I \tag{1}$$

where Ω is the angular velocity of rotation, d the distance to the source, and I the moment of inertia about the rotation axis. In this case the gravitational radiation is at 2Ω. The dimensionless number ϵ is often used to quantity the magnitude of the triaxiality:

$$\epsilon = \frac{\Delta I}{I}. \tag{2}$$

See Ushomirsky et al. (2000) and Haskell's contribution to these proceedings for theoretical estimates of upper limits on ϵ for neutron stars, and Owen (2005) for limits assuming some exotic equations of state.

Secondly, if the rotation axis is not aligned with a principal axis of the moment of inertia tensor, the star will undergo a free precessional type motion, emitting gravitational radiation in the process; see Jones and Andersson (2002). The exact spectrum of the radiation depends upon the relative sizes of the three principal moments of inertia and on the angular amplitude of the precession. In the physically realistic case of the small angle free precession of an approximately biaxial body the radiation is mainly at frequencies close to (but not exactly equal to) Ω and 2Ω. However, theory predicts that such precessional motion is likely to be highly damped, and it is not clear if there exist sufficiently strong excitation mechanisms to make free precession a good gravitational wave candidate; see Jones and Andersson (2002). To this end, a better understanding of the proposed free precession candidates discussed at this meeting would be invaluable (see e.g. contribution by Pons).

Finally, excitation of the normal modes of oscillation of the star would also lead to gravitational radiation emission. The frequency spectrum would depend upon the stellar structure, rotation rate, and upon which of the infinite zoo of oscillation modes was excited. As is the case for free precession, such emission is likely to be damped. Again, better understanding of proposed electromagnetic signatures of this sort of excitation (in this case the burst spectra of SGR, see the contributions of Israel and Watts) would help us quantify the likely gravitational wave detectability of mode oscillations for an isolated star.

The bulk of the discussion in this article will not depend upon which of the above mechanisms is operative. When it is necessary to be specific we will assume a rigidly rotating non-precessing star, as this is probably the most likely source on theoretical grounds.

One thing in common to all continuous sources is the parametrization of their gravitational wave signal by a set containing a *minimum* of 7 numbers:

$$\theta = \{h_0, f, \alpha_{\text{sky}}, \delta_{\text{sky}}, \phi_0, \iota, \psi\}. \tag{3}$$

Here h_0 is the signal amplitude, f the frequency, $\alpha_{\text{sky}}, \delta_{\text{sky}}$ two angles giving position on the sky (e.g. right ascension and declination), ϕ_0 a signal phase at some fiducial time (e.g. $t = 0$), and ι, ψ two more angles, giving the orientation of the star's spin axis. If the signal is exactly monochromatic, these 7 numbers would exactly parametrize the waveform. In reality, more parameters are likely to be required. The frequency is likely to be a slowly decreasing function of time, corresponding to an energy loss to gravitational and perhaps also electromagnetic radiation, requiring the addition of one or more frequency derivatives. If the star is in a binary system, the 5 Keplerian binary parameters must be added to remove the binary-induced Doppler motion. Any further non-smooth frequency variation (e.g. that due to timing noise) would complicate the parametrization further as described in Jones (2002b).

The size of the parameter space to be searched over is very important, as gravitational wave searches basically rely on match filtering over long duration (possibly of order years) data segments, with large numbers of source templates being required to catch all potential signals. For the foreseeable future, such searches will be computationally limited. A key theme of this article will be to describe ways in which astrophysical input could help reduce the massive computational burden by reducing the parameter space to be searched over. The reader is referred to Abbot et al. (2004) for information about the signal analysis procedure.

3 Upper bounds on the gravitational wave amplitude

We will first consider searches for gravitational waves from known sources, i.e. from stars where electromagnetic astronomy has already provided some information. The main such class of stars are the radio pulsars, whose sky locations and spin evolution are measured accurately; we will concentrate our discussion on them.

To gain a preliminary impression of the gravitational wave interest of the radio pulsars, Fig. 1 shows the *upper bound* on the gravitational wave amplitude from the known pulsar population, assuming that 100% of their spin-down energy is going into gravitational wave emission. The wave field is in fact given by (see e.g. Jones 2002b):

$$h_{\text{spindown}} = \frac{2}{d}\left(\frac{GI\dot{P}}{c^3 P}\right)^{1/2}, \tag{4}$$

so that radio astronomy supplies the spin period P, period derivative $\dot{P}$ and distance d, the latter being derived (in most

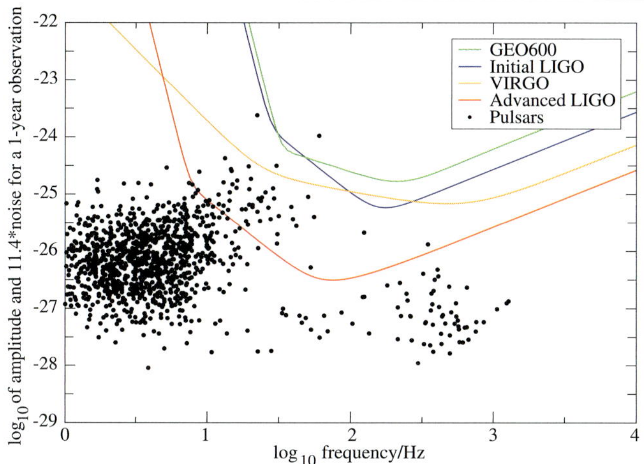

Fig. 1 Naive upper bounds on gravitational wave emission from pulsars, assuming 100% conversion of spin-down energy into gravitational weave energy

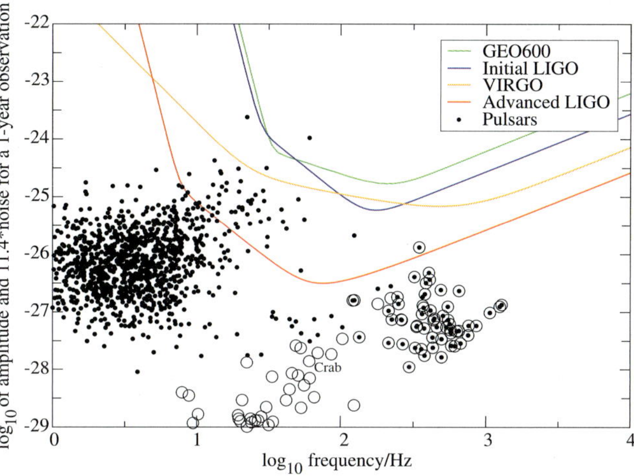

Fig. 2 More realistic upper bound on gravitational wave amplitudes

cases) by the dispersion of the radio signal. As can be seen, a number of pulsars lie above the Advanced LIGO noise-curves, some being young stars, some being millisecond pulsars.

However, for the young pulsars at least, there exists good evidence that in fact a large part of the spin-down energy is in fact going into the electromagnetic, not the gravitational wave channel. Furthermore, if one were to calculate the effective triaxiality ϵ defined previously, one finds that values as high as $\epsilon \sim 10^{-4}$ would be required. Theoretically, strains in the solid crust will probably limit ϵ to be no more than $\sim 10^{-7}$, and so such large values are completely implausible; see the contribution of Haskell for more details. More encouragingly, the ellipticities required for the millisecond pulsars are significantly smaller, of order $\epsilon \sim 10^{-8}$, and therefore perfectly plausible from the theoretical point of view. To convert Fig. 1 into a more realistic assessment of the likely upper bounds on gravitational wave emission from pulsars, we have produced Fig. 2, where in addition to the naive upper bound given above, we re-plot each pulsar as an open circle with the ellipticity set to *the minimum* of the ϵ value required to reproduce the observed spin-down and 10^{-7}, i.e. we set

$$\epsilon = \min\left(\epsilon_{\text{spin-down}}, 10^{-7}\right) \tag{5}$$

where

$$\epsilon_{\text{spin-down}} = \left[\frac{5c^5 \dot{P} P^3}{32(2\pi)^4 GI}\right]^{1/2}. \tag{6}$$

As is clear from Fig. 2, for the young pulsars, the gravitational wave amplitudes are reduced by orders of magnitude, as expected following the discussion above, while the millisecond pulsar estimates are either not affected at all or else only slightly reduced.

There a few rather obvious points that can be made at once. Firstly, it is clear from these figures that the older millisecond pulsars seem to be better bets for gravitational wave detection than the young stars. Secondly, there are precious few stars above the detectability curve of even advanced LIGO.

These points together lead to the equally obvious observation that one of the single most helpful inputs that conventional astronomy could make to gravitational wave searches is to supply new candidates for detection, with high spin frequencies, high spin-down rates and at small distances. Improvements to currently existing radio facilities and detection algorithms will surely help in this, but the large numbers of new pulsars that the high sensitivity Square Kilometer Array will provide may well prove crucial in identifying gravitationally-detectable pulsars.

4 Constraining the pulsar geometry

However, there is a slightly less obvious point that we will concentrate on here. There are a number of potential gravitational wave candidates that lie just below the detection threshold. To understand the significance of this, it is first necessary to understand the precise meaning of the noise-curves presented here. They mark out the threshold for a star to be detected at a signal-to-noise ration 11.4 in one year's worth of data. This number of 11.4 is chosen as it represents the signal-to-noise required for the probability of false detection to take the low value of 1%, while maintaining a 10% false dismissal probability. The value of 11.4 was determined by Monte Carlo simulation by the continuous source group of the LIGO Scientific Collaboration, running many 'fake' signals through their analysis pipeline; see Abbot et al. (2004). (Different analysis pipelines may give somewhat different values.) Crucially, the only information used from

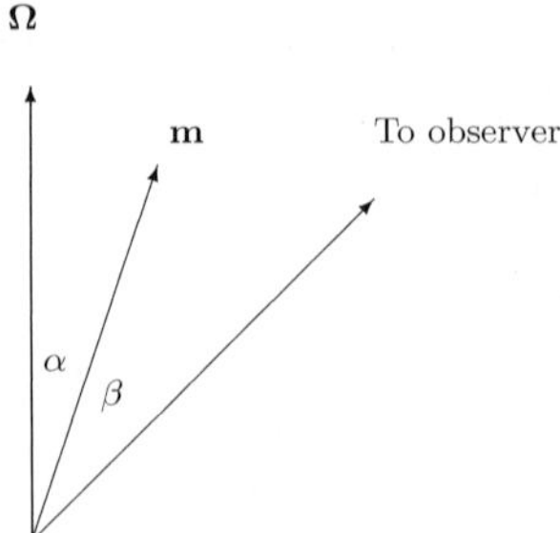

Fig. 3 Pulsar radio emission geometry: $\mathbf{\Omega}$ is the angular velocity vector, $\mathbf{m}$ is the rotating dipole, α is the (fixed) angle between $\mathbf{\Omega}$ and $\mathbf{m}$, and β is the impact parameter of $\mathbf{m}$ on line of sight

the radio data was the source phase evolution (basically the frequency and, if necessary, frequency corrections) and sky location. This left the amplitude h, arrival phase ϕ_0 and spin orientation angles ι, ψ to be extracted from the data.

However, as is well known in the matched filtering of noisy data, the larger the number of variables searched over, and the larger the allowable range of each variable, the larger the chances of the noise conspiring to mimic a signal, and therefore the larger one must set the signal-to-noise threshold for detection to maintain a given false detection rate. Therefore, the following question presents itself: is there any extra information, beyond the spin frequency and sky location, that could be extracted from the radio data to reduce the volume of parameter space over which the gravitational wave data analysts must search? If so, the critical signal-to-noise ratio threshold for detection could be set to some (to be determined) value lower than 11.4.

The answer to this question is a qualified 'yes,' and it is here that the expertise of the radio astronomy might be brought to bear to aid gravitational wave detection. Many pulsar astronomers have attempted to use their observations to probe the geometry of the radio emission by measuring the following angles (I'll follow their notation; see Fig. 3 for a diagram):

- α, the angle between the pulsar's spin axis and the line of sight,
- β, the impact parameter of the magnetic axis, i.e. the smallest angle that the magnetic axis makes with the observer's line of sight during each pulsar rotation.

The sum of these, $\iota = \alpha + \beta$, is the inclination angle of the pulsar's spin axis with respect to the observer, i.e. is an angle that appears explicitly in the gravitational wave signal; see (3). Astronomers are interested in this geometry as it can be used to probe the magnetic field outside the star and the radio-production mechanism. A crucial question from the gravitational wave point of view is how certain is the ι value so obtained? At the one extreme, if it seemed that the geometry as calculated from radio data was absolutely rock solid, then the angle ι could be used in the targeted gravitational wave searches, decreasing by one the number of unknowns to be extracted from the data, presumably increasing the accuracy with which the remaining unknowns could be extracted and lowering the critical signal-to-noise threshold for detection. If the geometry is less certain, then perhaps a gravitational wave-derived value of ι could be compared with the radio-derived one, testing the models that were used to produce the latter.

To gain a feeling of this, I will report on my literature search in this area, first describing the models used by radio astronomers to calculate α and β, and then give my impression of the usefulness of these in gravitational wave searches.

4.1 Methods of extracting ι

4.1.1 The rotating vector model (RVM)

This is the simplest and most widely used method, having been used by many authors. It involves fitting the radio linear polarization to the so-called *rotating vector model* first proposed by Radhakrishnan and Cooke (1969), whose key assumptions are:

- The magnetic field in the region where the radio pulsations are produced is purely dipolar.
- The observed radio polarization is along the projection of the magnetic field lines into a plane perpendicular to the direction to the observer.

With these assumptions, it is a matter of spherical geometry to show that the observed polarization angle ψ, as a function of pulsar rotation phase ϕ, is given by

$$\tan(\psi - \psi_0) = \frac{\sin\alpha \sin(\phi - \phi_0)}{\sin(\alpha+\beta)\cos\alpha - \cos(\alpha+\beta)\sin\alpha\cos(\phi-\phi_0)} \tag{7}$$

where ϕ_0 and ψ_0 are phase constants. If the polarization angle $\psi(\phi)$ can be measured to sufficient accuracy over a sufficiently large range in ϕ, a χ^2 fit to this formula can be carried out, yielding best-fit values for $\alpha, \beta, \phi_0, \psi_0$.

4.1.2 Rankin's method

One of the drawbacks of the RVM is that quite a wide interval of polarization data $\psi(\phi)$ is needed to attempt a fit. To bypass this difficulty, Rankin proposed a new method, based upon her work on classifying pulsar beams. In her method, some polarization information was still required, but only the relatively easily obtained value of the *maximum* rate of change polarization angle with rotation phase. In the RVM, this is given by differentiation of (7):

$$\left.\frac{d\psi}{d\phi}\right|_{\max} = \frac{\sin\alpha}{\sin\beta}. \tag{8}$$

This maximum occurs at the centre of the profile, where the signal-to-noise tends to be large.

Rankin has argued that the beams can be divided into central core regions and sets of outer concentric cones. The form of an individual pulsar's radio profile then depends upon which components the pulsar has in the first place (some have just a core, some a core and cones, some just cones), the spin-dipole angle α, and the impact parameter β. Rankin (1990) began by looking only at inter-pulsars, i.e. pulsars where it appears we see both north and south magnetic poles once per rotation. Such stars presumably have a near orthogonal geometry, i.e. $\alpha \approx 90^\circ$ and $\beta \ll \alpha$. In this case she found a surprising result, that over several orders of magnitude in spin period, the observed angular width of the core beam, $W_{\rm core}$ (in degrees) was related to the period P (in seconds) according to

$$W_{\rm core} = \frac{2.45^\circ}{P^{1/2}} \tag{9}$$

(see Figs. 1 and 3 of Rankin 1990). This is just what one would expect if the radio waves are beamed tangentially to the last open field lines from close to the star's surface, a theoretically appealing interpretation.

Encouraged by this result, Rankin then assumed that even in non-orthogonal rotators the width of the core beam was determined by the last open field lines (see Rankin 1993), leading to the straight-forward generalization

$$W_{\rm core} = \frac{2.45^\circ}{P^{1/2}\sin\alpha}. \tag{10}$$

Then a measurement of P in a pulsar with a clearly defined core component gives α, and if the maximum rate of change of polarization angle is measured, (8) allows β to be calculated, hence $\iota = \alpha + \beta$.

4.1.3 Lyne and Manchester's method

Lyne and Manchester (1988) also made use of (8), but obtained their extra piece of geometric information by making a fit to the width of the *conal* part of the pulsar beam. They obtained this fit by means of the following argument: They began by assuming an orthogonal geometry, i.e. $\alpha = 90^\circ$, and used (8) to calculate β. They then assumed that the beam was circular with opening angle ρ. It can be shown using spherical geometry that ρ is related to α, β and $\Delta\phi$, the rotational phase duration of the beam, according to

$$\sin^2\left(\frac{\rho}{2}\right) = \sin^2\left(\frac{\Delta\phi}{2}\right)\sin\alpha\sin(\alpha+\beta) + \sin^2\left(\frac{\beta}{2}\right). \tag{11}$$

They were then able to obtain ρ for many different pulsars. A plot of ρ verses P revealed a large scatter (Fig. 14 of their paper), but there was a fairly clear lower bound on ρ well fit by the line

$$\rho = \frac{6.5^\circ}{P^{1/3}}. \tag{12}$$

The above procedure overestimates ρ if $\alpha < 90^\circ$, and so Lyne and Manchester attributed the scatter in their plot to α being significantly less than 90° in most pulsars, and interpreted (12) as correctly giving the beam opening angle for all α.

To determine the geometry of a given pulsar, Lyne and Manchester then basically inverted the above argument, using observations of the period P, maximum rate of change of polarization angle $d\psi/d\phi$ and the rotational beam width $\Delta\phi$ to solve (8), (11) and (12) for α and β, hence ι.

Later authors have disputed the fit of (12), and have cited evidence for ρ scaling as $P^{-1/2}$ (Rankin 1993), but modulo such updating the method is still mentioned in the literature and was used relatively recently by Manchester et al. (1998).

4.1.4 Special relativistic refinement of the RVM

The RVM incorporates neither general nor special relativistic effects. Blaskiewicz et al. (1991) made an effort to correct for the latter. They argued that the main effect of special relativity was not to introduce aberration or contraction of the dipole, but to modify the motion of the plasma that produced the radio pulsar radiation. The upshot is they claim to have a refinement of the RVM, at the expense of having to include the pulse-producing plasma in the modelling.

4.2 Results in the literature

Over the years, many papers have appeared using one or other (or various combinations) of the above methods to investigate the geometry of pulsar emission. I've come across 30 or so such papers in my reading, and there are probably quite a few more. Some deal with just one or two pulsars, while some perform fits to many (e.g. Lyne and Manchester 1988; Blaskiewicz et al. 1991; Rankin 1993; von Hoensbroech and Xilouris 1997).

A few papers have compared the results of using different models to the same pulsars, e.g. Everett and Weisberg (2001). They found good agreement for some pulsars and poor agreement for others, making it difficult to draw clear conclusions as to which method is best. Having said that, the full χ^2 fit to the RVM is clearly safest, as by looking at how good the fit is you can judge how reliable the extracted angles are likely to be.

Unfortunately, most pulsars studied in the literature were slow rotators with $P \sim 1$ second, and of no interest to gravitational wave astronomers. It seems that the polarization angles of MSPs don't fit the RVM as well as those of slower rotators. Nevertheless, some authors have managed to produce fits for rapidly rotating stars, and so it certainly is possible, see e.g. Stairs et al. (1999), Lommen et al. (2000). The quoted random errors bars range from a few degrees to a few tens of degrees. I haven't attempted to trawl though a huge number of papers, looking for fits to all pulsars of gravitational wave interest, e.g. those lying above the 1-year spin-down curve, but it seems fits don't exist for most such pulsars.

4.3 Key points

Here are the key points as I see them:

- There is a small industry in the radio pulsar community attempting to model pulsar emission geometry. They are forced to assume a dipolar magnetic geometry, and, sometimes, further perhaps more questionable assumptions.
- Their results are probably too insecure to provide an *input* for a gravitational wave search. Even when the quoted *random* errors are small, there may be significant *systematic* errors, as highlighted in the comparisons carried out by Everett and Weisberg. Perhaps if a very high quality χ^2 fit were obtained one might be happy to use this, but I'm not aware of any for rapidly spinning stars.
- Because gravitational wave astronomers are confident that they understand gravitational wave generation, their extraction of ι is far less model dependent. So, it looks like gravitational wave results will be of use to radio astronomers, rather than vice versa. That's to say, a gravitationally-derived ι could be passed on to the radio astronomers. This would have two physical payoffs:

  1. If sufficiently good radio polarization data could be obtained, the RVM could be tested, telling us whether the magnetic field really is dipolar and of the polarization does indeed follow the field lines.
  2. With further assumptions (basically that the edges of the pulsar beam are emitted along the last open field lines) the altitude of radio emission could be calculated (see e.g. Kijak and Gill 1997).

  These things are of great interest to radio astronomers, and it is worth advertising the fact that a gravitational wave detection can help address them.

5 Supernova remnants

We will now extend our discussion beyond particular electromagnetic point sources to larger sky patches that seem to be good bets for containing unknown spinning neutron stars. The most obvious such locations are supernova remnants (SNRs) where a compact central object has not yet been identified. An important question from the gravitational wave point of view is to decide how large a solid angle to search over, as different directions require different templates in the matched filtering, caused by the directional dependence of the Doppler demodulation.

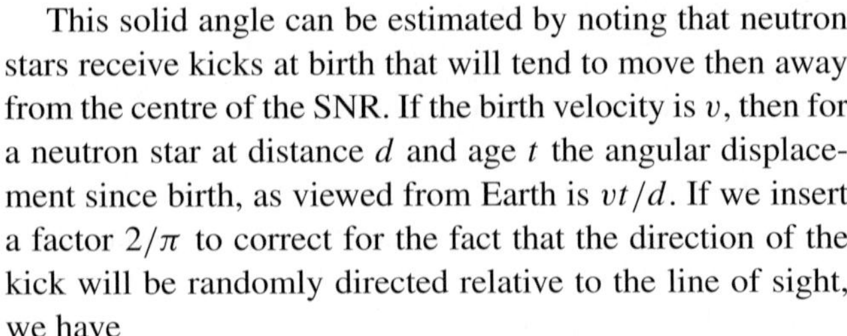
This solid angle can be estimated by noting that neutron stars receive kicks at birth that will tend to move then away from the centre of the SNR. If the birth velocity is v, then for a neutron star at distance d and age t the angular displacement since birth, as viewed from Earth is vt/d. If we insert a factor $2/\pi$ to correct for the fact that the direction of the kick will be randomly directed relative to the line of sight, we have

$$\theta = \frac{2}{\pi}\frac{vT}{d}. \tag{13}$$

Lyne and Lorimer (1994) calculate an average kick velocity of 450 km/s. Parameterizing in terms of this value:

$$\theta = 2.9 \times 10^{-3}\,\mathrm{rad}\left(\frac{v}{450\ \mathrm{km/s}}\right)\left(\frac{t}{10^4\ \mathrm{yrs}}\right)\left(\frac{\mathrm{kpc}}{d}\right). \tag{14}$$

It can be shown that this is much larger than the solid angle that can be Doppler demodulated with a single template for search durations of interest. This immediately suggests another way in which astronomical data can aid a gravitational wave search—maximally accurate age and distance estimates for SNRs can help gravitational wave observers decide how large a sky location to investigate.

Also, the fact that many sky patches must be search over means that even these relatively small-area searches (compared to a blind all-sky search) will be computationally costly, which begs the question: which supernova remnants are the best bet for gravitational wave searches? To gain insight into this, we can use a logic similar to that for pulsars, this time assuming a star born in SNR a time t ago, spinning at a rate $f_{\mathrm{GW},0}$ has spun-down purely by gravitational radiation emission. It is straight-forward to show that its frequency is now:

$$f_{\mathrm{GW}} = \frac{f_{\mathrm{GW},0}}{[1+t/\tau]^{1/4}} \tag{15}$$

where

$$\tau = \frac{5c^5}{8G\pi^4}\frac{1}{I\epsilon^2 f_{\mathrm{GW},0}^4}. \tag{16}$$

I is the moment of inertia and ϵ the (dimensionless) ellipticity. The corresponding gravitational wave amplitude, using the all-sky average of (55) of Kip's '300 Years of Gravitation' article in Misner et al. (1973), is

$$h = \frac{8G\pi^4}{c^4}\left(\frac{2}{15}\right)^{1/2}\frac{\epsilon I}{d}\frac{f_{\mathrm{GW},0}}{[1+t/\tau]^{1/2}}. \tag{17}$$

In the case where the star has spin-down a great deal since birth so that $f_{\mathrm{GW}} \ll f_{\mathrm{GW},0}$ these equations simplify to:

$$f_{\mathrm{GW}} \approx \frac{1}{\pi}\left[\frac{5c^5}{8G\pi^4}\frac{1}{I\epsilon^2 t}\right]^{1/4}, \tag{18}$$

Table 1 Galactic SNRs with known ages and distances

Id number	Common name	t/years	d/kpc
G4.5+6.8	(Kepler's)	402	2.9
G111.7-2.1	(Cas A)	333	3.4
G120.1+1.4	(Tycho)	434	2.4
G315.4-2.3	(185AD?)	1821	2.5
G327.6+14.6	(1006AD)	1001	1.8

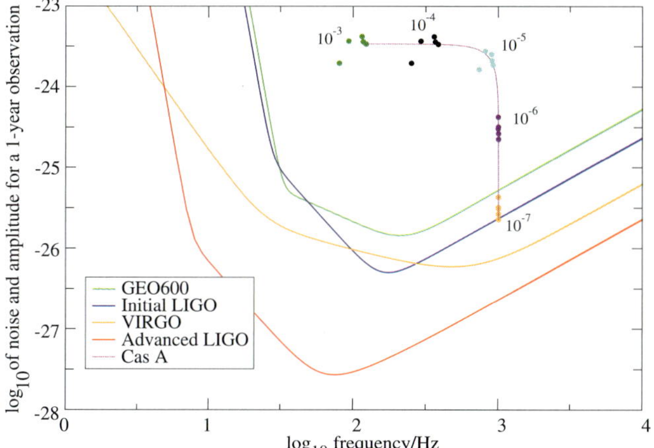

Fig. 4 Galactic SNRs, initial frequency 1 kHz

$$h \approx 4\left(\frac{G}{3c^3}\right)^{1/2}\frac{I^{1/2}}{dt^{1/2}}. \tag{19}$$

Note that some rather surprising cancellations have occurred: in this approximation h is independent of $f_{GW,0}$ and ϵ!

To make use of these formulae we need estimates of a number of SNR parameters. To this end we can make use of the on-line catalogue of SNRs by David Green: http://www.mrao.cam.ac.uk/surveys/snrs/. Distance estimates (sometimes only lower bounds) are given for a fairly large number of remnants, about 20% of the total population. However, ages are only given for the six or so historical supernovae. When the Crab is subtracted from the list of historical SNRs we are left with those listed in Table 1.

It would be useful to try and decide which remnant might be the best bet for gravitational wave observation. There's no completely obvious way of doing this, but a natural strategy would be to assume that all neutron stars are born with the same gravitational wave frequency and ellipticity and use (15–17) to calculate the current frequency and amplitude. Figure 4 shows the results for $f_{GW,0} = 1$ kHz and a variety of different ϵ values, as indicated. Each clump of 5 dots denotes the gravitational wave amplitude, as would be measured today, of the 5 remnants of Table 1 for a particular assumed value of ϵ. The continuous line is the set of amplitudes traced out by Cas A as its ellipticity is varied from

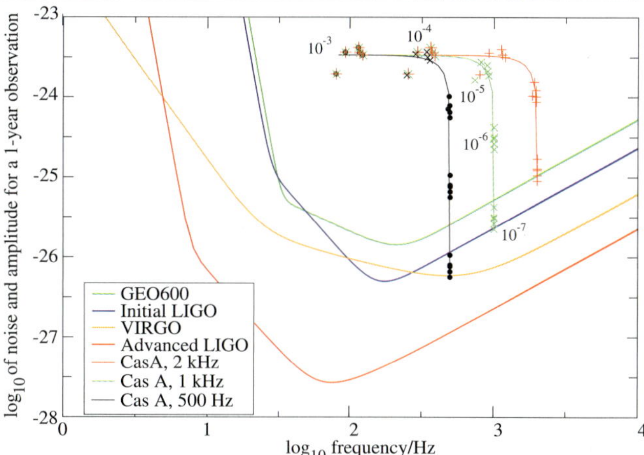

Fig. 5 Galactic SNRs, initial frequency 500 Hz, 1 kHz and 2 kHz

10^{-7} up to a value of 10^{-3}. This curve is included to guide the eye; it is *not* a trajectory in time! Note that the theoretical maximum for ϵ is somewhere around 10^{-8}, or 10^{-7} at a push, so all the points in the diagram would require something special to be going on. Figure 5 contains the same information as Fig. 4, with data for $f_{GW,0} = 500$ Hz and $f_{GW,0} = 2$ Hz superimposed.

For sufficiently small ellipticities the gravitational wave spin-down is negligible, and the wave amplitude is given to a good approximation by the $t = 0$ value of (17). The amplitudes then stand in the inverse ratio of the SNRs' distances. As ϵ is increased the wave amplitude at first increase in proportion, e.g. the $\epsilon = 10^{-6}$ amplitudes are 10 times those for $\epsilon = 10^{-7}$. The gravitational wave frequency hardly changes at all. As ϵ is increased further to values of around 10^{-5} spin-down gradually becomes important and the gravitational wave amplitude no longer increases in proportion to ϵ. For values of ϵ in excess of 10^{-5} the wave amplitude flattens out—increasing ϵ further only serves to decrease the gravitational wave frequency. Now the approximate (18) and (19) can be applied. The flattening-out of h is caused by two competing effects: for a given f_{GW}, h increases with ϵ, while, for a given age and $f_{GW,0}$, f_{GW} decreases with increasing ϵ; the two effects leave h unchanged.

Looking at Fig. 5, the effect of increasing the assumed initial spin frequency is significant for low ellipticity stars, but unimportant high ellipticity ones, as expected from the equations given above. Clearly, these astrophysically derived plots are of great use in deciding which systems to target, and accurate estimates of ages and distances for further systems would aid gravitational wave searches further.

6 Other targets: the compact object zoo

We will end by mentioning briefly one other area where astrophysical expertise could help inform gravitational wave search strategies.

Radio pulsars aren't the only point sources on the sky that might be worth targeting in gravitational wave searches. In recent years the zoo of possible compact objects has expanded greatly, with a sometimes bewildering array of new classes of object being invented. The author made a list of the classes of object discussed at the London meeting alone: isolated neutron stars, classical pulsars, rotating radio transients, X-ray dim neutron stars, compact central objects, anomalous X-ray pulsars, magnetars, and soft gamma-ray repeaters.

The question is, which of these stars are of gravitational wave interest? Central compact objects seem to be good bets, as they are probably hot young stars, which may still be rotating fast enough to radiate at accessible frequencies. However, there seems to be evidence of varying degrees of strength that the remaining classes of object are too slowly rotating to be detectable in the gravitational wave channel (the lowest LIGO/VIRGO frequencies of interest are of order a few tens of Hz). The question is, how certain can we be? For instance, all the isolated neutron stars that have had their spin frequencies measured are slow rotators, but how sure can we then be that all isolated neutron stars are slow rotors? If astronomers are able to supply convincing evolutionary arguments that all the isolated neutron stars (or any of the other classes of star above) are probably slow, then gravitational wave astronomers might decide to limit the time and resources invested in targeting such objects.

7 Summary

In this article we have described a number of areas where increased astrophysical input could influence gravitational wave searches. In the case of the radio pulsars, this input lies in selecting the most promising candidates and using radio data to help constrain the target's geometrical parameters, decreasing the parameter space to be searched over. In the case of supernova remnants this lay in using simple energetic arguments to try and choose the best remnants to target for as yet undiscovered stars.

There are many other areas where astrophysical modelling is important, e.g. deciding which areas of the Galaxy to target (Galactic centre verses spiral arms verses the Gould belt ...), understanding how timing noise might affect the gravitational wave stream, quantifying the possibility that there exist an electromagnetically invisible population of spinning down stars detectable in the gravitational wave channel only .... The list could go on. The key point of this contribution is promote debate on such issues, and to try and strengthen the links between traditional electromagnetic astronomy and the newly emerging field of gravitational wave astronomy, to the benefit of both. Hopefully future neutron star conferences will focus more on these issues, particularly when the gravitational wave participants are able to provide observations of their own!

References

Abbot, B., et al.: Phys. Rev. Lett. **94**, 1103 (2004)
Blaskiewicz, M., Cordes, J.M., Wasserman, I.: Astrophys. J. **370**, 643 (1991)
Everett, J.E., Weisberg, J.M.: Astrophys. J. **553**, 341 (2001)
Jones, D.I.: Class. Quant. Gravity **19**, 1255 (2002a)
Jones, D.I.: Phys. Rev. D **70**, 2002 (2002b)
Jones, D.I., Andersson, N.: Mon. Not. Roy. Astron. Soc. **331**, 203 (2002)
Kijak, J., Gill, J.: Mon. Not. Roy. Astron. Soc. **288**, 631 (1997)
Lommen, A.N., et al.: Astrophys. J. **545**, 1007 (2000)
Lyne, A.G., Lorimer, D.R.: Nature **369**, 127 (1994)
Lyne, A.G., Manchester, R.N.: Mon. Not. Roy. Astron. Soc. **234**, 477 (1988)
Manchester, R.N., Han, J.L., Qiao, G.J.: Mon. Not. Roy. Astron. Soc. **295**, 280 (1998)
Misner, C.W., Thorne, K.S., Wheeler, J.A.: Gravitation. Freeman, San Francisco (1973)
Owen, B.J.: Phys. Rev. Lett. **95**, 211101 (2005)
Radhakrishnan, V., Cooke, D.J.: Astrophys. Lett. **3**, 225 (1969)
Rankin, J.M.: Astrophys. J. **352**, 247 (1990)
Rankin, J.M.: Astrophys. J. **405**, 285 (1993)
Stairs, I.H., Thorsett, S.E., Camilo, F.: Astrophys. J. Suppl. Ser. **123**, 627 (1999)
Ushomirsky, G., Cutler, C., Bildsten, L.: Mon. Not. Roy. Astron. Soc. **319**, 902 (2000)
von Hoensbroech, A., Xilouris, K.M.: Astron. Astrophys. **324**, 981 (1997)

Astrophys Space Sci (2007) 308: 133–136
DOI 10.1007/s10509-007-9376-0

ORIGINAL ARTICLE

Dim isolated neutron stars, cooling and energy dissipation

M. Ali Alpar

Received: 11 August 2006 / Accepted: 15 November 2006 / Published online: 21 March 2007

Abstract The cooling and reheating histories of dim isolated neutron stars (DINs) are discussed. Energy dissipation due to dipole spindown with ordinary and magnetar fields, and due to torques from a fallback disk are considered as alternative sources of reheating which would set the temperature of the neutron star after the initial cooling era. Cooling or thermal ages are related to the numbers and formation rates of the DINs and therefore to their relations with other isolated neutron star populations. The possibility of energy dissipation at ages greater than about 10^6 yrs is a potentially important factor in determining the properties of the DIN population. Interaction with a fallback disk, higher multipole fields and activity of the neutron star are briefly discussed.

Keywords Isolated neutron stars · Cooling · Energy dissipation

PACS 97.60.Jd

1 Introduction

All "magnificent" seven ROSAT detected sources display X-ray fluxes consistent with thermal luminosities of the order of 10^{31}–10^{32} $\mathrm{ergs\,s^{-1}}$. In particular, RX J1856.5-3754, which is the only source in this class with a parallax distance determination and kinematic age, has luminosity 9.8×10^{31} $\mathrm{ergs\,s^{-1}}$ at distance 161 pc, according to the most recent parallax measurement quoted at this conference (Van Kerkwijk and Kaplan 2007). For RXJ1856.5-3754 and RXJ 0720.4-3125 spectra can be fit with two blackbodies to cover X-ray and optical data. This is probably an indicative representation of the actual temperature modulation due to the magnetic fields. Moreover, the soft to hard luminosity ratio is the same, 0.5, in both sources, possibly indicating similar physics on the surface. The soft blackbody seems to cover the entire neutron star surface. This suggests that the entire surface luminosity is of order 10^{31}–10^{32} $\mathrm{ergs\,s^{-1}}$ also in the latter two sources. Are luminosities of 10^{31}–10^{32} $\mathrm{ergs\,s^{-1}}$ standard for these sources?

Cooling provides the source of thermal X-ray luminosity for a young neutron star. Cooling proceeds first via neutrino emission from the interior of the neutron star, to give way to cooling by surface photon emission at an age of the order of 10^6 yrs. According to families of standard cooling curves (e.g. Tsuruta et al. 2002), luminosities in the range of 10^{31}–10^{32} $\mathrm{ergs\,s^{-1}}$ correspond to the end of the neutrino cooling era and the transition to photon cooling.

Why do we not detect a few sources at age 10^5 yr and luminosity 10^{33} $\mathrm{ergs\,s^{-1}}$? The space density of 10^5 yr old sources would be only 10 times less than the density of the 10^6 yr old sources, if the age distribution of neutron stars in the solar neighbourhood were uniform. With luminosities one order of magnitude larger than it is for the 10^6 yr old sources, according to the standard cooling curves, the detectable volume for the 10^5 yr old sources may not be significantly larger. The detectable neighbourhood is further limited by absorption. Nevertheless, within the present uncertainties in distance estimates some of the sources may have 10^{33} $\mathrm{ergs\,s^{-1}}$ luminosities. If there is a real deficiency of 10^5 yr old neutron stars with 10^{33} $\mathrm{ergs\,s^{-1}}$ luminosities, this may simply be due to exceptions to the standard cooling histories, as exemplified by the Vela pulsar. Thus some

M.A. Alpar (✉)
Sabancı University, Orhanlı, Tuzla, 34956 Istanbul, Turkey
e-mail: alpar@sabanciuniv.edu

of the local neutron stars at age 10^5 yr may be dimmer than expected by standard cooling curves.

We might expect an even larger number of older DINs, say at age 3×10^6 yrs. While the cooling luminosity drops sharply at ages of a few 10^6 yrs, such nearby sources would still be above XMM and Chandra detection limits, but of course almost impossible to identify unless the source is first discovered by other instruments. The lower thermal luminosities and effective temperatures of older sources means the spectra are shifted to lower photon energies and therefore detection may be more difficult because of absorption. With only 7 sources, distance and luminosity uncertainties and lack of reliable age estimates, it is at present too early to test whether the population of DINs is consistent with being powered by cooling luminosities alone. In particular it will be interesting to see if more DINs and age determinations yield sources at ages longer than 10^6 yrs without the drop-off in luminosity expected in the photon-cooling era. With typical distances of the order of 100 pc and if their galactic birthrate is of the order of the total neutron star birthrate, 10^{-2} yr^{-1}, the DINs may well be a population dominated by sources older than 10^6 yrs, whose luminosity is supplied by something else after the drop-off in the cooling luminosity. Discussion of their cooling histories and population syntheses (Popov et al. 2006, Popov et al. 2006) need to be extended to take into account the different possibilities of reheating for sources older than 10^6 yrs.

2 Energy dissipation

What else can happen after the initial 10^6 yrs of cooling? A star with viscous coupling between different internal components has to dissipate energy when it is spun up or down by an external torque. Reheating of neutron stars by energy dissipation was first considered by Alpar et al. (1984, 1986) in connection with the dynamical coupling between the pinned inner crust superfluid and the crust of the neutron star. The steady state thermal luminosity supplied by energy dissipation is of the form:

$$L_{\rm diss} = J|\dot{\Omega}| \qquad (1)$$

where J is a parameter describing the effective viscous coupling and $\dot{\Omega}$ is the spindown rate due to the external torque on the neutron star. In the case of the pinned inner crust superfluid-vortex creep model, $J \sim 10^{43}$ erg s.

Whether a detectable thermal luminosity can be provided by energy dissipation after the cooling luminosity drops off depends on the external torque on the star. The external torque due to a rotating magnetic dipole gives the spindown rate

$$I\dot{\Omega} = -2/3\mu^2\Omega^3/c^3 \qquad (2)$$

where I is the star's moment of inertia, μ is the dipole magnetic moment perpendicular to the rotation axis and Ω is the rotation rate. This yields the time dependence:

$$|\dot{\Omega}| = 4 \times 10^{-13}\,{\rm s}^{-2} t_6^{-3/2} I_{45}^{1/2} \mu_{30}^{-1} \qquad (3)$$

where t_6 is the age in units of 10^6 yrs, I_{45} the star's moment of inertia in units of 10^{45} gm cm^2, and μ_{30} the dipole magnetic moment in units of 10^{30} G cm^3. The expected energy dissipation rate is

$$L_{\rm diss} = 4 \times 10^{30}\,{\rm erg\,s}^{-1} t_6^{-3/2} I_{45}^{1/2} \mu_{30}^{-1}. \qquad (4)$$

A nearby ($d \sim$ a few 100 pc) isolated neutron star with an "ordinary" magnetic moment $\mu_{30} \sim 1$ might be detectable up to ages of about 10^7 yrs. For a constant dipole moment in the magnetar range, $\mu_{30} \sim 100$–1000, the dissipation luminosity is very low, and chances of detection after the 10^6 yrs of initial cooling are negligible. Field decay makes the prospects even worse.

An alternative source of external torque on an isolated neutron star is the torque from a fallback disk. Fallback disks were proposed for anomalous X-ray pulsars by Chatterjee et al. (2000). It was proposed by Alpar (2001) that the presence or absence, and properties, of a fallback disk is the initial condition, along with initial dipole moment and rotation rate, that determines the subsequent evolution of all classes of young neutron stars. In particular Alpar (2001) suggested that the DINs, being likely to be of ages of the order of 10^6 yrs or more, may be powered by energy dissipation due to the external torque supplied by a fallback disk (for those DINs whose cooling luminosity has already dropped off).

Torques from a fallback disk near rotational equilibrium with the neutron star can be estimated as

$$\begin{aligned} I|\dot{\Omega}| &\sim (\mu^2/r_A{}^3)(\Omega - \Omega_{\rm eq})/\Omega_{\rm eq} \\ &\sim (\mu^2/r_{\rm co}^3)(\Omega - \Omega_{\rm eq})/\Omega_{\rm eq} \\ &\sim (\mu^2\Omega^2/GM)(\Omega - \Omega_{\rm eq})/\Omega_{\rm eq} \end{aligned} \qquad (5)$$

where r_A and $r_{\rm co}$ are the Alfven and corotation radii in the disk, respectively and $\Omega_{\rm eq}$ is the equilibrium rotation period for the neutron star in interaction with the disk. This gives the energy dissipation rate estimate

$$\begin{aligned} L_{\rm diss} &\sim (J/I)(\mu^2\Omega^2/GM)(\Omega - \Omega_{\rm eq})/\Omega_{\rm eq} \\ &\sim 10^{31}\mu_{30}^2\Omega^2(M/M_\odot)^{-1}\,{\rm erg\,s}^{-1}, \end{aligned} \qquad (6)$$

taking $(\Omega - \Omega_{\rm eq})/\Omega_{\rm eq} \sim 0.1$ (comparable to the spread of DIN periods). Disks surviving beyond 10^6 yrs would keep most DINs (the oldest and most abundant, at ages beyond 10^6 yrs) at luminosities of 10^{31}–10^{32} ergs s^{-1} throughout the lifetime of the disk, which may be longer than 10^6 yrs.

Note that the energy dissipation rate with a fallback disk given in (6) is larger for large dipole magnetic moments, in contrast to the case of isolated dipole spindown, for which the energy dissipation rate is smaller for magnetars, as (4) shows.

3 Discussion

At typical separations of say 150 pc, the DINs would make up a galactic plane population of about 10 000. If the duration of the neutrino cooling epoch with $L_{\rm cool}$ larger than 10^{32} ergs s^{-1} is taken to be their representative age, this age is of the order of 10^6 yrs. The rate of formation is therefore of the order of 10^{-2} yr^{-1}. The DINs make up a substantial fraction of the supernova rate: they are a very abundant population of young neutron stars, comparable to isolated radio pulsars.

The main point of this note is that the turn-off age of standard (and non-standard) cooling, a few 10^5 yrs to 10^6 yrs, coincides with the typical age estimates of the DIN population. The X-ray luminosities of DINs are determined by their cooling histories at ages earlier than the cooling turn-off, which corresponds to the switch from neutrino cooling to surface photon cooling. Many of the DINs are likely to be at ages comparable to the cooling turn-off age. If reheating of the neutron star due to dynamical energy dissipation is large enough to sustain the X-ray luminosity beyond this critical age, the later luminosity evolution, and the population properties, will be effected by energy dissipation.

The rate and time evolution of energy dissipation depends on the external torque on the neutron star, and hence on the dipole magnetic moment, whether the star is a magnetar or a more ordinary neutron star, and also on whether a fallback disk is present. In principle these different options will effect the population properties and should be included in population syntheses. In practice, the present number of DINs is too small to allow any firm conclusions. Future identifications of ROSAT blank fields with more sources will improve the prospects (Treves et al. 2007 and references therein).

A fallback disk, if it remained active at 10^6 yrs, would provide a strong and steady energy dissipation rate. This model would lead to a DIN population dominated by old and X-ray luminous sources. We do not know, observationally, if such long lived disks are common. In AXPs the first fallback disk was recently discovered around the source 4U 0142+61 by Wang et al. (2006), and claimed to be a gas disk by Ertan et al. (2007a, 2007bb); see also Ertan and Çalışkan (2006).

For DINs, the dipole magnetic fields inferred from $\dot{P}$ and P, as well as surface fields inferred from absorption features for proton cyclotron lines are about 6×10^{13} G. This is of the order of or above the quantum critical field, and only an order of magnitude less than the inferred dipole magnetic fields of AXPs and SGRs. Periods are clustered in the same special narrow range as AXPs and SGRs. For 5 DINs observed periods range from 3.45 s to 11.37 s (Haberl 2005). This period clustering can be explained with fallback disks acting as a gyrostat around the neutron star, and determining its asymptotic equilibrium period (Alpar 2001). In this scheme DINs, AXPs and SGRs are supposed to be in a propeller interaction with the fallback disk. Most numerous, after the no-disk radio pulsars, must be the very small mass disk cases, where the disk would have a long lifetime. These are proposed to be the DINs. Near infrared H-band VLT archival observations for 5 DINs were analyzed by Mignani et al. (2007), who found that the present upper limits are not constraining regarding the presence of a fallback disk.

All these classes of objects are in the upper right hand part of the P–$\dot{P}$ diagram. All are close to but above the death line for radio pulsars. They share the P–$\dot{P}$ neighbourhood with the high dipole field radio pulsars. So why do the DINs (AXPs and SGRs also) not function as radio pulsars? This is more of a question for the DINs as their inferred dipole surface fields are of the same order as the highest dipole field radio pulsars: the AXPs and SGRs might short out the pair creation in the polar cap more easily by virtue of fields that are one order of magnitude larger than those of the radio pulsars.

Now there is a new enigma: RRATS (McLaughlin 2006), some of which are close to the same corner of the P–$\dot{P}$ diagram. RRATS also may have an intriguing connection with AXPs and SGRs suggested by, in addition to the neighbouring positions in the P–$\dot{P}$ diagram, also by the fact that they burst, though in the radio, and with very different signatures from the SGR and AXP bursts. What distinguishes between the different classes might be the presence or absence, and nature, of a fallback disk around the neutron star. The $\dot{M}$ history of the disk determines the equilibrium period to which the star approaches asymptotically. For $B = 10^{12}$–10^{13} G, a relatively wide range of $\dot{M}$ gives a narrow range of equilibrium periods: $P_{\rm eq}$ scales with $|\dot{M}|^{-4/7}$. An interesting example is that if the dipole magnetic field on the surface is the quantum critical field, and $\dot{M}$ is at the Eddington value, this leads to equilibrium periods of about 12 s.

This approach is not in conflict with magnetar models (Thompson and Duncan 1995; Woods and Thompson 2006). The magnetar model explains SGR and AXP bursts with processes involving very strong magnetic fields anchored in the neutron star crust. These magnetar fields store and release energy in local processes in the neutron star crust. The surface magnetar fields might be concentrated in the higher multipoles, while the dipole magnetic field on the surface of magnetars may be of order or less than the quantum critical field. One might speculate that when the quantum critical field is reached on the surface, leading to pair creation, near surface currents and surface heating, magnetic structure in the higher multipoles will be amplified, at the expense of the dipole (global) surface field, which might be limited to a value near the quantum critical field. As Zane and Turolla (2005) have shown, model fits to the light curves of DINs indicate the presence of multipole fields.

In the context of fallback disk models and hybrid models with magnetar fields in the higher multipoles, the fallback disk torques interacting with the dipole component of the magnetic field determine the spindown rate. In DINs, without accretion from the disk, these torques would lead to detectable thermal luminosities from the DINs if the disk lifetime extends beyond the 10^6 yr timescale of the cooling luminosity. As explored here, the energy dissipation rate for propeller disk torques near equilibrium is stronger than energy dissipation rates supplied by ordinary or magnetar dipole spindown, and may be roughly constant throughout the disk lifetime, unlike the dipole spindown case for which the energy dissipation rate decays rapidly with a power law time dependence. Furthermore the energy dissipation rate for fallback disk torques is larger for stronger dipole moments while the energy dissipation rate in the dipole spindown case is weaker for stronger dipole moments.

Acknowledgements I thank the Sabancı University Astrophysics and Space Forum and the Turkish Academy of Sciences for research support.

References

Alpar, M.A., Anderson, P.W., Pines, D., et al.: Astrophys. J. **276**, 325 (1984)
Alpar, M.A., Nandkumar, R., Pines, D.: Astrophys. J. **288**, 191 (1986)
Alpar, M.A.: Astrophys. J. **554**, 1245 (2001)
Chatterjee, B., Hernquist, L., Narayan, R.: Astrophys. J. **534**, 373 (2000)
Ertan, Ü., Çalışkan, Ş.: Astrophys. J. **649**, L87 (2006)
Ertan, Ü., Alpar, M.A., Erkut, M.H., Göğüş, E., et al.: Astrophys. Space Sci. **308**. DOI 10.1007/s10509-007-9328-8 (2007a)
Ertan, Ü., Erkut, M.H., Ekşi, Y.K., et al.: Astrophys. J. **657**, 441 (2007b), astro-ph/0606259
Haberl, F.: In: Proc. of the 2005 EPIC XMM-Newton Consortium Meeting, Schloss Ringberg, astro-ph/0510480 (2005)
McLaughlin, M.A., et al.: Nature **439**, 817 (2006)
Mignani, R.P., Bagnulo, S., De Luca, A., et al.: Astrophys. Space Sci. **308**. DOI 10.1007/s10509-007-9340-z, astro-ph/0608025 (2007)
Popov, S., Turolla, R., Possenti, A.: Mon. Not. Roy. Astron. Soc. **369**, L23 (2006)
Popov, S., Grigorian, H., Turolla, R., et al.: Astron. Astrophys. **448**, 327 (2006)
Thompson, C., Duncan, R.C.: Mon. Not. Roy. Astron. Soc. **275**, 255 (1995)
Treves, A., Campana, S., Chieregato, M., et al.: Astrphys. Space Sci. **308**. DOI 10.1007/s10509-007-9346-6, astro-ph 0609194 (2007)
Tsuruta, S., Teter, M.A., Takatsuka, T., et al.: Astrophys. J. **571**, L143 (2002)
Van Kerkwijk, M.H., Kaplan, D.L.: Astrophys. Space Sci. **308**. DOI 10.1007/s10509-007-9343-9, astro-ph/0607320 (2007)
Wang, Z., Chakrabarty, D., Kaplan, D.L.: Nature **440**, 772 (2006)
Woods, P.M., Thompson, C.: In: Lewin, W.H.G., van der Klis, M. (eds.) Compact Stellar X-ray Sources. Cambridge University Press, Cambridge, astro-ph/0406133 (2006)
Zane, S., Turolla, R.: Mon. Not. Roy. Astron. Soc. **366**, 727 (2005)

Astrophys Space Sci (2007) 308: 137–140
DOI 10.1007/s10509-007-9351-9

ORIGINAL ARTICLE

Accretion by isolated neutron stars

N.R. Ikhsanov

Received: 29 June 2006 / Accepted: 28 July 2006 / Published online: 20 March 2007

Abstract Accretion of interstellar material by an isolated neutron star is discussed. The point I address here is the interaction between the accretion flow and the stellar magnetosphere. I show that the interchange instabilities of the magnetospheric boundary under the conditions of interest are basically suppressed. The entry of the material into the magnetosphere is governed by diffusion. Due to this reason the persistent accretion luminosity of isolated neutron stars is limited to $<4\times10^{26}$ erg s^{-1}. These objects can also appear as X-ray bursters with the burst durations of $\sim$30 min and repetition time of $\sim10^5$ yr. This indicates that the number of the accreting isolated neutron stars which could be observed with recent and modern X-ray missions is a few orders of magnitude smaller than that previously estimated.

Keywords Accretion · Neutron stars · Magnetic field

PACS 97.10.Gz · 97.60.Jd · 98.38.Am

1 Introduction

As a neutron star moves through the interstellar medium it interacts in a time unit with the mass

$$\dot{\mathfrak{M}}_0 \simeq 10^9 \text{ g s}^{-1} n m^2 \left(\frac{V_{\rm rel}}{10^7 \text{ cm s}^{-1}}\right)^{-3}, \tag{1}$$

where m is the mass of the neutron star expressed in units of $1.4 M_{\odot}$, n is the number density of material situated beyond the accretion (Bondi) radius of the star expressed in units of 1 hydrogen atom cm^{-3} and $V_{\rm rel}$ is the relative velocity between the star and its environment, which is limited to the sound speed in the interstellar material as $V_{\rm rel} > V_{\rm s0}$. The mass capture rate by the star from its environment is therefore limited to $\dot{\mathfrak{M}}_{\rm c} \leq \dot{\mathfrak{M}}_0$.

A necessary condition for the captured material to reach the stellar surface is

$$r_{\rm m} < r_{\rm cor}, \tag{2}$$

where

$$r_{\rm m} = \left(\frac{\mu^2}{\dot{\mathfrak{M}}_{\rm c}\sqrt{2GM_{\rm ns}}}\right)^{2/7}, \tag{3}$$

is the magnetospheric radius of a neutron star, and

$$r_{\rm cor} = \left(\frac{GM_{\rm ns}P_{\rm s}^2}{4\pi^2}\right)^{1/3}, \tag{4}$$

is its corotational radius. Here μ, $M_{\rm ns}$ and $P_{\rm s}$ are the dipole magnetic moment, mass and spin period of the star, and G is the gravitational constant. Solving the inequality (2) for $P_{\rm s}$ one finds

$$P_{\rm s} > P_{\rm cd} \simeq 7000 \text{ s} \times \mu_{30}^{6/7} V_7^{9/7} n^{-3/7} m^{-11/7}, \tag{5}$$

where $\mu_{30} = \mu/10^{30}$ G cm^3. This implies that the spin-down rate of the neutron star in a previous epoch was

$$\dot{P} > 10^{-14}\left[\frac{P_{\rm s}}{7000 \text{ s}}\right]\left[\frac{t_{\rm sd}}{10^{10} \text{ yr}}\right]^{-1} \text{s s}^{-1}, \tag{6}$$

and therefore, suggests that only the stars whose initial dipole magnetic moment was in excess of 10^{29} G cm^3 could be a subject of further consideration (for a discussion see,

N.R. Ikhsanov (✉)
Institute of Astronomy, University of Cambridge,
Madingley Road, Cambridge, CB3 0HA, UK
e-mail: ikhsanov@ast.cam.ac.uk

e.g., Popov et al. 2000a). Here t_{sd} is the spin-down timescale of the neutron star.

Finally, a formation of an accretion disk around the magnetosphere of an isolated neutron star accreting material from the interstellar medium could be expected only if the relative velocity satisfies the condition $V_{rel} < V_0$, where

$$V_0 \simeq 10^5 \text{ cm s}^{-1} \mu_{30}^{-6/65} n^{3/65} m^{5/13} \times \left(\frac{V_t}{10^6 \text{ cm s}^{-1}}\right)^{21/65} \left(\frac{R_t}{10^{20} \text{ cm}}\right)^{-7/65}. \quad (7)$$

Here V_t is the velocity of turbulent motions of the interstellar material at a scale of R_t and the Kolmogorov spectrum of the turbulent motions is assumed (Prokhorov et al. 2002). This inequality, however, can unlikely be satisfied since V_0 is smaller than the speed of sound in the interstellar material, and therefore, is smaller than the lower limit to V_{rel}.

Thus, the accretion by old isolated neutron stars can be treated in terms of a spherical (Bondi) accretion onto a magnetized, slowly rotating neutron star. The accretion picture under these conditions has been reconstructed first by Arons and Lea (1976) and Elsner and Lamb (1976) and further developed by Lamb et al. (1977) and Elsner and Lamb (1984). An application of the results reported in these papers to the case of an isolated neutron star is discussed in the following sections.

2 Accretion flow at r_m

As shown by Arons and Lea (1976) and Elsner and Lamb (1976), the magnetosphere of a neutron star undergoing spherically symmetrical accretion is closed and, in the first approximation, prevents the accretion flow from reaching the stellar surface. The mass accretion rate onto the stellar surface is therefore limited to the rate of plasma entry into the magnetosphere. The fastest modes by which the material stored over the magnetospheric boundary can enter the stellar magnetic field are the Bohm diffusion and interchange instabilities (Elsner and Lamb 1984).

The rate of plasma diffusion in the considered case can be evaluated as (Ikhsanov 2003)

$$\dot{\mathfrak{M}}_B \le 2 \times 10^6 \text{ g s}^{-1} \zeta_{0.1}^{1/2} \mu_{30}^{-1/14} m^{15/7} n^{11/14} V_7^{33/14}, \quad (8)$$

where $\zeta_{0.1} = \zeta/0.1$ is the efficiency of the diffusion process normalized according to Gosling et al. (1991). This indicates that the luminosity of the diffusion-driven source is limited to

$$L_{x,dd} \le 4 \times 10^{26} \text{ erg s}^{-1} \times \zeta_{0.1}^{1/2} \mu_{30}^{-1/14} m^{22/7} n^{11/14} V_7^{33/14} r_6^{-1}. \quad (9)$$

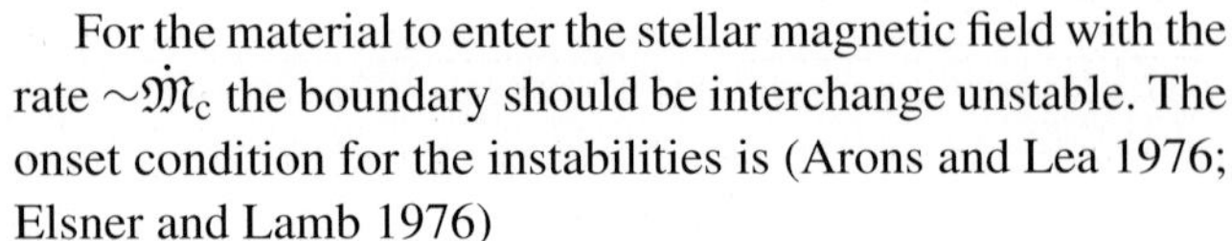

For the material to enter the stellar magnetic field with the rate $\sim\dot{\mathfrak{M}}_c$ the boundary should be interchange unstable. The onset condition for the instabilities is (Arons and Lea 1976; Elsner and Lamb 1976)

$$T_p(r_m) \le 0.1 T_{ff}(r_m), \quad (10)$$

where $T_p(r_m)$ and $T_{ff}(r_m)$ are the plasma temperature and the free-fall (adiabatic) temperature at the magnetospheric boundary, respectively. This indicates that a direct accretion of the captured material onto the stellar surface could occur only if the cooling of the plasma at the boundary dominates the heating.

The mechanism which is responsible for the cooling of plasma at the boundary is the bremsstrahlung emission. Indeed, the magnetospheric radius, free-fall temperature and number density of the material stored over the boundary are, respectively,

$$r_m \simeq 6 \times 10^{10} \text{ cm} \times \mu_{30}^{4/7} \dot{\mathfrak{M}}_9^{-2/7} m^{-1/7}, \quad (11)$$

$$T_{ff}(r_m) \simeq 10^7 \text{ K} \times \mu_{30}^{-4/7} \dot{\mathfrak{M}}_9^{2/7} m^{6/7}, \quad (12)$$

$$N_e(r_m) \simeq 300 \text{ cm}^{-3} \times \mu_{30}^{-6/7} \dot{\mathfrak{M}}_9^{10/7} m^{-2/7}. \quad (13)$$

Under these conditions both the cyclotron and Compton cooling time scale are significantly larger than the bremsstrahlung cooling time scale

$$t_{br}(r_m) \simeq 10^5 \text{ yr} \times T_7^{1/2} \left(\frac{N_e(r_m)}{300 \text{ cm}^{-3}}\right)^{-1}, \quad (14)$$

where $T_7 = T_{ff}(r_m)/10^7$ K.

The heating of the material at the magnetospheric boundary is governed by the following processes.

2.1 Adiabatic shock

As the captured material reaches the boundary it stops in an adiabatic shock. The temperature in the shock increases to $T_{ff}(r_m)$ on a dynamical time scale,

$$t_{ff}(r_m) \simeq 740 \text{ s} \times m^{-1/2} \left(\frac{r_m}{6 \times 10^{10} \text{ cm}}\right)^{3/2}. \quad (15)$$

Since $t_{ff}(r_m) \ll t_{br}(r_m)$ the height of the homogeneous atmosphere at the boundary proves to be $\sim r_m$. This prevents an accumulation of material over the boundary. Furthermore, as the condition $t_{ff}(r_G) < t_{br}(r_m)$ is satisfied throughout the gravitational radius of the neutron star a hot quasi-stationary envelope extended from r_m up to r_G forms (Davies and Pringle 1981). The formation of the envelope prevents the surrounding material from penetrating to within the gravitational radius of the neutron star. The mass of the envelope is, therefore, conserved. As the neutron star moves through the interstellar medium the surrounding material overflow the outer edge of the envelope with a rate $\dot{\mathfrak{M}}_{c}$.

Within an approximation of a non-rotating star whose "magnetic gate" at the boundary is closed completely the envelope remains in a stationary state on a time scale of $t_{br}(r_m)$. As the condition (10) is satisfied the boundary becomes unstable and material enters into the magnetic field and accretes onto the stellar surface with a rate of $\sim\dot{\mathfrak{M}}_{c}$. As shown by (Lamb et al. 1977), the time of the accretion event in this case is limited to $t_{burst} <$ a few $\times\, t_{ff}(r_m)$ during which the temperature of the envelope increases again to the adiabatic temperature (as the upper layers of the envelope comes to r_m). The corresponding source, therefore, would appear as an X-ray burster with the luminosity

$$L_{burst} \simeq 2 \times 10^{29} n V_7^{-3} m^3 r_6^{-1} \text{ erg s}^{-1}, \tag{16}$$

the typical outburst durations of $t_{burst} \le 30$ min and the repetition time of $t_{rep} \sim t_{br}(r_m) \sim 10^5$ yr.

2.2 Subsonic propeller

As shown by Davies and Pringle (1981), the rotation of a neutron star surrounded by the hot envelope can be neglected only if its spin period exceeds (Ikhsanov 2001)

$$P_{br} \simeq 10^5 \text{ s} \times \mu_{30}^{16/21} n^{-5/7} V_7^{15/7} m^{-34/21}. \tag{17}$$

Otherwise, the heating of plasma at the inner edge of the envelope due to propeller action by the star dominates cooling. The corresponding state of the neutron star is referred to as a subsonic propeller. The star remains in this state as long as its spin period satisfies the condition $P_{cd} < P_s < P_{br}$. The time during which the spin period increases from P_{cd} up to P_{br} is

$$\tau_{br} \simeq 2 \times 10^5 \text{ yr} \times \mu_{30}^{-2} I_{45} m \left(\frac{P_{br}}{10^5 \text{ yr}}\right), \tag{18}$$

where I_{45} is the moment of inertia of the neutron star expressed in units of 10^{45} g cm^2. This indicates that the spin periods of accreting isolated neutron stars are expected to be in excess of a day, and therefore, these objects can unlikely be recognized as pulsars.

2.3 Diffusion-driven accretor

As mentioned above, the "magnetic gate" at the magnetospheric boundary is not closed completely. The plasma flow through the interchange stable boundary is governed by the diffusion. As shown by Ikhsanov (2003), this leads to a drift of the envelope material towards the star and, as a result, to an additional energy source for heating of the envelope material. The heating due to the radial drift dominates the bremsstrahlung energy losses from the envelope if

$$\dot{\mathfrak{M}}_{c} \le 10^{14} \text{ g s}^{-1} \times \zeta_{0.1}^{7/17} \mu_{30}^{-1/17} V_7^{14/17} m^{16/17}. \tag{19}$$

This indicates that only the old isolated neutron stars which move slowly ($V_{rel} \ll 10^7$ cm s^{-1}) through a dense molecular cloud ($N_e > 10$ cm^{-3}) can be expected to be observed as the bursters. The rest of the population would appear as persistent X-ray sources with the luminosity of $L_x \le L_{x,dd}$ (see (9)).

3 Discussion

The results of this paper force us to reconsider previously made predictions about the number of old isolated neutron stars which could be observed with recent and current X-ray missions. In particular, the total flux of the persistent emission of these objects within the above presented accretion scenario is limited to $F < 10^{-16}$ erg cm^2 s^{-1} d_{100}^{-2}, where d_{100} is the distance to the source expressed in units of 100 pc. Furthermore, the mean energy of photons emitted by these objects within the blackbody approximation is close to 50 eV. This clearly shows that a detection of these sources by the Chandra and XMM-Newton is impossible.

The X-ray flux emitted by the accreting isolated neutron stars during the outbursts (see Sect. 2.1) is over the threshold of sensitivity of modern detectors. However, the probability to detect this event appears to be negligibly small. Indeed, the number of these sources which could be detected by Chandra and XMM-Newton is

$$N \le 10^{-5} \left(\frac{N(0)}{3 \times 10^4}\right) \left(\frac{t_{burst}}{30 \text{ min}}\right) \left(\frac{t_{rep}}{10^5 \text{ yr}}\right), \tag{20}$$

where $N(0)$ is the number of the sources which would be observed if the influence of the stellar magnetic field to the accretion flow at r_m were insignificant (Popov et al. 2000b).

Our results, therefore, naturally explain a lack of success in searching for the isolated neutron stars accreting material from the interstellar medium. They rather suggest that these objects can be considered as targets for coming space missions with more sensitive detectors in the soft X-ray part of the spectrum.

Acknowledgement I acknowledge the support of the European Commission under the Marie Curie Incoming Fellowship Program.

References

Arons, J., Lea, S.M.: Astrophys. J. **207**, 914 (1976)
Davies, R.E., Pringle, J.E.: Mon. Not. Roy. Astron. Soc. **196**, 209 (1981)
Elsner, R.F., Lamb, F.K.: Nature **262**, 356 (1976)
Elsner, R.F., Lamb, F.K.: Astrophys. J. **278**, 326 (1984)
Gosling, J.T., Thomsen, M.F., Bame, S.J., et al.: J. Geophys. Res. **96**, 14097 (1991)

Ikhsanov, N.R.: Astron. Astrophys. **368**, L5 (2001)
Ikhsanov, N.R.: Astron. Astrophys. **399**, 1147 (2003)
Lamb, F.K., Fabian, A.C., Pringle, J.E., Lamb, D.Q.: Astrophys. J. **217**, 197 (1977)
Popov, S.B., Colpi, M., Treves, A., et al.: Astrophys. J. **530**, 896 (2000a)
Popov, S.B., Colpi, M., Prokhorov, M.E., et al.: Astrophys. J. **544**, L53 (2000b)
Prokhorov, M.E., Popov, S.B., Khoperskov, A.V.: Astron. Astrophys. **381**, 1000 (2002)

Astrophys Space Sci (2007) 308: 141–149
DOI 10.1007/s10509-007-9349-3

ORIGINAL ARTICLE

Non-LTE modeling of supernova-fallback disks

Klaus Werner · Thorsten Nagel · Thomas Rauch

Received: 25 June 2006 / Accepted: 24 August 2006 / Published online: 22 March 2007

Abstract We present a first detailed spectrum synthesis calculation of a supernova-fallback disk composed of iron. We assume a geometrically thin disk with a radial structure described by the classical α-disk model. The disk is represented by concentric rings radiating as plane-parallel slabs. The vertical structure and emission spectrum of each ring is computed in a fully self-consistent manner by solving the structure equations simultaneously with the radiation transfer equations under non-LTE conditions. We describe the properties of a specific disk model and discuss various effects on the emergent UV/optical spectrum.

We find that strong iron-line blanketing causes broad absorption features over the whole spectral range. Limb darkening changes the spectral distribution up to a factor of four depending on the inclination angle. Consequently, such differences also occur between a blackbody spectrum and our model. The overall spectral shape is independent of the exact chemical composition as long as iron is the dominant species. A pure iron composition cannot be distinguished from silicon-burning ash. Non-LTE effects are small and restricted to few spectral features.

Keywords Radiative transfer; scattering · Neutron stars · Infall, accretion, and accretion disks

PACS 95.30.Jx · 97.60.Jd · 98.35.Mp

K. Werner (✉) · T. Nagel · T. Rauch
Institut für Astronomie und Astrophysik, Universität Tübingen, Tuebingen, Germany
e-mail: werner@astro.uni-tuebingen.de

1 Introduction

Anomalous X-ray pulsars (AXPs) are slowly rotating ($P_{\rm rot} =$ 5–12 s) young ($\leq$100 000 yr) isolated neutron stars. Their X-ray luminosities ($\approx 10^{36}$ erg/s) greatly exceed the rates of rotational energy loss ($\approx 10^{33}$ erg/s). It is now generally believed that AXPs are magnetars with magnetic field strengths greater than 10^{14} G and that their X-ray luminosity is powered by magnetic energy (Woods and Thompson 2006). As an alternative explanation the X-ray emission was attributed to accretion from a disk that is made up of supernova-fallback material (van Paradijs et al. 1995; Chatterjee et al. 2000; Alpar 2001). The fallback-disk model has difficulties to explain IR/optical emission properties of AXPs. When compared with disk models, the faint IR/optical flux suggests that any disk around AXPs must be very compact (e.g. Perna et al. 2000; Israel et al. 2004).

The discovery of optical pulsations in 4U 0142+61 which have the same period like the X-ray pulsations (Kern and Martin 2002) appears to be a strong argument against the disk model. It was argued that reprocessing of the pulsed NS X-ray emission in the disk cannot explain the high optical pulsed fraction, because disk radiation would be dominated by viscous dissipation and not by reprocessed NS irradiation (Kern and Martin 2002). Ertan and Cheng (2004), on the other hand, showed that these optical pulsations can be explained either by the magnetar outer gap model or by the disk-star dynamo model. Therefore, the observation of optical pulsations is not an argument against the disk model. A spectral break in the optical spectrum of 4U 0142+61 was discovered by Hulleman et al. (2004). This was also taken as an argument against the disk model because the authors do not expect such strong features from a thermally emitting disk. The recent discovery of mid-IR emission from this AXP (Wang et al. 2006), however, has strongly rekindled the interest in studies of fallback-disk emission properties. While this mid-IR emission was attributed to a cool, passive (X-ray irradiated) dust debris disk by Wang et al. (2006) it was shown by Ertan on this conference that it can be explained with a model for an active, dissipating gas disk. If true, then the disk emission properties allow to conclude on important quantities, e.g., the magnetic field strength of the neutron star can be derived from the inner disk radius.

Independent hints for the possible existence of fallback disks come from pulsars with particular spin-down properties. For example, the discrepancy between the characteristic age and the supernova age of the pulsar B1757-24 was explained by the combined action of magnetic dipole radiation and accretion torques (Marsden et al. 2001). Even more, the presence of jets from pulsars such like the Crab and Vela can possibly be explained by disk-wind outflow interacting with and collimating the pulsar wind (Blackman and Perna 2004).

A fallback-disk model was proposed in order to explain the X-ray enhancement following a giant flare of the Soft Gamma Repeater SGR 1900+14 (Ertan and Alpar 2003). The X-ray light curve is interpreted in terms of the relaxation of a fallback disk that has been pushed back by the gamma-ray flare. This model can also explain the long-term X-ray and IR enhancement light curves of the AXP 1E 2259+58 following a major bursting epoch (Ertan et al. 2006).

The presence of a fallback disk around the stellar remnant of SN 1987A has been invoked in order to explain its observed lightcurve which deviates from the theoretical one for pure radioactive decay (Meyer-Hofmeister 1992). From the non-detection of any UV/optical point source in the supernova remnant, however, tight constraints for the disk extension can be derived (Graves et al. 2005).

To our best knowledge, the emission from fallback disks in all studies was hitherto modeled with blackbody spectra. In view of the importance of disk models for the quantitative interpretation of observational data it is highly desirable to construct more realistic models by detailed radiation-transfer calculations.

2 Radial disk structure

For the modeling we employ our computer code AcDc (Nagel et al. 2004), that calculates disk spectra under the following assumptions. The radial disk structure is calculated assuming a stationary, Keplerian, geometrically thin α-disk (Shakura and Sunyaev 1973). As pointed out by Menou et al. (2001), for a comparison with observational data one probably has to use a more elaborate model, because near the outer disk edge the viscous dissipation and hence the surface mass density decline stronger with increasing radius than in an α-disk. However, the purpose of the present paper is to look for differential effects of various assumptions. Qualitatively, these effects can be expected to be independent of the detailed radial disk structure. In any case, it would be no problem to carry out the computations presented here with different radial structures.

The α-disk model is fixed by four global input parameters: Stellar mass $M_\star$ and radius $R_\star$ of the accretor, mass accretion rate $\dot{M}$, and the viscosity parameter α. For numerical treatment the disk is divided into a number of concentric rings. For each ring with radius R our code calculates the detailed vertical structure, assuming a plane-parallel radiating slab.

In contrast to a (planar) stellar atmosphere, which is characterized by $T_{\rm eff}$ and $\log g$, a particular disk ring with radius R is characterized by the following two parameters, which follow from the global disk parameters introduced above. The first parameter measures the dissipated and then radiated energy. It can be expressed in terms of an effective temperature $T_{\rm eff}$:

$$T^4_{\rm eff}(R) = [1 - (R_\star/R)^{1/2}]3GM_\star\dot{M}/8\sigma\pi R^3.$$

The second parameter is the half surface mass density Σ of the disk ring:

$$\Sigma(R) = [1 - (R_\star/R)^{1/2}]\dot{M}/3\pi\bar{w}.$$

σ and G are the Stefan-Boltzmann and gravitational constants, respectively. $\bar{w}$ is the depth mean of viscosity $w(z)$, where z is the height above the disk mid-plane. The viscosity is given by the standard α-parametrization as a function of the total (i.e. gas plus radiation) pressure, but numerous other modified versions are used in the literature. We use a formulation involving the Reynolds number Re, as proposed by Kriz and Hubeny (1986). We chose $Re = 15\,000$ which corresponds to $\alpha \approx 0.01$.

For the results presented here we selected the following input parameter values. The neutron star mass is 1.4 $M_\odot$. The radii of the inner and outer disk edges are 2000 and 200 000 km, respectively. The disk is represented by nine rings or, more precisely, by nine radial grid points. The radiation integrated over the whole disk is then computed by assigning a weight to each point's spectrum that resembles the area fraction that it represents. The main characteristics of the disk at the radial grid points are given in Table 1. The mass-accretion rate was set to $\dot{M} = 3 \times 10^{-9}$ $M_\odot$/yr. Figure 1 shows the radial run of $T_{\rm eff}$. We also display the Keplerian rotation velocity for the later discussion of our results. The radial distance from the neutron star is expressed in units of the NS radius which is set to $R_\star = 9.7$ km. But note from the above equations that the disk model is essentially independent of the stellar radius for large distances from the neutron star.

While $\Sigma(R)$ and $T_{\rm eff}(R)$ in columns 3 and 4 of Table 1 follow from the α-disk assumption, the quantities in the next

Table 1 Characteristics of the nine rings that compose the disk model. Surface mass density Σ and emergent flux (expressed as $T_{\rm eff}$) follow from the α disk prescription. The Rosseland optical depth $\tau_{\rm Ross}$ at the disk midplane, the mass density ρ and gravity $\log g$ at optical depth unity follow from our computations of the detailed vertical ring structure. The last column denotes the fraction of the disk area that is made up by each ring model in order to compute disk-integrated spectra

Ring	R [1000 km]	Σ [g/cm^2]	$T_{\rm eff}$ [1000 K]	$\log \tau_{\rm Ross}^{\rm midplane}$	$\log \rho(\tau_{\rm Ross}=1)$ [g/cm^3]	$\log g(\tau_{\rm Ross}=1)$ [cm/s^2]	% area fraction
1	2.0	2.9	305	3.1	−6.1	7.7	0.0025
2	2.5	2.8	258	3.2	−6.2	7.5	0.01
3	3.5	2.7	201	3.3	−6.4	7.2	0.034
4	6.0	2.6	135	3.5	−6.0	6.7	0.084
5	9.0	2.6	100	3.7	−5.9	6.4	0.19
6	14	2.5	72	3.3	−6.4	6.0	0.62
7	25	2.3	46	3.5	−6.9	5.5	1.70
8	40	2.2	33	3.9	−7.1	5.1	33
9	200	1.9	9.8	3.8	−7.6	3.6	64

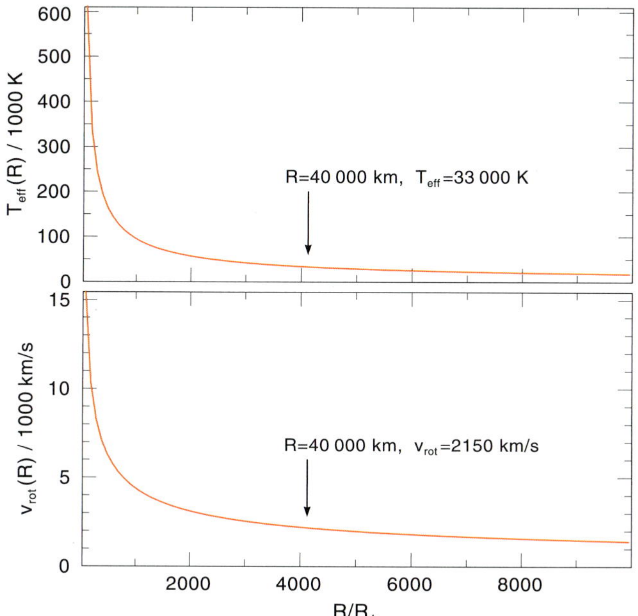

Fig. 1 Radial disk structure: Effective temperature (*top panel*) and Keplerian rotation velocity (*bottom panel*). *Arrows* mark a reference point at a distance of $R = 40\,000$ km (about 4000 stellar radii $R_\star$) from the neutron star as discussed in the text

three columns are the result from our detailed vertical structure calculations described below. It shows that the entire disk model is optically thick. The Rosseland optical depth at the disk midplane $\tau_{\rm Ross}^{\rm midplane}$ is >1000 at all radii. We also tabulate the mass density and the gravity at unity optical depth. That demonstrates that the conditions in the line forming regions of the disk resemble those in white dwarfs at the inner disk radii up to main sequence stars at the outer disk radii. The strength of Stark line broadening therefore strongly depends on the distance of the emitting region from the neutron star.

3 Vertical disk structure

The vertical structure of each disk ring is determined from the simultaneous solution of the radiation transfer equations plus the structure equations. The latter ones invoke radiative and hydrostatic equilibrium plus charge conservation. The structure equations also consist of the non-LTE rate equations for the atomic population densities. The solution of this set of highly non-linear integro-differential equations is performed using the Accelerated Lambda Iteration (ALI) technique (Werner and Husfeld 1985; Werner 1986; Werner et al. 2003).

The total observed disk spectrum, which depends on the inclination angle, is finally obtained by intensity integration over all rings accounting for rotational Doppler effects.

3.1 Radiation transfer, hydrostatic and radiative equilibrium

We consider the radiation transfer equation for the intensity I_ν at frequency ν:

$$\mu \frac{\partial I_\nu(\mu, z)}{\partial z} = -\kappa_\nu(z) I_\nu(\mu, z) + \eta_\nu(z)$$

with the opacity κ_ν and the emissivity η_ν. z measures the geometrical height above the disk midplane and μ is the cosine of the inclination angle i. The equation is solved using a short characteristics method. Opacities and emissivities are computed using atomic population densities that are obtained by solving the non-LTE rate equations. Our code allows for the irradiation of the disk by the central source, however, the results presented here are computed with zero incident intensity.

The radiation-transfer equations plus vertical structure equations are solved like in the stellar atmosphere case, but accounting for two basic differences. First, the gravity (entering the hydrostatic equation for the total, i.e. gas plus radiation, pressure) is not constant with depth, but increases with z. The gravity is the vertical component of the gravitational acceleration exerted by the central object (self-gravitation of the disk is negligible):

$$g = zGM_\star/R^3.$$

Second, the energy equation for radiative equilibrium balances the dissipated mechanical energy and the net radiative losses:

$$9/4\rho w G\Sigma/R^3 = 4\pi \int_0^\infty (\eta_\nu - \kappa_\nu J_\nu)\, d\nu,$$

where ρ and J_ν denote mass density and mean intensity, respectively. In the case of a stellar atmosphere the left-hand side of this equation vanishes and we get the usual radiative equilibrium equation. The solution is obtained by a generalized Unsöld-Lucy scheme and yields the vertical temperature structure.

Having calculated the vertical structures and spectra of the individual disk rings, the ring spectra are integrated to get the spectrum of the whole accretion disk:

$$I_\nu(i) = \cos(i) \int_{R_\mathrm{i}}^{R_\mathrm{o}} \int_0^{2\pi} I_\nu(i, \phi, r) r \, d\phi \, dr.$$

Here, R_i and R_o denote the inner and outer radius of the disk, and ϕ is the azimuthal angle. At this stage, the Keplerian rotation velocity v_rot is taken into account by assigning a Doppler shift of $\Delta\nu = \frac{\nu}{c}\mathrm{v}_\mathrm{rot} \sin\phi \sin i$ to the intensity emerging from a specific azimuthal ring sector.

3.2 Non-LTE rate equations

For each atomic level i the rate equation describes the equilibrium of rates into and rates out of this level and, thus, determines the occupation number n_i:

$$n_i \sum_{i \neq j} P_{ij} - \sum_{j \neq i} n_j P_{ji} = 0.$$

The rate coefficients P_{ij} have radiative and electron collisional components: $P_{ij} = R_{ij} + C_{ij}$. The radiative downward rate for example is given by:

$$R_{ji} = \left(\frac{n_i}{n_j}\right)^\star 4\pi \int_0^\infty \frac{\sigma_{ij}(\nu)}{h\nu} \left(\frac{2h\nu^3}{c^2} + J_\nu\right) e^{-h\nu/kT}\, d\nu.$$

$\sigma_{ij}(\nu)$ is the photon cross-section and $(n_i/n_j)^\star$ is the Boltzmann LTE population ratio.

The blanketing by millions of lines from iron arising from transitions between some 10^5 levels can only be attacked with the help of statistical methods (Anderson 1989; Dreizler and Werner 1993). At the outset, model atoms are constructed by combining many thousands of levels into a relatively small number of superlevels. The respective line transitions are grouped into superlines connecting these superlevels. In this case, the population densities of the superlevels are computed from the rate equations, in which the photon cross-sections $\sigma_{ij}(\nu)$ in the radiative rates R_{ij} do not contain only a single line profile but all individual lines that are combined into a superline. As an example we show such a cross-section in Fig. 2. The complete spectrum of our disk model ($\lambda = 4$–$300\,000$ Å) is sampled by 30 700 frequency points.

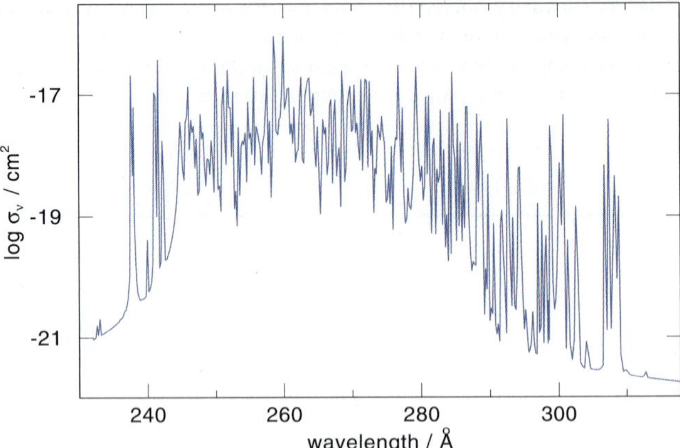

Fig. 2 Photon cross-section for a superline in the Fe IV ion (between superlevels number 1 and 7)

The model atoms that we have created for our disk calculations are summarized in Table 2. Most important is the iron model atom. It comprises the first eleven ionisation stages and a total number of more than 3 million lines. Atomic data are taken from Kurucz (1991) and the Opacity and Iron Projects (TIPTOPbase[1]).

3.3 Disk composition

The chemical composition of the supernova-fallback material in the disk is not exactly known. It depends on the amount of mass that goes into the disk. A disk with a small mass (say $\leq 0.001\ \mathrm{M}_\odot$) will be composed of silicon-burning ash (Menou et al. 2001). For simplicity, the results presented here are obtained by assuming a pure-iron composition. It turns out that the emergent spectrum is insensitive against the exact composition as long as iron is the dominant species (Sect. 4.1). For a respective test run for one specific ring we assumed a composition that represents a silicon-burning ash. It contains iron (80% mass fraction) as well as silicon and sulfur by 10% each.

[1] http://vizier.u-strasbg.fr/topbase/

Table 2 Summary of non-LTE model atoms for silicon, sulfur, and iron. The numbers in brackets at the iron ions give the number of individual lines summed up into superlines. Employed for a specific test run, the silicon and sulfur model atoms are tailored to the conditions encountered in disk ring number 8

Element	Ion	NLTE levels	Lines	
Si	III	6	4	
	IV	16	44	
	V	1	0	
S	III	1	0	
	IV	6	4	
	V	14	16	
	VI	1	0	
Fe	I	7	25	(141 821)
	II	7	25	(218 490)
	III	7	25	(301 981)
	IV	7	25	(1 027 793)
	V	7	25	(793 718)
	VI	8	33	(340 132)
	VII	7	22	(86 504)
	VIII	7	27	(8 724)
	IX	7	25	(36 843)
	X	7	28	(45 229)
	XI	1	0	

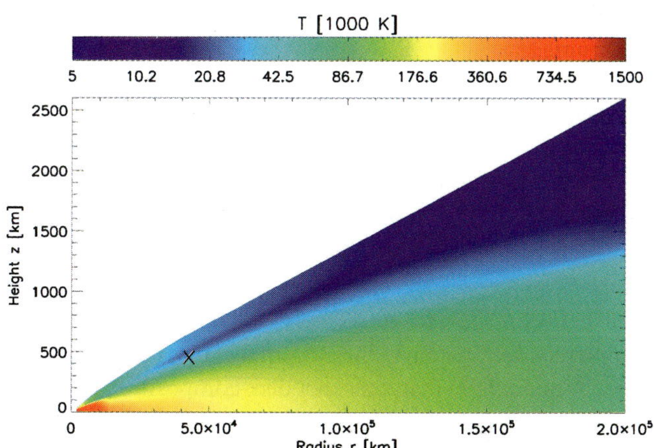

Fig. 3 This cut perpendicular to the midplane shows the temperature structure of the disk. Note that the *vertical scale* (height above the midplane) is expanded. The height-to-radius ratio is about 0.015. The *cross* marks the depth at $R = 40\,000$ km where $\tau_{\rm Ross} = 1$

3.4 Disk model properties

Figure 3 displays the temperature structure of the disk. The temperature varies between 1.5 million K at the midplane at the inner disk edge down to 6000 K in the upper layers at the outer disk edge. At all radii the vertical run of the temperature decreases almost monotonously with height above the midplane. A mild temperature reversal in the uppermost disk layers occurs. This turns out to be a non-LTE effect, because the respective LTE model has a strictly monotonous temperature run. We will discuss the consequences of this effect in Sect. 4.2.

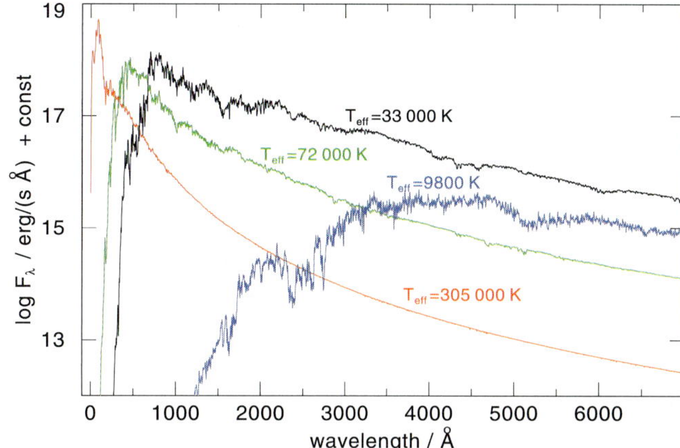

Fig. 4 Relative contribution of single disk rings to the total disk flux. For clarity, we only show the fluxes from rings number 1, 6, 8, and 9. The flux from ring 8 at $R = 40\,000$ km with $T_{\rm eff} = 33\,000$ K dominates the total disk spectrum at UV/optical wavelengths

Which disk regions contribute to the total disk spectrum and to what extent? In Fig. 4 we plot the emergent astrophysical flux from the area of four disk rings (rings 1, 6, 8, and 9, see Table 1), i.e., the computed flux per cm^2 is weighted with the ring area. The spectral flux distribution of the innermost ring with $T_{\rm eff} = 305\,000$ K has its peak value in the soft X-ray region. The contribution of this innermost region to the optical/UV spectrum is negligible. Cutting of the disk at this inner radius ($R = 2000$ km), therefore, is justified if this spectral range is of interest. The disk region that is dominating the UV/optical flux is represented by ring 8 with $T_{\rm eff} = 33\,000$ K. Its radius is 40 000 km, that is about 4000 neutron star radii. Its spectrum is dominated by strong blends of the numerous iron lines. Further out in the disk the effective temperature decreases and the flux contribution to the UV/optical spectrum decreases, too. Our outermost ring (number 9) has $T_{\rm eff} = 9800$ K, its flux maximum is at $\lambda = 4000$ Å and it is fainter than the inner neighbor ring 8 over the whole spectral range. Cutting off the disk at this outer radius ($R = 200\,000$ km) therefore does not affect the UV/optical spectral region.

4 Results

Because ring 8 is a representative disk region that determines the UV/optical spectrum, we will discuss its properties in more detail in the following three subsections, addressing the effects of chemical composition, non-LTE physics, and limb darkening on the spectrum. The radius of

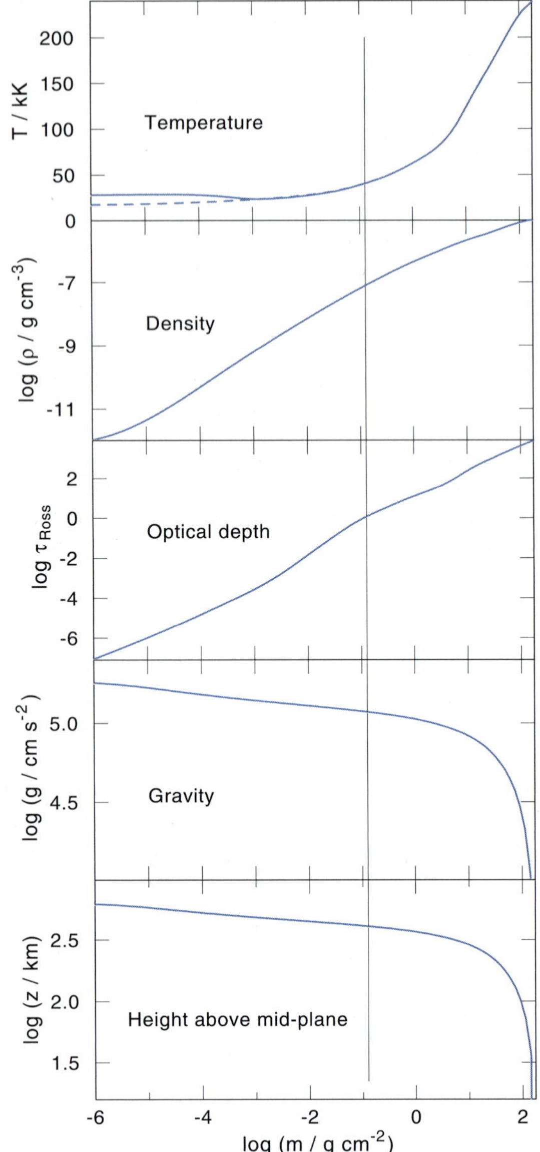

Fig. 5 Vertical stratification of the disk at $R = 40\,000$ km (ring 8, $T_{\rm eff} = 33\,000$ K). The quantities are plotted against the column mass density that is measured from the disk surface to the midplane (from *left* to *right*). The *vertical line* at $\log\, m = -0.9$ denotes the depth where $\tau_{\rm Ross} = 1$. The *dashed curve* in the *top panel* shows the temperature structure of a respective LTE model

this ring is marked by arrows in Fig. 1 and by a cross in Fig. 3.

In Fig. 5 we show the vertical structure of the disk at this radius ($R = 40\,000$ km). We plot the run of several quantities on a column-mass scale, measured inward from the surface toward the midplane of the disk. The vertical line at $\log m = -0.9$ marks the depth at which $\tau_{\rm Ross} = 1$, i.e., the region of spectral line formation. The temperature shows a strong increase towards the midplane and, as already mentioned, a mild temperature reversal in the optically thin surface layers. Together with the temperature run,

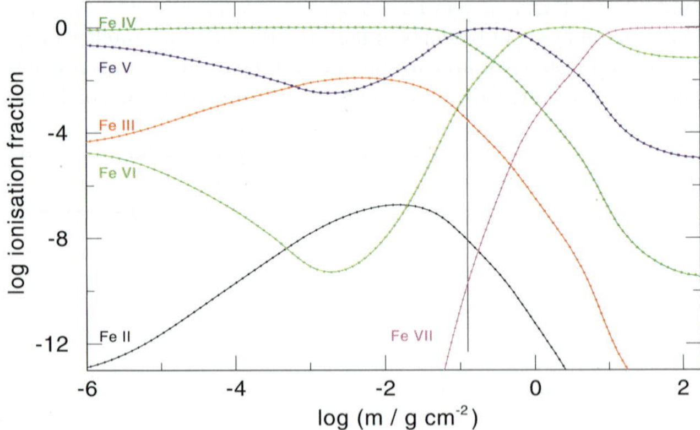

Fig. 6 Vertical iron ionisation structure at $R = 40\,000$ km (ring 8, $T_{\rm eff} = 33\,000$ K). The *vertical line* at $\log\, m = -0.9$ denotes the depth where $\tau_{\rm Ross} = 1$. The dominant iron opacities in the line forming regions are from Fe III–VI

the panels showing the mass density and gravity structure indicate that their values in the line forming region are comparable to those encountered in hot subdwarfs. The lowest panel shows that the disk height H at this distance from the NS is $\approx$600 km, i.e. $R/H = 0.015$.

In Fig. 6 we show the vertical ionisation structure of iron in the disk at $R = 40\,000$ km. The dominant ionisation stages in the line forming regions are Fe III–VI. At the midplane Fe VII is dominant. The temperature at any depth is so high that Fe I–II do not significantly contribute to the spectrum.

4.1 Effects of chemical composition

In Fig. 7 we show the flux spectrum of ring 8 in the wavelength range $\lambda = 6200$–6500 Å. It has been calculated for a pure iron composition as well as for a Fe/Si/S = 80/10/10 composition representing silicon-burning ash. The difference between the spectra is very small, because they are completely dominated by the extremely large number of iron lines. We conclude that the exact disk composition is not affecting the spectrum as long as iron is the dominant species.

While silicon and sulfur do not affect the overall spectrum by continuous background opacities, line features can be seen in the computed spectra, e.g. the Si IV resonance line in the UV. The line depth reaches about 50% of the continuum level but it would be difficult to detect even in medium resolution spectra when the disk inclination is high and the spectra are Doppler broadened by rotation.

4.2 Significance of non-LTE effects

For our particular disk model we expect that non-LTE effects are not very large. This is because of the relatively high gravities in the line forming regions, ranging between $\log g = 3.6$ in the outermost ring with $T_{\rm eff} = 9800$ K and

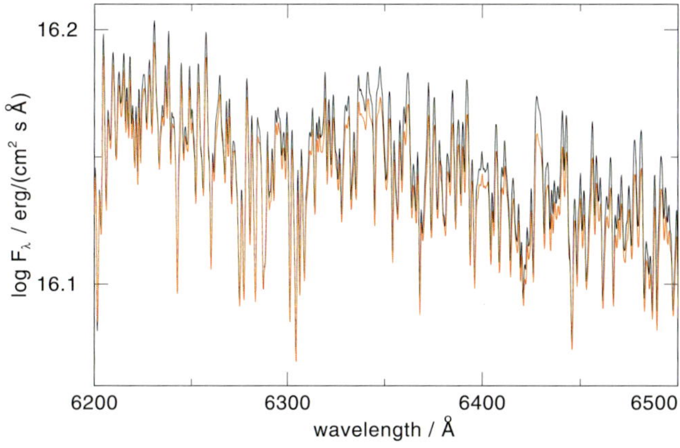

Fig. 7 Detail from the emergent disk spectrum at $R = 40\,000$ km (ring 8, $T_{\mathrm{eff}} = 33\,000$ K). We compare a pure iron composition (*black line*) with a silicon-burning ash composition (*red line*). The differences are marginal

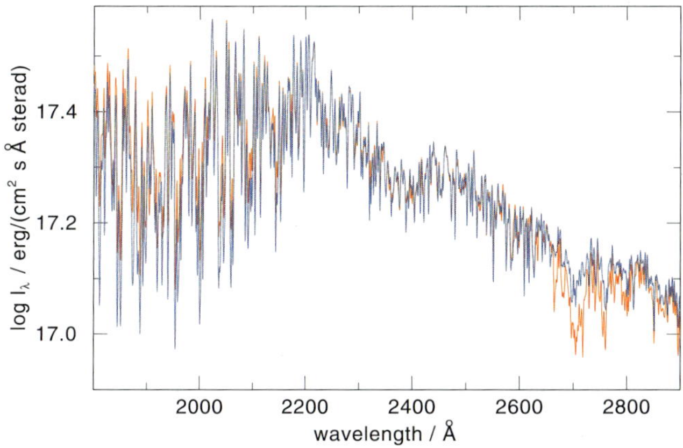

Fig. 8 Comparison of LTE and NLTE emergent disk intensities (*red and blue lines*, respectively) at $R = 40\,000$ km (ring 8, $i = 87°$, $T_{\mathrm{eff}} = 33\,000$ K)

$\log g = 7.7$ in the innermost region with 305 000 K. In Fig. 8 we compare the spectra of an LTE and a non-LTE model of disk ring 8 in the wavelength range $\lambda = 1800$–2900 Å, where the largest deviations were found. Indeed, non-LTE physics affects only narrow spectral regions. Only there, flux differences occur to an extent that the equivalent width of line blends changes by a factor of two. Accordingly, the temperature structures of both models deviate only in the uppermost layers of the disk (see top panel of Fig. 5) and, hence, only strong spectral lines that are still optically thick can be affected.

4.3 Limb-darkening effects

Our model spectra show distinct limb-darkening effects. The situation is similar to the stellar atmosphere case (center-to-limb variation of the specific intensity). Looking face-on we see into deeper and hotter (and thus "brighter") layers of the disk when compared to a more edge-on view. In Fig. 9 we compare the specific intensity emitted by ring 8 (per unit area) for a high and a low inclination angle. Overall, the "edge-on" spectrum is roughly a factor of two fainter than the "face-on" spectrum in the optical region. The difference increases towards the UV and amounts to a factor of about three. We conclude that limb darkening effects are important when disk dimensions are to be estimated from magnitude measurements.

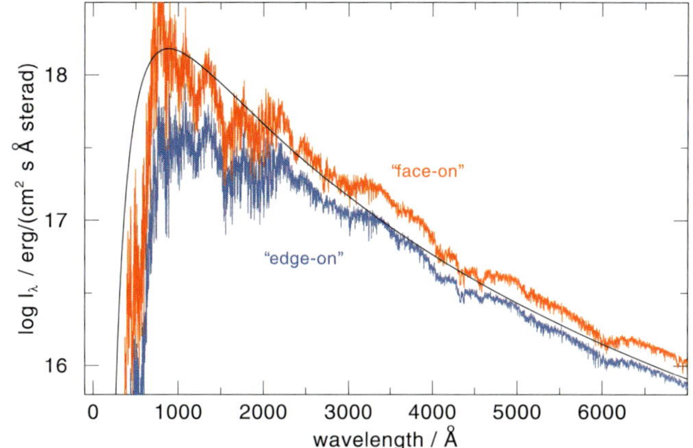

Fig. 9 Effect of limb darkening: Specific intensity of the disk at $R = 40\,000$ km (ring 8) seen under two inclination angles, namely 87° (i.e. almost edge-on, *blue*) and 18° (i.e. almost face-on, *red*). For comparison we also show a blackbody spectrum with $T = T_{\mathrm{eff}} = 33\,000$ K

It is also interesting to compare the intensities with a blackbody spectrum (Fig. 9). Depending on the wavelength band, the blackbody over- or underestimates the "real" spectrum up to a factor of two in the optical and a factor of four in the UV.

4.4 Rotational broadening

We have seen that the spectrum of an iron-dominated disk is characterized by strong blends of a large number of lines. At some wavelengths broad spectral features appear. It remains to be seen if Doppler effects from disk rotation smears out these features or if they could still be detectable. From the bottom panel of Fig. 1 we see that rotational broadening amounts to $\approx \sin i \cdot 2000$ km/s at $R = 40\,000$ km, corresponding to an orbital period of about two minutes. When seen almost edge-on, this rotational velocity is equivalent to a Doppler broadening of about $\Delta\lambda = 25$ Å at $\lambda 4000$ Å which clearly smears out any individual line profiles. In Fig. 10 we display the rotationally broadened spectrum of the entire disk model, seen under three different inclination angles. It is obvious that the broad line blends are so prominent that they do not disappear even for an almost edge-on view of the disk.

Among the strongest features is a 200 Å wide line blend at $\lambda 1500$ Å with an absorption depth of about 50% relative to

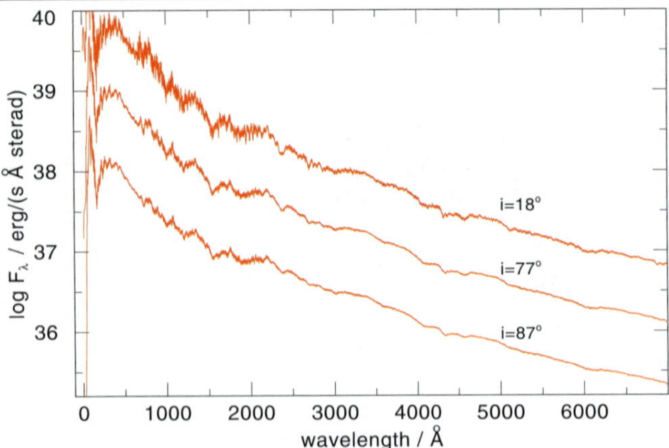

Fig. 10 Complete disk spectrum including Kepler rotation broadening, seen under three inclination angles. The broad iron-line blends are detectable even in the almost edge-on case

the continuum. Should the disk be cooler, then disk regions with $T_{\mathrm{eff}} \approx 9000$ K could dominate the optical emission and the disk spectrum might look more like that emitted by ring 9 in our model (Fig. 4). Strong iron-line blanketing could cause a spectral break. This contrasts with a statement in Hulleman et al. (2004), where the spectral break observed in the optical energy distribution of the AXP 4U 0142+61 is suggested as an argument against disk emission.

5 Summary and outlook

We have computed a model for a supernova-fallback disk in order to study its structure and optical/UV emission properties. We assumed an α-disk for the radial structure and performed detailed non-LTE radiation transfer calculations for the vertical structure and spectrum synthesis. The input parameters were:

Neutron star mass: 1.4 $M_{\odot}$
Inner and outer disk edge radii: $R = 2000$ and 200 000 km
Mass-accretion rate: $\dot{M} = 3 \times 10^{-9}\ M_{\odot}/\mathrm{yr}$.

We have identified that the disk region in the vicinity of $R = 40\,000$ km is the main contributor to the total disk spectrum at UV/optical wavelengths. We therefore investigated in some detail the disk properties at this radius.

We summarize our results as follows:

- The overall disk spectrum is independent of the detailed chemical composition as long as iron is the dominant species. In particular, a pure-iron composition is spectroscopically indistinguishable from a silicon-burning ash composition.
- The overall disk spectrum is hardly influenced by non-LTE effects, however, equivalent widths of individual line blends can change by a factor of two.
- Limb darkening affects the overall disk spectrum (in addition to the geometric foreshortening factor $\cos i$). Depending on inclination and spectral band, the disk intensity varies up to a factor of three.
- Depending on the inclination, the disk flux can be a factor of two higher or lower compared to a blackbody radiating disk.
- Strong iron line blanketing causes broad (>100 Å) spectral features that could be detectable even from almost edge-on disks. Disks that are cooler than our model (because of a lower mass-accretion rate) could even exhibit a spectral break in the optical band due to massive line blanketing.

We stress that these results hold strictly only for our particular disk model. In order to arrive at more general results a systematic parameter study (disk extent, accretion rate) of the disk emission is necessary. Also, it needs to be investigated in detail how fine the subdivision of the disk in a number of rings is necessary in order to achieve a computed spectrum with a certain accuracy. In addition, deviations from the α-disk model must be studied. Another important point will be the inclusion of disk irradiation by the X-ray emission from the neutron star. This will reveal the relative importance of viscous dissipation and reprocessed irradiation that is discussed in the context of simultaneous optical and X-ray pulsations in the AXP 4U 0142+61. At the moment we do not dare to make any prediction how this affects the results presented here.

The innermost disk ring has a very high effective temperature and its flux distribution peaks in the soft X-ray band. It needs to be investigated systematically under which conditions (inclination, inner disk radius, accretion rate) the innermost disk regions can contribute to the thermal spectrum of the magnetars.

Acknowledgements T.R. was supported by the German Ministry of Economy and Technology through the German Aerospace Center (DLR) under grant 50 OR 0201.

References

Alpar, M.A.: Astrophys. J. **554**, 1245 (2001)
Anderson, L.S.: Astrophys. J. **339**, 558 (1989)
Blackman, E.G., Perna, R.: Astrophys. J. **601**, L71 (2004)
Chatterjee, P., Hernquist, L., Narayan, R.: Astrophys. J. **534**, 373 (2000)
Dreizler, S., Werner, K.: Astron. Astrophys. **278**, 199 (1993)
Ertan, Ü., Alpar, M.A.: Astrophys. J. **593**, L93 (2003)
Ertan, Ü., Cheng, K.S.: Astrophys. J. **605**, 840 (2004)
Ertan, Ü., Göğüş, E., Alpar, M.A.: Astrophys. J. **640**, 435 (2006)
Graves, G.J.M., Challis, P.M., Chevalier, R.A., et al.: Astrophys. J. **629**, 944 (2005)
Hulleman, F., van Kerkwijk, M.H., Kulkarni, S.R.: Astron. Astrophys. **416**, 1037 (2004)
Israel, G.L., Rea, N., Mangano, V., et al.: Astrophys. J. **603**, L97 (2004)

Kern, B., Martin, C.: Nature **417**, 527 (2002)
Kriz, S., Hubeny, I.: Bull. Astron. Inst. Czechoslov. **37**, 129 (1986)
Kurucz, R.L.: In: Crivellari, L., Hubeny, I., Hummer, D.G. (eds.) Stellar Atmospheres: Beyond Classical Models. NATO ASI Series C, vol. 341, p. 441 (1991)
Marsden, D., Lingenfelter, R.E., Rothschild, R.E.: Astrophys. J. **547**, L45 (2001)
Menou, K., Perna, R., Hernquist, L.: Astrophys. J. **559**, 1032 (2001)
Meyer-Hofmeister, E.: Astron. Astrophys. **253**, 459 (1992)
Nagel, T., Dreizler, S., Rauch, T., Werner, K.: Astron. Astrophys. **428**, 109 (2004)
Perna, R., Hernquist, L.E., Narayan, R.: Astrophys. J. **541**, 344 (2000)
Shakura, N.I., Sunyaev, R.A.: Astron. Astrophys. **24**, 337 (1973)
van Paradijs, J., Taam, R.E., van den Heuvel, E.P.J.: Astron. Astrophys. **299**, L41 (1995)
Wang, Z., Chakrabarty, D., Kaplan, D.L.: Nature **440**, 772 (2006)
Werner, K., Husfeld, D.: Astron. Astrophys. **148**, 417 (1985)
Werner, K.: Astron. Astrophys. **161**, 177 (1986)
Werner, K., Deetjen, J.L., Dreizler, S., et al.: In: Hubeny, I., Mihalas, D., Werner, K. (eds.) Stellar Atmosphere Modeling. ASP Conf. Series, vol. 288, p. 31 (2003)
Woods, P.M., Thompson, C.: In: Lewin, W.H.G., van der Klis, M. (eds.) Compact Stellar X-Ray Sources, p. 547. Cambridge University Press (2006)

Astrophys Space Sci (2007) 308: 151–160
DOI 10.1007/s10509-007-9322-1

ORIGINAL ARTICLE

Chandra observations of neutron stars: an overview

M.C. Weisskopf · M. Karovska · G.G. Pavlov · V.E. Zavlin · T. Clarke

Received: 29 June 2006 / Accepted: 31 October 2006 / Published online: 20 March 2007

Abstract We present an overview of *Chandra* X-ray Observatory observations of neutron stars. The outstanding spatial and spectral resolution of this great observatory have allowed for observations of unprecedented clarity and accuracy. Many of these observations have provided new insights into neutron star physics. We present an admittedly biased and overly brief review of these observations, highlighting some new discoveries made possible by the Observatory's unique capabilities. This includes our analysis of recent multiwavelength observations of the putative pulsar and its pulsar-wind nebula in the IC443 SNR.

Keywords X-ray astronomy · Neutron stars · SNR · Crab pulsar · Vela pulsar · IC443 · B1509–58 · 1E 1207.4–5209 · SNR 292.0+1.8 · 3C58 · SNR 1987A · RX J1856.5–3754

PACS 97.60.Jd · 98.70.Qy

M.C. Weisskopf (✉)
Marshall Space Flight Center, VP62, AL 23812, USA
e-mail: martin@smoker.msfc.nasa.gov

M. Karovska
Smithsonian Astrophysical Observatory, 60 Garden St., Cambridge, MA 02138, USA
e-mail: karovska@head.cfa.harvard.edu

G.G. Pavlov
525 Davey Lab, Pennsylvania State University, University Park, PA 16802, USA
e-mail: pavlov@astro.psu.edu

V.E. Zavlin
National Space Science Technology Center, 320 Sparkman Drive, Huntsville, AL 35805, USA
e-mail: slava.zavlin@nsstc.nasa.gov

T. Clarke
Code 7213, Naval Research Laboratory, 4555 Overlook Ave SW, Washington, DC 20375, USA

T. Clarke
Interferometrics Inc., 13454 Sunrise Valley Dr. #240, Herndon, VA 20171, USA
e-mail: tracy.clarke@nrl.navy.mil

1 Introduction

For many users, the view of the *Chandra* X-Ray observatory is as pictured in Fig. 1. We therefore provide Fig. 2 to remind us that the Observatory, with Inertial Upper Stage (IUS) attached, was the largest and heaviest payload ever handled by NASA's Space Transportation system. The Observatory was launched on 1999 July 23 using the ill-fated orbiter Columbia. The IUS, a two-stage solid rocket booster, was subsequently fired and separated, sending *Chandra* toward a high elliptical orbit. On August 7, the fifth burn of *Chandra*'s integral propulsion system placed it into a 63.5 hour (80,800 km semi-major axis) orbit. On August 12, the Observatory's forward door opened, exposing the focal plane to celestial X-rays.

The Observatory was designed for 3 years of operation with a goal of five years. We are therefore extremely pleased that in July of 2006 the Observatory will have completed its seventh year of operation. Certain difficulties (slow degradation of the thermal shielding and contamination buildup on the Advanced CCD Imaging Spectrometer [ACIS] instrument's cold filters) notwithstanding, the Observatory continues to operate successfully. The gas supply, used to point the Observatory, has a lifetime of much more than 15 years. The

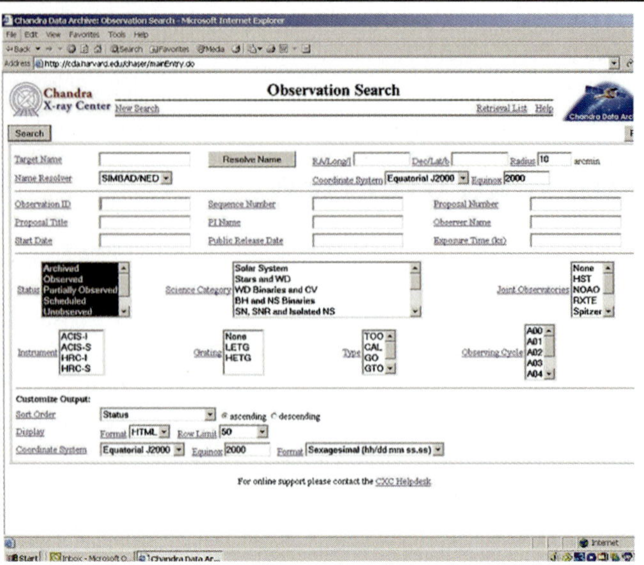

Fig. 1 *Chandra* as perceived by an observer today

Fig. 2 *Chandra* with IUS attached in Columbia's cargo bay

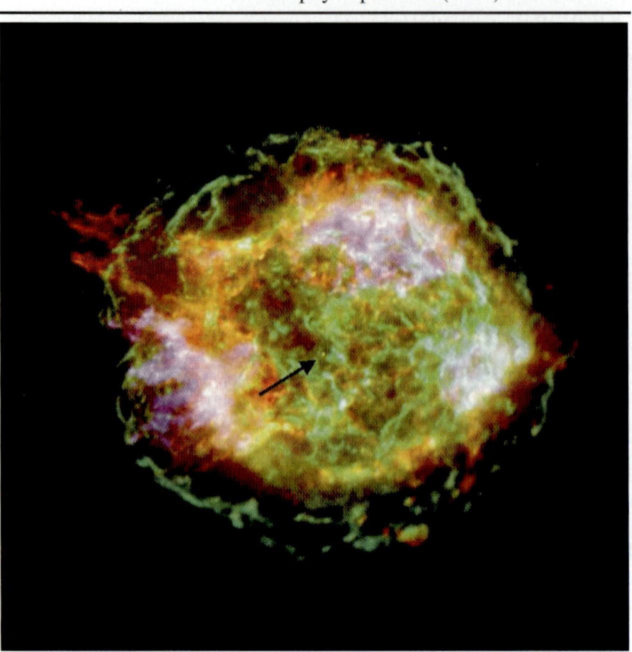

Fig. 3 *Chandra* image of Cas A with the point source at the *center*. Image is $8' \times 8'$. Courtesy *Chandra* X-Ray Center (CXC)

orbit will be stable for an even longer time period, and we anticipate many more years of usage.

Even the first observations with *Chandra* provided numerous startling insights concerning astrophysical systems in general, and neutron stars in particular. Two of the early images, those of the supernova remnant Cas A (Fig. 3) and that of the Crab Nebula and its pulsar (Fig. 4) have become two of the most spectacular and interesting observations made with *Chandra*.

2 The pulsar in the Crab Nebula

The *Chandra* X-Ray Observatory first observed the Crab Nebula and its pulsar during orbital calibration in 1999 (Weisskopf et al. 2000). That image showed a striking richness of X-ray spatial structures: an X-ray inner ring within the X-ray torus; the suggestion of a hollow-tube structure for the torus; X-ray knots along the inner ring and (perhaps) along the inward extension of the X-ray jet. The *Chandra* image also clearly resolved the X-ray torus and jet and counterjet which are all features that had been previously observed (Aschenbach and Brinkmann 1975; Brinkmann et al. 1985; Hester et al. 1995; Greiveldinger and Aschenbach 1999) but never with such clarity. On slightly larger scales, the image also showed a sharply bounded notch (WSW of the Pulsar) into the X-ray nebular emission, earlier associated with the "west bay" of the Nebula (Hester et al. 1995). Visible-light polarization maps of the Crab Nebula (Schmidt et al. 1979; Hickson and van den Bergh 1990) demonstrate

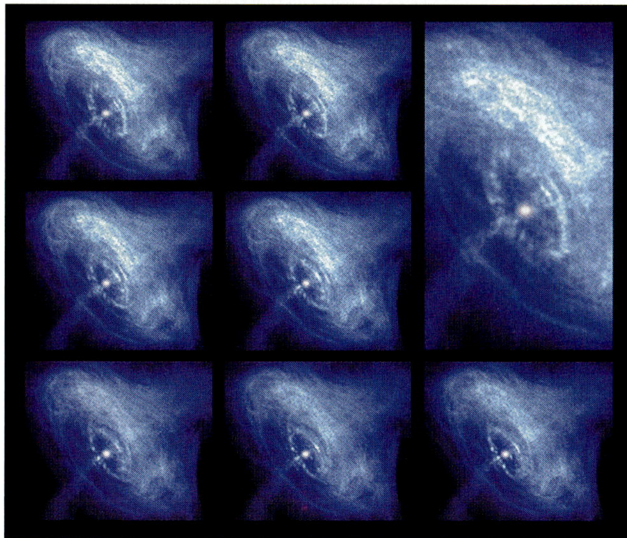

Fig. 4 Sequence of *Chandra* ACIS observations of the Crab Nebula and pulsar. The 7 images are $1.6' \times 1.6'$ and cover a 7 month time interval. The close up zooms in by a factor of 2. Courtesy CXC

that the magnetic field is parallel to the boundary of this notch, thus serving to exclude the X-ray-emitting relativistic electrons from the west bay.

The most striking feature of the X-ray image is, of course, the inner elliptical ring, lying between the pulsar and the torus, which can be interpreted as a termination shock in the pulsar wind. The existence of a termination shock had been predicted (Rees and Gunn 1974; Kennel and Coroniti 1984), but such a narrow ring, which can form only in a wind confined in the equatorial plane, had never been observed. On the ring reside a few compact knots (Fig. 4); one of them lying SE of the pulsar along the projected inward extension of the jet. The surface brightness of this knot is too high to be simply explained as the superposition of the ring's and jet's surface brightnesses. Ultimately the nature of these knots needs to be probed by means of high-resolution spectroscopy.

Subsequently, Tennant et al. (2001) observed the system with *Chandra*. They performed a most sensitive search for X-ray emission from the pulsar as a function of pulse phase, including pulse phases that had been traditionally referred to as "unpulsed". They found, as in the visible (Golden et al. 2000; Peterson et al. 1978), that the pulsar emits X-rays at *all* pulse phases.

Tennant et al. (2001) also confirmed prior observations (Pravdo et al. 1997; Massaro et al. 2000) which showed that the pulsar's power-law spectral index varied with pulse phase and extended the such measurements into the pulse minima. Finally, assuming that all of the flux from the pulsar at pulse minimum is attributable to thermal emission, the authors used these data to set a new upper limit to the blackbody temperature. As a representative case, they took the apparent angular radius of the neutron star $\theta_\infty = 2.1 \times 10^{-16}$ rad (corresponding to $R_\infty = 13$ km at $D = 2$ kpc) and $N_{\rm H} = 3 \times 10^{21}$ cm^{-2}. With these parameters, the blackbody temperature and luminosity that would account for all the flux observed at the pulse minimum were $T_\infty = 2.12$ MK ($kT_\infty = 183$ eV) and $L_\infty \approx 2.4 \times 10^{34}$ erg s^{-1}), which bounds the actual temperature and thermal luminosity of the neutron star. Subsequent *Chandra* LETGS observations and analyses of the spectrum as a function of pulse phase (Weisskopf et al. 2004) slightly improved this upper limit to $T_\infty < 1.76$ and 2.01 MK at 2σ and 3σ confidence levels.[1]

Weisskopf et al. (2004) also performed a detailed analysis of the phase-averaged spectrum. They were able to study the interstellar X-ray extinction due primarily to photoelectric absorption and secondarily to scattering by dust grains in the direction of the Crab Nebula. They confirmed the findings of Willingale et al. (2002) that the line-of-sight to the Crab is under-abundant in oxygen. Using the abundances and cross sections from Wilms et al. (2000), they found $[{\rm O/H}] = (3.33 \pm 0.25) \times 10^{-4}$. Spectral studies such as this, where the abundances are allowed to vary, are important as it is unlikely that standard abundances apply equally well to all lines of sight, especially those that intersect large quantities of SN debris (for more on this point see the discussion in Serafimovich et al. 2004).

In 2002, Hester et al. (2002) completed one phase of a set of coordinated observations of the Crab's pulsar wind nebula (PWN) using *Chandra* (ACIS-S in sub-array mode) and the *Hubble Space Telescope*. These spectacular observations revealed numerous dynamical features including wisps moving outward from the inner equatorial X-ray ring with a speed of about 0.5 c. These *Chandra* observations are summarized in Fig. 4.

Finally, *Chandra* (and *XMM-Newton*, see Willingale et al. 2002) has been used to study spectral variations as a function of position in the nebula. Weisskopf et al. (2000) first presented the variation of a hardness ratio (the ratio of flux in two energy bands) as a function of position as seen with *Chandra* using $5'' \times 5''$ pixels. Mori et al. (2004) followed this work with studies of the variation of the power-law spectral index as a function of position using $2.5'' \times 2.5''$ pixels. It is worth mentioning that the pileup effect impacted the data analysis and required corrections. These corrections were not accurate when bright spatial structure was present within an analysis pixel—dealing with that particular situation was noted by Mori et al. to be beyond the scope of their paper. One hopes this problem will be addressed by some enterprising expert in pileup in the future since *Chandra* is providing one with unique information as to these small features.

[1]These upper limits appear weaker than previous *ROSAT*-established upper limits set by Becker and Aschenbach (1995). The *ROSAT* limits were, however, too "optimistic" as discussed in Tennant et al. (2001).

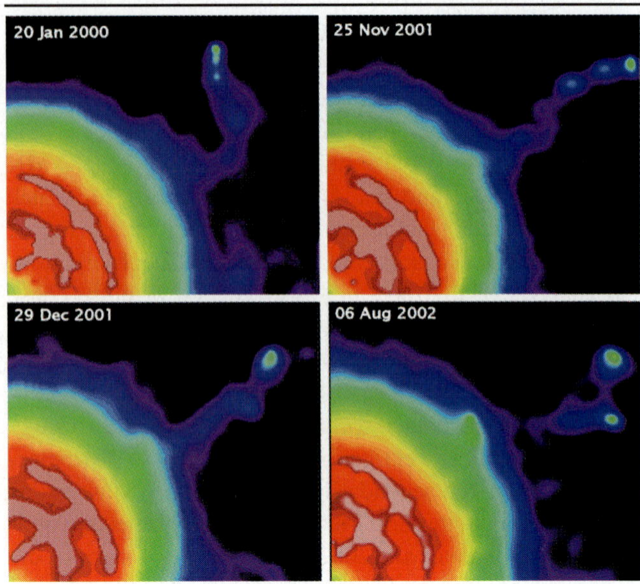

Fig. 5 Sequence of *Chandra* ACIS observations of the Vela pulsar and its jet (provided by O. Kargaltsev). Each image is $1.6' \times 1.2'$

3 The Vela pulsar

Chandra observations of the 89-ms period Vela pulsar and its surroundings (Helfand et al. 2001; Pavlov et al. 2001a, 2001b, 2003) have been most revealing. These observations showed the complex and time variable spatial structure of the region immediately surrounding the pulsar itself—a structure that includes two sets of arcs, a jet in the direction of the pulsar's proper motion and a counterjet. Furthermore, the *Chandra* images taken by Pavlov et al. (2003) also discovered that the continuation of the jet that extends to the NW is time-variable in both intensity and position on scales of days to weeks (Fig. 5). Pavlov et al. (2003) found bright blobs in the outer jet, moving away from the pulsar, and inferred flow velocities of 0.3–0.7 c. Finally, the apparent width of the outer jet appears to be approximately constant, despite large variations in appearance, indicating confinement. The analogy to a fire hose being held at its base is most appropriate.

4 IC 443

Historically, the first *Chandra* observation of the IC 443 SNR (Olbert et al. 2001) was a publicity tour-de-force as the first three authors were high school students at the time. The remnant has been the subject of many studies in different wavelength bands as there is a large variety of shocked molecular gas present due to the interaction with surrounding molecular clouds (see references in Olbert et al. 2001, and Bykov et al. 2005). IC 443 is also a candidate counterpart to the EGRET source 3EG J0617+2238. The original *Chandra* image clearly shows what appears to be a point source behind perhaps a bow shock and surrounded by a nebulosity that looks somewhat like a cometary tail. Olbert et al. (2001) also presented accompanying VLA observations which confirmed and complemented the X-ray spatial structure. No pulsations were reported either from the X-ray or the flat spectrum radio observations. Subsequent observations with both *Chandra* and *XMM* (Bocchino and Bykov 2001; Bykov et al. 2005) have also not detected pulsations making this a candidate radio-quiet neutron star with a PWN.

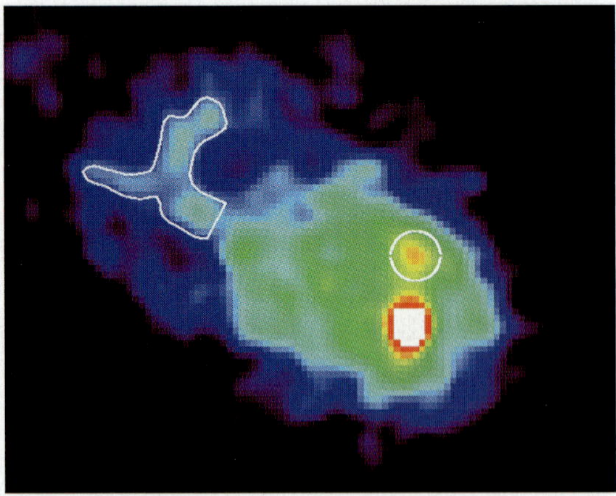

Fig. 6 IC443: an $50'' \times 40''$ image of the PWN core smoothed with a Gaussian of FWHM $= 2''$. The *white circle* indicates the second point-like source $6''$ northward from the central and brightest source (presumably a pulsar), while the other white contour marks the hook-like structure about $25''$ to the north-west from the pulsar

More recently, an additional *Chandra* observation was made by Gaensler et al. (2006); hereafter G2006. Based on this observation with better angular resolution (the target was much closer to the aimpoint), G2006 confirmed the previously reported X-ray structure and discussed additional morphological details of the presumed bow-shock PWN, which they described as a "tongue" of bright emission close to the neutron star, enveloped by a larger, fainter "tail" (a somewhat weird anatomical configuration!). G2006 also (reasonably) assumed that the brightest (point-like) enhancement in the X-ray image is the underlying rotation-powered pulsar (that has not been observed to pulse) that is creating the nebula. These authors further commented upon small extensions near this object, which might be evidence for polar jets such as those seen in the Crab.

We have been performing our own independent analysis of the *Chandra* data, supplemented by observations at other wavelengths and find a number of intriguing and puzzling morphological features.

Two compact sources in the X-ray PWN Our initial analysis of the early data showed the presence of not one, but two bright point-like enhancements in the X-ray data: the one

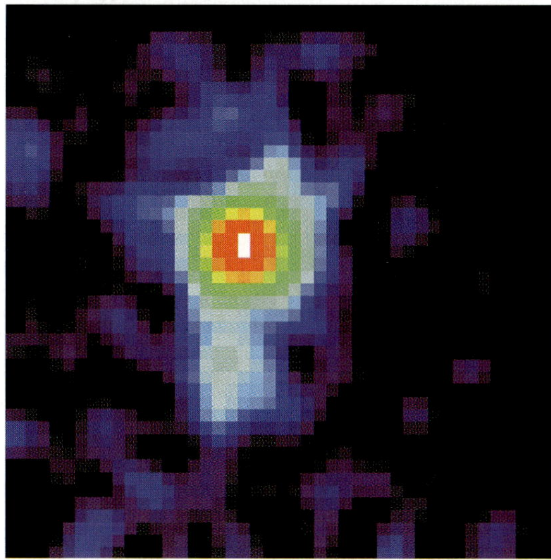

Fig. 7 IC443: an $8'' \times 8''$ image of the southern source (pulsar) binned in 1/9 ACIS pixels and smoothed with a Gaussian of FWHM $= 0.5''$. A $2''$-long structure extending southward from the pulsar is clearly visible

reported by Olbert et al. (2001) and an apparently harder source $6''$ north of the putative pulsar. Our conjecture was confirmed using the data from the longer more recent observation as shown in Fig. 6. It is not unreasonable to continue the assumption that the brighter enhancement to the south is the emission of the source powering the PWN, but it is worth investigating the northern feature more closely. Spectral analysis of the approximately 300 counts in the southern enhancement by G2006 is certainly consistent with the pulsar interpretation, but by no means definitive given the small number of counts. There are also not enough counts (between 20 and 50, depending on the background subtraction) from the northern source to determine spectral parameters and pursue the question as to the nature of this enhancement. Is it associated with the PWN or is it an interloper? Is it connected with the southern source by some faint "bridge" (e.g., a receding pulsar jet)? Although unlikely, is it possibly the pulsar?

"Jets" in the southern source G2006 noticed short extensions towards north and south of the southern point-like source and suggested these extensions might be the jets emanating from the pulsar. Such an interpretation means that the jets (and hence the pulsar's spin axis) are not coaligned with the symmetry axis of the PWN, contrary to many other cases. If so, this has important implications for our understanding of the "natal kicks" of neutron stars (the [mis]alignment constrains the kick timescale: if this timescale is larger than the pulsar period at birth, then the momentum imparted by the kick is rotationally averaged, resulting in alignment between velocity and spin vectors—see, e.g., Romani 2005). Our analysis of the same data con-

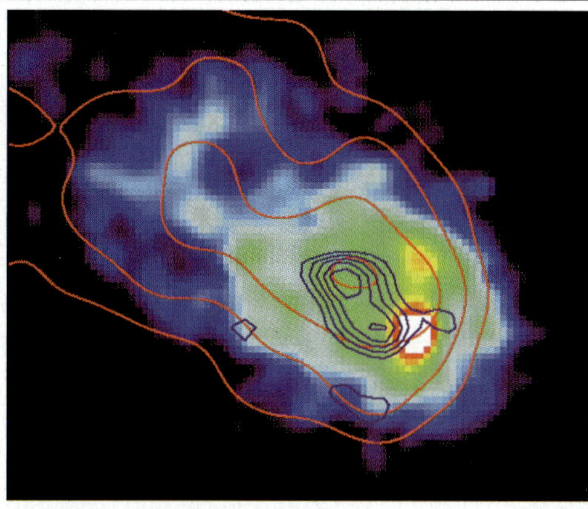

Fig. 8 IC443: same as Fig. 6 with overlaid contours of radio brightness measured at 8.6 (*red*) and 4.8 (*blue*) GHz

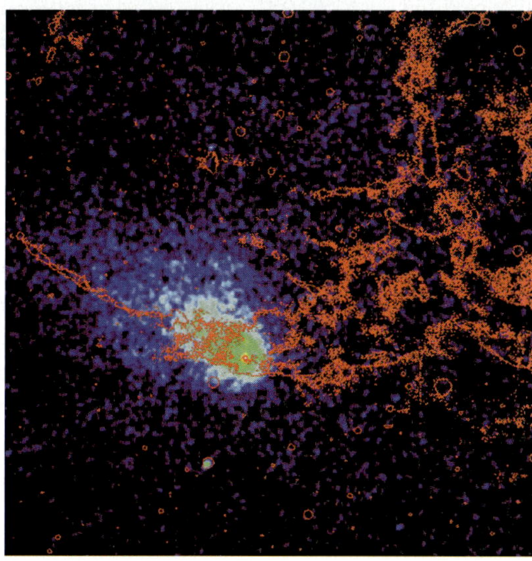

Fig. 9 IC443: X-ray (*green and blue*) and H_α (*red*) emission. The image size is approximately $5' \times 5'$

vincingly reveals only the southern extended feature, with an estimated number of counts of 56 ± 14 (in agreement with G2006), while the existence of a northern "jet" is questionable (Fig. 7).

The radio/X-ray surface brightness puzzle In a "classical" PWN, such as the Crab and Vela, the radio emission tends to peak at the outskirts of the nebula due to synchrotron cooling (e.g., Kargaltsev and Pavlov 2004). Yet for IC443, the radio emission appears contained within the boundaries defined by the X-rays (Fig. 8). This might be explained by different energy distributions of relativistic wind electrons (e.g., more electrons with lower energies in the IC443 PWN than in the Crab and Vela).

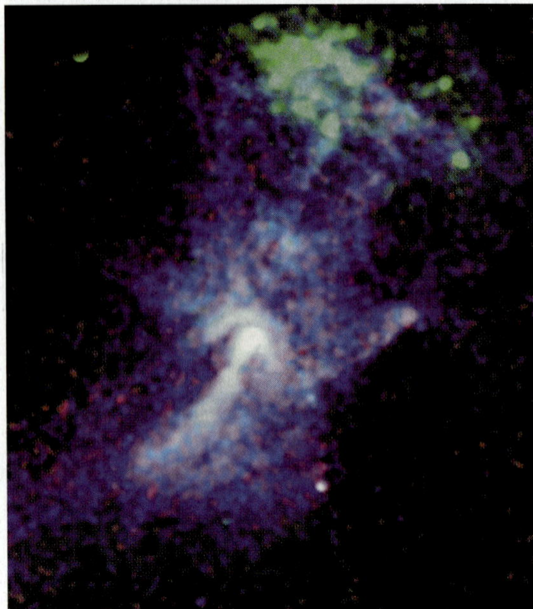

Fig. 10 *Chandra* observation of the PWN powered by PSR B1509–58 in SNR 320.4–1.2. The image is $10' \times 14'$. Courtesy CXC

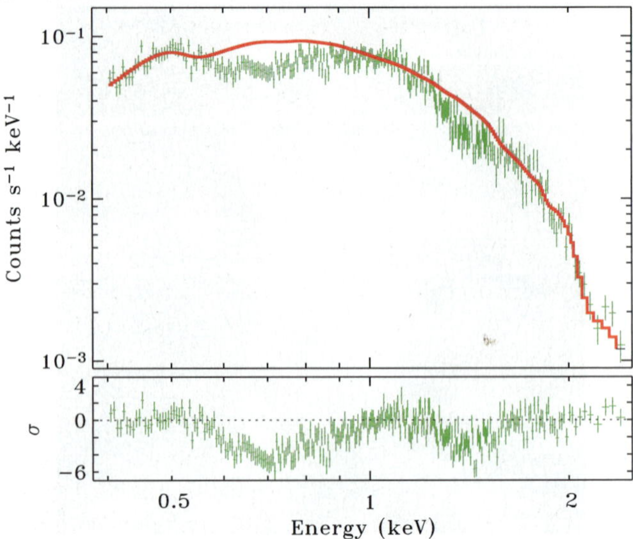

Fig. 11 The upper panel shows the data for 1E 1207.4–5209 and a featureless thermal spectrum. The lower panel displays the residuals to this spectrum and clearly indicates the presence of the two absorption features discussed in the text

The H_α *filament and the PWN* To our knowledge, we are the first to note the rather puzzling positional alignment between the X-ray PWN and the H_α filamentary structure shown in Fig. 9. Optical observations (performed by J. Thorstensen and R. Fesen) show no hint of this structure at other wavelengths. Observationally, it is extremely unlikely that the major axis of the PWN and the H_α filamentary structure would be so aligned. Trying to understand a possible astrophysical reason is, however, a challenge. One might speculate that the H_α structure shows the streamlines of the SNR gas flowing round the PWN, excited by interaction with the pulsar wind and PWN radiation. The long narrow H_α tail behind (i.e. to the north-east in Fig. 9) the X-ray PWN would then mean that the excited gas is moving in a narrow channel, suggesting some collimation mechanism. We will propose to investigate this mystery further in a deeper *Chandra* observation that would allow us to assess the correlation between the X-ray and H_α intensities at large distances behind the moving pulsar.

5 B1509–58

We complete our selection of *Chandra* images of neutron stars and their pulsar-driven wind nebulae with the image (Fig. 10) of the PWN powered by a young, 150 ms pulsar B1509–58 in SNR 320.4–1.2 (Gaensler et al. 2002). The pulsar is the bright source at the center of the nebula. A thin jet can be seen in the image to extend to the southeast. Just above the pulsar there is a small arc of X-ray emission, which seems to mark the location of the shock wave produced by the particles flowing away from the pulsar's equator. The cloud near the top of the image (known as RCW 89) may be due to high-temperature gas. This gas, possibly a remnant of the explosion associated with the creation of the pulsar, may have been heated by collisions with high-energy particles produced by the pulsar. Yatsu et al. (2005) provide a discussion of the interaction of the pulsar's jet with this material.

6 Finding pulsars

6.1 1E 1207.4–5209

Observations with *Chandra* have contributed at least two significant new insights into the source 1E 1207.4–5209, initially discovered with the *Einstein* Observatory (Helfand and Becker 1984) and located $6'$ from the center of SNR PKS 1209–51/52 (aka SNR 296.5+10.0). The first observation of this source with *Chandra*'s ACIS, taken in continuous clocking mode, resulted in the detection of a 424-ms period, which, of course, provided compelling evidence that the source is a neutron star (Zavlin et al. 2000). Since the source appears to be radio-quiet (Mereghetti et al. 1996; Kaspi et al. 1996), it may be either an active pulsar beamed out of our line of sight (or perhaps a rotating radio transient [RRAT]!?) or a truly radio-quiet neutron star, where the X-ray pulsations are caused perhaps by hot spots rotating in and out of our line of sight. A second *Chandra* observation allowed Pavlov et al. (2002a) to establish a preliminary estimate for the period derivative and hence, the spin-down power, characteristic age, and a dipole component of the magnetic field.

From the spectral analysis of the same two ACIS-S3 observations, Sanwal et al. (2002) found two significant absorption features centered at 0.7 and 1.4 keV with equivalent widths of about 0.1 keV (Fig. 11). These authors discussed several possible interpretations for the absorption, including cyclotron resonances and atomic features. They presented arguments favoring atomic transitions of once-ionized helium in the atmosphere of the neutron star assumed to be very strongly magnetized ($B \approx 10^{14}$ G). Other authors suggested different interpretations of the *Chandra*-discovered features. For example, Hailey and Mori (2002) argued that the absorption features were associated with He-like oxygen or neon in a field of $\approx 10^{12}$ G. More recent *XMM* observations (e.g., Mereghetti et al. 2002; Bignami et al. 2003; De Luca et al. 2004) not only confirmed the *Chandra*-detected absorption features at 0.7 and 1.4 keV, but also seemed to have detected an additional feature at 2.1 keV and evidence for a fourth one, at 2.8 keV. Taking all these latter data into account supports an explanation involving the fundamental and two, possibly three, harmonics of the electron cyclotron absorption in a field of order 10^{11} G. However, the two additional spectral features in the *XMM-Newton* data have not been universally accepted. Mori et al. (2005) have cast doubt as to the reality of the spectral features at 2.1 and 2.8 keV. Their arguments seem compelling, and it is thus unfortunate that the *Chandra* response is insufficient to weigh in on this question without expending significant amounts of observing time.

Zavlin et al. (2004) have continued to observe this target using both *Chandra* and *XMM-Newton*. They have detected significant non-monotonous variations in the spin period which they interpreted in light of three hypotheses: a glitching pulsar; variations in the accretion rate from a fall-back disk; and variations produced by orbital motion in a wide binary (see also the presentation by Woods et al. at this conference for new timing results).

We conclude this section by noting that the sequence of *Chandra* observations have provided important discoveries—the detection of the pulse period and firm detection of two absorption features. An important and unanswered question is what are the limits as to the presence of such spectral features for the other neutron stars in SNRs. A systematic comparison, if not already in progress, should be performed.

Finally we note that 1E 1207.4–5209 is a source that, in some critical respects, is similar to the central compact object (CCO) in the Cas A SNR (Pavlov et al. 2000): it is in a SNR, it is radio-quiet, and shows a thermal-like X-ray spectrum (albeit with somewhat lower temperature); however it pulses. Thus, on the one hand, the source might give us confidence that all CCOs will ultimately be found to pulse. On the other hand, it may happen that the *Chandra*-revealed characteristics of this source will remain unique among the CCOs and will be used to separate it from that class of objects. Time will tell.

Fig. 12 *Chandra* ACIS-S3 image of 3C58. Image is 130″ × 130″. Courtesy P. Slane

6.2 SNR 292.0+1.8

SNR 292.0+1.8 is, along with Cas A and Puppis A, an oxygen-rich supernova remnant. Hughes et al. (2001) used *Chandra* and detected a bright, spectrally hard, point source within an apparently extended region. This detection suggested the presence of a pulsar and its PWN. Radio observations (Camilo et al. 2002) then found a 135-ms radio pulsar. The subsequent detection by Hughes et al. (2003) of X-ray pulses at the expected period secured the identification. The X-ray spectrum is modeled with a simple power-law, although, as with Vela (and many other sources!), the fit to the data is not unique. From the motions of oxygen-rich optical knots and the size of the remnant, Ghavamian et al. (2005) estimated a kinematic age for SNR 292.0+1.8 of about 3200 years, assuming a distance of 6 kpc. This value is in good agreement with the pulsar's spin-down age of 2900 years.

6.3 3C58

The *Chandra* observations of 3C58 (SNR 130.7+3.1) were first performed by Murray et al. (2002) using the High Resolution Camera (HRC). These data imaged the previously detected X-ray point source (Becker et al. 1982). The early *Chandra* data also revealed an extended PWN and the presence of 66 ms pulsations from the central point source (J0205+6449). Deeper *Chandra* observations, using ACIS-S3, by Slane et al. (2004) produced images (Fig. 12) showing little similarity of this PWN with the Crab and Vela (in particular, the presence of numerous loop-like filaments), although one might speculate that we see a torus edge-on, projected in the north-south direction, and a one-sided jet westward of the pulsar.

One aspect of the *Chandra*-based research of 3C58 of special importance was the limit to any thermal emission from the surface of the young neutron star, J0205+6449. The search for thermal emission was first presented by Slane et al. (2002) and then refined by Slane et al. (2004) who found that the upper limits on surface temperature and thermal luminosity fall well below predictions of standard neutron star cooling. Yakovlev et al. (2002) discuss calculations of neutron star cooling in the context of J0205+6449 and concluded that the observations can be explained by the cooling of a superfluid neutron star where the direct Urca processes are forbidden.

Of course, neutron stars may be different, so that limits to the thermal components of 3C58 may not apply to all young neutron stars. In general, such analyses are not simple, requiring enhanced sensitivity for the detection of the putative thermal component often in the presence of a much stronger non-thermal flux from the magnetosphere of the pulsar, especially if one wants to measure the temperature—as opposed to setting an upper limit. *Chandra* is uniquely poised to provide the data for such studies due to its ability to maximally separate the pulsar from the surrounding nebulosity, yet often long observations are required.

7 Not finding neutron stars

Because of its tremendous sensitivity facilitated by its superb optics that have such low scatter and therefore provide high contrast, *Chandra* is ideally suited for searching for X-ray sources.

7.1 γ-Cygni (SNR 78.2+2.1)

Becker et al. (2004) used *Chandra* to search for the X-ray counterpart to the unidentified EGRET source 3EG J2020+4017. These authors investigated the possibility (Brazier et al. 1996) that RX J2020.2+4026 was the counterpart and concluded that it is associated with a K field star, excluding it from being counterpart of the bright γ-ray source.

The observation also demonstrated the difficulties one often encounters in searching for compact objects associated with a SNR. Thirty seven additional X-ray sources were detected in the field searched (which was only a fraction of the

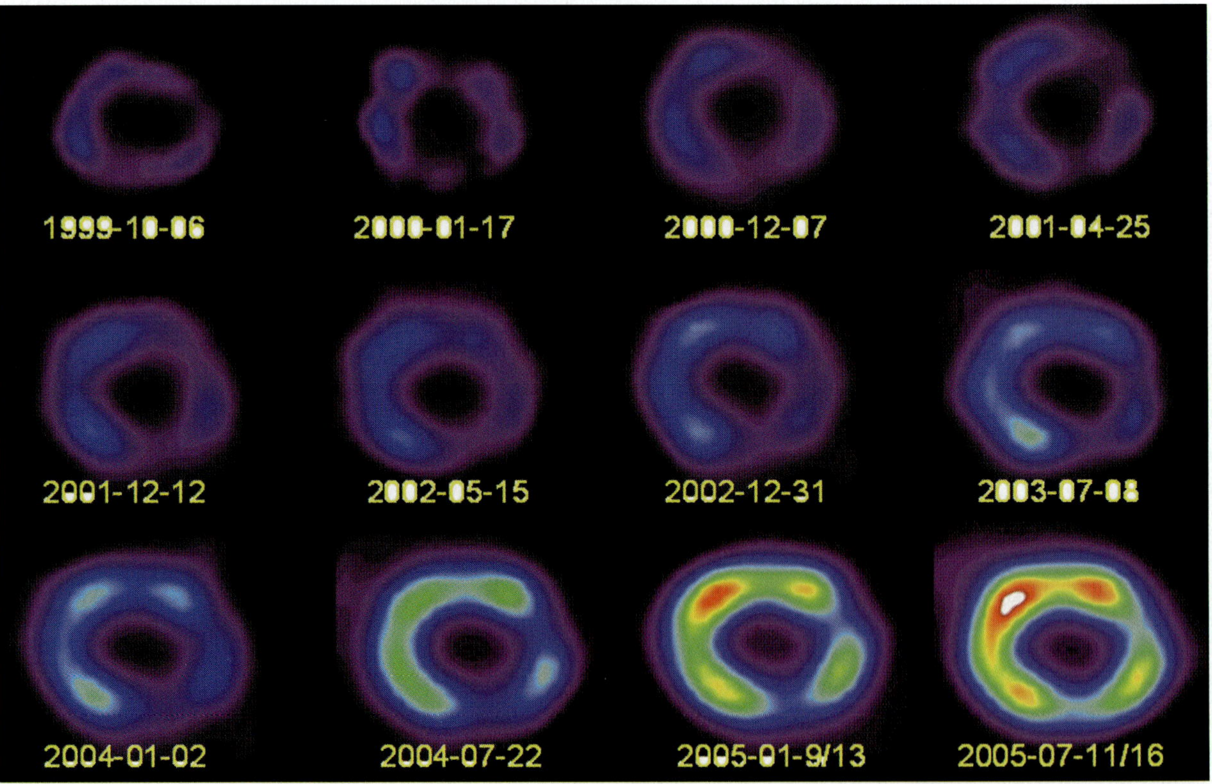

Fig. 13 Sequence of *Chandra* observations of SNR 1987A (provided by S. Park). Each image panel is approximately $2'' \times 2''$

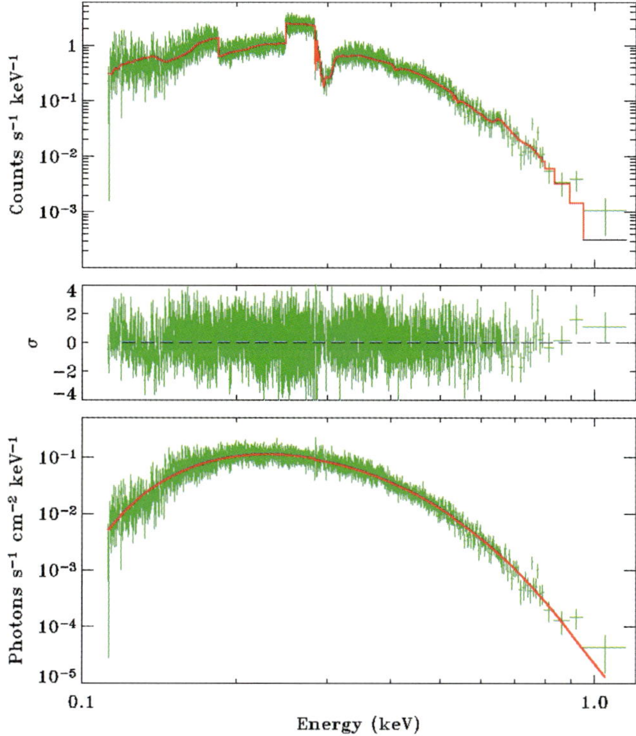

Fig. 14 *Chandra* LETGS spectrum of RX J1856.5–3754 based on 500 ksec of observation. The *upper panel* shows the raw data, and the best-fit blackbody model. The *middle panel* shows the residuals to the best fit and the *bottom panel* the incident flux

full size of the SNR). Radio observations reported by these authors, which covered the complete 99% EGRET likelihood contour of 3EG J2020+4017 with a sensitivity limit of $L_{820\ \mathrm{MHz}} = 0.09$ mJy kpc^2, were unable to find a pulsar. The absence of radio pulsations suggests that if there is a pulsar operating in γ-Cygni, the pulsar's emission geometry is such that the radio beam does not intersect with the line of sight. Alternatively, the X-ray counterpart might be a CCO which does not produce significant amounts of radio emission although CCOs are not known to show γ-ray emission. (Or perhaps the source is the counterpart to a RRAT, and the radio observations were not performed at the appropriate time.)

Without high-precision X-ray spectra of each of the candidate X-ray sources, and detailed follow up in other wavelength bands, there is essentially no satisfactory way in which to eliminate most of the candidates from consideration. This is especially true for neutron stars as the corresponding infrared-visible fluxes are very weak, making such observations especially difficult. In such cases, the principal and important *Chandra* contribution is to provide target lists with accurate positions as a basis for future studies.

7.2 SNR 1987A

Since the launch of *Chandra*, SN 1987A has been the focus of a series of repeated observations (see Park et al. 2005 and references therein). One goal of these observations is to detect the emergence of the X-ray flux from a newly born compact object. Figure 13 shows a sequence of such observations from 1999 through the middle of last year. These images are a textbook demonstration of the development of the various shocks resulting from a supernova event. The images also dramatically demonstrate that, thus far, no central compact object has yet appeared at X-ray wavelengths. Considering the spatial scale, these are observations that can only be performed with *Chandra*, now and for the foreseeable future.

8 RX J1856.5–3754

RX J1856.5–3754 is the brightest of a number of ROSAT-detected objects, which are thermal emitters with blackbody temperatures in the range of 50–100 eV. These sources appear to be isolated neutron stars, and therefore RX J1856.5–3754 was a promising candidate for a long *Chandra* observation in the hope that spectroscopy would provide a determination of parameters such as the radius, surface gravity, and gravitational redshift—all of which would lead one to the holy grail of the equation of state. The featureless, 500-ksec spectrum that resulted (Fig. 14, Burwitz et al. 2003) was therefore a surprise, although in retrospect it is worth noting that no features were detected in a much shorter (50-ksec; Burwitz et al. 2001) prior observation. (Ironically, spectral features were later found in at least 3 other, not so bright, sources of this class—Haberl 2006.) The absence of line features needs an explanation, and has triggered new and interesting theoretical work. At one point the object was even claimed to be a quark star (Drake et al. 2002). Their argument was based on the fact that the X-ray emitting area was much smaller than the area of a neutron star, but the optically emitting area does not support this. In our opinion, the quark star interpretation is no longer credible. Other suggestions include high ($\approx 10^{13}$ G) field strength (e.g., Trümper et al. 2003; see also Turolla et al. 2004) neutron stars, and a slowly spinning ($P > 100{,}000$ s) magnetar (Mori and Ruderman 2003). Braje and Romani (2002) and Pavlov et al. (2002b) noted that several effects (e.g., fast rotation) may act in suppressing spectral features from an atmosphere with heavy elements, but no period has been detected from this puzzling source. Although it seems quite certain that RX J1856.5–3754 is a very compact object, like a neutron star, its true nature still remains elusive.

Acknowledgements A number of the figures in this paper were provided by the *Chandra* X-Ray Center (CXC) and are publicly available via http://chandra.harvard.edu. MCW acknowledges support by the *Chandra Project*. MK is a member of the CXC, operated by the Smithsonian Astrophysical Observatory under contract NASA NAS8-39073. The work of GGP was supported by NASA grant NAG5-10865. VEZ is supported through a NASA Research Associateship Award. Basic research in radio astronomy performed by TC at NRL is supported by the Office of Naval Research.

References

Aschenbach, B., Brinkmann, W.: Astron. Astrophys. **41**, 147 (1975)
Becker, W., Aschenbach, B.: In: The Lives of Neutron Stars, vol. 47 Kluwer Academic, Dordrecht (1995)
Becker, R.H., Helfand, D.J., Szymkowiak, A.E.: Astrophys. J. **255**, 557 (1982)
Becker, W., Weisskopf, M.C., Arzoumanian, Z., et al.: Astrophys. J. **615**, 897 (2004)
Bignami, G.F., Caraveo, P.A., De Luca, A., et al.: Nature, **423**, 725–727 (2003)
Bocchino, F., Bykov, A.M.: Astron. Astrophys. **376**, 248 (2001)
Braje, T.M., Romani, R.W.: Astrophys. J. **580**, 1043 (2002)
Brazier, K.T.S., Kanbach, G., Carramiñana, A., Guichard, J., Merck, M.: Mon. Not. Roy. Astron. Soc. **281**, 1033 (1996)
Brinkmann, W., Aschenbach, B., Langmeier, A.: Nature **313**, 662 (1985)
Burwitz, V., Zavlin, V.E., Neuhäuser, R., et al.: Astron. Astrophys. **379**, 35 (2001)
Burwitz, V., Haberl, F., Neuhäuser, R. et al.: Astron. Astrophys. **399**, 1109 (2003)
Bykov, A.M., Bocchino, F., Pavlov, G.G.: Astrophys. J. **624**, L41 (2005)
Camilo, F., Manchester, R.N., Gaensler, B.M., et al.: Astrophys. J. **576**, L71 (2002)
De Luca, A., Mereghetti, S., Caraveo, P.A., et al.: Astron. Astrophys. **418**, 625 (2004)
Drake, J.J., Marshall, H.L., Dreizler, S., et al.: Astrophys. J. **572**, 996 (2002)
Gaensler, B.M., Arons, J., Kaspi, V.M., et al.: Astrophys. J. **569**, 878 (2002)
Gaensler, B.M., Chatterjee, S., Slane, P.O., et al.: Astrophys. J. **648**, 1037 (2006) astro-ph/0601081
Ghavamian, P., Hughes, J.P., Williams, T.B.: Astrophys. J. **635**, 365 (2005)
Golden, A., Shearer, A., Beskin, G.M.: Astrophys. J. **535**, 373 (2000)
Greiveldinger, C., Aschenbach, B.: Astrophys. J. **510**, 305 (1999)
Haberl, F.: These proceedings and astro-ph/0609066 (2006)
Hailey, C.J., Mori, K.: Astrophys. J. **578**, L133 (2002)
Helfand, D.J., Becker, R.H.: Nature **307**, 215 (1984)
Helfand, D.J., Gotthelf, E.V., Halpern, J.P.: Astrophys. J. **556**, 380 (2001)
Hester, J.J., Scowen, P.A., Sankrit, R., et al.: Astrophys. J. **448**, 240 (1995)
Hester, J.J., et al.: Astrophys. J. **577**, L49 (2002)
Hickson, P., van den Bergh, S.: Astrophys. J. **365**, 224 (1990)
Hughes, J.P., Slane, P.O., Burrows, D.N., et al.: Astrophys. J. **559**, L153 (2001)
Hughes, J.P., et al.: Astrophys. J. **591**, L139 (2003)
Kargaltsev, O., Pavlov, G.G.: In: Camilo, F., Gaensler, B.M. (eds.), Young Neutron Stars and their Environments, IAU Symposium, vol. 218, p. 195, Astronomical Society of the Pacific, San Francisko (2004)
Kaspi, V.M., et al.: Astron. J. **111**, 2028 (1996)
Kennel, C.F., Coroniti, F.V.: Astrophys. J. **283**, 694 (1984)
Massaro, E., Cusumano, G., Litterio, M., et al.: Astron. Astrophys. **361**, 695 (2000)
Mereghetti, S., Bignami, G.F., Caraveo, P.A.: Astrophys. J. **464**, 842 (1996)
Mereghetti, S., De Luca, A., Caraveo, P.A., et al.: Astrophys. J. **581**, 1280 (2002)
Mori, K., Chonko, J.C., Hailey, C.J.: Astrophys. J. **631**, 1082 (2005)
Mori, K., Burrows, D.N., Hester, J.J., et al.: Astrophys. J. **609**, 186 (2004)
Mori, K., Ruderman, M.A.: Astrophys. J. **592**, L75 (2003)
Murray, S.S., Slane, P.O., Seward, F.D., et al.: Astrophys. J. **568**, 226 (2002)
Olbert, C.M., Clearfield, C.R., Williams, N.E., et al.: Astrophys. J. **554**, L205 (2001)
Pavlov, G.G., Zavlin, V.E., Aschenbach, B., et al.: Astrophys. J. **531**, L53 (2000)
Pavlov, G.G., Zavlin, V.E., Sanwal, D., et al.: Astrophys. J. **552**, L129 (2001a)
Pavlov, G.G., Kargaltsev, O.Y., Sanwal, D., et al.: Astrophys. J. **554**, L189 (2001b)
Pavlov, G.G., Zavlin, V.E., Sanwal, D., et al.: Astrophys. J. **569**, L95 (2002a)
Pavlov, G.G., Zavlin, V.E., Sanwal, D.: In: Becker, W., Lesch, H., Trümper, J. (eds.) Proc. of 270th WE-Heraeus Seminar on Neutron Stars, Pulsars and Supernova Remnants, p. 273. Max-Planck-Institut, Garching (2002b) (Max-Planck-Institut Report 278)
Pavlov, G.G., Teter, M.A., Kargalstev, O., et al.: Astrophys. J. **591**, 1171 (2003)
Park, S., Zhekov, S.A., Burrows, D.N., et al.: Adv. Space Res. **35**, 991 (2005)
Peterson, B.A., Murdin, P., Wallace, P., et al.: Nature **276**, 475 (1978)
Pravdo, S.H., Angelini, L., Harding, A.K.: Astrophys. J. **491**, 808 (1997)
Rees, M.J., Gunn, J.E.: Mon. Not. Roy. Astron. Soc. **167**, 1 (1974)
Romani, R.W.: In: Rasio, F.A., Stairs, I.H. (eds.) Binary Radio Pulsars, ASP Conf. Ser., vol. 328, p. 337. ASP, San Francisco (2005)
Sanwal, D., Pavlov, G.G., Zavlin, V.E., et al.: Astrophys. J. **574**, L61 (2002)
Schmidt, G.D., Angel, J.R.P., Beaver, E.A.: Astrophys. J. **227**, 113 (1979)
Serafimovich, N.I., Shibanov, Yu.A., Lundqvist, P., et al.: Astron. Astrophys. **425**, 1041 (2004)
Slane, P., Helfand, D.J., Murray, S.S.: Astrophys. J. **571**, L45 (2002)
Slane, P., Helfand, D.J., van der Swaluw, E., et al.: Astrophys. J. **616**, 403 (2004)
Tennant, A.F., Becker, W., Juda, M., et al.: Astrophys. J. **554**, L173 (2001)
Trümper, J.E., et al.: In: The Restless High-Energy Universe, Proceedings of the Symposium Dedicated to Six Years of Successful BeppoSAX Operations, Amsterdam 5–8 May 2003
Turolla, R., Zane, S., Drake, J.J.: Astrophys. J. **603**, 265 (2004)
Weisskopf, M.C., Hester, J.J., Tennant, A.F., et al.: Astrophys. J. **536**, 81 (2000)
Weisskopf, M.C., O'Dell, S.L., Paerels, F., et al.: Astrophys. J. **601**, 1050 (2004)
Willingale, R., Bleeker, J.A.M., van der Heyden, K.J., et al.: Astron. Astrophys. **381**, 1039 (2002)
Wilms, J., Allen, A., McCray, R.: Astrophys. J. **542**, 914 (2000)
Yakovlev, D.G., et Kaminker, A.D., Haensel, P., et al.: Astron. Astrophys. **389**, L24 (2002)
Yatsu, Y., Kataoka, J., Kawai, N., et al.: Adv. Space Res. **35**, 1066 (2005)
Zavlin, V.E., Pavlov, G.G., Sanwal, D., et al.: Astrophys. J. **540**, L25 (2000)
Zavlin, V.E., Pavlov, G.G., Sanwal, D.: Astrophys. J. **606**, 444 (2004)

Astrophys Space Sci (2007) 308: 161–166
DOI 10.1007/s10509-007-9377-z

ORIGINAL ARTICLE

XMM-Newton observations of the isolated neutron star 1RXS J214303.7+065419/RBS1774

Mark Cropper · Silvia Zane · Roberto Turolla · Luca Zampieri · Matteo Chieregato · Jeremy Drake · Aldo Treves

Received: 3 July 2006 / Accepted: 5 September 2006 / Published online: 21 March 2007

Abstract We report *XMM-Newton* observations of the isolated neutron star RBS1774 and confirm its membership as an XDINS. The X-ray spectrum is best fit with an absorbed blackbody with temperature $kT = 101$ eV and absorption edge at 0.7 keV. No power law component is required. An absorption feature in the RGS data at 0.4 keV is not evident in the EPIC data, but it is not possible to resolve this inconsistency. The star is not seen in the UV OM data to $m_{AB} \sim 21$. There is a sinusoidal variation in the X-ray flux at a period of 9.437 s with an amplitude of 4%. The age as determined from cooling and magnetic field decay arguments is 10^5–10^6 yr for a neutron star mass of 1.35–1.5 $M_\odot$.

Keywords Stars: individual (1RXS J214303.7+065419, RBS1774) · Stars: neutron · X-rays: stars

PACS 97.60.Jd

M. Cropper (✉) · S. Zane
Mullard Space Science Laboratory, University College London, Holmbury St Mary, Dorking, Surrey RH5 6NT, UK
e-mail: msc@mssl.ucl.ac.uk

S. Zane
e-mail: sz@mssl.ucl.ac.uk

R. Turolla
Department of Physics, University of Padova, Via Marzolo 8, I-35131 Padua, Italy
e-mail: turolla@pd.infn.it

L. Zampieri
INAF-Osservatorio Astronomico di Padova, Vicolo dellOsservatorio 5, I-35122 Padua, Italy
e-mail: luca.zampieri@oapd.inaf.it

M. Chieregato · A. Treves
Department of Physics and Mathematics, Universita dellInsubria, Via Valleggio 11, I-22100 Como, Italy
e-mail: matteo.chieregato@pd.infn.it

A. Treves
e-mail: treves@mib.infn.it

J. Drake
6 Smithsonian Astrophysical Observatory, Mail Stop 3, 60 Garden Street, Cambridge, MA 02138, USA
e-mail: jdrake@head-cfa.harvard.edu

1 Introduction

RBS1774 (1RXS J214303.7+065419) is the seventh X-ray emitting, isolated neutron star (XDINS) to have been discovered. XDINS are X-ray sources with soft, thermal spectra and high X-ray/optical flux ratios, properties consistent with the surface emission from a neutron star. These objects therefore provide a unique opportunity to examine the properties of a neutron star surface, and, through spectral and temporal analysis, to investigate whether there is an atmosphere and of what nature, and to examine the thermal properties of the interior and surface layers and the configuration of the magnetic field. Further analysis also permits the age and the evolutionary status of the neutron star to be estimated, providing constraints on neutron star equations of state and models of the interior.

RBS1774 was identified as an XDINS by Zampieri et al. (2001) as a soft source in a *ROSAT* PSPC pointing of the BL Lac object MSS 2143.4+0704. Because it was observed off-axis, the positional accuracy was relatively poor. Zampieri et al. (2001) examined the optical field covered by the error circles for these data and the *ROSAT* all sky survey, and excluded brighter objects as the optical counterpart. They concluded that the counterpart must be fainter than $V \sim 23$, giving an X-ray:optical flux ratio >1000, thus indicating that the object is a neutron star. The *ROSAT* spectrum was acceptably fitted with a blackbody model with $kT \sim 92$ eV.

Further *XMM-Newton* X-ray observations of RBS1774 were reported in Zane et al. (2005). This paper summarises and updates the results provided in that paper, and the reader is referred there for a more detailed exposition.

2 Observations

The *XMM-Newton* observations of RBS1774 were made on 2004 May 31. Both EPIC cameras were in small window mode with the thin filter. After good-time-interval selection to excise data affected by elevated background, a total of 23 ks exposure time was available. The UVW1 ($\sim$280 nm) filter was selected for the Optical Monitor (OM) observation. The data were processed using the *XMM-Newton* SAS v6.0 with response files generated for the slightly off-axis position. The X-ray and UV images are shown in Fig. 1.

A more accurate source position and error circle were derived using the *XMM-Newton* data. The revised position is $\alpha = 21^{\rm h}43^{\rm m}3.3^{\rm s}$, $\delta = +6°54'17''$ with a 90% uncertainty radius of 3 arcsec.

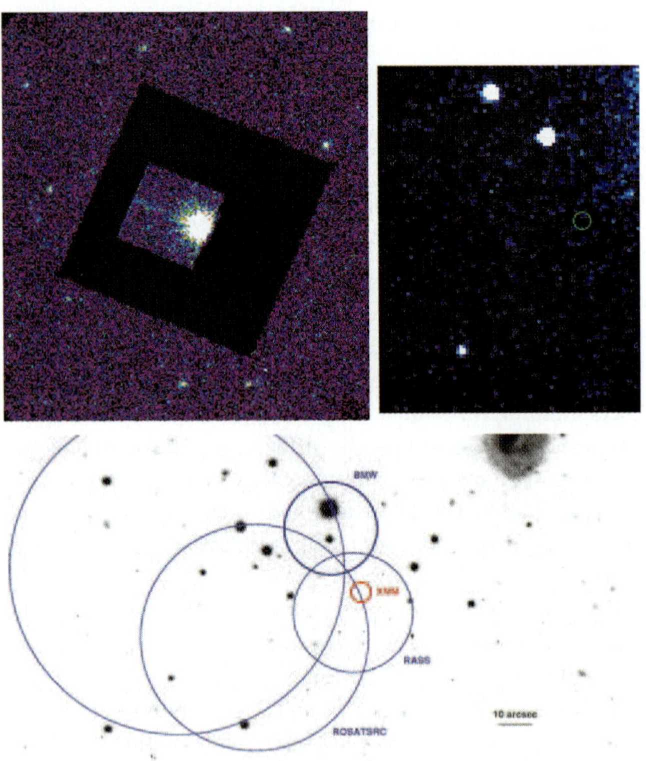

Fig. 1 The combined EPIC pn and MOS 0.2–10 keV X-ray image (*top left*) and OM 280 nm image (*top right*) of the RBS1774 field, together with an ESO NTT *R*-band image (*bottom*). The X-ray error circle is shown on the UV image in *green*. The UV and R-band image are on the same scale, while the EPIC image is on a larger scale (the central small window is 4.4 arcmin each side). North is to the *left* and East is *down* in these images

RBS1774 was well detected in the EPIC cameras and, at a low level, in the RGS. No source is evident in the OM image within the error circle. The limiting magnitude for the stacked OM exposures is $m_{AB} \sim 21$.

3 Spectral analysis

3.1 EPIC data

The EPIC pn and MOS data were simultaneously fitted with a blackbody spectrum modified with a cold absorber (the TBABS model in XSPEC using the 'angr' abundances). The fit was limited to energies between 0.2 and 1.2 keV: below 0.2 keV the EPIC calibrations are less reliable. The best fit temperature was 101.4 eV, similar to but slightly higher than that determined from the much poorer *ROSAT* data by Zampieri et al. (2001), and the n_H absorbing column was 3.65×10^{20} cm^{-2}. The fit is shown in Fig. 2a. The fit is poor, with $\chi^2_\nu = 1.36$, and there are significant residuals at $\sim$0.7 keV.

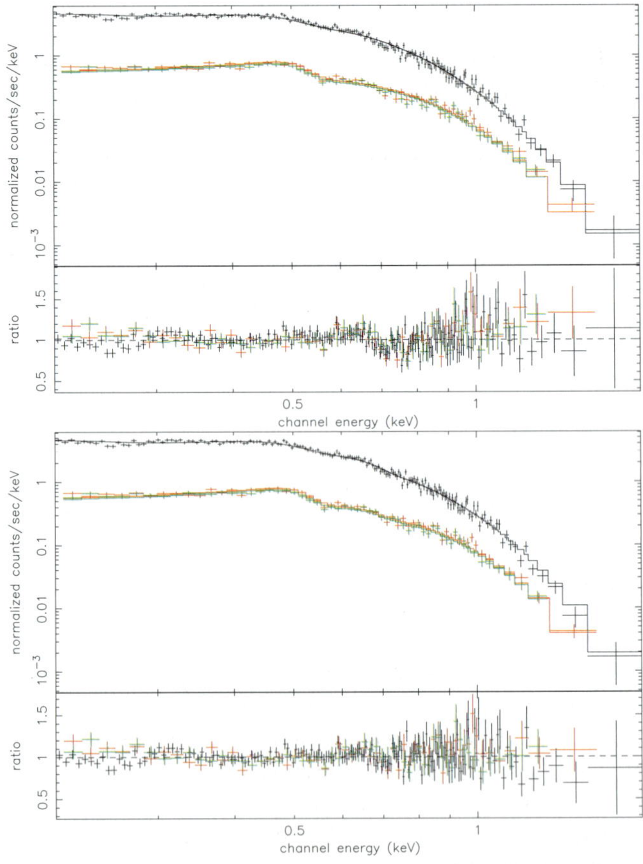

Fig. 2 (*Top*) The EPIC pn (*black*) and MOS 1 and 2 (*red* and *green* respectively) spectra, together with the best simultaneous blackbody and cold absorber fit. The residuals around 0.7 keV are significant. (*Bottom*) The same data with an additional absorption edge component. See Table 1 for details

Table 1 Model fit parameters for the blackbody (BB in XSPEC) plus cold absorber (TBABS) model obtained by simultaneously fitting data from all three EPIC cameras. The uncertainties are for the 68% confidence interval and the flux is the unabsorbed flux in the (0.2–2) keV band measured with EPIC-pn

Model	n_H 10^{20} cm^{-2}	kT_{bb}^{∞} eV	$E_{\text{edge/line}}$ eV	τ_{edge}; τ_{line}	σ_{line} eV	f_X^c erg cm^{-2} s^{-1}	χ^2/d.o.f.
bb	$3.65^{+0.16}_{-0.13}$	$101.4^{+0.5}_{-0.6}$				5.16×10^{-12}	1.36
bb+abs. edge	$3.60^{+0.21}_{-0.16}$	$104.0^{+0.6}_{-0.7}$	694^{+5}_{-11}	$0.25^{+0.03}_{-0.03}$		5.07×10^{-12}	1.17
bb+gauss. line	$3.74^{+0.14}_{-0.10}$	$102.1^{+0.5}_{-0.3}$	754^{+8}_{-9}	$4.8^{+1.0}_{-0.5}$	27^{+15}_{-8}	5.20×10^{-12}	1.20

Table 2 Model Fit Parameters for NSA atmospheric models in XSPEC absorbed by the TBABS cold absorber model. The star mass and radius are fixed at 1.4 $M_{\odot}$ and 10 km, respectively. The parameters are derived from simultaneously fitting the data from all three EPIC cameras. The uncertainties are for the 68% confidence interval

B 10^{12} G	n_H 10^{20} cm^{-2}	kT^{∞} eV	χ^2/d.o.f.
0	$8.01^{+0.38}_{-0.15}$	$24.5^{+0.2}_{-0.4}$	1.95
1	$8.61^{+0.16}_{-0.49}$	$37.0^{+0.6}_{-0.2}$	2.00
10	$8.86^{+0.19}_{-0.29}$	$39.7^{+3.9}_{-0.4}$	2.20

The addition of a power law component or a second blackbody was found to be not required or useful in improving the fit.

The NSA model for a magnetised neutron star H atmosphere (Zavlin et al. 1996) in XPSEC was also fitted to the data for different magnetic field strengths. The fits for this model were poorer than for the single-temperature blackbody fit.

Other XDINS have been found to have significant residuals when fitted to a blackbody spectrum (for example, Haberl 2005). As in those, it was found that the addition of an absorption line or edge at ~700 eV reduced the residuals significantly with a $\chi^2_\nu = 1.17$ (Fig. 2b).

The parameters derived from these fits are given in Table 1 and Table 2.

3.2 RGS data

RBS1774 is faint for the RGS but is detected with sufficient signal to allow a spectral analysis. The data from the two spectrometers were combined according to the prescription in Page et al. (2003) and are shown in Fig. 3. Also shown in Fig. 3 is the blackbody fit to the EPIC data (row 1 in Table 1) allowing for some renormalisation.

While broadly representative of the RGS data, there are significant discrepancies between the EPIC fit an the RGS data. The most obvious is a strong absorption feature at 0.4 keV, and a general slope in the residuals from 0.4–0.8 keV. Alternatively, the RGS response may be anomalously high below 0.35 keV. Depending on the location of

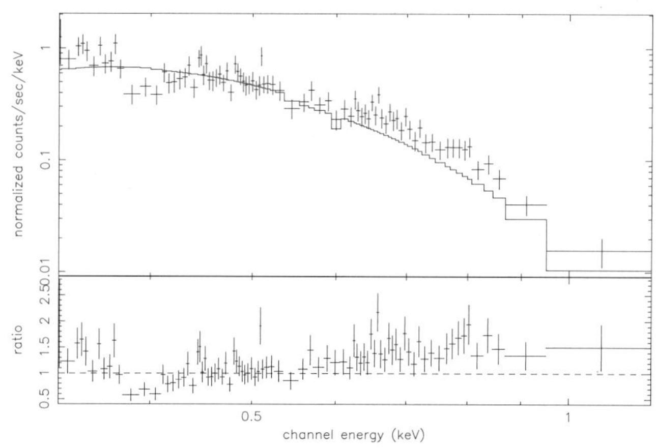

Fig. 3 The combined RGS spectrum of RBS1774, together with the renormalised EPIC model flux from the absorbed blackbody TBABS(BB) model from Table 1

the fitted model, it is unclear whether the 0.7 keV absorption in the EPIC data is seen in the RGS data.

3.3 Comparison

The main features of the X-ray spectrum of RBS1774 are clear: it can be described by a single absorbed blackbody with $kT \sim 101$ eV. Assuming a distance for the source of 300 pc, the effective radius of the blackbody is $R_{bb} \sim 2$ km. There is no significant power law component in the data.

The status of the spectral features is less clear. A line is not manifestly evident in the EPIC data at 0.4 keV. However, at these low energies, the energy resolution is poor, so that any line may be broadened to a degree that might be difficult to discern. In order to assess the possibility of a feature at 0.4 keV, an edge was introduced at this energy into the BB model in Table 1, but the optical depth of any feature was required to be insignificant by the fitting. This confirmed that a line absorbing 40% of the continuum at 0.4 keV as seen in the RGS data could not be hidden in the EPIC data as a result of the low energy resolution. On the other hand, however, the reduction in flux at 0.7 keV seen in the EPIC data, and which was fitted by an edge in Table 1, is at the 15% level in the residuals ratio, and could be accommodated in the RGS data, depending on the adopted continuum fit.

There are no known calibration features in the EPIC response matrices at 0.7 keV, and inspection of other soft well-exposed EPIC spectra (for example magnetic Cataclysmic Variables) showed no features at these energies. On the other hand, features with an absorption of 40% would easily be visible in RGS spectra of more luminous objects, but are not. It may be the case that the RGS calibration is intensity dependent.

Without further data it is not possible to resolve these inconsistencies. However, even though the presence or not of a line at 0.4 keV is unclear, it is reasonable to accept the reality of the 0.7 keV feature as assumed by Zane et al. (2005).

4 Timing analysis

The EPIC pn and MOS photon event files were analysed using a maximum likelihood periodogram technique (Zane et al. 2002; Cropper et al. 2004). The data were searched for periods in the range 10 ks down to 30 ms. Periods below 300 ms used only the EPIC pn data because of the time resolution of the EPIC MOS small window mode. A significant period (6σ with respect to the noise level) was found at 9.437 s. This is shown in Fig. 4.

In any large spectral range in a power spectrum, apparently significant peaks will occur by chance. The significance of the 9.437 s peak for periods in the 1–1000 s range is, however, secure at the 4σ level above any such peaks ($\Delta\chi^2 = 18.7$, 59638 independent periods). Moreover, although the 9.437 s period is present but not significant in the EPIC MOS data alone, when folded and averaged into phase intervals on the combined period, it can be seen (Fig. 5) that the phase and amplitude of the MOS and pn data are similar. The semi-amplitude of the 9.437 s variation is $3.6 \pm 0.6\%$ as determined from sinusoidal fits to the combined data.

The X-ray variation requires some temperature distribution over the surface of the star. This is consistent with the small polar cap inferred (for d = 300 pc) from the blackbody fit to the X-ray flux.

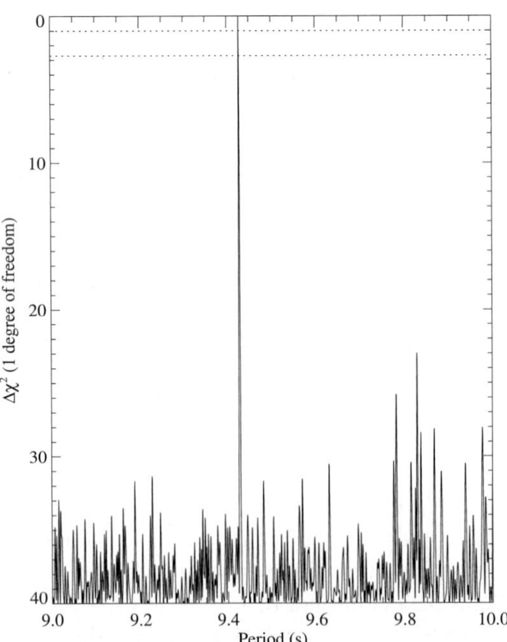

Fig. 4 The maximum likelihood periodogramme (MLP) of the combined EPIC data in the period range 9–10 s. The significance in $\Delta\chi^2$ can be read directly from the MLP, and the dotted lines near the peak show the period range for the 68% and 90% confidence levels

5 Discussion

5.1 Comparison with other XDINS

The main properties of the seven XDINS are shown in Table 3. This indicates that XDINS all have blackbody spectra with temperatures $kT \sim 50$–100 eV and no powerlaw component. All of those objects with detected periods are in the relatively narrow range of 3–12 s and the amplitude is <20%. Most XDINS have lines in their X-ray spectra in the range 0.2–0.8 keV. It is perhaps notable that RBS1774 is the hottest of the XDINS, as well as being the XDINS with the highest energy line feature.

5.2 Spectral features

Assuming the 0.7 keV line is real, there are several possibilities as to its origin. It could be a proton cyclotron line if $B \sim 1.4 \times 10^{14}$ G or a H ionisation edge (Ho et al.

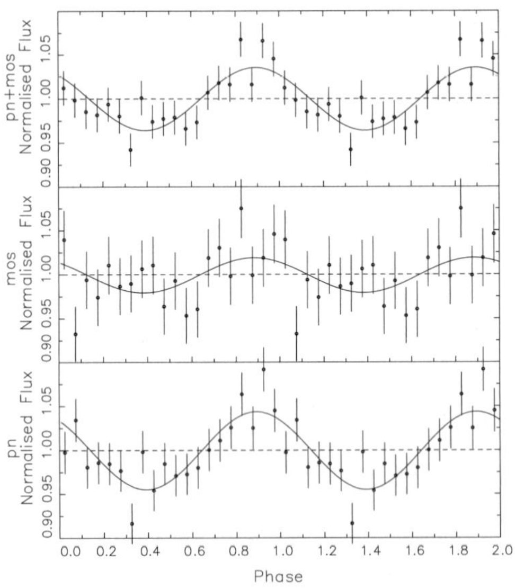

Fig. 5 The combined EPIC (*top*), MOS only (*middle*) and pn only (*bottom*) light curves folded on the 9.437 s period, together with the best fits for a sinusoidal variation in each case

Table 3 Summary of XDINSs properties. References are as follows: [1] Burwitz et al. (2003), [2] Haberl et al. (2004a), [3] Van Kerkwjik et al. (2004), [4] Haberl et al. (2003), [5] Haberl et al. (2004b), [6] Zane et al. (2005)

Source	kT_{bb}^{∞} eV	n_H 10^{20} cm^{-2}	$E_{\rm line}$ eV	P s	Semi-ampl.	Reference
RX J1856-3754	56.7	0.18	no	no	no	1
RX J0720-3125	85.2	1.38	270	8.39	11%	2
RX J1605.3+3249	94.1	0.68	493	no	no	3
RX J1308.6+2127	85.8	4.10	290	10.31	18%	4
RX J0420-5022	44.9	1.02	329	3.45	13%	5
RX J0806-4123	95.6	0.41	460	11.37	6%	5
RBS 1774	104.0	3.65	700	9.44	4%	6

2003) or He bound-bound transitions (Sanwal et al. 2002; Zavlin et al. 2004), again requiring $B \sim$ few $\times\, 10^{14}$ G. Otherwise, were the fields lower, it could be an electron cyclotron line if $B \sim 7.5 \times 10^{10}$ G or the result of atomic transitions of mid-Z H-like ion (such as CIV, NVII or OVIII) with $B \sim 10^{11}$ G. These latter fields are rather lower than almost all neutron star/ radio pulsar fields, so that the first two alternatives are to be preferred. If so, the field is magnetar-like, and similar to that in the XDINS RX J1605.3+3249 and RX J0806.4-4123. The spindown rate from dipole radiation for $B \sim$ few $\times\, 10^{14}$ G fields and $P = 10$ s is $\dot{P} \sim 10^{12}$ s s^{-1}.

5.3 Cooling age

As RBS1774 is the hottest XDINS, it is likely to be one of the youngest. The cooling age τ_c depends on the neutron star mass, and is somewhat model dependent, depending on what cooling effects are included. Using the Yakovlev et al. (2004) cooling model including proton superfluidity, the age for $kT \sim 100$ eV ranges from $\tau_c = 10^2$–10^5 years, depending on the mass and whether their 1p or 2p model is used. However, for masses ≤ 1.4 M$_\odot$, the cooling age is $\tau_c \sim 10^4$–10^5 yr.

5.4 Magnetic field decay age

If RBS1774 has a high magnetic field, its age can also be computed from the field decay time using the equations in Cropper et al. (2004):

$$B_0 = (P\dot{P})^{1/2}\left[b^{\frac{\alpha-2}{2}} - \frac{\alpha-2}{2}\frac{a}{b}(\dot{P})^{\frac{\alpha-2}{2}}(P^2 - P_o^2)\right]^{\frac{1}{2-\alpha}},$$

$$\tau_d = (a\alpha B_0^{\alpha})^{-1}\left[\left[1 - \frac{2-\alpha}{2}\frac{a}{bB_0^{2-\alpha}}(P^2 - P_o^2)\right]^{\frac{\alpha}{\alpha-2}} - 1\right]$$

where B_0 is in units of 10^{13} G, τ_d is the field decay age, P is in seconds, $\dot{P}$ in s/yr and the age is in units of 10^6 yr. P_0 is the period at the star birth, $b \approx 3$, and the parameters a, α discriminate between the three decay laws considered by Colpi et al. (2000). If $\dot{P} \sim 10^{-12}$ s s^{-1} is estimated from dipole radiation, and the current spin period of 9.437 s is used, then the age as determined by the magnetic field decay is $\tau_d \sim 10^4$ yr if the decay is by Hall cascade, or $\tau_d \sim 10^5$ yr if the decay is by ambipolar diffusion. The field decay τ_d and thermal τ_c ages are consistent for a neutron star mass of 1.35–1.5 M$_\odot$. More details are given in Zane et al. (2005).

6 Conclusions

The *XMM-Newton* observations show that RBS1774 has a thermal spectrum with $kT = 101$ eV and a low amplitude sinusoidal variation with a period of 9.437 s. A power law component can be excluded, and neutron star atmosphere models as in Zavlin et al. (1996) do not provide good fits.

The EPIC data require a line or edge at 0.7 keV. Such a feature can be accommodated in the RGS data for particular choices of continuum; however a significant feature at 0.4 keV in the RGS data cannot be accommodated within the EPIC fits. It is not clear how to resolve this discrepancy unless the EPIC or RGS calibration is incorrect at these energies, at least for the particular characteristics of this source. A line at 0.7 keV suggests magnetar-like magnetic fields for RBS1774, with an origin either as a proton cyclotron line, or from an H ionisation edge or He bound-bound transitions.

Cooling models (Yakovlev et al. 2004) predict ages of $\sim 10^5$ yr for a 1.35 M$_\odot$ neutron star down to 10^2 yr for a neutron star with mass 1.6 M$_\odot$. Ages from field decay arguments are determined to be 10^4–10^5 yr. The two mass determinations are consistent (for the models used here) for 1.35–1.5 M$_\odot$.

Acknowledgements We are grateful to Marten van Kerkwijk for discussions regarding the RGS data.

References

Burwitz, V., Haberl, F., Neuhäuser, R., et al.: Astron. Astrophys. **399**, 1109 (2003)

Colpi, M., Geppert, U., Page, D.: Astrophys. J. **529**, L29 (2000)

Cropper, M., Haberl, F., Zane, S., et al.: Mon. Not. Roy. Astron. Soc. **351**, 1099 (2004)

Haberl, F.: In: Proc. 2005 EPIC XMM-Newton Consortium Meeting—5 years of Science with XMM-Newton, Schloss Ringberg, 11–13 April 2005. MPE Report 288, p. 39. MPI, Garching (2005)

Haberl, F., Schwope, A.D., Hambaryan, V., et al.: Astron. Astrophys. **403**, L19 (2003)

Haberl, F., Zavlin, V.E., Trümper, J., et al.: Astron. Astrophys. **419**, 1077 (2004a)

Haberl, F., Motch, C., Zavlin, V.E., et al.: Astron. Astrophys. **424**, 635 (2004b)

Ho, W.C.G., Lai, D., Potekhin, A.Y., et al.: Astrophys. J. **599**, 1293 (2003)

Page, M.J., Soria, R., Wu, K., et al.: Mon. Not. Roy. Astron. Soc. **345**, 639 (2003)

Sanwal, D., Pavlov, G.G., Zavlin, V.E., et al.: Astrophys. J. **574**, L61 (2002)

Van Kerkwjik, M.H., Kaplan, D.L., Durant, M., et al.: Astrophys. J. **608**, 432 (2004)

Yakovlev, D.G., Gnedin, O.Y., Kaminker, A.D., et al.: Adv. Space Res. **33**, 523 (2004)

Zampieri, L., Campana, S., Turolla, R., et al.: Astron. Astrophys. **378**, L5 (2001)

Zane, S., Haberl, F., Cropper, M., et al.: Mon. Not. Roy. Astron. Soc. **334**, 345 (2002)

Zane, S., Cropper, M., Turolla, R., et al.: Astrophys. J. **637**, 397 (2005)

Zavlin, V.E., Pavlov, G.G., Shibanov, Yu.A.: Astron. Astrophys. **315**, 141 (1996)

Zavlin, V.E., Pavlov, G.G., Sanwal, D.: Astrophys. J. **606**, 444 (2004)

Astrophys Space Sci (2007) 308: 167–169
DOI 10.1007/s10509-007-9346-6

ORIGINAL ARTICLE

Persistent and transient blank field sources

A. Treves · S. Campana · M. Chieregato · A. Moretti ·
T. Nelson · M. Orio

Received: 29 June 2006 / Accepted: 11 September 2006 / Published online: 24 March 2007

Abstract Blank field sources (BFS) are good candidates for hosting dim isolated neutron stars (DINS). The results of a search of BFS in the ROSAT HRI images are revised. We then focus on transient BFS, arguing that they belong to a rather large population. The perspectives of future research on DINS are then discussed.

Keywords Stars: neutron

PACS 97.60.Jd

1 Introduction

Isolated neutron stars, which have overcome the pulsar phase are elusive sources. In principle they can shine from some residual internal energy (coolers), or because of their interaction with the interstellar medium (e.g. accretors). Their number in the Galaxy should be very high, about one percent of the total number of stars. Their significance as a population is of the utmost interest: they are the end-point of the evolution of a vast class of stars.

It is just because of these considerations that the discovery of dim isolated neutron stars by the ROSAT satellite has been a major achievement (see e.g. Treves et al. 2000; Haberl 2005; Zane et al. 2005).

DINS are one of the main attractions of this meeting, see in particular the presentations by Cropper and Popov. Their properties can be summarized as follows: softness $T \sim 100$ eV; closeness $d \sim 100$ pc; extremely dim optical counterparts ($V > 25$), periodicities of 5–10 s; absorption features below 1 keV. The seven DINS discovered thus far are probably a mixed bag, in the sense that the above properties may not appear all together. For instance the prototype of the class 1856-37, has a perfect black body spectrum in the X-ray band with neither absorption lines nor indications of pulsations. Most likely they are all coolers (Neuhäuser and Trümper 1999, Popov et al. 2000).

In order to further improve our knowledge of DINS it is mandatory to enlarge their sample. The procedure followed up to now to discover new DINS has been to search the ROSAT images for the so called "Blank Field Sources" (BFS, Cagnoni et al. 2002), i.e. X-ray sources without counterparts in other spectral bands, and use the properties listed above to argue that the candidate belongs to the class. This line was pursued by a number of authors, we mention in particular the recent paper by Agueros et al. (2006) where the ROSAT PSPC images (RASS) are compared with the Sloan Digital Sky Survey.

Here we focus on progresses of our search of BFS in the ROSAT HRI images (Chieregato et al. 2005), concentrating on the possible detection of transient BFS.

2 The ROSAT-HRI blank field sources

The ROSAT HRI fields cover ~3% of the sky but the advantage with respect to the PSPC is that the position of the

A. Treves (✉) · M. Chieregato
Insubria University, via Valleggio 11, Como, Italy
e-mail: aldo.treves@uninsubria.it

S. Campana · A. Moretti · M. Orio
INAF – Osservatorio Astronomico di Brera, Merate, Italy

M. Chieregato
Zürich University, Zuerich, Switzerland

T. Nelson
University of Wisconsin, Madison, WI, USA

Table 1 Blank Field Sources from the HRI Rosat catalogue, adapted from Chieregato et al. 2005

Source	Flux $10^{-13}\frac{\mathrm{erg}}{\mathrm{cm}^2\,\mathrm{s}}$	Prob. σ	Cts	f_X/f_{opt}	Opt. σ
0421-51	6.5	14.0	742	>141	5.3
1357+18	3.5	4.2	112	>47	6.3
2007-48	3.0	4.3	55	>65	5.2

source is much better determined, therefore the limit set by the absence of counterparts can be brought to a deeper level.

The ~30 000 sources of the ROSAT HRI Brera wavelet catalogue (Panzera et al. 2003) have been searched for objects (a) with extreme f_X/f_{opt}, (b) not too faint, (c) with total number of photon above a given threshold. Excluding known sources, three objects have been found which have a statistical significance $> 4\sigma$, and with the closest counterpart at $> 4\sigma$ (see Table 1). With respect to Chieregato et al. (2005) 0433+15 was excluded since it was recognized as spurious.

The brightest source is 0421-57. It is close to a bright star (see Chieregato et al. 2005, Fig. 2). It has been detected with the PSPC at essentially the same level revealed by the HRI. The source is soft (~0.2 keV, see also Sect. 4) but not as soft as other typical DINS. The two other sources are much weaker, and they have not been detected with HRI or PSPC when observed at different epochs. We will refer to them as transient BFS. Their light curve have been examined and we can exclude spike-like emission of duration of seconds or minutes.

3 New observations

A program of X-ray observation of the three sources with the Swift XRT is ongoing (P.I. Moretti). We observed all three BFS: 0421-57 (11 ks), 1357+18 (9 ks) and 2007-48 (8 ks). The last two sources were not detected, resulting in 0.3–10 keV 3σ upper limits of 1.8×10^{-3} counts s^{-1} and 2.2×10^{-3} counts s^{-1}, respectively. Assuming a power law spectrum with photon index 2 and a Galactic column density (2.1 and 5.1×10^{20} cm^{-2}) we obtain 5×10^{-14} and 8×10^{-14} erg cm^{-2} s^{-1} as upper limit on the unabsorbed flux in the 0.3–10 keV band. In the case of 0421-57 we detect the source at a rate of $(2.2 \pm 0.2) \times 10^{-2}$ counts s^{-1} (about 200 counts) but we are evaluating the contamination from a bright star closeby. The spectrum is very soft and consistent either with a black body (250 ± 30 eV) or a double Raymond-Smith model.

Several optical campaigns are now in progress. 1357+18 was observed with the I and R filters and the MiniMo camera with the 3.5 WIYN telescope in 10 minutes exposures on 2004 June (P.I.M. Orio). The seeing was about

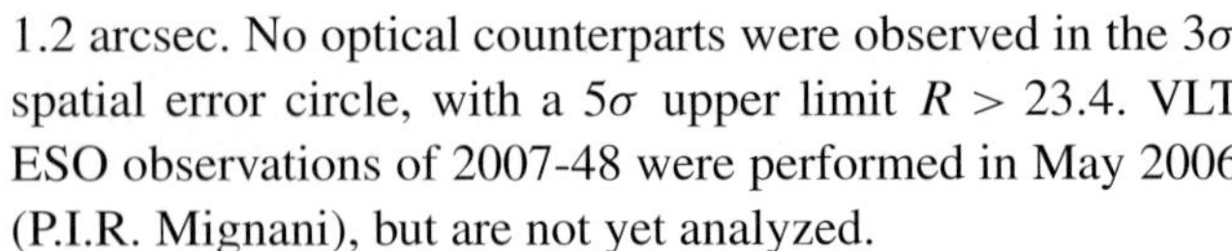

1.2 arcsec. No optical counterparts were observed in the 3σ spatial error circle, with a 5σ upper limit $R > 23.4$. VLT ESO observations of 2007-48 were performed in May 2006 (P.I.R. Mignani), but are not yet analyzed.

4 Discussion

We consider in particular the two transient BFS (1357+18, 2007-48). Note that the statistical significance is formally 4σ, and the total number of photons ~100. It is obvious that one must be extremely cautious about the reality of the sources, and because they are supposedly transient, one can't test with further observations.

In the following we suppose that the sources are real: the optical counterparts are dim indeed. What can transient BFS be? An extragalactic origin seems unlikely, because if they were some kind of BL Lac object, a persistent radio-emission would be expected. Gamma ray bursts are probably to be excluded too, because as noted above the light curves do not show short term variability. One should exclude also binaries with a non collapsed companion, because this should show up in the optical band. One is left to systems consisting of collapsed objects. The key point is that the population of which we have tentatively detected two members could be quite numerous. In fact the HRI field is ~0.2 deg^2, the total exposure time was $\sim 3 \times 10^7$ s. If the distribution of sources were isotropic this would translate in a rate of 10^5 transient BFS per year, otherwise the number should be scaled with the solid angle. The large parent population points to isolated neutron stars or white dwarfs. In particular one may wonder if transient BFS are related to a sudden release of the internal energy of a neutron star, a process which may be at work in the recently discovered transient radio pulsars (McLaughlin et al. 2006; Lyne, this conference).

5 Conclusions

In the fifteen years of research about DINS the progress was remarkable, yet there are two basic points that have not yet been achieved:

- There is still no example of a DINS which is convincingly powered by the accretion of the interstellar medium. While there are a number of arguments indicating that these objects should be much rarer than originally estimated (Treves and Colpi 1991; Blaes and Madau 1993; Perna et al. 2003 and references therein), these objects should finally show up, and their emission should be largely independent of the neutron star age. Neutron stars as old as the Galaxy could be a part of the lot.

- We have not yet any information on the cousins of DINS, i.e. isolated black holes (see e.g. Agol and Kamionkowski 2002).

The challenge for the future is obviously to explore the sky for DINS and their cousins, at a flux threshold which is an order of magnitude lower than that of ROSAT, and which is easily accessible to present generation X-ray telescopes. There is no doubt that the activity up to now has been rather slow, since rich X-ray and optical archives are already available.

Let us summarize the hopes for the future, which can derive from a thorough study of the existing data:

- discovery of accretion fed DINS;
- establishing or excluding the existence of transient BFS;
- discovery of isolated black holes;
- discovery of intermediate mass black holes possibly related to ultraluminous X-ray sources (e.g. Mapelli et al. 2006).

References

Agol, E., Kamionkowski, M.: Mon. Not. Roy. Astron. Soc. **334**, 553 (2002)
Agueros, M.A., Anderson, S.F., Margon, B., et al.: Astron. J. **131**, 1740 (2006)
Blaes, O., Madau, P.: Astrophys. J. **403**, 690 (1993)
Cagnoni, I., Elvis, M., Kim, D.W., et al.: Astrophys. J. **579**, 148 (2002)
Chieregato, M., Campana, S., Treves, A., et al.: Astron. Astrophys. **444**, 69 (2005)
Haberl, F.: In: Proceedings of the 2005 EPIC XMM-Newton Consortium Meeting. MPE Rep. **289**, 39 (2005)
Mapelli, M., Ferrara, A., Rea, N.: Mon. Not. Roy. Astron. Soc. **368**, 1340 (2006)
McLaughlin, M.A., Lyne, A.G., Lorimer, D.R., et al.: Nature **439**, 817 (2006)
Neuhäuser, R., Trümper, J.E.: Astron. Astrophys. **343**, 151 (1999)
Panzera, M.R., Campana, S., Covino, S., et al.: Astron. Astrophys. **399**, 351 (2003)
Perna, R., Narayan, R., Rybicki, G., et al.: Astrophys. J. **594**, 936 (2003)
Popov, S., Colpi, M., Prokhorov, M.E., et al.: Astrophys. J. **544**, L53 (2000)
Treves, A., Colpi, M.: Astron. Astrophys. **241**, 107 (1991)
Treves, A., Turolla, R., Zane, S., et al.: Publ. Astron. Soc. Pac. **112**, 297 (2000)
Zane, S., Cropper, M., Turolla, R., et al.: Astrophys. J. **627**, 397 (2005)

Astrophys Space Sci (2007) 308: 171–179
DOI 10.1007/s10509-007-9344-8

ORIGINAL ARTICLE

The Magnificent Seven in the dusty prairie

The role of interstellar absorption on the observed neutron star population

B. Posselt · S.B. Popov · F. Haberl · J. Trümper · R. Turolla · R. Neuhäuser

Received: 30 June 2006 / Accepted: 27 July 2006 / Published online: 23 March 2007

Abstract The Magnificent Seven have all been discovered by their exceptional soft X-ray spectra and high ratios of X-ray to optical flux. They all are considered to be nearby sources. Searching for similar objects with larger distances, one expects larger interstellar absorption resulting in harder X-ray counterparts. Current interstellar absorption treatment depends on chosen abundances and scattering cross-sections of the elements as well as on the 3D distribution of the interstellar medium. After a discussion of these factors we use the comprehensive 3D measurements of the Local Bubble by Lallement et al. (2003, Astron. Astrophys. **411**, 447) to construct two simple models of the 3D distribution of the hydrogen column density. We test these models by using a set of soft X-ray sources with known distances. Finally, we discuss possible applications for distance estimations and population synthesis studies.

Keywords Neutron stars · Absorption · ISM · X-ray:general

PACS 97.60.Jd · 98.38.-j

B. Posselt · F. Haberl · J. Trümper
Max-Planck-Institut für extraterrestrische Physik,
Giessenbachstrasse, 85748 Garching, Germany

B. Posselt · R. Neuhäuser
Astrophysikalisches Institut und Universitäts-Sternwarte,
Schillergäßchen 2-3, 07745 Jena, Germany

Present address:
B. Posselt (✉)
Observatoire Astronomique de Strasbourg, 11 rue de l'Université,
67000 Strasbourg, France
e-mail: posselt@quasar.u-strasbg.fr

S.B. Popov
Sternberg Astronomical Institute, Universitetskiy pr. 13,
Moscow 119992, Russia
e-mail: polar@sai.msu.ru

R. Turolla
INFN-Sezione di Padova, Dipartimento di Fisica Galileo Galilei,
Via Marzolo 8, 35131 Padova, Italy

1 Introduction

The Magnificent Seven (M7), as the ROSAT-discovered X-ray thermal isolated neutron stars are sometimes dubbed, are exceptional because of their soft blackbody-like spectra without non-thermal components; for reviews see e.g. Haberl (2007) or Haberl (2004). One of them has the second-best blackbody spectrum after the cosmic background radiation and this lead to one of the best neutron star radius estimates known (Trümper et al. 2004). Thought to be nearby, the M7 represent nearly half of the local young neutron star population. Despite several searches for new candidates of this radio-quiet neutron star class (e.g. Agüeros et al. 2006; Chieregato et al. 2005; Rutledge et al. 2003), no new M7-like object has been confirmed up to now. One aggravating circumstance these searches face is the hardening of the X-ray spectrum due to the interstellar absorption. Caused mainly by photo-electric absorption of photons by heavy elements, the interstellar absorption is usually described by the equivalent hydrogen column density assuming certain chemical abundances. Highly important is the clumpiness of the interstellar medium (ISM) causing (un)favourable lines of sights to search for new M7-like objects. The knowledge about the local ISM structure is also important in connection with the long-debated issue of isolated neutron stars accreting from the ISM.

1.1 Abundances and cross-sections

Most absorption models today presume element abundances independent of the line of sight. It was outlined by Wilms et al. (2000) that the total gas plus dust ISM abundances are lower than the local—Solar—abundances and the general ISM abundance uncertainties are still in the order of 0.1 dex because the measurements are very difficult. One can choose between several different abundance tables implemented e.g. in *XSPEC* for analysing X-ray spectra. We note here only two examples commonly used—the abundance tables by Anders and Grevesse (1989) ('angr' in *XSPEC*) and by Wilms et al. (2000) ('wilm' in *XSPEC*) considering newer measurements by e.g. Snow and Witt (1996); Cardelli et al. (1996); Meyer et al. (1997, 1998).

The individual abundances and photoionization cross-sections of the elements and their ions as well as dust grain properties (e.g. sizes, shapes) influence the total photoelectric absorption of X-rays. Again, we mention here only two exemplary works used very often for analysis of X-ray spectra. Morrison and McCammon (1983) did a polynomial fit of the effective absorption cross-section per hydrogen atom based on measured atomic absorption cross-sections by Henke et al. (1982) and abundances mostly from Anders and Ebihara (1982). With the exception of oxygen, Morrison and McCammon (1983) assume the elements to be either entirely in the gas phase or completely depleted into dust grains. The recent improved absorption treatment by Wilms et al. (2000) takes into account newer element abundances as noted above and more recent photoionization cross-section calculations by e.g. Verner et al. (1993); Verner and Yakovlev (1995). Wilms et al. (2000) additionally include an improved molecular cross-section for H_2. Besides this updated database Wilms et al. (2000) considered furthermore a simple spherical composite dust grain model in an MRN distribution (Mathis et al. 1977).

Both models are implemented in *XSPEC* and can be used via the routines '*wabs*' (Morrison and McCammon 1983) or '*tbabs*' (Wilms et al. 2000). The various absorption descriptions can lead to different results (see Fig. 1) or expectations of e.g. observable objects. The differences caused by the diverse treatments show our incomplete knowledge about the interstellar absorption and should be kept in mind especially when interpreting soft X-ray spectra.

1.2 The inhomogeneously distributed ISM

In X-ray astronomy with low spectroscopic resolution the chosen abundances and cross-sections are used to derive the hydrogen column density responsible for the instellar absorption. Cold and warm H-atoms (H I) as well as hydrogen molecules and ionized hydrogen (H II) contribute to the overall amount of hydrogen along a line of sight. These

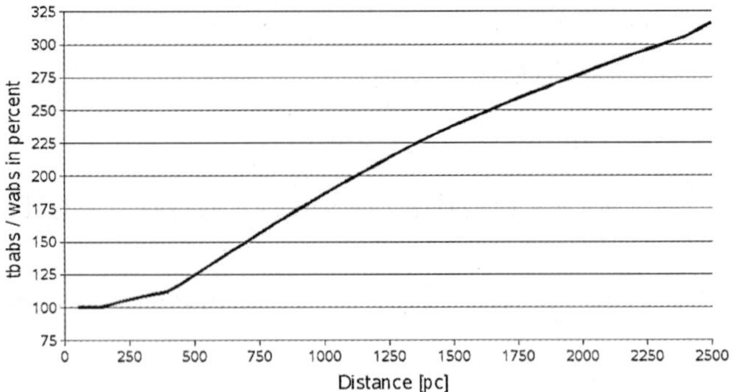

Fig. 1 The effect of different abundances and cross-sections on the absorption towards soft X-ray sources. For the same neutron star (blackbody of 90 eV, 11.67 km radius, and 1.48 Solar masses) the ratio of the X-ray fluxes (as would be expected to be observed by ROSAT) resulting from the two *XSPEC* routines '*tbabs*' and '*wabs*' are calculated for different distances. The galactic coordinate x varies from 50 to 2500 pc (direction: from the Sun towards the galactic center), y is always 0, z is 25 pc north. For the ISM distribution, we applied the same as used by Popov et al. (2000), which is described in more detail below. The effect of less absorption, resulting from the use of the newer routine '*tbabs*', is stronger at larger column densities, hence distances

different components of the hydrogen column density are estimated by different measurement methods, e.g. various extinction measurements as summarized by Knude (2002) or radio 21-cm observations of hydrogen as the extensive study by Dickey and Lockman (1990) and more recently by Kalberla et al. (2005). A wealth of theoretical models has been applied to these observations. However, most of them are only 2D with few more recent exceptions (see e.g. Drimmel et al. 2003; Amôres and Lépine 2005 or Marshall et al. 2006 for 3D models). The overall result is a highly inhomogeneously distributed interstellar medium within the Milky Way, present in bubbles with loops or dense rims, and shaped by molecular clouds, supernova explosions and stellar activity (for an illustrated overview see Henbest and Couper 1994). The ISM distribution is best known in the close Solar neighbourhood, especially in the Local Bubble (e.g. Breitschwerdt 1998). The most detailed view is provided by the study of Lallement et al. (2003), who measured NaI absorption towards 1005 sightlines with Hipparcos distances up to 350 pc from the Sun; for more details on the measurements and the applied density inversion method see Lallement et al. (2003), the precursor paper Sfeir et al. (1999), and Vergely et al. (2001). NaI is regarded as good tracer of the total amount of the neutral interstellar gas and the sodium column density is thought to be convertible into a hydrogen column density (e.g. Ferlet et al. (1985), but see also discussion in Sect. 2). For more distant regions the determination of a 3D distribution of the ISM is hampered mainly by the unknown or imprecise distances of the measured objects and the lack of a well-sampled grid of measurement points. This is true for all extinction-related methods to our knowledge and results in much more coarse sampling compared to the Solar neighbourhood. The results obtained by the various methods can differ for the same regions and a combination can be difficult. These difficulties result partly from the different observed astrophysical phenomena like measuring the dust emission or stellar magnitudes. They also result from different sets of stellar absolute magnitude calibrations and the underlying assumptions which can deviate from each other as interstellar extinction determination is an iterative process coupled with the identification of the statistical properties of stars (Hakkila et al. 1997).

2 Two simple 3D distribution model cubes of the absorbing ISM

Considering the absorption in the Solar neighbourhood, the result by Lallement et al. (2003) is a good starting point representing the currently best database of the local ISM distribution. R. Lallement kindly provided us with the NaI density cube derived using the inversion method developed by Vergely et al. (2001). Due to the smoothing length of 25 pc applied in this method by Lallement et al. (2003), structures smaller than 25 pc cannot be resolved. We note further that even if measurements go up to Hipparcos distances of 350 pc from the Sun, the sampling becomes coarser at larger distances. Thus the sodium density cube has a span of only 250 pc and starting from 200 pc one has to be careful dealing eventually with the a priori density information applied in the inversion method (R. Lallement, personal communication). From the NaI density cube we calculate the column density for the same grid (sampling 3.9 pc). This is then converted to H I column density applying the formula by Ferlet et al. (1985).

It has to be mentioned that there are on-going discussions about how well the sodium D-line absorption actually traces H I. The correlation found between the H I and Na column densities derived from Na-D absorption lines by Ferlet et al. (1985) was doubted by Welty et al. (1994), especially for low column densities ($\log N(\mathrm{NaI}) < 11$). The region of the Lallement cube that should not be used because it has such low column densities is indicated in Fig. 2.

However, the lowest known X-ray–measured hydrogen column density for one of the M7 converts to a sodium column density of $\log N(\mathrm{NaI}) > 11.5$ (applying the formula by Ferlet et al. 1985). Therefore this low-column density uncertainty is not important considering the M7 and negligible for the intended population synthesis with neutron stars (NSs) at usually larger distances.

Vergely et al. (2001) could not always find a correlation of the NaI density with the H I density. They concluded that the correlation is rather weak due to non-constant population ratios for interstellar medium species, thus different abundances. Lallement et al. (2003) noted the size of the

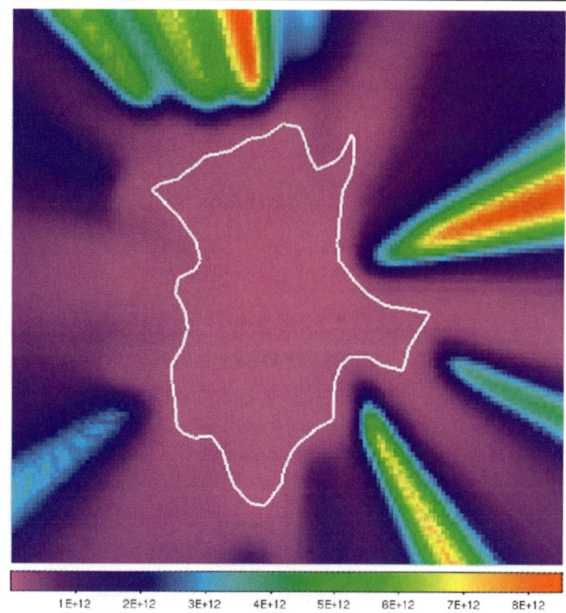

Fig. 2 The limits of using the Ferlet et al. (1985) formula. The sodium column density in units of cm^{-2} as derived from the density data by Lallement et al. (2003) is shown in galactic Cartesian coordinates at $z = 0$; x is towards the galactic center which is at the right, y towards $l = 90°$ is pointing up. The image spans 500 pc times 500 pc. The area within the contour has low column densities ($\log N(\mathrm{NaI}) < 11$) for which Welty et al. (1994) doubted whether the Ferlet et al. (1985) formula is applicable. This region of small sodium column densities is largest at $z = 0$ and becomes smaller for lower or larger z

Local Bubble revealed by H I is smaller than when derived by NaI. This was explained by the first ionization stage of NaI being below that of H I, resulting in a longer neutral phase of H I (Lallement et al. 2003). Recently, Hunter et al. (2006) presented new ultraviolet NaI observations towards 74 O- and B-type stars in the galactic disk. Where possible they also derived column densities of the NaI D-lines. Then they compared the correlation between the NaI and H I column densities. While there was an excellent agreement with Ferlet et al. (1985) for the D-lines, this was not the case for their ultraviolet absorption lines. Hunter et al. (2006) found a significant offset for the correlation they got from the NaI D-lines compared to the relation derived from the UV transitions, even after accounting for saturation effects in the D-lines. They argue that this offset is due to an underestimation of approximately 30% of $N(\mathrm{H})$ in the case of NaI D-lines observations. Recapitulating, one has to be aware of the still not completely solved problems concerning the conversion of NaI to H I column densities when it comes to interpreting individual results. However, we regard the measurements by Lallement et al. (2003) and the conversion to a hydrogen column density according to Ferlet et al. (1985) as the best possibility at hand to take into account the local inhomogeneity in the ISM for X-ray astronomy with low spectral resolution.

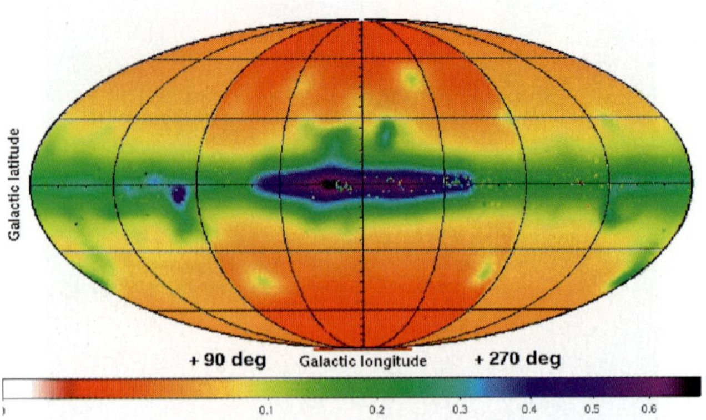

Fig. 3 The $N(\mathrm{H})$ predicted by the analytical model at 1 kpc. Positive galactic latitudes are up and $l = 0°$ is in the center. The $N(\mathrm{H})$ is in units of 10^{22} cm^{-2}. For description of the model and discussion see text

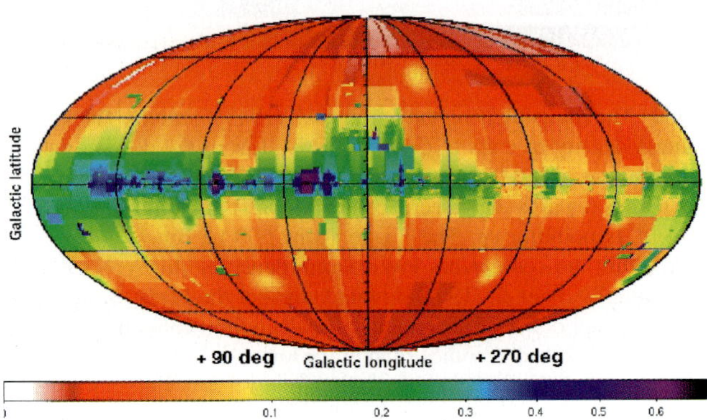

Fig. 4 The $N(\mathrm{H})$ predicted by the extinction model based on Hakkila et al. (1997) at 1 kpc. Positive galactic latitudes are up and $l = 0°$ is in the center. The $N(\mathrm{H})$ scale is in unities of 10^{22} cm^{-2}. For description of the model and discussion see text

At larger distances we consider two different models—one is based completely on extinction measurements and another one is described by analytical formulae. Hakkila et al. (1997) put several extinction studies carefully together in an easily accessible routine. All studies have been modified to statistically account for unsampled regions. Additionally, a correction method for the systematic underestimation between 1 and 5 kpc was developed. Errors were provided individually for each considered main survey and for their mean. The often large mean error values illustrate the disagreement among individual observations. The model by Hakkila et al. (1997) is a large scale model, capable of identifying e.g. molecular clouds at intermediate distances, but not sensitive to extinction variations of less than 1 degree. It is well suited for statistical studies.

The analytical model we apply is described by Popov et al. (2000). It is based on the formulae by Zane et al. (1995), Dickey and Lockman (1990), de Boer (1991). Popov et al. (2000) included additionally a galactocentric radius dependency for the number density of atomic and molecular hydrogen as estimated by Bochkarev (1992). The Local Bubble was taken into account by Popov et al. (2000) as a sphere of 140 pc radius having a constant low density of 0.1 particles cm^{-3}. As noted above, there are other more recent theoretical 3D extinction models. However, the analytical H density model by Popov et al. (2000) seems to be at least as good as e.g. the underlying model used by Amôres and Lépine (2005). The model of Drimmel et al. (2003) cannot be applied to distances less than a few hundred parsecs in the Solar neighbourhood and the data of Marshall et al. (2006) has a coarse sampling with distance bins of 1 kpc.

Both described models are applied only for distances larger than 230 pc in case they provide an $N(\mathrm{H})$ larger than derived from the sodium cube at 230 pc. Otherwise the hydrogen column density at 230 pc is taken since the column density can only increase with distance. We use spherical coordinates—the galactic coordinates l, b and distance d. The nominal sampling is one degree in l and b, and 10 pc for the distance. This is technically motivated and does not represent the actually achieved accuracy. Distances are covered up to 4500 pc.

2.1 Testing the models and further improvements

To test our models we first consider a relatively large number of test objects with good distances and extinction measurements, not necessarily determined from X-rays. Then we proceed to a few of the rare neutron stars with well known distances, having also small error bars for the absorbing $N(\mathrm{H})$ derived by X-ray observations.

As Hakkila et al. (1997) note, different studies do not agree precisely with each other due to the various applied methods or objects studied. The latter are usually very coarsely scattered and measurement errors can influence the extinction values of a large region. One possibility to overcome this problem is investigating open clusters. An open cluster has the advantage that one can measure a number of stars at approximately the same distance, which is usually well known. Therefore, open clusters are a good choice in aiming to minimize the extinction measurement errors along one line of sight. Usually one measures the reddening $E(B-V)$. We concentrate here on the recent comprehensive reddening measurements for 650 open clusters by Piskunov et al. (2006), representing a complete sample up to distances of about 850 pc from the Sun. Details about this open cluster study can be found in Piskunov et al. (2006); Kharchenko et al. (2005a, 2005b). To convert the reddening $E(B-V)$ we apply the formula by Paresce (1984): $N(\mathrm{H}) = 5.5 \times 10^{21} E(B-V)$ [cm^{-2}]. This is quite similar to the slope found in the X-ray study for the extinction A_V by Predehl and Schmitt (1995): $N(\mathrm{H}) = 1.79 \pm 0.03 \times A_V - 0.41$ [10^{21} cm^{-2}] if $A_V = 3.1 E(B-V)$. This is noted here because there is a deviation of the correlation by Predehl and Schmitt (1995) between A_V, indicator of the dust,

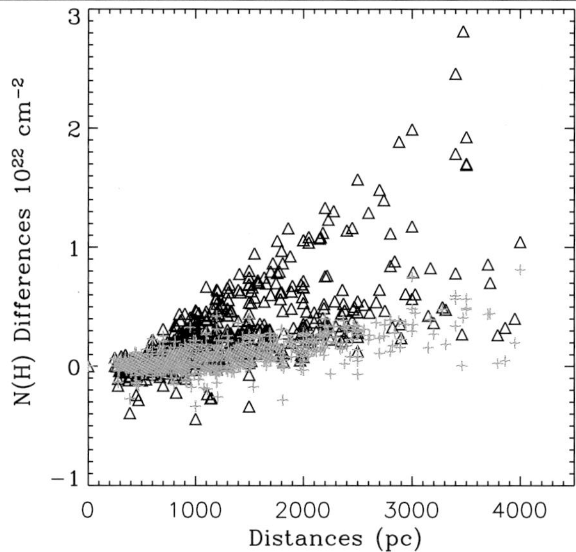

Fig. 5 $N(\mathrm{H})$ deviations at different distances. Shown are the differences between the model $N(\mathrm{H})$ at the distances of the open cluster sample and the $N(\mathrm{H})$ obtained directly from the reddening measurements of this sample. The results of the extinction model based on Hakkila et al. (1997) are plotted as grey crosses, those of the analytical model as black triangles. While the scatter at low distances is relatively small, the analytical model has a larger spread at larger distances

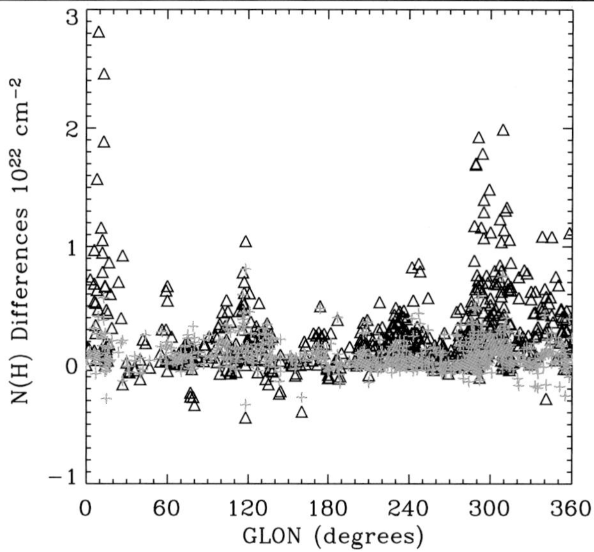

Fig. 6 $N(\mathrm{H})$ deviations at different galactic longitudes. The symbols for the models are the same as in Fig. 5. For both models a larger scatter is visible at GLON = 0°, 120°, and 300°

and $N(\mathrm{H})$, indicator for cold gas and dust, at low distances. This is probably due to a significantly lower amount of dust in the Local Bubble, as also found by Predehl and Schmitt (1995) for the source LMC X-1. Therefore, we consider only open clusters with distances larger than 230 pc.

We compared the hydrogen column densities we inferred from the open cluster reddenings with those obtained by the models. 628 out of the 650 open clusters by Piskunov et al. (2006) lie within the considered data cube with distances larger than 230 pc and smaller than 4.5 kpc. The scatter in the obtained extinction differences increases with distances for both models (see Fig. 5). For the model based on Hakkila et al. (1997) the mean deviation is around 9×10^{20} cm^{-2}, the corresponding mean value for the analytical model is 27×10^{20} cm^{-2}. Considering only distances up to 1 kpc these values are 3.3×10^{20} cm^{-2} and 8.1×10^{20} cm^{-2} respectively. Interestingly at some galactic longitudes ($l \approx 0°$, 120°, and 300°; see Fig. 6) and at the galactic latitude $b \approx 0°$ the scatter is more pronounced. The extinction routine by Hakkila et al. (1997) predicts the observed properties of the open clusters better than the analytical model. This can be partly explained by the use of (older) open cluster data (e.g. Fitzgerald 1968) in the routine by Hakkila et al. (1997). The analytical model tends to overpredict the extinction by an order of magnitude, especially at larger distances. Due to the convincing advantages of extinction data derived from open clusters, we include them finally as local inhomogeneities in our models for distances larger than 230 pc. Again, we take into account that the column density can only increase with distance. If $N(\mathrm{H})$ outside 230 pc is lower than the value at 230 pc in this direction we consequently changed the values based on the Lallement et al. (2003) to the lower value until an $N(\mathrm{H})$ increasing homogeneously with distance is reached. Aiming to apply our $N(\mathrm{H})$-models to X-ray detected neutron stars their performance for well known sources is interesting. We selected four of the rare NSs having at the same time known parallaxes and hydrogen column densities with small error bars obtained from X-ray observations. In Table 1 we present the model-derived distances for the measured $N(\mathrm{H})$ range as well as the model-derived $N(\mathrm{H})$ at the given parallactic distance. While the analytical model gives reasonable results for PSR B0656+14, Vela and RX J1856.5-37541 (RXJ1856 in the following), the extinction model based on Hakkila et al. (1997) underestimates the column density towards Vela significantly. As both ISM models are equal up to 230 pc, both give the same minimal distance of 220 pc towards Geminga which deviates from the claimed distance by Caraveo et al. (1996). Interestingly Faherty et al. (2007) reported recently a revised distance of 254^{+111}_{-59} pc towards Geminga. In the case of RXJ1856 the derived minimal and maximal distances are completely included in the $N(\mathrm{H})$ cube obtained by the Lallement et al. (2003) measurements. Here, an error of 25 pc applies while an error estimate is difficult for the analytical model at distances larger than 230 pc. The Hakkila et al. (1997) model has often a formal mean error of the same order as the obtained $N(\mathrm{H})$ values.

As noted in above it is still debated how good NaI D-lines are as tracer for hydrogen. In particular it is unclear whether molecular hydrogen is well traced. In Table 1 we include the results obtained for two further test objects situated in or

Table 1 Testing the models for individual objects

Name	Distance	Ref.	$N(\mathrm{H})$	Ref.	Analytical model			Extinction model		
					D_{MIN}	D_{MAX}	$N(\mathrm{H})$	D_{MIN}	D_{MAX}	$N(\mathrm{H})$
	[pc]		$[10^{20}\ \mathrm{cm}^{-2}]$		[pc]	[pc]	$[10^{20}\ \mathrm{cm}^{-2}]$	[pc]	[pc]	$[10^{20}\ \mathrm{cm}^{-2}]$
PSR B0656+14	288^{+33}_{-27}	1	1.73 ± 0.18	2	240	260	2.68	230	240	2.86
Vela	287^{+59}_{-34}	3	3.3 ± 0.3	4	300	330	2.73	860	910	1.07
Geminga	157^{+19}_{-17}	5	1.76 ± 0.95	6	220	300	0.09^{a}	220	240	0.09^{a}
RX J1856.5-37541	140 ± 40^{b}	7	0.74 ± 0.10	8	130	140	0.87	130	140	0.87
TY CrA	129 ± 11	9	130 ± 0.3^{c}	10	out	out	0.08	out	out	0.08
HD 18190	185	11	$> 7.21^{d}$	11	440	...	5.05	800	...	5.05

For each object the known distance and hydrogen column density $N(\mathrm{H})$ with references are given. Furthermore are listed for each model: the obtained minimal distance D_{MIN} for the lowest $N(\mathrm{H})$ value, the maximal distance D_{MAX} for the largest $N(\mathrm{H})$ value and the expected $N(\mathrm{H})$ for the known distance. The models are the same up to a distance of 230 pc (see text for details). The first 4 objects are neutron stars with VLBI or HST parallaxes and X-ray–obtained $N(\mathrm{H})$. TY CrA lies within the CrA star forming molecular cloud core close to RX J1856.5-37541. HD 18190 is thought to be situated behind MBM 12. 1: Golden et al. (2005); 2: Marshall and Schulz (2002); 3: Dodson et al. (2003); 4: Pavlov et al. (2001); 5: Caraveo et al. (1996); 6: Jackson and Halpern (2005); 7: Kaplan et al. (2002); 8: our EPIC-pn fit; 9: Casey et al. (1998); 10: Forbrich et al. (2006); 11: Hobbs et al. (1988)

[a] $\log(N(\mathrm{NaI})) = 10.6 < 11$ Ferlet et al. (1985) formula was doubted by Welty et al. (1994), see text

[b] Note the recently refined parallax by van Kerkwijk and Kaplan (2006) yielding a distance of 161^{+18}_{-14} pc

[c] Mean value

[d] Lower limit for $N(\mathrm{NaI})$ was converted to $N(\mathrm{H})$; *out*: needed $N(\mathrm{H})$ is not reached up to the here considered 1 kpc distance

close to molecular cloud cores—the star-forming site CrA close to RXJ1856 and MBM12 (Magnani et al. 1985). While towards MBM12 there is an enlarged but still underestimated amount of hydrogen (see also Lallement et al. 2003 for discussion of MBM12), the small CrA cloud is clearly missed. Both examples indicate that one has to be cautious in applying the $N(\mathrm{H})$ cube obtained from the Lallement et al. (2003) measurements in the direction of small molecular clouds. However, there are only few such density enhancements of molecular hydrogen in the close Solar neighbourhood. The extinction model as well as the analytical model are in principle sensitive to molecular hydrogen. At large distances >1 kpc the extinction model may miss molecular clouds due to their low angular size and bad sampling while the analytical model does not consider individual clouds.

3 Applications of the models

3.1 Distance estimates

A possible application is to estimate the distance to neutron stars without parallaxes or other distance measurements. This is the case of the majority of the M7 where despite for RXJ1856 only a preliminary parallactic distance of 330^{+170}_{-80} pc is reported for RX J0720.4-3125 by van Kerkwijk and Kaplan (2007). We obtain $N(\mathrm{H})$ by considering the best *XSPEC*-fits for blackbody and absorption lines of all currently available XMM EPIC-pn observations reduced with XMM SAS 6.5 for each neutron star. As noted by Haberl (2007), in this respect the XMM EPIC-pn is the best-suited instrument for relative measurements with reasonably small errors ($\approx 0.1 \times 10^{20}$ cm^{-2} for all objects). The model predictions for these column densities can be found in Table 2. As noted before, the technical sampling is 10 pc. If a $N(\mathrm{H})$ value lies between two sampling points we indicate this by noting e.g. 235. However, even errors in the best-studied regions ($\lesssim$230 pc) are around 25 pc, and can be much larger otherwise.

In case of the extinction model based on Hakkila et al. (1997) one cannot always derive a convincing value due to bad sampling with the (low) extinction not changing over a large scale of distances (e.g. up to 1 kpc). For the analytical model only the high latitude object RBS 1223 is a problem where at 1 kpc one arrives only at 3.2×10^{20} cm^{-2}. The limitations of this large-scale model appear here. However, absorption lines in the spectrum also influence $N(\mathrm{H})$. Schwope (2007) fitted the spectrum of RBS 1223 using recent new observations and obtained $N(\mathrm{H}) = 3.7 \times 10^{20}$ cm^{-2} for one absorption line. This value is comparable to our result in Table 2, taking into account the additional data. However, the fit was not very good and Schwope 2007 obtained better fits with two absorption lines $N(\mathrm{H})$, yielding values ranging from 1.2×10^{20} cm^{-2} to 1.8×10^{20} cm^{-2}. The lower value

Table 2 Distances obtained for the M7

Name	N(H) (#lines) $[10^{20}\ \text{cm}^{-2}]$	d_{ana} [pc]	d_{ext} [pc]	d_{ana130} [pc]
RX J1856.5-3754	0.7 (0L)	135	135	125
RX J0420.0-5022	1.6 (1L)	345	...	325
RX J0720.4-3125	1.2 (1L)	270	235	265
RX J0806.4-4123	1.0 (1L)	250	235	240
RBS 1223	4.3 (1L)	...	...	...
RX J1605.3+3249	2.0 (3L)	390	...	325
RBS 1774	2.4 (1L)	430	...	390

The N(H) was obtained by blackbody and absorption line (*tabs*) *XSPEC*-fits from XMM-*Newton* EPIC-pn observations (abundances and cross-sections are chosen according to Wilms et al. 2000). The number of lines used is indicated in parenthesis. The corresponding model prediction of the distance is given for the analytical model: d_{ana}, the extinction model based on Hakkila et al. (1997) d_{ext}, and the analytical model plus consideration of possible 30% underprediction by the sodium D-lines: d_{ana130}. See also text for discussion

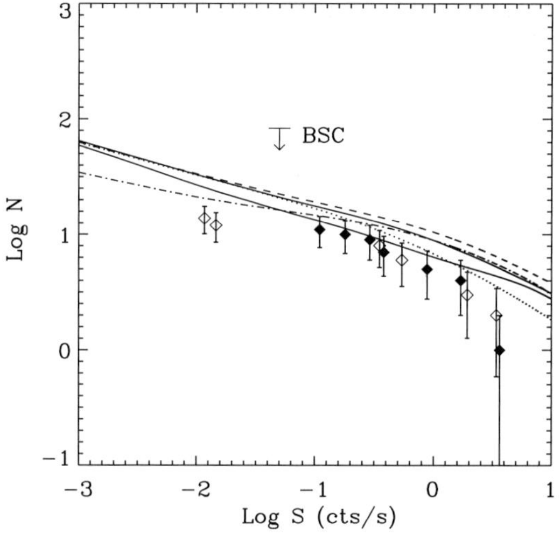

Fig. 7 The log N–log S curves for different models. Each curve shows the expected number of observable neutron stars at ROSAT PSPC count rates. The *solid line* represents the curve for the new progenitor distribution but old absorption model; the *dotted* and the *dashed line* are the results of the old simulations by e.g. Popov et al. (2003) for Gould Belt radii of $R_{\text{GB}} = 500$ pc and $R_{\text{GB}} = 300$ pc respectively. The only difference to the solid curve is the progenitor distribution. The results of taking into account the new progenitor distribution and additionally the new absorption models are shown by the *dashed-dotted line* for the refined analytical ISM model and the *dot-dot-dot-dashed line* for the extinction model based on Hakkila et al. (1997). Furthermore the measurement points of young NSs with thermal X-ray emission are plotted as in Popov et al. (2003). For discussion see text

would correspond to a distance of 525 pc when using the analytical model.

Overall, the distances derived from both models towards two neutron stars are in acceptable agreement: for RX J0720.4-3125 with 250 pc and for RX J0806.4-4123 with 240 pc. Since both NSs are close to the border of the sodium measurements, an error of $\approx$25 pc seems reasonable. In general, the values of the analytical model have to be used with caution as local clumpiness of the ISM is not considered by this model at distances >230 pc. Considering the open cluster test from above, the mean N(H) deviation of the analytical model up to 500 pc is $0.05 \times 10^{20}\ \text{cm}^{-2}$; the standard deviation, however, is $4.9 \times 10^{20}\ \text{cm}^{-2}$.

3.2 Population synthesis

The main motivation for us to take into account better absorption treatment is an improved population synthesis model based on that of Popov et al. (2000, 2003, 2005). A detailed discussion of the improvements is beyond the scope of this article and will be provided elsewhere (Popov, Posselt, in preparation). Therefore, we summarize here only the new developments and concentrate on illustrating the importance of the interstellar absorption.

While Popov et al. (2003, 2005) modeled the neutron star progenitor distribution as coming from infinitely thin disks, either from the galactic plane or the Gould Belt, the new initial distribution is more realistic. Up to the Hipparcos limit of 400 pc (ESA 1997) the known B2-O8 stars are considered individually as well as their affiliation to an OB association (de Zeeuw et al. 1999) resulting in birth properties depending on the age of the OB-association. The Gould Belt birth rate is applied here. For distances above 400 pc, the galactic disk is considered for few neutron stars randomly as well as 36 associations (Blaha and Humphreys 1989; Mel'nik and Efremov 1995) for most of the neutron stars. Due to the unknown ages of the associations, the birth probability is set proportional to the number of association members. In Fig. 7, the log N–log S curves are shown for the old thin-disk Gould Belt models with either 300 pc or 500 pc radius, the new progenitor distribution as well as for the new progenitor distribution considering both absorption models. All simulations were done for the same mass spectrum as in Popov et al. (2006), the cooling curve set labeled Model I in the same paper, the same abundances and cross-sections (in this case those of Morrison and McCammon (1983) for comparison) and identical remaining parameters. The results for the new progenitor distribution lie in-between those of the two Gould Belt sizes, a little apart from the measured integrated number of objects. This is also true for the curve considering the absorption by the analytical model. At the bright end the differences introduced by the analytical absorption model are only very small. Towards lower count rates, the curve lies significantly below the one without the refined 3D ISM model. Considering the absorption based on the extinction model by Hakkila et al. (1997) gives another picture. At the bright end the curve is much lower than that without the refined 3D ISM model, while the difference becomes smaller towards low count rates. We note here again that both absorption models are the same up to 230 pc, based on the measurements by Lallement et al. (2003). Both also consider the open cluster data, however slightly differently as the column density is only allowed to increase. The curve of the analytical model shows best the influence of an improved ISM distribution model compared to the old absorption treatment (e.g. Popov et al. 2003) where the analytical model without the data by Lallement et al. (2003) and Piskunov et al. (2006) was used. One might have expected that differences should be largest at low distances and hence for bright sources. However, averaging over the small Local Bubble gives apparently the same result as the simpler description before. The influence of the better model for the Solar vicinity is apparent in the expected location likelihoods for neutron stars. Going to larger distances the additive column densities are especially large around $l \approx 300^\circ$ to $l \approx 60^\circ$ in the galactic plane where many objects could be expected. Thus, the number of predicted neutron stars is reduced. The $N(\mathrm{H})$ model based on the extinction study by Hakkila et al. (1997) shows on average much higher $N(\mathrm{H})$ already at 240 pc compared to the analytical model. Given the nearly rectangular shape of the regions with relatively high values in the l, b plane these seem to be caused at least partly by bad sampling and the necessary statistical treatment. The higher $N(\mathrm{H})$ results in less observable sources bringing the corresponding $\log N$–$\log S$ much closer to the actually measured number of sources at the bright end.

We made similar test plots for the expected neutron star number per square degree like the one presented in Fig. 6 by Popov et al. (2005). We summarize here only the main results, for detailed discussion see Popov, Posselt 2007 (in prep.). Contrary to the old model the predicted neutron star number is enhanced in regions including the nearby Gould Belt OB associations (e.g. Upp Sco) with the new progenitor distribution. Sources are not anymore homogeneously distributed at all longitudes along the planes of the galactic disk or the Gould Belt. When it comes to consider the 3D ISM models, the influence of the absorption is on much smaller scales. Less sources would be expected at higher latitudes, and towards some small regions the high extinction hinders the possible detection of neutron stars. Differences between the two absorption models are pronounced only in small regions which to discuss in detail is beyond our intention here.

4 Conclusions

An understanding of the interstellar absorption is crucial for the interpretation of low-resolution soft X-ray spectra. Taking into account the 3D distribution of the ISM is important to determine the absorption of known sources as well as for searches of new soft X-ray sources towards a particular direction in the sky. We presented two simple models for 3D ISM distributions together with possible applications: distance estimation and the population synthesis for young isolated thermal, thus soft X-rays emitting neutron stars. Both models are large scale models and have to be used with caution especially considering distance estimations. Concerning population synthesis the two models influence the $\log N$–$\log S$ curve discriminatively, while the differences for the predicted spatial distribution of observable neutron stars are only small.

Acknowledgements We would like to thank R. Lallement providing us with the sodium density cube and for discussions as well as A. Piskunov for answering our questions regarding the open cluster study.

S.B. Popov also acknowledges the grants RFBR 04-02-16720 and 06-02-26586.

This research has made use of SAOimage ds9, developed by the Smithsonian Astrophysical Observatory.

References

Agüeros, M.A., Anderson, S.F., Margon, B., et al.: Astron. J. **131**, 1740 (2006)
Amôres, E.B., Lépine, J.R.D.: Astron. J. **130**, 659 (2005)
Anders, E., Ebihara, M.: Geochim. Cosmochim. Acta **46**, 2363 (1982)
Anders, E., Grevesse, N.: Geochim. Cosmochim. Acta **53**, 197 (1989)
Blaha, C., Humphreys, R.M.: Astron. J. **98**, 1598 (1989)
Bochkarev, N.G.: Basics of the ISM Physics. Moscow University Press, Moscow (1992)
Breitschwerdt, D.: In: IAU Colloq. 166: The Local Bubble and Beyond. LNP, vol. 506, p. 5 (1998)
Caraveo, P.A., Bignami, G.F., Mignani, R., et al.: Astrophys. J. **461**, L91 (1996)
Cardelli, J.A., Meyer, D.M., Jura, M., et al.: Astrophys. J. **467**, 334 (1996)
Casey, B.W., Mathieu, R.D., Vaz, L.P.R., et al.: Astron. J. **115**, 1617 (1998)
Chieregato, M., Campana, S., Treves, A., et al.: Astron. Astrophys. **444**, 69 (2005)
de Boer, H.: In: Bloemen H. (ed.) IAU Symp. 144: The Interstellar Disk-Halo Connection in Galaxies, vol. 333 (1991)
de Zeeuw, P.T., Hoogerwerf, R., de Bruijne, J.H.J., Brown, A.G.A., Blaauw, A.: Astron. J. **117**, 354 (1999)
Dickey, J.M., Lockman, F.J.: Annu. Rev. Astron. Astrophys. **28**, 215 (1990)
Dodson, R., Legge, D., Reynolds, J.E., et al.: Astrophys. J. **596**, 1137 (2003)
Drimmel, R., Cabrera-Lavers, A., López-Corredoira, M.: Astron. Astrophys. **409**, 205 (2003)
ESA. VizieR Online Data Catalog, 1239 (1997)
Faherty, J., Walter, F.M., Anderson, J.: Astrophys. Space Sci., DOI 10.1007/s10509-007-9368-0 (2007)
Ferlet, R., Vidal-Madjar, A., Gry, C.: Astrophys. J. **298**, 838 (1985)
Fitzgerald, M.P.: Astron. J. **73**, 983 (1968)
Forbrich, J., Preibisch, T., Menten, K.M.: Astron. Astrophys. **446**, 155 (2006)
Golden, A., Brisken, W.F., Thorsett, S.E., Benjamin, R.A., Goss, W.M.: Balt. Astron. **14**, 436 (2005)
Haberl, F.: Adv. Space Res. **33**, 638 (2004)
Haberl, F: Astrophys. Space Sci., DOI 10.1007/s10509-007-9342-x (2007)
Hakkila, J., Myers, J.M., Stidham, B.J., et al.: Astron. J. **114**, 2043 (1997)

Henbest, N., Couper, H.: The Guide to the Galaxy. Cambridge University Press, Cambridge (1994)
Henke, B.L., Lee, P., Tanaka, T.J., et al.: At. Data Nucl. Data Tables **27**, 1 (1982)
Hobbs, L.M., Penprase, B.E., Welty, D.E., et al.: Astrophys. J. **327**, 356 (1988)
Hunter, I., Smoker, J.V., Keenan, F.P., et al.: Mon. Not. Roy. Astron. Soc. **277**, (2006)
Jackson, M.S., Halpern, J.P.: Astrophys. J. **633**, 1114 (2005)
Kalberla, P.M.W., Burton, W.B., Hartmann, D., et al.: Astron. Astrophys. **440**, 775 (2005)
Kaplan, D.L., van Kerkwijk, M.H., Anderson, J.: Astrophys. J. **571**, 447 (2002)
Kharchenko, N.V., Piskunov, A.E., Röser, S., et al.: Astron. Astrophys. **440**, 403 (2005a)
Kharchenko, N.V., Piskunov, A.E., Röser, S., et al.: Astron. Astrophys. **438**, 1163 (2005b)
Knude, J.: Astrophys. Space Sci. **280**, 97 (2002)
Lallement, R., Welsh, B.Y., Vergely, J.L., Crifo, F., Sfeir, D.: Astron. Astrophys. **411**, 447 (2003)
Magnani, L., Blitz, L., Mundy, L.: Astrophys. J. **295**, 402 (1985)
Marshall, H.L., Schulz, N.S.: Astrophys. J. **574**, 377 (2002)
Marshall, D.J., Robin, A.C., Reyle, C., et al.: Astron. Astrophys. **453**, 635 (2006)
Mathis, J.S., Rumpl, W., Nordsieck, K.H.: Astrophys. J. **217**, 425 (1977)
Mel'nik, A.M., Efremov, Y.N.: Astron. Lett. **21**, 10 (1995)
Meyer, D.M., Cardelli, J.A., Sofia, U.J.: Astrophys. J. **490**, L103 (1997)
Meyer, D.M., Jura, M., Cardelli, J.A.: Astrophys. J. **493**, 222 (1998)
Morrison, R., McCammon, D.: Astrophys. J. **270**, 119 (1983)
Paresce, F.: Astron. J. **89**, 1022 (1984)
Pavlov, G.G., Zavlin, V.E., Sanwal, D., et al.: Astrophys. J. **552**, L129 (2001)
Piskunov, A.E., Kharchenko, N.V., Röser, S., et al.: Astron. Astrophys. **445**, 545 (2006)
Popov, S.B., Colpi, M., Treves, A., et al.: Astrophys. J. **530**, 896 (2000)
Popov, S.B., Colpi, M., Prokhorov, M.E., et al.: Astron. Astrophys. **406**, 111 (2003)
Popov, S.B., Turolla, R., Prokhorov, M.E., et al.: Astrophys. Space Sci. **299**, 117 (2005)
Popov, S., Grigorian, H., Turolla, R., Blaschke, D.: Astron. Astrophys. **448**, 327 (2006)
Predehl, P., Schmitt, J.H.M.M.: Astron. Astrophys. **293**, 889 (1995)
Rutledge, R.E., Fox, D.W., Bogosavljevic, M., et al.: Astrophys. J. **598**, 458 (2003)
Schwope, A.D., Hambaryan, V., Haberl, F., Motch, C.: Astrophys. Space Sci., DOI 10.1007/s10509-007-9375-1 (2007)
Sfeir, D.M., Lallement, R., Crifo, F., et al.: Astron. Astrophys. **346**, 785 (1999)
Snow, T.P., Witt, A.N.: Astrophys. J. **468**, L65 (1996)
Trümper, J.E., Burwitz, V., Haberl, F., et al.: Nucl. Phys. B Proc. Suppl. **132**, 560 (2004)
van Kerkwijk, M.H., Kaplan, D.L.: Astrophys. Space Sci., DOI 10.1007/s10509-007-9343-9 (2007)
Vergely, J.-L., Freire Ferrero, R., Siebert, A., et al.: Astron. Astrophys. **366**, 1016 (2001)
Verner, D.A., Yakovlev, D.G.: Astron. Astrophys. Suppl. Ser. **109**, 125 (1995)
Verner, D.A., Yakovlev, D.G., Band, I.M., Trzhaskovskaya, M.B.: At. Data Nucl. Data Tables **55**, 233 (1993)
Welty, D.E., Hobbs, L.M., Kulkarni, V.P.: Astrophys. J. **436**, 152 (1994)
Wilms, J., Allen, A., McCray, R.: Astrophys. J. **542**, 914 (2000)
Zane, S., Turolla, R., Zampieri, L., et al.: Astrophys. J. **451**, 739 (1995)

Astrophys Space Sci (2007) 308: 181–190
DOI 10.1007/s10509-007-9342-x

ORIGINAL ARTICLE

The magnificent seven: magnetic fields and surface temperature distributions

Frank Haberl

Received: 24 July 2006 / Accepted: 22 August 2006 / Published online: 23 March 2007

Abstract Presently seven nearby radio-quiet isolated neutron stars discovered in ROSAT data and characterized by thermal X-ray spectra are known. They exhibit very similar properties and despite intensive searches their number remained constant since 2001 which led to their name "The Magnificent Seven". Five of the stars exhibit pulsations in their X-ray flux with periods in the range of 3.4 s to 11.4 s. XMM-Newton observations revealed broad absorption lines in the X-ray spectra which are interpreted as cyclotron resonance absorption lines by protons or heavy ions and/or atomic transitions shifted to X-ray energies by strong magnetic fields of the order of 10^{13} G. New XMM-Newton observations indicate more complex X-ray spectra with multiple absorption lines. Pulse-phase spectroscopy of the best studied pulsars RX J0720.4-3125 and RBS 1223 reveals variations in derived emission temperature and absorption line depth with pulse phase. Moreover, RX J0720.4-3125 shows long-term spectral changes which are interpreted as due to free precession of the neutron star. Modeling of the pulse profiles of RX J0720.4-3125 and RBS 1223 provides information about the surface temperature distribution of the neutron stars indicating hot polar caps which have different temperatures, different sizes and are probably not located in antipodal positions.

Keywords Stars: neutron · Stars: magnetic fields · X-rays: stars

PACS 97.60.Jd · 98.70.Qy

F. Haberl (✉)
Max-Planck-Institut für extraterrestrische Physik,
Giessenbachstrasse, 85748 Garching, Germany
e-mail: fwh@mpe.mpg.de

1 Introduction

After the discovery of RX J1856.4-3754 as an isolated neutron star in ROSAT data (Walter et al. 1996; Walter and Matthews 1997), six further objects with very similar properties were found. Their X-ray emission is characterized by a soft, blackbody-like continuum which is little attenuated by photo-electric absorption by the interstellar medium, indicating small distances (of the order of a few 100 pc, Posselt et al. 2007). For the brightest object, RX J1856.4-3754, this was soon confirmed by the parallax measurement (Walter 2001; Kaplan et al. 2002c; Walter and Lattimer 2002). Including new HST observations the most recent revision of the parallax yields a distance of 161^{+18}_{-14} pc (van Kerkwijk and Kaplan 2007). A preliminary value for the parallax of RX J0720.4-3125 is presented by van Kerkwijk and Kaplan (2007) which corresponds to a distance of 330^{+170}_{-80} pc.

No indication for a hard, non-thermal X-ray emission component was found in the X-ray spectra. Compared to normal radio pulsars also no strong radio emission is detected: While for RX J0720.4-3125 and RX J0806.4-4123 deep radio observations failed to detect a radio counterpart (Johnston 2003; Kaplan et al. 2003b), for RBS 1223 and RBS 1774 the detection of weak radio emission was claimed (Malofeev et al. 2006, 2007). The constant X-ray flux on time scales of years, no obvious association with a supernova remnant and relatively high proper motion measurements for the three brightest objects suggest that we deal with a group of cooling neutron stars with ages of around 10^6 years. Recent reviews which summarize the observed properties of the seven thermally emitting INSs known today can be found in Treves et al. (2000), Motch (2001) and Haberl (2004, 2005).

The discovery of the seven neutron stars (which are often called the Magnificent Seven, hereafter M7) with

Table 1 X-ray and optical properties of the Magnificent Seven

Object	kT eV	Period s	Amplitude %	Optical mag	PM mas/year	Ref.
RX J0420.0-5022	44	3.45	13	$B = 26.6$		1
RX J0720.4-3125	85–95	8.39	8–15	$B = 26.6$	97	2, 3, 4, 5, 6
RX J0806.4-4123	96	11.37	6	$B > 24$		7, 1
RBS 1223[a]	86	10.31	18	$m_{50ccd} = 28.6$		8, 9, 10, 11
RX J1605.3+3249	96	6.88?	?	$B = 27.2$	145	12, 13, 14, 15
RX J1856.5-3754	62	–	<1.3	$B = 25.2$	332	16, 17, 18, 19
RBS 1774[b]	102	9.44	4	$B > 26$		20, 21, 22

[a]= 1RXS J130848.6+212708

[b]= 1RXS J214303.7+065419

References: (1) Haberl et al. (2004a); (2) Haberl et al. (1997); (3) Cropper et al. (2001); (4) Haberl et al. (2004b); (5) de Vries et al. (2004); (6) Motch et al. (2003); (7) Haberl and Zavlin (2002); (8) Schwope et al. (1999); (9) Hambaryan et al. (2002); (10) Kaplan et al. (2002a); (11) Haberl et al. (2003); (12) Motch et al. (1999); (13) Kaplan et al. (2003a); (14) van Kerkwijk et al. (2004); (15) Motch et al. (2005); (16) Walter and Matthews (1997); (17) Walter and Lattimer (2002); (18) Burwitz et al. (2003); (19) van Kerkwijk and Kulkarni (2001b); (20) Zampieri et al. (2001); (21) Zane et al. (2005); (22) Komarova (2006)

purely thermal X-ray spectra raised wide interest by theoreticians and observers as promising objects to learn about atmospheres and the internal structure of neutron stars (e.g. Paerels 1997). The apparent absence of non-thermal processes which would hamper the analysis of the X-ray emission allows a direct view onto the stellar surface. Limits on the radius of the star RX J1856.4-3754 were used to constrain the equation of state of neutron star matter (Pons et al. 2002; Trümper et al. 2004). The ROSAT PSPC spectra with low energy resolution were consistent with Planckian energy distributions with blackbody temperatures kT in the range 40–100 eV and little attenuation by interstellar absorption. More recent high resolution observations of the two brightest objects RX J1856.4-3754 (Burwitz et al. 2001) and RX J0720.4-3125 (Paerels et al. 2001; Pavlov et al. 2002; Kaplan et al. 2003b) were performed using the low energy transmission grating (LETG) aboard Chandra and the reflection grating spectrometers (RGS) of XMM-Newton. In particular the high signal to noise LETG spectrum from a 500 ks Chandra observation of RX J1856.4-3754 did not reveal any significant deviation from a blackbody spectrum (Burwitz et al. 2001; Burwitz et al. 2003). The statistical quality and energy band coverage of the RGS and LETG spectra of RX J0720.4-3125 are however insufficient to detect subtle narrow features in the spectrum (Paerels et al. 2001).

Four of the M7 exhibit clear pulsations in their X-ray flux, indicating the spin period of the neutron star (RX J0420.0-5022, RX J0720.4-3125, RX J0806.4-4123 and RBS 1223) and there is evidence that also RBS 1774 is a pulsar (Zane et al. 2005). A possible candidate period for RX J1605.3+3249 needs future confirmation. A summary with blackbody temperatures derived from (phase-averaged) EPIC-pn spectra, spin periods, amplitudes for the flux modulation, optical brightness and measured proper motions is given in Table 1 together with key references for each object. Despite extensive searches in ROSAT data (e.g. Rutledge et al. 2003; Chieregato et al. 2005; Agüeros et al. 2006) no further neutron star with similar characteristic properties was found.

The first significant deviations from a blackbody spectrum were reported by Haberl et al. (2003) from EPIC spectra of RBS 1223. A broad absorption feature near 300 eV was interpreted as proton cyclotron resonance absorption. Apart from this absorption line with an equivalent width of 160 eV the X-ray spectrum of RBS 1223 is blackbody-like. Haberl et al. (2003) also suggested that changes in the soft part of the X-ray spectrum of the pulsar RX J0720.4-3125 with pulse phase reported by Cropper et al. (2001) are caused by variable cyclotron absorption (Haberl et al. 2004b). Cyclotron resonance absorption features in the 0.1–1.0 keV band are expected in spectra from strongly magnetized neutron stars with field strengths in the range of 10^{10}–10^{11} G or 2×10^{13}–2×10^{14} G if caused by electrons or protons, respectively (see e.g. Zane et al. 2001; Zavlin and Pavlov 2002). Variation of the magnetic field strength over the neutron star surface leads to a broadening of the line (Ho and Lai 2004).

The relatively long spin periods in the range of 3.45–11.37 s (the majority of radio pulsars has periods less than 1 s) were the first indicator for strong magnetic fields. If the stars were born with millisecond spin periods, age estimates from neutron star cooling curves (Page et al. 2006) and proper motions (tracing the neutron star back to a likely birth place, Motch et al. 2006) of typically 10^6 years require B fields of the order of 10^{13} G to decelerate the rotation of the stars to their current periods. This was recently confirmed by the accurate determination of the pulse period derivative $\dot{P}$ of $0.698 \times 10^{-13}\ \mathrm{s\,s^{-1}}$ and $1.120 \times 10^{-13}\ \mathrm{s\,s^{-1}}$ for RX J0720.4-3125 and RBS 1223, respectively (Kaplan and van Kerkwijk 2005a, 2005b). In the magnetic dipole braking model this yields characteristic ages of 1.9×10^6 years and 1.5×10^6 years and B field strengths of 2.4×10^{13} G and 3.4×10^{13} G, respectively. Such strong fields supports the picture in which the broad absorption lines—if due to cyclotron resonance—at least for RX J0720.4-3125 and RBS 1223 are caused by protons.

2 Pulse-phase averaged X-ray spectra

Significant deviations from the absorbed blackbody spectrum were reported from XMM-Newton EPIC observations of RBS 1223 (Haberl et al. 2003) and RX J0720.4-3125 (Haberl et al. 2004b) and from XMM-Newton RGS spectra of RX J1605.3+3249 (van Kerkwijk et al. 2004). Also non-magnetic neutron star atmosphere models (e.g. Gänsicke et al. 2002; Zavlin and Pavlov 2002) fail to reproduce the spectra. Iron and solar mixture atmospheres cause too many absorption features in particular at energies between 0.5 and 1.0 keV which are not seen in the measured spectra. On the other hand the spectrum of a pure hydrogen model is similar in shape to that of a blackbody but results in a much lower effective temperature which would predict a far too high optical flux (see Pavlov et al. 1996). The modeling of the X-ray spectra was significantly improved by including a broad absorption line with Gaussian shape. Line energies around 300 eV were found for RX J0720.4-3125 and RBS 1223 from the medium energy resolution EPIC spectra, while a line with higher energy of 450–480 eV was discovered in the high energy resolution RGS spectrum of RX J1605.3+3249.

Spectral fits to the EPIC-pn spectra of the third brightest M7 star RX J1605.3+3249 obtained in 2003, January 17 and February 26 (satellite revolutions 569 and 589) also considerably improve when adding a Gaussian absorption line to the absorbed blackbody model (see Table 2 and Fig. 1). However, the quality of the fit with a reduced χ^2 of 2.39 (all XMM-Newton spectra presented here were obtained using the analysis software SAS version 6.5) is still not acceptable and worse than typical values around 1.5 one finds for joint fits to the available EPIC-pn spectra of RX J1856.4-3754 (see Fig. 2) or RX J0720.4-3125. Adding more lines further improves the fit quality and with three absorption lines an acceptable χ^2 is found. The best fit parameters for the models with different number of absorption lines are listed in Table 2. The width of the lines was fit as a single common parameter in the case of multiple lines ($\sigma = 87$ eV for the model with three lines). A remarkable result is the ratio of the line energies for the model with three lines. With $E_2/E_1 = 1.46 \pm 0.02$ and $E_3/E_1 = 1.94 \pm 0.06$ the line energy ratios are consistent with 1:1.5:2. It should also be noted that the depth of the lines decreases with line energy by a factor of about 5 (in terms of absorbed line photon flux) from one line to the next. A fit with the model with three absorption lines to the RGS spectra (only allowing a re-normalization factor between the instruments) shows that also the RGS spectra are consistent with that model (Fig. 3).

Similar to RX J1605.3+3249, new XMM-Newton observations also indicate a more complex X-ray spectrum for RBS 1223. With a total exposure time of more than 100 ks the increased statistical quality of the EPIC-pn spectra requires two absorption lines for an acceptable fit (Schwope et al. 2007). With 230 eV and 460 eV the line energies are a factor of two apart. Also, in the case of RX J0806.4-4123 a simple absorbed blackbody model yields an unacceptable fit to the EPIC-pn spectra (Haberl et al. 2004a). A model with one line results in $\chi^2_{\mathrm{red}} = 1.39$ which is formally acceptable. Adding another line with an energy twice that of the first one, further improves the fit to $\chi^2_{\mathrm{red}} = 1.05$, but more higher quality data is required to confirm the significance of this result.

Table 3 summarizes the magnetic field estimates for the M7 neutron stars utilizing either the magnetic dipole braking model for the pulsars with measured pulse period derivative $\dot{P}$ ($B = 3.2 \times 10^{19} (P \times \dot{P})^{1/2}$) or assuming that the absorption line is due to cyclotron resonance by protons ($B = 1.6 \times 10^{11} E(\mathrm{eV})/(1 - 2GM/c^2R)^{1/2}$). If multiple absorption lines were included in the spectral modeling, the lowest line energy was used for the estimate of the magnetic field.

Table 2 Model fits to the EPIC-pn spectra of RX J1605.3+3249

N_{H} $10^{20}\ \mathrm{cm^{-2}}$	kT eV	E_1 eV	E_2 eV	E_3 eV	F_1 10^{-4}	F_2 photons cm^{-2}	F_3 s^{-1}	EW$_1$ eV	EW$_2$ eV	EW$_3$ eV	χ^2_{red}
0.15	97	–	–	–	–	–	–				4.38
1.1	92	445	–	–	−67	–	–				2.39
1.4	92	440	1.5E$_1$	–	−75	−43	–				1.75
2.04 ± 0.04	91.6 ± 0.3	403 ± 2	589 ± 4	780 ± 24	-43 ± 1	-8.0 ± 0.8	-1.6 ± 0.4	96	76	67	1.39

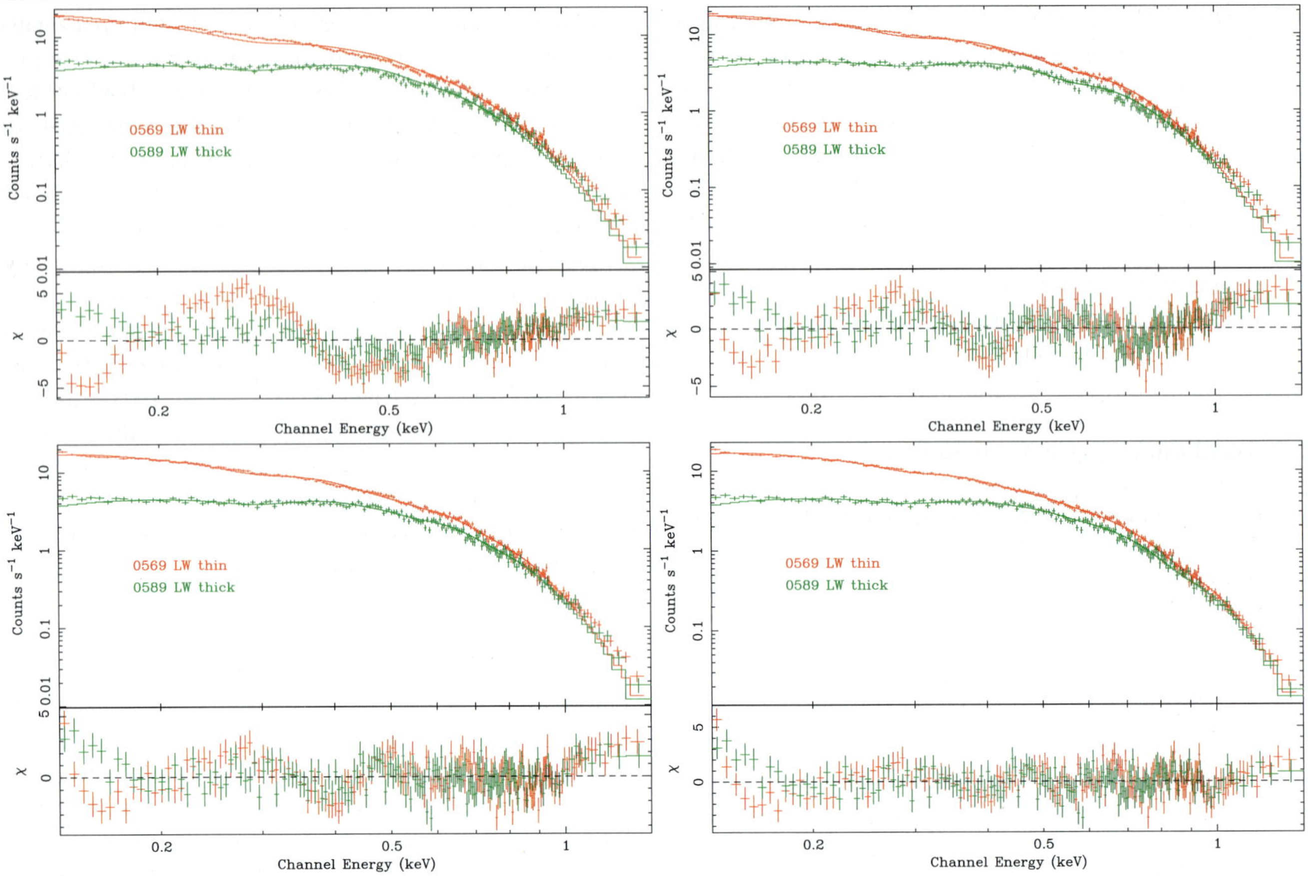

Fig. 1 Modeling of the EPIC-pn spectra of RX J1605.3+3249. From *top left* to *bottom right* the number of absorption lines increases from zero to three. As continuum a blackbody model with interstellar absorption is used

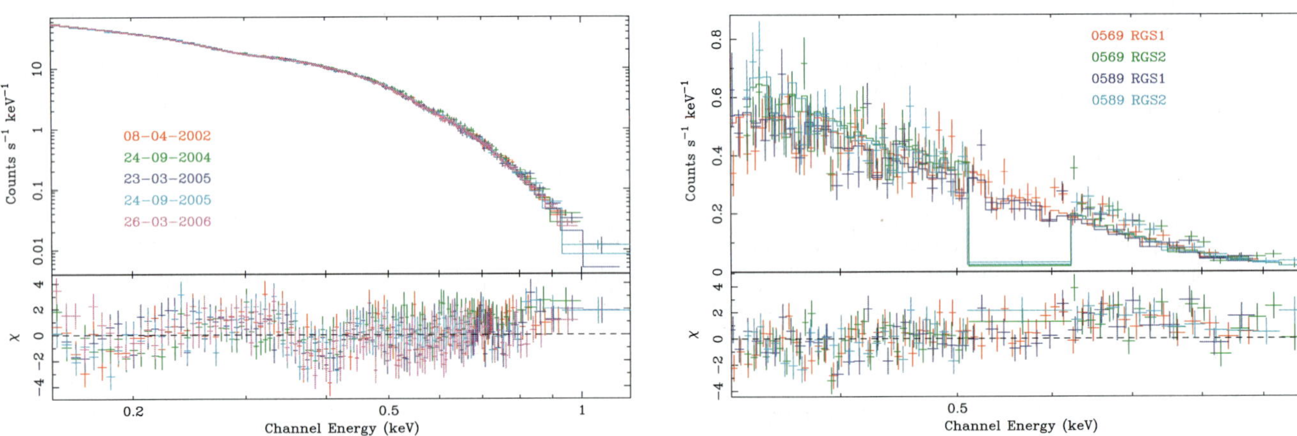

Fig. 2 EPIC-pn spectra of RX J1856.4-3754 obtained from five observations between 2002 and 2006 with the same instrumental setup (small window readout mode and thin filter). A joint fit with an absorbed blackbody model yields a reduced χ^2 of 1.55 and demonstrates the stability of detector and source to better than 1%. The dip in the residuals around 0.4 keV is a remaining calibration problem

Fig. 3 RGS spectra of RX J1605.3+3249 from the two observations in 2003, January 17 and February 26 (satellite revolutions 569 and 589). The *histograms* represent the best fit EPIC-pn model with three absorption lines allowing only a re-normalization to account for possible cross-calibration problems

The brightest M7 star, RX J1856.4-3754, does not show pulsations with a very low amplitude limit of 1.3% (Ransom et al. 2002; Burwitz et al. 2003). However, van Kerkwijk and Kulkarni (2001a) discovered the existence of a Hα nebula around the star with cometary-like morphology aligned with the direction of its proper motion. Assuming that the nebula

Table 3 Magnetic field estimates

Object	dP/dt 10^{-13} s s^{-1}	E_{cyc} eV	B_{db} 10^{13} G	B_{cyc} 10^{13} G
RX J0420.0-5022	<92	330	<18	6.6
RX J0720.4-3125	0.698(2)	280	2.4	5.6
RX J0806.4-4123	<18	430/306[a]	<14	8.6/6.1
RBS 1223	1.120(3)	300/230[a]	3.4	6.0/4.6
RX J1605.3 + 3249		450/400[b]		9/8
RX J1856.5-3754		–	∼1[c]	
RBS 1774	<60[d]	750	<24[d]	15

[a] Spectral fit with single/two lines

[b] With single line/three lines at 400 eV, 600 eV and 800 eV

[c] Estimate from Hα nebula assuming that it is powered by magnetic dipole breaking (Kaplan et al. 2002c; Braje and Romani 2002; Trümper et al. 2004)

[d] Radio detection: (Malofeev et al. 2006)

is powered by magnetic dipole braking and an age of the neutron star of $\sim 5 \times 10^5$ years which is inferred from its proper motion and the distance to the likely birth place, a magnetic field strength of $\sim 1 \times 10^{13}$ G is derived (Kaplan et al. 2002c; Braje and Romani 2002; Trümper et al. 2004). Such a field strength is consistent with the non-detection of a proton cyclotron feature which would have an energy below the sensitive range of X-ray instruments.

In two cases (RX J0720.4-3125 and RBS 1223) accurate $\dot{P}$ measurements exist. For both stars also absorption lines were found in their X-ray spectra which allows a comparison of B_{db} (field estimate from dipole braking) and B_{cyc} (field estimate from proton cyclotron resonance). The ratio B_{cyc}/B_{db} is 2.33 for RX J0720.4-3125 and 1.35 for RBS 1223 (for the spectral model with two lines). A similar low ratio for RX J0720.4-3125 would be obtained if a second (lower-energy) line around 140 eV exists. Unfortunately, the detection of a line at such low energies is outside the capability of the XMM-Newton instruments.

3 Neutron star surface temperature distributions

The first XMM-Newton observations of RX J0720.4-3125 revealed spectral changes with pulse phase expressed as hardness ratio variations (Cropper et al. 2001). Using polar cap models first constraints on the polar cap sizes and viewing geometries could be derived. The pulse profile in the 0.12 keV to 1.2 keV energy band was found to be approximately sinusoidal with a peak-to-peak amplitude of about 15% with the hardness ratio being softest slightly before flux maximum. Spectra from different pulse phases indicated a temperature variation and a change in the low-energy attenuation, which—when modeled with a simple absorbed blackbody spectrum—appears as variable absorption column density. Including more data Haberl et al. (2004b) interpreted this as broad absorption line with variable depth. Pulse phase spectroscopy of RBS 1223 (Schwope et al. 2005) also revealed that changes in temperature and absorption line depth are responsible for the spectral variations with pulse phase. The double-humped X-ray light curves of RBS 1223 in different energy bands were modeled by Schwope et al. (2005) assuming Planckian radiation from a neutron star surface with inhomogeneous temperature distribution. Again, a simple model with two hot spots with temperatures $T_1^\infty = 92$ eV and $T_2^\infty = 84$ eV and full angular sizes of 8° and 10° which are separated by ∼160° can reproduce the data. A more physical temperature distribution based on the crustal field models by Geppert et al. (2004) which can produce relatively strong temperature gradients from the magnetic poles to the equator was equally successful.

The neutron star RX J0720.4-3125 is unique among the M7 as it shows a gradual change of its X-ray spectrum over years which is accompanied by an energy-dependent change in the pulse profile (de Vries et al. 2004; Vink et al. 2004). Figure 4 shows folded light curves together with hardness ratios for six different epochs. They were derived from observations with identical instrumental setup to avoid systematic differences between different filters and CCD readout modes. As was found by de Vries et al. (2004) the pulse profile became deeper with time with a pulsed fraction of ∼8% in the year 2000, increasing to ∼15% by the end of 2003. Moreover, the spectra became considerable harder during these years as can be seen in the increase of the average hardness ratio in Fig. 4. Haberl et al. (2006) found first evidence for a trend reversal of the spectral evolution and interpret the involved period of 7.1±0.5 years as precession period of the neutron star.

The overall hardening of the spectra until 2004 and the trend-reversal can also be seen from the pulse-phase resolved spectra which are shown in Fig. 5 for four different XMM-Newton observations. Both, pulse-phase variations and long-term spectral changes can be modeled by variations in the blackbody temperature and absorption line depth (Haberl et al. 2006). In Fig. 6 the line equivalent width is plotted versus temperature kT. The temperature variation (∼2.5 eV) was smaller during the first observations and increased to ∼6 eV, almost as large as the long-term change of ∼8 eV seen in the phase-averaged spectra. This supports the idea that the temperature changes are caused by geometrical effects and different areas on the neutron star surface come into our view. In contrast, the amplitude in the equivalent width variation is ∼40 eV and did not change much over the years. The relatively large variation in the line depth with pulse phase and the strong increase over the years in-

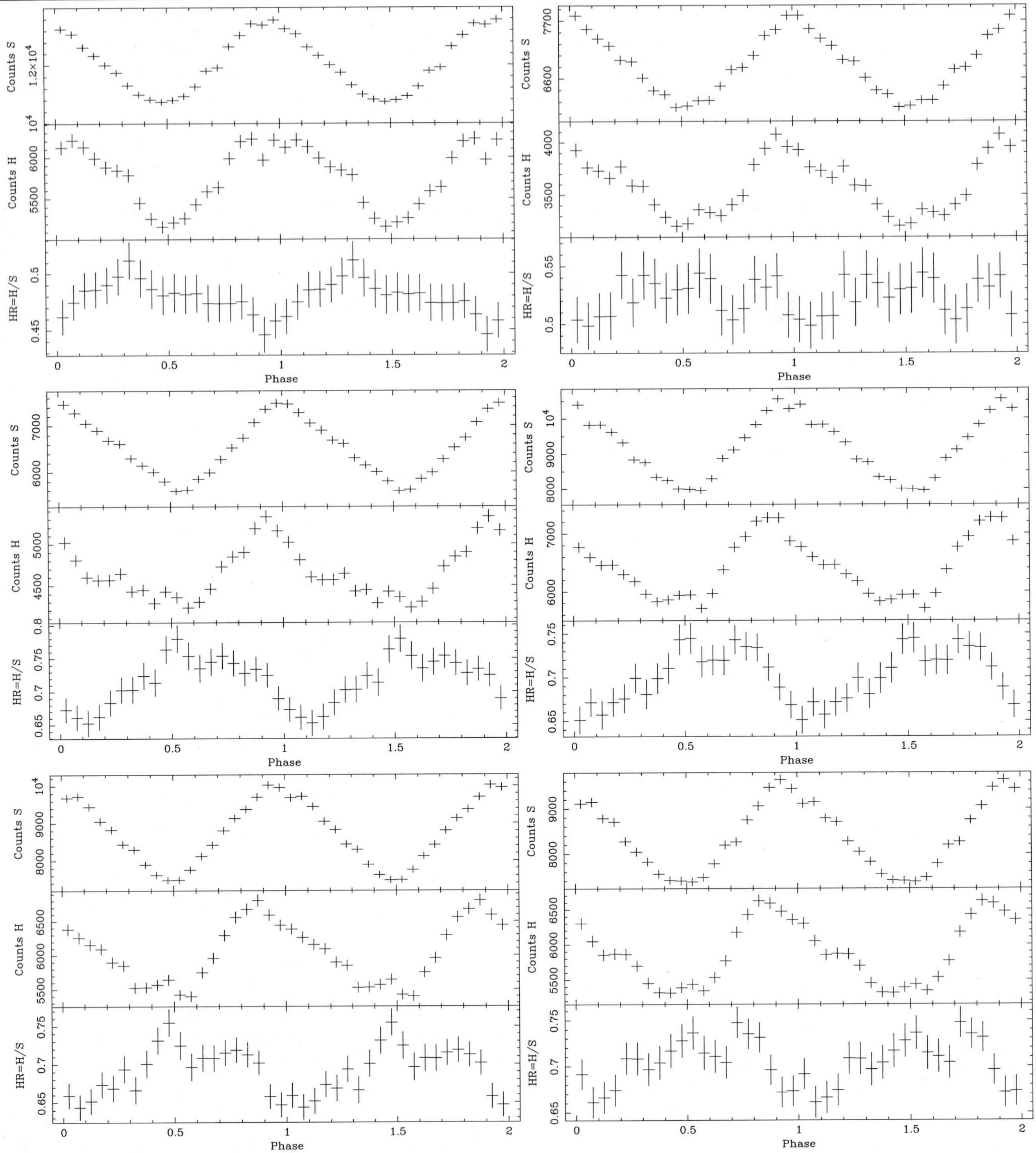

Fig. 4 Folded EPIC-pn light curves of RX J0720.4-3125 in two different energy bands (soft S: 120 eV–400 eV; hard H: 400 eV–1 keV) together with the hardness ratio. Different observations, all performed with the same instrumental setup (full-frame mode with thin filter), are shown from *top left* to *bottom right*: 2000 May 13, 2002 Nov. 6, 2004 May 22, 2005 April 28, 2005 Sep. 23 and 2005 Nov. 12. For a detailed summary of all XMM-Newton observation in the years 2000 to 2005 see Table 1 in Haberl et al. (2006). Pulse phases were calculated using the X-ray timing ephemerides ("All Data" solution) from Kaplan and van Kerkwijk (2005a).

dicates that the hotter polar cap which came into (better) view—possibly due to precession—plays the major role in the line absorption. More detailed analyses are required to see if this is mainly a temperature effect (for higher temperatures atomic line transitions are expected to be reduced because of a higher ionization degree) or a viewing effect

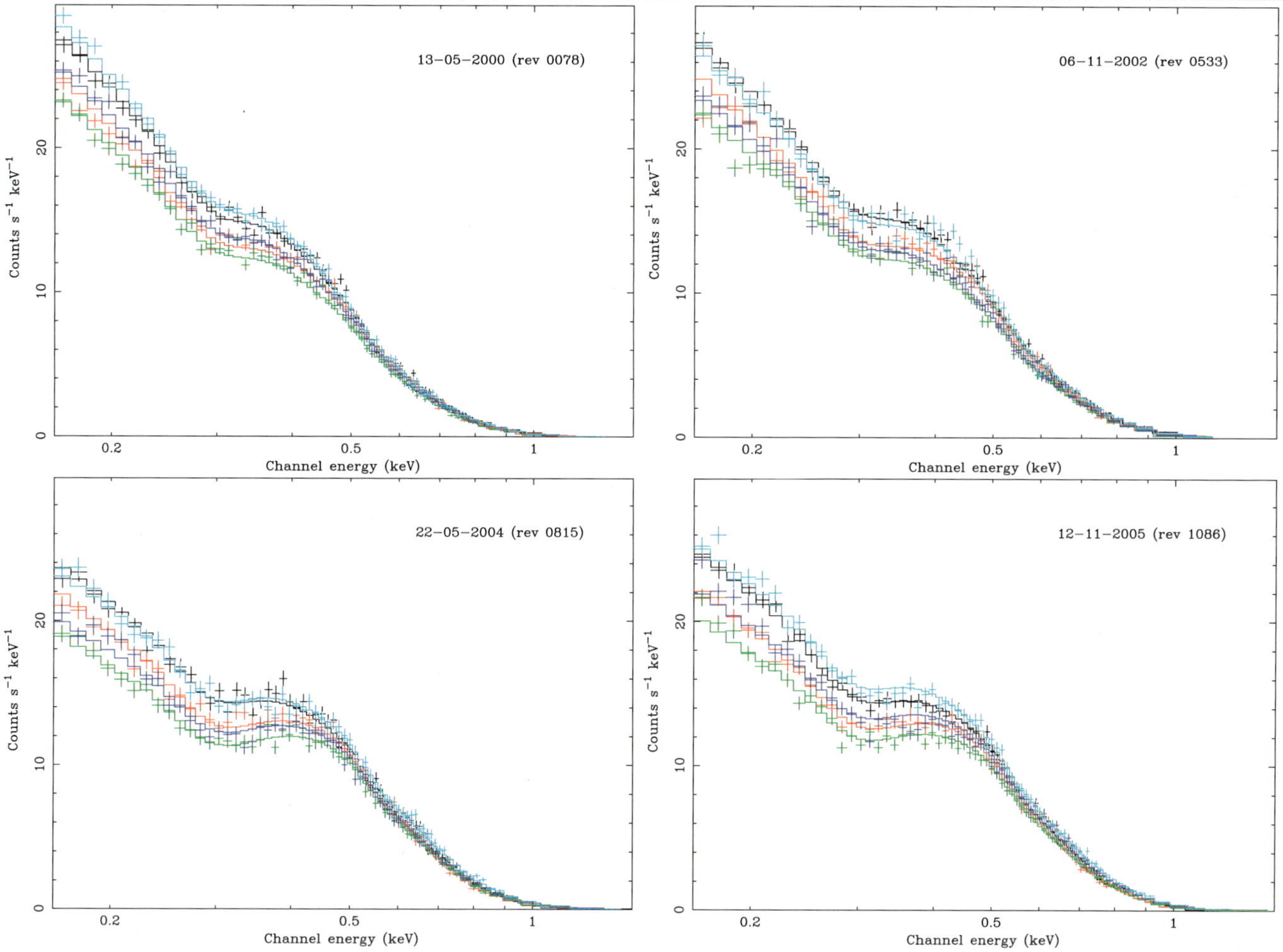

Fig. 5 Pulse-phase resolved EPIC-pn spectra of RX J0720.4-3125 from four different observations. During the pulse the spectra vary as is shown by the different colours used for different phases (0.0–0.2: *black*, 0.2–0.4: *red*, 0.4–0.6: *green*, 0.6–0.8: *blue*, 0.8–1.0: *light blue*; with phase 0.0 defined as intensity maximum as in Fig. 4). The intensity scale is linear to better resolve the changes at low energies

due to the dependence on the angle between the direction of radiation propagation and magnetic field lines.

Since the spectral observations do not yet cover a complete precession period Haberl et al. (2006) also performed a preliminary analysis of published and new pulse timing residuals (from three new XMM-Newton observations). This includes archival ROSAT, Chandra and XMM-Newton observations which are distributed over a total of $\sim$12 years (see Kaplan et al. 2002b; Zane et al. 2002; Cropper et al. 2004; Kaplan and van Kerkwijk 2005a). A sinusoidal fit to the phase residuals yields a period of 7.7 ± 0.6 years (Fig. 7) consistent with that derived from the spectral analysis. In a similar analysis van Kerkwijk and Kaplan (2007) infer a shorter period of 4.3 years. Apart from using new Chandra data instead of the new XMM-observations they also used different phase residuals for the early ROSAT observations (which are subject to cycle count ambiguity) compared to the original values published in Kaplan and van Kerkwijk (2005a). The reason for the different periods might be due to the energy dependence of the pulse profile which can lead to systematic differences between instruments with different spectral response and/or the deviations from a sinusoidal shape of the light curves (which moreover evolve with time) as it was assumed in the analysis of Kaplan and van Kerkwijk (2005a). Further monitoring of RX J0720.4-3125 and a more sophisticated analysis and modeling of the data are required to confirm the precession model and further constrain its parameters. This can provide independent information about the interior of the neutron star and its distortion from a spherical shape $\epsilon = (I_3 - I_1)/I_1 = P_{\rm spin}/P_{\rm prec} \approx 4 \times 10^{-8}$ which is larger than that reported for radio pulsars (e.g. Jones and Andersson 2001; Akgün et al. 2006) but smaller than that for Her X-1 (Trümper et al. 1986; Ketsaris et al. 2000).

Haberl et al. (2006) used a polar cap model similar to that applied by Schwope et al. (2005) to RBS 1223. Including free precession of the neutron star to explain the long-term spectral variations of RX J0720.4-3125 they can at least

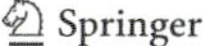

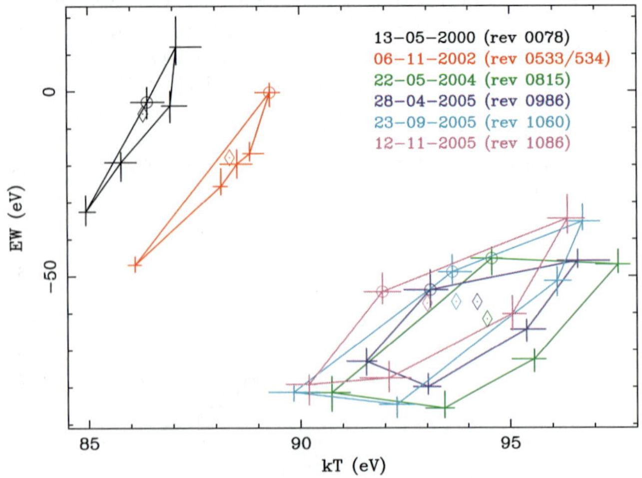

Fig. 6 Equivalent width of the absorption line vs. temperature kT derived from the EPIC-pn full-frame mode observations with thin filter. *Diamonds* denote the values derived from the phase-averaged spectra. During the pulse the parameters evolve counter-clockwise, the *circle* marks phase 0.0–0.2. This figure is a colour representation of Fig. 4 in Haberl et al. (2006)

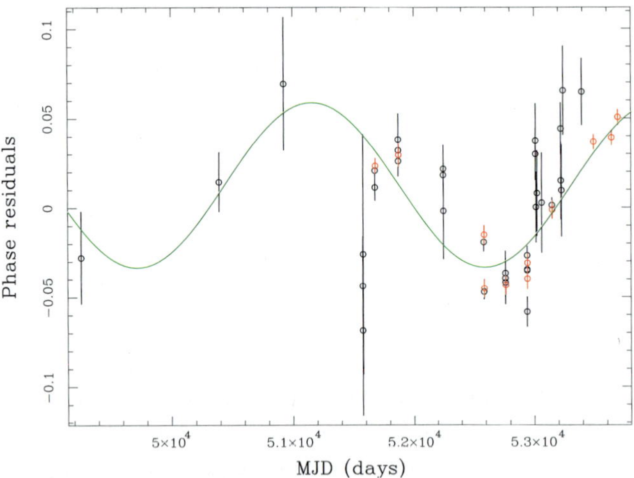

Fig. 7 Phase residuals for RX J0720.4-3125. The *black data points* are reproduced from Kaplan and van Kerkwijk (2005a) while *red points* are from the analysis of Haberl et al. (2006), which includes a re-analysis of published EPIC-pn results for a consistency check as well as three new data sets (the last three points)

qualitatively reproduce the variations in the X-ray spectra, changes in the pulsed fraction, shape of the light curve and the phase-lag between soft and hard energy bands. Like for RBS 1223, where the intensity peaks in the pulse profile are not separated by 0.5 in spin phase, the spots on the surface of RX J0720.4-3125 are probably also not located exactly in antipodal positions. This can explain the observed evolution of the light curve and hardness ratio with precession phase.

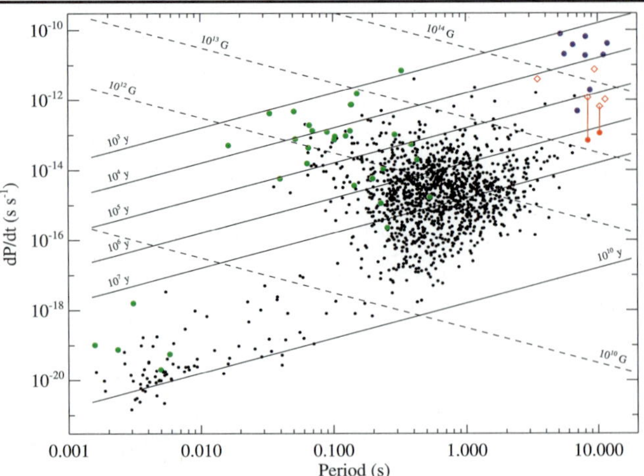

Fig. 8 P–$\dot{P}$ diagram for radio pulsars (*black dots*), magnetars (*blue dots*) and the Magnificent Seven. Red dots show the direct P and $\dot{P}$ measurements for RX J0720.4-3125 and RBS 1223, while the open diamonds mark the $\dot{P}$ values expected from the magnetic field estimates assuming proton resonance as origin of the absorption lines. For RX J0720.4-3125 and RBS 1223 both field estimates are connected with a vertical line. Marked in green are radio pulsars which are also detected at high energies (incomplete for the millisecond pulsars)

4 Discussion

The new results from optical and X-ray observations of the Magnificent Seven, thermally emitting isolated neutron stars, strongly support the model of cooling neutron stars with ages around a million years. Large proper motions measured for the three brightest objects make accretion from the interstellar medium to re-heat the neutron star very inefficient. Absorption features in the X-ray spectra of most of the Magnificent Seven and first measurements of the spin period derivative from the two pulsars RX J0720.4-3125 and RBS 1223 consistently point to magnetic field strengths of the order of 10^{13} G to 10^{14} G. This places them at the long spin period and high magnetic field end of the radio pulsar distribution. However, the $\dot{P}$ measurements still indicate field strengths below those of the magnetars (Fig. 8). The discovery of a few radio pulsars with similar magnetic field strengths and long periods (Camilo et al. 2000; Morris et al. 2002; McLaughlin et al. 2003) shows that radio emission can still occur at inferred field strengths close to the "quantum critical field" $B_{\rm cr} = m_e^2 c^3/e\hbar \simeq 4.4 \times 10^{13}$ G. Therefore it remains unclear if the M7 exhibit no radio emission at all or if we do not detect them because their radio beam is very narrow due to their large light cylinder radius and therefore does not cross the Earth.

In a $\sim 10^{13}$ G magnetic field cyclotron resonance lines at soft X-ray energies are expected to be caused by protons or highly ionized atoms of heavy elements. The ratios of line strengths in consecutive harmonic lines scales with $E_{\rm cyc}/(mc^2)$ (Pavlov et al. 1980) and for the involved high particle masses no harmonic lines should be detectable.

There is now more and more evidence for additional absorption lines in the X-ray spectra of the Magnificent Seven. An explanation for the additional (or even all) lines can be atomic bound-bound or bound-free transitions. In high B fields atomic orbitals are distorted into a cylindrical shape and the electron energy levels are similar to Landau states, with binding energies of atoms strongly increased. For hydrogen in a magnetic field of the order of 10^{13} G the strongest atomic transition is expected at energy $E/\text{eV} \approx 75(1 + 0.13\ln(B_{13})) + 63B_{13}$, with $B_{13} = B/10^{13}$ G (Zavlin and Pavlov 2002). For the line energies found in the spectra of thermal isolated neutron stars this requires similar field strengths to those derived assuming cyclotron absorption. Atomic line transitions are expected to be less prominent at higher temperatures because of a higher ionization degree (Zavlin and Pavlov 2002). For a more detailed discussion of expected line features that can be produced in a hydrogen atmosphere see van Kerkwijk and Kaplan (2007). It is also not clear how important the contribution of heavier elements in the atmosphere is because of the strong gravitational stratification forces (Mori and Hailey 2006). Further, the remarkable harmonic-like energy spacing of multiple lines needs to be understood. Studying their pulse phase dependence which can probably best be done for RBS 1223 might shed light on this.

The modeling of pulse profiles in different X-ray energy bands is a powerful tool to unveil the surface temperature distribution of the neutron stars (Zane and Turolla 2006). First studies of RX J0720.4-3125 and RBS 1223 yielded constraints on the polar cap geometry. It is remarkable that for both stars asymmetric configurations of the hot polar caps were inferred. In both cases different temperatures and sizes of the hot spots are found with the smaller spot being the hotter one. Probably, also in both stars the spots are not located in exactly antipodal positions. This could point to magnetic field configurations of an off-centered dipole or the involvement of higher-order multi-pole components. Studies of the effects of a strong magnetic field on the temperature distribution in the neutron star crust show that configurations which include dipolar poloidal and toroidal components can indeed reproduce the observed temperature distributions (Geppert et al. 2004, 2006).

Acknowledgements The XMM-Newton project is an ESA Science Mission with instruments and contributions directly funded by ESA Member States and the USA (NASA). The XMM-Newton project is supported by the Bundesministerium für Wirtschaft und Technologie/Deutsches Zentrum für Luft- und Raumfahrt (BMWI/DLR, FKZ 50 OX 0001), the Max-Planck Society and the Heidenhain-Stiftung.

References

Agüeros, M.A., Anderson, S.F., Margon, B., et al.: Astron. J. **131**, 1740 (2006)
Akgün, T., Link, B., Wasserman, I.: Mon. Not. Roy. Astron. Soc. **365**, 653 (2006)
Braje, T.M., Romani, R.W.: Astrophys. J. **580**, 1043 (2002)
Burwitz, V., Zavlin, V.E., Neuhäuser, R., et al.: Astron. Astrophys. **379**, L35 (2001)
Burwitz, V., Haberl, F., Neuhäuser, R., et al.: Astron. Astrophys. **399**, 1109 (2003)
Camilo, F., Kaspi, V.M., Lyne, A.G., et al.: Astrophys. J. **541**, 367 (2000)
Chieregato, M., Campana, S., Treves, A., et al.: Astron. Astrophys. **444**, 69 (2005)
Cropper, M., Zane, S., Ramsay, G., et al.: Astron. Astrophys. **365**, L302 (2001)
Cropper, M., Haberl, F., Zane, S., et al.: Mon. Not. Roy. Astron. Soc. **351**, 1099 (2004)
de Vries, C.P., Vink, J., Méndez, M., et al.: Astron. Astrophys. **415**, L31 (2004)
Gänsicke, B.T., Braje, T.M., Romani, R.W.: Astron. Astrophys. **386**, 1001 (2002)
Geppert, U., Küker, M., Page, D.: Astron. Astrophys. **426**, 267 (2004)
Geppert, U., Küker, M., Page, D.: Astron. Astrophys. **457**, 937 (2006)
Haberl, F.: Adv. Space Res. **33**, 638 (2004)
Haberl, F.: In: 5 Years of Science with XMM-Newton. MPE Report 288, pp. 39–44 (2005)
Haberl, F., Zavlin, V.E.: Astron. Astrophys. **391**, 571 (2002)
Haberl, F., Motch, C., Buckley, D.A.H., et al.: Astron. Astrophys. **326**, 662 (1997)
Haberl, F., Schwope, A.D., Hambaryan, V., et al.: Astron. Astrophys. **403**, L19 (2003)
Haberl, F., Motch, C., Zavlin, V.E., et al.: Astron. Astrophys. **424**, 635 (2004a)
Haberl, F., Zavlin, V.E., Trümper, J., et al.: Astron. Astrophys. **419**, 1077 (2004b)
Haberl, F., Turolla, R., de Vries, C.P., et al.: Astron. Astrophys. **451**, L17 (2006)
Hambaryan, V., Hasinger, G., Schwope, A.D., et al.: Astron. Astrophys. **381**, 98 (2002)
Ho, W.C.G., Lai, D.: Astrophys. J. **607**, 420 (2004)
Johnston, S.: Mon. Not. Roy. Astron. Soc. **340**, L43 (2003)
Jones, D.I., Andersson, N.: Mon. Not. Roy. Astron. Soc. **324**, 811 (2001)
Kaplan, D.L., van Kerkwijk, M.H.: Astrophys. J. **628**, L45 (2005a)
Kaplan, D.L., van Kerkwijk, M.H.: Astrophys. J. **635**, L65 (2005b)
Kaplan, D.L., Kulkarni, S.R., van Kerkwijk, M.H.: Astrophys. J. **579**, L29 (2002a)
Kaplan, D.L., Kulkarni, S.R., van Kerkwijk, M.H., et al.: Astrophys. J. **570**, L79 (2002b)
Kaplan, D.L., van Kerkwijk, M.H., Anderson, J.: Astrophys. J. **571**, 447 (2002c)
Kaplan, D.L., Kulkarni, S.R., van Kerkwijk, M.H.: Astrophys. J. **588**, L33 (2003a)
Kaplan, D.L., van Kerkwijk, M.H., Marshall, H.L., et al.: Astrophys. J. **590**, 1008 (2003b)
Ketsaris, N.A., Kuster, M., Postnov, K.A., et al: In: Proc. Intl. Workshop: Hot Points in Astrophysics, JINR, Dubna, Russia, pp. 192–205 (2000)
Komarova, V.: Optical search for counterparts to isolated neutron stars at the 6m telescope of SAO RAS. Poster presented at the Conference on High Energy Density Laboratory Astrophysics, Rice University, Houston, TX, 11–14 March 2006
Malofeev, V.M., Malov, O.I., Teplykh, D.A., et al.: Astron. Telegr. **798**, 1 (2006)
Malofeev, V.M., Malov, O.I., Teplykh, D.A.: Astrophys. Space Sci. **308**. DOI 10.1007/s10509-007-9341-y (2007)
McLaughlin, M.A., Stairs, I.H., Kaspi, V.M., et al.: Astrophys. J. **591**, L135 (2003)

Mori, K., Hailey, C.J.: Astrophys. J. **648**, 1139 (2006)
Morris, D.J., Hobbs, G., Lyne, A.G., et al.: Mon. Not. Roy. Astron. Soc. **335**, 275 (2002)
Motch, C.: In: X-ray Astronomy, Stellar Endpoints, AGN, and the Diffuse X-Ray Background, AIP Conference Proceedings, pp. 244–253 (2001)
Motch, C., Haberl, F., Zickgraf, F.J., et al.: Astron. Astrophys. **351**, 177 (1999)
Motch, C., Zavlin, V.E., Haberl, F.: Astron. Astrophys. **408**, 323 (2003)
Motch, C., Sekiguchi, K., Haberl, F., et al.: Astron. Astrophys. **429**, 257 (2005)
Motch, C., Pires, A.M., Haberl, F., et al.: Astrophys. Space Sci. **308**. DOI 10.1007/s10509-007-9338-6 (2007)
Paerels, F.: Astrophys. J. **476**, L47 (1997)
Paerels, F., Mori, K., Motch, C., et al.: Astron. Astrophys. **365**, L298 (2001)
Page, D., Geppert, U., Weber, F.: Nucl. Phys. A **777**, 497 (2006)
Pavlov, G.G., Shibanov, I.A., Iakovlev, D.G.: Astrophys. Space Sci. **73**, 33 (1980)
Pavlov, G.G., Zavlin, V.E., Trümper, J., et al.: Astrophys. J. **472**, L33 (1996)
Pavlov, G.G., Zavlin, V.E., Sanwal, D.: In: Becker, W., Lesch, H., Trümper, J. (eds.) Neutron Stars, Pulsars, and Supernova Remnants. MPE-Report 278, pp. 273–286 (2002)
Pons, J.A., Walter, F.M., Lattimer, J.M., et al.: Astrophys. J. **564**, 981 (2002)
Posselt, B., Popov, S.B., Haberl, F.: Astrophys. Space Sci. **308**. DOI 10.1007/s10509-007-9344-8 (2007)
Ransom, S.M., Gaensler, B.M., Slane, P.O.: Astrophys. J. **570**, L75 (2002)
Rutledge, R.E., Fox, D.W., Bogosavljevic, M., et al.: Astrophys. J. **598**, 458 (2003)
Schwope, A.D., Hasinger, G., Schwarz, R., et al.: Astron. Astrophys. **341**, L51 (1999)
Schwope, A.D., Hambaryan, V., Haberl, F., et al.: Astron. Astrophys. **441**, 597 (2005)
Schwope, A.D., Hambaryan, V., Haberl, F., et al.: Astrophys. Space Sci. **308**. DOI 10.1007/s10509-007-9375-1 (2007)
Treves, A., Turolla, R., Zane, S., et al.: Publ. Astron. Soc. Pac. **112**, 297 (2000)
Trümper, J., Kahabka, P., Oegelman, H., et al.: Astrophys. J. **300**, L63 (1986)
Trümper, J.E., Burwitz, V., Haberl, F., et al.: Nucl. Phys. B Proc. Suppl. **132**, 560 (2004)
van Kerkwijk, M.H., Kaplan, D.: Astrophys. Space Sci. **308**. DOI 10.1007/s10509-007-9343-9 (2007)
van Kerkwijk, M.H., Kulkarni, S.R.: Astron. Astrophys. **380**, 221 (2001a)
van Kerkwijk, M.H., Kulkarni, S.R.: Astron. Astrophys. **378**, 986 (2001b)
van Kerkwijk, M.H., Kaplan, D.L., Durant, M., et al.: Astrophys. J. **608**, 432 (2004)
Vink, J., de Vries, C.P., Méndez, M., et al.: Astrophys. J. **609**, L75 (2004)
Walter, F.M.: Astrophys. J. **549**, 433 (2001)
Walter, F.M., Lattimer, J.M.: Astrophys. J. **576**, L145 (2002)
Walter, F.M., Matthews, L.D.: Nature **389**, 358 (1997)
Walter, F.M., Wolk, S.J., Neuhäuser, R.: Nature **379**, 233 (1996)
Zampieri, L., Campana, S., Turolla, R., et al.: Astron. Astrophys. **378**, L5 (2001)
Zane, S., Turolla, R.: Mon. Not. Roy. Astron. Soc. **366**, 727 (2006)
Zane, S., Turolla, R., Stella, L., et al.: Astrophys. J. **560**, 384 (2001)
Zane, S., Haberl, F., Cropper, M., et al.: Mon. Not. Roy. Astron. Soc. **334**, 345 (2002)
Zane, S., Cropper, M., Turolla, R., et al.: Astrophys. J. **627**, 397 (2005)
Zavlin, V.E., Pavlov, G.G.: In: Becker, W., Lesch, H., Trümper, J. (eds.) Neutron Stars, Pulsars, and Supernova Remnants. MPE-Report 278, pp. 263–272 (2002)

Astrophys Space Sci (2007) 308: 191–201
DOI 10.1007/s10509-007-9343-9

ORIGINAL ARTICLE

Isolated neutron stars: magnetic fields, distances, and spectra

M.H. van Kerkwijk · D.L. Kaplan

Received: 27 July 2006 / Accepted: 31 August 2006 / Published online: 20 March 2007

Abstract We present timing measurements, astrometry, and high-resolution spectra of a number of nearby, thermally emitting, isolated neutron stars. We use these to infer magnetic field strengths and distances, but also encounter a number of puzzles. We discuss three specific ones in detail: (i) For RX J0720.4-3125 and RX J1308.6+2127, the characteristic ages are in excess of 1 Myr, while their temperatures and kinematic ages indicate that they are much younger; (ii) For RX J1856.5-3754, the brightness temperature for the optical emission is in excess of that measured at X-ray wavelengths for reasonable neutron-star radii; (iii) For RX J0720.4-3125, the spectrum changed from an initially featureless state to one with an absorption feature, yet there was only a relatively small change in $T_{\rm eff}$. Furthermore, we attempt to see whether the spectra of all seven sources, in six of which absorption features have now been found, can be understood in the context of strongly magnetised hydrogen atmospheres. We find that the energies of the absorption features can be reproduced, but that it remains puzzling that, for J0720.4-3125 specifically, the spectrum was featureless in one state, and that, generally, the spectra do not have high-energy tails that are harder than the Wien-like ones obseved.

M.H. van Kerkwijk (✉)
Department of Astronomy & Astrophysics, University of Toronto, 50 Saint George Street, Toronto, Ontario M5S 3H4, Canada
e-mail: mhvk@astro.utoronto.ca

D.L. Kaplan
Kavli Institute for Astrophysics and Space Research, Massachusetts Institute of Technology, 77 Massachusetts Ave, Room 37-664H, Cambridge, MA 02139, USA
e-mail: dlk@space.mit.edu

Keywords Atomic processes and interactions · Stellar atmospheres · Neutron stars

PACS 95.30.Dr · 97.10.Ex · 97.60.Jd

1 Introduction

One of the great benefits of the *ROSAT* All-Sky Survey (Voges et al. 1996) is that is has provides an unbiased sample of all classes of nearby neutron stars (limited only by their age and distribution of the local interstellar medium). Particularly interesting is the discovery of the group of seven nearby, thermally emitting, isolated neutron stars (INS; for a review, see Haberl, these proceedings).

The INS form the majority among the nearby neutron stars (typical distances are less than ~500 pc; Kaplan et al. 2002b; see also Posselt et al., Popov, these proceedings), yet are atypical of the neutron-star population represented by radio surveys: while pulsars detected by their thermal emission all have normal periods of less than a second, five out of the seven INS have periods about ten times longer (the remaining two appear to have no pulsations despite intensive searches; Ransom et al. 2002; van Kerkwijk et al. 2004, but see Haberl [these proceedings] for a possible periodicity in RX J1605.3+3249). A number of models – accretors (Wang 1997), middle-aged magnetars (Heyl and Kulkarni 1998; Heyl and Hernquist 1999), long-period pulsars (Kaplan et al. 2002a; Zane et al. 2002)—have been suggested to explain these objects.

A prime reason for studying the INS is the hope of constraining fundamental physics at very high densities: neutron stars are natural laboratories for quantum chromodynamics (Rho 2000). The overall goal is to determine the masses and radii of a number of neutron stars and hence constrain the equation of state (EOS) of ultra-dense matter (Lattimer and Prakash 2000; Lattimer, these proceedings). For the majority of known neutron stars (i.e., radio pulsars), this is complicated by the non-thermal emission that dominates the spectrum, but for the INS this is not the case: the X-ray spectra show thermal emission only. Hence, much effort has been spent trying to derive constraints from the INS (Burwitz et al. 2001, 2003; Drake et al. 2002; Pons et al. 2002). The constraints have not been very meaningful, however, because the data could not be interpreted properly: they just do not fit any current realistic models (Motch et al. 2003; Zane et al. 2004).

To make progress in understanding the thermal emission, we need first to know the basic ingredients: the elemental abundances, the temperature distribution, and the magnetic field strength. Furthermore, to use the thermal emission to infer radii, we need information about the distance. Fortunately, observational clues are now becoming available: broad absorption features at energies of 0.3–0.7 keV have been discovered in the spectra of six of the seven INS (Haberl et al. 2003, 2004a, 2004b; van Kerkwijk et al. 2004; Zane et al. 2005; see Haberl, these proceedings), and, as described below, magnetic field strengths have been inferred from timing solutions and new or improved parallaxes have been measured.

The outline of this contribution is as follows. First, in Sect. 2, we present timing solutions for RX J0720.4-3125 and RX J1308.6+2127, and discuss the resulting estimates of the magnetic field strengths and characteristic ages. Next, in Sect. 3, we describe new parallax distance measurements for RX J1856.5-3754 and RX J0720.4-3125. For the former, these resolve previous conflicting results, but also raise a puzzle: a rather large radius or high brightness temperature inferred for the optical emission. In Sect. 4, we turn to high-resolution X-ray spectra, comparing spectra of RX J0720.4-3125, before and after its spectral change, with those of RX J1308.6+2127. In Sect. 5, we attempt to interpret the observations assuming the sources have gaseous atmospheres, focussing on hydrogen, but also briefly discussing the possibility of helium. We summarise and discuss future work in Sect. 6.

From here on, we will refer to the INS in the text using abbreviated names: J0420 for RX J0420.0-5022, J0720 for RX J0720.4-3125, J0806 for RX J0806.4-4123, J1308 for RX J1308.6+2127 = RBS 1223, J1605 for RX J1605.3+3249, J1856 for RX J1856.5-3754, J2143 for RX J2143.0+0654 = RBS 1774 = RXS J214303.7+065419.

2 Timing solutions

Until recently, the typical magnetic field strength of the INS was just a guess, with a wide range of possibilities (10^{10}–10^{15} G)—similar to the wide uncertainty in magnetic field

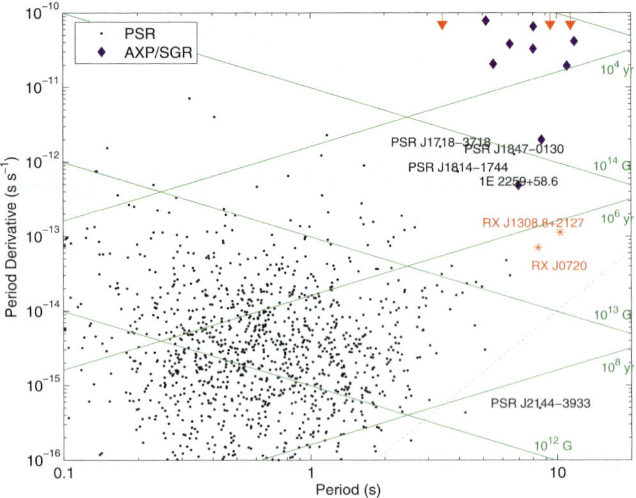

Fig. 1 P–$\dot{P}$ diagram, showing radio pulsars (*points*) and magnetars (*diamonds*); selected objects are labeled. Also shown are the five INS with periodicities: RX J0720.4-3125 and RX J1308.6+2127 are shown by the *stars*, while RX J0420.0-5022, RX J0806.4-4123, and RX J2143.0+0654 are the *arrows at the top* (since $\dot{P}$ is unknown). The *diagonal lines* show loci of constant dipole magnetic field and spin-down age, as labeled

strength implied for the different demographic models of the INS, although, based on the long spin periods, fields of a few 10^{13} G were considered most likely.

We have improved upon this situation using dedicated timing observations with *Chandra*, which, combined with *Chandra*, *XMM-Newton*, and *ROSAT* archival observations, allowed us to determine phase-coherent timing solutions for J0720 and J1308 stretching back at least 5 years (Kaplan and van Kerkwijk 2005a, 2005b). Using the periods and the period derivatives to place these objects on the classic P–$\dot{P}$ diagram (Fig. 1), one sees that they are intermediate between radio pulsars and magnetars (in line with the idea that they are long-period pulsars whose radio beams do not cross our line of sight; Kaplan et al. 2002a; Zane et al. 2002).

From the solutions, assuming the sources spin down by magnetic dipole radiation, one infers that they have similar magnetic fields: $B = 2.4 \times 10^{13}$ G and 3.4×10^{13} G for J0720 and J1308, respectively. As will become clear below (Sect. 5), this agrees quite well with the magnetic field strengths inferred from the absorption lines.

One also infers characteristic ages: $\tau_c \equiv P/2\dot{P} = 1.9$ Myr for J0720 and 1.5 Myr for J1308. These are puzzling, since they are substantially longer than expected based on cooling: From standard cooling curves (Page et al. 2004), the observed temperatures around 90 eV (10^6 K) correspond to ages of a few 10^5 yr. Even if one takes into account that the black-body temperature likely overestimates the effective temperature—as is clear from the fact that the extrapolation of a black-body fit to the X-ray data underpredicts the optical—one is very hard-pressed to find an age in excess of 10^6 yr; at 1.5 Myr, the effective temperature should be below 20 eV.

For J0720, there is an additional age estimate from kinematics (Motch et al. 2003; Kaplan 2004; also Motch, these proceedings): tracing back its proper motion, the most natural birthplace is the Trumpler 10 association; it would have left about 7×10^5 yr ago.[1]

Of course, the above discrepancy may simply mean that the characteristic age is a poor estimate of the true age. For breaking with $\dot{\nu} \propto \nu^n$, where ν is the spin frequency and n the so-called braking index (equal to 3 for magnetic dipole radiation), the true age is $t = (1 - P_0/P)^{n-1}(P/(n-1)\dot{P})$. Thus, one can obtain ages $t \ll \tau$ either if the initial spin period P_0 is close to the current one (i.e., the neutron star was born spinning slowly, but in no other systems is there evidence for $P_0 > 1$ s), or if n is substantially larger than 3. Lyne (these proceedings) presented evidence for values of n substantially different from 3, although most values were less, implying characteristic ages that are longer than the true age, contrary to what is required here.

As an alternative, we noted that one way of obtaining $P_0 \simeq P$, would be to have the neutron star undergo a phase in which it was accreting, either from a companion (which later disappeared, e.g., in a supernova explosion) or perhaps a fall-back disk such as that discovered around the AXP 4U 0142+61 (Wang et al. 2006). Intriguingly, the equilibrium spin period, $P_{\rm eq} \approx 5$ s $(B/10^{13}\ {\rm G})^{6/7}(\dot{M}/\dot{M}_{\rm Edd})^{-3/7}$, is roughly equal to the current observed periods for a magnetic field of a few 10^{13} G and an accretion rate $\dot{M}$ close to the Eddington rate $\dot{M}_{\rm Edd}$.

Finally, comparing the timing residuals, we find that for J0720, they are ~0.3 s, far larger than the measurement errors, while for J1308, they are consistent with the measurement errors, at ~0.01 s. The larger residuals for J0720 have been ascribed to precession (Haberl et al. 2006; also Haberl, these proceedings). To verify this, we tried including different terms in our timing model (see Fig. 2), and we indeed find that adding a periodic component improves the fit drastically, with the reduced χ^2 decreasing from $\chi^2_\nu = 11.9$ to 2.4. Trying different periods, however, the best period appears to be 4.3 yr (Fig. 3), and not ~7 yr as inferred from the spectral changes (at 7 yr, the fit is better than quadratic, but not that much different from a higher-order polynomial). This is in contrast to what is found by Haberl (these proceedings), who does find a best period of 7 yr. We think it is unlikely that this results from the different recent observations included (LETG versus XMM), but rather, as mentioned by Haberl, reflects ambiguities in the cycle count for the early ROSAT observations. Generally, we caution, though, that the timing residuals shown by J0720 are not exceptional: they are in line with trends seen for radio pulsars and similar apparent periodicities can be seen in the residuals of some of the Anomalous X-ray Pulsars (Kaspi, these proceedings).

[1] A possible origin in the Scorpius OB associations about 1.5 Myr ago has also been suggested, but the new parallax and proper motion we derived [partly described in Sect. 3 below] make this less likely. Furthermore, for J1856, which is cooler than J0720, the kinematic age of about 4×10^5 yr is not in doubt.

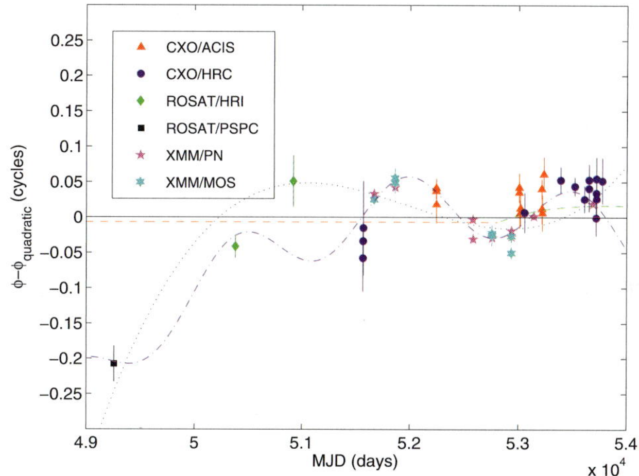

Fig. 2 Phase residuals for timing measurements for RX J0720.4-3125 relative to the best-fit quadratic ($\dddot{\nu} = 0$) model. The data include those from Kaplan and van Kerkwijk (2005a), plus a number of additional points from *Chandra* with LETG/HRC and one additional point from *XMM-Newton* with EPIC-PN (all at MJD > 53500); other EPIC-PN data exist (Haberl, these proceedings) but they are not yet public. The fit has $\chi^2_\nu = 11.93$ and the residuals have rms = 0.39 s. We also show alternate fits: the best-fit cubic model ($\dddot{\nu} \neq 0$; *dotted, black curve*), which has rms = 0.36 s and $\chi^2_\nu = 5.12$; the best-fit periodic model (*dot-dashed, blue*), which has a period of 4.3 yr and rms = 0.23 s and $\chi^2 = 2.35$; and a simple glitch model (*dashed, red*) with the glitch occurring at MJD 52821, which does not substantially improve the fit over the quadratic model. Note that there appears to be high-frequency residuals that are not fit by any of these models, as can be seen by the disagreement between simultaneous XMM observations with different instruments. Whether these are due to the energy dependence of the pulse profile or some instrumental effect, we cannot say, but it suggests that caution should be taken in interpreting the timing residuals.

3 Parallax measurements

A parallax measurement for the brightest INS, J1856, was first attempted by Walter (2001), using three observations with the Planetary Camera onboard the *Hubble Space Telescope* (*HST*); the resulting parallax implied a distance of ~60 pc. The measurement was tricky, however, and a much larger distance of 140 pc was derived from the same observations by Kaplan et al. (2002b); a larger distance, of 117 pc, was also found by Walter and Lattimer (2002), who redid their analysis and included a fourth PC observation.

In order to obtain more accurate distances, we have used the High-Resolution Camera (HRC) of the Advanced Camera for Surveys (ACS) onboard *HST*. This camera is more sensitive than the PC and has a smaller pixel scale, so that

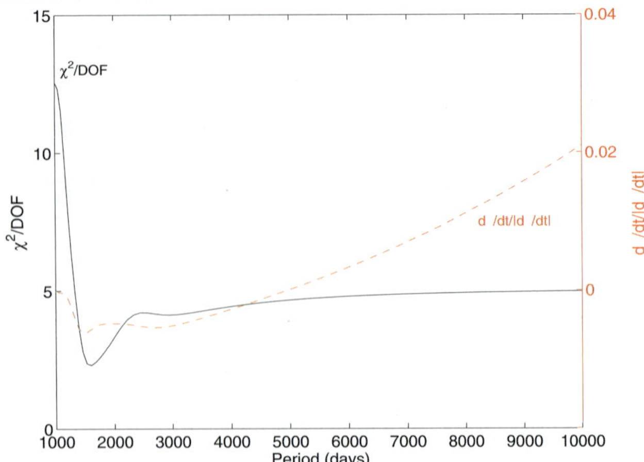

Fig. 3 Improvement in χ^2_ν for a timing model for RX J0720.4-3125 that includes a periodic component, as a function of period, fit to the data shown in Fig. 2. The χ^2_ν (*solid, black curve*) has its deepest minimum around 1580 days (4.3 yr); there may be a second minimum near 3000 days (8.2 yr), close to the ~7 yr period suggested by Haberl et al. (2006), but with these data it is clearly less significant. Also shown is the fractional change in $\dot{\nu}$ for the different fits (*dashed, red curve*); it changes by <2% for the full range of periods considered here, and thus does not influence the conclusions regarding the magnetic field or characteristic age

undersampling of the point-spread function is much less of an issue. Furthermore, across the pixels, the sensitivity is more uniform, reducing the variability in the point-spread function with pixel phase; as a result, much more accurate astrometry can be done with ACS/HRC (Anderson and King 2004). We obtained images of J1856 and J0720 in the blue F475W band, visiting each source eight times over two years.

For J1856, our analysis of the HRC data is virtually complete (Kaplan, van Kerkwijk, and Anderson, in preparation). In order to obtain as accurate a parallax as possible, we have taken into account the parallactic motion of the background reference stars, by determining their photometric parallaxes (assuming they are main-sequence stars; for the less distant stars—generally the brighter ones with strong weight—we confirm the photometric parallaxes astrometrically). With that, from the HRC data alone, we determine a parallax $\pi = 6.2 \pm 0.6$ mas, corresponding to a distance $d = 161^{+18}_{-14}$ pc. We are currently trying to improve the measurement further by including the PC data.

For J0720, a first analysis of the HRC data has just been completed. For this source, parallaxes of the background stars are much less important, since it is at low Galactic latitude and most objects are distant. Our preliminary parallax is $\pi = 3.0 \pm 1.0$ mas, corresponding to a distance $d = 330^{+170}_{-80}$ pc.

The factor two ratio in the distances to J1856 and J0720 is consistent with what was expected by Kaplan et al. (2002b) under the zeroth-order assumption that the optical flux for different sources scales as $f_\nu \propto T(R/d)^2$, that the radii R are similar, and that the temperature T in the region of the atmosphere emitting the optical emission scales with the temperature determined from fits to the X-ray spectrum. The distances also compare well with the distances of 135 ± 25 and 255 ± 25 pc inferred from the run of H I column density with distance (Posselt et al., these proceedings).

With our distances, we can estimate the radii for the two sources. We start by simply using the black-body fit to the X-ray spectra. For J1856, one finds $kT = 63$ eV and $R_\infty/d = 0.0364\ \mathrm{km\,pc^{-1}}$, which, with our new distance, implies a radiation radius $R_\infty \simeq 5.9$ km. This is smaller than a typical radius of a neutron star, but this is not unexpected, for two reasons. First, for the most likely atmospheric compositions, the opacity decreases with increasing frequency. As a result, at X-ray energies one sees relatively hot layers and a fit to the X-ray spectrum will thus overestimate the effective temperature and underestimate the radius (Pavlov et al. 1996). Second, the temperature distribution likely is not uniform, in which case the area inferred from the X-ray emission would simply correspond to that of the hotter parts.

In the above picture, one expects the optical emission to be in excess of the extrapolation from the black-body fit, since it arises from a cooler layer and from a larger area. And indeed, the spectral energy distribution, shown in Fig. 4, shows an excess. It poses a possible problem, however, since the optical excess is a factor 7, which implies a radiation radius of $R_{\infty,\mathrm{opt}} = 16\ (T_\mathrm{X}/T_\mathrm{opt})^2$ km (where T_opt and T_X are suitable averages of the temperatures of the optical and X-ray emitting regions, respectively). Given that one expects $T_\mathrm{opt} < T_\mathrm{X}$, the optical emission thus seems to imply that the radiation radius is quite a bit larger than 16 km. Yet, for most reasonable equations of state, a typical neutron star is expected to have a smaller radiation radius (e.g., Lattimer and Prakash 2001).

Of course, the above discrepancy may simply reflect our lack of understanding of neutron star atmospheres in strong magnetic fields: the temperature of the optical emission region may be larger than expected. In order to see what would be required, one can reverse the process: assume that the neutron star has a 'standard' mass and radius, and calculate the brightness temperature at each energy assuming that the emission originates from the whole surface. In Fig. 5, we show the result for $R_\infty = 14.7$ km (which is the value one obtains for $M = 1.35 M_\odot$ and $R = 12$ km). We see that this confirms the above reasoning: in order to produce the optical emission, the temperature in the emission region has to exceed 70 eV, i.e., be higher than that in the X-ray emitting region.

For J0720, a fit to the X-ray spectrum from 2000 (i.e., before the appearance of an absorption line) gives $kT = 85.7$ eV and $R_\infty/d = 0.0170\ \mathrm{km\,pc^{-1}}$. Taking the distance at face value, the implied radiation radius is 5.7 km, a little smaller than that of J1856, but easily consistent within

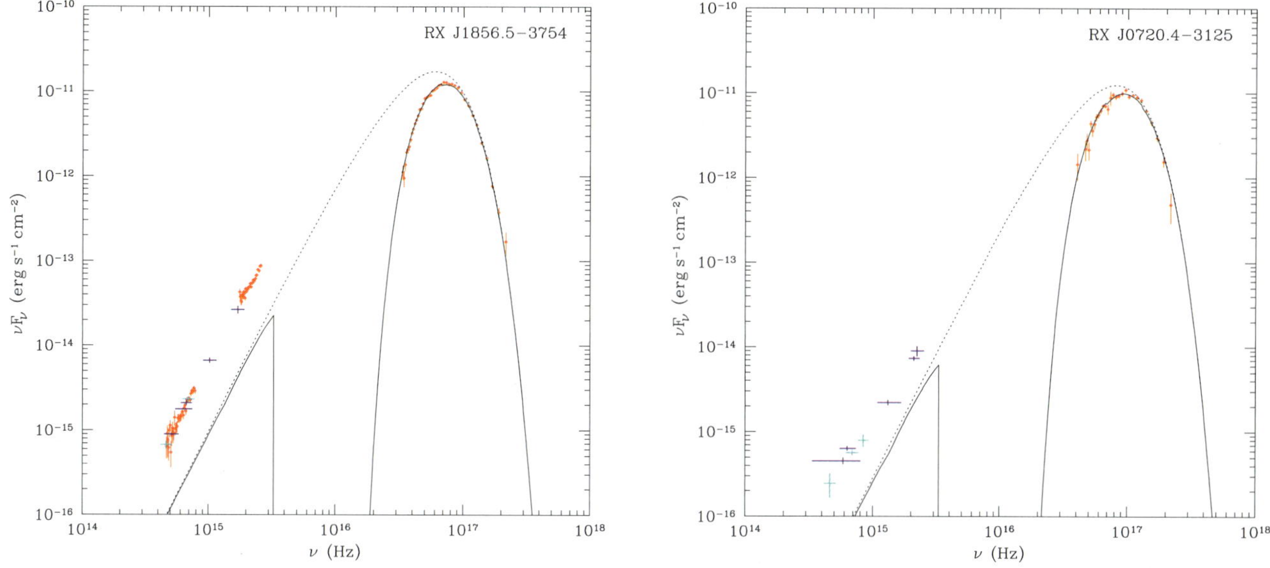

Fig. 4 Spectral energy distributions for RX J1856.5-3754 (*left*) and RX J0720.4-3125 (*right*; with the X-ray spectrum from before the appearance of an absorption feature). For both, the X-ray points are from LETG spectra, the *dark blue points* from *HST*, and the *cyan points* from ground-based observations. The optical and ultraviolet spectra for RX J1856.5-3754 are from VLT and *HST*, respectively. The *black, drawn curves* represent the best-fit black-body models to the X-ray data; the *dotted curves* are the same model without interstellar extinction

the 30% uncertainty due to the parallax measurement error. Since the optical excess is similar, the radiation radius for the optical emission is again large. In this case, however, the optical emission does not follow a Rayleigh-Jeans tail (Kaplan et al. 2003; Motch et al. 2003; Fig. 4), and hence it is not clear that the emission is from the surface. This can also be seen from the brightness temperatures (Fig. 5), which is not constant in the optical/ultraviolet range.

A further puzzle raised in comparing the sources, is that despite the fact that the X-ray emission areas are rather similar, the pulsation properties are very different: J0720 shows clear pulsations, with a pulsed fraction of 11% (Haberl et al. 1997), while J1856 shows no pulsations, to a limit of $\sim$1% (Ransom et al. 2002; Burwitz et al. 2003). This may reflect differences in geometry; for instance, for isotropic emission from two opposite magnetic poles, there is a fair range in parameters for which no pulsations would be observed (e.g., Beloborodov 2002). Of course, the presence of the pulsations constitutes a warning about the brightness temperatures shown in Fig. 5: the pulsations indicate the X-ray emission does not arise from the whole surface, and hence the temperatures of the X-ray emitting regions will be higher than those shown.

4 LETG spectra

The study of the X-ray spectra of the INS has made great strides with the advent of *Chandra* and *XMM-Newton*. Both the CCD instruments, in particular EPIC-PN, and the grating spectrometers LETG and RGS have been used extensively. Here, we focus on the grating instruments (for the exciting results from EPIC-PN, see Haberl, these proceedings). We will only discuss results from LETG, since that instrument covers the full range of energies at which INS emit and since the calibration of the RGS at longer wavelengths has been rather problematic.[2]

So far, three sources have been observed with LETG. By far the best spectrum is of J1856, taken using 500 ks of director's discretionary time. Unfortunately, and puzzlingly, the spectrum appears completely featureless, and is well described by a black-body model (Burwitz et al. 2001, 2003; Drake et al. 2002; Braje and Romani 2002). The second brightest source, J0720 has been studied extensively as well. A first spectrum was taken in 2000 (Kaplan et al. 2003), when the spectrum was featureless, and a second in 2004 (Vink et al. 2004), to confirm the change in spectrum discovered with RGS by de Vries et al. (2004). During 2005, the source was regularly monitored by us, for 300 ks total, in order to look for further spectral changes and to see if more than one spectral feature was present. Finally, a 100 ks observation was made of J1308 in guaranteed time by the MPE group (used for timing purposes in Schwope et al. 2005).

In Fig. 6, we show the LETG spectra of J0720 and J1308. For the former, we show separately the 'before' spectrum (from 2000, before the appearance of an absorption feature) and the 'after' spectrum (the average of all spectra taken af-

[2]The problems appear to be largely solved with the 2006 June 30 release of the Scientific Analysis Software.

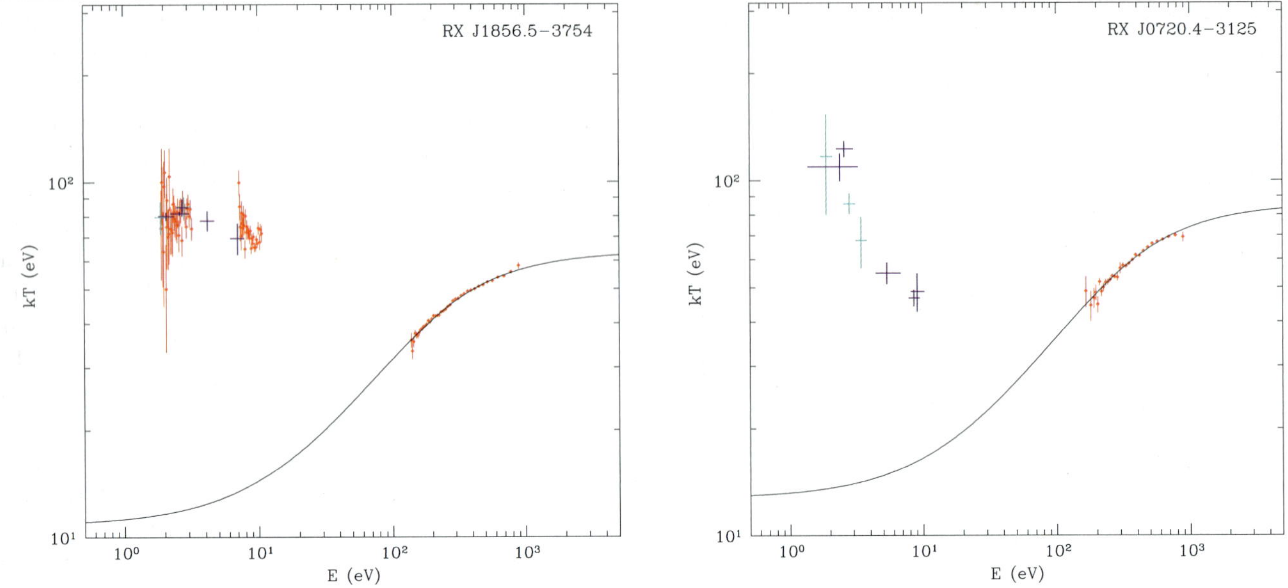

Fig. 5 Brightness temperatures for RX J1856.5-3754 (*left*) and RX J0720.4-3125 (*right*), assuming our parallax measurements are correct and that the emission arises from a neutron star with radiation radius $R_\infty = 14.7$ km (which is the value one obtains for $M = 1.35M_\odot$ and $R = 12$ km). The symbols and colours are as in Fig. 4 (the model curves are not at a constant temperature since the best-fit radiation radius is not equal to 14.7 km). One sees that the optical emission requires temperatures at least equal to those required for the X-ray emission. Note, however, that for RX J0720.4-3125 it is not clear the emission is thermal

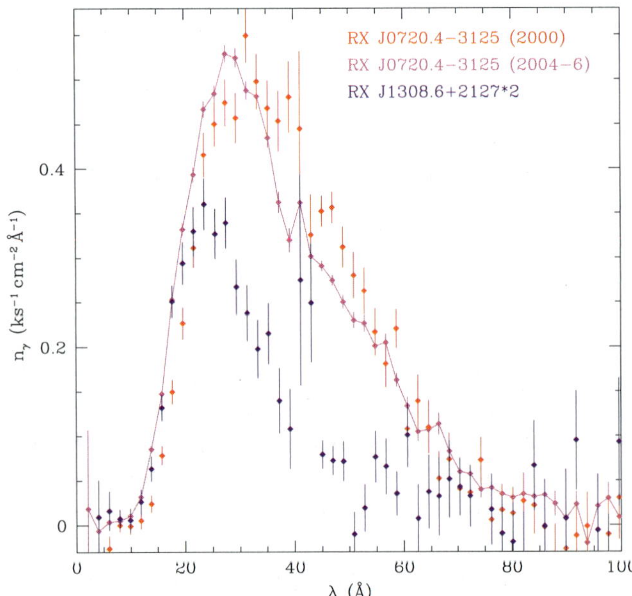

Fig. 6 Spectra of RX J0720.4-3125 and J1308.6+2127 taken with LETG. The *magenta points* connected with the line are the average of spectra of RX J0720.4-3125 taken after 2004 (about 300 ks). Compared to the earlier spectrum (*red points*), the spectrum is harder and has developed an absorption feature. On the other hand, compared to RX J1308.6+2127 (*blue points*, multiplied by two in order to match at the short-wavelength side), the feature is rather weak, even though the inferred temperature is similar (as is also clear from the good match at short wavelengths), and so is the magnetic field inferred from timing

ter the change; within our statistics, the individual spectra do not differ).

The comparison of the three spectra raises a number of questions. First, for J0720, how can a relatively small change in temperature (from ~80 to 90 eV) cause the appearance of a pronounced absorption feature? Second, independent of the state of J0720, why is the absorption feature much less strong than that in J1308, despite the fact that their temperatures and magnetic field strengths (from timing, Sect. 2) are rather similar?

Considering first the change in J0720, the simplest possibility would be that the increase in temperature corresponds to a relatively large change in the ionisation and/or dissociation balance. If so, this might give a clue to the corresponding energies of the matter in the atmosphere. The alternative would be that the region that heated up (or appeared in view) has either a different composition or a greatly different magnetic field strength compared to the regions that dominated the spectrum before the change. Neither possibility seems particularly appealing.

Whatever the physical reason for the appearance of the absorption feature, another more basic question is whether the change corresponds to a global change in the properties, or whether, instead, only a fraction of the surface changed or if ones viewpoint changed. A global change might occur if, e.g., heat was deposited deep inside the neutron star. In contrast, a change in a limited area would be expected if heat were deposited near the surface (with the affected area perhaps increasing in size with time), or if the source were precessing and different regions came into view (de Vries et al. 2004; Haberl et al. 2006; Haberl and Zane, these proceedings).

Of course, if only part of the area that we see changed, then the average 'after' spectrum we currently observe contains a contribution from the unchanged parts of the surface, i.e., from the cooler, featureless 'before' spectrum. Hence, the spectrum from the changed part should be hotter and should have a stronger line than one would infer from the average. We can set an upper limit to the contribution from the 'before' spectrum by requiring that it does not exceed the 'after' spectrum at any wavelength. From Fig. 6, one sees that the limit is about 70%, set by the 35–40 Å region.

The above could solve the second question: it might well be that after the change, the parts of the surface of J0720 that show an absorption feature in their spectrum, have a line as strong as that observed in J1308. It only appears weaker in the 'after' spectrum because it is diluted by the featureless emission from the unchanged parts of the surface. So, just a single question may be left: how can a neutron-star atmosphere, with presumably the same magnetic field strength and the same composition, and with only a modest, $\Delta T/T < 0.2$ temperature increase, emit such different spectra?

5 Strongly magnetised atmospheres

In interpreting the spectra, a major uncertainty is the composition. For a single source, this may be difficult to determine uniquely, but one can hope to make progress by treating the INS as an ensemble: ideally, it should be possible to understand the features (or lack thereof) in all INS with a single composition, appealing only to differences in temperature and magnetic field strength (constrained by observations where possible), which might lead to different ionisation states being dominant, and possibly the formation of molecules or even a condensate. Here, we discuss only the possibilities of hydrogen or helium atmospheres. For completeness, we note that gaseous atmospheres composed of heavier elements appear to be excluded by the lack of large numbers of features. Condensated from heavier elements are also being considered seriously (Pons, Ho, these proceedings), and detailed theoretical calculations are being carried out to determine at what magnetic field strength condensates can form (Medin and Lai 2006; Lai, these proceedings).

5.1 Hydrogen

The presence of a hydrogen atmosphere has often been considered by default, since if any hydrogen is present, gravitational settling will ensure it floats to the surface. Typically, it has been assumed the hydrogen is fully ionised, and spectral features have been interpreted as proton cyclotron lines. In strong magnetic fields, however, the binding energies of atoms increase (for a review, Lai 2001; Potekhin, these proceedings; see also Fig. 7), and for temperatures and fields appropriate for INS, a fraction of up to 10% of neutral hydrogen will be present (Potekhin et al. 1999). From initial model-atmosphere calculations that take the presence of neutral hydrogen into account (Ho et al. 2003, see their Fig. 3), it is clear that, e.g., at 10^6 K and 10^{13} G, the lines from neutral hydrogen have larger equivalent width than the proton cyclotron line (they are less deep but much wider, due to the so-called motional Stark effect; Pavlov and Meszaros 1993; Potekhin and Pavlov 1997); at lower temperatures or stronger magnetic fields, the fraction of neutral hydrogen increases and hence the difference should be even larger. In general, it is worth stressing that the features are very strong: they may not appear so on the logarithmic scale typically used, but they have depths depth often exceeding 50%, similar to what is observed for J1308.

Below, we first discuss whether the energies of the main features observed in the INS can be reproduced by a strongly magnetised atmosphere, and then turn to two possible problems: harmonically spaced lines found recently, and the featureless, black-body like spectra shown by some INS. We do not include the optical excess among these problems, since currently it is not clear any model makes reliable predictions for the optical emission: at a few 10^{13} G, the plasma frequency exceeds the frequencies of optical photons, and the models do not take into account the resulting significant deviations of the refractive index from unity (van Adelsberg and Lai 2006; see Kowalski and Saumon 2004 for a discussion of possible effects in the context of cool white-dwarf atmospheres composed of helium).

5.1.1 Line energies

In Fig. 7, we show the energies for features that might be produced in a hydrogen atmosphere: the electron and proton cyclotron lines, and the bound-bound and bound-free transitions of neutral hydrogen (relative to the ground state). Also shown are the approximate energies of the main features that have been detected in the various INS (corrected for an assumed gravitational redshift of 0.3)

In the figure, thick vertical lines indicate the two magnetic field strengths inferred from timing (Sect. 2). From those, it follows that if J0720 and J1308 have hydrogen atmospheres, the features are most likely due to the transition from the ground state to the first excited tightly bound state of neutral hydrogen, perhaps in combination with the proton cyclotron line. As argued in Sect. 4 above, the line in J0720 might be weaker than that in J1308 because the emission from part of its surface is featureless (which in itself is

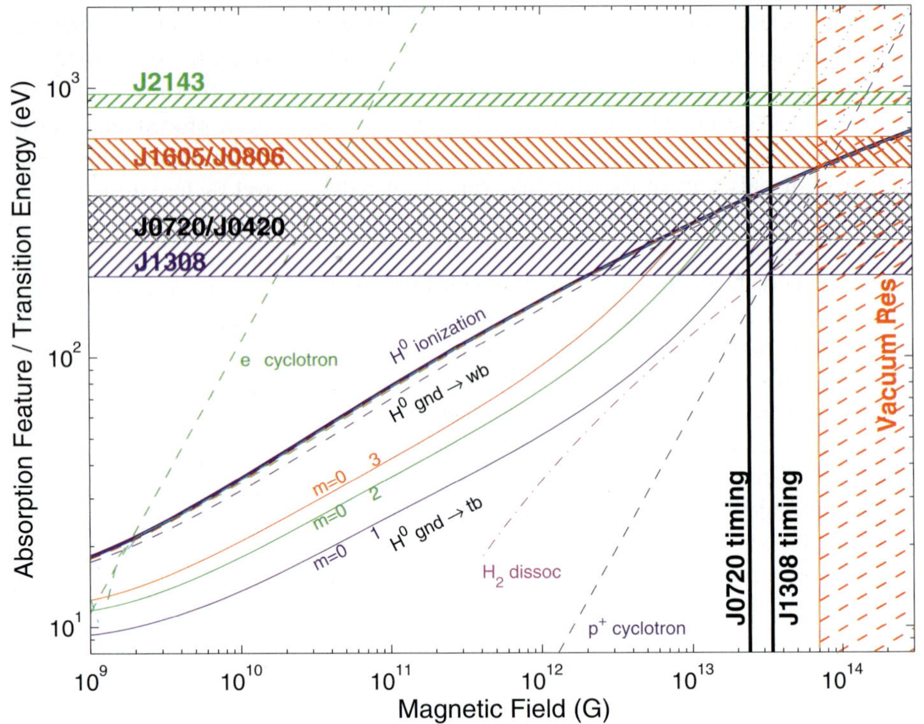

Fig. 7 Energy versus magnetic field for electron and proton cyclotron, ground state to tightly bound (tb) or weakly bound (wb) states in neutral hydrogen, hydrogen ionisation, and molecular H_2 dissociation (Ho et al. 2003; Lai 2001; Potekhin 1998). For fields above $\sim 10^{14}$ G (*hatched region* to the right), features may be washed out due to the effects of vacuum resonance mode conversion (Lai and Ho 2003). The *hatched bands* show the energies of the main absorption features (corrected for gravitational redshift) for the INS, and the *two vertical lines* indicate the dipole field strengths determined by timing. The sources are labeled with short-hand notation as in the text (see end of Sect. 1). References for the energies are: J0420, Haberl et al. (2004a); J0720, Haberl et al. (2004b); J0806, Haberl et al. (2004a); J1308, Haberl et al. (2003); J1605, van Kerkwijk et al. (2004); J2143, Zane et al. (2005)

problematic; we return to this below). Alternatively, the line in J0720 might be weaker because it is to the second excited tightly bound state (van Kerkwijk et al. 2004).

If the above is correct, the feature in J0420 likely has the same origin and thus its field should also be a few 10^{13} G. The features in J1605 and J0806 could result from the same transitions or from the ionisation edge, but in either case the implied magnetic field strength is higher, close to 10^{14} G. For J1605, van Kerkwijk et al. (2004) noted that the line was substantially weaker than that of J1308, and they suggested this might be due to the effect of vacuum resonance mode conversion, which for fields in excess of $\sim 7 \times 10^{13}$ G tends to weaken features (see Ho and Lai 2003; also Lai, these proceedings). Finally, for J2143, the line energy of 0.7 keV is substantially higher than what is observed for all other sources, and for any transition in neutral hydrogen, the upper state is auto-ionising: it is at an energy level that is higher than the continuum from the ground state. It is thus not clear whether the line could be due to neutral hydrogen. Instead, it might be due to the proton cyclotron line in a field of just over 10^{14} G. For these field strengths, the feature should be strongly weakened by vacuum resonance mode conversion (but not necessarily disappear; e.g., Ho and Lai 2004; van Adelsberg and Lai 2006); qualitatively, this is consistent with the rather modest observed strength Zane et al. (2005).

5.1.2 Possible problem 1: harmonically spaced lines

At the conference, evidence for harmonically spaced absorption lines was presented for three INS. For J1605, Haberl (these proceedings) found that apart from the line at 0.40 keV discovered by van Kerkwijk et al. (2004), the EPIC-PN data show a significant feature at 0.78 keV, i.e., at an energy that is in a 1:2 ratio with that of the stronger line. Furthermore, a third feature at 0.59 keV could be present, consistent with energies in a 2:3:4 ratio. For J1308, Schwope et al. (these proceedings) presented evidence that the single strong feature originally found at 0.3 keV or less by Haberl et al. (2003), could be composed of two features, at 0.23 and 0.46 keV, i.e., again harmonically spaced. Finally, for J0806, the single feature at 0.43 keV found by Haberl et al. (2004a) may again be better described by two features at 0.30 and 0.60 keV (Haberl, these proceedings).

It would appear tempting to interpret these features as cyclotron lines, since those naturally have harmonic energy ratios. It is difficult, however, to see how this could be possible for proton cyclotron lines, since the harmonics are expected to be exceedingly weak: the oscillator strength for the harmonic would be a factor $E/m_{\rm p}c^2$ weaker than that for the fundamental.

Instead, as mentioned in a discussion with George Pavlov, Joachim Trümper, and Frank Haberl at the meeting, a different solution may be suggested by the behaviour of the transitions of neutral hydrogen. As can be seen in Fig. 7, for any transition, above a certain magnetic field strength, the transition energy starts to become proportional to the proton cyclotron energy. As a result, at sufficiently strong magnetic field, the transitions become harmonically related. A possible problem, however, is that in this situation, the upper level of the transition is an auto-ionising state, i.e., it has an energy in excess of the continuum energy relative to the ground state. It will still lead to some additional opacity, but at present it is not clear whether this is sufficient. Fortunately, there is one prediction: for J1605, it would not be possible to explain the spectrum if there are really three features in a 2:3:4 ratio, without a strong corresponding '1'; thus, the prediction is that upon further analysis, the 0.59 keV feature will disappear.

Finally, we note in this context that it will be worth checking carefully that for J2143, the 0.7 keV feature observed is in fact not a 'harmonic.' From the present fits by Zane et al. (2005), a rather high $N_{\rm H}$ is inferred, and this could perhaps be an artefact of a strong absorption feature at $\sim$0.3 keV (From initial attempts, this appears unlikely; discussion with Mark Cropper at the meeting.)

5.1.3 Possible problem 2: featureless black-body spectra

Perhaps the most severe problem with the idea that the INS have pure hydrogen atmospheres is that the spectra of J1856 and J0720 (before the change) are featureless and well represented by black-body emission. For J1856, perhaps no features are expected, since its magnetic field strength, as inferred from the bow-shock shaped Hα nebula around the source (van Kerkwijk and Kulkarni 2001; Kaplan et al. 2002b), is below 10^{13} G, in which case all features may be below the observed band (Fig. 7), but for J0720 this explanation is not possible. Furthermore, for a mostly ionised atmosphere, the spectrum is expected to have a hard tail, unlike the observed exponential, Wien-like shape, since the free-free opacity decreases with increasing energy.

There are several possible solutions. First, there could be a reason for the opacity to be much greyer than currently estimated, so that the emission at all wavelengths originates from layers at similar depths and thus with similar temperature. The extreme version of this, discussed in detail by Pons and Ho (these proceedings), is that the two sources have a condensed surface (with possibly only a thin hydrogen layer on top). A less extreme version might be that the atmosphere does not contain just ionised and neutral hydrogen, but also molecules, which might have so many transitions that the opacity becomes effectively grey. Hydrogen molecules do indeed have a higher binding energy than hydrogen atoms, but the dissociation energy is only around 0.2 keV for a few 10^{13} G (Lai 2001; see Fig. 7). With temperatures only a factor two smaller, the abundance should be very small (as indeed found by, e.g., Potekhin et al. 1999; see their Fig. 7). Nevertheless, it may be worthwhile verifying this, making sure that the abundance and the resulting opacity are indeed negligible.

A possible alternative way to produce spectra resembling black bodies is by making the temperature profile in the atmosphere shallower, closer to isothermal. While this certainly appears ad hoc and likely would require significant fine-tuning, there is evidence for active magnetospheres: for J1856, the Hα nebula provides evidence of a pulsar wind, and for J0720, the optical emission appears to be partly non-thermal. If there is an active magnetosphere, some particles might hit the atmosphere, leading to additional heating; at the right locations, this could lead to rather different emergent spectra (e.g., Gänsicke et al. 2002).

At present, none of the above explanations seem satisfactory. Also, none provide an easy explanation for why some sources have featureless spectra while others have not (or why it would change). Perhaps the first parameter to consider would be the overall temperature, since J1856 is cool and the appearance of the absorption feature in J0720 was accompanied by a temperature increase. The increase was only small ($\Delta T/T < 0.2$), however, and furthermore, an absorption feature does appear to be present in the coolest INS, J0420 ($kT \simeq 45$ eV, Haberl et al. 2004a).

5.2 Helium

Above, we stated that if any hydrogen were present, it would float to the top. Recently, it has been questioned, however, whether an outer hydrogen envelope can survive (Chang and Bildsten 2004; Chang et al. 2004). The reason this is not certain is that some hydrogen will diffuse down and reach underlying Carbon or Oxygen layers, where, if the temperature is right, it will be burned. Indeed, Chang and Bildsten (2004) find that all of the hydrogen can be burned in the first 10^5 yr of a neutron star's life, in which case an atmosphere composed of helium might be left (unless hydrogen is replenished, as could happen due to spallation by relativistic particles from the magnetosphere, or very low levels of accretion).

Partly inspired by this possibility, Pavlov and Bezchastnov (2005) calculated properties of singly-ionised helium in strong magnetic fields. For a few 10^{13} G, the transition energies are again in the range in which features are observed in the INS, and hence it seems worthwhile to try to do a similar analysis as done above for hydrogen. From a very rough first attempt at producing model atmospheres (done by Kaya Mori and Wynn Ho), including neutral helium, it seems that, like for hydrogen, the features will be very strong. Typically, however, more than one very strong feature should be present, which appears to be in conflict with what is observed. The picture is currently incomplete, however, since molecules have not yet been considered, while for helium the binding energy of, e.g., He_2^+ is sufficiently high that it may well be present (a detailed calculation is tricky, since one has to have a decent estimate of the number of possible rotational and vibrational states).

6 Discussion and future prospects

Of the four main parameters mentioned in the introduction that determine the properties of the thermal emission from INS, we now appear to have reasonable handles on three: the shapes of the X-ray spectra indicate temperatures around 10^6 K, period derivatives imply magnetic field strengths of a few 10^{13} G, and parallax measurements show that a fair fraction of the surface is emitting X-ray radiation.

The main unknown appears to be the composition. We found that the energies of the observed absorption features can be matched fairly easily for hydrogen atmospheres. However, reproducing the smooth, featureless spectra of some INS, and the Wien-like high-energy side of the X-ray spectra in general, appears problematic, nor is it clear how the spectrum of J0720 could change from featureless to one that has an absorption line.

Fortunately, it should soon become clear whether these issues are real problems or not, since great progress is being made in constructing more reliable strongly magnetised hydrogen model atmospheres (Lai, Potekhin, these proceedings). From Sect. 5, it seems particularly important to include in full detail transitions to the auto-ionising levels, verify that all sources of opacity, including from (traces of) molecules are included, and check the influence, in particular on the temperature profile, of high-density effects and vacuum resonance mode conversion. At the same time, it would seem worthwhile to consider atmospheres of other elements; for the INS, He might be most relevant, but it would be good to check heavier elements as well, since these may cause the absorption features seen in 1E 1207.4-5209 (Hailey and Mori 2002).

From the observational side, the easiest route to further progress would appear to be timing. With further estimates of the magnetic fields, one can test the predictions based on hydrogen atmospheres, that J0420 has a field about as strong as that of J0720 and J1308, J0806 a stronger one, approaching 10^{14} G, and J2143 the strongest, in excess of 10^{14} G.

For the X-ray spectra, further monitoring is useful, but perhaps the largest advance will come from the unified analysis of all sources, which allows one to exclude instrumental effects. This is already well underway for the EPIC-PN data (Haberl, these proceedings), and similar studies of the LETG and RGS data should prove fruitful. As present, first steps are being taken in detailed modelling of the phase-resolved spectra (Haberl and Zane, these proceedings), and this should help obtain stronger constraints on the thermal distribution over the surface.

Finally, in the optical-ultraviolet regime, it would be good to complete the census of the sources, and obtain at least rough spectral energy distributions, to determine whether the emission is thermal, or whether there are non-thermal components. For sources that are sufficiently bright, proper motion measurements can help determine true ages and parallax measurements can help determine distances.

Acknowledgements It is a pleasure to thank Kaya Mori, Wynn Ho, George Pavlov, and Dong Lai for enlightening discussions about the physics of strongly magnetised, dense atmospheres, and Jay Anderson for his continuing patience and collaboration in trying to extract the best possible astrometry from *HST* images. We acknowledge financial support through a guest observer grant from NASA, as well as through individual grants from NSERC (M.H.v.K.) and a Pappalardo fellowship (D.L.K.).

References

Anderson, J., King I.: Multi-filter PSFs and Distortion Corrections for the HRC, ISR 04-15, Space Telescope Science Institute (2004)
Beloborodov, A.M.: Astrophys. J. **566**, L85 (2002)
Braje, T.M., Romani, R.W.: Astrophys. J. **580**, 1043 (2002)
Burwitz, V., Zavlin, V.E., Neuhäuser, R., et al.: Astron. Astrophys. **379**, L35 (2001)
Burwitz, V., Haberl, F., Neuhäuser, R., et al.: Astron. Astrophys. **399**, 1109 (2003)
Chang, P., Bildsten, L.: Astrophys. J. **605**, 830 (2004)
Chang, P., Arras, P., Bildsten, L.: Astrophys. J. **616**, L147 (2004)
de Vries, C.P., Vink, J., Méndez, M., et al.: Astron. Astrophys. **415**, L31 (2004)
Drake, J.J., et al.: Astrophys. J. **572**, 996 (2002)
Gänsicke, B.T., Braje, T.M., Romani, R.W.: Astron. Astrophys. **386**, 1001 (2002)
Haberl, F., Motch, C., Buckley, D.A. H., et al.: Astron. Astrophys. **326**, 662 (1997)
Haberl, F., Schwope, A.D., Hambaryan, V., et al.: Astron. Astrophys. **403**, L19 (2003)
Haberl, F., Motch, C., Zavlin, V.E., et al.: Astron. Astrophys. **424**, 635 (2004a)
Haberl, F., Zavlin, V.E., Trümper, J., et al.: Astron. Astrophys. **419**, 1077 (2004b)
Haberl, F., Turolla, R., de Vries, et al.: Astron. Astrophys. **451**, L17 (2006)
Hailey, C.J., Mori, K.: Astrophys. J. **578**, L133 (2002)
Heyl, J.S., Hernquist, L.: Phys. Rev. D **59**, 45005 (1999)
Heyl, J.S., Kulkarni, S.R.: Astrophys. J. **506**, L61 (1998)
Ho, W.C.G., Lai, D.: Mon. Not. Roy. Astron. Soc. **338**, 233 (2003)
Ho, W.C.G., Lai, D.: Astrophys. J. **607**, 420 (2004)
Ho, W.C.G., Lai, D., Potekhin, A.Y., Chabrier, G.: Astrophys. J. **599**, 1293 (2003)
Kaplan, D.L.: PhD thesis, California Institute of Technology (2004)
Kaplan, D.L., van Kerkwijk, M.H.: Astrophys. J. **628**, L45 (2005a)
Kaplan, D.L., van Kerkwijk, M.H.: Astrophys. J. **635**, L65 (2005b)
Kaplan, D.L., Kulkarni, S.R., van Kerkwijk, M.H., et al.: Astrophys. J. **570**, L79 (2002a)
Kaplan, D.L., van Kerkwijk, M.H., Anderson, J.: Astrophys. J. **571**, 447 (2002b)
Kaplan, D.L., van Kerkwijk, M.H., Marshall, H.L., et al.: Astrophys. J. **590**, 1008 (2003)

Kowalski, P.M., Saumon, D.: Astrophys. J. **607**, 970 (2004)
Lai, D.: Rev. Mod. Phys. **73**, 629 (2001)
Lai, D., Ho, W.C.G.: Astrophys. J. **588**, 962 (2003)
Lattimer, J.M., Prakash, M.: Phys. Rep. **333**, 121 (2000)
Lattimer, J.M., Prakash, M.: Astrophys. J. **550**, 426 (2001)
Medin, Z., Lai, D.: Phys. Rev. A **74**, 062508 (2006)
Motch, C., Zavlin, V.E., Haberl, F.: Astron. Astrophys. **408**, 323 (2003)
Page, D., Lattimer, J.M., Prakash, M., et al.: Astrophys. J. Suppl. Ser. **155**, 623 (2004)
Pavlov, G.G., Bezchastnov, V.G.: Astrophys. J. **635**, L61 (2005)
Pavlov, G.G., Meszaros, P.: Astrophys. J. **416**, 752 (1993)
Pavlov, G.G., Zavlin, V.E., Truemper, J., et al.: Astrophys. J. **472**, L33 (1996)
Pons, J.A., Walter, F.M., Lattimer, J.M., et al.: Astrophys. J. **564**, 981 (2002)
Potekhin, A.Y.: J. Phys. B **31**, 49 (1998)
Potekhin, A.Y., Pavlov, G.G.: Astrophys. J. **483**, 414 (1997)
Potekhin, A.Y., Chabrier, G., Shibanov, Y.A.: Phys. Rev. E **60**, 2193 (1999)
Ransom, S.M., Gaensler, B.M., Slane, P.O.: Astrophys. J. **570**, L75 (2002)
Rho, M.: nucl-th/0007073 (2000)
Schwope, A.D., Hambaryan, V., Haberl, F., et al.: Astron. Astrophys. **441**, 597 (2005)
van Adelsberg, M., Lai, D.: Mon. Not. Roy. Astron. Soc. **373**, 1495 (2006)
van Kerkwijk, M.H., Kulkarni, S.R.: Astron. Astrophys. **380**, 221 (2001)
van Kerkwijk, M.H., Kaplan, D.L., Durant, M., et al.: Astrophys. J. **608**, 432 (2004)
Vink, J., de Vries, C.P., Méndez, M., et al.: Astrophys. J. **609**, L75 (2004)
Voges, W. et al.: IAU Circ. **6420**, 2 (1996)
Walter, F.M.: Astrophys. J. **549**, 433 (2001)
Walter, F.M., Lattimer, J.M.: Astrophys. J. **576**, L145 (2002)
Wang, J.C.L.: Astrophys. J. **486**, L119 (1997)
Wang, Z., Chakrabarty, D., Kaplan, D.L.: Nature **440**, 772 (2006)
Zane, S., Cropper, M., Turolla, R., et al.: Astrophys. J. **627**, 397 (2005)
Zane, S., Haberl, F., Cropper, M., et al.: Mon. Not. Roy. Astron. Soc. **334**, 345 (2002)
Zane, S., Turolla, R., Drake, J.J.: Adv. Space Res. **33**, 531 (2004)

Astrophys Space Sci (2007) 308: 203–210
DOI 10.1007/s10509-007-9340-z

ORIGINAL ARTICLE

Studies of neutron stars at optical/IR wavelengths

R.P. Mignani · S. Bagnulo · A. De Luca · G.L. Israel · G. Lo Curto · C. Motch · R. Perna · N. Rea · R. Turolla · S. Zane

Received: 30 June 2006 / Accepted: 23 August 2006 / Published online: 28 March 2007

Abstract In the last years, optical studies of Isolated Neutron Stars (INSs) have expanded from the more classical rotation-powered ones to other categories, like the Anomalous X-ray Pulsars (AXPs) and the Soft Gamma-ray Repeaters (SGRs), which make up the class of the *magnetars*, the radio-quiet INSs with X-ray thermal emission and, more recently, the enigmatic Compact Central Objects (CCOs) in supernova remnants. Apart from 10 rotation-powered pulsars, so far optical/IR counterparts have been found for 5 *magnetars* and for 4 INSs. In this work we present some of the latest observational results obtained from optical/IR observations of different types of INSs.

Keywords Neutron stars · Optical · Infrared

PACS 97.60.Gb · 97.60.Jd

Based on observations with the NASA/ESA *Hubble Space Telescope*, obtained at the Space Telescope Science Institute, which is operated by AURA, Inc. under contract No NAS 5-26555.
Based on observations collected at the European Southern Observatory, Paranal, Chile under programme ID 63.P-0002(A), 71.C-0189(A)72.C-0051(A), 074.C-0596(A), 074.D-0729(A), 075.D-0333(A), 076.D-0613(A).

R.P. Mignani (✉) · S. Zane
MSSL-UCL, Holmbury St. Mary, Dorking RH56NT, UK
e-mail: rm2@mssl.ucl.ac.uk

S. Bagnulo · G. Lo Curto
ESO, A. de Cordova 3107, Vitacura Santiago 19001, Chile

A. De Luca
INAF-IASF, Via Bassini 15, 20133 Milan, Italy

G.L. Israel
INAF-OAR, Via di Frascati 33, 00040 Monte Porzio, Italy

C. Motch
OAS, rue de l'Université 11, 67000 Strasbourg, France

R. Perna
University of Colorado, 440 UCB, Boulder, 80309, USA

N. Rea
SRON, Sorbonnelaan 2, 3584 CA Utrecht, Netherlands

R. Turolla
Universitá di Padova, via Marzolo 8, I-35131 Padova, Italy
e-mail: turolla@pd.infn.it

1 Introduction

Being the first discovered Isolated Neutron Stars (INSs), rotation-powered pulsars (RPPs) were also the first ones identified in the optical. Recent summaries of the RPPs optical observations can be found in Mignani et al. (2004) and Mignani (2006). After the spectacular results of the 1990s, which yielded to seven of the ten present RPP identifications thanks to the ESO *NTT* (Mignani et al. 2000a) and to the *HST* telescopes (Mignani et al. 2000b), only PSR J0437-4715 (Kargaltsev et al. 2004) has been added to the record, despite several attempts carried out after the advent of the ESO *VLT* (e.g. Mignani et al. 1999, 2003a, 2003b, 2005; Mignani and Becker 2004). The optical emission properties of RPPs depend on the age, with the young ones featuring purely magnetospheric spectra and the middle-aged ones featuring composite spectra with an additional thermal component arising from the cooling neutron star surface. For older objects the situation is less clear although there is evidence for a dominant magnetospheric emission (Mignani et al. 2002; Zharikov et al. 2004), while only the very old PSR J0437-4715 features a purely thermal emission (Kargaltsev et al. 2004). Multi-wavelength observations carried out in the last decades have unveiled the existence of other groups of INSs, most of them radio-quiet, which have been later studied in the optical/IR. *ROSAT* observations lead to the identification of seven nearby ($\lesssim$300 pc) INSs dubbed "The Magnificent Seven" (S. Popov) with purely thermal X-ray emission (Haberl, this conference). Being no unanimous consensus on the acronime to use (J. Trümper, this conference) from now on I will personally refer to these objects as X-ray Thermal INSs (XTINSs). Four XTINS have optical counterparts, with the identification of three of them secured via proper motion measures. Their inferred velocities have also allowed to rule out surface heating from ISM accretion as the source of the thermal X-ray emission in favour of heating from the cooling neutron star core. The XTINS optical emission is mostly thermal and exceeds the extrapolation of the soft X-ray spectrum by a factor $\sim$10, which suggests that it arises from a cooler and larger area on the neutron star surface with respect to the X-ray one (e.g. Mignani et al. 2004). Other peculiar INSs discovered through their X-ray/γ-ray emission are the AXPs and the SGRs which are believed to be *magnetars*, neutron stars with hyper-strong magnetic fields ($\sim 10^{14-15}$ G). Out of the twelve *magnetars* so far identified (Woods and Thompson 2006), only four have been observed in the optical/IR. Very little is known on the optical/IR spectra of the *magnetars*, apart from the fact that they flatten with respect to the extrapolation of the soft X-ray spectrum. This flattening can be taken as an indication of either a turnover in the magnetar spectrum or of the presence of an additional emitting source (e.g. an X-ray irradiated fallback disk). Other very enigmatic, supposedly isolated, neutron stars are the so-called CCOs in SNRs (Pavlov et al. 2004). Out of the seven CCOs known, only two have proposed optical/IR counterparts, classified as low-mass *K* or *M* stars (Sanwal et al. 2002; Pavlov et al. 2004). This would suggest that CCOs are indeed binary rather than isolated neutron stars. Last entry in the INS family are the newly discovered Rapid Radio Transients or RRATs (Gaensler et al., this conference) for which optical/IR follow ups have just started.

In the following, we present some of the most recent observational results obtained with the *VLT* and the *HST* in the optical/IR studies of some groups of INSs, i.e. RPPs, XTINSs and CCOs. For the *magnetars*, recent results are reported by Israel et al. (this conference).

2 Optical observations of rotation-powered pulsars

2.1 The PSR J0537-6910 optical counterpart

PSR J0537-6910 is an X-ray pulsar (16 ms) in the LMC supernova remnant N157B. The measured period derivative $\dot{P} \approx 5 \times 10^{-14}$ s s^{-1} (Marshall et al. 1998; Cusumano et al. 1998)) implies a spin down age of $\sim$5 000 yrs and a rotational energy loss of $\dot{E} \approx 4.8 \times 10^{38}$ ergs s^{-1}, the highest among RPPs. As common of young pulsars, PSR J0537-6910 exhibits large glitches (Middleditch et al. 2006) and features a compact pulsar-wind nebula (Townsley et al. 2006). Although its large energy output makes PSR J0537-6910 a natural target for multi-wavelength observations, it has not yet been detected outside the X-ray band. In radio it is undetected down to $F_{1.4\ \mathrm{GHz}} \sim 0.01$ mJy (Crawford et al. 2005), which implies that it is significantly fainter than both the Crab and PSR B0540-69. First exploratory optical observations (Mignani et al. 2000c; Gouiffes and Ögelman 2000; Butler et al. 2002) also failed to identify the pulsar counterpart, mainly owing to the crowdedness of the field. More recently, deeper high-resolution observations were performed by Mignani et al. (2005) with the *Advanced Camera for Surveys* (*ACS*) aboard *HST* and three most likely counterparts were selected within the revised *CXO* position on the base of their spectral flux distributions and colors.

Follow-up timing observations of the candidate counterparts have been performed with the *Space Telescope Imaging Spectrometer* (*STIS*) aboard the *HST* with the *NUV-MAMA* (Mignani et al. 2007a). The instrument was used in its spectroscopic configuration with the PRISM disperser (1460-3270 Å) and in TIME-TAG mode (125 μs time resolution). The target was observed for five Continuous Viewing Zone orbits, yielding a total integration time of 25 200 s. All objects of Mignani et al. (2005), with the exception of 1, 4 and 8, fall in the 52″ × 2″ slit—see Fig. 1.

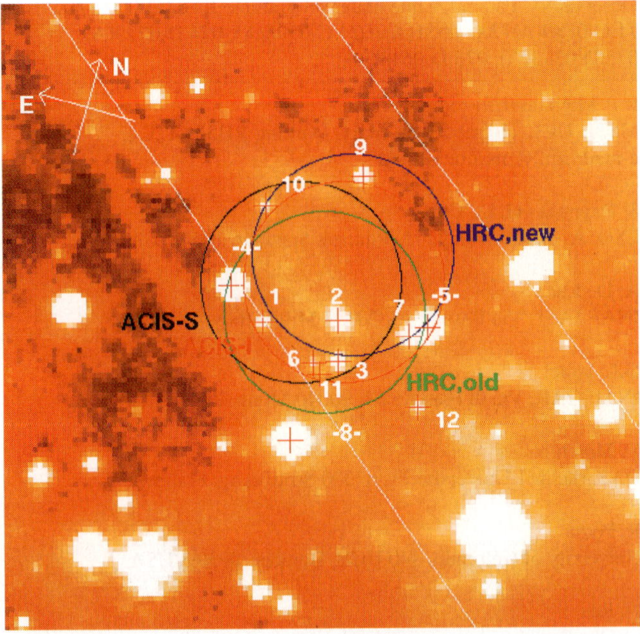

Fig. 1 *HST/ACS* image of the field of PSR J0537-6910 taken with the 814*W* filter with candidates labelled. The *circles* are the available *CXO* positions of the pulsar (Mignani et al. 2005). The projected orientation and width of the *STIS/NUV-MAMA* 52″ × 2″ slit are shown

Unfortunately, two objects only are detected in the two-dimensional spectrum, of which only one (object 5) is one of the candidates. The timing analysis does not reveil evidence for pulsations at the expected period, which definitely rules out object 5. At the same time, the extinction corrected [$E(B-V)=0.32$, Mignani et al. (2005)] near-UV flux upper limit (Log $F_\nu \sim -28.97$ ergs cm^{-2} s^{-1} Hz^{-1}) on the other candidates makes it unlikely that any of them be the pulsar counterpart, unless it has a very red spectrum. The optical counterpart of the more and more elusive PSR J0537-6910 is thus still unidentified. This implies that, as in the radio band, PSR J0537-6910 is intrinsically fainter than the Crab pulsar and PSR B0540-69. This suggests that the optical luminosity of RPPs decreases very fast, a scenario so far based only on the Vela pulsar case, which is about 10 times older than the Crab but four orders of magnitude fainter.

2.2 The optical polarization of the Vela pulsar

Besides the radio band, optical polarimetric observations of RPPs and of their synchrotron nebulae are uniquely able to provide deep insights into the highly magnetized relativistic environment of young rotating neutron stars. Being the first and the brightest ($V \sim 16.5$) RPP detected in the optical, polarization measures were first obtained for the Crab pulsar soon after its identification (Wampler et al. 1969). However, despite the substantial increase in the number of optically identified RPPs, the Crab is still the only one which has both precise and repeated polarization measures (e.g. Smith et al. 1988). Recently, Wagner and Seifert (2000) performed phase-averaged polarization observations of other three young pulsars with the *VLT*. For the Crab "twin" PSR B0540-69 ($V \sim 22.5$) they reported a polarization of ≈5% (with no quoted error bars), certainly contaminated by the contribution of the surrounding compact ($\sim 4''$ diameter) pulsar-wind nebula (Caraveo et al. 2001a). For PSR B1509-58 the value of the optical polarization is also very uncertain as the newly proposed counterpart is hidden in the PSF wings of the Caraveo et al. (1994) original candidate ($V=22$). Thus, both PSF subtraction problems and the object faintness ($R=26$), whose existence was never independently confirmed so far, make the reported polarization measure (≈10%, also quoted with no error bars) tentative. A polarization measurement of $8.5 \pm 0.8\%$ was finally reported for the Vela pulsar ($V \sim 23.6$).

In order to add information on the polarizaton properties of Vela, e.g. the angle of maximum polarization, we have undergone a careful reanalysis of the data set used by Wagner and Seifert. The details of the observations and data reduction are described elsewhere (Mignani et al. 2007b). We obtained a polarization of $9.4 \pm 4\%$, qualitatively in agreement with the one of Wagner and Seifert but with a much larger error. This is justified by the fact that, owing to the faintness of the target, the uncertainty on the background subtraction dominates the photometric errors on the polarized fluxes, and ultimately on the Stokes parameters. We are thus akin to conclude that this large difference is ascribed to an error underestimation on their side, likely to be related to the neglection of the background subtraction contribution. Thus, the value of the optical polarization of the Vela pulsar, contrary to previous claims, is still uncertain. We have also computed the angle on the plane of the sky corresponding to the direction of maximum polarization. Interestingly, we found that its value ($\theta = 145° \pm 14.7°$) is compatible, perspective wise, with the axis of the X-ray torus and jet observed by *CXO* (Pavlov et al. 2001a) and with the pulsar's proper motion vector (Caraveo et al. 2001b). Although we can not rule out a chance coincidence, it is tentalizing to speculate about the alignment between the polarization direction and the axis of symmetry of the X-ray structures as a tracer of the connection between the pulsar's magnetospheric activity and its interactions with the environment. More precise measures of the pulsar maximum polarization direction, possibly supported by still to come polarization measures in X-rays, will hopefully provide a more robust observational grounds for theoretical speculations.

3 Optical/IR observations of X-ray thermal isolated neutron stars

3.1 The proper motion of RXJ 1605.3+3249

An optical counterpart to RX J1605.3+3249 was identified by Kaplan et al. (2003) in an apparently blue object detected with *HST* at the *CXO* position. As in the case of other XTINS, e.g. RX J1856.5-3754 (van Kerkwijk and Kulkarni 2001) and RX J0720.4-3125 (Motch et al. 2003), the optical flux of the candidate counterpart was found in excess with respect to the Rayleigh-Jeans tail of the X-ray blackbody, adding weight to the proposed identification. This was confirmed by the measure of the object proper motion ($\mu = 144.5 \pm 13.2$ mas/year, position angle $\sim 350.14° \pm 5.65°$) by (Motch et al. 2005). We have obtained new observations of RX J1605.3+3249 with *ACS* to derive a more accurate proper motion measure (Zane et al. 2006). By comparing the target position measured in our 2005 *ACS* image with the 2001 *STIS* one of Kaplan et al. (2003) we measured a proper motion $\mu = 155.0 \pm 3.1$ mas/year with position angle $\sim 344° \pm 1°$ (see Fig. 2). This value confirms and updates the one of Motch et al. (2005) and settle the optical identification of RX J1605.3+3249. Furthermore, it strenghtens the identification of the neutron star birth place (for an age of $10^5 - 10^6$ years) with the Sco OB2 association, also suspected to be the birth place of other three XTINSs. The *ACS* photometry (filter 606*W*) has been compared with the one

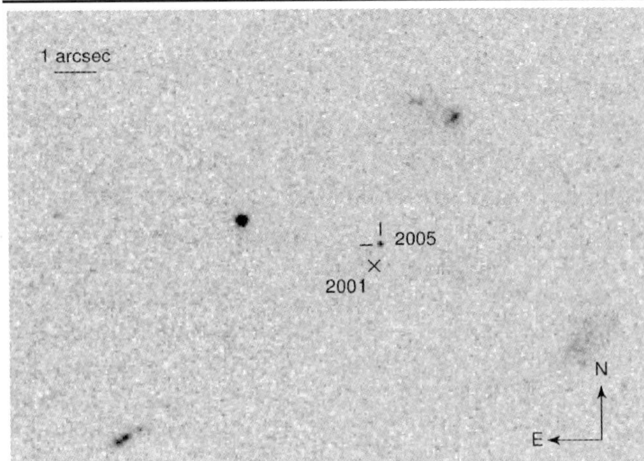

Fig. 2 *HST/ACS* image of RX J1605.3+3249 taken in January 2005 (Zane et al. 2006). The counterpart is marked by the two *ticks* while the cross indicates its position measured in the 2001 *HSTSTIS* image of Kaplan et al. (2003)

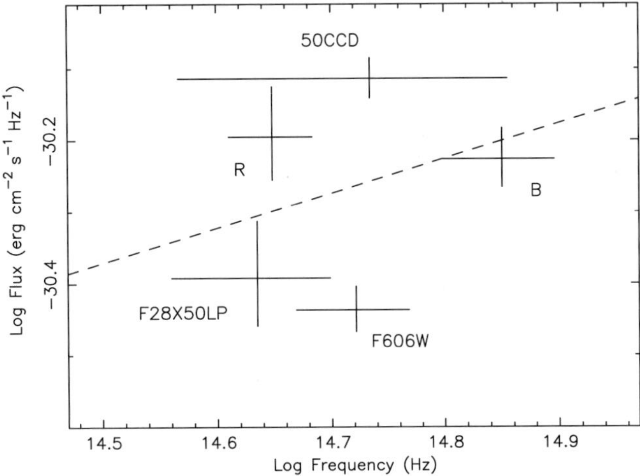

Fig. 3 Collection of the currently available (*HST* and *Subaru*) optical photometry of RX J1605.3+3249, corrected for a reddening $A_V = 0.11$ (Zane et al. 2006). The *dotted line* represents the best fit power-law ($\alpha = -0.5 \pm 0.5$)

of Motch et al. (2005) and Kaplan et al. (2003) to characterize the source optical spectrum. While Kaplan et al. (2003), on the base of two *HST* points, suggested a blackbody, Motch et al. (2005), on the base of the *Subaru B* and *R* points only, claimed a non-thermal spectrum ($\alpha \sim 1.5$). However, by using all available points we can not find any statistically acceptable fit (Fig. 3), not even by excluding the *ACS* point (Zane et al. 2006). Thus, the optical spectrum of RX J1605+3249 is virtually unconstrained. New observations taken with the same telescope and instrument set-up to provide a consistent photometry are required.

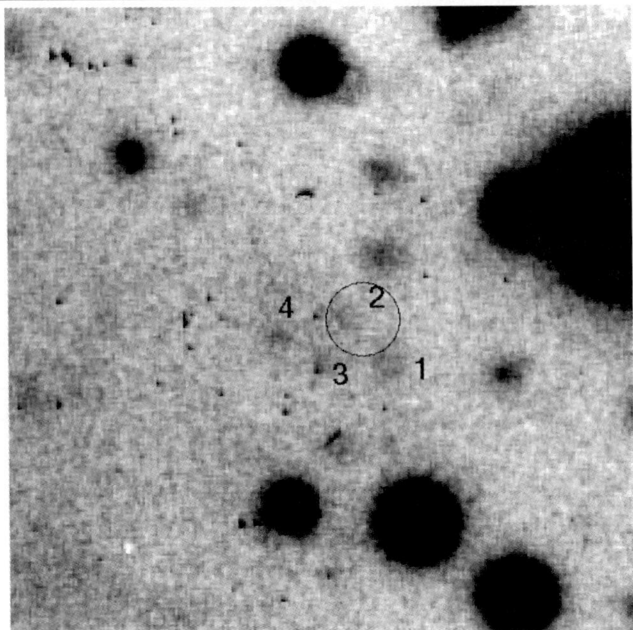

Fig. 4 *VLT/FORS1 V* band image of the RBS 1774 field (one hour integration time) after a Gaussian smoothing. The objects detected close to the *XMM* 3″ error circle are labelled

3.2 The search for the optical counterpart of RBS 1774

The X-ray source 1 RXS 214303.7+065419 (aka RBS 1774) is the last entry in the XTINS family (Zampieri et al. 2001) and one of the three which still wait for an optical identification. As a part of dedicated campaigns, we have carried out *VLT* observations of RBS 1774 with *FORS1*. Unfortunately, out of the 8 hours observing time (*B* and *V* bands) originally allocated in Service Mode, only one hour in *V* was actually executed. Besides, the quality of the observations was heavily affected by the very bad atmospheric conditions, with a seeing constantly above 1.5″. Figure 4 shows the *V* band image reduced through the *FORS1* pipeline, with the 3″ *XMM* error circle of RBS 1774 (Zampieri, private communication) overlaied. Although a few objects (1–4) are detected, none of them can be considered a realistic candidate counterpart to RBS 1774. First of all, they are at least a factor 10 brighter than the optically identified XTINS. Then, after comparing our photometry with the *B* band one of Komarava et al. (this conference) obtained with the *Subaru* they all turn out to be quite red, with a $B - V > 0.5$. This is confirmed by their detection in IR *VLT* images (see next section). No other object has been detected down to $V \sim 25.5$, which we set as the upper limit on the RBS 1774 flux.

3.3 Search for IR emission

It has been noted how the “Magnificent Seven” show intriguing similarities with the *magnetars*, which suggest a possible link between the two groups. Their spin periods are similar (3–12 s) and, through the observations of possible cyclotron X-ray absorption features, magnetic field of $B \sim 6-7 \times 10^{13}$ have been derived in three XTINSs. Although fainter than those of the *magnetars*, they are one order of magnitude stronger than those of the majority of radio pulsars. Also, for the two XTINSs with a measured period derivative the estimated X-ray luminosities turn out to be comparable or larger than the inferred spin-down energy. In *magnetars*, the X-ray luminosities indeed exceed their spin-down energy by at least 2 orders of magnitudes. Finding more similarities at other wavelengths is certainly of invaluable help to strengthen a possible link between the "Magnificent Seven" and the *magnetars*. Both SGRs and AXPs are known to have peculiar IR spectra, where the optical/IR spectrum flattens with respect to the extrapolation of the X-ray spectra (e.g. Israel et al. 2003, 2005). Whatever the origin of this spectral turnover, it is interesting to look for a similar behavior in the XTINSs. To this aim, the best targets are RXJ 0720.4-3125 and RXJ 1856-3754 since they are the only one with a rather accurate characterization of the optical spectrum (Motch et al. 2003; Kaplan et al. 2003). In both cases, a steep decline in the IR is expected, so that it would be easy to pinpoint a spectral flattening, if it is present.

IR observations of both RXJ 0720.4-3125 and RXJ 1856-3754, as well as for the three unidentified XTINSs RXJ 0420-5022, RXJ 0806-4122, RBS 1774 are available in the ESO archive. The observations were taken with the *VLT* between May 2004 and December 2005 using the *ISAAC* instrument with the H band filter. Integration times are varying between 4000 and 6000 s, split in shorter dithered exposures to enable for sky subtraction. The data were retrieved from the ESO archive together with the closest in time associated calibrations, and reduced. For RXJ 0720.4-3125 and RXJ 1856-3754 we have used the coordinates of their optical counterparts, while for RXJ 0420-5022, RXJ 0806-4122 and RBS 1774 we have used the available *CXO* and *XMM* coordinates. In all cases, no object was detected at the target position, with the only exception of RBS 1774 where we identified the objects already detected in our *FORS1 V* band image (see Sect. 3.2). We derived H band upper limits of 21.9 ± 0.15, 22.1 ± 0.1, 22.4 ± 0.1, 21.6 ± 0.2 and 21.7 ± 0.2 for RXJ 0420-5022, RXJ 0720-3125, RXJ 0806-4122, RXJ 1856-3754 and RBS 1774, respectively. Unfortunately, these upper limits are not very compelling. For RXJ 0720-3125 and RXJ 1856-3754, a spectral flattening redward of the R band would imply a H band magnitude of $\geq$24 and $\geq$23, respectively. Similar expectation values can be assumed also for RXJ 0420-5022, RXJ 0806-4122 and RBS 1774 assuming similar optical spectra and a factor $\sim$10 optical excess with respect to the extrapolation of their soft X-ray spectra. The derived constraints on a putative IR spectral flattening are not very compelling to constrain the presence of a fossil disk, either. By using the disk model of Perna et al. (2000) we were able only to exclude a disk extending at, or beyond, the light cylinder. Deeper IR observations to be performed with *NACO* at the *VLT* will allow us to improve these results.

4 Optical observations of compact central objects (CCOs)

4.1 1E 1207-5209 in G296.5+10.0

The CCO 1E 1207-5209 is one of the very few which pulsates in X-rays, with a period of 424 ms (Zavlin et al. 2000) and a period derivative $\dot{P} \sim 1.4 \times 10^{-14}$ (Pavlov et al. 2002; Mereghetti et al. 2002; De Luca et al. 2004). Strangely, the characteristic age ($\sim$470 000 years) is about two orders of magnitude higher that the age ($\sim$7000 years) of the associated SNR (Roger et al. 1988). In absolute, the most striking peculiarity of 1E 1207-5209 is the presence of three (possibly four) X-ray absorption features at regularly spaced energies. These features were interpreted in terms of electron or proton cyclotron absorption occurring in a magnetic field of $B \sim 8 \times 10^{10}$ G or $B \sim 1.6 \times 10^{14}$ G, respectively (Bignami et al. 2003; De Luca et al. 2004). These values are about two orders of magnitude lower/higher with respect to the value of the magnetic field inferred from the pulsar spin down ($B_d \sim 2 \times 10^{12}$ G). Different hypotheses to solve the age and B-field discrepancies have been proposed, including the possible influence of a fossil disk in the neutron star spin-down history. Recently, Zavlin et al. (2004) reported evidence for a non monotonous spin evolution which could imply that 1E 1207-5209 is either a strong glitcher or it is a binary. To understand the nature of the source, very deep optical observations have been performed both with the *VLT* and with the *HST*. De Luca et al. (2004) set upper limits of $R \sim 27.1$ and $V \sim 27.3$ on the optical brightness of the source. Soon after, optical *HST* and IR *VLT* observations unveiled a faint source ("star A", see Fig. 5), apparently compatible with the *CXO* position. The source colors, $m_{F555W} \sim 26.8$, $m_{F814W} \sim 23.4$, $J \sim 21.7$, $H \sim 21.2$ and $Ks \sim 20.7$, identify it as a late M star (Pavlov et al. 2004), thus implying that the 1E 1207-5209 is a binary. The fact that star A was not detected in the optical images of De Luca et al. (2004) would also imply that it is variable.

In order to investigate the proposed identification, we have carefully reassessed the *CXO* astrometry of 1E1207-5209 and we have compared its position with the one of star A. Our best *CXO* coordinates are $\alpha(J2000) = 12^{h}10^{m}0.826^{s}$, $\delta(J2000) = -52°26'28.43''$ with an associated uncertainty of $0.6''$. The *HST/ACS* images of the field have been retrieved from the ESO archive after on-the-fly reduction and recalibration. Single exposures have been combined using the *multidrizzle* task in IRAF, which also corrects for the CCD geometrical distortions. For each final

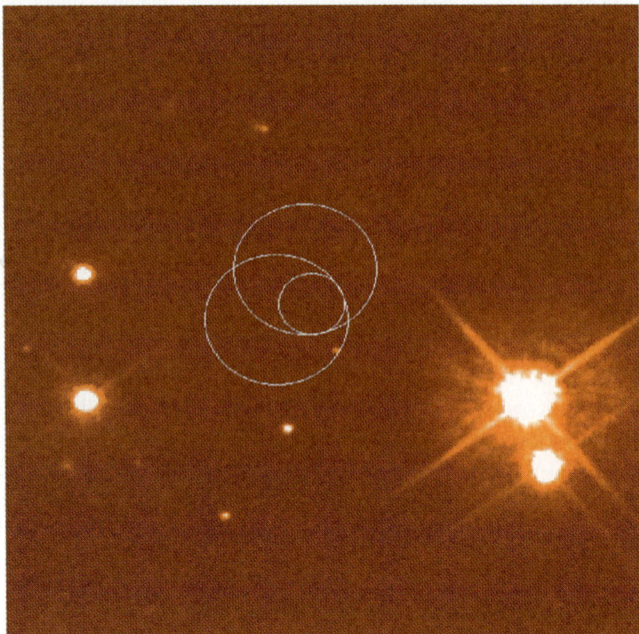

Fig. 5 *HST/ACS* 814*W* filter image of the 1E 1207-5209 field. The *upper and lower large circles* (1.5″ radius) are the *XMM* MOS1, MOS2 positions of the source, respectively (De Luca et al. 2004) while the *smaller* one (0.7″ radius) is the revised *CXO* position. Star A of Pavlov et al. (2004) is visible at the edge of the MOS2 error circle

image we have then recomputed the astrometry using as a reference a number of stars extracted from the GSC2 catalogue. The final error of the target position is 0.7″, inclusive of the accuracy of our astrometry (0.17″). The *CXO* position is shown in Fig. 5, overlaied on the *ACS* image taken with the 814*W* filter. The *CXO* error circle falls within the intersection of the MOS1 and MOS2 ones and it is significantly offset from the position of the candidate counterpart, which is right at the edge of the MOS1 error circle. We thus conclude that star A is unlikely to be the counterpart to 1E 1207-5209, although the ultimate piece of evidence should be obtained by its proper motion measurement, now in progress with the *HST*.

4.2 CXO J085201.4-461753 in Vela Jr

Vela Jr. (G266.1-1.2) is a very young (a few thousands years) and relatively nearby ($\leq$1 kpc) supernova remnant discovered in the ROSAT All Sky Survey (Aschenbach 1998). The CCO in Vela Jr was first studied with *ASCA* and *BeppoSax* (Mereghetti 2001) and later with *CXO* which also provided its sub-arcsec position (Pavlov et al. 2001b). The CXO J085201.4-461753 X-ray emission is characterized by a thermal-like spectrum, as in other CCOs, with no evidence of pulsations (Kargaltsev et al. 2002). First optical observations of CXO J085201.4-461753 were presented by Pellizzoni et al. (2002) using archived B and R observations taken with the ESO/MPG 2.2 m telescope. Although no counterpart was detected down to $B = 23$ and $R = 22$, the digitized H_α plates taken with the UK Schmidt telescope unveiled the presence of an extended emission blob (~6″ diameter) which was interpreted as a bow-shock nebula seen face-on. We have performed deeper observations of the Vela Jr. CCO with *FORS1* at the *VLT*. To minimize the light pollution from an object ("star Z" of Pavlov et al. 2001b) located ~1.5″ away from our target, we split the integration time in 20 exposures of 260 s each. In order to achieve the best possible spatial resolution, *FORS1* was used in its High Resolution mode with a corresponding pixel size of 0.1″. A very bright star located 40″ away from the target was masked using the *FORS1* occulting bars. Observations were collected with good seeing (~0.9″) and airmass (~1.3) conditions. A 17″ × 17″ zoom of the *FORS1* R band image of the field is shown in Fig. 6, after pipeline reduction and average combination of the single exposures. While no point-like source appears at the *CXO* position down to $R \sim 26$, a compact optical nebula is detected. We exclude that this nebula is an artifact due to a PSF anomaly in star Z, to a defect in the image flat fielding or to any instrumental effect. Both its position and extent though are consistent with the one of the putative H_α nebula seen by Pellizzoni et al. (2002), which clearly indicates that they are the same object. Unfortunately, the available B band upper limit is to shallow to constraint the nebula spectrum. It is thus unclear weather it is indeed a bow-shock or it is some kind structure similar to the pulsar-wind nebulae seen around RPPs. Follow-up *VLT* observations, carried out at the time of writing, will hopefully help to unveil both the nature of this nebula and of the CCO.

5 Infrared observations of high magnetic field pulsars

5.1 PSR J1119-6127

About 40 radio pulsars have been detected by the Parkes Pulsar Survey with magnetic fields larger than 10^{13} G (Camilo et al. 2000). In particular, five of them have magnetar-like magnetic fields larger than the quantum critical field $B_c = 4.33 \times 10^{13}$ G above which radio emission is expected to be suppressed, meaning that they are not expected to be radio pulsars at all. Despite having such high magnetic fields, these high-magnetic field radio pulsars (HBRPs) do not behave as *magnetars*. First of all, they are radio pulsars, while pulsed radio emission has been discovered so far only in the transient AXP XTE J1810-197 (Camilo et al. 2006). Second, only two HBRPs, PSR J1119-6127 (Gonzales and Safi-Harb 2003) and PSR J1718-3718 (Kaspi and McLaughlin 2005) have been detected in X-rays, with luminosities $L_X \sim 10^{32-33}$ ergs s^{-1} lower than those of the

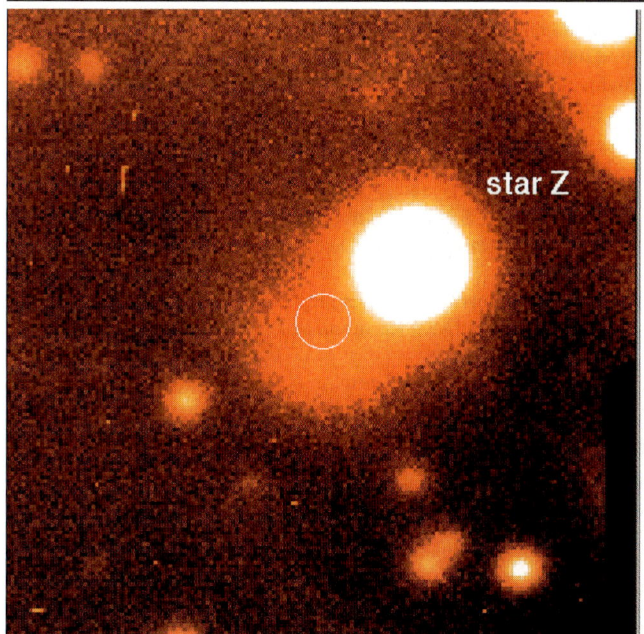

Fig. 6 $17'' \times 17''$ section of the *VLT/FORS1 R* band image of the Vela Jr. CCO. The *circle* ($0.6''$ radius) corresponds to the *CXO* position uncertainty. Star Z of Pavlov et al. (2001b) is labelled

magnetars. Finally, they do not show bursting emission, either in X-rays or in γ-rays, as AXPs and SGRs instead do. These differences might be explained assuming, e.g., that HBPSs are dormant transients, that their lower X-ray luminosities are a consequence of their lower magnetic fields, or simply assuming that different evolutionary paths or stages account for the different phenomenologies. Of course, one possibility is that these HBRPs are not genuine *magnetars* because the spin-derived magnetic field values are polluted by the torques produced by a fossil disk. The possible existence of fossil disks around INSs has been demonstrated by the recent *Spitzer* discovery of a disk around the AXP 4U 0142+61 (Wang et al. 2006). Thus, if HBRPs do have fossil disks, they should be detectable through deep, high-resolution, IR observations. To this aim, we have started a program of IR observations of HBRPs with the *VLT*. Since the IR luminosity of a disk scales with the X-ray one (Perna et al. 2000), our primary candidates are those HBRPs detected in X-rays.

The field of PSR J1119-6127 was observed in Service Mode between January and February 2006 with NAos-COnica (*NACO*), an adaptive optics image and spectrometer at the *VLT*. In order to provide the best combination between angular resolution and sensitivity, *NACO* was operated in its S27 mode with a corresponding field of view of $28'' \times 28''$ and a pixel scale of $0.027''$. Observations were performed in the J, H and K_s bands for a total integration time of 2 hours each, dithered and split in short exposures of 55 s for sky subtraction requirements. The seeing conditions ($\sim 0.6''$) allowed for an optimal use of the adaptive optics. The data were reduced independently using the ESO *NACO* pipeline and procedures run under the *eclipse* package. The pulsar is undetected in any of the three observing passbands down to limiting magnitudes of ~ 24, ~ 23 and ~ 22 in the J, H and K_s passbands, respectively. We have then compared these upper limits with the disk models of Perna et al. (2000). With the due caution that the data analysis is still in progress, the results of our simulations do not presently allow to rule out the presence of a disk extending down to the magnetospheric radius. Further theoretical and simulation work will allows to better constrain weather, and how, the putative disk interact with the neutron star.

Acknowledgements Roberto Mignani warmly thanks the ESO/Chile Scientific Visitors Programme for supporting his visit at the ESO Santiago Offices (Vitacura), where part of this work was finalised. A special thank goes to A. Micol (ST-ECF) for the support in the reduction of the *HST/ACS* data.

References

Aschenbach, B.: Nature **396**, 141 (1998)

Bignami, G.F., Caraveo, P.A., De Luca, A., et al.: Nature **423**, 725 (2003)

Butler, R.F., Golden, A., Shearer, A., et al.: In: Becker W., et al. (eds.) Proceedings WE-Heraeus Seminar on Neutron Stars, Pulsars, and Supernova Remnants, Bonn, Germany, January 21–25, 54. MPE Report 278, Garching (2002)

Camilo, F., Kaspi, V.M., Lyne, A.G., et al.: Astrophys. J. **541**, 367 (2000)

Camilo, F., Ransom, S.M., Halpern, J.P., Reynolds, J., Helfand, D.J., et al.: Transient pulsed radio emission from a magnetar. Nature **442**, 892–895 (2006)

Caraveo, P.A., Mereghetti, S., Bignami, G.F.: Astrophys. J. **423**, L125 (1994)

Caraveo, P.A., Mignani, R.P., De Luca, A., et al.: In: Livio, M., et al. (eds.) Proceedings A decade of HST Science, Baltimore, USA, 11–14 April 2000, p. 9. Space Telescope Science Institute, Baltimore (2001a)

Caraveo, P.A., De Luca, A., Mignani, R.P., et al.: Astrophys. J. **561**, 930 (2001b)

Crawford, F., McLaughlin, M., Johnston, S., et al.: Adv. Space Res. **35**, 1181 (2005)

Cusumano, G., Maccarone, M.C., Mineo, T., et al.: Astron. Astrophys. **333**, L55–L58 (1998)

De Luca, A., Mereghetti, S., Caraveo, P.A., et al.: Astron. Astrophys. **418**, 625 (2004)

Gonzales, M., Safi-Harb, S.: Astrophys. J. **591**, L143 (2003)

Gouiffes, K., Ögelman, H.: In: Kramer, M., et al. (eds.) Proceedings Pulsar Astronomy—2000 and Beyond, Bonn, Germany, 30 August–3 September 1999. ASP Conference Series, p. 301. San Francisco (2000)

Israel, G.L., Covino, S., Perna, R., et al.: Astrophys. J. **589**, L93–L96 (2003)

Israel, G.L., Covino, S., Mignani, R., et al.: Astron. Astrophys. **438**, L1 (2005)

Kaplan, D.L., Kulkarni, S.R., van Kerkwijk, M.H.: Astrophys. J. **588**, L33 (2003)

Kargaltsev, O., Pavlov, G.G., Sanwal, D., et al.: Astrophys. J. **580**, 1060 (2002)

Kargaltsev, O., Pavlov, G.G., Romani, R.W.: Astrophys. J. **602**, 327 (2004)

Kaspi, V., McLaughlin, M.: Astrophys. J., **618**, L41 (2005)
Marshall, F.E., Gotthelf, E.V., Zhang, W., et al.: Astrophys. J. **499**, L179 (1998)
Mereghetti, S.: Astrophys. J. **548**, 213 (2001)
Mereghetti, S., De Luca, A., Caraveo, P.A., et al.: Astrophys. J. **581**, 1280 (2002)
Middleditch, J., Marshall, F.E., Wang, Q.D., Gotthelf, E.V., Zhang, W.: Predicting the starquakes in PSR J0537-6910. Astrophys. J. **652**, 1531–1546 (2006)
Mignani, R.P., Becker, W.: Adv. Space Res. **33**, 616 (2004)
Mignani R., Caraveo, P.A., Bignami, G.F.: Astron. Astrophys. **343**, L5 (1999)
Mignani, R.P., Caraveo, P.A., Bignami, G.F.: Messenger **99**, 22 (2000a)
Mignani, R.P., Caraveo, P.A., Bignami, G.F.: STScI Newsl. **17**, 1 (2000b)
Mignani, R., Pulone, L., Marconi, G., et al.: Astron. Astrophys. **355**, 603 (2000c)
Mignani, R.P., De Luca, A., Caraveo, P.A., et al.: Astrophys. J. **580**, L147 (2002)
Mignani, R.P., Manchester, R., Pavlov, G.G.: Astrophys. J. **582**, 978 (2003a)
Mignani, R.P., De Luca, A., Kargaltzev, O., et al.: Astrophys. J. **594**, 419 (2003b)
Mignani, R.P., De Luca, A., Caraveo, P.A.: In: Camilo, F., Gaensler, B. (eds.) Proceedings Young Neutron Stars and Their Environments, Sydney, Australia, 14–17 July 2003, p. 391. Astronomical Society of the Pacific, San Francisco (2004)
Mignani, R.P., Pulone, L., Iannicola, G., et al.: Astron. Astrophys. **431**, 659 (2005)
Mignani, R.P.: In: Baykal, A., et al. (eds.) Proceedings The Electromagnetic Spectrum of Neutron Stars, Marmaris, Turkey, 7–18 June 2004, p. 133. Springer, Berlin (2006)
Mignani, R.P., Kargaltzev, O.Y., Pavlov, G.G.: The more and more elusive optical counterpart of PSR J0537-6910. Astron. Astrophys. (2007a, in preparation)
Mignani, R.P., Bagnulo, S., Dyks, Y., Lo Curto, G., Slowikowska, A.: The optical polarization of the Vela pulsar revisited. Astron. Astrophys. (2007b, in press)
Motch, C., Zavlin, V.E., Haberl, F.: Astron. Astrophys. **408**, 323 (2003)
Motch, C., Sekiguchi, K., Haberl, F., et al.: Astron. Astrophys. **429**, 257 (2005)
Pavlov, G.G., Kargaltsev, O.Y., Sanwal, D., et al.: Astrophys. J. **554**, L189 (2001a)
Pavlov, G.G., Sanwal, D., Kiziltan, B., et al.: Astrophys. J. **559**, L131 (2001b)
Pavlov, G.G., Zavlin, V.E., Sanwal, D., et al.: Astrophys. J. **569**, L95 (2002)
Pavlov, G.G., Sanwal, D., Teter, M.A.: In: Camilo, F., Gaensler, B. (eds.) Proceedings Young Neutron Stars and Their Environments, Sydney, Australia, 14–17 July 2003, p. 239. Astronomical Society of the Pacific, San Francisco (2004)
Pellizzoni, A., Mereghetti, S., De Luca, A.: Astron. Astrophys. **393**, L65 (2002)
Perna, R., Hernquist, L., Narayan, R.: Astrophys. J. **541**, 344 (2000)
Roger, R.S., Milne, D.K., Kesteven, M.J., et al.: Astrophys. J. **332**, 940 (1988)
Sanwal, D., Garmire, G.P., Garmire, A., et al.: Bull. Am. Astron. Soc. **34**, 74 (2002)
Smith, F.G., Jones, D.H.P., Dick, J.S.B., et al.: Mon. Not. Roy. Astron. Soc. **233**, 305 (1988)
Townsley, L.K., Broos, P.S., Feigelson, E.D., et al.: Astron. J. **131**, 2140 (2006)
van Kerkwijk, M.H., Kulkarni, S.R.: Astron. Astrophys. **378**, 986 (2001)
Wagner, S.J., Seifert, W.: In: Kramer, M., et al. (eds.) Proceedings Pulsar Astronomy—2000 and Beyond, Bonn, Germany, 30 August–3 September 1999. ASP Conference Series, p. 315. San Francisco (2000)
Wampler, E.J., Scargle, J.D. and Miller, J.S.: Astrophys. J. **157**, L1 (1969)
Wang, Z., Chakrabarty, D., Kaplan, D.L.: A debris disk around an isolated young neutron star. Nature **440**, 772–775 (2006)
Woods, P.M., Thompson, C.: Soft gamma repeaters and anomalous X-ray pulsars: magnetar candidates. In: Lewin, W.H.G., van der Klis, M. (eds.) Compact Stellar X-ray Sources. Cambridge Astrophysics Series, vol. 39, pp. 547–586. Cambridge University Press, Cambridge (2006)
Zane, S., De Luca, A., Mignani, R.P., et al.: Astron. Astrophys. **457**, 619 (2006)
Zampieri L., Campana, S., Turolla, R., et al.: Astron. Astrophys. **378**, L5–L9 (2001)
Zharikov, S.V., Shibanov, Yu.A., Mennickent, R.E., et al.: Astron. Astrophys. **417**, 1017 (2004)
Zavlin, V.E., Pavlov, G.G., Sanwal, D., et al.: Astrophys. J. **540**, L25 (2000)
Zavlin, V.E., Pavlov, G.G., Sanwal, D.: Astrophys. J. **606**, 444 (2004)

Astrophys Space Sci (2007) 308: 211–216
DOI 10.1007/s10509-007-9341-y

ORIGINAL ARTICLE

Radio emission from AXP and XDINS

V.M. Malofeev · O.I. Malov · D.A. Teplykh

Received: 30 June 2006 / Accepted: 20 September 2006 / Published online: 22 March 2007

Abstract The result of the search for, and the observations of radio emission from two groups of isolated neutron stars: AXP 1E 2259+586 and XDINS 1RXS J1308.6+212708 and 1RXS J214303.7+065419 are reported. The observations were carried out on two sensitive transit radio telescopes at a few frequencies in the range 42–112 MHz. The flux densities, mean pulse profiles, as well as, the estimation of the dispersion measures, distances and integrated radio luminosities of all objects are presented. Comparison with X-ray data shows large differences in the mean pulse widths and luminosities.

Keywords Neutron stars—radio emission

PACS 97.60.Gb · 97.60.Jd

1 Introduction

Recent gamma-ray and X-ray observations have led to the discovery of a few interesting groups of pulsars: anomalous X-ray pulsars (AXPs) (for example, Mereghetti 2001), soft gamma-ray repeaters (SGRs) (Gaensler et al. 2001) and nearby dim isolated neutron stars with strong magnetic fields (XDINS) or the magnificent seven (Treves et al. 2000). These isolated neutron stars have different parameters than the larger group of "normal" radio pulsars and ordinary X-ray pulsars do. These objects have long periods (5–12 s) and large period derivatives, 10^{-11}–10^{-13} s/s. They are young objects with characteristic ages of up to several million years. Most of AXPs and SGRs objects are located close to the plane of the Galaxy, and nearly half of them are inside supernova remnants. The main problem connected with these objects is the source of their energies, which sometimes imply luminosities that are two to three orders of magnitude higher than can be provided by the rotational kinetic-energy losses. All the models proposed until recently, including the most popular magnetar model of Duncan and Thompson (1992), which proposes the superstrong magnetic fields 10^{14}–10^{15} G to explain this energy source encounters serious difficulties (Malov et al. 2003). XDINSs are not associated with supernova remnants and maybe in these compact stars we can have a clean view of the star surface in X-ray, without contamination from magnetospheric emission (Zane et al. 2005). The presence of the absorption feature in the X-ray spectra of a few XDINSs (Haberl et al. 2003) give the possibility to measure the magnetic field strength independently from spin-down measurements.

One of the key arguments in favor of the magnetar model was the absence of radio emission from AXPs, SGRs and XDINSs: this could be naturally explained as a consequence of the absence the electron-positron cascades that are responsible for radio emission in some models in the presence of such strong fields (Baring and Harding 1998). The situation was changed when pulsed radio emission was detected from the SGR 1900+14 (Shitov et al. 2000) and the AXP 1E2259+586 (Malofeev et al. 2001, 2004, 2005). This led to searches for mechanism of pair production, such as two-photon process for the generation of radio emission in the

V.M. Malofeev (✉) · O.I. Malov · D.A. Teplykh
P.N.Lebedev Physical Institute, Russian Academy of Sciences, Pushchino, Moscow region, PRAO FIAN, 142290, Russia
e-mail: malofeev@prao.ru

O.I. Malov
e-mail: malov_jr@prao.ru

D.A. Teplykh
e-mail: teplykh@prao.ru

framework of the magnetar model by Zhang (2001, 2003), as well as completely new models, such as the action of drift waves at the periphery of the magnetosphere (Malov et al. 2003). The importance of searches for the radio emission of AXPs, SGRs and XDINs is obvious and recently the radio emission from two XDINSs: 1RXS J1308.6+212708 (Malofeev et al. 2004, 2005) and 1RXS J2143.7+065419 (Malofeev et al. 2006) and the transient AXP XTE J1810-197 (Camilo et al. 2006) has been detected. In this report we present briefly data on the radio emission of the AXP 1E 2259+586 and the XDINS J1308.6+212708, and more detail we shall report about the radio emission of XDINS 1RXS J2143.7+065419.

The X-ray object 1E 2259+586 (G109.1.1.0) was detected in the direction toward the supernova remnant CTB 109 in 1980 by Gregory and Fahlman, who also detected pulsed emission (1E 2259+586). The second object, 1RXS J1308+21, was discovered in 2001 by Hambaryan et al. (2002) and the rotational period for this pulsar was recently redetermined $P = 10.32$ s (Haberl et al. 2003). The detection of pulsed X-ray emission of XDINS 1RXS J2143.7+065419 was made by Zane et al. in 2005.

2 Observations

Observations of the AXP 1E 2259+586 began on March 1999. Regular observations every one-two months began in February 2001. The second object, 1RXS J1308+21, has been observed in the same mode since December 2001. We consider here data obtained for both pulsars until April 2005. The observations of third pulsar 1RXS J2143+06 was started in October 2005 and we present data obtained until April 2006. Most of the observations were carried out on the high-sensitivity Large Phased Array (LPA) of the Lebedev Physical Institute (Pushchino) at a frequency of 111 MHz. A few simultaneous observations have been made at 87, 61 or 42 MHz on the East–West arm of the DKR-1000. This antenna operates at 30–110 MHz. The effective area is $\sim$20 000 m^2 and $\sim$7000 m^2 for LPA and the East–West arm of the DKR-1000, respectively. Both radio telescopes are transit instruments; the durations of each observational session on the LPA were 6.2 (1E 2259+586) and 3.3 min for 1RXS J1308+21 and 1RXS J2143+06, respectively, and those on the DKR-1000 were longer by factor of three for each pulsar. A filter-bank receiver with a bandwidth of 20 kHz and 64 channels at 111 MHz and 32 or 64 channels at 87, 61 and 42 MHz was used. As a rule, the data-sampling interval was 25.6 or 51.0976 ms and the receiver time constant was 30 or 100 ms. These parameters were used for integrating a signal with a known period. In addition, we used a method to search for pulsed radio emission with unknown period (Malofeev et al. 2005). To increase reliability we carried out numerous observations, using double the period. We have used a technique that has been tested with observations of faint pulsars, as well as the Geminga pulsar (Malofeev and Malov 1997). Some observations were calibrated using reference radio sources with known flux densities.

3 Mean profiles

We detected weak periodic pulsed radio emission from the AXP 1E 2259+586 at the 111 MHz and 87 MHz. This result have been reported (Malofeev and Malov 2001; Malofeev et al. 2004, 2005). The observations of 1RXS J1308+21 began in 2001 and after a few month of observations, we detected periodic pulsed radio emission at 111, 87 and 61 MHz (Malofeev et al. 2004, 2005). In this paper we present the radio emission at 42 MHz for this pulsar. In total, we had more than 400 days of observations at 111 MHz, and a few dozen days of them at 87 and 61 MHz for the AXP 1E 2259+586 over the entire observational interval. About one-third of the observations were corrupted by strong interference, and the pulsar signal did not exceed 4σ on more than one-third of the days. We obtained more than 80 and more than 10 records useful for analysis for 1RXS J1308+21 at 111 MHz, and at every lower frequency (87, 61 and 42 MHz), respectively. About 30 good records of 1RXS J2143+06 have been obtained during Oct. 2005–March 2006 at the frequency 111 MHz.

A search programme on the base of Fourier-amplitude spectra has also been pursued. Tests of this method using observations of known pulsars demonstrate that pulsars with fluxes $>$70 mJy can be confidently detected. We were able to obtain good spectra for AXP 1E 2259+586 and XDINS 1RXS J1308+21, some examples of which were presented by Malofeev et al. (2004, 2005).

3.1 AXP 1E 2259+586

Since the integrations for individual days were carried out over only 53 pulses, the signal-to-noise ratio of the mean pulse rarely reached five. Accordingly, we improved the signal-to-noise ratio by summing the data for a number of different days. Since we did not have precise timing for this pulsar, we summed days when the observations were carried out using the double period, and the pulses were aligned using visible pulses. In this case, we should observe two pulses separated by precisely one pulsar period in the summed profile, as is shown in Fig. 1a. Folding these data using the pulsar period (Fig. 1b) yielded a very narrow profile with a mean duration of 120 ± 20 ms, or 1.7% of the period. This is one of the narrowest pulses observed for radio pulsars. Furthermore, in contrast to the X-ray profile, we do not observe

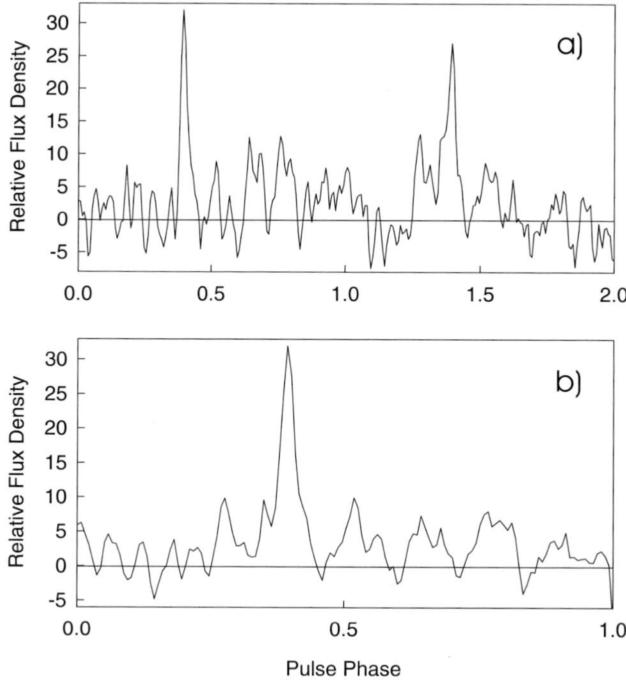

Fig. 1 **a** Integrated profile of the AXP 1E 2259+586 at 111.2 MHz (in arbitrary units) obtained by summing 16 days, when the observing window equals twice the apparent period (416 doubled periods); **b** Integrating profile obtained by the folding of data with one period, i.e., the sum of 832 periods

the interpulse with an amplitude 20% of the main pulse. We were also able to detect pulsed periodic emission from this pulsar at 87.5 MHz (Malofeev et al. 2005).

3.2 XDINS 1RXS J1308+21

We observe a narrow pulse at 111 MHz, which has a duration of 140 ± 20 ms, or 1.35% of the period ($P = 10.32$ s). Figure 2 shows the mean profiles at 111, 87, 61 and 42 MHz. In addition to a narrow pulse, this pulsar displays an interpulse at a phase of ~0.5 of period, which is clearly visible in Fig. 2 at all frequencies.

3.3 XDINS 1RXS J2143+06

The mean profiles are shown of Figs. 3, 4. This pulsar demonstrates a wider and more complex profile, than two other objects. Very probable the profile has three components separated on 400 ms and total width of the integration profile at 50 percent of the intensity maximum is about 1000 ms. Some days we observed the interpulse at the phase 0.5 of period (Fig. 4). To confirm the presence of periodic pulsed radio emission we used the search programme. The sum of 11 amplitude spectra is shown on Fig. 5. There is few first harmonics at the frequencies suitable to the pulsar period. A few features are shown by the arrows.

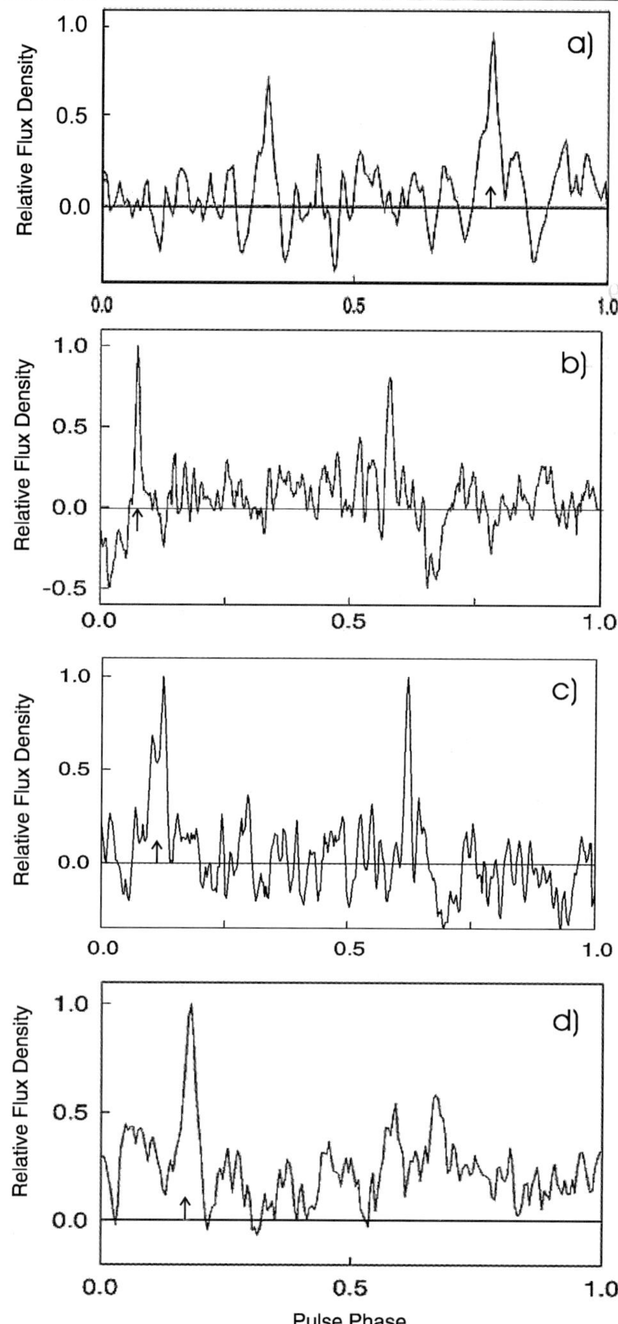

Fig. 2 Mean profiles of 1RXS J1308+21 **a** 111.2 MHz (in arbitrary units), the sum of 50 periods; **b** 87 MHz, sum of 14 periods; **c** 62 MHz, sum of 100 periods, and **d** sum of 206 periods. The phases of main pulses are shown by the arrows

4 Dispersion measure, flux density and period

Determination of the dispersion measures is extremely important, since it makes it possible to obtain independent estimates of the distances to the pulsars. We estimated the dispersion measure using the best data in the 111.22–110.58 MHz frequency band, which was covered by 64 receiver channels. The signal-to-noise ratio and pulse du-

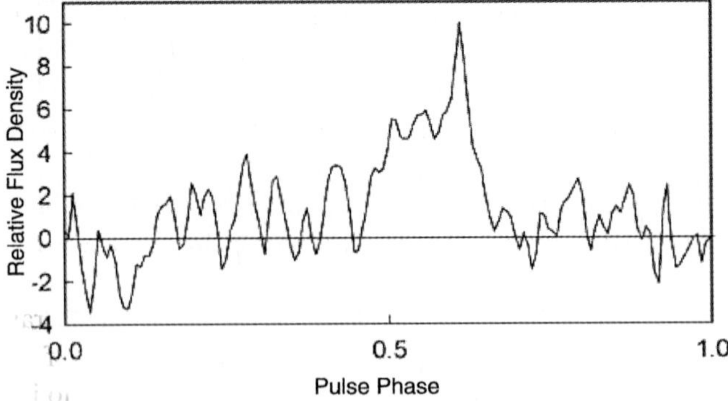

Fig. 3 Integrated profile of the 1RXS J2143+06 at 111.2 MHz (in arbitrary units) obtained by integrating of 20 periods on 16 Feb. 2006

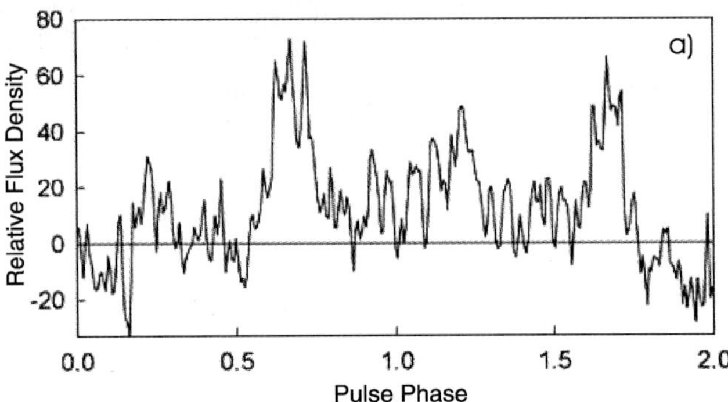

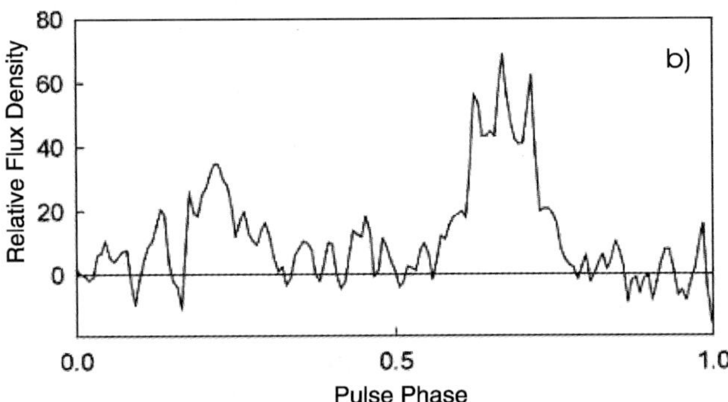

Fig. 4 **a** Integrated profile of the 1RXS J2143+06 at 111.2 MHz (in arbitrary units) obtained by summing 23 days during four month of 2004, when the observing window equals twice the apparent period (225 doubled periods); **b** Integrating profile obtained by the folding of data with one period, i.e., the sum of 450 periods

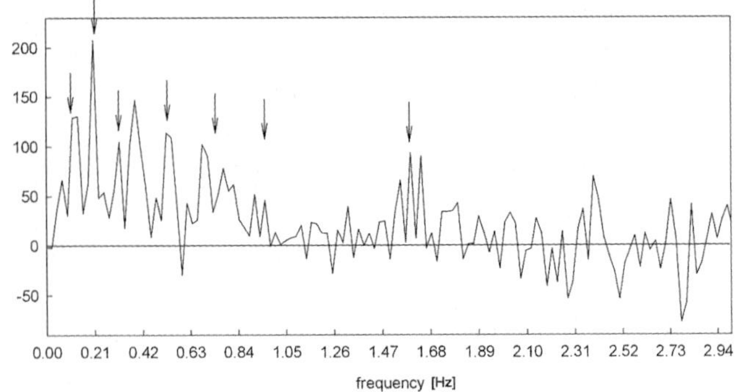

Fig. 5 Amplitude spectrum the 1RXS J2143+06. A few first harmonics are shown by the *arrows*

ration as a function of the dispersion measure for pulsars 1E 2259+586 and J1308+21 was shown by Malofeev et al. (2004, 2005). We obtained that the mean profile had the highest signal-to-noise ratio for $DM = 8 \pm 5$ pc/cm^3 in the case of XDINS J2143+06. All values of the dispersion measures are presented in Table 1.

We measured the fluxes at 111 MHz via calibration using the radio sources with known fluxes. The flux of 1E 2259+586 was measured by averaging of 30 days of observations over 3.5 years, and that of 1RXS J1308+21 on 10 days of observations over 1.5 years. For the J2143+06 the flux density was obtained as mean value during 6 days of observations (Table 1).

Table 1 Measured and calculated parameters of the three pulsars

Parameter	1E 2259+586	1RXS J1308+21	1RXS J2143+06
DM, pc/cm^3	79 ± 4	5.7 ± 0.5	8 ± 5
S_{111}, mJy	35 ± 25	50 ± 20	60 ± 25
w_{50} (111 MHz), ms	120 ± 20	140 ± 20	990 ± 60
D, kpc	3.6 ± 0.2	0.25 ± 0.02	0.4 ± 0.2
L_R, erg/s	3×10^{28}	3×10^{26}	9×10^{26}
L_x, erg/s	7.9×10^{34}	3×10^{31}	
E, erg/s	5.6×10^{31}	0.4×10^{31}	

The period and period derivative for pulsars 1E 2259+586 and J1308+21 (Malofeev et al. 2004) is in agreement with X-ray measurements by Gavriil and Kaspi (2002) and Hambaryan et al. (2002). New data presented by Kaplan and van Kerkwijk (2005) for J1308+21 gives a much smaller value of the period derivative, that requires us to redetermine our value using all data obtained during 2001–2005. We measured the period and its derivative ($P = 9.43707(10)$ s and negative $\dot{P} = -15(22) \times 10^{-13}$ s/s) during the MJD range 53657.7-53799.3 for J2143+06.

5 Discussion and summary

Table 1 lists the main parameters of the radio emission from the three pulsars. Comparison with the X-ray data shows that the radio measurements both extend our knowledge about these objects and carry fundamentally new information. We have detected strongly differing durations of the mean pulses, and derived independent estimates of the distances to sources based on their dispersion measures. The existence of the radio emission itself represents a fundamentally new fact, which raises doubts about either the magnetar model or our understanding of radio emission in superstrong magnetic fields.

Distance estimates for the pulsar 1E 2259+586 (or rather for the supernova remnant CTB 109; the pulsar is almost at its center) in the literature are between 3.5 and 4.5 kpc (see, e.g., Gregory and Fahlman 1980). Our measurements of the dispersion measure, 79 ± 4 pc/cm^3, yield a distance of 3.6 ± 0.7 kpc for the model electron density distribution in the Galaxy of Taylor and Cordes (1993). Estimates of the distance to 1RXS J1308+21, or, more precisely, the star RBS 1223, obtained using several methods lie in the broader interval from 0.1 to 1.5 kpc (Hambaryan et al. 2002). Our dispersion measure (Table 1) and the model for the Galaxy yield an estimated distance to the pulsar of $0.25^{+0.2}_{-0.1}$ kpc. This suggests that the pulsar is close to us. In the case of J2143+06 we can estimate the distance as 0.4 kpc, in agreement with the estimate 0.28 kpc given by Zane et al. (2005).

The X-ray luminosity of 1E 2259+586 for a distance of 3.6 kpc $\log L_x(\mathrm{erg/s}) = 34.9$ (Malov et al. 2003) remains three orders of magnitude higher than the rate at which this star is losing rotational kinetic energy. The X-ray luminosity of 1RXS J1308+21 for a distance of 0.25 kpc is $L_x = 2.6 \times 10^{31}$ erg/s. If $\dot{P} = 1.110^{-13}$ s/s (Kaplan and van Kerkwijk 2005) we obtain $E = 0.4 \times 10^{31}$ erg/s, that is less than the X-ray luminosity. To estimate the total radio luminosity, we must know the spectrum of the pulsar, or at least the spectral index. Using our flux density measurement at 111 MHz and the upper limits at 87 MHz and 600 MHz and 1500 MHz: $S_{600} < 2.3$ mJy (Lorimer et al. 1998) and $S_{1500} < 0.05$ mJy (Coe et al. 1994) for 1E2259+586 and $S_{1400} < 0.94$ mJy (White et al. 1997) for J1308+21, we estimate the spectral index of 1E2259+586 to be $\alpha > 2.5$ and for J1308+21 $\alpha > 1.7$.

Given that the spectra of all three pulsars are probably steep, we estimated the integral radio luminosities of the pulsars by adopting the value $\alpha = 2.5$; this yielded $L_R = 3 \times 10^{28}$ erg/s for 1E 2259+586, $L_R = 3 \times 10^{26}$ erg/s for 1RXS J1308+21 and $L_R = 9 \times 10^{26}$ erg/s for J2143+06. Thus, these pulsars do not have extremely high luminosities in the radio. While 1E 2259+586 has a typical radio luminosity, both XDINSs J1308+21 and J2143+06 have among the lowest radio luminosities.

Comparison of the radio and X-ray data suggests large differences in two observed parameters. First, the radio and X-ray pulse durations differ by a factor of 16-18 for two objects and by a factor of only 5 for J2143+06. Second, while an interpulse is observed in both the radio and X-ray for J1308+21, 1E 2259+586 displays an interpulse only in the X-ray, and J2143+06 shows an irregular interpulse only in the radio band. In addition, there is a huge difference in the radio and X-ray luminosities (five and six orders of magnitude for two first pulsars). This result is very important.

1. We have detected periodic pulsed emission from the AXP 1E 2259+586 and two XDINSs 1RXS J1308+21 and J2143+06 in observations carried out on two radio telescopes of the Pushchino Radio Astronomy Observatory. The pulsars parameters are listed in Table 1.
2. We have obtained independent estimates of the distances to pulsars, which are within the intervals of distances determined using other methods.
3. The main difference between the radio from the X-ray pulsed emission is that the integrated radio pulses of all objects are much narrower. In addition, there is radio interpulse for XDINSs 1RXS J1308+21 and J2143+06, but no appreciable one for AXP 1E 2259+586.
4. The presence of weak radio emission in AXP 1E 2259+586 and SGR 1900+14 (Shitov et al. 2000) and strong one in XTE J1810-197 (Camilo et al. 2006), together with the recent detection of a radio pulsar (J1847-0130) with a long period ($P = 6.7$ s) and period derivative ($\dot{P} = 1.3 \times 10^{-12}$ s/s) (McLaughlin et al. 2003), similar to those of AXPs and SGRs, suggests the need to re-examine radio emission mechanisms in the magnetar model, or to consider other AXP and SGR models that do not involve superstrong magnetic fields.

Acknowledgements The authors are grateful to the staff of radio telescopes of the Pushchino Radio Astronomy Observatory for help with the observations. This work and report was partially supported by the Russian Foundation for Basic Research (projects no. 06-02-16888, 06-02-16810) and the National Science Foundation (project no. 00-098685).

References

Baring, M.G., Harding, A.K.: Astrophys. J. **507**, L55 (1998)
Camilo, F., Ransom, S., Halpern, J.: Nature **442**, 892 (2006)
Coe, M.J., Jones, Z.R., Lehto, H.: Mon. Not. Roy. Astron. Soc. **270**, 178 (1994)
Duncan, R.C., Thompson, C.: Astrophys. J. **392**, L9 (1992)
Gaensler, B.M, Slane, P.O., Gotthelf, E.V., Vasisht, G.: Astrophys. J. **559**, 963 (2001)
Gavriil, F.P., Kaspi, V.M.: Astrophys. J. **567**, 1067 (2002)
Gregory, P.C., Fahlman, G.G.: Nature **287**, 805 (1980)
Haberl, F., Schwope, A.D., Hambaryan, V., et al.: Astron. Astrophys. **403**, L19 (2003)
Hambaryan, V., Hasinger, G., Schwope, A.D., Schulz, N.S.: Astron. Astrophys. **381**, 98 (2002)
Kaplan, D.L., van Kerkwijk, M.H.: Astrophys. J. **635**, L65 (2005)
Lorimer, D.R., Lyne, A., Camilo, F.: Astrophys. J. **331**, L1002 (1998)
Malofeev, V.M., Malov, O.I.: Nature **389**, 697 (1997)
Malofeev, V.M., Malov, O.I.: astro-ph/0106435 (2001)
Malofeev, V.M., Malov, O.I., Teplykh, D.A.: In: Camilo, F., Gaensler, B.M. (eds.) IAU Symp. 218: Young Neutron Stars and Their Environments. Ast. Soc. Pac. Conf. Ser., vol. 218, p. 261 (2004)
Malofeev, V.M., Malov, O.I., Teplykh, D.A., et al: Astron. Rep. **49**, 242 (2005)
Malofeev, V.M., Malov, O.I., Teplykh, D.A., et al.: ATel#798 (2006)
Malov, I.F., Machabeli, G.Z., Malofeev, V.M.: Astron. Rep. **47**, 232 (2003)

McLaughlin, M.A., Stairs, I.H., Kaspi, V.M., et al.: Astrophys. J. **591**, L135 (2003)

Mereghetti, S.: In: Kouveliotou, C., Ventura, J., van den Heuvel, E. (eds.) Conf.: The Neutron Star–Black Hole Connection. NATO Science Series C: Mathematical and Physical Sciences, vol. 567, p. 351. Kluwer, Dordrecht (2001), astro-ph/9911252 (1999)

Shitov, Yu.P., Pugachev, V.D., Kutuzov, S.M.: In: Kramer, M., Wex, N., Wielebinski, R. (eds.) IAU Colloq. 177: Pulsar Astronomy—2000 and Beyond, Bonn. Ast. Soc. Pac. Conf. Ser., vol. 202, p. 685 (2000)

Taylor, J.H., Cordes, J.M.: Astrophys. J. **411**, 674 (1993)

Treves, A., Turolla, R., Zane, S., Colpi, M.: Publ. Astron. Soc. Pac. **112**, 297 (2000)

White, R.L., Becker, R.H., Helfand, D.J., Gregg, M.D.: Astrophys. J. **475**, 479 (1997)

Zane, S., Cropper, M., Turolla, et al.: Astrophys. J. **627**, 397 (2005)

Zhang, B.: Astrophys. J. **562**, L59 (2001)

Zhang, B.: Acta Astron. Sinic. E **44**, 215 (2003)

Astrophys Space Sci (2007) 308: 217–224
DOI 10.1007/s10509-007-9338-6

ORIGINAL ARTICLE

Measuring proper motions of isolated neutron stars with Chandra

Constraints on the nature and origin of the ROSAT discovered isolated neutron stars

Christian Motch · Adriana M. Pires · Frank Haberl · Axel Schwope

Received: 14 July 2006 / Accepted: 24 August 2006 / Published online: 20 March 2007

Abstract The excellent spatial resolution of the Chandra observatory offers the unprecedented possibility to measure proper motions at X-ray wavelength with relatively high accuracy using as reference the background of extragalactic or remote galactic X-ray sources. We took advantage of this capability to constrain the proper motion of RX J0806.4-4123 and RX J0420.0-5022, two X-ray bright and radio quiet isolated neutron stars (INSs) discovered by ROSAT and lacking an optical counterpart. In this paper, we present results from a preliminary analysis from which we derive 2σ upper limits of 76 mas/yr and 138 mas/yr on the proper motions of RX J0806.4-4123 and RX J0420.0-5022 respectively. We use these values together with those of other ROSAT discovered INSs to constrain the origin, distance and evolutionary status of this particular group of objects. We find that the tangential velocities of radio quiet ROSAT neutron stars are probably consistent with those of 'normal' pulsars. Their distribution on the sky and, for those having accurate proper motion vectors, their possible birth places, all point to a local population, probably created in the part of the Gould Belt nearest to the earth.

C. Motch (✉)
CNRS, UMR 7550, Observatoire Astronomique de Strasbourg, 11 rue de l'Université, 67000 Strasbourg, France
e-mail: motch@astro.u-strasbg.fr

A.M. Pires
Instituto de Astronomia, Geofísica e Ciências Atmosféricas, Universidade de São Paulo, R. do Matão 1226, 05508-090 Sao Paulo, Brazil

F. Haberl
Max-Plank Institut für Extraterrestrische Physik, Giessenbachstr., 85748 Garching, Germany

A. Schwope
Astrophysikalisches Institut Potsdam, An der Sternwarte 16, 14482 Potsdam, Germany

Keywords Stars: neutron · X-rays: individuals (RX J0806.4-4123, RX J0420.0-5022)

PACS 97.60.Jd · 98.70.Qy

1 Introduction

The vast majority of isolated neutron stars (INS) are being discovered as radio pulsars. However, recent X-ray surveys have revealed that a number of these objects manifest themselves at X-ray and γ-ray wavelengths. A surprising result of the ROSAT all sky survey has been the discovery of a small group of INS exhibiting properties at variance from those of the more familiar radio-pulsars (see Haberl 2005 for a recent review). First, these neutron stars do not exhibit radio emission, although there are recent claims of the detection at a very low radio flux for two of these objects (Malofeev et al. 2007). Most of them display X-ray pulsations with periods close to 10 s. Their X-ray spectra are essentially thermal with temperatures in the range of 40 to 100 eV. Many ROSAT INSs display shallow broad absorption lines on the top of the thermal continuum, which are interpreted as proton cyclotron lines or as atomic transitions in high magnetic field conditions ($\geq$few 10^{13} G). At present, it is still unclear whether or not ROSAT INSs constitute a homogeneous group, representative of a neutron star population different from standard radio-pulsars. The absence of strong radio emission can be due either to a very strong magnetic field (in excess of the quantum field) or to the fact that the radio pencil beam (which narrows at long spin periods) does not sweep the earth. Taking into account that only the closest sources have been detected by ROSAT because of their intrinsic X-ray faintness, their spatial density may well be comparable to that of normal radio-pulsars. ROSAT INSs may be therefore just the tip of the iceberg, a formerly hidden large population of stellar remnants, possibly related to the growing number of radio-quiet and X-ray bright compact central objects in SNRs (see Pavlov et al. 2004 for a review).

The generally low absorption ($N_H \sim 10^{20}$ cm^{-2}) derived from the soft X-ray observations suggest relatively small distances of the order of a few hundred parsecs. Small distances are indeed inferred from the HST parallax of the two X-ray brightest members, RX J1856.5-3754 ($d = 178^{+22}_{-17}$ pc, Kaplan 2003) and RX J0720.4-3125 ($d = 330^{+170}_{-80}$ pc, van Kerkwijk and Kaplan 2007). Proper motions in the range of 100 to 330 mas yr^{-1} are now measured for the three X-ray brightest ROSAT INSs (Walter 2001; Motch et al. 2003, 2005). These high velocities imply very low Bondi-Hoyle accretion rates and thus exclude with high confidence the possibility that the neutron star is re-heated by accretion of matter from the interstellar medium. The thermal emission of ROSAT INSs is most probably due to the progressive cooling of a middle-aged $\approx 10^5$–10^6 yr neutron star.

In this presentation we report on proper motion measurements of two ROSAT discovered INSs, RX J0806.4-4123 and RX J0420.0-5022. These two neutron stars do not have established optical counterparts which could be used to measure their proper motions with optical telescopes. Instead we take advantage of the outstanding imaging quality of the Chandra observatory to constrain their proper motion using two observations obtained in 2002 and 2005. The first part of the paper describes the observations and the method used to analyse them and to detect the displacement on the sky. In a second section, we use extensive MARX simulations to validate our results. We finally discuss the implication of our findings on the origin and distance of the group of thermally X-ray emitting but radio-quiet isolated neutron stars discovered by ROSAT.

2 X-ray and optical properties of RX J0806.4-4123 and RX J0420.0-5022

RX J0806.4-4123 was discovered by Haberl et al. (1998) in the ROSAT all-sky survey with a PSPC count rate of 0.38 cts/s. Follow-up observations with the XMM-Newton satellite revealed pulsations at a period of 11.4 s (Haberl and Zavlin 2002; Haberl et al. 2004). Its thermal X-ray spectrum ($kT = 96$ eV) lies among the hottest of this group. As for other ROSAT INSs, the quality of the spectral fit is significantly improved by adding a shallow absorption line centered around 460 eV. The source undergoes very little interstellar absorption ($N_H = 4 \times 10^{19}$ cm^{-2} or 1.1×10^{20} cm^{-2} depending on the inclusion or not of the low energy broad absorption line). Optical studies have been so far hampered by the low galactic latitude of RX J0806.4-4123 ($b = -4.98°$) and no strong constraint exists on the optical brightness of the optical counterpart. ESO WFC imaging do not reveal optical counterparts brighter than an estimated R magnitude of 22 in the Chandra error circle. Deep radio observations failed to detect any source at a position consistent with that of RX J0806.4-4123 (Johnston 2003).

RX J0420.0-5022 is a slightly fainter INS, also discovered in the ROSAT all-sky survey (Haberl et al. 1999). Its PSPC count rate is 0.11 cts/s. XMM-Newton spectra show that the source is the coolest of its group with $kT = 45$ eV and display evidences of a broad absorption line at $E \sim 330$ eV (Haberl et al. 2004). The interstellar absorption toward RX J0420.0-5022 is $N_H = 1.0 \times 10^{20}$ cm^{-2} or 2.0×10^{20} cm^{-2}, again depending on the inclusion or not of a broad line in the spectral model, and is on average about twice that of RX J0806.4-4123. Repeated XMM-Newton EPIC observations have now confirmed a pulsation period of 3.45 s, the shortest of all ROSAT discovered INSs. A one hour long ESO-VLT exposure suggests the presence of a $B = 26.6 \pm 0.3$ mag object in the Chandra error circle, which owing to its faintness could well be the optical counterpart of the X-ray source.

3 Chandra observations

3.1 Detecting proper motions

We started in Cycle 3 a long term observing program aiming at measuring the motion of several ROSAT discovered INSs, either lacking optical counterparts, or with optical counterparts too faint to be measured from the ground in a reasonable amount of time.

A reference astrometric frame can be built from the background of AGNs with a relatively high accuracy. At low galactic latitudes, a fraction of the field sources are identified with remote galactic objects, mostly active stars, which typically have proper motions one or two orders of magnitude below that of the neutron star and can thus be used to build the reference frame. In order to minimize systematic effects, the 2002 and 2005 observations were acquired with the same instrumental setting (aim point, readout mode) and at similar times of the year and roll angles (see Table 1). In both cases, the neutron star was located at the aim point of the instrument. The duration of the Chandra observation is determined by the need to acquire a large enough number of well positioned background extragalactic sources. In spite of the X-ray softness of RX J0806.4-4123, we preferred to

Table 1 Journal of observations

Object	Observation date	Exposure time (ks)	Roll angle (deg)
RX J0806.4-4123	2002-02-23	17.7	322.4
RX J0806.4-4123	2005-02-18	19.7	325.6
RX J0420.0-5022	2002-11-11	19.4	18.6
RX J0420.0-5022	2005-11-07	19.7	23.6

use ACIS-I instead of a BI chip of ACIS-S to avoid strong pile up and eventual severe degradation of the positional accuracy of the INS as a result of the distortion of the PSF. For RX J0420.0-5022, which is significantly fainter and softer than RX J0806.4-4123, we chose to use ACIS-S to obtain the best position on the INS at the expense of a slightly smaller field of view.

3.2 Source detection

Data reduction and source detection was performed using CIAO 3.3.0.1 and the latest calibration database. The 2002 and 2005 observations were reprocessed in an homogeneous way and the best attitude corrections applied. The level-1 event files were corrected for known processing offsets and were reprocessed to the level-2 stage. The standard Chandra data reduction pipeline randomizes the X-ray event positions detected within a given pixel in order to remove the "gridded" appearance of the images and to avoid any aliasing effects. In principle, this randomization process could slightly degrade the source centering accuracy. We thus had to reprocess the raw data in order to remove the pixel randomization following the science thread available at the Chandra X-ray center. Source detection was run on both the normal standard processing data and de-randomized data. The *wavdetect* algorithm (wavelet transform) was used with pixel scales 1, 2 and 4 well suited to the detection of unresolved sources at moderate off-axis distances. We only considered sources detected in the central CCDs, chips 0 to 3 and 6 and 7 for ACIS-I and ACIS-S respectively.

The final relative astrometric quality of the reference frames also depends on the number of sources common to the two observations and on the quality of the determination of their positions. We thus tried five different source detection thresholds, defined in *wavdetect* as the probability of a false detection at any given pixel. These thresholds ranged from 10^{-8} up to a level of 5×10^{-5} in order to find the best compromise between the number of sources and the mean positional errors, both increasing with increasing threshold. The largest value implies that about 50 false sources per CCD chip enter the *wavdetect* source list. However, since we only consider sources detected in both the 2002 and 2005 observations and located at a maximum distance of one to three arcsec, the actual probability that the common source is spurious is very low. Best positions are obtained in the energy range in which the contrast between the source spectra and the unresolved extragalactic and instrumental backgrounds is optimal. For the neutron star, measured in the same conditions as the reference sources, the choice of the energy range is much less critical since the object is brighter than any of the background sources. We decided to test energy bands 0.3–10 keV, 0.3–5 keV, 0.3–3 keV, 0.5–5 keV and 0.5–2 keV.

3.3 Matching method

Our relative astrometric frame determination uses the reconstructed equatorial positions of the X-ray sources. We allowed for translations of the right ascension and declination coordinates as well as for rotation around the aim point, resulting from possible errors in the attitude solution. A first algorithm simply uses a least square method ignoring the source positional errors. In a second step we implemented a maximum likelihood method consisting in maximizing the quantity;

$$L = \sum_{i,j} \exp\left(-\frac{1}{2}\left(\frac{d_{ij}}{\sigma_{ij}}\right)^2\right)$$

where d_{ij} are the distances between sources i and j after transformation and σ_{ij} the error on their distances, both quantities being computed independently for right ascension and declination.

3.4 Simulations

To our knowledge, the Chandra data base does not contain repeated ACIS observations of high proper motion isolated neutron stars with suitable properties to be used as test data. In order to check our detection chain and constrain the resulting positional errors, we thus created several simulated data sets using the CXC ray-tracer MARX 4.2.1.

A total of 26 and 12 sources were simulated for the fields of RX J0806.4-4123 and RX J0420.0-5022, respectively. These sources are those common (within 3″) to the actual 2002 and 2005 observations as detected in the 0.5–5.0 keV band and with a threshold of 10^{-6} for RX J0806.4-4123 and in the 0.5–2.0 keV band with a threshold of 10^{-5} for RX J0420.0-5022. We assumed a power law spectral energy distribution, typical of that of the extragalactic population which is known to be dominant on the X-ray background. Theoretical spectra of an absorbed power law of spectral index $\Gamma = 1.7$, undergoing a hydrogen column density of $N_H = 5.21 \times 10^{21}$ cm^{-2} (for the field of RX J0806.4-4123) and $N_H = 1.07 \times 10^{20}$ cm^{-2} (for the field of RX J0420.0-5022), were created using XSPEC 12.2.1. The column density values were derived from the colour excess in the direction of the sources as provided by the maps of Schlegel et

al. (1998) available at the NASA/IPAC Extragalactic Database – respectively, $E_{(B-V)} = 0.971$ and $E_{(B-V)} = 0.020$ – and applying the Predehl and Schmitt (1995) relation between the optical extinction and the X-ray absorption, $N_H = (1.79 \pm 0.03) A_V \times 10^{21}$ cm^{-2}. On the other hand, the neutron stars were assumed to radiate as blackbodies with $kT = 96$ eV and $kT = 44$ eV (Haberl 2005).

Each source was individually simulated giving as input parameters its equatorial coordinates, photon flux and spectral energy distribution, as well as the observation exposure and start time (in order to apply the proper degree of ACIS hydrocarbon contamination that is continuously degrading the quantum efficiency of the CCDs, specially at energies below 1 keV), the detector type at the focal plane and the coordinates of the aim-point. The flux normalization of each source was computed converting the ACIS count rates into flux in units of photons s^{-1} cm^{-2} in the 0.1–10.0 keV energy band. Events generated for each source were then merged into a single event file, which mimics the spatial and flux distribution of the actual sources present in the fields of RX J0806.4-4123 and RX J0420.0-5022. Background X-ray events were extracted from experimental blank fields provided by the calibration database. These blank fields were reprojected onto the tangent plane of the observations and, for each simulation, the events were randomly picked from the map so as to reproduce the level of background noise of the observation in the 0.5–5.0 keV band. Finally, sources and background were merged into a single event file.

The count distribution of the artificial sources is in good agreement with a Poisson distribution. We had to adjust the input flux of each source simulation to correct for the exposure map, vignetting and discrepant counts for the few sources located close to CCD edges or bad pixels and columns. It also seems that the *wavdetect* algorithm systematically underestimates by a few percents or more the number of X-ray photons and the input source parameters in MARX had to be increased accordingly to recover a distribution in source counts similar to that of the actual observations.

A total of 14 pairs of simulated event lists were generated for each ACIS I and S detector configurations. The main parameters changed were (i) the displacement applied to the central source, 0, 0.34, 0.75 and 1″, (ii) the offset applied to the 2005 reference coordinate system with respect to the 2002 one, 0 and 0.29″ and (iii) the position of the aim point which was moved by the equivalent of 1/4, 1/2 and 2/3 of a pixel in order to produce different distributions of the source PSF on the CCD pixel grid, although the Lissajous shaped dithering should in principle wash out a large part of the effect. In each of the 14 configurations, both event lists with and without pixel randomization applied were considered.

We discuss here preliminary results of the ACIS-I simulations corresponding to the RX J0806.4-4123 observations.

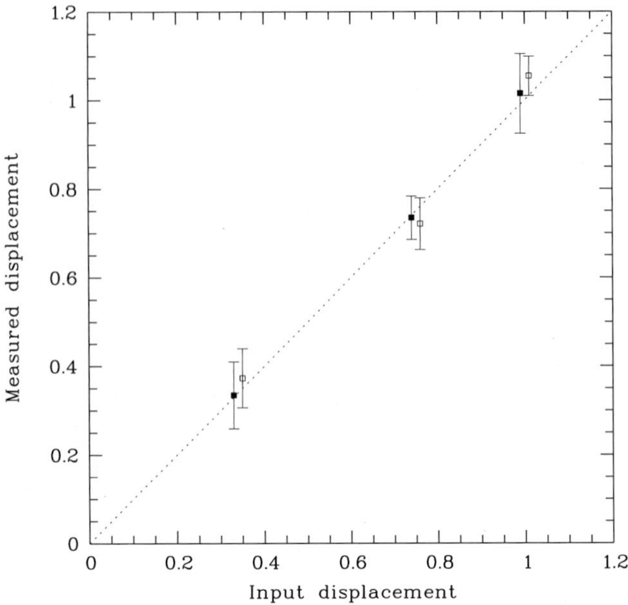

Fig. 1 Measured displacement as a function of the displacement entered in the simulations. Energy band 0.3–5 keV and 0.5–5 keV. *Filled squares* and *open squares* represent simulations with and without pixel randomization (*small offsets* applied to the input values for clarity)

A detailed analysis of all ACIS-I and S simulations will be presented in a forthcoming paper.

Owing to its large number of photons, the formal errors on the position of the isolated neutron star delivered by *wavdetect* are very small, of the order of 0.02″. In actual data, however, one should also take into account a small additional error which results from uncertainties in the transformation between the detector pixel and sky coordinates. Deep ACIS-I observation of a field in Orion containing over a thousand of bright X-ray sources show that this additional error is of the order of ∼0.07″ (Chandra Proposer's Observatory Guide). Therefore, most of the uncertainty on the proper motion probably comes from the accuracy with which we can link the reference frames of the two epochs. This accuracy is given by $\sigma^2_{\rm frame} = \frac{1}{n(n-1)} \sum_{i=1}^{n} (d_i^{12})^2$ with d_i^{12} being the distance between source i as observed in the first and second observation. Most of the configurations yield a $\sigma_{\rm frame}$ close to 0.10″. Best results are obtained when restricting the energy band in which source detection is done to 0.3–5.0 keV or 0.5–5.0 keV.

In our simulations, the pixel randomization process has basically no influence on the accuracy of the frame matching and there is no strong dependency either on the threshold used for source detection. We show in Fig. 1 the relation between the input displacements and their values measured on the simulated data. On the average, the motion is recovered with an rms accuracy of the order of 0.07 to 0.10″. In a similar manner, the shifts in right ascension and declination are recovered with a rms accuracy of 0.07″.

3.5 Results

As for the MARX 4.2.1 simulations, the $\sigma_{\rm frame}$ for real data are minimum for the energy ranges 0.3–5.0 keV and 0.5–5.0 keV for ACIS-I and 0.3–3.0 and 0.5–5.0 keV for ACIS-S. The number of sources common to both observations varies with detection threshold, energy range and maximum allowed distance between the 2002 and 2005 positions to include the source in the computation of the transformation. It ranges from 12 to 39 sources for RX J0806.4-4123 and from 6 to 16 sources for RX J0420.0-5022. Introducing sources more distant than $1''$ (at large off-axis angles) somewhat worsens the quality of the relative astrometry. Detection thresholds smaller than 10^{-6} give results of lower quality as a consequence of the decreasing number of real sources detected at the two epochs. Using the best combination of energy bands and detection thresholds, the mean distance between reference sources after transformation is $0.46''$ for ACIS-I and $0.57''$ for ACIS-S, again with no strong dependency on the fact that pixel randomization is applied or not. These mean distances are slightly larger by $\sim$30% than those expected from the formal errors given by *wavdetect*. A small offset of $0.14''$ in right ascension and $0.39''$ in declination is found between the 2002 and 2005 coordinate systems of the field of RX J0806.4-4123, while no really significant translation or rotation is needed between the equatorial systems of the two observations of RX J0420.0-5022.

For RX J0806.4-4123, the measured displacement averaged over the best energy bands and detection thresholds is $\sim 0.1''$, corresponding to a $\sim 1\sigma$ effect. The error on the transformation, $\sigma_{\rm frame}$ probably dominates the error budget. With a best value of $\sigma_{\rm frame} = 0.09''$ and assuming an additional systematic error of $0.07''$, the 3σ upper limit on the total displacement of RX J0806.4-4123 over the 3 years time interval is $0.34''$ implying $\mu \leq 114$ mas/yr (or $\mu \leq 76$ mas/yr at the 2σ level).

Because of the smaller central field of ACIS-S and accordingly smaller number of reference sources, the mean $\sigma_{\rm frame}$ is $0.19''$ for the field of RX J0420.0-5022. The mean displacement measured is $0.09''$ corresponding to a $\sim 0.4\sigma$ effect. For RX J0420.0-5022 too, we do not measure any statistically significant motion and can put a 3σ upper limit of $\mu = 207$ mas/yr (or $\mu \leq 138$ mas/yr at the 2σ level).

Our observing programme has thus established the possibility to measure proper motions at X-ray wavelengths with relatively good accuracy. In particular, the proper motions of RX J1856.5-3754, RX J1605.3+3249 and possibly RX J0720.4-3125 would have been detected with ACIS-I and a three year time interval. If Chandra lives long enough, very significant constraints or detections of the proper motions of the ROSAT discovered INSs that are too faint to be observed with optical telescopes will be possible.

4 Discussion

The upper limit on the proper motions of RX J0806.4-4123 and RX J0420.0-5022 derived from our Chandra observations provide significant constraints on the space velocities of these two objects. In particular, the low value of 76 mas/yr obtained for RX J0806.4-4123 is well below those measured so far for the three X-ray brightest objects but comparable to those of nearby radio pulsars (see Table 2). The likely small distance of the source ($d \leq 240$ pc, Posselt et al. 2007) imply low transverse velocities of less than 83 km s^{-1}.

On Fig. 2 we compare the transverse velocities of ROSAT discovered INSs with those of nonrecycled radio pulsars, total and young (age $\leq 3 \times 10^6$ yr) populations. Radio pulsar data are taken from Hobbs et al. (2005). The distribution of the inferred 2-D velocities of the ROSAT INSs appears consistent with that of the radio pulsars. The absence of fast moving ROSAT INSs is intriguing but not really statistically significant, since the probability of such a combination remains above the 10% level. According to Hobbs et al. (2005), young radio pulsars have a mean 2-D velocity slightly larger than old ones. However, the two distributions overlap, as seen on Fig. 2, and the significance of this difference remains uncertain. The ATNF catalogue (Manchester et al. 2005) contains a total of 6 radio pulsars with $d \leq 1$ kpc, younger than 4.25 Myr and with measured proper motion. They have 2-D velocity in the range of 60 to 354 km s^{-1} with a mean of 177 km s^{-1}, clearly consistent with that observed for the ROSAT INSs.

The measurement of proper motions can potentially provide a wealth of important informations on the evolutionary status, origin and age of the neutron star. A large spatial velocity, typically in excess of 30 km s^{-1} rules out the possibility that the star is reheated by accretion of interstellar material.

Table 2 Proper motions, distances and transverse velocities of nearby young neutron stars

Name	Proper motion (mas/yr)	Distance (pc)	V_T (km s^{-1})
RX J1856.5-3754	332 ± 1	178^{+22}_{-17}	283
RX J0720.4-3125	97 ± 12	250	114
RX J1605.3+3249	144.5 ± 13.2	<410	<280
RX J0806.4-4123	≤ 76	240 ± 25	<83
RX J0420.0-5022	≤ 138	≤ 340	<241
B0656+14	44.1 ± 0.7	288^{+33}_{-27}	60.1
B1929+10	103.4 ± 0.2	361^{+10}_{-8}	176
Vela Pulsar	58.0 ± 0.1	287^{+19}_{-17}	79
Geminga	170 ± 4	273 ± 84	219

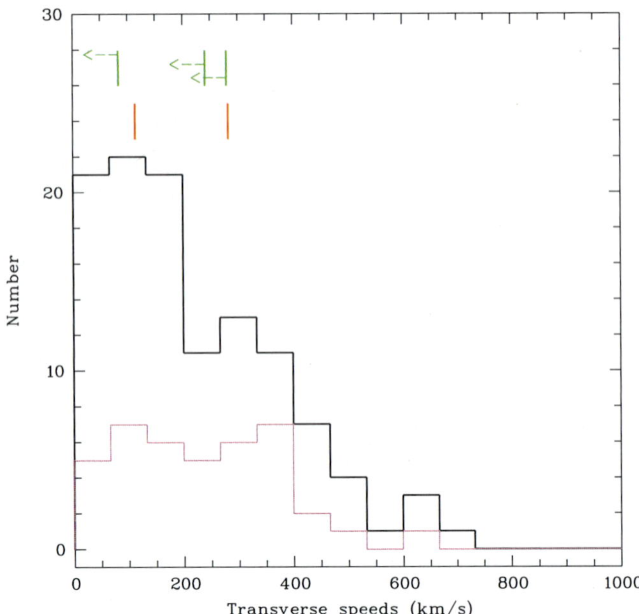

Fig. 2 Transverse speeds of ROSAT discovered INSs compared to those of nonrecycled radio pulsars. Data taken from Hobbs et al. (2005). The *black thick line* and the *thin magenta line* show the histogrammes of the entire population and of the young (age $\leq 3 \times 10^6$ yr) respectively. Measured values and upper limits for the ROSAT INSs are shown on the *top*

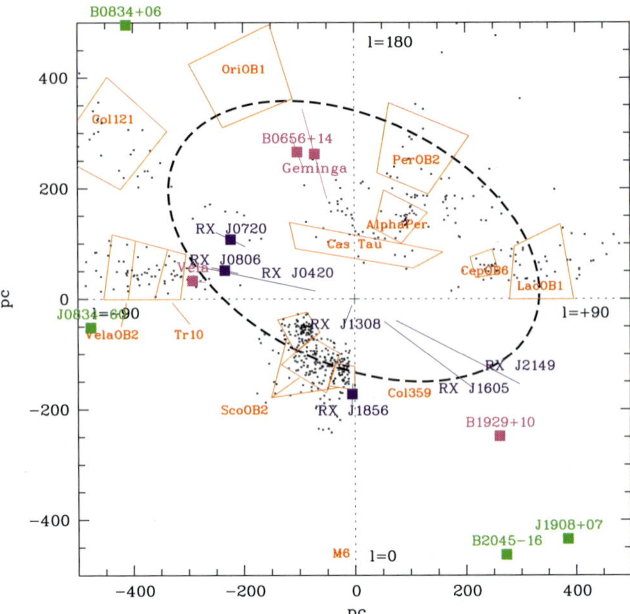

Fig. 3 Positions of nearby neutron stars and OB associations projected on the Galactic plane. *Red boxes* show the OB associations boundaries. Hipparcos stars with a probability higher than 75% to be linked to the OB associations are represented by *black dots*. *Blue lines* and *filled squares* show the possible positions of the ROSAT discovered INS, assuming a distance range of 100 to 400 pc for those which do not have distance estimates. Radio or γ-ray pulsars younger than 4.25 Myr and located within 1 kpc are shown as *magenta symbols* (when a parallax distance exists) or *green symbols* when the distance is estimated from dispersion measurements. The Gould Belt is shown as a *thick dashed line*

In this respect, the upper limits on the 2-D velocities of RX J0806.4-4123 and RX J0420.0-5022 are still large and the possibility that they accrete from the interstellar medium cannot yet be seriously considered. The fact that these two INSs share many properties such as X-ray energy distribution and pulsation periods with high spatial velocity ROSAT INSs also argues for a similar nature.

If measured with enough accuracy, the proper motion vector of the neutron star can also be used to compute possible past trajectories assuming a range of present distances and radial velocities. OB stars progenitors of neutron stars are not distributed randomly on the sky but rather concentrate in relatively large associations sharing a common space motion. The exact fraction of early type stars found in OB associations remains in many cases uncertain due to the errors on the photometric distances and the generally fuzzy boundaries attributed to the associations. Garmany (1994) shows that the relative fraction of O and B stars found in associations is larger than 75% and 58% respectively. The fact that a group of backwards trajectories crosses the past position of a known OB association is thus a hint, no definite proof since other trajectories are possible, that the neutron star progenitor was located there when it underwent supernova explosion. It becomes then possible to compute possible travel times which can be compared to theoretical cooling curves, with the spin down time and, if associated to a SNR, with the expansion time. For seven nearby and young isolated neutron stars listed in Table 3 a tentative birth place has been proposed in the recent literature.

We show on Fig. 3 the positions projected on the galactic plane of neutron stars younger than 4.25 Myr and located at distances less than 1 kpc. Radio pulsar data were extracted from the pulsar catalogue maintained by the Australian Telescope National Facility (Manchester et al. 2005). We also show the boundaries of the classical OB associations as defined in Humphreys (1978) and de Zeeuw et al. (1999). For the Gould Belt, we use the determination of Perrot and Grenier (2003) based on HI and H_2 clouds and Hipparcos distances to the nearby OB associations.

Neutron star travel times from their supposed birth place to their present location are difficult to determine with great accuracy. The large extents of the possible parent association and the unknown radial velocities usually yield rather large uncertainties. In the case of the two youngest nearby radio pulsars, the Vela pulsar (10^4 yr) and B0656+14 (10^5 yr), there is a good agreement between the estimated travel time, the pulsar age computed assuming magnetic dipole braking and the age of the remnant of the parent supernova (Hoogerwerf et al. 2001; Thorsett et al. 2003). The absence of obvious supernova remnants at the possible birth place of B1929+10 and Geminga is consistent with their older ages.

Table 3 Possible birth places of nearby young neutron stars

Name	Spin down age (yr)	Travel time (yr)	Birth place	Reference
RX J1856.5-3754	?	4×10^5–1×10^6	Upper Sco OB2	1, 2
RX J0720.4-3125	2×10^6	6×10^5–3×10^6	Tr 10 + Vela OB2 or Lower Sco OB2	3, 11
RX J1605.3+3249	?	$\sim 10^6$	upper Sco OB2	4
B0656+14	10^5	10^5	Monogem Ring	5, 6
B1929+10	3×10^6	1×10^6	Upper Sco OB2	7, 8
Vela Pulsar	10^4	10^4	Vela OB2	9, 8
Geminga	3.4×10^5	3.4×10^5	Cas-Tau or Ori OB1	10

(1) Kaplan (2003); (2) Walter (2001); (3) Motch et al. (2003); (4) Motch et al. (2005); (5) Brisken et al. (2003); (6) Thorsett et al. (2003); (7) Chatterjee et al. (2004); (8)Hoogerwerf et al. (2001); (9) Dodson et al. (2003); (10) Pellizza et al. (2005); (11) Kaplan and van Kerkwijk (2005)

In this respect, the lack of conspicuous remnants in the surroundings of the hypothetical birth locations of the ROSAT discovered INSs is also consistent with the relatively large ages ($\geq 3 \times 10^5$ yr) derived from the flight times (see Table 3).

According to Perrot and Grenier (2003), the Gould Belt has an elliptical shape with major and minor semi-axes of 373 ± 5 pc and 233 ± 5 pc respectively. The Sun lies inside the Belt and is located about 104 ± 4 pc offset from its center. It is currently closer to the region of the Gould Belt populated by the Sco OB2 and Vela OB2 + Trumpler 10 associations (see Fig. 3). Backwards trajectories suggest an origin in the closest region of the Belt for RX J1856.5-3754, RX J0720.4-3125, RX J1605.3+3249 and B1929+10. Two INSs, Geminga and B0656+14 were possibly born in or close to the Orion complex, on the opposite side of the Gould Belt. As visible in Fig. 3, all ROSAT discovered INSs lie in the half sky centered on Sco OB2 and containing the closest part of the Gould Belt.

No such sensitivity anisotropy exists in the ROSAT all sky survey at the level of 0.14 cts/s, the PSPC count rate of RX J0420.0-5022 which is the faintest of the ROSAT discovered INSs (Voges et al. 1999). In order to detect and recognise the extreme softness of such a source a minimum of $\sim$200 s exposure time is needed. Only 10% of the whole sky had exposures shorter than this value. Furthermore, the distribution of the low exposed regions are not specifically directed towards the far side of the Gould Belt where ROSAT INSs are apparently not detected.

The very soft thermal-like energy distributions of the ROSAT INSs render their detectability very sensitive to the amount of interstellar absorption on the line of sight. The distribution of the local interstellar medium has been the subject of extensive studies, mostly based on the measurement of the NaID doublet at 5890 Å (Welsh et al. 1994; Sfeir et al. 1999). Our current environment is rather unusual since we appear to be located in a local bubble cavity of very low interstellar gas density. The bubble is surrounded by a "wall" of interstellar matter. Its shape is very irregular with the closest distance to the edge of the bubble being at $\sim$60 pc in the direction of the Galactic center. Being based on more than a thousand of directions, the NaI absorption map of Lallement et al. (2003) provides a good description of the local interstellar medium up to distances of about 350 pc for absorptions up to a few 10^{20} cm^{-2}. Their map shows no evidence for any asymmetry which could explain the absence of detection of INSs in the half sky where they apparently miss.

Therefore, although based on small number statistics, the distribution of ROSAT discovered INSs on the sky might well reveal the major role played by the local Gould Belt in their generation. Popov et al. (2003) have indeed highlighted the importance of the Gould Belt for understanding the bright part of the observed X-ray $\log N(>S)$-$\log S$ INS curve.

5 Conclusions

The high astrometric quality of the Chandra Observatory offers the possibility to measure proper motions of X-ray sources with an unprecedented accuracy. Our observations and simulations show that displacements as small as $\sim 0.3''$ can be detected. Applying this method to two ROSAT discovered INSs, we derive an upper limit on the transverse speed of RX J0806.4-4123 at the lower end of the observed distribution for radio pulsars while less constraining limits are obtained for RX J0420.0-5022. We show that the spatial distribution and possible birth places of this particular group of INSs argue in favour of a local population and for a production dominated by the closest part of the Gould Belt.

Acknowledgements We acknowledge the use of the ATNF Pulsar Catalogue available at http://www.atnf.csiro.au/research/pulsar/psrcat.

References

Brisken, W.F., Thorsett, S.E., Golden, A., et al.: Astrophys. J. **593**, L89 (2003)

Chatterjee, S., Cordes, J.M., Vlemmings, W., et al.: Astrophys. J. **604**, 339 (2004)

de Zeeuw, P.T., Hoogerwerf, R., de Bruijne, J.H.J., et al.: Astron. J. **117**, 354 (1999)

Dodson, R., Legge, D., Reynolds, J.E., et al.: Astrophys. J. **596**, 1137 (2003)

Garmany, C.D.: Publ. Astron. Soc. Pac. **106**, 25 (1994)

Haberl, F.: 5 years of Science with XMM-Newton. MPE Report 288, p. 39 (2005)

Haberl, F., Zavlin, V.E.: Astron. Astrophys. **391**, 571 (2002)

Haberl, F., Motch, C., Pietsch, W.: Astron. Nachr. **319**, 97 (1998)

Haberl, F., Pietsch, W., Motch, C.: Astron. Astrophys. **351**, L53 (1999)

Haberl, F., Motch, C., Zavlin, V.E., et al.: Astron. Astrophys. **424**, 635 (2004)

Hobbs, G., Lorimer, D.R., Lyne, A.G., et al.: Mon. Not. Roy. Astron. Soc. **360**, 974 (2005)

Hoogerwerf, R., de Bruijne, J.H.J., de Zeeuw, P.T.: Astron. Astrophys. **365**, 49 (2001)

Humphreys, R.M.: Astrophys. J. Suppl. Ser. **38**, 309 (1978)

Johnston, S.: Mon. Not. Roy. Astron. Soc. **340**, L43 (2003)

Kaplan, D.: In: Proceedings of the Workshop Physics and Astrophysics of Neutron Stars, Santa Fe, New Mexico, 28 July–1 August 2003 (2003)

Kaplan, D.L., van Kerkwijk, M.H.: Astrophys. J. **628**, L45 (2005)

Lallement, R., Welsh, B.Y., Vergely, J.L., et al.: Astron. Astrophys. **411**, 447 (2003)

Malofeev, V., Malov, O., Teplykh, D.: Astrophys. Space Sci., DOI 10.1007/s10509-007-9341-y (2007)

Manchester, R.N., Hobbs, G.B., Teoh, A., et al.: Astron. J. **129**, 1993 (2005)

Motch, C., Zavlin, V.E., Haberl, F.: Astron. Astrophys. **408**, 323 (2003)

Motch, C., Sekiguchi, K., Haberl, F., et al.: Astron. Astrophys. **429**, 257 (2005)

Pavlov, G.G., Sanwal, D., Teter, M.A.: In: IAU Symposium, vol. 218, p. 239 (2004)

Pellizza, L.J., Mignani, R.P., Grenier, I.A., et al.: Astron. Astrophys. **435**, 625 (2005)

Perrot, C.A., Grenier, I.A.: Astron. Astrophys. **404**, 519 (2003)

Popov, S.B., Colpi, M., Prokhorov, M.E., et al.: Astron. Astrophys. **406**, 111 (2003)

Posselt, et al.: Astrophys. Space Sci., DOI 10.1007/s10509-007-9344-8 (2007)

Predehl, P., Schmitt, J.H.M.M.: Astron. Astrophys. **293**, 889 (1995)

Schlegel, D.J., Finkbeiner, D.P., Davis, M.: Astrophys. J. **500**, 525 (1998)

Sfeir, D.M., Lallement, R., Crifo, F., et al.: Astron. Astrophys. **346**, 78 (1999)

Thorsett, S.E., Benjamin, R.A., Brisken, W.F., Golden, A., Goss, W.M.: Astrophys. J. **592**, L71 (2003)

van Kerkwijk, M.H., Kaplan, D.L.: Astrophys. Space Sci., DOI 10.1007/s10509-007-9343-9 (2007)

Voges, W., et al.: Astron. Astrophys. **349**, 389 (1999)

Walter, F.M.: Astrophys. J. **549**, 433 (2001)

Welsh, B.Y., Craig, N., Vedder, P.W., et al.: Astrophys. J. **437**, 638 (1994)

Astrophys Space Sci (2007) 308: 225–230
DOI 10.1007/s10509-007-9368-0

ORIGINAL ARTICLE

The trigonometric parallax of the neutron star Geminga

Jacqueline Faherty · Frederick M. Walter · Jay Anderson

Received: 30 June 2006 / Accepted: 25 August 2006 / Published online: 20 March 2007

Abstract We obtained a series of four observations of the isolated neutron star Geminga over an 18 month period using the Advanced Camera for Surveys (ACS) Wide Field Camera (WFC) on the Hubble Space Telescope in order to determine its trigonometric parallax. We find the parallax $\pi = 4.0 \pm 1.3$ mas, corresponding to a distance to Geminga of 250^{+120}_{-62} pc, a result 60% larger than the previously published value. The proper motion is 178.2 ± 1.8 mas/year. In this paper, we describe the analysis techniques in detail since the amplitude of the parallactic shift is smaller than the camera's pixel size. We fit each star in the images with an appropriate effective PSF and applied a distortion correction to generate stellar positions accurate to 0.01 pixels ($\sim$0.5 mas). The 134 stars common to all images serve to establish a reference frame for alignment of the image series. Our observations were made around the times of maximum parallactic shift. We discuss the implications of this new distance measurement for the inferred radius of Geminga, and the neutron star equation of state.

Keywords Astrometry · Geminga · Parallax · Neutron star

PACS 95.10.Jk · 97.60.Jd

J. Faherty (✉) · F.M. Walter
Stony Brook University, Stony Brook, NY, USA
e-mail: jfaherty@amnh.org; fwalter@astro.sunysb.edu

J. Anderson
Rice University, Houston, TX, USA
e-mail: jayowl@rice.edu

1 Introduction

Geminga (1E0630+17.8; PSR B0630+18) is the prototypical isolated multi-wavelength pulsar. It is one of the brightest persistent gamma ray sources in the sky. It pulses at frequencies from γ rays up to (possibly) the MHz radio (Caraveo et al. 1996). The small characteristic spindown age ($P/2\dot{P}$) of 340 000 years may account for Geminga's energetic characteristics. The flux and the spectral energy distribution of Geminga have been measured, and the spectrum is predominantly thermal. Model fitting (generally with black bodies) provides an angular diameter and a surface flux. Lacking an accurately known distance, the luminosity and the true transverse motions—fundamental physical parameters—are poorly known.

Knowledge of the distance is of fundamental importance, as many physical parameters scale from observational measures as the first or second powers of the distance. But because of the sheer scale of the universe, the parallax remains one of the most difficult measurements that astronomers attempt. The plate scale of the Hubble Space Telescope (HST) Advanced Camera for Surveys Wide Field Camera (ACS/WFC) is $\sim$0.05 arcsec/pixel, therefore objects more distant than 40 pc have parallactic displacements of less than one pixel. Measurements of such small relative motions are challenging, as we shall show below.

As one of the closest of the isolated neutron stars, and as the seminal γ-ray pulsar, the measurement of the distance to Geminga will provide good constraints on the energetics of young pulsars. For an assumed mass (Geminga is a single star), a well-constrained radius will limit the possible interior equation of state of a neutron star. Astrophysical observations of the equation of state of compact objects serve two purposes: to provide bounds on the symmetry energy of nuclear matter in the high neutron-fraction limit unattain-

able in the laboratory, and to probe whether or not strange matter stars exist in our universe.

The stellar radius is the product of the angular diameter and its distance. The angular diameter is a model-dependent quantity, but is specified uniquely if we know the surface temperature (for a black body), or the temperature and composition for a stellar atmosphere, or the surface emissivity if the surface is solid (Pavlov et al. 1996; Pons et al. 2002; Gänsicke et al. 2002; Rajagopal and Romani 1996; Turolla et al. 2004).

The most direct astronomical measure of distance is the trigonometric parallax. With current instrumentation, only a few compact objects are close enough and bright enough that one can reasonably expect to measure a significant geometric parallax. It is only for those objects whose spectrum is dominated by thermal emission that one can hope to determine a meaningful radius. The nearby thermal compact objects include:

1. RX J1856-3754, the brightest of the isolated thermal neutron stars. The X-ray spectrum is indistinguishable from a black body; the optical data lie on a Rayleigh-Jeans tail, but about a factor 7 above the extrapolation of the X-ray black body. The flux is constant to $<1.3\%$ at frequencies from 10^{-3} to 50 Hz (Burwitz et al. 2003). Its distance has been determined to be 117 ± 12 pc by Walter and Lattimer (2002) using the HST WFPC2 camera. This distance constrains the radius to lie between 14 and 17 km. Kaplan et al. (2002) have reported a revised distance of 140–180 pc.
2. RX J0720-3125, an isolated neutron star with a spectrum similar to that of RXJ 1856-3754. However, it exhibits a spectral feature in the X-rays and has an 8.4 second period (Habert et al. 2006). The differences between it and RX J1856-3754 may lie in its inclination to our line of sight. The distance is estimated to be <410 pc (Kulkarni and van Kerkwijk 1998). A program to determine its parallax has begun using ACS HRC images and is close to completion.
3. Geminga, the isolated 237 msec pulsar, whose parallax was reported by Caraveo et al. (1996). The distance they report is 157^{+59}_{-34} pc based on three WFPC2 images. We undertook the program we report here in part because of an inability to reproduce that result.

In this paper we report on our efforts to determine the trigonometric parallax of Geminga. We focus on the technique, but conclude with some thoughts on the astrophysical implications of the revised distance.

2 Previous measurements

Caraveo et al. (1996) observed the Geminga field 3 times with the Wide Field Planetary Camera 2 (WFPC2) between March 1994 and March 1995. They reported a distance to Geminga of 157^{+59}_{-34} pc, with a proper motion of $\mu = 170 \pm 6.0$ mas/year at a position angle of $55.0 \pm 2.0°$. The distance has been questioned for several reasons:

1. The data consisted of only three images. This is the absolute minimum required to solve jointly for parallax and proper motion. A fourth image was later obtained, but the results have never been published.
2. Caraveo et al. (1996) employed what were the accepted geometric corrections for the WFPC2 at that time. Subsequently, Anderson and King (1999) found and characterized the 34th row error which potentially has a large effect on the measurement.
3. The experiment design, though usable, was less than ideal. Geminga is located at high galactic latitude. The small field of view and low sensitivity of the WFPC2/PC yields less than 10 objects in the PC field appropriate for the background astrometric grid. Since the WFPC2 images are not all centered at the same coordinates, not all objects are present in the three images, which compounds the difficulty of creating an accurate astrometric grid.

In light of our inability to reproduce the measurement, we decided to re-visit Geminga with a different instrument and longer observations.

3 WFC data

3.1 Observational design

The parallactic displacement is an ellipse whose orientation and shape are fixed by the ecliptic coordinates of the targets. At Geminga's ecliptic longitude of 98°, the ellipse is oriented essentially parallel to the celestial equator (the inclination is 1.9°). At Geminga's ecliptic latitude of $-5°$, the ellipse is highly elongated E-W, with an eccentricity of about 0.995 (on the ecliptic the eccentricity is 1). Essentially all the parallactic displacement will be east-west.

We planned our observations to occur near the times of maximum parallactic displacement, at 6 month intervals (Table 1). We made four successful observations. There was a tradeoff between the wider field and higher sensitivity of the WFC, which yields a better detection of Geminga ($V = 26.5$ mag) and a more robust astrometric grid, and the smaller pixels of the High Resolution Camera (HRC). We selected the WFC. We used the F555W (broadband V) filter, as a compromise between the relatively blue color of the target and the relatively red colors of the distant stars that make up the astrometric grid. Each observation required three orbits. In each orbit we acquired 4 images, with standard half-pixel dithering and CR-split images. Each visit

Table 1 Epochs with dates the images were taken and the number of stars in each image. Maximum parallactic shift was on March 27 and September 30. The total number of stars common to all images was 134

Epoch	1	2	3	4
Date	10-07-03	03-18-04	09-21-04	03-22-05
# of stars	310	311	303	286
Stars common with epoch 1		200	225	191

therefore consists of 6 images at each of two dithered positions. The total integration time is about 8 ksec at each epoch.

Aside from the target, there are 134 stars common to all 4 epochs (see Table 1).

3.2 Distortion corrected positions

As these observations were obtained from above the atmosphere, the primary distortions in our images arise in the HST optics and the detector. We considered working with the Drizzled (_drz) images, which have been distortion corrected and chip-combined by the STSDAS package Multi-Drizzle. However, the positions on the _drz images were not internally self-consistent at our required accuracy because of the re-sampling by Multi-Drizzle. We therefore went back to the basics and used the Flattened (_flt) images which are only bias-subtracted and flat-fielded.

Removing the instrumental distortion is the most important step in this measurement, and its accuracy dominates the error in our result. In an Instrument Science Report (ISR), Anderson (2006) provides a Fortran routine for determining distortion corrected positions for ACS WFC images. That ISR describes a publicly available program *img2xym_WFC.09x10*. The program uses a library of effective point-spread-functions (ePSF) that cover nearly all areas of the two WFC chips. We fit each of the point sources in all 48 images using this software, yielding a table of distortion-corrected X, Y pixel positions.

We rotated the resultant stellar positions for epoch 1 using the ORIENTAT angle in the _drz FITS header for each image, so that the X, Y axes aligned with right ascension (RA), and declination (DEC). We chose to use the angle in the _drz header because it is distortion corrected and well understood. The _flt header is affected by distortion and is not as reliable. We needed to rotate only epoch 1 by this angle because all of the other epochs would be transformed into these positions when we solved for the epoch-to-epoch transformations (see Sect. 4.1). The uncertainty in this angle is a negligible contribution to the positional error. This is the only instance when we used the FITS header information for a calculation.

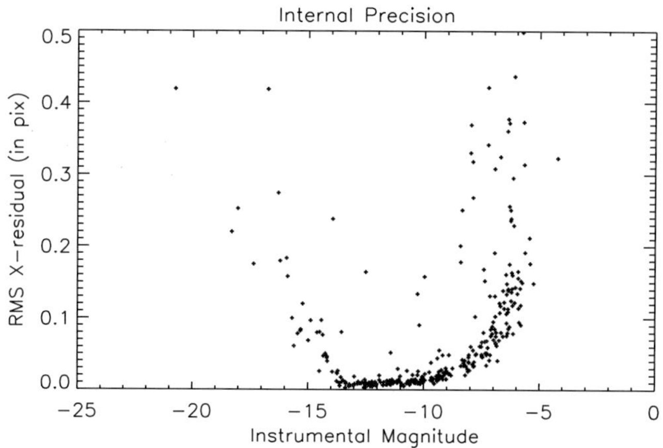

Fig. 1 The Root Mean Square (RMS) residuals for the X positions on the chip vs. instrumental magnitude for epoch 1 to epoch 1 prior to any star selection cut-off applications. The "saddle"-like shape of the internal precision shows that there is a range of magnitudes (we select $-14 < M < -8$) that will provide the best residuals for our reference stars

4 WFC DATA analysis

4.1 Star selection

We began with an internal comparison of all the stars in the 12 images at a single epoch. Over the approximately 5 hour interval, there should be no detectable stellar motions. We arbitrarily chose one of the 12 images as the standard plate. We then transformed the other 11 images to this standard frame by fitting the relative offsets, plate scales, and image rotation. Since the stars should show no external motion within each visit, these transformations indicate the reproducibility of the positions. Figure 1 shows that the brightest and the faintest stars do not transform well.

Therefore, we implemented two star selection cutoffs:

1. We selected only stars in the instrumental magnitude range $-14 < M < -8$, as those yielded the lowest residuals (See Fig. 1). Geminga was just beyond that range with an instrumental magnitude of $M = -6.8$. However from close examination of the stars at this magnitude, Geminga actually transforms very well, with residuals of $\sim$0.09 pixels.
2. We required that the reference stars transform with residuals <0.05 pixels. There was no point in forcing the solution to be orders of magnitude better than what we could do for our target object, therefore we did not attempt to obtain a solution down to the astrometric limit of 0.01 pixels.

We used a basic six parameter, least-squares, linear transformation to solve for the plate constants. As stated above, within any visit the external stellar motions are negligible.

The transformation solved the following two equations:

$$X = x_{2o} + A(X_1 - X_0) + B(Y_1 - Y_0), \quad (1)$$

$$Y = y_{2o} + C(X_1 - X_0) + D(Y_1 - Y_0), \quad (2)$$

where x_{2o} and y_{2o} set an arbitrary origin and were therefore set to the center of the field, A, B, C, D solve for displacement, rotation, and an arbitrary stretch in the two coordinates, and (X_0, Y_0), (X_1, Y_1) are transformed positions in the reference image (0) and the other image (1). X and Y are the transformed coordinates.

We iterated over all the stars in each image and examined the residuals at each step removing stars prior to the next iteration that did not fit the above criteria. The process converged on a solution that had reference stars with residuals <0.05 pixels (see Fig. 3). The root mean square (RMS) of both the X and the Y positions were nearly identical (see Figs. 1 and 2). This was expected as it would otherwise in-

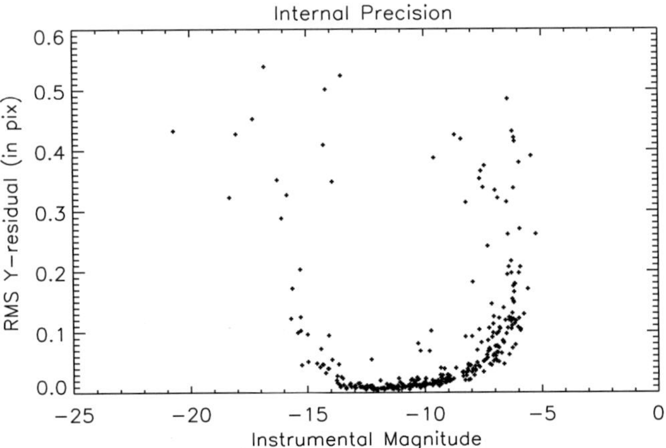

Fig. 2 As Fig. 1, but for the Y positions

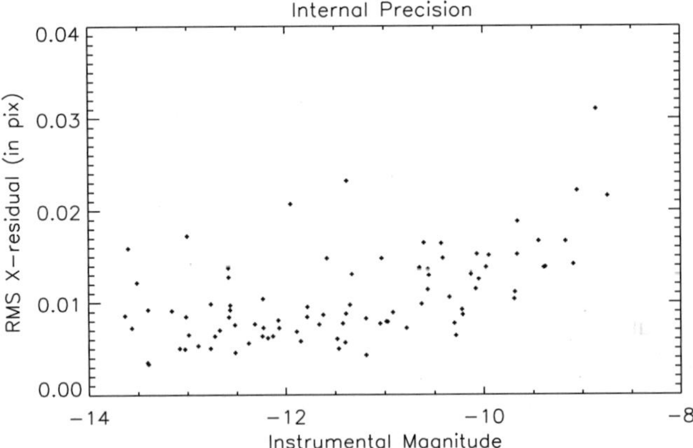

Fig. 3 The stars that transformed within the star selection cut-off limitations for epoch 3 into epoch 1. Compared to the above figure this one shows that with the two star selection cut-offs ($-14 < M < -8$; residuals <0.05 pixels), we obtain a group of stars that have residuals <0.04 pixels

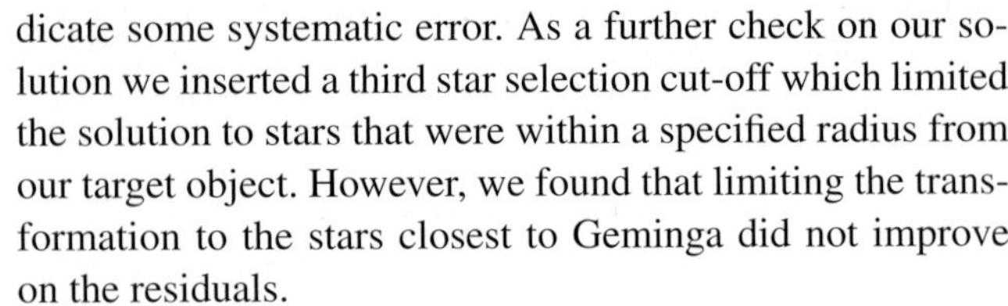
dicate some systematic error. As a further check on our solution we inserted a third star selection cut-off which limited the solution to stars that were within a specified radius from our target object. However, we found that limiting the transformation to the stars closest to Geminga did not improve on the residuals.

4.2 Removing the outliers: justification

Some stars did not transform well for several reasons. The PSF becomes distorted as stars approach saturation, and at the faint end counting statistics dominate the uncertainties. There is a middle range that proved optimal for the transformations. Stars that were outside that range had poorly understood positions. However, from Fig. 1 it is clear that even in the well-defined range there are a few outliers. This type of deviation implies that there is a problem with the image of the star on the chip. Most likely, the stars were lying on a bad pixel or were too close to a cosmic ray and therefore were poorly understood.

4.3 Transforming between epochs

Once we had aligned the positions at a given epoch, we averaged the 12 solutions. This reduced the positional uncertainty by $\sqrt{12}$. We then transformed the later epoch images to the average epoch 1 positions, and produced average positions at each epoch. We implemented the same cut-offs that we had for the initial epoch 1 internal transformation: we eliminated stars that had an instrumental magnitude outside the range of $-14 < M < -8$, and those that transformed with residuals >0.05 pixels.

Our first and our third epoch images were closely aligned, as were our second and fourth epoch images, but the two sets were rotated ~180° relative to each other. The transformations for stars at corners of the WFC chips between the rotated images have fairly large residuals. This cross chip transformation is a known issue (Anderson 2006). The distortion solution holds well for the main part of the chips but breathing can have a large effect at the edges.

We initially assumed that all reference stars had negligible proper motion and parallax. Non-negligible proper motion and parallax will cause stars to fail the transformation with large residuals. Of the 134 stars common to the four epochs, 45 transform well-enough to define our astrometric grid.

As a check, we transformed each image at each epoch into itself (again arbitrarily choosing one image as standard) as we had done for epoch 1 and then transform those averaged positions back into our first. We found that the internal residuals at each epoch were very similar. We found that the solution converged on similar stars in both techniques. There was no statistically significant difference between these two approaches to the data. We took this as further evidence that we fully understood the uncertainties in the transformations between epochs.

4.4 Least squares solution for the parallax and proper motion

With the corrected positions for all stars at all four epochs in hand, we could solve for the proper motion and parallax of all the stars in the field. The solution that we obtained in the previous steps was applied to each star in the field, even those that had been eliminated earlier for poor transformations. Here we are only concerned with our target star Geminga so we report only the parallax and proper motion of that object. We used a least squares robust estimator called Gaussfit (Jefferys et al. 1987) to solve the equations for the proper motion and parallax of our target star.

Parallax and proper motion are coupled, and, since we have four epochs with proper motion in both X and Y, we are over-constrained for these three components. We used the transformed positions of Geminga (obtained using the plate constants solved for above) and solved the following equations:

$$X = X_1 + \mu_x * tc + pi_x, \tag{3}$$

$$Y = Y_1 + \mu_y * tc + pi_y. \tag{4}$$

X_1 and Y_1 are the epoch 1 positions, X and Y are the positions at later epochs. tc is the time between epoch 1 and a subsequent epoch, pi_x and pi_y contain the parallactic factors, and μ_x and μ_y are the proper motion components. We used JPL's DE405 software to obtain the parallactic factor for each epoch.

Gaussfit, a program written to do this type of astrometric solution, solves for the proper motion and parallax of our target star Geminga. We kept all of our numbers in pixels for the transformations as we did not want to introduce further error by converting to RA and DEC values. We use a plate scale of the WFC determined by Anderson's field solution of 0.04972 ± 0.00002 arcsec/pixel. The uncertainty of the position of Geminga is ± 0.025 pixels (1.2 mas) from the internal scatter of the 12 positions at each epoch. The uncertainty of the transformation is ± 0.01 pixels (0.5 mas). The net uncertainty is 0.027 pixels (1.3 mas). The measured parallactic displacement of Geminga is 0.080 ± 0.027 pixels, for a parallax of $\pi = 4.0 \pm 1.3$ mas, and a distance of 250^{+120}_{-62} pc. The proper motion $\mu = 178.2 \pm 1.8$ mas/year at a position angle of 52.9 ± 0.4 deg. The components of the proper motion are $\mu_x = 2.860 \pm 0.025$ pix/year and $\mu_y = 2.160 \pm 0.025$ pix/year. Using the above plate scale this corresponds to:

$$\mu_\alpha = 142.2 \pm 1.2 \text{ mas/year} \tag{5}$$

and

$$\mu_\delta = 107.4 \pm 1.2 \text{ mas/year}. \tag{6}$$

Caraveo et al. (1996) reported a comparable proper motion but at a position angle of $55.0 \pm 2.0°$. The μ_α component of proper motion is within 1.0 σ of the Caraveo result and the μ_δ result is within 2.5σ.

5 Conclusions

The radius is usually estimated by modeling the spectral energy distribution with two blackbodies to yield an angular radius, and then multiplying by some assumed distance. A stellar atmosphere would modify the emission characteristics of the neutron star and yield a larger radius for a given luminosity however we assume a black body model as that is the standard approach for this particular object. Most authors have either used the Caraveo et al. (1996) 157 pc distance, or have assumed a distance of 200 pc. Caraveo et al. (1996) report a neutron star radius of 10 km with no quoted uncertainty using their calculated distance and modeling Geminga. De Luca et al. (2005) report a radius of 8.6 ± 1.0 km at 157 pc. At our revised distance of 250 pc and including the error on our measurement, the De Luca result would then increase to $13.7^{+6.8}_{-3.7}$ km. Kargaltsev et al. (2005) report a neutron star radius of 12.9 ± 1.0 for a distance of 200 pc, based on fits to the XMM-Newton X-ray spectra. Using our 250 pc distance that radius would increase to $16.1^{+7.8}_{-4.2}$ km.

The $\pm 30\%$ uncertainty on the parallax translates into a minimum $^{+48}_{-25}\%$ uncertainty in the radius. Unfortunately this is not quite good enough to usefully constrain the Equation of State. For an assumed mass of $1.4 M_\odot$, the range of acceptable radii include most equations of state, including those incorporating strange quark matter (Lattimer and Prakash 2005).

Geminga is thought to be a runaway from the Orion OB1a association (Smith et al. 1994), at a distance of 330 pc and an age of 10 Myr. Our derived distance is consistent with this. The observed transverse velocity V_T corresponds to a projected velocity of 205^{+90}_{-47} km/s. The three dimensional space velocity, assuming a point of origin, carries Geminga into the heart of the Orion OB1a association in about 1.5 Myr. An origin in the more distant but younger Orion OB1b association requires a longer trek. This is yet another case where the spindown age of a pulsar seems to be significantly shorter than its kinematic age.

The 30% uncertainty is disappointing, in that we cannot usefully constrain the astrophysics of Geminga. But we are not yet done. The results reported herein are based solely on the four ACS/WFC images. We are currently re-analyzing the earlier WFPC2 images to see if we can successfully

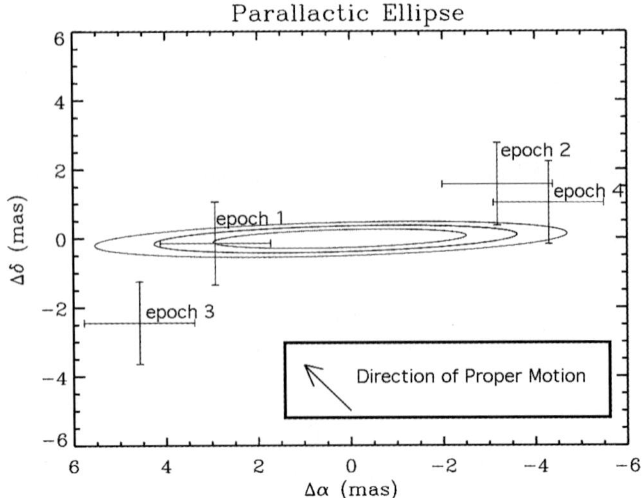

Fig. 4 The parallactic ellipse, $\pi = 4.0 \pm 1.2$ mas. The proper motion $\mu = 178.2 \pm 1.0$ mas/year has been subtracted. The three ellipses are the quoted solution (250 pc) and the $\pm 1\sigma$ limits. The direction of the proper motion is shown at the position angle of 52.9°. The length of the arrow is ~2 mas which at the given proper motion would correspond to 4.1 days of motion

transform them into our reference frames. If we can, we will have just over a 10 year baseline for the proper motion, which will significantly reduce its uncertainty. Reducing the uncertainty on the proper motion will yield an improvement in the measurement of the parallax of Geminga.

Acknowledgements This research has profited from discussions with J. Lattimer, M. Prakash, and G. Pavlov. J.F. learned a great deal about astrometry at the Michelson summer school held at Yale, and is grateful for the support offered by the organizers. This research has been supported by a grant from the Space Telescope Science Institute.

References

Anderson, J.: Instrument Science Report ACS ISR 2006-01 (2006)
Anderson, J., King, I.R.: Publ. Astron. Soc. Pac. **111**, 1095 (1999)
Burwitz, V., Haberl, F., Neuhäuser, R., et al.: Astron. Astrophys. **399**, 1109 (2003)
Caraveo, P.A., Bignami, G.F., Mignani, R., et al.: Astrophys. J. **461**, L91 (1996)
De Luca, A., Caraveo, P.A., Mereghetti, S., et al.: Astrophys. J. **623**, 1051 (2005)
Gänsicke, B.T., Braje, T.M., Romani, R.W.: Astron. Astrophys. **386**, 1001 (2002)
Haberl, F., Turolla, R., de Vries, C.P., et al.: Astron. Astrophys. **451**, L17 (2006)
Jefferys, W.H., Fitzpatrick, M.J., McArthur, B.E.: Celest. Mech. **41**, 39 (1987)
Kaplan, D.L., van Kerkwijk, M.H., Anderson, J.: Astrophys. J. **571**, 447 (2002)
Kargaltsev, O.Y., Pavlov, G.G., Zavlin, V.E., et al.: Astrophys. J. **625**, 307 (2005)
Kulkarni, S.R., van Kerkwijk, M.H.: Astrophys. J. **507**, L49 (1998)
Lattimer, J.M., Prakash, M.: Phys. Rev. Lett. **94**(11), 111, 101 (2005)
Pavlov, G.G., Zavlin, V.E., Truemper, J., et al.: Astrophys. J. **472**, L33 (1996)
Pons, J.A., Walter, F.M., Lattimer, J.M., et al.: Astrophys. J. **564**, 981 (2002)
Rajagopal, M., Romani, R.W.: Astrophys. J. **461**, 327 (1996)
Smith, W., Cunha, K., Plez, B.: Astron. Astrophys. **281**, L41 (1994)
Turolla, R., Zane, S., Drake, J.J.: Astrophys. J. **603**, 265 (2004)
Walter, F.M., Lattimer, J.M.: Astrophys. J. **576**, L145 (2002)

Astrophys Space Sci (2007) 308: 231–238
DOI 10.1007/s10509-007-9369-z

ORIGINAL ARTICLE

The puzzling X-ray source in RCW103

A. De Luca · P.A. Caraveo · S. Mereghetti · A. Tiengo · G.F. Bignami

Received: 14 July 2006 / Accepted: 14 September 2006 / Published online: 24 March 2007

Abstract 1E 161348-5055 (1E) is a compact object lying at the center of the 2000 year old Supernova Remnant (SNR) RCW103. Its original identification as an isolated, radio-quiet neutron star has been questioned in recent years by the observation of a significant long-term variability, as well as by reports of a possible periodicity at ~6 hours. Here we report conclusive evidence for a strong (nearly 50%) periodic modulation of 1E at 6.67 ± 0.03 hours, discovered during a long (90 ks) XMM-Newton observation performed in August 2005, when the source was in a "low state". The source spectrum varies along the 6.67 hr cycle. No fast pulsations are seen. 1E could be a very young binary system, possibly composed of a compact object and a low-mass star in an eccentric orbit. This would be the first example of a low-mass X-ray binary (LMXB) associated with a SNR, and thus the first LMXB for which we know the precise birth date, just 2000 years ago. Alternatively, if it is an isolated neutron star, the unprecedented combination of age, period and variability may only fit in a very unusual scenario, featuring a peculiar magnetar, dramatically slowed-down over 2000 years, possibly by a supernova debris disc.

Keywords X-rays: stars · Stars: neutron · X-rays: binaries

PACS 98.38.Mz · 97.60.Jd · 97.80.Jp

A. De Luca (✉) · P.A. Caraveo · S. Mereghetti · A. Tiengo · G.F. Bignami
INAF/IASF Milano, Via Bassini 15, 20133 Milano, Italy
e-mail: deluca@iasf-milano.inaf.it

G.F. Bignami
Istituto Universitario di Studi Superiori (IUSS) c/o Collegio Giasone del Maino, Via Luino 4, 27100 Pavia, Italy

1 Introduction

The X-ray point source 1E was discovered by Tuohy and Garmire (1980) with the Einstein observatory close to the geometrical center of the very young (~2000 yr, Carter et al. 1997) shell-type SNR RCW 103. The association of the point source with the SNR is very robust on the basis of the good positional coincidence (1E lies within ~20″ of the SNR center); furthermore, HI observations of this region (Reynoso et al. 2004) pointed to a spatial correlation of the two objects in view of their similar distance (~3.3 kpc). Historically, 1E was the first radio-quiet neutron star candidate found in a young SNR, characterized by a thermal X-ray spectrum, no counterparts at other wavelengths, no pulsations, no non-thermal extended emission. Since then, a handful of similar sources have been discovered inside young SNRs and dubbed, as a class, "Central Compact Objects" (CCOs: Pavlov et al. 2002).

What makes 1E unique among CCOs is its very peculiar temporal behaviour. First, the source shows significant long-term flux variations. A factor 10 variability was already evident within the combined Einstein/ROSAT/ASCA dataset (Gotthelf et al. 1999). More recently, Chandra ACIS observations showed an increase of the flux by a factor ~60 (from $\sim 8 \times 10^{-13}$ erg cm^{-2} s^{-1} to $\sim 5 \times 10^{-11}$ erg cm^{-2} s^{-1} in 0.5–8 keV) between September 1999 and February 2000 (Garmire et al. 2000a); subsequent monitoring showed a fading to the level of $\sim 10^{-11}$ erg cm^{-2} s^{-1} (Sanwal et al. 2002). Moreover, the first Chandra observation (~16 ks long) hinted a ~6 hours periodicity (Garmire et al. 2000b). However, such periodicity was not clearly recognized during a second observation (23 ks) in February 2000, when the source was found in very high state, but it was possibly seen again in a subsequent, longer 50 ks observation which showed a very complex light curve including "dips" (Sanwal et al. 2002). A first observation with XMM-Newton (19 ks long) was performed on September 2001. Using such XMM data, Becker and Aschenbach (2002) did not find any periodicity, but confirmed the remarkable variability of the source and its peculiar light curve, with a shallow intensity dip.

On the optical side, deep observations of the field of 1E have been recently performed in the near-infrared band with VLT/FORS1, VLT/ISAAC as well as with HST/NICMOS. Pavlov et al. (2004) reported three faint potential IR counterparts, with $H \sim 22$–23. An upper limit of $R > 25.6$ was set by Wang and Chakrabarty (2002), based on Magellan-1 6.5 m telescope observations. No firm conclusions about the association of the IR sources with 1E could be drawn.

In order to settle the periodicity issue, as well as to better assess the phenomenology of 1E, we performed a deep (90 ks), uninterrupted observation with XMM-Newton in 2005. We report here on the outcome of such observation. The phenomenology of 1E as observed in 2005 is described in Sect. 2. In Sect. 3 we give the results of our re-analysis of the previous XMM-Newton observation performed in 2001 to allow for a consistent assessment of the source evolution. In Sect. 4 the long-term flux history of 1E is studied, using public Chandra monitoring observations as well as historical measurements with the Einstein, ASCA and ROSAT satellites. Finally, in Sect. 5, possible interpretations for 1E are considered. A more complete discussion of the nature of the source may be found in De Luca et al. (2006).

2 1E as observed with XMM-Newton: the 2005 low state

Our new, deep XMM-Newton observation of 1E was performed on 2005, August 23 (07:11 UT). The total observing time was of 89 ks. The EPIC/pn detector (Strüder et al. 2001) was operated in Small Window mode ($4' \times 4'$ field of view, 5.6 ms time resolution) with the thin optical filter; the EPIC/MOS cameras (Turner et al. 2001) were set in their Full Frame mode ($15'$ radius field of view, 2.6 s time resolution) with the medium optical filter.

Observation Data Files were retrieved from the XMM-Newton archive and were processed with the XMM-Newton Science Analysis Software (SASv6.5.0) using standard pipeline tasks (*epproc* and *epproc* for the pn and MOS, respectively). Photon time of arrivals were converted to the solar system barycenter. The observation was affected by a few particle background episodes (soft proton flares) of rather low intensity. Thus, no time-screening was performed, but a stringent spatial selection was applied in order to select source events. An extraction radius of $15''$ (corresponding to an encircled energy fraction of $\sim$0.65 at 2 keV), allows to minimize the contamination both from particle background and from the diffuse emission of the bright host supernova remnant RCW103. Background events were extracted from an appropriate $25''$ radius circular region[1] located at the same distance from the readout node as the source region on the pn detector plane. With the adopted selection, the source background-subtracted count rates in the 0.5–8.0 keV range are 0.382 ± 0.003 cts s^{-1}, 0.132 ± 0.001 cts s^{-1} and 0.139 ± 0.001 cts s^{-1} in the pn, MOS1 and MOS2, respectively, background accounting for $\sim$35% (pn) and $\sim$30% (MOS) of the total counts in the source region.

2.1 Timing phenomenology

Background-subtracted light curves in different energy ranges were extracted for each detector separately, selecting source and background events as explained above.

The combined pn/MOS light curve in the 0.5–8 keV range is shown in Fig. 1 (lower curve), where the count rates have been converted to flux units using appropriate conversion factors according to the source time-averaged spectrum (see next section). Our long and uninterrupted observation shows unambiguously that the source is indeed periodical. A nearly sinusoidal pulse shape and a very large flux modulation between the peak and the minimum are apparent in Fig. 1. A fit to the light curve with a sin function ($\chi^2_\nu = 2.67$, 84 d.o.f.) allowed to measure the source period to be $P = 24\,000 \pm 100$ s ($\sim$6.7 hours). We note that a better fit to the pulse shape may be obtained adding a second harmonic to the fundamental sin function ($\chi^2_\nu = 1.63$, 81 d.o.f.), which in any case does not alter the estimate of the period ($P = 24\,000 \pm 100$ s). The Pulsed Fraction (PF) in 0.5–8 keV, defined as $\mathrm{PF} = F_{\mathrm{max}} - F_{\mathrm{min}}/F_{\mathrm{max}} + F_{\mathrm{min}}$, where F_{max} and F_{min} are the observed background-subtracted count rates at maximun and minimum, respectively, is $\mathrm{PF} = 43.5 \pm 1.8\%$. While the pulse shape does not change significantly (see the folded profiles in Fig. 2, right panel), a remarkable increase of the pulsed function as a function of the energy is observed: $\mathrm{PF} = 37.1 \pm 2.8\%$ in the 0.5–2 keV range, $\mathrm{PF} = 56.8 \pm 2.9\%$ in the 2–8 keV range. The hardness ratio, also plotted in Fig. 2 (right panel) shows clearly that the source emission is softer at the minimum and harder at the maximum of the pulse. Using pn data (time resolution of

[1] The bright and patchy supernova remnant emission requires a particular care to be devoted in choosing the background region. The choice is also constrained by the limited ($4' \times 4'$) field of view of the pn camera. We selected a circular region located at the same distance from the readout node as the source region, centered on 16:17:42.45, −51:02:38.3, having a surface brightness comparable to the one of the region surrounding the target, as estimated from Chandra/ACIS images. The same region was used for all the EPIC cameras in both the 2005 and the 2001 observations. We note that with different background selections, the soft ($E < 1.5$ keV) part of the spectrum of the source may slightly change, which yields somewhat different best fit spectral parameters (e.g. N_{H}). In any case, the main results reported in this work do not depend on the choice of the background region.

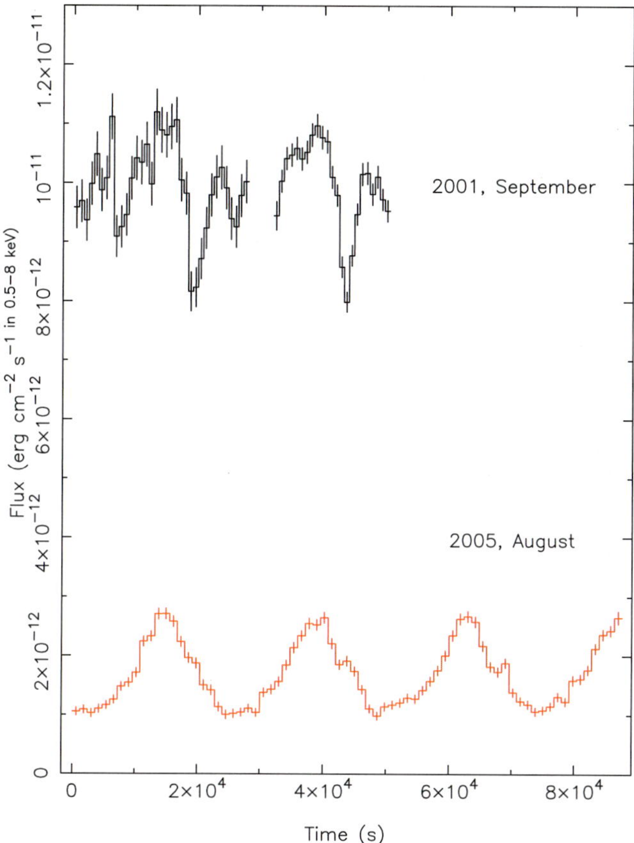

Fig. 1 Background-subtracted light curves of 1E as observed with XMM-Newton in 2005 and in 2001. Time is measured since the start of the observations. Observation starting times have been artificially aligned; no folding has been performed. The long 90 ks observation of 2005 shows unambiguously the source 6.67 hr (24 ks) periodicity. The same periodicity may be seen in the 2001 data, in spite of the ~3 ks interruption due to the satellite slew. Note the large variation in the source phenomenology between the active state and the low state. Adapted from De Luca et al. (2006)

5.6 ms) we performed a deep search for faster periodicities down to periods as small as 12 ms. We did not find any significant signal. Assuming a sinusoidal pulse shape, we may set an upper limit of ~10% on the pulsed fraction (99% confidence level).

2.2 Spectral phenomenology

Spectra for source and background were extracted for each detector using the regions described above and rebinned in order to have at least 30 counts per channel and to oversample the energy spectral resolution of the instruments by a factor 3. Ad-hoc response matrices and effective area files were generated using the SAS tasks *rmfgen* and *arfgen*. The spectral analysis was performed using Xspec v11.3. The pn and MOS spectra were fitted simultaneously in the 0.7–8.0 keV range. Errors are quoted at the 90% confidence level for one parameter of interest unless otherwise specified. Single-component models (including the effects of interstellar absorption) do not yield acceptable fits: while a power law is totally inadequate ($\chi^2_\nu > 2.3$, 493 d.o.f.), a blackbody curve yields a better description ($\chi^2_\nu = 1.45$, 493 d.o.f.), although significant deviations from the model are seen at $E > 3$ keV. The fit improves adding a second component to the blackbody curve, either a second, hotter blackbody ($\chi^2_\nu = 1.18$, 490 d.o.f.), or a power law ($\chi^2_\nu = 1.19$, 490 d.o.f.). The best fit parameters are reported in Table 1. The observed flux in the 0.5–8 keV range is of $\sim 1.7 \times 10^{-12}$ erg cm^{-2} s^{-1}.

In order to investigate the spectral variability along the 6.7 hours period, we extracted phase-resolved spectra, selecting the phase intervals corresponding to the minimum, the rise, the peak and the decline of the pulse. Focusing to the minimum and peak spectra, assuming the blackbody plus power law model, as a first step we kept all parameters fixed to their best fit values for the phase-averaged spectrum and we allowed only an overall normalization factor to vary. This yielded unacceptable fits to the data ($\chi^2_\nu > 3.75$, 632 d.o.f.), residuals clearly showing that the spectrum of the source is softer at the minimum and harder at the peak.

The spectral variation turned out to be rather complex: a better fit could be obtained allowing the normalizations of the two spectral component to vary independently ($\chi^2_\nu \sim 1.4$, 628 d.o.f.); the fit further improves ($\chi^2_\nu \sim 1.2$, 626 d.o.f.) allowing either the BB temperature or the PL photon index to vary. Using the double blackbody model, e.g., we found at minimum $N_{\rm H} = (5 \pm 1) \times 10^{21}$ cm^{-2}, $kT_{\rm BB1} = 0.47 \pm 0.03$ keV, $R_{\rm BB1} = 560 \pm 60$ m, $kT_{\rm BB2} = 1.4 \pm 0.7$ keV, $R_{\rm BB2} < 80$ m; at the peak $N_{\rm H} = (9 \pm 1) \times 10^{21}$ cm^{-2}, $kT_{\rm BB1} = 0.54 \pm 0.02$ keV, $R_{\rm BB1} = 750 \pm 40$ m, $kT_{\rm BB2} = 1.3 \pm 0.4$ keV, $R_{\rm BB2} = 40 \pm 20$ m.

3 1E as observed with XMM-Newton: the 2001 active state

The previous XMM-Newton observation of 1E started on 2001, September 3 (18:45 UT) and lasted for 19.9 ks. The instrumental setup was the same selected for the 2005 observation. Additional data on 1E were serendipitously collected during a 29.0 ks long observation devoted to the energetic pulsar PSR J1617-5055, located ~7′ North of RCW103: our target was imaged by the MOS cameras (operated in Full frame mode with the Medium optical filter), while no useful data on 1E were recorded by the pn detector, operated in Timing mode. Such observation (started on September 3, 10:01 UT) was performed immediately before the one devoted to 1E. Thus, the MOS data on 1E cover a ~50 ks time span, with an interruption of ~3 ks due to the satellite slew. The data reduction and analysis was performed as described for the 2005 observation. We used the same selections to extract source and background events. The resulting background-subtracted count rates (0.5–8 keV) are

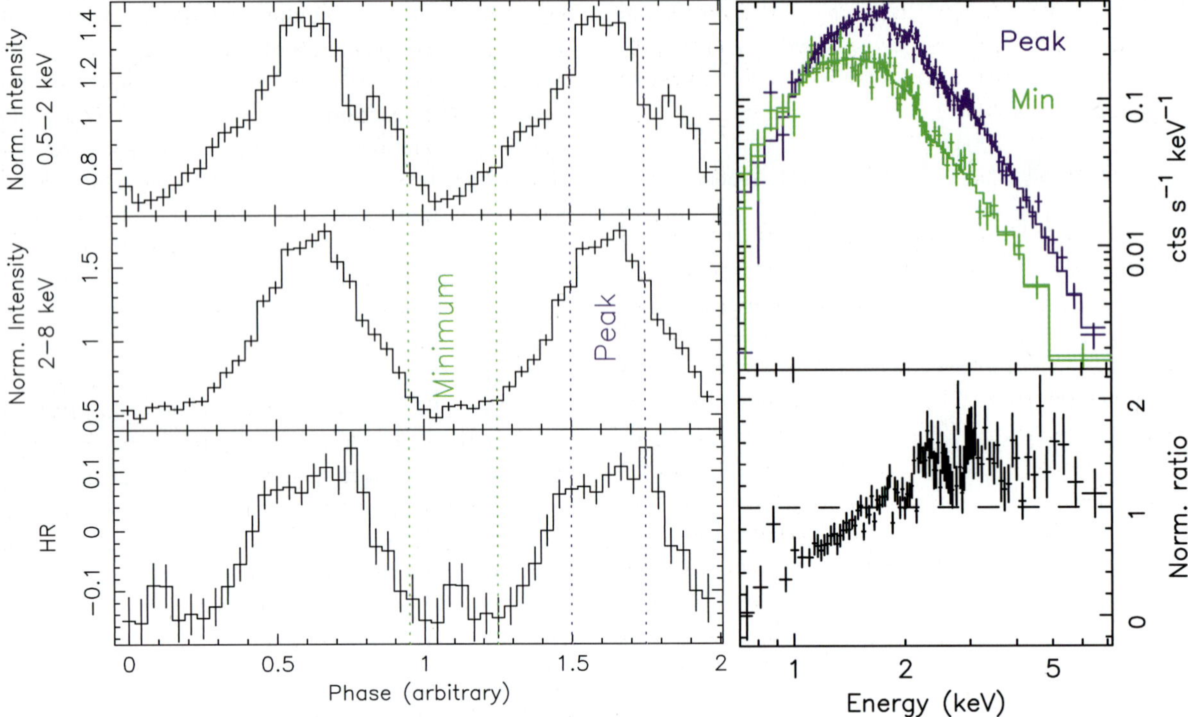

Fig. 2 *Left panel*: Background-subtracted, folded ($P = 24$ ks) light curves in the Soft (0.5–2 keV) and Hard (2–8 keV) energy ranges for 1E as observed in 2005, together with the corresponding hardness ratio, defined as (Hard–Soft)/(Hard+Soft). *Right panel*: Phase-resolved energy spectra of 1E extracted from phase intervals corresponding to the peak and to the minimum of the light curve, as well as ratio between the peak spectrum and the renormalized best fit model for the minimum. From De Luca et al. (2006)

Table 1 Best fit parameters for the time-averaged spectra. Two spectral models are used, consisting of the sum of two blackbody curves (BB+BB) and the sum of a blackbody curve and of a power law (BB+PL). The values obtained for the low state (2005) and for the high state (2001) are shown. Errors are at 90% confidence level for one parameter of interest. A distance of 3.3 kpc has been assumed

	BB+BB (2005)	BB+PL (2005)	BB+BB (2001)	BB+PL (2001)
N_H (10^{22} cm^{-2})	0.65 ± 0.04	0.85 ± 0.20	1.08 ± 0.05	1.35 ± 0.15
kT_1 (keV)	0.511 ± 0.015	0.514 ± 0.009	0.503 ± 0.020	0.540 ± 0.008
R_1 (m)	610 ± 35	570 ± 30	1560 ± 120	1310 ± 50
kT_2 (keV)	$1.0^{+0.4}_{-0.2}$	–	$0.93^{+0.13}_{-0.09}$	–
R_2 (m)	35^{+22}_{-12}	–	220^{+100}_{-60}	–
Γ	–	$2.9^{+0.4}_{-0.9}$	–	$3.0^{+0.3}_{-0.5}$
$F^{obs}_{0.5-8}$ [a] (erg cm^{-2} s^{-1})	$(1.7 \pm 0.1) \times 10^{-12}$	$(1.7 \pm 0.1) \times 10^{-12}$	$(9.9 \pm 0.6) \times 10^{-12}$	$(9.9 \pm 0.6) \times 10^{-12}$
L_{BB1} [b] (erg s^{-1})	3.0×10^{33}	2.9×10^{33}	2.0×10^{34}	1.9×10^{34}
L_{BB2} [c] (erg s^{-1})	3.8×10^{32}	–	4.6×10^{33}	–
L_{PL} [d] (erg s^{-1})	–	1.6×10^{33}	–	1.7×10^{34}
χ^2_ν	1.18	1.19	1.09	1.14
d.o.f.	490	490	517	517

[a]Observed flux, 0.5–8 keV (statistical error at 68% confidence level)

[b]Bolometric luminosity of first blackbody

[c]Bolometric luminosity of second blackbody

[d]Luminosity of power law component, 0.5–8 keV

1.936 ± 0.013, 0.692 ± 0.006 and 0.701 ± 0.006 for the pn, MOS1 and MOS2, respectively (background accounting for $\sim$7% of the counts in the source regions); such values are $\sim$5 time higher than in 2005.

3.1 Timing phenomenology

Background-subtracted light curves were generated for each detector in different energy ranges. In order to combine the data collected in the two observations into a unique, 50 ks long light curve, we computed count-rate to flux conversion factors according to the observed source spectrum (see below) and accounting for the significant vignetting effects affecting the first section of data. The resulting light curve (0.5–8 keV) is shown in Fig. 1 (upper curve). In spite of the 3 ks interruption, the source periodicity may be clearly recognized, the light curve covering $\sim$2 periods. The pulse profile is very complex, with a narrow dip (also noted by Becker and Aschenbach 2002) $\sim$2500 s long, occurring after the main intensity peak and a peak-to-dip modulation of $\sim$25%. The pulsed fraction, defined as in Sect. 2.2, is PF = 11.7 ± 1.4%, remarkably lower than in 2005. We note in any case that the modulation in flux has a semi-amplitude of $\sim 10^{-12}$ erg cm^{-2} s^{-1}, which is very similar to the value observed in 2005. The central times of the two dips (evaluated by fitting a Gaussian curve to their profile) were used to compute the period, which turned out to be of $24\,200 \pm 300$ s, consistent with the value measured in 2005. A secondary, less pronounced dip is also apparent in MOS data, separated by $\sim$0.5 in phase with respect to the deepest one. The two dips identify two broad peaks per period which show in their turn some substructure. Energy-resolved light curves folded at the 24 ks period show the deepest dip to be shallower and broader in the 2–8 keV range with respect to the 0.5–2 keV range. A hardness ratio study gives some evidence for a spectral hardening during dip. No faster (down to 12 ms) modulation was found with an upper limit of $\sim$10% on the pulsed fraction.

3.2 Spectral phenomenology

As for the 2005 observation, we obtained the best fit to the data using double-component models, either the sum of two blackbody curves ($\chi^2_\nu = 1.09$, 517 d.o.f.), or the sum of a blackbody curve to a power law ($\chi^2_\nu = 1.14$, 517 d.o.f.). The best fit parameters are given in Table 1.

The time-averaged observed flux is $\sim 10^{-11}$ erg cm^{-2} s^{-1}, $\sim$6 times higher than in 2005. The spectrum of the source in its "high state" of 2001 is remarkably harder than in the "low state" of 2005 (see Fig. 3). To reproduce the complex spectral changes, a variation of both spectral components is required, both using the double blackbody model and using the blackbody plus power law model. The spectra of the SNR extracted in the two observations (using the

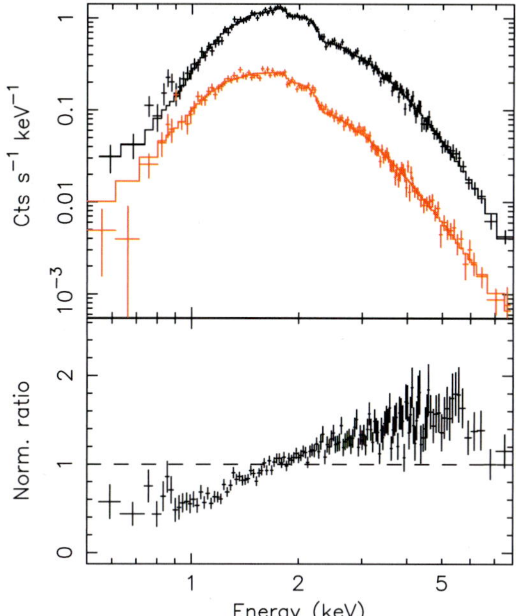

Fig. 3 Spectrum of 1E as observed in 2001 (*black data points*), compared to the spectrum as observed in 2005 (*red data points*), as well as ratio between the 2001 spectrum and the renormalized 2005 best fit model. From De Luca et al. (2006)

background region described above) were also compared, but no changes were found.

In order to investigate the spectral variation along the pulse phase in the high state, we extracted time-resolved spectra selecting two 2500 s time intervals centered on the deepest dip and on the peak, respectively. The time-integrated best fit model was assumed as a template. A simple renormalization of such model does not fit the data ($\chi^2_\nu \sim 1.6$, 350 d.o.f.). The spectrum corresponding to the dip is markedly harder. The spectral evolution cannot be described by extra absorption; using the double blackbody model, e.g., a variation in the main blackbody emitting surface or temperature is required, hinting for a complex spectral variation. In any case, a more detailed modelling is hampered by the limited statistics.

4 The long-term flux evolution

In order to investigate the flux variation of 1E on the years time scale, we have analyzed public Chandra/ACIS monitoring ($\sim$4 ks) observations performed between January 2001 and January 2005 (see also Sanwal et al. 2002), together with the $\sim$20 ks ACIS observations of September 1999 and February 2000. "Level 1" event files were retrieved from the public Chandra archive and processed using standard pipelines (*acis_process_events*) within the Chandra Interactive Analysis of Observations (CIAO v3.2.1) software

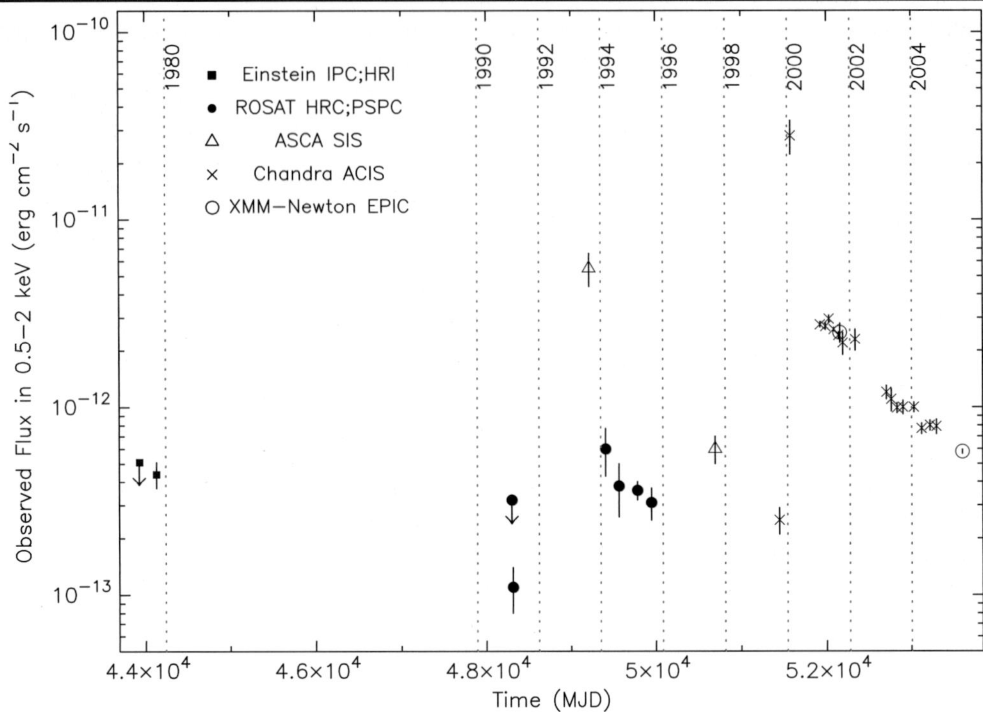

Fig. 4 Secular flux evolution of 1E along more than 25 years, as derived from Einstein, ROSAT, ASCA, Chandra and XMM-Newton observations. Adapted from De Luca et al. (2006)

package. Source events were selected from appropriate circular regions corresponding to encircled fractions of ~0.9. In February 2000 the source was found in a very high state and data are heavily affected by pile-up. To overcome the problem, we excluded the PSF core, extracting source events from an annulus with inner and outer radii of 2.5″ and 6″, respectively. Background events were extracted in all cases from 20″ radius regions located close to the target.

We extracted spectra for source and background and we generated appropriate response matrices and effective area files using the *psextract* script. Source spectra were rebinned in order to have at least 30 counts per channel. Each spectrum was fitted in Xspec in the 0.8–8 keV range using a double blackbody curve modified by interstellar absorption. The observed flux in the 0.5–2 keV range was then computed using the best fit model.[2] The resulting fluxes have been plotted in Fig. 4 as a function of the time. We reported in the same figure the XMM-Newton values presented in the previous sections, as well as the "historical" measurements performed with the Einstein, ROSAT and ASCA satellites (Gotthelf et al. 1999) since the discovery of the source in 1979.

[2]In the case of the "very high state" of February 2000, a significant error (>10%) is possibly introduced by the energy dependence of the Chandra Point Spread Function (PSF), since the source spectrum was extracted by the PSF wings. To account for this problem, we assumed a conservative, 20% error on the February 2000 flux. A more sophisticated analysis, using the model described by Davis (2001), is beyond the scope of this work.

The very large flux increase (~2 orders of magnitude) occurred between September 1999 and February 2000 (Garmire et al. 2000a) is clearly visible in the plot. A continuous decline in flux over a 5.5 years time span is also apparent. In August 2005 the source is approaching the flux level of the very low 1999 state. At least another episode of major variation in flux is seen in historical data (Gotthelf et al. 1999), starting a few months before the ASCA observation of August 1993. Thus, the source outbursts are possibly recurrent on a few year time scale.

5 Conclusions

The long 90 ks XMM-Newton observation of 1E of August 2005 caught the source in a low-state with respect to its secular flux evolution and provided conclusive evidence of the source periodicity, with $P = 6.67$ hr. The flux modulation is very strong (pulsed fraction of ~45%), with a single-peaked, nearly sinusoidal pulse profile. A remarkable spectral variation is seen along the 6.67 hr cycle, emission being harder at the peak and softer at the minimum. The source behaviour in the active state of 2001 was remarkably different, with a ~6 times higher average flux and a harder spectrum. The pulse shape was much more complex, including 2 dips per period, with a smaller pulsed fraction (~12%), but a similar pulsed flux. Such properties, together with the very young age of the source and with the underluminous nature of its optical/IR counterpart, settle the case for a unique phenomenology. 1E defies an easy identification with any currently known class of X-ray sources, forcing to consider unusual scenarii (see De Luca et al. 2006, for a more detailed discussion).

As a first possibility, the 6.67 hr periodicity could be the orbital period of a binary system featuring a compact object generated by the recent SN explosion, as well as a normal companion star, for which the optical/IR data set stringent constraints: only a M4 dwarf with $M < 0.5 M_{\odot}$ is allowed. 1E would be the first case of a low-mass X-ray binary (LMXB) associated with a SNR, and thus it would be by far the youngest known member of such a class of sources. Indeed, the very unusual phenomenology, remarkably different from that of standard LMXBs (e.g., 1E has a much lower luminosity, stronger flux modulation and spectral variation along the orbit, as well as a very unusual secular variability) could be explained by the very young age. 1E could be in an early and possibly short-lived phase of a LMXB life, dominated by the effects of an expected (Kalogera 1996) initial orbital eccentricity. This could drive a "double accretion process". First, the compact object should capture a fraction of the companion star's wind. This effect would be strongly modulated along an eccentric orbit, as a result of the variation in the relative velocity of the companion star wind and of the compact object. A simple computation shows that the accretion rate would vary by a factor ranging from $\sim$3 to $\sim$9, assuming an eccentricity in the 0.2–0.5 range, for a system featuring a 1.4 $M_{\odot}$ neutron star and a dwarf star of 0.4–$0.2 M_{\odot}$. The modulation in the accretion rate has a single peak occurring in the receding part of the orbit, away from periastron (anomaly $\sim$120°). This effect could account for the nearly sinusoidal flux modulation seen in the low state (which is also possibly present in the active state). Second, under the same assumptions for the binary system configuration, the first Lagrangian point could come very close to the companion star surface in a narrow range of orbital phases near periastron. This could drive a non-steady Roche-lobe overflow, possibly triggered by instabilities in the companion star, resulting in the formation of an accretion disc. The presence of a disc could account for the long-term variations in flux (disc instabilities? mass ejections from the companion?), as well as for the dips in the light curve of the high state (occultations by disc structures?). Much work would be needed in order to test the binary picture in its details. We note that, if the compact object is a neutron star, accretion may occur only if the ejector and propeller barriers may be overcome (Illarionov and Sunyaev 1975), which implies a very low magnetic field and/or a slow spin period for the 2 kyr old neutron star. The observed luminosities imply (Davies and Pringle 1981) $P \sim (0.35\text{–}2.5) B_{10}^{6/7}$ s, where B_{10} is the magnetic field in units of 10^{10} G. Such properties are somewhat different with respect to standard expectations for a 2000 year old NS. On the other hand, X-ray production by a black hole accreting at low rate could be problematic (see e.g. Narayan 2005).

As an alternative, 1E could be an isolated compact object. In this case, a very unusual scenario is required (see De Luca et al. 2006, for a more complete discussion) to explain its unique phenomenology: 1E could be a neutron star with an ultra-high magnetic field, i.e. a magnetar (Woods and Thompson 2006), now rotating at 6.67 hr. Indeed, most of the observed properties of 1E could easily fit in a magnetar scenario. The spectrum of 1E and its luminosity are very similar to those of most AXPs. The secular flux variability of 1E is comparable to the long lasting outburst of the transient AXP XTE J1810–197 (Ibrahim et al. 2004; Gotthelf and Halpern 2005). Moreover, magnetars show long-term variations in spectrum, pulse shape and pulsed fraction similar to the ones observed for 1E (see, e.g., the cases of the AXP 1E 1048.1-5937 and of SGR 1900+14, Tiengo et al. 2005; Woods and Thompson 2006). All known magnetars, however, spin $\sim$1000 times faster than 1E, with periods clustered in the 5–12 s range. A very efficient slowing-down mechanism is required to quench the rotation of 1E over a 2000 y time scale. A viable possibility could be the propeller effect due to the material of a supernova fallback disc. Indeed, using the models by Chatterjee et al. (2000) and Francischelli et al. (2002), we estimated that a magnetar featuring a 5×10^{15} G magnetic field could be slowed down to a spin period of $\sim$6.7 hr in $\sim$2000 yr by a fallback disc with a mass of $3 \times 10^{-5} M_{\odot}$. This requires a birth rotation period greater than 300 ms, to avoid an early ejector phase which would have pushed the disc away from the neutron star. A birth period $>$300 ms seems too long to fit in the most popular theory for the origin of the huge magnetar magnetic fields: the dynamo effect in the protoneutron star proposed by Duncan and Thompson (1992) implies birth periods of order 1 ms. However, recent evidences (Vink and Kuiper 2006) suggest that not all magnetar are born as hyper-fast rotating NSs. 1E could thus be the first example of a previously unknown class of slowly-rotating, isolated neutron stars, braked by an early interaction with supernova fallback material.

Future observations, both in X-rays as well as in the optical/IR, will help to provide a conclusive interpretation to the unique phenomenology of 1E.

References

Becker, W., Aschenbach, B.: In: Becker, W., Lesch, H., Trümper, J. (eds.) Proc. of the 270. WE-Heraeus Seminar on Neutron Stars, Pulsars, and Supernova Remnants. MPE Report 278, p. 64. Max-Plank-Institut für extraterrestrische Physik, Garching bei München (2002)

Carter, L.M., Dickel, J.R., Bomans, D.J.: Publ. Astron. Soc. Pac. **109**, 990 (1997)

Chatterjee, P., Hernquist, L., Narayan, R.: Astrophys. J. **534**, 373 (200)

Davies, R.E., Pringle, J.E.: Mon. Not. Roy. Astron. Soc. **196**, 209 (1981)

Davis, J.E.: Astrophys. J. **562**, 575 (2001)
De Luca, A., Caraveo, P.A., Mereghetti, S., Tiengo, A., Bignami, G.F.: Science **313**, 814 (2006)
Duncan, R.C., Thompson, C.: Astrophys. J. **392**, L9 (1992)
Francischelli, G.J., Wijers, R.A.M.J., Brown, G.E.: Astrophys. J. **565**, 471 (2002)
Garmire, G.P., Pavlov, G., Garmire, A.B., Zavlin, V.E.: IAU Circular 7350 (2000)
Garmire, G.P., Garmire, A.B., Pavlov, G., Burrows, D.N.: Bull. Am. Astron. Soc. **32**, 1237 (2002)
Gotthelf, E.V., Halpern, J.P.: Astrophys. J. **632**, 1075 (2005)
Gotthelf, E.V., Petre, R., Vasisht, G.: Astrophys. J. **514**, L107 (1999)
Ibrahim, A.I., Markwardt, C.B., Swank, J.H.: Astrophys. J. **609**, L21 (2004)
Illarionov, A.F., Sunyaev, R.A.: Astron. Astrophys. **39**, 185 (1975)
Kalogera, V.: Astrophys. J. **471**, 352 (1996)
Narayan, R.: Astrophys. Space Sci. **300**, 177 (2005)
Pavlov, G.G., Sanwal, D., Garmire, G.P., Zavlin, V.E.: In: Slane, P.O., Gaensler, B.M. (eds.) Neutron Stars in Supernova Remnants. ASP Conference Series, p. 247. ASP, San Francisco (2002)
Pavlov, G.G., Sanwal, D., Teter, M.A.: In: Camilo, F., Gaensler, B.M. (eds.) Young Neutron Stars and their Environments, IAU Symp. No. 218, p. 297. ASP, San Francisco (2004)
Reynoso, E.M., Green, A.J., Johnston, S., et al.: Proc. Astron. Soc. Aust. **21**, 82 (2004)
Sanwal, D., Garmire, G.P., Garmire, A., Pavlov, G.G., Mignani, R.: Bull. Am. Astron. Soc. **34**, 764 (2002)
Strüder, L., Briel, U., Dennerl, K., et al.: Astron. Astrophys. **365**, L18 (2001)
Tiengo, A., Mereghetti, S., Turolla, R., et al.: Astron. Astrophys. **437**, 997 (2005)
Tuohy, I., Garmire, G.: Astrophys. J. **239**, L107 (198)
Turner, M.J.L., Abbey, A., Arnaud, M., et al.: Astron. Astrophys. **365**, L27 (2001)
Vink, J., Kuiper, L.: Mon. Not. Roy. Astron. Soc. **370**, L14 (2006)
Wang, Z., Chakrabarty, D.: In: Slane, P.O., Gaensler, B.M. (eds.) Neutron Stars in Supernova Remnants. ASP Conference Series, p. 297. ASP, San Francisco (2002)
Woods, P.M., Thompson, C.: In: Lewin, W., van der Klis, M. (eds.) Compact Stellar X-ray Sources, p. 547. Cambridge University Press, Cambridge (2006)

Astrophys Space Sci (2007) 308: 239–246
DOI 10.1007/s10509-007-9339-5

ORIGINAL ARTICLE

Evidence for a binary companion to the central compact object 1E 1207.4-5209

Peter M. Woods · Vyacheslav E. Zavlin · George G. Pavlov

Received: 31 July 2006 / Accepted: 7 November 2006 / Published online: 27 March 2007

Abstract Unique among neutron stars, 1E 1207.4-5209 is an X-ray pulsar with a spin period of 424 ms that contains at least two strong absorption features in its energy spectrum. This neutron star is positionally coincident with the supernova remnant PKS 1209-51/52 and has been identified as a member of the growing class of radio-quiet compact central objects in supernova remnants. From previous observations with *Chandra* and *XMM-Newton*, it has been found that the 1E 1207.4-5209 is not spinning down monotonically as is common for young, isolated pulsars. The spin frequency history requires either strong, frequent glitches, the presence of a fall-back disk, or a binary companion. Here, we report on a sequence of seven *XMM-Newton* observations of 1E 1207.4-5209 performed during a 40 day window between 2005 June 22 and July 31. Due to unanticipated variance in the phase measurements during the observation period that was beyond the statistical uncertainties, we could not identify a unique phase-coherent timing solution. The three most probable timing solutions give frequency time derivatives of $+0.9$, -2.6, and $+1.6 \times 10^{-12}$ Hz s^{-1} (listed in descending order of significance). We conclude that the local frequency derivative during our *XMM-Newton* observing campaign differs from the long-term spin-down rate by more than an order of magnitude. This measurement effectively rules out glitch models for 1E 1207.4-5209. If the long-term spin frequency variations are caused by timing noise, the strength of the timing noise in 1E 1207.4-5209 is much stronger than in other pulsars with similar period derivatives. Therefore, it is highly unlikely that the spin variations are caused by the same physical process that causes timing noise in other isolated pulsars. The most plausible scenario for the observed spin irregularities is the presence of a binary companion to 1E 1207.4-5209. We identified a family of orbital solutions that are consistent with our phase-connected timing solution, archival frequency measurements, and constraints on the companions mass imposed by deep IR and optical observations.

Keywords X-rays · Neutron stars: individual: (1E 1207.4-5209) · Supernovae: individual (PKS 1209-51/52)

PACS 97.60.Jd · 97.60.Bw

P.M. Woods (✉)
Dynetics, Inc., 1000 Explorer Blvd., Huntsville, AL 35806, USA
e-mail: peter.woods@dynetics.com

V.E. Zavlin
Space Science Laboratory, NASA MSFC SD50, Huntsville, AL 35805, USA
e-mail: vyacheslav.zavlin@msfc.nasa.gov

G.G. Pavlov
Pennsylvania State University, 525 Davey Lab, University Park, PA 16802, USA
e-mail: pavlov@astro.psu.edu

1 Introduction

There exist a handful of enigmatic X-ray point sources, very likely young neutron stars, positionally coincident with supernova remnants (SNRs) whose nature remains uncertain. These objects are commonly referred to as central compact objects (CCOs) that are characterized by soft, thermal X-ray spectra and an absence of ordinary pulsar activity such as radio pulsations, γ-ray emission and pulsar wind nebulae (Pavlov et al. 2002a, 2004).

The CCO 1E 1207.4-5209 (1E1207 hereafter) in the PKS 1209-51/52 SNR is a particularly interesting member of this class in that it is both an X-ray pulsar (with a 0.424 s period; Zavlin et al. 2000) and the only CCO found to possess prominent absorption lines in its spectrum (Sanwal et al. 2002). The X-ray energy spectrum is best modeled with a continuum blackbody component of temperature $kT \approx 0.14$ keV and at least two broad absorption lines centered at 0.7 and 1.4 keV. The strength of these lines depend upon the rotational phase of the pulsar (Mereghetti et al. 2002). Two additional features, at 2.1 and 2.8 keV, have been reported (Bignami et al. 2003); however, their validity has been questioned (Mori et al. 2005).

The physical origin for the spectral lines remains unknown. Sanwal et al. (2002) have concluded that these lines cannot be associated with transitions in Hydrogen atoms and argued that neither electron nor proton cyclotron resonance could cause these features. These authors suggest that the lines could be due to absorption by once-ionized Helium in a magnetic field $B \sim 2 \times 10^{14}$ G (see also Pavlov and Bezchastnov 2005), while Hailey and Mori (2002) and Mori and Hailey (2006) argue that the lines could be formed in an Oxygen atmosphere with $B \sim 10^{11}$–10^{12} G. As different interpretation imply very different magnetic field strengths, measuring the field strength of 1E1207 would be most important for understanding the nature of the spectral lines. For isolated pulsars, the most straightforward method for estimating dipole magnetic field strengths is to measure the spin frequency ν and its time derivative (spin-down rate) $\dot{\nu}$.

Zavlin et al. (2004) studied the spin evolution of 1E1207 using a compilation of *Chandra* and *XMM-Newton* observations covering 3.5 years. They found that the spin frequency of this pulsar is not steadily decreasing as one would expect for a magnetically-braking dipole. Instead, the frequency evolution was quite erratic, leading Zavlin et al. to consider three possible explanations: (i) the star is undergoing frequent glitches, (ii) the star is surrounded by a debris disk that influences its spin evolution through accretion and propeller torques, or (iii) the star is a member of a (non-accreting) binary system. Another possibility is that the spin-down of the star is influenced by timing noise, a ubiquitous property of isolated neutron stars whose physical origin is unclear. The sparse frequency history of 1E1207 could not distinguish between these models.

Here, we report on a sequence of seven *XMM-Newton* observations of 1E1207 that were designed to perform phase-coherent timing in order to precisely measure the pulse frequency and frequency derivative of the source. A measurement of the local frequency derivative would help us to distinguish between the possible scenarios. Below, we describe the observation (Sect. 2) and our timing analysis of the *XMM-Newton* data set (Sect. 3), and discuss the resulting constraints on the physical mechanisms for the spin-frequency evolution in 1E1207 (Sect. 4).

2 XMM-Newton observations

During a 40 day interval between 2005 June 22 and July 31, *XMM-Newton* observed 1E1207 seven times, with the EPIC PN camera as the primary instrument. The first three and last two pointings had effective exposures of 10–15 ks, while the central (fourth) exposure was about 45 ks. The fifth pointing of about 7 ks exposure was shorter than planned because of strong background contamination. The spacing between consecutive observations was approximately 15, 5, 1, 1, 5, and 15 days. The spacing and durations of the *XMM-Newton* pointings were planned in such a way as to allow phase-coherent timing of the pulsar over the full time span of 40 days (see Sect. 3). The exact exposures, observing epochs, and other observational details for these observations are listed in Table 1.

Table 1 *XMM-Newton* observation log for 1E 1207.4-5209

#	ObsID	Date	Central epoch (MJD)	Span (ks)	PN Exp.[a] (ks)	r[b] (arcsec)	Counts[c]	max Z_1^2
1	0304531501	2005 Jun 22	53543.600542	15.1	10.6	35	13233	41.1
2	0304531601	2005 Jul 05	53556.141927	18.2	12.7	35	12858	37.0
3	0304531701	2005 Jul 10	53561.404084	20.5	14.3	20	17651	43.1
4	0304531801	2005 Jul 11	53562.456311	63.4	44.4	35	56804	120.4
5	0304531901	2005 Jul 12	53563.335586	9.6	6.7	20	8559	19.1
6	0304532001	2005 Jul 17	53568.112822	16.5	11.5	35	14696	81.2
7	0304532101	2005 Jul 31	53582.691485	17.7	12.4	20	15524	26.5
	Sum	...	...	161.0	112.6	...	139325	...

[a]Effective source exposure times after filtering. See text for details

[b]Source extraction radius used for event selection

[c]Number of counts used for timing analysis

For each observation, the PN camera was operated in small window mode, with 5.6 ms time resolution. Starting from the observation data files, all data were processed using *XMMSAS* version 6.5.0. After running the tool `epchain`, we extracted light curves from the observed field of view minus a circular region that included 1E1207. These light curves were used to identify and filter out periods of high background. Source (plus background) counts for timing analysis were extracted from a circular region for each observation and filtered using standard criteria and the good time intervals we determined. The radii of the extraction regions, 35″ or 20″, are listed in Table 1. A smaller radius of 20″ was used to improve the signal-to-noise ratio for the three observations where the background rate within the good time intervals was elevated. The filtered event lists were barycentered to the location R.A. = $12^{\rm h}10^{\rm m}0.^{\rm s}80$, Decl. = $-52°26'25.''1$ using the *XMMSAS* tool `barycen`. Finally, we selected counts within the energy range 0.4–2.5 keV to maximize the signal-to-noise ratio of the pulsed signal before beginning our timing analysis. The numbers of selected counts are given in Table 1 (background was estimated to contribute less than 15% in each dataset).

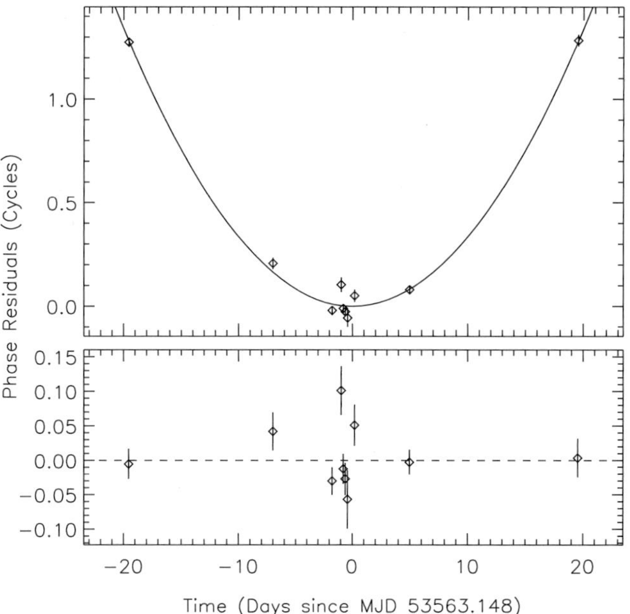

Fig. 1 Pulse phase residuals from *XMM-Newton* observations of 1E1207 during the 2005 observing campaign for model MOD1 (Table 2). *Top*: Phase residuals minus a linear trend. *Bottom*: Phase residuals minus a quadratic trend. Note that the central (longest) observation is split into four segments of equal spans

3 Phase-coherent timing analysis

Phase-coherent timing analysis requires careful spacing of individual observations such that an extrapolation of the measured phase model, $\phi(t) = \phi(t_0) + \nu(t - t_0) + \frac{1}{2}\dot{\nu}(t - t_0)^2 + \cdots$, for a given observation or set of observations is precise enough to predict the phase to the next observation to much better than a pulse cycle. The advantage of this approach is that one can achieve far more precise measurements of the pulse frequency and higher derivatives than by using independent pulse frequency measurements with the same total exposure. This approach is commonly applied to all types of pulsars, including Anomalous X-ray Pulsars (e.g. Gavriil and Kaspi 2004), Soft Gamma Repeaters (e.g. Woods et al. 2002), and radio-quiet Isolated Neutron Stars (Kaplan and van Kerkwijk 2005a, 2005b).

3.1 Pulse phase fitting technique

As the frequency error ($\delta\nu$) in an individual observation is inversely proportional to its duration, $\delta\nu_j \propto T_j^{-1}$, the longer central exposure served as our reference point. We measured the pulse frequency during this observation first via a Z_1^2 search (see Sect. 3.2) and then refined this measurement as follows. We split the observation into 4 segments and folded these segments on the measured frequency to generate pulse profiles for each segment. Next, we cross-correlated each pulse profile with a high signal-to-noise pulse template and measured phase offsets. The pulse template is first derived from the central observation folded at the initial frequency. The phase offsets for the 4 segments were fitted to a straight line and the slope of this line was added to the initial frequency to produce our refined frequency. The short gaps between the central exposure and the adjacent exposures were expected to preserve the phase information, i.e. the propagated phase error (e.g., between the 4-th and 5-th observations, $\delta\phi = \delta\nu(t_5 - t_4)$) was expected to be $\ll$1 cycle, which would mean that no pulse cycles are missed in the phase model. As one incorporates more and more data over a wider time span, the precision of the phase model improves, and one can tolerate larger gaps between observations. Note that the template pulse profile is updated as more data are included until the full data set is utilized. By the time we incorporated the measured phases from the first and final observations into our fit, it became clear that the phase offsets did not conform to a simple linear trend, and a quadratic term ($\propto\dot{\nu}$) was added to the phase model, $\phi(t) = \phi(t_0) + \nu(t - t_0) + \frac{1}{2}\dot{\nu}(t - t_0)^2$. However, even the inclusion of the quadratic term did not reduce the variance of the phase residuals to the point where we obtained an acceptable fit ($\chi^2 = 19.2$ for 8 degrees of freedom; see Fig. 1).

The poor fit to the quadratic phase model indicated that we either converged on an alias solution or 1E1207 exhibits significant "phase noise"[1] on a time scale of weeks. An alias

[1] In this context, we refer to phase noise simply as deviations from our simple quadratic phase model beyond statistical errors. Note that an

timing solution would be when there are an incorrect number of cycle counts between consecutive observations. Phase noise can be characterized in many ways such as the presence of a strong cubic term ($\propto\ddot{\nu}$), white noise, periodic variations, etc. To ensure that the poor χ^2 value in the solution we found is *not* the consequence of misidentified cycle counts, we employed a technique used for timing noisy rotators such as Soft Gamma Repeaters (Woods et al. 2007). In this technique, we measure the pulse phase and frequency at each of the 7 observing epochs. The phase for each observation was measured by folding the data from each observation on a pulse ephemeris of constant frequency determined by the central observation and computing the phase difference between this profile and a template pulse profile. The pulse frequencies for the short observations were measured by splitting the data into two segments of equal duration, folding each segment on the pulse frequency measured for the central observation, measuring phase shifts for each pulse profile relative to the pulse template, and fitting these two pulse phases to a line to determine the local pulse frequency. Finally, we perform a least-squares fit to the set of 7 phases and 7 frequencies, where we vary the number of cycles between consecutive observations by integer increments (Woods et al. 2007). This provides a family of solutions to the full data set which define the pulse phase evolution according to a quadratic model covering the 40-day time interval. None of the solutions (for a quadratic phase model) provide a statistically acceptable fit to the data. For all possible solutions, the "null hypothesis probability" (i.e. the probability of measuring the large χ^2 values by chance, assuming that the model is correct) is very small. Fit parameters for the top three solutions (ranked in order of increasing χ^2) are given in Table 2. We chose to consider a limited number of solutions; therefore, we selected only the solutions that had a probability of getting the measured χ^2 by chance of 10^{-5} or larger (three solutions). We found that the best-fit model is equivalent to the solution we identified via our bootstrap phase-fitting method described earlier. Phase residuals for this model are shown in Fig. 1. Since none of the identified solutions provide a statistically-acceptable fit to the data, we conclude that 1E1207 does, in fact, exhibit significant phase noise on a time scale of weeks. It seems unlikely that the excess noise we observed is due to underestimating our phase errors. Analysis of *XMM-Newton* data from other pulsars using the same software, although covering time spans shorter than 40 days, have consistently yielded reduced χ^2 values of $\sim$1 (e.g Woods et al. 2004).

The presence of the phase noise does not allow us to unambiguously phase-connect the complete data set and thus measure a unique frequency and frequency derivative for the full 40-day time span. To place some constraints on the frequency derivative during our observing sequence, we employed a Monte-Carlo simulation to estimate statistical significance of the multiple solutions. For this simulation, we first had to choose a model for the phase noise. We selected two models: (i) a cubic phase term and (ii) white noise. In both cases, the amplitude of the model noise variance was equal to the total variance in the top three fits to the data minus the statistical variance. In our simulation, we generated phases for each observing epoch which included three components: the model phase (including ν and $\dot{\nu}$ terms), Gaussian measurement noise, and the model phase noise. In addition, we simulated frequency measurements at each epoch assuming Gaussian measurement noise (i.e. we neglect phase noise on the time scale of the observation duration). For each phase model, we generated 10^5 realizations and fit for the cycle counts between consecutive epochs as we did for the measured data to identify all possible timing solutions for each realization. In each realization, we identified the rank of the true timing solution in terms of χ^2. The most constraining results were obtained from the white noise model for the phase noise. For this model, we found that the true timing solution was among the top three solutions (ranked in order of χ^2) 90% of the time and was the top solution 65% of the time. Assuming white phase noise, our simulation suggests that we can be 90% confident that the true pulse ephemeris for 1E1207 is MOD1, MOD2 or MOD3 given in Table 2. Similarly, these results suggest there is a 65% chance that MOD1 defines the appropriate cycle counts between observing epochs, and hence, reflects the correct pulse ephemeris.

The differences between the three pulse ephemerides listed in Table 2 amount to small differences in the cycle counts between the four outer observations in our observing sequence (i.e. a few additional or less cycles between observations 1 and 2, 2 and 3, 5 and 6, and 6 and 7). In fact, we can only be sure of the cycle count accuracy between the three central observations (observations 3, 4 and 5 in Table 1). To show this explicitly, we fit for the cycle counts between the central three observations as we did for the full data set, only we limited the order of the phase model to be first order on account of the short time span (2 days). We measure a difference in χ^2 of 91 for 3 degrees of freedom between the best-fit ephemeris identified in our search and the next closest. Clearly, we were able to phase-connect this subset of the data and unambiguously identify the local pulse frequency ($\nu = 2.35776187(31)$ Hz over the time range 53561.328 to 53563.347 MJD TDB).

Although the method described in this section is very efficient, it has some limitations. For large cycle count corrections between consecutive observations, the local pulse ephemeris will change considerably as will the folded pulse profile. In turn, the pulse phase measurement will likely also

alternative definition of phase noise has specific meaning in the context of pulsar timing noise (e.g. Cordes and Helfand 1980).

Table 2 Candidate pulse ephemerides for 1E 1207.4-5209 for 2005 June through July

Model	Epoch[a] (MJD TDB)	ν[b] (Hz)	$\dot{\nu}$ (10^{-12} Hz s^{-1})	χ^2/dof	Null hypothesis probability	Z_1^2	Z_2^2
MOD1	53563.148	2.357761722(16)	+0.890(22)	25.2/11	2.9×10^{-3}	323.0	333.5
MOD2	53563.148	2.357762311(16)	−2.651(23)	32.6/11	2.3×10^{-4}	314.9	325.8
MOD3	53563.148	2.357761720(17)	+1.607(24)	37.6/11	3.6×10^{-5}	318.5	325.2

[a]Pulse ephemerides are valid over the time range 53543.547 to 53582.746 MJD TDB

[b]Numbers given in parentheses indicate the 1σ error in the least significant digit(s). The statistical errors are inflated by a factor

be affected. In practice, the differences in the pulse shapes of 1E1207 for the three pulse ephemerides reported here are insignificant. For very large cycle count corrections, where this effect becomes important, the χ^2 contribution from the frequency measurements begin to dominate the total χ^2, and these peaks are effectively suppressed. Even so, this method is relatively new and not extensively tested. To verify the results obtained with this technique, we employ the Z_n^2 test, a traditional approach to X-ray timing.

3.2 The Z_n^2 test

The Z_n^2 statistic (e.g., Buccheri et al. 1983) is defined as follows:

$$Z_n^2 = \frac{2}{N}\sum_{k=1}^{n}\left[\left(\sum_{i=1}^{N}\cos 2\pi k\phi_i\right)^2 + \left(\sum_{i=1}^{N}\sin 2\pi k\phi_i\right)^2\right], \quad (1)$$

where $\phi_i = \nu(t_i - t_0) + \dot{\nu}(t_i - t_0)^2/2 + \cdots$ is the phase of i-th event, $t_i - t_0$ is the event arrival time counted from an epoch t_0 of zero phase, n is the number of harmonics involved in the test, and N is the number of events. For a signal with a nearly sinusoidal pulse profile, such as observed from 1E1207, the Z_1^2 (Rayleigh) test is known to give excellent results. For a sinusoidal signal, the expected peak value of Z_1^2 is $Nf_{\rm p}^2/2$, where $f_{\rm p}$ is the pulsed fraction. For 1E1207, the pulsed fraction was measured to be $f_{\rm p} = 8$–12% (Zavlin et al. 2000; Pavlov et al. 2002b). The peak Z_1^2 values found in the individual data sets (Table 1) are in a reasonable agreement with those given by this estimate.

The Z_n^2 test has been used for a phase-coherent timing analysis of several observations spread over a large time span by Mattox et al. (1996) and Zavlin et al. (1999), and we follow the approach described by those authors. To account for the phase connection, we apply the Z_n^2 test (for $n = 1$ and 2) to the whole data set of seven observations. To determine the parameters ν and $\dot{\nu}$ of the quadratic phase model, we calculated the Z_n^2 on a dense two-dimensional grid [$\nu - 2.3577$ Hz = 41–81 μHz, $|\dot{\nu}| < 1 \times 10^{-11}$ Hz s^{-1}], with ν and $\dot{\nu}$ spacings of 0.02 μHz and 2×10^{-14} Hz s^{-1}, respectively. A contour map obtained with the Z_1^2 statistic is shown in Fig. 2. Because of the cycle-count ambiguities during the gaps between the consecutive observations, the map shows multiple peaks, one of them corresponding to the true ν, $\dot{\nu}$ solution and the others being aliases. The first, third and fourth highest peaks in this map correspond to MOD1, MOD3 and MOD2, respectively (see Table 2). The top three peaks in a similar Z_2^2 map are at the same ν, $\dot{\nu}$ as MOD1, MOD2 and MOD3, respectively. If the phase connection between separate data sets were perfect, then the peak corresponding to the true solution would be much higher than the aliases. However, in our case the difference between the heights of the peaks turned out to be too small to single out a unique solution. For instance, in addition to the highest peak in the Z_1^2 map, $Z_{1,\max}^2 =$ 323.0 at $\nu = 2,357,761.72$ μHz, $\dot{\nu} = +0.90 \times 10^{-12}$ Hz s^{-1}, we see four peaks with $310 < Z_1^2 < 320$ in Fig. 2, at different ν, $\dot{\nu}$ values. Similar to the method described in Sect. 3.1, the differences in peak values of ν, $\dot{\nu}$ correspond to different (integer) numbers of cycles ($\sim 8 \times 10^6$) during the full observational time span $T = 3393.8$ ks. We are not aware of statistical criteria to estimate significance of separate peaks in this approach, and we can only assume that the solutions corresponding to several highest Z_n^2 peaks cannot be ruled out on statistical grounds. We also note that the lack of perfect phase-coherence is supported by the fact that the largest Z_1^2 is much smaller than $\sum_{j=1}^{7} Z_{1,j}^2 = 346.8$ (at the same ν, $\dot{\nu}$), the value we would expect to obtain for perfect phase connection. Thus, the results of the Z_n^2 search for the 1E1207 frequency and frequency derivative are generally consistent with the results reported in Sect. 3.1.

4 Origin of the erratic spin behavior

The deviations from monotonic spin-down in 1E1207 are substantial, and they manifest on timescales of years to as short as possibly weeks as evidenced by the phase noise detected here. We now consider four possibilities for both the erratic long-term spin behavior and short-term phase noise in 1E1207: (i) frequent glitching, (ii) accretion and propeller

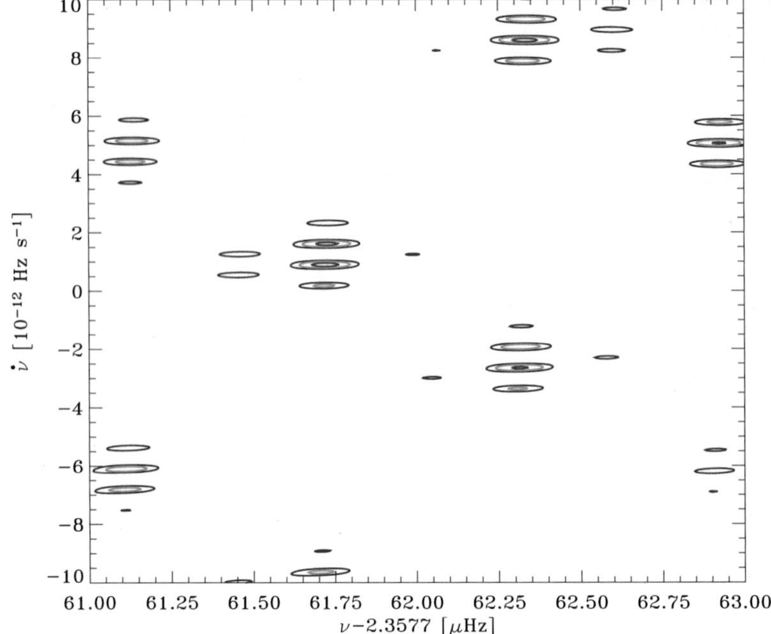

Fig. 2 Contour plot of the Z_1^2 power on the ν−$\dot{\nu}$ grid. The *purple, green* and *red contours* correspond to $Z_1^2 = 250$, 280, and 310, respectively

torques from a circumstellar debris disk, (iii) timing noise in an isolated neutron star, and (iv) orbital Doppler shifts caused by the presence of a binary companion.

For any glitch model, the glitch frequency and amplitude would have to be very high to account for the observed spin variability (see Zavlin et al. 2004). Moreover, the most viable timing solutions indicating spin down over the 40-day observing span differ from the long-term spin-down of 1E1207 over the last 5.5 years ($\sim -4 \times 10^{-14}$ Hz s^{-1}) by an order of magnitude. For example, the most likely spin-down solution (MOD2) has a frequency derivative more than one order of magnitude larger. Because only a very contrived glitch model could account for the long-term frequency history, and this model would provide no explanation for the short-term phase noise, the glitch model is effectively excluded by these observations.

Debris disks left over from the supernova explosions that produce neutron stars could alter the spin evolution of the central neutron star via accretion and propeller torques (Zavlin et al. 2004). If the spin-up rate of 1E1207 during the 40-day interval were equal to the values measured for MOD1 or MOD3, then the mass accretion rate would have to be very large ($\dot{m} > 3 \times 10^{16}$ g s^{-1}). Such a large accretion rate would require a large increase in X-ray luminosity which is not observed. Even the spin-down solutions would require significant optical and IR emission from the disk. Deep IR observations of 1E1207 have shown no indication of even a cool, passive debris disk (Wang et al. 2007). Thus, it appears unlikely that a debris disk is the cause of the spin variability in 1E1207.

Timing noise (irregular evolution of the pulse phase with time) is a ubiquitous phenomenon in isolated neutron stars. This variability is in addition to the usual variation caused by magnetic braking. It has been demonstrated that the magnitude of these irregular variations depends upon the spin-down rate of the pulsar (e.g., Cordes and Helfand 1980). Millisecond pulsars show the smallest timing noise while magnetars exhibit very strong timing noise. In the case of magnetars, these variations manifest as changes in the effective spin-down rate of up to factors of 5 on a time scale of years. For a convenient (albeit crude) description of timing noise, Arzoumanian et al. (1994) introduced a "stability parameter" defined by the following equation: $\Delta_{\log t} = \log(|\ddot{\nu}|t^3/6\nu)$, where t is the time during which the pulse phase has been monitored ($t = 10^8$ s is a commonly used characteristic time), and $\ddot{\nu}$ is the formal value of the second frequency derivative obtained from fitting a cubic model to pulse phases (it is much larger in magnitude than the actual $\ddot{\nu}$ for noisy pulsars). Third-order polynomial fits to the 1E1207 phase residuals of the top three candidate timing solutions yielded insignificant measurements of $\ddot{\nu}$. The timing observations of 1E1207 during 5.5 years were too sparse to fit the pulse phases with any model. Therefore, to estimate $\ddot{\nu}$ and Δ_8, we fitted the dependence $\nu(t) = \nu_0 + \dot{\nu}(t - t_0) + \ddot{\nu}(t - t_0)^2/2$ to the frequency history covering the last 5.5 years. (The same exercise was performed by Heyl and Hernquist (1999) for some Anomalous X-ray Pulsars, and it was shown to be reasonably accurate by Gavriil and Kaspi (2004).) Choosing $t_0 = 52700$ MJD TDB, we found $\nu_0 = 2{,}357{,}762.8 \pm 0.2$ μHz, $\dot{\nu} = (-3.4 \pm 0.6) \times 10^{-14}$ Hz s^{-1}, and $\ddot{\nu} = (5.9 \pm 1.9) \times 10^{-22}$ Hz s^{-2}, which translates to a timing noise level of $\Delta_8 = 1.6$. In Fig. 3, we show the period derivative versus the timing noise parameter Δ_8 for

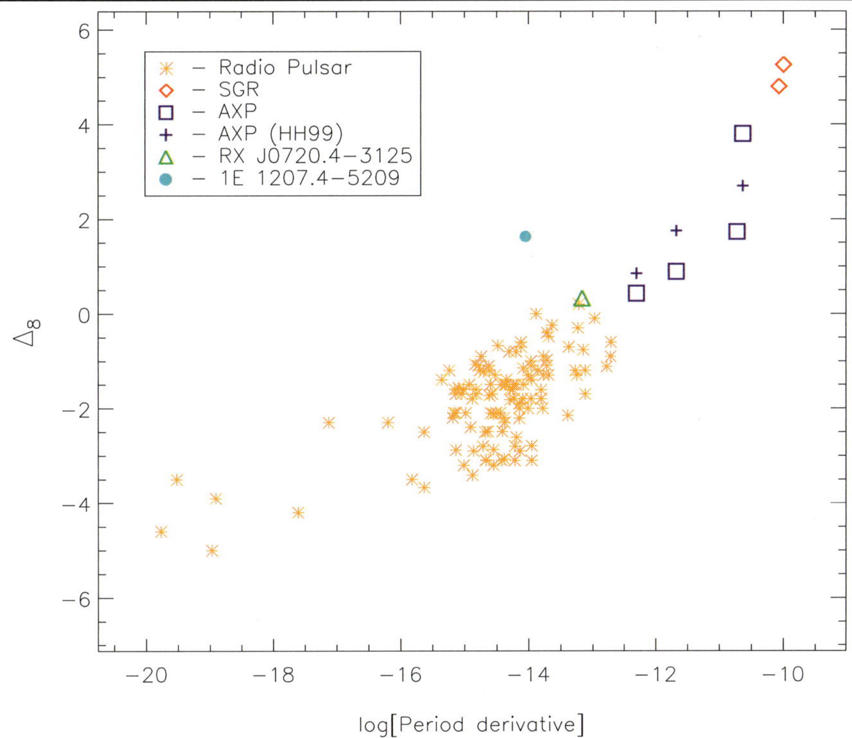

Fig. 3 The timing noise parameter Δ_8 after Arzoumanian et al. (1994) for 119 radio pulsars, the isolated neutron star RX J0720.4-3125 (Kaplan and van Kerkwijk 2005b), four Anomalous X-ray Pulsars (*blue squares*—(Gavriil and Kaspi 2004); *blue plus signs*—(Heyl and Hernquist 1999)), two Soft Gamma Repeaters (Woods et al. 2002) and the CCO 1E 1207.4-5209

126 isolated pulsars of various flavors as well as for the CCO 1E1207. The isolated pulsars fall along a relatively well-defined locus, from the quiet millisecond pulsars to the noisy Soft Gamma Repeaters, while 1E1207 stands out from this trend with an anomalously large timing noise strength, some 2–4 orders of magnitude higher than isolated pulsars at similar spin-down rates. Although such an estimate for the timing noise parameter is, by necessity, very crude, its enormously high magnitude, together with the gross inconsistency of the local (June–July 2005) spin-down rate with the long-term average, suggest that the erratic frequency behavior in this source is not due to the same effect that causes timing noise in other isolated neutron stars.

The most straightforward explanation for the long-term spin variations in 1E1207 is the presence of a binary companion. Note that this model cannot explain the observed short-term phase noise. Current IR and optical limits for 1E1207 exclude main sequence companions earlier than M5 and even white dwarfs with effective temperatures greater than $\sim 10^4$ K (Fesen et al. 2006; Wang et al. 2007). Allowable companion masses are less than 0.2 $M_\odot$ for late-type stars. Even with such low-mass companions, the resulting Doppler shifts are large enough to account for the frequency variations in 1E1207. Using the archival spin frequencies in combination with the phases from our three candidate timing solutions, we fit the data to a circular orbital model whose phase evolution is defined by the following equation: $\phi(t) = \phi(t_0) + \nu(t - t_0) + \frac{1}{2}\dot{\nu}(t - t_0)^2 + A \sin \omega(t - t_0)$. This is the same equation as given in Sect. 3.1 with an additional sinusoidal term to account for the orbital Doppler shifts. We identified a family of acceptable orbits for each of the three timing solutions listed in Table 2. The full set of allowable timing solutions are too numerous to list. We can place only very crude constraints on the orbital periods to fall between 120 and 600 days. The mass functions range between 1×10^{-7} and 5×10^{-5} $M_\odot$ for acceptable orbital solutions. For a 90° inclination and a 1.4 $M_\odot$ neutron star, the corresponding companion mass range is 0.007 to 0.05 $M_\odot$, well within the existing limits on companion masses from IR and optical observations. For illustrative purposes, we show two example orbital solutions that are consistent with the existing timing data for 1E1207 (Fig. 4).

5 Conclusions

We observed 1E1207 with *XMM-Newton* seven times during the course of a 40 day interval in an effort to measure the local pulse frequency and frequency derivative with high precision. Due to unanticipated phase noise, we were unable to phase-connect the full data set. From systematic pulse ephemeris searches, we identified a number of possible timing solutions, none of which had spin-down rates close to the average spin-down rate of 1E1207 over the last 5 years. The interpretation of the erratic long-term timing behavior of 1E1207 in terms of the usual pulsar timing noise would require timing noise levels much higher than seen in isolated neutron stars with comparable spin-down rates. The

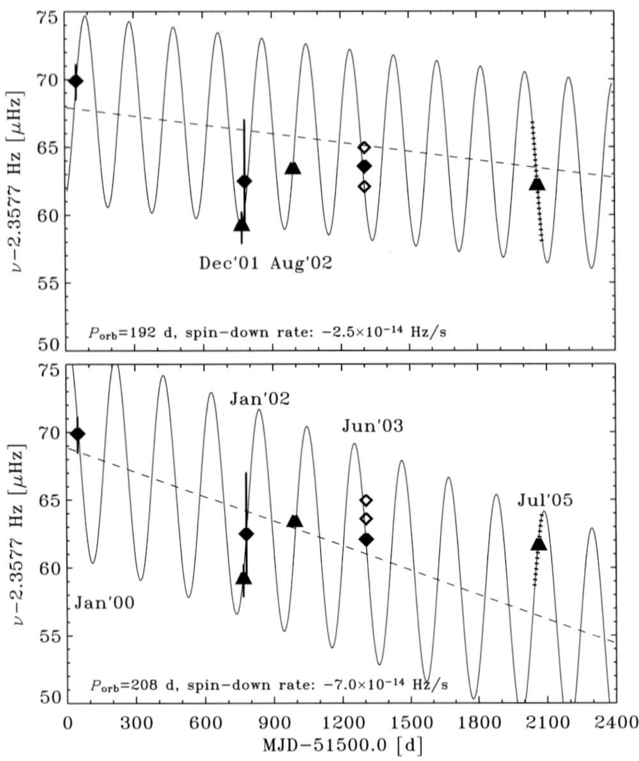

Fig. 4 Candidate orbital solutions for 1E1207 consistent with our MOD2 (*top*) timing solution and MOD1 (*bottom*) timing solution

most plausible explanation for the erratic long-term pulse frequency evolution of 1E1207 is the presence of a binary companion, although the current data do not allow us to place strong constraints on system parameters. Further efforts at phase-coherent pulse timing observations of 1E1207 are required to (i) unambiguously identify the nature of the long-term pulse frequency variations, and (ii) confirm and further investigate the observed short-term phase noise. Just one additional 40-day observing sequence with higher sampling would yield far better constraints on orbital parameters should 1E1207, in fact, possess a binary companion.

Acknowledgements This work was supported by NASA grants NNG05GN91G and NAG5-10865. V.E.Z. is supported by a NASA Research Associateship Award at NASA Marshall Space Flight Center.

References

Arzoumanian, Z., Nice, D.J., Taylor, J.H., et al.: Astrophys. J. **422**, 671 (1994)

Bignami, G.F., Caraveo, P.A., De Luca, A., et al.: Nature **423**, 725 (2003)

Buccheri, R., Bennett, K., Bignami, G.F., et al.: Astron. Astrophys. **128**, 245 (1983)

Cordes, J.M., Helfand, D.J.: Astrophys. J. **239**, 640 (1980)

Fesen, R.A., Pavlov, G.G., Sanwal, D.: Astrophys. J. **636**, 848 (2006)

Gavriil, F.P., Kaspi, V.M.: Astrophys. J. Lett. **609**, 67 (2004)

Hailey, C.J., Mori, K.: Astrophys. J. Lett. **578**, 133 (2002)

Heyl, J.S., Hernquist, L.: Mon. Not. Roy. Astron. Soc. **340**, L37 (1999)

Kaplan, D.L, van Kerkwijk, M.H.: Astrophys. J. Lett. **635**, 65 (2005a)

Kaplan, D.L., van Kerkwijk, M.H.: Astrophys. J. Lett. **628**, 45 (2005b)

Mattox, J.R., Halpern, J.P., Caraveo, P.A.: Astron. Astrophys. **120**, 77 (1996)

Mereghetti, S., De Luca, A., Caraveo, P.A., et al.: Astrophys. J. **581**, 1280 (2002)

Mori, K., Hailey, C.: astro-ph/0301161 v2, Astrophys. J. (2006, submitted)

Mori, K., Chonco, J.C., Hailey, C.J.: Astrophys. J. **631**, 1082 (2005)

Pavlov, G.G., Bezchastnov, V.G.: Astrophys. J. Lett. **635**, L61 (2005)

Pavlov, G.G., Sanwal, D., Garmire, G.P., et al.: In: Slane, P.O., Gaensler, B.M. (eds.) Neutron Stars in Supernova Remnants. ASP Conf. Ser. **271**, 247 (2002a)

Pavlov, G.G., Zavlin, V.E., Sanwal, D., et al.: Astrophys. J. Lett. **569**, 95 (2002b)

Pavlov, G.G., Sanwal, D., Teter, M.A.: In: Camilo, F., Gaensler, B.M. (eds.), IAU Symp. 218, Young Neutron Stars and Their Enviroments, p. 239. ASP, San Francisco (2004)

Sanwal, D., Pavlov, G.G., Zavlin, V.E., et al.: Astrophys. J. Lett. **574**, 61 (2002)

Wang, Z., Kaplan, D.L., Chakrabarty, D.: Astrophys. J. **655**, 261 (2007)

Woods, P.M., Kaspi, V.M., Thompson, C., et al.: Astrophys. J. **605**, 378 (2004)

Woods, P.M., Kouveliotou, C., Gögüs, E., et al.: Astrophys. J. **576**, 381 (2002)

Woods, P.M., Kouveliotou, C., Finger, M.H., et al.: Astrophys. J. **654**, 470 (2007)

Zavlin, V.E., Pavlov, G.G., Sanwal, D., et al.: Astrophys. J. Lett. **540**, 25 (2000)

Zavlin, V.E., Pavlov, G.G., Sanwal, D.: Astrophys. J. **606**, 444 (2004)

Zavlin, V.E., Trümper, J., Pavlov, G.G.: Astrophys. J. Lett. **525**, 959 (1999)

Astrophys Space Sci (2007) 308: 247–257
DOI 10.1007/s10509-007-9336-8

ORIGINAL ARTICLE

Towards self-consistent models of isolated neutron stars

The case of the X-ray pulsar RX J0720-3125

J.A. Pons · J.F. Pérez-Azorín · J.A. Miralles · G. Miniutti

Received: 30 June 2006 / Accepted: 11 August 2006 / Published online: 21 March 2007

Abstract We propose a self–consistent model to explain all observational properties reported so far on the isolated neutron star (INS) RX J0720-3125 with the aim of giving a step forward towards our understanding of INSs. For a given magnetic field structure, which is mostly confined to the crust and outer layers, we obtain theoretical models and spectra which account for the broadband spectral energy distribution (including the apparent optical excess), the X-ray pulsations, and for the spectral feature seen in the soft X-ray spectrum of RX J0720-3125 around 0.3 keV. By fitting our models to existing archival X-ray data from 6 different XMM–Newton observations and available optical data, we show that the observed properties are fully consistent with a *normal* neutron star, with a proper radius of about 12 km, a temperature at the magnetic pole of about 100 eV, and a magnetic field strength of 2–3 $\times$ 10^{13} G. Moreover, we are able to reproduce the observed long–term spectral evolution in terms of free precession which induces changes in the orientation angles of about 40 degrees with a periodicity of 7 years. In addition to the evidence of internal toroidal components, we also find strong evidence of non–dipolar magnetic fields, since all spectral properties are better reproduced with models with strong quadrupolar components.

Keywords Stars: neutron · Stars: magnetic fields · Radiation mechanisms: thermal

PACS 97.60.Jd · 98.38.Am

1 Introduction

A decade after the discovery of the soft X-ray source RX J185635-3754 (Walter et al. 1996) with the *ROSAT* PSPC, the thermal component associated with the direct emission from the neutron star's surface has been detected in more than 20 X-ray sources. In many cases the thermal component is superimposed on a power–law tail, but seven of these objects are well characterized in the X-ray band as simple blackbodies with temperatures ranging between 60 and 100 eV. The apparent small emitting surface of RX J185635-3754, as inferred from the best blackbody fit and its parallax (Walter and Lattimer 2002), has led to speculation about its nature and has been considered as evidence for the existence of strange stars, i.e. a self–bound objects made of up, down and strange quarks (Pons et al. 2002; Drake et al. 2002). Although isolated compact stars probably represent a real opportunity to place stringent constraints on the equation of state (EOS) of dense matter from astrophysical measurements, one must be cautious before concluding that an apparently small X-ray emitting surface provides evidence for a quark star. Indeed, the data could well be explained by assuming a standard misaligned X-ray pulsar and reliable conclusions on the nature of the source can be reached only by considering all available information, including optical data. Using the Hubble Space Telescope (HST), Walter and Matthews (1997) subsequently identified the optical counterpart of RX J185635-3754 which turns

J.A. Pons (✉) · J.F. Pérez-Azorín · J.A. Miralles
Departament de Física Aplicada, Universitat d'Alacant,
Ap. Correus 99, 03080 Alacant, Spain
e-mail: jose.pons@ua.es

G. Miniutti
Institute of Astronomy, University of Cambridge, Madingley
Road, Cambridge CB3 0HA, UK

out to be 7 times brighter than the extrapolation of the X-ray–detected blackbody into the optical V band, most likely indicating the presence of large temperature anisotropies on the star's surface. The observed optical fluxes have been confirmed by subsequent observations (Pons et al. 2002). Remarkably, the other three isolated compact X-ray sources that have been detected in the optical band (RX J0720-3125, RX J1308+2127, and RX J1605+3249) also exhibit a significant optical excess over the extrapolation of the X-ray blackbody (a factor 5 to 14). None of them is a known radio source.

In this paper we shall focus on one member of the family of INSs, RX J0720-3125. The source was discovered with *ROSAT* (Haberl et al. 1997), and its X-ray spectrum was soon found to be well described by a blackbody with a temperature of $kT \sim 82$ eV. It is a nearby object (≈ 300 pc, Kaplan et al. 2003) and shows low interstellar absorption ($n_H = 1$–1.5×10^{20} cm^{-2}), as the other members of the INS family. More interestingly, it is a confirmed X-ray pulsar with a period of 8.391 s (Haberl et al. 1997) and it is one of the two INSs with a reliable measure of the period derivative $\dot{P} = 7 \times 10^{-14}$ s s^{-1} (Kaplan and van Kerkwijk 2005a), which implies a magnetic field of about $B = 2.4 \times 10^{13}$ G.[1] Another important observational property, common to other INSs, is that the observed optical flux is a factor of 6 higher than the extrapolation of the best–fitting blackbody (BB) model of the X-ray spectrum to the optical band. This apparent optical excess can be explained by the existence of large temperature anisotropies over the surface (Pons et al. 2002). In the case of RX J0720-3125, the presence of an anisotropic temperature distribution is also strongly supported by the observed X-ray pulsations which have a relatively large amplitude ($\sim$11%).

More recently, the study of INSs has suffered a new twist when observations with XMM-Newton have revealed deviations from a pure BB X-ray spectrum in the form of absorption features observed in the 0.1–1.0 keV band. In the case of RX J0720-3125, a phase dependent absorption line around 270 eV has been recently reported (Haberl et al. 2004). This feature has been associated with proton cyclotron resonant absorption and/or bound–bound H or H-like He transitions (Haberl et al., van Kerkwijk et al. 2003, 2004). Both require a magnetic field of $\sim 5 \times 10^{13}$ G, consistent with the dipole braking estimate (2.4×10^{13} G) within a factor of 2. This estimate only reflects the large scale magnetic field structure, whereas the presence of smaller scale inner magnetic fields (i.e. strong toroidal components) seems very likely. MHD core–collapse simulations show indeed that toroidal magnetic fields are quickly generated by

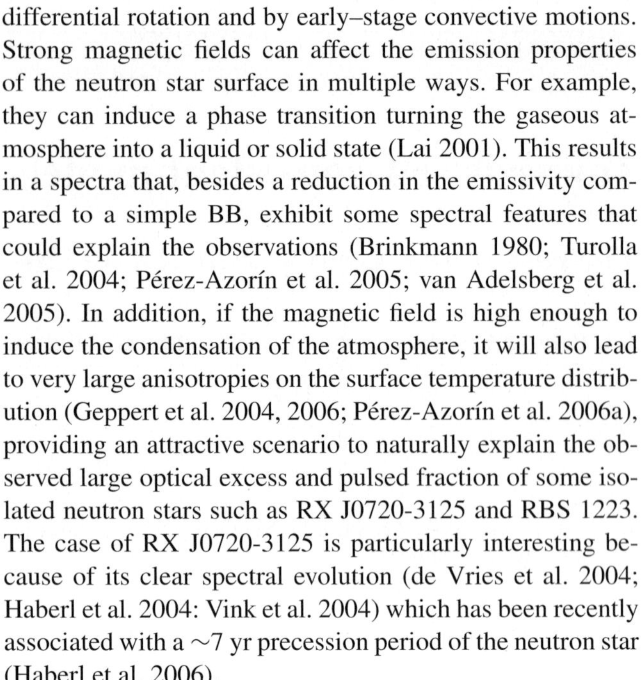

differential rotation and by early–stage convective motions. Strong magnetic fields can affect the emission properties of the neutron star surface in multiple ways. For example, they can induce a phase transition turning the gaseous atmosphere into a liquid or solid state (Lai 2001). This results in a spectra that, besides a reduction in the emissivity compared to a simple BB, exhibit some spectral features that could explain the observations (Brinkmann 1980; Turolla et al. 2004; Pérez-Azorín et al. 2005; van Adelsberg et al. 2005). In addition, if the magnetic field is high enough to induce the condensation of the atmosphere, it will also lead to very large anisotropies on the surface temperature distribution (Geppert et al. 2004, 2006; Pérez-Azorín et al. 2006a), providing an attractive scenario to naturally explain the observed large optical excess and pulsed fraction of some isolated neutron stars such as RX J0720-3125 and RBS 1223. The case of RX J0720-3125 is particularly interesting because of its clear spectral evolution (de Vries et al. 2004; Haberl et al. 2004: Vink et al. 2004) which has been recently associated with a $\sim$7 yr precession period of the neutron star (Haberl et al. 2006).

The paper is organized as follows. In Sect. 2 we review the main properties of our theoretical model (magnetic field configuration, temperature anisotropy, etc.). In Sect. 3 we summarize the X-ray observations used in this work. In Sect. 4, we discuss and motivate the choices made to limit the parameter–space of our theoretical models, and we present our results obtained by applying our realistic and physically motivated models to the available X-ray data. In Sects. 5 and 6 we discuss (i) the observational evidence for precession, and (ii) the interpretation of the excess optical flux, respectively. Finally, in Sect. 7, we summarize our results and discuss open questions and caveats.

2 The theoretical model

In a previous paper (Pérez-Azorín et al. 2006a) (PMP06 hereafter) we have presented detailed calculations of the temperature distribution in the crust and condensed envelope of neutron stars in the presence of strong magnetic fields, obtained through axisymmetric, stationary solutions of the heat diffusion equation with anisotropic thermal conductivities. Having explored a variety of magnetic field strengths and configurations, we concluded that variations in the surface temperature by factors 2–10 are easily obtained with $B \approx 10^{13}$–10^{14} G whereas the average luminosity (and therefore the inferred effective temperature) depends only weakly on the strength of the magnetic field.

The main effect of the magnetic field on the temperature distribution can be understood by looking at the expression for the heat flux:

$$\mathbf{F} = -\kappa_\perp[\nabla\tilde{T} + x_B^2(\mathbf{b}\cdot\nabla\tilde{T})\mathbf{b} + x_B(\mathbf{b}\times\nabla\tilde{T})] \quad (1)$$

[1] The other object with a measure of the period derivative is RBS 1223, with $\dot{P} = 1.12 \times 10^{-13}$ s s^{-1}, which implies $B = 3.4 \times 10^{13}$ G (Kaplan and van Kerkwijk 2005b).

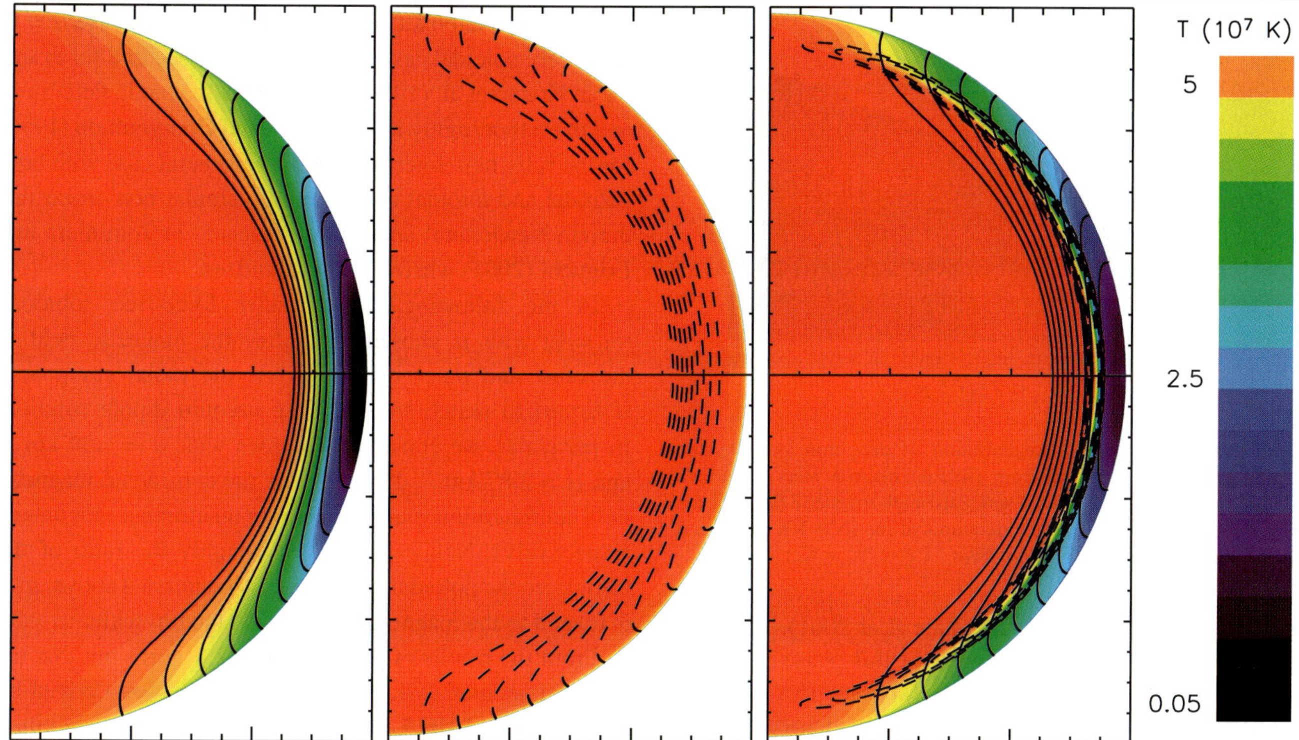

Fig. 1 Temperature distribution in the crust and envelope of a neutron star with different magnetic field toroidal components. The poloidal component (*solid lines*) is the same in all models ($B_p = 10^{13}$ G) and it is confined to the outer layers. The *left panel* shows results for a purely poloidal field, the *central panel* a force-free configuration, and the *right panel* corresponds to a toroidal component confined to a narrow region in the crust. In the two latter cases the *dashed lines* show contours of constant B_ϕ. The scale has been stretched by about a factor 2 to emphasize the crustal region. Notice that in the *central panel* most of the gradients are generated in the envelope, but the final surface temperature is similar (PMP06)

where $\mathbf{b}$ is the unit vector in the direction of the magnetic field and $\tilde{T}$ is the redshifted temperature. For simplicity, we have considered the classical relaxation time approximation in which only electrons carry heat. The ratio between the parallel conductivity ($\kappa_\parallel$) and the perpendicular one ($\kappa_\perp$) is related to the magnetization parameter (x_B) as $\kappa_\parallel/\kappa_\perp = 1 + x_B^2$ (Urpin and Yakovlev 1980).

The Hall contribution to the heat flux is given by the last term on the right hand side of (1). If the magnetic field geometry has only poloidal components and the temperature distribution does not depend on the azimuthal angle (ϕ); the divergence of the Hall term vanishes (Geppert et al. 2004) and it does not affect the energy balance equation. However, for a magnetic field structure with a toroidal component, this term contributes to the heat flux, even in axial symmetry. When the magnetization parameter is large ($x_B \gg 1$), the dominant contribution to the flux is proportional to $x_B^2(\mathbf{b}\cdot\nabla T)$. Therefore, in order to reach the stationary configuration the temperature distribution must be such that the constant–temperature surfaces are nearly aligned with the magnetic field lines. This is shown explicitly in the left panel of Fig. 1, where we show the stationary solution for a purely poloidal configuration confined to the crust and the outer layers. The alignment is enforced in most of the crust and envelope and strong radial gradients are generated only close to the surface. When a toroidal component is introduced, the situation changes because the Hall term in (1) induces large meridional fluxes ($\sim x_B$) resulting in an almost isothermal crust. This is clearly seen in the central panel which shows the temperature distribution for a force-free magnetic field with a toroidal component in the outer layers (crust and envelope). For comparison, we also consider another non–force-free model (right panel) which has a toroidal component confined to a thin crustal region (toroidal confined, TC in the following), with a maximum value of 2×10^{15} G. It acts as an insulator keeping a different temperature at both sides of the toroidal field. In the region external to the toroidal field, only the poloidal component is present and the isothermal surfaces are aligned with the magnetic field lines. This does not happen if the B_ϕ component extends all the way up to the surface, as in the central panel. We point out again that the poloidal field is the same in all three models (solid lines, $B_p = 10^{13}$ G) even if the field lines have been omitted in the central panel for clarity.

As stated in the Introduction, a strong enough magnetic field can induce a phase transition turning the gaseous at-

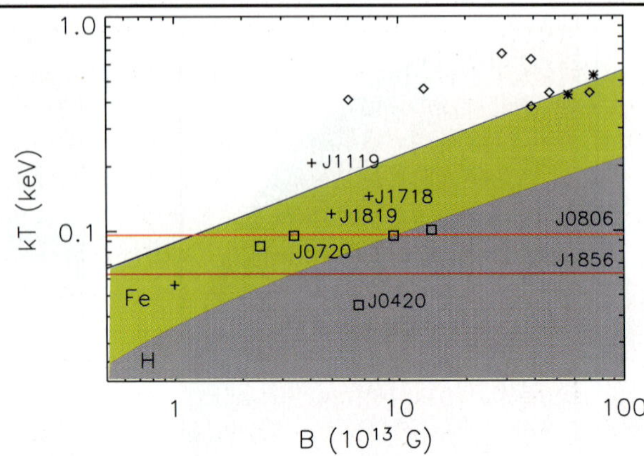

Fig. 2 Phase diagram for condensation of pure iron (*yellow*) and pure hydrogen (*gray*). Data from isolated neutron stars (*squares*), SGRs (*stars*), AXPs (*diamonds*) and some high magnetic field pulsars (*crosses*) are shown. The BB temperatures of the INSs RX J1856 and RX J0806 are shown as *horizontal lines*

mosphere into a liquid or solid state (Lai 2001). In Fig. 2 we show the kT–B phase diagram for this phase transition together with the observational data for some INSs, pulsars and magnetars. We also point out that condensed–surface models predict the existence of a spectral edge which, for $B \approx 10^{13}$–10^{14} G, falls in the range 0.2–0.5 keV. The absorption feature is a promising possible explanation for the absorption line recently reported in RX J0720-3125 (Haberl et al. 2004). We refer the interested reader to a previous work (PMP06) for details about the calculations. Here, we briefly describe the main features of the magnetic field geometry and surface temperature distribution of the models which are later applied to the available X-ray observations of RX J0720-3125.

The structure of relativistic stars with both poloidal and toroidal magnetic field components has been studied e.g. in (Ioka and Sasaki 2004). It was shown that all relevant quantities can be determined from a stream function that satisfies the relativistic Grad-Shafranov equation. In the linear regime (weak magnetic field, in the sense that induced deformations are small), the Grad-Shafranov equation becomes simpler and, under the assumptions of axisymmetry and stationarity, the general interior solution has the form (Ioka and Sasaki 2004)

$$\mathbf{B} = B_0\left(2\frac{\cos\theta}{r^2}\Psi(r), -\frac{\sin\theta}{r}\frac{\partial\Psi(r)}{\partial r}, \mu\frac{\sin\theta}{r}\Psi(r)\right) \qquad (2)$$

where μ is a constant related to the wavenumber of the magnetic field. In general, the l multipole of the stream function must satisfy the differential equation

$$\frac{d^2\Psi(r)}{dr^2} + \left(\mu^2 - \frac{l(l+1)}{r^2}\right)\Psi(r) = 4\pi r^2 \rho a_0, \qquad (3)$$

where ρ is the energy density, $a_0 = a_0(r)$ is a function that depends on the boundary conditions, and where we have omitted the relativistic factors for clarity. In the case $\mu = 0$, the purely poloidal case is recovered, as discussed for example in Konno et al. (1999) for the weak magnetic field limit, and more thoroughly in Bonazzola and Gourgoulhon (1996) for the fully non–linear case. For the general case with both poloidal and toroidal components, a similar non–linear (although Newtonian) analysis is presented in Tomimura and Eriguchi (2005) where (2) is generalized.

In the following, we consider force-free solutions ($a_0 = 0$) with μ chosen to confine the magnetic field to the crust and outer regions, while the radial component is forced to match smoothly the vacuum dipole solution. In particular, we consider two cases with $\mu = 1.34\ \mathrm{km}^{-1}$ and $\mu = 3.87\ \mathrm{km}^{-1}$ depending on the considered magnetic field configuration (purely dipolar or quadrupole–dominated respectively). Notice that μR is roughly the ratio of the toroidal to the poloidal field while, as mentioned above, μ^{-1} is related to the magnetic field typical length–scale.

From the results of our calculations, it turns out that the classical semi-analytic temperature distribution derived by Greenstein and Hartke (1983) remains a good approximation. A simple but accurate analytical approximation to the temperature distribution for most models, which is more realistic that discontinuous two temperature models, is

$$T^4(\theta) = T_p^4 \frac{\cos^2\theta}{\cos^2\theta + \frac{a+1}{4}\sin^2\theta} + T_{\min}^4, \qquad (4)$$

where T_p is the temperature at the magnetic pole, $T_{\min}$ is the minimum temperature reached on the surface of the star (typically $T_{\min} < 10T_p$) and the parameter a takes into account the relative strength between poloidal and toroidal components. For a dipolar field $a = 0$, while for force-free models with magnetic field confined to the crust and envelope $a \approx 250$. We have found that the approximate expression above is accurate at the 3% level. Example of temperature distributions and more details can be found in Pérez–Azorín et al. (2006b).

The main features of the considered models are: (i) purely dipolar models show nearly equal antipodal hot polar caps (with only minor differences due to the Hall term); (ii) when a quadrupolar component is present, the north/south symmetry is broken by the different temperatures and sizes of the polar caps; (iii) a dominant quadrupolar component implies also the existence of a hot "equatorial" belt, that can however be displaced from the equator depending on the relative strength of the dipolar and quadrupolar components. More complex geometries could be obtained by including higher order multipoles. In general, smaller values of μ correspond to larger angular sizes of polar caps and/or belts.

Table 1 Summary of the XMM–Newton observations. All observations are performed in Full Frame with the Thin Optical filter applied except in Rev. 175 where the Medium filter was used

Rev.	Instrument	Epoch	Exposure
78	mos1	2000 May 13	48 ks
175	pn	2000 Nov. 21–22	23 ks
533	pn	2002 Nov. 6–7	26 ks
533	mos1	2002 Nov. 6–7	30 ks
534	pn	2002 Nov. 8–9	27 ks
534	mos1	2002 Nov. 8–9	32 ks
622	mos1	2003 May 2–3	24 ks
711	mos1	2003 Oct. 27–28	14 ks
815	pn	2004 May 22–23	22 ks
815	mos1	2004 May 22–23	26 ks
1086	pn	2005 Nov. 12–13	34 ks

3 Summary of the XMM-Newton observations

RX J0720-3125 has been observed several times by *XMM–Newton* and we focus here on EPIC data (pn and MOS 1) from the 8 publicly available observations. The *XMM–Newton* observations span a period of about 5.5 years starting from May 2000 (Rev. 78) to November 2005 (Rev. 1086). Here we ignore the observations performed with the cameras operated in Small Window because of the worse calibration with respect to Full Frame science modes. The pn data of Rev. 078 are also excluded from the analysis due to problems in the SAS 6.5.0 reduction pipeline (see http://xmm.vilspa.esa.es/ for details). The remaining data provide a homogeneous set and have all been collected with the cameras operated in Full Frame (FF) mode with the Thin or Medium filter applied, and a summary of the observations is presented in Table 1. The MOS 2 data are consistent with MOS 1 and do not add relevant information to our analysis. Photon pile–up has been minimised by following standard procedures (see e.g. Haberl et al. 2004). Since the softest energies suffer from calibration uncertainties, we consider the 0.18–1.2 keV band only, after having checked that the inclusion of data down to 0.13 keV does not change our results in any noticeable way.

4 Spectral analysis with realistic, self–consistent emission models with a condensed surface

Simple BB plus absorption line fits to the X-ray spectra of RX J0720-3125 provide a very good description of the X-ray data, but they cannot explain in a self–consistent way the Optical flux which is observed to exceed by about a factor 6 the predicted one. In the following, we present an attempt to describe the X-ray data with synthetic spectra for the emission from a condensed iron surface, obtained through detailed numerical simulations.

We have considered a fiducial NS model fixing the mass at $1.4M_{\odot}$ and the radius at 12.27 km. While the first choice is well motivated by observations of NSs in binary systems the radius of NSs is less certain. In the present case, however, we are interested in the X-ray (and optical) emission from the star surface which, in absence of spectral features that can reveal the redshift at the surface, will depend very little on the assumed radius. Indeed, our models are not significantly affected if the NS radius is varied in a sensible range (e.g. 10–15 km). This motivates and justifies the choice of a fiducial model which greatly simplifies the numerical task of computing large grids of models if all parameters are allowed to vary. The magnetic field intensity is defined at the pole and we consider variations in the range of $B_p = 0.5\text{–}6 \times 10^{13}$ G, as suggested by the value of 2.4×10^{13} G estimated from the period decay. As for the magnetic field configuration, we explore two different cases: a purely dipolar magnetic field with $\mu = 1.34$ km^{-1}, and a quadrupole–dominated one ($\beta_d = 0.05$) with $\mu = 3.87$ km^{-1}. This is only a small sub–set of the possible configurations and our choice is based on a relatively time–consuming assessment of which configuration provides a better description of the X-ray data. In particular, (very) different values for the radial length–scale of the magnetic field (μ) resulted in inconsistent results in which the radius of the NS, as inferred from the X-ray spectral fits, was different from that of the input model.

Having set up a baseline model, we have built tabular XSPEC models[2] as a function of the pole temperature and magnetic field strength for several orientations within the range allowed by the pulsation profiles (as discussed in the next section). We then consider all available *XMM–Newton* observations of RX J0720-3125 and fit our realistic (and limited in parameter–space) models to the data, with the addition of photo-electric absorption.

4.1 Constraining the parameter space with the pulsed fraction

In Fig. 3 we show contour plots of the pulsed fraction, for two of our calculations, as a function of the angle between spin and magnetic axis ($\mathcal{B}$) and between spin and observer direction ($\mathcal{O}$). In general, we can classify the pulsation profile (i.e. the shape of the light curve once folded at the observed period) in two groups. The models with large $\mathcal{O}+\mathcal{B}$ are characterized by a non–sinusoidal profile, or even two visible maxima; this is because we actually can see both poles in each period if $\mathcal{O}+\mathcal{B} > 90°$. Therefore, INSs which

[2]Tabular models are available upon request to the authors.

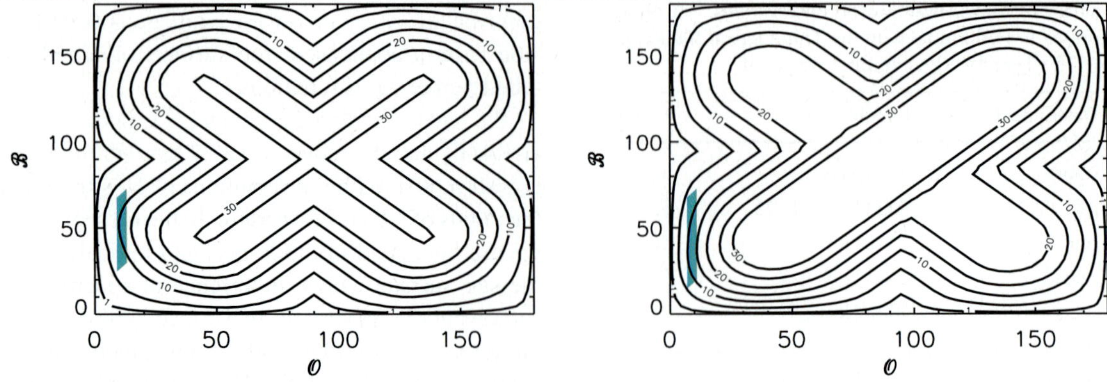

Fig. 3 *Left*: Lines of constant pulsed fraction for a purely dipolar magnetic field, with $n_H = 1.33 \times 10^{20}$ cm^{-2}, $T_p = 115$ eV and $B_p = 1.8 \times 10^{13}$ G. $\mathcal{B}$ is the angle between rotation and magnetic axis, and $\mathcal{O}$ the angle between the rotation axis and the observer direction. The flux has been obtained by integration over the whole surface taking into account General Relativistic light bending effects. *Right*: same as in the left panel but for a quadrupole–dominated magnetic field ($\beta_d = 0.05$) with the same T_p, B_p as before and with $n_H = 1.2 \times 10^{20}$ cm^{-2}

exhibit non–sinusoidal pulse profiles do lie in this region. On the other hand, models with small $\mathcal{O} + \mathcal{B}$ always show a single peaked profile which is very close to sinusoidal when either one of the angles is small. This could be the case of RX J0720-3125, that shows a very regular and almost perfectly sinusoidal pulsation profile. Next, we can reduce the range of angles to those consistent with the observed value of the pulsed fraction. For a nearly spherical neutron star, the rotation axis is essentially aligned with the (vector) total angular momentum of the star, which is conserved. Therefore, the variation of the angle $\mathcal{O}$ with time is too small to be observable, but the star can wobble around its symmetry axis (in general different from both, rotation and magnetic axis) with a free precession timescale of a few years for oblateness of the order of 10^{-7} (Jones and Andersson 2001). We have tried different orientations in the range of $\mathcal{O}$ and $\mathcal{B}$ allowed by the observed pulsed fraction. We find that for purely dipolar models, the observed pulsed fraction values are best reproduced if $\mathcal{O} = 12°$ and $\mathcal{B}$ varies in the range 30°–60° (i.e. 180° $- \mathcal{B}$ in the range 120°–150°). With this choice of parameters, the pulsed fraction of our models always lies in the range between 9% and 13%, consistent with the observed values. Therefore, we will focus on the vertical shaded region, fixing $\mathcal{O}$ and allowing for variations in $\mathcal{B}$.

The right panel of Fig. 3 is the same as in the left panel, but for the model with a dominant quadrupolar component ($\beta_d = 0.05$). The very different surface temperature distribution (two hot spots plus a hot belt close to the equator, as discussed in Sect. 2) produces significant differences with respect to the purely dipolar case. First, the allowed maximum pulsed fraction is much higher, up to 45%. Second, the north/south symmetry has been broken, and the results are different when changing either $\mathcal{B}$ or $\mathcal{O}$ by $\pi - \mathcal{B}$ or $\pi - \mathcal{O}$ (the symmetry with respect the simultaneous interchange of both angles by their supplementary is kept). Third, a portion of the hot belt is now visible, which will modify the spectrum. We proceed as in the previous case, and we first localize the region in which we are allowed to vary the angles $\mathcal{O}$ and $\mathcal{B}$, keeping the pulsed fraction at about 9–11% and close to a sinusoidal shape, as observed. This corresponds to $\mathcal{O} = 11°$ and $\mathcal{B} = 20°$–65° (see Fig. 3, right). As mentioned, the choice $\mathcal{O} = 169°$ and $\mathcal{B} = 115°$–160° produces the same results.

4.2 Dipolar vs. quadrupolar magnetic fields

We begin our spectral analysis with a purely dipolar magnetic field configuration, suggested by the regular sinusoidal shape of the light curve of RX J0720-3125 (when folded at the spin period of 8.391 s). If one lets all parameters to vary, the statistical quality of the fits obtained with the realistic models is comparable to that obtained with BB plus Gaussian models (Pérez–Azorín et al. 2006b), but the accuracy on the most relevant parameters ($\mathcal{B}$, $kT_{\rm pole}$, and B_p) is limited by the fact that the inferred variations of $n_{\rm H}$ drive the fitting results, strongly affecting the soft energy band where most of the photons are collected. Given the high proper motion of the source, a variation in the absorbing column cannot be excluded, but the observed variation seem to occur almost randomly and the pn and MOS 1 data often give inconsistent results for the same observation. It is in our opinion more realistic to assume that the column density is the same in all observations.

To overcome the inter–calibration uncertainties, we here consider the pn observations only and we force the absorbing column density to be the same in all observations (see Pérez–Azorín et al. (2006b) for more details and results). We also force the model normalization (directly related to the NS radius and distance) and the magnetic field intensity to be the same in all observations. Our results are presented in Table 2. The best–fitting magnetic field intensity B_p is

Table 2 Joint fits to EPIC–pn observations for a purely dipolar field ($\beta_d = 1$) with $\mu = 1.34$ and $\mathcal{O} = 12°$, forcing the hydrogen column density, the normalization constant and the magnetic field intensity to be the same for all observations. The best fit gives $n_H = 1.33^{+0.04}_{-0.02} \times 10^{20}$ cm^{-2}, $R = (D/300\ \text{pc})11.3^{+1.1}_{-1.8}$ km, and $B_p = 14.1^{+0.6}_{-0.2} \times 10^{12}$ G. The overall χ^2/d.o.f. is 1570/819

Rev.	$\mathcal{B}$	kT_{pole} (eV)
175-pn	$20.5^{+6.5}_{-10.5}$	$120.6^{+0.2}_{-0.6}$
533-pn	$29.7^{+1.9}_{-2.2}$	$123.3^{+0.3}_{-0.4}$
534-pn	$31.3^{+1.7}_{-2.4}$	$123.5^{+0.4}_{-0.4}$
815-pn	$56.0^{+0.7}_{-0.7}$	$132.3^{+0.5}_{-0.4}$
1086-pn	$52.2^{+1.5}_{-1.5}$	$131.2^{+0.3}_{-0.7}$

Table 3 Same as Table 2 for a quadrupole dominated magnetic field ($\beta_d = 0.05$) with $\mu = 3.87$ and $\mathcal{O} = 11°$. The best fit gives $n_H = 1.23^{+0.04}_{-0.04} \times 10^{20}$ cm^{-2}, $R = (D/300\ \text{pc})12.7^{+1.0}_{-2.1}$ km, and $B_p = 18.2^{+1.6}_{-1.5} \times 10^{12}$ G. The overall χ^2/d.o.f. is 1192/819

Rev.	$\mathcal{B}$	kT_{pole} (eV)
175	$73.9^{+6.0}_{-14.1}$	$104.6^{+0.4}_{-0.4}$
533	$58.6^{+2.0}_{-2.4}$	$107.5^{+0.2}_{-0.3}$
534	$57.6^{+2.0}_{-2.4}$	$107.5^{+0.4}_{-0.3}$
815	$26.9^{+1.7}_{-0.6}$	$118.2^{+0.5}_{-0.2}$
1086	$31.8^{+0.9}_{-1.1}$	$116.3^{+0.5}_{-0.3}$

of the order of 1.4×10^{13} G, about a factor 2 smaller than that inferred from the observed period decay, but compatible with the uncertainties due to the lack of accurate knowledge of the magnetic field geometry. Nevertheless, a clear long–term evolution of the magnetic field orientation $\mathcal{B}$ emerges from our results. $\mathcal{B}$ increases with time up to Rev. 815, while it decreases in the last observation (Rev. 1086). By considering the vertical shaded area in Fig. 3 (or its equivalent complementary at $\pi - \mathcal{B}$), this solution suggests that the pulsed fraction must also exhibit a long–term evolution, with a maximum pulsed fraction around $\mathcal{B} \sim 55°$, corresponding to the latest observations. We point out that the $\mathcal{B}$ long–term evolution has the same behaviour as the BB temperature and the absorption line equivalent width (Haberl et al. 2006) and (see below) pulsed fraction. Our realistic model is thus trying to reproduce the observed long–term spectral variability with changes in the magnetic field orientation, which is what is expected if precession is responsible for the long–term variability (as suggested by Haberl et al. (2006) on the basis of BB fits to the data). The dipolar model explored so far seems able to describe the bulk of the observed long–term variability in terms of precession and is therefore very promising. However, the quality of the fits is not very satisfactory and we thus explore in the following a case in which a strong quadrupolar component is added to the magnetic field configuration.

We proceed as above, considering the sub–set of pn data only, and forcing the absorbing column density, the normalization, and the magnetic field intensity to be one and the same in all observations. Our results are presented in Table 3. We obtain a very significant statistical improvement with respect to the purely dipolar model described above and our results now approach the statistical quality of the phenomenological fits (see Pérez–Azorín et al. (2006b) for a comparison). With the addition of the quadrupolar component, the best–fitting magnetic field intensity raises to $1.8 \pm 0.2 \times 10^{13}$ G which is more in line with the value inferred from period decay. As for the dipolar model discussed above, the magnetic field orientation $\mathcal{B}$ clearly shows a long–term evolution similar to the $\pi - \mathcal{B}$ evolution of the dipolar model. The issue of the long–term evolution of the magnetic axis orientation will be discussed in Sect. 5.

4.3 The hardness ratio anti–correlation

The better statistical quality of the fits to the X-ray data and the more consistent value of the magnetic field intensity are, up to some extent, an indication of the presence of higher order multipolar components. However, as recently discussed in Zane and Turolla (2006) there is a more conclusive observational fact that allows us to distinguish between the purely dipolar and the multipolar case. A striking feature of the X-ray light curves of RX J0720-3125 is the clear anti–correlation of the hardness ratio with the pulse profiles in both the hard and the soft band. This feature cannot be explained by purely dipolar models. A possible explanation that has been proposed is a model with two hot spots, with one of them displaced by about 20° from the south pole (Haberl et al. 2006).

Motivated by the well established hardness ratio anti–correlation, we explored whether our models can reproduce the observed behaviour. In Fig. 4 we show the folded light curves in the hard and soft bands, and the hardness ratio for our best–fitting quadrupole–dominated model (top panel) and for the data (bottom). The observed anti–correlation of the hardness ratio with the pulse profiles is qualitatively well reproduced by our model. We point out that such good agreement is impossible to obtain with pure dipolar fields in which the North/South symmetry is not broken. We have not attempted at this stage to find the best–fitting model to both the light curves and the phase averaged spectra because this is a formidable numerical task, especially when dealing with realistic models that need a serious computational effort to be produced. However, the good qualitative agreement seems to indicate that we could be seeing a neutron star with a *hot spot and a hot belt*, as an alternative to the case of displaced hot spots.

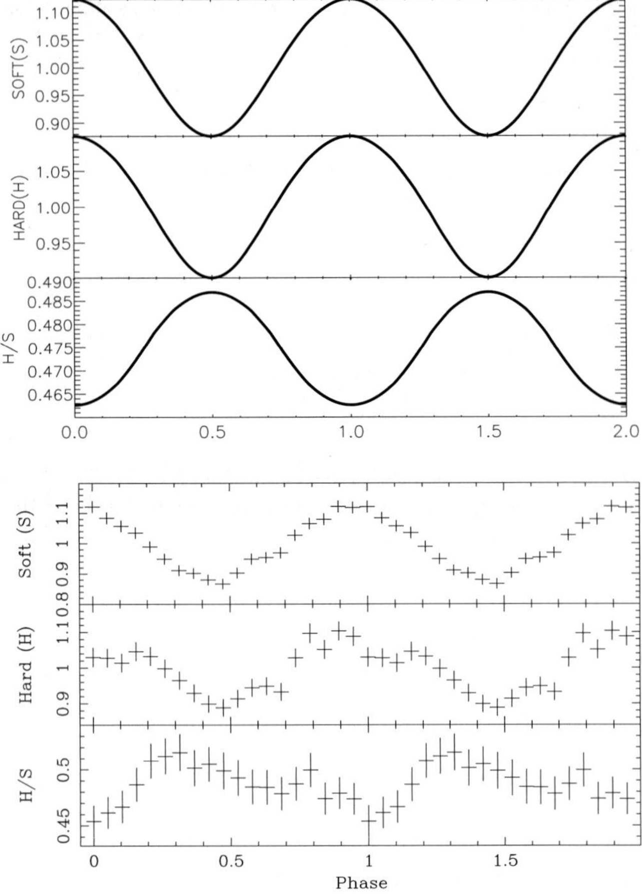

Fig. 4 Pulse profiles in two energy bands (soft: 0.12–0.4 keV, hard: 0.4–1.0 keV) for a realistic model with $\mathcal{O} = 11$, $\mathcal{B} = 54$, $T_p = 115$ eV, $B_p = 1.8 \times 10^{13}$ G, $n_H = 1.2 \times 10^{20}$ cm^{-2} and $\beta_d = 0.05$. The *third panel* is the hardness ratio (HARD/SOFT). The *bottom panel* shows the observational data corresponding to observation rev. 534. Notice that we have not attempted to perform a fit of the pulse profiles, it is simply a comparison with a realistic model

5 Precession

Evidence for precession of RX J0720-3125 has recently been reported (Haberl et al. 2006). The evidence was based on results from BB plus absorption line fits to the available X-ray data in which the BB temperature shows a long–term evolution best described by a sine wave with a period of 7.1 ± 0.5 yr. If the variability is due to free precession of the neutron star, one should be able to find a unique model (temperature distribution, magnetic field configuration, etc.) that explains all observational properties with (periodic) changes in the relative orientation only. We then consider our results for the dipolar and quadrupole–dominated models (see Tables 2 and 3) and fit the time evolution of the relevant angle ($\mathcal{B}$) with a periodic function. The results are shown in Fig. 5. Notice that the right panel shows $\pi - \mathcal{B}$ (this is just a matter of the arbitrary choice of the origin of coordinates). Both models (dipole and quadrupole–dominated) are consistent with the interpretation of a precessing neutron star with a precession period of about 7 years and a relatively large wobbling angle. Both cases are very well described by a sine wave but, as mentioned earlier, the hardness ratio anti–correlation strongly favors models with a significant quadrupole component. Our results point in the direction that most of the spectral variation can be explained by precession of the neutron star. Notice that a precession timescale of a few years has also already been reported for some pulsars (Link and Epstein 2001).

6 Optical flux

An issue under debate in the condensed–surface models is the emissivity in the optical band. When the effect of the motion of ions is completely neglected the optical flux is depressed (Turolla et al. 2004; Pérez-Azorín et al. 2005). However, ignoring the motion of ions at low energies may not be justified. Indeed, a simple treatment considering the

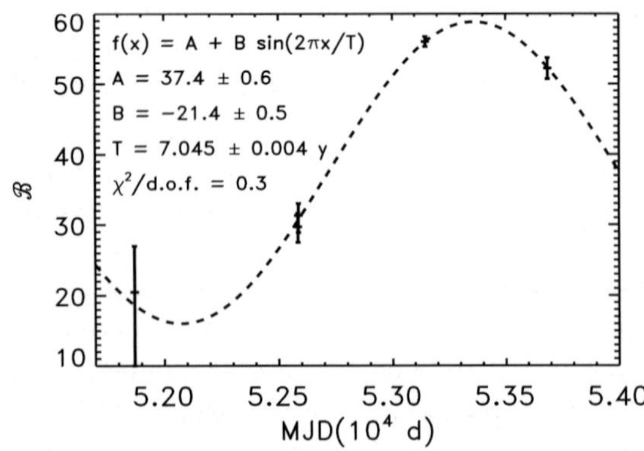

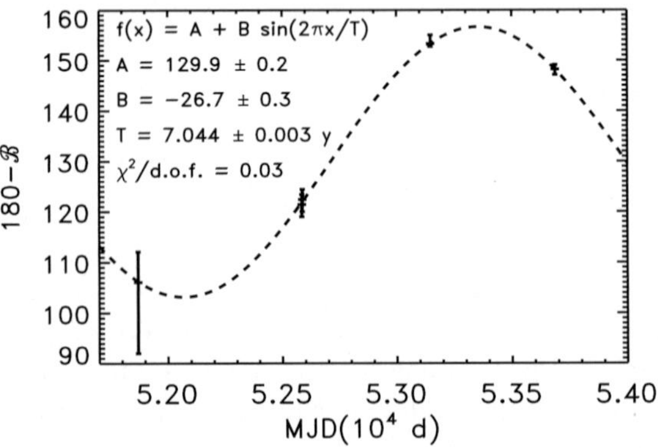

Fig. 5 Time variation of the orientation of the magnetic axis, $\mathcal{B}$, from the results shown in Tables 2 (*left panel*, $\beta_d = 1$) and 3 (*right panel*, $\beta_d = 0.05$). The *solid line* shows the best fit to the data (see text in figure)

ions as free particles leads to very different results (Ginzburg 1970). This simple approach (to consider them as free particles) could be approximately correct if the condensed surface is in a liquid state, but it is clearly largely inaccurate in the case of a solid surface where the ions are placed in a lattice. However, even in the presence of a solid lattice, the ions do actually move and resonant plasmon or phonon excitations are generally important. Moreover, strong magnetic fields can also modify the motion of the ions (the ion cyclotron radius is much smaller than the separation between ions in the lattice).

It is clearly out of the scope of this paper to discuss the microphysical behaviour of this condensed (liquid or solid) atmospheres, and we do not know much about the characteristic frequencies of the lattice and how it couples to the magnetic field. We just want to point out that several important energy scales (the ion cyclotron energy $\hbar\omega_{B,i}$, kT, and $\hbar\omega_{p,\mathrm{ion}}$) are similar ($\sim$0.1 keV) under our typical conditions and, consequently, a careful analysis of the emissivity in the optical band is needed before we can establish severe constraints on the models. As a compromise, since the *free ion* treatment predicts an emissivity in the optical band ($\alpha = 1$ corresponds to a blackbody) of about $\alpha = 0.7$ and the *frozen ions* model predicts approximately $\alpha = 0.2$ (Pérez-Azorín et al. 2005; van Adelsberg et al. 2005), we will show both limits in the following figure. Naively, one expects that the real emissivity is confined between the two values.

In Fig. 6, we show the unfolded spectrum and best–fitting models of RX J0720-3125 (Rev. 534) together with the available optical-UV data (Kaplan et al. 2003). We have not attempted to fit simultaneously the optical and X-ray data, but the good agreement is evident. We also show (dotted lines) the best–fitting BB plus Gaussian model to the same data to illustrate its similarity with the realistic model in the X-ray band and to highlight the well known problem of under–predicting the optical flux.

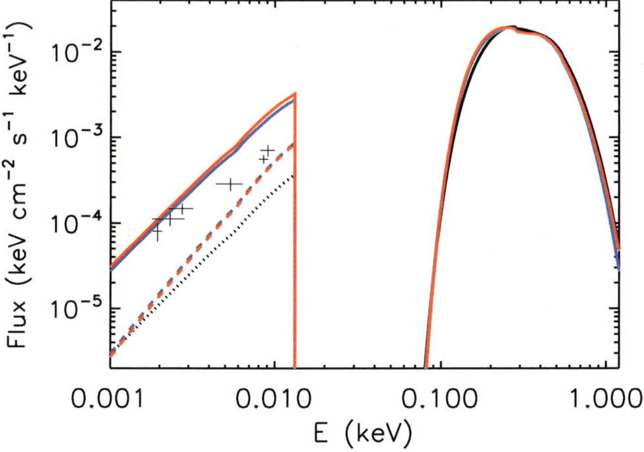

Fig. 6 Spectral energy distribution and best–fitting models for the XMM–Newton observation in revolution 534. We show the phenomenological model (BB+Gaussian, *black solid*), our realistic quadrupole–dominated model (Table 3, *blue solid*), and our realistic dipole–dominated one (Table 2, *red solid*). The dotted line is the optical tail of the BB model. Solid and dashed lines in the optical band correspond to models with free and frozen ions, respectively. Available optical and UV data are also shown (*crosses*)

The earlier (lower energy) optical observations (Jul 2001) show a somewhat higher flux than the latter ones (Feb 2002). The fact the optical observations are not consistent with a pure Rayleigh-Jeans tail has been discussed by Kaplan et al. (2003), who found that the best fit consists of two BBs plus a power law. However, they also comment that the spectrum can be consistent with a Rayleigh-Jeans tail if the deviations have a temporal nature, which is thought to be unlikely because the X-ray flux is constant. Interestingly, all observations fall in between the two lines predicted by the *free ions* and *frozen ions* models, suggesting that more effort must be placed in understanding the physics of the emission processes.

7 Summary and conclusions

We have shown that our models of emission from the condensed surface of neutron stars with strong magnetic fields are consistent with the observed X-ray and optical spectra, the observed deviation from a pure thermal spectrum in the X-ray band, and the long–term spectral variability of RX J0720-3125. Although we do not exclude the presence of a resonant proton cyclotron absorption or bound-bound transitions in H, we do not need to invoke any of the above mechanisms to explain the observed absorption feature in the soft X-ray band which is naturally produced in our models. We remind here that our model only considers the surface thermal emission, which could be modified by the interaction of the radiation with the plasma in the magnetosphere.

Once the magnetic field configuration is given, we are able to obtain a self-consistently calculated thermal spectrum that reproduces reasonably well all available observations. Our analysis indicates that most of the long–term spectral variation of the source can be explained in terms of neutron star precession. Similar conclusions have been reached independently by other groups (Haberl et al. 2006). Notice, however, that our results still show some variability of the pole temperature correlated to the variability of the orientation angle. In theory, a *perfect* model should provide the surface temperature distribution explaining the long–term variation by changes in the orientation angle only. We have provided a step forward in that direction, since our fits are obtained keeping fixed n_H, normalization (i.e. distance), magnetic field intensity, and the star model (mass, radius, temperature distribution geometry), but we still need to allow for small variations of the pole temperature. It is of

course possible that the temperature and/or size of the hot regions really changes on a timescale of 7 years, but we are not aware of any mechanism able to do this in such a short timescale. Moreover, if this was the case, no long–term evolution periodicity would be expected.

Since we have only explored a few magnetic field configurations, an extensive study including a more complete grid of models for spectral fitting is likely to improve the statistical quality of the fits presented here. Nevertheless, we are confident that some qualitative features we have discussed above are robust: (i) the models that reproduce the observational data have magnetic fields confined to the crust and the outer regions, and strong toroidal components (about an order of magnitude larger than the poloidal one). This naturally leads to large surface temperature anisotropies which can explain also the observed pulsations. (ii) Models with strong quadrupolar components are in general favored. Higher order multipoles (octupole, etc.) can in principle be added, but high order multipoles are probably in conflict with large pulsation amplitudes. (iii) The anti-correlation of the hardness ratio also requires a temperature distribution with some degree of asymmetry. In particular, models with two hot spots (maybe with different temperatures and sizes) and a hot belt close to the equator can also explain the observations.

The condensed–surface model under consideration implies the presence of a spectral edge around 0.3 keV and the absence of other spectral lines expected from gaseous atmospheres with heavy elements. We think that this part of our results is not as robust as the strong anisotropy in the temperature distribution produced by toroidal fields and non–dipolar components. There are interesting alternatives that can explain the observed spectral absorption feature equally well, but the physical motivation of the condensed surface is that the magnetic field is high enough to produce the condensation, if the composition is primarily iron or similar heavy elements. Strongly magnetized H atmospheres (or other light elements) are still an alternative, but to our knowledge there is no fully consistent treatment for arbitrary magnetic field orientations and structure. More work in both the condensed–surface and strongly magnetized atmospheres frameworks will be certainly worth to pursue in the near future and is likely to significantly improve our current understanding of INS emission mechanisms.

If future observations confirm, as discussed here, that RX J0720-3125 (and other INSs) is subject to precession with a relatively large wobbling angle combined with a relatively short period of the order of years, it would have very interesting implications. As pointed out by Link (2003), a large wobbling angle with a precession timescale of the order of years is in conflict with the standard picture of a type II superconducting core. This scenario favors either type I superconductivity or a situation in which the magnetic field does not penetrate into the core and is instead mainly confined to the crust and outer regions. We point out that this was exactly the starting assumption of our models: a magnetic field living in the crust that produces the large temperature anisotropies needed to reproduce the observed properties. Alternatively, Jones (2004) suggests that long-term precession leads to the conclusion that nuclei and superfluid neutrons do not coexist, which also would establish severe constraints on the equation of state. At present, the uncertainties in our knowledge of vortex dynamics in the inner crust can lead to different conclusions, all of them interesting and based on the observational confirmation of long-lived under-damped precession.

We are also working to study other NSs that we expect to be well described by the condensed–surface models. Preliminary results on RBS1223 are promising, since we could find good fits to the very irregular light curve and the large pulsed fraction, again with quadrupolar magnetic fields. Another interesting NS is the pulsar PSR J1119-6127, that has a very high pulsed fraction (70%). This large value cannot be reproduced with purely dipolar components but we found some (still axisymmetric) magnetic field configurations that can explain this extremely large variability. Results about this objects will be reported in future work.

Acknowledgements This work has been supported by the Spanish Ministerio de Ciencia y Tecnología grant AYA 2004-08067-C03-02. J.A.P. is supported by a *Ramón y Cajal* contract from the Spanish MEC. G.M. thanks the UK PPARC for support.

References

Bonazzola, S., Gourgoulhon, E.: Astron. Astrophys. **312**, 675 (1996)
Brinkmann, W.: Astron. Astrophys. **82**, 352 (1980)
de Vries, C.P., Vink, J., Méndez, M., et al.: Astron. Astrophys. **415**, L34 (2004)
Drake, J.J., Marshall, H.L., Dreizler, S., et al.: Astrophys. J. **572**, 996 (2002)
Geppert, U., Küker, M., Page, D.: Astron. Astrophys. **426**, 267 (2004)
Geppert, U., Küker, M., Page, D.: Astron. Astrophys. **457**, 937 (2006)
Ginzburg, V.L., Propagation of Electromagnetic Waves in Plasmas, 2nd edn. Pergamon Press, Oxford (1970)
Greenstein, G., Hartke, G.J.: Astrophys. J. **271**, 283 (1983)
Haberl, F., Motch, C., Buckley, D.A.H., et al.: Astron. Astrophys. **326**, 662 (1997)
Haberl, F., Schwope, A.D., Hambaryan, V., et al.: Astron. Astrophys. **403**, L19 (2003)
Haberl, F., Zavlin, V.E., Trümper, J., et al.: Astron. Astrophys. **419**, 1077 (2004)
Haberl, F., Turolla, R., de Vries, C.P., et al.: Astron. Astrophys. **451**, L17 (2006)
Ioka, K., Sasaki, M.: Astrophys. J. **600**, 296 (2004)
Jones, D.I., Andersson, N.: Mon. Not. Roy. Astron. Soc. **324**, 811 (2001)
Jones, P.B.: Phys. Rev. Lett. **92**, 149001 (2004)
Link, B.: Phys. Rev. Lett. **91**, 101101 (2003)
Link, B., Epstein, R.I.: Astrophys. J. **556**, 392 (2001)
Kaplan, D.L., van Kerkwijk, M.H., Marshall, H.L., et al.: Astrophys. J. **590**, 1008 (2003)

Kaplan, D.L., van Kerkwijk, M.H.: Astrophys. J. **628**, L45 (2005)

Kaplan, D.L., van Kerkwijk, M.H.: Astrophys. J. **635**, L65 (2005)

Konno, K., Obata, T., Kojima, Y.: Astron. Astrophys. **352**, 211 (1999)

Lai, D.: Rev. Mod. Phys. **73**, 629 (2001)

Pérez-Azorín, J.F., Miralles, J.A., Pons , J.A.: Astron. Astrophys. **433**, 275 (2005)

Pérez-Azorín, J.F., Miralles, J.A., Pons, J.A.: Astron. Astrophys. **451**, 1009 (2006)

Pérez-Azorín, J.F., Miralles, J.A., Pons, J.A., et al.: Astron. Astrophys. **459**, 175 (2006)

Pons, J.A., Walter, F.M., Lattimer, J.M., et al.: Astron. Astrophys. **564**, 981 (2002)

Turolla, R., Zane, S., Drake, J.J.: Astrophys. J. **603**, 265 (2004)

Tomimura, Y., Eriguchi, Y.: Mon. Not. Roy. Astron. Soc. **359**, 1117 (2005)

Urpin, V., Yakovlev, D.G.: Sov. Astron. **24**, 425 (1980)

van Adelsberg, M., Lai, D., Potekhin, A.: Astrophys. J. **628**, 902 (2005)

van Kerkwijk, M.H., Kaplan, D.L., Durant, M., et al.: Astrophys. J. **608**, 432 (2004)

Vink, J., de Vries, C.P., Méndez, M., et al.: Astrophys. J. **609**, L75 (2004)

Walter, F.M., Lattimer, J.M.: Astrophys. J. **575**, L145 (2002)

Walter, F.M., Matthews, L.D.: Nature **389**, 358 (1997)

Walter, F.M., Wolk, S.J., Neuhäuser, R.: Nature **379**, 233 (1996)

Zane, S., Turolla, R.: Mon. Not. Roy. Astron. Soc. **366**, 727 (2006)

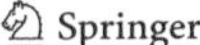

Astrophys Space Sci (2007) 308: 259–265
DOI 10.1007/s10509-007-9299-9

ORIGINAL ARTICLE

Neutron star surface emission: Beyond the dipole model

Silvia Zane

Received: 3 July 2006 / Accepted: 28 July 2006 / Published online: 21 March 2007

Abstract Recent Chandra and XMM-Newton observations of a number of X-ray "dim" pulsating neutron stars revealed quite unexpected features in the emission from these sources. Their soft thermal spectrum, believed to originate directly from the star surface, shows evidence for a phase-varying absorption line at some hundred eVs. The pulse modulation is relatively large (pulsed fractions in the range ~ 8–35% in amplitude), the pulse shape is often non-sinusoidal, and the hard X-ray color appears to be anti-correlated in phase with the total emission. Moreover, the prototype of this class, RX J0720.4-3125, has been found to undergo rather sensible changes both in its spectral and timing properties over a timescale of a few years. By modeling the light curves of two sources, RBS 1223 and RX J0720.4-3125, it has been found evidence for two hot regions located at a slightly non antipodal direction. All these new findings are difficult to reconcile with the standard picture of a cooling neutron star endowed with a purely dipolar magnetic field. Here we present more realistic models of surface emission, where the effects of different neutron star thermal and magnetic surface distributions are accounted for. We show how a star-centered field made of a dipolar and a quadrupolar component can influence the properties of the observed light curves and we present results that account self-consistently for toroidal and poloidal crustal field configurations.

Keywords Stars: neutron · X-rays · Magnetic fields

PACS 95.85.Nv · 97.10.Qh · 97.60.Jd

S. Zane (✉)
Mullard Space Science Laboratory, UCL, Holmbury St. Mary, Dorking, Surrey, RH5 6NT, UK
e-mail: sz@mssl.ucl.ac.uk

1 Introduction

Over the last few years a number of high resolution spectral and timing observations of thermally emitting neutron stars (NSs) become available thanks to new generation X-ray satellites (both *Chandra* and *XMM-Newton*), opening new perspectives in the study of these sources. In this respect, particularly interesting has been the outcome of deep observations of a small group of dim X-ray sources, originally discovered with *ROSAT* (hereafter XDINSs,[1] see e.g. Treves et al. 2000 and Haberl 2004 for reviews; Zane et al. 2005). In a sense, one may claim that these are the only "genuinely isolated" NSs: their soft thermal emission is unmarred by (non-thermal) magnetospheric activity nor by the presence of a supernova remnant or a binary companion. XDINSs play a key role in compact objects astrophysics: these are the only sources in which we can have a clean view of the compact star surface, and as such offer an unprecedented opportunity to confront theoretical models of neutron star surface emission with observations. In particular, when X-ray pulsations are detected, it is possible to study the shape and evolution of the pulse profile of the thermal emission, obtaining information about the thermal and magnetic map of the star surface.

The main observational characteristics of the seven XDINSs discovered so far are summarized in Table 1. The XDINSs X-ray spectrum is blackbody-like with temperatures in the range ~ 40–100 eV and, with no exception, it cannot be reproduced by available atmospheric models

[1] Although XDINSs is rather widely used, the debate on the acronym to designate these neutron stars is going on; see the contribution by J. Trümper in this conference, http://www.mssl.ucl.ac.uk/~sz/Conference_files/pres/Truemper_London.pdf

Table 1 Spectral and timing parameters of XDINSs. The last entry is the semi-amplitude of the pulsed fraction

Source	kT (eV)	$E_{\rm line}$ (eV)	P (s)	Semi-Ampl.
RX J0420.0-5022	44	329	3.45	13%
RX J0806.4-4123	96	460	11.37	6%
RBS 1223	86	290	10.31	18%
RX J0720.4-3125[a]	85	270	8.39	11%
RX J0806.4-4123	60			<1%
RBS 1774	100	700	9.44	4%
RBS 1556	96	493		

[a]Note that this is a varying source, see text for details

(which, on the other hand, are all computed under the simplifying assumption that temperature and magnetic field are both constant over the star surface). The need for a more complex field geometry and thermal map is also evident from the study of the X-ray light curve. So far, pulsations have been detected in five sources (with periods in the range 3–11 s), and in each case the pulsed fraction is relatively large ($\sim$ 8–35% in amplitude). Quite surprisingly, and contrary to what one would expect in a simple dipolar geometry, often the hardness ratio is minimum at the pulse maximum (Cropper et al. 2001; Haberl et al. 2003). Broad absorption features have been detected around $\sim$ 300–700 eV in all pulsating XDINSs and the line strength appears to vary with the pulse phase. In addition, the X-ray light curves exhibit a certain asymmetry, with marked deviations from a pure sinusoidal shape at least in the case of RBS 1223 (Haberl et al. 2003; Schwope et al. 2005).

In the standard picture, emission from an isolated, cooling NS arises when thermal radiation originating in the outermost surface layers traverses the atmosphere which covers the star crust. Although the emerging spectrum is thermal, it is not a blackbody because of radiative transfer in the magnetized atmosphere and the inhomogeneous surface temperature distribution. The latter is controlled by the crustal magnetic field, since thermal conductivity across the field is highly suppressed, and bears the imprint of the field topology.

The currently proposed one-dimensional magnetic atmosphere models represent a substantial improvement with respect to a simple blackbody representation, inasmuch they include most of the relevant physics, but a large amount of work remains to be done before they can be claimed to be fully satisfactory. In fact, models available so far only account for a single temperature and a single value of the magnetic field strength and inclination. The radiative transfer computations are basically plane-parallel and refer to a single surface patch with fixed B and T; the star surface is then assumed to consist of patches all equal to each other. Beside other things, changes in the magnetic inclination with respect to the line of sight are not accounted for.

The high quality data now available for thermally emitting NSs, and XDINSs in particular, demand for a detailed modeling of surface emission to be exploited to a full extent. Such a treatment should combine both an accurate formulation of radiation transport in the magnetized atmosphere and a quite general description of the thermal and magnetic surface distributions, which, necessary, must go beyond the simple dipole approximation. The ultimate goal is to produce a completely self-consistent model, capable to reproduce simultaneously both the spectral and timing properties. Here we review the first steps taken in this direction and present a systematic study of X-ray light curves from XDINSs.

2 Beyond the dipolar model

2.1 Observational motivations

Besides the observational motivations listed above, the evidence for a patchy and complex thermal and magnetic distribution at the star surface also arises from the study of two particular XDINSs. The first example is RBS 1223, that shows a peculiar double-peaked light curve (see e.g. Fig. 1). The separation between the two maxima is 0.47 phase units, different from that between the two minima of 0.43 phase units. As discussed by Schwope et al. (2005), this can be explained if the emission originates from two hot caps *which are not exactly antipodal*. By assuming that the hot regions are simple blackbody emitters, these authors were able to constrain the minimum spot separation to be $\sim 130°$. The NS viewing geometry is described in terms of two angles χ and ξ which give the inclination of the line of sight (LOS) and of the dipole axis with respect to the star spin axis; in the case at hand Schwope et al. (2005) found that the source should be a nearly orthogonal rotator, with $\chi \sim \xi \sim 80°$.

A further and possibly stronger evidence for the existence of hot spots not exactly antipodal comes from the spectacular example of RX J0720.4-3125. Although XDINSs were unanimously believed to be steady sources, recently *XMM-Newton* observations have unexpectedly revealed a substantial change in the spectral shape and pulse profile of this source over a timescale of a few years (De Vries et al. 2004; Vink et al. 2004; Haberl et al. 2006). In particular Haberl et al. (2006) have recently shown that precession of a neutron star with two hot spots of different T and size may account for the variations in X-ray spectra, changes in the pulsed fraction, shape of the light curve and phase-lag between soft and hard energy bands, provided that the two spots are not antipodal, but located at $\sim 160°$ from each other (see Fig. 2). The inferred precession period is 7.1 ± 0.5 yrs, consistent within the errors with that found by an independent sinusoidal fit to pulse timing residuals from a coherent analysis covering $\sim$ 12 yrs ($\sim 7.7 \pm 0.6$ yrs).

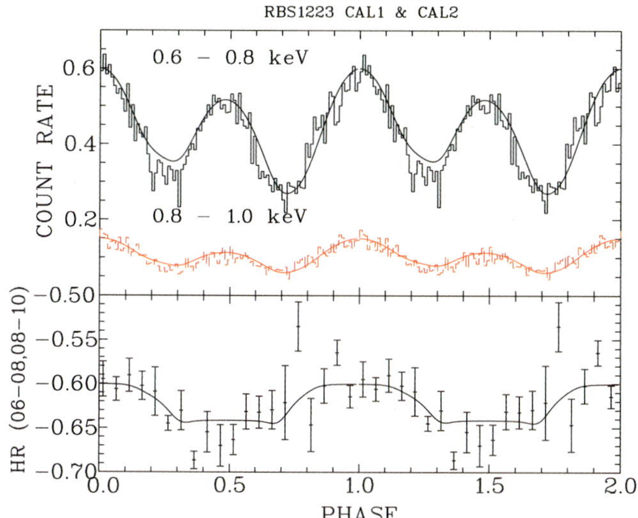

Fig. 1 Soft X-ray *light curve* of RBS 1223 in the 0.6–0.8 keV and 0.8–1.0 keV energy bands and corresponding hardness ratio. *Solid lines* are predicted based on crustal field temperature profiles (see text for details). Taken from Schwope et al. (2005), with permission of the authors

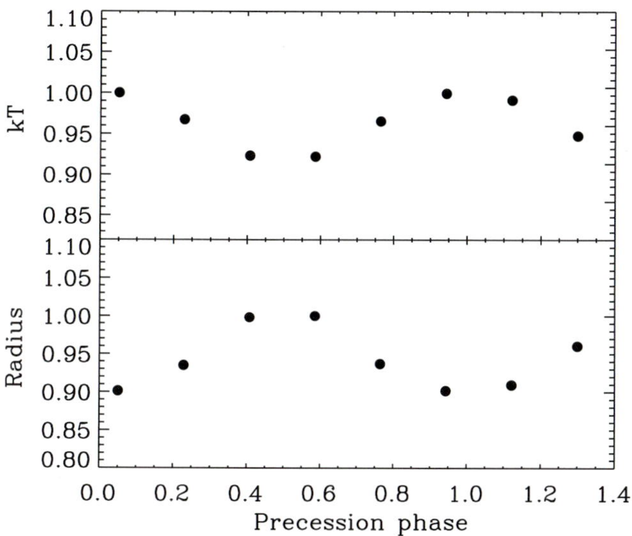

Fig. 2 The variation of the blackbody temperature and radius as a function of the precession phase, as predicted by a two cap model. Isotropic (*blackbody*) emission has been assumed. The computed variations are consistent with those observed in RX J0720.4-3125. Here $T_1 = 80$ eV, $T_2 = 100$ eV (T_i being the temperature of the ith spot), $\xi = 75°$, $\chi = 5°$ and the spot separation is 160°. See Haberl et al. (2006) for all details

2.2 Theoretical models

The quite large pulsed fraction, pulse asymmetry, long-term variations and the possible existence of hot spots, possibly in non-symmetrical positions, seem difficult to explain by assuming that thermal emission originates at the NS surface, at least if the thermal surface distribution is that induced by a simple core-centered dipolar magnetic field. In this case, the temperature monotonically decreases from the poles to the equator, as shown long ago by Greenstein and Hartke (1983). The corresponding pulse profiles have been investigated in detail by Page (1995), under the assumption that each surface patch emits (isotropic) blackbody radiation. Because of gravitational effects and of the smooth temperature distribution, the pulse modulation is quite modest (pulsed fraction $\lesssim 10\%$) for any reasonable value of the star radius. Moreover, being the temperature distribution symmetrical about the magnetic equator, the pulse shape itself is always symmetrical, regardless of the viewing geometry. Larger pulsed fractions may be reached by the proper inclusion of an atmosphere (because of magnetic beaming, e.g. Pavlov et al. 1994) or by considering a star-centered dipolar + quadrupolar field Page and Sarmiento (1996).

On the other hand, the appearance of non-antipodal caps with different properties requires a complex and non-axisymmetric thermal surface distribution. Theoretically, this can be obtained in (at least) two ways: either by complicating the field geometry in the exterior of the NS (e.g. Zane and Turolla 2006a; see Sect. 3), or by assuming that the external field is a dipole but by allowing for toroidal components in the crust (e.g. Geppert et al. 2004, 2005). A first step in investigating the latter scenario has been presented by Geppert et al. (2004, 2005), who considered field geometries in which currents associated to the dipolar field are distributed in both the crust and core of the star, giving rise to strong toroidal components in the crustal field. They have found that, if the meridional component of the crustal field dominates over the radial one, the non-uniformity of the temperature is not restricted to the crust, but it extends over the entire envelope. As a consequence, the surface temperature map turns out very different from that predicted by the Greenstein and Hartke (1983) model and it is relatively easy to generate small hot regions strongly confined toward the poles with different size and temperature. The main limitation of the model is that so far is computed by assuming axisymmetry, therefore it can not mimic the non-antipodality which seems to be required by observations. Besides, radiative beaming and, in general, the presence of an atmospheric layer have not been accounted for. Despite this, Schwope et al. (2005) presented an attempt to apply this model to the observed light curve of RBS 1223 (see Fig. 1, solid line). In order to mimic the relatively large difference in the two spots required by the data, the star has been artificially divided into two hemispheres and two theoretical models with different field parameters have been used. Also, to simulate the required non-antipodality, the two hemispheres have been slightly inclined with respect to each other. The solid line shown in Fig. 1 is the result of this process, and suggests that a (still unavailable) 3-D computation may provide a consistent satisfactory answer to the problem.

3 Toward a self consistent study of neutron star surface emission: atmospheric effects

As mentioned before, an alternative way to produce a more complex surface T distribution is to consider a star-centered dipolar + quadrupolar field. The presence of multipolar components induces large temperature variations even between nearby regions and this results in larger pulsed fractions and asymmetric pulse profiles. In Zane and Turolla (2006a) we explored this possibility, and we developed a tool for computing the phase-resolved spectrum accounting for both radiative transfer in a magnetized atmosphere and general relativistic ray-bending. The former effect has been included by making use of an archive in which the specific intensity relative to pure H atmospheric models has been computed for a set of values of effective temperature, photon energy and direction, strength and inclination of the local magnetic field. With no exception we have found that the broad characteristics of all the XDINSs light curves observed so far can be successfully reproduced for a suitable combination of quadrupolar magnetic field components and viewing angles (although, due to the multidimensionality of the problem, the fit is not necessarily unique; see Fig. 3 for an example and Zane and Turolla (2006a) for all details and a discussion of other XDINSs light curves).

This is also confirmed by the result of a principal component analysis of a set of more than 78000 model light curves, computed varying the quadrupolar components and the geometrical angles. By performing this study, we find that the 3-D space defined by the first three principal components provides a quite accurate description of the dataset since they account for a large fraction (72%) of the total variance of our model population. Also, profiles close to each other in this 3-D space must exhibit a similar shape. By comparing (in this 3-D space) the principal component representation of our theoretical and observed XDINSs light curves, we find that the observed pulse profiles never fall in the small region occupied by purely dipolar models. However, they all lie close to the quadrupolar models, indicating that a quadrupolar configuration capable to reproduce the observed features must exist (see Fig. 3 in Zane and Turolla 2006a).

One of the most interesting features of our code is that it allows for a study of the beaming effects induced by the presence of the NS atmosphere. The recent detection of absorption features at $\approx$ 300–700 eV in the spectrum of XDINSs and their interpretation in terms of proton cyclotron resonances or bound-bound transitions in H or H-like He, may indicate that these sources possess strong magnetic fields, up to $B \sim 9 \times 10^{13}$ G (Van Kerkwjik et al. 2004; Haberl et al. 2004; Zane et al. 2005). A (slightly) more direct measurement, based on the spin-down, has been obtained so far only in two sources (i.e. RX J0720.4-3125, e.g. Cropper et al. 2004, Kaplan and van Kerkwijk 2005a and RBS 1223, Kaplan and van Kerkwijk 2005b). In these cases the inferred field is again as high as a few $\times 10^{13}$ G. This has important implications, since in a strongly magnetized medium photon propagation is anisotropic and occurs preferentially along the field (magnetic beaming, e.g. Pavlov et al. 1994).

This effect has been studied in detail by Ozel (2001, 2002). Figure 4, taken from Ozel (2001), shows the angular dependence of the radiation emerging from a pure H,

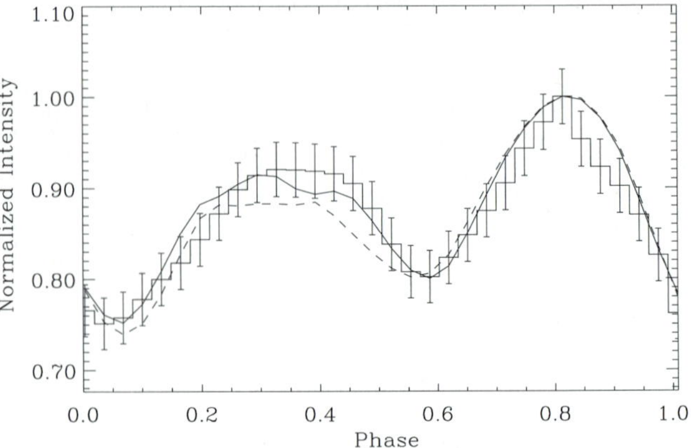

Fig. 3 Fit to the EPIC-PN (0.12–0.5 keV) *light curve* of RBS 1223 detected by Haberl et al. (2003) during rev. 561. The model fit is as in Zane and Turolla (2006a), i.e. it accounts for a NS magnetic field up to a quadrupolar order, radiative transfer effects and gravitational bending. Data point refer to the smoothed observed *light curve*; *dashed line*: trial solution; *solid line*: best fit solution. Here the best fit viewing angles are $\xi = 0.0°$, $\chi = 95.1°$. See Zane and Turolla (2006a) for details

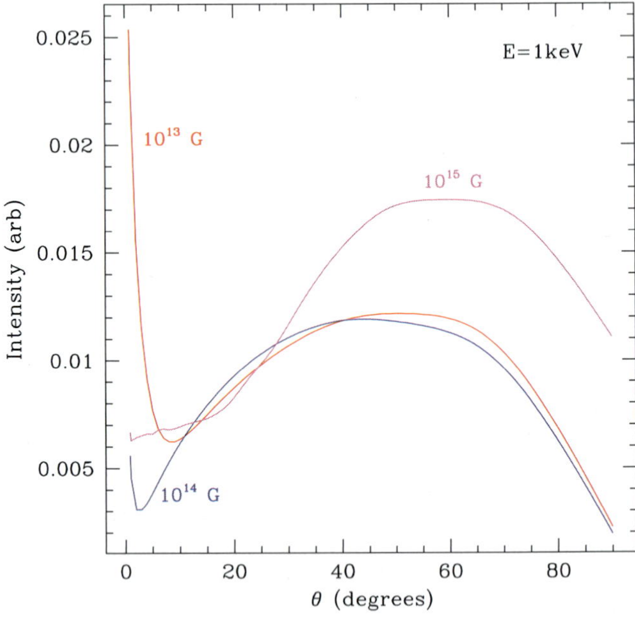

Fig. 4 Angle dependence of the intensity at 1 keV emergent from a pure H, plane parallel atmospheric layer with effective temperature $T_{\rm eff} = 0.3$ keV and different values of the magnetic field. Taken from Ozel (2001), reproduced by permission of the AAS

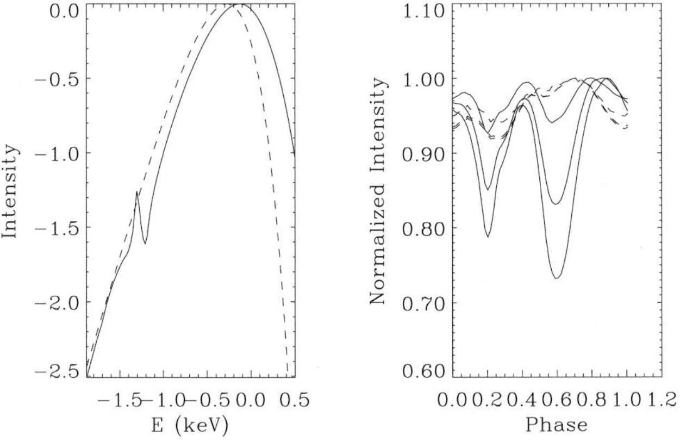

Fig. 5 Effects of Radiative beaming. Phase average spectrum (*left*) and *light curves* in the 0.1–2 keV, 0.1–0.5 keV and 0.5–2 keV bands (*right*, arbitrary normalization). *In both panels solid lines* refer to atmospheric models, while *dashed lines* to blackbody emission. With reference to Zane and Turolla (2006a) for the notation, the model parameters are: $B_{\rm dip} = 6 \times 10^{12}$ G, $T_{\rm pol} = 2.5 \times 10^6$ K, $b_0 = 0.5$, $b_1 = b_3 = b_4 = 0$, $b_2 = 0.9$ (for the components of the quadrupolar field), $\chi = 90°$, $\xi = 60°$. Taken from Zane and Turolla (2006a)

magnetized atmospheric layer for different values of the (assumed to be constant) magnetic field. As we can see, the radiation pattern is made of two components: a narrow pencil beam that is concentrated along the field direction (i.e. at $\lesssim 5°$–$10°$ for $B = 10^{13}$ G and at even smaller angles for higher field strengths) and a much broader fan beam pointing toward an intermediate direction (~ 20–$40°$). It is important to stress that this pattern is quite independent on the details of radiative transfer (similar results have been found with the code used here) and reflects an intrinsic characteristic of the angular dependence of the magnetized cross section. In particular, extraordinary photons contribute to both the pencil and the fan beams, while ordinary photons (whose emissivity/opacity drops sharply away from the poles) give a $\sim 50\%$ contribution to the emission in the narrow beam.

Effects of radiative beaming are illustrated in Fig. 5, were we compare phase average spectra and light curves computed by using radiative atmosphere model with those obtained under the assumption of isotropic blackbody emission. In both cases, a quadrupolar magnetic field has been assumed. As we can see, the pulse profiles are substantially different in the two cases. In particular, when radiative beaming is accounted for the spectrum is harder, the pulse fraction is relatively large ($\sim 20\%$), and the pulse shape exhibits a marked energy dependence.

Figures 6, 7 and 9 illustrate again the same effect, but in these cases we compare the light curves computed: (i) by assuming the thermal distribution expected in case of a simple dipole, (ii) by assuming a thermal distribution as in Geppert et al. (2005), (iii) by assuming isotropic (blackbody) emission and (iv) by considering atmospheric effects. These re-

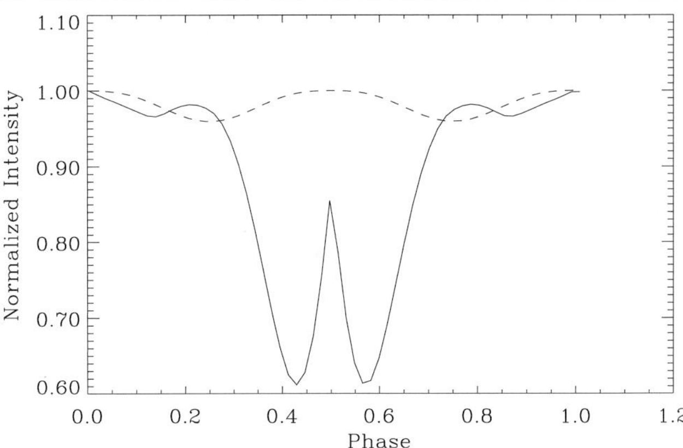

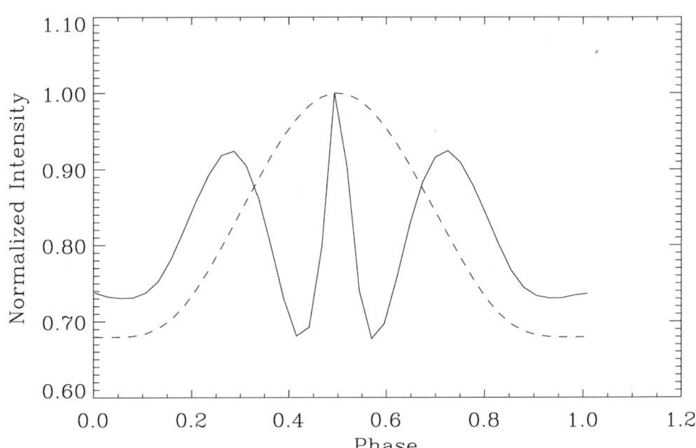

Fig. 6 *Top:* X-ray *light curve* computed by assuming isotropic emission (*dashed line*) or including radiative transfer effects (*solid line*). A pure dipolar configuration has been assumed for the magnetic field and the thermal map is as in Greenstein and Hartke (1983). *Bottom:* same as *in the top panel*, but the thermal map is as in Fig. 8. *In both panels* the viewing angles are $\chi = 90°$, $\xi = 90°$

sults have been obtained by readapting the code developed in Zane and Turolla (2006a, 2007).

In the examples presented in the bottom panels of Figs. 6, 7 and in Fig. 9, we used a thermal distribution as in Geppert et al. (2005), which is computed by assuming that toroidal and poloidal magnetic fields permeate the NS crust and envelope. The external magnetic field is assumed to be a dipole of polar strength 6×10^{12} G, while for the internal one, with reference to the notation used in Geppert et al. (2005), we used $B^0_{\rm crust} = 3 \times 10^{12}$ G, $B^0_{\rm core} = 3 \times 10^{12}$ G, $B^0_{\rm tor} = 10^{15}$ G as parameters for the crustal, core, and toroidal magnetic components respectively. The maximum of the toroidal field is located at $x_{\rm core} + 0.2(x_{\rm crust} - x_{\rm core})$, and the neutron star surface temperature is $T_{\rm surf} = 1.2 \times 10^6$ K (see again Geppert et al. 2005 for notation and details). The resulting temperature distribution is shown in Fig. 8.

Let us first consider the case illustrated in Fig. 6. As we can see, the strong angular dependence of the magnetic cross section gives rise to strong interpulses in the light curve.

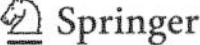

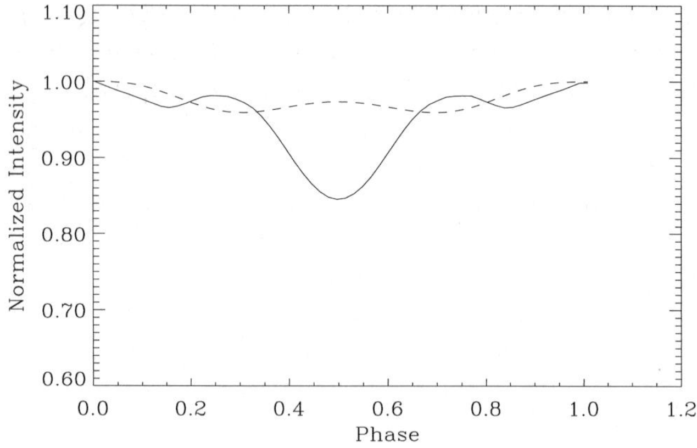

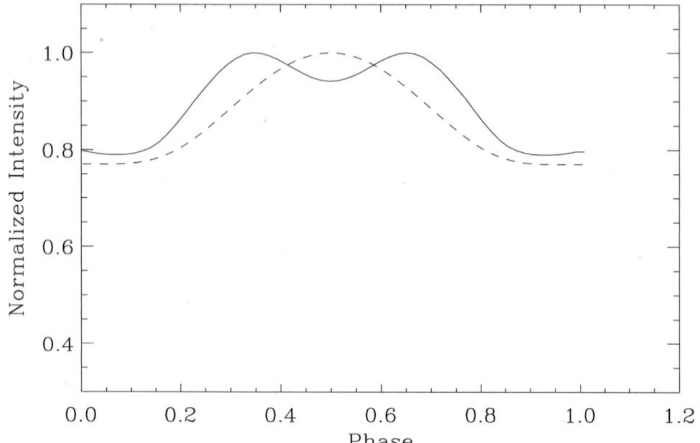

Fig. 7 *Top:* same as in the top panel of Fig. 6 (i.e. pure dipolar configuration and *light curve* computed with or without radiative beaming effects accounted for) but for $\chi = 60°$, $\xi = 60°$. *Bottom:* Same as in the bottom panel of Fig. 6 (i.e. thermal map as in Geppert et al. (2005) and *light curve* computed with or without radiative beaming effects accounted for) but for $\chi = 60°$, $\xi = 60°$

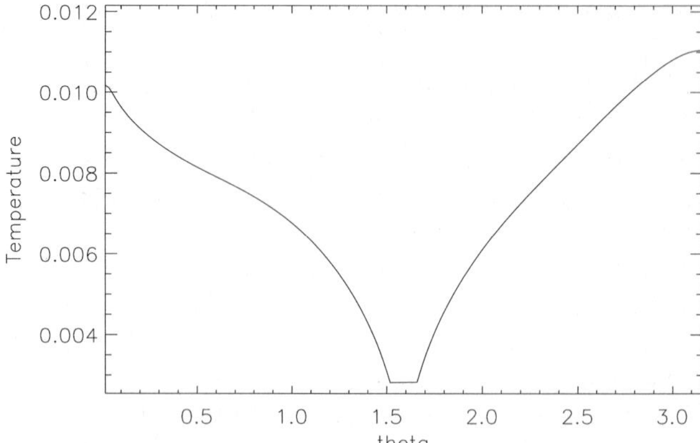

Fig. 8 The surface thermal distribution used for the light curve in the bottom panels of Figs. 6, 7 and in 9. The model and the corresponding magnetic field configurations are described in Geppert et al. (2005). Note the steep temperature gradient away from the polar regions and the difference in temperature (of a factor $\sim 10\%$) between the two caps. See text for details and model parameters

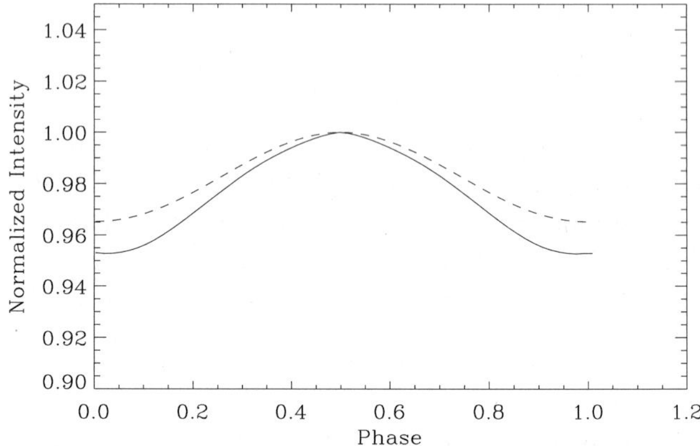

Fig. 9 Same as in the bottom panel of Fig. 6 (i.e. thermal map as in Geppert et al. (2005) and *light curve* computed with or without radiative beaming effects accounted for) but for $\chi = 75°$, $\xi = 5°$

In particular, due to the narrow pencil beam component in the angular dependence of the magnetic cross section, light curves of nearly orthogonal rotators must exhibit a sharp peak (visible in both top and bottom panels of Fig. 6 at phase 0.5), while the two broad fan beams components give rise to the two symmetric "bumps" visible at phase ~ 0.2, 0.6. Moreover, the pulse shape presented in the bottom panel differs from that of the top panel in that the two hot caps are more concentrated toward the poles and have a different temperature, with one (here into view at phases 0.1) about 10% colder with respect to the other (here visible at phase 0.5; note that the phase is arbitrary). Correspondingly, the emitted intensity at phase 0 is a factor $\sim 30\%$ lower than that at phase 0.5 (this effect is particularly clear from the comparison of the dashed lines in the panels, since in these profiles a pure blackbody emitter has been assumed). Overall, this means that when accounting for a thermal map à la Geppert et al. (2005) results in the interpulses being even more evident.

All results are consistent with those by Ozel (2002), which however concentrated on magnetar-like field strengths and concluded that the pencil beam is observationally insignificant, because of its confinement at small angles. As shown here, for field strengths as low as $\sim 10^{13}$ G, the narrow pencil beam is less sharply confined and becomes observationally important provided the source is a nearly orthogonal rotator. As shown in Fig. 7, at larger values of the inclination between LOS and spin axis, its effect is no more visible and only the contributions of the wide fan beams remain dominant. Moreover, all effects due to radiative bending are far less crucial if the source is a nearly aligned rotator (see Fig. 9, a viewing geometry which is relevant for observations of RX J0720.4-3125, see Haberl et al. 2006). In this case the assumption of isotropic emission provides a reliable approximation.

4 Summary

There is an increasing observational evidence that the magnetic field of XDINS is far more complicated than a simple dipole. Very often the observed lightcurves present a certain degree of asymmetry (skewness), and in at least two cases there is evidence for the presence of two hot surface regions with different temperature and size, which are not exactly in antipodal positions. These findings suggest that the star magnetic distribution, which in turns dictates the surface thermal map, is complex and non axisymmetric. Theoretically, two possible scenarios have been considered so far:

1. The presence of multipolar (quadrupolar) components in the magnetic field *outside* the star (see e.g. Zane and Turolla 2006a).
2. The existence of both toroidal and poloidal magnetic fields in the NS crust and envelope (Geppert et al. 2004, 2005). In this case, the assumption of a simple dipolar field *outside* the star has been retained.

Here we presented the current status-of-the-art of both approaches, and discuss their capabilities (and limitations) in reproducing the observed data. We also presented realistic atmospheric models, where the effects of different neutron star thermal and magnetic surface distributions on the observed spectra and lightcurves are accounted for. In the case 1, we find that the main characteristics of the observed XDINSs lightcurves can be reproduced. We readapted the numerical code described in (Zane and Turolla 2006a) to account for a scenario of the second kind and to include the effects of an atmospheric H layer above the condensed envelope (this is part of a work in preparation by Zane et al. 2007). Here we presented a preliminary study of the radiative beaming effects introduced by the presence of an atmosphere. We find that radiative beaming substantially affects the properties of the lightcurve of nearly orthogonal rotators, giving rise to interpulses in the pulse shape. These features are even more evident when the temperature distribution is patchy (as in the case in which the poloidal and toroidal crustal fields are present). The effects are less dramatic in the case of nearly aligned rotators, for which the assumption of isotropic emission gives acceptable results.

As far as other observational implications are concerned, we notice that, in both cases 1 and 2, the measurement of the magnetic field strength as inferred from timing studies is not affected by the details of the magnetic configuration in none. In fact, timing studies are only sensitive to the large scale dipolar component. However, in the first scenario variations of cyclotron lines or atomic atmospheric lines (which depend on the local value of the magnetic field) with the spin phase are expected to be larger than in the second case, when the atmosphere is permeated by a simple dipolar field. On the other hand, spectral features do not originate necessarily in the atmospheric layers. Recently, Pons et al. (2007) considered a magnetic configuration as in Geppert et al. (2005) and computed the emissivity from the neutron star crust and condensed envelope, predicting the appearance of broad spectral edges at energies between 0.2 and 0.6 keV (presented at this meeting).

Overall, these studies shows that deep observations of XDINSs spread over the star spin period (or, in the case of RX J0720.4-3125, complemented by studies of spectral variations over the precession period) are crucial and provide a powerful tool to unveil the complex thermal and magnetic surface map of Galactic neutron stars.

Acknowledgements We thank A. Schwope for the permission to reprint Fig. 1 and we thank F. Ozel for the permission to reprint Fig. 4. S.Z. thanks the UK Particle Physics and Astronomy Research Council (PPARC) for its support through a PPARC Advanced Fellowship.

References

Cropper, M., Zane, S., Ramsay, G., et al.: Astron. Astrophys. **365**, L302 (2001)
Cropper, M., Haberl, F., Zane, S., et al.: Mon. Not. Roy. Astron. Soc. **351**, 1099 (2004)
De Vries, C.P., Vink, J., Méndez, M., et al.: Astron. Astrophys. **415**, L31 (2004)
Geppert, U., Kueker, M., Page, D.: Astron. Astrophys. **426**, 267 (2004)
Geppert, U., Kueker, M., Page, D.: Astron. Astrophys. (2005, submitted), astro-ph/0512530
Greenstein, G., Hartke, G.J.: Astrophys. J. **271**, 283 (1983)
Haberl, F., Schwope, A.D., Hambaryan, V., et al.: Astron. Astrophys. **403**, L19 (2003)
Haberl, F.: Adv. Space Res. **33**, 638 (2004)
Haberl, F., Motch, C., Zavlin, V.E., et al.: Astron. Astrophys. **424**, 635 (2004)
Haberl, F., Turolla, R., De Vries, C., et al.: Astron. Astrophys. **451**, L17 (2006)
Kaplan, D., Van Kerkwijk, M.H.: Astrophys. J. Lett. **628**, L45 (2005a)
Kaplan, D., Van Kerkwijk, M.H.: Astrophys. J. Lett. **635**, L65 (2005b)
Ozel, F.: Astrophys. J. **563**, 276 (2001)
Ozel, F.: Astrophys. J. **575**, 397 (2002)
Page, D.: Astrophys. J. **442**, 273 (1995)
Page, D., Sarmiento, A.: Astrophys. J. **473**, 1067 (1996)
Pavlov, G.G., Shibanov, Yu.A., Ventura, J., et al.: Astron. Astrophys. **289**, 837 (1994)
Pons, J.A., Pérez-Azorin, J.F., Miralles, J.A., et al.: Towards self-consistent models of isolated neutron stars: The case of the X-ray pulsar RX J07020-3125. Astrophys. Space Sci. **308**. DOI 10.1007/s10509-007-9336-8 (2007)
Schwope, A.D., Hambaryan, V., Haberl, F., et al.: Astron. Astrophys. **441**, 597 (2005)
Treves, A., Turolla, R., Zane, S., et al.: Publ. Astron. Soc. Pac. **112**, 297 (2000)
Van Kerkwjik, M.H., Kaplan, D.L., Durant, M., et al.: Astrophys. J. **608**, 432 (2004)
Vink, J., De Vries, C.P., Méndez, M., et al.: Astrophys. J. Lett. **609**, L75 (2004)
Zane, S., Cropper, M., Turolla, R., et al.: Astrophys. J. **627**, 397 (2005)
Zane, S., Turolla, R.: Mon. Not. Roy. Astron. Soc. **366**, 727 (2006a)
Zane S., Turolla, R., Geppert, U., et al.: (2007, in preparation)

Astrophys Space Sci (2007) 308: 267–277
DOI 10.1007/s10509-007-9337-7

ORIGINAL ARTICLE

Molecular systems in a strong magnetic field

How atomic–molecular physics in a strong magnetic field might look like

Alexander V. Turbiner

Received: 10 July 2006 / Accepted: 13 October 2006 / Published online: 21 March 2007

Abstract Brief overview of one-two electron molecular systems made out of protons and/or α-particles in a strong magnetic field $B \le 4.414 \times 10^{13}$ G is presented. A particular emphasis is given to the one-electron exotic ions $H_3^{++}(pppe)$, $He_2^{3+}(\alpha\alpha e)$ and to two-electron ions $H_3^{+}(pppee)$, $He_2^{++}(\alpha\alpha ee)$.

Quantitative studies in a strong magnetic field are very complicated technically. Novel approach to the few-electron Coulomb systems in magnetic field, which provides accurate results, based on variational calculus with physically relevant trial functions is briefly described.

Keywords Molecular ions (traditional and exotic) · Strong magnetic field

PACS 31.15.Pf · 31.15-p · 31.10.+z · 32.60.+i · 97.10.Ld

Our goal is to present a brief review of the properties of $1e$-atomic-molecular systems (both traditional and exotic (marked by bold)) in a magnetic field:

$$\mathrm{H},\ \mathrm{H}_2^{+},\ \mathbf{H}_3^{2+},\ \mathbf{H}_4^{3+},$$

$$(\mathbf{HeH})^{2+},\ (\mathbf{H{-}He{-}H})^{3+},\ (\mathbf{He{-}H{-}He})^{4+},$$

$$\mathrm{He}^{+},\ \mathbf{He}_2^{3+},$$

(the list is complete for all magnetic fields in $0 \le B \le 4.414 \times 10^{13}$ G) and $2e$-systems

$$\mathrm{H}^{-},\ \mathrm{H}_2,\ \mathrm{H}_3^{+},\ \mathbf{H}_4^{2+},$$

$$(\mathbf{HeH})^{+},$$

$$\mathrm{He},\ \mathbf{He}_2^{2+},$$

(this list is incomplete so far) made from protons and/or α-particles.

In general, a relevance of exploration of atomic-molecular physics in a strong magnetic field is motivated by a discovery of neutron stars which is characterized from one side by a strong surface magnetic field and from another side by the atmosphere possibly made from atomic-molecular compounds. Recent discovery by Chandra X-ray spatial observatory in the soft X-ray spectrum of the isolated neutron star 1E1027.4-5209 of two absorption features (Sanwal et al. 2002) pushed ahead drastically these studies. A present status of the $1e$-molecular systems in a strong surface magnetic field is reviewed in Turbiner and Lopez-Vieyra (2006).

What does it mean 'strong magnetic field'? A magnetic field of order or higher than the characteristic atomic magnetic field for which the Larmor radius is equal to the Bohr radius

$$B = B_0 \equiv \frac{m_e^2 e^3 c}{\hbar^3} = 2.3505 \times 10^9 \text{ G}.$$

Where a non-relativistic consideration can be used? For magnetic fields which are smaller than the Schwinger limit $B_{\rm rel}$—the magnetic field for which the electron cyclotron energy is equal to the electron mass

$$B \le B_{\rm rel} = \frac{m_e^2 c^3}{\hbar e} = 4.414 \times 10^{13} \text{ G}.$$

A.V. Turbiner (✉)
Instituto de Ciencias Nucleares, UNAM, Mexico DF 04510, Mexico
e-mail: turbiner@nucleares.unam.mx

Hence, for magnetic fields $0 \leq B \lesssim B_{\text{rel}}$ the Schroedinger equation can be used.

What is the Born–Oppenheimer (BO) approximation (of zero order)? Protons and/or α-particles are assumed infinitely massive. Corrections to BO due to finite nuclear mass(es) can be estimated by the lowest vibrational and rotational energies with a probable contribution coming from the cyclotron energies of the nuclei. Estimates show that these corrections appear to be of the same order of magnitude for all studied magnetic fields. They grow as a magnetic field increase but do not exceed 10–20% for binding energies for the highest $B \leq B_{\text{rel}}$. Thus, the numerical results obtained using the Schroedinger equation in BO approximation and presented below, in physics applications should be slightly modified by taking into account the relativistic corrections and the finite-mass effects. However, the existing accuracies in observational data do not require such a modification.

Why the problem is so difficult? Mostly, it is due to the following reasons:

- Highly-non-uniform asymptotics of the potential at large distances: in some spacial directions it grows $\propto r^2$, in others it vanishes.
- It is a problem of several centers.
- A posteriori we know that we deal with a problem of weakly-bound states,

  $$E_{\text{binding}}/E_{\text{total}} \ll 1,$$

  (e.g. for H_2^+ at $B = 10^{13}$ G this ratio is $\lesssim 10^{-2}$).

In general, atomic units are used throughout the article if different is not stated. The energies and potentials are measured in Rydbergs.

Method

- Variational Calculation.
- *Simple* and *unique* trial function applicable for the whole range of accessible magnetic fields (0–4.414×10^{13} G which can lead to a sufficiently high accuracy in total energy.

How to choose trial functions? (see e.g. Turbiner 1984)

- Physical relevance (as many as possible physics properties should be encoded even for a cost of simplicity).
- Mathematical (computational) simplicity should *not* be a guiding principle.
- Resulting perturbation theory should be convergent (see below).

Variational calculus in a framework of perturbation theory (see Turbiner 1984)

For chosen square-integrable Ψ_{trial} let us find a trial potential

$$V_{\text{trial}} = \frac{\nabla^2 \Psi_{\text{trial}}}{\Psi_{\text{trial}}}, \qquad E_{\text{trial}} = 0. \tag{1}$$

Hence, we know the Hamiltonian for which the normalized Ψ_{trial} is an eigenfunction

$$H_{\text{trial}} \Psi_{\text{trial}} = [p^2 + V_{\text{trial}}] \Psi_{\text{trial}} = 0, \tag{2}$$

then

$$\begin{aligned} E_{\text{var}} &= \int \Psi^*_{\text{trial}} H \Psi_{\text{trial}} \\ &= \int \psi^*_{\text{trial}} \underbrace{H_{\text{trial}} \Psi_{\text{trial}}}_{=0} + \int \Psi^*_{\text{trial}} (H - H_{\text{trial}}) \Psi_{\text{trial}} \\ &= 0 + \int \Psi^*_{\text{trial}} (V - V_{\text{trial}}) \Psi_{\text{trial}} = E_0 + E_1. \end{aligned} \tag{3}$$

Hence,

- The variational energy is a sum of the first two terms of a certain perturbation series with perturbation potential $(V - V_{\text{trial}})$.
- Choosing ψ_{trial} appropriately we can obtain the *convergence* of the perturbation series (in order to get convergence the ratio $|\frac{V - V_{\text{trial}}}{V_{\text{trial}}}|$ should be bounded if in addition this ratio tends to zero at large distances it leads to faster convergence) and if it is so the *rate* of convergence can be made as fast as possible (sometimes, it is reached by making a minimization of the energy functional with respect of parameters of ψ_{trial}, however, it is not always the case that minimization leads to an increase the rate of convergence.).
- How to calculate E_2 *in practice*?—in general, it is unsolved problem yet, the only exception is one-dimensional case.

Example Hydrogen atom in a magnetic field (ground state) In the excellent review by Garstang (1977) the physics of the problem is described, while the most accurate calculations at present are done in Kravchenko et al. (1996).

If the magnetic field **B** is directed along z-axis the problem is characterized by the potential

$$V = -\frac{2}{r} + \frac{B^2}{4} \rho^2, \quad \rho^2 = x^2 + y^2. \tag{4}$$

Let us choose a trial function

$$\psi_0 = \exp(-\alpha r - \beta B \rho^2 / 4), \tag{5}$$

where α, β are physically meaningful variational parameters: α measures the charge (anti)-screening due to a magnetic field presence, β "measures" a magnetic permittivity. The trial potential is

$$V_0 = \frac{\Delta \psi_0}{\psi_0} = -\frac{2\alpha}{r} + \frac{\beta^2 B^2}{4}\rho^2 + \frac{\alpha\beta B}{2}\frac{\rho^2}{r}, \tag{6}$$

and the reference point for energy is

$$E_0 = -\alpha^2 + \beta B. \tag{7}$$

Variational energy gives *relative accuracy* $\sim 10^{-4}$ *in total energy for* $0 < B < 4.414 \times 10^{13}$ **G** comparing to an accurate calculation!

Remark (see Potekhin and Turbiner 2001) The trial function

$$\psi_0 = \exp(-\varphi), \tag{8}$$

with

$$\varphi = \sqrt{\alpha^2 r^2 + (\gamma_1 r^3 + \gamma_2 r^2 \rho + \gamma_3 r \rho^2 + \gamma_4 \rho^3) + \beta^2 B^2 \rho^4/16}$$

where $\alpha, \gamma_{1\ldots4}, \beta$ are parameters, gives *relative accuracy* $\sim 10^{-7}$ in total energy for magnetic fields $0 < B < 4.414 \times 10^{13}$ G.

The binding (ionization) energy $E_b = B - E_T$ grows with a magnetic field increase in quite drastic way. For instance, E_b (found variationally using ψ_0) is equal to

H: $\quad E_b(10\,000 \text{ a.u.}) = 27.95\,Ry$,

He^+: $\quad E_b(10\,000 \text{ a.u.}) = 78.43\,Ry$

which is many times larger than in the field-free case.

PART I

1 One-electron molecular systems

(i) (ppe) *system*

H_2^+ molecular ion (Parallel Configuration)

The potential which describes the (ppe) system in a magnetic field with infinitely massive protons on the same magnetic line (see Fig. 1) is given by

$$V = -\frac{2}{r_1} - \frac{2}{r_2} + \frac{2}{R} + Bm_\ell + \frac{B^2\rho^2}{4}, \tag{9}$$

where $\rho^2 = x^2 + y^2$ and m_ℓ is magnetic quantum number.

Trial Functions for H_2^+ *(ground state,* $m_\ell = 0$*)*

I.

$$\psi_1 = \underbrace{e^{-\alpha_1(r_1+r_2)}}_{\text{Heitler–London}}\ \underbrace{e^{-\beta_1 B\rho^2/4}}_{\text{Landau}}, \tag{10}$$

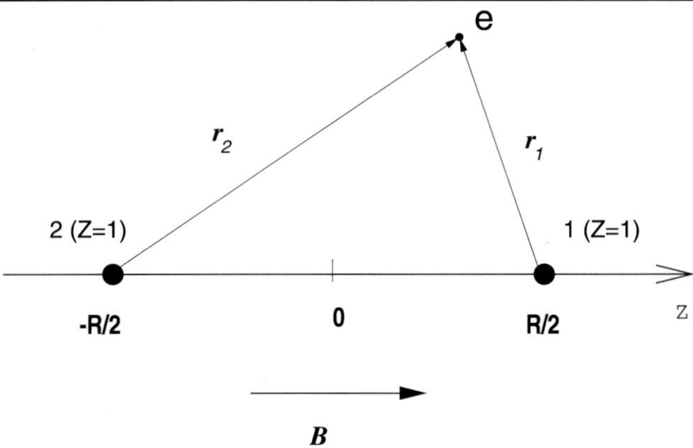

Fig. 1 The geometrical setting for the (ppe) system

where α_1, β_1, R are variational parameters. It is a product of the Heitler–London function and the lowest Landau orbital (modified by β_1). This function is the exact eigenfunction of the lowest energy in the potential

$$V_1^{\text{trial}} = -2\alpha_1\left(\frac{1}{r_1} + \frac{1}{r_2}\right) + {\beta_1}^2\frac{B^2\rho^2}{4} + 2{\alpha_1}^2 - \beta_1 B \\ + \underbrace{2{\alpha_1}^2 \mathbf{n}_1\cdot\mathbf{n}_2 + \alpha_1\beta_1 B\rho^2\left(\frac{1}{r_1} + \frac{1}{r_2}\right)}_{V - V_{\text{trial}}}. \tag{11}$$

At $\alpha_1 = \beta_1 = 1$ the Coulomb singularities and Harmonic Oscillator behavior at large distances are reproduced exactly. At $\alpha_1, \beta_1 \neq 1$ (anti)screening of the nuclear charges and of the magnetic permittivity appear, respectively. We assume that the modified Heitler–London approximation can give a significant contribution for internuclear distances around the equilibrium. It will be verified a posteriori. Such a trial function describes a situation when the electron is attached to both charged centers assuming a type of coherent interaction. One can call it a 'covalent' coupling of the system.

II.

$$\psi_2 = \underbrace{(e^{-\alpha_2 r_1} + \sigma e^{-\alpha_2 r_2})}_{\text{Hund–Mulliken}}\, e^{-\beta_2 B\rho^2/4}, \tag{12}$$

where α_2, β_2, R are variational parameters. Such a trial function describes a situation when the electron is attached to either one charged center or another assuming a type of incoherent interaction. One can call it a 'ionic' coupling of the system, $H + p$. We assume that this function can give a significant contribution for large internuclear distances. It will be verified a posteriori.

In order to describe intermediate region between two domains $R \simeq R_{\text{eq}}$ and $R \gg R_{\text{eq}}$ we use two types of *interpolations*.

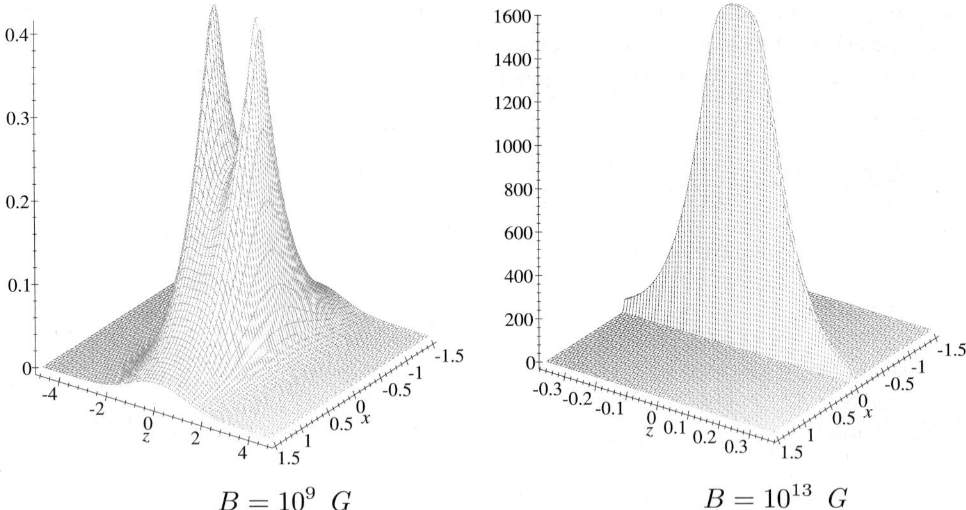

Fig. 2 Electronic density $|\Psi|^2$ for H_2^+ in parallel configuration

III-1. Non-linear Interpolation (simplest)

$$\psi_{3_1} = \underbrace{(e^{-\alpha_3 r_1 - \alpha_4 r_2} + \sigma e^{-\alpha_3 r_2 - \alpha_4 r_1})}_{\text{Guillemin-Zener}} e^{-\beta_3 B\rho^2/4}, \qquad (13)$$

where $\alpha_3, \alpha_4, \beta_3, R$ are variational parameters. If $\alpha_3 = \alpha_4$ then $\psi_{3_1} \to \psi_1$, if $\alpha_4 = 0$ then $\psi_{3_1} \to \psi_2$. It realizes "ionic $\leftrightarrow$ covalent coupling" interpolation which is verified a posteriori.

III-2. Linear Interpolation

$$\psi_{3_2} = A_1\psi_1 + A_2\psi_2. \qquad (14)$$

IV. Superposition of the two kinds of the interpolation

$$\psi_4 = A_{3_1}\psi_{3_1} + A_{3_2}\psi_{3_2}. \qquad (15)$$

Conclusion:

- This single function describes the system in the range of magnetic fields up to $B = 4.414 \times 10^{13}$ G.
- Till very recently, this 10-parametric trial function ψ_4 led to the *lowest* total energies for the ground state compared with numerous previous calculations for $B > 10^{10}$ G up to $B = 4.414 \times 10^{13}$ G. Now the most accurate calculations are due to Vincke and Baye (2006).
- For $B \lesssim 10^{10}$ G: Relative accuracy $\sim 10^{-5}$ in binding energy (by making a comparison with Vincke and Baye 2006).
- A simple modification of ψ_4 allows to study excited states $1\sigma_u$, $1\pi_{g,u}$, $1\delta_{g,u}$ etc.—the lowest states of different magnetic numbers and parities.

Physical phenomenon: For $B \sim 5 \times 10^{11}$ G the coupling changes from 'ionic' (incoherent) type to 'covalent' (coherent) type (see Fig. 2) (for further details see Turbiner and Lopez-Vieyra 2006).

(ii) *(ppe) system*

H_2^+ molecular ion (Inclined Configuration, see Fig. 3)

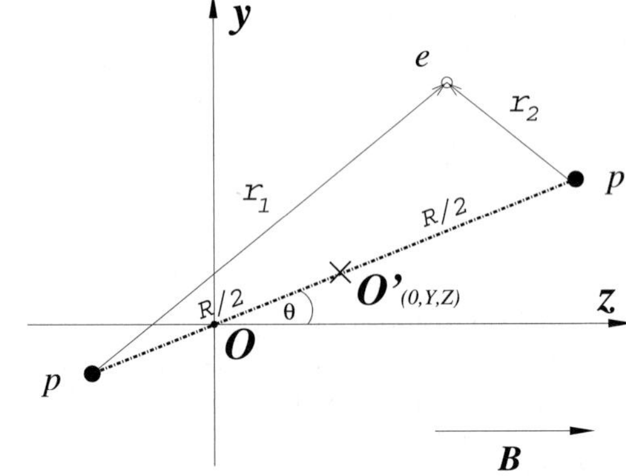

Fig. 3 The geometrical setting for the (ppe) system (inclined configuration)

Physics (see for references and details a review Turbiner and Lopez-Vieyra 2006):

- Parallel configuration $\Theta = 0$ is optimal for all studied magnetic fields (total energy takes a minimal value), see e.g. Fig. 4.
- H_2^+ does not exist for large inclinations at $B > 10^{11}$ G (total energy curve has no minimum at finite R), see Fig. 5.
- H_2^+ is stable for all B and the most bound $1e$-system made from protons for $B \lesssim 10^{13}$ G: $H_2^+ \nrightarrow H + p$.

Binding (ionization) energy of H_2^+ grows with a magnetic field increase. For example,

$$E_b(10\,000 \text{ a.u.}) = 45.80\,Ry,$$

at $R_{eq} = 0.118$ a.u. is very large being almost twice larger than one for the H-atom (see above). Transition energy between the ground state and the lowest excited state also grows with a magnetic field increase reaching at 10 000 a.u.

$$\Delta E(1\sigma_g \to 1\pi_u) = 11.73\,Ry.$$

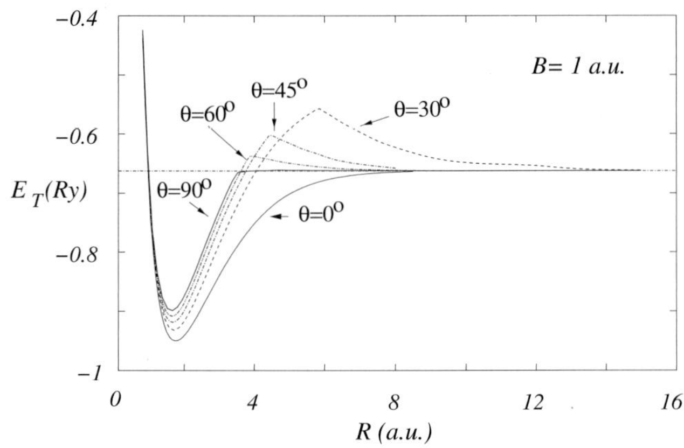

Fig. 4 Total energy E_T of the (ppe)-system as function R for different inclinations at $B = 1$ a.u.

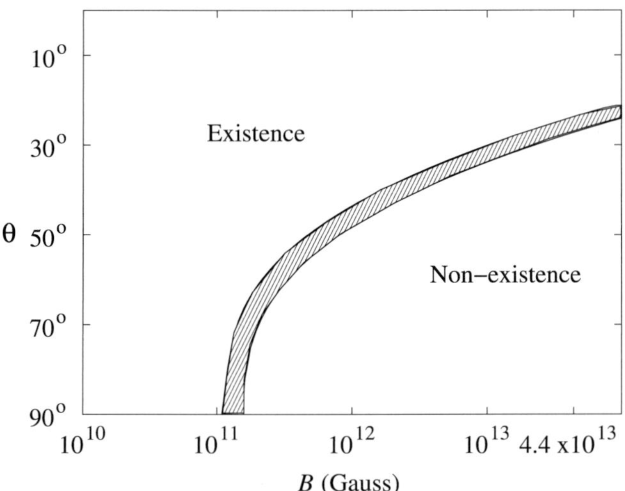

Fig. 5 H_2^+-ion: domains of existence ↔ non-existence

(iii) *(pppe) system*
H_3^{2+}: linear parallel configuration
(The molecular axis and magnetic line coincide)
The potential has a form

$$V = \frac{2}{R_-} + \frac{2}{R_+} + \frac{2}{R_- + R_+} - \frac{2}{r_1} - \frac{2}{r_2} - \frac{2}{r_3} + \frac{B^2\rho^2}{4}, \tag{16}$$

(for a geometrical settings see Fig. 6).

For $B < 10^{11}$ G the total energy $E(R_+, R_-)$ of the $(pppe)$ system displays no minimum for finite $R_{+/-}$, at most, a slight irregularity. However, at $B \gtrsim 10^{11}$ G the total energy $E(R_+, R_-)$ has explicitly pronounced minimum for finite $R_+ = R_-$ which is stable towards small deviations from linearity. Hence, *the parallel configuration is optimal*. Thus, at $B \gtrsim 10^{11}$ G the system $(pppe)$ has a bound state, which manifests existence of the exotic molecular ion H_3^{2+} (Turbiner et al. 1999). Furthermore, the excited states

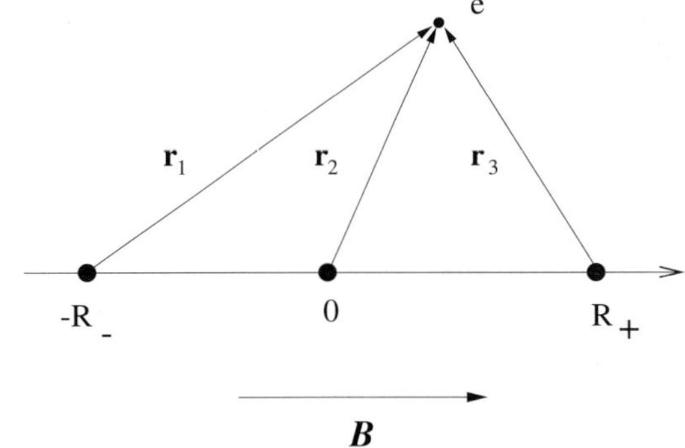

Fig. 6 The geometrical setting for the $(pppe)$ system (linear parallel configuration)

of the positive parity $1\pi_u$, $1\delta_g$ of the H_3^{2+} ion can exist for $B \gtrsim 10^{11}$ G. In general, the ion has a finite lifetime—it can decay

$$H_3^{2+} \to H_2^+ + p,$$

however, a process

$$H_3^{2+} \nrightarrow H + p + p,$$

is always prohibited. For $B \gtrsim 3 \times 10^{13}$ G the first decay also becomes prohibited, H_3^{2+} is stable system which is even the most bound $1e$-system made from protons.

Binding (ionization) energy of H_3^{2+} is

$$E_b(10\,000 \text{ a.u.}) = 45.41\,Ry,$$

at $R_\pm^{eq} = 0.130$ a.u., it is very large in comparison with ionization energy for the H atom but it is comparable with H_2^+. Transition energy between the ground state and the lowest excited state at 10 000 a.u. is

$$\Delta E(1\sigma_g \to 1\pi_u) = 12.78\,Ry,$$

being comparable with one for H_2^+.

(iv) *(pppe) system*
H_3^{2+}: triangular configuration
(The magnetic field is perpendicular to the equilateral triangle formed by protons)

At $B < 10^8$ G the total energy of $(pppe)$ in a spacial configuration has no minimum (or even does not display any irregularity) at finite interproton distances. However, at $10^8\,\text{G} \lesssim B \lesssim 10^{11}$ G the total energy $E(R)$ has a well-pronounced minimum with protons in a triangular equilateral configuration at finite size R of a triangle side (Lopez-Vieyra and Turbiner 2002) (see Fig. 7). This state of H_3^{2+} is metastable (it has a finite lifetime) if the protons are externally supported in the plane perpendicular to the magnetic field direction. Otherwise, it is unstable. The minimum disappears at $B > 10^{11}$ G, hence H_3^{2+} in triangular configuration does not exist anymore. This state can be considered

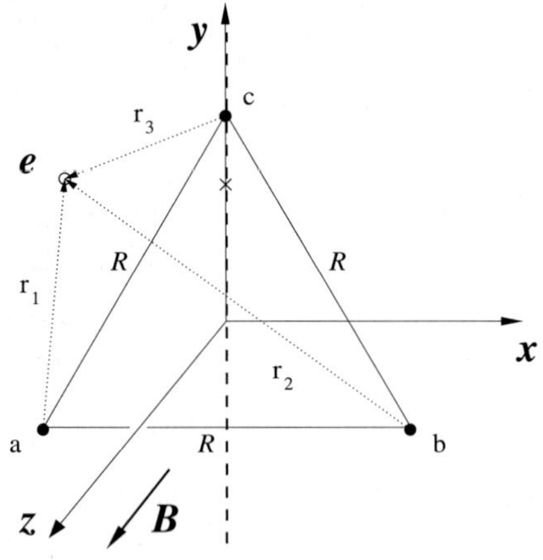

Fig. 7 The geometrical setting for the $(pppe)$ system (triangular configuration)

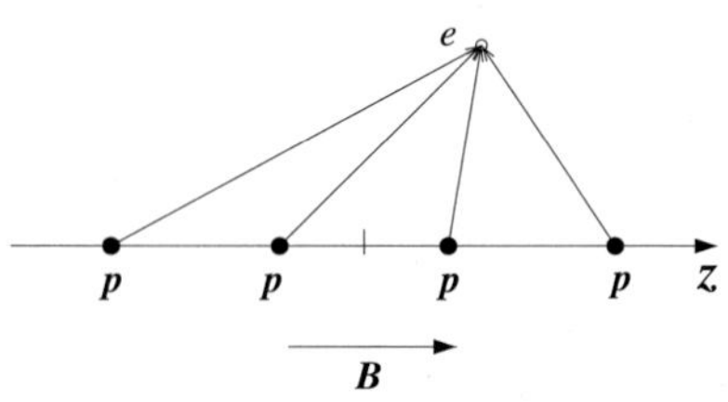

Fig. 8 $(ppppe)$ in linear parallel configuration

as a *precursor to linear configuration.* No more stable or metastable spatial configurations of protons are found!

(v) *(ppppe) system*

H_4^{3+} molecular ion (Parallel Configuration)

Surprisingly, at $B \gtrsim 10^{13}$ G the system $(ppppe)$ has a minimum in total energy in linear parallel configuration at finite internuclear distances (see Fig. 8). It manifests the existence of the molecular ion $H_4^{(3+)}$ as metastable state. It can decay

$$H_4^{3+} \to H_2^+ + p + p, \qquad H_4^{3+} \to H_3^{2+} + p,$$

but

$$H_4^{3+} \nrightarrow H + p + p + p.$$

Binding (ionization) energy (see Turbiner and Lopez-Vieyra 2006)

$$E_b(3 \times 10^{13}\,\mathrm{G}) = 38.42\,Ry,$$

is sufficiently large. Its excited state $1\pi_u$ also exists but no more excited states are found.

(vi) *(αpe) system*

$(HeH)^{2+}$ molecular ion (Parallel Configuration)

- For $B < 10^{12}$ G the system (αpe) is not bound.
- For $B \gtrsim 10^{12}$ G the system (αpe) has a bound state manifesting the possible existence of the molecular ion $(HeH)^{(2+)}$. Also this systems can exist in two excited states $1\pi, 1\delta$.
- For $B \gtrsim 10^{13}$ G the ion $(HeH)^{(2+)}$ becomes stable:

  $$(HeH)^{(2+)} \nrightarrow He^+ + p.$$

- Parallel Configuration is always optimal (see Fig. 9).

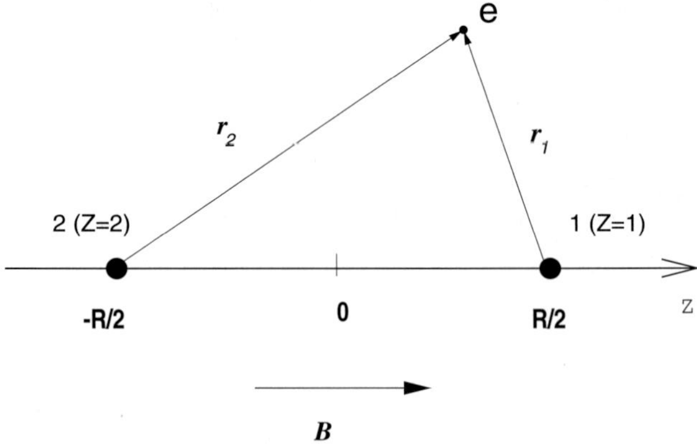

Fig. 9 The geometrical setting for the (αpe) system

Binding (ionization) energy is equal to

$$E_b(10\,000\,\mathrm{a.u.}) = 77.30\,Ry,$$

at $R_{eq} = 0.142$ a.u., it is almost twice larger then one for the H-atom and the hydrogenic ions H_2^+, H_3^{2+} being also larger than one for He^+. Transition energy between the ground state and the lowest excited state at 10 000 a.u. is

$$\Delta E(1\sigma \to 1\pi) = 20.80\,Ry,$$

(for details see Turbiner and Lopez-Vieyra 2006, 2007).

(vii) *(ααe) system*

He_2^{3+} molecular ion (Parallel Configuration)

- For $B < 2 \times 10^{11}$ G the system $(\alpha\alpha e)$ does not exist
- For $B \gtrsim 2 \times 10^{11}$ G the system $(\alpha\alpha e)$ is bound manifesting the existence of the molecular ion $He_2^{(3+)}$, and even two excited states of positive parity $1\pi_u$, $1\delta_g$ do also exist
- For $B \gtrsim 10^{12}$ G ion $He_2^{(3+)}$ becomes *stable*:

  $$He_2^{(3+)} \nrightarrow He^+ + \alpha,$$

- Parallel Configuration is always optimal.

Binding (ionization) energy

$$E_b(10\,000\,\mathrm{a.u.}) = 86.23\,Ry,$$

(cf. $[E_b^{He^+}(10\,000\,\mathrm{a.u.}) = 78.43\,Ry]$) at $R_{eq} = 0.150$ a.u. It is almost twice larger in comparison with one for the atom H and hydrogenic ions H_2^+, H_3^{2+}, and also larger than He^+.

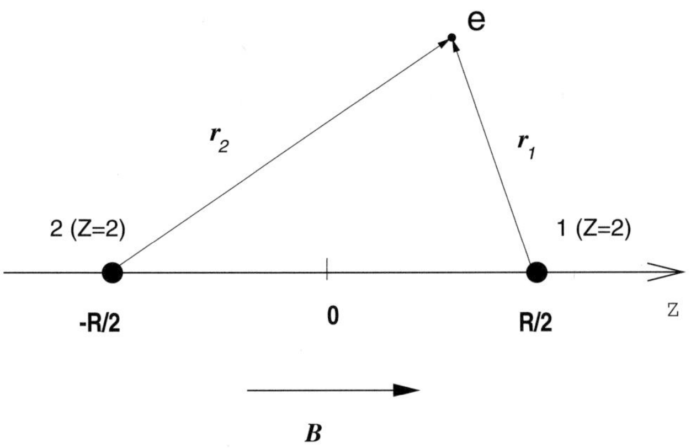

Fig. 10 The geometrical setting for the $(\alpha\alpha e)$ system

Transition energy between the ground state and the lowest excited state at 10 000 a.u. is equal to

$$\Delta E(1\sigma_g \to 1\pi_u) = 24.69\,Ry,$$

(for details see Turbiner and Lopez-Vieyra 2006, 2007), it is the largest among transition energies for one-electron systems.

It is discovered *the striking* approximate relation between the binding energies of the most bound one-electron systems made from α-particles (i) and made from protons (ii):

$$E_b^{\mathrm{He}^+,\mathrm{He}_2^{(3+)}} \approx 2E_b^{\mathrm{H}_2^+,\mathrm{H}_3^{2+}}, \tag{17}$$

which holds for 10^{11} G $< B < 10^{14}$ G.

- For 10^{11} G $< B < 10^{12}$ G in l.h.s. E_b is the binding energy of the He^+ atomic ion, otherwise E_b is the binding energy of the exotic ion He_2^{3+}.
- For 10^{11} G $< B < 10^{13}$ G in r.h.s. E_b is the binding energy of H_2^+, otherwise E_b is the binding energy of the exotic ion H_3^{2+}.

2 Summary

One-electron linear systems (for details see Turbiner and Lopez-Vieyra 2006) Optimal configuration for all linear molecular ions H_2^+, H_3^{2+}, $H_4^{(3+)}$, $(HeH)^{2+}$ and $He_2^{(3+)}$ is parallel: heavy centers are situated along magnetic field line (when exist).

When magnetic field *grows*:

- Binding energy of H, H_2^+, H_3^{2+}, H_4^{3+}, $(HeH)^{2+}$ and He_2^{3+} *grows* (whenever the given system exists).
- Natural size of the systems H_2^+, H_3^{2+}, H_4^{3+}, $(HeH)^{2+}$ and He_2^{3+} *decreases*.
- H_2^+ has the *lowest* E_{total} for $0 < B \lesssim 10^{13}$ G among $1e$ systems made from protons.
- H_3^{2+} has the *lowest* E_{total} for $B \gtrsim 10^{13} G$ among $1e$ systems made from protons.
- The Hydrogen atom has the highest total energy being the *least* bound $1e$ system made from protons for $0 < B \lesssim 4.414 \times 10^{13}$ G.
- Possible existence of the system $H_5^{(4+)}$ for $B > 4.4 \times 10^{13}$ G (at $\sim 10^{13}$ G); but a reliable statement requires a consideration of relativistic corrections and finite-mass effects. It would be the longest $1e$ hydrogenic chain.
- For $B \gtrsim 10^{12}$ G the exotic He_2^{3+} has the *lowest total energy* among systems made from protons and/or α-particles.
- At $B \sim 3 \times 10^{13}$ G for the ions H_2^+ and H_3^{2+} in optimal configuration the binding energies $\equiv$ ionization energies coincide, both are equal to $\sim$700 eV, while for the He_2^{3+}-ion in parallel configuration the ionization energy is $\sim$1400 eV. It corresponds to the energies of absorption features observed in the radiation from 1E1027.4-5209 Sanwal et al. (2002).
- Many more exotic $1e$ systems may occur at the Schwinger limit $B \sim 4.414 \times 10^{13}$ G and beyond.

For molecular systems we are not aware about any quantitative reliable studies of radiation transitions, both bound-bound and bound-free ones.

• *Technical observation*:
Surprisingly, many quite sophisticated methods allow to find 1, 2, 3 significant digits in binding energy only. For example, the Hartree–Fock method used by E. Salpeter and collaborators (see Lai et al. 1992) for H_2^+ at 10^{11} G gives a single significant digit only. Usually, serious difficulties occur when attempting to go beyond those 1–3 significant digits, to higher accuracy. It may need narrow-specialized methods to employ and much efforts. It was implicitly demonstrated in the numerous papers by P. Schmelcher and co-authors and, finally, in Vincke and Baye (2006).

3 Physics of $1e$-systems in a magnetic field

Any quantal charged particle moves in a magnetic field almost freely inside of a channel situated along a magnetic line of the average transverse size which is defined by the Larmor radius, $\sim B^{-1/2}$. The walls of this channel are "soft", the potential grows quadratically in transverse size ρ. Placing several charged particles inside of this channel one can check the electrostatic stability of the system calculating the Coulomb energy (see Fig. 11). One of the simplest systems of such a type is one when there are two heavy charges Z and the electron situated in the middle of them. It is easy to find that the Coulomb energy of this system is negative for charges $Z = 1, 2, 3$ and for any distance between heavy charges which implies a stability. A natural question is for what size of the channel the transverse fluctuations do not

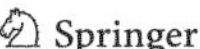

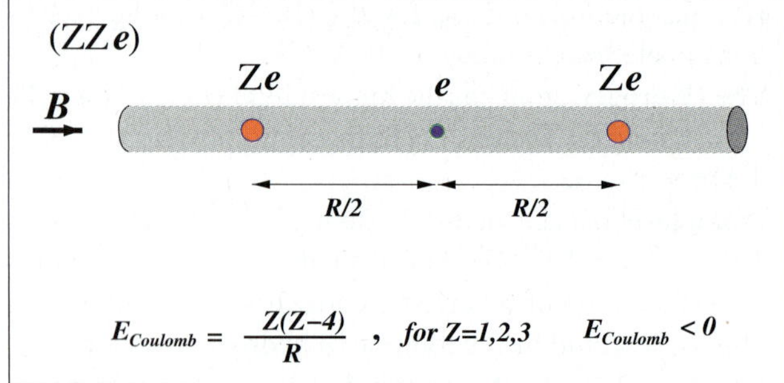

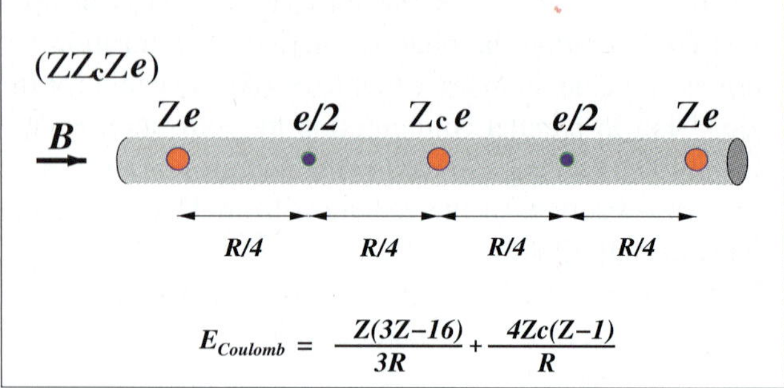

Fig. 11 $1e$ molecular systems: a qualitative picture

ruin such a picture. As for traditional ion H_2^+ this picture holds for any transverse size—the system (ppe) is bound for any magnetic field strength. A first critical transverse size is $\sim$0.1 a.u.—for this transverse size three exotic two- and three-center systems begin to occur: $(HeH)^{2+}$, He_2^{3+}, H_3^{2+} and then they continue to exist for smaller transverse sizes. They appear at first as metastable systems but for the transverse size $\gtrsim$0.01 a.u. become stable. Surprisingly, this transverse size is also the next critical transverse size. At the transverse size $\sim$0.01 a.u. five more exotic two-, three- and four-center systems begin to occur: $(Li\,H)^{3+}$, Li_2^{5+}, $(H\text{–}He\text{–}H)^{3+}$, $(He\text{–}H\text{–}He)^{4+}$, H_4^{3+}, then they continue to exist for smaller transverse sizes.

PART II

4 Two-electron molecular systems

Two-electron systems in a magnetic field is a subject which was explored in a very little extend. It is related with a fact that the studies of these systems are quite complicated technically. To our opinion even the simplest atomic systems H^-, He are not fully understood (see e.g. Al-Hujaj and Schmelcher 2003 and references therein) although it is undoubted that both systems exist for all accessible magnetic fields. One can say that the simplest molecule H_2 only was a subject of quite intense studies. However, a situation is far to be clear.

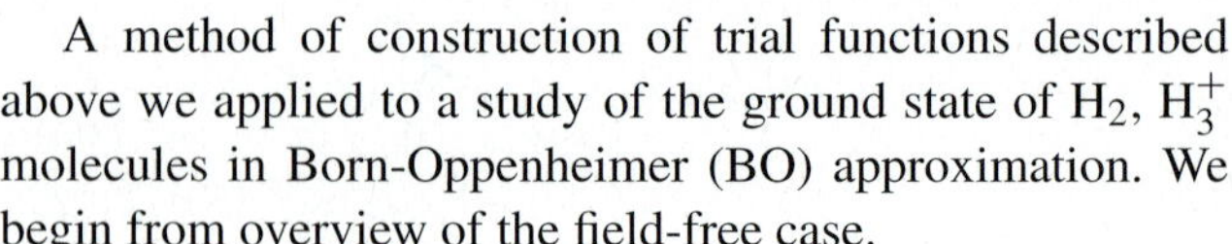
A method of construction of trial functions described above we applied to a study of the ground state of H_2, H_3^+ molecules in Born-Oppenheimer (BO) approximation. We begin from overview of the field-free case.

5 H_2 (ground state)

$E_{BO} = -2.3469\,Ry$ (James and Coolidge 1933, 15 parameters).

$E_{BO} = -2.3478\,Ry$ (Heidelberg group '01, >200 non-centered Gaussian orbitals).

$E_{BO} = -2.3484\,Ry$ (see Turbiner and Guevara 2007, 14 parameters).

$E_{BO} = -2.3489\,Ry$ (*record calculations*, $\gtrsim$1000 James and Coolidge type functions).

6 H_3^+ (lowest linear spin-triplet state)

$E_{BO} = -2.2284\,Ry$ (Schaad et al.'74, Configuration Interaction (CI) method (see Turbiner et al. 2006a for reference)).

$E_{BO} = -2.2297\,Ry$ (see Turbiner et al. 2006a, 23 parameters).

$E_{BO} = -2.2322\,Ry$ (Preiskorn et al. 91, CI + James and Coolidge type functions (see Turbiner et al. 2006a for reference)).

In Turbiner et al. (2006a, 2006b) electronic correlation appears in explicit form $\exp(\gamma r_{12})$ in trial functions, where γ is a parameter. A difficult technical problem of the multidimensional integration with high precision which in the past always was tried to avoid by everybody was successfully resolved.

7 Physics of $2e$-systems in a magnetic field

It is well-known that in absence of magnetic field the state of the lowest energy is the spin-singlet, the electron spins are antiparallel. It seems absolutely natural that for a sufficiently strong magnetic field the state of the lowest energy (if bound) is the spin-triplet, with both electron spins are antiparallel to the magnetic field direction. Hence, for a certain magnetic field there is a transition from spin-singlet to spin-triplet state. Therefore quantum numbers of the ground state are changed. However, a situation might be more complicated than described. If we neglect inter-electron interaction then for spin-singlet case the electrons can have the same quantum numbers except for the spin projection. A transition to spin-triplet state can occur without a change of the electron quantum numbers except for the spin projection if electrons are situated sufficiently far away from each other and the Pauli repulsion can be non-essential. Of course, if

the electrons are close to each other a transition singlet—triplet is accompanied by a change of quantum numbers of electron(s). It is exactly what happens in the case of the Helium atom—for zero and small magnetic fields the ground state is $^1\sigma$ and then for a certain magnetic field the state $^3\pi$ becomes the ground state Al-Hujaj and Schmelcher (2003). It might be different for the case of molecules—it might exist a domain in magnetic fields where the lowest energy state is $^3\Sigma$. It is known that at very large magnetic fields the transverse size of the electronic cloud is very small, of the order of the Larmor radius. The longitudinal size also shrinks as a magnetic field grows, $\sim 1/\log B$. It leads naturally to a conclusion at a sufficiently large magnetic field for any two-electron Coulomb system the lowest energy state should be $^3\Pi$—the spin-triplet state with the total magnetic quantum number equals to -1 Kadomtsev and Kudryavtsev (1971); Ruderman (1971). What remains to be answered is two quantitative questions: from what magnetic field strength the state $^3\Pi$ is the ground state and does a domain in B exist where the state $^3\Sigma$ is the state of the lowest energy. As for the latter question it is found that for two two-electron atomic-like systems: H^- and He this domain is absent (see Al-Hujaj and Schmelcher 2003 and references therein).

We follow a physics picture of the channel which surrounds the magnetic line. Let us "place" inside of the channel several heavy, positively charged particles and two electrons. One can check the electrostatic stability of the system by calculating the Coulomb energy. The simplest systems of such a type are made of several protons and two electrons. It is easy to check that the Coulomb energy is negative for the case of 1, 2, 3, 4, 5 protons and it remains true for any distances between protons (see Fig. 12). It assumes the electrostatic stability of these systems. Natural questions are (i) for what size of the channel the transverse fluctuations do not ruin such a qualitative picture and (ii) how to estimate equilibrium distances. We do not know how to answer these questions qualitatively although quantitatively it can be answered by solving the corresponding Schroedinger equation.

(i) *(pppee) system*

H_3^+ (Turbiner et al. 2006a, first detailed study)

The Hamiltonian is

$$\mathcal{H} = \sum_{\ell=1}^{2}\left(-\nabla_\ell^2 + \frac{B^2}{4}\rho_\ell^2\right) - \sum_{\substack{\ell=1,2\\ \kappa=a,b,c}} \frac{2}{r_{\ell\kappa}} + \frac{2}{r_{12}} + \frac{2}{R_+} + \frac{2}{R_-} + \frac{2}{R_+ + R_-} + B(\hat{L}_z + 2\hat{S}_z), \tag{18}$$

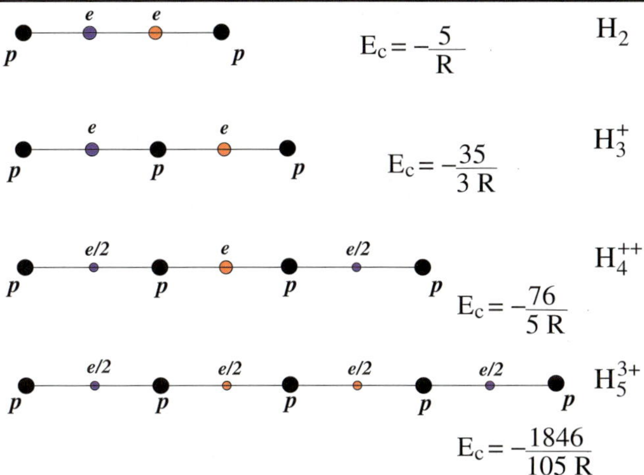

Fig. 12 2e-molecular systems: a qualitative picture

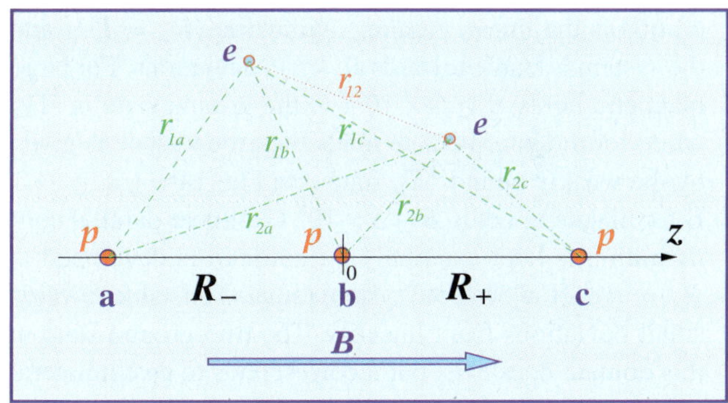

Fig. 13 The H_3^+ molecular ion in parallel configuration in a uniform constant magnetic field $\mathbf{B} = (0, 0, B)$

(for a geometrical setting see Fig. 13), where $\hat{L}_z = \hat{l}_{z_1} + \hat{l}_{z_2}$ and $\hat{S}_z = \hat{s}_{z_1} + \hat{s}_{z_2}$ are the z-components of the total angular momentum and total spin, respectively, and $\rho_\ell = \sqrt{x_\ell^2 + y_\ell^2}$.

Trial function:

$$\begin{aligned}\psi^{(\text{trial})} &= (1 + \sigma_e P_{12}) \\ &\quad \times (1 + \sigma_N P_{ac})(1 + \sigma_{N_a} P_{ab} + \sigma_{N_a} P_{bc}) \\ &\quad \times \rho_1^{|m|} e^{i m \phi_1} e^{\gamma r_{12} - \beta_1 B \rho_1^2/4 - \beta_2 B \rho_2^2/4} \\ &\quad \times e^{-\alpha_1 r_{1a} - \alpha_2 r_{1b} - \alpha_3 r_{1c} - \alpha_4 r_{2a} - \alpha_5 r_{2b} - \alpha_6 r_{2c}},\end{aligned} \tag{19}$$

where $\sigma_e = \pm 1$ stands for spin singlet ($S = 0$) and triplet states ($S = 1$), respectively.

For S_3-permutationally symmetric case $\sigma_N = \sigma_{N_a} = \pm 1$. P_{ac} interchanges the two extreme protons a and c, and α_{1-6}, β_{1-2} and γ are variational parameters.

The operators P_{12} interchanges electrons ($1 \leftrightarrow 2$). One can consider possible degenerations of $\psi^{(\text{trial})}$ which appear when some α's are made equal and/or vanish. Then to take a linear superposition of $\psi^{(\text{trial})}$ and its degenerations as a new trial function.

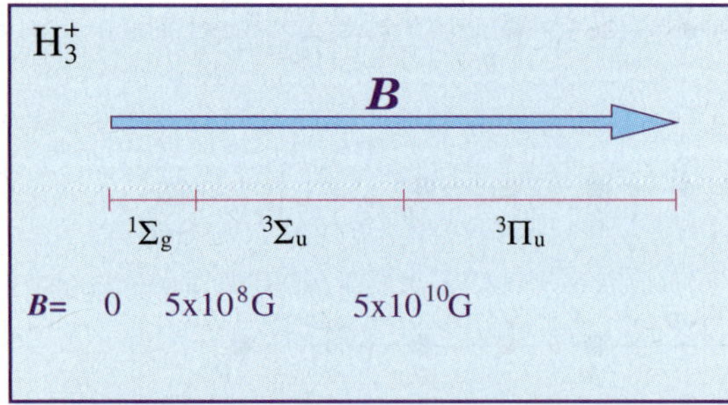

Fig. 14 Ground state evolution for the H_3^+-ion in parallel configuration as a function of the magnetic field strength

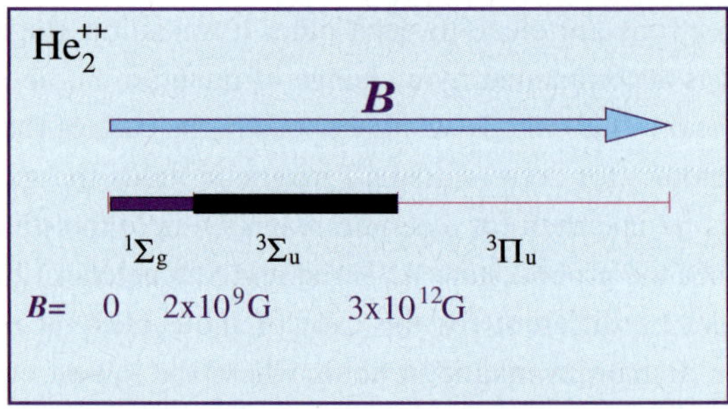

Fig. 15 Ground state evolution for the He_2^{2+}-ion in parallel configuration as a function of the magnetic field strength where $^1\Sigma_g$ is metastable, $^3\Sigma_u$ is unbound, $^3\Pi_u$ is strongly bound

The results for H_3^+:

- For any magnetic field the H_3^+ ion is stable.
- For magnetic fields $B \geq 5\times 10^8$ G the optimal configuration is the linear, parallel, symmetric ($R_+ = R_-$) and the system is stable towards all small deviations. For large magnetic fields $B \geq 5\times 10^{10}$ G the ground state is $^3\Pi_u$ while for the intermediate fields the ground state is given by the weakly bound $^3\Sigma_u$ state (see Fig. 14).
- If for magnetic fields $B \leq 5\times 10^8$ G a linear parallel configuration is kept externally a minimum is developed at $R_+ = R_-$. However, the system remains unstable towards small deviations from linearity. The true ground state in this domain does exist but it corresponds to an equilateral triangular configuration.

At $B = 10\,000$ a.u. the total energy of the ground state of H_3^+ is equal to

$$E_T(H_3^+(^3\Pi_u)) = -95.21\,Ry,$$

at $R_\pm^{eq} = 0.093$ a.u. It coincides with the minus double ionization energy and is significantly smaller than the total energy of the ground state of the H_2 molecule (see below)

$$E_T(H_2(^3\Pi_u)) = -71.39\,Ry,$$

as well as the sum of the total energies of $H_2^+(1\pi_u)$ and H(1s)

$$E_T(H_2^+(1\pi_u) + H(1s)) = -62.02\,Ry.$$

It is worth noting that the dissociation energy for $H_3^+ \to H_2 + p$ is quite large, 23.82 Ry. The transition energy from the ground state to the lowest excited state $^3\Delta_g$ is equal to

$$\Delta E(^3\Pi_u \to {}^3\Delta_g) = 7.76\,Ry.$$

(ii) *($\alpha\alpha ee$) system*
He_2^{2+} (see Turbiner and Guevara 2006, the first study) (parallel configuration, the lowest excited states)
The results for He_2^{2+} (see Fig. 15):

- For all magnetic fields where the He_2^{2+} ion exists the parallel configuration is optimal,
- He_2^{2+} is metastable at $B \leq 0.85$ a.u.: $He_2^{2+} \to He^+ + He^+$
- He_2^{2+} is stable and strongly bound at $B \geq 1100$ a.u. with $^3\Pi_u$ as the ground state
- He_2^{2+} *does not exist at surprisingly large domain* $0.85 \leq B \leq 1100$ a.u.

At $B = 10000$ a.u. the total energy of the ground state $^3\Pi_u$ of He_2^{2+} is

$$E_T(^3\Pi_u) = -174.51\,Ry,$$

with $R_{eq} = 0.106$ a.u. while the total energy of two Helium ions with the parallel electron spins is

$$E_T(He^+ + He^+) = \begin{cases} -156.85\,Ry & \text{for the state } (1s1s), \\ -137.26\,Ry & \text{for the state } (1s2p_{-1}), \end{cases}$$

where in brackets the quantum state of the first and second Helium ion, respectively, are indicated. It is worth noting that $E_T(^3\Pi_u)$ is almost twice less than the total energy of $He_2^{3+}(1\sigma_g) + e$:

$$E_T(He_2^{3+}(1\sigma_g) + e) = -86.23\,Ry.$$

The transition energy from the ground state $^3\Pi_u$ to the lowest excited state $^3\Delta_g$ is equal to

$$\Delta E(^3\Pi_u \to {}^3\Delta_g) = 13.87\,Ry,$$

it is small comparing to ionization-dissociation energies.

(iii) *(ppee) system*
H_2 (ground state (Turbiner 1983 (first calculation), ... Heidelberg group '90–'03, in particular, Detmer et al. 1998, for a review and references see e.g. Al-Hujaj and Schmelcher 2003).

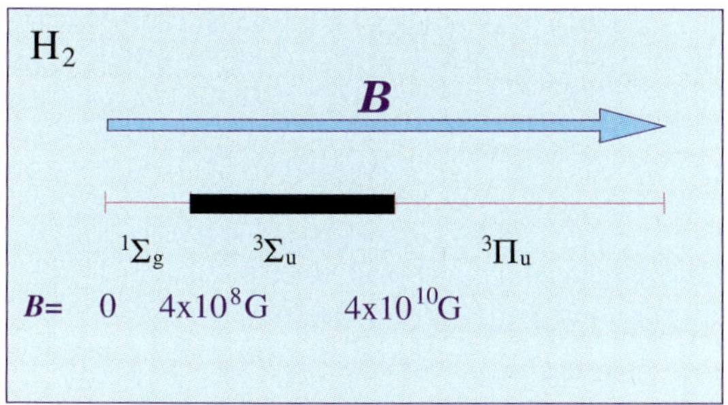

Fig. 16 Ground state evolution for the H_2 molecule in parallel configuration as a function of the magnetic field strength where $^1\Sigma_g$ is stable, $^3\Sigma_u$ is unbound, $^3\Pi_u$ is strongly bound

The results:

- When H_2 exists the parallel configuration is optimal.
- The ground state evolves from $^1\Sigma_g$ at $B \lesssim 0.18$ a.u. to unbound $^3\Sigma_u$ ($R_{\rm eq}$ tends to infinity, the system appears as two separated H-atoms) and then at $B \gtrsim 15.6$ a.u. to strongly bound $^3\Pi_u$ state (see Fig. 16 and Al-Hujaj and Schmelcher 2003, the numbers from Turbiner and Guevara 2007).
- When H_2 exists the system is always stable, however, for all magnetic fields the total energy of the ground state is $E_T(H_2) > E_T(H_3^+)$.

Due to a presumed importance of this system in a strong magnetic field several research teams made studies, many striking qualitative effects are predicted. However, it must be emphasized that due to the technical complexity of the problem the accuracy of obtained results is hard to be estimated. For example, our analysis says that the total energy of the $^3\Pi_u$ state at $B = 10^{11}$ G obtained in the Hartree–Fock method Lai et al. (1992) gives correctly not more that two significant digits. As for the $^1\Sigma_g$ state at $B \le 0.2$ a.u. not more than three significant digits in the binding energy are found correctly in the Gaussian orbital method employed by the Heidelberg group although it was claimed the six digit accuracy.

Many more $2e$ systems should be studied: H_4^{2+}, H_5^{3+} ... ($2e$ hydrogenic linear chains?), $(HeH)^+$, $(H\text{–}He\text{–}H)^{++}$, $(He\text{–}H\text{–}He)^{3+}$, He_3^{4+}, etc. We do not think it is a moment to draw a conclusion about physics of $2e$ molecular systems in a strong magnetic field. So far any concrete applications are obstructed by absence of analysis of radiation transitions.

Acknowledgements The author brings his sincere gratitude to J.C. López Vieyra and N.L. Guevara with whom he shared a pleasure to study this beautiful physics, and to G.G. Pavlov for the encouragement to write a present mini-review. This work was supported in part by FENOMEC and PAPIIT grant IN121106.

References

Al-Hujaj, O.A., Schmelcher, P.: Phys. Rev. A **67**, 023403 (2003)

Detmer, T., Schmelcher, P., Cederbaum, L.: Phys. Rev. A **57**, 1767 (1998)

Garstang, R.H.: Rep. Prog. Phys. **40**, 105 (1977)

James, H.M., Coolidge, A.S.: J. Chem. Phys. **1**, 825 (1933)

Kadomtsev, B.B., Kudryavtsev, V.S.: Pis'ma Zh. Eksp. Teor. Fiz. **13**, 15, 61 (1971)

Kravchenko, Yu.P., Liberman, M.A., Johansson, B.: Phys. Rev. A **54**, 287 (1996)

Lai, D., Salpeter, E., Shapiro, S.L.: Phys. Rev. A **45**, 4832 (1992)

Lopez-Vieyra, J.C., Turbiner, A.V.: Phys. Rev. A **66**, 023409 (2002)

Potekhin, A.Y., Turbiner, A.V.: Phys. Rev. A **63**, 065402 (2001)

Sanwal, D., Pavlov, G.G., Zavlin, V.E.: Astrophys. J. **574**, L61 (2002)

Ruderman, M.: Phys. Rev. Lett. **27**, 1306 (1971)

Turbiner, A.V.: Pis'ma Zh. Eksp. Teor. Fiz. **38**, 510 (1983)

Turbiner, A.V.: Usp. Fiz. Nauk **144**, 35 (1984), Yad. Fiz. **46**, 204 (1987)

Turbiner, A.V., Guevara, N.L.: Phys. Rev. A **74**, 063419 (2006), preprint astro-th/0610928

Turbiner, A.V., Guevara, N.L.: Collect. Czechoslov. Chem. Commun., a special issue dedicated to the memory of Professor Koutecky (2007, in press), preprint physics/0606120

Turbiner, A.V., Lopez-Vieyra, J.C.: Phys. Rep. **424**, 309 (2006)

Turbiner, A.V., Lopez-Vieyra, J.C.: Int. J. Mod. Phys. (2007, in press), preprint astro-th/0412399

Turbiner, A.V., Lopez-Vieyra, J.C., Solis, H.: Pis'ma Zh. Eksp. Teor. Fiz. **69**, 800 (1999)

Turbiner, A.V., Lopez-Vieyra, J.C., Guevara, N.L.: Phys. Rev. A **71**, 023403 (2005)

Turbiner, A.V., Guevara, N.L., Lopez-Vieyra, J.C.: preprint physics/0606083 (2006a)

Turbiner, A.V., Guevara, N.L., Lopez-Vieyra, J.C.: in preparation (2006b)

Vincke, M., Baye, D.: J. Phys. B **39**, 2605 (2006)

Astrophys Space Sci (2007) 308: 279–286
DOI 10.1007/s10509-007-9366-2

ORIGINAL ARTICLE

Thin magnetic hydrogen atmospheres and the neutron star RX J1856.5–3754

Wynn C.G. Ho · David L. Kaplan · Philip Chang · Matthew van Adelsberg · Alexander Y. Potekhin

Received: 1 July 2006 / Accepted: 31 August 2006 / Published online: 21 March 2007

Abstract RX J1856.5–3754 is one of the brightest nearby isolated neutron stars, and considerable observational resources have been devoted to it. However, current models are unable to satisfactorily explain the data. We show that our latest models of a thin, magnetic, partially ionized hydrogen atmosphere on top of a condensed surface can fit the entire spectrum, from X-rays to optical, of RX J1856.5–3754, within the uncertainties. In our simplest model, the best-fit parameters are an interstellar column density $N_H \approx 1 \times 10^{20}$ cm^{-2} and an emitting area with $R^\infty \approx 17$ km (assuming a distance to RX J1856.5–3754 of 140 pc), temperature $T^\infty \approx 4.3 \times 10^5$ K, gravitational redshift $z_g \sim 0.22$, atmospheric hydrogen column $y_H \approx 1$ g cm^{-2}, and magnetic field $B \approx (3\text{–}4) \times 10^{12}$ G; the values for the temperature and magnetic field indicate an effective average over the surface.

W.C.G. Ho (✉) · D.L. Kaplan
Kavli Institute for Astrophysics and Space Research, Massachusetts Institute of Technology, Cambridge, MA 02139, USA
e-mail: wynnho@slac.stanford.edu

P. Chang
Department of Astronomy, University of California, 601 Campbell Hall, Berkeley, CA 94720, USA

M. van Adelsberg
Center for Radiophysics and Space Research, Department of Astronomy, Cornell University, Ithaca, NY 14853, USA

A.Y. Potekhin
Ioffe Physico-Technical Institute, Politekhnicheskaya 26, 194021 St Petersburg, Russia

Keywords Stars: atmospheres · Stars: individual (RX J1856.5–3754) · Stars: neutron · X-rays: stars

PACS 97.60.Jd · 97.10.Ex

1 Introduction

Seven candidate isolated, cooling neutron stars (INSs) have been identified by the ROSAT All-Sky Survey, of which the two brightest are RX J1856.5–3754 and RX J0720.4–3125 (see Treves et al. 2000; Pavlov et al. 2002; Kaspi et al. 2006 for a review). These objects share the following properties: (1) high X-ray to optical flux ratios of $\log(f_X/f_{\rm optical}) \sim$ 4–5.5, (2) soft X-ray spectra that are well described by blackbodies with $kT \sim 50$–100 eV, (3) relatively steady X-ray flux over long timescales, and (4) lack of radio pulsations.

For the particular INS RX J1856.5–3754, single temperature blackbody fits to the X-ray spectra underpredict the optical flux by a factor of ~6–7 (see Fig. 5). X-ray and optical/UV data can best be fit by two-temperature blackbody models with $kT_X^\infty = 63$ eV, emission size $R_X^\infty = 5.1\,(d/140\,{\rm pc})$ km,[1] $kT_{\rm opt}^\infty = 26$ eV, and $R_{\rm opt}^\infty = 21.2\,(d/140\,{\rm pc})$ km (Burwitz et al. 2001, 2003; van Kerkwijk and Kulkarni 2001a; Braje and Romani 2002; Drake et al. 2002; Pons et al. 2002; see also Pavlov et al. 2002;

[1]The most recent determination of the distance to RX J1856.5–3754 is ≈160 pc (Kaplan et al., in preparation). However, the uncertainties in this determination are still being examined. Therefore, we continue to use the previous estimate of 140(±40) pc from Kaplan et al. (2002) since the uncertainty in this previous value encompasses both the alternative estimate of 120 pc from Walter and Lattimer (2002) and the new value.

Trümper et al. 2004), where $T^\infty = T_{\rm eff}/(1+z_g)$, $R^\infty = R^{\rm em}(1+z_g)$, and $R^{\rm em}$ is the physical size of the emission region. The gravitational redshift z_g is given by $(1+z_g) = (1-2GM/Rc^2)^{-1/2}$, where M and R are the mass and radius of the NS, respectively. However, the lack of X-ray pulsations (down to the 1.3% level) puts severe constraints on such two-temperature models (Drake et al. 2002; Ransom et al. 2002; Burwitz et al. 2003). It is possible that the magnetic axis is aligned with the spin axis or the hot magnetic pole does not cross our line of sight (Braje and Romani 2002). Alternatively, RX J1856.5–3754 may possess a superstrong magnetic field ($B \gtrsim 10^{14}$ G) and has spun down to a period $>10^4$ s (Mori and Ruderman 2003), though Toropina et al. (2006) argue that this last case cannot explain the Hα nebula found around RX J1856.5–3754 (van Kerkwijk and Kulkarni 2001b). On the other hand, a single uniform temperature is possible if the field is not dipolar but small-scale (perhaps due to turbulence at the birth of the NS; see, e.g., Bonanno et al. 2005, and references therein).

Even though blackbody spectra fit the data, one expects NSs to possess atmospheres of either heavy elements (due to debris from the progenitor) or light elements (due to gravitational settling or accretion); we note that a magnetized hydrogen atmosphere may provide a consistent explanation for the broad spectral feature seen in the atmosphere of RX J0720.4–3125 (Haberl et al. 2004; Kaplan and van Kerkwijk 2005). The lack of any significant spectral features in the X-ray spectrum argues against a heavy element atmosphere (Burwitz et al. 2001, 2003), whereas single temperature hydrogen atmosphere fits overpredict the optical flux by a factor of $\sim$100 (Pavlov et al. 1996; Pons et al. 2002; Burwitz et al. 2003). However, these hydrogen atmosphere results are derived using non-magnetic atmosphere models. Only a few magnetic (fully ionized) hydrogen or iron atmospheres have been considered (e.g., Burwitz et al. 2001, 2003), and even these models are not adequate. Since $kT \sim$ tens of eV for RX J1856.5–3754 and the ionization energy of hydrogen at $B = 10^{12}$ G is 160 eV, the presence of neutral atoms must be accounted for in the magnetic hydrogen atmosphere models; the opacities are sufficiently different from the fully ionized opacities that they can change the atmosphere structure and continuum flux (Ho et al. 2003; Potekhin et al. 2004), which can affect fitting of the observed spectra.

Another complication in fitting the observational data of RX J1856.5–3754 (and RX J0720.4–3125) with hydrogen atmosphere models is that the model spectra are harder at high X-ray energies. On the other hand, observations of RX J0720.4–3125 suggest it possesses a dipole magnetic field $B \approx 2\times10^{13}$ G (Kaplan and van Kerkwijk 2005). It is probable then that RX J0720.4−3125 (and possibly RX J1856.5–3754) is strongly magnetized with $B \sim 10^{13}$–10^{14} G, and its high energy emission is softened by the effect of vacuum polarization, which can show steeper high energy tails (Ho and Lai 2003). Rather than resorting to a superstrong magnetic field, an alternate possibility is that there exists a "suppression" of the high energy emission. One such mechanism is examined in Motch et al. (2003; see also Zane et al. 2004), specifically, a geometrically thin hydrogen atmosphere at the surface that is optically thick to low energy photons and optically thin to high energy photons. The high energy photons that emerge then bear the signature of the lower temperature (compared to atmospheres that are optically thick at all energies) at the inner boundary layer (usually taken to be a blackbody) of the atmosphere model; this leads to a softer high energy tail. Motch et al. (2003) find a good fit to RX J0720.4–3125 in this case by using a non-magnetic atmosphere model with $kT_{\rm eff} = 57$ eV, a hydrogen column density $y_{\rm H} = 0.16$ g cm^{-2}, and distance of 204 pc, and assuming a $M = 1.4M_\odot$, $R = 10$ km NS.

We examine this last possibility by fitting the entire spectrum of RX J1856.5–3754 with the latest partially ionized hydrogen atmosphere models (constructed using the opacity and equation of state tables from Potekhin and Chabrier 2003) and condensed matter in strong magnetic fields (see van Adelsberg et al. 2005). The goal of the paper is to provide a self-consistent picture of RX J1856.5–3754 that resolves the major observational and theoretical inconsistencies: (1) blackbodies fit the spectrum much better than realistic atmosphere models, (2) strong upper limits on X-ray pulsations suggest RX J1856.5–3754 may have a largely uniform temperature (and hence magnetic field) over the entire NS surface, (3) the inferred emission size from blackbody fits are either much smaller or much larger than the canonical NS radius of 10–12 km. Because of observational uncertainties (see Sect. 3) and the computationally tedious task of constructing a complete grid of models, we do not attempt to prove the uniqueness of our results; rather we try only to reproduce the overall spectral energy distribution and argue for the plausibility of our model.

2 Models of neutron star atmospheres

Thermal radiation from a NS is mediated by the atmosphere (with scaleheight $\sim$1 cm). In the presence of magnetic fields typical of INSs ($B \gtrsim 10^{12}$ G), radiation propagates in two polarization modes (see, e.g., Mészáros 1992). Therefore, to determine the emission properties of a magnetic atmosphere, the radiative transfer equations for the two coupled photon polarization modes are solved (see Ho et al. 2003; Potekhin et al. 2004, and references therein for details on the construction of the atmosphere models). We note that the atmosphere models formally have a dependence, through hydrostatic balance, on the surface gravity g $[= (1+z_g)GM/R^2]$ and thus the NS radius R; however, the emergent spectra do not vary much using different values of g around 2×10^{14} cm s^{-2} (Pavlov et al. 1995). Nevertheless, we construct models using a surface gravity that is consistent with the inferred radius obtained from the spectral fits in Sect. 4 ($g = 1.1 \times 10^{14}$ cm s^{-2} with $M = 1.4M_{\odot}$, $R = 14$ km, and $z_g = 0.2$). Also, though our atmosphere models can have a magnetic field at an arbitrary angle Θ_B relative to the surface normal, the models considered here have the magnetic field aligned perpendicular to the stellar surface. Since the magnetic field and temperature distributions over the NS surface are unknown, synthetic spectra from the whole surface are necessarily model-dependent. However, if the temperature variation is large (so that a small surface area emits at high $T_{\rm eff}$ and a large area emits at lower $T_{\rm eff}$) and because spectra for various Θ_B can be qualitatively similar (see, e.g., Zavlin et al. 1995; Lloyd 2003), then a single magnetic field and temperature model can approximately describe the overall spectrum (see Ho et al. 2007 for more detailed calculations, including surface B and T variations). We describe other elements of our atmosphere models below.

2.1 Partially ionized atmospheres

As discussed in Sect. 1, previous works that attempted to fit the spectra of RX J1856.5–3754 and other INSs with magnetic hydrogen atmosphere models assume the hydrogen is fully ionized. The temperature obtained using these models (or simple blackbodies) are in the range $kT^{\infty} \approx 40 - 110$ eV. Contrast this with the atomic hydrogen binding energies of 160 eV and 310 eV at $B = 10^{12}$ G and 10^{13} G, respectively. Therefore the atmospheric plasma must be partially ionized.

Figure 1 illustrates the spectral differences between a fully ionized and a partially ionized hydrogen atmosphere. The atomic fraction is $<10\%$ throughout the atmosphere, where the atomic fraction is the number of H atoms with non-destroyed energy levels divided by the total number of protons (see Potekhin and Chabrier 2003; Ho et al. 2003). Besides the proton cyclotron line at $\lambda_{Bp} = 1966(B/10^{12}\ {\rm G})^{-1}(1 + z_g)$ Å, the other features are due to bound-bound and bound-free transitions. In particular, these are the $s = 0$ to $s = 1$ transition at (redshifted) wavelength $\lambda = 170$ Å, the $s = 0$ to $s = 2$ transition at $\lambda = 110$ Å, and the bound-free transition at $\lambda = 61$ Å. The quantum number s measures the B-projection of the relative proton-to-electron angular momentum (see, e.g., Lai 2001). For a moving atom, this projection is not an integral of motion, but nonetheless the quantum number s (or $m = -s$) remains unambiguous and convenient for numbering discrete states of the atom (see Potekhin 1994). Because of magnetic broadening, the features resemble dips rather than ordinary spectral lines (see Ho et al. 2003).

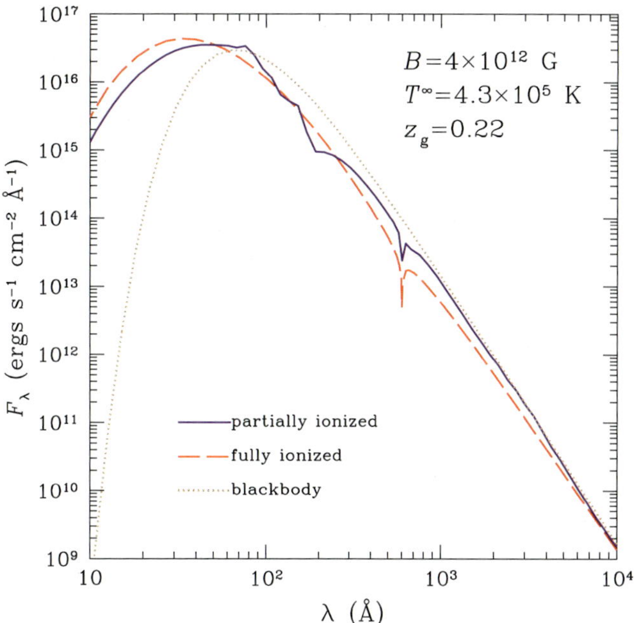

Fig. 1 Spectra of hydrogen atmospheres with $B = 4 \times 10^{12}$ G and $T^{\infty} = 4.3 \times 10^5$ K. The *solid line* is for a partially ionized atmosphere, the *dashed line* is for a fully ionized atmosphere, and the *dotted line* is for a blackbody. All spectra are redshifted by $z_g = 0.22$

2.2 Thin atmospheres

Conventional NS atmosphere models assume the atmosphere is geometrically thick enough so that it is optically thick at all photon energies (optical depth $\tau_\lambda \gg 1$ for all wavelengths λ); thus the observed photons are all created within the atmosphere layer. The input spectrum (usually taken to be a blackbody) at the bottom of the atmosphere is not particularly important in determining the spectrum seen above the atmosphere since photons produced at this innermost layer undergo many absorptions/emissions. The observed spectrum is determined by the temperature profile and opacities of the atmosphere. For example, atmosphere spectra are harder than a blackbody (at the same temperature) at high energies as a result of the non-grey opacities (see Fig. 1); the opacities decline with energy so that high energy photons emerge from deeper, hotter layers in the atmosphere than low energy photons.

On the other hand, consider an atmosphere that is geometrically thinner than described above, such that the atmosphere is optically thin at high energies but is still optically thick at low energies ($\tau_\lambda < 1$ for $\lambda < \lambda_{\rm thin}$ and $\tau_\lambda > 1$ for $\lambda > \lambda_{\rm thin}$). Thus photons with wavelength $\lambda < \lambda_{\rm thin}$ pass through the atmosphere without much attenuation (and their contribution to thermal balance is small since most of the energy is emitted at $\lambda > \lambda_{\rm thin}$ in the case of RX J1856.5–3754). This is illustrated in Fig. 2. Thus if the innermost atmosphere layer (at temperature $T_{\rm thin}$) emits as a black-

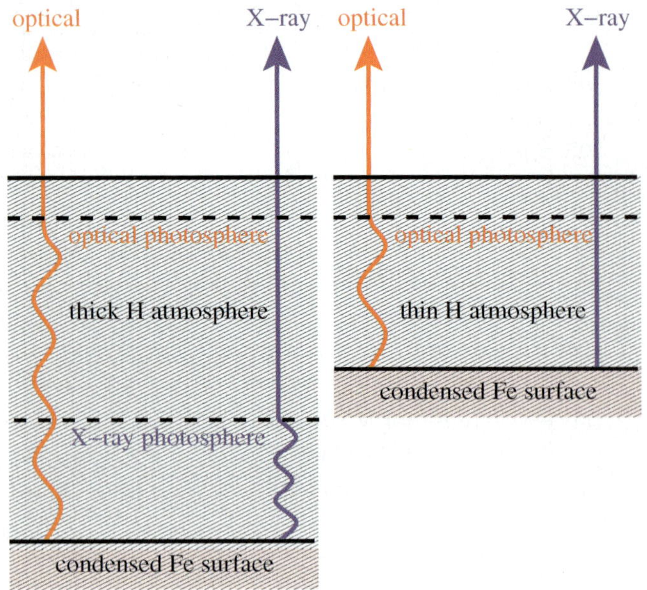

Fig. 2 Schematic diagram illustrating the difference between a "thick" atmosphere (*left*) that is optically thick to photons of all wavelengths versus a "thin" atmosphere (*right*) that is optically thick to long wavelength photons but optically thin to short wavelength photons

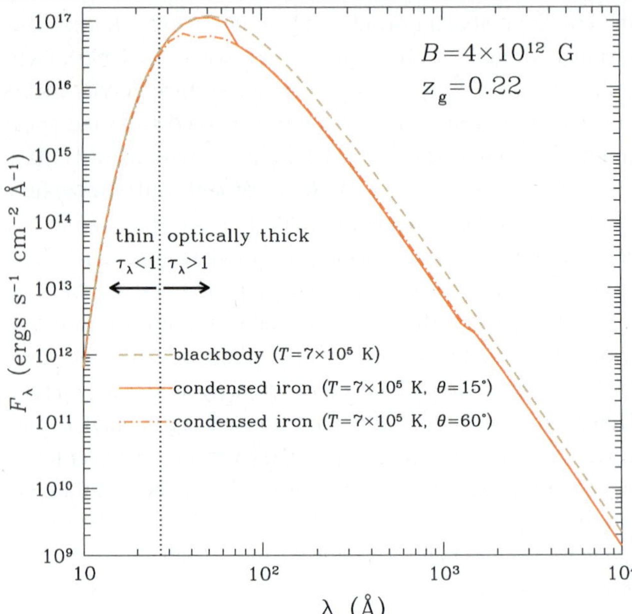

Fig. 3 Condensed iron spectrum (*solid line* for photon propagation direction $\theta = 15^\circ$ and *dot-dashed* line for $\theta = 60^\circ$) with $B = 4 \times 10^{12}$ G, $T = 7 \times 10^5$ K, and $\rho = 2.35 \times 10^4$ g cm^{-3}, compared to a blackbody (*dashed line*) with the same temperature. All spectra are redshifted by $z_g = 0.22$. The *vertical line* separates the wavelength ranges where the atmosphere is optically thin ($\tau_\lambda < 1$) and optically thick ($\tau_\lambda > 1$)

body, then the observed spectrum at $\lambda < \lambda_{\rm thin}$ will just be a blackbody spectrum at temperature $T = T_{\rm thin}$. Motch et al. (2003) showed that a "thin" atmosphere can yield a softer high energy spectrum than a "thick" atmosphere and used a thin atmosphere spectrum to fit the observations of RX J0720.4−3125. How such thin hydrogen atmospheres may be created is discussed in Ho et al. (2007).

2.3 Condensed iron versus blackbody emission

In addition to atmosphere models in which the deepest layer of the atmosphere is assumed to be a blackbody, we construct (more realistic and self-consistent) models in which this layer undergoes a transition from a gaseous atmosphere to a condensed surface. A surface composed of iron is a likely end-product of NS formation, and Fe condenses at $\rho \approx 561\, AZ^{-3/5} B_{12}^{6/5}$ g cm^{-3} $\approx 2.35 \times 10^4$ g cm^{-3} and $T \lesssim 10^{5.5} B_{12}^{2/5}$ K $\approx 5.5 \times 10^5$ K for the case considered here (Lai 2001); note that there is several tens of percent uncertainty in the condensation temperature (Medin and Lai, private communication; see also Medin and Lai 2006a, 2006b). Hydrogen condenses at $T \sim 2 \times 10^4 B_{12}^{0.65}$ K and thus requires much lower temperatures than is relevant for our case. The condensed matter surface possesses different emission properties than a pure blackbody (Brinkmann 1980; Turolla et al. 2004; van Adelsberg et al. 2005; Pérez-Azorín et al. 2005); in particular, features can appear at the plasma and proton cyclotron frequencies.[2]

We use the calculations of van Adelsberg et al. (2005) to determine the input spectrum in our radiative transfer calculations of the atmosphere. However, at the temperature ($T_{\rm thin} \approx 7 \times 10^5$ K) of the condensed layer relevant to our thin atmosphere models that fit the spectrum of RX J1856.5–3754, the input spectrum (where $\tau_\lambda \lesssim 1$) is effectively unchanged from a blackbody (since the temperature profile is nearly identical, with $\Delta T_{\rm thin} \sim 3\%$). Thus there are only slight differences in the resulting surface spectrum. This is illustrated in Fig. 3, where we show the emission spectrum from a $B = 4 \times 10^{12}$ G condensed iron surface at $T = 7 \times 10^5$ K and $\rho = 2.35 \times 10^4$ g cm^{-3} and compare this to a blackbody at the same temperature. The deviation from a blackbody is smaller at low angles of photon propagation θ and increases for increasing θ, as illustrated by the two angles $\theta = 15^\circ$ and 60° (see van Adelsberg et al. 2005). Thus for most angles θ, the condensed surface spectra at short

[2] van Adelsberg et al. (2005) consider an approximation in their treatment of the ion contribution to the dielectric tensor which leads to a spectral feature at the proton cyclotron frequency. However, because of the uncertainty in this approximation, the strength of the feature is not well-determined. Nevertheless, our results are not at all strongly dependent on this feature (or the input spectrum at these low energies) because the optical depth of the atmosphere $\tau_\lambda \gg 1$ at the proton cyclotron (and plasma) frequency (see text for discussion).

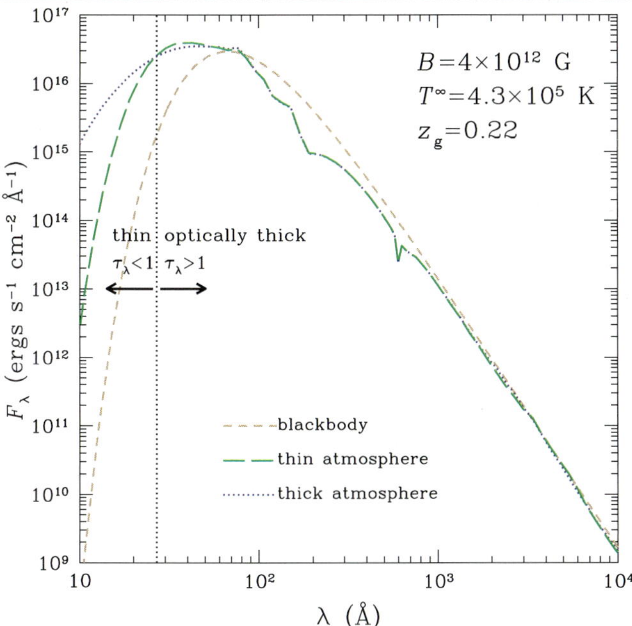

Fig. 4 Spectra of hydrogen atmospheres with $B = 4 \times 10^{12}$ G and $T^{\infty} = 4.3 \times 10^5$ K. The *dotted* and *long-dashed lines* are the model spectra using the "thick" atmosphere and "thin" atmosphere with $y_{\rm H} = 1.2$ g cm^{-2}, respectively (see text for details). The *short-dashed* line is for a blackbody with the same temperature. All spectra are redshifted by $z_g = 0.22$. The *vertical line* separates the wavelength ranges where the atmosphere is optically thin ($\tau_\lambda < 1$) and optically thick ($\tau_\lambda > 1$)

wavelengths (where $\tau_\lambda \ll 1$, so that this surface is visible to an observer above the atmosphere) are virtually identical to a blackbody. On the other hand, the atmosphere is optically thick at longer wavelengths, where the condensed surface spectra deviate from a blackbody; thus the condensed surface and the spectral features are not visible. The resulting atmosphere spectra seen by a distant observer are shown in Fig. 4. The harder spectrum at high energies in the "thick" atmosphere becomes much softer in the "thin" atmosphere and takes on a blackbody shape. In contrast, there is a negligible difference where the atmosphere is optically thick.

3 Observations and analysis

We collect publically available optical, UV, and X-ray data on RX J1856.5–3754. These data have been discussed elsewhere so our treatment will be brief. First, we assemble the optical (B- and R-band) photometry from the Very Large Telescope (VLT) from van Kerkwijk and Kulkarni (2001a) and the *Hubble Space Telescope* (*HST*) WFPC2 F170W, F300W, F450W, and F606W photometry (Walter 2001; Pons et al. 2002) as analyzed by van Kerkwijk and Kulkarni (2001a). We also take the optical VLT spectrum from van Kerkwijk and Kulkarni (2001a) and a STIS far-UV spectrum.[3] The spectra are entirely consistent with the photometry as calibrated by van Kerkwijk and Kulkarni (2001a), although given the limited signal-to-noise ratio of the spectra, we rely primarily on the photometry in what follows. We then use the *Extreme Ultraviolet Explorer* (*EUVE*; Haisch et al. 1993) data as discussed and reduced by Pons et al. (2002). Finally, we take the RGS spectrum from the 57-ks *XMM-Newton* observation and the 505-ks *Chandra* LETG spectrum that are discussed by Burwitz et al. (2003).

A source of uncertainty is that, as mentioned by Burwitz et al. (2003), the RGS and LETG data are not entirely consistent in terms of flux calibration: while they have very similar shapes (and hence implied temperatures and absorptions) the radii inferred from blackbody fits differ by as much as 10% and the overall flux by as much as 20%. Since the LETG fits in Burwitz et al. (2003) are more consistent with those of the CCD instruments on *XMM-Newton* (EPIC-pn and EPIC-MOS2) and in our opinion the current low-energy calibration of LETG is more reliable, we adjust the flux of the RGS data upward by 17% to force agreement with the *Chandra* data. We do not know for certain which calibration (if either) is entirely accurate, so some care must be taken when interpreting the results at the 10–20% level. Fully reliable calibration or even cross-calibration at the low-energy ends of the *Chandra* and *XMM-Newton* responses (<0.2 keV) is not currently available (see, e.g., Kargaltsev et al. 2005), and the detailed response of *EUVE* compared to those of *Chandra* and *XMM-Newton* is also unknown. Therefore, for accuracy in doing the EUV/X-ray fits, we concentrate on the LETG data, which are consistent and have high-quality calibration.

We follow the HRC-S/LETG analysis threads "Obtain Grating Spectra from LETG/HRC-S Data,"[4] "Creating Higher-order Responses for HRC-S/LETG Spectra,"[5] "Create Grating RMFs for HRC Observations,"[6] and "Compute LETG/HRC-S Grating ARFs"[7] and use CIAO[8] version 3.2.2 and CALDB version 3.2.2. We extracted the dispersed events and generated response files for orders ±1, ±2, and ±3. After fitting the LETG data, we compare the fit results qualitatively with the RGS and *EUVE* data; the general agreement is good, but we do not use them quantitatively.

[3] Datasets: O5G702010-O5G702050, O5G703010-O5G703050, O5G704010-O5G704050, O5G705010-O5G705050, O5G751010-O5G751050, O5G752010-O5G752050. O5G701010-O5G701050, and

[4] See http://asc.harvard.edu/ciao/threads/spectra_letghrcs/.

[5] See http://asc.harvard.edu/ciao/threads/hrcsletg_orders/.

[6] See http://asc.harvard.edu/ciao/threads/mkgrmf_hrcs/.

[7] See http://asc.harvard.edu/ciao/threads/mkgarf_letghrcs/.

[8] http://cxc.harvard.edu/ciao/.

Table 1 Fits to the X-ray data. Numbers in parentheses are 68% confidence limits in the last digit(s). The formal fit uncertainty for $R^{\infty} < 10\%$; however, since the radius determination depends on the distance, we conservatively adopt the current ~30% distance uncertainty as our radius uncertainty

	Atmosphere	Blackbody
	Model parameters	
B (10^{12} G)	4	
$y_{\rm H}$ (g cm^{-2})	1.2	
	Fit results	
$N_{\rm H}$ (10^{20} cm^{-2})	1.30(2)	0.91(1)
T^{∞} (10^5 K)	4.34(2)	7.36(1)
R^{∞} (d_{140} km)	17	5.0
z_g	0.22(2)	
$\chi_{\rm r}^2$/dof	0.86/4268	0.86/4269

4 Atmosphere model fitting

Because of data reduction and cross-calibration issues (see Sect. 3) and possible variations in the interstellar absorption abundances (standard abundances are assumed), we do not feel that a full fit of the data is justified at this time. Therefore we do not fit for all of the parameters in a proper sense nor do we perform a rigorous search of parameter space. Instead, we fit for a limited subset of parameters while varying others manually. This allows us to control the fits in detail and reduce the computational burden of preparing a continuous distribution (in B, $T_{\rm eff}$, and $y_{\rm H}$) of models, yet still determine whether our model qualitatively fits the data.

For a given magnetic field and atmosphere thickness, we generate partially ionized atmosphere models for a range of effective temperatures. (Note that, since the continuum opacity of the dominant photon polarization mode decreases for increasing magnetic fields, the required thickness $y_{\rm H}$ of the atmosphere increases for increasing B.) We then perform a χ^2 fit in `CIAO` to the LETG data over the 10–100 Å range (0.12–1.2 keV) for the absorption column density $N_{\rm H}$ [using the `TBabs` absorption model from `Xspec` (Wilms et al. 2000), although other models such as `phabs` give similar results], the temperature T^{∞}, the normalization (parameterized by R^{∞}), and the redshift z_g, where we interpolate over T^{∞}. We obtain a good fit, and Table 1 lists the best-fit parameters and model; the radius is given assuming a distance $d_{140} = d/(140 \text{ pc}) = 1$ (see footnote 1). While we find a range of magnetic fields [$B \approx (3-4) \times 10^{12}$ G] that give acceptable fits, changes in the magnetic field outside this range (but still within $B = 10^{12}-10^{13}$ G) and atmosphere thickness lead to worse fits. At $B > 10^{13}$ G, spectral features due to proton cyclotron and bound species appear in the observable soft X-ray range (though they are likely to be broadened due to magnetic field variations over the surface of the NS), which are not seen.

To further evaluate the quality of this fit, we fit the same LETG data with a blackbody. The blackbody fit yields parameters (see Table 1) that are very similar to those derived by Burwitz et al. (2003; see Sect. 1). Given the comparable $\chi_{\rm r}^2$ ($\approx$1) we achieve from our blackbody and atmosphere model fits (along with the low-energy calibration uncertainties), we are confident that the atmosphere model describes the observations just as well as a blackbody.

Next, we examine the quality of the fit to the optical/UV data. van Kerkwijk and Kulkarni (2001a) showed that these data are well fit by a Rayleigh–Jeans power law ($F_\lambda \propto \lambda^{-4}$) with an extinction $A_V = 0.12 \pm 0.05$ mag. In our fitting, we try using both the optical extinction implied by the X-ray absorption ($A_V = N_{\rm H}/(1.79 \times 10^{21} \text{ cm}^{-2})$ mag; Predehl and Schmitt 1995) and fitting freely for A_V, but we find that these fits are too unconstrained and that the value of A_V is not sensitively determined by the data (indeed, this is reflected in the large uncertainties found by van Kerkwijk and Kulkarni 2001a). As a result, we fix A_V to 0.12 mag. We also assume a single value for the reddening ($R_V = 3.1$) and use the reddening model of Cardelli et al. (1989) with corrections from O'Donnell (1994). We find that our best-fit model to the X-ray data also produces a λ^{-4} power law but underpredicts the optical/UV data by a factor of 15–20%.[9] Looking in detail at the highest quality optical data point (the *HST* F606W measurement), we find that it is 15% above our model spectra. However, the error budget is 3% (photometric uncertainty), 5% (A_V uncertainty), 10% (uncertainty in the optical model flux), and 5% (uncertainty in the fitted optical flux due to the X-ray normalization), and therefore the 15% disagreement can easily be explained by known sources of uncertainty.

Figure 5 shows the observations of RX J1856.5–3754. We also overlay the blackbody and our $B = 4 \times 10^{12}$ G atmosphere model spectra with the parameters given in Table 1. As discussed in Sect. 1, the data are generally featureless, while the models show spectral features; at $B \approx 4 \times 10^{12}$ G, the features due to bound species lie in the extreme UV to very soft X-ray range and are thus hidden by interstellar absorption. Overall, we see that our atmosphere model spectra are generally consistent with the X-ray and optical/UV data, while a blackbody underpredicts the optical/UV by a factor of 6–7.

[9]Comparing monochromatic fluxes derived from photometry with the models is not sufficiently accurate for a detailed, quantitative analysis. A more accurate method would involve convolving the models with the filter bandpasses and predicting monochromatic count-rates to compare with the data (e.g., van Kerkwijk and Kulkarni 2001a; Kaplan et al. 2003). However, given the assumptions about extinction and reddening and the level of accuracy of this analysis, the first approach will suffice here.

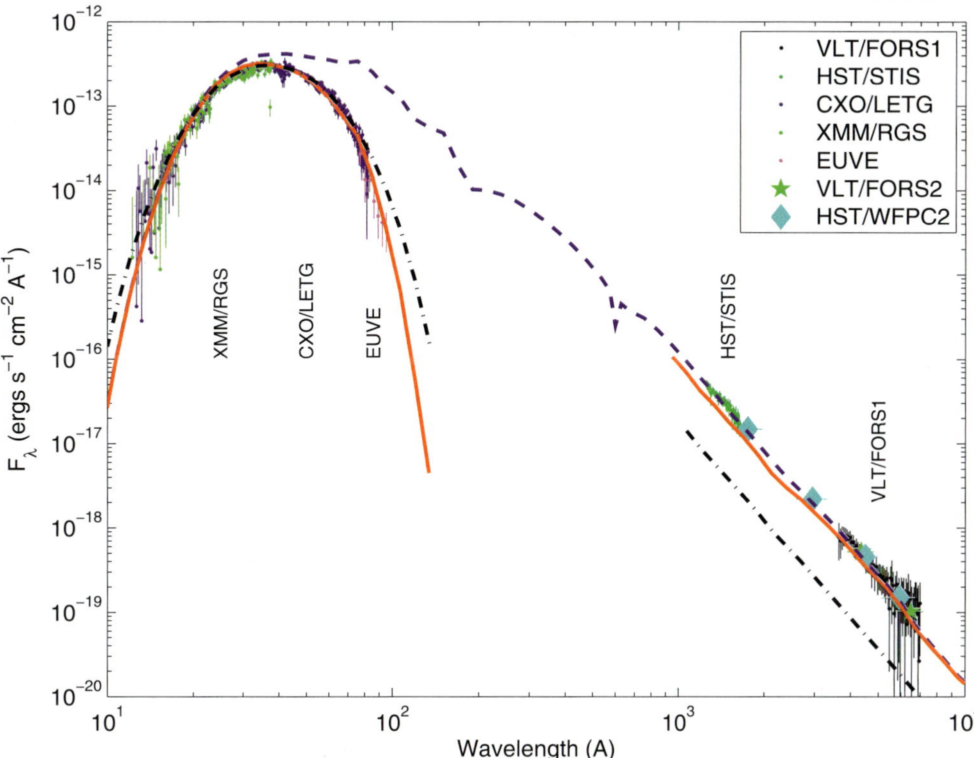

Fig. 5 Spectrum of RX J1856.5–3754 from optical to X-ray wavelengths. The data points are observations taken from various sources. Error bars are one-sigma uncertainties. Optical spectra are binned for clarity: STIS data into 30 bins at a resolution of 12 Å and VLT data into 60 bins at 55 Å resolution. The *solid line* is the absorbed (and redshifted by $z_g = 0.22$) atmosphere model spectrum with $B = 4 \times 10^{12}$ G, $y_H = 1.2$ g cm^{-2}, $T^\infty = 4.3 \times 10^5$ K, and $R^\infty = 17$ km. The *dashed line* is the unabsorbed atmosphere model spectrum. The *dash-dotted line* is the (absorbed) blackbody fit to the X-ray spectrum with $R^\infty = 5$ km

5 Summary and conclusions

We have gathered together observations of the isolated neutron star RX J1856.5–3754 and compared them to our latest magnetic, partially ionized hydrogen atmosphere models. Prior works showed that the observations were well-fit by blackbody spectra. Here we obtain good fits to the overall multiwavelength spectrum of RX J1856.5–3754 using the more realistic atmosphere model. In particular, we do not overpredict the optical flux obtained by previous works and require only a single temperature atmosphere. [Note that this single temperature (and magnetic field) serves as an average value for the entire surface.] In addition to the neutron star orientation and viewing geometry, the single temperature helps to explain the non-detection of pulsations thus far. At high X-ray energies, where the atmosphere is optically thin, the model spectrum has a "blackbody-like" shape due to the emission spectrum of a magnetized, condensed surface beneath the atmosphere. The atmosphere is optically thick at lower energies; thus features in the emission spectrum of the condensed surface are not visible when viewed from above the atmosphere. The "thinness" of the atmosphere helps to produce the featureless, blackbody-like spectrum seen in the observations.

Based on a possible origin within the Upper Scorpius OB association, the age of RX J1856.5–3754 is estimated to be about 5×10^5 yr (Walter 2001; Walter and Lattimer 2002; Kaplan et al. 2002). Our surface temperature determination ($kT^\infty = 37$ eV) is only a factor of 1.7 below the blackbody temperature ($kT^\infty = 63$ eV) obtained by previous works and therefore does not place much stronger constraints on theories of neutron star cooling (see, e.g., Page et al. 2004; Yakovlev and Pethick 2004). It may also be noteworthy that RX J1856.5–3754 is one of the cooler isolated neutron stars and possibly possesses the lowest magnetic field [$B \approx (3\text{–}4) \times 10^{12}$ G]; the lower magnetic field implies a more uniform surface temperature distribution and weaker radiation beaming.

Finally, the emission radius we derive from our atmosphere model fits is $R^\infty \approx 17\,(d/140\text{ pc})$ km (although recall the distance and flux uncertainties discussed in footnote 1 and Sect. 3, respectively). Accounting for gravitational redshift ($z_g \sim 0.22$), this yields $R^{\rm em} \approx 14$ km. This is much larger than the inferred radius obtained by just fitting the X-ray data with a blackbody ($R^\infty_{\rm X} \approx 5$ km). As a result, there does not appear to be a need to resort to more exotic explanations such as quark or strange stars (e.g., Drake et al. 2002; Xu 2003; see Walter 2004; Weber 2005 for a review), at least for the case of RX J1856.5–3754. On the other hand, our radius is small compared to the radius derived from fitting the optical/UV data ($R^\infty_{\rm opt} \approx 21$ km). For a $1.4\,M_\odot$ neutron star, the latter implies a low redshift ($z_g \approx 0.12$) and very large intrinsic radius ($R^{\rm em}_{\rm opt} \approx 19$ km); this is ruled out by neutron star equations of state, while our radius $R^{\rm em} \approx 14$ km only requires a standard, stiff equation of state (see, e.g., Lattimer and Prakash 2001).

Acknowledgements We greatly appreciate assistance from Marten van Kerkwijk, Gilles Chabrier, Chris Thompson, Dong Lai, and Lars Bildsten. We thank the anonymous referee for useful comments. W.H. is supported by NASA through Hubble Fellowship grant HF-01161.01-A awarded by STScI, which is operated by AURA, Inc., for NASA, under contract NAS 5-26555. D.K. was partially supported by a fellowship from the Fannie and John Hertz Foundation and by HST grant GO-09364.01-A. Support for the HST grant GO-09364.01-A was provided by NASA through a grant from STScI, which is operated by AURA, Inc., under NASA contract NAS 5-26555. P.C. is supported by the NSF under PHY 99-07949, by JINA through NSF grant PHY 02-16783, and by NASA through Chandra Award Number GO4-5045C issued by CXC, which is operated by the SAO for and on behalf of NASA under contract NAS 8-03060. P.C. acknowledges support from the Miller Institute for Basic Research in Science, University of California Berkeley. A.P. is supported by FASI through grant NSh-9879.2006.2 and by RFBR through grants 05-02-16245 and 05-02-22003.

References

Bonanno, A., Urpin, V., Belvedere, G.: Astron. Astrophys. **440**, 199 (2005)

Braje, T.M., Romani, R.W.: Astrophys. J. **580**, 1043 (2002)

Brinkmann, W.: Astron. Astrophys. **82**, 352 (1980)

Burwitz, V., Zavlin, V.E., Neuhäuser, R., et al.: Astron. Astrophys. **379**, L35 (2001)

Burwitz, V., Haberl, F., Neuhäuser, R., et al.: Astron. Astrophys. **399**, 1109 (2003)

Cardelli, J.A., Clayton, G.C., Mathis, J.S.: Astrophys. J. **345**, 245 (1989)

Drake, J.J., et al.: Astrophys. J. **572**, 996 (2002)

Haberl, F., Zavlin, V.E., Trümper, J., et al.: Astron. Astrophys. **419**, 1077 (2004)

Haisch, B., Bowyer, S., Malina, R.F.: J. Br. Interplanet. Soc. **46**, 331 (1993)

Ho, W.C.G., Lai, D.: Mon. Not. Roy. Astron. Soc. **338**, 233 (2003)

Ho, W.C.G., Lai, D., Potekhin, A.Y., et al.: Astrophys. J. **599**, 1293 (2003)

Ho, W.C.G., Kaplan, D.L., Chang, P., et al.: Mon. Not. Roy. Astron. Soc. **375**, 821 (2007)

Kaplan, D.L., van Kerkwijk, M.H.: Astrophys. J. Lett. **628**, L45 (2005)

Kaplan, D.L., van Kerkwijk, M.H., Anderson, J.: Astrophys. J. **571**, 447 (2002)

Kaplan, D.L., van Kerkwijk, M.H., Marshall, H.L., et al.: Astrophys. J. **590**, 1008 (2003)

Kargaltsev, O.Y., Pavlov, G.G., Zavlin, V.E., et al.: Astrophys. J. **625**, 307 (2005)

Kaspi, V.M., Roberts, M.S.E., Harding, A.K.: In: Lewin, W.H.G., van der Klis, M. (eds.) Compact Stellar X-ray Sources. Cambridge University Press, Cambridge (2006, in press). Preprint: astro-ph/0402136

Lai, D.: Rev. Mod. Phys. **73**, 629 (2001)

Lattimer, J.M., Prakash, M.: Astrophys. J. **550**, 426 (2001)

Lloyd, D.A.: Mon. Not. Roy. Astron. Soc. (2003, submitted) (astro-ph/0303561)

Medin, Z., Lai, D.: Phys. Rev. A **74**, 062507 (2006a) (astro-ph/0607166)

Medin, Z., Lai, D.: Phys. Rev. A **74**, 062508 (2006b) (astro-ph/0607277)

Mészáros, P.: High-Energy Radiation from Magnetized Neutron Stars. University of Chicago Press, Chicago (1992)

Mori, K., Ruderman, M.A.: Astrophys. J. Lett. **592**, L75 (2003)

Motch, C., Zavlin, V.E., Haberl, F.: Astron. Astrophys. **408**, 323 (2003)

O'Donnell, J.E.: Astrophys. J. **422**, 158 (1994)

Page, D., Lattimer, J.M., Prakash, M., et al.: Astrophys. J. Suppl. Ser. **155**, 623 (2004)

Pavlov, G.G., Zavlin, V.E., Sanwal, D.: In: Becker, W., Lesch, H., Trümper, J. (eds.) Proc. 270 WE-Heraeus Seminar on Neutron Stars, Pulsars, and Supernova Remnants, p. 273. MPI, Garching (2002) (MPE Rep. 278)

Pavlov, G.G., Shibanov, Yu.A., Zavlin, V.E., Meyer, R.D.: In: Alpar, M.A., Kiziloğlu, Ü., van Paradijs, J. (eds.) Lives of the Neutron Stars, p. 71. Kluwer, Boston (1995)

Pavlov, G.G., Zavlin, V.E., Trümper, J., et al.: Astrophys. J. Lett. **472**, L33 (1996)

Pérez-Azorín, J.F., Miralles, J.A., Pons, J.A.: Astron. Astrophys. **433**, 275 (2005)

Pons, J.A., Walter, F.M., Lattimer, J.M., et al.: Astrophys. J. **564**, 981 (2002)

Potekhin, A.Y.: J. Phys. B. **27**, 1073 (1994)

Potekhin, A.Y., Chabrier, G.: Astrophys. J. **585**, 955 (2003)

Potekhin, A.Y., Lai, D., Chabrier, G., et al.: Astrophys. J. **612**, 1034 (2004)

Predehl, P., Schmitt, J.H.M.M.: Astron. Astrophys. **293**, 889 (1995)

Ransom, S.M., Gaensler, B.M., Slane, P.O.: Astrophys. J. Lett. **570**, L75 (2002)

Toropina, O.D., Romanova, M.M., Lovelace, R.V.E.: Mon. Not. Roy. Astron. Soc. **371**, 569 (2006). Preprint: astro-ph/0606254

Treves, A., Turolla, R., Zane, S., et al.: PASP **112**, 297 (2000)

Trümper, J.E., Burwitz, V., Haberl, F., et al.: Nucl. Phys. B Proc. Suppl. **132**, 560 (2004)

Turolla, R., Zane, S., Drake, J.J.: Astrophys. J. **603**, 265 (2004)

van Adelsberg, M., Lai, D., Potekhin, A.Y., et al.: Astrophys. J. **628**, 902 (2005)

van Kerkwijk, M.H., Kulkarni, S.R.: Astron. Astrophys. **378**, 986 (2001a)

van Kerkwijk, M.H., Kulkarni, S.R.: Astron. Astrophys. **380**, 221 (2001b)

Walter, F.M.: Astrophys. J. **549**, 433 (2001)

Walter, F.M.: J. Phys. G **30**, S461 (2004)

Walter, F.M., Lattimer, J.M.: Astrophys. J. Lett. **576**, L145 (2002)

Weber, F.: Prog. Part. Nucl. Phys. **54**, 193 (2005)

Wilms, J., Allen, A., McCray, R.: Astrophys. J. **542**, 914 (2000)

Xu, R.X.: Astrophys. J. Lett. **596**, L59 (2003)

Yakovlev, D.G., Pethick, C.J.: Annu. Rev. Astron. Astrophys. **42**, 169 (2004)

Zane, S., Turolla, R., Drake, J.J.: Adv. Space Res. **33**, 531 (2004)

Zavlin, V.E., Pavlov, G.G., Shibanov, Yu.A., et al.: Astron. Astrophys. **297**, 441 (1995)

Astrophys Space Sci (2007) 308: 287–296
DOI 10.1007/s10509-007-9383-1

ORIGINAL ARTICLE

Ultraviolet emission from young and middle-aged pulsars

Connecting X-rays with the optical

Oleg Kargaltsev · George Pavlov

Received: 21 July 2006 / Accepted: 19 September 2006 / Published online: 21 March 2007

Abstract We present the UV spectroscopy and timing of three nearby pulsars (Vela, B0656+14 and Geminga) recently observed with the Space Telescope Imaging Spectrograph. We also review the optical and X-ray properties of these pulsars and establish their connection with the UV properties. We show that the multiwavelengths properties of neutron stars (NSs) vary significantly within the sample of middle-aged pulsars. Even larger differences are found between the thermal components of Geminga and B0656+14 as compared to those of radio-quiet isolated NSs. These differences could be attributed to different properties of the NS surface layers.

Keywords Neutron Stars · Pulsars · Geminga · PSR B0656+14 · Vela pulsar · RX J1856.5-3754 · RX J0720.4-3125 · RX J1308.6+2127

PACS 97.60.Gb · 97.60.Jd

1 Introduction

Optical through X-rays radiation from a typical isolated neutron star (NS) is expected to exhibit two components: thermal radiation from the NS surface and non-thermal radiation from the NS magnetosphere. The spectrum of *non-thermal radiation* can be described by a power-law (PL) model with a spectral index $-1 \lesssim \alpha \lesssim 0$ ($F_\nu \propto \nu^\alpha$). This radiation is commonly interpreted as synchrotron emission from relativistic electrons/positrons accelerated by the electric fields near the NS surface and from secondary particles produced in the pair cascades. The magnetospheric radiation is intrinsically anisotropic and, therefore, it shows strong pulsations. The non-thermal emission completely dominates the multiwavelength spectrum in very young pulsars (e.g., Crab and B0540-69).

As the magnetospheric emission becomes fainter with increasing pulsar age, *thermal emission* becomes detectable at $\tau \equiv P/2\dot{P} \gtrsim 10$ kyrs, being seen as a "thermal hump" on top of the flat non-thermal spectrum. This thermal radiation is emitted from the NS surface which can be non-uniformly heated (e.g., due to anisotropic heat conductivity of the crust and bombardment by relativistic particles). Indeed, in several cases at least two thermal components with different temperatures are needed to fit the X-ray spectrum (e.g., Pavlov et al. 2002; Zavlin and Pavlov 2004; De Luca et al. 2005). The cooler *thermal soft* (TS) component is commonly interpreted as emission from the bulk of the NS surface. The hotter *thermal hard* (TH) component is often attributed to the NS polar caps (PCs) which can be additionally heated by the magnetospheric particles. As expected, the TS temperatures generally decrease with pulsar age. However, they do not fall onto a monotonically decreasing cooling curve, which suggests that NSs may have different masses (e.g., Yakovlev and Pethick 2004). In X-rays, the TS component usually shows weaker and broader pulsations than the TH component. The TS pulsations can be caused by non-uniformities of the NS surface temperature and by anisotropy of local emissivity in the strong magnetic field.

This work was supported by STScI grants GO-9182 and GO-9797 and NASA grant NAG5-10865.

O. Kargaltsev (✉) · G. Pavlov
Department of Astronomy & Astrophysics, Pennsylvania State University, 525 Davey Lab, University Park, PA 16802, USA
e-mail: oyk100@psu.edu

G. Pavlov
e-mail: pavlov@astro.psu.edu

Spectra of middle-aged pulsars ($10 \lesssim \tau \lesssim 500$ kyr) are particularly interesting because they often exhibit all the three emission components. The nearby middle-aged pulsars, Vela ($\tau \approx 11$ kyrs), B0656+14 (hereafter B0656; $\tau \approx 110$ kyrs) and Geminga ($\tau \approx 340$ kyrs), are relatively bright and have been extensively studied in both the optical and X-rays. Given their different ages, the three pulsars provide a representative cut through the population of middle-aged pulsars, allowing one to study the evolution of the pulsar properties.

The X-ray spectra of these three pulsars can be described by a three-component, TS+TH+PL, model, with the thermal components modeled as BBs. However, such fits result in substantially different model parameters in each case.

The youngest Vela pulsar shows the highest $T_{\rm TS} \approx 1.2$ MK emitted from a region with $R_{\rm TS} \approx 3.7$ km. This $R_{\rm TS}$ is substantially smaller than a typical NS radius, which might indicate that a large fraction of the NS surface is too cold to be seen in X-rays.[1] A factor of ten older B0656 has a lower temperature, $T_{\rm TS} \approx 0.7$ MK, and a much larger emitting area, $R_{\rm TS} \approx 12$ km, close to that of a typical NS. The oldest of the three pulsars, Geminga, is even colder, $T_{\rm TS} \approx 0.5$ MK, and has the largest emitting area ($R_{\rm TS} \approx 13$ km).

The PL component, with $\alpha_{\rm X} \approx -1$, is the strongest in the Vela pulsar, $L_{X,{\rm PL}} \approx 3 \times 10^{31}$ erg s^{-1} (in 0.2–10 keV), becoming weaker in B0656 ($\alpha_{\rm X} \approx -0.5$; $L_{X,{\rm PL}} \approx 1.4 \times 10^{31}$ erg s^{-1}), and much weaker in Geminga ($\alpha_{\rm X} \sim -0.6$; $L_{X,{\rm PL}} \approx 2.2 \times 10^{30}$ erg s^{-1}). This PL component is typically interpreted as synchrotron emission produced by relativistic particles in the pulsar magnetosphere.

The strength of the TH component also appears to decrease with pulsar age. Being relatively strong in the Vela pulsar ($L_{\rm TH} \approx 6.3 \times 10^{31}$ erg s^{-1}), it becomes weaker in B0656 ($L_{\rm TH} \approx 3.5 \times 10^{31}$ erg s^{-1}) and decreases dramatically in Geminga ($L_{\rm TH} \approx 4.4 \times 10^{29}$ erg s^{-1}). The sizes of the TH emission regions are close to the conventional PC radii [$r_{\rm pc} \equiv (2\pi R_{\rm NS}^3/cP)^{1/2}$; see Table 1] in Vela and B0656, while in Geminga, $R_{\rm TH} \approx 46$ m, is surprisingly small when compared to its $r_{\rm pc} = 440$ m. This may suggest that the surface of Geminga is more uniformly heated and the "TH component" simply mimics a phase-dependent PL component.[2] However, regardless of the exact nature of the TH and PL components, the soft X-ray (0.3–2 keV) spectra of these three pulsars are clearly dominated by the thermal emission.

[1] Alternatively, it may mean that the BB description of the thermal spectrum is inadequate, and a more realistic atmosphere model should be used. For instance, the Vela pulsar spectrum fits equally well by a two-component, NS atmosphere plus PL, model, which gives $T \approx 0.7$ MK at $R = 13$ km (Pavlov et al. 2001; see also Kargaltsev 2004).

[2] From a phase-resolved analysis of the XMM data, Jackson and Halpern (2005) conclude that the X-ray spectrum of Geminga can be alternatively described by a BB+PL model with a phase-dependent PL slope.

Table 1 Basic properties of three middle-aged pulsars and three RQINSs with measured spin-down parameters and/or parallaxes (Kaplan and van Kerkwijk 2005a, 2005b; see also van Kerkwijk & Kaplan, these proceedings)

Name	Dist. pc	τ kyr	P s	$\log \dot{E}$ ergs s^{-1}	B TG	$r_{\rm pc}$ m
Geminga	$\sim$200	340	0.24	34.52	1.6	440
B0656+14	288^{+33}_{-27}	110	0.38	34.58	4.7	346
Vela	293^{+19}_{-17}	11	0.09	36.84	3.8	717
RX J0720	333^{+167}_{-83}	1900	8.4	30.67	24	7
RX J1308	$\sim$700	1500	10.3	30.60	34	7
RX J1856	161^{+17}_{-14}	$\sim$400	...	...	...	...

Observing the TS radiation over a wide range of frequencies is particularly important. X-ray observations can only probe the Wein tail of the TS spectrum emitted from the NS surface. This information is insufficient to reconstruct the overall shape of the spectrum which can deviate from the simple black-body due to the presence of an atmosphere or nonuniformity of surface temperature. Chemical composition of the atmosphere can dramatically affect the X-ray spectrum (Romani 1987; Pavlov et al. 1995; Zavlin et al. 1996; Zavlin and Pavlov 2002), with light element atmospheres leading to a large Wien excess. Any surface temperature inhomogeneities will also complicate the spectrum, with hotter regions being increasingly important at higher frequencies. For these reasons, comparison of thermal X-ray emission with optical-UV emission on the Rayleigh-Jeans (R-J) side of the thermal hump is particularly valuable. The challenge here is that the non-thermal magnetospheric emission becomes increasingly dominant at longer wavelengths.

The observations with ground-based telescopes and the Hubble Space Telescope (HST) have provided multi-band photometry from near-IR (NIR) to near-UV (NUV) and showed that the NIR-optical spectra of the three pulsars are non-thermal, with a hint of a R-J component seen in Geminga and B0656+14 at $\lambda \lesssim 3000$ Å. The observations in UBVRIHJ filters show that the optical spectrum of the Vela pulsar remains very flat from NIR to UV, with $\alpha_{\rm O} \approx -0.1$ (Shibanov et al. 2003). For the other two pulsars, the NIR-optical spectra are steeper, $\alpha_{\rm O} \approx -0.4$ (Koptsevich et al. 2001; Shibanov et al. 2005).

For all the three pulsars optical pulse profile measurements have been carried out. The Vela and B0656, show strong non-sinusoidal pulsations (Gouiffés 1998; Kern et al. 2003). Optical pulsations of Geminga were only marginally detected in the B band (Shearer et al. 1998).

Despite the extensive coverage in the optical and X-rays, very few NUV observations had been carried out until recently, and the FUV parts of the spectra have remained completely unexplored. To fill this gap, we have undertaken an observational campaign with the MAMA-NUV and MAMA-FUV detectors of the Space Telescope Imaging Spectrograph (STIS) aboard the HST.

2 Observations with STIS in far- and near-UV

We have observed all the there pulsars in the NUV (1800–3000 Å) and FUV (1150–1700 Å) passbands (see Table 2). The data reduction and analysis are described in detail by Kargaltsev et al. (2005), Romani et al. (2005), and Kargaltsev and Pavlov (2007). Here we summarize the main results of these observations.

2.1 Geminga

In the FUV passband we obtained a low-resolution spectrum of Geminga using MAMA-FUV with the grating G140L. The observed flux in the 1155–1702 Å range is $\mathcal{F}_{\rm FUV} = (3.72 \pm 0.24) \times 10^{-15}$ ergs s^{-1} cm^{-2}, corresponding to the luminosity $L_{\rm FUV} = 4\pi d^2 \mathcal{F}_{\rm FUV} = (1.78 \pm 0.11) \times 10^{28} d_{200}^2$ ergs s^{-1}. Fitting the spectrum with the absorbed PL model gives $\alpha_{\rm FUV} = 1.43 \pm 0.53$ for $E(B-V) = 0.03$. The PL slope is close to that of the R-J spectrum, $F_\nu \propto \nu^2$, suggesting that the observed radiation is dominated by thermal emission from the NS surface. To estimate the NS surface temperature, we fit the absorbed BB model to the spectrum. Since the FUV frequencies are in the R-J part of the spectrum, the temperature values are strongly correlated with the radius-to-distance ratio (approximately, $T \propto d^2/R^2$). For a typical NS radius, $R = 13$ km, and the assumed distance $d = 200$ pc, the inferred temperatures are 0.31 ± 0.01 and 0.41 ± 0.02 MK, for $E(B-V) = 0.03$ and 0.07, respectively.

The Geminga pulsar was imaged with MAMA-NUV using the broad-band filter F25SRF2. The NUV flux depends on the assumed spectral slope and extinction. For a plausible $E(B-V) = 0.03$ and $\alpha_{\rm NUV} = 1$, the unabsorbed NUV flux is $\mathcal{F}_{\rm NUV} = 2.2 \times 10^{-15}$ ergs cm^{-2} s^{-1} in 1800–3000 Å.

We used the 125 μs time resolution of STIS for timing analysis of both NUV and FUV data. We found statistically significant pulsations with frequencies lying within (−0.3, +0.7) μHz around the $f = 4,217,608.6953$ μHz frequency estimated from multi-epoch timing observations in γ-rays and X-rays (Jackson et al. 2002). The folded (source plus background) NUV light curve, plotted in Fig. 2, shows one broad (FWHM ≈0.8 of the period), flat-top peak per period, centered at the phase $\phi \approx 1.0$. The most notable feature of the pulse profile is the narrow dip at $\phi \approx 0.45$. The pulsed fraction, defined as the ratio of the number of counts above the minimum level to the total number of counts in the light curve, is about 28%, which corresponds to the intrinsic source pulsed fraction $f_{\rm p} \approx 40\%$.

The FUV light curve, plotted in the same Fig. 2, also shows a sharp, asymmetric dip at $\phi \approx 0.45$ with an additional shallower dip at $\phi \approx 0.95$. The pulsed fraction in the observed (source + background) radiation is about 40% (intrinsic pulsed fraction 60–70%.).

2.2 B0656+14

The FUV spectrum of B0656 has been measured with MAMA-FUV using the G140L grating. Unfortunately, only two of the eight planned orbits of data were collected because of the STIS failure on August 3, 2004. The observed flux in the 1153–1700 Å band is $\mathcal{F}_{\rm FUV} = (4.3 \pm 0.3) \times 10^{-15}$ ergs s^{-1} cm^{-2}, corresponding to $L_{\rm FUV} = (4.2 \pm 0.3) \times 10^{28} d_{288}^2$ ergs s^{-1}. The fit with the absorbed PL model results in a rather steep spectral slope $\alpha_{\rm FUV} = 1.51 \pm 0.62$ for a plausible $E(B-V) = 0.03$. Similar to Geminga, this suggests that the FUV radiation could be dominated by thermal emission from the NS surface. The absorbed BB fit gives the surface temperature of 0.71 ± 0.03 MK for

Table 2 STIS observations

Pulsar	Date	Instrument	Exposure, s	$\Delta\lambda$,[a] Å	$\mathcal{F}$,[b] cgs	f_p,[c] %
Geminga	2002 Feb 27	NUV-MAMA/F25SRF2	11367	1800–3000	1.7×10^{-15}	40
	2002 Feb 26	FUV-MAMA/G140L	10674	1155–1702	3.7×10^{-15}	65
B0656+14	2001 Sep 1	NUV-MAMA/PRISM	6791	1790–2950	2.6×10^{-15}	67
	2001 Nov 16	NUV-MAMA/PRISM	12761	1790–2950	2.6×10^{-15}	89
	2004 Jan 20	FUV-MAMA/G140L	4950	1153–1700	4.2×10^{-15}	64
Vela	2002 May 28	NUV-MAMA/F25SRF2	2895	1800–3000	7.2×10^{-15}	87
	2002 May 28	FUV-MAMA/G140L	3060	1153–1701	8.2×10^{-15}	73

[a] Instrument + filter passband

[b] Observed flux in the corresponding passband (for the NUV observations of B0656, the flux has been measured from the two observations combined)

[c] Intrinsic (corrected for the background) pulsed fraction

$E(B-V)=0.03$, $R=13$ km, and the distance of 288 pc (Brisken et al. 2003). The corresponding unabsorbed bolometric luminosity is $L_{\rm bol}=(3.1\pm0.5)\times10^{32}$ ergs s^{-1}. Thus, the measured brightness temperature is substantially higher than that of Geminga.

The NUV spectrum and light curve of B0656 have been obtained using MAMA-NUV with PRISM (Shibanov et al. 2005). We re-analyzed these data to facilitate the direct comparison with the FUV and other multiwavelength data. The observed flux in the 1790–2950 Å range is $\mathcal{F}_{\rm NUV}=(2.63\pm0.38)\times10^{-15}$ erg s^{-1} cm^{-2}, corresponding to $L_{\rm NUV}=(2.61\pm0.38)\times10^{28}d_{288}^2$ erg s^{-1}. Fitting the spectrum with the absorbed PL model, we find $\alpha_{\rm NUV}=1.09\pm0.41$ for a plausible $E(B-V)=0.03$. The PL slope is somewhat flatter than that of the FUV spectrum, which can be explained by a larger contribution of the non-thermal component.

Making use of the event time-tags, we created FUV and NUV pulse profiles folded with the radio ephemeris provided by Kramer (2005; priv. comm.). The FUV pulse profile (Fig. 2) shows two peaks of approximately equal strengths. The minimum at $\phi\approx0.55$ is shallower than the other minimum at $\phi\approx1.05$. The observed pulsed fraction, ≈36%, corresponds to the intrinsic pulsed fraction of ≈64%. We also folded the light curves for each of the NUV visits and co-added them into the combined NUV pulse profile (Fig. 2). The NUV pulse shape resembles the FUV one but shows a higher pulsed fraction (the observed and intrinsic pulsed fractions are 22% and 80%, respectively). The lower pulsed fraction in FUV is evidence of a larger contribution of thermal emission at the FUV wavelengths (thermal emission is intrinsically less pulsed than the non-thermal emission). The NUV and FUV light curves can be interpreted as two non-thermal peaks on top of a single low-amplitude thermal pulse with the maximum at $\phi\approx0.5$. This interpretation would imply a substantially softer spectrum at the pulse minima than at the peaks.

To verify this assumption, we produced the phase-resolved NUV through FUV spectrum for three phase intervals [0.5; 0.7], [0.9; 1.2], and [0.2; 0.5] + [0.7; 0.9], corresponding to each of the two peaks and two minima combined. The difference in the slopes of the PL model fitted to the peaks spectra does not exceed the spectral index uncertainties. However, the spectrum at the pulse minima is substantially steeper ($\alpha=1.5\pm0.2$) than the peak spectra ($\alpha=0.6\pm0.2$). This behavior is in line with the expected larger contribution of thermal emission at the minima phases.

2.3 Vela pulsar

The spectroscopic (FUV) and imaging (NUV) data has been acquired using exactly the same instrumental setup as in the Geminga observations (see Table 1) and reduced in a similar way.

For a plausible PL slope, $\alpha_{\rm NUV}=0$ (see Sect. 3.3) and color index $E(B-V)=0.05$ (estimated from the hydrogen column density found from the X-ray fits; Sanwal et al. 2002), the unabsorbed NUV flux is $\mathcal{F}_{\rm NUV}\approx1.1\times10^{-14}$ ergs cm^{-2} s^{-1} in 1800–3000 Å.

The FUV spectrum of the Vela pulsar is significantly flatter than those of Geminga and B0656, which reflects its predominantly non-thermal nature even throughout the FUV passband. The total observed flux in the 1153–1701 Å range is $\mathcal{F}_{\rm FUV}=(8.19\pm0.36)\times10^{-15}$ ergs s^{-1} cm^{-2}, corresponding to $L_{\rm FUV}=(8.78\pm0.39)\times10^{28}$ ergs s^{-1} at $d=300$ pc (Dodson et al. 2003). Fitting the spectrum with the absorbed power-law model, we found $\alpha_{\rm FUV}=0.06\pm0.39$ at $E(B-V)=0.05$.

Figure 2 shows the NUV and FUV light curves folded with appropriate radio ephemeris (courtesy of R.N. Manchester). At least four narrow peaks can be identified in each of the light curves. The higher signal-to-noise (S/N) NUV light curve shows some substructure in the two main peaks. The observed (source plus background) pulsed fractions are 52% and 73% for the FUV and NUV light curves, respectively. The corresponding intrinsic pulse fractions are 73% and 87%.

3 Multiwavelength spectra and light curves

To compare the UV emission with the NIR, optical and X-ray emission, we plotted multiwavelength spectra of the three pulsars in Fig. 1 (NIR to FUV) and Fig. 3 (NIR to X-rays). The FUV, NUV and X-ray light curves are shown in Fig. 2. Below we briefly discuss the multiwavelength properties for each of the pulsars.

3.1 Geminga

The top panel of Fig. 1 shows the UV fluxes together with the multi-band NIR-optical photometry adopted from Kargaltsev et al. (2005) and Shibanov et al. (2005).

As the NIR through FUV spectrum cannot be described by a simple PL model, we fit this spectrum with a BB+PL model. For the fixed $R/d=13$ km/200 pc, we obtain $T=0.30\pm0.02$ MK, $\alpha_{\rm O}=-0.46\pm0.12$, $F_0=0.11\pm0.02$ μJy for $E(B-V)=0.03$, where $\alpha_{\rm O}$ and F_0 are the parameters of the PL component: $F_\nu=F_0\,(\nu/1\times10^{15}\ {\rm Hz})^{\alpha_{\rm O}}$. Notice that the parameters of the BB component are virtually the same as obtained from the FUV-MAMA spectrum alone. We see that the BB emission dominates at $\lambda\lesssim3000$ Å, while the magnetospheric PL emission dominates at longer wavelengths.

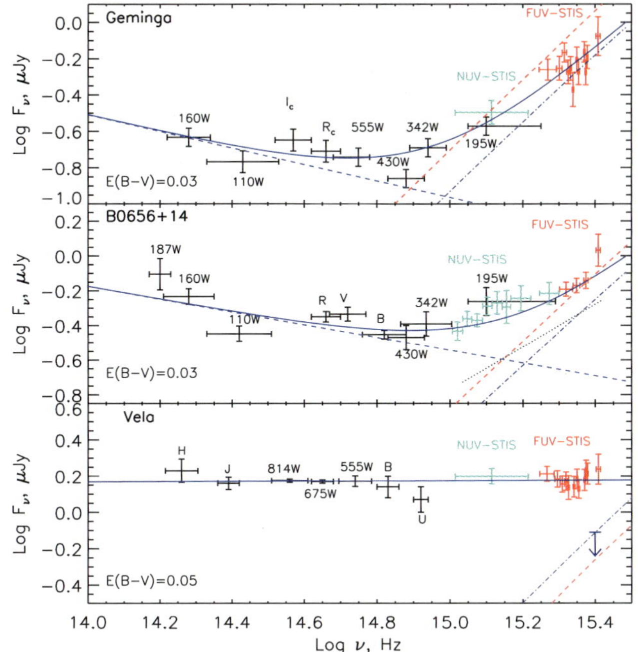

Fig. 1 Unabsorbed NIR through FUV spectra of the Geminga, B0656 and Vela pulsars (*top* to *bottom*). The PL+BB fits are shown by the *solid blue lines*. The contributions of the individual components are also shown (*blue dashed line* for PL, *blue dash-dotted line* for BB). The dashed red line shows the extrapolation of the TS component fitted to the X-ray spectrum (see also Fig. 3). For B0656 we have also plotted the fit to the NUV+FUV spectrum at the pulse minima (*black dotted line*)

To investigate the connection between the optical/UV and X-ray properties, we use the data obtained with XMM-Newton. The X-ray spectrum can be reasonably well described by a three-component TS+TH+PL model (see Sect. 1 and Kargaltsev et al. 2005 for details), shown in the top panel of Fig. 3. One can see that the extrapolation of the TS component *overpredicts* the FUV fluxes by a factor of 1.6 [for $E(B-V)=0.03$; see also Fig. 1]. This FUV *deficit* can also be demonstrated by plotting together the temperature-radius confidence contours for the TS component and for the BB fit to the FUV-MAMA spectrum (Fig. 4). We see that at plausible values of interstellar extinction, $E(B-V)\lesssim 0.07$, the FUV contours lie at smaller radii (or much lower temperatures) than the X-ray contours, in contrast to some other neutron stars (see Sect. 4.2). The extrapolation of the X-ray PL component suffers from large uncertainties and is marginally consistent with the optical fluxes.

The background-subtracted X-ray light curves are compared with the UV and optical light curves in Fig. 2. In the 0.2–0.5 and 2–8 keV bands, the radiation is dominated by the TS and PL components, respectively. The PL light curve (pulsed fraction $f_p = 34\% \pm 8\%$) shows two pronounced peaks per period, resembling the γ-ray light curve (Jackson et al. 2002), albeit with a smaller distance between the peaks, and a hint of a third peak, at $\phi \approx 0.2$. On the contrary, the 0.2–0.5 keV light curve ($f_p = 30\% \pm 2\%$) is characterized by one broad peak per period (with small "ripples", perhaps due to contribution from the PL and TH components). The minimum of the TS light curve is approximately aligned in phase with one of the minima of the PL light curve, being shifted by $\Delta\phi \approx 0.1$ from the sharp dips of the NUV and FUV light curves.

As the thermal soft emission dominates the spectrum from 4 eV to 1 keV, the non-thermal pulsations can only be observed in the optical or hard X-rays. The double-peaked shape of the hard X-ray (2–8 keV) pulsations does not correlate with the TS pulsations. The optical pulsations, marginally detected in the B band (Shearer et al. 1998), also show a double-peaked structure, with a larger separation between the peaks ($\Delta\phi \approx 0.6$ in the optical vs. $\Delta\phi \approx 0.4$ in the hard X-rays). The phases of the optical peaks differ from those of the hard X-ray peaks. To understand and resolve these apparent inconsistencies, a higher quality optical light curve should be obtained.

3.2 B0656+14

Similar to Geminga, we fit the NIR through FUV spectrum of B0656 (middle panel in Fig. 1; NIR and optical data are from Koptsevich et al. 2001 and Shibanov et al. 2005) with a two-component, BB+PL, model. For the fixed $R/d = 13$ km/288 pc, we found the temperature $T = 0.50 \pm 0.05$ MK, spectral index $\alpha_O = -0.41 \pm 0.08$, and PL normalization $F_0 = 0.27 \pm 0.02$ μJy for $E(B-V) = 0.02$. One can see that the BB emission dominates at $\lambda \lesssim 2000$ Å. The fit also confirms that, in the NUV, the fraction of thermal photons is smaller than the fraction of nonthermal photons by a factor of 1.5, which is compatible with the deeper minima and higher pulsed fraction in the NUV light curve.

We plot the NIR to X-rays spectra for B0656 in the middle panel of Fig. 3. The X-ray spectrum is again fitted with the TS+TH+PL model. We see that the extrapolation of the TS component from X-rays to lower frequencies roughly coincides with the FUV spectral flux and goes slightly above (by a factor of 1.3) the UV thermal component. This excess, however, is not statistically significant (see Fig. 4).

The double-peaked structure seen in the optical (Kern et al. 2003) and NUV/FUV light curves changes over to a single peak in the X-ray light curves (Fig. 2). It is possible that the asymmetric 0.25–0.7 keV pulse profile consists of two components: a broad one, peaked at about $\phi \approx 0.60$–0.65, and a narrower component centered at $\phi \approx 0.90$–0.95, close to the radio pulse phase, $\phi = 0$. In this case the maximum of the broader component would approximately coincide with the shallower minimum in the NUV and FUV light curves, which is located at the phase of a presumable maximum in the thermal UV component (see Sect. 2.2). The peak of the

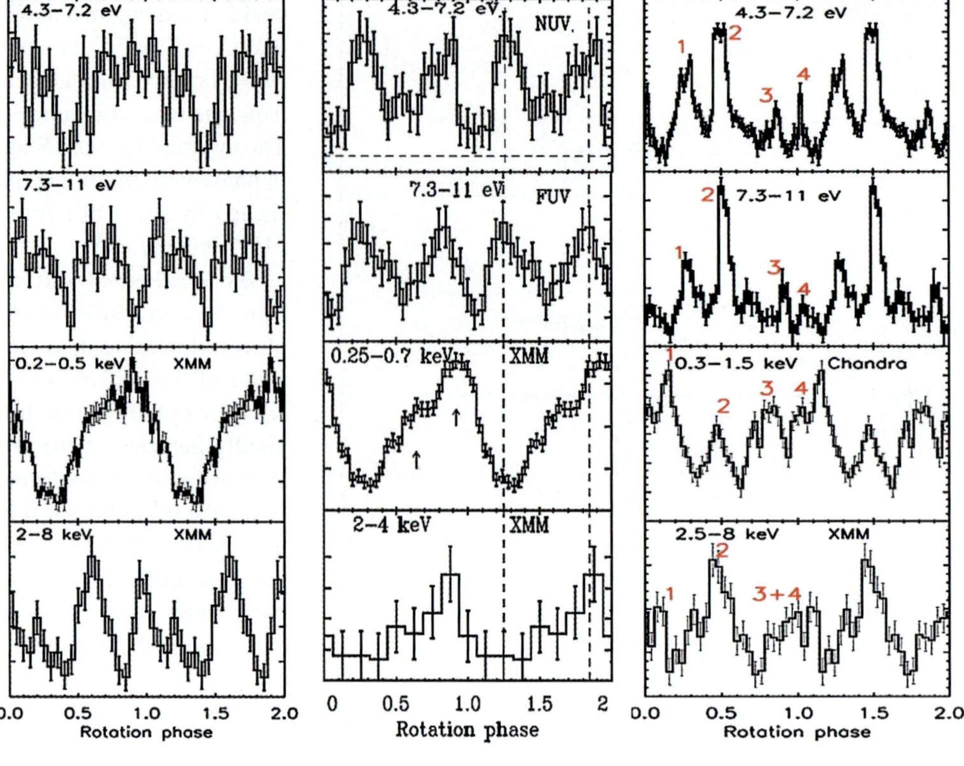

Fig. 2 NUV, FUV and X-ray (*top* to *bottom*) *light curves* of the Geminga, B0656+14 and Vela pulsars (*left* to *right*)

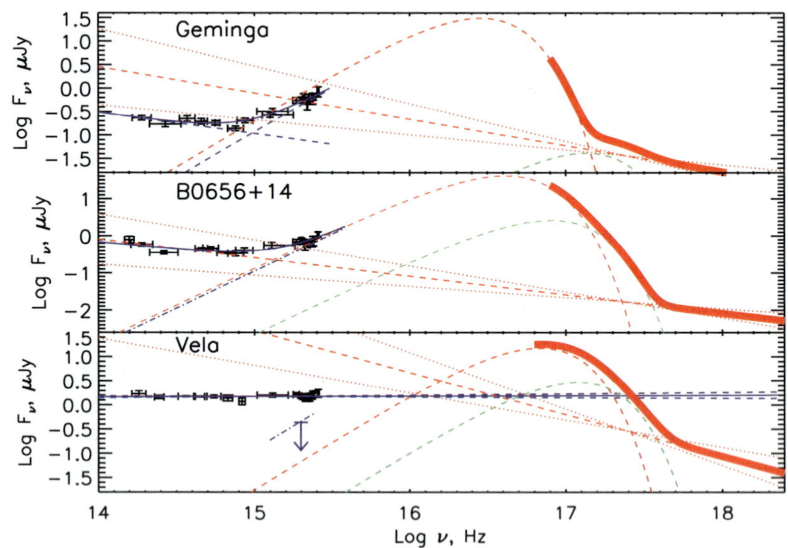

Fig. 3 Unabsorbed multiwavelength spectra of the Geminga, B0656+14 and Vela pulsars (*top* to *bottom*). In all cases the X-ray spectra (*thick red lines*) are fitted with the TS+TH+PL model. The TS and PL X-ray components are shown with *dashed red lines* (the uncertainties of the PL fit are *dotted lines*). The TH component is shown by the *green dashed line*. The *solid blue lines* show PL+BB fits to the optical through FUV spectra; the *blue dashed* and *dash-dotted lines* are the contributions of the PL and BB components, respectively

narrow component is approximately aligned with the single peak of the hard (2–4 keV) X-ray light curve.

Given the non-thermal interpretation of the peaks in the optical-UV light curves and the close match between the optical and X-ray PL components (Fig. 3), it is surprising that the 2–4 keV pulse profile of B0656 shows only one peak (at the phase of the second optical/UV peak; $\phi \approx 0.85$). Since the phase-resolved spectroscopy shows similar PL slopes in both UV peaks, they are likely to be of a similar origin. The absence of the second peak at the X-ray frequencies could be explained assuming that the X-rays are emitted closer to the NS in a more narrow cone while the UV emission is generated higher in the magnetosphere and spread over a broader beam. Alternatively, it could be that the "X-ray PL" is not magnetospheric emission (see Sect. 4.3), and the match between the X-ray and optical spectra is just a coincidence.

3.3 Vela pulsar

Figure 1 shows the spectrum of the Vela pulsar from NIR to UV, with the optical/NIR data points adopted from Shibanov et al. (2003). This very flat spectrum nicely fits a PL model.

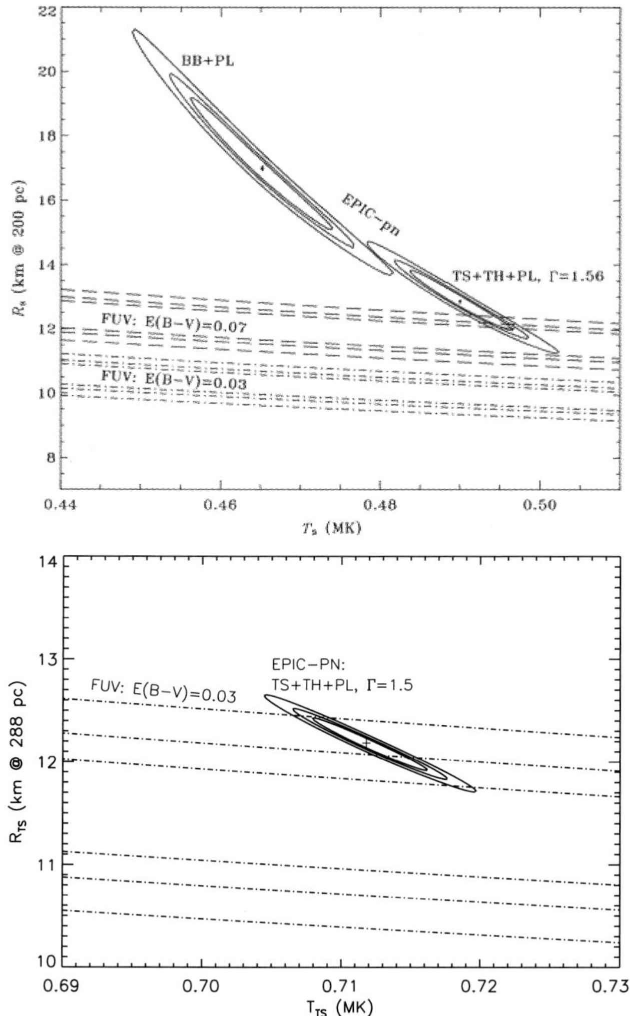

Fig. 4 *Top:* Confidence contours (68%, 90%, and 99%) in the (*soft*) temperature-radius plane obtained from fitting the EPIC-pn spectra (*solid lines*) with the TS+TH+PL and BB+PL models (*labels near the contours*). The *dashed* and *dash-dotted lines* show the confidence contours obtained from fitting the FUV spectrum with a BB model for two values of $E(B-V)$. See Kargaltsev et al. (2005) for other details. *Bottom:* Similar confidence contours but for B0656+14

The best-fit spectral index and normalization are $\alpha_O = 0.01 \pm 0.02$ and $F_\nu = 1.50 \pm 0.03$ μJy (at $\nu = 10^{15}$ Hz), respectively. The spectral index is larger than that obtained by Shibanov et al. (2003), $\alpha_O = -0.12$, from fitting only IR and optical data. On the other hand, the optical-UV spectrum is substantially flatter than the non-thermal X-ray spectrum, $\alpha_X \approx -1$.

The UV spectrum of the Vela pulsar allows one to obtain a restrictive upper limit on the NS surface temperature (see Romani et al. 2005 for details). The spectrum remains largely nonthermal even at the pulse minima, which constrains the NS temperature to $\lesssim 0.4(d/300\ \mathrm{pc})^2 (R/13\ \mathrm{km})^{-2}$ MK, assuming the BB model for the thermal emission and $E(B-V) = 0.05$. The limit becomes less restrictive, $\lesssim 0.6(d/300\ \mathrm{pc})^2(R/13\ \mathrm{km})^{-2}$ MK, if the hydrogen NS atmosphere model is used to describe the thermal spectrum. The inferred upper limit is surprisingly low compared to the temperature of the older B0656 (Table 3).

In the UV, the Vela pulse profile is dominated by multicomponent nonthermal emission. In Fig. 2 we show the NUV, FUV and X-ray light curves together. The X-ray light curves obtained with Chandra (Sanwal et al. 2002) and RXTE (Harding et al. 2002) are extremely complex, with multiple peaks likely originating from different regions of magnetosphere or corresponding to physically different spectral components. In agreement with Harding et al. (2002), we identify at least five components in the multiwavelength pulse profile. No peaks are seen in the UV at the phases $\phi = 0.12$ and 0.56 of the two γ-ray peaks (Kanbach et al. 1994). Instead, in the FUV they are replaced by tighter spaced peaks 1 and 2, which continue to the optical. We also find that the UV peaks 2 and 4 have counterparts in the soft X-ray light curve while peaks 1 and 3 appear to be shifted by $\Delta\phi \approx 0.1$ with respect to the other two soft X-ray peaks. The UV light curves show strong correlation with the hard X-ray light curves obtained with RXTE above 10 keV (Harding et al. 2002), i.e. peaks 2, 3 and 4 are almost at the same phases in both UV and RXTE bands. At the same time, the correlation between the UV and 2.5–8 keV light curves is less obvious. The spectral indices of the individual UV components are rather uncertain and show no correlation with the RXTE components (Romani et al. 2005). Should a better NUV/FUV data be ever obtained, a joint phase-resolved analysis of the UV and hard X-ray data may be more revealing.

4 Discussion

4.1 Thermal emission from NS surface

The spectra of all the three pulsars show thermal emission, but the strength and position of the thermal hump vary. In the Vela pulsar, thermal emission is seen only in soft (0.3–2 keV) X-rays while in Geminga and B0656 the FUV spectra are also predominantly thermal. The observed FUV and soft X-ray spectra represent the R-J and Wien tails of the TS component. The TS BB temperatures measured in X-rays decrease from ≈1.2 MK in the Vela pulsar to ≈0.7 MK in B0656 and ≈ 05 MK in the oldest Geminga. At the same time, the bolometric TS luminosity of the Vela pulsar is surprisingly low, which can be formally explained by a smaller emitting area of the TS component. The actual reason for this difference could be different properties (particularly, chemical composition and degree of ionization) of the NS atmosphere. In both Geminga and B0656 the brightness temperatures derived from the PL+BB fit to the combined NIR through UV data are lower than the temperatures of the

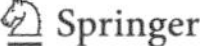

Table 3 X-ray, UV and optical properties of middle-aged pulsars and RQINSs

Name	$T_{\rm UV}$ MK	$\log L_{\rm FUV}$ ergs/s	$\log L_{\rm PL,O}$ ergs/s	$\alpha_{\rm O}$	$T_{\rm TS}$ MK	$R_{\rm TS}$ km	$\log L_{\rm TS}$ ergs/s	$T_{\rm TH}$ MK	$R_{\rm TH}$ km	$\log L_{\rm TH}$ ergs/s	$\log L_{{\rm PL},X}$ erg/s	α_X
Vela	$\lesssim$0.4	29.15	28.57	−0.01	1.17	3.7	32.26	2.1	0.7	31.80	31.49	−1.0
B0656+14	0.53	28.74	27.87	−0.41	0.71	12.2	32.43	1.4	1.1	31.54	31.15	−0.5
Geminga	0.30	28.33	27.17	−0.46	0.49	12.9	31.83	2.3	0.05	29.64	30.34	−0.6
RX J0720	0.37	28.44	27.78?	0.3?	1.0	5.1	32.58	...	...	...	...	...
RX J1856	0.80	28.76	...	...	0.72	8.7	32.65	...	...	...	...	...

Note. First column is the brightness temperature measured from the TS+PL fit to the NIR through FUV spectra (see Fig. 1) for a 13 km NS radius. Second column gives the unabsorbed FUV luminosity in the 1150–1700 Å passband [$E(B-V)=0.05$, 0.03 and 0.03 for Vela, B0656 and Geminga, respectively]. Third and forth columns are the PL component slope and luminosity (in 4000–9000 Å), respectively. Columns three through ten list the parameters of the TS+TH+PL fits to the X-ray spectra. Thermal component luminosities are the bolometric luminosities while the unabsorbed PL luminosity, $L_{{\rm PL},X}=4\pi d^2\mathcal{F}_{{\rm PL},X}$, is in the 0.2–10 keV band. The luminosities and radii are calculated for the distances listed in Table 1

X-ray TS component (see Table 3). The observed thermal UV spectrum of Geminga lies a factor of 1.5–2 below the continuation of the X-ray TS component (for a plausible extinction), whereas the UV deficit is substantially smaller, if present at all, in B0656. In the case of Vela pulsar, we could only estimate an upper limit on the UV brightness temperature, which lies above the extrapolation of the X-ray TS component.

Turning to the light curves (Fig. 2), one can see that in Geminga and B0656 thermal soft X-ray light curves show broad maxima with some non-thermal "bumps" on top of them. The bump (centered at $\phi\approx 0.9$) is relatively large in B0656, making the soft X-ray light curve asymmetric. The soft X-ray pulse profile of Vela is much more complex, reflecting the larger non-thermal contribution than in the other two pulsars. However, one still can argue that there is a single broad thermal maximum in the $\phi=0.7$–1.2 range.

The FUV light curves of B0656 and, especially, Geminga are also expected to be largely thermal. Yet, Geminga shows strong double-peaked FUV pulsations, different from both the NUV and the soft X-ray pulsations (see Fig. 3). Neither FUV nor soft X-ray pulsations of Geminga can be produced by locally isotropic blackbody emission. Although the soft X-ray pulsations can be explained by a magnetized atmosphere model (e.g., Zavlin and Pavlov 2002), the high pulsed fraction and the peculiar shape of the FUV pulsations are not predicted by this model (notice, however, that the current atmosphere models are not applicable at the optical-UV frequencies). A possible explanation of the strong thermal pulsations invokes a "screen" of absorbing plasma suspended in the NS magnetosphere, which may partially eclipse the surface emission at certain rotation phases.

4.2 UV emission of Geminga and B0656 versus RQINSs

Over the last decade seven soft X-ray sources with thermal spectra have been discovered by ROSAT. Very large X-ray-to-optical flux ratios/limits and a lack of radio emission allow one to conclude that these objects are radio-quiet isolated neutron stars (RQINSs). Extensive X-ray observations of RQINSs with Chandra and XMM-Newton have provided high-quality spectra and period measurements for most of them (F. Haberl, these proceedings). Recent period derivative measurements by Kaplan and van Kerkwijk (2005a, 2005b) in RX J0720.4-3125 and RX J1308.6+2127 (J0720 and J1308 hereafter) place these RQINSs near the anomalous X-ray pulsars (AXPs) in the P–$\dot{P}$ diagram. However, the X-ray properties of RQINSs are quite different from those of AXPs (the BB temperatures of RQINSs are much lower, and no PL tails are seen). On the other hand, the X-ray spectra of RQINSs are also very different from those of ordinary old radio pulsars ($\tau\gtrsim 1$ Myr; $B\sim 10^{12}$ G; e.g., Zavlin and Pavlov 2004; Kargaltsev et al. 2006). In terms of their thermal X-ray properties, RQINSs resemble middle-aged pulsars (see Table 3); however, the optical-UV properties of RQINSs show important differences.

Out of the seven RQINSs, five have been detected in the optical, and UV spectra have been obtained for two of the five, RX J1856.5-3754 (J1856) and J0720 (van Kerkwijk and Kulkarni 2001; Kaplan et al. 2003). The UV spectra of these RQINSs look thermal ($F_\nu\propto\nu^2$), and even their optical fluxes seem to follow the same R-J curves, although Kaplan et al. (2003) report a faint non-thermal PL component, with $\alpha_{\rm O}\approx 0.3$, $L_{\rm PL,O}\sim 6\times 10^{27}$ ergs s^{-1} for J0720 (for the other three RQINSs, the spectral slopes in the optical are uncertain). We note that no detectable optical magnetospheric component is expected at least in J0720 and J1308 (and likely in other RQINSs, whose spin-down powers, $\dot{E}$, have not been measured yet) if the nonthermal optical efficiency of these objects is similar to those of radio pulsars, $L_{\rm PL,O}/\dot{E}\sim 10^{-7}$–$10^{-6}$ (Zavlin and Pavlov 2004). For instance, one would expect $L_{\rm PL,O}\sim 10^{24}$–10^{25} ergs s^{-1} for J0720, i.e. a factor of $>10^3$ higher efficiency is needed to explain the result of Kaplan et al. (2003). More importantly, RQINSs exhibit significant *excess in thermal UV emission*, with the UV fluxes exceeding the continuations of the X-ray thermal spectra by a factor of 5–9. The nature of this UV excess is not firmly established yet. It could occur if the UV and soft X-ray thermal components are emitted from different regions (with different areas) on a non-uniformly heated NS surface (Pavlov et al. 2002). Alternatively, these NSs may have a condensed (solid) surface, possibly covered by a tenuous atmosphere that is optically thin in X-rays but optically thick in the optical-UV (e.g., Motch et al. 2003). Whatever is the nature of the large UV excess in RQINSs, we do not see it in B0656 and see a *UV deficit* in Geminga.

The apparently smaller UV-emitting area of Geminga, as compared to the X-ray-emitting area, cannot be explained by a nonuniform temperature distribution. We might speculate that the temperature distribution over the bulk of Geminga's surface is more uniform than in J1856 and J0720, e.g., because of different geometry and strength of the magnetic field that affects the surface temperature distribution. However, to explain why the more uniformly heated Geminga exhibits quite substantial pulsations of its thermal X-ray radiation while no pulsations have been detected from J1856, one has to assume very special orientations of the J1856's spin and magnetic axes. The lack of UV excess in Geminga can also be attributed to different chemical composition or lower temperature of the Geminga's surface (Kargaltsev et al. 2005). However, the latter argument does not apply to B0656 whose UV and soft X-ray thermal components show higher temperatures, comparable to those of J1856 and other RQINSs. One could also speculate that the surface of RQINSs is solid while in B0656 and Geminga it is in a gaseous state. Indeed, the magnetic fields of Geminga and B0656 are 5–10 times smaller than those of J0720 and J1308, implying lower condensation temperatures (van Adelsberg et al. 2005). In addition, irradiation of the NS surface by energetic particles and photons generated in the magnetosphere may ablate the NS surface and facilitate formation of thick atmospheres in active pulsars while we see no indications of magnetospheric activity in RQINSs. The presence of thick atmospheres (lack of solid surface) in pulsars is supported by the fact that a hydrogen atmosphere model provides a reasonably good description of the TS spectrum in the Vela pulsar while the BB model gives a very small NS radius (Pavlov et al. 2001). (Such fully ionized atmosphere models, however, are not directly applicable to colder Geminga and B0656 because their atmospheres are not fully ionized.) In RQINSs, the solid surface, directly seen in X-rays, would emit the spectrum that resembles a BB but with the emissivity reduced by a factor of $\lesssim 2$ (van Adelsberg et al. 2005), which would explain the apparently smaller X-ray radii of RQINSs (R_{TS} in Table 3). However, even a tenuous atmosphere on top of the solid surface can be opaque in UV, resulting in a larger UV emitting area. Within this interpretation, one still needs to explain the absorption features observed in the spectra of several RQINSs. Although it was suggested that these features can be formed in a hydrogen or helium atmosphere (see van Kerkwijk & Kaplan, these proceedings), they could also be due to the resonances at various hybrid frequencies in the dense magnetized condensate (van Adelsberg et al. 2005).

4.3 Nonthermal emission

In Geminga and B0656, the PL components fitted to the hard X-ray and optical spectra show similar slopes $\alpha_X \approx \alpha_O \approx -0.5$. However, in B0656 the extrapolation of the X-ray PL matches the optical points much better that in Geminga (see Fig. 3). For the Vela pulsar, the slope of the PL component is much steeper in X-rays, $\alpha_X \approx -1$, than in the UV-optical, $\alpha_O \approx 0.0$. However, one should keep in mind that the slopes of the X-ray PL components are measured in the narrow 2–8 keV band (Fig. 3) and hence are very uncertain because of a large background at these energies and, possibly, because the spectrum deviates from a simple PL. For instance, a comptonized tail of thermal radiation could mimic the nonthermal PL component in the narrow 2–8 keV band. Measuring hard X-ray spectra above $\gtrsim 10$ keV will help to better understand the nature of the nonthermal emission in Geminga and B0656.

The 2–8 keV light curves show rather sharp and strong peaks for all the three pulsars. If the 2–8 keV emission is indeed due to the comptonization of thermal emission, one needs to explain its high anisotropy. The number of hard X-ray peaks varies from a single peak in B0656's light curve to a double-peaked structure in Geminga, and even more complex multi-peak light curve in the Vela pulsar.

In the UV-optical, the Vela pulsar shows the strongest pulsations, with at least four narrow peaks. In contrast to the Vela pulsar, the UV light curves of B0656 have a large thermal contribution ($\approx$70% in FUV and $\approx$50% in NUV). Its NUV light curve shows two large peaks with a much wider separation than the main peaks in the NUV pulse profile of the Vela pulsar. Due to a much lower S/N in the B0656 NUV light curve, smaller and narrower non-thermal peaks may remain undetected. Similar to the Vela pulsar, the NUV light curve of B0656 is noticeably different from its 2–4 keV light curve. No obvious non-thermal contribution to the UV light curves is seen in Geminga, which is consistent with a large thermal fraction in the UV. However, non-thermal pulsations should be better seen in the optical ($\lambda \lesssim 4300$ Å).

5 Conclusions

We have observed three pulsars (Geminga, B0656+14, and Vela) at the far-UV and near-UV wavelengths with the STIS

MAMA detectors and measured the pulsar spectra and pulse profiles. We have also analyzed the X-ray data and found that optical through X-ray spectra consist of thermal and non-thermal components in all the three pulsars. In particular, we found the following:

1. Thermal contribution to the FUV passband increases with pulsar age. The thermal component is not seen in the younger Vela pulsar ($\tau \approx 11$ kyrs), but it dominates the FUV spectrum in the Geminga pulsar ($\tau \approx 340$ kyrs). As the FUV spectrum grows more thermal, the "thermal hump", whose maximum is located in soft X-rays/EUV, shifts toward lower frequencies.
2. The X-ray spectra of the two older pulsars, Geminga and B0656, resemble those of RQINSs. However, their UV properties are noticeably different. In sharp contrast to RQINSs, whose spectra show large *UV/optical excess*, Geminga shows *UV deficit* while the continuation of the X-ray TS component approximately matches the R-J component in B0656. Likely, this reflects the differences in the magnetic field and phase state of the NS surface layers.
3. While no UV pulsations have been detected in RQINSs, Geminga shows strong, non-sinusoidal pulsations in the FUV range, where the spectrum is dominated by the thermal component. To explain the strong FUV pulsations, one may need to invoke magnetospheric absorption at certain rotation phases.

Acknowledgements We thank Slava Zavlin for the help with the X-ray data analysis and David Kaplan for providing the most recent parallax measurements for RQINSs.

References

Brisken, W.F., Thorsett, S.E., Golden, A., Goss, W.M.: Astrophys. J. **593**, 89 (2003)
De Luca, A., Caraveo, P.A., Mereghetti, S., et al.: Astrophys. J. **623**, 1051 (2005)
Dodson, R., Legge, D., Reynolds, J.E., McCulloch, P.M.: Astrophys. J. **596**, 1137 (2003)
Gouiffés, C.: In: Shibazaki, N., Kawai, N., Shibata, S., Kifune, T. (eds.) Neutron Stars and Pulsars, p. 363. Univ. Acad. Press, Tokyo (1998)
Harding, A.K., Strickman, M.S., Gwinn, C.: Astrophys. J. **576**, 376 (2002)
Jackson, M.S., Halpern, J.P.: Astrophys. J. **633**, 1114 (2005)
Jackson, M.S., Halpern, J.P., Gotthelf, E.V., Mattox, J.R.: Astrophys. J. **578**, 935 (2002)
Kanbach, G., Arzoumanian, Z., Bertsch, D.L., et al.: Astron. Astrophys. **289**, 855 (1994)
Kaplan, D.L., van Kerkwijk, M.H.: Astrophys. J. **628**, 45 (2005a)
Kaplan, D.L., van Kerkwijk, M.H.: Astrophys. J. **635**, 65 (2005b)
Kaplan, D.L., van Kerkwijk, M.H., Marshall, H.L., et al.: Astrophys. J. **590**, 1008 (2003)
Kargaltsev, O.: Ph.D. Thesis, Pennsylvania State University (2004)
Kargaltsev, O., Pavlov, G.G.: Astrophys. J. (2007, in preparation)
Kargaltsev, O., Pavlov, G.G., Zavlin, V.E., Romani, R.W.: Astrophys. J. **625**, 307 (2005)
Kargaltsev, O., Pavlov, G.G., Garmire, G.P.: Astrophys. J. **636**, 406 (2006)
Kern, B., Martin, C., Mazin, B., Halpern, J.P.: Astrophys. J. **597**, 1049 (2003)
Koptsevich, A.B., Pavlov, G.G., Zharikov, S.V., et al.: Astrophys. J. **370**, 1004 (2001)
Motch, C., Zavlin, V.E., Haberl, F.: Astron. Astrophys. **408**, 323 (2003)
Pavlov, G.G., Shibanov, Yu.A., Zavlin, V.E., Meyer, R.D.: In: Alpar, M.A., Kiziloglu, Ü., van Paradijs, J. (eds.) The Lives of the Neutron Stars, p. 71. Kluwer Academic, Dordrecht (1995)
Pavlov, G.G., Zavlin, V.E., Sanwal, D., et al.: Astrophys. J. **552**, 129 (2001)
Pavlov, G.G., Zavlin, V.E., Sanwal, D.: In: Becker, W., Lesch, H., Trümper, J. (eds.) Proc. 270 WE-Heraeus Seminar on Neutron Stars, Pulsars and Supernova Remnants, MPE-Reports, vol. 278, p. 273 (2002)
Romani, R.W.: Astrophys. J. **313**, 718 (2005)
Romani, R.W., Kargaltsev, O., Pavlov, G.G.: Astrophys. J. **627**, 383 (2005)
Sanwal, D., Pavlov, G.G., Kargaltsev, O., et al.: In: Neutron Stars in Supernova Remnants. ASP Conf., vol. 271, p. 353 (2002)
Shearer, A., Golden, A., Harfst, S., et al.: Astron. Astrophys. **335**, 21 (1998)
Shibanov, Yu.A., Koptsevich, A.B., Sollerman, J., Lundqvist, P.: Astron. Astrophys. **406**, 645 (2003)
Shibanov, Yu.A., Sollerman, J., Lundqvist, P., et al.: Astron. Astrophys. **440**, 693 (2005)
Shibanov, Yu.A., Zharikov, S.V., Komarova, V.N., et al.: Astron. Astrophys. **448**, 313 (2006)
van Adelsberg, M., Lai, D., Potekhin, A.Y., et al.: Astrophys. J. **628**, 902 (2005)
van Kerkwijk, M.H., Kulkarni, S.R.: Astron. Astrophys. **378**, 986 (2001)
Yakovlev, D.G., Pethick, C.J.: Annu. Rev. Astron. Astrophys. **42**, 169 (2004)
Zavlin, V.E., Pavlov, G.G.: In: Becker, W., Lesch, H., Trümper, J. (eds.) Proc. 270 Heraeus Seminar on Neutron Stars, Pulsars and Supernova Remnants, MPE Reports, vol. 278, p. 263 (2002)
Zavlin, V.E., Pavlov, G.G.: In: Proc. XMM-Newton EPIC Consortium. Mem. Soc. Astron. It., vol. 75, p. 458 (2004a)
Zavlin, V.E., Pavlov, G.G.: Astrophys. J. **616**, 452 (2004b)
Zavlin, V.E., Pavlov, G.G., Shibanov, Yu.A.: Astron. Astrophys. **315**, 141 (1996)

Astrophys Space Sci (2007) 308: 297–307
DOI 10.1007/s10509-007-9297-y

ORIGINAL ARTICLE

Studying millisecond pulsars in X-rays

Vyacheslav E. Zavlin

Received: 7 April 2006 / Accepted: 21 August 2006 / Published online: 16 March 2007

Abstract Millisecond pulsars represent an evolutionarily distinct group among rotation-powered pulsars. Outside the radio band, the soft X-ray range ($\sim$0.1–10 keV) is most suitable for studying radiative mechanisms operating in these fascinating objects. X-ray observations revealed diverse properties of emission from millisecond pulsars. For the most of them, the bulk of radiation is of a thermal origin, emitted from small spots (polar caps) on the neutron star surface heated by relativistic particles produced in pulsar acceleration zones. On the other hand, a few other very fast rotating pulsars exhibit almost pure nonthermal emission generated, most probably, in pulsar magnetospheres. There are also examples of nonthermal emission detected from X-ray nebulae powered by millisecond pulsars, as well as from pulsar winds shocked in binary systems with millisecond pulsars as companions. These and other most important results obtained from X-ray observations of millisecond pulsars are reviewed in this paper, as well as results from the search for millisecond pulsations in X-ray flux of the radio-quite neutron star RX J1856.5-3754.

Keywords X-rays · Neutron stars · Millisecond pulsars

PACS 95.85.Nv · 97.10.Qh · 97.60.Jd · 97.60.Gb

1 Introduction

Millisecond pulsars (MSPs) significantly differ in properties from other (ordinary) radio pulsars. First of all, MSPs possess very short and stable spin periods, $P \lesssim 50$ ms, with extremely small period derivatives, $\dot{P} \lesssim 10^{-18}$ s s^{-1}. These two main parameters separate MSPs from the majority of other pulsars, as illustrated in the P–$\dot{P}$ diagram[1] shown in Fig. 1. According to the conventional pulsar magnetic-braking model, MSPs are very old neutron stars, with characteristic ages $\tau = P/2\dot{P} \sim 0.1$–10 Gyr, and low surface magnetic fields $B_{\rm surf} \simeq 3.2 \times 10^{19}(P\dot{P})^{1/2} \lesssim 10^{10}$ G (see Fig. 1). They were presumably spun up by angular momentum transfer during a mass accretion phase in binary systems[2] (e.g., Alpar et al. 1982), and their low values of $B_{\rm surf}$ could be explained by the Ohmic and/or accretion-induced decay of magnetic field (see Cumming 2005 and references therein). Discoveries of several accretion-driven binary MSPs, first of all the famous pulsar SAX J1808.4-3658 with $P \simeq 2.5$ ms, support this hypothesis (see, e.g., Wijnands 2004 for a review on accreting MSPs).

Since the discovery of the first fast rotating pulsar B1937+21 by Backer et al. (1982), MSPs have been extensively searched for and studied in radio domain. Currently, about 130 MSPs are known (Manchester et al. 2005). Outside the radio band, as MSPs are intrinsically faint at optical wavelengths and most of them ($\sim$80%) reside in binary systems with optically brighter white dwarf companions, the soft X-ray energy range ($\sim$0.1–10 keV) is the main source of information on these pulsars. The detection of pulsed X-ray emission from the brightest (and nearest) MSP J0437-4715 with *ROSAT* (Becker and Trümper 1993) initiated a series of dedicated X-ray observations of these intriguing objects in 90's with this satellite and also *ASCA*,

V.E. Zavlin (✉)
Space Science Laboratory, NASA MSFC SD50, Huntsville, AL 35805, USA
e-mail: vyacheslav.zavlin@msfc.nasa.gov

[1] Based on the pulsar catalog provided by the Australia Telescope National Facility (Manchester et al. 2005) and available at http://www.atnf.csiro.au/research/pulsar.

[2] It is why MSPs are often called "recycled" pulsars.

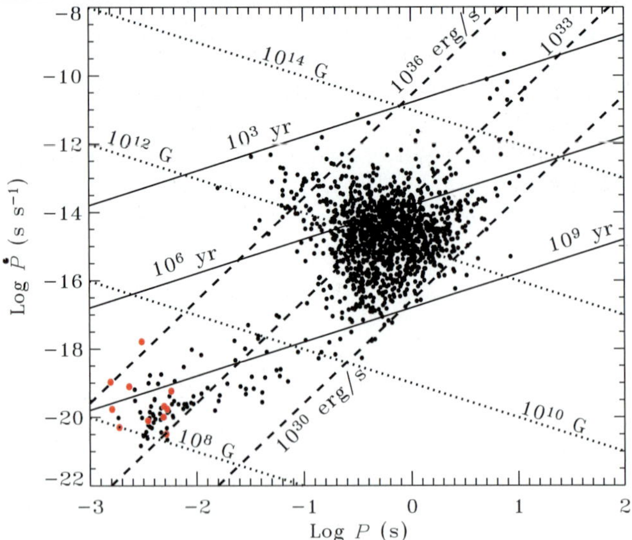

Fig. 1 P–$\dot{P}$ diagram for about 1500 radio pulsars (*dots*). Millisecond pulsars are located in the *lower-left corner* of the diagram. The pulsars from Tables 1 and 3 are shown with *red dots*. *Straight lines* correspond to constant values of pulsar characteristic age $\tau = 10^3$, 10^6 and 10^9 yr, surface magnetic field $B_{\rm surf} = 10^8$, 10^{10}, 10^{12} and 10^{14} G, and spin-down energy $\dot{E} = 10^{30}$, 10^{33} and 10^{36} erg s^{-1}

Table 1 Main parameters of eight MSPs

PSR	P (ms)	d (kpc)	τ (Gyr)	$\log \dot{E}$ (erg s^{-1})
B1937+21	1.56	3.57	0.24	36.0
B1957+20	1.61	2.49	2.24	35.0
J0218+4232	2.32	2.67	0.48	35.4
B1821-24	3.05	3.09	0.03	36.3
J0030+0451	4.87	0.32	7.71	33.5
J2124-3358	4.93	0.27	6.01	33.6
J1024-0719	5.16	0.39	>27.25	<32.9
J0437-4715	5.76	0.14	6.51	33.5

Beppo*SAX* and *RXTE*. Later on this observational "relay" has been continued with *Chandra* and *XMM*-Newton. So far, firm X-ray detections have been reported for about three dozens of isolated (solitary, or non-accreting if in binaries) MSPs. The majority of these objects are located in globular clusters (mostly 47 Tuc—see Bogdanov et al. 2006). This paper mainly concentrates on eight MSPs, listed in Table 1, with available detailed information on properties of detected X-ray emission. Besides P and τ, Table 1 gives estimates on the distances to these objects, d (inferred from either pulsar parallaxes or dispersion measures and the NE2001 galactic electron density model by Cordes and Lazio 2003),[3] and on their spin-down energies, $\dot{E} = 4\pi^2 I P^{-3}\dot{P}$ (assuming a standard neutron star moment of inertia, $I = 10^{45}$ g cm^2). Note that for PSR J1024-0719 the intrinsic period derivative is not well determined, $\dot{P} < 3 \times 10^{-21}$ s s^{-1} (Hotan et al. 2006), that results in the lower and upper limits on τ and $\dot{E}$, respectively.

Generally, X-ray emission from radio pulsars consists of two different components, thermal and nonthermal, generated on the neutron star surface or in its vicinity. The nonthermal component is usually described by a power-law (PL) spectral model and attributed to radiation produced by synchrotron and/or inverse Compton processes in the pulsar magnetosphere, whereas the thermal emission can originate from either the whole surface of a cooling neutron star or small hot spots around the magnetic poles (polar caps; PCs) on the star surface, or both. As predicted by virtually all pulsar models, these PCs can be heated up to X-ray temperatures ($\sim$1 MK) by relativistic particles generated in pulsar acceleration zones. A conventional assumption about the PC radius is that it is close to the radius within which open magnetic field lines originate from the pulsar surface, $\sim[2\pi R_{\rm NS}^3/cP]^{1/2} \simeq 2\,[P/5\ {\rm ms}]^{-1/2}$ km (for a neutron star radius $R_{\rm NS} = 10$ km). In case of MSPs, the entire surface at a neutron star age of $\sim$1 Gyr is too cold, $\lesssim 0.1$ MK, to be detectable in X-rays (although it may be seen in the UV/FUV band—see Sect. 6). Therefore, only nonthermal (magnetospheric) and/or thermal PC radiation is expected to be observed in X-rays from these objects. In addition to the radiation produced by MSPs themselves, nonthermal emission from pulsar-wind nebulae (PWNe) associated with MSPs moving at supersonic velocities ($\gtrsim$100 km s^{-1}) through interstellar medium may be detected. Another source of nonthermal X-ray radiation generated in binary systems can be an intrabinary shock formed where the pulsar wind and matter from the stellar component collide (Arons and Tavani 1993), although such a component would be hardly separated from radiation produced by the pulsar itself (unless properties of the nonthermal emission varies with orbital phase).

The MSPs in Table 1 belong to two distinct groups: those which emit almost pure nonthermal radiation and those with a predominantly thermal PC component. The next two sections provides details on the X-ray properties of these objects. Section 4 is devoted to three X-ray emitting MSPs whose properties remain uncertain. X-ray PWNe powered by MSPs are briefly discussed in Sect. 5. A summary is given in Sect. 6.

[3] For the MSPs of the second group in Table 1 (with $d < 1$ kpc), the distance estimates obtained from the pulsar parallaxes and those inferred from their dispersion measures agree well with each other (see Hotan et al. 2006 and Lommen et al. 2006).

2 Nonthermally emitting MSPs

Of the MSPs listed in Table 1, these are PSRs B1937+21, B1957+20, J0218+4232, and B1821-24. They are characterized by large values of spin-down energy, $\dot{E} > 10^{35}$ erg s^{-1}, and shorter spin periods, $P < 3.1$ ms. PSR J0218+4232 is also one of few γ-pulsars known (Kuiper et al. 2000). X-ray observations of these MSPs revealed their nonthermal spectra, albeit with different photon indices Γ. Figs. 2, 3, 4 and 5 show the spectra detected with *Chandra* and/or *XMM*-Newton from these MSPs. Although the spectral data on these pulsars do not formally require any other component (in addition to a PL), one cannot exclude that there is also a contribution of thermal PC emission in the detected X-ray fluxes, as indicated in Figs. 2, 3, 4 and 5. For PSRs B1821-24, J0218+4232, and B1937+21 the mea-

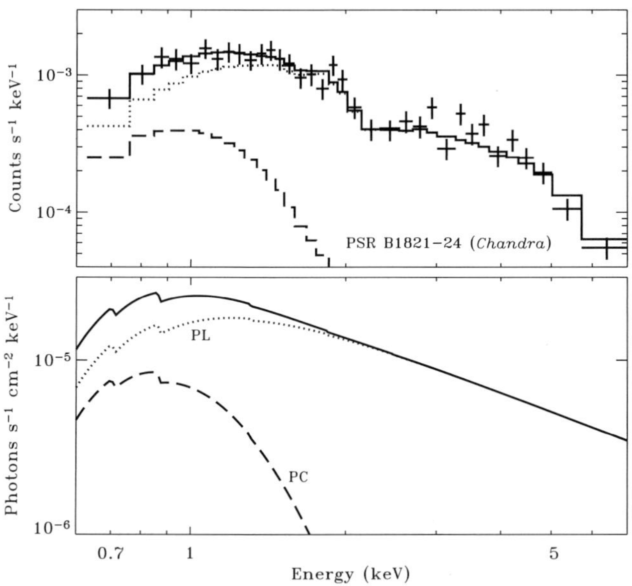

Fig. 2 X-ray spectrum of PSR B1821-24 as detected with the *Chandra* ACIS-S instrument (*crosses*) fitted with a PL model of $\Gamma = 1.2$ (*dotted curves*) plus a possible PC component (*dashes*). See also Becker et al. (2003)

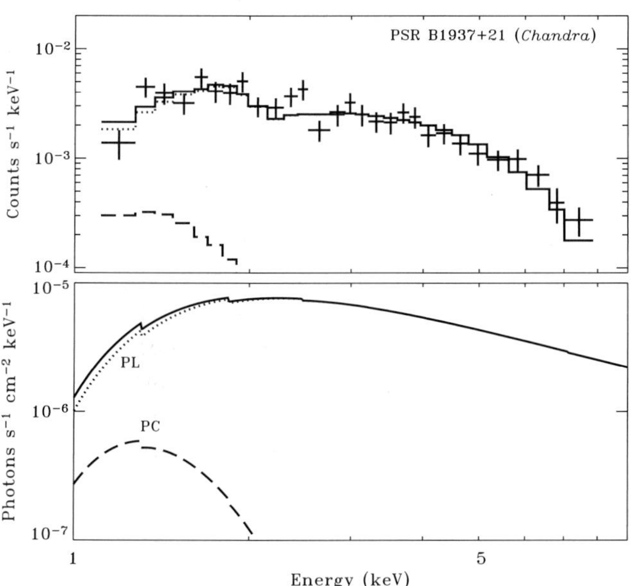

Fig. 3 Same as in Fig. 2 for PSR B1937+21 and a PL model of $\Gamma = 1.2$

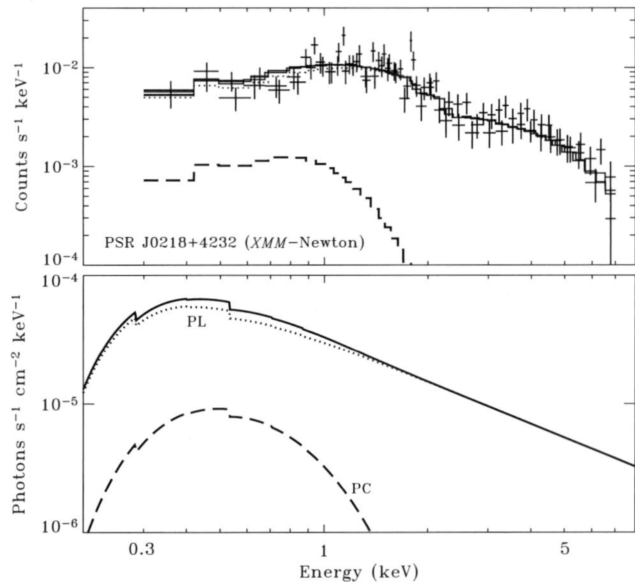

Fig. 4 Same as in Fig. 2 for PSR J0218+4232 (as detected with the *XMM*-Newton EPIC-MOS instruments) and a PL model of $\Gamma = 1.1$ (see also Webb et al. 2004)

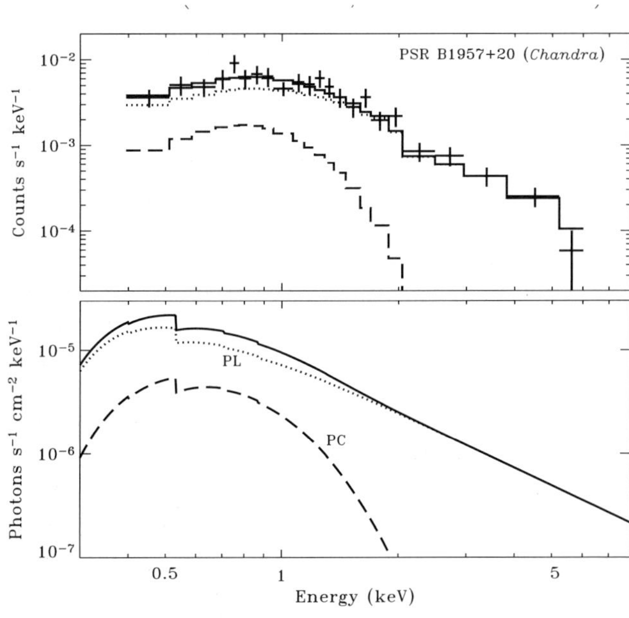

Fig. 5 Same as in Fig. 2 for PSR B1957+20 and a PL model of $\Gamma = 1.9$. (Note that the detected flux can contain a contribution from a possible intrabinary shock—see Sect. 2 and Stappers et al. 2003 for more details)

Table 2 X-ray luminosities of eight MSPs

PSR	Log $L^{\rm nonth}$ (erg s^{-1})	Log $L^{\rm pc}_{\rm bol}$ (erg s^{-1})	Log $\eta^{\rm nonth}$	Log $\eta^{\rm pc}$
B1937+21	32.8	<31.5	−3.2	<−4.5
B1957+20	31.5	<30.4	−3.5	<−4.6
J0218+4232	32.6	<31.0	−2.8	<−4.4
B1821-24	32.7	<32.0	−3.6	<−4.3
J0030+0451	29.8	30.3	−3.7	−3.2
J2124-3358	<29.0	30.0	<−4.6	−3.6
J1024-0719	<28.9	29.3	...	>−3.6
J0437-4715	29.7	30.2	−3.8	−3.3

sured photon indices are $\Gamma \simeq 1.1$–1.2. The spectrum of PSR B1957+20, was found to be much steeper, with $\Gamma \simeq 1.9$ (although the X-ray radiation of this pulsar can consist of two component—see below). Note that these estimates on Γ are derived for the phase-integrated fluxes, whereas the observational data on PSRs J0218+4232 and B1937+21 indicate that the spectral slope may change with pulsar rotational phase (Webb et al. 2004; Nicastro et al. 2004). However, much more sensitive observations are required to confirm this effect. Regarding the X-ray spectrum of PSR B1821-24 (located in the globular cluster M28), Becker et al. (2003) speculated that there is a marginal evidence for an emission line around 3.3 keV (see also Fig. 2). If this feature is real,[4] it could be interpreted as cyclotron emission from an optically thin corona above the pulsar provided its magnetic field is strongly different from a centered dipole. The estimated nonthermal (isotropic) luminosities[5] of these four objects in the 0.2–10 keV range, $L^{\rm nonth}$, and the corresponding "nonthermal" efficiencies, $\eta^{\rm nonth} = L^{\rm nonth}/\dot{E}$, are given in Table 2. Note that a fraction of the X-ray emission detected from the eclipsing pulsar B1957+20, which is thought to ablate its companion in a close binary system with a 9.2-hr orbital period, may be due to an intrabinary shock between the pulsar wind and that of the companion star. The *Chandra* data on this MSP showed an apparent (at a 99% confidence level) modulation of the X-ray emission detected at the pulsar's position with orbital phase (Stappers et al. 2003), with lowest and highest fluxes during and immediately after eclipse (respectively). Assuming that this flux modulation is genuine, Stappers et al. (2003) obtained a 50% estimate on the contribution of X-ray emission from the intrabinary shock in the total flux detected from the pulsar.

[4]There is no indication of a feature at this photon energy in the spectra of PSRs B1957+20 and B1937+21 detected with the same *Chandra* ACIS-S instrument (Figs. 3 and 5).

[5]X-ray luminosities of the pulsars discussed in Sects. 2 and 3 are derived for the distances given in Table 1.

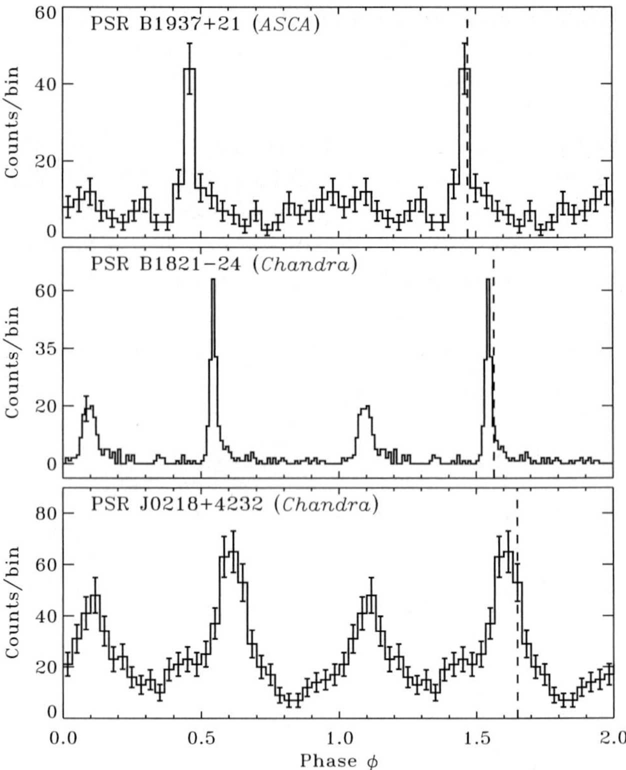

Fig. 6 Pulsed profiles of three nonthermally emitting MSPs. Vertical dashed lines indicate phases of main radio pulses. See Takahashi et al. (2001), Rutledge et al. (2004), and Kuiper et al. (2002) for more details on timing analysis of X-rays from PSRs B1937+21, B1821-24, and J0218+4232 (respectively)

This fraction is accounted for in deriving the nonthermal luminosity and efficiency for PSR B1957+20 given in Table 2.

Another prominent feature of the nonthermal emission from three of these MSPs is the shape of the X-ray pulsed profiles, with strong and narrow main pulses and large pulsed fractions $f_{\rm p}$ ranging from about 65% for PSR B0218-4232 up to nearly 100% for PSRs B1937+21 and B1821-24. Figure 6 presents the pulsars' light curves. Nonthermal pulsed emission from these MSPs was also detected up to about 20 keV with *RXTE* (Rots et al. 1998; Kuiper et al. 2004; Cusumano et al. 2003). The main X-ray and radio peaks of these pulsars were found to be nearly aligned in phase (based on both *Chandra* and *RXTE* data).[6] This suggests that nonthermal photons emitted from these MSPs in the radio and X-ray bands are generated in the same (or close) zones, although it still remains unclear where those zones are located: closer to the neutron star surface (as suggested by the polar-cap model of Harding et al. 2005) or near the pulsar light cylinder (according to the outer-gap model—

[6]Although Takahashi et al. (2001) found from *ASCA* data that the main X-ray and radio pulses of PSR B1937+21 are separated by $\Delta\phi \simeq 0.5$ in phase.

e.g., Cheng et al. 1986; Romani 1996). Pulsations of the X-ray flux emitted by PSR B1957+20 have to be detected yet.[7]

3 Thermally emitting MSPs

This is the second group of objects in Table 1 which consists of PSRs J0030+0451, J2124-3358, J1024-0719, and J0437-4715. These MSPs are less energetic in terms of $\dot{E}$, have longer spin periods and are much closer to Earth. X-ray spectra of the two brightest members of this group, PSRs J0030+0451 and J0437-4715 cannot be fitted with a single PL model, whereas for the other two pulsars such a fit yields too large photon indices, $\Gamma \simeq 3$–4, and the obtained estimates on hydrogen column density, $n_{\rm H} \simeq (1\text{–}2) \times 10^{21}$ cm^{-2}, towards these objects greatly exceed those inferred from independent measurements (Zavlin 2006). A broken-PL model, suggested by Becker and Aschenbach (2002) to interpret the spectra of PSRs J0030+0451 and J0437-4715, results in unrealistic estimates on $n_{\rm H}$. A model involving a PL and a simple one-temperature thermal component also faces the same problem.

Analyzing *ROSAT* data on PSR J0437-4715, Zavlin and Pavlov (1998) suggested a thermal model implying a nonuniform temperature distribution over PCs. These authors discussed that relativistic particles bombarding magnetic poles of an MSP could heat a region larger than the conventional PC size (see Sect. 1) because the low magnetic field of the pulsar does not prevent the released heat from propagating along the neutron star surface. The applied model assumes two identical PCs on magnetic poles covered with a weakly magnetized hydrogen atmosphere (Zavlin et al. 1996). It also takes into account the GR effects (redshift and bending of photon trajectories near the neutron star surface). In this model the thermal radiation depends on PC temperature and radius, the neutron star mass-to-radius ratio, which determines the GR effects, and the star geometry (the viewing and magnetic angles, ζ and α, respectively). The sketch shown in Fig. 7 illustrates the neutron star geometry and the effect of light bending (see also, e.g., Zavlin et al. 1995 for more details on the GR effects on pulsar PC emission). The nonuniform temperature was approximated with a step-like function, referred as PC "core" and "rim". Observations of PSR J0437-4715 with *Chandra* (Zavlin et al. 2002) and *XMM*-Newton (Zavlin 2006) showed that such a model, supplemented with a PL component of $\Gamma \simeq 2.0$, fits well the spectrum of PSR J0437-4715 up to 10 keV and yields reasonable pulsar parameters as well as $n_{\rm H}$. The thermal model provides the bulk of the pulsar's X-ray flux,

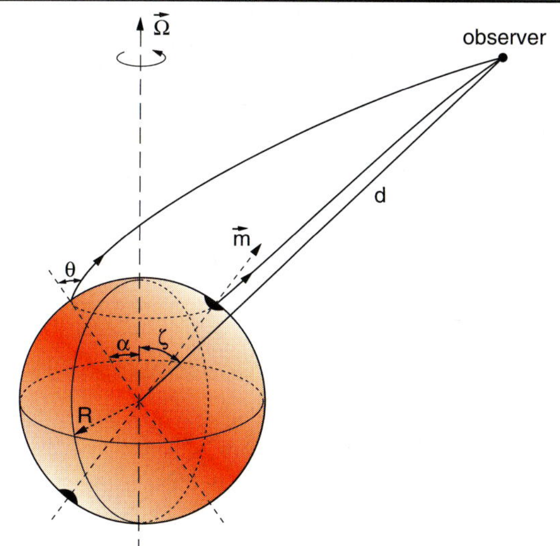

Fig. 7 Sketch illustrating bending of photon trajectories in a strong gravitational field near the surface of a neutron star with rotational and magnetic axes $\boldsymbol{\Omega}$ and $\boldsymbol{m}$, and viewing and magnetic angles ζ and α (respectively)

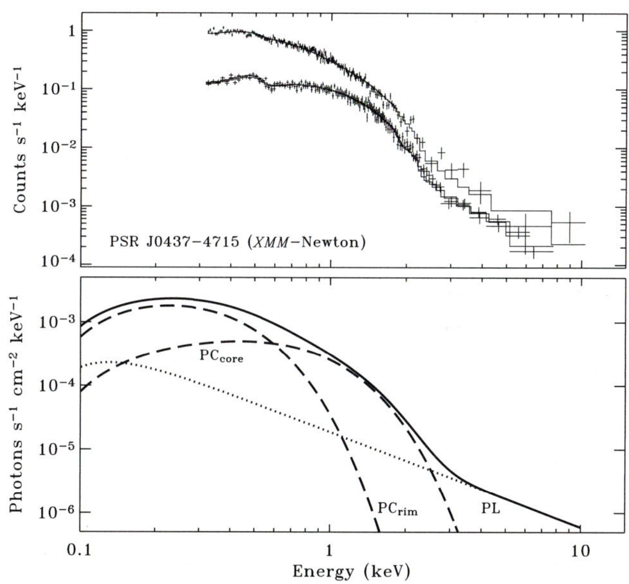

Fig. 8 X-ray spectra of PSR J0437-4715 as detected with the *XMM*-Newton EPIC-pn and MOS instruments (*crosses*). The solid curves show a best fitting model, PL (*dots*) plus a two-temperature PC component (*dashes*), "core" and "rim" (see Sect. 3)

whereas the PL component prevails only at photon energies $E \gtrsim 3$ keV. A similar (nonuniform PCs plus PL) model works also well on the spectral data of PSR J0030+0451. Figures 8 and 9 demonstrate this. The inferred parameters of the thermal components are: $T_{\rm pc}^{\rm core} \simeq 1.4$ and 2.1 MK, $T_{\rm pc}^{\rm rim} \simeq 0.5$ and 0.8 MK, $R_{\rm pc}^{\rm core} \simeq 0.4$ and 0.1 km, $R_{\rm pc}^{\rm rim} \simeq 2.6$ and 1.4 km, for PSRs J0437-4715 and J0030+0451, respectively (in the latter case the PL index was fixed at $\Gamma = 1.5$ because of a poorer data statistics at higher energies). For

[7] No results from the *XMM*-Newton observation of this MSP conducted in October 2004 have been reported yet.

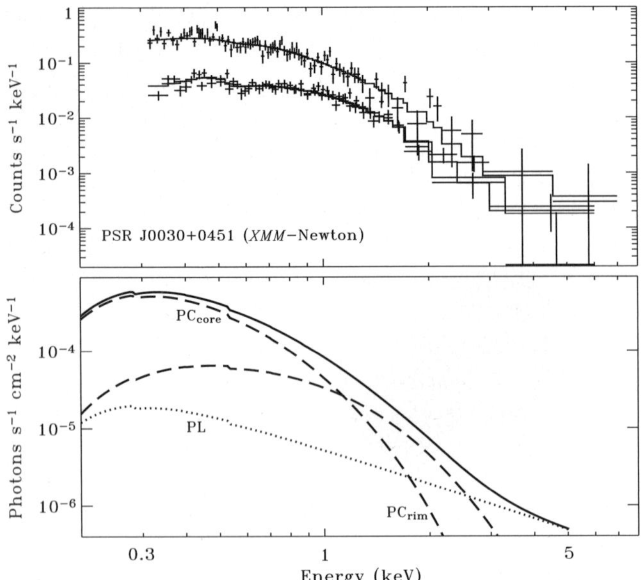

Fig. 9 Same as in Fig. 8 for PSR J0030+0451

PSRs J2124-3358 and J1024-0719 the same model yields PC parameters close to those mentioned above, although low quality of the observational data on these two objects at $E > 3$ keV allows one to put only upper limits on intensity of the nonthermal component (Zavlin 2006). The bolometric luminosities of one PC, $L_{\rm bol}^{\rm pc}$, and "PC efficiencies", $\eta^{\rm pc} = L_{\rm bol}^{\rm pc}/\dot{E}$, can be found in Table 2, together with the estimates on $L^{\rm nonth}$ and $\eta^{\rm nonth}$ for the nonthermal component (or corresponding 1σ upper limits). Upper limits on the luminosity of a possible thermal PC component in the X-ray fluxes of the four nonthermally emitting pulsars (Sect. 2) are also given in Table 2. Note that a more detailed study of possible thermal emission from these four pulsars is hampered by large distances (a few kpc) and, hence, strong interstellar absorption ($n_{\rm H} \sim [2–4] \times 10^{21}$ cm^{-2}) towards these objects.

As alternative to the PC-plus-PL model of the X-ray emission from PSR J0437-4715, Bogdanov et al. (2006) suggested a purely thermal interpretation in which the harder X-ray spectral tail (at $E \gtrsim 3$ keV) is a result of the inverse Compton scattering of soft thermal PC photons by energetic electrons/positrons in an optically thin thermal layer of a temperature $kT_{\rm e} \sim 150$ keV located presumably in the pulsar's magnetosphere. However, as existence of such a thermal layer in a pulsar magnetosphere does not seem to be physically supported, this Comptonization interpretation remains rather questionable.

X-ray emission from all these four thermally emitting MSPs is pulsed, with pulsed fraction ranging from about 35% up to 50% (Becker and Aschenbach 2002; Zavlin 2006). The pulsed profiles of PSRs J0437-4715, J2124-3358, and J1024-0719 are rather similar in shape, with single broad pulses, whereas the light curve of PSR J0030+0451 exhibits two pulses per period separated by $\Delta\phi \simeq 0.5$ in phase. This indicates that the geometry of PSR J0030+0451 (the angles ζ and α) is different from those of the three others. For example, in the framework of the conventional pulsar model with the magnetic dipole at the neutron star center, PSR J0030+0451 can be a nearly orthogonal rotator (i.e., $\zeta \simeq \alpha \simeq 90°$) with two pulses in its light curve being due to contributions from two PCs seen during the pulsar's rotation. For the other MSPs, the bulk of the detected X-ray fluxes is expected to come mostly from one PC. However, the X-ray pulses of PSRs J0437-4715, J1024-0719, and J2124-3358 are clearly asymmetric, with a longer rise and a faster decay for the former two MSPs and with the opposite behavior for the latter pulsar. None of these shapes can be explained by a simple axisymmetric temperature distribution. A feasible interpretation is that the observed asymmetry in the pulsed profiles of these three MSPs is caused by contribution of the nonthermal component whose peak is shifted in phase with respect to the pulse of the thermal emission. Results of the energy-resolved timing and phase-resolved spectroscopy on the *XMM*-Newton data on PSR J0437-4715 support this explanation (Zavlin 2006). For PSR J2124-3358, there may be an alternative interpretation of the shape of the pulsar's light curve: the steeper rise and longer trail could be caused by relativistic effects (in particular, the Doppler boost) in fast rotating pulsars (Braje et al. 2000), although it then should be understood why these effects are not seen in the pulse profiles of the other three MSPs whose the spin periods close to that of PSR J2124-3358 (Table 1). Another way to attempt to explain the observed light curves is to involve a model with a decentered magnetic dipole (or another magnetic field configuration—see the presentation by S. Zane at this conference), what has not been done yet for interpreting pulsed X-ray emission from MSPs. In any case, to produce a pulsed fraction at a level of $f_{\rm p} \gtrsim 30\%$, thermal emission has to be anisotropic, as predicted by neutron star atmosphere models, to counteract the effect of light bending near the star surface on pulsations of the thermal flux (Zavlin et al. 1995). So far, the difference in phases between radio and X-ray pulses was determined only for PSR J0437-4715 based on *Chandra* data (Zavlin et al. 2002; see also Fig. 10). As the bulk of the X-ray flux from this pulsar is of the thermal PC origin, the small difference in phases of the radio and X-ray peaks suggests that the pulsar's radio emission is generated close to the neutron star surface, unless the effects of field-line sweepback and/or aberration cancel the travel-time difference.

4 More X-ray emitting MSPs

Three pulsars, J0034-0534, J0751+1807, and J1012+5307, are examples of MSPs firmly detected in X-rays but whose

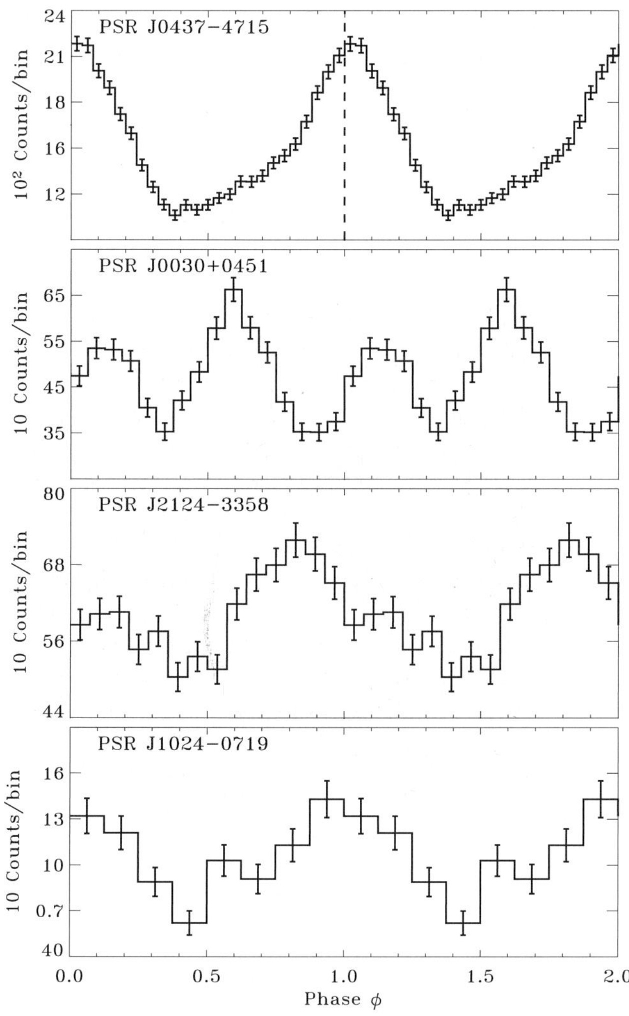

Fig. 10 Pulsed profiles of thermally emitting MSPs obtained from *XMM*-Newton observations (Zavlin 2006). The vertical dashed line in the upper panel indicates phase of a radio pulse of PSR J0437-4715 (see Sect. 3 for details)

radiative properties remain uncertain. X-ray observations of these objects were not long enough to provide data suitable for discriminating among various spectral models. Those data could be equally well fitted with a PL model of $\Gamma \sim 1.7$ or a thermal (blackbody) model of a temperature of ~3 MK (see Webb et al. 2004 and Zavlin 2006). Estimates on the total flux are the only rather reliable X-ray characteristics available for these objects. Table 3 presents the main pulsar parameters[8] together with the estimated X-ray luminosities in the 0.2–10 keV range, L_X, and corresponding efficiencies, $\eta_X = L_X/\dot{E}$. In terms of the spin-down power $\dot{E}$, PSRs J0751+1807 and J1012+5307 join the four pulsars exhibiting thermal PC emission (Sect. 3), while PSR J0034-0534 seems to be more energetic and closer to the first group of the nonthermally emitting MSPs (Sect. 2). On the other hand, PSR J0034-0534 is most "underluminous", its luminosity, $L_X \simeq 0.5 \times 10^{30}$ erg s^{-1}, is as low as that estimated for PSR J1024-0719, the least energetic pulsar among the MSPs discussed in Sects. 2–4. The X-ray efficiency of PSR J0034-0534, $\eta_X \simeq 1.6 \times 10^{-5}$, is smaller at least by an order of magnitude than those measured for the other MSPs (Tables 2 and 3). However, it is worth mentioning that the estimate on $\dot{E}$ given in Table 2 for PSR J0034-0534 assumes that the intrinsic period derivative of the pulsar is as directly measured in radio observations, i.e., without accounting for the Shklovskii effect on $\dot{P}$ due to pulsar proper motion (Shklovskii 1970). The proper motion of PSR J0034-0534 has not been determined yet. Still, it is reasonable to assume that the transverse component of the pulsar's velocity may be about 100 $(d/0.54\ \mathrm{kpc})^{1/2}$ km s^{-1} (what is not unusual for a binary MSP—see Lommen et al. 2006). Then, its intrinsic $\dot{P}$ and, hence, $\dot{E}$ would be reduced by a factor of 10, making the pulsar's X-ray efficiency similar to those found for the MSPs with $\dot{E} < 10^{34}$ erg s^{-1}.

5 MSPs and their nebulae

Relativistic pulsar winds, which carry away pulsar rotational energy, interact with ambient medium, that is expected to form PWNe detectable at different wavelengths. In X-rays, about 30 PWNe are currently known (see Kaspi et al. 2006; Gaensler and Slane 2006 and Kargaltsev et al. 2007 for reviews, as well as electronic PWN catalogs),[9] thanks mainly to the superb spatial resolution of *Chandra* (see the presentation by Weisskopf et al. at this conference for a review). The observed X-ray PWNe are diverse in properties. Some of them are of a torus-like structure with jets along the symmetry axis (which apparently coincides with the pulsar's spin axis, as suggested for the young Crab and Vela pulsars—Weisskopf et al. 2000; Pavlov et al. 2003). Several others have a cometary-like shape caused by the pulsar motion. The "Mouse" PWN powered by PSR J1747-2958 (Gaensler et al. 2004) is confined within a bow-shaped boundary without a shell-like structure. X-ray PWNe associated with PSR B1757-24 (Kaspi et al. 2001), J1509-5859 and J1809-1917 (Sanwal et al. 2005), and the Geminga pulsar (Caraveo et al. 2003; De Luca et al. 2006; Pavlov et al. 2006) have elongated structures that look like "trails" (or "wakes") behind the moving pulsars. These tails could be pulsar jets confined by toroidal magnetic fields, or they could be associated with shocked relativistic wind confined by the ram pressure of the surrounding interstellar medium. Two X-ray nebulae

[8] The distance to PSR J0751+1807 is derived from the pulsar's parallax (Nice et al. 2005), whereas the other two estimates are obtained from the pulsars' dispersion measures and the NE2001 model.

[9] http://www.astro.psu.edu/users/green/psrdatabase/psrcat.htm, http://www.physics.mcgill.ca/~pulsar/pwncat.html.

Table 3 Three MSPs with estimates on X-ray flux

PSR	P (ms)	d (kpc)	τ (Gyr)	$\log \dot{E}$ (erg s^{-1})	$\log L_X$ (erg s^{-1})	$\log \eta_X$
J0034-0534	1.88	0.54	6.0	34.5	29.7	−4.8
J0751+1807	3.48	0.63	7.7	33.8	30.3	−3.5
J1012+5307	5.26	0.41	8.6	33.4	30.4	−3.0

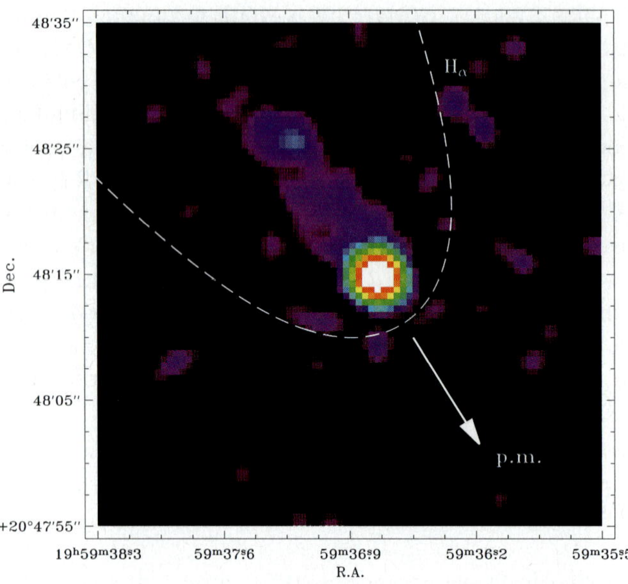

Fig. 11 *Chandra* image of PSR B1957+20 and its tail-like X-ray PWN. The dashed curve sketches the H_α nebula associated with the pulsar. The arrow indicates the direction of the pulsar's proper motion

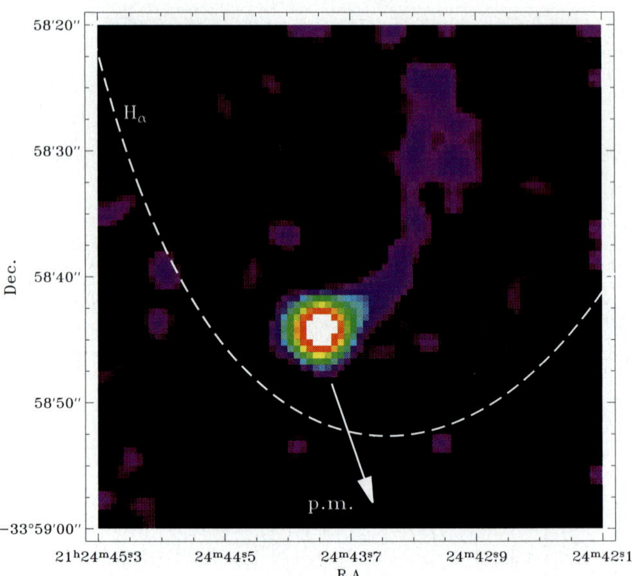

Fig. 12 Same as in Fig. 11 for PSR J2124-3358

produced by the MSPs, B1957+20 (Stappers et al. 2003) and J2124-3358 (Hui and Becker 2006), belong to the latter group of PWNe. These are shown in Figs. 11 and 12. Besides, optical observations revealed H_α bow-shocks surrounding these two pulsars (sketched with dashed curves in Figs. 11 and 12). Both X-ray PWNe have a form of a 20″-long tail, although there is a striking difference between their shapes. The H_α bow-shock and X-ray tail of PSR B1957+20 are fairly symmetric, with the symmetry axis being nearly aligned with the direction of the pulsar's proper motion (Fig. 11). For PSR J2124-3358, both the bow-shock and the X-ray tail are bent and highly asymmetric with respect to the proper motion vector (Fig. 2). Gaensler et al. (2002) argued that such an unusual shape of a PWN could be caused by a combination of effects associated with anisotropy of the pulsar wind and density nonuniformity of the surrounding medium. Analyzing the properties of the X-ray PWN of PSR B1957+20, Stappers et al. (2003) speculated that the efficiency with which relativistic particles are accelerated in the pulsar's wind can be as high as those inferred for young pulsars with much stronger surface magnetic fields. Comparing ratios of X-ray luminosities of tail-like PWNe to pulsar spin-down energies may be considered as an additional evidence in favor of this conclusion: the ratio, $L_X/\dot{E} \sim 10^{-4}$, estimated for the tail associated with PSR B1957+20 is very close to that found for the nebula produced by the 16-kyr-old PSR B1757-24 (Kaspi et al. 2001) and greatly exceeds those inferred for the Geminga's tail, $\sim 4 \times 10^{-6}$ (Pavlov et al. 2006) and the Vela's southeast jet, $\sim 1 \times 10^{-6}$ (Pavlov et al. 2003). Contrary to these two objects, the third MSP whose supersonic motion through the ambient medium causes an H_α bow-shock, J0437-4715, has not been found to power an X-ray nebula, suggesting a very low magnetic field in the expected PWN region (Zavlin et al. 2002). No an X-ray PWN associated with the most energetic MSPs B1937+21 has been detected in a deep *Chandra* observation of this pulsar.

6 Summarizing remarks

X-ray observations of the MSPs discussed in this paper and those located in the globular clusters (Bogdanov et al. 2006) revealed that most of them emit predominantly thermal PC emission with a luminosity of $L_{\rm bol}^{\rm pc} \sim 10^{30}$–$10^{31}$ erg s^{-1} and "PC efficiency" of $\eta^{\rm pc} \sim 10^{-4}$–$10^{-3}$. It is worth noting that these estimates on $\eta^{\rm pc}$ are close to that derived for the old ordinary pulsar B0950+08 ($\tau \simeq 17$ Myr, $P \simeq 253$ ms, $\dot{E} \simeq 6 \times 10^{32}$ erg s^{-1}), $\eta^{\rm pc} \simeq 2.4 \times 10^{-4}$ (Zavlin and Pavlov 2004), suggesting that heating mechanisms in ordinary and rapidly spinning pulsars may be quite similar. Comparing pulsar parameters of the MSPs with detected X-ray radiation shows that the four MSPs with nonthermal radiation possess much large values of $\dot{E}$, by at least a factor of 50, than those estimated for the thermally emitting pulsars. The numbers presented in Table 2 indicate that the larger is $\dot{E}$, the higher is the "nonthermal" efficiency $\eta^{\rm nonth}$, whereas the opposite tendency is apparent for $\eta^{\rm pc}$. A similar behavior of $\eta^{\rm pc}$ follows from results of the theoretical modeling of the PC heating by returning positrons produced through curvature radiation and inverse Compton scattering (see Figs. 7 and 8 in Harding and Muslimov 2001 and 2002, respectively, with account for the relation $\dot{E} \sim \tau^{-[n+1]/[n-1]} = \tau^{-2}$, for the magneto-dipole braking index $n = 3$). In addition, high luminosity of the nonthermal emission may be associated with large values of the magnetic field at the pulsar light cylinder (Saito et al. 1997), $B_{\rm lc} = B_{\rm surf}[2\pi R_{\rm NS}/cP]^3$. Indeed, the nonthermally emitting MSPs possess magnetic fields, $B_{\rm lc} \sim (0.3\text{–}1) \times 10^6$ G, close to that of the Crab pulsar ($\sim 1 \times 10^6$ G) and exceeding those in the MSPs with thermal emission (as well as in PSRs J0751+1807 and J1012+5307), $B_{\rm lc} \sim 0.03 \times 10^6$ G, at least by an order of magnitude. This may indicate that the nonthermal radiation of MSPs is generated in emission zone(s) close to the light cylinder (as predicted by the outer-gap pulsar models) and that $B_{\rm lc}$ is an important parameter governing the magnetospheric activity. On the other hand, a model by Harding et al. (2005) for acceleration and pair cascades on open field lines close to pulsar PCs describes rather well the spectrum of PSR J0218+4232 observed from X-rays through γ-rays. Hence, at the present stage, neither of the two models (outer-gap or polar-cap) of the nonthermal emission of MSPs could be regarded as more favorable. Besides, one cannot rule out that the pulsar spin period also affects $\eta^{\rm nonth}$—the MSPs of the first group have shorter periods than those of the other four pulsars (Table 1). Definitely, a bigger sample of X-ray emitting MSPs is required to draw more conclusive speculations.

The results of Sect. 3 point out that PCs of MSPs are likely nonuniform, with temperatures decreasing by a factor of 2–3 from the PC center down to its edge. There have not been yet reliable calculations of the temperature distribution around the pulsar magnetic poles. Hence, sophisticated theoretical models of PC heating and temperature distribution are required for further investigation of thermal X-ray emission from MSPs.

As shown by Pavlov and Zavlin (1997), modeling of an X-ray pulsed profile can yield constraints on the pulsar mass-to-radius ratio as well as its geometry. To do that, more elaborated models of magnetospheric pulsed emission and data of a better quality at higher energies are required to disentangle nonthermal and thermal components.

Chandra and *XMM*-Newton provided remarkable detections of two double-neutron-star binary systems with MSPs as companions, J0737-3039 and J1537+1155 (Campana et al. 2004; McLaughlin et al. 2004; Pellizzoni et al. 2004; Kargaltsev et al. 2006). Although the collected data are of a rather scanty statistics, one could conclude that the inferred X-ray spectra and luminosities of these two system are similar, suggesting the same mechanisms generating the detected radiation. Among various interpretations proposed, the most plausible one is that the observed emission consists of a thermal PC component emitted by the MSPs plus nonthermal X-rays from the interaction of the pulsar winds with the neutron star companions. This interpretation explains a gap observed in the orbital dependence of the X-ray flux emitted by the highly eccentric J1537+1155 system (Kargaltsev et al. 2006). To investigate these systems in more detail, a phase-resolved spectroscopy on X-ray data of a better quality is required to separate possible components in X-rays from these objects. Further observations are expected to constrain both the properties of the pulsar winds and those of the pulsars themselves in these (and other) double-neutron-star systems.

So far, only one MSP, J0437-4715, has been observed and detected in the UV/FUV band (Kargaltsev et al. 2004). The shape of the inferred spectrum suggests thermal emission from the whole neutron star surface of a surprisingly high temperature of about 0.1 MK. A heating mechanism should be operating in a Gyr-old neutron star to keep its surface at such temperature. As discussed by Kargaltsev et al. (2004), among several possible mechanisms (both internal and external), chemical heating (Reisenegger 1995) and frictional heating (e.g., Cheng et al. 1992) of the neutron star core are the most plausible options. To understand thermal evolution of neutron stars, more MSPs (e.g., the other three pulsars discussed in Sect. 3) and close ordinary old pulsars (PSR B0950+08 is one of the best candidates—Zavlin and Pavlov 2004) should be observed in the UV/FUV band. The example of PSR J0437-4715 shows that such an observational program would be feasible.

As a final remark, it is worth mentioning that so far all known isolated (non-accreting) MSPs were first discovered in radio. In X-rays a first attempt to find millisecond pulsations from an isolated compact object was done under assumption that the most famous (and intriguing) member of the group of the "dim" isolated, radio-quite, and thermally emitting neutron stars discovered with *ROSAT* (see the presentation by Haberl at this conference), RX J1856.5-3754, could be an MSP (see Pavlov and Zavlin 2003 for a discussion). One of the reasons to look for millisecond pulsations was the fact that the X-ray flux detected from this object in deep observations with *Chandra* and *XMM*-Newton revealed no pulsations at periods above 20 ms (the period range to which those observational data were sensitive), with a stringent upper limit on the pulsed fraction $f_p < 1.3\%$. To check the MSP hypothesis, a deep *XMM*-Newton observation of RX J1856.5-3754 in an instrumental model with a temporal resolution of 0.03 ms was conducted in April 2004. Despite a very large number of source counts collected in this observation ($\sim 2 \times 10^5$), no significant pulsations were found at periods in the range $P = 1$–20 ms. The derived 1σ upper limit on the pulsed flux for this period range is $f_p < 2.1\%$. This nondetection virtually excludes the MSP interpretation for RX J1856.5-3754 and supports alternative hypotheses of a very special neutron star's geometry and/or a very long spin period ($P \gtrsim 10$ hr).

Acknowledgements The author thanks George Pavlov for useful discussions. This work is supported by a NASA Research Associateship Award at NASA Marshall Space Flight Center. The study of RX J1856.5-3754 with *XMM*-Newton was made possible through the NASA grant NNG04GI801G.

References

Alpar, M.A., Cheng, A.F., Ruderman, M.A., et al.: Nature **300**, 728 (1982)
Arons, J., Tavani, M.: Astrophys. J. **403**, 249 (1993)
Backer, D.C., Kulkarni, S.R., Heiles, C., et al.: Nature **300**, 615 (1982)
Becker, W., Aschenbach, B.: In: Becker, W., Lesch, H., Trümper, J. (eds.) Neutron Stars, Pulsars and Supernova Remnants. MPE Report 278, p. 64 (2002)
Becker, W., Trümper, J.: Nature **365**, 528 (1993)
Becker, W., Schwarz, D.A., Pavlov, G.G., et al.: Astrophys. J. **594**, 798 (2003)
Bogdanov, S., Grindlay, J.E., Heinke, C.O., et al.: Astrophys. J. **646**, 1104 (2006)
Bogdanov, S., Grindlay, J.E., Rybicki, G.B.: Astrophys. J. **648**, L55 (2006)
Braje, T.M., Romani, R.W., Rauch, K.P.: Astrophys. J. **531**, 447 (2000)
Campana, S., Possentti, A., Burgay, M.: Astrophys. J. Lett. **613**, 53 (2004)
Caraveo, P.A., Bignami, G.F., De Luca, A., et al.: Science **301**, 1345 (2003)
Cheng, K.S., Ho, C., Ruderman, M.: Astrophys. J. **300**, 500 (1986)
Cheng, K.S., Chau, W.Y., Zhang, J.L., et al.: Astrophys. J. **396**, 135 (1992)
Cordes, J.M., Lazio, T.J.W., preprint, astro-ph/0301598 (2003)
Cumming, A.: In: Rasio, F.A., Stairs, I.H. (eds.) Binary Radio Pulsars. ASP Conference Series, vol. 238, pp. 311. ASP, San Francisco (2005)
Cusumano, G., Hermsen, W., Kramer, M., et al.: Astron. Astrophys. **410**, L9 (2003)
De Luca, A., Caraveo, P.A., Mattana, F., et al.: Astron. Astrophys. **445**, L9 (2006)
Gaensler, B.M., Jones, D.H., Stappers, B.W.: Astrophys. J. **580**, L137 (2002)
Gaensler, B.M., van der Swaluw, E., Camilo, F., et al.: Astrophys. J. **616**, 383 (2004)
Gaensler, B.M., Slane, P.O.: Annu. Rev. Astron. Astrophys. **44**, 17 (2006)
Harding, A.K., Muslimov, A.G.: Astrophys. J. **556**, 987 (2001)
Harding, A.K., Muslimov, A.G.: Astrophys. J. **568**, 862 (2002)
Harding, A.K., Usov, V.V., Muslimov, A.G.: Astrophys. J. **622**, 531 (2005)
Hotan, A.W., Bailes, M., Ord, S.M.: Mon. Not. Roy. Astron. Soc. **369**, 1502 (2006)
Hui, C.Y., Becker, W.: Astron. Astrophys. **448**, L13 (2006)
Kargaltsev, O.Y., Pavlov, G.G., Romani, R.W.: Astrophys. J. **602**, 327 (2004)
Kargaltsev, O.Y., Pavlov, G.G., Garmire, G.P.: Astrophys. J. **646**, 1139 (2006)
Kargaltsev, O.Y., Pavlov, G.G., Garmire, G.P.: Astrophys. J., in press, astro-ph/0611599 (2007)
Kaspi, V.M., Gotthelf, E.V., Gaensler, B.M., et al.: Astrophys. J. Lett. **562**, 163 (2001)
Kaspi, V.M., Roberts, M.S.E., Harding, A.K.: In: Lewin, W.H.G., van der Klis, M. (eds.) Compact Stellar X-ray Sources, p. 279. Cambridge University Press, Cambridge (2006)
Kuiper, L., Hermsen, W., Verbunt, F., et al.: Astron. Astrophys. **359**, 615 (2000)
Kuiper, L., Hermsen, W., Verbunt, F., et al.: Astrophys. J. **577**, 917 (2002)
Kuiper, L., Hermsen, W., Stappers, B.: Adv. Space Res. **33**, 507 (2004)
Lommen, A.N., Kipphorn, R., Nice, D.J., et al.: Astrophys. J. **642**, 1012 (2006)
Manchester, R.N., Hobbs, G.B., Teoh, A., et al.: Astron. J. **129**, 1993 (2005)
McLaughlin, M.A., Camilo, F., Burgay, M., et al.: Astrophys. J. Lett. **605**, 41 (2004)
Nicastro, L., Cusumano, G., Löhmer, O., et al.: Astron. Astrophys. **413**, 1065 (2004)
Nice, D.J., Splaver, E.M., Stairs, I.H., et al.: Astrophys. J. **634**, 1242 (2005)
Pavlov, G.G., Zavlin, V.E.: Astrophys. J. Lett. **490**, 91 (1997)
Pavlov, G.G., Zavlin, V.E.: In: Bandiera, R., Maiolino, R., Mannucci, F. (eds.) The XXI Texas Symposium on Relativistic Astrophysics, p. 319. World Science (2003)
Pavlov, G.G., Teter, M.A., Kargaltsev, O.Y., et al.: Astrophys. J. **591**, 1157 (2003)
Pavlov, G.G., Sanwal, D., Zavlin, V.E.: Astrophys. J. **643**, 1146 (2006)
Pellizzoni, A., De Luca, A., Mereghetti, S., et al.: Astrophys. J. Lett. **612**, 49 (2004)
Reisenegger, A.: Astrophys. J. **442**, 749 (1995)
Romani, R.W.: Astrophys. J. **470**, 469 (1996)
Rots, A.H., Jahoda, K., Makomb, D.J., et al.: Astrophys. J. **501**, 749 (1998)
Rutledge, R.E., Fox, D.W., Kulkarni, S.R., et al.: Astrophys. J. **613**, 522 (2004)
Saito, Y., Kawai, N., Kamae, T., et al.: Astrophys. J. Lett. **447**, 37 (1997)
Sanwal, D., Pavlov, G.G., Garmire, G.P.: Am. Astron. Soc. **206**, 43.02 (2005)
Shklovskii, I.S.: Sov. Astron. **13**, 562 (1970)
Stappers, B.W., Gaensler, B.M., Kaspi, V.M., et al.: Science **299**, 1372 (2003)
Takahashi, M., Shibata, S., Torii, K., et al.: Astrophys. J. **554**, 316 (2001)
Webb, N.A., Olive, J.-F., Barret, D.: Astron. Astrophys. **417**, 181 (2004a)
Webb, N.A., Olive, J.-F., Barret, D., et al.: Astron. Astrophys. **419**, 269 (2004b)

Weisskopf, M.C., Hester, J.J., Tennant, A.F., et al.: Astrophys. J. Lett. **536**, 81 (2000)

Wijnands, R.: Nucl. Phys. B **132**, 486 (2004)

Zavlin, V.E.: Astrophys. J. **638**, 951 (2006)

Zavlin, V.E., Pavlov, G.G.: Astron. Astrophys. **329**, 583 (1998)

Zavlin, V.E., Pavlov, G.G.: Astrophys. J. **616**, 452 (2004)

Zavlin, V.E., Shibanov, Yu.A., Pavlov, G.G.: Astron. Lett. **21**, 149 (1995)

Zavlin, V.E., Pavlov, G.G., Shibanov, Yu.A.: Astron. Astrophys. **315**, 141 (1996)

Zavlin, V.E., Pavlov, G.G., Sanwal, D., et al.: Astrophys. J. **569**, 894 (2002)

Astrophys Space Sci (2007) 308: 309–316
DOI 10.1007/s10509-007-9319-9

ORIGINAL ARTICLE

X-ray observations of PSR B0355+54 and its pulsar wind nebula

Katherine E. McGowan · W. Thomas Vestrand · Jamie A. Kennea · Silvia Zane · Mark Cropper · France A. Córdova

Received: 26 June 2006 / Accepted: 13 October 2006 / Published online: 21 March 2007

Abstract We present X-ray data of the middle-aged radio pulsar PSR B0355+54. The XMM-Newton and Chandra observations show not only emission from the pulsar itself, but also compact diffuse emission extending $\sim 50''$ in the opposite direction to the pulsar's proper motion. Our analysis also indicates the presence of fainter diffuse emission extending $\sim 5'$ from the point source. The morphology of the diffuse component is similar to the ram-pressure confined pulsar wind nebulae detected for other sources. We find that the compact diffuse component is well-fitted with a power-law, with an index that is consistent with the values found for other pulsar wind nebulae. The core emission from the pulsar can be characterized with a thermal plus power-law fit, with the thermal emission most likely originating in a hot polar cap.

Keywords Pulsars · Individual (PSR B0355+54) · X-rays

K.E. McGowan (✉)
School of Physics and Astronomy, University of Southampton, Highfield, Southampton, SO17 1BJ, UK
e-mail: kem@astro.soton.ac.uk

W.T. Vestrand
Los Alamos National Laboratory, Los Alamos, NM 87545, USA

J.A. Kennea
525 Davey Laboratory, Pennsylvania State University, University Park, PA 16802, USA

S. Zane · M. Cropper
Mullard Space Science Laboratory, Holmbury St. Mary, Dorking, RH5 6NT, UK

F.A. Córdova
Chancellor's Office, University of California, Riverside, CA 92521, USA

1 Introduction

Isolated pulsars constitute one of the most powerful laboratories for studying particle acceleration in astrophysics. A significant fraction of the energy from rotation-powered pulsars is converted into a wind (Rees and Gunn 1974), which travels at a velocity close to the speed of light. The interaction of this pulsar wind with the ambient medium produces a shock and acceleration of the relativistic particles at the shock generates synchrotron emission. This non-thermal diffuse emission manifests itself as a pulsar wind nebulae (PWNe) at radio and X-ray energies (e.g. Rees and Gunn 1974; Gaensler 2001). Due to the short synchrotron lifetimes of high energy electrons, X-ray emission from a PWN directly traces the current energetics of the pulsar. The spectral and morphological characteristics of an X-ray PWN therefore reveal the structure and composition of the pulsar wind and the orientation of the pulsar's spin axis and/or velocity vector.

The middle-aged 156 ms radio pulsar PSR B0355+54 is a known X-ray emitter (Helfand 1983; Seward and Wang 1988; Slane 1994). Helfand (1983) reported the first detection in X-rays of the source using data from Einstein, stating that emission extended $5'$ from the pulsar. However, Seward and Wang (1988) analyzed the Einstein data and concluded that while there was evidence for weak emission $1.7'$ from the source position, emission from the pulsar itself was not detected. Nevertheless, they did not rule out the possibility that the emission could be associated with a PWN. Slane (1994) detected PSR B0355+54 in a 20 ks ROSAT observation, but owing to the lack of counts it was not feasible to perform a spectral analysis. The analysis of the ROSAT data also led to the detection of faint extended emission $\sim 1.6'$ from the pulsar position, but Slane (1994) did not believe there was enough evidence to support a link between the source and the extended emission.

In this paper, we present XMM-Newton and Chandra observations of PSR B0355+54 which we use to investigate the presence of diffuse emission that can be attributed to a PWN.

2 Observations and data reduction

XMM-Newton observed PSR B0355+54 for 29 ks on 2002 February 10. The EPIC pn was operated in Small Window mode with the thin filter, and the EPIC MOS in Full Frame mode with the medium filter. We reduced the data using the XMM-Newton Science Analysis System (SAS version 6.1.0). The data were filtered to include only single, double, triple and quadruple photon events for MOS and only single and double photon events for pn. We excluded events that may have incorrect energies, for example those next to the edges of the CCDs and next to bad pixels, and only included photons with energies in the range 0.3–10 keV.

PSR B0355+54 was also observed for 66 ks on 2004 July 16 with the ACIS-S array on Chandra in the very faint timed exposure imaging mode. We performed standard data processing using CIAO version 3.2. The data were filtered to restrict the energy range to 0.3–10 keV and to exclude times of high background.

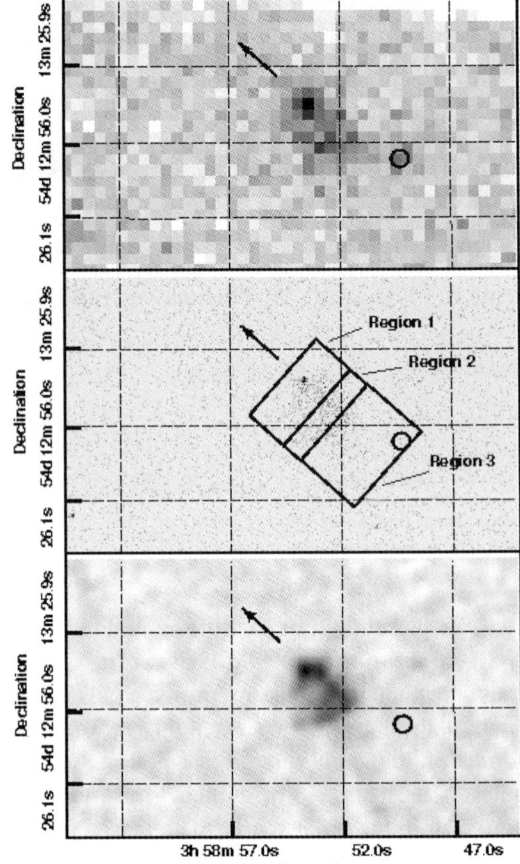

Fig. 1 X-ray detection of PSR B0355+54 and its diffuse emission. *Top panel*: *Gray-scale plot* of the 0.3–10 keV XMM-Newton EPIC image. *Middle panel*: *Gray-scale plot* of the 0.3–10 keV Chandra ACIS image. *The boxes* define the regions used to extract spectra for the diffuse emission. *Bottom panel*: The same image as *the middle panel* with the contribution from the X-ray point source removed. The image is smoothed with a Gaussian of width $\sim 2''$. *The arrow* shows the direction of the proper motion of the pulsar and has a length of $20''$

3 Spatial analysis

The images created from the EPIC data show relatively strong emission at the pulsar position (RA $= 03^{\rm h}58^{\rm m}53.72^{\rm s}$, Dec $= +54^\circ 13' 13.7''$) and evidence for extended emission near to PSR B0355+54 (see Fig. 1, top panel). To confirm the presence and examine the extent of the diffuse emission in the XMM-Newton data we have compared the detected pn emission with that for a point source. We calculated the intensity for the pulsar by using bilinear interpolation at regularly spaced points along the direction of proper motion of PSR B0355+54 (Chatterjee et al. 2004). We compared this profile with the XMM-Newton point-spread function (PSF) for the pn at 1.5 keV, which we generated using the King profile parameters included in the XMM-Newton calibration file "XRT3_XPSF_0006.CCF.plt".[1] In Fig. 2 (top panel) we show the profiles for the pulsar and the pn PSF.

The Chandra ACIS image also reveals a faint tail of emission in the opposite direction to the pulsar's proper motion (see Fig. 1, middle and bottom panels). Again we determined the net counts from the source and diffuse emission at regularly spaced intervals along the direction of the pulsar's proper motion. We generated a PSF for PSR B0355+54 using the Chandra PSF library evaluated at 1.5 keV and the location relevant to our source. The PSF was normalized to the total counts in PSR B0355+54. We calculated the net counts for the PSF in the same intervals as for PSR B0355+54. The source and PSF profiles are shown in Fig. 2 (bottom panel).

We smoothed the Chandra ACIS image with a Gaussian of width $\sim 2''$ (Fig. 1, bottom panel), the resulting image suggests that there are two regions of enhanced diffuse emission—one near to the pulsar and the other $\sim 10''$ away. The intensity profiles for the XMM-Newton and Chandra data indicate that the core of the X-ray emission lies within $5''$ of the pulsar position. Both emission profiles indicate that the diffuse emission extends out to $\sim 50''$, with the bulk of the emission lying within 20–30$''$ of PSR B0355+54. The profiles also show evidence for a dip in the emission at $\sim 10''$ agreeing with the smoothed ACIS image.

[1] See http://xmm.vilspa.esa.es/docs/documents/CAL-SRN-0100-0-0.ps.gz for more information.

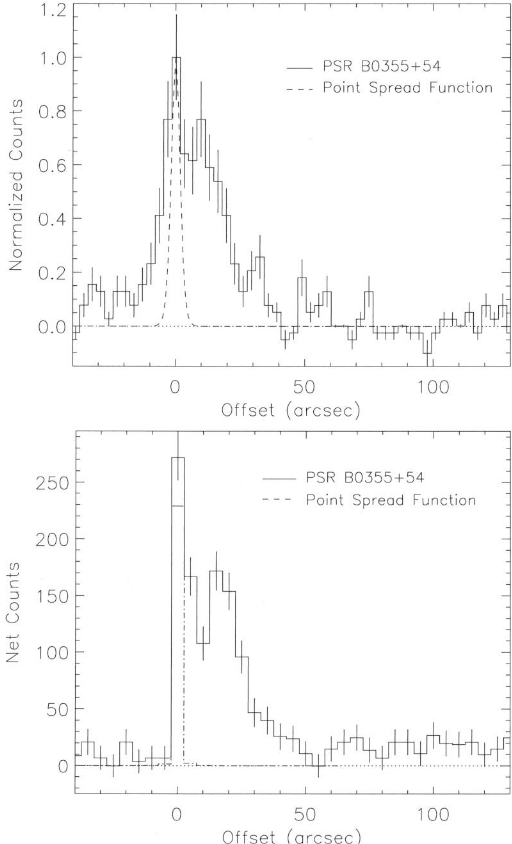

Fig. 2 X-ray emission from PSR B0355+54 as a function of distance from the point source along the direction of the proper motion of the pulsar (*solid line*), compared to the instrument PSF determined at 1.5 keV and the location of the pulsar (*dashed line*). *Top panel*: XMM-Newton pn. *Bottom panel*: Chandra ACIS

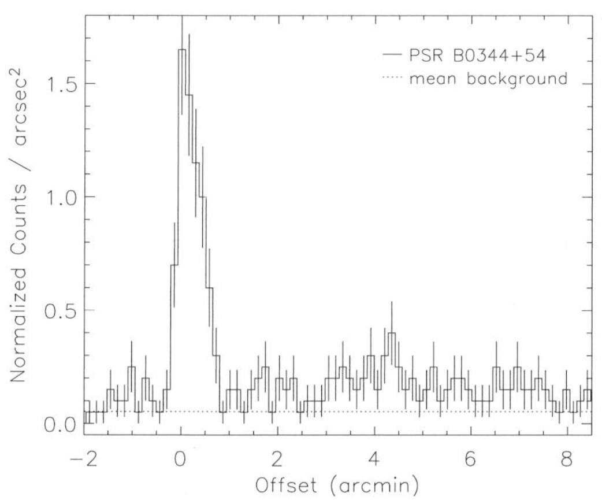

Fig. 3 XMM-Newton MOS1 X-ray emission from PSR B0355+54 as a function of distance from the point source along the direction of the pulsar's proper motion (*solid line*), compared to the mean background (*dotted line*)

Detection of emission from PSR B0355+54 at distances of 1.6′ to 5′ from the source position i.e. on a larger scale than shown in Fig. 1, have been reported by Helfand (1983), Seward and Wang (1988) and Slane (1994), see also Tepedelenlioğlu and Ögelman (2005). Visual inspection of a smoothed version of the MOS1 image indicates that there is a region of enhanced emission extending a few arcminutes south of the pulsar. The intensity of the MOS1 emission in this direction was determined using bilinear interpolation at regularly spaced intervals from the source. We show in Fig. 3 the distribution of counts as a function of distance from the source compared to the mean background. Our analysis suggests that there is an excess of counts at 1′ and ∼3–5′ from the point source.

4 Spectral analysis

In order to investigate the properties of the X-ray emission from PSR B0355+54 and the compact ($\leq 50''$) diffuse nebula, we compared the spectra extracted from different spatial regions. Our results from the spatial analysis suggest that the core of the pulsar emission lies within 5″ of the pulsar's position. However, in the case of the XMM-Newton data, this size of aperture does not contain enough counts for a meaningful analysis.

The pulsar spectrum has been extracted from the XMM-Newton observation using a circular region of radius 30″, centered on the pulsar's radio position that was determined from VLBI measurements (Chatterjee et al. 2004). The background was extracted from a region of similar size offset from the pulsar position. The total counts contained in the source region is 1143 with an estimated 562 from background. The spectrum was regrouped by requiring at least 50 counts per spectral bin. We created a photon redistribution matrix (RMF) and ancillary region file (ARF) for the spectrum. The subsequent spectral fitting and analysis was performed using XSPEC, version 11.3.1.

We modelled the spectrum in the 0.5–9.0 keV range. Initially we fitted the spectrum with single-component models including absorbed power-law, blackbody and magnetized, pure H atmospheric (Pavlov et al. 1995) models. The spectrum is best-fitted with a power-law with index $\Gamma = 1.5^{+0.5}_{-0.3}$ and column density $N_H = (0.50^{+0.36}_{-0.20}) \times 10^{22}$ cm^{-2}, with $\chi^2_\nu = 0.7$ for 16 degrees of freedom. This value for the power-law index is similar to the values found for other PWNe (e.g. Kaspi et al. 2005). The Galactic hydrogen column in the direction of PSR B0355+54 is $N_H = 0.88 \times 10^{22}$ cm^{-2}. The fit results in an unabsorbed 0.3–10 keV energy flux of $(2.3^{+1.0}_{-0.7}) \times 10^{-13}$ ergs cm^{-2} s^{-1}. We also fitted the spectrum with blackbody plus power-law and atmospheric plus power-law models, both modified by photoelectric absorption. The multi-component models give similar values for reduced χ^2, however the temperatures implied by the fits are poorly constrained. It is likely that the presence of

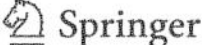

the pulsar wind nebula, and being unable to separate the pulsar core and diffuse emission, effects our ability to constrain the thermal component in the spectral fits of the XMM-Newton data. The XMM-Newton spectrum with the best-fitting power-law model is shown in Fig. 4.

For the Chandra data we extracted a spectrum for the core of the pulsar emission from a circular region of radius 5″ centered on the radio position. An annulus centered on the pulsar position was used to extract the background, with inner and outer radii of 6″ and 10″, respectively. We find a total of 244 counts contained in the source region, with 29 counts attributed to background. We created the RMF and ARF files using standard CIAO tools. Before fitting the spectrum we regrouped the data, requiring a minimum of 15 counts per spectral bin.

We fitted the spectrum in the energy range 0.5–7.0 keV using the same models as for the XMM-Newton data. In the first instance we let the neutral hydrogen column density be a free parameter; however this led to unreasonably small values for N_H. Subsequently we fixed the column density at the value found from the power-law fit to the XMM-Newton spectrum of PSR B0355+54. We find that the Chandra spectrum can also be characterized by a power-law. The model has a power-law index of $\Gamma = 1.9^{+0.4}_{-0.3}$ which is consistent within the 90% uncertainties to the value found from the XMM-Newton data. However, in the case of the Chandra data we find that a thermal plus power-law model provides a better fit statistically. The data are equally well-fitted by a blackbody plus power-law and a magnetized, pure H atmospheric (Pavlov et al. 1995, "nsa" model in XSPEC) plus power-law model. The results of the Chandra spectral fitting are given in Table 1.

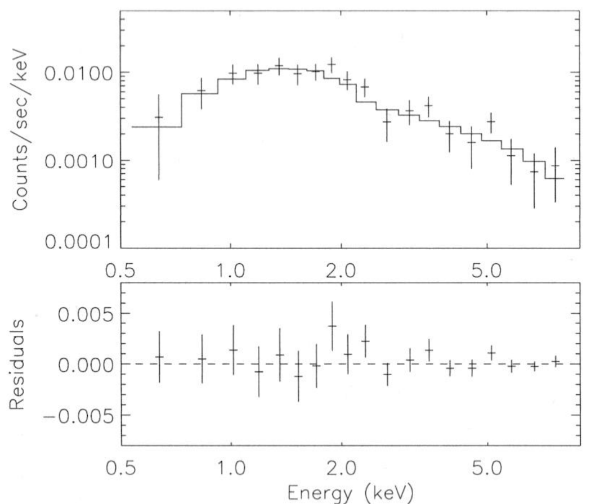

Fig. 4 The XMM-Newton spectrum of PSR B0355+54 with best-fit power-law model. Also shown are the residuals from comparison of the data to the model

For the blackbody plus power-law model the best-fit parameters are a power-law index of $\Gamma = 1.0^{+0.2}_{-1.0}$ and temperature of $T = (2.32^{+1.16}_{-0.81}) \times 10^6$ K. Using the parallax determined distance to the source of $D = 1.04^{+0.21}_{-0.16}$ kpc (Chatterjee et al. 2004) implies a blackbody emitting radius of $0.12^{+0.16}_{-0.07}$ km. This value is too small to be reconciled with the radius of the neutron star and would indicate that the origin of the emission is a hot polar cap. We find an unabsorbed 0.3–10 keV energy flux of $(6.4^{+19.3}_{-1.1}) \times 10^{-14}$ ergs cm^{-2} s^{-1}. The magnetized, pure H atmospheric plus power-law model best-fit parameters are a power-law index of $\Gamma = 1.5^{+0.5}_{-0.4}$, temperature of $T^{\infty}_{\rm eff} = (0.45^{+0.20}_{-0.22}) \times 10^6$ K, and a radius for the neutron star of $R_{\rm NS} = 7.2^{+7.2}_{-2.2}$ km. For this fit the distance to the source and the mass of the neutron star were fixed at $D = 1.04$ kpc and $M_{\rm NS} = 1.4 M_{\odot}$, respectively. The magnetic field of the neutron star was fixed at $B = 10^{12}$ G (this is a good approximation since the pulsar magnetic field as inferred from radio timing properties is $B = 8.4 \times 10^{11}$ G, Hobbs et al. 2004; Manchester et al. 2005). The unabsorbed 0.3–10 keV energy flux for this fit is $(1.5^{+54.0}_{-0.7}) \times 10^{-13}$ ergs cm^{-2} s^{-1}. The Chandra spectrum of the core emission from PSR B0355+54 is shown in Fig. 5 with the best-fitting blackbody plus power-law model (top) and magnetized, pure H atmospheric plus power-law model (bottom).

To analyze the compact diffuse emission we created a new events file in which the emission from the pulsar core was removed. We extracted a spectrum for the diffuse component from a rectangular region of 40″ × 55″,

Table 1 Best-fit parameters for the Chandra X-ray emission of PSR B0355+54

Region	Model	Γ	$T/T^{\infty}_{\rm eff}$ ($\times 10^6$ K)	$R_{\rm BB}/R_{\rm NS}$ (km)	χ^2_ν (dof)	$F_X^{\rm unabs}$ (0.3–10 keV) (erg cm^{-2} s^{-1})
Core	PL	$1.9^{+0.4}_{-0.3}$	...	...	0.5 (34)	$(4.9^{+1.6}_{-0.7}) \times 10^{-14}$
	BB + PL	$1.0^{+0.2}_{-1.0}$	$2.32^{+1.16}_{-0.81}$	$0.12^{+0.16}_{-0.07}$	0.3 (32)	$(6.4^{+19.3}_{-1.1}) \times 10^{-14}$
	NSA + PL	$1.5^{+0.5}_{-0.4}$	$0.45^{+0.20}_{-0.22}$	$7.2^{+7.2}_{-2.2}$	0.4 (32)	$(1.5^{+54.0}_{-0.7}) \times 10^{-13}$
Diffuse—all	PL	$1.4^{+0.3}_{-0.3}$	...	...	1.0 (50)	$(1.7^{+0.8}_{-0.5}) \times 10^{-13}$
Diffuse—1	PL	$1.4^{+0.4}_{-0.4}$	...	...	1.0 (16)	$(5.7^{+3.3}_{-1.8}) \times 10^{-14}$
Diffuse—2	PL	$1.5^{+0.3}_{-0.3}$	...	...	1.4 (16)	$(7.6^{+3.4}_{-1.9}) \times 10^{-14}$
Diffuse—3	PL	$1.2^{+0.5}_{-0.4}$	...	...	1.2 (17)	$(5.3^{+4.4}_{-2.0}) \times 10^{-14}$

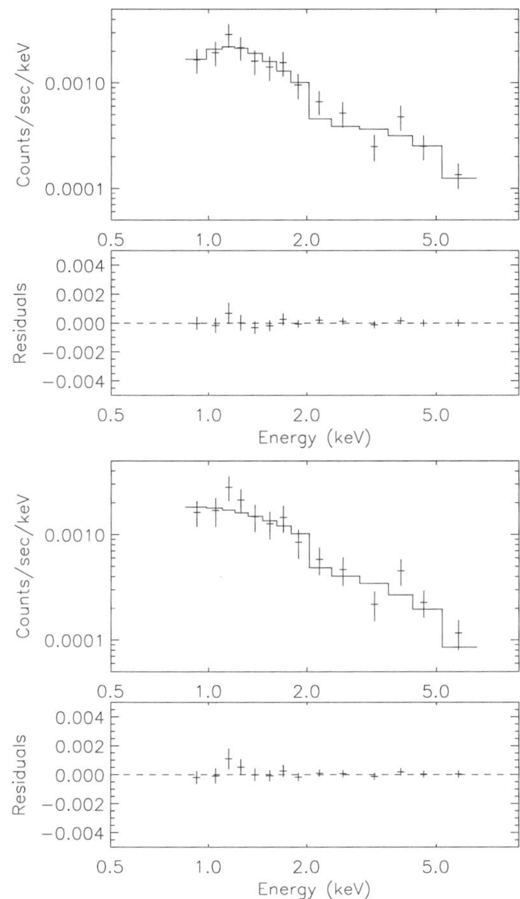

Fig. 5 The Chandra spectrum of the core emission of PSR B0355+54 with best-fit blackbody plus power-law model (*top*) and magnetized, pure H atmospheric plus power-law model (*bottom*). Also shown are the residuals from comparison of the data to the model in each case

centered on the emission and orientated along the direction of the pulsar's proper motion. The background was extracted from a region of similar size offset from the diffuse emission. The diffuse component extraction region contains 1207 counts, with an estimated 414 counts due to background. We created the RMF and ARF files using standard CIAO tools. Before fitting the spectrum we regrouped the data, requiring a minimum of 15 counts per spectral bin. We modelled the spectrum over 0.5–7.0 keV with an absorbed power-law, keeping the column density fixed at $N_H = 0.50 \times 10^{22}$ cm^{-2}. The best-fit has a power-law index of $\Gamma = 1.4 \pm 0.3$ and unabsorbed 0.3–10 keV energy flux of $(1.7^{+0.8}_{-0.5}) \times 10^{-13}$ ergs cm^{-2} s^{-1}.

In order to investigate the possibility of spectral evolution along the extended X-ray emission we created spectra for three regions of the compact diffuse emission. The sizes of the regions were chosen with the aim of having similar numbers of counts in each region. The three extraction regions, orientated along the direction of proper motion, are as follows, region 1: $40'' \times 18''$, contains a total of 396 counts with 130 attributed to background, region 2:

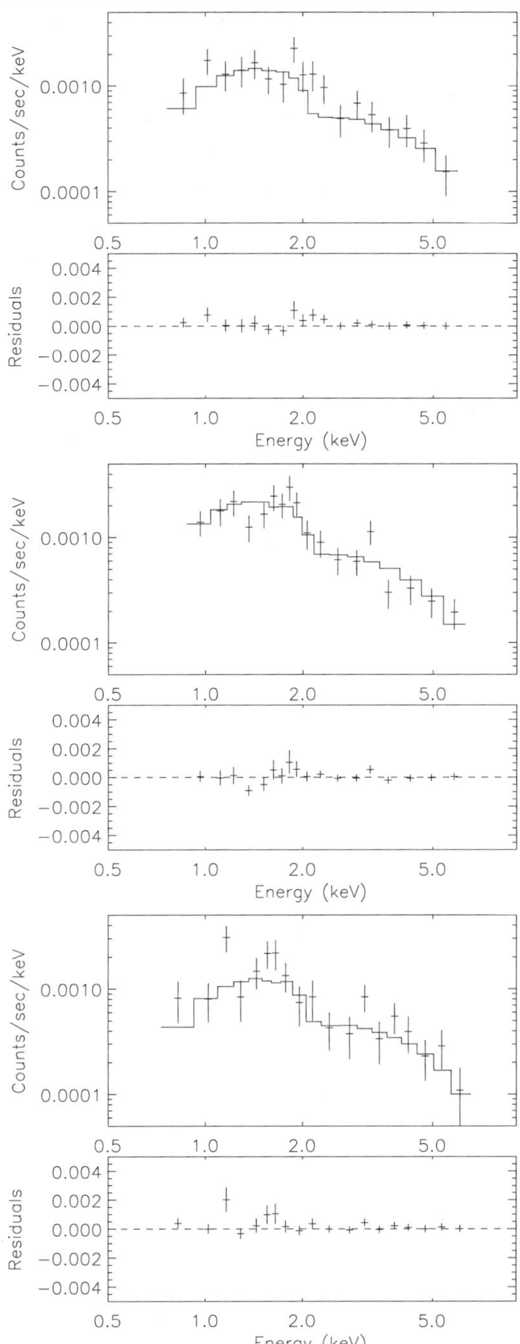

Fig. 6 Spectral fitting to Chandra data of PSR B0355+54. *First–third panels*: Diffuse emission from regions 1–3, respectively, with best fit power-law models. Also shown are the residuals from comparison of the data to the model in each case

$40'' \times 9''$, contains a total of 356 counts with 69 attributed to background, region 3: $40'' \times 28''$, contains a total of 454 counts with 214 attributed to background (see Fig. 1 (middle panel)). The background was extracted from the same region as above. We created response files for each region and regrouped the spectra, requiring a minimum of 15 counts per spectral bin. The spectra were fitted in the 0.5–7.0 keV energy range with a power-law and fixed column density of $N_H = 0.50 \times 10^{22}$ cm^{-2}. We find the best-fit power-law indices for regions 1–3 are $\Gamma = 1.4 \pm 0.4$, 1.5 ± 0.3 and $1.2^{+0.5}_{-0.4}$, respectively. Due to the uncertainties on the indices the presence of any spectral variability remains unclear. The unabsorbed 0.3–10 keV energy fluxes are $(5.7^{+3.3}_{-1.8}) \times 10^{-14}$, $(7.6^{+3.4}_{-1.9}) \times 10^{-14}$ and $(5.3^{+4.4}_{-2.0}) \times 10^{-14}$ ergs cm^{-2} s^{-1}, respectively. The Chandra spectra of the diffuse emission from regions 1–3 are shown in Fig. 6 (first–third panel) with the best-fitting power-law models.

5 Timing analysis

We barycentrically corrected the photon arrival times in the XMM-Newton pn event file before performing the temporal analysis. We extracted data for the source from a circular region of 15″ centered on the pulsar position. The total counts encompassed in this region is 391, with the background contributing 151 counts.

In order to search for an X-ray modulation at the PSR B0355+54 spin frequency, we first determined a predicted pulse frequency at the epoch of our XMM-Newton observations, assuming a linear spin-down rate and using the radio measurements (Hobbs et al. 2004; Manchester et al. 2005). We calculate $f = 6.3945388$ Hz at the midpoint of our observation (MJD 52315.7). As glitches and/or deviations from a linear spin-down may alter the period evolution, we then searched for a pulsed signal over a wider frequency range centered on $f = 6.39454$ Hz. We searched for pulsed emission using two methods. In the first method we implement the Z_n^2 test (Buccheri et al. 1983), with the number of harmonics n being varied from 1 to 5. In the second method we calculate the Rayleigh statistic (de Jager 1991; Mardia 1972) and then calculate a maximum likelihood periodogram (MLP) using the C statistic (Cash 1979) to determine significant periodicities in the data sets.

The most significant Z_n^2-statistic occurs for $n = 1$. With the number of harmonics equal to one, the Z_n^2-statistic corresponds to the well known Rayleigh statistic. We find three peaks with >90% significance in the MLP, all with corresponding peaks from the Z_1^2-test. The dominant peak from the Z_1^2-test occurs at $6.3945447^{+0.0000167}_{-0.0000107}$ Hz, with the corresponding peak in the MLP occurring at $6.3945467^{+0.00000821}_{-0.00000818}$ Hz. The uncertainties quoted are the 68% confidence limits on the position of the peak. Both frequencies are consistent, within the 68% contour, with the predicted pulse frequency, and with each other within the 90% contour. The second most prominent peak from the Z_1^2-test, and the corresponding peak in the MLP, are not consistent with the predicted pulse frequency.

While we have detected a frequency that is in agreement with the predicted pulse frequency for PSR B0355+54 we

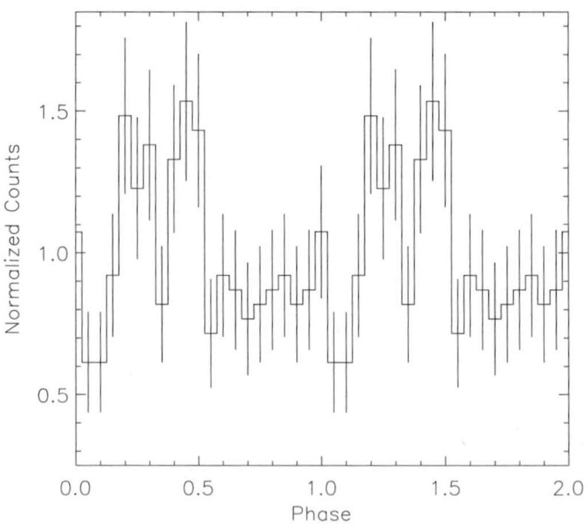

Fig. 7 pn data in the 0.3–10 keV energy range for PSR B0355+54 folded on the frequency found from the Z_1^2-test. The data are folded using the radio ephemeris

caution that the Z_1^2 peak has a probability of chance occurrence of 3×10^{-3}. Further observations of the source are needed to show whether the modulation detected is in fact pulsed X-ray emission from PSR B0355+54. We have folded the data on the frequency found from the Z^2-test (see Fig. 7); by fitting the profile with a sinusoid we find that the modulation amplitude is $21 \pm 8\%$. This value for the pulsed fraction is reasonable if the emission originates in a hot polar cap, while a much larger pulsed fraction would be expected from magnetospheric emission.

6 Discussion

Our spatial analysis of the XMM-Newton and Chandra observations of PSR B0355+54 have not only revealed X-rays from the pulsar, but have provided definitive proof of diffuse emission extending in the opposite direction to the pulsar's proper motion. Similar detections of extended emission have been seen for other sources (e.g. N157B, Wang and Gotthelf 1998; PSR B1757-24, Frail and Kulkarni 1991; Kaspi et al. 2001; PSR B1957+20, Stappers et al. 2003; PSR B1951+32, Li et al. 2005), and have been interpreted as emission from a ram-pressure confined PWN.

We cannot separate the core and diffuse emission components for the XMM-Newton data and find that the spectrum can be well-fitted with a power-law model with index of 1.5. The nebular emission is most likely dominating the spectrum. The core emission from the Chandra data can be well-fitted by a thermal plus power-law model with $T = 2.32 \times 10^6$ K and index of 1.0. The size of the emitting region resulting from the fit implies that the flux originates from a hot polar cap. We can also fit the spectrum with a pure H, magnetized atmospheric plus power-law model with $T_{\rm eff}^{\infty} = 0.45 \times 10^6$ K, $\Gamma = 1.5$ and $R_{\rm NS} = 7.2$ km. Taking into account the possible detection of pulsed emission from PSR B0355+54 it is likely that the emitting region is a hot polar cap.

It is suggested that the presence of a PWN is related to the spin-down power of the pulsar, and for sources with $\log \dot{E} \leq 36$ the PWN emission efficiency is significantly reduced (Frail and Scharringhausen 1997; Gaensler et al. 2000; Gotthelf 2003). For PSR B0355+54 $\log \dot{E} = 34.6$, making it one of a handful of sources with spin-down power below this limit with a detectable PWN (cf. Geminga, Caraveo et al. 2003). Using the results from the blackbody plus power-law fit to the core emission detected with Chandra we determine an isotropic unabsorbed luminosity in the 0.3–10 keV band of 8.3×10^{30} erg s^{-1}. With $\dot{E} = 4.5 \times 10^{34}$ erg s^{-1} for PSR B0355+54, this leads to a conversion efficiency of 2×10^{-4}. So in fact, we find that the conversion efficiency of the point source is similar to the values found for other pulsars (see e.g. Becker and Trümper 1997; Gaensler et al. 2004). Our analysis also indicates that the compact diffuse component is more luminous than the point source, with a conversion efficiency of 5×10^{-4} in the 0.3–10 keV range. This result is again consistent with other sources (Becker and Trümper 1997). In addition, it is reported that when the pulsar spin-down energy is $\log \dot{E} \leq 36.5$ the morphology of the PWN seems to transition from toroidal to a jet/tail (Kaspi et al. 2005). Our measurements of PSR B0355+54 appear to agree with this trend.

The morphology of the diffuse emission depends on how the interaction with the interstellar medium (ISM) or supernova remnant constrains the flow of particles (e.g. Reynolds and Chevalier 1984). For a pulsar that is moving with a supersonic space velocity, the interaction of the supersonic flow with the ambient medium causes the speed of the flow to decrease sharply, while the density increases, forming a bow shock. In addition to the bow shock, which is at some distance ahead of the pulsar, a reverse shock is formed nearer to the source which terminates the pulsar wind.

The results of the spatial analysis of the XMM-Newton pn and Chandra ACIS data of PSR B0355+54 indicate that the bulk of the diffuse emission extends $\sim 50''$ [$0.25(d/1.04$ kpc) pc] downstream from the pulsar. Using the measurements of the pulsar's proper motion (Chatterjee et al. 2004) we find that the transverse velocity of PSR B0355+54 is $v_t = 61$ km s^{-1}. This implies that the time taken for the pulsar to have traversed the length of the diffuse emission is >4000 yr. In addition, by considering the analysis of the XMM-Newton MOS1 data we find that the diffuse emission may extend as far as $\sim 5'$ [$1.51(d/1.04$ kpc) pc] from the point source. This results in a travel time of >24000 yr for the pulsar. Following the work of Wang and Gotthelf (1998) (see also Kaspi et al. 2001) the synchrotron lifetime of an electron of energy E (in keV) can be defined as $t_s \sim 40 E^{-1/2} B_{-4}^{-3/2}$ yr, where $B_{-4}^{-3/2}$ is the magnetic field in units of 10^{-4} G. Assuming that the dominant loss mechanism is synchrotron emission, i.e. $B > 3.2$ μG, and that the energy of the photon is $E \sim 5$ keV, then $t_s \sim 3000$ yr. This indicates that the diffuse emission that we detect is not due to particles deposited by the pulsar as it travelled through space. Hence, there must be a constant supply of wind particles travelling at velocities greater than the space velocity of the pulsar. In addition, the particle flow velocity must be high enough such that the time for the flow to cross the length of the diffuse emission is less than the radiative lifetime of the particles.

Using the Chandra data we have modelled the spectrum of the compact diffuse emission, excluding the contribution from the pulsar, finding that the data can be well-fitted with a power-law. In other sources the power-law is seen to soften as one moves away from the pulsar position (see e.g. Slane et al. 2002; Li et al. 2005; Kaspi et al. 2005). An increase in the spectral index is expected as the particles will be cooler, i.e. older, at greater distance from the pulsar. Our results indicate that we are detecting relatively hard emission, but due to the uncertainties, we are unable to comment on any changes in the spectral slope. To measure cooling the PWN must be of an adequate size, it may be that for PSR B0355+54 the compact diffuse region is not large enough for a substantial change in power-law index to be measured, and there are too few counts in the more extended diffuse region to perform a spectral analysis. It is noted however that by comparing the spectral indices from the blackbody plus power-law fit to the core emission and the power-law fit to the compact diffuse emission we do detect an increase in Γ of ~ 0.5.

Gaensler et al. (2004) have presented a detailed analysis of the diffuse X-ray emission associated with the radio source G359.23-0.82, also known as "the Mouse". Their hydrodynamic simulations show that there are a number of regions that can be defined in a pulsar bow shock. These include a pulsar wind cavity, shocked pulsar wind material, contact discontinuity (CD) and shocked ISM.

The energetic shocked particles from the pulsar are confined by the CD, the position of which denotes the transition to the shocked ISM. Following the method of Gaensler et al. (2004) we have estimated the distance between the peak of the emission from PSR B0355+54 and the sharp cut-off in brightness ahead of the pulsar. Using the same limit as Gaensler et al. (2004) i.e. where the X-ray surface brightness falls by $1/e^2 = 0.14$, we find a distance of $0.9'' \pm 0.2''$, giving the CD a projected radius of $r_{\rm CD} = 0.004 \pm 0.001$ pc. Here, and in the following, we have used a distance to the pulsar of 1.04 kpc (Chatterjee et al. 2004). From (1) of Gaensler et al. (2004) we can estimate the radius of the forward termination shock (TS), $r_{\rm TS}^F \sim 0.003$ pc. This corresponds to an angular distance of $\theta = 0.59''$. Comparing our values to those for the Mouse implies that the emission in front of the pulsar is more compact in PSR B0355+54 than for the Mouse. In both cases the close proximity of the forward TS to the peak X-ray emission renders the TS undetectable. Using our results and (2) of Gaensler et al. (2004) we find that PSR B0355+54 produces a ram pressure of $\rho v_t^2 \sim 1.4 \times 10^{-9}$ ergs cm^{-3}. Assuming cosmic abundances and equating the inferred velocity to that measured from the proper motion, this gives $v_t \sim 247 n_0^{-1/2}$ km s^{-1}, where n_0 is the number density of the ambient medium. For PSR B0355+54 we determine $n_0 \approx 0.06$ cm^{-3}, which is not unrealistic.

Additional information can be obtained by equating the pressure of the pulsar wind (assumed isotropic), to that of the ambient medium. By introducing the Mach number $M = v_t/c_s$, where c_s is the adiabatic sound speed in the ambient medium, and using the same prescription as Gaensler et al. (2004) for a representative ISM pressure (i.e. $P_{\rm ISM} = 2400 k P_0$ erg cm^{-3}, with $0.5 \le P_0 \le 5$ and k is the Boltzmann's constant), this gives:

$$\dot{E}/[4\pi (r_{\rm TS}^F)^2 c] = 2400 k \gamma_{\rm ISM} P_0 M^2, \quad (1)$$

from which we can obtain an estimate of the Mach number. We find that for PSR B0355+54 the sound speed c_s of the medium lies in the range 1–30 km s^{-1}. The three principal phases of the ISM are generally named cold, warm and hot and are characterized by typical sound speed values of 1, 10 or 100 km s^{-1}; according to this denomination our result implies that the pulsar is moving in either a cold or mildly warm ambient gas. For comparison, in the case of the Mouse, Gaensler et al. (2004) found that the most probable pulsar velocity requires that the pulsar is moving through a warm phase of the ISM.

Gaensler et al. (2004) also discuss the possible detection of the backward TS in their data. Their simulations show that this feature has a closed surface, while the CD and bow shock are unrestricted. The backward TS should lie much further away from the pulsar than the forward TS, i.e. $r_{\rm TS}^B \gg r_{\rm TS}^F$. In principle this means that the backward TS may be detectable. The possible dip we see in the profiles for the PSR B0355+54 data could indicate the presence of the backward TS. The angular separation of the dip in our data is $\sim 10''$, a value consistent with that for the Mouse. For a backward TS, Gaensler et al. (2004) predict that there would be a lack of spectral evolution, a result we have found for the diffuse emission of PSR B0355+54. However, we note that the feature in the Mouse data (and simulations) is quite compact in the north-south direction in comparison to the PSR B0355+54 feature. In addition, the number of counts we detect for PSR B0355+54 may hinder our investigation of the presence of such a feature. Deeper observations are needed to probe further the PWN of PSR B0355+54.

Acknowledgements This work is based on observations obtained with *XMM-Newton*, an ESA science mission with instruments and contributions directly funded by ESA Member States and NASA. Support for this work was provided by the National Aeronautics and Space Administration through Chandra Award Number NNG04EF62I issued by the Chandra X-ray Observatory Center, which is operated by the Smithsonian Astrophysical Observatory for and on behalf of the National Aeronautics Space Administration under contract NAS8-03060.

References

Becker, W., Trümper, J.: Astron. Astrophys. **326**, 682 (1997)
Buccheri, R., et al.: Astron. Astrophys. **128**, 245 (1983)
Caraveo, P.A., Bignami, G.F., DeLuca, A., et al.: Science **301**, 1345 (2003)
Cash, W.: Astrophys. J. **228**, 939 (1979)
Chatterjee, S., Cordes, J.M., Vlemmings, W.H.T., et al.: Astrophys. J. **604**, 339 (2004)
de Jager, O.C.: Astrophys. J. **378**, 286 (1991)
Frail, D.A., Kulkarni, S.R.: Nature **352**, 785 (1991)
Frail, D.A., Scharringhausen, B.R.: Astrophys. J. **480**, 364 (1997)
Gaensler, B.M.: In: Holt, S.S., Hwang, U. (eds.) Young Supernova Remnants, AIP Conference Proceedings, vol. 565, p. 295. American Institute of Physics, New York (2001)
Gaensler, B.M., Stappers, B.W., Frail, D.A., et al.: Mon. Not. Roy. Astron. Soc. **318**, 58 (2000)
Gaensler, B.M., van der Swaluw, E., Camilo, F., et al.: Astrophys. J. **616**, 383 (2004)
Gotthelf, E.V.: Astrophys. J. **591**, 361 (2003)
Helfand, D.J.: In: Danzinger, J., Gorenstein, P. (eds.) IAU Symp. 101, Supernova Remnants and Their X-ray Emission, vol. 101, p. 471. Reidel, Dordrecht (1983)
Hobbs, G., Lyne, A.G., Kramer, M.: Mon. Not. Roy. Astron. Soc. **353**, 1311 (2004)
Kaspi, V.M., Gotthelf, E.V., Gaensler, B.M., et al.: Astrophys. J. **562**, 163 (2001)
Kaspi, V.M., Roberts, M.S.E., Harding, A.K.: In: Lewin, W.H.G., van der Klis, M. (eds.) Compact Stellar X-ray Sources, vol. 39, p. 279. Cambridge Univ. Press, Cambridge (2005)
Li, X.H., Lu, F.J., Li, T.P.: Astrophys. J. **628**, 931 (2005)
Manchester, R.N., Hobbs, G.B., Teoh, A., et al.: Astrophys. J. **129**, 1993 (2005)
Mardia, K.V.: Statistics of Directional Data. Academic, London (1972)
Pavlov, G.G., Shibanov, Y.A., Zavlin, V.E., et al.: In: Alpar, A., Kiliźoglu, U., van Paradijs, J. (eds.) The Lives of Neutron Stars, p. 71. Kluwer Academic, Dordrecht (1995)
Rees, M.J., Gunn, J.E.: Mon. Not. Roy. Astron. Soc. **167**, 1 (1974)
Reynolds, S.P., Chevalier, R.A.: Astrophys. J. **278**, 630 (1984)
Seward, F.D., Wang, Z.-R.: Astrophys. J. **332**, 199 (1988)
Slane, P.: Astrophys. J. **437**, 458 (1994)
Slane, P.O., Helfand, D.J., Murray, S.S.: Astrophys. J. **571**, 45 (2002)
Stappers, B.W., Gaensler, B.M., Kaspi, V.M., et al.: Science **299**, 1372 (2003)
Tepedelenlioğlu, E., Ögelman, H.: Astrophys. J. Lett. (2005, submitted), astro-ph/0512209
Wang, Q.D., Gotthelf, E.V.: Astrophys. J. **494**, 623 (1998)

Astrophys Space Sci (2007) 308: 317–323
DOI 10.1007/s10509-007-9320-3

ORIGINAL ARTICLE

New phase-coherent measurements of pulsar braking indices

Margaret A. Livingstone · Victoria M. Kaspi · Fotis P. Gavriil · Richard N. Manchester · E.V.G. Gotthelf · Lucien Kuiper

Received: 10 July 2006 / Accepted: 11 October 2006 / Published online: 20 March 2007

Abstract Pulsar braking indices offer insight into the physics that underlies pulsar spin-down. Only five braking indices have been measured via phase-coherent timing; all measured values are less than 3, the value expected from magnetic dipole radiation. Here we present new measurements for three of the five pulsar braking indices, obtained with phase-coherent timing for PSRs J1846-0258 ($n = 2.65 \pm 0.01$), B1509-58 ($n = 2.839 \pm 0.001$) and B0540-69 ($n = 2.140 \pm 0.009$). We discuss the implications of these results and possible physical explanations for them.

Keywords Pulsars · Timing

PACS 95.85.Nv · 97.60.Gb

M.A. Livingstone (✉) · V.M. Kaspi
Physics Department, McGill University, Rutherford Physics Building, 3600 University Street, Montreal, Qc H3A 2T8, Canada
e-mail: maggie@physics.mcgill.ca

F.P. Gavriil
NASA Goddard Space Flight Center Code 662,
X-ray Astrophysics Laboratory, Greenbelt, MD 20771, USA

R.N. Manchester
Australia National Telescope Facility, CSIRO, P.O. Box 76, Epping, NSW 1710, Australia

E.V.G. Gotthelf
Columbia Astrophysics Laboratory, Columbia University, 550 West 120th Street, New York, NY 10027-6601, USA

L. Kuiper
Netherlands Institute for Space Research, Sorbonnelaan 2, 3584 CA, Utrecht, Netherlands

1 Introduction

A very commonly assumed model for pulsar spin-down posits that

$$\dot{\nu} = -K\nu^n, \tag{1}$$

where ν is the pulse frequency, $\dot{\nu}$ is the frequency derivative, K is a constant and n is the braking index. The braking index is then given by

$$n = \frac{\nu\ddot{\nu}}{\dot{\nu}^2}. \tag{2}$$

The braking index provides insight into the physics that drives pulsar spin-down. Typically, it is assumed that magnetic dipole radiation underlies pulsar evolution, resulting in $n = 3$ (e.g. Manchester and Taylor 1977). However, other processes could, in principle, cause the pulsar to radiate and would result in different values for n and K. For example, a pulsar spun down entirely by the loss of relativistic particles would have $n = 1$ (Michel and Tucker 1969). A pulsar losing energy via gravitational radiation or quadrupole magnetic radiation would spin down with $n = 5$ (Blandford and Romani 1988).

Braking indices have proven difficult to measure. To date, only six have been reported even though more than 1600 pulsars are known. Evidently, the pulsar properties necessary for a measurement of n are rare; the pulsar must: spin down quickly; experience few, small, and relatively infrequent glitches; and be relatively uncontaminated by timing noise, a low-frequency stochastic process superposed on the deterministic spin-down of the pulsar. The youngest pulsars, of which the Crab pulsar is the most famous example, uniquely possess these three qualities. The reason then for the paucity of measured values of n is a direct consequence of the relative rarity of very young pulsars, i.e. those with characteristic ages on the order of 1 kyr, where characteristic age is defined as

$$\tau_c \equiv \frac{P}{2\dot{P}} = \frac{\nu}{2\dot{\nu}}. \tag{3}$$

Of the six pulsars with measured n, five were obtained via phase-coherent timing. All five of these pulsars: PSRs J1846-0258, B0531+21 (the Crab pulsar, $n = 2.51 \pm 0.01$), B1509-58, J1119-6127 ($n = 2.91 \pm 0.05$), and B0540-69, have characteristic ages less than 2 kyr (Livingstone et al. 2005a, 2005b, 2006; Lyne et al. 1993; Camilo et al. 2000). The sixth measurement, that of the Vela pulsar ($n = 1.4 \pm 0.2$), could not be obtained with phase-coherent timing due to large glitches (Lyne et al. 1996). Timing noise and large glitches begin to seriously contaminate measurements of n when pulsars have characteristic ages $\sim$5 kyr (McKenna and Lyne 1990; Marshall et al. 2004).

In this paper we report on long-term *Rossi X-ray Timing Explorer* observations of three young pulsars, PSRs J1846-0258, B1509-58 and B0540-69. We present braking index measurements for each of these pulsars obtained via phase-coherent timing.

2 Phase-coherent pulsar timing

The most accurate method of extracting pulsar timing parameters is phase-coherent timing, that is, accounting for every turn of the pulsar. Pulse times of arrival (TOAs) are measured and fitted to a Taylor expansion of pulse phase, ϕ at time t given by

$$\phi(t) = \phi(t_0) + \nu_0(t - t_0) + \frac{1}{2}\dot{\nu}_0(t - t_0)^2 + \frac{1}{6}\ddot{\nu}_0(t - t_0)^3 + \cdots, \tag{4}$$

where subscript 0 denotes a parameter at the reference epoch, t_0. TOAs and initial spin parameters are input to pulse timing software (e.g. TEMPO[1]) and refined spin parameters and timing residuals are output.

The existence of timing noise and glitches in young pulsars is well known to contaminate the measurement of deterministic spin parameters. Though powerful, a fully phase-coherent timing solution can be sensitive to these contaminants. In such cases, a partially coherent method may be employed. In this case, local phase-coherent measurements of ν, $\dot{\nu}$ and possibly $\ddot{\nu}$ are made. Though the effects of timing noise cannot be eliminated, the noise component is more readily identified and separated from the deterministic component of the spin-down, as shown in Sect. 5. In addition, glitches can be easier to identify with this method, as will be shown in Sect. 6 of this paper.

[1] http://www.atnf.csiro.au/research/pulsar/tempo.

3 Observations

In this paper we describe observations of three young pulsars PSRs J1846-0258, B1509-58, and B0540-69 taken with the Proportional Counter Array (PCA) on board the *Rossi X-ray Timing Explorer* (RXTE). The PCA consists of five collimated xenon/methane multianode proportional counter units (PCUs). The PCA operates in the 2–60 keV energy range, has an effective area of $\sim$6500 cm^2 and has a 1 degree field of view. While RXTE has no imaging capability, it has excellent time resolution of $\sim$1 μs (Jahoda et al. 1996). This makes RXTE ideal for observing young, rapidly rotating pulsars.

Observations of PSR B1509-58 were taken in "GoodXenonWithPropane" mode, while observations of the other two sources were taken in "GoodXenon" mode. Both modes record the photon arrival time with 1 μs-resolution and photon energy with 256-channel resolution. The number of PCUs active during an observation varies, but is typically three. For PSRs B1509-58 and J1846-0258, which have relatively hard spectra, all three Xenon layers and photons with energies ranging from 2–60 keV were used, while for the softer spectrum source, PSR B0540-69, only the top Xenon layer and photons with energies ranging from 2–18 keV were used. Further details of X-ray and radio observations of PSR B1509-58 are given in (Livingstone et al. 2005b) and references therein. Details of RXTE observations of PSR B0540-69 can be found in (Livingstone et al. 2005a) while details of PSR J1846-0258 observations can be found in (Livingstone et al. 2006) and references therein.

Data were reduced using standard FITS tools as well as in-house software developed for analyzing RXTE data for pulsar timing. Data from different PCUs were merged and binned at (1/1024) s resolution. Photon arrival times were corrected to barycentric dynamical time (TDB) at the solar system barycenter using the J2000 source positions and the JPL DE200 solar system ephemeris.

For PSRs B0540-69 and J1846-0258, initial ephemerides were found by performing periodograms on observations to determine values of ν. Several values of ν were fitted with a linear least squares fit to determine an initial value of $\dot{\nu}$. These initial values were then used as input to a Taylor expansion of TOAs to determine more accurate parameters (4). PSR B1509-58 has a previously determined ephemeris from radio timing data obtained with the Molonglo Observatory Synthesis Telescope and the Parkes Radio Telescope (Kaspi et al. 1994). We were able to extend that fit with 7.6 yr of RXTE data, removing a constant, but not well determined, offset between radio and X-ray TOAs.

4 PSR J1846-0258

PSR J1846-0258 is a very young pulsar ($\tau_c = 723$ yr) located at the center of the supernova remnant Kesteven 75. It has a relatively long spin period of 324 ms and a large magnetic field[2] of $B \sim 5 \times 10^{13}$ G. PSR J1846-0258 has been observed with RXTE for 6.3 yr since its discovery in 1999 (Gotthelf et al. 2000).

Using our initial ephemeris we obtained a phase-coherent timing solution valid over a 3.5 yr interval in the range MJD 51574-52837. Three spin parameters (ν, $\dot{\nu}$ and $\ddot{\nu}$) were required by the fit. In addition, we discovered a small glitch near MJD 52210±10. The fitted glitch parameters are $\Delta\nu/\nu = 2.5(2) \times 10^{-9}$ and $\Delta\dot{\nu}/\dot{\nu} = 9.3(1) \times 10^{-4}$. Note that these and all other quoted uncertainties are 68% confidence intervals, unless otherwise indicated. The wide spacing of data near the glitch prevent the detection of any short-timescale glitch recovery. Timing residuals are shown in Fig. 1. The top panel of Fig. 1 shows residuals with ν, $\dot{\nu}$, $\ddot{\nu}$ and glitch parameters fitted. The residuals clearly show systematics due to timing noise and possibly unmodeled glitch recovery. In order to minimize contamination of long-term timing parameters, we fitted additional frequency derivatives to render the residuals consistent with Gaussian distributed residuals (a process known as 'whitening' residuals. See, for example, Kaspi et al. 1994). For this timing solution, a total of eight frequency derivatives were fitted, shown in the bottom panel of Fig. 1. The braking index resulting from this 'whitened' timing solution is $n = 2.64 \pm 0.01$. Complete spin-down parameters are given in Table 1.

Phase was lost over a 78-day gap in the data near MJD 52837, indicated by the fact that a timing solution attempting to connect over this gap fails to predict the pulse frequency at previous epochs. The loss of phase is likely due to timing noise or a second glitch. However, these two possibilities cannot be distinguished due to the relatively long gap in the data set.

A second phase-coherent solution was obtained for 1.8 yr from MJD 52915-53579 with ν, $\dot{\nu}$ and $\ddot{\nu}$. Timing residuals with these three parameters fit are shown in the top panel of Fig. 2. Again, systematics due to timing noise and/or glitch recovery remain in these residuals. To 'whiten' timing residuals, five total frequency derivatives were fitted from the data, shown in the bottom panel of Fig. 2. Complete spin parameters for this timing solution are given in Table 1.

Quoting the average value of n from the two independent timing solutions, which are in agreement, gives $n = 2.65 \pm 0.01$. As is the case for all measured values of n, this value is significantly less than 3, the value consistent with spin-down via magnetic dipole radiation. This implies that some other

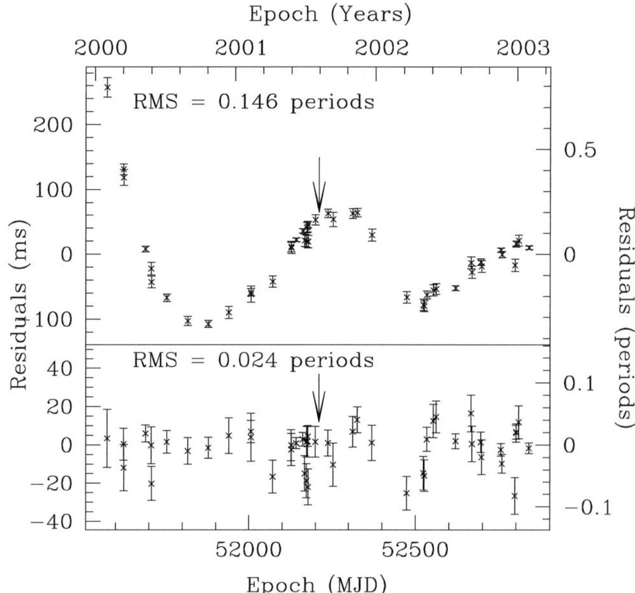

Fig. 1 Phase-coherent X-ray timing analysis of the young pulsar PSR J1846-0258 spanning a 3.5-yr interval in the range MJD 51574-52837 (after Livingstone et al. 2006). *Top panel*: Residuals with ν, $\dot{\nu}$, $\ddot{\nu}$ as well as glitch parameters $\Delta\nu$ and $\Delta\dot{\nu}$ fitted. The glitch epoch, MJD 52210 is indicated by the arrow. *Bottom panel*: Residuals with glitch parameters and eight frequency derivatives in total fitted to render the residuals consistent with Gaussian noise

Table 1 Spin parameters for PSR J1846-0258

Parameter	First solution	Second solution
Dates (MJD)	51574.2–52837.4	52915.8–53578.6
Epoch (MJD)	52064.0	53404.0
ν (s^{-1})	3.0782148166(9)	3.070458592(1)
$\dot{\nu}$ (10^{-11} s^{-2})	−6.71563(1)	−6.67793(5)
$\ddot{\nu}$ (10^{-21} s^{-3})	3.87(2)	3.89(4)
Braking Index (n)	2.64(1)	2.68(3)
Glitch epoch (MJD)	52210(10)	
$\Delta\nu/\nu$	$2.5(2) \times 10^{-9}$	
$\Delta\dot{\nu}/\dot{\nu}$	$9.3(1) \times 10^{-4}$	

physical process must contribute to the spin-down of all of these pulsars.

This measurement of n for PSR J1846-0258 increases its age estimate (Livingstone et al. 2006). The commonly known characteristic age (3) implicitly assumes that $n = 3$. A more physical estimate can be made once n is known. The age then can be estimated as

$$\tau = \frac{1}{n-1}\tau_c \leq 884 \text{ yr}. \tag{5}$$

The estimate is an upper limit since the initial spin frequency of the pulsar is not known. The upper limit approaches an equality when the pulsar is born spinning much faster than

[2] $B \equiv 3.2 \times 10^{19}(P\dot{P})^{1/2}$ G.

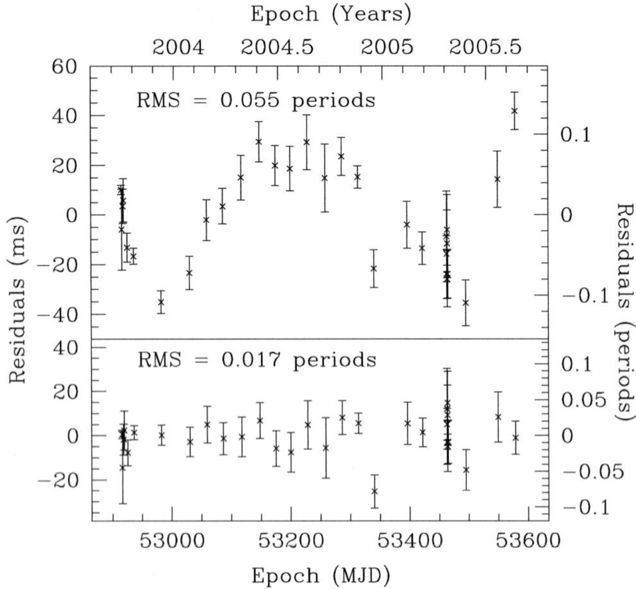

Fig. 2 Phase-coherent X-ray timing analysis of PSR J1846-0258 spanning an 1.8-yr interval in the range MJD 52915-53579 (after Livingstone et al. 2006). *Top panel*: Residuals with ν, $\dot{\nu}$, $\ddot{\nu}$ fitted. *Bottom panel*: Residuals with five frequency derivatives total fitted to render the residuals consistent with Gaussian noise

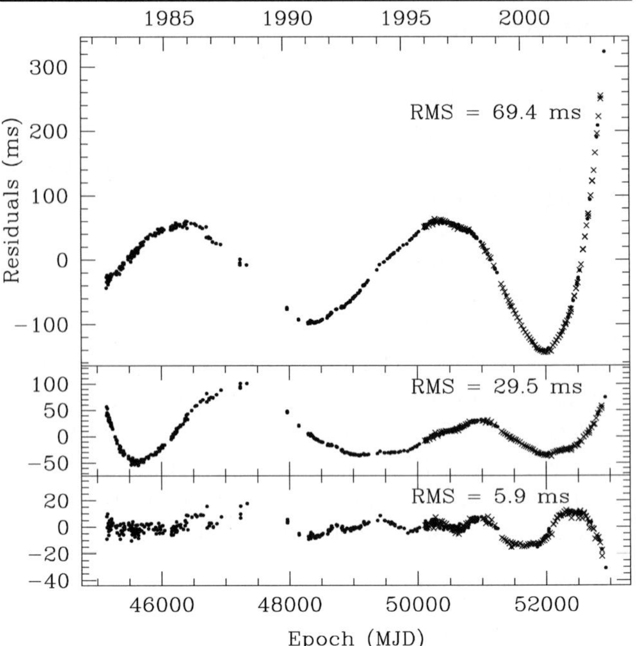

Fig. 3 Timing residuals for PSR B1509-58. Radio TOAs are shown as *dots* and the X-ray TOAs are shown as *crosses* (after Livingstone et al. 2005b). *The top panel* has pulse frequency and three frequency derivatives removed, *the middle panel* also has the fourth frequency derivative removed, and *the bottom panel* shows residuals after the removal of five frequency derivatives

its present spin frequency. Given the long period of the pulsar and the estimated initial spin period distribution (e.g. Faucher-Giguère and Kaspi 2006), the latter is likely to have occurred. This age estimate for PSR J1846-0258 is less than the known age of the Crab pulsar of 952 yr.

5 PSR B1509-58

The young pulsar PSR B1509-58 was discovered in 1982 and has been observed regularly ever since, first with radio telescopes such as the Molonglo Observatory Synthesis Telescope (Manchester et al. 1985) and the Parkes Radio Observatory (Kaspi et al. 1994), and more recently with RXTE (Rots 2004; Livingstone et al. 2005b). We phase-connected all 21.3 yr of available radio and X-ray timing data to determine the braking index. Timing residuals are shown in Fig. 3. The top panel shows the timing residuals with ν and three frequency derivatives fitted; the middle panel shows residuals with the fourth frequency derivative also fitted; the bottom panel shows timing residuals with five frequency derivatives fitted. Remarkably for such a young pulsar, *no* glitches were detected in this time period. Spin parameters from this phase-coherent analysis are $\nu = 6.633598804(3)\ \mathrm{s}^{-1}$, $\dot{\nu} = -6.75801754(4) \times 10^{-11}\ \mathrm{s}^{-2}$, $\ddot{\nu} = 1.95671(2) \times 10^{-21}\ \mathrm{s}^{-1}$ at epoch MJD 49034.5. These parameters imply a braking index of $n = 2.84209(3)$, though timing noise that could not be completely removed by fitting additional frequency derivatives contributes to a systematic uncertainty that is not included in the formal uncertainty quoted here. To solve this problem, we performed a partially phase-coherent analysis by making independent measurements of n.

Due to the large value of $\dot{\nu}$ for this pulsar, a significant measurement of n can be made in approximately 2 yr, without noticeable contamination of the measured spin parameters from timing noise. Thus, having over 20 years of data allows 10 independent measurements of n, which are shown in Fig. 4. No secular variation of n over 21.3 yr is seen, however, there is significant deviation from the average value of $n = 2.839 \pm 0.003$. This uncertainty was determined by a 'bootstrap' analysis which is a robust method of determining uncertainties when the formal uncertainties are thought to underestimate the true values, i.e. due to the presence of timing noise (Efron 1979). Note that this value is in agreement with that obtained with the fully phase-coherent timing solution, as well as the previously reported value of $n = 2.837 \pm 0.001$ (Kaspi et al. 1994). The reduced χ^2 is 15 for 9 degrees of freedom. This variation is likely due to the same timing noise process that can be observed in timing residuals. Here, the variation is at the $\sim$1.5% level. A similar analysis has been performed for PSR J1846-0258 where variations are seen to be on the order of $\sim$5%, though only at the 2σ level (Livingstone et al. 2006) and for the Crab pulsar where variations are on the order of 0.5% (Lyne et al. 1993).

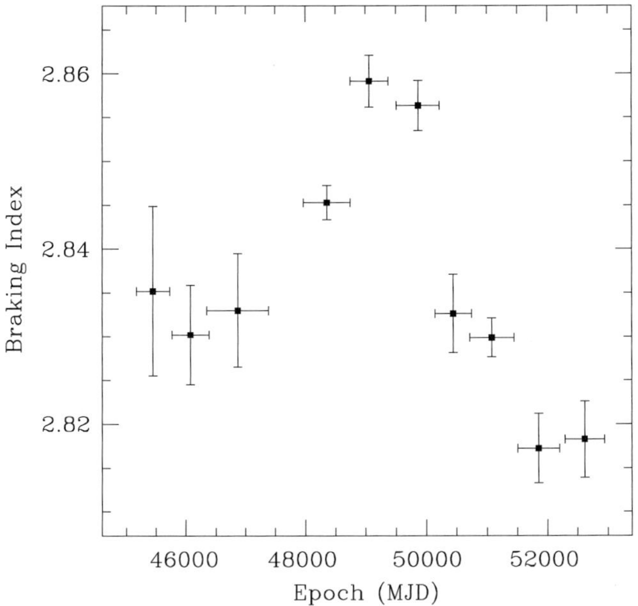

Fig. 4 Braking index calculated at 10 epochs of ~2 yr in length (after Livingstone et al. 2005b). There is no statistically significant secular change of 21.3 yr of data. The average value is 2.839 ± 0.003, in agreement with the previously reported value (Kaspi et al. 1994) and the value obtained from a phase-coherent analysis. The reduced χ^2 value is 15 for 9 degrees of freedom, suggesting contamination by timing noise

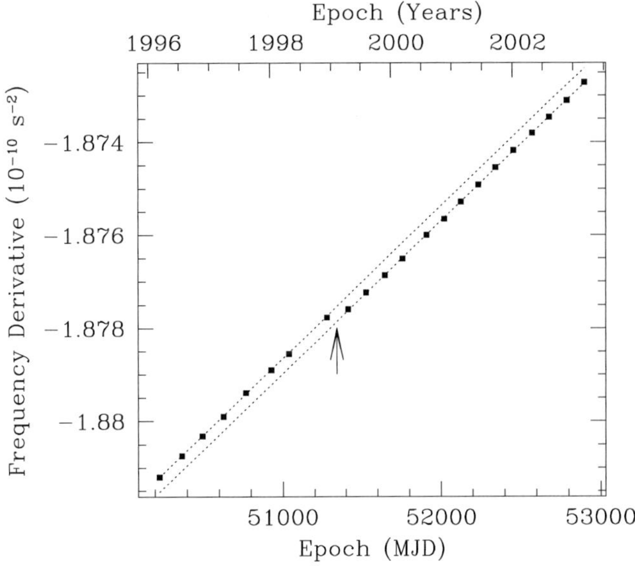

Fig. 5 Measurements of $\dot{\nu}$; the slope is $\ddot{\nu}$ (after Livingstone et al. 2005a). The glitch occurring near MJD 51342 is shown with an arrow. The pre-glitch slope is $\ddot{\nu} = 3.81(3) \times 10^{-21}$ s^{-3}, while the post-glitch slope is $\ddot{\nu} = 3.81(1) \times 10^{-21}$ s^{-3}. The average of pre- and post-glitch n is 2.140 ± 0.009. Measurement uncertainties are smaller than the points

6 PSR B0540-69

Located in the Large Magellenic Cloud, PSR B0540-69 is commonly known as the 'Crab Twin', due to its similar spin and nebular properties. For instance, its period of 50 ms and magnetic field $B \sim 5 \times 10^{12}$ G are nearer to those of the Crab pulsar ($P = 33$ ms, $B \sim 4 \times 10^{12}$ G) than for any other pulsar. Due to its large distance, PSR B0540-69 is very difficult to detect in the radio waveband (Manchester et al. 1993), hence regular radio timing of this source is not practical.

The lack of regular, long-term timing observations for this pulsar has led to conflicting values of n in the literature; reported values range from $n = 1.81 \pm 0.07$ to $n = 2.74 \pm 0.01$ (Zhang et al. 2001; Ögelman and Hasinger 1990). Widely spaced timing observations greatly increases the risk of losing phase if a phase-coherent solution is attempted. If instead of a phase-coherent timing solution, measurements of frequency are obtained over widely spaced intervals, small glitches can easily be missed and the effects of timing noise are difficult to discern. Two conflicting values of n are of particular interest since they are based on overlapping data fro RXTE.

Zhang et al. (2001) reported on 1.2 years of regular timing observations and found a small magnitude glitch at MJD 51325 ± 45 with parameters $\Delta\nu/\nu = (1.90 \pm 0.04) \times 10^{-9}$ and $\Delta\dot{\nu}/\dot{\nu} = (8.5 \pm 0.5) \times 10^{-5}$. They used the 300 days of data available after the glitch to measure a braking index of $n = 1.81 \pm 0.07$. Cusumano et al. (2003) extended the data set to 4.6 yr and reported that no glitch occurred. In contrast to the previous value, they measured $n = 2.125 \pm 0.001$.

We re-examined all previously reported RXTE data and extended the data set by 3 yr in order to resolve the discrepant timing solutions and measure the true braking index for this source. We phase connected a total of 7.6 yr of data and found a small glitch near MJD 51335 with parameters $\Delta\nu/\nu \sim 1.4 \times 10^{-9}$ and $\Delta\dot{\nu}/\dot{\nu} \sim 1.33 \times 10^{-4}$, in agreement with those reported by Zhang et al. (2001). This glitch is very small, and is most easily seen by the change in $\dot{\nu}$ at the glitch epoch. Figure 5 shows 22 measurements of $\dot{\nu}$ obtained from individual phase-coherent analyses, with the fitted glitch epoch indicated by an arrow. The slope of the line, that is, the second frequency derivative $\ddot{\nu}$, does not change significantly after the glitch (before $\ddot{\nu} = 3.81(3) \times 10^{-21}$ s^{-3}, after $\ddot{\nu} = 3.81(1) \times 10^{-21}$ s^{-3}). Uncertainties on $\ddot{\nu}$ were determined by a bootstrap analysis (Efron 1979). We use the average to determine the braking index, found to be $n = 2.140 \pm 0.009$.

In agreement with Zhang et al. (2001), we report a small glitch near MJD 51335, though our value of n is significantly larger. By phase-connecting only the same 300 day subset of data that they used to measure n, we find $n = 1.82 \pm 0.01$, in agreement with their result. The low value of n in this case appears to be the result of timing noise and/or glitch recovery contaminating the relatively short time baseline used to measure n.

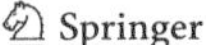

Our measured value of n is 1.7σ from that reported by Cusumano et al. (2003, $n = 2.125 \pm 0.001$) though they do not report a glitch and their uncertainty does not account for the effects of timing noise. The reason for the agreements between our measured values is that their value of n was determined by two phase-coherent fits to the data, before and after the glitch epoch reported by Zhang et al. (2001), despite the fact that Cusumano et al. (2003) report no glitch.

7 Implications and physical explanations for $n < 3$

All measured values of n are less than 3, the value consistent with spin-down due solely to magnetic dipole radiation. This implies that an additional torque is contributing to the spin-down of young pulsars. Also intriguing is the relatively wide range of measured values of n, shown in Table 2. A measurement of n immediately provides a correction to the age estimate of the pulsar given by the characteristic age, as shown with (5). Comparisons of the characteristic age and age estimated with n are also given in Table 2. Although the age estimate is always increased with a measurement of $n < 3$, it should be noted that the age estimates given are upper limits since the initial spin frequencies are not known. The calculation of magnetic fields are also affected by a measurement of $n < 3$, since these are obtained assuming pure magnetic dipole radiation. Unfortunately, there is no simple formula to estimate the correction to the dipole magnetic field as there is for the age. Specific details of the spin-down torque are required to uncover the true magnetic field of pulsars.

There are several theories that attempt to explain the measurements of $n < 3$. One explanation is that the pulsar's magnetic field grows or counter-aligns with the spin axis. This is equivalent to allowing the 'constant', K, in the simple model of pulsar spin down (1) to vary with time (Blandford and Romani 1988). An advantage of this model is that it can be tested if precision measurements of the third frequency derivative can be made (Blandford 1994). To date, the third frequency derivative has been measured only for the Crab pulsar (Lyne et al. 1993) and PSR B1509-58 (Livingstone et al. 2005b), though neither is known with sufficient precision to rule out the null hypothesis of a constant value of K. Timing noise and in the case of the Crab pulsar, glitches, may prevent a sufficiently precise measurement from ever being made.

Table 2 Braking index measurements via phase-coherent timing. Also given are the characteristic age, τ_c and age estimate using n, τ

Pulsar	n	τ_c	τ
J1846-0258	2.65(1)	723	884
B0531+21	2.51(1)	1240	1640
B1509-58	2.839(3)	1550	1690
J1119-6127	2.91(5)	1610	1680
B0540-69	2.140(9)	1670	2940

Another suggestion is that a fall-back disk formed from supernova material modulates the spin-down of young pulsars, providing a propeller torque in addition to the torque from magnetic dipole radiation. This would cause the pulsar to lose energy more quickly leading to a measured value $2 < n < 3$ (Alpar et al. 2001). A difficulty in this model is that the disk must not suppress the pulsed radio emission during the propeller phase (Menou et al. 2001).

In recent years, much work has been done on modeling the pulsar magnetosphere. Fully physical, three dimensional, time-dependent models of the pulsar magnetosphere are still some time away, however, significant progress has been made and there is some suggestion that $n < 3$ may be a natural result of a plasma filled magnetosphere (see, for example Spitkovsky 2005; Timokhin 2006; Contopoulos and Spitkovsky 2006).

The idea that plasma in the magnetosphere affects the torque acting on a pulsar is gaining acceptance with the first observational evidence for this having recently being presented. Kramer et al. (2006) show that PSR B1931+24, which has curious quasi-periodic nulling behavior, spins down at different rates when it is 'on' or 'off'. Specifically, the pulsar has a faster rate of spin down when it is observed in the radio waveband than when it goes undetected. This implies a connection between the radio emission mechanism and the spin-down torque. The interpretation presented by the authors is that the radio emission mechanism is only active when sufficient plasma is present in the magnetosphere, and that this plasma exerts a torque on the pulsar, spinning it down faster than in the absence of plasma. If this is indeed the case, then all observable pulsars should have $n < 3$.

Melatos (1997) suggested that the solution to the $n < 3$ problem is related to the angle between the spin and magnetic axes, α, as well as to currents in the magnetosphere. Melatos postulates that the magnetosphere can be considered to be split into two sections, an inner and outer magnetosphere. The division occurs at the 'vacuum' radius, the location where particles are no longer confined to field lines. The inner magnetosphere will then corotate with the neutron star and can be considered part of the radius of the rotating dipole. However, since this radius is less than, but comparable in size to, that of the light cylinder, the dipole can no longer be treated as a point, but has some finite size. As a result, $2 < n < 3$ and n approaches 3 as a pulsar ages. This model is especially attractive because it provides an explanation for the large scatter in observed values of n, and provides a prediction for n given measured values of ν, $\dot{\nu}$ and α. Given the large uncertainties on known values of α, the model roughly agrees with measurements of n for

PSRs B1509-58, B0540-69 and the Crab pulsar. PSR J1119-6127 does not appear to have a well determinable α. However, following the Melatos model, the measured value of $n = 2.91 \pm 0.05$ predicts a range of $10^\circ \leq \alpha \leq 32^\circ$ (Crawford and Keim 2003). Our recent measurement of n for PSR J1846-0258 allows a prediction of $\alpha = 8.1-9.6^\circ$ (95% confidence). At present, there is no reported radio detection of this source (Kaspi et al. 1996), however, were it one day detected, radio polarimetric observations could in principle constrain α.

8 Conclusions

The five very young pulsars with values of n measured via phase-coherent timing (Table 2) show a wide range of spin properties and behaviors. The glitch behavior exhibited by these pulsars is widely varied, ranging from PSR B1509-58, which has not glitched in 21.3 yr of continuous timing observations, to the Crab pulsar, which experiences a glitch on average, every $\sim$2 yr. The measured values of n for these five pulsars span the relatively wide range between $2.140(9) < n < 2.91(5)$. With the exception of the value of $n = 2.91 \pm 0.05$ for PSR J1119-6127, which is nearly compatible with $n = 3$, the measured values of n are significantly less than 3. The physical cause of the spin-down of pulsars remains one of the outstanding problems in pulsar astronomy.

Acknowledgements This research made use of data obtained from the High Energy Astrophysics Science Archive Research Center Online Service, provided by the NASA-Goddard Space Flight Center. The Molonglo Radio Observatory is operated by the University of Sydney. The Parkes radio telescope is part of the Australia Telescope which is funded by the Commonwealth of Australia for operation as a National Facility managed by CSIRO. M.A.L. is an NSERC PSGS-D Fellow. V.M.K. is a Canada Research Chair. F.P.G. is a NASA Postdoctoral Program (NPP) Fellow. E.V.G. acknowledges NASA ADP grant ADP04-0000-0069. Funding for this work was provided by NSERC, FQRNT, CIAR and CFI. Funding has also been provided by NASA RXTE grants over the course of this study.

References

Alpar, M.A., Ankay, A., Yazgan, E.: Astrophys. J. Lett. **557**, L61 (2001)
Blandford, R.: Mon. Not. Roy. Astron. Soc. **267**, L7 (1994)
Blandford, R.D., Romani, R.W.: Mon. Not. Roy. Astron. Soc. **234**, 57 (1988)
Camilo, F., Kaspi, V.M., Lyne, A.G., et al.: Astrophys. J. **541**, 367 (2000)
Contopoulos, I., Spitkovsky, A.: Astrophys. J. **643**, 1139 (2006)
Crawford, F., Keim, N.C.: Astrophys. J. **590**, 1020 (2003)
Cusumano, G., Massaro, E., Mineo, T.: Astron. Astrophys. **402**, 647 (2003)
Efron, B.: Ann. Stat. **7**, 1 (1979)
Faucher-Giguère, C.-A., Kaspi, V.M.: Astrophys. J. **643**, 332 (2006)
Gotthelf, E.V., Vasisht, G., Boylan-Kolchin, M., Torii, K.: Astrophys. J. **542**, L37 (2000)
Jahoda, K., Swank, J.H., Giles, A.B., et al.: Proc. SPIE **2808**, 59 (1996)
Kaspi, V.M., Manchester, R.N., Siegman, B., et al.: Astrophys. J. **422**, L83 (1994)
Kaspi, V.M., Manchester, R.N., Johnston, S., et al.: Astron. J. **111**, 2028 (1996)
Kramer, M., Lyne, A.G., O'Brien, J.T., et al.: Science **312**, 549 (2006)
Livingstone, M.A., Kaspi, V.M., Gavriil, F.P.: Astrophys. J. **633**, 1095 (2005a)
Livingstone, M.A., Kaspi, V.M., Gavriil, F.P., Manchester, R.N.: Astrophys. J. **619**, 1046 (2005b)
Livingstone, M.A., Kaspi, V.M., Gotthelf, E.V.G., Kuiper, L.: Astrophys. J. **647**, 1286 (2006)
Lyne, A.G., Pritchard, R.S., Smith, F.G.: Mon. Not. Roy. Astron. Soc. **265**, 1003 (1993)
Lyne, A.G., Pritchard, R.S., Graham-Smith, F., Camilo, F.: Nature **381**, 497 (1996)
Manchester, R.N., Durdin, J.M., Newton, L.M.: Nature **313**, 374 (1985)
Manchester, R.N., Mar, D., Lyne, A.G., et al.: Astrophys. J. **403**, L29 (1993)
Manchester, R.N., Taylor, J.H.: Pulsars. Freeman, San Francisco (1977)
Marshall, F.E., Gotthelf, E.V., Middleditch, J., et al.: Astrophys. J. **603**, 682 (2004)
McKenna, J., Lyne, A.G.: Nature **343**, 349 (1990)
Melatos, A.: Mon. Not. Roy. Astron. Soc. **288**, 1049 (1997)
Menou, K., Perna, R., Hernquist, L.: Astrophys. J. **559**, 1032 (2001)
Michel, F.C., Tucker, W.H.: Nature **223**, 277 (1969)
Ögelman, H., Hasinger, G.: Astrophys. J. Lett. **353**, L21 (1990)
Rots, A.: In: Kaaret, P., Lamb, F.K., Swank, J.H. (eds.) Proceedings of X-ray Timing 2003. Rossi and Beyond, vol. 714, p. 309. Melville (2004)
Spitkovsky, A.: In: Bulik, T., Rudak, B., Madejski, G. (eds.) Proceedings of Astrophysical Sources of High Energy Particles and Radiation, vol. 801, p. 253 (2005)
Timokhin, A.N.: Mon. Not. Roy. Astron. Soc. **368**, 1055 (2006)
Zhang, W., Marshall, F.E., Gotthelf, E.V., et al.: Astrophys. J. **554**, L177 (2001)

Astrophys Space Sci (2007) 308: 325–333
DOI 10.1007/s10509-007-9364-4

ORIGINAL ARTICLE

Thermal X-ray emission from hot polar cap in drifting subpulse pulsars

Janusz Gil · George Melikidze · Bing Zhang

Received: 24 June 2006 / Accepted: 7 September 2006 / Published online: 27 March 2007

Abstract Within the framework of the partially screened inner acceleration region the relationship between the X-ray luminosity and the circulational periodicity of drifting subpulses is derived. This relationship is quite well satisfied in pulsars for which an appropriate radio and X-ray measurements exist. A special case of PSR B0943+10 is presented and discussed. The problem of formation of a partially screened inner acceleration region for all pulsars with drifting subpulses is also considered. It is argued that an efficient inner acceleration region just above the polar cap can be formed in a very strong and curved non-dipolar surface magnetic field.

Keywords Neutron stars · Pulsars · Radio emission · X-ray

PACS 97.60.Jd · 97.60.Gb · 96.60.tg · 95.85.Nv

1 Introduction

Drifting subpulses is a puzzling phenomenon widely regarded as a powerful tool for the investigation of mechanisms of pulsar radio emission. Weltevrede et al. (2006) presented recently the results of a systematic, unbiased search for subpulse modulation in 187 pulsars and found that the fraction of pulsars showing drifting subpulse phenomenon is likely to be larger than 55% (see also these Proceedings). The authors concluded that the conditions required for drifting mechanism to work cannot be very different from the emission mechanism of radio pulsars. Weltevrede et al. (2006) suggest that the subpulse drifting phenomenon in an intrinsic property of the emission mechanism, although drifting could in some cases be very difficult or even impossible to detect due to low signal to noise ratio.

The classical vacuum gap model of Ruderman and Sutherland (1975), in which sparks-associated sub-beams of subpulse emission circulate around the magnetic axis due to $\mathbf{E} \times \mathbf{B}$ drift of spark plasma filaments, provides the most natural and plausible explanation of drifting subpulse phenomenon. However, despite its popularity, it suffers from the well known binding energy problem. Recently Gil and Mitra (2001) revisited this problem and found that the formation of vacuum gap (VG) is in principle possible, although it requires a very strong surface magnetic fields, much stronger than a dipolar components inferred from the observed spin-down rate. In other words, Gil and Mitra (2001) concluded that in general, the surface magnetic field at the polar cap must be significantly nondipolar.

Gil et al. (2003) developed further the idea of the inner acceleration region above the polar cap by including the partial screening by thermal flow from the surface heated by sparks. They succeeded to construct a self-consistent model, which they applied to a number of well known drifters. In this paper we apply the partially screened inner acceleration model to a new set of 102 pulsars from Weltevrede et al. (2006) and show that it can work in every case, provided that the non-dipolar surface magnetic field is sufficiently strong.

We acknowledge the support of the Polish State Committee for scientific research under Grant P03D 029 26. G.M. was partially supported by Georgian NSF grant ST06/4-096.

J. Gil (✉) · G. Melikidze
Institute of Astronomy, University of Zielona Góra, Lubuska 2, 65-265, Zielona Góra, Poland
e-mail: jag@astro.ia.uz.zgora.pl

G. Melikidze
Abastumani Astrophysical Observatory, Al. Kazbegi ave. 2a, 0160, Tbilisi, Georgia

B. Zhang · J. Gil
Physics University of Nevada Las Vegas, Las Vegas, NV, USA

The thermostatic regulation sets the surface temperature at the level of few MK, which is consistent with recent XMM-Newton observations of drifting subpulse pulsar B0943+10 performed and analyzed by Zhang et al. (2005).

2 Charge depleted inner acceleration region

The inner acceleration region above the polar cap results from the deviation of a local charge density ρ from the co-rotational charge density Goldreich and Julian (1969) $\rho_{\mathrm{GJ}} = -\mathbf{\Omega} \cdot \mathbf{B}_s/2\pi c \approx \pm B_s/cP$, where the positive/negative sign corresponds to $^{56}_{26}\mathrm{Fe}$ ions/electrons, for antiparallel and parallel relative orientation of magnetic and spin axes, respectively. However, due to lack of space we will discuss only the positive sign in this paper. As already mentioned in the Introduction, more and more evidence exists both observational and theoretical that the actual surface magnetic field B_s is highly non-dipolar. Its magnitude can be described in the form $B_s = bB_d$ (Gil and Sendyk 2000), where the enhancement coefficient $b > 1$ and $B_d = 2 \times 10^{12}(P\dot{P}_{-15})^{1/2}$ G is the canonical, star centered dipolar magnetic field, P is the pulsar period and $\dot{P}_{-15} = \dot{P}/10^{-15}$ is the period derivative.

The polar cap is defined as the locus of magnetic field lines that penetrate the so-called light cylinder (Goldreich and Julian 1969). In general case, the polar cap radius can be written as $r_p = b^{-0.5} 1.45 \times 10^4 P^{-0.5}$ cm (Gil and Sendyk 2000), where the factor $b^{-0.5}$ describes squeezing of the dipolar polar cap area due to the magnetic flux conservation. One should realize, however, that the polar cap radius r_p expressed by the above equation is only a characteristic dimension of the region above which a high accelerating potential drop can develop. In fact, the actual polar cap can be largely deformed from circularity in the presence of strong nondipolar surface magnetic field.

As mentioned in the Introduction, the charge depletion above the polar cap, traditionally called vacuum gap, can result from bounding of the positive $^{56}_{26}\mathrm{Fe}$ ions (at least partially) in the neutron star surface. If this is possible, then positive charges cannot be supplied at the rate that would compensate the inertial outflow through the light cylinder. As a result, a significant part of the unipolar potential drop (Goldreich and Julian 1969; Ruderman and Sutherland 1975) develops above the polar cap, which can accelerate charge particles to relativistic energies and power the pulsar radiation mechanism. The characteristic height h of such an acceleration region is determined by the mean free path of pair producing high energy photons. In other words, the growth of the accelerating potential drop is limited by the avalanche production of electron-positron plasma (e.g. Ruderman and Sutherland 1975; Cheng and Ruderman 1980). The accelerated positrons will leave the acceleration region, while the electrons will bombard the polar cap surface, causing a thermal ejection of ions. This thermal ejection will cause partial screening of the acceleration potential drop ΔV corresponding to a shielding factor $\eta = 1 - \rho_i/\rho_{\mathrm{GJ}}$ (Cheng and Ruderman 1980), where ρ_i is charge density of ejected ions,

$$\Delta V = \eta(2\pi/cP)B_s h^2 \tag{1}$$

is the potential drop and h is the height of the acceleration region. The gap potential drop is completely screened when the total charge density $\rho = \rho_i + \rho_+ = \rho_{\mathrm{GJ}}$ reaches the co-rotational value.

Two VG models can be considered under the near-threshold (NT) condition valid for strong surface magnetic field $B_s > 0.1B_q$, where $B_q = 4.414 \times 10^{13}$ G is the so-called quantum magnetic field (Ruderman and Sutherland 1975 and references therein): Curvature Radiation dominated Near Threshold Partially Screened Gap model and Inverse Compton Scattering dominated Near Threshold Partially Screened Gap model, in which the potential drop is limited by the curvature radiation (CR) and the resonant inverse Compton scattering (ICS) seed photons, respectively (for details see Gil and Mitra 2001; Gil and Melikidze 2002; Gil et al. 2003 and references therein). The latter will not be discussed in this paper due to lack of space and the former will just be called Partially Screened Gap (PSG henceforth). In the strong surface magnetic field $B_s > 5 \cdot 10^{12}$ G the high energy photons with energy $E_f = \hbar\omega$ produce electron-positron pairs at or near the kinematic threshold $\hbar\omega = 2mc^2/\sin\theta$, where $\sin\theta = l_{ph}/\mathcal{R}$, l_{ph} is the photon mean free path for pair formation and $\mathcal{R} = \mathcal{R}_6 \cdot 10^6$ cm is the radius of curvature of magnetic field lines.

Let us consider PSG model with the cascading e^-e^+ pair plasma production driven (or at least dominated) by the curvature radiation photons with typical energy

$$\hbar\omega = (3/2)\hbar\gamma^3 c/\mathcal{R} \tag{2}$$

where γ is the typical Lorentz factor of electrons/positrons moving relativistically along surface magnetic field lines with a radius of curvature $\mathcal{R}$. In the quasi-steady conditions the height h of the acceleration region is determined by the mean free path l_{ph} for pair production by energetic CR photon in strong and curved magnetic field. Following Gil and Mitra (2001) and Gil and Melikidze (2002) and including partial screening (Gil et al. 2003) we obtain

$$h \approx h^{\mathrm{CR}} = (3.1 \times 10^3)\mathcal{R}_6^{2/7}(\eta b)^{-3/7}(P/\dot{P}_{-15})^{3/14}\ \mathrm{cm}. \tag{3}$$

3 Binding energy problem

If the binding energy of $^{56}_{26}\mathrm{Fe}$ ions is large enough to prevent thermionic emission, a charge depleted acceleration region can form just above the polar cap. Normally, at the solid-vacuum interface, the charge density of outflowing ions is roughly comparable with density of the solid at the surface temperature $kT_s = \varepsilon_c$, where ε_c is the cohesive (binding) energy and $k = 8.6 \times 10^{-8}$ keV K^{-1} is the Boltzman constant. However, in case of pulsars, only the corotational charge density ρ_{GJ} can be reached, and the $^{56}_{24}$Fe ion number density corresponding to ρ_{GJ} is about $\exp(-30)$ times lower than in the neutron star crust. Since the density of outflowing ions ρ_i decreases in proposition to $\exp(-\varepsilon_c/kT_s)$, then one can write $\rho_i/\rho_{GJ} \approx \exp(30-\varepsilon/kT_s)$. At the critical temperature $T_i = \varepsilon_c/30k$ the ion outflow reaches the maximum value $\rho = \rho_{GJ}$, and the numerical coefficient 30 is determined from the tail of the exponential function with an accuracy of about 10%.

Calculations of binding energies are difficult and uncertain (see Lai 2001, for review and critical discussion). In this paper we will use the results of Jones (1986), which were recommended by Lai (2001) in his review paper. Jones (1986) obtained $\varepsilon_c = 0.29, 0.60$ and 0.92 keV for $B_s = 2, 5$ and 10×10^{12} G, respectively. These values can be approximately represented by the function $\varepsilon_c \simeq (0.18 \text{ keV})(B_s/10^{12})^{0.7}$ G and converted into critical ion temperature

$$T_i \simeq (0.7 \times 10^5 \text{ K})(B_s/10^{12} \text{ G})^{0.7} \simeq (1.2 \times 10^5 \text{ K}) b^{0.7} (P\dot{P}_{-15})^{0.36} \simeq (10^6 \text{ K})(B_s/B_q)^{0.7}, \quad (4)$$

above which the thermionic ion flow reaches the maximum GJ density, thus screening completely any acceleration potential drop above the polar cap, where b, B_s and B_q are explained earlier in the paper. Below this temperature the charge-depleted acceleration region will form, that should most likely discharge in a quasi-steady manner by a number of sparks, as originally proposed by Ruderman and Sutherland (1975) (see more detailed discussion in Gil et al. 2003). The electron–positron plasma produced by sparking discharges co-exists with thermally ejected ions, whose charge density can be characterized by the shielding factor in the form $\eta = 1 - \exp[30(1 - T_i/T_s)]$. At the temperature $T_i = T_s$ the shielding factor $\eta = 0$ (corresponding to fully developed space–charge limited flow with $\rho_i = \rho_{GJ}$), but even a very small drop of T_s below T_i, much smaller than 10%, corresponds to creation of the pure vacuum gap with $\eta = 1$ ($\rho_i = 0$). Thus, the condition for PSG can be written in the form $T_s \lesssim T_i$, meaning that the actual surface temperature T_s should be slightly lower (few percent) than the critical ion temperature T_i, which for a given pulsar is determined purely by the surface magnetic field B_s. Although the pure vacuum gap of Ruderman and Sutherland (1975) is inconsistent with X-ray observations of drifting subpulse pulsar B0943+10 (see Zhang et al. 2005), the condition for existence of the inner acceleration region can be used in a practical form

$$T_s = T_i, \quad (5)$$

remembering that in reality T_s cannot be exactly equal to T_i but slightly (few percent) lower. We introduced the parameter B_s/B_q for a convenience of presenting results in a graphical form (Fig. 1).

4 Thermostatic regulation of surface temperature

Following the original suggestion by Cheng and Ruderman (1980), Gil et al. (2003) argued that because of the exponential sensitivity of the accelerating potential drop ΔV to the surface temperature T_s, the actual potential drop should be thermostatically regulated at a temperature close to the critical temperature T_i or. In fact, when ΔV is large enough to ignite the avalanche pair production, then the backflow of relativistic charges will deposit their kinetic energy in the polar cap surface and heat it at a predictable rate. This heating will induce thermionic emission from the surface, which will in turn decrease the potential drop that caused the thermionic emission in the first place. As a result of these two oppositely directed tendencies, the quasi-equilibrium state should be established, in which heating due to electron bombardment is balanced by cooling due to thermal radiation (see Gil et al. 2003 for more details). This should occur at a temperature T_s slightly lower than T_i.

The quasi-equilibrium condition is $Q_{\text{cool}} = Q_{\text{heat}}$, where $Q_{\text{cool}} = \sigma T_s^4$ is a cooling power surface density by thermal radiation from the polar cap surface and $Q_{\text{heat}} = \gamma m_e c^3 n$ is heating power surface density due to back-flow bombardment, $\gamma = e\Delta V/m_e c^2$ is the Lorentz factor and ΔV is the accelerating potential drop, $n = n_{GJ} - n_i = \eta n_{GJ}$ is the charge number density of back-flow particles that actually heat the polar cap surface, η is the shielding factor, n_i is the charge number density of thermally ejected flow and $n_{GJ} = \rho_{GJ}/e = 1.4 \times 10^{11} b \dot{P}_{-15}^{0.5} P^{-0.5}$ cm^{-3} is the corotational charge number density. It is straightforward to obtain an expression for the quasi-equilibrium surface temperature in the form

$$T_s = (1.98 \times 10^6 \text{ K})(\dot{P}_{-15}/P)^{1/4} \eta^{1/2} b^{1/2} h_3^{1/2} \quad (6)$$

where $h_3 = h/10^3$ cm is the normalized height of the acceleration region (3). As a result of thermostatic regulation described above, this temperature should be very close to but slightly lower than the critical ion T_i or electron T_e temperature (4), (5), (6)).

The X-ray thermal luminosity L_x can be written as

$$L_x = \sigma T_s^4 \pi r_p^2 = 1.2 \times 10^{32} (\dot{P}_{-15}/P^3)(\eta h/r_p)^2 \text{ erg/s} \quad (7)$$

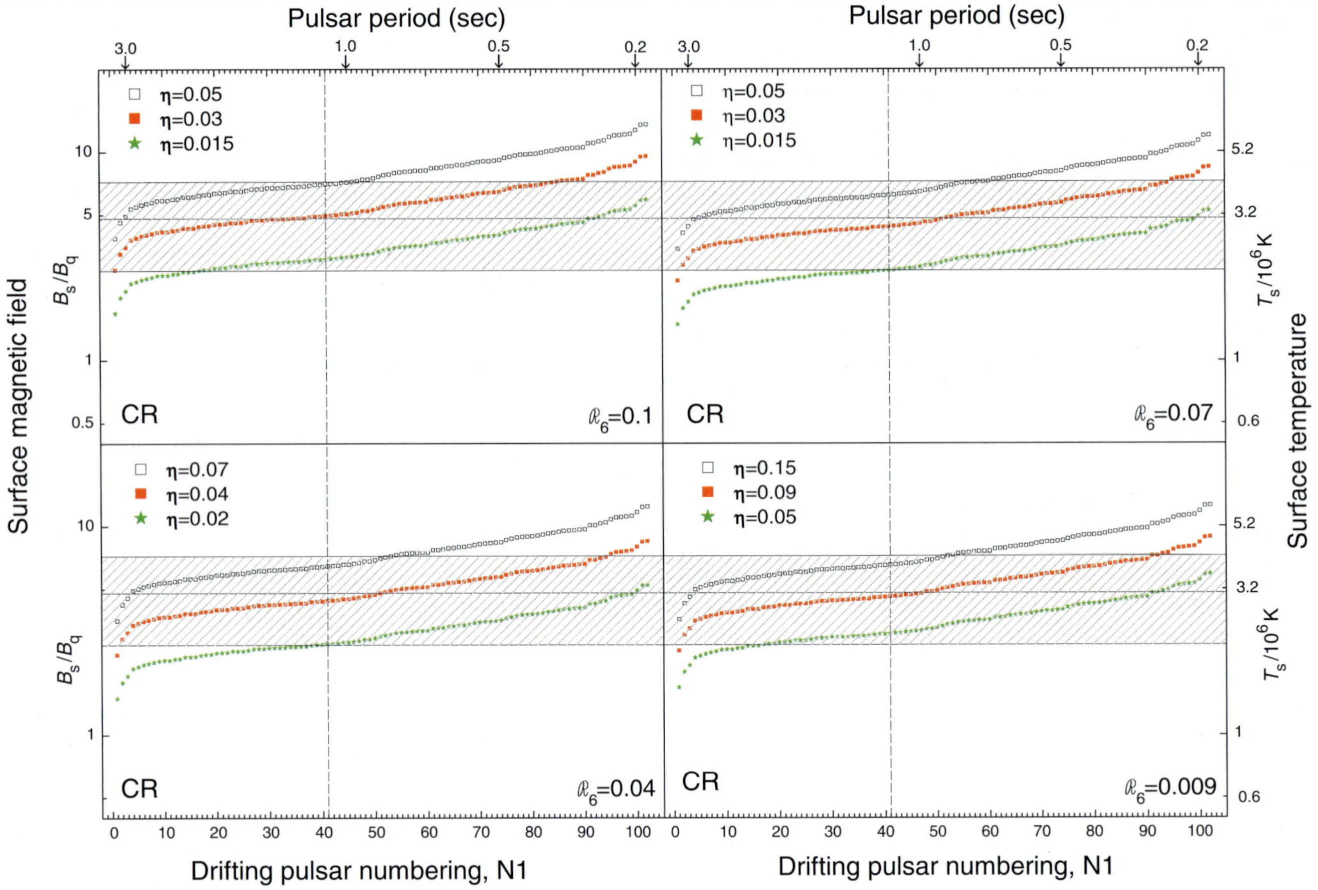

Fig. 1 Models of partially screened inner acceleration regions (PSG) driven by curvature radiation seed photons above positively charged (ion case) polar cap for 102 pulsars from Table 1, sorted according to pulsar period. The horizontal axes correspond to the pulsar number N1 (*bottom*) or pulsar period (*top*). The vertical axes correspond to the surface magnetic field B_s/B_q (*left-hand side*), or surface temperature $T_s/10^6$ K (*right-hand side*). The calculations were made for conditions corresponding to very strong ($B_s > 5\times10^{12}$ G) and curved ($0.1 > \mathcal{R}_6 > 0.005$) surface magnetic field. The vertical lines correspond to the case of PSR B0943+10 (N1 = 41) and the hatched area encompassing three horizontal lines correspond to the range of surface magnetic field and temperature inferred for this pulsar from the XMM-Newton observation (Zhang et al. 2005). These models allow to read off the physical conditions existing in the acceleration region above the polar cap of a particular pulsar in the form of parameters such as B_s, $\mathcal{R}_6$ and η

where $r_p = 1.45\times10^4 b^{-1/2}P^{-1/2}$ cm is the actual polar cap radius. This X-ray luminosity can be compared with the spin-down power

$$\dot{E} = I\Omega\dot{\Omega} = 3.95\times10^{31}\,\dot{P}_{-15}/P^3 \text{ erg/s} \tag{8}$$

where $I = 10^{45}$ g cm^2. For the pulsar in which the tertiary circulational periodicity of drifting subpulses $\hat{P}_3 = NP_3$ (Ruderman and Sutherland 1975) is known, where P_3 is a basic subpulse drift periodicity and N is a number of circulating sub-beams (Deshpande and Rankin 1999, 2001; Gil and Sendyk 2003), one can derive a relationship between observable and physical parameters of PSG model in the form

$$\eta h/r_p \approx P/\hat{P}_3 \tag{9}$$

(Gil et al. 2003; Zhang et al. 2005). We can now write the equation for thermal X-ray luminosity from the heated polar cap

$$L_x \approx 2.5\times10^{31}\left(\dot{P}_{-15}/P^3\right)\left(\hat{P}_3/P\right)^{-2} \tag{10}$$

or for the efficiency

$$L_x/\dot{E} = 0.63\left(\hat{P}_3/P\right)^{-2} \tag{11}$$

where $\dot{E}$ is defined by (8). It is particularly interesting and important that this equation does not depend on details of the sparking gap model (η, b, h), which results from 9 relating physical parameters of the acceleration region with the observational properties of drifting subpulses. Thus, (10) and (11) describe a very simple relationship between properties of spark-associated drifting subpulses observed in radio band and characteristics of X-ray thermal radiation from the polar cap heated by sparks.

Table 1 List of 102 pulsars with drifting subpulses compiled from Weltevrede et al. (2006)

Name	N	Name	N	Name	N	Name	N	Name	N
B0011+47	31	B0751+32	22	B1642-03	83	J1901-0906	12	B2043-04	18
B0031-07	49	B0809+74	27	J1650-1654	13	B1911-04	52	B2021+51	72
B0037+56	40	J0815+0939	63	B1702-19	91	B1914+13	93	B2044+15	38
B0052+51	6	B0818-13	32	B1717-29	64	J1916+0748	70	B2045-16	8
B0136+57	94	B0820+02	51	B1737+13	53	B1917+00	29	B2053+36	99
B0138+59	33	B0823+26	71	B1738-08	7	B1919+21	26	B2106+44	81
B0144+59	100	B0826-34	10	B1753+52	4	B1923+04	43	B2110+27	35
B0148-06	20	B0834+06	28	B1804-08	102	B1924+16	68	B2111+46	46
B0301+19	23	B0919+06	78	B1818-04	66	B1929+10	98	B2148+63	84
B0320+39	3	B0940+16	42	B1819-22	9	B1933+16	86	B2154+40	19
B0329+54	59	B0943+10	41	B1822-09	54	B1937-26	82	B2255+58	85
B0450+55	90	B1039-19	24	J1830-1135	1	B1942-00	44	B2303+30	17
B0523+11	87	B1112+50	14	B1839+56	15	B1944+17	77	B2310+42	88
B0525+21	2	B1133+16	36	B1839-04	11	B1946+35	58	B2319+60	5
B0540+23	95	B1237+25	25	B1841-04	47	B1952+29	79	B2324+60	96
B0609+37	92	B1508+55	56	B1844-04	67	B1953+50	73	B2327-20	16
B0621-04	45	B1530+27	39	B1845-01	61	B2000+40	50	J2346-0609	37
B0626+24	75	B1540-06	60	B1846-06	21	J2007+0912	76	B2351+61	48
B0628-28	30	B1541+09	55	B1857-26	65	B2011+38	97		
B0727-18	74	B1604-00	80	B1859+03	62	B2016+28	69		
B0740-28	101	B1612+07	34	B1900+01	57	B2020+28	89		

The comparison of theoretical predictions and observational data is summarized in Table 2. For PSR B0943+10 with $P = 1.097$ s and $\dot{P}_{-15} = 3.72$, which is the only pulsar for which both $\hat{P}_3 = 37.4$ P (Deshpande and Rankin 1999, 2001) and L_x (Zhang et al. 2005) is measured, the above equation is well satisfied. Naturally, the efficiency $L_x/\dot{E} = 4.5 \times 10^{-4}$ is also satisfied in PSR B0943+10, for which the observational value $L_x/\dot{E} = 5 \times 10^{-4}$ (Zhang et al. 2005). It should be mentioned that the thermal fit suggests that the emitting area is about $1^{+4}_{-0.4} \times 10^7$ cm^2. Even the largest area allowed by the fit is much smaller than the conventional polar cap area for a dipole magnetic field 6×10^8 cm^2 (Zhang et al. 2005). Consequently $b \sim 60$ for this pulsar.

PSR B1133+16 with $P = 1.19$ s, $\dot{P}_{-15} = 3.7$, and $\dot{E} = 9 \times 10^{31}$ erg s^{-1} is almost a twin of PSR B0943+10. Kargaltsev et al. (2006) observed this pulsar with Chandra and found an acceptable BB fit $L_x/\dot{E} = (0.77^{+0.13}_{-0.15}) \times 10^{-3}$, $A_{\rm bol} = (0.5^{+0.5}_{-0.3}) \times 10^7$ cm^2 and $T_s \approx 2.8 \times 10^6$ K. These values are also very close to those of PSR B0943+10, as should be expected for twins. Using (10) we can predict $\hat{P}_3/P = 27^{+5}_{-2}$ for B1133+16. Interestingly, Nowakowski (1996) obtained fluctuation spectrum for this pulsar with clearly detected long period feature corresponding to about $32P$. Most recently, Weltevrede et al. (2006) found $P_3/P = 3 \pm 2$ and long period feature corresponding to $(33 \pm 3)P$ in the fluctuation spectrum of PSR B1133+16. The latter value seem to coincide with that of Nowakowski (1996), as well as with our predicted range of $\hat{P}_3$. We therefore claim that this is the actual tertiary periodicity in PSR B1133+16. The two above pulsars are summarized in Table 2.

PSR B0826-34 with $P = 1.85$ s, $\dot{P}_{-15} = 1.0$, $\dot{E} = 6 \times 10^{30}$ erg s^{-1}. This pulsar has $P_3 \approx 1P$ (highly aliased) and $\hat{P}_3 \approx 14P$, i.e. $N = 14$ (Gupta et al. 2004). With these values we can predict from (10) that $L_x = 2 \times 10^{28}$ erg s^{-1}. This pulsar should be as bright as PSR B0943+10 and thus it is worth observing with *XMM-Newton*. The shielding parameter $\eta \approx 0.16$.

PSR B0834+06 with $P = 1.27$ s, $\dot{P}_{-15} = 6.8$, and $\dot{E} = 1.3 \times 10^{32}$ erg s^{-1}. This pulsar has $\hat{P}_3 = 15P$, and $P_3 = 2.16$ P (Asgekar and Deshpande 2001), implying the number of sparks $N \approx 15/2.16 \approx 7$. From (10) we obtain $L_x = 37 \times 10^{28}$ erg s^{-1}. This pulsar should be almost 8 times more luminous than PSR B0943+10, and we strongly recommend to observe it with *XMM-Newton*. The shielding parameter $\eta \approx 0.07$.

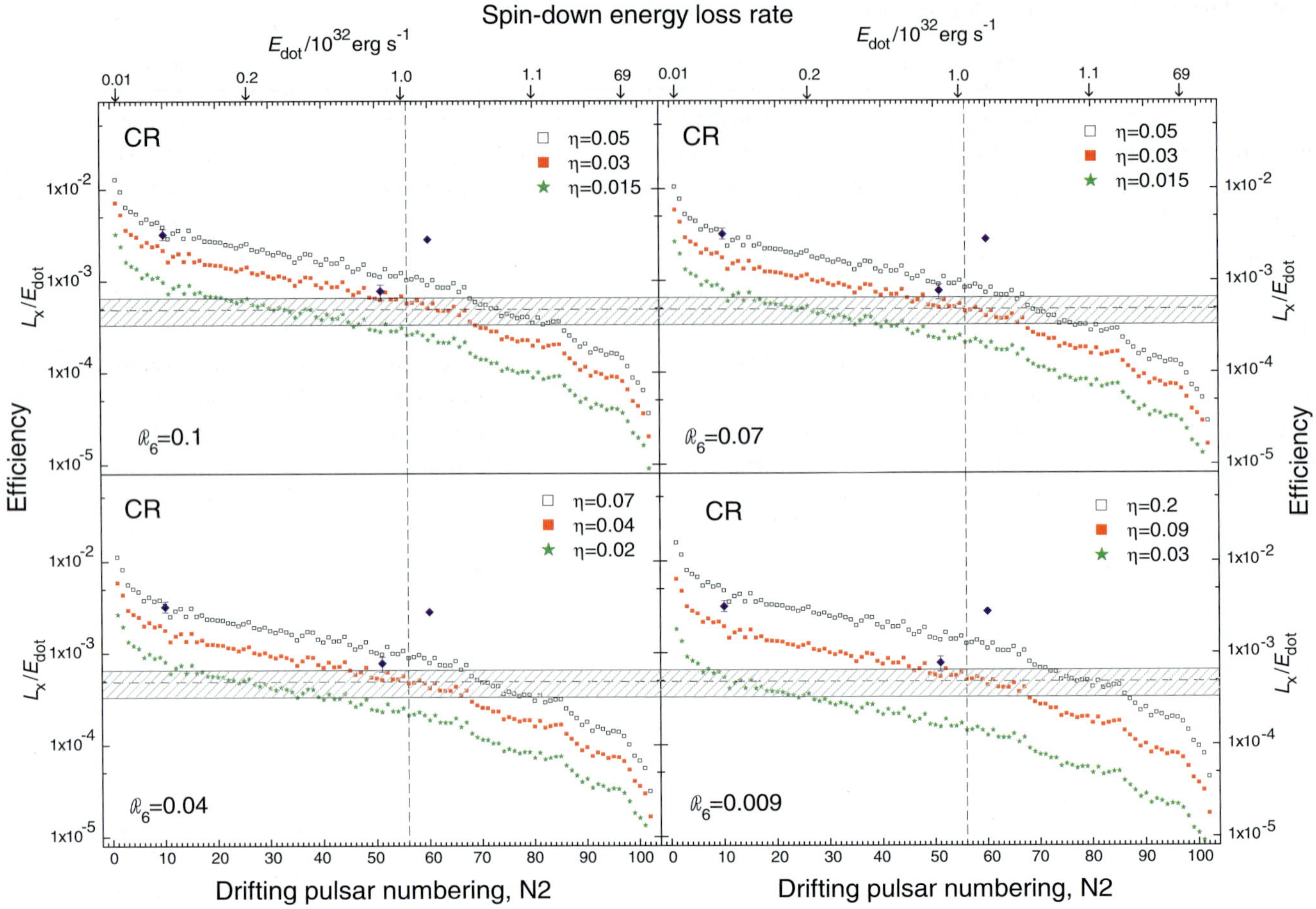

Fig. 2 The thermal radiation efficiency from the hot polar cap for 102 pulsars from Table 1 for different parameters of the PSG marked in the legends. Pulsars are sorted according to the pulsar spin-down luminosity (N2 in Table 1). For clarity of presentation the surface field is fixed at $B_s/B_q = 4.8$. The *vertical dashed lines* (N2 = 56) and *three horizontal dashed lines* ($L_x = 5 \pm 2 \times 10^{28}$ erg/s) within the hatched area correspond to the special case of PSR B0943+10 (Tables 1 and 2). Three other pulsars for which values of $\hat{P}_3/P$ is directly determined from single pulse radio data are included as diamond symbols: B0826-34 (N2 = 10), B1133+16 (N2 = 51) and B0834+06 (N2 = 60)

5 PSG models for drifting subpulse pulsars

The results of model calculations for 102 pulsars with drifting subpulses (see Table 1) are presented in Fig. 1. Both panels correspond to quasi-steady discharge of the accelerating potential drop by cascading production of the electron–positron pairs driven (or at least dominated) by the curvature radiation seed photons. Pulsars were sorted according to decreasing value of the pulsar period P. We have then plotted B_s/B_q (left-hand side vertical axes) versus the pulsar number (which corresponds to a particular pulsar according to Table 1), where $B_s = bB_d$. The actual value of $B_s/B_q = 0.045b(P \cdot \dot{P}_{-15})^{0.5}$ for a given pulsar was computed from the condition $T_s = T_i$, where T_s is the actual surface temperature. The vertical axes on the right hand side are expressed in terms of the surface temperature T_s computed from (4) for the upper panel (ion case) and lower panel (electron case), respectively. Different plots correspond to different normalized radius of curvature ranging from $\mathcal{R}_6 = 1$ to $\mathcal{R}_6 = 0.01$. Different symbols (described in the left upper panel) used to plot exemplary curves correspond to an arbitrarily chosen values of shielding factor $\eta = 1, 0.1, 0.05$ and 0.01, respectively (each curve thus represents the same shielding conditions η for all 102 pulsars considered). Note that the curve corresponding to pure vacuum gap ($\eta = 1$) does not fit into any panel for B_s/B_q below 10, or T_s below 5 MK. This means that in realistic pulsars the shielding parameter $\eta \ll 1$. As for the surface magnetic field value $B_s \sim B_q$, which suggests that $b \gg 1$ for almost all cases. The observations support such an assumption (Zhang et al. 2005).

The vertical lines in Fig. 1 correspond to PSR B0943+10 (number 41), which was observed using XMM-Newton by Zhang et al. (2005). Three horizontal lines correspond to T_s equal to 2, 3 and 4 MK (from bottom to the top), respectively, calculated from (4) for the ion case. Thus the hatched area encompassing these lines corresponds to the range of surface temperatures $T_s = (3 \pm 1) \times 10^6$ K deduced ob-

Table 2 Comparison of theoretical and observational data

Name	$\hat{P}_3/P$		$L_x/\dot{E} \times 10^{-3}$		$L_x \times 10^{28}$		b	$T_s^{(\mathrm{obs})}$	$T_s^{(\mathrm{pred})}$	B_d	B_s
PSR B	Obs.	Pred.	Obs.	Pred.	Obs.	Pred.	$A_{\mathrm{pc}}/A_{\mathrm{bol}}$	10^6 K	10^6 K	10^{12} G	10^{14} G
0943+10	37.4	36	$0.49^{+0.06}_{-0.16}$	0.45	$5.1^{+0.6}_{-1.7}$	4.7	60^{+140}_{-48}	$3.1^{+0.9}_{-1.1}$	$3.3^{+1.2}_{-1.1}$	3.95	$2.37^{+5.53}_{-1.90}$
1133+16	(33^{+3}_{-3})	27^{+5}_{-2}	$0.77^{+0.13}_{-0.15}$	$0.58^{+0.12}_{-0.09}$	$6.8^{+1.1}_{-1.3}$	$5.1^{+1.0}_{-0.8}$	$11.1^{+16.6}_{-5.6}$	$2.8^{+1.2}_{-1.2}$	$2.1^{+0.5}_{-0.4}$	4.25	$0.47^{+0.71}_{-0.24}$
0826-34	14^{+1}_{-1}			$3.2^{+0.5}_{-0.4}$		$2.0^{+0.33}_{-0.25}$				2.74	
0834+06	15			2.8		37				5.94	

servationally for the polar cap of B0943+10 from XMM-Newton observations (Zhang et al. 2005, see their Fig. 1). Although, for a given pulsar the charge depleted acceleration region can form at any B_s allowed by the condition $T_s = T_i$, we know that in B0943+10 the surface temperature T_s is about 3 MK (Zhang et al. 2005), which implies $B_s = bB_d \sim 2 \times 10^{14}$ G.

Analysis of Fig. 1 shows that in the CR dominated ion case, a partially screened charge depleted acceleration region (PSG) can form for all pulsars from Table 1. As one can see, the shielding factor η should be much smaller than 1.0 (perhaps even well below 0.1), otherwise the required surface magnetic field would be unrealistically high (exceeding 10^{15} G). This means that more than 90% of the vacuum (Ruderman and Sutherland 1975) potential drop above the heated polar cap is shielded due to thermionic emission. Also, the larger values of the radius of curvature $\mathcal{R}_6$ correspond to smaller values of the shielding factor η. The values $\mathcal{R}_6 \gtrsim 0.5$ seem disfavored. Realistically, the required surface magnetic field extends from about $0.1B_q$ to about $10B_q$, which is the range used on the vertical scale. Although the upper range values exceeding 10^{14} G may seem very high, one should realize that there are at least three pulsars which have dipolar magnetic fields above 10^{14} G (McLaughlin et al. 2003, and references therein) and the non-dipolar surface fields could be even stronger.

Figure 2 present calculations of the temperature T_s and the efficiency $L_x/\dot{E}$ using (6) and (7), respectively, for 102 pulsars from Table 1. One should emphasize that these calculations did not assume a priori an existence of the PSG in form of the condition $T_s = T_c$ (5). Thus, these figures should be considered in association with Fig. 1, which does take into account this condition. The vertical solid lines correspond to PSR B0943+10 for which $T_s = (3 \pm 1) \times 10^6$ K (corresponding to $B_s/B_q = 4.8$ in Fig. 1) and $L_x/\dot{E} = (5 \pm 2) \times 10^{-4}$ (Zhang et al. 2005). Due to weak dependence of the efficiency on the magnetic field we decided to use only one value of $B_s/B_q = 4.8$ in Fig. 2 in order to avoid overlapping of too many lines. The observational range of $L_x/\dot{E}$ and T_s are marked by the hatched belts. We have sorted pulsars in according to increasing value of $\dot{E}$, which are marked on the top of the figure (N2 in Table 1). If $L_x/\dot{E} \sim 10^{-3} \div 10^{-4}$ for all pulsars, then one can say that the shielding factor η should be larger for pulsars with larger $\dot{E}$. Also, it follows from Fig. 4 that if $T_s = (2 \div 4) \times 10^6$ K in all pulsars, then the shielding factor η should be larger for longer period pulsars.

6 Discussion and conclusions

The phenomenon of drifting subpulses has been widely regarded as a powerful tool for understanding the mechanism of coherent pulsar radio emission. Ruderman and Sutherland (1975) first proposed that drifting subpulses are related to $\mathbf{E} \times \mathbf{B}$ drifting sparks discharging the high potential drop within the inner acceleration region above the polar cap. The subpulse-associated streams of secondary electron–positron plasma created by sparks were penetrated by much faster primary beam. This system was supposed to undergo a two-stream instability, which should lead to generation of the coherent radiation at radio wave lengths. In this paper we applied the partially screened gap model proposed by Gil et al. (2003) to PSR B0943+10, a famous drifter for which Zhang et al. (2005) successfully attempted to measure the thermal X-ray from hot polar cap. We derived a simple relationship between the X-ray luminosity L_x from the polar cap heated by sparks and the tertiary periodicity $\hat{P}_3$ of the spark-associated subpulse drift observed in radio band. This relationship reflects the fact that both the drifting (radio) and the heating (X-rays) rates are determined by the same value of the electric field in the partially screened gap. As a consequence of this coupling (10) and (11) are independent of details of the acceleration region. In PSRs B0943+10 and B1133+16, the only two pulsars for which both L_x and $\hat{P}_3$ are measured, the predicted relationship between observational quantities holds very well. We note that $\hat{P}_3$ is very difficult to measure and its value is known only for four pulsars: B0943+10, B1133+16, B0826-34 and B0834+06. The successful confrontation of the predicted X-ray luminosity with the observations in PSRs B0943+10 and B1133+16 encourages further tests of the model with future X-ray observations of other drifting pulsars. This is particularly relevant to PSR B0834+06, whose predicted X-ray luminosity is much higher than in PSR B0943+10, while the distance to both pulsars is almost identical. It is worth mentioning that due

to relatively poor photon statistics it is still not absolutely clear whether the X-ray radiation associated with polar cap of PSR B0943+10 is thermal or magnetospheric (or both) in origin. Observations of PSR B0834+06 with *XMM-Newton* should help to resolve this question ultimately.

In both the steady SCLF model and the pure vacuum gap model, the potential increases with height quadratically at lower altitudes. However the growth rate is significantly different—the latter is faster by a factor of R/r_p. It is well known that in the pure vacuum case (Ruderman and Sutherland 1975), the potential grows so fast that a primary particle could quickly generate pairs with a high multiplicity, and that some backward returning electrons generate more pairs and soon a "pair avalanche" occurs and the potential is short out by a spark. In the PSG model we are advocating, the potential drops by a factor of η. For essentially all the cases we are discussing, this η value is much larger than the r_p/R value required for the steady SCLF to operate, although it is much less than unity. The steady state condition is not satisfied, and the gap is more analogous to a vacuum gap, i.e. the pair discharging happening intermittently. In fact, the partially screened potential drop is still above the threshold for the magnetic pair production, which in strong and curved surface magnetic field is a condition necessary and sufficient for the sparking breakdown.

The original Ruderman and Sutherland (1975) pure VG model predicts much too high a subpulse drift rate and an X-ray luminosity to explain the case of PSR B0943+10 and other similar cases. Other available acceleration models predict too low a luminosity and the explanation of drifting subpulse phenomenon is generally not clear at all (see Zhang et al. 2005 for more detailed discussion). In summary, the bolometric X-ray luminosity for the space charge limited flow (Arons and Scharlemann 1979) pure vacuum gap (Ruderman and Sutherland 1975) and partially screened gap (Gil et al. 2003) is $(10^{-4}-10^{-5})\dot{E}$ (Harding and Muslimov 2002), $(10^{-1}-10^{-2})\dot{E}$ (Zhang et al. 2005), and $\sim 10^{-3}\dot{E}$ (this paper), respectively. The latter model also predicts right $\mathbf{E}\times\mathbf{B}$ plasma drift rate. Thus, combined radio and X-ray data are consistent only with the partially screened VG model, which requires very strong (generally non-dipolar) surface magnetic fields. Observations of the hot-spot thermal radiation almost always indicate bolometric polar cap radius much smaller than the canonical Goldreich–Julian value. Most probably such a significant reduction of the polar cap size is caused by the flux conservation of the non-dipolar surface magnetic fields connecting with the open dipolar magnetic field lines at distances much larger than the neutron star radius.

Our analysis suggests the following pulsar picture: In the strong magnetic fields approaching 10^{14} G at the neutron star surface, the binding energy is high enough to prevent a full thermionic flow from the hot polar cap at the corotation limited level. A partially screened vacuum gap develops with the acceleration potential drop exceeding the threshold for the magnetic pair formation. The growth of this potential drop should be naturally limited by a number of isolated electron–positron spark discharges. As a consequence, the polar cap surface is heated by back-flowing particles to temperatures $T_s \sim 10^6$ K, just below the critical temperature T_c at which the thermionic flow screens the gap completely. The typical radii of curvature of the field lines $\mathcal{R}$ is of the order of polar cap radii $r_p \sim 10^3$–10^4 cm. The only parameter that is thermostatically adjusted in a given pulsar is the shielding parameter $\eta = 10^{-3}(B_s/B_q)\mathcal{R}_6^{-0.5}P \sim 0.001T_6^{1.43}\mathcal{R}_6^{-0.5}P$, which determines the actual level of charge depletion with respect to the pure vacuum case ($\eta = 1$), and in consequence the polar cap heating rate as well as the spark drifting rate. It is worth to emphasize that $\eta \sim 0.1$ for longer period pulsars $P \sim 1$ s and $\eta \sim 0.01$ for shorter period pulsars. Our calculations are consistent with PSR B0943+10 and few other drifting pulsars, for which the signatures of X-ray emission from the hot polar cap were detected (Table 2).

The sparks operating at the polar cap generate streams of secondary electron–positron plasma flowing through the magnetosphere. These streams are likely to generate beams of coherent radio emission that can be observed in the form of subpulses. Usov (1987) first pointed out that the non-stationarity associated with sparking discharges naturally leads to a two-stream instability as the result of mutual penetration between the slower and the faster plasma components. Asseo and Melikidze (1998) developed this idea further, calculated the growth rates, and demonstrated that the instability is very efficient in generating Langmuir plasma waves at distances of many stellar radii $r_{\rm ins} \sim 10^{4-5}h^{\rm CR} = 10^{7-8}$ cm, where $h^{\rm CR}$ is the height of the acceleration region described by (21) (Melikidze et al. 2000). This is exactly where pulsar's radio emission is supposed to originate (e.g. Kijak and Gil 1998). Conversion of these waves into coherent electromagnetic radiation escaping from the pulsar magnetosphere was considered and discussed by Melikidze et al. (2000) and Gil et al. (2004). These authors demonstrated that the nonlinear evolution of Langmuir oscillations developing in pulsar's magnetosphere leads to formation of charged, relativistic solitons, able to emit coherent curvature radiation at radio wavelengths. The component of this radiation that is polarized perpendicularly to the planes of dipolar magnetic field can escape from the magnetosphere (see Lai et al. 2001, for observational evidence of such polarization of pulsar radiation). The observed pulsar radiation in this picture is an indirect consequence of sparking discharges within the inner acceleration region just above the polar cap. In light of this paper we can therefore argue that the coherent pulsar radio emission is conditional on the presence of strong nondipolar surface magnetic fields at the polar cap, with a strength about 10^{13-14} G and radius of curvature of the order 10^4 cm.

In the very strong surface magnetic field assumed within the accelerator, processes such as photon splitting (Baring and Harding 2001) and bound pair creation (Usov and Melrose 1995) may become important. It has been suggested that such processes could potentially delay pair creation and thus increase the height and voltage of the accelerator. For the photon splitting case, the delay is significant only if photons with both polarization modes split — a hypothesis in strong magnetic fields (Baring and Harding 2001). It could be possible that only one mode split (Usov 2002). In such a case the gap height and potential of a PSG would not be significantly affected due to the high pair multiplicity in strong, curved magnetic fields. For bound pairs (Usov and Melrose 1995, 1996), in the very hot environment near the neutron star surface (with temperatures exceeding 1 MK), it is possible that bound pairs could not survive long from photoionization. Following Bhatia et al. (1992) and Usov and Melrose (1996) one can roughly estimate the mean free path for bound pair photo-ionization. It turns out that for temperatures around (2–3) MK this mean free path is of the order of few meters, which is considerably less than the height of CR-driven PSG. In such a case, even if the bound pairs are initially produced, they would not significantly delay the pair formation. One could address this potential problem by referring to the detailed case study of PSR B0943+10 analyzed in this paper. This is the only pulsar in which we have full information concerning $\mathbf{E} \times \mathbf{B}$ drift rate and the polar cap heating rate. In the analysis we have used two methods. The first method is independent on details of accelerating region (such as height, potential drop, etc.) but is based only on the subpulse drift radio-data and X-ray luminosity (10) and (11). The second one includes the detailed treatments of model parameters without considering the delaying effect by photon splitting and bound pairs. Both methods give consistent results, as illustrated in Figs. 1 and 2. We therefore conclude that the delaying processes, if any, are not significant in the PSGs given the physical conditions we invoke.

References

Arons, J., Scharlemann, E.T.: Astrophys. J. **231**, 854 (1979)

Asgekar, A., Deshpande, A.: Mon. Not. Roy. Astron. Soc. **326**, 1249 (2001)

Asseo, E., Melikidze, G.I.: Mon. Not. Roy. Astron. Soc. **301**, 59 (1998)

Baring, M.G., Harding, A.K.: Astrophys. J. **547**, 929 (2001)

Bhatia, V.B., Chopra, N., Panchapakesan, N.: Astrophys. J. **388**, 131 (1992)

Cheng, A.F., Ruderman, M.A.: Astrophys. J. **235**, 576 (1980)

Deshpande, A.A., Rankin, J.M.: Astrophys. J. **524**, 1008 (1999)

Deshpande, A.A., Rankin, J.M.: Mon. Not. Roy. Astron. Soc. **322**, 438 (2001)

Gil, J., Melikidze, G.I.; Astrophys. J. **577**, 909 (2002)

Gil, J., Mitra, D.: Astrophys. J. **550**, 383 (2001)

Gil, J., Sendyk, M.: Astrophys. J. **541**, 351 (2000)

Gil, J., Sendyk, M.: Astrophys. J. **585** (2003)

Gil, J., Melikidze, G.I., Geppert, U.: Astron. Astrophys. **407**, 315 (2003)

Gil, J., Lyubarski, Yu., Melikidze, G.I.: Astrophys. J. **600**, 872 (2004)

Goldreich, P., Julian, H.: Astrophys. J. **157**, 869 (1969)

Gupta, Y., Gil, J., Kijak, J., et al.: Astron. Astrophys. **426**, 229 (2004)

Harding, A.K., Muslimov, A.G.: Astrophys. J. **568**, 862 (2002)

Jones, P.B.: Mon. Not. Roy. Astron. Soc. **218**, 477 (1986)

Kargaltsev, O., Pavlov, G.G., Garmire, G.P.: Astrophys. J. **636**, 406 (2006)

Kijak, J., Gil, J.: Mon. Not. Roy. Astron. Soc. **299**, 855 (1998)

Lai, D.: Rev. Mod. Phys. **73**, 629 (2001)

Lai, D., Chernoff, D.F., Cordes, J.M.: Astrophys. J. **549**, 1111 (2001)

McLaughlin, M.A., Stairs, I.H., Kaspi, V.M., et al.: Astrophys. J. **591**, L135 (2003)

Melikidze, G.I., Gil, J.A., Pataraya, A.D.: Astrophys. J. **544**, 1081 (2000)

Nowakowski, L.: Astrophys. J. **457**, 868 (1996)

Usov, V.V.: Astrophys. J. **320**, 333 (1987)

Usov, V.V.: Astrophys. J. **572**, L87 (2002)

Usov, V.V., Melrose, D.B.: Aust. J. Phys. **48**, 571 (1995)

Usov, V.V., Melrose, D.B.: Astrophys. J. **464**, 306 (1996)

Ruderman, M.A., Sutherland, P.G.: Astrophys. J. **196**, 51 (1975)

Weltevrede, P., Edwards, R.I., Stappers, B.A.: Astron. Astrophys. **445**, 243 (2006) (astro-ph/0507282)

Zhang, B., Sanwal, D., Pavlov, G.G.: Astrophys. J. **624**, L109 (2005)

Astrophys Space Sci (2007) 308: 335–343
DOI 10.1007/s10509-007-9365-3

ORIGINAL ARTICLE

The example of effective plasma acceleration in a magnetosphere

V.S. Beskin · E.E. Nokhrina

Received: 24 June 2006 / Accepted: 11 October 2006 / Published online: 17 March 2007

Abstract The problem of effective transform of Poynting flux energy into the kinetic energy of relativistic plasma outflow in a magnetosphere is considered. In this article we present an example of such acceleration. In order to perform it, we use the approach of ideal axisymmetric magnetohydrodynamics (MHD). For highly magnetized plasma outflow we show that a linear growth of Lorentz factor with a cylindrical distance from the rotational axis is a general result for any field configuration in the sub-magnetosonic flow. In the far region the full magnetohydrodynamics problem for one-dimensional flow is considered. It turns out that the effective plasma outflow acceleration is possible in the paraboloidal magnetic field. It is shown that such an acceleration is due to the drift of charged particles in the crossed electric and magnetic field. The clear explanation of the absence of acceleration in the monopole magnetic field if given.

Keywords Neutron stars · Magnetosphere · Plasma acceleration · MHD and plasmas

PACS 94.30.-d · 94.20.wc · 95.30.Qd

V.S. Beskin (✉)
P.N. Lebedev Physical Institute, RAS, Leninsky pr. 53, Moscow, 119991, Russia
e-mail: beskin@lpi.ru

E.E. Nokhrina
Moscow Institute of Physics and Technology, Institutsky per. 9, Dolgoprudny, 141700, Russia
e-mail: nokhrinaelena@gmail.com

1 Introduction

We regard a problem of acceleration of plasma outflow in a pulsar magnetosphere. Convenient way to characterize MHD flows in a vicinity of a pulsar is to introduce the magnetization parameter σ, as was first done by Michel (1969):

$$\sigma = \frac{e\Omega\Psi_{\rm tot}}{4\lambda m_e c^3}. \qquad (1)$$

Here $\Psi_{\rm tot}$ is the total magnetic flux, and $\lambda = n/n_{\rm GJ}$ is the multiplication parameter of plasma ($|e|n_{\rm GJ} = |\Omega \mathbf{B}|/2\pi c$ is the Goldreich–Julian charge density). The magnetization parameter characterizes the quotient of electro-magnetic flux to particle rest mass energy flux near the surface of an object. The so-called σ-problem, i.e., the problem of transformation of electro-magnetic energy into particle kinetic energy, appears while one tries to explain an effective particle acceleration in the magnetic field.

Indeed, the theoretical modeling predicts the flow to be strongly magnetized at its origin in the vicinity of radio pulsars, i.e. $\sigma \gg 1$ (Michel 1991; Beskin et al. 1993). On the other hand, at large distance from a pulsar, observations and modeling allow us to determine the magnetization parameter $\sigma \approx 10^{-3}$ (Kennel and Coroniti 1984).

Up to now the axisymmetric stationary MHD approach gave an inefficient particle acceleration beyond the fast magnetosonic surface (Beskin et al. 1998; Bogovalov 1997; Begelman and Li 1994). This seemed to be a general conclusion for any structure of a flow, however a lack of acceleration in the supersonic region was rather a consequence of monopole field (Spitkovsky and Arons 2004; Thomson et al. 2004). Different approaches have been made in order to explain such an appealing disagreement between the observations and theory. The reconnection in a striped wind was discussed as one of possible acceleration scenarios (Coroniti 1990; Lyubarsky and Kirk 2001; Kirk and Lyubarsky 2001); however, this mechanism can account only for the non-axisymmetric part of a Poynting flux. Besides, for the Crab pulsar, it was shown by Lyubarsky and Kirk (2001) that effective acceleration may take place only beyond the standing shock. Another acceleration process can be connected with possible restriction of the longitudinal current, and thus the appearance of a light surface $|\mathbf{E}| = |\mathbf{B}|$ at the finite distance from a central object. In this case the effective energy conversion and the current closure takes place in the boundary layer near the light surface (Beskin et al. 1993; Chiueh et al. 1998; Beskin and Rafikov 2000). The self-similar approach is used to get semi-analytic solutions to the quasi-linear transfield equation, however such approach has its own disadvantages (see, e.g. Bogovalov (1995)). The effective energy transform has been got by Vlahakis and Königl (2003) in the r self-similar model.

In this work we show that in the sub-magnetosonic region any highly magnetized flow is accelerating linearly with the non-dimensional distance from a symmetry axis x as

$$\gamma = x, \tag{2}$$

with a maximal Lorentz factor at the fast magnetosonic surface (FMS) being equal to the classical value

$$\gamma_{\mathrm{F}} \leq \sigma^{1/3}. \tag{3}$$

We find the solution for MHD flow in the parabolic magnetic field—the force-free solution, first found by Blanford (1976). We solve the problem of particle acceleration self-consistently within the approach of stationary axisymmetric magnetohydrodynamics. In the sub-magnetosonic region the general results are applicable. However, in the super-magnetosonic region we treat the problem as one-dimensional and solve the full MHD problem. We find, that a linear acceleration $\gamma = x$ continues until the half of the Poynting flux energy is transformed into the particle kinetic energy. In the work by Beskin et al. (1998) the Michel's monopole solution was taken as the zero approximation to the flow. For this structure of the magnetic field the Lorentz factor was found to be $\sigma^{1/3}$ on the fast magnetosonic surface, which was located at the finite distance unlike the force-free limit. Beyond that singular surface the acceleration turned out to be ineffective. Treating the problem numerically, an ineffective acceleration and collimation have been found by Bogovalov (1997) and later by Komissarov (2004) for the monopole outflow. Now we took the flow near another force-free solution, i.e. Blandford's paraboloidal magnetic field (Blanford 1976). The obvious difference of this zero-approximation in comparison with monopole solution is a well-collimated flow even in the force-free limit. The physical reason for ineffective acceleration in the monopole magnetic field and an effective acceleration in the parabolic magnetic field is discussed.

We must emphasize that the model presented in the article is an example of possible magnetic field configuration where an effective plasma acceleration can be achieved.

2 Basic equations

Let us consider a stationary axisymmetric MHD flow of cold plasma in a flat space. Within this approach magnetic field is expressed by

$$\mathbf{B} = \frac{\nabla\Psi \times \mathbf{e}_{\hat{\varphi}}}{2\pi\varpi} - \frac{2I}{\varpi}\mathbf{e}_{\hat{\varphi}}. \tag{4}$$

Here $\Psi(r,\theta)$ is the stream function, $I(r,\theta)$ is the total electric current inside magnetic tube $\Psi(r,\theta) = \mathrm{const}$, and $\varpi = r\sin\theta$ is the distance from the rotational axis. In this paper we put $c = 1$. Owing to the condition of zero longitudinal electric field, one can write down the electric field as

$$\mathbf{E} = -\frac{\Omega_{\mathrm{F}}}{2\pi}\nabla\Psi \tag{5}$$

where the angular velocity Ω_{F} is constant on the magnetic surfaces: $\Omega_{\mathrm{F}} = \Omega_{\mathrm{F}}(\Psi)$. The frozen-in condition $\mathbf{E} + \mathbf{v} \times \mathbf{B} = 0$ gives us

$$\mathbf{u} = \frac{\eta}{n}\mathbf{B} + \gamma\Omega_{\mathrm{F}}\varpi\mathbf{e}_{\hat{\varphi}} \tag{6}$$

where $\mathbf{u}$ is four-velocity of a flow, and n is a concentration in the comoving reference frame. Function η is the ratio of particle flux to the magnetic field flux. Using the continuity equation $\nabla(n\mathbf{u}) = 0$, one gets that η is constant on magnetic surfaces as well: $\eta = \eta(\Psi)$.

Two extra integrals of motion are the total energy flux, conserved due to stationarity,

$$E(\Psi) = \frac{\Omega_{\mathrm{F}} I}{2\pi} + \mu\eta\gamma, \tag{7}$$

and the z-component of the angular momentum, conserved due to axial symmetry,

$$L(\Psi) = \frac{I}{2\pi} + \mu\eta\varpi u_{\hat{\varphi}}. \tag{8}$$

Here μ is the relativistic enthalpy which includes the rest energy. It is constant for the cold flow. The fifth integral of motion is the entropy $s(\Psi)$, which is equal to zero for the cold flow under consideration.

If the flux function Ψ and the integrals of motion are given, all other physical parameters of the flow can be determined using the following algebraic relations (Camenzind 1986; Beskin 1997):

$$\frac{I}{2\pi} = \frac{L - \Omega_{\rm F}\varpi^2 E}{1 - \Omega_{\rm F}^2\varpi^2 - \mathcal{M}^2}, \tag{9}$$

$$\gamma = \frac{1}{\mu\eta} \cdot \frac{E - \Omega_{\rm F}L - \mathcal{M}^2 E}{1 - \Omega_{\rm F}^2\varpi^2 - \mathcal{M}^2}, \tag{10}$$

$$u_{\hat\varphi} = \frac{1}{\varpi\mu\eta} \cdot \frac{(E - \Omega_{\rm F}L)\Omega_{\rm F}\varpi^2 - \mathcal{M}^2 L}{1 - \Omega_{\rm F}^2\varpi^2 - \mathcal{M}^2} \tag{11}$$

where the Alfvénic Mach number $\mathcal{M}$ is

$$\mathcal{M}^2 = \frac{4\pi\eta^2\mu}{n}. \tag{12}$$

To determine $\mathcal{M}^2$ one should use the definition of Lorentz factor $\gamma^2 - \mathbf{u}^2 = 1$ which gives the Bernoulli equation in the form

$$\frac{K}{\varpi^2 A^2} = \frac{1}{64\pi^4} \cdot \frac{\mathcal{M}^4(\nabla\Psi)^2}{\varpi^2} + \mu^2\eta^2. \tag{13}$$

Here

$$A = 1 - \Omega_{\rm F}^2\varpi^2 - \mathcal{M}^2, \tag{14}$$

$$K = \varpi^2(E - \Omega_{\rm F}L)^2(1 - \Omega_{\rm F}^2\varpi^2 - 2\mathcal{M}^2) + \mathcal{M}^4(\varpi^2E^2 - L^2). \tag{15}$$

The cold transonic flow is characterized by two singular surfaces: the Alfvénic surface and the fast magnetosonic surface (FMS). The first is determined by the condition of nulling the denominator A in the relations (9–11). FMS can be defined as the singularity of gradient of the Mach number. Writing (13) in the form

$$(\nabla\Psi)^2 = F = \frac{64\pi^4}{\mathcal{M}^4} \cdot \frac{K}{A^2} - \frac{64\pi^4}{\mathcal{M}^4}\varpi^2\mu^2\eta^2,$$

and taking gradient of it, we can get

$$\nabla_k\mathcal{M}^2 = \frac{N_k}{D}.$$

Here

$$D = \frac{A}{\mathcal{M}^2} + \frac{1}{\mathcal{M}^2} \cdot \frac{B_\varphi^2}{B_{\rm P}^2},$$

$$N_k = -\frac{A\nabla^i\Psi \cdot \nabla_k\nabla_i\Psi}{(\nabla\Psi)^2} + \frac{A}{2(\nabla\Psi)^2}\nabla_k' F,$$

and the operator ∇_k' acts on all quantities but $\mathcal{M}^2$. The regularity conditions

$$D = 0; \; N_r = 0; \; N_\theta = 0$$

define the position of FMS and the relation between the integrals of motion on it.

Finally, the stream equation on the function $\Psi(r, \theta)$ can be written as (Beskin 1997)

$$\nabla_k\left[\frac{1}{\varpi^2}(1 - \Omega_{\rm F}^2\varpi^2 - \mathcal{M}^2)\nabla^k\Psi\right] + \Omega_{\rm F}(\nabla\Psi)^2\frac{d\Omega_{\rm F}}{d\Psi} + \frac{64\pi^4}{\varpi^2}\frac{1}{2\mathcal{M}^2}\frac{\partial}{\partial\Psi}\left(\frac{G}{A}\right) = 0 \tag{16}$$

where

$$G = \varpi^2(E - \Omega_{\rm F}L)^2 + \mathcal{M}^2L^2 - \mathcal{M}^2\varpi^2E^2. \tag{17}$$

Operator $\partial/\partial\Psi$ acts only on the integrals of motion. The stream equation (16) contains the magnetic flux function Ψ and four integrals of motion: $E(\Psi)$, $L(\Psi)$, $\eta(\Psi)$, and $\Omega_{\rm F}(\Psi)$, i.e., it has the Grad–Shafranov form.

3 Sub-sonic flow and the fast magnetosonic surface—a case of an arbitrary field configuration

In this section we will show that for highly magnetized flow $W_{\rm part} \ll W_{\rm field}$ for any field configuration the following always holds:

1. In the sub-sonic flow $\gamma = x$.
2. On the FMS $\gamma_{\rm F} \le \sigma^{1/3}$.
3. The FMS is situated inside the cylinder with a radius $R_{\rm L}\sigma^{1/3}$.

In a general case it is natural to define Michel's magnetization parameter as

$$\sigma = \frac{E_{\rm A}}{\mu\eta} = \max_{\Psi\in[0;\Psi_0]}\frac{E(\Psi)}{\mu\eta}. \tag{18}$$

Here $E_{\rm A}$ is the total energy flux amplitude. In this case the energy integral can be written as $E(\Psi) = \sigma\mu\eta h(\Psi)$, where $h(\Psi)$ is a known function for a particular force-free flow configuration, $h \in [0; 1]$.

The position of the fast magnetosonic surface can be found without knowing the particular force-free field configuration. In order to find the position of FMS one can rewrite the Bernoulli equation (13) in the form

$$q^4 + 2q^3 - \left(\xi + \frac{1}{\Omega_{\rm F}^2\varpi^2}\right)q^2 + (2q+1)\left(\frac{\mu^2\eta^2}{E^2} + \frac{(e')^2}{\Omega_{\rm F}^2\varpi^2}\right) = 0. \tag{19}$$

Here by definition

$$q = \mathcal{M}^2/\Omega_{\rm F}^2\varpi^2,$$

$$e' = E(\Psi) - \Omega_{\rm F} L(\Psi)$$
$$= \mu\eta\gamma(1 - \Omega_{\rm F}\varpi v_{\hat{\varphi}}) \approx \gamma_{\rm in}\mu\eta = \text{const},$$

and

$$\xi = 1 - \frac{\Omega_{\rm F}^4\varpi^2(\nabla\Psi)^2}{64\pi^4 E^2}. \quad (20)$$

It is easy to check that for the force-free solution

$$\xi = 1 - \frac{|\mathbf{E}_{\rm p}|^2}{|\mathbf{B}_\varphi|^2} \ll 1. \quad (21)$$

For example, $\xi \equiv 0$ for Michel force-free monopole outflow.

Let us find the position of FMS and the Lorentz factor on it. In order to show that $q \ll 1$ on the FMS we write (10) assuming $\sigma \gg 1$ and $\Omega_{\rm F}^2\varpi^2 \gg 1$. Solving this equation for q, we get

$$q = \frac{\gamma\mu\eta}{E} \approx \frac{\gamma}{\sigma}. \quad (22)$$

Thus, q is approximately equal to the ratio of particle kinetic energy to the full energy of the flow, i.e., $q \ll 1$ for the magnetically dominated flow, since $\gamma = \sigma$ corresponds to the particle dominated flow.

As a result, one can rewrite (19) in the form

$$2q^3 - \left(\xi + \frac{1}{\Omega_{\rm F}^2\varpi^2}\right)q^2 + \frac{\mu^2\eta^2}{E^2} + \frac{(e')^2}{\Omega_{\rm F}^2\varpi^2} = 0. \quad (23)$$

Here the terms q^4 and q were omitted due to their smallness in comparison with $2q^3$ and 1 respectively.

Fast magnetosonic surface corresponds to the intersection of two roots of (23), or to the condition of discriminant Q being equal to zero (Beskin et al. 1998). The regularity conditions can be equivalently written as $\nabla Q = 0$. For (23) discriminant Q is expressed by

$$Q = \frac{1}{16}\frac{\mu^4\eta^4}{E^4} - \frac{1}{16}\frac{1}{27}\frac{\mu^2\eta^2}{E^2}\left(\xi + \frac{1}{\Omega_{\rm F}^2\varpi^2}\right)^3. \quad (24)$$

Here we neglected the $(e')^2/\Omega_{\rm F}^2\varpi^2$ in comparison with $\mu^2\eta^2/E^2$ as we seek the position of the FMS in the region where $W_{\rm part} \ll W_{\rm field}$, i.e. not too close to the axis, where the flow is always dominated by the particles. The condition of root intersection $Q = 0$ can be rewritten as

$$\xi(\mathbf{r}_{\rm F}) + \frac{1}{\Omega_{\rm F}^2\varpi_{\rm F}^2} = 3\left(\frac{\mu\eta}{E}\right)^{2/3}. \quad (25)$$

Values of $q(\mathbf{r}_{\rm F})$ and $\gamma_{\rm F} = \gamma(\mathbf{r}_{\rm F})$ due to condition $Q = 0$ do not depend on the sum $(\xi + 1/\Omega_{\rm F}^2\varpi^2)$ and on the FMS are equal to

$$q(\mathbf{r}_{\rm F}) = \frac{1}{\sigma^{2/3}h^{2/3}(\Psi)}, \quad (26)$$

$$\gamma_{\rm F} = \frac{E}{\mu\eta}q\Big|_{\rm F} = \sigma^{1/3}h^{1/3}(\Psi). \quad (27)$$

As $\max_{\Psi\in[0;\Psi_0]} h(\Psi) = 1$ by definition, the maximal Lorentz factor at the FMS is equal to $\sigma^{1/3}$.

To obtain the position of FMS, we need one of the criticality conditions, that we will write as $\partial Q/\partial\alpha = 0$, where α is one coordinate of the orthogonal coordinate system $\{\Psi, \alpha, \varphi\}$, $\nabla\Psi \cdot \nabla\alpha = 0$. This condition gives

$$\frac{\partial\xi}{\partial\alpha} = \frac{2}{\Omega_{\rm F}^2\varpi^2}\frac{\partial\varpi}{\partial\alpha}. \quad (28)$$

Taking approximately $\partial\xi/\partial\alpha \approx \xi/\alpha$, we get from (25) and (28)

$$r\sin\theta|_F = R_{\rm L}\sigma^{1/3}h^{1/3}(\Psi), \quad (29)$$

which leads to the result

$$r\sin\theta|_F \le R_{\rm L}\sigma^{1/3}. \quad (30)$$

To obtain q, and, consequently, γ, in the sub-magnetosonic region, we shall make a natural assumption that q and ξ grow monotonically from correspondingly σ^{-1} and 0 near the origin of the flow to $(\mu\eta/E)^{2/3}$ and $1/\Omega_{\rm F}^2\varpi^2$ on the fast magnetosonic surface. Thus, in (23) one can neglect the terms q^3 and ξq^2 in comparison with $(\mu\eta/E)^2$ and $q^2/(\Omega_{\rm F}\varpi)^2$ correspondingly. After that the solution of (23) is expressed by

$$q = \frac{\Omega_{\rm F}}{\sigma}\frac{\varpi}{h(\Psi)}. \quad (31)$$

For the Lorentz factor we obtain

$$\gamma = \frac{E}{\mu\eta}q = \frac{r}{R_{\rm L}}\sin\theta. \quad (32)$$

Thus, we conclude, that for highly magnetized flow the Lorentz factor of a plasma outflow grows linearly with the non-dimensional distance from the rotational axis.

4 The problem in the parabolic magnetic field

As we just have showed, in the vicinity of a central object a particle energy flux is much smaller than that of electromagnetic field. In this case it is possible to consider the contribution of particle inertia as a small disturbance to the quantities of the force-free flow. It can be done in the sub-magnetosonic region, since on the fast magnetosonic critical surface $\gamma \leq \sigma^{1/3}$, and consequently the flow is still highly magnetized: $W_{\rm part}/W_{\rm field} \leq \sigma^{-2/3}$.

However, in the far region the possibility of using this method is questionable, and it is not applicable beyond the fast magnetosonic surface in the problem under consideration. We shall show that in the far region the flow becomes one-dimensional, and we shall be able to elaborate the full MHD problem.

In this section we will state the problem in the parabolic magnetic field and the boundary conditions.

4.1 The zero approximation

Our goal is to determine the physical parameters of a flow in the paraboloidal magnetic field. For this reason it is convenient to use the following orthogonal coordinates:

$$X = r(1-\cos\theta); \quad Y = r(1+\cos\theta); \quad \varphi.$$

Here X stands for the certain magnetic surface in the force-free Blandford's paraboloidal solution, Y is the distance along it, and φ is an azimuthal angle. The latter does not appear in the equations because of the axial symmetry of the problem. The flat metric in this coordinates is

$$g_{\rm XX} = \frac{X+Y}{4X}; \quad g_{\rm YY} = \frac{X+Y}{4Y}; \quad g_{\varphi\varphi} = XY.$$

Then Blandford's force-free solution can be written as (Lee and Park 2004):

$$\frac{d\Psi}{dX} = \frac{\pi C}{\sqrt{1+\Omega_{\rm F}^2(X)X^2}}, \tag{33}$$

$$I_0(\Psi) = \frac{C\Omega_{\rm F}(X)X}{2\sqrt{1+\Omega_{\rm F}^2 X^2}}, \tag{34}$$

where $\Omega_{\rm F}$ is an arbitrary function of Ψ, and C is a constant. In particular, for $\Omega_{\rm F} = {\rm const}$ we have

$$\Psi_0(X) = \frac{\pi C}{\Omega_{\rm F}} \ln\left(\Omega_{\rm F}X + \sqrt{(\Omega_{\rm F}X)^2+1}\right). \tag{35}$$

4.2 The boundary conditions

As was previously emphasized, for cold plasma the problem is characterized by two singular surfaces: Alfvénic and fast magnetosonic surfaces. Consequently, we need to specify four boundary conditions on the disc surface D (Beskin 1997). For simplicity, we consider the case

$$\Omega_{\rm F}(\Psi)|_{\rm D} = \begin{cases} \Omega_{\rm F} = {\rm const}, & \Omega_{\rm F}X|_{\rm D} < 1, \\ 0, & \Omega_{\rm F}X|_{\rm D} \geq 1, \end{cases}$$

$$\gamma|_{\rm D} = {\rm const} = \gamma_{\rm in} \ll \sigma^{1/3},$$

$$\eta(\Psi)|_{\rm D} = {\rm const} = \eta,$$

$$\Psi|_{\rm D} = \frac{\pi C}{\Omega_{\rm F}} \ln\left(\Omega_{\rm F}\varpi + \sqrt{(\Omega_{\rm F}\varpi)^2+1}\right).$$

The condition $\Omega_{\rm F} = {\rm const}$ naturally restricts the region of a flow under consideration. Since we assume that the magnetic field is frozen in the disc, we must consider the flow only when $\Omega_{\rm F}X|_D = \Omega_{\rm F}\varpi < 1$. In fact, we should use the inequality $\Omega_{\rm F}X \ll 1$. For $\Omega_{\rm F}\varpi \geq 1$ the angular velocity must drop rapidly, so we assume that it becomes equal to zero beyond the field line $\Omega_{\rm F}X = 1$.

5 Sub-magnetosonic flow and fast magnetosonic surface—the parabolic magnetic field

Let us use the results of Sect. 3 for parabolic magnetic field. Using (29) and taking into account that $h(\Psi) = \Omega_{\rm F}X$, we obtain the exact form of the FMS:

$$r_{\rm F}(\theta) = R_{\rm L}\left(\frac{\sigma}{\theta}\right)^{1/2} \tag{36}$$

where $R_{\rm L} = \Omega_{\rm F}^{-1}$ is the radius of a light cylinder. Thus, the Lorentz factor on the FMS is

$$\gamma = \sqrt{\sigma\theta}. \tag{37}$$

The small disturbance to the flux function was found by Beskin and Nokhrina (2006).

6 Super-magnetosonic flow

In the super-magnetosonic region we are able to elaborate the full MHD problem since the flow becomes one-dimensional. In order to do it we need to emphasize some features of the paraboloidal configuration of magnetic field:

1. The character of the flow may change in the vicinity of the singular surface.
2. It was shown by Beskin and Nokhrina (2006) that for paraboloidal magnetic field the curvature of field lines plays no role in the force balance inside the working volume $\Omega_{\rm F}X < 1$. This allows us to consider the flow as one-dimensional.
3. Positions of the fast magnetosonic surfaces in the paraboloidal field and in the cylindrical field coincide.

Let us clarify the last point. Although the stream equation written for one-dimensional cylindrical flow has no singularity that can be associated with the fast magnetosonic surface, it was shown by Beskin and Malyshkin (2000) that for $\mathcal{M}_0^2 < 1$ there must be sub-magnetosonic flow inside the surface defined by

$$\Omega_F \varpi \sim \sigma \frac{B_{ext}}{\Omega_F^2 \Psi_{jet}} \qquad (38)$$

where $\varpi \approx r\theta$ is the distance from the rotational axis, and Ψ_{jet} is the total flux inside the jet. This we shall call fast magnetosonic surface for the magnetized cylindrical jet immersed into the external magnetic field B_{ext}.

We choose the vacuum paraboloidal magnetic field in the region, where $\Omega_F = 0$,

$$B_z = \frac{C}{2r(1+\Omega_F^2 X^2)^{1/2}} \approx \frac{C}{2z} \qquad (39)$$

to be the external field for our problem.

We need to define C in terms of the total flux Ψ_{jet}. According to our definition of σ (18) for paraboloidal flow,

$$E = \mu\eta\sigma\Omega_F X = \frac{\Omega_F^2 \Psi}{4\pi^2}. \qquad (40)$$

Please, keep in mind that this definition of E, consistent with (18), differs from the definition by Beskin and Malyshkin (2000) and Beskin and Nokhrina (2006) by a factor 2. As a result, we shall define the magnetic flux inside the region $\Omega_F X < 1$ as Ψ_{jet}, and thus C is expressed by

$$C = \frac{\Psi_{jet}\Omega_F}{\pi}. \qquad (41)$$

Combining (38), (39), and (41) one can get the position of the FMS to be

$$r_F \approx R_L \left(\frac{\sigma}{\sin\theta\,\cos\theta}\right)^{1/2} \approx R_L \left(\frac{\sigma}{\theta}\right)^{1/2} \qquad (42)$$

which coincides with (36).

Hence, the flow becomes actually 1D in the vicinity of the FMS, not to say about the supersonic region. For this reason we can consider the supersonic flow as one-dimensional. But unlike Beskin and Malyshkin (2000) we would use the paraboloidal magnetic field $B_p(z)$ (39) outside the working volume as an external one. It gives us the z-dependence of all the characteristics of the flow.

Let us formulate the one-dimensional problem. For the cylindrical flow after making use of (41) we find that the integrals of motion near the axis from the work by Beskin and Malyshkin (2000) are the same as the integrals of motion in our paraboloidal problem:

$$L(\Psi) = \frac{\Omega_F \Psi}{4\pi^2}, \qquad (43)$$

$$\Omega_F(\Psi) = \text{const}, \qquad (44)$$

$$\eta(\Psi) = \text{const}, \qquad (45)$$

$$E(\Psi) = \gamma_{in}\mu\eta + \Omega_F L = e' + \Omega_F L \qquad (46)$$

where $e' = \text{const}$. Introducing non-dimensional variables

$$y = \sigma \frac{\Psi}{\Psi_0}, \qquad (47)$$

$$x = \Omega_F \varpi, \qquad (48)$$

one can rewrite (13, 16) as a set of ordinary differential equations for y and M^2 (Beskin and Malyshkin 2000):

$$\begin{aligned}(1 - x^2 - \mathcal{M}^2)^2 \left(\frac{dy}{dx}\right)^2 &= \frac{\gamma_{in}^2 x^2}{\mathcal{M}^4}(1 - x^2 - 2\mathcal{M}^2) \\ &\quad + x^2(\gamma_{in} + y)^2 - y^2 - \frac{x^2}{\mathcal{M}^4}(1 - x^2 - \mathcal{M}^2)^2, \end{aligned} \qquad (49)$$

$$\begin{aligned}(\gamma_{in}^2 + x^2 - 1)\frac{d\mathcal{M}^2}{dx} &= 2x\mathcal{M}^2 - \frac{\gamma_{in}^2 x \mathcal{M}^2}{(1 - x^2 - \mathcal{M}^2)} + \frac{y^2\mathcal{M}^6}{x^3(1 - x^2 - \mathcal{M}^2)}.\end{aligned} \qquad (50)$$

Here we have already used the particular form of the integrals of motion (43–46).

Analytically, from the set of (49–50) one can get the following results:

1. $x \ll \gamma_{in}$

$$\mathcal{M}^2 = \mathcal{M}_0^2 = \text{const}, \qquad y = \frac{\gamma_{in}}{\mathcal{M}_0^2}x^2, \qquad (51)$$

i.e., the poloidal magnetic field is approximately constant.

2. $x \gg \gamma_{in}$

(a) $\mathcal{M}_0^2 \gg \gamma_{in}^2$,

$$\mathcal{M}^2 = \frac{\mathcal{M}_0^2}{\gamma_{in}^2}x^2, \qquad y \propto \ln\frac{x}{\gamma_{in}}, \qquad (52)$$

i.e., the poloidal magnetic field decreases as $B_p \propto \varpi^{-2}$ (Chiueh et al. 1991; Eichler 1993; Bogovalov 1995).

(b) $\mathcal{M}_0^2 \ll \gamma_{in}^2$,

$$\mathcal{M}^2 = \frac{\mathcal{M}_0^2}{\gamma_{in}}x, \qquad y = \frac{\gamma_{in}}{\mathcal{M}_0^2}x^2, \qquad (53)$$

i.e., again $B_p \approx \text{const}$.

Using the equality $q = \gamma\mu\eta/E$, one can get for the Lorentz factor

$$\gamma = \frac{qE}{\mu\eta} = \mathcal{M}^2\frac{y}{x^2}. \qquad (54)$$

Then, for $x \gg \gamma_{\rm in}$ and $\mathcal{M}_0^2 \ll \gamma_{\rm in}^2$ the following linear dependence is valid (Beskin and Malyshkin 2000):

$$\gamma = x. \tag{55}$$

Let us find the distance along the axis until which the linear growth of the Lorentz factor continues. In order to do this one should write the constant poloidal magnetic field which defines $\mathcal{M}_0^2$:

$$B_z = \frac{\Psi_0 \Omega_{\rm F}^2}{2\pi\sigma x}\frac{dy}{dx} = \frac{\Psi_0 \Omega_{\rm F}^2 \gamma_{\rm in}}{2\pi\sigma \mathcal{M}_0^2}. \tag{56}$$

As this magnetic field should be equal to the outer one, we get

$$z = \sigma\gamma_{\rm in} R_{\rm L}, \tag{57}$$

and the greatest Lorentz factor near the boundary of the working volume is

$$\gamma = (\sigma\gamma_{\rm in})^{1/2}. \tag{58}$$

For $z > \sigma\gamma_{\rm in} R_{\rm L}$ we shall perform numerical calculation in the intermediate region between two limits $\mathcal{M}_0 \ll \gamma_{\rm in}$ and $\mathcal{M}_0 \gg \gamma_{\rm in}$. The boundary condition is the equality of the magnetic flux and of magnetic field. Thus, for every z we need to find the proper value of $\mathcal{M}_0^2$ that would allow us to make internal poloidal magnetic field on the border of the working volume be equal to the external paraboloidal magnetic filed, as the internal flux becomes equal to the total flux of a jet $\Psi_{\rm jet}$. In this case the border of the working volume $x_{\rm jet}$ is defined during the numerical integration. The integration shows that the border remains almost paraboloidal $x_{\rm jet}(z) \approx z^{1/2}$ (see Beskin and Nokhrina (2006)). Such an integration gives the linear growth of Lorentz factor until about $\sigma/2$ (see Beskin and Nokhrina (2006)).

Thus, the Lorentz factor γ grows linearly with the distance from the axis reaching the value σ near the border of the working volume for $z = \sigma^2 R_{\rm L}/4$. This corresponds to the transformation of about a half of electro-magnetic energy into the kinetic one (compare it with the result of Vlahakis and Königl 2003). The maximal Lorentz factor in this problem is σ:

$$\frac{E}{\mu\eta} = \gamma_{\rm in} + \sigma \tag{59}$$

which follows from the definition (10). Thus, in the paraboloidal magnetic field the effective particle acceleration can be realized.

Here we should point out the main difference of our problem from the work in which the Michel's monopole solution was taken as the first approximation:

1. Even the force-free flow is well collimated already.
2. The curvature term of the paraboloidal problem does not play any role in the force balance, which allowed us to treat the flow as one-dimensional in the supersonic domain.
3. The considered flow is supported from outside by the much greater flux than the one which is contained in the inner region. This allows the flow to widen and, consequently, to accelerate outflowing particles inside the working volume.

In the next section we will discuss the reason for the effective acceleration in the parabolic magnetic field and the ineffective acceleration for the monopole field configuration.

7 Comparison of an outflow acceleration for monopole and paraboloidal magnetic fields

Why is the outflow acceleration is effective for one solution and extremely ineffective for the other? Particles in the flow are moving in the crossed electric and magnetic field. For $|\mathbf{E}| < |\mathbf{B}|$ there always exists such a reference frame where the particles are moving in the pure magnetic field. Let us write the well known formulas for the field transform

$$E_y = \frac{E'_y + \frac{v}{c}H'_z}{\sqrt{1-\left(\frac{v}{c}\right)^2}}, \tag{60}$$

$$H_z = \frac{H'_z + \frac{v}{c}E'_y}{\sqrt{1-\left(\frac{v}{c}\right)^2}} \tag{61}$$

where v is the drift velocity perpendicular both to electric and magnetic field. To screen the electric filed, i.e. to make $E'_y = 0$, a particle must have velocity

$$\frac{v}{c} = \frac{E_y}{H_z}, \tag{62}$$

or, for the appropriate fields in the MHD flow,

$$\frac{v}{c} = \frac{|\mathbf{E}|}{|\mathbf{B}|} = \frac{|\mathbf{E}_{\rm p}|}{\sqrt{|\mathbf{B}_{\rm p}|^2 + |\mathbf{B}_\varphi|^2}}. \tag{63}$$

Rewriting this equation in terms of a flux function and integrals of motion, we obtain

$$\frac{v}{c} = \left(\sqrt{\frac{1}{\Omega_{\rm F}^2\varpi^2} + \left(\frac{4\pi I}{\Omega_{\rm F}\varpi|\nabla\Psi|}\right)^2}\right)^{-1}. \tag{64}$$

Let us denote

$$A = 4\pi I/\Omega_{\rm F}\varpi|\nabla\Psi|. \tag{65}$$

For Lorentz factor we get

$$\gamma = \sqrt{\frac{1 + x^2 A^2}{1 + x^2 (A^2 - 1)}}. \tag{66}$$

We see that for $A \approx 1$, i.e. for $|\mathbf{E}_\mathrm{p}| \approx |\mathbf{B}_\varphi|$, the value of a Lorentz factor depends on ratio of 1 and $x^2(A^2 - 1)$:

1. For $1 \gg x^2(A^2 - 1)$ we have $\gamma \approx x$.
2. For $1 \ll x^2(A^2 - 1)$ we have $\gamma \approx 1/\sqrt{A^2 - 1}$.

Thus, when the discrepancy between values of electric and magnetic fields is so small that $1 \gg x^2(A^2 - 1)$, the condition $E' = 0$ forces particle to have drift velocity extremely close to c, and the Lorentz factor scaling $\gamma = x$ holds. On the other hand, if the value of toroidal magnetic field is slightly greater, than the value of electric field, the drift velocity needed to screen the electric field in the particles rest frame must be smaller.

As we have showed in Sect. 3, in the sub-magnetosonic region the condition $1 \gg x^2(A^2 - 1)$ always holds, and so the Lorentz factor growth as

$$\gamma = x \tag{67}$$

is a general result for any field configuration. Let us check this for the solutions in the monopole and paraboloidal magnetic field.

7.1 Monopole magnetic field. Super-magnetosonic flow

Taking the flux function found by Beskin et al. (1998)

$$\Psi = \Psi_0\big(1 - \cos\theta + k(r)\sin^2\theta\sigma^{-2/3}\big) \tag{68}$$

where the function $k(r)$ is of order of unity, we obtain

$$A^2 = 1 + \frac{k(r)}{\sigma^{2/3}}. \tag{69}$$

As we regard the super-magnetosonic flow, $r \gg \sigma^{1/3}$. Thus, $x^2\sigma^{-2/3} \gg 1$, and

$$\gamma \approx \sigma^{1/3}. \tag{70}$$

This is the result in the work by Beskin et al. (1998), meaning that there is no particle acceleration in the super-sonic flow.

7.2 Parabolic magnetic field. Super-magnetosonic flow

For one-dimensional flow we write (65) as

$$A = 2\frac{y}{x}\frac{1}{dy/dx}. \tag{71}$$

Taking the analytical solution for the one-dimensional flow $y \propto x^{2/(1+M_0^2/\gamma_\mathrm{in}^2)}$ from (Beskin and Nokhrina 2006), we obtain

$$A^2 - 1 = \frac{M_0^2}{\gamma_\mathrm{in}^2} \ll 1. \tag{72}$$

As we use $x \gg \gamma_\mathrm{in}^2$ and $M_0^2 > 1$,

$$x^2(A^2 - 1) \gg 1. \tag{73}$$

So

$$\gamma = x. \tag{74}$$

We see that the drift velocity considerations gives us clear physical picture of effective/ineffective outflow acceleration. The reason for ineffective particle acceleration in the super-magnetosonic region of monopole outflow is considerable discrepancy in magnitudes of electric and toroidal magnetic field. Thus, smaller drift velocity is needed to screen the electric field in the outflow rest frame.

8 Conclusions

We have showed that for any highly magnetized sub-manetosonic flow the following is true:

1. $\gamma = \dfrac{r\sin\theta}{R_\mathrm{L}}$, for $\mathbf{r} < \mathbf{r}_\mathrm{L}$.
2. On the FMS $\gamma = \sigma^{1/3}h^{1/3}(\Psi)$, where $h(\Psi) \in [0; 1]$, so the maximal Lorentz factor on FMS is $\sigma^{1/3}$.

We have found that in the super-magnetosonic region in the parabolic magnetic field the effective plasma acceleration continues until

$$\gamma = \frac{1}{2}\sigma, \tag{75}$$

and we explain such an acceleration as the drift velocity of particles in the crossed electric and magnetic fields.

We must emphasize that this result can not be described by the self-similar solution, since in the self-similar approach all the critical surfaces have a conical shape. Thus, self-similar solution is valid only far enough from the axis. On the other hand, our solution regularly approaches the symmetry axis (see details in Beskin and Nokhrina 2006).

Although the model with the parabolic magnetic field seems to be unlikely for radio pulsars since it needs a disk around a central object, but it presents an example that the effective particle acceleration is possible for some field configurations. In particular, we can be sure that the effective acceleration can be achieved if

$$(\Omega_\mathrm{F} r\sin\theta)^2\left(\frac{|\mathbf{B}_\varphi|^2}{|\mathbf{E}_\mathrm{p}|^2} - 1\right) \ll 1, \tag{76}$$

when the drift velocity in the crossed electric and magnetic fields corresponds to

$$\gamma = x. \tag{77}$$

Acknowledgements This work was partially supported by the Russian Foundation for Basic Research (Grant No. 05-02-17700) and Dynasty fund. Elena Nokhrina thanks the Conference Organizing Committee for the PPARC/Padova University Grant, and the RFFR for the travel grant (No. 06-02-26645).

References

Begelman, M.C., Li, Zh.-Yu.: Astrophys. J. **426**, 269 (1994)
Beskin, V.S., Gurevich, A.V., Istomin, Ya.N.: Physics of the Pulsar Magnetosphere. Cambridge University Press, Cambridge (1993)
Beskin, V.S., Phys. Uspekhi **40**, 659 (1997)
Beskin, V.S., Kuznetsova, I.V., Rafikov, R.R.: Mon. Not. Roy. Astron. Soc. **299**, 341 (1998)
Beskin, V.S., Malyshkin, L.M.: Astron. Lett. **26**, 208 (2000)
Beskin, V.S., Rafikov, R.R.: Mon. Not. Roy. Astron. Soc. **313**, 433 (2000)
Beskin, V.S., Nokhrina, E.E.: Mon. Not. Roy. Astron. Soc. **367**, 375 (2006)
Blandford, R.D.: Mon. Not. Roy. Astron. Soc. **176**, 465 (1976)
Bogovalov, S.V.: Astron. Lett. **21**, 565 (1995)
Bogovalov, S.V.: Astron. Astrophys. **327**, 662 (1997)
Camenzind, M.: Astron. Astrophys. **162**, 32 (1986)
Chiueh, T., Li, Zh.-Yu., Begelman, M.C.: Astrophys. J. **377**, 462 (1991)
Chiueh, T., Li, Zh.-Yu., Begelman, M.C.: Astrophys. J. **505**, 835 (1998)
Coroniti, V.F.: Astrophys. J. **349**, 538 (1990)
Eichler, D.: Astrophys. J. **419**, 111 (1993)
Kennel, C.F., Coroniti, F.V.: Astrophys. J. **283**, 694 (1984)
Kirk, J.G., Lyubarsky, Y.: J. Astron. Soc. Aust. **18**, 415 (2001)
Komissarov, S.S.: Mon. Not. Roy. Astron. Soc. **350**, 1431 (2004)
Lee, H.K., Park, J.: Phys. Rev. D **70**, 063001 (2004)
Lyubarsky, Y., Kirk, J.G.: Astrophys. J. **547**, 437 (2001)
Michel, F.C.: Astrophys. J. **158**, 727 (1969)
Michel, F.: Theory of Neutron Star Magnetosphere. The University of Chicago Press, Chicago (1991)
Spitkovsky, A., Arons, J.: Astrophys. J. **603**, 669 (2004)
Thomson, T.A., Chang, P., Quataert, E.: Astrophys. J. **611**, 380 (2004)
Vlahakis, N., Königl, A.: Astrophys. J. **596**, 1080 (2003)

Astrophys Space Sci (2007) 308: 345–351
DOI 10.1007/s10509-007-9334-x

ORIGINAL ARTICLE

Impact of neutron star oscillations on the accelerating electric field in the polar cap of pulsar

or could we see oscillations of the neutron star after the glitch in pulsar?

A.N. Timokhin

Received: 2 July 2006 / Accepted: 31 August 2006 / Published online: 20 March 2007

Abstract Pulsar "standard model", that considers a pulsar as a rotating magnetized conducting sphere surrounded by plasma, is generalized to the case of oscillating star. We developed an algorithm for calculation of the Goldreich-Julian charge density for this case. We consider distortion of the accelerating zone in the polar cap of pulsar by neutron star oscillations. It is shown that for oscillation modes with high harmonic numbers (l, m) changes in the Goldreich-Julian charge density caused by pulsations of neutron star could lead to significant altering of an accelerating electric field in the polar cap of pulsar. In the moderately optimistic scenario, that assumes excitation of the neutron star oscillations by glitches, it could be possible to detect altering of the pulsar radioemission due to modulation of the accelerating field.

Keywords Stars: neutron, oscillations, magnetic fields · Pulsars: general

PACS 97.60.Jd · 97.60.Gb

1 Introduction

Neutron stars (NS) are probably the most dense objects in the Universe. There are extreme physical conditions inside NS, i.e. the magnetic field is close to the quantum limit, the pressure is of the order of the nuclear one and the typical radius of a NS is only about 2–3 times larger than its gravitational radius. Knowledge of properties of matter under such extreme conditions is very important for fundamental physics. It is impossible to reconstruct such physical circumstances in terrestrial laboratories, therefore, study of the NS's internal structure would give an unique opportunity for experimental verification of several fundamental physical theories.

To study interiors of a celestial body one have to perform some kind of seismological study, by comparing observed frequencies of eigenmodes with frequencies inferred from theoretical considerations. Eigenfrequencies of oscillations in the crust of NS as well as in its interiors were calculated in several papers (see e.g. McDermott et al. 1988; Chugunov 2006, and references there). However, for seismological study there should exist both (i) a mechanism for excitation of oscillations, (ii) a mechanism modulating radiation of the celestial object.

There are two types of known NSs: member of binary systems and isolated ones. The former radiate due to accretion of the matter from the companion. For these stars there are many possibilities to excite oscillations, for example by instabilities in the accretion flow. However, in this case it would be difficult to distinguish whether a particular feature in the power spectrum of the object is due to oscillations of the NS or it is caused by some processes in the accretion disc/column. Because of this ambiguity we think that the study of isolated NS should be more promising in regard of the seismology.

The vast majority of known isolated neutron stars are radiopulsars. The glitch (sudden change of the rotational period) is probably the only possible mechanism for excitation of oscillations for isolated pulsars. Radiation of radiopulsars is produced mostly in the magnetosphere. In or-

This work was partially supported by RFBR grant 04-02-16720, and by the grants N.Sh.-5218.2006.2 and RNP-2.1.1.5940.

A.N. Timokhin (✉)
Sternberg Astronomical Institute, Universitetskij pr. 13, 119992, Moscow, Russia
e-mail: atim@sai.msu.ru

der to judge whether oscillations of the NS could produce detectable changes in pulsar radiation, the impact of the oscillations on the magnetosphere must be considered. There is a widely accepted model of radiopulsar as a highly magnetized NS surrounded by non-neutral plasma (Goldreich and Julian 1969). Although, there is still no self-consistent theory of radiopulsars, there is a general agreement regarding basic picture for the processes in the magnetosphere. Oscillations of the star can generate electric field as it happens in the case of rotation. Generalization of the formalism developed for rotating NS to the case of oscillating star should help to obtain the desired information.

The first attempt to generalize Goldreich-Julian (GJ) formalism to the case of oscillating NS was made in Timokhin et al. (2000). It was developed a general algorithm for calculation of the GJ charge density in the near zone of an oscillating NS. Using this algorithm GJ charge density and electromagnetic energy losses were calculated for the case of toroidal oscillations of the NS. Here we apply this formalism to the case of spheroidal oscillation modes, representing wide class of stellar modes (r-, g-, p- modes). We consider also impact of stellar oscillations on the acceleration mechanism in the polar cap of pulsar and discuss the possibility of observation of the NS oscillations.

2 Main results

2.1 General formalism

Let us start by considering the case of a non-rotating oscillating NS. Motion of the conducting NS surface in the strong magnetic field of the star generates electric field as in the case of rotation. Only oscillation modes with non-vanishing velocity $\mathbf{V}^{\mathrm{osc}}$ at the surface will disturb the magnetosphere. For the same reason as it is in the pulsar "standard model", the electric field in the magnetosphere of an oscillating star should be perpendicular to the magnetic field. Otherwise charged particles will be accelerated by a longitudinal (parallel to $\mathbf{B}$) electric field and their radiation will give rise to electron-positron cascades producing enough particles to screen the accelerating electric field (Sturrock 1971). As in the case of rotating stars we will define the Goldreich-Julian electric field $\mathbf{E}_{\mathrm{GJ}}$ as the field which is everywhere perpendicular to the magnetic field of the star, $\mathbf{E}_{\mathrm{GJ}} \perp \mathbf{B}$, and the GJ charge density as the charge density, which supports this field

$$\rho_{\mathrm{GJ}} \equiv \frac{1}{4\pi} \nabla \cdot \mathbf{E}_{\mathrm{GJ}}. \tag{1}$$

For simplicity we consider only a zone near the NS, at the distances $r \ll 2\pi c/\omega$, where ω is the frequency of NS oscillations. For many global oscillation modes (see e.g. McDermott et al. 1988) the polar cap accelerating zone is well withing this distance, therefore, we can study the changes of the accelerating electric field in the polar cap caused by oscillation. In the near zone all physical quantities change harmonically with time, i.e. the time dependence enters only through the term $e^{-\mathrm{i}\omega t}$.

We make an additional assumption, that changes of the magnetic field induced by currents in the NS crust are much larger than the distortion caused by currents flowing in the magnetosphere. This assumption is considered as a first order approximation according to the small parameter (ξ / R_{NS}), where ξ is the amplitude of oscillation and R_{NS}—the NS radius. In other words, outside of the NS

$$\nabla \times \mathbf{B} = 0 \tag{2}$$

in the first order in (ξ / R_{NS}). This assumption can be rewritten in terms of a condition on the current density in the magnetosphere as

$$j \ll \rho_{\mathrm{GJ}} c \left(\frac{c}{\omega r} \right) \equiv j_{\mathrm{GJ}} \left(\frac{c}{\omega r} \right). \tag{3}$$

Condition (3) implies that the current density in the near zone of the magnetosphere is less that the GJ current density connected with oscillations multiplied by a large factor $c/(\omega r)$. So, if the current density in the magnetosphere is of the order of the GJ current density, assumption (2) is valid. Under these assumption it is possible to solve the problem analytically in general case, i.e. to develop an algorithm for finding an analytical solution for the GJ charge density for arbitrary configuration of the magnetic field and arbitrary velocity field on the NS surface.

Under assumption (2) the magnetic field can be expressed through a scalar function P as

$$\mathbf{B} = \nabla \times \nabla \times (P \mathbf{e_r}). \tag{4}$$

The GJ electric field depends also on a scalar potential Ψ_{GJ} through the relation

$$\mathbf{E}_{\mathrm{GJ}} = -\frac{1}{c} \nabla \times (\partial_t P \mathbf{e_r}) - \nabla \Psi_{\mathrm{GJ}}. \tag{5}$$

The GJ charge density is then expressed as

$$\rho_{\mathrm{GJ}} = -\frac{1}{4\pi} \Delta \Psi_{\mathrm{GJ}}. \tag{6}$$

An equation for Ψ_{GJ} is

$$\Delta_\Omega P \partial_r \Psi_{\mathrm{GJ}} - \partial_r \partial_\theta P \partial_\theta \Psi_{\mathrm{GJ}} - \frac{1}{\sin^2\theta} \partial_r \partial_\phi P \partial_\phi \Psi_{\mathrm{GJ}} + \frac{1}{c \sin\theta} (\partial_r \partial_\phi P \partial_\theta \partial_t P - \partial_r \partial_\theta P \partial_\phi \partial_t P) = 0, \tag{7}$$

where Δ_Ω is an angular part of the Laplace operator. This is the first order linear partial differential equation for the GJ electric potential Ψ_{GJ}. As the equation for Ψ_{GJ} is linear, each oscillation mode can be treated separately. This equation is valid for arbitrary configuration of the magnetic field and for any amplitude of the surface oscillation, provided that condition (2) is satisfied. Dependence on oscillation mode appears in the boundary conditions and also through the time derivative $\partial_t P$. For different oscillation modes both the equation and boundary conditions for Ψ_{GJ} are different. Derivation of (7), boundary condition for functions Ψ_{GJ} as well as more detailed discussion of used approximations can be found in Timokhin et al. (2000).

2.2 Goldreich Julian charge density

In Timokhin et al. (2000) (7) was solved for the case of small-amplitude toroidal oscillations and dipolar configuration of unperturbed magnetic field. Solutions had been obtained with a code written in computer algebra language MATHEMATICA. Now we have developed a new version of this code, which allows to obtain analytical solutions of (7) for a more complicated case of spheroidal modes. Any vector field on a sphere can be represented as a composition of toroidal ($\nabla \cdot \mathbf{V}^{\mathrm{osc}} = 0$) and spheroidal ($\nabla \times \mathbf{V}^{\mathrm{osc}} = 0$) vector fields (Unno et al. 1979). So, now we are able to calculate GJ electric field and charge density for arbitrary oscillations of a NS with dipole magnetic field.

Similar to the case of toroidal oscillations the small current approximation turned to be valid for a half of all oscillation modes. For oscillation modes with velocity field, which is symmetric relative to the equatorial plane (see an example of such mode in Fig. 1 (left)), solution of (7) is smooth everywhere (see Fig. 2 (left)). For modes with antisymmetrical velocity field (an example of such mode is shown in Fig. 1 (right)), Ψ_{GJ} is discontinuous at the equatorial plane (see Fig. 2 (right)). There is the following reason for such behavior. Dipolar magnetic field is antisymmetric relative to the equatorial plane. Antisymmetric motion of the field line footpoints give rise to a twisted configuration of the magnetic field, which cannot be curl-free. So, for such modes a strong current will flow along closed magnetic field lines, $j \gg j_{GJ}$. However, there is no physical reason why a smooth solution for GJ electric field can not exists also for such modes. An argument supporting this hypothesis is a solution for twisted force-free magnetic field found by Wolfson (1995). In his solution, which corresponds to the toroidal mode $(2, 0)$, the configuration of force-free twisted magnetic field is supported by a strong current flowing along magnetic field lines.

For the modes with smooth solutions our approximation should be valid. On the other hand, a strong electric current will flow only along closed magnetic field lines. The current density along open magnetic field lines should be close to j_{GJ} and condition for the small current approximation (see (3)), will be satisfied in the open field line domain. Then, Ψ_{GJ} in the polar cap could be obtained by solving (7). Ψ_{GJ} in the polar cap will differ from the solutions obtained here, because the boundary conditions should be set at the polar cap boundaries and not on the whole surface on the NS. The main properties of Ψ_{GJ} regarding its qualitative dependence on the coordinates, used in our discussion, would be, however, similar to the properties obtained from our solutions.

As expected, the GJ charge density distribution follows the distribution of the velocity field (see e.g. Figs. 3, 4, 5). With increasing of the harmonic numbers, ρ_{GJ} falls more rapidly with the distance, what is also expected for multipolar solutions. A remarkable property of GJ charge density distribution near oscillating star is that the local maxima of ρ_{GJ} increases with increasing of both l and m (see Figs. 6, 7). The reasons for this are as follows. The electric field induced by oscillations is of the order

$$E_{GJ} \sim \frac{V^{\mathrm{osc}}}{c} B. \tag{8}$$

The charge density supporting this electric field is of the order of $E/\Delta x$, where Δx is a characteristic distance of electric field variation. For a mode with harmonic number l this size is of the order of R_{NS}/l. Hence, for ρ_{GJ} we have an estimate

$$\rho_{GJ} \sim l \frac{E}{4\pi R_{NS}} \sim l \frac{V^{\mathrm{osc}}}{c} \frac{B}{4\pi R_{NS}}. \tag{9}$$

So, for the same amplitude of velocity of oscillation the amplitude of variation of the GJ charge density is larger for higher harmonics. Equation (9) is in a good agreement with the exact results shown in Figs. 6, 7.

2.3 Particle acceleration in the polar cap of pulsar

The GJ charge density induced by NS oscillations influences particle acceleration mechanism in the polar cap of pulsar. As we will show, oscillations will have the strongest impact on the accelerating electric field in the polar cap of pulsar for models with free particle escape from the NS surface (Scharlemann et al. 1978; Muslimov and Tsygan 1992). The accelerating electric field in pulsars arises due to deviation of the charge density of the plasma from the local GJ charge density. For pulsar models with Space Charge Limited Flow (SCLF) the charge density of the flow ρ at the NS surface is equal to the local value of the GJ charge density, $\rho(R_{NS}) = \rho_{GJ}(R_{NS})$. Magnetic field lines diverge and the charge density of the flow decreases with increasing of the distance from the star. However, the local GJ charge density decreases in a different way and at some distance r from the NS $\rho_{GJ}(r) \neq \rho(r)$. The discrepancy between these charge densities gives rise to

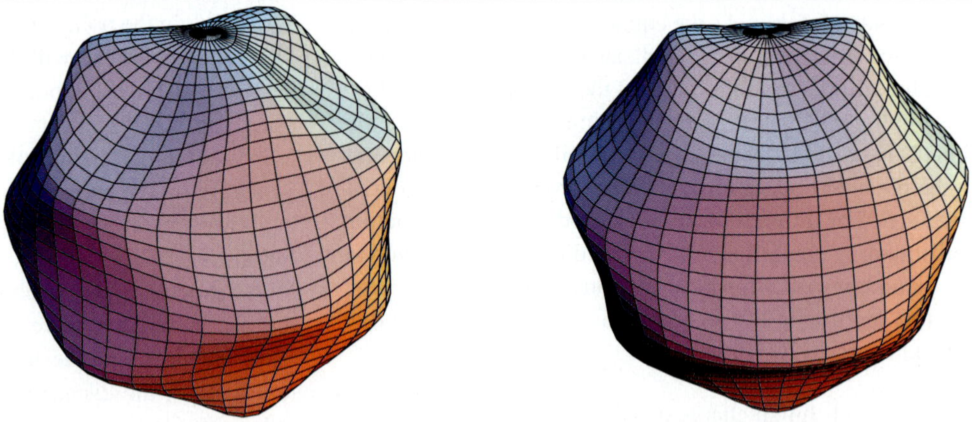

Fig. 1 The shape of the star surface during oscillations *left* for spheroidal mode (7, 3), *right* for spheroidal mode (7, 2). The shape changes harmonically with time

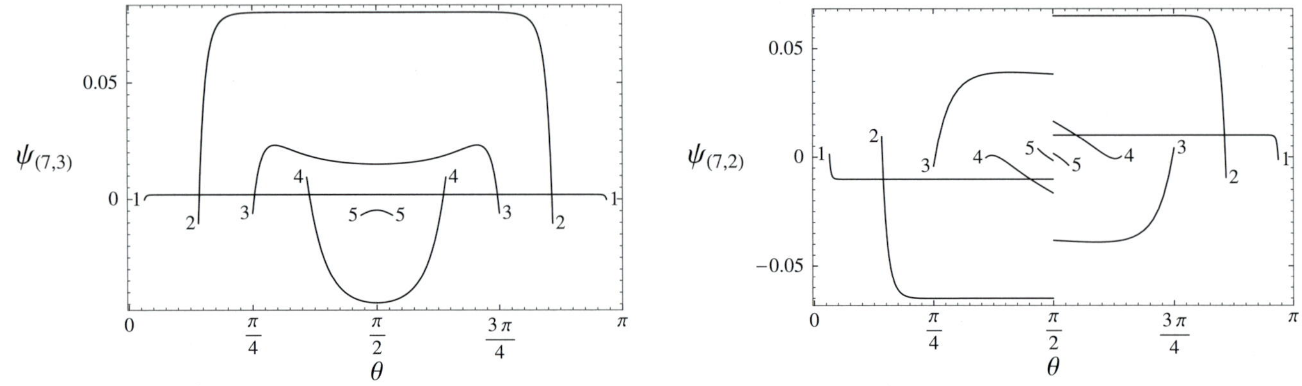

Fig. 2 Electric potential Ψ_{GJ} along a dipolar magnetic field line as a function of the polar angle θ is shown for 5 field lines with azimuthal angle $\phi = 0$. *Left* for spheroidal mode (7, 3), *right* for spheroidal mode (7, 2). The quantity changes harmonically with time

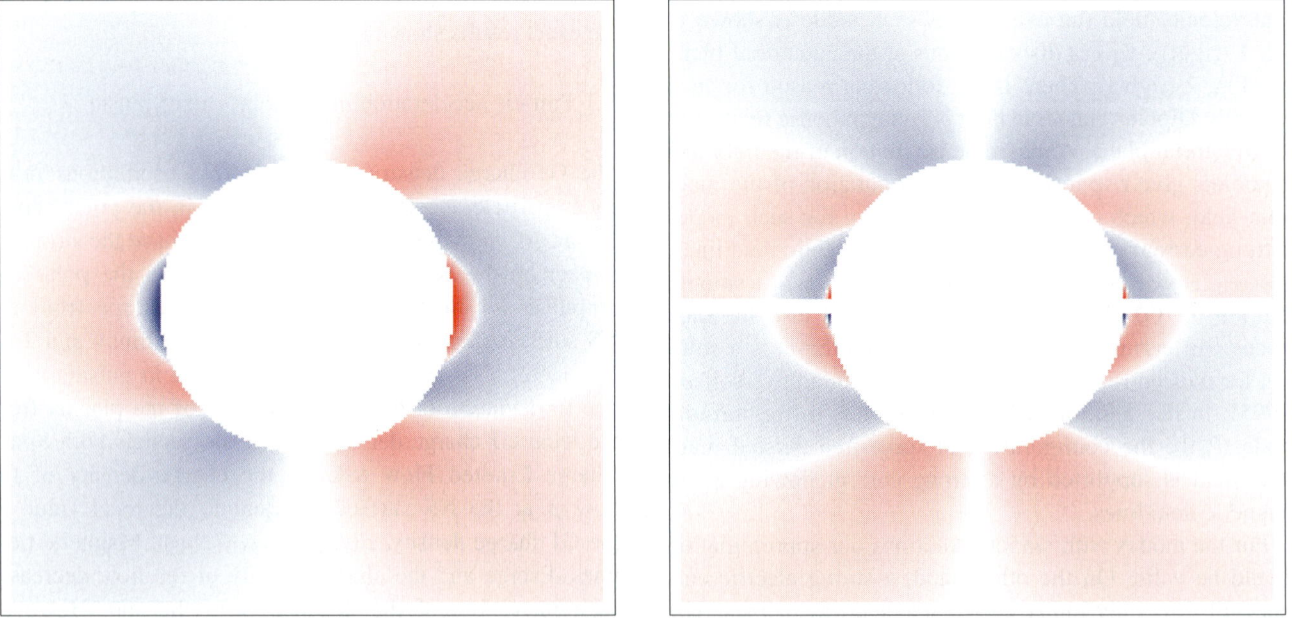

Fig. 3 Charge density ρ_{GJ} near the NS for azimuthal angle $\phi = 0$. Positive values of the charge density are shown in *red* and negative ones in *blue*. *Left*: ρ_{GJ} for the spheroidal mode (7, 3); *right*: ρ_{GJ} for the spheroidal mode (7, 2). The quantity changes harmonically with time. NB: at the equatorial plane ($\theta = \pi/2$) ρ_{GJ}^{72} is infinite

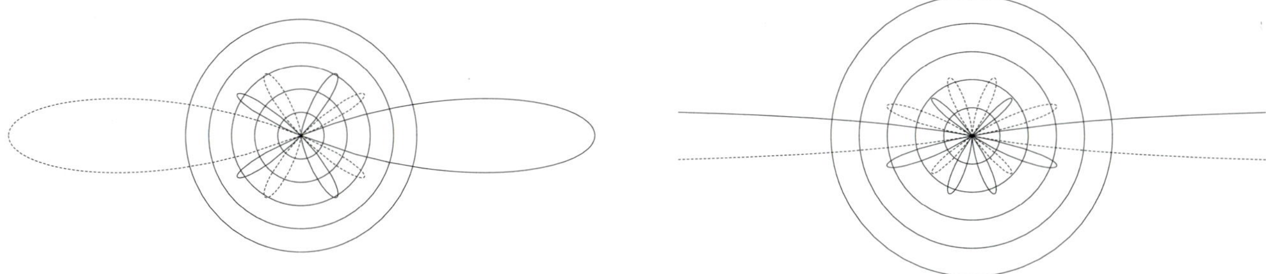

Fig. 4 Charge density $\rho_{\rm GJ}(r = R_{\rm NS}, \theta, \phi = 0)$ on the NS surface is shown in a polar coordinate system $(|\rho_{\rm GJ}^{lm}(R_{\rm NS}, \theta, 0)|, \theta)$. Positive values of charge density are shown by the *solid line* and negative ones by the *dashed line*. *Left*: $\rho_{\rm GJ}$ for the spheroidal mode $(7, 3)$; *right*: $\rho_{\rm GJ}$ for the spheroidal mode $(7, 2)$. The quantity changes harmonically with time. *Circles* correspond to the values of $\rho_{\rm GJ} = (0.1, 0.2, 0.3, 0.4, 0.5)$ in normalized units. NB: on the equatorial plane $(\theta = \pi/2)$ $\rho_{\rm GJ}^{72}$ is infinite

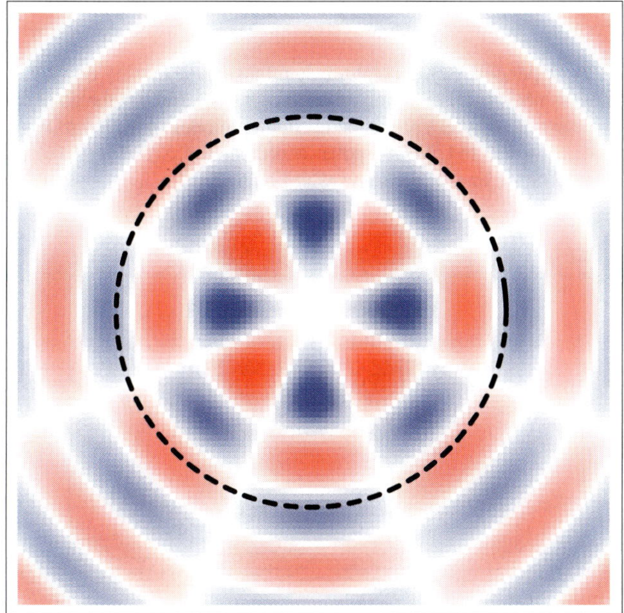

Fig. 5 View of the polar cap of pulsar from the top. Goldreich-Julian charge density for oscillation mode (44, 4) is shown by the *color map* (positive values in *red*, negative—in *blue*). The quantity changes harmonically with time. The polar cap boundary for a pulsar with period 3 ms is shown by the *dashed line*

a longitudinal electric field (see Scharlemann et al. 1978; Muslimov and Tsygan 1992).

The GJ charge density for oscillation modes with large l, m falls very rapidly with the distance. Hence, the charge density of a charge-separated flow for oscillating NS will exceed the local GJ charge density at some distance from the star. This produces a *decelerating* electric field. Therefore, in the case of rotating and oscillating NS, if $\rho_{\rm GJ}^{\rm osc} < \rho_{\rm GJ}^{\rm rot}$, the effective accelerating electric field will be reduced periodically due to superposition of accelerating and decelerating electric fields.

The oscillational GJ charge density $\rho_{\rm GJ}^{\rm osc}$ for modes with large l, m decreases practically to zero already at 2–3 NS radii and the whole oscillational GJ charge density contributes to the decelerating electric field. While in the case of rotation only ~15% of the GJ charge density contributes to the accelerating electric field for SCLF in the polar cap (Muslimov and Tsygan 1992). The most important factor increasing modification of the accelerating electric field is that the amplitude of $\rho_{\rm GJ}^{\rm osc}$ increases with increasing of the harmonic numbers of the mode. In Fig. 8 we show the difference between the charge density of SCLF and the local GJ charge density for dipolar magnetic field as a function of the distance for rotating and oscillating stars. In order to demonstrate the importance of the discussed effects we consider a rather grotesque case when the linear velocity of rotation is equal to the maximum velocity of oscillations. It is evident from this plot, that for large enough l and m the decelerating electric field caused by stellar oscillations could be of comparable strength with the accelerating electric field even if $V^{\rm osc} \ll V^{\rm rot}$.

Let us estimate the harmonic number of the oscillation mode where decelerating electric field would have a given impact on the accelerating electric field induced by NS rotation. The decelerating electric will be κ times less that the rotational accelerating electric field,

$$\kappa \equiv \frac{E_{\rm dec}^{\rm osc}}{E_{\rm acc}^{\rm rot}} \sim \frac{\rho_{\rm GJ}^{\rm osc}}{0.15\rho_{\rm GJ}^{\rm rot}}, \tag{10}$$

if

$$l \sim 0.15\kappa \frac{V^{\rm rot}}{V^{\rm osc}}, \tag{11}$$

here $V^{\rm rot}$ is the linear velocity of NS rotation at the equator. For example for $l \sim 100$, from this equation we can see, that for canceling of the accelerating electric field it is sufficient that the velocity amplitude is only $\sim 10^{-3}$ of rotational velocity at the NS equator.

For pulsar operating in Ruderman and Sutherland (1975) regime the impact of stellar oscillations on the accelerating field in the polar cap will be reduced by the factor of ~10.

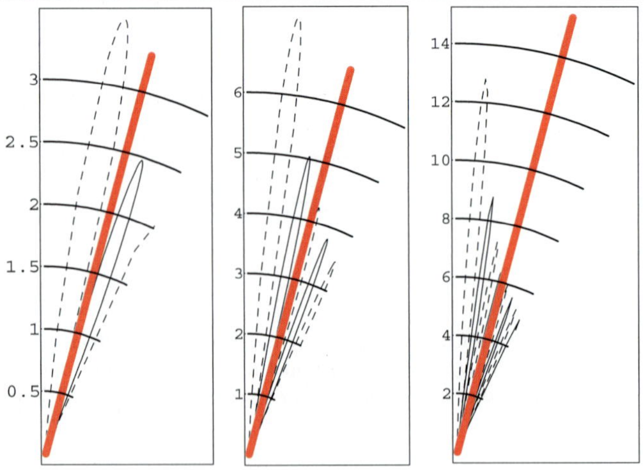

Fig. 6 Dependency of the Goldreich-Julian charge density on the harmonic number l. Distribution of ρ_{GJ} on the NS surface in the polar cap of pulsar is shown for oscillation modes (*from left to right*) (28, 4), (44, 4), (64, 4) in polar coordinates $(|\rho_{\mathrm{GJ}}^{lm}(R_{\mathrm{NS}}, \theta, 0)|, \theta)$. Positive values of charge density are shown by the *solid line* and negative ones by the *dashed line*. The *red line* shows the angle at which the last closed field line intersect the NS surface for a pulsar with period 3 ms. *Circular segments* correspond to the values of ρ_{GJ} in normalized units shown at the left side of each plot. The same normalization for ρ_{GJ} is used as in Fig. 4

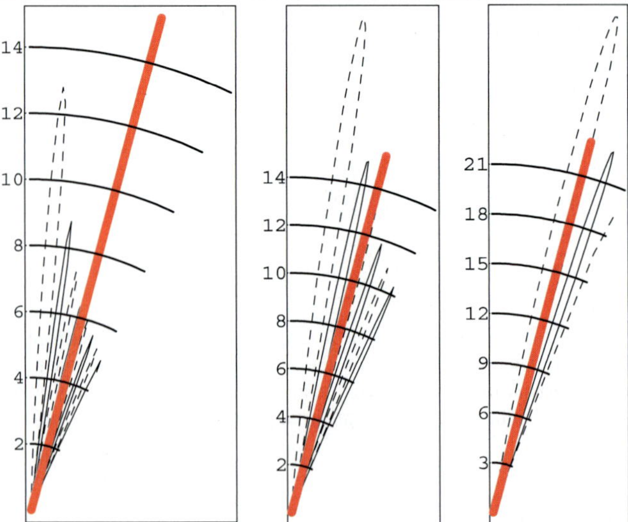

Fig. 7 Dependency of the Goldreich-Julian charge density on the harmonic number m. Distribution of ρ_{GJ} on the NS surface in the polar cap of pulsar is shown for spheroidal oscillation modes (*from left to right*) (64, 4), (64, 8), (64, 14) in polar coordinates $(|\rho_{\mathrm{GJ}}^{lm}(R_{\mathrm{NS}}, \theta, 0)|, \theta)$. Meaning of the lines is the same as in Fig. 6

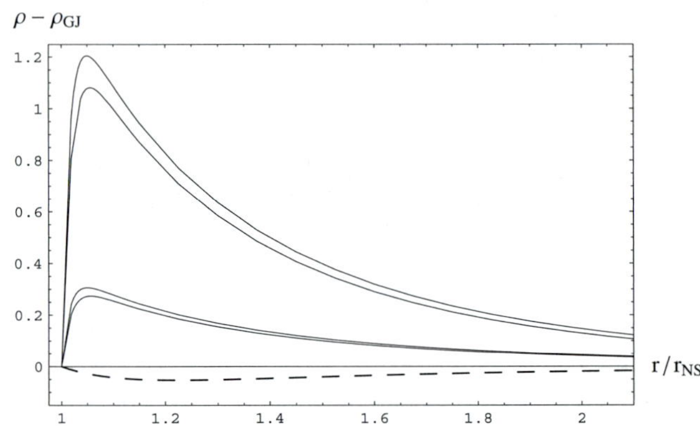

Fig. 8 $\Delta\rho \equiv \rho - \rho_{\mathrm{GJ}}$ – difference between the charge density of a space charge limited flow and the local Goldreich-Julian charge density along magnetic field lines in the polar cap of pulsar. $\Delta\rho$ is shown (in arbitrary units) by *solid lines* for spheroidal modes with different (l, m), from top to bottom: (64, 14), (54, 14), (64, 2), (54, 2). The same relation for an aligned rotator is shown by the *dashed line*. Negative values of $\Delta\rho$ give rise to an accelerating electric field, positive ones—to a decelerating field

In this model the accelerating electric field is generated in a vacuum gap, so the whole rotational GJ charge density contribute to the accelerating electric field and in this case $\kappa \sim \rho_{\mathrm{GJ}}^{\mathrm{osc}}/\rho_{\mathrm{GJ}}^{\mathrm{rot}}$.

3 Discussion

We have shown, that oscillations of the NS can induce changes in the accelerating electric field, which are more stronger than a naive estimation $(V^{\mathrm{osc}}/c)B$. Indeed, for high harmonics the induced electric field will be $\sim l$ times stronger. In order to make definitive predictions about observable parameters is necessary to study the polar cap acceleration zone more detailed. An accelerating electric potential and the height of the pair formation front should be calculated.

However, we can do simple estimations using (11). Let us estimate the harmonic number of the mode which can cancel the accelerating electric field, assuming the mode is excited by a glitch. As we pointed in Sect. 2.1 only modes with non-zero amplitude on the NS surface can produce changes in the magnetosphere. The distribution of oscillational motion plays crucial role. If oscillations are trapped in the NS crust a rather small energy will be required to pump the oscillation amplitude to the level high enough for strong disturbance of the accelerating electric field. If a fraction ϵ of the NS mass M_{NS} is involved in the oscillations the amplitude of the oscillational velocity is of the order

$$V^{\mathrm{osc}} \sim \sqrt{\frac{2W^{\mathrm{osc}}}{\epsilon M_{\mathrm{NS}}}}, \tag{12}$$

where W^{osc} is the total energy of the mode. The energy transferred during the glitch of the amplitude $\Delta\Omega$ is

$$W^{\mathrm{glitch}} = i I_{\mathrm{NS}} \Omega \Delta\Omega = i I_{\mathrm{NS}} \left(\frac{2\pi}{P}\right)^2 \frac{\Delta\Omega}{\Omega}, \tag{13}$$

where i is the fraction of the total momentum of inertia of the NS I_{NS} coupled to the crust. Let us assume that some fraction η of this energy goes into excitation of oscillations. Using (13), (12), (11) we get conditions for harmonic number of modes which would periodically cancel the accelerating electric field, as

$$l > 300\eta_{\%}^{-1/2}\epsilon^{1/2}i^{-1/2}\left(\frac{\Delta\Omega}{\Omega}\right)_6^{-1/2}, \tag{14}$$

for SCLF model, and for Ruderman and Sutherland (1975) model:

$$l > 2000\eta_{\%}^{-1/2}\epsilon^{1/2}i^{-1/2}\left(\frac{\Delta\Omega}{\Omega}\right)_6^{-1/2}. \tag{15}$$

Here $\eta_{\%}$ is measured in per cents and the relative magnitude of the glitch $(\Delta\Omega/\Omega)_6$ is normalized to 10^{-6}. It is widely accepted, that the origin of pulsar glitches is angular momentum transfer from the NS core to the crust. In the frame of this model the fraction of the energy which can go into excitation of NS oscillation η is of the order of $\Delta\Omega/\Omega$, i.e. it is very small, of the order of $\sim 10^{-6}$. We may speculate however, that excited oscillations are trapped in the NS crust, i.e. ϵ is also very small. Such global oscillation modes ($l \sim$ several hundreds) could induce substantial changes in the accelerating electric field.

As we mentioned in Sect. 2.1 all physical quantities in the solutions obtained here oscillate with the frequency of the star oscillations. The accelerating electric field close to the local geometrical maxima of the oscillational GJ charge density will be weakened periodically by the decelerating effect due to stellar pulsations. The field oscillation will influence the particle distribution in the open field line zone of the pulsar magnetosphere and it should produce some observable effects. Depending on oscillation mode and position of the line of sight a complicated pattern will appear periodically in individual pulse profiles. Although individual pulses are highly variable, the presence of periodical features should be possible to discover in the power spectra of pulsars, provided the oscillations are excited to a high enough level and observations have been made with hight temporal resolution. If one observes some feature, which appears just after the glitch, then decreases and disappears after some time, and it never appears in the normal pulsar emission, then one can undoubtedly attribute this feature to the NS oscillations.

Acknowledgements I wish to thank M. Ali Alpar for discussion.

References

Chugunov, A.I.: Mon. Not. Roy. Astron. Soc. **369**, 349 (2006)
Goldreich, P., Julian, W.H.: Astrophys. J. **157**, 869 (1969)
McDermott, P.N., van Horn, H.M., Hansen, C.J.: Astrophys. J. **325**, 725 (1988)
Muslimov, A.G., Tsygan, A.I.: Mon. Not. Roy. Astron. Soc. **255**, 61 (1992)
Ruderman, M.A., Sutherland, P.G.: Astrophys. J. **196**, 51 (1975)
Scharlemann, E.T., Arons, J., Fawley, W.M.: Astrophys. J. **222**, 297 (1978)
Sturrock, P.A.: Astrophys. J. **164**, 529 (1971)
Timokhin, A.N., Bisnovatyi-Kogan, G.S., Spruit, H.C.: Mon. Not. Roy. Astron. Soc. **316**, 734 (2000)
Unno, W., Osaki, Y., Ando, H., Shibahashi, H.: Forest Grove. University of Tokyo Press, Tokyo (1979)
Wolfson, R.: Astrophys. J. **443**, 810 (1995)

Astrophys Space Sci (2007) 308: 353–361
DOI 10.1007/s10509-007-9362-6

ORIGINAL ARTICLE

Heat blanketing envelopes and thermal radiation of strongly magnetized neutron stars

Alexander Y. Potekhin · Gilles Chabrier · Dmitry G. Yakovlev

Received: 3 July 2006 / Accepted: 1 November 2006 / Published online: 20 March 2007

Abstract Strong ($B \gg 10^9$ G) and superstrong ($B \gtrsim 10^{14}$ G) magnetic fields profoundly affect many thermodynamic and kinetic characteristics of dense plasmas in neutron star envelopes. In particular, they produce strongly anisotropic thermal conductivity in the neutron star crust and modify the equation of state and radiative opacities in the atmosphere, which are major ingredients of the cooling theory and spectral atmosphere models. As a result, both the radiation spectrum and the thermal luminosity of a neutron star can be affected by the magnetic field. We briefly review these effects and demonstrate the influence of magnetic field strength on the thermal structure of an isolated neutron star, putting emphasis on the differences brought about by the superstrong fields and high temperatures of magnetars. For the latter objects, it is important to take proper account of a combined effect of the magnetic field on thermal conduction and neutrino emission at densities $\rho \gtrsim 10^{10}$ g cm^{-3}. We show that the neutrino emission puts a B-dependent upper limit on the effective surface temperature of a cooling neutron star.

Work supported in parts by RFBR (Grants 05-02-16245 and 05-02-22003), FASI (Grant NSh-9879.2006.2), and CNRS French–Russian program (Grant PICS 3202).

A.Y. Potekhin (✉) · D.G. Yakovlev
Ioffe Physico-Technical Institute, Politekhnicheskaya 26, 194021 St. Petersburg, Russia
e-mail: palex@astro.ioffe.ru

D.G. Yakovlev
e-mail: yak@astro.ioffe.ru

G. Chabrier
Ecole Normale Supérieure de Lyon, CRAL (UMR 5574 CNRS), 46 allée d'Italie, 69364 Lyon, France
e-mail: chabrier@ens-lyon.fr

Keywords Neutron stars · Dense plasmas · Magnetic fields

PACS 97.60.Jd · 52.25.Xz

1 Introduction

Thermal emission from neutron stars can be used to measure the magnetic field, temperature, and composition of neutron-star envelopes, and to constrain the properties of matter under extreme conditions (see, e.g., Yakovlev and Pethick 2004; Yakovlev et al. 2005, and references therein). To achieve these goals, one should use reliable models of the atmosphere or condensed surface, where the thermal spectrum is formed, and of deeper layers, which provide thermal insulation of hot stellar interiors. In these layers, the effects of strong magnetic fields can be important. In recent years, significant progress has been achieved in the theoretical description of neutron-star envelopes with strong magnetic fields, but new challenges are put forward by observations of magnetars. In Sect. 2 we briefly overview recent work on the construction of models of neutron star atmospheres with strong magnetic fields and on the modeling of spectra of thermal radiation formed in an atmosphere or at a condensed surface. We list important unsolved theoretical problems which arise in this modeling. In Sect. 3, after a brief review of the effects of strong magnetic fields on the thermal structure and effective temperature of neutron stars, we describe our new calculations of the thermal structure. Compared to the previous results (Potekhin and Yakovlev 2001; Potekhin et al. 2003), we have taken into account neutrino energy losses in the outer crust of the star. We show that neutrino emission strongly affects the temperature profile in a sufficiently hot neutron star and places an upper limit on its surface temperature T_s and photon luminosity L_γ.

2 Thermal emission from magnetized surface layers of a neutron star

2.1 Neutron star atmosphere

It was realized long ago (Pavlov et al. 1995) that a neutron star atmosphere model should properly include the effects of a strong magnetic field and partial ionization. Models of *fully ionized* neutron star atmospheres with strong magnetic fields were constructed by several research groups (e.g., Shibanov et al. 1992; Zane et al. 2000; Ho and Lai 2003, and references therein). The most recent papers highlighted the effects that can be important for atmospheres of magnetars: the ion cyclotron feature (Ho and Lai 2001; Zane et al. 2001) and vacuum polarization, including a conversion of normal radiation modes propagating in the magnetized atmosphere (Ho and Lai 2003; Lai and Ho 2003).

Early studies of *partial ionization* in the magnetized neutron star atmospheres (e.g., Rajagopal et al. 1997; reviewed by Zavlin and Pavlov 2002) were based on an oversimplified treatment of atomic physics and nonideal plasma effects in strong magnetic fields. At typical parameters, the effects of thermal motion of bound species are important. So far these effects have been taken into account only for hydrogen plasmas. Thermodynamic functions, absorption coefficients, the dielectric tensor and polarization vectors of normal radiation modes in a strongly magnetized, partially ionized hydrogen plasma have been obtained and used to calculate radiative opacities and thermal radiation spectra (see Potekhin et al. 2004, and references therein).

The summary of the magnetic hydrogen atmosphere models and the list of references is given by Potekhin et al. (2006). The model is sufficiently reliable at 10^{12} G $\lesssim B \lesssim 10^{13.5}$ G, i.e., in the field range typical of isolated radio pulsars. It provides realistic spectra of thermal X-ray radiation (Potekhin et al. 2004). Potekhin and Chabrier (2004) extended this model to higher B. However, there remain the following unsolved theoretical problems that prevent to obtain reliable results beyond the indicated field range:

- The calculated spectra at $B \gtrsim 10^{14}$ G depend on the adopted model of mode conversion owing to the vacuum resonance and on the description of the propagation of photons with frequencies below the plasma frequency. Neither of these problems has been definitely solved. Their solution is also important for modeling the low-frequency (UV and optical) tail of the spectrum.
- At low T or high B, hydrogen atoms recombine in H_n molecules and eventually form a condensed phase (see Sect. 2.2). Corresponding quantum-mechanical data are very incomplete.
- At 10^9 G $\lesssim B \lesssim 10^{11}$ G, transition rates of moving H atoms have not been calculated because of their complexity. There is the only one calculation of the energy spectrum of bound states appropriate to this range of B (Lozovik and Volkov 2004).
- A more rigorous treatment of radiative transfer in the atmosphere requires solving the transfer equations for the Stokes parameters which has not been done so far for partially ionized atmospheres (see, e.g., Lai and Ho 2003; van Adelsberg and Lai 2006 for the cases of fully ionized atmospheres).

Finally, we note that it is still not possible to calculate accurate atmospheric spectra at $B \gtrsim 10^{12}$ G for chemical elements other than hydrogen, because of the importance of the effects of motion of atomic nuclei in the strong magnetic fields. Apart from the H atom, these effects have been calculated only for the He atom (Al-Hujaj and Schmelcher 2003a, 2003b), which *rests* as a whole, but has a moving nucleus, and for the He^+ ion (Bezchastnov et al. 1998; Pavlov and Bezchastnov 2005). The data of astrophysical relevance for He^+ are partly published and partly in preparation (see Pavlov and Bezchastnov 2005); one expects to have a He/He^+ magnetic atmosphere model available in the near future.

2.2 Condensed surface and thin atmosphere

The notion that an isolated magnetic neutron star has a condensed surface was first put forward by Ruderman (1971), who considered the iron surface. Lai and Salpeter (1997) and Lai (2001) studied the phase diagram of strongly magnetized hydrogen and showed that, when the surface temperature T_s falls below some critical value (dependent on B), the atmosphere can undergo a phase transition into a condensed state. A similar phase transition occurs for the equation of state of partially ionized, nonideal, strongly magnetized hydrogen plasma, constructed by Potekhin et al. (1999) for $B \lesssim 10^{13.5}$ G and extended by Potekhin and Chabrier (2004) to the magnetar field strengths. It is analogous to the "plasma phase transition" suggested in plasma physics at $B = 0$ (see, e.g., Chabrier et al. 2006 for discussion and references). According to Potekhin et al. (1999), the critical point for the phase transition in the hydrogen plasma is located at the density $\rho_c \approx 143 B_{12}^{1.18}$ g cm^{-3} and temperature $T_c \approx 3 \times 10^5 B_{12}^{0.39}$ K, where $B_{12} = B/10^{12}$ G. At $T < T_c$ the density ρ_{cond} of the condensed phase increases up to a few times of ρ_c. On the other hand, according to Lai (2001), the surface density of a condensed phase for heavy elements is $\rho_{cond} \approx 560 A Z^{-0.6} B_{12}^{1.2}$ g cm^{-3}, where Z and A are the charge and mass numbers of the ions. These two estimates of ρ_{cond} are in qualitative agreement. Lai (2001) estimates the critical temperature of hydrogen as $T_c \lesssim 0.1 E_s \lesssim 10^{5.5} B_{12}^{0.4}$ K (E_s being the cohesive energy), also in agreement with the above estimate. Jones (1986) calculated the cohesive energy for Ne and Fe at 10^{12} G $\lesssim B \le 10^{13}$ G, using the density functional theory

(DFT) and obtained $E_s \sim 0.1B_{12}$ keV. Recently, Medin and Lai (2006) performed DFT calculations of the cohesive energies for zero-pressure condensed hydrogen, helium, carbon, and iron at 10^{12} G $\leq B \lesssim 10^{15}$ G. For instance, they found that the cohesive energy per carbon atom ranges from ~50 eV at $B = 10^{12}$ G to 20 keV at 10^{15} G. The cohesive energy per iron atom varies from ~0.8 keV at $B = 10^{13}$ G to 33 keV at 10^{15} G. These calculations suggest T_c of the same order of magnitude as the above estimate for hydrogen.

Note that the models of Potekhin et al. (1999) and Lai and Salpeter (1997) are constructed in the framework of the "chemical picture" of plasmas, whose validity near the plasma phase transition can be questionable (Chabrier et al. 2006). Thus the position (and the very existence) of the condensed surface requires further theoretical investigation and experimental or observational verification. Hopefully, this can be done by analyzing observations of thermal emission from neutron stars.

The thermal emission from the magnetized surface was studied by Brinkmann (1980); Turolla et al. (2004); Pérez-Azorín et al. (2005), and van Adelsberg et al. (2005). The spectrum exhibits modest deviations from blackbody across a wide energy range, and shows mild absorption features associated with the ion cyclotron frequency (energy $\hbar\omega_{ci} = 6.3B_{12}Z/A$ eV) and the electron plasma frequency (energy $\hbar\omega_p = 28\sqrt{\rho Z/A}$ eV, where ρ is in g cm^{-3}). However, the predictions of the ion cyclotron feature and the spectrum at lower frequencies are not firm. The uncertainty arises from motion of the ions in the electromagnetic field around their equilibrium lattice positions. Most of the models treat the ions as fixed (non-moving). Only van Adelsberg et al. (2005) considered two limits of fixed and free ions. In reality, however, the ions are neither fixed nor completely free (see van Adelsberg et al. 2005 for estimates of possible uncertainties).

In addition, the condensed surface of a neutron star can be surrounded by a "thin" atmosphere, which is transparent to X-rays, but optically thick at lower wavelengths. Such a hypothesis has been first invoked by Motch et al. (2003) for explaining the spectrum of the isolated neutron star RX J0720.4−3125. Recently, the hydrogen atmosphere model, described in Sect. 2.1, together with the condensed surface emission model of van Adelsberg et al. (2005) have been successfully used for fitting the spectrum of the isolated neutron star RX J1856.5–3754 (Ho et al. 2007) (assuming the atmosphere to be "thin" as defined above).

3 Heat transport through magnetized envelopes

3.1 Overview of previous work

The link between the magnetized atmosphere and stellar interior is provided by a *heat blanketing (insulating) envelope*. The solution of heat transport problem relates the effective surface temperature T_s to the temperature T_b at the inner boundary of the blanketing envelope. Without a magnetic field, it is conventional to place the inner boundary at $\rho = \rho_b = 10^{10}$ g cm^{-3}. In this case, it can be treated in the quasi-Newtonian approximation with fractional errors $\lesssim 10^{-3}$ (Gudmundsson et al. 1983). A strong magnetic field, however, greatly affects heat transport and, consequently, the thermal structure of the envelope. The thermal structure of neutron star envelopes with radial magnetic fields (normal to the surface) was studied by Van Riper (1988) (also see Van Riper 1988 for references to earlier work). His principal conclusion was that the field reduces the thermal insulation of the heat blanketing envelope due to the Landau quantization of electron motion. The thermal structure of the envelope with magnetic fields normal and tangential to the surface was analyzed by Hernquist (1985) and Schaaf (1990a). The tangential field increases the thermal insulation of the envelope, because the Larmor rotation of electrons reduces the transverse electron thermal conductivity.

The case of arbitrary angle θ_B between the field lines and the normal to the surface was studied by Greenstein and Hartke (1983) in the approximation of constant (density and temperature independent) longitudinal and transverse thermal conductivities. The authors proposed a simple formula which expresses T_s at arbitrary θ_B through two values of T_s calculated at $\theta_B = 0$ and 90°. The case of arbitrary θ_B was studied also by Schaaf (1990b) and Heyl and Hernquist (1998, 2001).

Potekhin and Yakovlev (2001) reconsidered the thermal structure of blanketing iron envelopes for any θ_B, using improved thermal conductivities (Potekhin 1999). Potekhin et al. (2003) analyzed accreted blanketing envelopes composed of light elements. In agreement with an earlier conjecture of Hernquist (1985) and simplified treatments of Page (1995) and Shibanov and Yakovlev (1996), they demonstrated that the dipole magnetic field (unlike the radial one) does not necessarily increase the total stellar luminosity L_γ at a given T_b. On the contrary, the field $B \sim 10^{11}$–10^{13} G lowers L_γ, and only the fields $B \gtrsim 10^{14}$ G significantly increase it. Potekhin et al. (2003) shifted the inner boundary of the blanketing envelope to the neutron drip, $\rho_b = 4 \times 10^{11}$ g cm^{-3}, because in some cases they found a non-negligible temperature drop at $\rho > 10^{10}$ g cm^{-3}. They obtained that magnetized accreted envelopes are generally more heat-transparent than non-accreted ones (the same is true in the field-free case, studied by Potekhin et al. 1997). However, this heat transparency enhancement is less significant, when the transparency is already enhanced by a superstrong magnetic field.

Recently Potekhin et al. (2005) showed that qualitatively the same dependence of L_γ on B and on the chemical composition holds not only for dipole, but also for small-scale field configurations.

Geppert et al. (2004, 2006) studied heat-blanketing envelopes with a magnetic field anchored either in the core or in the inner crust of the star (with dipole and toroidal field components of different strengths). They showed that a superstrong field in the inner crust of a not too hot star can significantly affect the surface temperature distribution and make it nonuniform and even asymmetric, with hot spots having different temperatures. Similar results were obtained by Pérez-Azorín et al. (2006), who also evaluated pulsed fractions and phase-dependent spectra of neutron stars with strong magnetic fields anchored in the inner crust.

3.2 Heat-blanketing envelopes of magnetars: the effect of neutrino emission

Magnetars differ from ordinary pulsars in two respects: they possess superstrong surface magnetic fields, and they are generally younger and hotter. The first circumstance suggests to extend the heat-blanketing layer to deeper layers (at least to the neutron drip density as was done by Potekhin et al. 2003). The second indicates that neutrino emission can be important in the heat-blanketing envelopes. Accordingly, we have modified our computer code, which calculates the thermal structure (Potekhin and Yakovlev 2001; Potekhin et al. 2003), to make it fully relativistic and to allow for energy sinks during heat diffusion.

3.2.1 Basic equations

A complete set of equations for mechanical and thermal structure of a spherically symmetric star in hydrostatic equilibrium has been derived by Thorne (1977). They can easily be transformed to the form valid in an envelope of a star with radial heat transport, anisotropic (slowly varying) temperature distribution over any spherical layer, and a force-free magnetic field. Assuming quasistationary heat transport and neutrino emission, these equations reduce to the following system of ordinary differential equations for the metric function (gravitational potential) Φ, the local heat flux F_r, temperature T, and gravitational mass m, contained within a sphere of circumferential radius r, as functions of pressure P:

$$\frac{d\Phi}{d\ln P} = -\frac{1}{\mathcal{K}_h}\frac{P}{\rho c^2}, \tag{1}$$

$$\frac{1}{r^2}\frac{d(r^2 F_r)}{d\ln P} = \frac{P}{\rho g}\frac{Q}{\mathcal{K}_r^2\mathcal{K}_h\mathcal{K}_g} - 2F_r\frac{d\Phi}{d\ln P}, \tag{2}$$

$$\frac{d\ln T}{d\ln P} = \frac{3}{16\,\sigma T^4}\frac{F_r}{g}\frac{K'P}{\mathcal{K}_h\mathcal{K}_g} - \frac{d\Phi}{d\ln P}, \tag{3}$$

$$\frac{dr}{d\ln P} = -\frac{P}{\rho g}\frac{1}{\mathcal{K}_r\mathcal{K}_h\mathcal{K}_g}, \tag{4}$$

$$\frac{dm}{d\ln P} = -\frac{4\pi r^2 P}{g\mathcal{K}_r\mathcal{K}_h\mathcal{K}_g}. \tag{5}$$

Here ρ is the mass density (equivalent energy density of the matter), Q is the net energy loss per unit volume ($Q = Q_\nu$ is the neutrino emissivity in our case, although generally $Q = Q_\nu - Q_\mathrm{h}$, Q_h being the heat deposition rate, e.g., due to nuclear reactions), $K' = K\rho_\mathrm{bar}/\rho$, K is the opacity, $\rho_\mathrm{bar} = n_\mathrm{bar} m_\mathrm{H}$ is the so-called "baryon mass density," n_bar is the baryon number density, m_H is the mass of the hydrogen atom, σ is the Stefan–Boltzmann constant, $g = Gm/(r^2\mathcal{K}_r)$ is the local gravity, and G is the gravitational constant. Furthermore,

$$\mathcal{K}_r = (1 - 2Gm/rc^2)^{1/2}, \tag{6}$$

$$\mathcal{K}_h = 1 + P/\rho c^2, \tag{7}$$

$$\mathcal{K}_g = 1 + 4\pi r^3 P/mc^2, \tag{8}$$

are general relativistic corrections to radius, enthalpy and surface gravity, respectively. We adopt the conventional definition of the opacity K, used also by Thorne (1977). In our notations $K' = 16\sigma T^3/(3\kappa\rho)$, where κ is the total (electron plus radiative) thermal conductivity of the plasma. Note that $\rho \approx \rho_\mathrm{bar}$ and $K' \approx K$ in the entire neutron star envelope. We use the same thermal conductivities as in Potekhin and Yakovlev (2001). The effective radial thermal conductivity in a local part of the surface equals $\kappa = \kappa_\parallel \cos^2\theta_B + \kappa_\perp \sin^2\theta_B$, where $\kappa_\parallel$ and $\kappa_\perp$ are the components of the conductivity tensor responsible for heat transport along and across field lines, respectively.

The local (non-neutrino) luminosity equals the integral of the flux over the sphere of radius r,

$$L_r = \int \sin\theta\, d\theta\, d\varphi r^2 F_r(\theta, \varphi), \tag{9}$$

where (θ, φ) are the polar and azimuthal angles. For a magnetic dipole model (Ginzburg and Ozernoy 1964), $\tan\theta = 2\tan\theta_B$.

The boundary conditions to (1–5) are

$$\Phi_\mathrm{s} = \frac{\mathcal{K}_{g\mathrm{s}}}{2}, \qquad F_{r\mathrm{s}} = F_R = \sigma T_\mathrm{s}^4, \qquad r_\mathrm{s} = R, \qquad m_\mathrm{s} = M,$$

and the surface pressure is determined, following Gudmundsson et al. (1983), by the condition $K_{\mathrm{rad,s}} P_\mathrm{s}/g_\mathrm{s} = 2/3$, where K_rad is the radiative opacity. The subscript s refers to surface values; M is the gravitational stellar mass.

General relativity correction factors $\mathcal{K}_r$, $\mathcal{K}_h$, and $\mathcal{K}_g$ in (6–8) are nearly constant because $(M - m)/M \sim 10^{-5}$ and $P/\rho c^2 \sim 10^{-2}$ at the bottom of the outer crust. However, they are taken into account in our code, in order to extend calculations to deeper neutron star layers, when required.

Equations (2), (3) are one-dimensional, which implies that the mean temperature gradient along stellar radius is large compared to the tangential temperature gradient, i.e., $\epsilon \equiv |\partial T/\partial x|/|\partial T/\partial r| \ll 1$, where x is a coordinate along the stellar surface. Let us roughly estimate the mean value of ϵ for a large-scale (e.g., dipole) magnetic field, following Greenstein and Hartke (1983). In this case, $\epsilon \sim (\overline{T}_s/T_0)(l_0/R)$, where l_0 is the depth at which the temperature distribution becomes nearly isotropic ($T = T_0$ at the depth l_0) and $\overline{T}_s$ is the mean surface temperature. At the bottom of the outer crust (i.e., assuming $T_0 = T_b$) we have $l_0/R \lesssim 0.1$ (e.g., $l_0 \approx 0.6$ km for $M = 1.4M_\odot$ and $R = 10$ km) and $\overline{T}_s/T_0 \sim 10^{-2}$ (see Sect. 3.2.2). This estimate gives $\epsilon \lesssim 10^{-3}$.

Nevertheless, the one-dimensional approximation is inaccurate at those loci where magnetic field lines are tangential to the surface, because in this case one cannot neglect their curvature, $\partial\theta_B/\partial x$. For large-scale magnetic field ($\partial\theta_B/\partial x \sim R^{-1}$), the maximum size a of such sites can be estimated as the distance at which an initially tangential field line crosses the depth l_0 (which assumes that heat flows along field lines, i.e., $\kappa_\perp/\kappa_\| \ll l_0/a$). Thus, $a \lesssim \sqrt{Rl_0}$. Since T_s is minimal at such sites, their contribution to the total stellar luminosity can be neglected in the first approximation.

3.2.2 Results

We have solved (1–5) using a straightforward generalization of the Runge–Kutta method employed in our previous papers (Potekhin et al. 1997; Potekhin and Yakovlev 2001; Potekhin et al. 2003). Temperature profiles have been calculated within a local part of the blanketing envelope with a locally constant magnetic field. At every value of P, a corresponding value of ρ was found from the equation of state of magnetized electron–ion relativistic plasma (e.g., Potekhin and Yakovlev 1996), using approximations of Fermi–Dirac integrals presented in Potekhin and Chabrier (2000). In all examples shown in the figures we chose the "canonical" neutron star mass $M = 1.4M_\odot$ and radius $R = 10$ km. The neutrino emissivity is calculated as $Q_\nu = Q_{\rm pair} + Q_{\rm pl} + Q_{\rm syn} + Q_{\rm brems}$, where the contributions due to electron-positron pair annihilation $Q_{\rm pair}$, plasmon decay $Q_{\rm pl}$, synchrotron radiation of neutrino pairs by electrons $Q_{\rm syn}$, and bremsstrahlung in electron-nucleus collisions $Q_{\rm brems}$ are given, respectively, by (22), (38), (56), and (76) of Yakovlev et al. (2001). According to the results of Itoh et al. (1996) and Yakovlev et al. (2001), other neutrino emission mechanisms are unimportant in the outer crust of neutron stars.

Our calculations show that neutrino emission is crucially important for the thermal structure of neutron stars with internal temperature $T_b \gtrsim 10^9$ K.

Figure 1 shows temperature profiles at $T_b = 10^8$ K, 10^9 K, and 2×10^9 K for magnetic fields $B = 10^{12}$ G and 10^{15} G, directed perpendicular to the stellar surface. The present results (*solid* and *dot-dashed lines*) are compared with the profiles calculated neglecting neutrino emission (dashed lines). At the lowest temperature $T_b = 10^8$ K there is virtually no difference (all the lines coincide). At $T_b = 10^9$ K, the difference is noticeable, and at $T_b = 2\times10^9$ K it is large. Because of the growth of neutrino emission with increasing temperature at $\rho \gtrsim 10^{10}$ g cm^{-3}, T_s is nearly independent of T_b at $T_b \gtrsim 10^9$ K, but depends on the magnetic field.

In this figure we also compare temperature profiles for the different heavy-element compositions of the outer envelope: iron (*dot-dashed lines*) and ground-state matter (Negele and Vautherin 1973). The effect of composition is

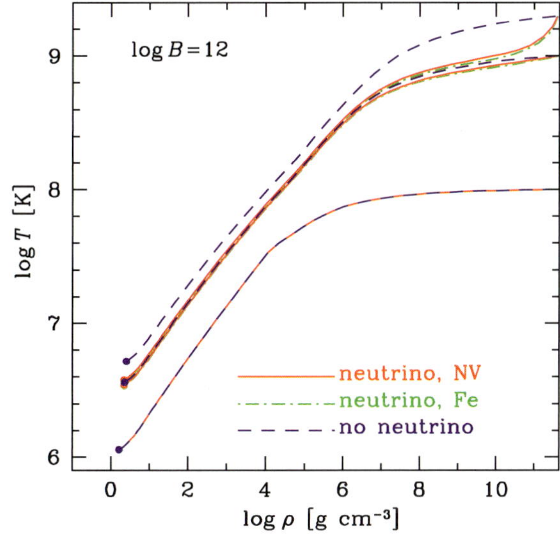

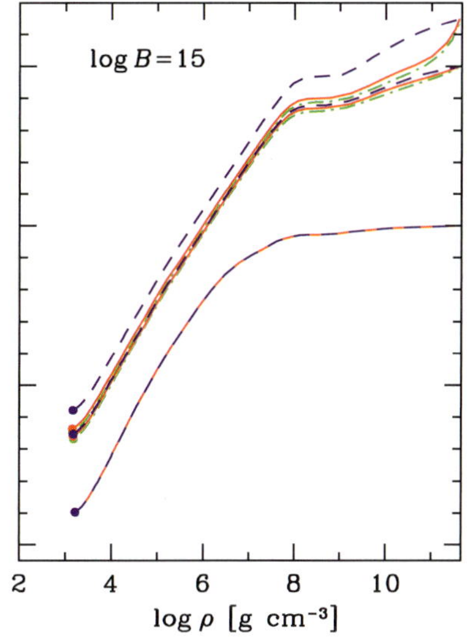

Fig. 1 Temperature profiles in the outer crust of a neutron star with magnetic field $B = 10^{12}$ G (*left panel*) or $B = 10^{15}$ G (*right panel*), directed perpendicular to the stellar surface ($\theta_B = 0$). *Solid lines* (NV)—ground-state (Negele–Vautherin) nuclear composition of the envelope, *dot-dashed lines*—^{56}Fe envelope. For comparison, *dashed lines* show the temperature profiles without allowance for neutrino emission (for ground-state matter). For every family of curves, temperature at the neutron drip density is fixed to $T_b = 10^8$ K, 10^9 K, and 2×10^9 K. The *dots at the left end* of the profiles correspond to the radiative surface, where the optical depth equals 2/3 and $\sigma T^4 = F_R$

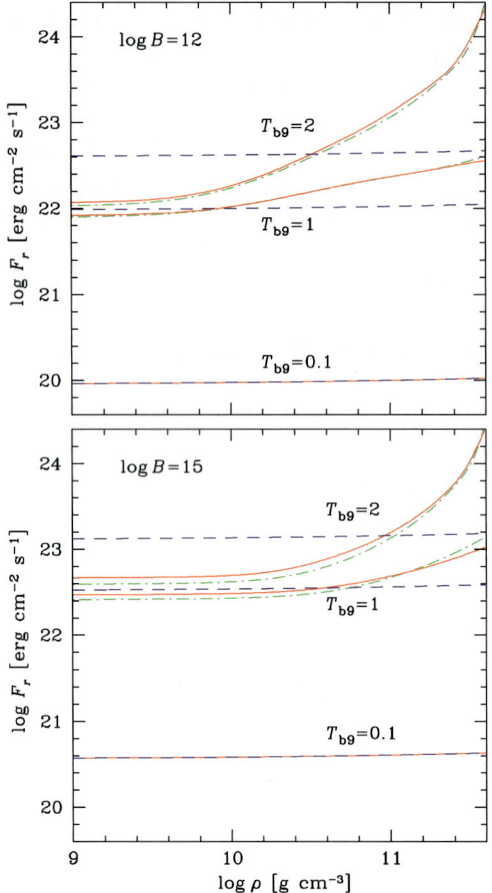

Fig. 2 Profiles of local radial heat flux F_r for the cases shown in Fig. 1. *Top panel*: $B = 10^{12}$ G, *bottom panel*: $B = 10^{15}$ G; *solid lines*—ground-state matter, *dot-dashed lines*—^{56}Fe, *dashed lines*—without neutrino emission for ground-state matter

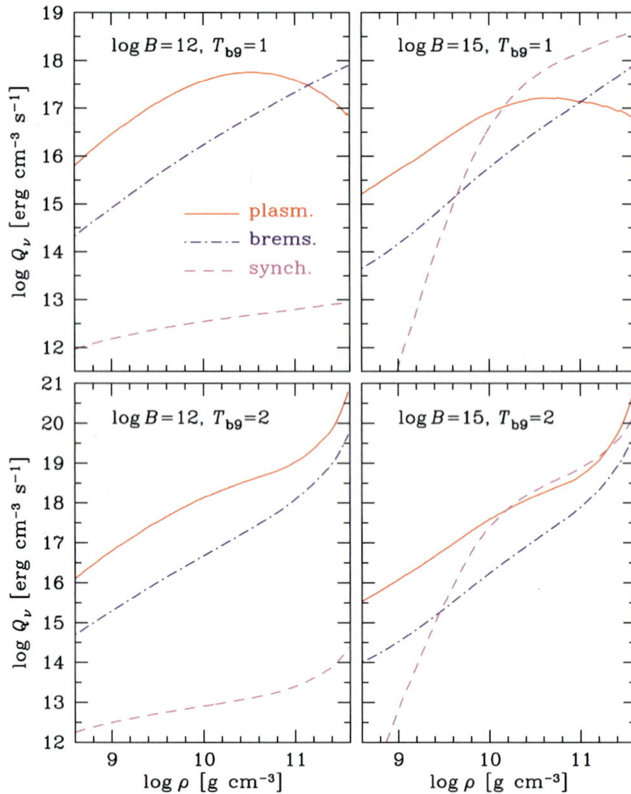

Fig. 3 Contributions to the total neutrino emissivity Q_ν from plasmon decay (*solid lines*), electron neutrino bremsstrahlung (*dot-dashed lines*), and electron synchrotron radiation (*dashed lines*). *Left panels*: $B = 10^{12}$ G (radial field); *right panels*: $B = 10^{15}$ G; *top panels*: $T = 10^9$ K; *bottom panels*: $T = 2 \times 10^9$ K

not strong, but noticeable when the neutrino emission is important, because Q_ν depends on the electron number density that is a function of composition for a given ρ.

Figure 2 demonstrates the profile of the local heat flux F_r, for the same cases as in Fig. 1, plotted by the same line types. Without neutrino emission (*dashed lines*), F_r would be nearly constant, with only $\approx$2% increase towards the inner crust due to the General Relativity effects (associated with the variation of the metric function Φ) and $\approx$9% increase because of the spherical geometry (the r^2 factor). The neutrino emission leads to a strong dependence of the flux on ρ and violates the familiar relation between F_r and T_s derived in the absence of energy sinks.

Figure 3 shows which neutrino emission mechanism dominates at given ρ, T_b, and B. For superstrong magnetic fields, the neutrino synchrotron mechanism dominates in certain density ranges, which does not happen at "ordinary pulsar" $B \sim 10^{12}$ G. It is natural because of the strong B-dependence of the synchrotron neutrino emissivity. Pair annihilation neutrino emissivity is not seen in the figure, because it is too small. Notice that the emissivity of plasmon decay and bremsstrahlung processes can be affected by superstrong magnetic fields which has not been explored so far (and we present the emissivities in the field-free case). A slow dependence of these emissivities on B, seen in Fig. 3, is indirect (caused by the dependence of temperature profiles on B).

Figures 4 and 5 give the $T_s(T_b)$ relation for the magnetic fields $B = 10^{12}$ G and $B = 10^{15}$ G perpendicular and parallel to the radial direction. The relations obtained with and without allowance for neutrino emission are plotted by solid and dashed lines, respectively. We see that at $T_b \lesssim 10^8$ K the neutrino emission does not affect T_s. At higher $T_b \gtrsim 10^9$ K, in contrast, this emission is crucial: if $Q_\nu = 0$, then T_s continues to grow up with increasing T_b, whereas with realistic Q_ν the surface temperature tends to a constant limit, which depends on θ_B and B. In most cases this limit is reached when $T_b \sim 10^9$ K, but for a superstrong field (*right panel*) and transverse heat propagation, it is reached at still smaller $T_b \sim 3 \times 10^8$ K.

In Fig. 6 we explore joint effects of magnetic field, neutrino emission, and the shift of the inner boundary from 10^{10} g cm^{-3} to the neutron drip. Here magnetic field lines are directed along the surface (perpendicular to the direc-

tion of heat transport). Therefore, these effects are most pronounced. The solid lines are the results of accurate calculations; the *dashed lines*, as before, show the model with $Q_\nu = 0$; and *dot-dashed* lines are obtained by solving (1–5) in the whole domain $\rho_s \le \rho \le 4 \times 10^{11}$ g cm^{-3}, but with the magnetic field artificially "switched off" at $\rho > 10^{10}$ g cm^{-3}. This is a simulation of the model, where the heat transport in the magnetized plasma is solved accurately up to $\rho_b = 10^{10}$ g cm^{-3}, while after this boundary nonmagnetic heat balance and transport equations are solved. In the absence of the results reported here, the latter model was used by Kaminker et al. (2007) to study the thermal structure and evolution of magnetars as cooling neutron stars with a phenomenological heat source in a spherical internal layer. In the *left panel* of Fig. 6, the *solid* and *dot-dashed lines* almost coincide, indicating that in this case $\rho_b = 10^{10}$ g cm^{-3} may provide a sufficient accuracy. In the *right panel*, in contrast, the effect of the shift of the inner boundary is quite visible. Therefore, we conclude that the development of our thermal structure code, reported here, will allow us to study the thermal history of magnetars at a higher accuracy level.

Meanwhile, a comparison of the profiles shown in Figs. 1 and 6 prompts that at $T \gtrsim 10^9$ K and $B \sim 10^{15}$ G the magnetic effects on the conductivity could be important at still higher densities in the inner crust. Investigation of this possibility requires taking into account the effects of free (possibly superfluid) neutrons on thermal conduction and neutrino emission. We are planning to perform such study in the future.

In the above figures we have shown the results of calculations where the surface of a neutron star was assumed diffuse, i.e., without the phase transition discussed in Sect. 2.2.

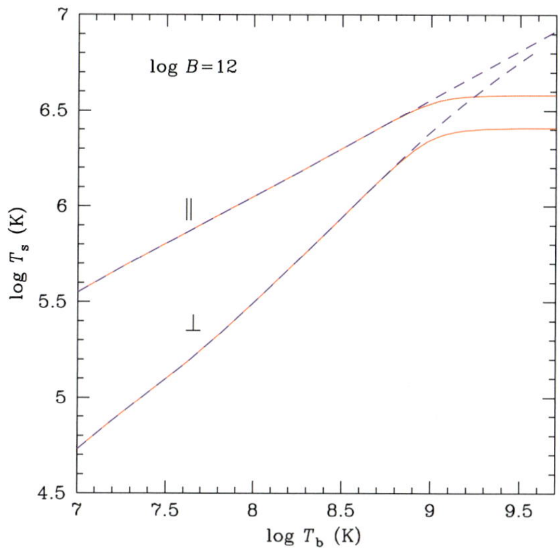

Fig. 4 Surface temperature T_s as a function of temperature T_b at the neutron drip point for a neutron star with the magnetic field $B = 10^{12}$ G, directed along stellar radius ($\theta_B = 0$, parallel heat transport, sign ∥) or along the surface ($\theta_B = 90°$, transverse transport, ⊥). Ground-state composition is assumed. *Solid lines* present calculation, *dashed lines*—neutrino emission is neglected

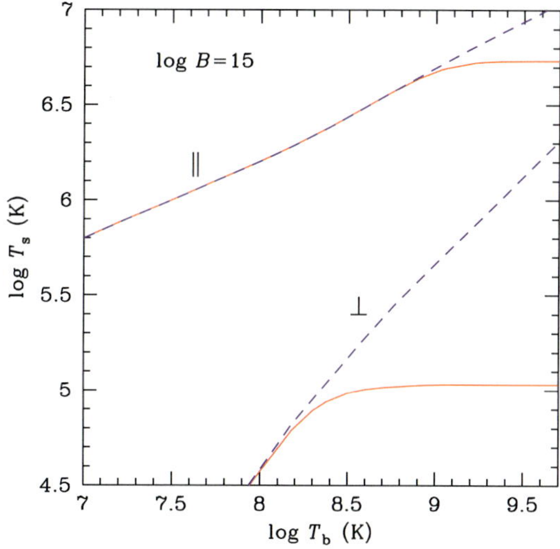

Fig. 5 The same as in Fig. 4, but for $B = 10^{15}$ G

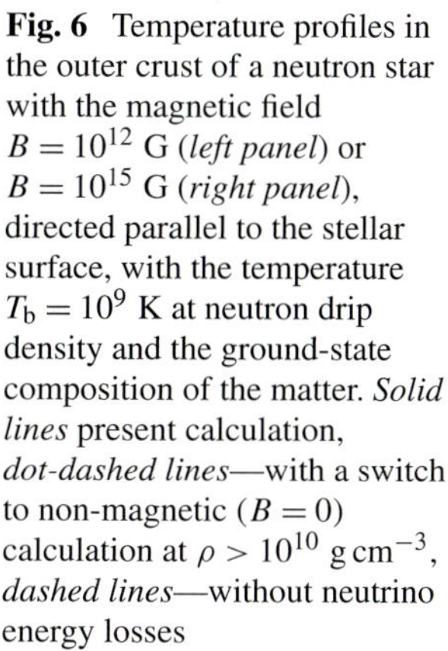

Fig. 6 Temperature profiles in the outer crust of a neutron star with the magnetic field $B = 10^{12}$ G (*left panel*) or $B = 10^{15}$ G (*right panel*), directed parallel to the stellar surface, with the temperature $T_b = 10^9$ K at neutron drip density and the ground-state composition of the matter. *Solid lines* present calculation, *dot-dashed lines*—with a switch to non-magnetic ($B = 0$) calculation at $\rho > 10^{10}$ g cm^{-3}, *dashed lines*—without neutrino energy losses

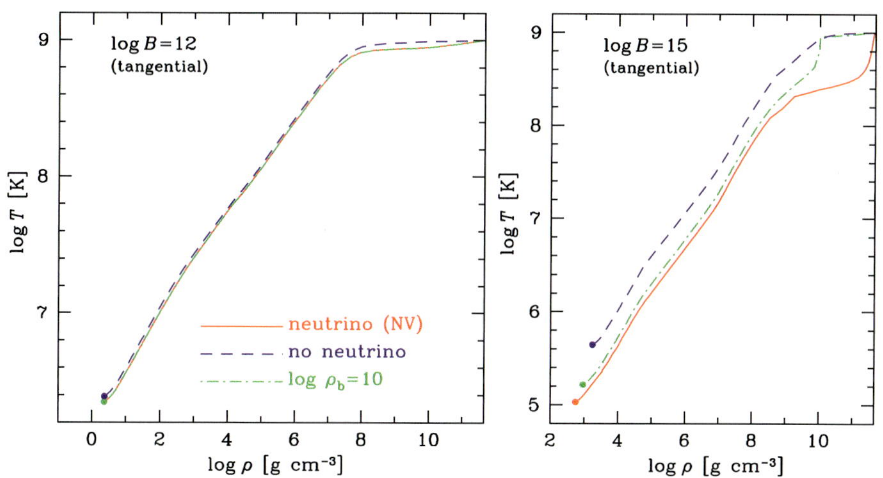

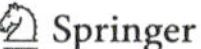

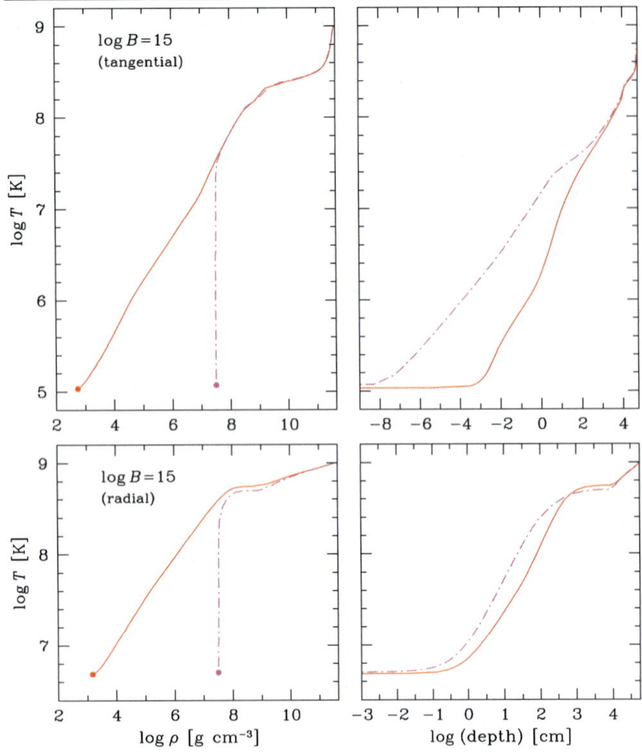

Fig. 7 Temperature in the outer crust of a neutron star as a function of density (*left panels*) or proper depth behind the radiative surface (*right panels*). Magnetic field $B = 10^{15}$ G is directed parallel (*top panels*) or perpendicular (*bottom panels*) to the stellar surface; $T_b = 10^9$ K. *Solid lines*—the model of diffuse surface (no phase transition), *dot-dashed lines*—condensed surface

However, for $B = 10^{15}$ G and $T_b \lesssim 10^9$ K the surface temperature T_s is below the estimates of the critical temperature T_c given in Sect. 2.2. This possibility is explored in Fig. 7. Here we use the equation of state containing non-ideal terms for strongly magnetized fully ionized plasma (Sect. IIIB of Potekhin et al. 1999), which enforce the phase transition. At $B = 10^{15}$ G the surface density is very high, $\rho_{\rm cond} \approx 3.1 \times 10^7$ g cm^{-3}, in our model (and the analytic estimate in Sect. 2.2 gives $\rho_{\rm cond} \sim 2 \times 10^7$ g cm^{-3}). The *solid lines* in the *left panels* reproduce the profiles shown in Figs. 1 and 6, whereas the *dot-dashed lines* display the case of magnetic condensation. In the *right panels*, we show the same temperature profiles as a function of local proper depth l ($\mathrm{d}l = -\mathcal{K}_r\,\mathrm{d}r$), measured from the radiative surface. Although the temperature profiles with and without magnetic condensation are drastically different in the surface layers, the effective temperature remains almost the same. Thus, the magnetic condensation does not significantly affect the $T_s(T_b)$ relation. This should not be surprising, because, as explained by Gudmundsson et al. (1983), the main regulator of the T_b–T_s relation is the "sensitivity strip" where κ has a minimum. This domain, a real bottleneck for the heat outflow, lies at $\rho > \rho_{\rm cond}$ (except for low T_s, not considered here), and therefore it is not affected by the condensation.

4 Summary

We have described the main effects of strong magnetic fields on the properties of neutron star atmospheres and heat blanketing envelopes, with the emphasis on the difference between ordinary neutron stars and magnetars. Observations of magnetars pose new theoretical problems and challenges because of magnetars' superstrong magnetic fields and high temperatures. We also report a solution to one of such problems, which consists in taking into account neutrino energy losses in the outer crust of hot, strongly magnetized neutron stars. We have demonstrated that, because of these losses, the effective surface temperature T_s almost ceases to depend on the temperature T_b in the inner crust as soon as T_b exceeds 10^9 K. A direct consequence of this observation is that in the absence of powerful energy sources in outer envelopes, the stationary (time-averaged) effective temperature cannot be raised above the value that it would have at $T_b \approx 10^9$ K, irrespectively of the energy release in the deeper layers.

Acknowledgements We are grateful to the anonymous referee for useful comments. A.Y.P. thanks the organizers of the conference "Isolated Neutron Stars: from the Interior to the Surface" (London, April 24–28, 2006), especially Silvia Zane and Roberto Turolla, for perfect organization, attention and support. Fruitful discussions with participants of this conference have significantly contributed to the developments partly reported in the present paper.

References

Al-Hujaj, O.-A., Schmelcher, P.: Phys. Rev. A **67**, 023403 (2003a)
Al-Hujaj, O.-A., Schmelcher, P.: Phys. Rev. A **68**, 053403 (2003b)
Bezchastnov, V.G., Pavlov, G.G., Ventura, J.: Phys. Rev. A **58**, 180 (1998)
Brinkmann, W.: Astron. Astrophys. **82**, 352 (1980)
Chabrier, G., Saumon, D., Potekhin, A.Y.: J. Phys. A Math. Gen. **39**, 4411 (2006)
Geppert, U., Küker, M., Page, D.: Astron. Astrophys. **426**, 267 (2004)
Geppert, U., Küker, M., Page, D.: Astron. Astrophys. **457**, 957 (2006)
Ginzburg, V.L., Ozernoy, L.M.: Zh. Eksp. Teor. Fiz. **47**, 1030 (1964) (English translation: Sov. Phys. JETP **20**, 689)
Greenstein, G., Hartke, G.J.: Astrophys. J. **271**, 283 (1983)
Gudmundsson, E.H., Pethick, C.J., Epstein, R.I.: Astrophys. J. **272**, 286 (1983)
Hernquist, L.: Mon. Not. Roy. Astron. Soc. **213**, 313 (1985)
Heyl, J.S., Hernquist, L.: Mon. Not. Roy. Astron. Soc. **300**, 599 (1998)
Heyl, J.S., Hernquist, L.: Mon. Not. Roy. Astron. Soc. **324**, 292 (2001)
Ho, W.C.G., Lai, D.: Mon. Not. Roy. Astron. Soc. **327**, 1081 (2001)
Ho, W.C.G., Lai, D.: Mon. Not. Roy. Astron. Soc. **338**, 233 (2003)
Ho, W.C.G., Kaplan, D.L., Chang, P., van Adelsberg, M., Potekhin, A.Y.: Astrophys. Space Sci. DOI 10.1007/s10509-007-9366-2
Itoh, N., Hayashi, H., Nishikawa, A., Kohyama, Y.: Astrophys. J. Suppl. Ser. **102**, 411 (1996)
Jones, P.B.: Mon. Not. Roy. Astron. Soc. **218**, 477 (1986)
Kaminker, A.D., Yakovlev, D.G., Potekhin, A.Y., et al.: Astrophys. Space Sci. DOI 10.1007/s10509-007-9358-2
Lai, D.: Rev. Mod. Phys. **73**, 629 (2001)
Lai, D., Ho, W.C.G.: Astrophys. J. **588**, 962 (2003)

Lai, D., Salpeter, E.E.: Astrophys. J. **491**, 270 (1997)
Lozovik, Yu.E., Volkov, S.Yu.: Phys. Rev. A **70**, 023410 (2004)
Medin, Z., Lai, D.: Phys. Rev A **74**, 062508 (2006)
Motch, C., Zavlin, V.E., Haberl, F.: Astron. Astrophys. **408**, 323 (2003)
Negele, J.W., Vautherin, D.: Nucl. Phys. A **207**, 298 (1973)
Page, D.: Astrophys. J. **442**, 273 (1995)
Pavlov, G.G., Bezchastnov, V.G.: Astrophys. J. **635**, L61 (2005)
Pavlov, G.G., Shibanov, Yu.A., Zavlin, V.E., et al.: Neutron star atmospheres. In: Alpar, M.A., Kiziloğlu, Ü., van Paradijs, J. (eds.) The Lives of the Neutron Stars, Proceedings of NATO Advanced Study Institute Ser. C, vol. 450, p. 71. Kluwer, Dordrecht (1995)
Pérez-Azorín, J.F., Miralles, J.A., Pons, J.A.: Astron. Astrophys. **433**, 275 (2005)
Pérez-Azorín, J.F., Miralles, J.A., Pons, J.A.: Astron. Astrophys. **451**, 1009 (2006)
Potekhin, A.Y.: Astron. Astrophys. **351**, 787 (1999)
Potekhin, A.Y., Chabrier, G.: Phys. Rev. E **62**, 8554 (2000)
Potekhin, A.Y., Chabrier, G.: Astrophys. J. **600**, 317 (2004)
Potekhin, A.Y., Yakovlev, D.G.: Astron. Astrophys. **314**, 341 (1996)
Potekhin, A.Y., Yakovlev, D.G.: Astron. Astrophys. **374**, 213 (2001)
Potekhin, A.Y., Chabrier, G., Yakovlev, D.G.: Astron. Astrophys. **323**, 415 (1997)
Potekhin, A.Y., Chabrier, G., Shibanov, Yu.A.: Phys. Rev. E **60**, 2193 (1999)
Potekhin, A.Y., Yakovlev, D.G., Chabrier, G., et al.: Astrophys. J. **594**, 404 (2003)
Potekhin, A.Y., Lai, D., Chabrier, G., et al.: Astrophys. J. **612**, 1034 (2004)
Potekhin, A.Y., Urpin, V.A., Chabrier, G.: Astron. Astrophys. **443**, 1025 (2005)
Potekhin, A.Y., Chabrier, G., Lai, D., Ho, W.C.G.: J. Phys. A Math. Gen. **39**, 4453 (2006)
Rajagopal, M., Romani, R., Miller, M.C.: Astrophys. J. **479**, 347 (1997)
Ruderman, M.: Phys. Rev. Lett. **27**, 1306 (1971)
Schaaf, M.E.: Astron. Astrophys. **227**, 61 (1990a)
Schaaf, M.E.: Astron. Astrophys. **235**, 499 (1990b)
Shibanov, Yu.A., Yakovlev, D.G.: Astron. Astrophys. **309**, 171 (1996)
Shibanov, Yu.A., Zavlin, V.E., Pavlov, G.G., Ventura, J.: Astron. Astrophys. **266**, 313 (1992)
Thorne, K.S.: Astrophys. J. **212**, 825 (1977)
Turolla, R., Zane, S., Drake, J.J.: Astrophys. J. **603**, 265 (2004)
van Adelsberg, M., Lai, D.: Mon. Not. Roy. Astron. Soc. **373**, 1495 (2006)
van Adelsberg, M., Lai, D., Potekhin, A.Y., et al.: Astrophys. J. **628**, 902 (2005)
Van Riper, K.A.: Astrophys. J. **329**, 339 (1988)
Yakovlev, D.G., Pethick, C.: Annu. Rev. Astron. Astrophys. **42**, 169 (2004)
Yakovlev, D.G., Kaminker, A.D., Gnedin, O.Y., et al.: Phys. Rep. **354**, 1 (2001)
Yakovlev, D.G., Gnedin, O.Y., Gusakov, M.E., et al.: Nucl. Phys. A **752**, 590 (2005)
Zane, S., Turolla, R., Treves, A.: Astrophys. J. **537**, 387 (2000)
Zane, S., Turolla, R., Stella, L., et al.: Astrophys. J. **560**, 384 (2001)
Zavlin, V.E., Pavlov, G.G.: Modeling neutron star atmospheres. In: Becker, W., Lesch, H., Trümper, J. (eds.) Proceedings of the 270. WE-Heraeus Seminar on Neutron Stars, Pulsars, and Supernova Remnants, p. 263. MPE, Garching (2002) (MPE Report 278)

Astrophys Space Sci (2007) 308: 363–369
DOI 10.1007/s10509-007-9363-5

ORIGINAL ARTICLE

Equation of state of neutron star cores and spin down of isolated pulsars

P. Haensel · J.L. Zdunik

Received: 30 June 2006 / Accepted: 9 October 2006 / Published online: 16 March 2007

Abstract We study possible impact of a softening of the equation of state by a phase transition, or appearance of hyperons, on the spin evolution of isolated pulsars. Numerical simulations are performed using exact 2-D simulations in general relativity. The equation of state of dense matter at supranuclear densities is poorly known. Therefore, the accent is put on the general correlations between evolution and equation of state, and mathematical strictness. General conjectures referring to the structure of the one-parameter families of stationary configurations are formulated. The interplay of the back bending phenomenon and stability with respect to axisymmetric perturbations is described. Changes of pulsar parameters in a corequake following instability are discussed, for a broad choice of phase transitions predicted by different theories of dense matter. The energy release in a corequake, at a given initial pressure, is shown to be independent of the angular momentum of collapsing configuration. This result holds for various types of phases transition, with and without metastability. We critically review observations of pulsars that could be relevant for the detection of the signatures of the phase transition in neutron star cores.

Keywords Dense matter · Equation of state · Pulsars: general · Stars: neutron · Stars: rotation

PACS 26.60.+c · 97.60.Gb · 97.60.Jd · 45.20.dc

This work was partially supported by the Polish MNiI Grant no. 1P03D-008-27.

P. Haensel (✉) · J.L. Zdunik
N. Copernicus Astronomical Center, Polish Academy of Sciences, Bartycka 18, PL-00-716 Warsaw, Poland
e-mail: haensel@camk.edu.pl

J.L. Zdunik
e-mail: jlz@camk.edu.pl

1 Introduction

The equation of state (EOS) above a few times nuclear density is a main mystery of neutron stars. Our degree of ignorance concerning EOS of neutron star core is therefore high. In particular, different theories of dense matter predict various types of softening of the EOS above some density, due to phase transition to exotic phases of hadronic matter (meson condensates, quark matter), or presence of new baryons, such as hyperons (for a review, see Glendenning 2000).

Since their discovery in 1967, thanks to the exceptional precision of their timing, isolated radio pulsars are cosmic laboratories to study physics of ultradense matter. Such a pulsar looses its angular momentum, J, due to radiation. In response to the angular momentum loss, $\dot{J} < 0$, the pulsar changes its rotation frequency $f = 1/\text{period} = \Omega/2\pi$ (measured in pulsar timing) and increases its central density and pressure ρ_c, P_c. The response of the stellar structure to $\dot{J}$ depends on the equation of state (EOS) of the neutron star core, and is sensitive to appearance of new particles or of a new phase at the neutron star center. In particular, crossing the phase-transition region by ρ_c is reflected by specific "non-standard" behavior of f as a function of time. In the present paper we analyze the relation between the pulsar spin down and the EOS. This problem was studied previously by many authors, who considered various types of EOSs of dense matter predicted by different theories (Weber and Glendenning 1991, 1992; Glendenning et al. 1997; Heiselberg and Hjorth-Jensen 1998; Chubarian et al. 2000; Spyrou and Stergioulas 2002). In the present paper we summarize and review results obtained using a different approach. Namely, through a flexible parametrization of dense matter EOS, we are able to study the problem of spin down evolution, including stability and the back bending phenomenon (spin-up accompanying angular momentum loss), for a very broad, con-

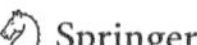

tinuous family of EOSs with and without phase transitions, and select those features which are genuine properties of the EOSs. On the other hand, the use of parametrized EOSs allows to perform very precise 2-D calculations, crucial for reliable analysis of stability, and avoiding usual severe limitations resulting from the lack of precision.

The plan of the present paper is as follows. In Sect. 2 we review effects of phase transitions on the equation of state (EOS) of dense matter under the assumption of full thermodynamic equilibrium. Then in Sect. 3 we remind stability criteria for relativistic rotating stars. Importance of stability for the back bending phenomenon is discussed in Sect. 4. In Sect. 5 various cases of evolutionary tracks of isolated pulsars, corresponding to different EOSs of dense matter, are studied, and general properties of the structure of stable families of stationary configurations are deduced. Corequakes resulting from instability of rotating configurations are discussed in Sect. 6. In Sect. 7 we show how a phase transition in neutron star core can affect the time evolution of the pulsar rotation period. Some remarkable features of changes in pulsar parameters resulting from a corequake are discussed in Sects. 8–9. Finally, Sect. 10 contains discussion and conclusions.

Results reviewed here were obtained in last few years within a fruitful collaboration with M. Bejger and E. Gourgoulhon (Zdunik et al. 2004, 2006a, 2006b; Bejger et al. 2005). We present these results in the most general terms, and summarize them in the form of conjectures (which in the future may become strict theorems), which on the one hand are general (i.e., do not depend on a specific dense matter model), and on the other hand seem to us useful for further numerical studies of the intimate relation between EOS and neutron star dynamics.

2 First order phase transition in thermodynamic equilibrium

Two possible cases of the EOS softening by a 1st order phase transition are shown schematically in Fig. 1. Simplest case is that of a first order phase transition from a pure N (normal) phase to a pure S (superdense) phase. It occurs at a well defined pressure P_0, and is accompanied by a density jump: $\rho_{\rm N} < \rho_{\rm S}$. Such a type of phase transition occurs for sufficiently strong pion or kaon condensation, and is also characteristic for many models of quark deconfinement. For a sufficiently small surface tension at the N–S interface, $\sigma < \sigma_{\rm crit}$, one has to contemplate a more complicated pattern of the N–S phase coexistence. Above some $P_{\rm N}^{\rm m}$ a mixture NS is preferred over the pure N phase, with fraction of the S phase increasing from zero to one at $P = P_{\rm S}^{\rm m}$, and above $P_{\rm S}^{\rm m}$ pure S phase is preferred. Such a situation might be possible for meson (kaon or pion) condensations or quark matter, provided $\sigma < \sigma_{\rm crit}$. Alas, our knowledge of physics of superdense matter is not sufficient to decide which case (pure or mixed) type of first order phase transition is actually realized in dense neutron star cores.

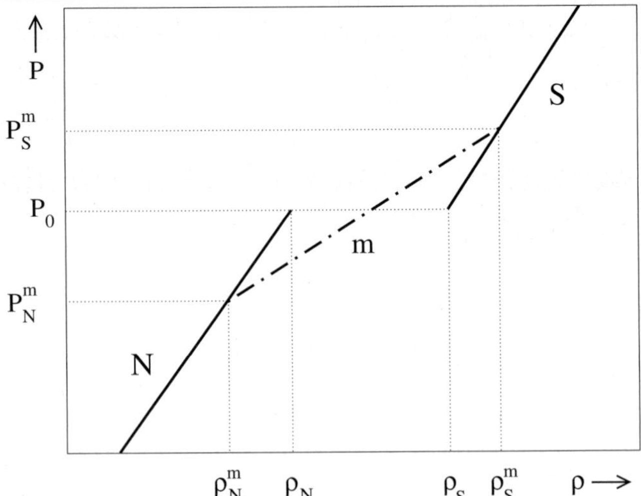

Fig. 1 Two types of the EOS with 1st order phase transition. *Thick solid lines*: stable pure N and S phases. Mixed phase is represented by a *dash-dot line*. For further explanation see the text. From Bejger et al. (2005)

3 Stability of hydrostatic equilibria

Neutron stars are assumed to be built of an ideal fluid. Therefore, stresses resulting from elastic shear, electromagnetic field, viscosity, etc. are neglected.

In our case, stability criteria are:

Stability criteria—nonrotating stars Nonrotating configuration form a one-parameter family $\mathcal{C}(x)$, and are labeled by $x = \rho_{\rm c}$ or $x = P_{\rm c}$. One considers stability of these 1-D configurations with respect to radial perturbations and finds that a configuration is stable if $\mathrm{d}M/\mathrm{d}x > 0$ and unstable if $\mathrm{d}M/\mathrm{d}x < 0$ (Harrison et al. 1965).

Stability criteria—rotating stars They were precisely formulated by Friedman et al. (1988). Stationary rotating configurations form a two-parameter family: $\mathcal{C}(x, \Omega)$. Their stability with respect to axially symmetric perturbations can be checked using any of the following criteria, involving baryon (rest) mass $M_{\rm b}$ or stellar angular momentum J:

(1) Configuration is stable if $(\partial M/\partial x)_{J=\rm const.} > 0$ and unstable if $(\partial M/\partial x)_{J=\rm const.} < 0$, or
(2) Configuration is stable if $(\partial J/\partial x)_{M_{\rm b}=\rm const.} < 0$ and unstable if $(\partial J/\partial x)_{M_{\rm b}=\rm const.} > 0$

4 Back-bending and stability of rotating configurations

Zdunik et al. (2006a) have shown examples of the EOS with 1st order phase transitions between the pure N and S phases, as well as mixed-phase transitions, for which one of the two situations occurs:

1. All configurations are stable but the back bending exists.
2. The phase transition results in the instability region for rotating stars.

To distinguish between these two cases we can look at the behavior either of the curves $M_b(\rho_c)_J$ or $J(f)_{M_b}$; the quantity fixed along a sequence is indicated by the lower index. For such curves the instability criterion directly applies. However, is not so easy to detect the instability using the $I(f)_{M_b}$ plot, where $I \equiv J/\Omega$ is stellar moment of inertia. The unstable case may look similar to that corresponding to a fully stable sequence. An example is presented in Fig. 2, where we plot two functions: $J(f)_{M_b}$ and $I(f)_{M_b}$ for the two models of EOS (MSt and MUn) with softening due a mixed-phase segment. Upper panels correspond to the stable model MSt and lower to the MUn model for which exist region of unstable configurations. The difference between the MSt and MUn cases is clearly visible in left panels ($J(f)_{M_b}$ curves), where we can easily recognize the instability region for MUn model, by applying the condition $(\partial J/\partial \rho_c)_{M_b} > 0$, and keeping in mind that ρ_c is monotonic along this curve. However, the upper and lower right panels ($I(f)_{M_b}$) are quite similar, without any qualitative difference. Simultaneously, the back bending phenomenon in the MUn case is very large (almost by 200 Hz), and obviously the phase transition results in an instability (increase of J for increasing ρ_c). An yet, $I(\rho_c)$ is a monotonic decreasing function all the time.

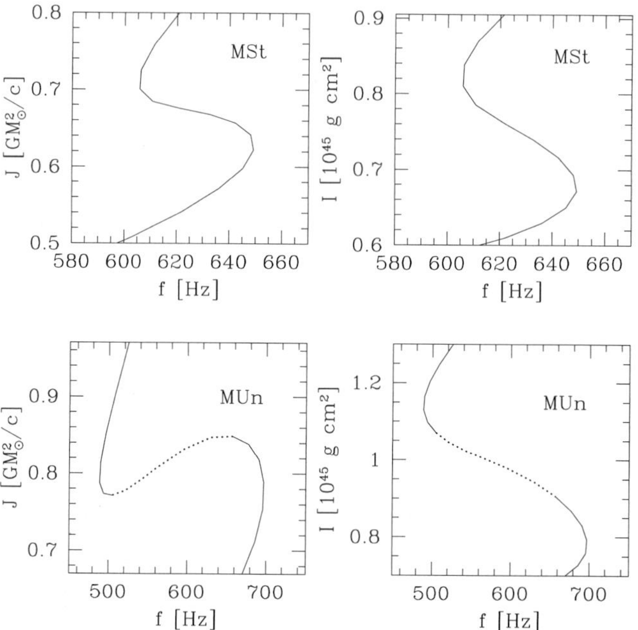

Fig. 2 Total angular momentum versus rotation frequency f (*left panels*), and moment of inertia $I \equiv J/\Omega$ versus f (*right panels*), for EOSs MSt and MUn of Zdunik et al. (2006a). The stability criterion is easily applied to left panels. It is clear that for the MSt EOS back bending feature is not associated with an instability, with all configurations being stable. On the contrary, the MUn EOS produces back bending with a large segment of unstable configurations. Simultaneously, the $I(f)$ curves for both EOSs are very similar and apparently show very similar back-bending shapes

5 Invariance of structure of (one-parameter) families $\{\mathcal{C}_X\}$

We select a one-parameter family (represented by a curve) from a two-parameter set of stationary configuration by fixing one of parameters, denoted by X, so that within this family $X = \text{const.}$ Here $X = M_b, J, f$. Examples of such one-parameter families are shown in Figs. 3, 4. As shown by Zdunik et al. (2006a), studying stability of these one parameter families reveals a very interesting invariance property. This property can be generalized to include also the special case of marginally unstable configurations, which correspond to the inflection point $(\partial M/\partial x)_{J=\text{const.}} = 0$ and $(\partial^2 M/\partial x^2)_{J=\text{const.}} = 0$. For the $M_b = \text{const.}$ families the marginal instability corresponds to $(\partial J/\partial x)_{M_b=\text{const.}} = 0$ and $(\partial^2 J/\partial x^2)_{M_b=\text{const.}} = 0$.

Only stable configurations are astrophysically relevant. In what follows, we use the term "unstable segment" of one parameter family in the restricted sense: we will always mean a segment composed of unstable configurations which

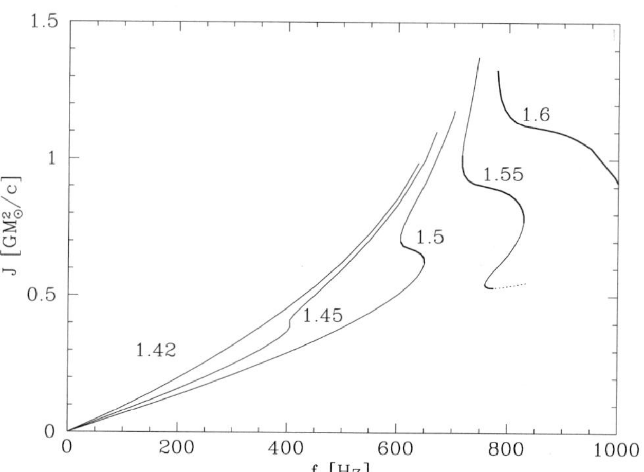

Fig. 3 Examples of the evolutionary tracks in the J–f plane for an EOS which produces back bending but does not lead to splitting of the tracks into disjoint stable branches, separated by unstable segments. Curves are labeled by baryon mass (constant along a track) in solar masses. *Thick segments* correspond to back bending phenomenon. A *small dotted* termination at the lower end of the 1.55 curve corresponds to a collapse into black hole after reaching the minimum J allowed for stationary configurations. From Zdunik et al. (2006a)

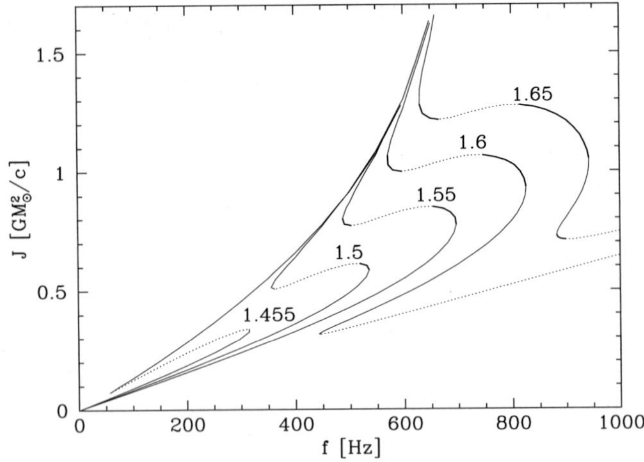

Fig. 4 Examples of the evolutionary tracks in the J–f plane for an EOS which produces unstable segments. Curves are labeled by baryon mass (constant along a track) in solar masses. *Thick segments* correspond to back bending phenomenon. From Zdunik et al. 2006a

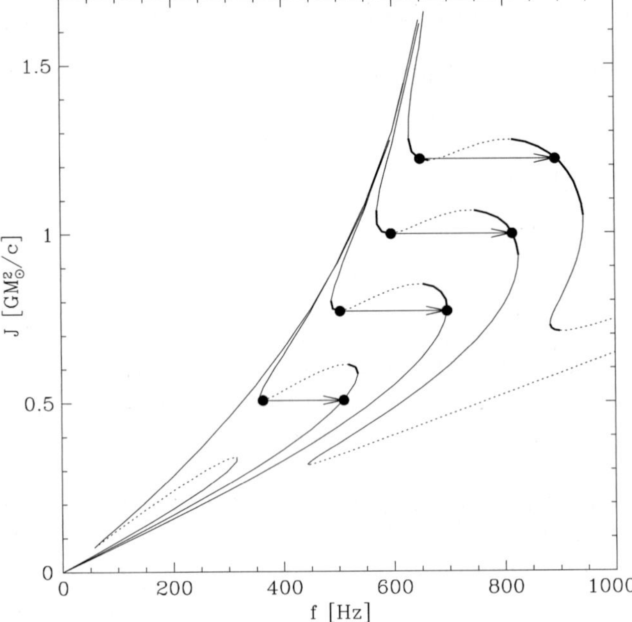

Fig. 5 Evolutionary tracks of an isolated pulsar loosing J in the J–f plane, when a softening of the EOS due to a phase transition imply instability regions (*dotted segments*). *Arrows* lead from unstable configuration to a stable collapsed one, with the same baryon mass and angular momentum. From Zdunik et al. (2006a)

is bounded on both sides by stable configurations. A remarkably general invariance property can be formulated as three conjectures:

(1) All stable $\{\mathcal{C}\}_{\rm stat}$ remains all stable $\{\mathcal{C}_X\}_{\rm rot}$.
(2) If $\{\mathcal{C}\}_{\rm stat}$ contains unstable segment then every $\{\mathcal{C}_X\}_{\rm rot}$ contains unstable segment too.
(3) If $\{\mathcal{C}\}_{\rm stat}$ contains a marginally unstable $\mathcal{C}$ then each $\{\mathcal{C}_X\}_{\rm rot}$ contains a marginally unstable $\mathcal{C}_X$.

These conjectures are based on very precise numerical 2-D simulations performed for hundreds of EOSs with softening due to phase transitions. They can also be rephrased in a more compact form:

1′ Single family of stable static configurations ⇔ single family of stable rotating configurations.
2′ Two disjoint families of stable static configurations separated by a family of unstable configurations ⇔ two disjoint families of stable rotating configurations separated by a family of unstable configurations (constant $M_{\rm b}$, or constant J, or constant f).

Therefore, topology of the one-parameter families of hydrostatic stationary equilibrium configurations is a genuine feature of the EOS, not altered by rigid rotation.

6 Instabilities and starquakes

Entering the unstable segment of the evolutionary track, during pulsar spin down, leads to a discontinuous transition (starquake) to a stationary configuration on another stable segment of the evolutionary track. This is illustrated by examples of evolutionary tracks in the J–f plane in Fig. 5. It is assumed that there is no matter ejection during the quake and that the timescale of it is so short that the angular momentum loss (by radiation of electromagnetic and gravitational waves) can be neglected. Therefore, a spinning down pulsar reaches instability point and collapses (with spinning up!) into a new stable configuration $\mathcal{C}$ with the same baryon mass and angular momentum, and then continues its evolution moving in the J–f plane (Fig. 5). The unavoidability of a corequake is best seen by showing the track in the M–J plane in Fig. 6. The only way out from $\mathcal{C}_{\rm i}$ is collapse to $\mathcal{C}_{\rm f}$, keeping $M_{\rm b}$ and J constant. The energy released during the corequake is $\Delta E = (M_{\rm i} - M_{\rm f})c^2$. Let us stress, that the corequake is accompanied by differential rotation, redistribution of the angular momentum, and breaking of the axial symmetry. It is a nonstationary phenomenon which is to be studied using methods beyond those applied to describe 2-D rigid rotation in the present paper.

7 Back bending and pulsar timing

A spinning down isolated pulsar looses its angular momentum and energy. The energy balance is described by the standard formula

$$\dot{M}c^2 = -\kappa \Omega^\alpha \quad (1)$$

where the right-hand side describes the energy loss due to radiation. The standard assumption is that the Ω-dependence

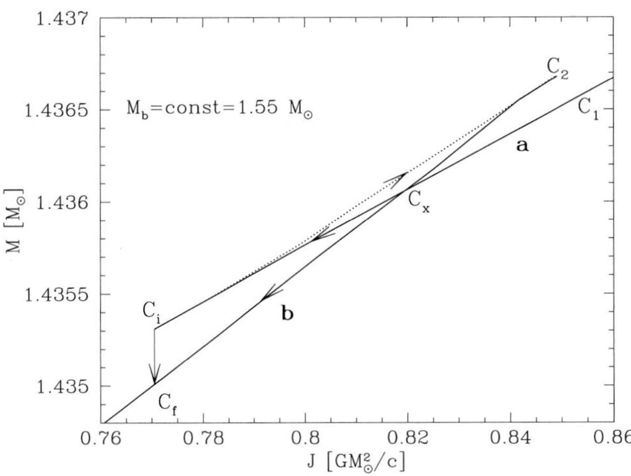

Fig. 6 Gravitational mass of the star as a function of its angular momentum, for a fixed baryon mass of the star. Central density is increasing along the curve as indicated by the *arrows*. The *upper* (*dotted*) *segment* corresponds to the unstable configurations. From Zdunik et al. (2006a)

of the moment of inertia I can be neglected, so that during spin down $I = I(\Omega = 0) = \text{const}$. However, in the phase transition epoch the Ω dependence of I is crucial for the evolution track, timing, and stability. Standard approximation $I = I(\Omega = 0) = \text{const}$. leads to a severe *overestimate* of the spin-down rate and an *underestimate* of pulsar age. An example visualizing this effect is shown in Fig. 7. We assume that a $P = 10$ ms pulsar is observed at time $= 0$. As we see, evolution back in time using $I = \text{const}$. (*dotted lines*, corresponding to different M_{b}) can be very misleading if a phase transition region was crossed in the past. Let us notice that in that "phase transition epoch" the breaking index $n = \Omega\ddot{\Omega}/\dot{\Omega}^2$ was reaching huge values. Breaking index $\gg 1$ is a signature of a phase transition taking place at the pulsar center.

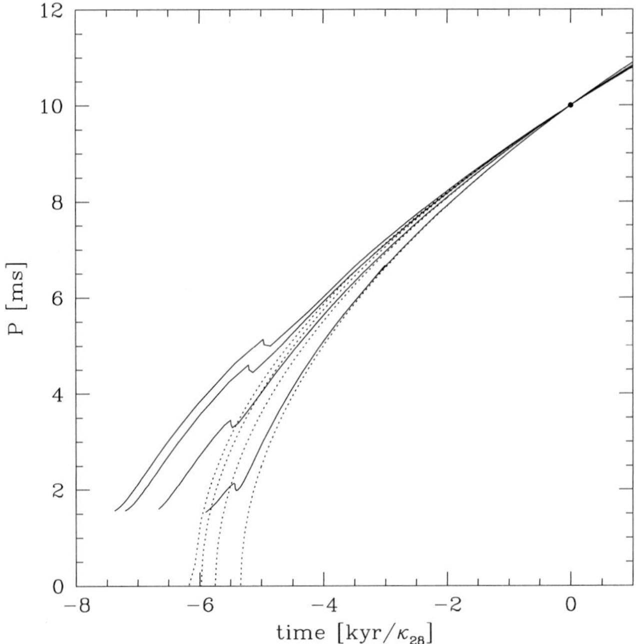

Fig. 7 The evolution of the pulsar period P when the energy loss is described by the magnetic dipole breaking with $\alpha = 4$. *Solid curves*—results obtained for an EOS with a phase transition, for different baryon masses. *Dotted lines* correspond to a standard model with $I = I(0) = \text{const}$. The unit of time is 1000 yr$/\kappa_{28}$, where $\kappa_{28} = \kappa/10^{28}$ (in the c.g.s. units), and κ is a parameter in the right-hand side of (1). From Zdunik et al. (2006a)

8 Spin up, shrinking of radius, and energy release in starquakes in pulsars

The changes in f, equatorial radius R_{eq}, and the energy release can be evaluated by comparing the relevant quantities in the stationary, rigidly rotating states (initial and final), $\mathcal{C}_{\text{i}} \longrightarrow \mathcal{C}_{\text{f}}$. Assumed conditions are $M_{\text{b,i}} = M_{\text{b,f}}$, $J_i = J_{\text{f}}$. The energy release is then given by $\Delta E = -\Delta M c^2$.

Examples of numerical results obtained in 2-D simulations are shown in Fig. 8. A remarkable feature is a very weak dependence of ΔE on J_i. Therefore ΔE can be calculated using 1-D code for non-rotating stars and this gives excellent prediction (within better than 20%) even for high J_{i}, when the stars are strongly flattened by rotation! This has important practical consequence: there is no need for 2-D simulations to get reliable estimate of the energy release, 1-D calculation for non-rotating stars is sufficient.

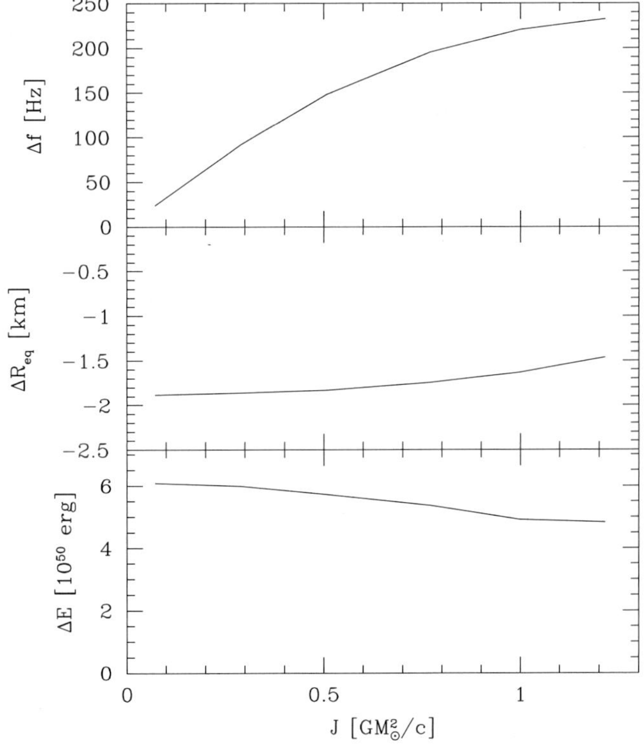

Fig. 8 Changes of stellar parameters of a rotating isolated neutron stars, due to a corequake which occurs after the star reaches an unstable configuration. An example from Zdunik et al. (2006a)

9 Metastability and two types of corequakes

The case of spherical, non rotating neutron stars was studied long time ago (Haensel et al. 1986; Zdunik et al. 1987). Corequakes in rotating neutron stars were recently studied by Zdunik et al. (2006a, 2006b).

First order phase transition allows for metastability of the N phase at $P > P_0$ (see Fig. 9). Depending on the magnitude of the density jump at $P = P_0$ between stable N and S phases, one has to distinguish two types of 1st order phase transitions.

Weak and moderate 1st order phase transitions characterized by $\rho_S/\rho_N < \frac{3}{2} + \frac{3}{2}P_0/\rho_N c^2$. A starquake is then triggered by nucleation of a droplet of the S phase in a metastable core of N phase (see Fig. 10).

Strong 1st order phase transitions characterized by $\rho_S/\rho_N > \frac{3}{2} + \frac{3}{2}P_0/\rho_N c^2$. Configurations with $P_c > P_0$ with small S cores are then *unstable* even under full equilibrium conditions (i.e., for no metastability). They collapse into configurations with a large S phase core.

One finds that in all cases ΔE is, to a very good approximation, independent of J of collapsing configuration, provided the energy release is calculated at fixed overcompression in the center of the metastable N-star core, $\delta \overline{P} \equiv (P_c - P_0)/P_0$. Generally, a starquake is triggered for $P_c = P_{nucl}$, when the S phase nucleates, which initiates a transition depicted in Fig. 10. Independence of $\Delta E(\delta \overline{P})$ from J of collapsing configuration is clearly seen in Fig. 11. All points lie on the same curve, which coincides with that obtained for nonrotating case.

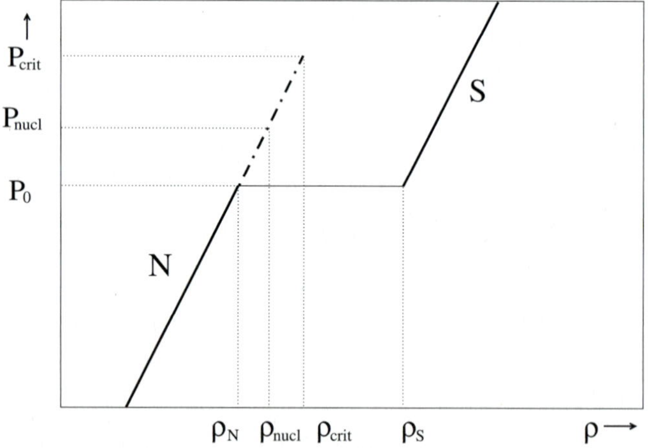

Fig. 9 A schematic representation, in the $\rho - P$ plane, of an EOS with a 1st order phase transition. *Solid segments*: stable N and S phases in thermodynamic equilibrium. *Dash-dot segment*: metastable N phase. The S phase nucleates at P_{nucl}. At P_{crit} nucleation of the S phase is instantaneous, because the energy barrier separating the N phase from the S one vanishes

Independence of the energy release of J_i is valid for both weak, moderate, and strong 1st order phase transition. It has a paramount practical importance: simple 1-D calculations for non-rotating stars are sufficient to get correct value of the energy release for even rapidly rotating stars.

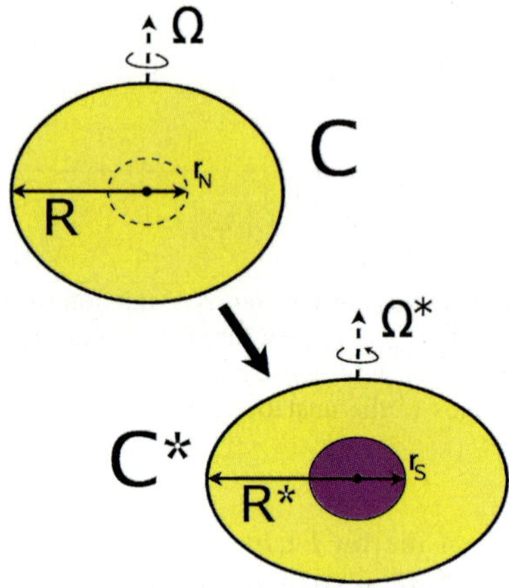

Fig. 10 Transition from a one-phase configuration $\mathcal{C}$ with a metastable core of radius r_N to a two-phase configuration $\mathcal{C}^\star$ with a S-phase core of radius r_S. The two configurations have the same baryon number (baryon mass) and the same angular momentum. From Bejger et al. (2005)

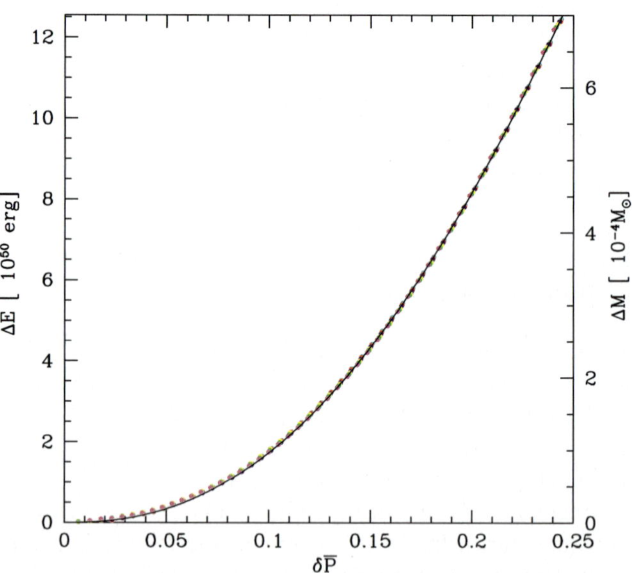

Fig. 11 The energy release due to a corequake of rotating neutron star as a function of the overpressure $\delta \overline{P}$ at the center of the metastable N phase core, for a model of the EOS with a moderate 1st order phase transition. The points of different color correspond to different values of the total angular momentum of rotating star, but they all lie on a same line (Zdunik et al. 2006b, to be published). Notice, that rotation is very rapid, with kinetic/gravitational energy ratio up to $T/W \simeq 0.1$

10 Conclusions

For isolated pulsar without phase transition in the center the evolution of the period during spin-down is smooth and monotonic. For stars with baryon mass below the maximum allowable one for static stars, $M_{\rm b,max}^{\rm (stat)}$, the stellar angular momentum J is monotonically decreasing with decreasing rotation frequency f, $\mathrm{d}J/\mathrm{d}f > 0$. For supramassive stars, with fixed baryon mass $M_{\rm b} > M_{\rm b,max}^{\rm (stat)}$, the derivative $\mathrm{d}J/\mathrm{d}f$ changes sign very close to the Keplerian (mass shedding) limit. However, in this region the effect of spin-up by the angular momentum loss results directly from effects of general relativity, and has nothing to do with the equation of state (Cook et al. 1992; Zdunik et al. 2004). The relevant value of f is then so close to the mass shedding limit, that the star is susceptible to many other instabilities and hence it is unlikely to be observed as a radio pulsar.

A softening of the EOS of dense matter due to a phase transition or appearance of hyperons can affect spin evolution tracks of isolated pulsars. In particular, back bending epochs and instabilities followed by corequakes may appear. One parameter families (sequences) of rigidly rotating configurations loosing angular momentum have a stability/instability structure identical with that of spherical non-rotating family. Energy release associated with a corequake in rotating pulsar depends only on the overpressure in the center of its metastable core and does not depend on the rotation rate of collapsing configuration. This independence on the initial rotation rate holds universally, for weak, moderate, and strong phase transitions, and with or without metastability. This means that the energy release in a corequake in a rotating star can be calculated, with a very good precision, in 1-D nonrotating model with the same pressure at the core center.

It should be mentioned that there is still no direct observational evidence of the back bending phenomenon. Although there exist many isolated pulsars with a measured *decreasing* period, they are located in globular clusters and the effect of $\dot{P} < 0$ is usually explained by the acceleration in the globular cluster gravitational field (for a recent review see Camilo and Rasio 2005). The effect of period clustering (which would be another consequence of the phase-transition softening of EOS) is not clear from observational point of view. There seems to be some evidence for such effect (e.g. Chakrabarty 2005) for accreting neutron stars, but not for isolated neutron stars considered in the present paper. However, for accreting neutron stars the effect of back bending is expected to be strongly suppressed due to the simultaneous increase of the stellar mass during accretion (Zdunik et al. 2005). There might be some indirect evidence for back bending hided in the evolutionary history of pulsars. Some evolutionary scenarios applied to the observed pulsars result in unrealistically short initial period if one assumes standard magnetic dipole braking. This paradox could be explained by the existence of the back-bending epoch during pulsar's life (Spyrou and Stergioulas 2002; Zdunik et al. 2006a, 2006b). Finally, measurements of very large values of braking indices (Johnston and Galloway 1999) could be an indication of the closeness to $\dot{f} = 0$ state associated with back bending. However, this might be explained in a more standard way, by the behavior connected with the glitch phenomena (Johnston and Galloway 1999; Alpar and Baykal 2006).

References

Alpar, M., Baykal, A.: Mon. Not. Roy. Astron. Soc. **372**, 489 (2006)
Bejger, M., Haensel, P., Zdunik, J.L.: Mon. Not. Roy. Astron. Soc. **359**, 699 (2005)
Camilo, F., Rasio, F.A.: In: Rasio, F.A., Stairs, I.H. (eds.) Binary Radio Pulsars. ASP Conference Series vol. 328, p. 147. ASP, San Francisco (2005)
Chakrabarty, D.: In: Interacting Binaries. Accretion, Evolution, and Outcomes. AIP Conference Proceedings, vol. 797, p. 71. (2005)
Chubarian, E., Grigorian, H., Poghosyan, G., Blaschke, D.: Astron. Astrophys. **357**, 968 (2000)
Cook, G.B., Shapiro, S.L., Teukolsky, S.A.: Astrophys. J. **398**, 203 (1992)
Friedman, J.L., Ipser, J.R., Sorkin, R.D.: Astrophys. J. **325**, 722 (1988)
Glendenning, N.K.: Compact Stars, Nuclear Physics, Particle Physics and General Relativity. Springer, New York (2000)
Glendenning, N.K., Pei, S., Weber, F.: Phys. Rev. Lett. **79**, 1603 (1997)
Haensel, P., Zdunik, J.L., Schaeffer, R.: Astron. Astrophys. **160**, 251 (1986)
Harrison, B.K., Thorne, K.S., Wakano, M., Wheeler, J.A.: Gravitation Theory and Gravitational Collapse. The University of Chicago Press, Chicago (1965)
Heiselberg, H., Hjorth-Jensen, M.: Phys. Rev. Lett. **80**, 5485 (1998)
Johnston, S., Galloway, D.: Mon. Not. Roy. Astron. Soc. **306**, L50 (1999)
Spyrou, N.K., Stergioulas, N.: Astron. Astrophys. **395**, 151 (2002)
Weber, F., Glendenning, N.K.: Phys. Lett. B **265**, 1 (1991)
Weber, F., Glendenning, N.K.: Astrophys. J. **390**, 541 (1992)
Zdunik, J.L., Haensel, P., Schaeffer, R.: Astron. Astrophys. **172**, 95 (1987)
Zdunik, J.L., Haensel, P., Gourgoulhon, E., Bejger, M.: Astron. Astrophys. **416**, 1013 (2004)
Zdunik, J.L., Haensel, P., Bejger, M.: Astron. Astrophys. **441**, 207 (2005)
Zdunik, J.L., Bejger, M., Haensel, P., Gourgoulhon, E.: Astron. Astrophys. **450**, 746 (2006a)
Zdunik, J.L., Bejger, M., Haensel, P., Gourgoulhon, E.: Energy release associated with a first-order phase transition in a rotating neutron star core (2006b) astro-ph/0610188. Astron. Astrophys. (2007, in press)

Astrophys Space Sci (2007) 308: 371–379
DOI 10.1007/s10509-007-9381-3

ORIGINAL ARTICLE

Equation of state constraints from neutron stars

James M. Lattimer

Received: 21 June 2006 / Accepted: 9 November 2006 / Published online: 20 March 2007

Abstract Recent measurements of thermal radiation from neutron stars have suggested a rather broad range of radiation radii ($R_\infty = R/\sqrt{1-2GM/Rc^2}$). Sources in M13 and Omega Cen imply $R_\infty \sim 12$–14 km, but X7 in 47 Tuc implies $R_\infty \sim 16$–20 km and RX J1856-3754 $R_\infty > 17$ km. If these measurements are all correct, only a limited selection of EOS's could be consistent with them, but a broad range of neutron star masses (up to 2 $M_\odot$) would also be necessary. The surviving equations of state are incompatible with significant softening above nuclear saturation densities, such as would occur with Boson condensates, a low-density quark-hadron transition, or hyperons. Other potential constraints, such as from QPO's, radio pulsar mass and moment of inertia measurements, and neutron star cooling, are compared.

Keywords Neutron stars · Equation of state · Pulsars

PACS 26.60.+c · 97.60.Jd · 7.60.Gb

1 Introduction

Neutron stars are manifestations of the densest known objects and therefore serve as laboratories for dense matter physics. However, the two most important properties of neutron stars—their maximum masses and typical radii—are not well known. The maximum mass is controlled by the stiffness of the nuclear force at densities in excess of five times n_s, the standard density inside nuclei ($n_s \simeq 0.16$ baryons fm^{-3} is also referred to as the saturation density of symmetric nuclear matter). Interestingly, the neutron star radius is instead controlled by properties of the nuclear force in the immediate vicinity of n_s, in particular by the density dependence of the nuclear symmetry energy (Lattimer and Prakash 2001) (the symmetry energy is the difference, at a given density, between the baryon energies of pure neutron matter and symmetric nucleonic matter).

The neutron star maximum mass, which is a consequence of general relativity and does not have meaning in Newtonian gravity, was considered by (Rhoades and Ruffini 1974), who established that it was about 2–3 $M_\odot$. They assumed the equation of state is known up to a fiducial energy density ρ_f, presumably 1–2 times $\rho_s = m_b c^2 n_s$ where m_b is the baryon mass, and is also assumed to be causal at higher densities (i.e., $dP/d\rho \le 1$). With these assumptions, neutron star masses cannot exceed $M_{RR} = 4.2\sqrt{\rho_f/\rho_s}\,M_\odot$. Causality considerations also set a limit to the maximum compactness for a neutron star of mass M (Lindblom 1984; Glendenning 1992)

$$R_{\min} \simeq 2.83GM/c^2, \tag{1}$$

assuming that $\rho_f \ge \rho_s$. A minimum compactness, or maximum radius, can be determined from considerations of rotational stability. In Newtonian gravity, the absolute maximum rotation rate is set by the "mass-shedding" limit $2\pi\nu_{\max} = \sqrt{GM/R^3}$, or

$$\nu_{\max}^{\text{rigid}} = 1838\left(\frac{10\text{ km}}{R}\right)^{3/2}\left(\frac{M}{M_\odot}\right)^{1/2}\text{ Hz} \tag{2}$$

for a rigid sphere of radius R. However, the actual limiting rate differs from this because rotation increases the equatorial radius and general relativity introduces frame-dragging

US DOE Grant DE-FG02-87ER-40317.

J.M. Lattimer (✉)
Department of Physics and Astronomy, Stony Brook University, Stony Brook, NY 11794 3800, USA
e-mail: lattimer@astro.sunysb.edu

among other effects. Using the results of fully relativistic, axially symmetric, calculations (Friedman et al. 1986; Lattimer et al. 1990; Salgado et al. 1994), Lattimer and Prakash (2004) showed that the minimum spin period for a star of a given non-rotating mass M can still be approximated by (2), but with a coefficient of 1045. This results in a maximum radius

$$R_{\max} = 10.3\left(\frac{1000\ \text{Hz}}{\nu}\right)^{2/3}\left(\frac{M}{\mathrm{M}_\odot}\right)^{1/3}\ \text{km}. \tag{3}$$

Note that this expression refers to the radius of the star when it is not rotating. Near the mass-shedding limit, the equatorial radius of the star would be about 50% larger.

Recently, the pulsar PSR J1748-2446ad with a spin rate of 716 Hz was discovered (Hessels et al. 2006). With this value, (3) suggests that for a 1.4 $\mathrm{M}_\odot$ star, the non-rotating radius would be limited by $R < 14.4$ km. However, there is some likelihood that this star spun up through the accretion of a few tenths of a solar mass. If this star is, for example, 1.7 $\mathrm{M}_\odot$, the upper limit to the non-rotating radius would be 15.4 km. Unfortunately, neither limit is very restrictive at present.

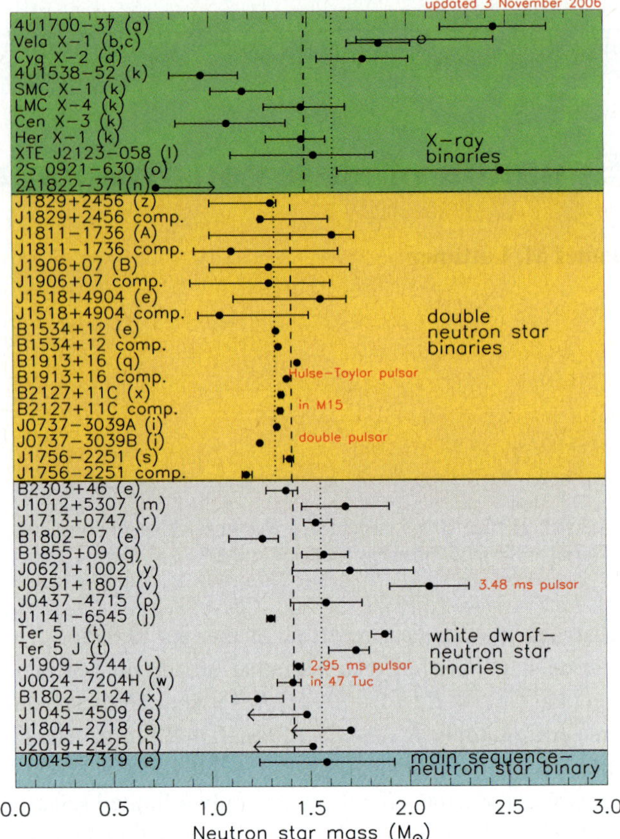

Fig. 1 Measured and estimated masses of neutron stars in radio binary pulsars (*gold, silver* and *blue regions*) and in X-ray accreting binaries (*green*). Letters in parentheses refer to citations: (a) Clark et al. (2002); (b) Barziv et al. (2001); (c) Quaintrell et al. (2003); (d) Orosz and Kuulkers (1999); (e) Thorsett and Chakrabarty (2001); (f) Nice et al. (2004); (g) Nice et al. (2003); (h) Nice et al. (2001); (i) Lyne et al. (2004); (j) Bailes et al. (2003); (k) van Kerkwijk et al. (1995); (l) Gelino et al. (2003); (m) Langer et al. (2001); (n) Jonker et al. (2003); (o) Jonker et al. (2005); (p) van Stratan et al. (2001); (q) Weisberg and Taylor (2005); (r) Splaver et al. (2005); (s) Faulkner et al. (2004); (t) Ransom et al. (2005); (u) Jacoby et al. (2005); (v) Nice et al. (2005); (w) Freire et al. (2003); (x) Jacoby (2005); (y) Splaver et al. (2002); (z) Champion et al. (2005); (A) Corongiu et al. (2004); (B) Lorimer et al. (2006)

2 Recent mass measurements

Several recent observations of neutron stars have direct bearing on the determination of the maximum mass. The most accurately measured masses are from timing observations of the radio binary pulsars. As shown in Fig. 1, which represents the latest compilation of measured neutron star masses including references, these include pulsars orbiting another neutron star, a white dwarf or a main-sequence star. The compact nature of several binary pulsars permits detection of relativistic effects, such as Shapiro delay or orbit shrinkage due to gravitational radiation reaction, which constrains the inclination angle and permits measurement of each mass in the binary. A sufficiently well-observed system can have masses determined to impressive accuracy. The textbook case is the binary pulsar PSR 1913 + 16, in which the masses are 1.3867 ± 0.0002 and 1.4414 ± 0.0002 $\mathrm{M}_\odot$, respectively (Weisberg and Taylor 2005).

One significant development concerns mass determinations in binaries with white dwarf companions, which show a broader neutron star mass range than binary neutron star pulsars. Perhaps a rather narrow set of evolutionary circumstances conspire to form double neutron star binaries, leading to a restricted range of neutron star masses. This restriction is relaxed for other neutron star binaries. Evidence is accumulating that a few of the white dwarf binaries may contain neutron stars larger than the canonical 1.4 $\mathrm{M}_\odot$ value, including the fascinating case (Nice et al. 2005) of PSR J0751+1807 in which the estimated mass with 1σ error bars is 2.1 ± 0.2 $\mathrm{M}_\odot$. For this neutron star, a mass of 1.4 $\mathrm{M}_\odot$ is about 4σ from the optimum value. In addition, to 95% confidence, one of the two pulsars Ter 5 I and J has a claimed mass larger than 1.68 $\mathrm{M}_\odot$. While the *mean* observed value of the white dwarf-neutron star binaries exceeds that of the double neutron star binaries by 0.25 $\mathrm{M}_\odot$, it is nevertheless the case that the 1σ errors of all but one of these systems extends into the range below 1.45 $\mathrm{M}_\odot$. Continued observations guarantee that these errors will be reduced. Raising the limit for the neutron star maximum mass could eliminate entire families of EOS's, especially those in which substantial softening begins around 2 to $3n_s$. This could be extremely significant, since exotica (hyperons, Bose condensates, or quarks) generally reduce the maximum mass appreciably.

Masses can also be estimated for another handful of binaries which contain an accreting neutron star emitting X-rays, as shown in Fig. 1. Some of these systems are characterized by relatively large masses, but the estimated errors are also large. The system of Vela X-1 is noteworthy because its lower mass limit (1.6 to 1.7 $M_\odot$) is at least mildly constrained by geometry (Quaintrell et al. 2003).

2.1 QPO's

Accreting neutron stars often display subtle periodicities in their X-ray fluxes. These quasi-periodic oscillations (QPO's) involve frequencies related to the star's rotation as well as higher frequencies connected with the inner edge of the accretion disc. For neutron stars that satisfy $R < 6GM/c^2$, there is a gap between the surface and the last stable circular orbit. Presumably, the accretion disc extends inward to a position close to the last stable orbit. Matter forced inwards plunges onto the neutron star. A leading model of QPO's suggests that among the high frequencies observed is a frequency equal to the Keplerian orbital frequency of the inner edge of the accretion disc. If this frequency is that of the innermost stable circular orbit, $\nu_{\rm ISCO}$, the mass of the star is constrained in general relativity to be (Miller et al. 1998)

$$\frac{M}{M_\odot} = \frac{\nu_{\rm ISCO}}{2200\ \rm Hz}\left(1 + 0.75\frac{cJ}{GM^2}\right), \tag{4}$$

where J is the angular momentum of the neutron star. Typically, $cJ/GM^2 \simeq 0.1$ and measured values of $\nu_{\rm ISCO} \simeq$ 1200–1300 Hz imply $M \sim 1.8$–$2.0\ M_\odot$. These values are consistent with the large masses inferred from some white dwarf-neutron star binaries, which also evolved through significant accretion.

2.2 Implications

Assuming that the hyperon-nucleon couplings are comparable to the nucleon-nucleon couplings typically results in the appearance of Σ^- and Λ hyperons around 2–$3n_s$ in neutron star matter. Neutron star matter is in beta equilibrium, such that $\mu_n - \mu_p = \mu_e = \mu_{\Sigma^-}$ and $\mu_n = \mu_\Lambda$. As a consequence, the proton fraction in such matter is quite small, of order 5–10%. Little is known about the symmetry dependence of the hyperon-nucleon couplings, i.e., hyperon-nucleon couplings are chiefly constrained by hyperon-nucleus masses, but these are more or less symmetric nuclei. If hyperons indeed appear as early as 2–$3n_s$, the maximum neutron star mass becomes relatively small, less than 1.6 $M_\odot$. Ironically, it is possible to increase the neutron star maximum mass in this case if a different form of strangeness can appear prior to hyperons, for example, deconfined uds quark matter. Maximum masses up to approximately 2.0 $M_\odot$ are then possible (Alford et al. 2005) but require special fine-tuning of the nucleonic and quark matter parameters. Therefore a confirmation of a neutron star mass in excess of 2 $M_\odot$ would be especially interesting.

A further consequence of measured neutron star masses is the determination of the largest possible density in stars. Consider, first, a constant density fluid. Coupled with the causality constraint, (1), the central density of the resulting star is

$$\rho_{c,u} = \frac{3M}{4\pi R_{\rm min}^3} \simeq 6.1 \times 10^{15}\left(\frac{M_\odot}{M}\right)^2\ \rm g\,cm^{-3}. \tag{5}$$

However, such a star is unrealistic since it has infinite sound speed and high densities at the surface. It was discovered (Lattimer and Prakash 2005) that the density distribution $\rho = \rho_c[1 - (r/R)^2]$ provided an upper limit to the density for a given mass for all known equations of state. This density distribution corresponds is the analytic solution of Einstein's equations known as Tolman VII (Tolman 1939). Again using the causality limit from (1), one sees that

$$\rho_{c,\rm VII} = \frac{15M}{8\pi R_{\rm min}^3} \simeq 15.2 \times 10^{15}\left(\frac{M_\odot}{M}\right)^2\ \rm g\,cm^{-3}. \tag{6}$$

It should be emphasized that since this represents the upper limit to the central density of the measured star, the actual central density of the maximum mass star must be even smaller. With a mass measurement of 2.1 $M_\odot$, for example, the limiting energy density is $3.4 \times 10^{15}\ \rm g\,cm^{-3}$, or about $8n_s$. If the maximum mass was in fact about 10% larger, the limiting energy density becomes $2.8 \times 10^{15}\ \rm g\,cm^{-3}$, or about $7n_s$. This could be small enough to rule out the possible appearance of a deconfined quark phase in neutron star cores.

3 Radius constraints

As previously mentioned, the radius is primarily determined by the density dependence of the symmetry energy. This arises through the relation between the neutron star radius and the matter pressure at intermediate densities. That such a relation exists can be seen most clearly by considering Newtonian polytropes. For the pressure-density relation

$$P \simeq K\rho^{1+1/n}, \tag{7}$$

where K is a constant and n is the polytropic index, hydrostatic equilibrium implies that

$$R \propto K^{n/(3-n)} M^{(1-n)/(3-n)}. \tag{8}$$

Realistic equations of state typically have $n \simeq 1$, but K is uncertain by a factor of 5 or 6. For some average density ρ_*,

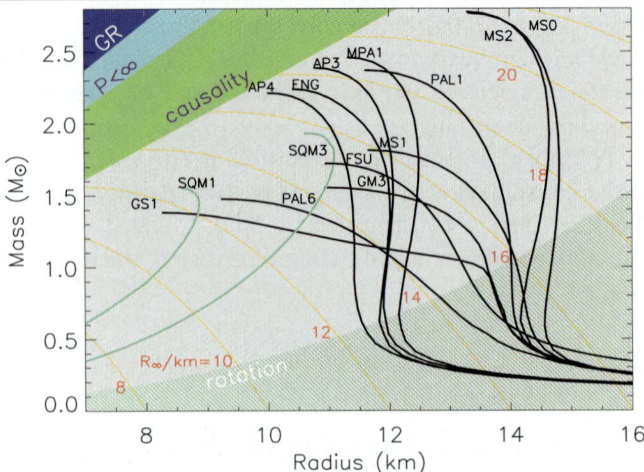

Fig. 2 Mass-radius trajectories for typical equations of state (see Lattimer and Prakash 2001 for notation and descriptions) are shown as black curves. Green curves are self-bound quark stars. The orange lines are constant values of the radiation radius, $R_\infty = R/\sqrt{1-2GM/Rc^2}$. The *dark blue region* is excluded by the GR constraint $R > 2GM/c^2$, the *light blue region* is excluded by the finite pressure constraint $R > (9/4)GM/c^2$, and the *green region* is excluded by causality, $R > 2.9GM/c^2$. The *light green region* shows the region $R > R_{\max}$ excluded by the pulsar J1748-2446ad

in the vicinity of $1 - 2m_b c^2 n_s$, suppose the pressure is P_*. For the case $n = 1$, one therefore has

$$R \propto P_*^{1/2} \rho_*^{-1} M^0. \tag{9}$$

The independence of R from M is a characteristic result of mass-radius trajectories for equations of state without extreme low-density softening, as seen in Fig. 2. Equation (9) suggests that the radius scales with the square root of the fiducial pressure P_*. However, general relativity plays an important role, and in practice the scaling is such that $R \propto P_*^{1/4}$ (Lattimer and Prakash 2001).

There is only one analytic solution of Einstein's equations that explicitly relates the radius, mass and pressure (Buchdahl 1967). This solution assumes the equation of state

$$\rho = 12\sqrt{p_* p} - 5p, \tag{10}$$

where p_* is a parameter. For low densities, one sees that this equation of state is that of an $n = 1$ polytrope, so this solution is a reasonable approximation for a neutron star. For this solution, the radius is an implicit function of M and p_*:

$$R = (1-\beta)\sqrt{\frac{\pi}{288 p_*(1-2\beta)}}, \tag{11}$$

where $\beta = GM/Rc^2$. Computation of $d\ln R/d\ln p$ for fixed M and n leads to a result of about $1/4$.

This correlation is significant because the pressure of degenerate neutron-star matter near the nuclear saturation density n_s is, in large part, determined by the symmetry properties of the EOS. For the present discussion, we introduce the incompressibility K and the skewness K', and expand the energy per particle about its values at n_s and $x = 1.2$, where x is the proton fraction:

$$E(n,x) = -16 + \frac{K}{18}\left(1 - \frac{n}{n_s}\right)^2 + \frac{K'}{27}\left(1 - \frac{n}{n_s}\right)^3 + E_{\rm sym}(n)(1-2x)^2 \cdots. \tag{12}$$

Here, $E_{\rm sym}$ is the symmetry energy function, approximately the energy difference at a given density between symmetric and pure neutron matter. The symmetry energy parameter is defined as $S_v \equiv E_{\rm sym}(n_s)$. Leptonic contributions $E_e = (3/4)\hbar c x(3\pi^2 n x^4)^{1/3}$ must be added. Matter in neutron stars is in beta equilibrium, i.e., $\mu_e = \mu_n - \mu_p = -\partial E/\partial x$, so the equilibrium proton fraction at n_s is $x_s \simeq (3\pi^2 n_s)^{-1}(4S_v/\hbar c)^3 \simeq 0.04$. The pressure at n_s is

$$P(n_s, x_s) = n_s(1-2x_s)\left[n_s S_v'(1-2x_s) + S_v x_s\right] \simeq n_s^2 S_v', \tag{13}$$

due the small value of x_s; $S_v' \equiv (\partial E_{\rm sym}/\partial n)_{n_s}$. The pressure depends primarily upon S_v'. The equilibrium pressure at moderately larger densities similarly is insensitive to K and K'. Experimental constraints to the compression modulus K, most importantly from analyses of giant monopole resonances, give $K \cong 220$ MeV. The skewness parameter K' has been estimated to lie in the range 1780–2380 MeV. Evaluating the pressure for $n = 1.5n_s$,

$$P(1.5n_s) = 2.25n_s\left[K/18 - K'/216 + n_s(1-2x)^2(\partial E_{\rm sym}/\partial n)_{1.5n_s}\right]. \tag{14}$$

Note that the contributions from K and K' largely cancel and, once again, the pressure is dominated by the symmetry energy derivative term.

While it might become possible to constrain the properties of the nuclear symmetry energy in the laboratory through the measurements of neutron skin thicknesses of neutron-rich nuclei, significant difficulties will be encountered in obtaining the quantity S_v' in the necessary density range. First of all, nuclei sample nucleonic matter in the density range $n < n_s$ alone, while we need the symmetry energy in the range 1–$2n_s$. Secondly, the neutron skin thickness is sensitive to the integral quantity $\int (S_v/E_{\rm sym}(n) - 1)d^3r$ which is not uniquely connected to $dE_{\rm sym}/dn$.

The main classes of equations of state in common use in nuclear astrophysics include non-relativistic potential models, relativistic field-theoretical models and self-bound models usually representing deconfined quark matter. The main features of each are as follows:

- Non-relativistic potential models
  - Momentum- and density-dependent potentials
  - Power-series density expansion
  - Density-dependent effective nucleon masses
  - Generally have relatively slowly varying symmetry energies, smaller neutron star radii
  - Can become acausal
  - Can be constrained to fit low-density matter properties
- Relativistic field-theoretical models
  - Interactions mediated by bosons (ω, σ, ρ)
  - Implicitly causal
  - Generally have linearly increasing symmetry energies, larger neutron star radii
  - Not easily constrained to fit low-density matter properties
- Self-bound models
  - Strange quark matter with a lower binding energy than hadronic matter (iron) at zero pressure
  - Radii of high mass stars not necessarily smaller than hadronic stars

A recently proposed Skyrmion force model (Ouyed and Butler 1999; Jaikumar and Ouyed 2006) which is a variation of relativistic field theoretical models designed to maximize neutron star radii, has somewhat bizarre properties near nuclear saturation density. Even when fit to reasonable symmetry energies and incompressibilities at saturation, these properties rapidly deviate at densities only slightly larger or lower than n_s. As a result, it is doubtful this model can be well-fit to nuclear properties.

4 Thermal emission observations

The quantity inferred from thermal observations of a neutron star's surface is the so-called radiation radius $R_\infty = R/\sqrt{1-2GM/Rc^2}$. It results from a combination of flux and temperature measurements, which have to be redshifted at the Earth from the neutron star's surface. The unredshifted Kirchoff's law for a black body, for example,

$$F = \left(\frac{R}{d}\right)^2 \sigma T^4, \tag{15}$$

becomes

$$F_\infty = \left(\frac{R_\infty}{d}\right)^2 \sigma T_\infty^4. \tag{16}$$

Contours of the quantity R_∞ are displayed in Fig. 2. A measured value of R_∞ sets upper limits to both R and M, but without another estimate of mass or radius only limited constraints are possible. The major uncertainties involved in its determination include the distance, interstellar H absorption that is important near the peak and at lower energies of the spectrum,and details concerning the composition of the atmosphere and its magnetic field strength and structure. The most reliable measurements will originate from sources where one or more of these uncertainties can be controlled. For example, the nearby sources RX J1856-3754 (Walter and Lattimer 2002) and Geminga (Walter and Faherty 2006) have measured parallaxes, and quiescent X-ray binaries in globular clusters for which reliable distances are or will soon be available. In addition, the X-ray binaries in globular clusters, having undergone recent accretion episodes, are expected to have low magnetic field H-rich atmospheres, the simplest of all situations.

A characteristic of these X-ray sources, in the cases in which the distances are small enough to allow detection of optical emission, is that the optical fluxes are a factor $f \simeq 5$–7 times the amount expected from extrapolating the X-ray blackbody fits onto the Rayleigh-Jeans tail. This is one consequence of the neutron star atmosphere and its redistribution of the flux from a simple blackbody. The slope of the Rayleigh-Jeans tail is a measure of the optical temperature, and typically $T_{\rm opt} \sim T_X/2$. Therefore, in order to properly fit the overall flux distribution, a larger radius is needed than the X-ray fit alone would imply. The optical flux may be written as

$$F_{\rm opt} \propto 4\pi \left(\frac{R_{\rm opt}}{d}\right)^2 T_{\rm opt} = 4\pi f \left(\frac{R_X}{d}\right)^2 T_X \tag{17}$$

while the total stellar radius is a function of the effective optical and X-ray radii:

$$R = \sqrt{R_{\rm opt}^2 + R_X^2} = R_X \sqrt{1 + \frac{T_X}{T_{\rm opt}} f} = R_X \sqrt{1+2f}. \tag{18}$$

This explicitly shows that a factor $f \sim 5$–7 results in an increase in inferred radius of about a factor 3–4 over that inferred from X-rays alone. Radii inferred from X-ray data alone are therefore suspect, and often much too small.

Results from atmospheric fitting of the data for RX J1856-3754 and various globular cluster sources are displayed in Fig. 3. (Whether or not a correction for non-uniform surface temperatures has to be made in the case of globular cluster sources is not clear.) The interesting feature of Fig. 3 is that if one accepts the present results at face value, only equation of state curves that pass through all the permitted regions can be accepted. This would eliminate several relativistic field theoretical equations of state such as GM3, FSU, MS0, MS1 and MS2, while still permitting non-relativistic potential models like AP3 and AP4. Other relativistic field models like ENG and MPA1 are also allowed. In addition, note that the surviving equations of state all support large masses: crucial if the 2.1 $M_\odot$ value for PSR J0751+1807 survives.

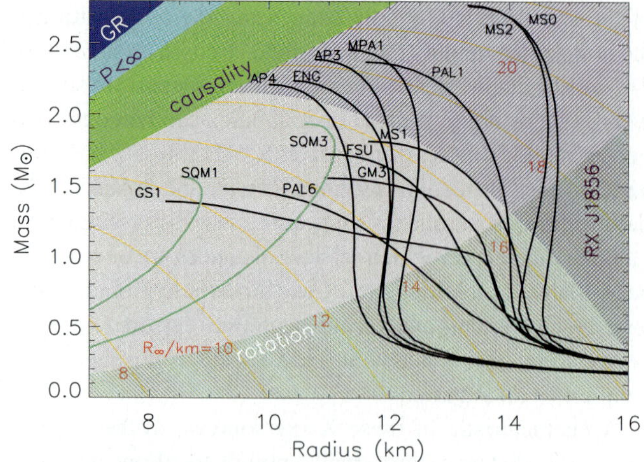

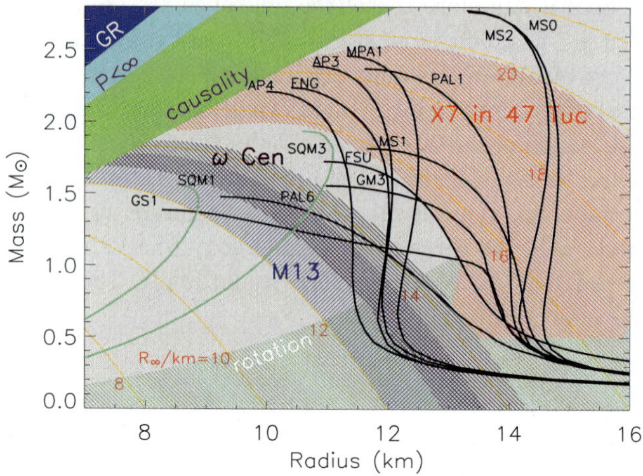

Fig. 3 Results for allowed mass and radius regions for thermally-emitting neutron stars. *Top figure* is for RX J1856-3754 (Walter and Lattimer 2002), while the *bottom figure* shows globular cluster sources in M13 (Gendre et al. 2003b), ω Cen (Gendre et al. 2003a) and 47 Tuc (Heinke et al. 2006)

5 Other radius observables

Other limits to the radius could be set by redshifts, pulsar glitches, quasi-periodic oscillations (QPO's), Eddington-limited burst fluxes, neutron star seismology (surface oscillations), and the moment of inertia.

5.1 Redshift

The simplest of these is the redshift, which is defined as

$$z = \left(1 - \frac{2GM}{Rc^2}\right)^{-1/2} - 1. \tag{19}$$

Possibly the best case is that of the active X-ray burster EXO 0748-676 (Cottam et al. 2002) for which a pair of lines consistent with He-like Fe imply $z \simeq 0.35$. This interpretation is not universally accepted, however. Other sources with lines have been observed, notably XTE J1814-338 with $z < 0.38$ and 4U1820-30 for which it is claimed $0.2 < z < 0.3$. Note that a simultaneous measurement of R_∞ and z determine both M and R:

$$R = R_\infty(1+z)^{-1},$$
$$M = R_\infty(1+z)^{-1}\left[1-(1+z)^{-2}\right]\frac{c^2}{2G}. \tag{20}$$

5.2 Eddington limit

Burst sources have long been known to have peak fluxes that are to order of magnitude comparable to the Eddington limited flux,

$$F_{\mathrm{edd},\infty} = \frac{cGMd^2}{\kappa}\sqrt{1-2GM/Rc^2}. \tag{21}$$

For $M = 1.4\,\mathrm{M}_\odot$ and $R = 10$ km, and assuming a hydrogen-poor atmosphere, $\kappa = \sigma_T = 0.2(1+X)$ cm^2 g^{-1}, where X is the hydrogen mass fraction and σ_T is the Thompson opacity. This is difficult to ascertain to much precision because the distances to the sources are rather uncertain, in general. Nevertheless, statistical studies (van Paradijs 1981) indicate that average maximum luminosity of X-ray bursts at infinity are about 60% larger than (21) predicts. Various suggestions have been proposed to explain the super-Eddington luminosities in these bursts. If one can be certain, however, that a particular burst is limited by the Eddington limit (or some known multiplicative factor), if R_∞/d can be measured, and a redshift z is also observed from a single source, then all three unknowns d, M and R can be simultaneously found (Ozel 2006). This might be possible for the source EXO 0748-676.

5.3 Glitches

Pulsar glitches, which are sudden discontinuities in the spin-down of pulsars, seem to involve the transfer of angular momentum from an isolated component to the entire star. A leading model for glitches supposes that the isolated component consists of superfluid neutrons in the crust. As the star spins down due to pulsar electromagnetic emissions, the superfluid component's angular velocity differs to a greater and greater degree from the star's. When this differential becomes large enough, somehow coupling increases to the point that angular momentum is suddenly transferred from the superfluid component to the star. In the case of the Vela pulsar, some 30 years of observations indicate a steady overall angular momentum transfer rate that indicates at least 1.4% of the total moment of inertia of the star is involved with the isolated component. It is possible to relate the moment of inertia fraction in the crust to stellar properties:

$$I_{\mathrm{crust}}/I_{\mathrm{star}} > 0.014 \propto P_t R^4 M^{-2}, \tag{22}$$

where P_t is the pressure at the core-crust interface. Depending on the equation of state, $0.20\ \text{MeV fm}^{-4} < P_t < 0.65\ \text{MeV fm}^{-3}$, once again indicating our lack of knowledge of the nuclear symmetry energy. Using the upper limit for P_t allows one to set a minimum R for a given M for Vela (Link et al. 1999),

$$\frac{M}{M_\odot} < 0.018\left(\frac{R}{10\ \text{km}}\right)^2. \tag{23}$$

Better understanding of the nuclear symmetry energy would permit a lower upper limit to P_t to be determined, which would make this constraint more restrictive.

5.4 Oscillations

A similar kind of limitation on radii can be obtained from observations of oscillations during X-ray bursts, presumably due to vibrations of the star's crust. This neutron star seismology has been exploited to estimate the thickness of the crust in the soft gamma-ray repeaters SGR 1806-20 and SGT 1900+14, $\Delta R/R \simeq 0.10$ (Watts and Strohmayer 2006). This result can be used to set a mass-radius constraint as follows: Begin with hydrostatic equilibrium in the crust, setting $M(r) = M$, ignoring P relative to ρc^2 and to $Mc^2/4\pi R^3$, and ignoring the internal energy compared to nc^2:

$$\frac{dP(r)}{m_b c^2 n} = \frac{d\mu}{m_b c^2} - \frac{GM}{r^2 - 2GMr/c^2}dr, \tag{24}$$

where $\mu = \mu_n$ is the baryon chemical potential in beta equilibrium (cold, catalyzed matter). Upon integration from the base of the crust (t) to the surface (0), (24) becomes

$$\frac{\mu_t - \mu_0}{m_b c^2} = \frac{1}{2}\ln\frac{r_t(R - 2GM/c^2)}{R(r_t - 2GM/c^2)}. \tag{25}$$

Note that $r_0 = R$ and $r_t = R - \Delta R$. This can be manipulated, upon defining $\ln\mathcal{H} \equiv 2(\mu_t - \mu_0)/m_b$, to

$$\frac{GM}{Rc^2} = \frac{1}{2} - \frac{\mathcal{H}}{2}\frac{\Delta R}{R}(\mathcal{H} - 1 + \Delta R/R)^{-1}. \tag{26}$$

A heuristic determination of $\mathcal{H}$ can be made as follows. The core-crust interface corresponds to the phase transition between nuclei and uniform matter. Near the transition, the uniform matter is nearly pure neutron matter, with a proton fraction of just a few percent. The value of $\mathcal{H}$ is determined by the baryon density of the phase transition, n_t, and by certain EOS parameters. It is customary to write the energy E of nuclear matter (at zero temperature) as an expansion around its value for symmetric matter at the saturation density $n_s \simeq 0.16\ \text{fm}^{-3}$:

$$E(n,x) \simeq -16 + \frac{K}{18}(1-u)^2 + S_v(u)[1-2x]^2, \tag{27}$$

where x is the proton fraction and $u \equiv n/n_s$. Energies are expressed in units of MeV. The quantity $K \simeq 225$ MeV is the incompressibility parameter and $S_v(u)$ is the symmetry energy. To a good approximation, $S_v(u) \simeq S_{v0}u^i$, where $S_{v0} \simeq 30$ MeV is the usual volume symmetry energy parameter. The exponent i is in the range of 2/3–1 depending on the nuclear force model. Using (27), the neutron chemical potential is

$$\begin{aligned}\mu_n &= \left.\frac{\partial[nE(n,x)]}{\partial n(1-x)}\right|_{nx} = E(n,x) + n\frac{\partial E(n,x)}{\partial n} - x\frac{\partial E(n,x)}{\partial x}\\ &= -16 + \frac{K}{18}(1-u)(1-3u)\\ &\quad + S_{v0}u^i(1-2x)\big[1+2x+i(1-2x)\big]\\ &\simeq -16 + \frac{K}{18}(1-u)(1-3u) + S_{v0}(1+i)u^i.\end{aligned} \tag{28}$$

In the last line, we approximated $x \simeq 0$. Zero density matter at the surface has $\mu_n = \mu_0 \approx -9$ MeV. Therefore, for the four cases defined by $i = 1, 2/3$ and $u = u_t = 1/2, 2/3$, one finds $\mu_t - \mu_0 = 19.9, 21.4, 28.8$, and 27.0 MeV, respectively, so that $\mathcal{H} = 1.043$ and 1.047 for $u_t = 1/2$, and 1.063 and 1.059 for $u_t = 2/3$. For a value of $\Delta R/R = 0.1$, (26) then yields $GM/Rc^2 = 0.136$ and 0.143 for $u_t = 1/2$ and 0.174 and 0.167 for $u_t = 2/3$, which translate into $z = 0.172, 0.184, 0.239$ and 0.226, respectively. The result apparently depends more on the value of u_t than on i. The value of u_t is a complicated interplay among the compression modulus, the symmetry energy and the surface energy of nuclei (Lamb et al. 1983). Note that for $M = 1.4\ M_\odot$, a radius of 12 and 14 km yields $GM/Rc^2 = 0.172$ and 0.147, respectively.

5.5 Moment of inertia

Spin orbit coupling is of the same order of magnitude as post-post-Newtonian effects and is potentially observable in compact enough binary pulsars (Barker and O'Connell 1975; Damour and Schaeffer 1988). Precession of the spins in a compact binary leads to two observable effects: the inclination angle is altered and their is additional perihelion advance. The recently discovered pulsar binary, PSR J073703039 offers an opportunity to measure these quantities, possibly within a few years (Lattimer and Schutz 2005). In this system, changes in the inclination angle will be nearly impossible to measure, since the effect is proportional to the cosine of the inclination angle, which is estimated to be almost 90 degrees. However, the periastron advance produces a timing residual proportional to the sine of the inclination angle. A measurement of the moment of inertia to within 10% would allow a 6% radius determination, since, roughly, $I \propto MR^2$. A relatively equation of state independent result for the moment of inertia in terms of the mass and radius of a neutron star is

$$I \simeq (0.237 \pm 0.008) M R^2 \times \left[1 + 4.2 \frac{M\ \mathrm{km}}{R\mathrm{M}_\odot} + 90 \left(\frac{M\ \mathrm{km}}{R\mathrm{M}_\odot}\right)^4\right] \quad (29)$$

which is valid for equations of state capable of supporting more than about 1.65 $\mathrm{M}_\odot$ (Lattimer and Schutz 2005). Since the estimated error of this general expression is less than 3%, and M will be determined to extremely high accuracy, a moment of inertia measurement of about 10% accuracy translates into a 6% error in R.

6 Neutron star cooling

While it is outside the scope of this review to develop the details of neutron star cooling (the reader is referred to articles by Page and Reisenegger in this volume), it should be mentioned that how rapidly neutron stars cool is a sensitive function of the internal composition. Inasmuch as the internal composition depends on the overall mass of the neutron star, an indirect measure of the mass of the star is possible. This is especially pertinent for those supernova remnants which have yet to yield detectable sources of thermal emission, such as 3C58 (Slane et al. 2004), and supernova remnants G084.2-0.8, G093.3+6.9, G127.1+0.5 and G315.4-2.3 (Kaplan et al. 2004). If these remnants in fact contain neutron stars, the stars may have cooled rapidly due to one of the several direct Urca processes, which is much more efficient than standard cooling due to the modified Urca process. If any part of the star, however miniscule, cools via a direct Urca process, the entire star will follow due to the efficient transport of heat to this region (Page et al. 2004). Since the direct Urca process is triggered at the density at which the proton fraction (in nucleonic matter) is 1/9, and at slightly lower densities when muons are considered, rapid cooling depends upon the density dependence of the symmetry energy and the star's central density. The central density critically depends upon the star's mass, so a mass estimate could follow. A radiation radius determination for the same star would then allow an extraction of the radius itself.

References

Alford, M., Braby, M., Paris, M., Reddy, S.: Astrophys. J. **629**, 969 (2005)
Bailes, M., Ord, S.M., Knight, H.S., Hotan, A.W.: Astrophys. J. **595**, L49 (2003)
Barker, B.M., O'Connell, R.F.: Phys. Rev. D **12**, 329 (1975)
Barziv, O., Karper, L., van Kerkwijk, M.H., Telging, J.H., van Paradijs, J.: Astron. Astrophys. **377**, 925 (2001)
Buchdahl, H.-A.: Astrophys. J. **147**, 310 (1967)
Champion, D.J., Lorimer, D.R., McLaughlin, M.A., Xilouris, K.M., Arzoumanian, Z., Freire, P.C.C., Lommen, A.N., Cordes, J.M., Camilo, F.: Mon. Not. Roy. Astron. Soc. **363**, 929 (2005)
Clark, J.S. et al.: Astron. Astrophys. **392**, 909 (2002)
Cottam, J., Paerels, F., Mendez, M.: Nature **420**, 51 (2002)
Corongiu, A., Kramer, M., Lyne, A.G., Löhmer, O., D'Amico, N., Possenti, A.: Mem. Soc. Astron. Ital. Suppl. **5**, 188 (2004)
Damour, T., Schaeffer, G.: Nuovo Cimento B **101**, 127 (1988)
Faulkner, A.J. et al.: Astrophys. J. Lett. **618**, L119 (2004)
Freire, P.C., Camilo, F., Kramer, M., Lorimer, D.R., Lyne, A.G., Manchester, R.N., D'Amico, N.: Mon. Not. Roy. Astron. Soc. **340**, 1359 (2003)
Friedman, J.L., Parker, L., Ipser, J.R.: Astrophys. J. **304**, 115 (1986)
Gelino, D.M., Tomsick, J.A., Heindl, W.A.: Bull. Am. Astron. Soc. **34**, 1199 (2003); Tomsick, J.A., private communication
Gendre, B., Barret, B., Webb, N.A.: Astron. Astrophys. **400**, 521 (2003a)
Gendre, B., Barret, B., Webb, N.A.: Astron. Astrophys. **403**, L11 (2003b)
Glendenning, N.K.: Phys. Rev. D **46**, 4161 (1992)
Heinke, C.O., Rybicki, G.B., Narayan, R., Grindlay, J.E.: Astrophys. J. **644**, 1090 (2006)
Hessels, J.W.T. et al.: Science **311**, 1901 (2006)
Jacoby, B.A.: PhD dissertation, CalTech (2005)
Jacoby, B.A., Hotan, A., Bailes, M., Ord, S., Kulkarni, S.R.: Astrophys. J. **629**, 113 (2005)
Jaikumar, P., Ouyed, R.: Astrophys. J. **639**, 354 (1999)
Jonker, P.G., van der Klis, M., Groot, P.J.: Mon. Not. Roy. Astron. Soc. **339**, 663 (2003)
Jonker, P.G., Steeghs, D., Nelemans, G., van der Klis, M.: Mon. Not. Roy. Astron. Soc. **356**, 621 (2005)
Kaplan, D.L. et al.: Astrophys. J. Suppl. Ser. **153**, 269 (2004)
Lamb, D.Q., Lattimer, J.M., Pethick, C.J., Ravenhall, D.G.: Nucl. Phys. A **411**, 449 (1983)
Lange, Ch. et al.: Mon. Not. Roy. Astron. Soc. **326**, 274 (2001)
Lattimer, J.M., Prakash, M.: Astrophys. J. **550**, 426 (2001)
Lattimer, J.M., Prakash, M.: Science **304**, 536 (2004)
Lattimer, J.M., Prakash, M.: Phys. Rev. Lett. **94**, 1101 (2005)
Lattimer, J.M., Schutz, B.F.: Astrophys. J. **629**, 979 (2005)
Lattimer, J.M., Prakash, M., Masak, D., Yahil, A.: Astrophys. J. Lett. **355**, L241 (1990)
Lindblom, L.: Astrophys. J. **278**, 364 (1984)
Link, B., Epstein, R.I., Lattimer, J.M.: Phys. Rev. Lett. **83**, 3362 (1999)
Lorimer, D.R. et al.: Astrophys. J. **640**, 428 (2006)
Lyne, A.G. et al.: Science **303**, 1153 (2004)
Miller, M.C., Lamb, F.K., Psaltis, D.: Astrophys. J. **508**, 791 (1998)
Nice, D.J., Splaver, E.M., Stairs, I.H.: Astrophys. J. **549**, 516 (2001)
Nice, D.J., Splaver, E.M., Stairs, I.H.: In: Bailes, M., Nice, D.J., Thorsett, S.E. (eds.) Radio Pulsars. ASP Conf. Ser., vol. 302, p. 75. Astron. Soc. Pac., San Francisco (2003), astro-ph/0210637
Nice, D.J., Splaver, E.M., Stairs, I.H.: In: Camilo, F., Gaensler, B.-M. (eds.) IAU Symposium 218, p. 49 (2004), astro-ph/0311296, also private communication
Nice, D.J., Splaver, E.M., Stairs, I.H., Löhmer, L., Jessner, A., Kramer, M., Cordes, J.M.: Astrophys. J. **634**, 1242 (2005)
Orosz, J.A., Kuulkers, E.: Mon. Not. Roy. Astron. Soc. **305**, 132 (1999)
Ouyed, R., Butler, M.: Astrophys. J. **522**, 453 (1999)
Ozel, F.: Nature **441**, 1115 (2006)
Page, D., Lattimer, J.M., Prakash, M., Steiner, A.W.: Astrophys. J. Suppl. Ser. **155**, 623 (2004)
Quaintrell, H., Norton, A.J., Ash, T.D.C., Roche, P., Willems, B., Bedding, T.R., Baldry, I.K., Fender, R.P.: Astron. Astrophys. **401**, 303 (2003)
Ransom, S.M., Hessels, J.W.T., Stairs, I.H., Freire, P.C., Camilo, F., Kaspi, V.M., Kaplan, D.L.: Science **307**, 892 (2005)

Rhoades, C.E., Jr., Ruffini, R.: Phys. Rev. Lett. **32**, 324 (1974)
Salgado, M., Bonazzola, S., Gourgoulhon, E., Haensel, P.: Astron. Astrophys. **291**, 155 (1994)
Slane, P., Helfand, D.J., Van der Swaluw, E., Murray, S.S.: Astrophys. J. **616**, 403 (2004)
Splaver, E.M., Nice, D.J., Arzoumanian, Z., Camilo, F., Lyne, A.G., Stairs, I.H.: Astrophys. J. **581**, 509 (2002)
Splaver, E.M., Nice, D.J., Stairs, I.H., Lommen, A.N., Backer, D.C.: Astrophys. J. **620**, 405 (2005)
Thorsett, S.E., Chakrabarty, D.: Astrophys. J. **512**, 288 (1999)
Tolman, R.C.: Phys. Rev. **55**, 364 (1939)
van Kerkwijk, M.H., van Paradijs, J., Zuiderwijk, E.J.: Astron. Astrophys. **303**, 497 (1995)
van Paradijs, J.: Astron. Astrophys. **101**, 174 (1981)
van Straten, W. et al.: Nature **412**, 158 (2001)
Walter, F.M., Faherty, J.K.: In: Page, D., Turolla, R., Zane, S. (eds.), Isolated Neutron Stars. Springer, Berlin (2006)
Walter, F.M., Lattimer, J.M.: Astrophys. J. **576**, 145 (2002)
Watts, A.L., Strohmayer, T.E.: Astrophys. J. **637**, 117 (2006)
Weisberg, J.M., Taylor, J.H.: In: Rasio, F.A., Stairs, I.H. (eds.) Binary Radio Pulsars. ASP Conf. Ser., vol. 328, p. 25. Astron. Soc. Pac., San Francisco (2005), astro-ph/0407149

Astrophys Space Sci (2007) 308: 381–385
DOI 10.1007/s10509-007-9335-9

ORIGINAL ARTICLE

Neutron star masses: dwarfs, giants and neighbors

Sergei Popov · David Blaschke · Hovik Grigorian · Mikhail Prokhorov

Received: 27 June 2006 / Accepted: 13 October 2006 / Published online: 20 March 2007

Abstract We discuss three topics related to the neutron star (NS) mass spectrum. At first we discuss the possibility to form low-mass ($M \lesssim 1M_{\odot}$) objects. In our opinion this and suggest this is possible only due to fragmentation of rapidly rotating proto-NSs. Such low-mass NSs should have very high spatial velocities which could allow identification. A critical assessment of this scenario is given. However, the mechanism has its own problems, and so formation of such objects is not very probable. Secondly, we discuss mass growth due to accretion for NSs in close binary systems. With the help of numerical population synthesis calculations we derive the mass spectrum of massive ($M > 1.8M_{\odot}$) NSs. Finally, we discuss the role of the mass spectrum in population studies of young cooling NSs. We formulate a kind of *mass constraint* which can be helpful, in our opinion, in discussing different competitive models of the thermal evolution of NSs.

Keywords Neutron stars · Pulsars · Binary systems

PACS 97.60.Jd · 97.60.Gb · 97.80.Jp

S.B.P. wants to thank the Organizers for support and hospitality. The work of S.B.P. was supported by the RFBR grant 06-02-16025 and by the "Dynasty" Foundation (Russia). The work of M.E.P.—by the RFBR grant 04-02-16720 and that of H.G. by DFG grant 436 ARM 17/4/05.

S. Popov (✉) · M. Prokhorov
Sternberg Astronomical Institute, Universitetski pr. 13,
119992 Moscow, Russia
e-mail: polar@sai.msu.ru

M. Prokhorov
e-mail: mike@sai.msu.ru

D. Blaschke
Gesellschaft für Schwerionenforschung mbH (GSI), Planck str. 1,
D-64291 Darmstadt, Germany
e-mail: Blaschke@theory.gsi.de

D. Blaschke
Bogoliubov Laboratory for Theoretical Physics, JINR Dubna,
141980 Dubna, Russia

H. Grigorian
Institut für Physik, Universität Rostock, D-18051 Rostock,
Germany
e-mail: hovik.grigorian@uni-rostock.de

H. Grigorian
Department of Physics, Yerevan State University,
375049 Yerevan, Armenia

1 Introduction

Mass is one of the key parameters for neutron star (NS) physics and astrophysics. It can be measured with high precision in binary radio pulsar systems. Until very recently all estimates were obtained in the very narrow region of 1.35–$1.45M_{\odot}$ (Thorsett and Chakrabarty 1999). These values lie very close to the Chandrasekhar limit for white dwarfs. Thus, $M = 1.4M_{\odot}$ was considered to be the standard value of the NS mass. Recently, the range widened towards lower masses after the discovery of the double pulsar J0737-3039 (Burgay et al. 2003). One of the NSs in this system has $M = 1.25M_{\odot}$ (Lyne et al. 2004). Several other examples are known (see Sect. 4).

The mass range is extended also towards higher masses, although these results are less certain. There is one NS in a binary radio pulsar system with mass significantly higher than the canonical value $1.4M_{\odot}$. It is the pulsar J0751+1807 with the mass $2.1^{+0.4}_{-0.5}$ (95% confidence level) (Nice et al.

2005). A recent reanalysis of RXTE data suggests that in low-mass X-ray binaries there are several candidates for neutron stars with masses between 1.9 and $2.1M_{\odot}$ (Barret et al. 2005) as, e.g., 4U 1636-536. The small number of massive radio pulsars, however, can be a result of selection effects.

Cooling curves of NSs are strongly dependent on the star mass, the mass spectrum is one of the most important ingredients of the population synthesis of these sources and, unfortunately, one of the less known. Still, theoretical considerations (Woosley et al. 2002) can shed light on the general properties of the mass spectrum of NSs. This properties have to be taken into account when confronting theoretical models of the thermal evolution of NSs with observational data.

In this paper we discuss three issues: the formation of low-mass NSs, the mass growth of NSs in binaries, and the role of NS masses in the population synthesis of young cooling objects.

2 Formation of low-mass neutron stars

The results presented in this section are based on the research note (Popov 2004).

In many models of thermal evolution of compact objects (neutron stars—NS, hybrid stars—HyS, strange stars—SS) low-mass sources with $0.8M_{\odot} < M < 1M_{\odot}$[1] remain hot for a relatively long time (about few million years).[2] During all that time they remain hotter than more massive stars (Blaschke et al. 2004 and references therein). In that sense they are promising candidates to be observed as *coolers*, and their detection is of great interest for the physics of dense matter (Carriere et al. 2003). However, in most of models of NS formation (Woosley et al. 2002; Fryer and Kalogera 2001; Timmes et al. 1996) no objects with $M < 1$–$1.2M_{\odot}$ are formed. It is likely that just because masses of stellar cores are always heavier than $1.2M_{\odot}$ even for the solar metallicity (and heavier for lower metallicity, see Woosley et al. 2002).

In our opinion the only way to form a low-mass object from a (relatively) high-mass core is fragmentation (see however a discussion in (Xu 2005), where the author discusses the formation of low-mass SSs from white dwarfs via accretion induced collapse).

The fragmentation of a rapidly rotating proto-NS due to dynamical instabilities as part of a two-stage supernova (SN) explosion mechanism was suggested by Berezinskii et al. (1988), later the mechanism was developed by Imshennik and Nadezhin (1992). This mechanism can explain several particular features of SN explosions (for example the delay in the neutrino signal from SN1987A). In this scenario a compact object inevitably obtains a high kick velocity. Recently, the mechanism was studied in some details by Colpi and Wasserman (2002) in application to kick phenomenon. Because of high temperature the exploding (lighter) companion can be significantly more massive than the minimum mass for cold stars ($\sim 0.1M_{\odot}$), up to $\sim 0.7M_{\odot}$. In that case from the initial object of, say, 1.2–1.3 solar masses due to fragmentation and explosion of the lighter part, we can finally obtain a high-velocity NS with a mass of about 0.8–$1M_{\odot}$ or lower if the mass of the initial object was smaller.

Even lower masses can appear if in the fragmentation three bodies are formed. In such a case the lightest or an intermediate mass fragment can be dynamically ejected from the system (again with significant velocity about thousands km s^{-1}). Such ejected compact objects can have masses about 0.2–0.5 solar masses. In the remaining pair the lighter one can start to accrete onto the second companion because of the orbit shrinking due to gravitational waves emission, and after reaching the minimum mass ($\sim 0.1M_{\odot}$) it explodes. So, the remaining compact object would also have relatively low mass ($\sim 1M_{\odot}$) and high velocity.

Objects formed after fragmentation have particular predictable properties: high spatial velocity, high surface temperature, velocity vector nearly perpendicular to the spin axis (since in this mechanism the kick is always obtained in the orbital plane which coincides with the equator of the initial proto-NS).

Due to high kicks low-mass compact objects are not expected to appear in binaries (at least they should be rarely found in binary systems). To find a low-mass compact object one has to search for a hot, young, high velocity NS.

It is reasonable to expect that mass and kick velocity are anti-correlated, as far as a higher mass of the remaining object corresponds to the lighter exploded component, which means to a wider orbit, and to lower orbital velocity of the remaining more massive component. Also higher kicks lead to smaller fall-back (Colpi and Wasserman 2002).

To reach fragmentation conditions (the dynamical instability) it is necessary that the progenitor core is rapidly rotating. Rotation of isolated progenitors and its influence on properties of newborn NSs was studied in several papers (see, for example, Heger et al. 2003 and references therein). To obtain a rapidly rotating compact object it is necessary to avoid spin-down influence of the magnetic field, so probably compact objects born after fragmentation should be low magnetized. It means that low-mass neutron stars are not expected to be normal radio pulsars. Because of the same reason they are not expected to show any kind of magnetar activity.

[1] Here and below speaking about compact objects we mean the gravitational mass.

[2] Objects with even lower mass, $\sim 0.5M_{\odot}$, also are relatively hot (Blaschke et al. 2004).

We have to note, that the mechanism of SN explosion suggested by Berezinskii et al. (1988) and Imshennik (1992) has its internal problems. There is the possibility, that a rapidly rotating proto-NS can just loose part of its mass in the form of an outflow in the equatorial plane. In that case two spiral arms appear, no second (or third) component is formed, and the kick can be relatively small. The fraction of lost mass is very small, about 4% (Houser et al. 1994), so that the final mass cannot be much lower than the initial one. If the fragmentation in the process of NS formation never happens in nature, then, in our opinion, it is very improbable, that low-mass compact objects can exist. The discovery of a high velocity low-mass NS, HyS or SS will be a strong argument in favour of the mechanism.

To conclude: the fragmentation of a proto-NS can be a unique mechanism of the formation of low-mass compact objects, which are expected to have several peculiar characteristics that can help to distinguish them among possible candidates. However, the realization of this mechanism in Nature is not very promising.

3 Massive neutron stars in binaries

In this section we present our recent calculations of the mass growth of NSs in close binary systems (Popov and Prokhorov 2005).

Observationally, high masses of NSs are mainly supported by data on X-ray binaries, where recently new results have been extracted from a reanalysis of RXTE data which has suggested that sharp and reproducible changes in QPO properties are related to the innermost stable circular orbit. A mass estimate of 1.9–2.1$M_\odot$ has been given for 4U 1636-536 (Barret et al. 2005). Estimates for several systems give also very high mass values with rather large uncertainties: 1.8–2.4$M_\odot$ for Vela X-1[3] (Quaintrell et al. 2003), $2.4 \pm 0.27 M_\odot$ for 4U 1700-37 (Clark et al. 2002; see also Heineke et al. 2003; van Kerkwijk 2004). Shahbaz et al. (2004) presented observations of a low-mass X-ray binary 2S 0921-630/V395 Car for which the 1σ mass range for the compact object is 2.0–4.3$M_\odot$.

The existence of NSs with $M \sim$ 1.8–2.4$M_\odot$ is not in contradiction with the present day theory of NS interiors. There are several models with stiff equation of state which allow the existence of NSs with masses larger than $2M_\odot$ (see a review and references in Haensel 2003; Lattimer and Prakash 2005 and in the contribution of J. Lattimer in these proceedings). Here we will assume that masses of NSs with extreme rotation can reach the maximum value $M_{\rm max} = 3.45 M_\odot$ according to the estimate by Ouyed and Butler (1999). If, however, in nature $M_{\rm max}$ is lower than this value, the part of the obtained mass spectrum exceeding it should be attributed to black holes.

3.1 Model

The model we use is discussed in detail in Popov and Prokhorov (2005). For our calculations we use the "Scenario Machine" code developed at the Sternberg Astronomical Institute.[4] A description of most of the parameters of the code can be found in Lipunov et al. (1996). The main assumptions of the scenario are:

- All NSs are born with $M = 1.4 M_\odot$.
- At the common envelope stage a hypercritical accretion (with $\dot{M}$ much larger than the Eddington value) is possible.
- During accretion the magnetic field of a NS decays to a value that cannot prevent rapid (maximum) rotation of the NS.
- The Oppenheimer–Volkoff mass of a rapidly rotating NS (the critical mass of a BH formation) is assumed to be $3.45 M_\odot$ according to Ouyed (2004).

3.2 Mass spectrum

As we are interested here in systems with a high mass ratio it is necessary to consider three different situations after NS formation when the secondary fills its Roche lobe: (i) a normal star can fill its Roche lobe without common envelope formation; (ii) a normal star can fill its Roche lobe with common envelope formation; (iii) a WD fills its Roche lobe.

To fill the Roche lobe a normal secondary star has to evolve further than the main sequence stage. During its evolution prior to the Roche lobe overflow the mass of the star is nearly constant. A common envelope is not formed if the normal star is not significantly heavier than the NS. In this regime, mass is not lost from the binary system. For more massive secondaries, formation of a common envelope is inevitable, mass transfer is unstable. In this regime, a significant fraction of the mass flow is lost from the system, so the mass of the NS grows less effectively. After the common envelope stage the orbital separation becomes smaller, so later on even a degenerate core of the secondary—a WD—can fill the Roche lobe.

In Fig. 1 we present our calculations of the mass spectrum of massive NSs. About 25% of accreting massive NSs have normal stars as secondaries, the rest have WD companions. The formation rate of massive NSs was found to be 6.7×10^{-7} yrs^{-1}. This corresponds to ~10 000 of these compact stars in the Galaxy.

[3]This range is based on the two estimates given in Quaintrell et al. (2003): 1.88 ± 0.13 and $2.27 \pm 0.17 M_\odot$.

[4]http://xray.sai.msu.ru/sciwork/scenario.html and http://xray.sai.msu.ru/~mystery/articles/review/.

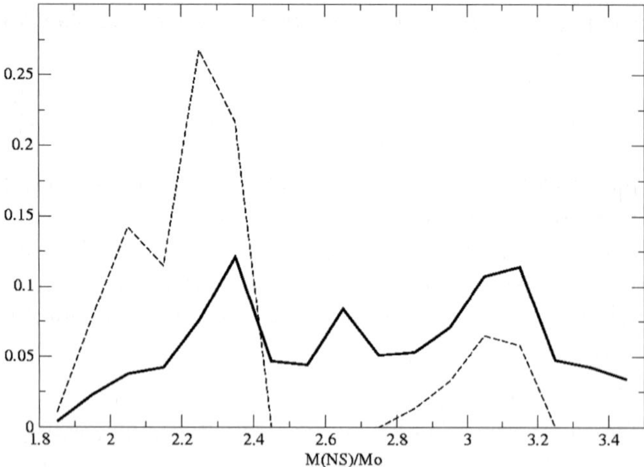

Fig. 1 Mass distribution of NSs. As we are interested only in the massive population we do not show the results for compact objects with $M < 1.8M_{\odot}$. The upper mass limit corresponds to SkyS with maximum rotation (Ouyed 2004). The *dashed line* represents results for the scenario with zero kick. The *solid line* is the non-zero kick. Left peaks for both distributions correspond to NSs with a single episode of accretion. Right peaks are formed by NSs which also increased their masses via accretion from WDs. Distributions were normalized to unity, i.e. the area below each line is equal to one

4 Mass constraint

The discussion below is related to our recent study of criteria to test theoretical cooling curves (Popov et al. 2006b) based on an approach for hybrid stars with color superconducting quark matter cores (Grigorian et al. 2005).

Cooling curves of NS are strongly mass dependent (see contributions by D. Page, A. Kaminker and others in this volume). Unfortunately, a mass determination with high precision is available only for NSs in binary systems. Compact objects in X-ray binaries could accrete a significant amount of matter. For some of the radio pulsars observed in binaries, accretion also played an important rôle. Without any doubts masses of millisecond pulsars do not represent their initial values. However, there is a small number of NSs with well determined masses, for which it is highly possible that these masses did not change significantly since these NSs were born (data on NS masses can be found, for example, in Lorimer 2005 and references therein). These are secondary (younger) components of double NS systems. According to standard evolutionary scenarios these compact objects never accreted a significant amount of mass. Their masses lie in the narrow range $1.18–1.39M_{\odot}$. Now there are nine double-NS systems with well-estimated masses of the secondary components. This set of data is a very good evidence in favour of the mass spectrum used in our population synthesis calculations (Popov et al. 2006a). Of course, some effects of binary evolution can be important, and so for isolated stars (or stars in very wide binaries) the situation can be slightly different. However, with these observational estimates of initial masses of NSs we feel more confident using the spectrum with a small number of NSs with $M \gtrsim 1.4–1.5M_{\odot}$.

Brighter sources are easier to discover. So, among known cooling NSs the fraction of NSs with *typical* masses, i.e. in the range $1.1M_{\odot} \lesssim M \lesssim 1.5M_{\odot}$, should be even higher than in the original mass spectrum. So, we have the impression that it is necessary to try to explain even cold (may be with an exception of 1–2 coldest) sources with $M \lesssim 1.4–1.5M_{\odot}$. Especially, the Magnificent seven and other young close-by compact objects should be explained as most typical representatives of the whole NS population. We want to underline that, even being selected by their observability in soft X-rays, these sources form one of the most uniform samples of young isolated NSs. In this sense, the situation where a significant number of sources are explained by cooling curves corresponding to $1.5M_{\odot} \lesssim M \lesssim 1.7M_{\odot}$ should be considered as a disadvantage of the model. Particularly, Vela, Geminga and RX J1856-3754 should not be explained as massive NSs.

All the above gives us the opportunity to formulate the conjecture of a *mass spectrum constraint*: data points should be explained mostly by NSs with *typical* masses.

References

Barret, D., Olive, J.-F., Miller, M.C.: Mon. Not. Roy. Astron. Soc. **361**, 855 (2005)
Berezinskii, V.S., Castagnoli, C., Dokuchaev, V.I., et al.: Nuovo Cimento C **11**, 287 (1988)
Blaschke, D., Grigorian, H., Voskresenky, D.N.: Astron. Astrophys. **424**, 979 (2004)
Burgay, M., D'Amico, N., Possenti, A., et al.: Nature **426**, 531 (2003)
Carriere, J., Horowitz, C.J., Piekarewicz, J.: Astrophys. J. **593**, 463 (2003)
Clark, J.S., Goodwin, S.P., Crowther, P.A., et al.: Astron. Astrophys. **392**, 909 (2002)
Colpi, M., Wasserman, I.: Astrophys. J. **581**, 1271 (2002)
Fryer, C.L., Kalogera, V.: Astrophys. J. **554**, 548 (2001)
Grigorian, H., Blaschke, D., Voskresensky, D.N.: Phys. Rev. C **71**, 045801 (2005)
Haensel, P.: In: Motch, C., Hameury, J.-M. (eds.) The Proceedings of the Conference Final Stages of Stellar Evolution. EAS Publications Series, vol. 7, p. 249 (2003)
Heger, A., Woosley, S.E., Langer, N., et al.: In: Maeder, A., Eenens, P. (eds.) The Proceedings of the IAU Symposium, vol. 215, p. 591. Astronomical Society of the Pacific (2003)
Heinke, C.O., Grindlay, J.E., Lloyd, D.A., et al.: Astrophys. J. **588**, 452 (2003)
Houser, J.L., Centrella, J.M., Smith, S.C.: Phys. Rev. Lett. **72**, 1314 (1994)
Imshennik, V.S.: Pis'ma Astron. Zh. **18**, 489 (1992)
Imshennik, V.S., Nadezhin, D.K.: Pis'ma Astron. Zh. **18**, 79 (1992)
Lattimer, J.M., Prakash, M.: Phys. Rev. Lett. **94**, 111101 (2005)
Lipunov, V.M., Postnov, K.A., Prokhorov, M.E.: Astrophys. Space Sci. **9**, 1 (1996)
Lorimer, D.R.: Living Rev. Rel. **8**, 7 (2005)
Lyne, A., Burgay, M., Kramer, M., et al.: Science **303**, 1153 (2004)
Nice, D.J., Splaver, E.M., Stairs, I.H.: In: Rasio, F.A., Stairs, I.H. (eds.) The Proceedings of the Conference Binary Radio Pulsar ASP Conference Series, vol. 328, p. 371. Astronomical Society of the Pacific (2005)

Ouyed, R.: astro-ph/0402122 (2004)
Ouyed, R., Butler, M.: Astrophys. J. **522**, 453 (1999)
Popov, S.B.: astro-ph/0403710 (2004)
Popov, S.B., Prokhorov, M.E.: Astron. Astrophys. **434**, 649 (2005)
Popov, S.B., Grigorian, H., Turolla, R., et al.: Astron. Astrophys. **448**, 327 (2006a)
Popov, S.B., Grigorian, H., Blaschke, D.: Phys. Rev. C, **74**, 025803 (2006b)
Quaintrell, H., Norton, A.J., Ash, T.D.C., et al.: Astron. Astrophys. **401**, 313 (2003)
Shahbaz, T., Casares, J., Watson, C., et al.: Astrophys. J. **616**, L123 (2004)
Timmes, F.X., Woosley, S.E., Weaver, T.A.: Astrophys. J. **457**, 834 (1996)
Thorsett, S.E., Chakrabarty, D.: Astrophys. J. **512**, 288 (1999)
van Kerkwijk, M.H.: In: Camilo F., Gaensler, B.M. (eds.) The Proceeding of the IAU Symposium, vol. 218, p. 283. Astronomical Society of the Pacific (2004)
Woosley, S.E., Heger, A., Weaver, T.A.: Rev. Mod. Phys. **74**, 1015 (2002)
Xu, R.X.: Mon. Not. Roy. Astron. Soc. **356**, 359 (2005)

Astrophys Space Sci (2007) 308: 387–394
DOI 10.1007/s10509-007-9360-8

ORIGINAL ARTICLE

A microscopic equation of state for protoneutron stars

G.F. Burgio · M. Baldo · O.E. Nicotra · H.-J. Schulze

Received: 27 June 2006 / Accepted: 25 October 2006 / Published online: 15 March 2007

Abstract We study the structure of protoneutron stars within the finite-temperature Brueckner–Bethe–Goldstone many-body theory. If nucleons, hyperons, and leptons are present in the stellar core, we find that neutrino trapping stiffens considerably the equation of state, because hyperon onsets are shifted to larger baryon density. However, the value of the critical mass turns out to be smaller than the "canonical" value $1.44 M_{\odot}$. We find that the inclusion of a hadron-quark phase transition increases the critical mass and stabilizes it at about $1.5–1.6 M_{\odot}$.

Keywords Dense matter · Equation of State · Neutron Stars

PACS 26.60.+c · 21.65.+f · 24.10.Cn · 97.60.Jd

1 Introduction

After a protoneutron star (PNS) is successfully formed in a supernova explosion, neutrinos are temporarily trapped within the star (Prakash et al. 1997). The subsequent evolution of the PNS is strongly dependent on the stellar composition, which is mainly determined by the number of trapped neutrinos, and by thermal effects with values of temperatures up to 30–40 MeV (Burrows and Lattimer 1986; Pons et al. 1999). Hence, the equation of state (EOS) of dense matter at finite temperature is crucial for studying the macrophysical evolution of PNS.

G.F. Burgio (✉) · M. Baldo · O.E. Nicotra · H.-J. Schulze
INFN Sezione di Catania, Via S. Sofia 64, 95123 Catania, Italy
e-mail: fiorella.burgio@ct.infn.it

Only a few microscopic calculations of the nuclear EOS at finite temperature are available so far. The variational calculation by Friedman and Pandharipande (1981) was one of the first semi-microscopic investigations of the finite-temperature EOS. The results predict a Van der Waals behavior for symmetric matter, which leads to a liquid-gas phase transition with a critical temperature $T_c \approx 18–20$ MeV. Later, Brueckner calculations (Lejeune et al. 1986; Baldo and Ferreira 1999) and chiral perturbation theory at finite temperature (Kaiser et al. 2002) confirmed these findings with very similar values of T_c. The Van der Waals behavior was also found in the finite-temperature relativistic Dirac–Brueckner calculations of Ter Haar and Malfliet (1986, 1987) and Huber et al. (1999), although at a lower temperature.

We have developed a microscopic EOS in the framework of the Brueckner–Bethe–Goldstone (BBG) many-body approach including nucleons and hyperons and extended to finite temperature. This EOS has been successfully applied to the study of the limiting temperature in nuclei (Baldo and Ferreira 1999; Baldo et al. 2004). Recently, this EOS has been extended for including neutrino trapping, and results have been presented in Nicotra et al. (2006a).

The scope of this work is to discuss composition and structure of these newly born stars with the EOS previously mentioned, also including a possible transition to quark matter. In fact, superdense matter in PNS cores may consist of weakly interacting quarks rather than hadrons, due to the asymptotic freedom. The appearance of quarks can alter substantially the chemical composition of a PNS, with observable consequences on the evolution, like onset of metastability and abrupt cessation of the neutrino signal (Pons et al. 2001).

We have studied the effects of a hadron-quark phase transition within the MIT bag model, and found that the presence of quark matter increases the value of the maximum mass of a PNS, and stabilizes it at about 1.5–1.6$M_\odot$, no matter the value of the temperature.

This paper is organized as follows. In Sect. 2 we briefly review the BBG theory of nuclear matter at finite temperature, including both nucleons and hyperons. In Sect. 3 we discuss the chemical composition of a PNS, whereas in Sect. 4 we study its structure, even including a possible transition to a quark phase. Finally, in Sect. 5, we draw our conclusions.

2 The BBG EOS at finite temperature

In the recent years, the BBG perturbative theory has made much progress, since its convergence has been firmly established (Day 1981; Song et al. 1998; Baldo et al. 2000b; Sartor 2006), and has been extended in a fully microscopic and self-consistent way to the hyperonic sector (Schulze et al. 1995, 1998, 2006; Baldo et al. 1998, 2000a). Moreover, the addition of phenomenological three-body forces (TBF) based on the Urbana model (Carlson et al. 1983; Schiavilla et al. 1986), permitted to improve to a large extent the agreement with the empirical saturation properties (Baldo et al. 1997; Zhou et al. 2004).

The finite-temperature formalism which is closest to the BBG expansion, and actually reduces to it in the zero-temperature limit, is the one formulated by Bloch and De Dominicis (1958, 1959a, 1959b). In this approach the essential ingredient is the two-body scattering matrix K, which, along with the single-particle potential U, satisfies the self-consistent equations

$$\begin{aligned}\langle k_1k_2|K(W)|k_3k_4\rangle &= \langle k_1k_2|V|k_3k_4\rangle \\ &\quad + \mathrm{Re}\sum_{k_3'k_4'}\langle k_1k_2|V|k_3'k_4'\rangle \frac{[1-n(k_3')][1-n(k_4')]}{W-E_{k_3'}-E_{k_4'}+i\epsilon} \\ &\quad \times \langle k_3'k_4'|K(W)|k_3k_4\rangle \end{aligned} \tag{1}$$

and

$$U(k_1) = \sum_{k_2} n(k_2)\langle k_1k_2|K(W)|k_1k_2\rangle_A, \tag{2}$$

where k_i generally denote momentum, spin, and isospin. Here V is the two-body interaction, and we choose the Argonne V_{18} nucleon-nucleon potential (Wiringa et al. 1995). $W = E_{k_1} + E_{k_2}$ represents the starting energy, and $E_k = k^2/2m + U(k)$ the single-particle energy. Equation (1) coincides with the Brueckner equation for the K matrix at zero temperature, if the single-particle occupation numbers $n(k)$ are taken at $T = 0$. At finite temperature $n(k)$ is a Fermi distribution. For a given density and temperature, (1) and (2) have to be solved self-consistently along with the equation for the grand-canonical potential density ω and the free energy density, which has the following simplified expression

$$f = \sum_i \left[\sum_k n_i(k)\left(\frac{k^2}{2m_i} + \frac{1}{2}U_i(k)\right) - Ts_i\right], \tag{3}$$

where

$$s_i = -\sum_k \left(n_i(k)\ln n_i(k) + [1-n_i(k)]\ln[1-n_i(k)]\right) \tag{4}$$

is the entropy density for component i treated as a free gas with spectrum $E_i(k)$. For a more extensive discussion of this topic, the reader is referred to (Baldo 1999) and references therein.

In deriving (3), we have introduced the so-called *Frozen Correlations Approximation*, i.e., the correlations at $T \neq 0$ are assumed to be essentially the same as at $T = 0$. This means that the single-particle potential $U_i(k)$ for the component i can be approximated by the one calculated at $T = 0$. This allows to save computational time and simplify the numerical procedure. It turns out that the assumed independence is valid to a good accuracy, at least for not too high temperature (Baldo and Ferreira 1999, Fig. 12).

In our many-body approach, we have also introduced TBF among nucleons, in order to reproduce correctly the nuclear matter saturation point $\rho_0 \approx 0.17$ fm^{-3}, $E/A \approx -16$ MeV. Since a complete microscopic theory of TBF is not available yet, we have adopted the phenomenological Urbana model (Carlson et al. 1983; Schiavilla et al. 1986), which consists of an attractive term due to two-pion exchange with excitation of an intermediate Δ resonance, and a repulsive phenomenological central term. In the BBG approach, the TBF is reduced to a density-de pendent two-body force by averaging over the position of the third particle, assuming that the probability of having two particles at a given distance is reduced according to the two-body correlation function. The corresponding EOS reproduces correctly the nuclear matter saturation point (Baldo et al. 1997; Zhou et al. 2004), and gives values of incompressibility and symmetry energy at saturation compatible with those extracted from phenomenology (Myers and Swiatecki 1996).

In Fig. 1 we display the free energy (upper panels) and the internal energy (lower panels) obtained following the above discussed procedure, both for symmetric and purely neutron matter, as a function of the nucleon density. Calculations are reported for several values of the temperature between 0 and 50 MeV. We notice that the free energy of symmetric matter shows a typical Van der Waals behavior (with $T_c \approx 16$ MeV) and is a monotonically decreasing function of the temperature. On the contrary, the internal energy is an increasing function of the temperature. At $T = 0$ the free energy coincides with the total energy and the corresponding curve is just the usual nuclear matter saturation curve.

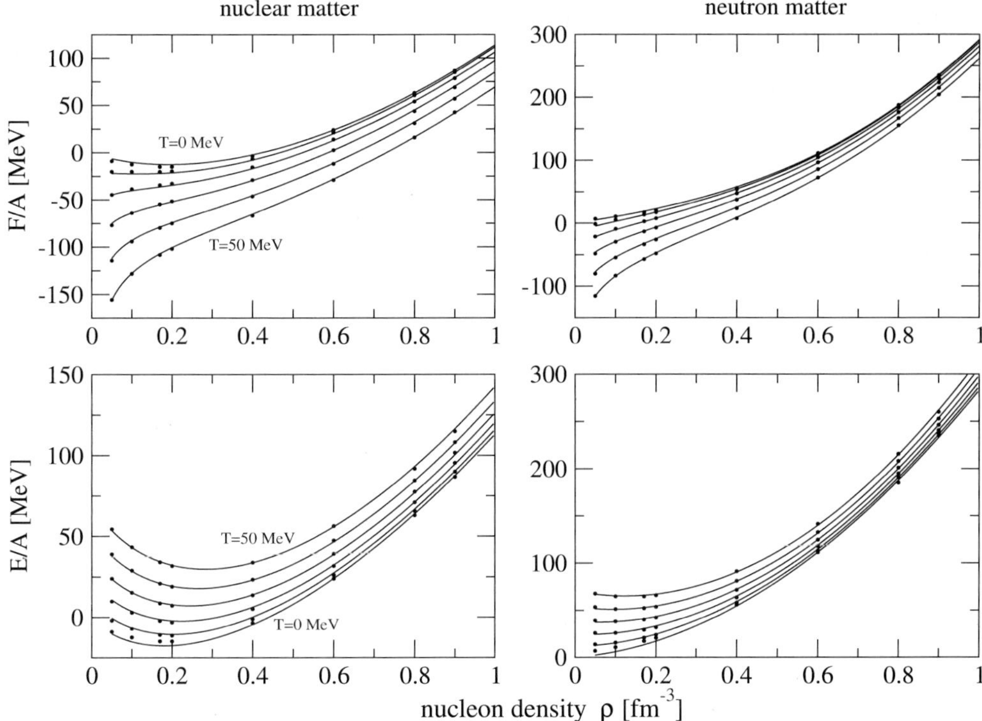

Fig. 1 Finite-temperature BBG EOS for symmetric (*left-hand panels*) and purely neutron (*right-hand panels*) matter. The upper panels show the free energy and the lower panels the internal energy per particle as a function of the nucleon density. The temperatures vary from 0 to 50 MeV in steps of 10 MeV

2.1 Inclusion of hyperons

The fast rise of the baryon chemical potentials with density in neutron star cores (Glendenning 1982, 1985) may trigger the appearance of strange baryonic species, i.e., hyperons. For this purpose we have extended the BBG approach in order to include the Σ^- and Λ hyperons (Schulze et al. 1998, 2006; Baldo et al. 1998, 2000a). The inclusion of hyperons requires the knowledge of the nucleon-hyperon (NH) and hyperon-hyperon (HH) interactions. In our past papers, we have shown results obtained at $T = 0$, and we have used the Nijmegen soft-core NH potential (Maessen et al. 1989), and neglected the HH interactions, since no reliable HH potentials are available yet. For these reasons, we present in this article finite-temperature calculations using free hyperons. A more complete set of calculations obtained with the inclusion of the NH interaction at finite temperature will be published elsewhere.

We have found that due to its negative charge the Σ^- hyperon is the first strange baryon appearing in the reaction $n + n \to p + \Sigma^-$, in spite of its substantially larger mass compared to the neutral Λ hyperon ($M_{\Sigma^-} = 1197$ MeV, $M_\Lambda = 1116$ MeV). The presence of hyperons strongly softens the EOS, mainly due to the larger number of baryonic degrees of freedom. This EOS produces a maximum neutron star mass that lies slightly below the canonical value of $1.44 M_\odot$ (Taylor and Weisberg 1989), as confirmed by Schulze et al. (2006) also with the NSC97 hyperon potentials (Stoks and Rijken 1999). This could indicate the presence of non-baryonic (quark) matter in the interior of heavy neutron stars (Burgio et al. 2002a, 2002b; Baldo et al. 2003; Maieron et al. 2004). This point is discussed more extensively below.

3 Composition and EOS of hot stellar matter

For stars in which the strongly interacting particles are only baryons, the composition is determined by the requirements of charge neutrality and equilibrium under the weak processes

$$B_1 \to B_2 + l + \bar{\nu}_l, \qquad B_2 + l \to B_1 + \nu_l, \tag{5}$$

where B_1 and B_2 are baryons and l is a lepton, either an electron or a muon. When the neutrinos are trapped, these two requirements imply that the relations

$$\sum_i q_i x_i + \sum_l q_l x_l = 0 \tag{6}$$

and

$$\mu_i = b_i \mu_n - q_i(\mu_l - \mu_{\nu_l}) \tag{7}$$

are satisfied. In the expression above, $x_i = \rho_i/\rho_B$ represents the baryon fraction for the species i, μ_i the chemical potential, b_i the baryon number, and q_i the electric charge. Equivalent quantities are defined for the leptons l.

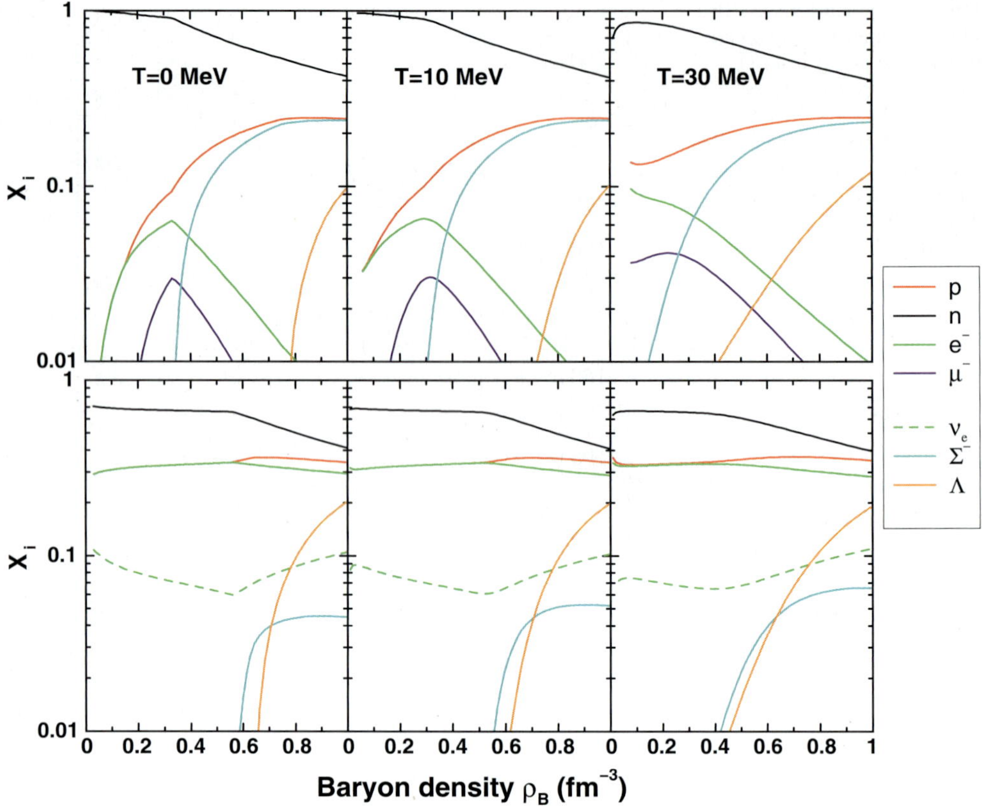

Fig. 2 Relative populations for neutrino-free (*upper panels*) and neutrino-trapped (*lower panels*) matter as a function of the baryon density for several values of the temperature

For stellar matter containing nucleons and hyperons as relevant degrees of freedom, the chemical equilibrium conditions read explicitly

$$\begin{aligned} &\mu_n - \mu_p = \mu_e - \mu_{\nu_e} = \mu_\mu + \mu_{\bar{\nu}_\mu}, \\ &\mu_{\Sigma^-} = 2\mu_n - \mu_p, \\ &\mu_\Lambda = \mu_n. \end{aligned} \tag{8}$$

The nucleon chemical potentials are calculated starting from the free energy and its partial derivatives with respect to the total baryon density ρ and proton fraction x_p, i.e.,

$$\mu_n(\rho, x_p) = \left[1 + \rho\frac{\partial}{\partial\rho} - x_p\frac{\partial}{\partial x_p}\right]\frac{f}{\rho}, \tag{9}$$

$$\mu_p(\rho, x_p) = \left[1 + \rho\frac{\partial}{\partial\rho} + (1 - x_p)\frac{\partial}{\partial x_p}\right]\frac{f}{\rho}, \tag{10}$$

whereas the chemical potentials of the noninteracting leptons and hyperons are obtained by solving numerically the free Fermi gas model at finite temperature. More details are given in Nicotra et al. (2006a).

Because of trapping, the numbers of leptons per baryon of each flavor $l = e, \mu$,

$$Y_l = x_l - x_{\bar{l}} + x_{\nu_l} - x_{\bar{\nu}_l}, \tag{11}$$

are conserved on dynamical time scales. Gravitational collapse calculations of the white-dwarf core of massive stars indicate that at the onset of trapping, the electron lepton number $Y_e = x_e + x_{\nu_e} \approx 0.4$, the precise value depending on the efficiency of electron capture reactions during the initial collapse stage. Moreover, since no muons are present when neutrinos become trapped, the constraint $Y_\mu = x_\mu - x_{\bar{\nu}_\mu} = 0$ can be imposed. We fix the Y_l at these values in our calculations for neutrino-trapped matter.

Let us now discuss first the populations of beta-equilibrated stellar matter, by solving the chemical equilibrium conditions given by (8), supplemented by electrical charge neutrality and baryon number conservation. In Fig. 2 we show the particle fractions as a function of baryon density, for different values of the temperature. The upper panels show the particle fractions when stellar matter does not contain neutrinos, whereas the lower panels show the populations in neutrino-trapped matter. We observe that the electron fraction is larger in neutrino-trapped than in neutrino-free matter, and, as a consequence, the proton population is larger.

Neutrino trapping strongly influences the onset of hyperons. In fact, the threshold density of the Σ^- is shifted to high density, whereas Λ's appear at slightly smaller density. This is due to the fact that the Σ^- onset depends on the neutron and lepton chemical potentials, i.e., $\mu_n + \mu_e - \mu_{\nu_e}$, which stays at larger values in neutrino-trapped matter than in the neutrino-free case because of the larger fraction of electrons, thus delaying the appearance of the Σ^- to higher baryon

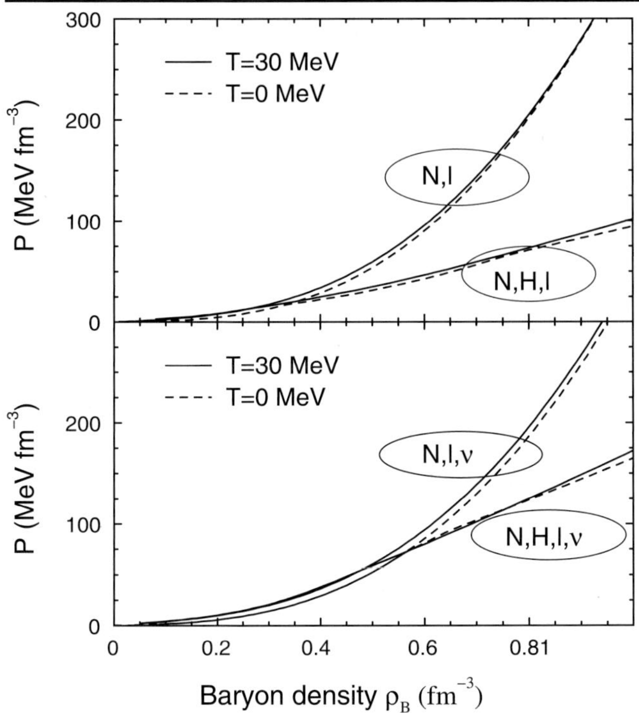

Fig. 3 Pressure as a function of baryon density for beta-equilibrated matter at temperatures $T = 0$ and 30 MeV. The *upper* (*lower*) *panel* shows the EOS in neutrino-free (neutrino-trapped) matter, with nucleons only (*upper curves*) and nucleons plus hyperons (*lower curves*)

density and limiting its population to a few percent. On the other hand, the Λ onset depends on the neutron chemical potential only, which keeps at lower values in the neutrino-trapped case. When the temperature increases, more and more hyperons are present also at low densities, but they represent only a small fraction of the total baryon density in this region of the PNS. Altogether, the hyperon fractions are much smaller than in the neutrino-free matter. Therefore the corresponding EOS will be stiffer than in the neutrino-free case.

Once the composition of the β-stable stellar matter is known, one can proceed to calculate the free energy density f and then the pressure p through the usual thermodynamical relation

$$p = \rho^2 \frac{\partial (f/\rho)}{\partial \rho}. \tag{12}$$

The resulting EOS is displayed in Fig. 3, where the pressure for beta-stable asymmetric matter, without (upper panel) and with (lower panel) neutrinos, is plotted as a function of the baryon density at temperatures $T = 0$ and 30 MeV. Let us begin with discussing the case of neutrino-free matter, shown without (upper curves) and with hyperons (lower curves). We notice that thermal effects produce a slightly stiffer EOS with respect to the cold case, and that at very high densities they almost play no role.

In the case that only nucleons are present, the EOS gets softer with increasing temperature at high baryon density. This behavior is at variance with the results obtained by Prakash et al. (1997). In this regard, we should notice that in our calculations we are considering an isothermal profile, whereas in Prakash et al. (1997) the profile is isentropic. Another difference between the two approaches is in the many-body method. A complete comparison can be made only by adopting an isentropic description within our BHF approach. This matter is left to further investigations.

The inclusion of hyperons produces a dramatic effect, because the EOS gets much softer, no matter the value of the temperature. In this case, thermal effects dominate over the whole density range since on the average the Fermi energies are smaller. A similar behavior was found in Prakash et al. (1997).

In the lower panel wc show the corresponding neutrino-trapped case. The EOS is slightly softer than in the neutrino-free case if only nucleons and leptons are present in the stellar matter. Again, the presence of hyperons introduces a strong softening of the EOS, but less than in the neutrino-free case, because now the hyperons appear later in the matter and their concentration is lower. Thermal effects are rather small also in this case, except for the disappearance of the hyperon onsets.

4 (Proto)neutron star structure

The stable configurations of a (proto)neutron star can be obtained from the well-known hydrostatic equilibrium equations of Tolman, Oppenheimer, and Volkov (Shapiro and Teukolsky 1983) for the pressure p and the enclosed mass m,

$$\frac{dp(r)}{dr} = -\frac{Gm(r)\epsilon(r)}{r^2} \frac{\left[1 + \frac{p(r)}{\epsilon(r)}\right]\left[1 + \frac{4\pi r^3 p(r)}{m(r)}\right]}{1 - \frac{2Gm(r)}{r}}, \tag{13}$$

$$\frac{dm(r)}{dr} = 4\pi r^2 \epsilon(r), \tag{14}$$

once the EOS $p(\epsilon)$ is specified, being ϵ the total energy density (G is the gravitational constant). For a chosen central value of the energy density, the numerical integration of (13) and (14) provides the mass-radius relation.

For the description of the (proto)neutron star crust, we have joined the hadronic EOS described above with the ones by Negele and Vautherin (1973) in the medium-density regime (0.001 fm^{-3} $< \rho <$ 0.08 fm^{-3}), and the ones by Feynman et al. (1949) and Baym et al. (1971) for the outer crust ($\rho <$ 0.001 fm^{-3}).

Simulations of supernovae explosions (Burrows and Lattimer 1986; Pons et al. 1999) show that the PNS has neither an isentropic nor an isothermal profile. For simplicity

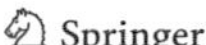

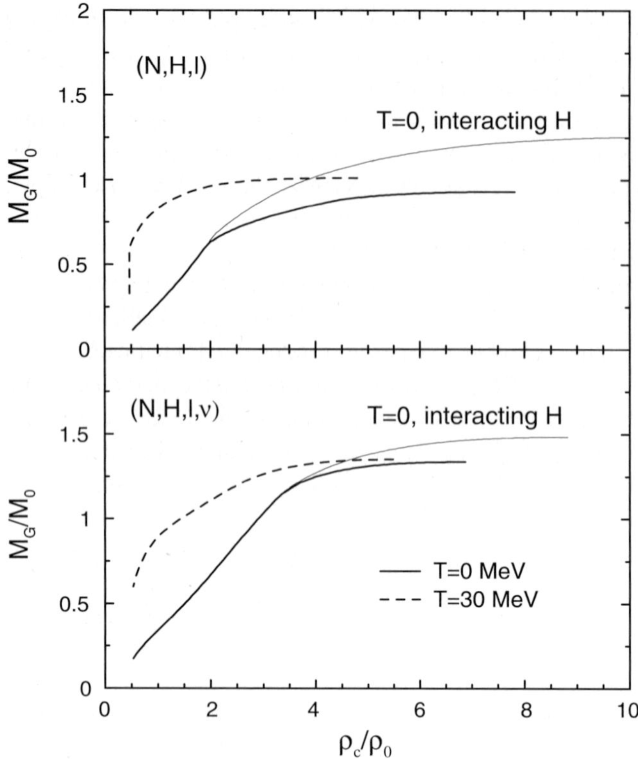

Fig. 4 The gravitational mass (in units of the solar mass) as a function of the central baryon density (normalized with respect to the saturation value $\rho_0 = 0.17\ \mathrm{fm}^{-3}$) at temperatures $T = 0$ and 30 MeV. The *upper* (*lower*) *plot* regards neutrino-free (neutrino-trapped) matter. The *thin solid curves* denote configurations of cold stars employing interacting hyperons, whereas the *remaining curves* show results obtained using free hyperons

we will assume a constant temperature inside the star and attach a cold crust for the outer part. This schematizes the temperature profile of the PNS. More realistic temperature profiles can be obtained by modelling the neutrinosphere both in the interior and in the external outer envelope, which is expected to be much cooler. A proper treatment of the transition from the hot interior to the cold outer part can have a dramatic influence on the mass-central density relation in the region of low central density and low stellar masses. In particular, the "minimal mass" region, typical of cold neutron stars (Zel'dovich and Novikov 1971; Shapiro and Teukolsky 1983), can be shifted in PNS to much higher values of central density and masses. A detailed analysis of this point can be found in (Gondek et al. 1997), where a model of the transition region between the interior and the external envelope is developed. However, as discussed in the next section, the maximum mass region is not affected by the structure of this low-density transition region.

In Fig. 4 we show the gravitational mass (in units of the solar mass $M_\odot = 1.98 \times 10^{33}$ g) as a function of the central baryon density for stars containing hyperons. We observed in Fig. 3 that the EOS softens considerably when hyperons are included, both in neutrino-free and neutrino-trapped matter. As a consequence the mass—central density relation is also significantly altered by the presence of hyperons and the value of the critical mass is about $1.3M_\odot$ for neutron stars (upper plot, thin curve) and $1.5M_\odot$ for protostars in the ν-trapped stage (lower plot, thin curve). Those results are obtained employing interacting hyperons at zero temperature, choosing the Nijmegen potential as nucleon-hyperon potential (Maessen et al. 1989), which is well adapted to the available experimental NH scattering data. However, since the value of the critical mass for cold stars falls below the mass of the best observed pulsar, i.e., $1.44M_\odot$ (Taylor and Weisberg 1989), the EOS of high-density nuclear matter comprising only baryons (nucleons and hyperons) is probably unrealistic (even taking into account the present uncertainty of hyperonic two-body and three-body forces), and must be supplemented by a transition to quark matter. This has been discussed extensively in (Burgio et al. 2002a, 2002b; Baldo et al. 2003; Maieron et al. 2004), and will be briefly recalled in the next subsection.

Nevertheless, it is interesting to observe that the maximum mass of hyperonic protostars is larger (by about $0.3M_\odot$) than the one of cold neutron stars. The reason is the minor importance of hyperons in the neutrino-saturated matter, which leads to a stiffer EOS, see Fig. 3, and to a larger maximum mass. This feature could lead to metastable stars suffering a delayed collapse while cooling down, as discussed in (Prakash et al. 1997; Pons et al. 1999). Metastable stars occur within a range of masses near the maximum mass of the initial configuration and remain stable only for several seconds after formation. In order to study metastability, it is useful to calculate the baryonic mass M_B, which is proportional to the number of baryons in the system and is constant during the evolution of the isolated star, if no accretion is assumed during the entire PNS evolution. For that, (13) and (14) must be supplemented with

$$\frac{dM_B(r)}{dr} = \frac{4\pi r^2 \rho_B m_N}{\sqrt{1 - 2Gm(r)/r}}, \tag{15}$$

where $m_N = 1.67 \times 10^{-24}$ g is the nucleon mass. As one can see from Fig. 5, if hyperons are present (lines ending with a dot), then deleptonization, that is the transition from $Y_\nu = 0.4$ to $Y_\nu = 0$, lowers the range of baryonic masses that can be supported by the EOS from about $1.15M_\odot$ to about $0.84M_\odot$. The window in the baryonic mass in which neutron stars are metastable is thus about $0.31M_\odot$. On the other hand, if hyperons are absent (lines ending with a diamond), the maximum baryonic mass increases during deleptonization, and no metastability occurs.

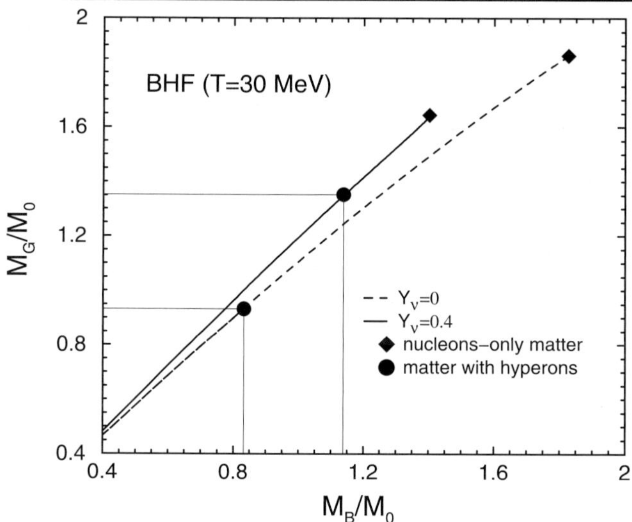

Fig. 5 Gravitational mass as a function of the baryon mass for neutrino-trapped matter at temperature $T = 30$ MeV (*solid curve*) and for cold neutrino-free matter (*dashed curve*). A dot at the end of the curves indicates matter with noninteracting hyperons, a diamond indicates purely nucleonic matter

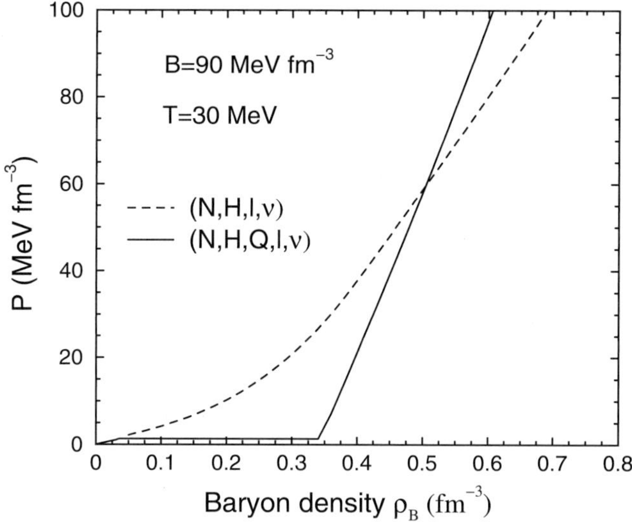

Fig. 6 Pressure as a function of the baryon density at temperature $T = 30$ MeV, and a bag constant $B = 90$ MeV fm^{-3}. The *solid* (*dashed*) *curve* is for neutrino-trapped matter containing baryons and quark matter (only baryons)

4.1 Including quark matter

The appearance of quark matter in the interior of massive neutron stars is one of the main issues in the physics of these compact objects. Calculations of neutron star structure, based on microscopic nucleonic EOS, indicate that the central particle density may reach values larger than $1/\text{fm}^3$. In this density range the nucleon cores (dimension $\approx$0.5 fm) start to touch each other, and it is hard to imagine that only nucleonic degrees of freedom can play a role. Rather, it can be expected that even before reaching these density values, the nucleons start to loose their identity, and quark degrees of freedom are excited at a macroscopic level. Unfortunately, while the microscopic theory of the nucleonic EOS has reached a high degree of sophistication, the quark matter EOS is poorly known at zero temperature and at the high baryonic density appropriate for NS.

Here we will briefly discuss some results obtained by using the MIT bag model (Chodos et al. 1974) at finite temperature, with a bag constant $B = 90$ MeV fm^{-3}. In order to study the hadron-quark phase transition, we have performed a Maxwell construction between the baryon phase and the quark phase. A more realistic model is the Glendenning construction (Glendenning 1992), which determines the range of baryon density where both phases coexist, yielding an EOS containing a pure hadronic phase, a mixed phase, and a pure quark matter region. Using the Maxwell construction implies that the phase transition is sharp, and no mixed phase exists. The pressure is characterized by a plateau extended over a range of baryon density, which separates the purely hadron phase at low density from the pure quark phase at higher density.

Table 1 The values of the maximum mass and corresponding central density (normalized with respect to the saturation density $\rho_0 = 0.17$ fm^{-3}) for different stellar configurations

	T (MeV)	$M_G/M_\odot$	ρ_c/ρ_0
PNS	30	1.53	8.2
PNS	50	1.53	7.9
NS	0	1.5	9.5

This is shown in Fig. 6, which displays the case of hot, neutrino-trapped matter. The solid line represents the EOS with the hadron-quark phase transition, whereas the dashed line is the purely hadronic case. As we can see, the EOS which comprises the hadron and the quark phases is stiffer at large baryon density than the purely hadronic case. A stiffer EOS causes larger values of the critical mass for a neutron star, as we found in (Burgio et al. 2002b).

In particular, the critical mass may reach values up to 1.5–1.6$M_\odot$, depending on the value of the bag constant. The maximum mass increases with decreasing value of B, but the latter cannot be too small if stability of symmetric nuclear matter at saturation has to be ensured. We have also checked that these results are quite general, and do not change appreciably if a density-dependent bag constant is introduced (Nicotra et al. 2006b).

In Table 1, we show the values of the critical mass for PNS with trapped neutrinos at temperatures $T = 30$ and 50 MeV and compare with the value for a cold NS. We find that the hadron-quark phase transition stabilizes the value of the maximum mass, which turns out to be independent of the temperature and equal to 1.53$M_\odot$ for the adopted value of the bag constant. In this case we do not find any metastable stars, at variance with the findings of Prakash et al. (1997). More detailed calculations will be presented elsewhere (Nicotra et al. 2006b).

5 Conclusions

In this paper we have studied the structure of (proto)neutron stars on the basis of a microscopically derived EOS for baryonic matter at finite temperature, in the framework of the Brueckner–Hartree–Fock many-body theory. Configurations with or without trapped neutrinos were considered. We found that the maximum mass of a hyperonic PNS is substantially larger than the one of the cold star, because both neutrino trapping and finite temperature tend to stiffen the EOS. Trapping shifts the onset of hyperons, in particular the Σ^-, to considerably higher density and reduces their concentrations.

However, as in the case of cold neutron stars, the addition of hyperons demands for the inclusion of quark degrees of freedom in order to obtain a maximum mass larger than the observational lower limit. For this purpose, we have studied the hadron-quark phase transition within the MIT bag model, and performed a Maxwell construction between the two phases. We found that the inclusion of quark matter stabilizes the value of the critical mass of a PNS at about 1.5–$1.6 M_\odot$, no matter the value of the temperature.

References

Baldo, M.: Nuclear Methods and the Nuclear Equation of State. World Scientific, Singapore (1999)

Baldo, M., Ferreira, L.S.: Phys. Rev. C **59**, 682 (1999)

Baldo, M., Bombaci, I., Burgio, G.F.: Astron. Astrophys. **328**, 274 (1997)

Baldo, M., Burgio, G.F., Schulze, H.-J.: Phys. Rev. C **58**, 3688 (1998)

Baldo, M., Burgio, G.F., Schulze, H.-J.: Phys. Rev. C **61**, 055801 (2000a)

Baldo, M., Song, H.Q., Giansiracusa, G., Lombardo, U.: Phys. Lett. B **473**, 1 (2000b)

Baldo, M., Buballa, M., Burgio, G.F., Neumann, F., Oertel, M., Schulze, H.-J.: Phys. Lett. B **562**, 153 (2003)

Baldo, M., Ferreira, L.S., Nicotra, O.E.: Phys. Rev. C **69**, 034321 (2004)

Baym, G., Pethick, C., Sutherland, D.: Astrophys. J. **170**, 299 (1971)

Bloch, C., De Dominicis, C.: Nucl. Phys. **7**, 459 (1958)

Bloch, C., De Dominicis, C.: Nucl. Phys. **10**, 181 (1959a)

Bloch, C., De Dominicis, C.: Nucl. Phys. **10**, 509 (1959b)

Burgio, G.F., Baldo, M., Sahu, P.K., Santra, A.B., Schulze, H.-J.: Phys. Lett. B **526**, 19 (2002a)

Burgio, G.F., Baldo, M., Sahu, P.K., Schulze, H.-J.: Phys. Rev. C **66**, 025802 (2002b)

Burrows, A., Lattimer, J.M.: Astrophys. J. **178**, 307 (1986)

Carlson, J., Pandharipande, V.R., Wiringa, R.B.: Nucl. Phys. A **401**, 59 (1983)

Chodos, A., Jaffe, R.L., Johnson, K., Thorn, C.B., Weisskopf, V.F.: Phys. Rev. D **9**, 3471 (1974)

Day, B.D.: Phys. Rev. C **24**, 1203 (1981)

Feynman, R., Metropolis, F., Teller, E.: Phys. Rev. **75**, 1561 (1949)

Friedman, B., Pandharipande, V.R.: Nucl. Phys. A **361**, 502 (1981)

Glendenning, N.K.: Phys. Lett. B **114**, 391 (1982)

Glendenning, N.K.: Astrophys. J. **293**, 470 (1985)

Glendenning, N.K.: Phys. Rev. D **46**, 1274 (1992)

Gondek, D., Haensel, P., Zdunik, J.L.: Astron. Astrophys. **325**, 217 (1997)

Huber, H., Weber, F., Weigel, M.K.: Phys. Rev. C **57**, 3484 (1999)

Kaiser, N., Fritsch, S., Weise, W.: Nucl. Phys. A **697**, 255 (2002)

Lejeune, A., Grangé, P., Martzolff, M., Cugnon, J.: Nucl. Phys. A **453**, 189 (1986)

Maessen, P., Rijken, Th., de Swart, J.: Phys. Rev. C **40**, 2226 (1989)

Maieron, C., Baldo, M., Burgio, G.F., Schulze, H.-J.: Phys. Rev. D **70**, 043010 (2004)

Myers, W.D., Swiatecki, W.J.: Nucl. Phys. A **601**, 141 (1996)

Negele, J.W., Vautherin, D.: Nucl. Phys. A **207**, 298 (1973)

Nicotra, O.E., Baldo, M., Burgio, G.F., Schulze, H.-J.: Astron. Astrophys. **451**, 213 (2006a)

Nicotra, O.E., Baldo, M., Burgio, G.F., Schulze, H.-J.: Phys. Rev. D **74**, 123001 (2006b)

Pons, J.A., Reddy, S., Prakash, M., Lattimer, J.M., Miralles, J.A.: Astrophys. J. **513**, 780 (1999)

Pons, J.A., Steiner, A.W., Prakash, M., Lattimer, J.M.: Phys. Rev. Lett. **86**, 5223 (2001)

Prakash, M., Bombaci, I., Prakash, M., Ellis, P.J., Lattimer, J.M., Knorren, R.: Phys. Rep. **280**, 1 (1997)

Sartor, R.: Phys. Rev. C **73**, 034307 (2006)

Schiavilla, R., Pandharipande, V.R., Wiringa, R.B.: Nucl. Phys. A **449**, 219 (1986)

Schulze, H.-J., Lejeune, A., Cugnon, J., Baldo, M., Lombardo, U.: Phys. Lett. B **355**, 21 (1995)

Schulze, H.-J., Baldo, M., Lombardo, U., Cugnon, J., Lejeune, A.: Phys. Rev. C **57**, 704 (1998)

Schulze, H.-J., Polls, A., Ramos, A., Vidaña, I.: Phys. Rev. C **73**, 058801 (2006)

Shapiro, S.L., Teukolsky, S.A.: Black Holes, White Dwarfs, and Neutron Stars. Wiley, New York (1983)

Song, H.Q., Baldo, M., Giansiracusa, G., Lombardo, U.: Phys. Rev. Lett. **81**, 1584 (1998)

Stoks, V.G.J., Rijken, Th.A.: Phys. Rev. C **59**, 3009 (1999)

Taylor, J.H., Weisberg, J.M.: Astrophys. J. **345**, 434 (1989)

Ter Haar, B., Malfliet, R.: Phys. Rev. Lett. **56**, 1237 (1986)

Ter Haar, B., Malfliet, R.: Phys. Rep. **149**, 207 (1987)

Wiringa, R.B., Stoks, V.G.J., Schiavilla, R.: Phys. Rev. C **51**, 38 (1995)

Zel'dovich, Y.B., Novikov, I.D.: Stars and Relativity. University of Chicago Press, Chicago (1971)

Zhou, X.R., Burgio, G.F., Lombardo, U., Schulze, H.-J., Zuo, W.: Phys. Rev. C **69**, 018801 (2004)

Astrophys Space Sci (2007) 308: 395–402
DOI 10.1007/s10509-007-9361-7

ORIGINAL ARTICLE

Modelling the dynamics of superfluid neutron stars

N. Andersson

Received: 7 July 2006 / Accepted: 10 October 2006 / Published online: 21 March 2007

Abstract In this brief summary I describe our recent work on superfluid neutron star dynamics. I review results on shear viscosity, hyperon bulk viscosity, vortex mediated mutual friction and the modelling of multifluid systems in general. For each problem I provide a set of questions that need to be addressed by future work.

Keywords Neutron stars · Superfluids · Dissipation

1 Introduction

As the temperature of a system decreases towards absolute zero, matter either freezes to a solid or becomes superfluid. The first case is dominated by a structured lattice. In the second case, the system acts as a macroscopic quantum system. Regardless of the outcome, the rules that apply in an extremely cold system may be very different from those that describe the system at higher temperatures. This is, of course, a well-known fact and the different possible phases of matter have been studied in great detail. In particular, increasingly sophisticated laboratory experiments continue to provide insights into the details of superfluid/superconducting systems, ranging from the standard Bose–Einstein condensates to superfluid Helium (Donnelly 1991) and atomic systems exhibiting fermion Cooper pairing. This area of research is highly relevant for those that are fascinated by neutron star physics. In fact, mature neutron stars may provide the ultimate testing ground for theoretical physics. They are expected to contain an elastic nuclear crust permeated by superfluid neutrons. An outer core where superfluid neutrons coexist with superconducting protons transitions to the deeper core which may contain exotica like superfluid hyperons and deconfined quarks in a colour superconducting state. As if this was not enough, the presence of a "solid" core remains a possibility (even though the relevant lattice parameters are unknown).

Trying to understand the details of the various neutron star phases and their possible impact on astronomical observations is an exciting challenge. First of all, it requires a working knowledge of much of modern theoretical physics. Secondly, one will need some understanding of the observations. This includes decades of data for radio pulsars, in particular associated with the glitch events that provide the strongest evidence of the presence of (at least partially) decoupled superfluid components. X-ray observations of accreting neutron stars in binary systems, and isolated neutron stars closer to the Earth, are common and the data for the existence of magnetars is now very convincing. In fact, the recent observations of likely crustal oscillations following the giant flares in SGR1806-20 and SGR1900+14 (Israel 2007; Watts and Strohmayer 2007; Samuelsson and Andersson 2007, this volume) may be the harbingers of neutron star asteroseismology. Future gravitational-wave observations should add another observational window, although it will likely require an advanced generation of detectors sensitive at high frequencies.

As the observations continue to improve we will have more precise quantitative opportunities to test our theoretical models. To meet these tests the current models have to be improved. In fact, many of our "favourite" models are far too rudimentary to pass any closer scrutiny. This is not surprising given that the relevant theory problems are difficult. Consider for example the possibility of superfluid vortex pinning in the neutron star crust. At the phenomenolog-

N. Andersson (✉)
School of Mathematics, University of Southampton, Southampton, UK
e-mail: na@maths.soton.ac.uk

ical level, the association between the observed spin-up of the crust in a glitch and the transfer of angular momentum from an initially faster spinning (superfluid) component is natural. To invoke vortex unpinning as the key agent that motivates the event is also natural. Yet it is clear from the effort that has gone into glitch modelling that it is difficult to describe the process in a truly quantitative way.

The dynamics of a superfluid neutron star is associated with a number of similar problems that need to be better understood. The main aim of the research described in this brief review is to arrive at such an understanding. Right now we are quite far away from this goal. In some cases, e.g. problems concerning basic shear viscosity, we may have a clear picture already, but in others, e.g. how the calculated shear viscosity coefficients are used in the case of a superfluid system, we are only beginning to be able to pose (what might be) the right questions.

This brief review is a summary of my presentation at the London neutron star meeting. It is in no sense a complete review of the various problems. Rather it is a self-centered view of my thinking at the time of writing. The interested reader will find references to the relevant literature in the cited papers.

Before I discuss our recent work, let me provide a list of problems that I find particularly interesting:

How do pulsar glitches really work? As already alluded to above, this remains a vexing issue. In my (somewhat pessimistic view) view, the present models are not truly quantitative. They likely contain the right elements, but the details of the underlying physics are not yet agreed upon. It is also worth noting that most models predict how the glitch proceeds and how the system relaxes back to quasi-equilibrium. Very few models provide a real mechanism to explain why the glitch happened in the first place.

How do we understand neutron star free precession? The observational indications that some neutron stars are wobbling now seem relatively strong. The obvious question is why this behaviour is so rare. After all, free precession is the most natural mode of motion for any rotating body. The answer will provide insight into the damping of fluid motion in the star, and thus may constrain our models for internal dissipation. There could also be implications for vortex pinning or vortex-flux tube interactions in the superfluid/superconducting core (see Link's contribution). It would also be nice to have a model for the excitation of the precession in the first place.

Do the superfluid degrees of freedom affect the oscillation properties of the star? This is an important question that impacts on future attempts to probe neutron star physics via asteroseismology. To provide a "useful" answer we need to study the problem using realistic models for the superfluid pairing gaps (which, of course, are not agreed upon by theorists) within general relativity. The latter is key in order to make the results accurate enough that it is meaningful to compare to observations. In principle, I don't think the mode calculation would be very difficult. At least not for non-rotating stars, and as long as one is prepared to accept that the pairing gaps are likely to remain unknown up to perhaps a factor of a few. However, if one wants to account for the presence of the crust (which should be penetrated by superfluid neutrons) and the magnetic field, then the modelling becomes much more challenging. In the case of the crust, we have not (until very recently: Carter and Samuelsson 2006) had any theoretical formulations that allow for the presence of a superfluid component, while in the magnetic field case we do not have a good understanding of the internal field structure.

Neutron stars may radiate detectable gravitational waves through a number of scenarios. One possibility is that the inertial r-modes are driven unstable as they radiate gravitationally. The basic mechanism behind this instability is well understood, and a large number of damping mechanisms that counteract the instability have been considered. Thus it has become clear that the key deciding factors relate to superfluidity. The presence of hyperons in the deep stellar core may prevent the instability from happening, but if the hyperons are superfluid then the chemical reactions that lead to bulk viscosity are suppressed and the effect on the instability may not be considerable. Another important mechanism is the so-called mutual friction, which in this context relates to the damping due to the scattering of electrons off of magnetic fields associated with the superfluid neutron vortices. So far all studies of mode-damping due to this effect have been based on straight vortices (as in a rigidly rotating star). But this may not be the appropriate model! If there is a significant oscillation in the star, which has some vorticity associated with it, then it seems likely that the vortices will get tangled up. The system would then be in a "turbulent" state, and from the analogous problem in superfluid Helium (Donnelly 1991) we know that the mutual friction force is then rather different.

Finally, I would like to emphasise the multi-fluid aspects of these problems. In most available studies, concerning for example viscosity, it is implicitly assumed that we can apply the standard single fluid equations of motion. Yet a multi-fluid system has extra degrees of freedom (the second sound in Helium and the analogous "superfluid modes" in an oscillating neutron star). These are unlikely to be "passive". In fact, we know that the equations that are used to model simple superfluid neutron stars admit a so-called two-stream instability (Andersson et al. 2004). One can speculate that this instability becomes relevant when the two components of the star rotate at different rates, as in the case of a spinning down neutron star with a pinned superfluid component. The instability mechanism is very simple and familiar from other physical systems, and it would be interesting to know whether it can operate in a neutron star as well.

2 Shear viscosity

We have recently investigated the effects of superfluidity on the shear viscosity in a neutron star core (Andersson et al. 2005). We were motivated to do this by a puzzling result from the literature. The available results suggest that the shear viscosity is stronger in a superfluid neutron star than it is when the star forms a normal-fluid system (Cutler and Lindblom 1987). This result contradicts our experience from other superfluid systems like Helium, and since the shear viscosity affects mode damping, free precession and relaxation after spin-up events in a crucial way, it is important that we understand it.

To address the problem we combined existing theoretical results for the viscosity coefficients with data for the various superfluid energy gaps into a "complete" description. This leads to a simple model for the electron viscosity which is relevant both when the protons form a normal fluid and when they become superconducting. This turns out to be the key distinguishing factor.

Below the neutron superfluid transition temperature the dominant contribution to the shear viscosity comes from the scattering of relativistic electrons. To model the shear viscosity we need to consider two contributions. Above the transition temperature at which the protons become superconducting, electron–proton scattering leads to a viscosity coefficient

$$\eta_{ep} \approx 1.8 \times 10^{18} \left(\frac{x_p}{0.01} \right)^{13/6} \rho_{15}^{13/6} T_8^{-2} \text{ g/cm s} \quad (1)$$

where x_p is the proton (electron) fraction, $T_8 = T/10^8$ K and $\rho_{15} = \rho/10^{15}$ g/cm^3. Meanwhile, when the protons are superconducting the dominant effect is due to electrons scattering off of each other. Then we have

$$\eta_{ee} \approx 4.4 \times 10^{19} \left(\frac{x_p}{0.01} \right)^{3/2} \rho_{15}^{3/2} T_8^{-2} \text{ g/cm s}. \quad (2)$$

The protons play the key role since individual scattering processes add like "parallel resistors". That is, we have

$$\tau = \left[\frac{1}{\tau_{ee}} + \frac{1}{\tau_{ep}} \right]^{-1} \quad \text{where } \tau_{ee} \gg \tau_{ep} \quad (3)$$

and it is clear that the most important contribution comes from the most frequent scattering process.

When the protons are superconducting the electron–proton scattering will be suppressed, essentially because there will be fewer states available for the protons to scatter into. In order to allow for the transition to proton superconductivity, we can introduce a suppression factor $\mathcal{R}_p$ such that

$$\tau_{ep} \longrightarrow \frac{\tau_{ep}}{\mathcal{R}_p}. \quad (4)$$

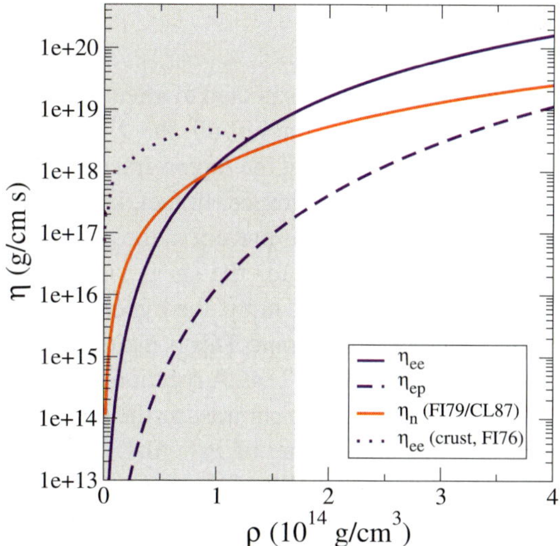

Fig. 1 Various shear viscosity estimates for a neutron star at temperature 10^8 K. The *solid blue line* shows our estimate η_{ee} while the *dashed blue curve* corresponds to η_{ep}. These estimates are compared to (i) the total shear viscosity in a normal fluid neutron star, which is dominated by neutron–neutron scattering (FI79/CL87, *solid red line*) taken from Cutler and Lindblom (1987), and (ii) results obtained by Flowers and Itoh for the electron–electron shear viscosity in the crust region (crust, FI76, *dotted blue curve*). The crust is indicated by the *gray region*. (See Andersson et al. (2005) for the relevant references.)

Far below the critical transition temperature at which the protons become superconducting we should have $\mathcal{R}_p \rightarrow 0$, and we see from (3) that the electron–electron scattering then dominates the shear viscosity.

Our results, which are illustrated in Fig. 1, explain in a clear way why proton superconductivity leads to a significant strengthening of the shear viscosity. It should be noted that the superfluidity of the neutrons is not the factor which leads to electron–electron scattering becoming the main shear viscosity agent. Rather, it is the fact that the onset of superconductivity suppresses the electron–proton scattering. As can be seen from Fig. 1, the electron–electron shear viscosity is not too different from the result for neutrons scattering off of each other. This means that, in the temperature range where shear viscosity dominates, the damping of neutron star oscillations will be quite similar (modulo multifluid effects) in the extreme cases when (i) the neutrons and protons are both normal, and (ii) when the neutrons and protons are superfluid/superconducting, respectively. The contrast with the case when the neutrons are superfluid and the protons normal (and viscosity is dominated by η_{ep}) is clear from Fig. 1. This is an interesting observation because it shows that proton superconductivity (or rather absence thereof) could have a significant effect on the dynamics of a neutron star core.

Future work needs to (i) *implement the different degrees of freedom of a multifluid system, and* (ii) *determine the required suppression factors (like* $\mathcal{R}_p$*).*

3 Hyperon viscosity

The presence of hyperons is expected to affect a neutron star in several important ways. First of all, the Σ^- carries negative charge which means that the lepton fractions drop significantly following its appearance. In fact, in some models there are virtually no electrons present in the core of the star. This may have implications for the shear viscosity results that we discussed earlier. Secondly, the hyperons can act as an extremely efficient refrigerant. This is mainly because the hyperons may undergo direct URCA reactions essentially as soon as they appear, in clear contrast to the protons which must exceed a threshold value of $x_p \sim 0.1$. Studies of the effect of this enhanced cooling have shown that a neutron star with a sizeable hyperon core would cool extremely fast. In fact, it would cool so fast that it would be much colder than observational data suggests. This could be taken as evidence against an exotic core, but more likely it suggests that the hyperons are (at least partly) superfluid. This is a natural explanation since superfluidity leads to a reduction of the relevant reaction rates.

Analogous non-leptonic reactions have recently been invoked as a mechanism for producing a very strong bulk viscosity that may suppress the gravitational-wave driven instability of the r-modes (Lindblom and Owen 2002) (see also the contribution by Chatterjee and Bandyopadhyay). This mechanism is particularly important because, in contrast to the standard bulk viscosity associated with β-reactions it is relevant at low temperatures (the coefficient scales as T^{-2} rather than T^6). For a while it was argued that the presence of hyperons would make the r-mode instability completely irrelevant. This is now understood not to be the case (Nayyar and Owen 2006). Basically, the hyperons must be superfluid in order not to lead to conflicts with cooling data. Then the nuclear reactions that lead to the bulk viscosity must also be suppressed, and the effect on the r-mode instability may not be so great after all, see Fig. 2. In fact, the most recent estimates suggest that the r-modes may be unstable in accreting neutron stars in LMXBs, possibly regulating the spin of these stars. Since the associated gravitational-wave signal may be detectable by advanced detectors, it is important that we understand this problem better.

Future work needs to (i) *improve our understanding of the hyperon pairing gaps and the associated suppression factors for bulk viscosity, and* (ii) *develop a multifluid description of bulk viscosity that can be used below the relevant superfluid transition temperature.*

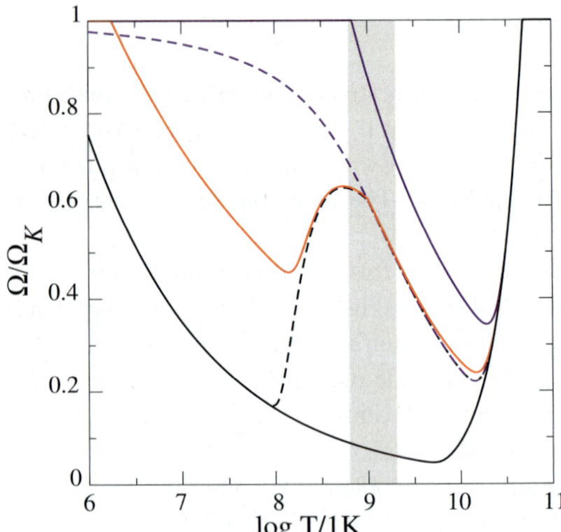

Fig. 2 Various estimates of the r-mode instability for a simple neutron star model with a hyperon core. The figure shows the critical rotation rate Ω/Ω_K as a function of the core temperature T. The *solid blue curve* represents the case of normal fluid hyperons. The r-modes are unstable above the curve, and even though there appears to be a sizeable instability window the hyperon viscosity provides a strong suppression (compare to the *solid black curve* which represents the case when there are no hyperons in the star). Basically, the remaining instability window is at such high temperatures that the star would cool through the instability regime in a minute or so. Far too fast to allow the modes to grow to a large amplitude. The dashed blue curve accounts for rotational effects, and shows that the modes are unstable in the fastest spinning stars also at much lower temperatures. The *solid red* and *dashed black curves* assume that the hyperon viscosity is suppressed below a given critical temperature (here taken to be low, at 10^9 K), and correspond to the case when there is a crust (and a viscous Ekman layer is in operation) and the star is entirely fluid, respectively.

4 Modelling multifluids

The equations that govern a general conservative multifluid system can be derived from a constrained variational principle (Prix 2004; Andersson and Comer 2006a). The fundamental variables in this framework are the number densities n_x, where x is a "constituent" index that identifies the different particle species, and the associated transport velocities v_x^i. To complete the model one must provide an energy functional which represents the equation of state. In the isotropic (meaning that there are no preferred directions) two-fluid problem this functional takes the form $E(n_n, n_p, w^2)$ where $w_i^{yx} \equiv v_i^y - v_i^x$ (x $\neq$ y are constituent indices). This immediately leads to

$$dE = \sum_{x=n,p} \mu_x \, dn_x + \alpha \, dw^2 \tag{5}$$

where $w^2 = w_i^{yx} w_{yx}^i$ and

$$\tilde{\mu}_x = \frac{\mu_x}{m_B} = \frac{1}{m_B}\frac{\partial E}{\partial n_x} \quad \text{and} \quad \alpha = \frac{\partial E}{\partial w^2} \tag{6}$$

(with m_B the baryon mass) defines the chemical potential per unit mass $\tilde{\mu}_x$ and the entrainment parameter α.

The equations of motion for the system can be written

$$\partial_t n_x + \nabla_i (n_x v_x^i) = 0 \tag{7}$$

and

$$n_\mathrm{x}(\partial_t + \mathcal{L}_{v_\mathrm{x}})p_i^\mathrm{x} + n_\mathrm{x}\nabla_i\left(\mu_\mathrm{x} - \frac{1}{2}m_\mathrm{B}v_\mathrm{x}^2\right) = 0 \tag{8}$$

where $\mathcal{L}_{v_\mathrm{x}}$ represents the Lie derivative along v_x^i (see Andersson and Comer (2006b) for an explanation why the Lie derivative is natural in this context). In (8) the momentum (per particle) is given by $p_i^\mathrm{x} = m_\mathrm{B}(v_i^\mathrm{x} + \varepsilon_\mathrm{x} w_i^\mathrm{yx})$, where $\varepsilon_\mathrm{x} = 2\alpha/m_\mathrm{B}n_\mathrm{x}$. From this we can understand the nature of the entrainment effect. It is such that the velocity and momentum of each constituent in the multifluid system are no longer parallel. An alternative, perhaps more intuitive, description would represent this mechanism in terms of an effective mass m_x^* (Andersson et al. 2006). In the case of the outer core of a neutron star, where superfluid neutrons coexist with superconducting protons the entrainment arises because of the strong interaction. Each proton/neutron is endowed with a cloud of particles of the opposite species, and when it flows it drags some of the other fluid along with it thus affecting its momentum.

The entrainment is known to play an important role in neutron star dynamics. In particular, it has been shown to affect the spectrum of stellar pulsation (Prix et al. 2004; Gusakov and Andersson 2006). Moreover, it is one of the key ingredients in the prescription for the vortex mediated mutual friction.

Future work needs to (i) *provide the entrainment parameters at finite temperatures (as in* Gusakov and Haensel (2005)) *(essentially doing equation of state calculations "out of equilibrium"), and* (ii) *develop fluid models that allow for the presence of both superfluid condensates and the quasiparticle excitations that will be present when the star is no longer at* $T = 0$.

5 Mutual friction

A superfluid mimics bulk rotation by forming an array of vortices. The dynamics of these vortices, and their interaction with the different particle species may significantly affect the evolution of the system. In a neutron star the presence of vortices leads to "mutual friction" between interpenetrating superfluids (neutrons and protons). This couples the two fluids on a relatively short timescale, which has implications for both glitch recovery estimates and the damping of pulsation modes. In fact, an estimate by Lindblom and Mendell indicates that the gravitational-wave driven instability of the fundamental f-modes will be prohibited by the mutual friction (Lindblom and Mendell 1995). The same effect is also relevant, albeit not quite in such a dramatic fashion, for the r-modes (Lindblom and Mendell 2000).

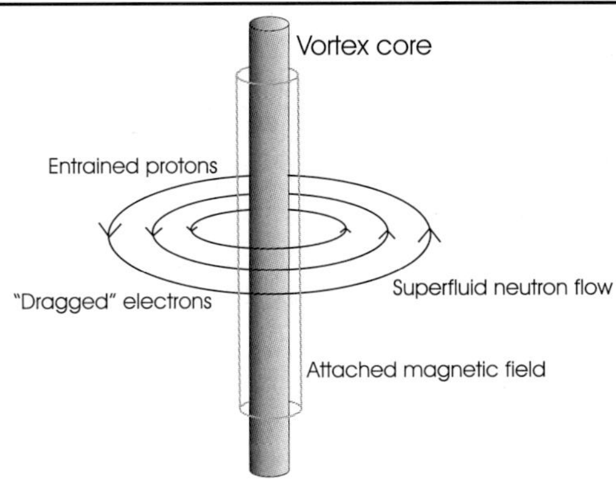

Fig. 3 A schematic illustration of the flow around a superfluid neutron vortex. Because of the entrainment effect, the vortex induces flow also in the protons. This leads to a magnetic field forming on the vortex, and the scattering of electrons off of this magnetic field leads to a dissipative mechanism known as mutual friction.

In a recent study, we have revisited the problem of the mutual friction force for neutron stars (Andersson et al. 2006). It is known that the entrainment plays a key role in determining the strength of the mutual friction, and to make progress and study various astrophysical scenarios we needed to develop a description of this mechanism within our multifluid framework. Our results provide useful checks on the classic work by Alpar et al. (1984) (who were the first to discuss the problem) and Mendell (1991) (whose mutual friction coefficients were used in the studies of mode-instabilities in superfluid neutron stars).

Briefly, the mutual friction that is expected to be the most important in the outer core of a mature neutron star originates as follows. The superfluid neutrons form vortices, which represent the quantisation of the momentum circulation. That is, we have

$$\kappa^i = \frac{h}{2m_\mathrm{n}}\hat{\kappa}^i = \epsilon^{ijk}\nabla_j p_k^\mathrm{n} \tag{9}$$

where $\hat{\kappa}^i$ is a normalised tangent vector to the vortex. Here it is important to understand that it is the circulation of momentum that is quantised, not that of the velocity. At first sight this distinction may seem to suggest that our model differs from the standard picture. However, this would be wrong. One must remember that in the orthodox description for superfluid Helium, the so-called superfluid "velocity" is in fact a rescaled momentum (Prix 2004). Hence, the two pictures are consistent, the only difference being that we make the true dynamical role of the variables clearer. Now, the flow of neutrons associated with the vortex induces flow in part of the protons because of the entrainment. This leads to the generation of a magnetic field, see Fig. 3, the strength of which is estimated as

$$B \approx 2\times 10^{14}\mathrm{G}\left(\frac{x_\mathrm{p}}{0.05}\right)\rho_{14}\left|\frac{m_p - m_p^*}{m_p^*}\right| \tag{10}$$

where $\rho_{14} = \rho/10^{14}$ g/cm^3. Finally, electrons scatter dissipatively off of this magnetic field, leading to a coupling of the two fluids in the system (the superfluid neutrons and a conglomerate of all the charged particles).

To arrive at the standard expression for the mutual friction force, we balance the Magnus "force", which acts on a vortex as the superfluid flows past it, and a resistive force due to electrons scattering off of the vortex. It should be noted that the Magnus effect is already present in the smooth averaged equations of motion (8), which means that in practice we only add resistivity to the right-hand side of the equations. Anyway, the force-balance leads to the mutual friction force acting on the neutrons being

$$f^i_{\mathrm{mf}} = \mathcal{B}\rho_n n_v \underbrace{\epsilon^{ijk}\hat{\kappa}_j\epsilon_{klm}\kappa^l w^m_{\mathrm{np}}}_{\mathrm{projects}\perp\kappa^i} + \mathcal{B}'\rho_n n_v \underbrace{\epsilon^{ijk}\kappa_j w^{\mathrm{np}}_k}_{\mathrm{acts}\perp w^{\mathrm{np}}_k}. \qquad (11)$$

This force has been used to model a number of astrophysical scenarios. In particular, the coupling of the two fluids following a glitch event, see the following section, and the damping of oscillations in a superfluid neutron star. It is, however, important to understand that this is only a first approximation to the real mutual friction interaction. There are presently two alternatives that should be taken seriously. The first, which has been advocated by Sedrakian and collaborators for over 20 years (see Sedrakian and Sedrakian 1997 and references given therein), is based on the notion of vortex clusters. The idea is that each neutron vortex is associated with a large collection of proton fluxtubes. This increases the scattering cross section for the electrons, and as a result the mutual friction force is many orders of magnitude stronger than given above. The upshot of this is (somewhat counter-intuitively) that the coupling between the fluids take place on a much longer timescale (months rather than seconds). The other alternative accounts for the curvature of the vortices, and the possibility that they will get tangled up in a state that in many ways is reminiscent of turbulence. The lack of a preferred direction of the vortex array obviously affects the form of the mutual friction force. This is well known from work on superfluid Helium (Donnelly 1991). Nevertheless, this alternative form for the mutual friction has only recently been discussed for neutron stars. Peralta et al. (2005) are leading the way by carrying out numerical simulations of a turbulent two-fluid system. This work has a number of interesting implications, and it will be exciting to see where this will take us.

Future work needs to investigate the alternative manifestations for the mutual friction in detail. In my opinion the most pressing issue concerns the turbulence. Since it is known to be the key effect in experiments on superfluid Helium one has every reason to expect that it will be relevant for neutron stars as well.

6 Dynamical coupling

One of the most important problems that can be addressed once we know the form of the mutual friction force regards the relaxation timescale following a pulsar glitch. The idea is that in the initial state the vortices are pinned. This leads to a build-up of a rotational lag between the two fluids as the charged component (the crust and the core protons/electrons) spins down. At some critical relative rotation rate, the pinning force will no longer be able to hold the vortices in place. Global unpinning then leads to evolution described by the equations that we have derived (provided that the vortices remain straight during the evolution, a debatable assumption). All we need in order to be able to estimate the relaxation timescale following the glitch are the coefficients in (11). As we have recently shown (Andersson et al. 2006), the dimensionless coefficients are $\mathcal{B}' = \mathcal{B}^2$ and

$$\mathcal{B} = \frac{\mathcal{R}}{\rho_{\mathrm{n}}\kappa}$$
$$\approx 4\times 10^{-4}\left(\frac{\delta m_{\mathrm{p}}}{m_{\mathrm{p}}}\right)^2\left(\frac{m_{\mathrm{p}}}{m^*_{\mathrm{p}}}\right)^{1/2}\left(\frac{x_{\mathrm{p}}}{0.05}\right)^{7/6}\rho_{14}^{1/6} \qquad (12)$$

where $\delta m_{\mathrm{p}} = m_{\mathrm{p}} - m^*_{\mathrm{p}}$. Working out the dynamical coupling timescale we easily find

$$\left.\begin{array}{l} n_{\mathrm{n}}\partial_t p^{\mathrm{n}}_i + \cdots = f^{\mathrm{mf}}_i \\ n_{\mathrm{p}}\partial_t p^{\mathrm{p}}_i + \cdots = -f^{\mathrm{mf}}_i \end{array}\right\} \rightarrow \frac{m^*_{\mathrm{p}}}{m_{\mathrm{p}}}\partial_t w^{\mathrm{np}}_i + \cdots \approx -\frac{\mathcal{B}\kappa n_v}{x_{\mathrm{p}}} w^{\mathrm{np}}_i. \qquad (13)$$

That is, the timescale on which the two fluids are (locally) coupled can be estimated as

$$\tau_d \approx \frac{m^*_{\mathrm{p}}}{m_{\mathrm{p}}}\frac{x_{\mathrm{p}}}{\mathcal{B}\kappa n_v}$$
$$\approx 10P(s)\left(\frac{m^*_{\mathrm{p}}}{\delta m_{\mathrm{p}}}\right)^2\left(\frac{x_{\mathrm{p}}}{0.05}\right)^{-1/6}\rho_{14}^{-1/6}. \qquad (14)$$

This suggests that the two core fluids are coupled on a timescale of about 10 rotation periods (for typical parameters). This estimate should be compared to the classic result of Alpar and Sauls (1988) who suggest that the coupling timescale is 400–10^4 periods. This relatively rapid coupling is usually taken as evidence that glitches must originate in the crust superfluid, not in the core. Our calculations, with a coupling that is about 1 order of magnitude faster than the old result, obviously support this conclusion. However, we need to understand why our estimate differs from that of Alpar and Sauls. The result may seem surprising since the parameters in the mutual friction force are similar in the two studies. The difference comes from the set-up of the coupling timescale problem. Our estimate followed immediately from the two-fluid equations, and represents the timescale on which mutual friction couples the two components in the system. In contrast, Alpar and Sauls consider the equation of motion for the vortices. Setting the system into relative rotation, they work out the timescale for the vortices to relax to a new equilibrium position. The key difference between the two estimates is that in the Alpar–Sauls scenario the rotational lag between the fluids is maintained while the vortices relax. This means that their model does not conserve angular momentum during the relaxation, which explains why the two results are different.

Fig. 4 An illustration of the different kinds of motion that lead to dissipation via particle scattering in a two-fluid problem (*blue* and *red*). The "*red*" fluid is assumed to be at rest, and the "*blue*" fluid (i) flows linearly, (ii) contracts and expands, and (iii) undergoes shearing motion.

7 Assignments

At the end of my presentation, I gave the audience a number of "homework assignments." These represent problems that I feel need more attention, and which I hope to consider in the near future. It seems appropriate to conclude this brief write-up by outlining these problems.

As we are beginning to understand the basic dynamics of the two-fluid model for neutron stars, we should look ahead and develop a framework that would allow us to make our models more realistic. Most importantly, we need to account for finite temperature effects and viscosity. This is a difficult problem area, but I feel that we cannot shirk away from it. Recently, Greg Comer and I have developed a flux-conservative formalism for multi-fluid systems, including both dissipation and entrainment (Andersson and Comer 2006a). Our formulation has much in common with work on the mechanics of mixtures and multiphase flows. Our first study of the problem is quite formal, and should be relevant for any multifluid situation in Newtonian physics. Of course, we want to apply the framework to neutron stars. To do this, we have considered two model systems. The results make it painfully clear that more realistic models will be significantly more complex than the systems we have considered so far. Our analysis shows that the simplest "reasonable" model for a neutron star core, starting with four fluids (neutrons, protons, electrons and entropy) and reducing to three degrees of freedom (superfluid neutrons, entropy and everything else), requires no less than 19 dissipation coefficients.

It will be a challenge to understand the role and values of these different coefficients. Yet it may not be too bad a problem, because the meaning of many of the new coefficients is quite clear. Consider for example dissipation due to particle scattering. In a single fluid problem scattering leads to viscosity due to shearing motion in the flow. In a situation where a relative flow is possible, there are more degrees of freedom that lead to particle scattering. Consider the schematic illustration in Fig. 4, where the two fluids are represented by blue and red particles. There are three different ways that relative flow can induce scattering, and result in dissipation. Taking the "red" fluid to be at rest, the blue fluid can (i) flow linearly, (ii) contract and expand, or (iii) undergo shearing motion. The viscosity coefficients associated with these different degrees of freedom are related and should only differ by "geometric factors." They should certainly be calculable, and might actually be known from kinetic theory already.

As described in other contributions to these proceedings, it is likely that we now have the first observations of neutron star oscillations. This is fantastic news, but it provides us with a challenge that must be met in the near future. We need to develop accurate theoretical models to put constraints on e.g. the crust physics via these observations. However, if we want to do this we must consider fully relativistic models. This is the only way to obtain results that have the precision required to make a comparison with observations meaningful. In fact, we might even have to account for the presence of superfluid neutrons in the crust, a very difficult problem. Moreover, the observed systems are all magnetars and one might expect the magnetic field to be important. It seems likely (at least to me) that the magnetic field will couple any motion in the crust to the core, thus altering the nature of the pulsation modes and making any analysis of "pure crust oscillations" less relevant (Glampedakis et al. 2006). This would force us to study global oscillations of magnetised fully relativistic stars, certainly not a simple problem.

As I have already indicated, we need to worry about superfluid "turbulence." I would not be at all surprised if studies of this problem lead to results that change our understanding of neutron star dynamics significantly. One advantage is that there has been a lot of work on the analogous problem for superfluid Helium (Donnelly 1991), and we can hope to benefit from these results. Most importantly we need to understand whether a turbulent description is relevant for neutron stars. If it is, how does it manifest itself? What is the effect on, for example, glitch relaxation and mutual friction damping of neutron star oscillations?

In addition to these problems, I can think of a number of issues concerning multifluid aspects of exotic phases like hyperons and deconfined quarks. Certainly the neutron star community is bustling with exciting ideas, but we have a lot to do before we truly understand the dynamics of superfluid neutron stars.

Acknowledgements Much of the work described in this article was done in collaboration with Greg Comer. Without his contribution the results discussed here would not have been the same.

References

Alpar, M.A., Sauls, J.A.: Astrophys. J. **327**, 723–725 (1988)
Alpar, M.A., Langer, S.A., Sauls, J.A.: Astrophys. J. **282**, 533–541 (1984)
Andersson, N., Comer, G.L.: Class. Quantum Grav. **23**, 5505 (2006a)
Andersson, N., Comer, G.L.: Relativistic fluid dynamics: physics for many different scales. Preprint gr-qc/0605010. Living Rev. Relativ. (2006b, to appear)
Andersson, N., Comer, G.L., Prix, R.: Mon. Not. Roy. Astron. Soc. **354**, 101–110 (2004)
Andersson, N., Comer, G.L., Glampedakis, K.: Nucl. Phys. A **763**, 212–229 (2005)
Andersson, N., Sidery, T., Comer, G.L.: Mon. Not. Roy. Astron. Soc. **368**, 162–170 (2006)
Carter, B., Samuelsson, L.: Class. Quantum Grav. **23**, 5367 (2006)
Cutler, C., Lindblom, L.: Astrophys. J. **314**, 234–241 (1987)
Donnelly, R.J.: Quantized Vortices in Helium II, Cambridge (1991)
Glampedakis, K., Samuelsson, L., Andersson, N.: Mon. Not. Roy. Astron. Soc. Lett. **371**, L74 (2006)
Gusakov, M.E., Andersson, N.: Temperature dependent pulsations of superfluid neutron stars. Mon. Not. Roy. Astron. Soc. online early 10.1111/j.1365-2966.2006.10982.x. Preprint astro-ph/0602282
Gusakov, M.E., Haensel, P.: Nucl. Phys. A **761**, 333–348 (2005)
Israel, G.: Astrophys. Space Sci., DOI 10.1007/s10509-007-9321-2 (2007)
Lindblom, L., Mendell, G.: Astrophys. J. **444**, 804–809 (1995)
Lindblom, L., Mendell, G.: Phys. Rev. D **61**, 104003 (2000)
Lindblom, L., Owen, B.J.: Phys. Rev. D **65**, 063006 (2002)
Mendell, G.: Astrophys. J. **380**, 530 (1991)
Nayyar, M., Owen, B.J.: Phys. Rev. D **73**, 084001 (2006)
Peralta, C., Melatos, A., Giacobello, M., Ooi, A.: Astrophys. J. **635**, 1224–1232 (2005)
Prix, R.: Phys. Rev. D **69**, 043001 (2004)
Prix, R., Comer, G.L., Andersson, N.: Mon. Not. Roy. Astron. Soc. **348**, 625–637 (2004)
Samuelsson, L., Andersson, N.: Astrophys. Space Sci., DOI 10.1007/s10509-007-9301-6 (2007)
Sedrakian, A.D., Sedrakian, D.M.: Astrophys. J. **447**, 305 (1997)
Watts, A.L., Strohmayer, T.E.: Astrophys. Space Sci., DOI 10.1007/s10509-007-9296-z (2007)

Astrophys Space Sci (2007) 308: 403–412
DOI 10.1007/s10509-007-9316-z

ORIGINAL ARTICLE

Cooling of neutron stars with strong toroidal magnetic fields

Dany Page · Ulrich Geppert · Manfred Küker

Received: 24 November 2006 / Accepted: 28 November 2006 / Published online: 28 March 2007

Abstract We present models of temperature distribution in the crust of a neutron star in the presence of a strong toroidal component superposed to the poloidal component of the magnetic field. The presence of such a toroidal field hinders heat flow toward the surface in a large part of the crust. As a result, the neutron star surface presents two warm regions surrounded by extended cold regions and has a thermal luminosity much lower than in the case the magnetic field is purely poloidal. We apply these models to calculate the thermal evolution of such neutron stars and show that the lowered photon luminosity naturally extends their life-time as detectable thermal X-ray sources.

Keywords Neutron star · Magnetic field

PACS 97.60.Jd · 26.60.+c · 95.30.Tg

Work partially supported by UNAM-DGAPA grant #IN119306.

D. Page (✉)
Departamento de Astrofísica Teórica, Instituto de Astronomía, Universidad Nacional Autónoma de México, Mexico, D.F. 04510, Mexico
e-mail: page@astroscu.unam.mx

U. Geppert
Departament de Física Aplicada, Universitat d'Alacant, 03080 Alacant, Spain
e-mail: geppert@ua.es

M. Küker
Astrophysikalisches Institut Potsdam, An der Sternwarte 16, 14482 Potsdam, Germany
e-mail: mkueker@aip.de

1 Introduction

It is generally considered that, within a few decades after its birth in a core collapse supernova, a neutron star reaches a state of "isothermality" characterized by a uniform high interior temperature. The stellar surface is much colder and protected from the hot interior by a thin ($\sim$100 meters) layer, the envelope, which acts as a heat blanket. As was argued by Greenstein and Hartke (1983), in the presence of a sufficiently strong magnetic field, $>10^{10}$ G, the surface temperature of the neutron star will not be uniform as is expected in the unmagnetized case since the magnetic field severely limits the ability of electrons to transport heat in directions perpendicular to itself. As a result, the regions around the magnetic poles, where the magnetic field is almost radial, are expected to be significantly warmer than the regions around the magnetic equator, where the field is almost tangent to the surface. Since then much work has been dedicated to study the effects of the magnetic field on the properties of the neutron star envelope and crust (see Potekhin and Yakovlev 2001 and Potekhin et al. 2003 for recent works). In the presence of a sufficiently strong magnetic field, $\geq 10^{12}$–10^{13} G, the anisotropy of heat transport actually extends to much higher densities and can even be present within the whole crust. Recently, we have shown (Geppert et al. 2004, 2006) that, in cases where the field geometry in the crust is such that the meridional component of the field dominates over its radial component in a large part of the crust, the non-uniformity of the temperature, previously considered to be restricted to the envelope, may actually extend to the whole crust. The largest effect was found when a strong toroidal component was included in the crust, superposed to the poloidal component. This result, that the geometry of the magnetic field *in the interior* of the neutron star leaves an observable imprint at the surface, potentially allows us to study the internal structure of the magnetic field through modelling of the spectra and pulse profiles of thermally emitting neutron stars.

There exist growing observational evidence that the anisotropy of heat transport in the envelope alone, assuming an otherwise isothermal crust, can not explain the surface temperature distributions of some observed neutron stars. In the case of several of the "Magnificent Seven" (see, e.g., Haberl 2004, 2007) optical broad band photometric detections can be interpreted as being due to the Rayleigh-Jeans tail of a blackbody. However, these optical data are well above the Rayleigh-Jeans tail of the blackbody detected in the X-ray ("optical excess") and indicate the presence of an extended cold component of much larger area than the warm component observed in X-ray, the latter having an emitting radius ($\sim$3–5 km) much smaller than the usually assumed radius of a neutron star ($\sim$10–15 km). Schwope et al. (2005) tried to fit the lightcurve of RBS 1223 and concluded that only a surface temperature profile with relatively small, about 4–5 km across, hot polar regions may explain the observations. Pons et al. (2002) and Trümper et al. (2004) had arrived qualitatively at the same conclusion when they fitted the combined X-ray and optical spectrum of RX J1856.5-3754. In both cases, the smallness of the hot region is much below what can be reached by considering anisotropic heat transport limited to only a thin envelope.

Little is known about the magnetic field structure in neutron stars which is very likely determined by processes during the proto-neutron star phase and/or in a relatively short period after that epoch. A proto-neutron star dynamo (Thompson and Duncan 1993) is unlikely to generate purely poloidal fields while differential rotation will easily wrap any poloidal field and generate strong toroidal components (Kluźniak and Ruderman 1998; Wheeler et al. 2002). The magneto-rotational instability (Balbus and Hawley 1991) also most certainly acts in proto-neutron stars (Akiyama et al. 2003) and results in toroidal fields from differential rotation (Balbus and Hawley 1998). Thus, it seems realistic to consider the effect of magnetic field configurations which consist of poloidal and toroidal components.

Besides their possible relevance for modelling the observed thermal radiation of the "Magnificent Seven", and probably other strong field isolated neutron stars, toroidal magnetic fields may have a strong effect on the thermal evolution of the stars. The highly non-uniform surface temperatures they can induce result in reduced thermal luminosities and hence reduced energy losses during the late photon cooling era. It is our purpose in this work to explore this issue and we will present preliminary models of cooling of neutron stars with strong toroidal fields.

The structure of this paper is the following. In Sect. 2 we briefly review the basic ingredients and concepts involved in modelling the cooling of a neutron star and in Sect. 3 we present in some detail the simple mathematical formalism which describes dipolar fields, both poloidal and toroidal. The next section, Sect. 4, presents our results on the effect of strong magnetic fields on the neutron star crust temperature distribution, and these results are applied to model the cooling of the star in Sect. 5. Finally, we discuss our results in Sect. 6 and offer some tentative conclusions and prospects for future work.

2 Basic mechanisms of neutron star cooling

The essentials of neutron star cooling are expressed in the energy balance equation. In its Newtonian formulation this equation reads

$$\frac{dE_{\rm th}}{dt} = C_{\rm v}\frac{dT}{dt} = -L_\nu - L_\gamma + H, \tag{1}$$

where $E_{\rm th}$ is the thermal energy content of the star, $C_{\rm v}$ its total specific heat and T its internal temperature, which we assume to be uniform here. The energy sinks are the neutrino luminosity L_ν and the surface photon luminosity L_γ. The source term H includes all possible "heating mechanisms" which will be neglected in the present work; however the decay of a strong magnetic field can result in significant heating (Kaminker et al. 2006, 2007). Only the basic points will be presented here and we refer the reader to the reviews Yakovlev and Pethick (2004) and Page et al. (2006), or Page et al. (2004), for more details.

The specific heat $C_{\rm v}$ receives its dominant contribution from the baryons in the core of the star and about 10% contribution from the leptons. The crustal lattice, free neutrons in the inner crust and electrons also provide a small contribution. When nucleons become superfluid (neutrons) or superconducting (protons) their contribution to $C_{\rm v}$ is severely reduced and may even practically vanish.

The neutrino luminosity L_ν is usually dominated by the core. We consider the slow emission processes of the modified Urca family and the similar bremsstrahlung ones. Once nucleon pairing turns on, neutrino emission from the formation and breaking of Cooper pairs is also included.

2.1 Neutron star envelopes (without magnetic fields)

Our main interest here is the photon luminosity L_γ. In the absence of a magnetic field one expects the surface to have a uniform temperature $T_{\rm e}$, the *effective temperature*, and can express L_γ as

$$L_\gamma = 4\pi R^2 \sigma_{\rm SB} T_{\rm e}^4 \quad \text{or} \quad L_{\gamma\infty} = 4\pi R_\infty^2 \sigma_{\rm SB} T_{\rm e\infty}^4, \tag{2}$$

where R is the star's radius and $\sigma_{\rm SB}$ the Stefan–Boltzmann constant (the quantities "at infinity" are defined as $L_{\gamma\infty} =$

$e^{2\Phi}L_\gamma$, $R_\infty = e^{-\Phi}R$, and $T_{e\infty} = e^{\Phi}T_e$, where $e^{\Phi} \equiv (1 - 2GM/Rc^2)^{1/2}$ is the redshift factor). In order to integrate (1) one needs a relationship between T and T_e. The assumption of uniform interior temperature T is reasonable for most of the interior, given the huge thermal conductivity of degenerate matter and once the star is old enough to have relaxed from the initially complicated temperature structure produced at its birth, but is certainly not possible in the upper layers of the star where density is low enough for the electrons not to be fully degenerate. One traditionally separates out these upper layers from the cooling calculation and treats them as an *envelope*. A typical cut density is 10^{10} g cm^{-3} and the resulting envelope, with a depth of the order of 100 m, can be studied separately in a plane parallel approximation. Gundmundsson et al. (1982, 1983) presented detailed models of neutron star envelopes and showed that T_e is related to the temperature at the bottom of the envelope, T_b at density $\rho_b = 10^{10}$ g cm^{-3}, through the simple relation

$$T_e = 0.87 \times 10^6 \text{ K } g_{s\,14}^{1/4} T_b^{0.55}, \tag{3}$$

where g_{s14} is the surface gravity in units of 10^{14} cm s^{-2}. A relation such as (3) is usually called a "T_b–T_e relationship". The models leading to (3) assumed the envelope is formed of iron-like nuclei, and it was shown by Chabrier et al. (1997) (see also Potekhin et al. 1997) that if light elements, such as H, He, C, O, are present deep enough in the envelope, the increase in thermal conductivity (which is roughly proportional to Z^{-1} in liquid matter, Z being the element's charge) results in much higher luminosities, by up to one order of magnitude. Since the temperature gradient penetrates deeper in hotter stars, larger amounts of light elements are necessary to alter heat transport at high T_b than at lower T_b.

2.2 Heat transport with magnetic fields

In the absence of a magnetic field the thermal conductivity κ is conveniently written as

$$\kappa_0 = \frac{1}{3} c_v \bar{v}^2 \tau = \frac{\pi^2 k_B^2 T n_e}{3 m_e^*} \tau, \tag{4}$$

where $\bar{v}$ is the mean velocity of the heat carriers, c_v their specific heat per unit volume, and τ their collisional time; the second expression is particularized to relativistic electrons, the dominant heat carriers in neutron star crusts (Yakovlev and Urpin 1980). In the presence of a magnetic field, due to the classical Larmor rotation of electrons, heat flow may be anisotropic and κ becomes a tensor

$$\kappa = \begin{pmatrix} \kappa_\perp & \kappa_\wedge & 0 \\ -\kappa_\wedge & \kappa_\perp & 0 \\ 0 & 0 & \kappa_\parallel \end{pmatrix} \tag{5}$$

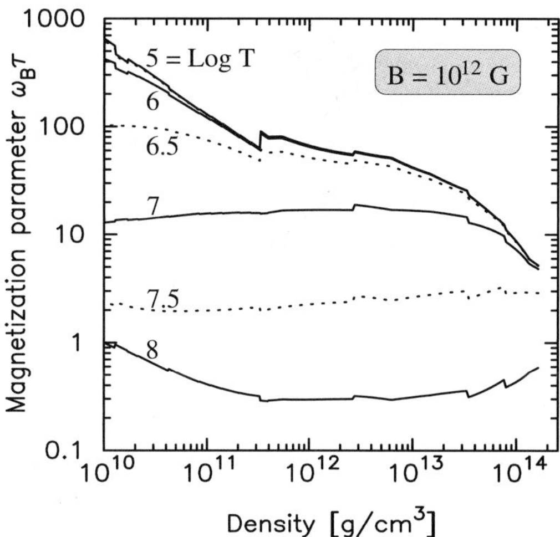

Fig. 1 Magnetization parameter $\omega_B\tau$ vs. density at six different temperatures (as labeled on *the curves*) assuming a uniform magnetic field of strength $B = 10^{12}$ G. Its value for different field strengths scales linearly in B. (Figure from Geppert et al. 2004)

(assuming the field **B** oriented along the z-axis) whose components have the form

$$\kappa_\parallel = \kappa_0,$$
$$\kappa_\perp = \frac{\kappa_0}{1 + (\omega_B\tau)^2}, \tag{6}$$
$$\kappa_\wedge = \frac{\kappa_0\,\omega_B\tau}{1 + (\omega_B\tau)^2},$$

where $\omega_B = eB/m_e^* c$ is the electron cyclotron frequency. The condition $\omega_B\tau \gg 1$, which implies strong anisotropy, is easily realized in a neutron star envelope, and also possibly in the whole crust (Geppert et al. 2004). Values of the magnetization parameter $\omega_B\tau$ are plotted in Fig. 1. Notice that τ receives contribution from both electron-phonon and electron-impurity scattering, with frequencies $\nu_{\text{e-ph}}$ and $\nu_{\text{e-imp}}$ resp., and is given by $\tau = (\nu_{\text{e-ph}} + \nu_{\text{e-imp}})^{-1}$. Electron-impurity scattering dominates at low temperatures, and/or high densities, and is T-independent while $\nu_{\text{e-ph}}$ goes roughly as T^2: combination of these two different T-dependencies is the reason for the complex behavior of $\omega_B\tau$ seen in Fig. 1.

In case the field is strong enough to be quantizing, the expressions (6) have to be modified (in particular τ also becomes anisotropic) but the essential result that $\kappa_\perp \ll \kappa_\parallel$, when $\omega_B\tau \gg 1$, remains (Potekhin 1999).

2.3 Magnetized neutron star envelopes

In considering magnetic field effects on heat transport in a neutron star the simplest case to handle is the envelope: since its thickness, ~100 m, is much smaller than the length scale over which the field is expect to vary significantly, ~km, one

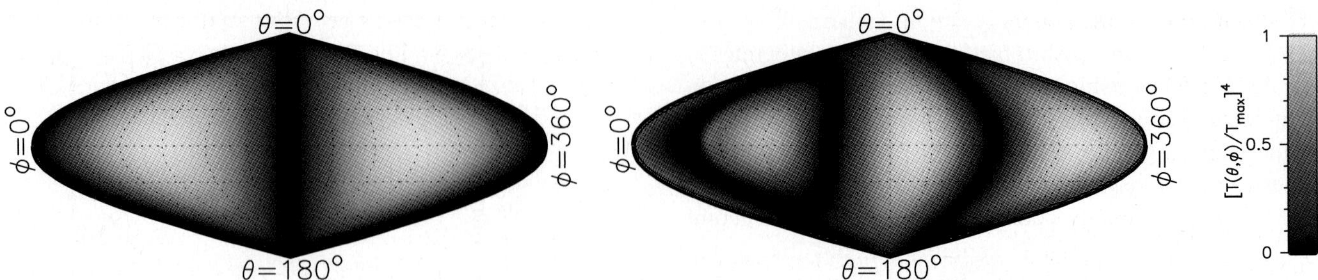

Fig. 2 Two examples of surface temperature distribution obtained from (8). The maps cover the whole neutron star surface in an area preserving projection. For better viewing, the magnetic field symmetry axis is located in the equatorial plane, oriented from $\phi = 90°$ to $\phi = 270°$. *The left panel* has a dipolar field only while *the right panel* also contains a quadrupolar field superposed to the same dipole. (Figure from Page et al. 2006)

can consider heat transport on a given small patch on the surface to be independent of the rest of the surface. Moreover, the field strength B and its angle with respect to the radial direction, Θ_B, can be considered as uniform in the patch and the heat flux considered as essentially radial. The thermal conductivity in a direction making an angle Θ_B with the field, considering the κ tensor of (5), is then given by

$$\kappa(\Theta_B) = \cos^2\Theta_B \times \kappa_\parallel + \sin^2\Theta_B \times \kappa_\perp. \tag{7}$$

With this form of $\kappa(\Theta_B)$ and a radial flux, heat transport in the envelope at this surface patch is a one dimensional problem. The solution is a "T_b–T_e" relationship which depends on B and Θ_B, and the obtained effective temperature T_e is a local one which we will write as $T_s(\theta, \phi)$ in the sense that the flux emerging from this patch is $\sigma_B T_s^4$. Given the form of $\kappa(\Theta_B)$, (7), Greenstein and Hartke (1983) proposed the simple interpolation formula

$$\begin{aligned} T_s(\theta,\phi)^4 &\equiv T_s(T_b; B, \Theta_B)^4 \\ &\approx \cos^2(\Theta_B) \times T_s(T_b; B, \Theta_B = 0)^4 \\ &\quad + \sin^2(\Theta_B) \times T_s(T_b; B, \Theta_B = 90°)^4 \end{aligned} \tag{8}$$

for arbitrary angle Θ_B in terms of the two cases of radial ($\Theta_B = 0$) and tangential ($\Theta_B = 90°$) field. Recently, Potekhin and Yakovlev (2001) and Potekhin et al. (2003) have presented detailed calculations and fitted their results by an expression similar to (8).

For a given geometry of the magnetic field (not necessarily dipolar) and assuming that the temperature at the bottom of the envelope, T_b, is not affected by the magnetic field and hence is uniform around the star, one can generate the expected surface temperature distribution $T_s(\theta, \phi)$ by piecing together envelope models through applying relationships of the form of (8) at each point of the surface (Page 1995; Page and Sarmiento 1996). Two examples of such expected surface temperature distributions are shown in Fig. 2. The photon luminosity is then given by

$$L_\gamma = 4\pi R^2 \cdot \iint \frac{d\Omega}{4\pi}\, \sigma_B\, T_s(\theta,\phi)^4 \equiv 4\pi R^2 \cdot \sigma_B T_e^4, \tag{9}$$

where T_e is again defined as in (2).

By this method one can easily generate surface temperature distributions corresponding to the geometry of the magnetic field at the surface (assuming that the same field geometry is maintained throughout the underlying thin envelope). However, the results of Fig. 1 show that the anisotropy in heat transport is likely to extend much deeper into the crust than just the envelope. To handle such cases one is hence forced to define the field geometry in the whole crust, which is what we do in the next section.

3 The internal magnetic field

We will only consider here axisymmetric configurations (more general ones will hopefully be considered in the future once a 3D transport code is available). In this case it is convenient to separate the magnetic field in two components

$$\mathbf{B} = \mathbf{B}^{\rm pol} + \mathbf{B}^{\rm tor}, \tag{10}$$

the poloidal and toroidal components, respectively, where $\mathbf{B}^{\rm pol}$ only has $\mathbf{e}_r$ and $\mathbf{e}_\theta$ components and $\mathbf{B}^{\rm tor}$ only an $\mathbf{e}_\phi$ component[1] (the $\mathbf{e}_{r,\theta,\phi}$ being units vectors in the spherical coordinate directions, with the $\theta = 0$ axis coinciding with the field symmetry axis). Thus, the magnetic field lines of $\mathbf{B}^{\rm tor}$ are simply circles centered on the symmetry axis but the field lines of $\mathbf{B}^{\rm pol}$ are more complicated. Let us expand B_r in Legendre polynomials as $B_r = \sum F_l(r)\, P_l(\cos\theta)$. We will only consider the first ($l = 1$) term, the dipole, and write F_1 as S/r^2 so that

$$B_r = \frac{S(r)}{r^2}\cos\theta. \tag{11}$$

Then Maxwell's equation $\nabla \cdot \mathbf{B} = 0$ implies

$$B_\theta = -\frac{1}{2r}\frac{\partial S}{\partial r}\sin\theta. \tag{12}$$

[1]Such decomposition is also possible for non-axisymmetric fields but is more involved (Rädler 2000).

We also need the three boundary conditions

$$S(r=0)=0,\quad S(r=R)=B_0R^2 \quad \text{and} \quad \left.\frac{\partial S}{\partial r}\right|_{r=R} = -\frac{S(R)}{R} = -B_0R, \tag{13}$$

which ensure regularity at the star's center and smooth matching with an external vacuum dipolar field, of strength B_0 at the magnetic poles, at the stellar surface. Besides these boundary conditions, the Stoke function S is totally arbitrary but a choice of it is equivalent to choosing the location of the currents sustaining the poloidal field since the latter, having only a j_ϕ component, are given by

$$j_\phi = \frac{c}{8\pi}\frac{\sin\theta}{r}\left(\frac{\partial^2 S}{\partial r^2} - \frac{2S}{r^2}\right). \tag{14}$$

Notice also a simple physical interpretation of the Stoke function: the magnetic flux through the star's equatorial plane in a circle of radius r is simply given by $\pi S(r)$, and the boundary condition gives a total flux $\Phi = \pi R^2 B_0$, as it should be. In vacuum, S simply reduces to B_0R^3/r.

The locations of the currents in the stellar interior is unknown but it is natural to separate them in two components, located in the core and in the crust and accordingly separate $\mathbf{B}^{\rm pol}$ as

$$\mathbf{B}^{\rm pol} = \mathbf{B}^{\rm core} + \mathbf{B}^{\rm crust}. \tag{15}$$

The core protons are expected to become a superconductor, with critical temperatures $T_c \sim 10^9$ K (see, e.g., Page et al. 2004), soon after the star's birth: the magnetic field is then confined into flux tubes and maintained by proton supercurrents. Since, by definition, $\mathbf{B}^{\rm core}$ corresponds to currents located in the core, within the crust it is described by a vacuum dipolar poloidal field, and we will parametrize it by its surface strength at the magnetic pole $B_0^{\rm core}$. We do not need to specify the distribution of the supercurrents sustaining $\mathbf{B}^{\rm core}$ and the geometry of this component in the core since the field, being confined to fluxoids which occupy only a very small volume, is not expected to alter heat transport in the core.

For the $\mathbf{B}^{\rm crust}$ component, we need to specify its geometry in the crust, i.e., the corresponding Stoke function $S^{\rm crust}$. The arbitrariness involved in such specification can be somewhat relieved by considering models of the time evolution of this component: currents spontaneously migrate toward the highest density region of the crust, where the electrical resistivity is smallest, until they reach the crust-core boundary where their migration is stopped by the proton superconductor (see, e.g., Fig. 4 in Page et al. 2000). We use an $S^{\rm crust}$ function resulting from such evolutionary calculations and scale it to vary the overall strength of $\mathbf{B}^{\rm crust}$, which we parametrize by $B_0^{\rm crust}$ defined as its strength at the magnetic pole.

For the toroidal component $\mathbf{B}^{\rm tor}$, we also expand it in Legendre polynomials $P_l(\sin\theta)$ and keep only the $l=1$ dipolar term

$$\mathbf{B}^{\rm tor} = T(r)\sin\theta. \tag{16}$$

We do not have to consider the part of $\mathbf{B}^{\rm tor}$ confined to the core, because of proton superconductivity, and specify only the part confined to the crust. The only restrictions are the boundary conditions

$$T(r=R_{\rm core}) = 0 \quad \text{and} \quad T(r=R)=0. \tag{17}$$

We are not aware of evolutionary calculations of toroidal field in neutron star crust and hence are left with a guess about the possible shape and size of the T function: following Geppert et al. (2006) we see it as a free parameter and consider several choices. We parametrize the strength of $\mathbf{B}^{\rm tor}$, for each choice of T, by $B_0^{\rm tor}$ defined as the maximum value of $\|\mathbf{B}^{\rm tor}\|$ in the crust.

Pérez-Azorín et al. (2006) have presented similar models of field structure and the resulting temperature distributions in the crust and chose force-free field configurations to specify the field geometry. It is reasonable to assume the field could evolve into a force-free configuration during the early neutron star life, but there is no physical reason why the later evolution, driven by Ohmic decay and Hall interactions, will conserve the force-free condition and this motivates our considering the function T as rather arbitrary. However, our choice of $S^{\rm crust}$, inspite of being based on evolutionary calculations of purely poloidal fields, may not be too realistic because of the coupling between poloidal and toroidal components from the Hall term. We show in Fig. 3 a sketch of the field geometry we consider and in Fig. 4 our choices of the functions S and T.

4 Magnetic field effects in the crust

Given the magnetic field geometries described in the previous sections, Geppert et al. (2006) calculated the resulting

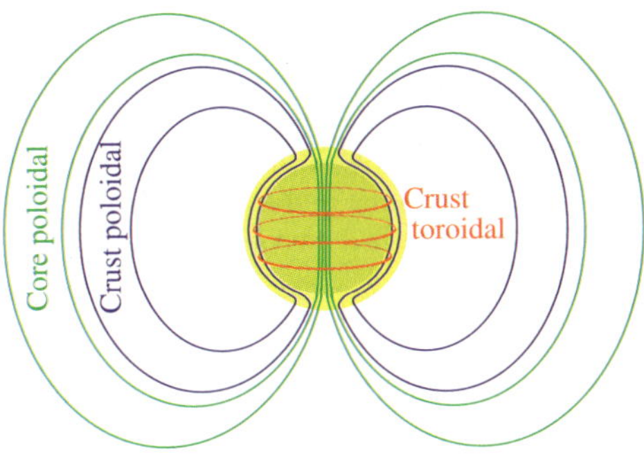

Fig. 3 The three components of the magnetic field considered in this work

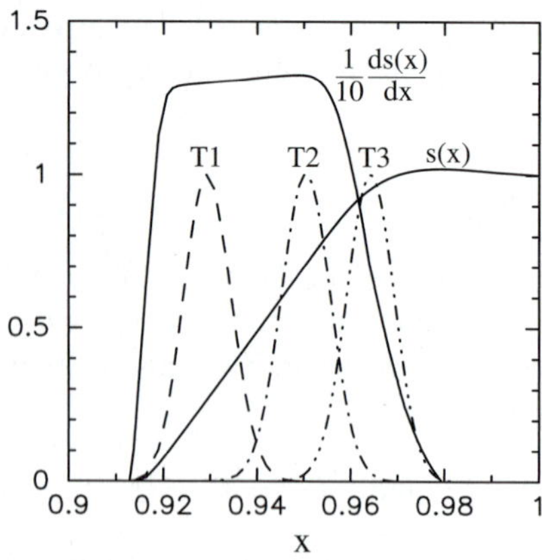

Fig. 4 *Continuous lines*: normalized Stokes function $s(x) = S/R^2 B_0^{\rm crust}$ and its derivative (scaled by a factor 10) used in this work, which generate B_r (11) and B_θ (12) resp. *Discontinuous lines*: the three different normalized functions $t(x) = T/B_0^{\rm tor}$ we consider for the toroidal field, B_φ (16), which we label as *T1*, *T2*, and *T3*. *On the horizontal axis* $x \equiv r/R$. (Figure from Geppert et al. 2006)

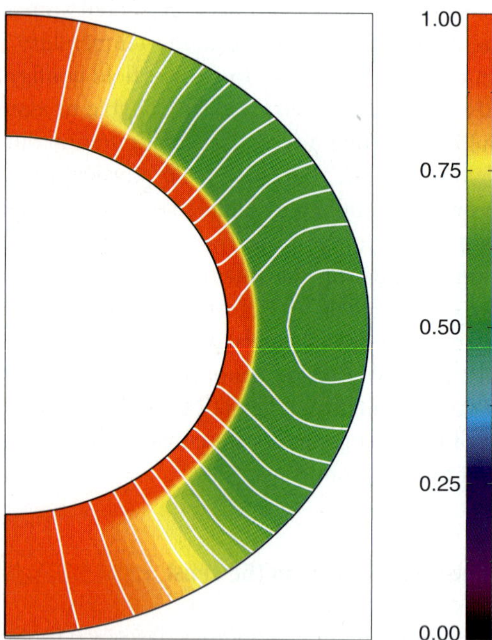

Fig. 5 Temperature distribution in a strongly magnetized neutron star crust (whose thickness has been stretched by a factor five for easier reading). The chosen field scale parameters are $B_0^{\rm core} = 7.5 \times 10^{12}$ G, $B_0^{\rm crust} = 2.5 \times 10^{12}$ G, $B_0^{\rm tor} = 3 \times 10^{15}$ G, and the toroidal component's generating functions T is the model "T1" of Fig. 4. The color code maps the relative temperature, i.e., $T(r,\theta)/T_{\rm core}$, with a core temperature $T_{\rm core} = 6 \times 10^7$ K. *White lines* show field lines of $\mathbf{B}^{\rm pol}$, the *field lines* of $\mathbf{B}^{\rm tor}$ being perpendicular to the plane of the figure. The heat blanketing effect of the toroidal component is clearly visible. (From Geppert et al. 2006)

temperature distributions in the neutron star crust, in a stationary state. In theses calculations, heat transport is solved in the crust for densities from $\rho = \rho_{\rm core} = 1.6 \times 10^{14}\ {\rm g\,cm^{-3}}$ down to $\rho = \rho_{\rm b} \equiv 10^{10}\ {\rm g\,cm^{-3}}$. In the core, at $\rho > \rho_{\rm core}$,

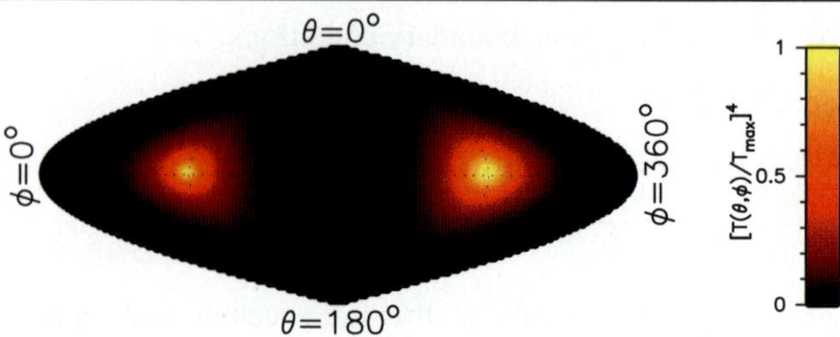

Fig. 6 Surface temperature distribution corresponding to the crust temperature shown in Fig. 5. For better viewing the magnetic symmetry axis is in the equator ($\theta = 90°$) pointing at $\phi = 90°$. (From Geppert et al. 2006)

the temperature $T_{\rm core}$ is assumed to be constant and uniform while at the outer zone, at $\rho = \rho_{\rm b}$, the models are matched with magnetized envelopes (treated as the outer boundary condition). Since the 2D transport code we use does not yet include correct specific heat it does not have the capability to perform realistic time dependent calculations and only stationary results can be obtained, i.e., the thermal evolution is followed until the temperature profile would not change anymore with time (this stationary limit is independent of the specific heat). Moreover, neutrino energy losses are also not included (neutrino emission, for high temperature, and also strong magnetic fields, is not negligible and should be included for accurate calculations, see, e.g., Potekhin et al. 2007). Figure 5 shows an example of the resulting temperature distribution in the crust and Fig. 6 shows the resulting surface temperature distribution: this latter figure should be compared with the surface temperature distribution shown in Fig. 2 where isothermal crusts were considered and magnetic field affected heat transport only within the thin envelope.

For a fixed magnetic field configuration and several values of $T_{\rm core}$, we calculate the crustal temperature distribution and the resulting surface temperature $T_{\rm s}(\theta)$, which is ϕ independent because of axial symmetry, and obtain $T_{\rm e}$ from (9). A set of results is displayed in Fig. 7: we show models which, within our selection of field configurations, maximize the magnetic field effects, with 75% of the poloidal flux coming from crustal currents and the toroidal component located in the middle of the crust. With a purely poloidal field, i.e., $B_0^{\rm tor} = 0$, one sees little effect and the crust stays close to isothermality, except at high temperature, $T_{\rm core} = 10^9$ K, where the envelope's temperature gradient extends deeper into the crust than the $\rho = 10^{10}\ {\rm g\,cm^{-3}}$ arbitrary cut. Significant effects appear when $B_0^{\rm tor} \gg 10^{14}$ G and when $B_0^{\rm tor} \geq 10^{15}$ G the crust is highly non-isothermal even at the highest temperature $T_{\rm core} = 10^9$ since then $\omega_{\rm B}\tau \gg 1$ in the whole crust. There is an intriguing change in the shape of the T-surface when going from $T_{\rm core} > 10^8$ K to $T_{\rm core} < 10^8$ K, which is most evident at the highest value $B_0^{\rm tor} = 3 \times 10^{15}$ G: it is most probably due to the shift in the dominant scattering process from electron-phonon, at high T where τ

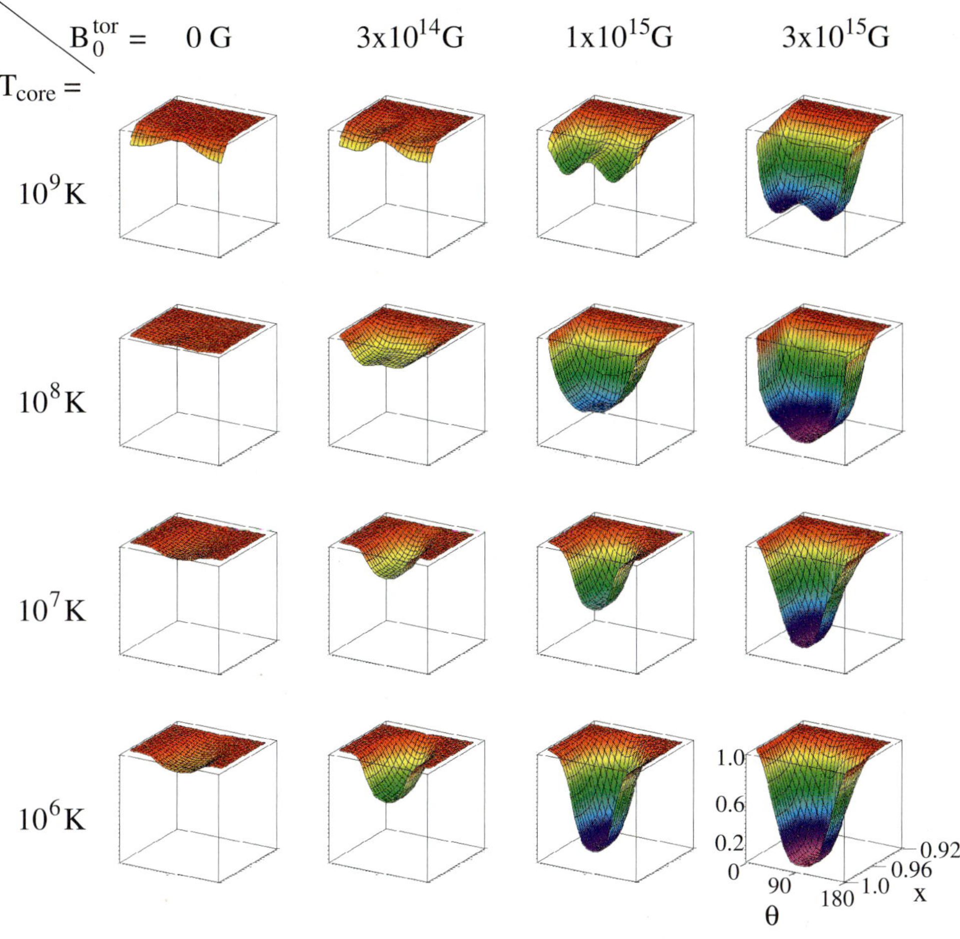

Fig. 7 3D plots of the temperature distribution in a strongly magnetized neutron star crust. *The horizontal axes* show radial coordinate $x = r/R$ and polar angle θ while *the vertical scale* is T/T_{core}. Four values of T_{core} are considered and the chosen field scale parameters for the poloidal field are $B_0^{\text{core}} = 2.5 \times 10^{12}$ G and $B_0^{\text{crust}} = 7.5 \times 10^{12}$ G in all cases while the toroidal component scale B_0^{tor} is varied, the overall shape of the T function (16), being the model "*T2*" of Fig. 4

is T-dependent, to electron-impurity, at low T where τ is T-independent, as mentioned in Sect. 4 (we hope to study this effect in more detail in a future work.).

5 Cooling of strongly magnetized neutron stars

In order to perform (time dependent) cooling calculations we will here use the models described in the previous section to produce a set of outer boundary conditions for our 1D cooling code (see, e.g., Page et al. 2004). We fix the outer boundary at a density $\rho_{\text{b}} = 10^{14}$ g cm^{-3}, instead of 10^{10} g cm^{-3} when envelope models are used as boundary conditions. As is obvious from Fig. 6, compared to Fig. 2, the photon luminosity, and hence T_{e}, is much lower in presence of a strong toroidal field than when the crust is considered as isothermal and Fig. 8 shows the resulting T_{b}–T_{e} relationships obtained for nine different magnetic field configurations. For the most extreme case, T_{e} can be reduced by a factor of 2.5, and hence L_γ by a factor 40, compared to the isothermal crust case. Notice that we chose $\rho_{\text{b}} = 10^{14}$ g cm^{-3} inspite of our crustal models starting at $\rho_{\text{core}} = 1.6 \times 10^{14}$ g cm^{-3} because we found a very small temperature gradient in the range 1.0–1.6 $\times 10^{14}$ g cm^{-3} and hence prefer to treat this density range with the cooling code. The 1D cooling code solves the energy balance and heat transport equations in their full general relativistic forms.

The cooling models we will consider here are within the minimal cooling paradigm of Page et al. (2004). In short, this means neutrino emission in the stellar core is from the modified Urca and the similar nucleon bremsstrahlung processes, neutron and proton pairing is taken into account with the resulting neutrino emission from the Cooper pair breaking and formation process and the alteration of the specific heat. We use pairing critical temperatures for proton $^1S_0$ from Takatsuka (1973), for neutron $^1S_0$ from Schwenk et al. (2003), and for neutron $^3P_2$ the model "a" from Page et al. (2004). The star model is a 1.4 M$_\odot$ neutron star built with the equation of state of Akmal et al. (1998) and hence, as part of the minimal cooling paradigm, charged meson condensates,[2] hyperons, and deconfined quark matter are not present in the star.

[2]The APR equation of state shows the presence of a π^0 condensate, but it has no noticeable effect on the cooling.

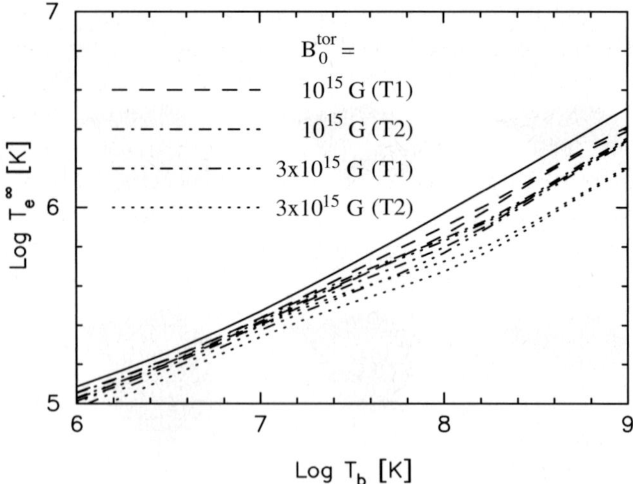

Fig. 8 Possible T_b–T_e relationships for a neutron star with a strong toroidal magnetic field. The various sets of curves correspond to two different values of B_0^{tor} and two different locations of the toroidal field, labelled as "*T1*" and "*T2*" as in Fig. 4. For each case we also consider two different splitting of the poloidal field: $B_0^{crust} = 7.5 \times 10^{12}$ G and $B_0^{core} = 2.5 \times 10^{12}$ G, or $B_0^{crust} = 2.5 \times 10^{12}$ G and $B_0^{core} = 7.5 \times 10^{12}$ G, the cases with the larger B_0^{core} resulting in higher T_e. *The continuous curve* shows the T_b–T_e relationships for an isothermal crust with the same poloidal field $B_0^{pol} = 10^{13}$ G and magnetic field effects included only in the envelope (Potekhin and Yakovlev 2001)

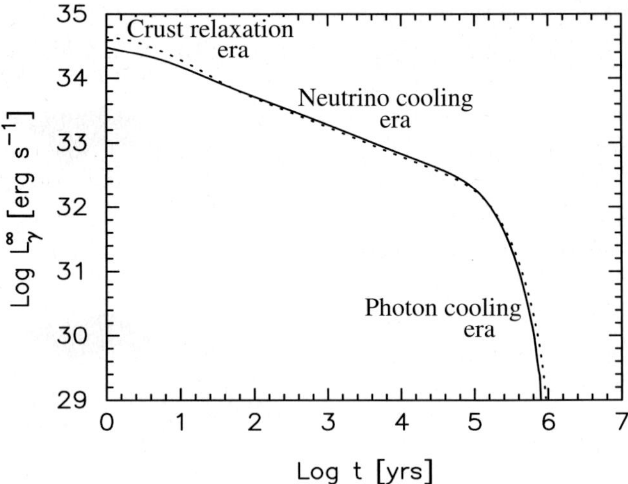

Fig. 9 Comparison of cooling trajectories, i.e., red-shifted surface photon luminosity vs. time, obtained with our truncated models using $\rho_b = 10^{14}$ g cm^{-3} (*continuous curve*) vs. a "full" model having $\rho_b = 10^{10}$ g cm^{-3} (*dotted curve*). Both models consider a poloidal dipolar field of strength $B_0^{pol} = 10^{13}$ G. The truncated model uses the T_b–T_e relationship for an isothermal crust shown in Fig. 8 while the full model uses directly the T_b–T_e relationship from Potekhin and Yakovlev (2001)

First we assess the effect of the approximation of truncating the star at $\rho_b = 10^{14}$ g cm^{-3}, and treating most of the crust through our stationary solutions as part of the outer boundary condition, instead of using the traditional outer boundary at 10^{10} g cm^{-3}. A comparison of cooling trajectories with these two different boundary densities, and in the absence of a toroidal field component, is shown in Fig. 9. The major difference appears at early times during the crust relaxation era: a "full" (i.e., using $\rho_b = 10^{10}$ g cm^{-3}) model shows strong radial temperature gradients in the crust at this stage (see, e.g., Gnedin et al. 2001) while the truncated (i.e., using $\rho_b = 10^{14}$ g cm^{-3}) model assumes an isothermal crust. Another way of seeing it is that our 2D crustal models only considered stationary states and are hence only applicable when the cooling time scale is much longer than the thermal relaxation time of the crust, a condition which is certainly not fulfilled during this early cooling phase. Later, during the neutrino cooling era, the truncated model is slightly warmer than the "full" model because neutrino emission from its truncated crust is missing. During the photon cooling era the difference between the two models becomes larger, now due to the smaller specific heat of the truncated model which consequently cools faster. Overall, moving the outer boundary from 10^{10} g cm^{-3} to 10^{14} g cm^{-3} only has a small, and quite negligible, effect except at early ages which we will hence not show in our next results.

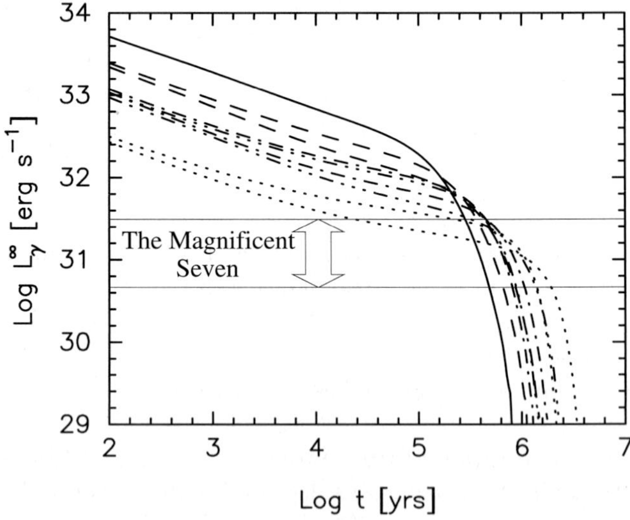

Fig. 10 Cooling of neutron stars with strong toroidal magnetic fields: red-shifted surface photon luminosity vs. time. Each curve corresponds to one of the T_b–T_e relationships displayed in Fig. 8, with the same line-style

Having now the confidence that stripping the star of most of its crust, and properly including the stripped part into the outer boundary condition, introduces an acceptably small error, we can proceed with models having strong toroidal crustal fields. We show in Fig. 10 our results for the nine field configurations of the T_b–T_e relationships presented in the Fig. 8. During the neutrino cooling era, the star's core evolution is driven by L_ν and the surface temperature, and L_γ, simply follows the evolution of the core. So during this phase all nine models have exactly the same evolution but they look very different at the surface: models with a higher T_e for a given T_b, as shown in Fig. 8, have a higher photon luminosity. During the photon cooling era the results are inverted since L_γ drives the cooling and models with a higher T_e for a given T_b result in a larger L_γ and consequently they undergo faster cooling.

We have preferred to plot the cooling curves as L_γ vs t instead of T_e vs. t: given the highly non-uniform $T_s(\theta, \phi)$ the effective temperature T_e looses any observational meaning. More detailed models taking this into account will be presented in a future work (Page et al. 2007).

6 Discussion and conclusions

We have extended our previous results about temperature distribution in a strongly magnetized neutron star crust (Geppert et al. 2004, 2006) to a broader range of temperatures and applied them to study the cooling of neutron stars with strong toroidal magnetic fields. Geppert et al. (2004) had shown that with a purely poloidal field entirely confined to the crust strongly non-uniform temperature distributions develop in the crust. Then Geppert et al. (2006) showed that allowing part of this poloidal field to permeate the core significantly reduced the temperature non-uniformity but that the inclusion of a strong toroidal field component can result in crustal temperature gradients even stronger than in the first case (similar results have been obtained by Pérez-Azorín et al. 2006). We have shown here that, due to the strength of the field, the magnetization parameter $\omega_B \tau$ is very large in most of the crust even at core temperatures as high as 10^9 K and the large crustal temperature gradients are still present in such hot stars. Considering these results we have been able to perform cooling calculations of such neutron stars with a strong toroidal crustal magnetic field, with some approximations which, as we have shown, introduce only very small errors in the results.

Our final results are displayed in Fig. 10 and compared with the estimated luminosity range of the "Magnificent Seven" (Haberl 2004). This comparison indicates most probable ages between 0.5 and 1.5 Myrs for these stars, depending on the magnetic field structure, but this range could be extended down to 0.05 Myrs for the brightest ones and up to 3 Myrs for the dimmest ones if the most extreme field geometry is considered. Besides the uncertainty in the field structure, which in itself results in the range of predicted ages shown in Fig. 10, there are two other key ingredients in cooling models which can have similar effects: nucleon pairing in the core and the chemical composition of the envelope. The former significantly affects the specific heat during the photon cooling era and different assumptions about the values of T_c can introduce a factor of a few in the predicted cooling ages while the latter can significantly affect the photon luminosity (see discussion in Sect. 2.1) and introduce another uncertainty of a few (we refer the reader to the detailled presentation of Page et al. 2004). Considering these three theoretical uncertainties, magnetic field geometry, nucleon pairing, and envelope chemical composition, it is certainly possible to extend the theoretical cooling ages, for photon luminosities between $\sim 3 \times 10^{30}$ and $\sim 3 \times 10^{31}$, to cover the range from 10^5 to almost 10^7 years!

Can we reasonably expect to reduce this enormous theoretical age uncertainty? Precession of some neutron stars (Link 2007; Pons 2007) may imply that neutron pairing critical temperatures in the core are very low (Link 2003), a possibility which moreover has a strong theoretical fundament (Schwenk and Friman 2004): this would imply a large specific heat and favor long cooling times (within the minimal cooling paradigm). Interpretation of the absorption lines detected in most of the "Magnificent Seven" (see, e.g., Haberl 2007 and van Kerkwijk and Kaplan 2007) may provide crucial information about the surface chemical composition. However, atomic and molecular physics in strong field still has many secrets to be unveiled (Turbiner 2007) and, moreover, what is needed is the chemical composition of the deeper layers of the envelope, i.e., several tens of meters below the surface. Finally, the structure and evolution of the magnetic field with toroidal components, in the neutron star context as considered in this paper, is an almost uncharted territory and much progress can be expected.

Acknowledgements We thank José Pons and many other participants of the meeting "Isolated Neutron Stars" for discussions, as well as Jillian A. Henderson for careful and critical reading of this paper.

References

Akiyama, S., Wheeler, J.C., Meier, D.L., et al.: Astrophys. J. **584**, 954 (2003)
Akmal, A., Pandharipande, V.R., Ravenhall, D.G.: Phys. Rev. C **58**, 1804 (1998)
Balbus, S.A., Hawley, J.F.: Astrophys. J. **376**, 214 (1991)
Balbus, S.A., Hawley, J.F.: Rev. Mod. Phys. **70**, 1 (1998)
Chabrier, G., Potekhin, A.Y., Yakovlev, D.G.: Astrophys. J. **477**, L99 (1997)
Geppert, U., Küker, M., Page, D.: Astron. Astrophys. **426**, 267 (2004)
Geppert, U., Küker, M., Page, D.: Astron. Astrophys. **457**, 937 (2006)
Gnedin, O.Y., Yakovlev, D.G., Potekhin, A.Y.: Mon. Not. Roy. Astron. Soc. **324**, 725 (2001)
Greenstein, G., Hartke, G.J.: Astrophys. J. **271**, 283 (1983)
Gudmundsson, E.H., Pethick, C.J., Epstein, R.I.: Astrophys. J. **259**, L19 (1982)
Gudmundsson, E.H., Pethick, C.J., Epstein, R.I.: Astrophys. J. **272**, 286 (1983)
Haberl, F.: Adv. Space Res. **33**, 638 (2004)
Haberl, F.: Astrophys. Space Sci., DOI:10.1007/s10509-007-9342-x (2007)
Kaminker, A.D., Yakovlev, D.G., Potekhin, A.Y., et al.: Mon. Not. Roy. Astron. Soc. **371**, 477 (2006)

Kaminker, A.D., Yakovlev, D.G., Potekhin, A.Y.: Astrophys. Space Sci., DOI:10.1007/s10509-007-9358-2 (2007)

Kluźniak, W., Ruderman, M.: Astrophys. J. **505**, L113 (1998)

Link, B.: Phys. Rev. Lett. **91**, 101101 (2003)

Link, B.: Astrophys. Space Sci., DOI:10.1007/s10509-007-9315-0 (2007)

Page, D.: Astrophys. J. **442**, 273 (1995)

Page, D., Sarmiento, A.: Astrophys. J. **473**, 1067 (1996)

Page, D., Geppert, U., Zannias, T.: Astron. Astrophys. **360**, 1052 (2000)

Page, D., Lattimer, J.M., Prakash, M., et al.: Astrophys. J. Suppl. **155**, 623 (2004)

Page, D., Geppert, U., Weber, F.: Nucl. Phys. A **777**, 497 (2006)

Page, D., Henderson, J.A., Geppert, U.: (2007, in preparation)

Pérez-Azorín, J.F., Miralles, J.A., Pons, J.A.: Astron. Astrophys. **452**, 1009 (2006)

Pons, J.A.: Astrophys. Space Sci., DOI:10.1007/s10509-007-9336-8 (2007)

Pons, J.A., Walter, F.M., Lattimer, J.M., et al.: Astrophys. J. **564**, 981 (2002)

Potekhin, A.Y., Chabrier, G., Yakovlev, D.G.: Astron. Astrophys. **323**, 415 (1997)

Potekhin, A.Y.: Astron. Astrophys. **351**, 797 (1999)

Potekhin, A.Y., Yakovlev, D.G.: Astron. Astrophys. **374**, 213 (2001)

Potekhin, A.Y., Yakovlev, D.G., Charbier, G., et al.: Astrophys. J. **594**, 404 (2003)

Potekhin, A.Y., Chabrier, G., Yakovlev, D.G.: DOI:10.1007/s10509-007-9362-6 (2007)

Rädler, K.-H.: The generation of cosmic magnetic fields. In: Page, D., Hirsch, J. (eds) Lect. Notes in Phys., vol. 556, p. 101 (2000)

Schwenk, A., Friman, B.: Phys. Rev. Lett. **92**, 082501 (2004)

Schwenk, A., Friman, B., Brown, G.E.: Nucl. Phys. A **713**, 191 (2003)

Schwope, A.D., Hambaryan, V., Haberl, F., et al.: Astron. Astrophys. **441**, 597 (2005)

Takatsuka, T.: Prog. Theor. Phys. **50**, 1754 (1973)

Thompson, C., Duncan, R.C.: Astrophys. J. **408**, 194 (1993)

Trümper, J., Burwitz, V., Haberl, F., et al.: Nucl. Phys. B Proc. Suppl. **132**, 560 (2004)

Turbiner, A.V.: Astrophys. Space Sci., DOI:10.1007/s10509-007-9337-7 (2007)

van Kerkwijk, M.H., Kaplan, D.L.: Astrophys. Space Sci., DOI:10.1007/s10509-007-9343-9 (2007)

Wheeler, J.C., Meier, D.L., Wilson, J.R.: Astrophys. J. **568**, 807 (2002)

Yakovlev, D.G., Pethick, C.J.: Annu. Rev. Astron. Astrophys. **42**, 169 (2004)

Yakovlev, D.G., Urpin, V.A.: Sov. Astron. **24**, 303 (1980)

Astrophys Space Sci (2007) 308: 413–418
DOI 10.1007/s10509-007-9331-0

ORIGINAL ARTICLE

Internal heating and thermal emission from old neutron stars

Constraints on dense-matter and gravitational physics

Andreas Reisenegger · Rodrigo Fernández · Paula Jofré

Received: 21 July 2006 / Accepted: 9 November 2006 / Published online: 21 March 2007

Abstract The equilibrium composition of neutron star matter is achieved through weak interactions (direct and inverse beta decays), which proceed on relatively long time scales. If the density of a matter element is perturbed, it will relax to the new chemical equilibrium through non-equilibrium reactions, which produce entropy that is partly released through neutrino emission, while a similar fraction heats the matter and is eventually radiated as thermal photons. We examined two possible mechanisms causing such density perturbations: (1) the reduction in centrifugal force caused by spin-down (particularly in millisecond pulsars), leading to *rotochemical heating*, and (2) a hypothetical time-variation of the gravitational constant, as predicted by some theories of gravity and current cosmological models, leading to *gravitochemical heating*. If only slow weak interactions are allowed in the neutron star (modified Urca reactions, with or without Cooper pairing), rotochemical heating can account for the observed ultraviolet emission from the closest millisecond pulsar, PSR J0437-4715, which also provides a constraint on $|dG/dt|$ of the same order as the best available in the literature.

Keywords Stars: neutron · Dense matter · Relativity · Stars: rotation · Pulsars: general · Pulsars: individual (PSR B0950+08, PSR J0108-1431, PSR J0437-4715)

PACS 91.10.Op · 06.20.Jr · 97.60.Jd

This work made use of NASA's Astrophysics Data System Service, and received financial support from FONDECYT through regular grants 1020840 and 1060644.

A. Reisenegger · P. Jofré
Departamento de Astronomía y Astrofísica, Pontificia Universidad Católica de Chile, Casilla 306, Santiago 22, Chile

A. Reisenegger
e-mail: areisene@astro.puc.cl

R. Fernández
Department of Astronomy & Astrophysics, University of Toronto, Toronto, ON, M5S 3H8, Canada
e-mail: fernandez@astro.utoronto.ca

Present address:
P. Jofré
Max-Planck-Institut für Astrophysik, Karl-Schwarzschild-Str. 1, 85741 Garching bei München, Germany
e-mail: jofre@MPA-Garching.MPG.de

1 Introduction

Neutron star matter is composed of degenerate fermions of various kinds: neutrons (n), protons (p), electrons (e), probably muons (μ) and possibly other, more exotic particles. (We refer to electrons and muons collectively as leptons, l.) Neutrons are stabilized by the presence of other, stable fermions that block (through the Pauli exclusion principle) most of the final states of the beta decay reaction $n \rightarrow p + l + \bar{\nu}$. The large chemical potentials μ_i ($\approx$Fermi energies) for all particle species i also make inverse beta decays, $p + l \rightarrow n + \nu$, possible. The neutrinos (ν) and antineutrinos ($\bar{\nu}$) leave the star without further interactions, contributing to its cooling (e.g., Shapiro and Teukolsky 1983; Yakovlev and Pethick 2004). The two reactions mentioned tend to drive the matter into a chemical equilibrium state, defined by $\eta_{npl} \equiv \mu_n - \mu_p - \mu_l = 0$.

If a matter element is in some way driven away from chemical equilibrium ($\eta_{npl} \neq 0$), free energy is stored, which is released by an excess rate of one reaction over the other. This energy is partly lost to neutrinos and antineutrinos (undetectable at present), and partly used to heat the matter. The heat is eventually lost as thermal (ultraviolet) photons emitted from the stellar surface.

The chemical imbalance can be caused by a change in the density of the stellar matter. This can in turn be produced in different ways. The first to be considered (by Finzi 1965; Finzi and Wolf 1968) was stellar pulsation; however, so far no clear evidence for this process has been seen. Gravitational collapse (Haensel 1992; Gourgoulhon and Haensel 1993) and mass accretion are also possible mechanisms, but in these contexts the non-equilibrium heating is probably overwhelmed by the energy released through other channels.

Here we review our work on neutron star heating through beta processes in two other contexts, which we consider to be the most promising in revealing information about the physics of dense matter and gravitation. One is *rotochemical heating* (Reisenegger 1995, 1997; Fernández and Reisenegger 2005, hereafter FR05; Reisenegger et al. 2006), in which the precisely measurable decrease of the stellar rotation rate, through the related reduction of the centrifugal force, makes the star contract progressively, keeping it away from chemical equilibrium. The more speculative *gravitochemical heating* is based on the hypothesis that the gravitational "constant" may in fact vary in time (Dirac 1937; Brans and Dicke 1961), causing a similar contraction or expansion of the neutron star (Jofré et al. 2006, hereafter JRF06). We refer to our published papers for a detailed discussion of our methods and formalism (see also Flores-Tulián and Reisenegger 2006). Here we restrict ourselves to a general, unified discussion of these two processes and their main implications.

2 Time evolution

The formalism for calculating the evolution of the temperature and chemical imbalances for the case of rotochemical heating is described in Sect. 2 of FR05. Here, we just outline the fundamental equations and the modifications required in order to treat the gravitochemical case as well. The evolution of the internal temperature, T, taken to be uniform inside the star, is given by the thermal balance equation,

$$\dot{T} = \frac{1}{C(T)}\left[L_H(\eta_{npl}, T) - L_\nu(\eta_{npl}, T) - L_\gamma(T)\right], \tag{1}$$

where C is the total heat capacity of the star, L_H is the total power released by the heating mechanism, L_ν the total power emitted as neutrinos, and L_γ the power released as thermal photons. Here and in what follows (including all figures), all temperatures, chemical imbalances, stellar radii, and luminosities are "redshifted" to the reference frame of a distant observer at rest with respect to the star.

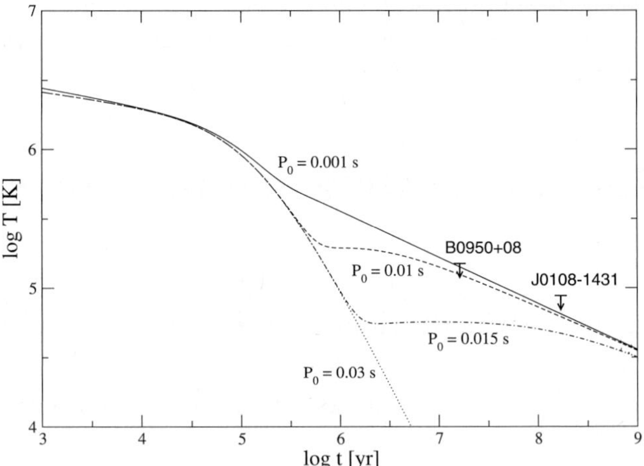

Fig. 1 Predicted time-evolution of the surface temperature, T_s, of a neutron star with rotochemical heating. All *curves* correspond to stars with mass $M = 1.4 M_\odot$, equation of state A18 + δv + UIX* (Akmal et al. 1998), which allows only modified Urca processes to occur, and magnetic dipole spin-down with $B_{\rm dipole} = 2.5 \times 10^{11}$ G. Each *curve* is labeled by the assumed initial rotation period. The upper limits correspond to observational constraints for pulsars B0950+08 (Zavlin and Pavlov 2004) and J0108-1431 (Mignani et al. 2003, as interpreted by Kargaltsev et al. 2004), both of which have magnetic dipole field strengths very close to the assumed value

The evolution of the chemical imbalances is given by

$$\begin{aligned}\dot{\eta}_{npl} = &-\left[A_{D,l}(\eta_{npe}, T) + A_{M,l}(\eta_{npe}, T)\right] \\ &-\left[B_{D,l}(\eta_{np\mu}, T) + B_{M,l}(\eta_{np\mu}, T)\right] \\ &- R_{npl}\Omega\dot{\Omega} + C_{npl}\dot{G}.\end{aligned} \tag{2}$$

The functions A and B quantify the effect of reactions towards restoring chemical equilibrium, and thus have the same sign of η_{npl} (FR05). The subscripts D and M refer to direct Urca reactions,

$$\begin{aligned}&n \to p + l + \overline{\nu}, \\ &p + l \to n + \nu,\end{aligned} \tag{3}$$

which are possibly forbidden by momentum conservation, and modified Urca,

$$\begin{aligned}&n + N \to p + N + l + \overline{\nu}, \\ &p + l + N \to n + N + \nu,\end{aligned} \tag{4}$$

where an additional nucleon N must participate in order to conserve momentum (e.g., Shapiro and Teukolsky 1983; Yakovlev and Pethick 2004). The scalars R_{npl} and C_{npl} quantify the departure from equilibrium due to the changes in the centrifugal force ($\propto \Omega\dot{\Omega}$) and the gravitational constant ($\dot{G}$), being positive and depending on the stellar model and equation of state (FR05; Reisenegger et al. 2006; JRF06).

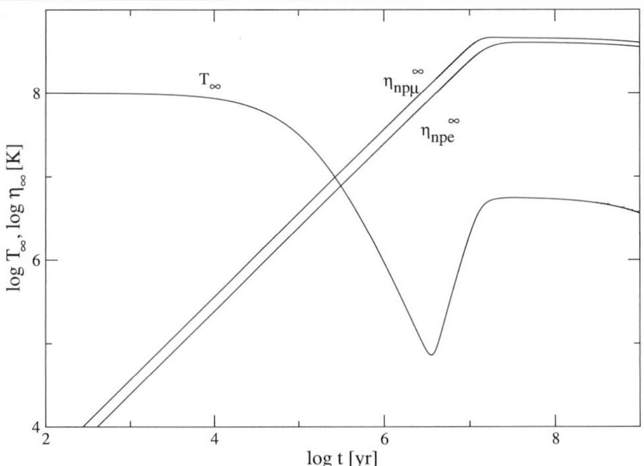

Fig. 2 Evolution of the internal temperature and chemical imbalances under the rotochemical heating effect for a $1.4M_\odot$ star calculated with the A18 + δv + UIX* equation of state (Akmal et al. 1998), with initial temperature $T = 10^8$ K, null initial chemical imbalances, and magnetic dipole spin-down with field strength $B = 10^8$ G and initial period $P_0 = 1$ ms (taken from FR05)

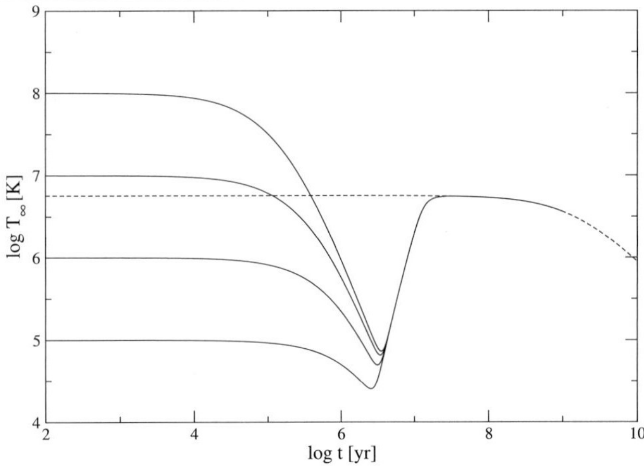

Fig. 3 Evolution of the internal temperature under rotochemical heating for different initial temperatures. We set $\eta_{npe} = \eta_{np\mu} = 0$. The *short-dashed line* is the quasi-equilibrium solution, obtained by solving $\dot T = 0$ and $\dot\eta_{npe} = \dot\eta_{np\mu} = 0$. The stellar model and spin-down parameters are the same as in Fig. 2 (from FR05)

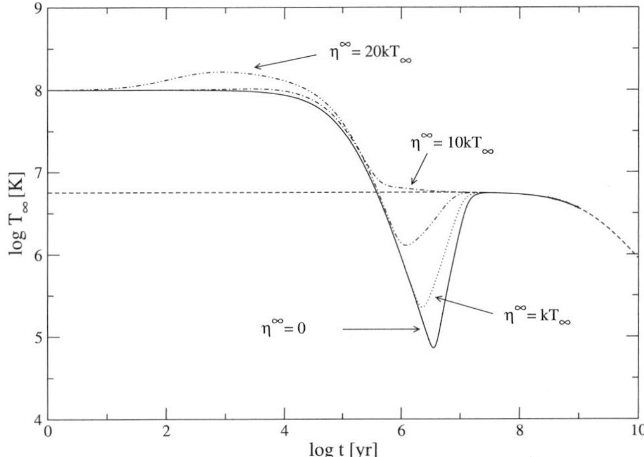

Fig. 4 Evolution of the internal temperature under rotochemical heating for different initial chemical imbalances $\eta_{npe} = \eta_{np\mu} \equiv \eta$ and the same initial temperature $T = 10^8$ K at $t = 0$. The line styles, the stellar model, and the spin-down parameters are the same as in Fig. 3 (from FR05)

Figure 1 shows the solution of the coupled differential equations (1) and (2) for the evolution of a classical pulsar with a moderate magnetic field and different assumed initial rotation periods under pure rotochemical heating ($\Omega\dot\Omega < 0$, $\dot G = 0$). It can be seen that, for very fast initial rotation, the pulsar can be kept warm beyond the standard cooling time of $\sim 10^7$ yr, at a level that is close to current observational constraints.

The case of a "millisecond pulsar" (a neutron star with fast rotation and a weak magnetic dipole field) is illustrated in Fig. 2. It first cools down from its high birth temperature, while the chemical potential imbalances η_{npl} slowly increase due to the decreasing rotation rate, until rotochemical heating increases the temperature again, and the reactions stop the rise of the chemical potential imbalances.

3 Stationary state

If the relevant forcing ($\Omega\dot\Omega$ or $\dot G$) changes slowly with time, the star eventually arrives at a stationary state, where the rate at which the equilibrium concentrations are modified by this forcing is the same as that at which the reactions drive the system toward the new equilibrium configuration, with heating and cooling balancing each other (Reisenegger 1995). The evolution to this state for pure rotochemical heating is illustrated in Fig. 2. Figures 3 and 4 show that the state reached is independent on the assumed initial conditions.

The properties of this stationary state can be obtained by the simultaneous solution of (1) and (2) with $\dot T = \dot\eta_{npl} = 0$. The existence of the stationary state makes it unnecessary to model the full evolution of the temperature and chemical imbalances of the star in order to calculate the final temperature, since the stationary state is independent of the initial conditions (see FR05 for a detailed analysis of the rotochemical heating case). For given values of $\Omega\dot\Omega$ and $\dot G$, it is thus possible to calculate the temperature of an old pulsar that has reached the stationary state, without knowing its exact age.

When only modified Urca reactions operate, it is possible to solve analytically for the stationary values of the photon luminosity L_γ^{st} and chemical imbalances η_{npl}^{st}, as a function of stellar model and current value of $\Omega\dot\Omega$ and $\dot G$. The reason for this is that the longer equilibration timescale given by the slower modified Urca reactions yields stationary chemical imbalances satisfying $\eta_{npl} \gg kT$. In this limit, the term

$L_H - L_\nu$ in the thermal balance equation can be written as $K_{Le}\eta_{npe}^8 + K_{L\mu}\eta_{np\mu}^8$, where $K_{L,l}$ are positive constants that depend only on stellar mass and equation of state (FR05; JRF06). For typical equations of state, the photon luminosity in the stationary state is

$$L_\gamma^{\rm st} \simeq 10^{30\text{–}31}\left|\frac{\dot{P}_{-20}}{P_{\rm ms}^3} + \frac{\dot{G}/G}{3\times 10^{-11}\ {\rm yr}^{-1}}\right|^{8/7} {\rm erg\ s}^{-1}, \quad (5)$$

where $P_{\rm ms}$ is the rotation period in milliseconds, and $\dot{P}_{-20}$ is its time derivative in units of 10^{-20} (dimensionless). The effective surface temperature of the star in the stationary state is

$$T_s^{\rm st} \simeq (2\text{–}3)\times 10^5\left|\frac{\dot{P}_{-20}}{P_{\rm ms}^3} + \frac{\dot{G}/G}{3\times 10^{-11}\ {\rm yr}^{-1}}\right|^{2/7} {\rm K}. \quad (6)$$

Finally, the timescale for the system to reach the stationary state is

$$\tau_{\rm st} \simeq 2\times 10^7\left|\frac{\dot{P}_{-20}}{P_{\rm ms}^3} + \frac{\dot{G}/G}{3\times 10^{-11}\ {\rm yr}^{-1}}\right|^{-6/7} {\rm yr}. \quad (7)$$

4 Comparison to observations

In order to verify this model and constrain the value of $|\dot{G}/G|$, we need a neutron star that (1) has a measured surface temperature (or at least a good enough upper limit on the latter), and (2) is confidently known to be older than the timescale to reach the stationary state. So far, the only object satisfying both conditions is the millisecond pulsar closest to the Solar System, PSR J0437-4715 (hereafter J0437), whose surface temperature was inferred from an HST-STIS ultraviolet observation by Kargaltsev et al. (2004). Its spin-down age, $\tau_{\rm sd} \simeq 5\times 10^9$ yr (e.g., van Straten et al. 2001), and the cooling age of its white dwarf companion, $\tau_{\rm WD} \simeq (2.5\text{–}5.3)\times 10^9$ yr (Hansen and Phinney 1998), are much longer than the time required to reach the steady state for both rotochemical and gravitochemical heating, in the latter case under the condition that $|\dot{G}/G| \geq 10^{-13}\ {\rm yr}^{-1}$.

Consequently, we consider stellar models constructed from different equations of state, with masses satisfying the constraint obtained for J0437 by van Straten et al. (2001), $M_{\rm psr} = 1.58 \pm 0.18 M_\odot$, and calculate the stationary temperature for each. Figure 5 compares the predictions for the case of pure rotochemical heating with the measured spin-down parameters of J0437 (for various equations of state and neutron star masses) to the temperature inferred from the observation of Kargaltsev et al. (2004).

In Fig. 6, we compare the same observational constraints on the temperature of this pulsar to the theoretical predictions for pure gravitochemical heating, assuming $|\dot{G}/G| = 2\times 10^{-10}\ {\rm yr}^{-1}$. As can be seen, this value is such that

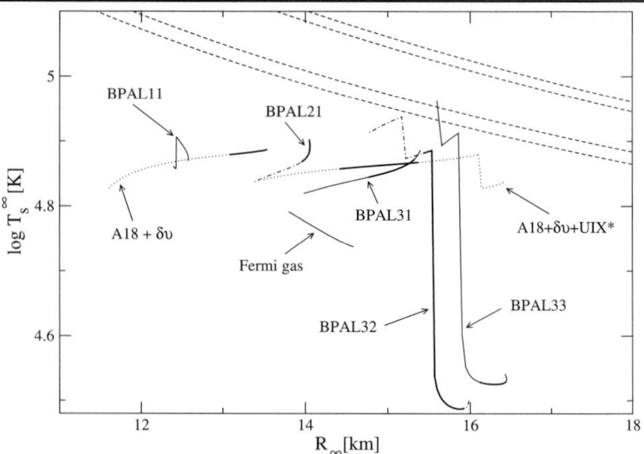

Fig. 5 Surface temperature due to rotochemical heating in the stationary state as function of stellar radius for different equations of state, shown as *solid lines* (APR from Akmal et al. 1998 and BPAL from Prakash et al. 1988), for the spin parameters of PSR J0437-4715. *Dashed lines* are 68% and 90% confidence contours of the blackbody fit to the emission from this pulsar (Kargaltsev et al. 2004). *Bold lines* indicate, for each equation of state, the mass range allowed by the constraint of van Straten et al. (2001), $M_{\rm PSR} = 1.58 \pm 0.18 M_\odot$. BPAL32 and BPAL33 allow direct Urca reactions in the observed mass range of PSR J0437 (taken from FR05)

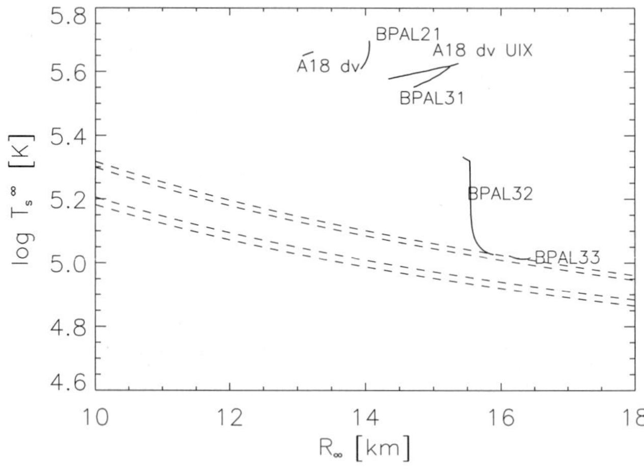

Fig. 6 Surface temperature due to gravitochemical heating in the stationary state as function of stellar radius for different equations of state. The value of $|\dot{G}/G| = 2\times 10^{-10}\ {\rm yr}^{-1}$ is chosen so that all stationary temperature curves lie above the observational constraints. Otherwise, the meanings of lines and symbols are as in Fig. 5 (from JRF06)

the stationary temperatures of all stellar models lie just above the 90% confidence contour, and therefore represents a rather safe and general upper limit.

When the stellar mass becomes large enough for the central pressure to cross the threshold for direct Urca reactions, T_s drops abruptly, due to the faster relaxation towards chemical equilibrium. This occurs in two steps, as electron and muon direct Urca processes have different threshold densities (see, e.g., FR05). Conventional neutron star cooling models reproduce observed temperatures better when only

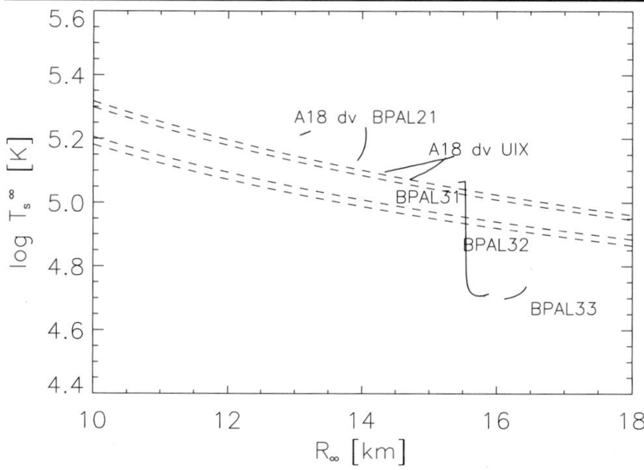

Fig. 7 Same as Fig. 6, but now the value of $|\dot{G}/G| = 4 \times 10^{-12}$ yr^{-1} is chosen such that only the stationary temperature curves with modified Urca reactions are above the observational constraints (from JRF06)

modified Urca reactions are considered (e.g., Yakovlev and Pethick 2004; Page et al. 2007). Restricting our sample to the equations of state that allow only modified Urca reactions in the mass range considered here, namely A18 + δv, A18 + δv + UIX, BPAL21, and BPAL31, we obtain a more restrictive upper limit on $|\dot{G}|$, as shown in Fig. 7, yielding $|\dot{G}/G| < 4 \times 10^{-12}$ yr^{-1}.

5 Discussion and conclusions

5.1 Rotochemical heating

Using the equations of state that allow only modified Urca reactions within the allowed mass range for PSR J0437-4715, rotochemical heating predicts an effective temperature in the narrow range $T_{s,\mathrm{eq}} = (6.9\text{–}7.9) \times 10^4\ K$, about 20% lower than the blackbody fit of Kargaltsev et al. (2004). There are three possible reasons why the prediction does not quite match the observation:

1. We are not taking superfluidity into account. This would reduce Urca reaction rates, lengthening the equilibration timescale and raising the stationary-state temperature (Reisenegger 1997).
2. We are neglecting other heating mechanisms (some of them directly related to superfluidity), which could further raise the temperature at any stage in the thermal evolution. Nonetheless, in millisecond pulsars, all proposed mechanisms appear to be less important than rotochemical heating (Schaab et al. 1999; Kargaltsev et al. 2004).
3. The thermal spectrum could deviate from a blackbody, as for the isolated neutron star RX J1856-3754, which has a well-determined blackbody X-ray spectrum that underpredicts the optical flux (Walter and Matthews 2001), indicating a more complex spectral shape of its thermal emission.

Kargaltsev et al. (2004) stress that PSR J0437-4715 has a higher surface temperature than the upper limit for the younger, "classical" pulsar J0108-1431, $T_s < 8.8 \times 10^4$ K, inferred from the optical non-detection by Mignani et al. (2003) and shown in our Fig. 1. In the rotochemical heating model, these two pulsars are in very different regimes: J0437 is in the stationary state in which its temperature can be predicted from its spin-down parameters, whereas J0108 has a 680 times smaller spin-down power ($\propto \Omega\dot{\Omega}$), and will therefore not reach a detectable stationary state. Its equilibration timescale, according to (7), is 2×10^{11} yr, longer than the age of the Universe and certainly much longer than the spin-down age of the pulsar. Thus, its heat content (if any) is due to its earlier, faster rotation, which may have built up a significant chemical imbalance that is currently being decreased by ongoing reactions in its interior (see Fig. 1). Depending on its initial rotation period, its surface temperature may be substantially smaller than both J0437's observed temperature and its own current upper limit.

5.2 Gravitochemical heating

Table 1 lists some of the many experiments performed so far to test the constancy of G (see Uzan 2003 and Will 2006 for recent reviews). The second column contains the upper limits on its time variation, most usefully expressed as $|\dot{G}/G|$, and the third is a rough time scale over which each experiment is averaging this variation. Based on the latter, the experiments can be separated into three classes. The first two experiments on the list measure the variation of G from the early Universe to the present time, and the constraint on the present-day value of $|\dot{G}/G|$ is based on assuming a time dependence $G(t) \propto t^{-\alpha}$, where t is the time since the Big Bang, and α is a constant constrained by these experiments. The next four are sensitive to variations over long timescales, 10^{9-10} yr, but without reaching into the very early Universe. The last four experiments measure the change of G directly over short, "human" timescales of years or few decades. Even though results from the first category are nominally the most restrictive on a long-term variation of G, they depend crucially on the assumed form of the variation of G near the Big Bang. Thus, it is still useful to consider measurements of the second and third categories, which could directly detect variations of G in more recent times.

The new method advocated here, namely gravitochemical heating of neutron stars, falls closest to the second category, as its timescale is much longer than human, but does not reach into the early Universe. However, it probes somewhat shorter timescales than the other methods in this category. In the most general case, when direct Urca reactions are allowed to operate, we obtain an upper limit $|\dot{G}/G| < 2 \times 10^{-10}$ yr^{-1}. Restricting the sample of equations of state

Table 1 Previous upper bounds on $|\dot{G}/G|$

Method	$\|\dot{G}/G\|_{\max}$ $[10^{-12}\ \mathrm{yr}^{-1}]$	Time scale [yr]	Reference
Big Bang nucleosynthesis	0.4	1.4×10^{10}	Copi et al. (2004)
Microwave background	0.7	1.4×10^{10}	Nagata et al. (2004)
Globular cluster isochrones	35	10^{10}	Degl'Innocenti et al. (1996)
Binary neutron star masses	2.6	10^{10}	Thorsett (1996)
Helioseismology	1.6	4×10^{9}	Guenther et al. (1998)
Paleontology	20	4×10^{9}	Eichendorf and Reinhardt (1977)
Lunar laser ranging	1.3	24	Williams et al. (2004)
Binary pulsar orbits	9	8	Kaspi et al. (1994)
White dwarf oscillations	250	25	Benvenuto et al. (2004)

to those that allow only modified Urca reactions, we obtain a much more restrictive upper limit, $|\dot{G}/G| < 4 \times 10^{-12}\ \mathrm{yr}^{-1}$ on a time scale $\sim 10^8$ yr (the time for the neutron star to reach its quasi-stationary state), competitive with constraints obtained from the other methods probing similar timescales. However, since the composition of matter above nuclear densities is uncertain and millisecond pulsars are generally expected to be more massive than classical pulsars, we cannot rule out the result for the direct Urca regime.

Further progress in our knowledge of neutron star matter will allow this method to become more effective at constraining variations in G. The method can also be improved with an increased sample of objects with measured thermal emission or good upper limits.

Acknowledgements We thank G. Pavlov and O. Kargaltsev for letting us know about their work in advance of publication and for kindly providing their data in electronic form. The authors are also grateful to C. Dib, O. Espinosa, S. Flores-Tulián, M. Gusakov, E. Kantor, R. Mignani, D. Page, M. Taghizadeh, and M. van Kerkwijk for discussions that benefited the present paper.

References

Akmal, A., Pandharipande, V.R., Ravenhall, D.G.: Phys. Rev. C **58**, 1804 (1998)
Benvenuto, O., García-Berro, E., Isern, J.: Phys. Rev. D **69**, 2002 (2004)
Brans, C., Dicke, R.H.: Phys. Rev. **124**, 925 (1961)
Copi, C., Davis, A., Krauss, L.: Phys. Rev. Lett. **92**, 17 (2004)
Degl'Innocenti, S., et al.: Astron. Astrophys. **312**, 345 (1996)
Dirac, P.: Nature **139**, 323 (1937)
Eichendorf, W., Reinhardt, M.: Mitt. Astron. Ges. Hamburg **42**, 89 (1977) (result cited in Uzan 2003)
Fernández, R., Reisenegger, A.: Astrophys. J. **625**, 291 (2005): FR05
Finzi, A.: Phys. Rev. Lett. **15**, 599 (1965)
Finzi, A., Wolf, R.A.: Astrophys. J. **153**, 835 (1968)
Flores-Tulián, S., Reisenegger, A.: Mon. Not. Roy. Astron. Soc. **372**, 276 (2006)
Gourgoulhon, E., Haensel, P.: Astron. Astrophys. **271**, 187 (1993)
Guenther, D.B., Krauss, L.M., Demarque, P.: Astrophys. J. **498**, 871 (1998)
Haensel, P.: Astron. Astrophys. **262**, 131 (1992)
Jofré, P., Reisenegger, A., Fernández, R.: Phys. Rev. Lett. **97**, 131102 (2006): JRF06
Hansen, B.M.S., Phinney, E.S.: Mon. Not. Roy. Astron. Soc. **294**, 569 (1998)
Kargaltsev, O., Pavlov, G.G., Romani, R.: Astrophys. J. **602**, 327 (2004)
Kaspi, V.M., Taylor, J.H., Ryba, M.F.: Astrophys. J. **428**, 713 (1994)
Mignani, R.P., Manchester, R.N., Pavlov, G.G.: Astrophys. J. **582**, 978 (2003)
Nagata, R., Chiba, T., Sugiyama, N.: Phys. Rev. D **69**, 3512 (2004)
Page, D., Geppert, U., Küker, M.: Astrophys. Space Sci. **308**. DOI 10.1007/s10509-007-9316-z (2007)
Prakash, M., Ainsworth, T.L., Lattimer, J.M.: Phys. Rev. Lett. **61**, 2518 (1988)
Reisenegger, A.: Astrophys. J. **442**, 749 (1995)
Reisenegger, A.: Astrophys. J. **485**, 313 (1997)
Reisenegger, A., Jofré, P., Fernández, R., Kantor, E.: Astrophys. J. **653**, 568 (2006)
Schaab, Ch., Sedrakian, A., Weber, F., Weigel, M.K.: Astron. Astrophys. **346**, 465 (1999)
Shapiro, S.L., Teukolsky, S.A.: Black Holes, White Dwarfs, and Neutron Stars. Wiley, New York (1996)
Thorsett, S.E.: Phys. Rev. Lett. **77**, 1432 (1996)
Uzan, J.: Rev. Mod. Phys. **75**, 403 (2003)
van Straten, W., et al.: Nature **412**, 158 (2001)
Walter, F.M., Matthews, L.D.: Nature **389**, 358 (1997)
Will, C.: Living Rev. Relativ. **9**, 3 (2006)
Williams, J.G., Turyshev, S.G., Boggs, D.H.: Phys. Rev. Lett. **93**, 261101 (2004)
Yakovlev, D.G., Pethick, C.J., Boggs, D.H.: Annu. Rev. Astron. Astrophys. **42**, 169 (2004)
Zavlin, V.E., Pavlov, G.G.: Astrophys. J. **616**, 452 (2004)

Astrophys Space Sci (2007) 308: 419–422
DOI 10.1007/s10509-007-9332-z

ORIGINAL ARTICLE

Thermal emission areas of heated neutron star polar caps

M. Ruderman · A.M. Beloborodov

Received: 28 July 2006 / Accepted: 16 November 2006 / Published online: 27 March 2007

Abstract Observed hot spots on neutron stars are often associated with polar caps heated by the backflow of energetic electrons or positrons from accelerators on bundles of open magnetic field lines. Three effects are discussed that may be relevant to formation of hot spots and their areas. (1) The area of a polar cap is proportional to the ratio of the star's surface dipole field to the local field at the polar cap. Because the field is coupled to the evolving spin of the superfluid core of the star, this ratio can depend on the stellar spin and its history. (2) The hot emission area may appear smaller to a distant observer when emitted X-rays propagate through electron-positron plasma created in the magnetosphere. The X-rays then change their energy spectrum because of cyclotron resonant scattering by pairs. (3) Hot spots may form on the star's surface as a result of crust motions that are driven by the pull of core flux tubes pinned to the crust. Such motions twist the footprints of closed magnetic loops of the magnetosphere and induce an electric current in the loop, which will heat those footprints.

Keywords Neutron stars · Polar caps · Magnetic fields

PACS 97.60.Jd

A.M.B. was supported by NASA grant NNG-06-G107G.

M. Ruderman (✉) · A.M. Beloborodov
Physics Department and Columbia Astrophysics Laboratory, Columbia University, 538 W 120th Street, New York, NY 10027, USA
e-mail: mar@astro.columbia.edu

A.M. Beloborodov
e-mail: amb@phys.columbia.edu

1 Introduction

A neutron star's (NS's) polar caps are the surface footprints of the "open" magnetic field lines which connect the surface to the light cylinder. In the simplest models the magnetic field is generally approximated as that from a spinning central dipole (or a uniformly magnetized stellar interior). The open field lines come from two polar caps at opposite locations on the NS's crust. These have canonical polar cap areas,

$$A_c \sim \frac{\pi \Omega R^3}{c} \tag{1}$$

for a NS with radius R and spin rate Ω. The surface polar caps are directly heated by backflow of very energetic electrons (or positrons) from accelerators on the open field line bundles. Hot polar cap areas can be inferred from X-ray observations of the polar cap temperatures and luminosities. However, for many reasons this cannot yet be accomplished with confidence in the quantitative results. There is, for example, uncertainty about the emissivity of the polar cap surface (composition; effects of long $e^{\pm}$ bombardment, excavation by particle outflow, strong electric and magnetic fields; etc.) It does appear, however, that heated polar cap areas are typically much less than those of (1) (see e.g. Caraveo et al. 2004). This could be for several reasons:

(a) NS magnetic fields are formed with a structure different from the central dipole, and retain it as the star cools and evolves.
(b) Accelerator backflow fills only a small part of the polar cap ("sparks") and thermal conductivity does not merge the many tiny hot spot emission areas (Gil, these proceedings).

(c) Polar caps areas change with NS spin-down and spin-up in a way not described by (1), even for the simplest initial magnetic field configurations.
(d) Thermal X-ray emission from the surface of some NSs is processed during its passage through the inner magnetosphere in a way which strongly diminishes inferred radiation areas.

Below we shall discuss only (c) and present some ongoing considerations of (d). We also briefly discuss another possibility of forming small hot spots: the heating of footprints of twisted closed magnetic loops in the magnetosphere.

The spin-down (up) torque on a solitary NS is the stellar moment of inertia I times the observed $\dot{\Omega}$. Generally,

$$I\dot{\Omega} \sim B_{\mathrm{lc}}^2 r_{\mathrm{lc}}^3, \tag{2}$$

where B_{lc} is the magnetic field at the light cylinder of radius $r_{\mathrm{lc}} = c/\Omega$. Because the total magnetic flux carried by the open field-line bundle $\Phi \sim r_{\mathrm{lc}}^2 B_{\mathrm{lc}}$ is conserved,

$$\Phi = A_{\mathrm{pc}} B_{\mathrm{pc},r}, \tag{3}$$

with A_{pc} the polar cap area and $B_{\mathrm{pc},r}$ the radial component of the polar cap magnetic field. The "observed" surface dipole field is defined as

$$B_d \equiv \left(\frac{r_{\mathrm{lc}}}{R}\right)^3 B_{\mathrm{lc}}. \tag{4}$$

Then,

$$A_{\mathrm{pc}} = A_c \frac{B_d}{B_{\mathrm{pc},r}}. \tag{5}$$

Significant changes in A_{pc} from the conventional A_c of (1) occur whenever $B_d/B_{\mathrm{pc},r}$ changes from unity.

2 Evolution of A_{pc}

Within about 10^4 seconds after its birth, a canonical NS is cooled enough by neutrino emission to allow the formation of a solid crust of thickness $\Delta \sim 10^5$ cm. The crust encloses a core of neutrons together with a few percent of protons and electrons. Within a few years the core protons form a superconducting fluid. The magnetic field through it will become organized into a twisted array of quantized flux tubes (type II superconductor) or, perhaps, but less likely, an array of non-superconducting strips with magnetic field $B \sim 10^{15}$ G separated by superconducting regions in which $B = 0$ (type I superconductor). After about 10^3 years the core neutrons become a superfluid in which vorticity is organized into a parallel array of quantized vortex lines with number density per unit area $n_v = (2m_n/\pi\hbar)\Omega \sim (10^4\ \mathrm{s}/P)\ \mathrm{cm}^{-2}$.

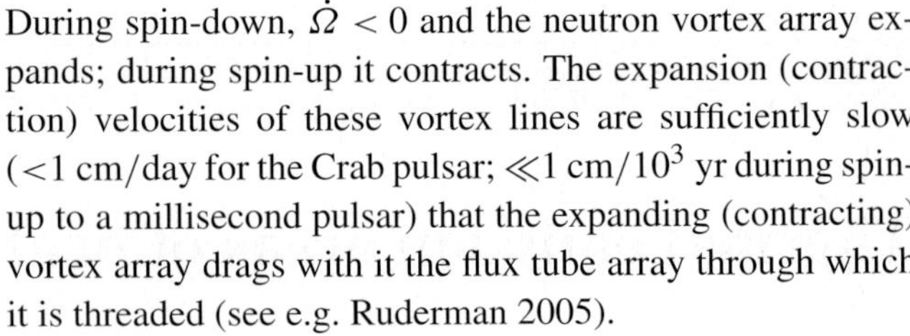
During spin-down, $\dot{\Omega} < 0$ and the neutron vortex array expands; during spin-up it contracts. The expansion (contraction) velocities of these vortex lines are sufficiently slow ($<$1 cm/day for the Crab pulsar; $\ll$1 cm/10^3 yr during spin-up to a millisecond pulsar) that the expanding (contracting) vortex array drags with it the flux tube array through which it is threaded (see e.g. Ruderman 2005).

The conducting solid crust of the star interferes with this evolution because it tries to keep the initial configuration of the magnetic field. The crust is not a superconductor and flux tubes expand and merge as they cross the core-crust boundary. If the crust were infinitely strong and had infinite electrical conductivity, it would not allow any changes in the surface magnetic field: the field would remain frozen forever in the static crust.

The real crust has a finite strength and can be broken by the magnetic pull from the core flux tubes. Then it can rearrange itself through horizontal motions of crustal plates. Such motions are, however, incompressible and do not change the radial component of the surface magnetic field B_r, i.e., one can assume $B_r = \mathrm{const}$ on a moving plate.

The real crust also has a finite electrical conductivity. B_r remains frozen in the crust as long as the timescale of field diffusion is longer than the star's age t. Calculations of electrical conductivity of the crust (Jones 2004) imply that this "frozen-in" stage lasts $\sim 10^6$ yr. During this stage, constant $B_{\mathrm{pc},r}$ implies that[1]

$$A_{\mathrm{pc}} \propto A_c B_d, \qquad t \ll 10^6\ \mathrm{yr}. \tag{6}$$

When the star becomes much older than 10^6 yr, the field evolution is dictated by the spin-down of the core, as if the star had no crust. Then the component of magnetic field parallel to $\mathbf{\Omega}$, $B(\|)$, changes $\propto \Omega$ following the expansion of the vortex array. In particular, the polar-cap magnetic field $B_{\mathrm{pc}}(\|) \propto \Omega$. Assuming, for a rough estimate, $B_{\mathrm{pc},r} \sim B_{\mathrm{pc}}(\|)$, one finds

$$A_{\mathrm{pc}} \propto B_d, \qquad t \gg 10^6\ \mathrm{yr}. \tag{7}$$

Note also that there is a relatively unexplored regime, lasting perhaps $t < 10^3$ yr, after core proton superconductivity has been established but before the neutron superfluid vortex array has formed. During it the flux-tubes pull on the crust may act to reduce B_d greatly. (One indicator of such a regime would be an observed pulsar spin-down age $P/2\dot{P}$ much greater than the true pulsar age inferred from the supernova remnant.)

[1] For simplicity, this estimate assumes that the bundle of open field lines remains pinned to the same crustal plate. This assumption may not hold in general for moving plates.

3 Millisecond pulsars

Slow spin-up of a NS to a millisecond pulsar squeezes all the core's flux-tubes, together with the NS polar caps, toward the spin axis. Then $A_{\rm pc}$ decreases. Spin-up from a spin-period $P \sim 10$ s to the minimum $P \sim 10^{-3}$ s by a companion-fed accretion disk gives a polar cap radius at the core-crust interface which is $\ll$ the crust thickness Δ. Then the polar cap area at the crust surface becomes

$$A_{\rm pc} \sim \pi \frac{\Omega R}{c} \Delta^2 \sim 10^{-2}\ {\rm km}^2. \tag{8}$$

This is about two orders of magnitude less than the conventionally assumed polar cap area: $A_c \sim \pi \Omega R^3/c \sim 1$ km^2 when $P \sim$ several ms. Such very small $A_{\rm pc}$ may be consistent with those reported here by Zavlin (these proceedings) for the almost aligned rotator J0437, and the two orthogonal rotators J0030 and J2124. There are, however, still considerable uncertainties in inferring observed $A_{\rm pc}$ from X-ray data.

4 Young neutron stars: other considerations

Several problems complicate the comparison of $A_{\rm pc}$ inferred from observations and the predictions of (6). One, of course, is identifying the emission from a hot polar cap in the presence of larger emissions from the stellar cooling and the power-law emission by $e^\pm$ continually created in the NS magnetosphere. There is also a large uncertainty in the expected spectrum of polar-cap emission: it is not a simple blackbody and it depends on composition and possible anomalies because of local fields and current flows. The inferred emission area increases if the observed spectrum is fitted by a hydrogen-atmosphere model (Pavlov et al. 2001).

Another problem may be the processing of surface emission by $e^\pm$ pairs continually created in the inner magnetosphere of a young neutron star. Pairs are created by curvature γ-rays from the inner- or outer-gap accelerators. Depending on the field geometry, the γ-rays may or may not propagate to the closed magnetosphere and $e^\pm$-populate the closed field lines. A photon with energy $\hbar\omega_X \sim 1$ keV will pass through an $e^\pm$ cyclotron resonance at a radius r such that $\omega_X \sim (eB_d/m_e c)(R/r)^3$, which gives r of several NS radii for typical $B_d \sim 10^{13}$ G. If the inner magnetosphere is efficiently populated with $e^\pm$, it may become opaque to surface X-rays because of resonant scattering.

The scattering greatly reduces the spin modulation of the observed X-rays. It would have little effect on their spectrum if $\gamma_\parallel$, the Lorentz factor of $e^\pm$ along $\mathbf{B}$, equals 1. One expects, however, $\gamma_\parallel > 1$ because the energy $E_\parallel = 2(\gamma_\parallel - 1)m_e c^2$ is left in the motion of a created pair after its fast synchrotron cooling. This parallel energy is transferred to the ambient X-rays through resonant scattering.

The number flux of surface X-rays $\dot{N}_X$ is not altered by the scattering process, but their luminosity L_X is increased: $L_X \to L_X({\rm surface}) + L_\parallel(e^\pm)$, where $L_\parallel(e^\pm)$ is the power deposited in the magnetosphere in the form of $E_\parallel$. If the processed X-rays are fit to a blackbody spectrum with given fixed $\dot{N}_X$ and the new L_X, the inferred emission area would scale as L_X^{-3}, i.e.

$$\frac{A_{\rm BB}({\rm inferred})}{A_{\rm BB}({\rm surface})} \sim \left[\frac{L_X({\rm surface})}{L_X({\rm surface}) + L_\parallel(e^\pm)}\right]^3. \tag{9}$$

For example, consider the processing of the cooling radiation from the entire star with area $A_{\rm NS} = 4\pi R^2$, neglecting the polar-cap contribution. The surface luminosity is given by

$$L_X({\rm surface}) \approx 10^{32}\left(\frac{t}{10^4\ {\rm yr}}\right)^{-0.3}\ {\rm erg\ s^{-1}}, \tag{10}$$

where t is the age of the star (Yakovlev et al. 2002). Suppose that $L_\parallel(e^\pm)$ scales with Goldreich-Julian current as $L_\parallel \propto I_{\rm GJ} \propto (B_d t)^{-1}$. In the simplest model with constant B_d, this would give

$$L_\parallel(e^\pm) = 10^{32} a \left(\frac{t}{10^4\ {\rm yr}}\right)^{-1}\ {\rm erg\ s^{-1}}, \tag{11}$$

where a is a dimensionless normalization parameter expected to be $\sim$1. Then (9) gives the inferred ratio $A_{\rm BB}/A_{\rm NS}$ as a function of star's age t, which seems consistent with the trend in the data on $A_{\rm BB}$ (Fig. 1) for $a \sim 2$. However, it can hardly explain all data. The predicted thermal luminosity $L_X = (t_4^{-0.3} + a t_4^{-1}) \times 10^{32}$ erg/s does not agree with the observed L_X. We also note that objects with smallest $A_{\rm BB}$ show weak or no nonthermal components, i.e. there is no sign of synchrotron radiation and pair creation in these objects.[2] Young neutron stars show significant diversity, and a single model predicting certain $A_{\rm BB}(t)$ and $L_X(t)$ cannot fit them all. The possibility of cyclotron scattering in some objects may contribute to this diversity. If it does, one might expect reductions in $A_{\rm pc}$ similar to (9).

Finally, we outline another possibility for forming hot spots on a young neutron star. The pull of the evolving core flux tubes on the crust can break it and create a shear motion of crustal plates. This motion generates $\nabla \times \mathbf{B}$ in the surface field, thereby initiating currents in the magnetosphere. The magnetosphere remains force-free, $\nabla \times \mathbf{B} \parallel \mathbf{B}$, so currents flow along $\mathbf{B}$, but it becomes non-potential. Ohmic dissipation of the currents at the footprints of a twisted magnetic loop can create hot spots.

[2] This does not exclude, however, significant pair production and synchrotron emission since this emission may be beamed away from our line of sight.

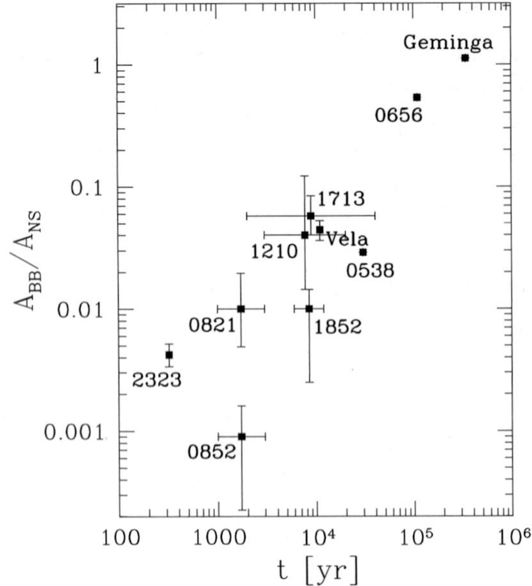

Fig. 1 The ratio of inferred blackbody emission areas (A_{BB}) to surface area $A_{NS} = 10^{13}$ cm^2 versus NS age t. The data are taken from Pavlov et al. (2001), Pavlov et al. (2004), and Zavlin and Pavlov (2004)

A similar model explains the coronal activity and surface heating of magnetars, with the induced voltage between the footprints of a loop $e\Phi_e \sim$ GeV (Beloborodov and Thompson 2007). Such voltage is required to maintain the creation of pairs that can conduct the current $\mathbf{j} = (c/4\pi)\nabla \times \mathbf{B}$ in the twisted magnetosphere. The magnetospheric particles deposit their energy $e\Phi_e$ in the surface layer at the loop footprint. For normal young NS, Φ_e will be different, especially if the closed magnetosphere is efficiently $e^{\pm}$-populated by curvature γ-rays coming from the open field lines. The resistivity of the upper crust may also contribute to dissipation of the twist current and ohmic heating of the footprint.

Acknowledgements A.M.B. is supported by NASA grant NNG-06-G107G.

References

Beloborodov, A.M., Thompson, C.: Astrophys. J. **657**, 967 (2007)
Caraveo, P.A., et al.: Science **305**, 376 (2004)
Jones, P.: Phys. Rev. Lett. **93**, 271101 (2004)
Pavlov, G.G., et al.: Astrophys. J. **552**, L129 (2001)
Pavlov, G.G., Sanwal, D., Teter, M.: In: Camilo, F., Gaensler, B. (eds.) Young Neutron Stars and Their Environments, IAU Symposium 218, p. 239. ASP, San Francisco (2004)
Ruderman, M.: Causes and consequences of magnetic field changes in neutron stars. astro-ph/0510623 (2005)
Yakovlev, D.G., et al.: In: Becker, W., Lesch, H., Trümper, J. (eds.) Proceedings of the 270 WE-Heraeus Seminar on Neutron Stars, Pulsars and Supernova Remnants, p. 287. Max-Plank-Institut für extraterrestrische Physik, Garching bei München (2002)
Zavlin, V.E., Pavlov, G.G.: Mem. Soc. Astron. Ital. **75**, 458 (2004)

Astrophys Space Sci (2007) 308: 423–430
DOI 10.1007/s10509-007-9358-2

ORIGINAL ARTICLE

Cooling of magnetars with internal layer heating

A.D. Kaminker · D.G. Yakovlev · A.Y. Potekhin · N. Shibazaki · P.S. Shternin · O.Y. Gnedin

Received: 1 July 2006 / Accepted: 13 October 2006 / Published online: 21 March 2007

Abstract We model thermal evolution of magnetars with a phenomenological heat source in a spherical internal layer and compare the results with observations of persistent thermal radiation from magnetars. We show that the heat source should be located in the outer magnetar's crust, at densities $\rho \lesssim 5 \times 10^{11}$ g cm^{-3}, and the heating rate should be $\sim 10^{20}$ erg cm^{-3} s^{-1}. Heating deeper layers is extremely inefficient because the thermal energy is mainly radiated away by neutrinos and does not warm up the surface to the magnetar's level. This deep heating requires too much energy; it is inconsistent with the energy budget of neutron stars.

Keywords Dense matter—stars · Magnetic fields—stars · Neutron—neutrinos

PACS 97.60.Jd · 95.85.Sz · 26.60.+c

A.D. Kaminker (✉) · D.G. Yakovlev · A.Y. Potekhin · P.S. Shternin
Ioffe Physico-Technical Institute, Politekhnicheskaya 26, 194021 Saint-Petersburg, Russia
e-mail: kam@astro.ioffe.ru

D.G. Yakovlev
e-mail: yak@astro.ioffe.ru

A.Y. Potekhin
e-mail: palex@astro.ioffe.ru

P.S. Shternin
e-mail: peter@astro.ioffe.ru

N. Shibazaki
Rikkyo University, 171-8501 Tokyo, Japan
e-mail: shibazak@rikkyo.ac.jp

O.Y. Gnedin
University of Michigan, 500 Church Street 949, Ann Arbor, MI 48109-1042, USA
e-mail: ognedin@umich.edu

1 Introduction

It is likely that soft gamma repeaters (SGRs) and anomalous X-ray pulsars (AXPs) belong to the same class of objects, *magnetars*, which are relatively hot, isolated slowly rotating neutron stars of age $t \lesssim 10^5$ yr with unusually strong magnetic fields, $B \gtrsim 10^{14}$ G (see, e.g., Woods and Thompson 2006, for a recent review). There have been numerous attempts to explain the activity of these sources and the high level of their X-ray emission by the release of the magnetic energy in their interiors, but a reliable theory is still absent.

In this paper we study the thermal evolution of magnetars as cooling isolated neutron stars. We do not develop a self-consistent theory of the magnetar thermal structure and evolution but use a simplified phenomenological approach which allows us to draw definite model-independent conclusions. We show that magnetars are too warm and cannot be purely cooling neutron stars; they should have some additional heating source, which we assume to operate in magnetar interiors. We determine the parameters of the heating source (its location and power) which are consistent with the high level of observed thermal radiation of SGRs and AXPs and with the energy budget of an isolated neutron star. We will mainly follow the consideration of our recent paper (Kaminker et al. 2006).

2 Observational data

For comparing theoretical calculations with observations, we select the same seven magnetar sources—two SGRs and five AXPs—as in Kaminker et al. (2006). We take the estimates of their spindown ages t, surface magnetic fields B and the blackbody surface temperatures T_s^∞ (redshifted for

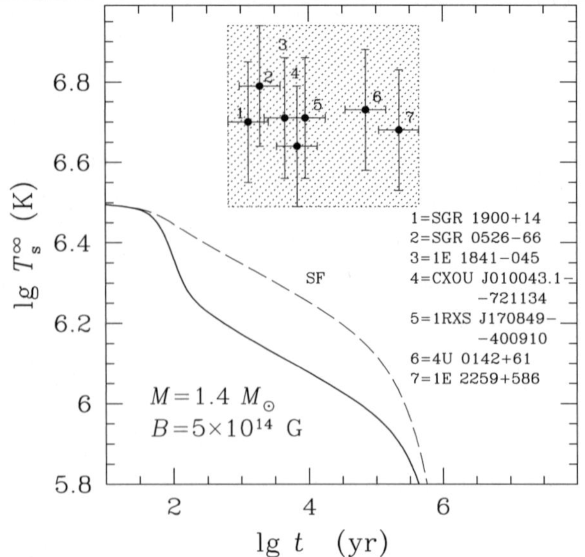

Fig. 1 Blackbody surface temperatures T_s^∞ and ages t of two SXTs and five AXPs (from Kaminker et al. 2006). The shaded rectangle is the "magnetar box". The observational data are compared to theoretical cooling curves of the $1.4 M_\odot$ neutron star with the magnetic field $B = 5 \times 10^{14}$ G and no internal heating sources; the star is either nonsuperfluid (the *solid line*) or has strong proton superfluidity in the core (the *dashed line SF*)

a distant observer) from Tables 14.1 and 14.2 of the review paper by Woods and Thompson (2006) and from the original paper by McGarry et al. (2005). The selected data are displayed in Fig. 1. The original publications which report these data are:

1. SGR 1900+14—Woods et al. (2001, 2002)
2. SGR 0526-66—Kulkarni et al. (2003)
3. 1E 1841-045—Gotthelf et al. (2002) and Morii et al. (2003)
4. CXOU J010043.1-721134—McGarry et al. (2005)
5. 1RXS J170849-400910—Gavriil and Kaspi (2002) and Rea et al. (2003)
6. 4U 0142+61—Gavriil and Kaspi (2002) and Patel et al. (2003)
7. 1E 2259+586—Gavriil and Kaspi (2002) and Woods et al. (2004)

Because the estimates of T_s^∞ for SGR 1627-41 are absent we do not include this SGR in our data set. Also, we do not include SGR 1806-20 and several AXPs whose thermal emission component and characteristic age are less certain. The pulsed fraction of radiation from the selected sources is $\lesssim$20%, and the pulsed fraction from some of them is even lower ($\lesssim$10%). This may mean that the thermal radiation can be emitted from a substantial part of the surface, although the pulsed fraction can be lowered by the gravitational lensing effect.

Figure 1 shows the blackbody surface temperatures T_s^∞ of the selected SGRs and AXPs versus their spindown ages t. Woods and Thompson (2006) present the values of T_s^∞ and t without formal errors, whereas the actual errors are certainly large. We introduce, somewhat arbitrarily, the 30% errorbars to the values of T_s^∞ and a factor of 2 uncertainties into the values of t. The data are too uncertain and our theoretical models are too simplified to study every source separately. Instead, we attempt to understand the existence of magnetars as cooling neutron stars that belong to the "magnetar box", the shaded rectangle in Fig. 1. We assume that the data reflect an average persistent thermal emission from magnetars (excluding bursting states). Two theoretical cooling curves in Fig. 1 are explained in Sect. 4.

3 Physics input

Calculations have been performed with our general relativistic cooling code (Gnedin et al. 2001). It models the thermal evolution of an isolated neutron star by solving the equations of heat diffusion within the star. The code takes into account heat outflow via neutrino emission from the entire stellar body and via energy transport within the star, leading to thermal photon emission from the surface. To simplify calculations, the star is artificially divided (e.g., Gudmundsson et al. 1983) into a thin outer heat blanketing envelope and the bulk interior. The blanketing envelope extends from the surface to the layer of density $\rho = \rho_b \sim 10^{10}$–$10^{11}$ g cm^{-3}; its thickness is of a few hundred meters. In the blanketing envelope the code uses the solution of the stationary thermal diffusion problem obtained in the approximation of a thin plane-parallel layer for a dipole magnetic field configuration (neglecting neutrino emission). This solution relates temperature T_b at the bottom of the blanketing envelope ($\rho = \rho_b$) to the effective surface temperature T_s averaged over the neutron star surface (e.g., Potekhin and Yakovlev 2001; Potekhin et al. 2003). In the bulk interior ($\rho > \rho_b$) the code solves the full set of equations of thermal diffusion in the spherically symmetric approximation, neglecting the effects of magnetic fields on thermal conduction and neutrino emission.

In the calculations we have mainly used the T_b–T_s relation obtained specifically for simulating the magnetar evolution. The relation has been derived assuming $\rho_b = 10^{10}$ g cm^{-3} and the magnetized blanketing envelope made of iron. (In principle, the envelope may contain lighter elements provided by accretion; we will consider this case elsewhere but do not expect to obtain qualitatively different results.) The calculations show (Sect. 4) that magnetars are hot inside and have large temperature gradients extending deeply within the heat blanketing envelope. Accordingly, even high magnetar magnetic fields do not drastically influence the average thermal flux emergent through the blanketing envelope and the T_b–T_s relation (see, e.g., Potekhin et al. 2003). For certainty, we mainly assume the dipole magnetic field in the blanketing envelope with $B = 5 \times 10^{14}$ G at the magnetic poles. Some variations of B will not change our principal conclusions (Kaminker et al. 2006).

We expect that an anisotropy of heat transport, induced by the magnetic field in a warm magnetar crust at $\rho > \rho_b$, is not too high to dramatically modify the temperature distribution in the bulk of the star and to produce large deviations of this distribution from spherical symmetry (Kaminker et al. 2006). At lower temperatures the anisotropy would be much stronger and the effects of the magnetic field in the stellar bulk could be much more significant (Geppert et al. 2004, 2006).

Our standard cooling code takes into account the effects of magnetic fields only in the heat blanketing envelope. In this paper we have also included, in a phenomenological way, the effects of magnetic fields on the thermal evolution in the bulk of the star. For this purpose we have introduced a heat source located within a spherical layer at $\rho > \rho_b$ and associated possibly with the magnetic field (Sect. 5). The heating rate H [$\mathrm{erg\,cm^{-3}\,s^{-1}}$] has been taken in the form

$$H = H_0 \Theta(\rho_1, \rho_2) \exp(-t/\tau), \tag{1}$$

where H_0 is the maximum rate, $\Theta(\rho_1, \rho_2)$ is a step-like function ($\Theta \approx 1$ in the heating layer, $\rho_1 < \rho < \rho_2$; and $\Theta \approx 0$ outside this layer, with a sharp but continuous transitions at the boundaries of the layer), t is the stellar age, and τ is the duration of the heating. A specific form of H as a function of ρ and t is unimportant for our main conclusions. We do not specify the nature of the heat source (although we discuss possible models in Sect. 5). In the majority of calculations we set $\tau = 5 \times 10^4$ years to explain high observed thermal states of all selected SGRs and AXPs (Sect. 4). We vary H_0, ρ_1 and ρ_2 (and we vary additionally τ in some runs), in order to understand which rate, location and duration of heat release are consistent with observations and with the energy budget of an isolated neutron star.

To quantify the energy generation within the star, we introduce the total heat power W^∞ [$\mathrm{erg\,s^{-1}}$], redshifted for a distant observer,

$$W^\infty(t) = \int \mathrm{d}V\, \mathrm{e}^{2\Phi} H, \tag{2}$$

where $\mathrm{d}V$ is the proper volume element; Φ is the metric function which determines gravitational redshift; and $H(\rho, t)$ is given by (1).

In the neutron star core we employ the equation of state of stellar matter proposed by Akmal et al. (1998) (their model Argonne V18 + δv + UIX*). This equation of state is currently thought to be the most elaborated equation of state of neutron-star matter. We use a convenient parameterization of this equation of state suggested by Heiselberg and Hjorth-Jensen (1999) and extended by Gusakov et al. (2005) (their version denoted as APR III). In this model, neutron star cores are composed of neutrons, protons, electrons, and muons. The maximum gravitational mass of stable neutron stars is $M = 1.929 M_\odot$. The powerful direct Urca process of neutrino emission (Lattimer et al. 1991) operates only in the inner cores of massive neutron stars with $M > 1.685 M_\odot$ (at densities $\rho > 1.275 \times 10^{15}\ \mathrm{g\,cm^{-3}}$).

We use neutron star models of two masses, $M = 1.4 M_\odot$ and $M = 1.9 M_\odot$. The $1.4 M_\odot$ model is an example of a neutron star with standard (not too strong) neutrino emission in the core; in a non-superfluid neutron star this neutrino emission is mainly produced by the modified Urca process. The circumferential radius of the $1.4 M_\odot$ star is $R = 12.27$ km, and the central density is $\rho_c = 9.280 \times 10^{14}\ \mathrm{g\,cm^{-3}}$. The $1.9 M_\odot$ model is an example of a neutron star with the neutrino emission greatly enhanced by the direct Urca process in the inner stellar core. For this star, $R = 10.95$ km, and $\rho_c = 2.050 \times 10^{15}\ \mathrm{g\,cm^{-3}}$.

4 Cooling calculations

Figure 1 shows the theoretical cooling curves $T_s^\infty(t)$ for the $1.4 M_\odot$ isolated magnetized neutron star without internal heating. The solid line is for a nonsuperfluid neutron star, and the dashed line SF is for a star with strong proton superfluidity in the core. This superfluidity greatly suppresses neutrino emission in the core which noticeably increases T_s^∞ at the neutrino cooling stage (e.g., Yakovlev and Pethick 2004). The surface temperature of these stars is highly nonuniform (the magnetic poles are much warmer than the equator); the figures show the average effective surface temperature (e.g., Potekhin et al. 2003).

The key problem is that the magnetars are much warmer than ordinary cooling neutron stars. The observations of ordinary neutron stars can be explained (within many different scenarios) by the cooling theory of neutron stars without additional heating sources (e.g., Yakovlev and Pethick 2004; Page et al. 2006). In contrast, the observations of magnetars imply that the magnetars have additional heating sources. We assume that these sources are located in the bulk of magnetars, at $\rho > \rho_b$. Note that there are alternative models which suggest that powerful energy sources operate in magnetar magnetospheres (Beloborodov and Thompson 2006).

Thus, to explain the observations of magnetars we have calculated a series of cooling models of neutron stars with $B = 5 \times 10^{14}$ G and internal heating sources given by (1). The results are presented in Figs. 2, 3, 4, 5 and 6. Our simulations with a powerful internal heating show that after a short initial relaxation ($t \lesssim 10$ years) the star reaches a quasi-stationary state. These states are governed by the heating source; the generated energy is mainly carried away by

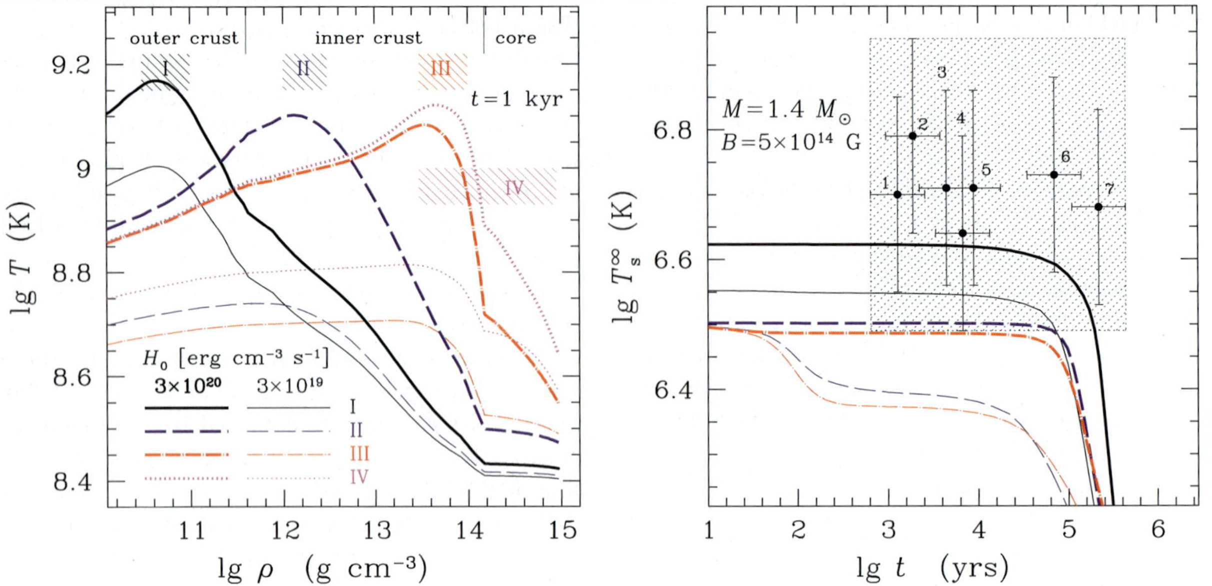

Fig. 2 *Left:* Temperature profiles within the neutron star of the mass $M = 1.4M_\odot$ and age $t = 1000$ years with four different positions *I–IV* of the heating layer (see text) shown by hatched rectangles and with two levels of the heating rate $H_0 = 3 \times 10^{19}$ and 3×10^{20} erg cm^{-3} s^{-1} (*thin* and *thick lines*, respectively). The magnetic field is $B = 5 \times 10^{14}$ G. *Right:* Cooling curves for the models with the heating layers *I–III* compared with the observations. From Kaminker et al. (2006)

neutrinos, although some fraction diffuses to the surface and radiates away by photons. The interior of these cooling neutron stars is essentially non-isothermal.

Our results are sufficiently insensitive to the neutron star mass, and we will mainly employ the neutron star model with $M = 1.4M_\odot$ (but present some results for the $1.9M_\odot$ star in Fig. 3). In addition, the results are not too sensitive to superfluidity in stellar interiors and we will mostly consider non-superfluid neutron stars. We assume the duration of the energy release $\tau = 5 \times 10^4$ years in all calculations excluding those presented in Fig. 4.

The left panel of Fig. 2 shows the temperature distributions inside the star with $M = 1.4M_\odot$ and $t = 1000$ years. The age $t = 1000$ years is taken as an example; the results are similar for all values of $t \lesssim \tau$. We have considered four locations of the heat layer, $\rho_1 - \rho_2$ (see (1)):

(I) 3×10^{10}–10^{11} g cm^{-3}(in the outer crust, just below the heat blanketing envelope)
(II) 10^{12}–3×10^{12} g cm^{-3}(at the top of the inner crust)
(III) 3×10^{13}–10^{14} g cm^{-3}(at the bottom of the inner crust) and
(IV) 3×10^{13}–9×10^{14} g cm^{-3}(at the bottom of the inner crust and in the entire core)

These locations are marked by hatched rectangles. We take two different heat rates, $H_0 = 3 \times 10^{19}$ and 3×10^{20} erg cm^{-3} s^{-1}, and the duration of the heat release $\tau = 5 \times 10^4$ yr.

As seen from the left panel of Fig. 2, the neutron star core is much colder than the crust because the core quickly cools down via neutrino emission (via the modified Urca process in our case). Placing the heat sources far from the surface does not allow us to maintain a high surface temperature. The heating layer can be hot, but the thermal energy is carried away by neutrinos and does not flow to the surface. For the deep heating layers (cases II, III, or IV), the heat rate $H_0 = 3 \times 10^{19}$ erg cm^{-3} s^{-1}is insufficient to warm the surface to the magnetar level. The higher rate 3×10^{20} erg cm^{-3} s^{-1}in these layers is more efficient but, nevertheless, it is much less efficient than in the layer I. For $H_0 = 3 \times 10^{20}$ erg cm^{-3} s^{-1}the heating of the crust bottom (case III) and the heating of the entire core (case IV) lead to the same surface temperature of the star. Therefore, the best way to warm the surface is to release the energy in the outer crust, close to the surface.

The right panel of Fig. 2 shows cooling curves of the stars with $M = 1.4M_\odot$ for the same models of the heating layer as in the left panel (without curves for the case IV; they are not presented to simplify the figure). The cooling curves of the star of age $t \lesssim \tau$ are almost horizontal. This indicates that the star is kept warm owing to the internal heating alone.

Figure 3 compares the thermal evolution of neutron stars with the two masses, $M = 1.4M_\odot$ and $M = 1.9M_\odot$. We consider the same three locations I–III of the heating layer and one heating rate $H_0 = 3 \times 10^{20}$ erg cm^{-3} s^{-1}. The thick lines correspond to the $1.4M_\odot$ star (they are the same as in the right panel of Fig. 2). The thin lines refer to the massive $1.9M_\odot$ star. Such a star has an extremely large neutrino luminosity produced by the direct Urca process operating in the inner core (Sect. 3). If the heating layer is

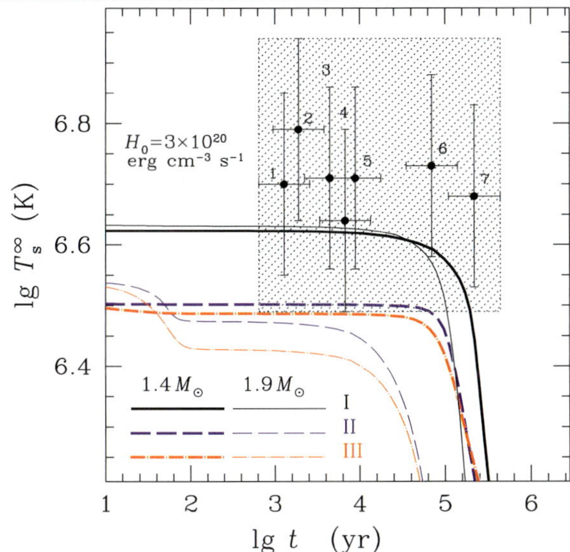

Fig. 3 Same as in the right panel of Fig. 2 for the three positions of the heating layer (*I–III*), one value of the heat rate and two neutron star masses, $1.4M_\odot$ (*thick lines*) and $1.9M_\odot$ (*thin lines*). From Kaminker et al. (2006)

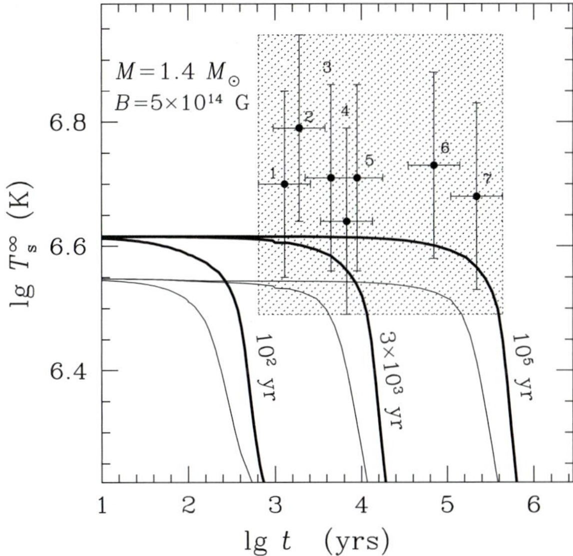

Fig. 4 Cooling curves of $1.4M_\odot$ neutron star with the heating layer in the outer crust (case I) for the same two heating rates $H_0 = 3 \times 10^{19}$ erg cm^{-3} s^{-1}(*thin lines*) and 3×10^{20} erg cm^{-3} s^{-1}(*thick lines*) as in Fig. 2 but for three values of the heat-release duration, $\tau = 10^2$, 3×10^3, and 10^5 years

located in the inner crust (cases II and III), the enhanced neutrino emission of the massive star induces rapid neutrino cooling and noticeably decreases the surface temperature of the star. If, in contrast, the heating layer is located in the outer crust (case I), a strong neutrino emission from the core weakly affects the surface temperature. The surface temperature becomes nearly insensitive to the physics of the stellar core and the inner crust, in particular, to the neutrino emission processes and superfluidity of matter. This means that the surface layers are thermally decoupled from the inner crust and the core. The same happens in ordinary young and hot cooling stars, before their internal thermal relaxation is over (Lattimer et al. 1994; Gnedin et al. 2001). The thermal decoupling in magnetars justifies our consideration of non-superfluid neutron stars. Notice, however, that the effects of superfluidity can be crucial for the cooling of ordinary neutron stars; e.g., Yakovlev and Pethick (2004), Page et al. (2006).

Our results demonstrate that for explaining the observations of magnetars one should locate the heating source in the outer crust and assume the heating rate $H_0 \gtrsim 10^{20}$ erg cm^{-3} s^{-1}. According to (1), the heating rate decays exponentially at $t \gtrsim \tau$, and the surface temperature drops down in response. This means the end of the magnetar stage—the star transforms into an ordinary neutron star cooling predominantly via the surface photon emission (e.g., Potekhin et al. 2003).

Figure 4 demonstrates the dependence of the cooling curves on τ, the duration of the heating process. The heating source is located in the outer crust. Since our magnetar models are wholly supported by the heating, they become cool as soon as the heating is switched off. The models with a short heating stage, $\tau = 100$ and 3000 years, cannot explain all the sources from the "magnetar box", while the models with $\tau \gtrsim 10^4$–10^5 years can explain them. Longer τ would require too large energy, in contradiction with the energy budget of neutron stars (Sect. 5). Therefore, the value $\tau \sim 5 \times 10^4$ years accepted for the majority of our cooling models seems to be optimal.

Figure 5 illustrates the sensitivity of magnetar cooling to the values of the thermal conductivity in the inner neutron-star crust. We present temperature profiles in the star with $M = 1.4M_\odot$ at $t = 1000$ years and cooling curves of this star for the heating layer that is located in the outer crust (case I) and has the heating rate $H_0 = 3 \times 10^{20}$ erg cm^{-3} s^{-1}. The thick lines are the same as in Fig. 2; they are calculated using our standard cooling code, which includes only the electron thermal conductivity in the crust (Gnedin et al. 2001). However, the inner crust contains not only electrons and atomic nuclei, but also free neutrons. As a consequence, thermal energy can also be transported by free neutrons. This neutron transport can be efficient, especially if neutrons are superfluid. The effect may be similar to the well known effect in superfluid ^{4}He, where no temperature gradients can be created in laboratory experiments because these gradients are immediately smoothed out by convective flows (e.g., Tilley and Tilley 1990). To simulate such an effect we have artificially introduced the layer of high thermal conduction in the inner crust, at densities from 3×10^{12} g cm^{-3} to 10^{14} g cm^{-3}, where we amplified the thermal conductivity (arbitrarily, for illustration) by a factor

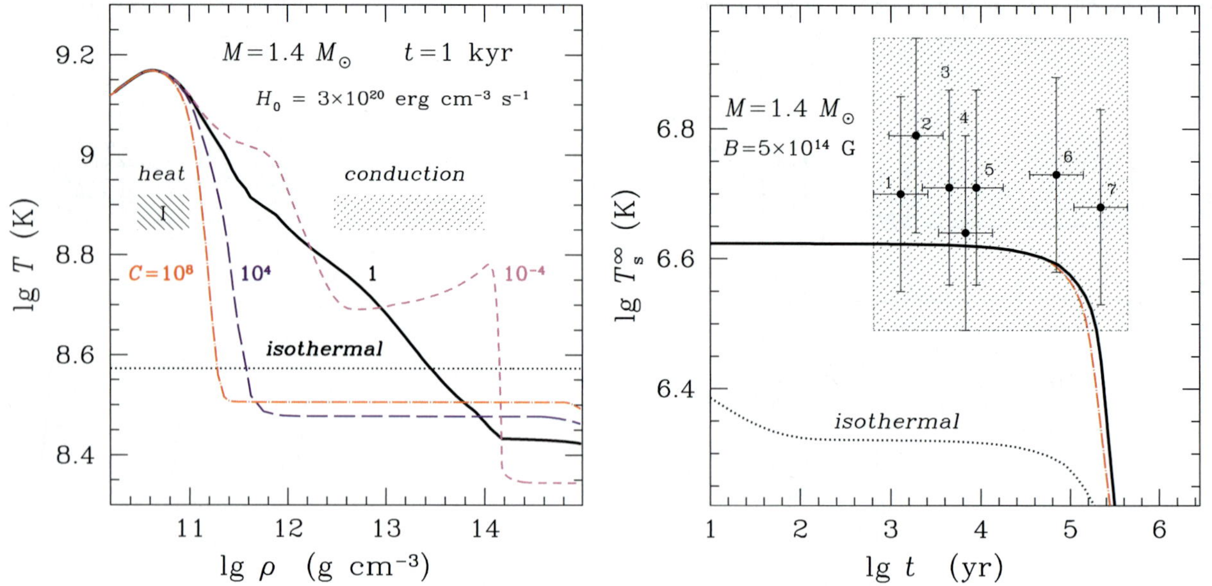

Fig. 5 The influence of thermal conduction in the inner crust on the thermal structure and evolution of the 1.4 $M_\odot$ magnetar with the heating layer in the outer crust (case I) and the heating rate $H_0 = 3 \times 10^{20}$ erg cm^{-3} s^{-1}. *Left:* The temperature profiles in the magnetar at $t = 1000$ years. The hatched rectangles show the positions of the heating layer and the layer, where the thermal conductivity was modified. *Right:* The cooling curves. The *thick solid lines* are the same as in Fig. 2. *Thinner long-dash, dot-and-dash,* and *short-dash* lines are for the star with the thermal conductivity modified by a factor of $C = 10^4$, 10^8, and 10^{-4}, respectively. The *dotted line* (marked *isothermal*) is for an infinite thermal conductivity at $\rho > 10^{10}$ g cm^{-3}. From Kaminker et al. (2006)

of $C = 10^4$ or 10^8. The amplification drastically changes the thermal structure of the inner crust, making it much cooler and almost isothermal. In addition, we have made a test run reducing artificially the thermal conductivity in the same layer by a factor of 10^4. This corresponds to $C = 10^{-4}$ and it mainly warms up the inner crust. Nevertheless, all these significant changes of the thermal state of the inner crust have almost no effect on the surface temperature and the cooling curves (which give another manifestation of the thermal decoupling of the outer crust from the inner stellar regions). Finally, we have simulated the neutron star cooling in the approximation of infinite thermal conductivity in the star bulk ($\rho > \rho_b = 10^{10}$ g cm^{-3}). Then the released heat is instantly spread over the star bulk and makes the stellar surface much cooler than in the case of finite conduction.

Figure 6 shows the integrated heating rate W^∞ given by (2) and the photon thermal surface luminosity of the star L_γ^∞ versus parameters of the heating layers. In the left panel we take three locations of the heating layer (I, II, III) and vary the heating rate H_0. One can see that only the heating of the layer I can produce $L_\gamma^\infty \gtrsim 3 \times 10^{35}$ erg s^{-1}, typical for magnetars. Moreover, the surface luminosity increases with H_0 much slower than W^∞. For $H_0 \gtrsim 10^{20}$ erg cm^{-3} s^{-1}and the layers II and III, the luminosity clearly saturates, so that pumping more energy into the heating layer does not affect L_γ^∞. The efficiency of converting the input heat into the surface photon emission ($L_\gamma^\infty / W^\infty$) is generally small. The highest efficiency is achieved if we warm up the outer crust (the layer I) at a low rate.

In the right panel of Fig. 6 we present L_γ^∞ and W^∞ as a function of the maximum density ρ_2 of the heating layer, for one value of the heating rate $H_0 = 10^{20}$ erg cm^{-3} s^{-1}and three fixed minimum densities of the heating layer ($\rho_1 = 3 \times 10^{10}$, 10^{12}, and 3×10^{13} g cm^{-3}). One can see the saturation of L_γ^∞ with growing ρ_2. If the heating layer extends into the stellar core, the integrated heating rate W^∞ is enormous, but this enormous energy is almost completely emitted by neutrinos.

Figure 6 compares theoretical values of L_γ^∞ with the thermal surface luminosities from the "magnetar box" (the lower shaded strip, estimated using the adopted values of T_s^∞ and the 1.4$M_\odot$ neutron star model). On the one hand, the heating should be intense to raise L_γ^∞ to the magnetar values. On the other hand, it is tacitly assumed that the energy W^∞ is persistently deposited into the heating layer during magnetar's life. As a consequence, the total deposited energy $E_{\rm tot}$ has to be restricted (cannot exceed the energy that can be stored in a neutron star). Let us assume further that the maximum energy of the internal heating is $E_{\rm max} \sim 10^{50}$ erg. Then the maximum energy generation rate is $W_{\rm max} \sim E_{\rm max}/\tau \sim 3 \times 10^{37}$ erg s^{-1}, which is plotted by the upper horizontal solid line in Fig. 6. In this case the upper shaded space above this line is prohibited by the neutron-star energy budget.

A successful theory of magnetars as cooling neutron stars with internal heating has to satisfy two principal require-

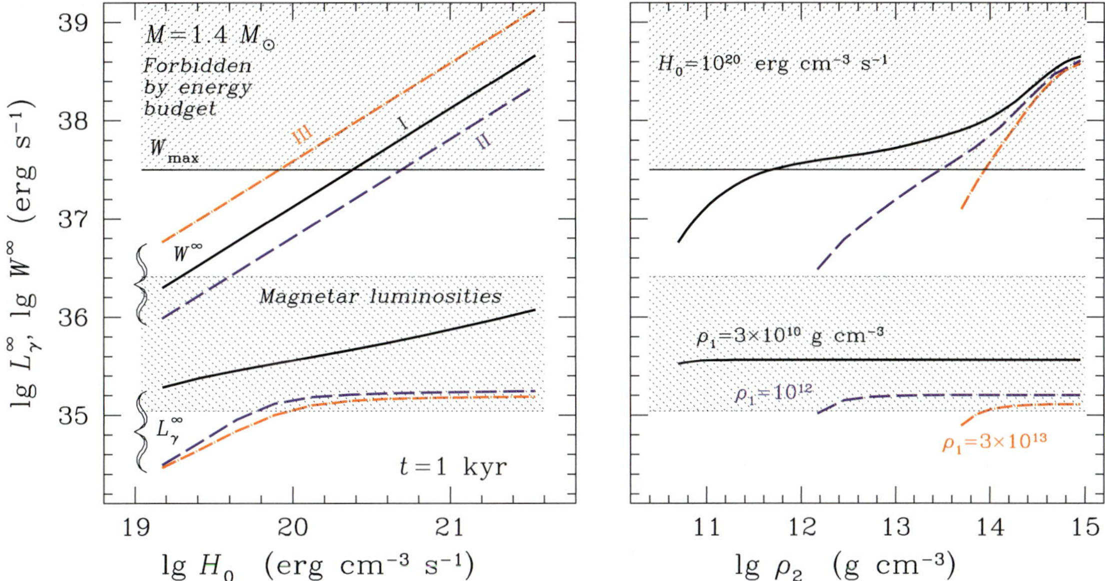

Fig. 6 The total heating power W^∞ (*higher curves*) and the surface photon luminosity L_γ^∞ (*lower curves*) versus parameters of the heating layer compared to the values of L_γ^∞ from the "magnetar box" (the *lower shaded strip*) and to the values of W^∞ forbidden by energy budget (the *upper shading*) for the neutron star with $M = 1.4M_\odot$ and $t = 1000$ years. *Left:* Three fixed positions of the heating layer of variable heating rate H_0. *Right:* Three fixed minimum densities ρ_1 of the heating layer, the fixed heating rate $H_0 = 10^{20}$ erg cm^{-3} s^{-1}, and variable maximum density ρ_2. From Kaminker et al. (2006)

ments. First, L_γ^∞ has to be sufficiently large to reach the "magnetar box"; and second, W^∞ has to be sufficiently low to avoid the prohibited region. These requirements can be reconciled for the heating source which is located in the outer stellar crust and has the heating rate H_0 between 3×10^{19} and 3×10^{20} erg cm^{-3} s^{-1}. A typical efficiency of heat conversion into the surface emission under these conditions is $L_\gamma^\infty / W^\infty \sim 10^{-2}$.

We have tested the sensitivity of our theoretical cooling curves to the level of neutrino emission in the magnetar core and crust. Variations of the neutrino emissivity in the inner crust and the core of a magnetar heated in the outer crust can strongly modify the internal thermal structure of the star but have almost no effect on the surface temperature and photon luminosity L_γ^∞. In contrast, L_γ^∞ is sensitive to the neutrino emission in the outer crust. In a warm outer crust, the neutrino emission is mainly produced by plasmon decay and electron-nucleus bremsstrahlung (see, e.g., Yakovlev et al. 2001). Very strong magnetic fields in magnetar crusts can greatly influence the plasmon decay process (which has not been studied in detail).

5 Discussion and conclusions

We have calculated the thermal structure and evolution of magnetars—SGRs and AXPs—in attempt to explain high surface temperatures and energy budget of these highly magnetized neutron stars. We have tried to present robust results independent of any specific theoretical model of the internal heating (available models are reviewed, e.g., by Woods and Thompson 2006 and Heyl 2006). We expect that our results place stringent constraints on possible models. Our main conclusions are:

(A) If the heating source is located inside the neutron star, it must be close to the surface, in the outer crust, at densities $\rho \lesssim 5 \times 10^{11}$ g cm^{-3}, and the heating rate should range from $\sim 3 \times 10^{19}$ to 3×10^{20} erg cm^{-3} s^{-1}. Were the heating source located deeper in the star, the heating energy would be radiated away by neutrinos, and it would be impossible to warm up the surface. This deeper heating would be very inefficient and would require more energy than a neutron star can have.

(B) Heating sources in the outer crust create a strongly heterogeneous temperature distribution within the neutron star. The temperature in the heating layer is higher than 10^9 K, but the deeper interior stays much colder. The thermal structure of the heating layer and the temperature of the magnetar surface are nearly insensitive to such physical parameters of the core and the inner crust as the equation of state, neutrino emission, thermal conductivity, superfluidity of baryons. This means thermal decoupling of the outer crust from the deeper layers. The total energy released in the heating layer during the magnetar lifetime ($\sim 10^4$–10^5 years) cannot be lower than 10^{49}–10^{50} erg; maximum 1% of this energy can be radiated by photons from the surface. This does not necessarily mean that the energy is stored in the outer crust—it must be converted into heat there.

Our results support the widely spread suggestion that the magnetars are not powered by their rotation, or by accretion, or by thermal energy accumulated in a cooling neutron star, or by strain energy accumulated in the crust. All these energy sources contain much less than 10^{49} erg required to explain the existence of the magnetars.

However, the necessary energy can be stored in the magnetic field if, for instance, the neutron star has a field $B \sim (1–3) \times 10^{16}$ G in its core. The Ohmic decay of this magnetic field can be accompanied by a strong energy release in the outer magnetar crust, where the electric conductivity is especially low and the Ohmic dissipation is strong. The magnetic field structure in the star can be much more complicated than magnetic dipole, which can further enhance the Ohmic dissipation in the outer crust. There could be other mechanisms of the magnetic energy release in the outer crust. For instance, rearrangements of the internal magnetic field in the course of the magnetar evolution can lead to the generation of waves (perturbations). These waves can propagate toward the stellar surface, decay in the outer crust and warm up this crust. Some mechanisms of wave generation are outlined by Thompson and Duncan (1996).

It is possible that the thermal radiation of magnetars is emitted from a smaller fraction of the magnetar surface, for instance, from hot spots around magnetic poles. Then the total heating energy could be lower. However, observed thermal X-ray luminosities of magnetars $\sim 10^{34}$–10^{36} erg s^{-1} (e.g., Mereghetti et al. 2002; Kaspi and Gavriil 2004; Woods and Thompson 2006) are in a reasonable agreement with the range of thermal luminosities in Fig. 6, which are calculated assuming the emission from the entire surface.

There is still no solid theory of the internal magnetar heating. Our basic results are that the heating energy released in magnetars must be at least two orders of magnitude higher than the photon thermal energy emitted through their surface, and the energy release should occur in the outer stellar crust. These conclusions are model-independent and stem from the well-known principle that hot stellar objects are strong sources of neutrino emission.

Also, let us point out alternative theories of magnetars, which assume (e.g., Beloborodov and Thompson 2006) that the main energy release takes place in the magnetar's magnetosphere, and the radiation spectrum is formed there (owing to comptonization and reprocession of the quasi-thermal spectrum).

Acknowledgements A.D.K. and A.Y.P. are grateful to the organizers of the conference "Isolated Neutron Stars: from the Interior to the Surface" for careful attention and financial support. Work supported in part by the Russian Foundation for Basic Research (grants 05-02-16245, 05-02-22003), by the Federal Agency for Science and Innovations (grant NSh 9879.2006.2), by the Rikkyo University Invitee Research Associate Program, and by the Dynasty Foundation.

References

Akmal, A., Pandharipande, V.R., Ravenhall, D.G.: Rhys. Rev. C **58**, 1804 (1998)

Beloborodov, A.M., Thompson, C.: Astrophys. J. (2006, submitted), astro-ph/0602417

Gavriil, F.P., Kaspi, V.M.: Astrophys. J. **567**, 1067 (2002)

Geppert, U., Küker, M., Page, D.: Astron. Astrophys. **426**, 267 (2004)

Geppert, U., Küker, M., Page, D.: Astron. Astrophys. **457**, 937 (2006)

Gnedin, O.Y., Yakovlev, D.G., Potekhin, A.Y.: Mon. Not. Roy. Astron. Soc. **324**, 725 (2001)

Gotthelf, E.V., Gavriil, F.P., Kaspi, V.M., Vasisht, G., Chakrabarty, D.: Astrophys. J. **564**, L31 (2002)

Gudmundsson, E.H., Pethick, C.J., Epstein, R.I.: Astrophys. J. **272**, 286 (1983)

Gusakov, M.E., Kaminker, A.D., Yakovlev, D.G., Gnedin, O.Y.: Mon. Not. Roy. Astron. Soc. **363**, 555 (2005)

Heiselberg, H., Hjorth-Jensen, M.: Astrophys. J. **525**, L45 (1999)

Heyl, J.: In: Proc. XXII Texas Symposium on Relativistic Astrophysics (2006, in press), astro-ph/0504077

Kaminker, A.D., Yakovlev, D.G., Potekhin, A.Y., Shibazaki, N., Shternin, P.S., Gnedin, O.Y.: Mon. Not. Roy. Astron. Soc. **371**, 477 (2006)

Kaspi, V.M., Gavriil, F.P.: Nuc. Phys. B (Proc. Suppl.) **132**, 456 (2004)

Kulkarni, S.R., Kaplan, D.L., Marshall, H.L., Frail, D.A., Murakami, T., Yonetoku, D.: Astrophys. J. **585**, 948 (2003)

Lattimer, J.M., Pethick, C.J., Prakash, M., Haensel, P.: Phys. Rev. Lett. **66**, 2701 (1991)

Lattimer, J.M., Van Riper, K.A., Prakash, M., Prakash, M.: Astrophys. J. **425**, 802 (1994)

McGarry, M.B., Gaensler, B.M., Ransom, S.M., Kaspi, V.M., Veljkovik, S.: Astrophys. J. **627**, L137 (2005)

Mereghetti, S., Chiarlone, L., Israel, G.L., Stella, L.: In: Becker, W., Lesch, H., Trümper, J. (eds.) 270 WE-Heraeus Seminar on Neutron Stars, Pulsars and Supernova Remnants, MPE Report 278, Garching, p. 29 (2002)

Morii, M., Sato, R., Kataoka, J., Kawai, N.: Publ. Astron. Soc. Jpn. **55**, L45 (2003)

Page, D., Geppert, U., Weber, F.: Nucl. Phys. A **777**, 497 (2006)

Patel, S.K., Kouveliotou, C., Woods, P.M., Tennant, A.F., Weisskopf, M.C., Finger, M.H., Wilson, C.A., Göğüs, E., van der Klis, M., Belloni, T.: Astrophys. J. **587**, 367 (2003)

Potekhin, A.Y., Yakovlev, D.G.: Astron. Astrophys. **374**, 213 (2001)

Potekhin, A.Y., Yakovlev, D.G., Chabrier, G., Gnedin, O.Y.: Astrophys. J. **594**, 404 (2003)

Rea, N., Israel, G.L., Stella, L., Oosterbroek, T., Mereghetti, S., Angelini, L., Campana, S., Covino, S.: Astrophys. J. **586**, L65 (2003)

Thompson, C., Duncan, R.C.: Astrophys. J. **473**, 322 (1996)

Tilley, D.R., Tilley, J.: Superfluidity and Superconductivity. IOP, Bristol (1990)

Woods, P.M., Thompson, C.: In: Lewin, W.H.G, van der Klis, M. (eds.) Compact Stellar X-ray Sources, pp. 547. Cambridge University Press, Cambridge (2006)

Woods, P.M., Kouveliotou, C., Göğüs, E., Finger, M.H., Swank, J., Smith, D.A., Hurley, K., Thompson, C.: Astrophys. J. **552**, 748 (2001)

Woods, P.M., Kouveliotou, C., Göğüs, E., Finger, M.H., Swank, J., Markwardt, C.B., Hurley, K., van der Klis, M.: Astrophys. J. **576**, 381 (2002)

Woods, P.M., Kaspi, V.M., Thompson, C., Gavriil, F.P., Marshall, H.L., Chakrabarty, D., Flanagan, K., Heyl, J., Hernquist, L.: Astrophys. J. **605**, 378 (2004)

Yakovlev, D.G., Pethick, C.J.: Annu. Rev. Astron. Astrophys. **42**, 169 (2004)

Yakovlev, D.G., Kaminker, A.D., Gnedin, O.Y., Haense, P.: Phys. Rep. **354**, 1 (2001)

Astrophys Space Sci (2007) 308: 431–434
DOI 10.1007/s10509-007-9314-1

ORIGINAL ARTICLE

What do exotic equations of state have to offer?

J.E. Horvath

Received: 12 July 2006 / Accepted: 11 September 2006 / Published online: 21 March 2007

Abstract We present a short general overview of the main features of exotic models of neutron stars, focusing on the structural and dynamical predictions derived from them. In particular, we discuss the presence of "normal" quark matter and Color-Flavor Locked (CFL) states, including their possible self-bound versions, and mention some different proposals emerging from the study of QCD microphysics. A connection with actual observed data is the main goal to be addressed at this talk and along the meeting. It is demonstrated that exotic equations of state are *not* soft if the vacuum contributions are large enough, and argued that recent measurements of high pulsar masses ($M \geq 2M_{\odot}$) create problems for *hadronic* models in which hyperons should be present.

Keywords · Pulsars · Internal structure · Quarks

PACS 97.60.Jd · 12.38.Mh

1 Introduction

The quest of the internal composition of compact stars has been going on for decades, in close connection with the work in nuclear and particle physics. Although deemed in some sense "simpler" than magnetospheric phenomena, dealing with matter at the extreme conditions inside pulsars/neutron stars has never been easy. As 2006, the consensus about the nature of matter at several times the nuclear saturation density is weak, if anything. Several phases/components of the nuclear fluid have been proposed and studied, but decisive evidence for or against them is hard to obtain. In fact, and as discussed several times during this meeting, advances on the observational side has allowed one for the first time to probe key macroscopic properties of compact stars (masses, radii and a few others) that reflect the internal composition indirectly. However, a few remarks on this last statement are in order: on the one hand the precision attained by measurements has improved greatly but *not* to the point far beyond any suspicion, and on the other hand, supranuclear components with poorly constrained parameters (coupling constants, vacuum expectation values, etc.) will not be ruled out, even by very precise observations. Therefore, work is needed on the theoretical and terrestrial laboratories as well.

Overall, I believe it is fair to state that the bulk structural properties are well-known in the subnuclear domain (more strictly, below the neutron drip density $\sim 10^{11}$ g cm^{-3}). However, important questions involving magnetic fields still remain (van Adelsberg et al. 2005). This is quite important to settle since it is where the star surface is seen by experiments due to electromagnetic radiation (see discussions about spectral lines and related topics in this meeting). There is also important information in timing irregularities, most notably glitches, believed to originate at the inner crust (although the "conventional wisdom" has been recently challenged, see the contribution by B. Link in this meeting and references therein). However, deep below the stellar crust it is increasingly difficult to construct a clear picture of the composition and therefore of the stellar structure. Condensates (π^-, K^-, etc.) are still possible depending on microphysics and drastically alter the thermal and dynamical behavior of the star. Fundamental degrees of free-

The author would like to acknowledge the financial support of CNPq (Brazil).

J.E. Horvath (✉)
IAG, Department of Astronomy, Universidade de São Paulo, Rua do Matão 1226, 05508-900 Sao Paulo, SP, Brazil
e-mail: foton@astro.iag.usp.br

dom (i.e. quarks) may also constitute the main part of the core, and even most of the star if they happen to be of the "self-bound" type (i.e. do not decay back to ordinary nuclear species once formed). Recently, a lot of attention has been paid to paired quark matter in a variety of phases still being studied. A recent summary of these matters can be found in Weber (2005). Hereafter we shall mainly concentrate on the self-bound phases because they bring the greatest modifications to the structure and they remain viable alternatives to the supranuclear matter composition.

2 Strange quark matter and its paired version

Strange quark matter (SQM) is an extreme version of a cold quark plasma in which, by hypothesis, the energy per baryon number unit is selected (with the chosen parameters) to fall below the mass of the nucleon. This possibility of having matter so strongly bound that it does not wish to return to the normal hadronic state was first discussed by Bodmer (1971), rediscovered by Terazawa (1979) and finally relaunched colorfully by Witten (1984) more than 20 years ago. Many studies, both experimental and theoretical, devoted to the SQM hypothesis have produced interesting results and some controversial arguments *against* its existence, but with loopholes in them. A few candidates have appeared to SQM in cosmic rays (Björken and McLerran 1979; Ichimura et al. 1993; Choutko 2003; Madsen 2005) with fluxes consistent with astrophysical injection scenarios (i.e. merging of compact stars, supernovae). SQM formation on $\tau \leq 1$ s timescale has been studied (Benvenuto and Horvath 1989; Lugones et al. 1994; Dai et al. 1995) and tentatively related to core-collapse supernovae, perhaps driven by photons (Chen and Xu 2006) instead of mechanical energy transfer or neutrinos.

The existence of SQM would be important for compact objects, since within this picture *all* of them should be "strange stars" instead of neutron stars. However, there is still the issue of the timescale for the conversion, since while in supernovae models the latter is quite short, it could be stretched by several orders of magnitude depending on microphysical details (see, for example, Lugones and Bombaci 2005). For the static, non-rotating structure, strange stars are constructed by integrating the Tolman–Oppenheimer–Volkoff equation with an equation of state of the form

$$P = \frac{1}{3}(\rho - 4B) \tag{1}$$

which has been extensively used because of its proximity with more detailed calculations including the finite s-quark mass and quark–quark interactions. The importance of the vacuum term, here written as $4B$ in the spirit of the well-known MIT bag model (Degrand et al. 1975) which produces a zero point pressure at finite (and large) energy density can not be overstated: the Bodmer–Witten–Terazawa hypothesis would preclude a "normal" matter crust in contact with it (thus limiting its total mass) unless a structured form of the quark matter itself is present (Benvenuto et al. 1990; Heiselberg et al. 1993; Alford et al. 2006a) and may be responsible for phenomena commonly attributed to the inner crust (i.e. glitches). The alternative is a "floating" normal crust supported by electrostatic forces, and then necessarily quite light, perhaps too light to produce the observed phenomenology.

Quite independently of these considerations, there is a widespread belief that, because of its underlying free quark derivation, an equation of state (EOS) like (1) must be very soft. This is far from being true: the issue of the softness/stiffness is rather related to the vacuum energy term, the real agent which determines the hardness of the EOS. We shall give examples of this behavior when discussing the stellar models.

A lot of activity has recently been seen on the effects of pairing interactions in dense quark matter. The issue is not new, since in the early '80s a few works addressed the superfluid/superconducting properties of paired quarks (Bailin and Love 1984 and references therein). However, those approaches were based on perturbative schemes, and therefore obtained (quite consistently) gaps of the order of 1 MeV or so, much smaller than the natural scales of the problem (say, the quark chemical potential). The recent works (Alford and Cowan 2006; Gómez Dumm et al. 2006 and references therein) have tried to calculate the phase diagram more directly, without resorting to perturbative schemes. As a result, several pairing possibilities (u and d quarks only, 2SC phase; all u, d and s quarks at a common Fermi momentum—not energy!—, the CFL state) were found with gaps as large as 100 MeV. Other possibilities, like a gapless CFL phase or the solid-like Larkin–Ovchinnikov–Fulde–Ferrell (LOFF) state are being considered and reflect a high complexity of the QCD phase diagram that might be important for compact stars (Ruster et al. 2004).

While this task continues, it is perhaps worthwhile to remark the importance of pairing energies for the stability issue discussed above: it is found in simple models (Lugones and Horvath 2003) that the stability window (i.e. the place in parameter space inside which paired CFL matter would be stable) is greatly enhanced when compared with the same parameters for unpaired matter. Figure 1 displays the situation for a model with constant gaps which were varied within the expected range. Perhaps pairing is a big clue to the ground state of matter relevant to astrophysics after all. This is the meaning of the (unimaginative) name "CFL strange matter".

3 Effects on stellar models

The general trend of stellar models calculated with self-bound equations of state is well-known: in sharp contrast

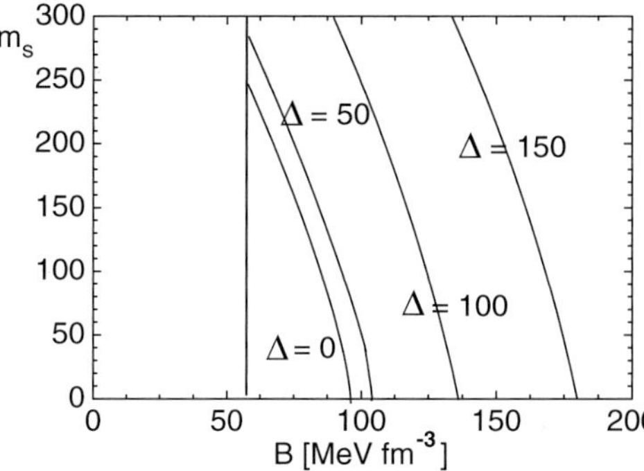

Fig. 1 The stability windows for CFL strange matter. If the strange quark mass m_s and the bag constant B lie inside the bounded region the CFL state is absolutely stable. Each window is plotted for $P=0$, the stability window is the region between the *vertical line* (obtained by requiring instability of two-flavor quark matter) and the *curve* with a given value of the gap Δ as indicated by the label. Note the enlargement of the window with increasing Δ. See Lugones and Horvath (2002) for details

with neutron matter calculations, for which R grows to $\sim$100 km for finite small baryonic mass values, self-bound stars can be found, in principle, down to tennis-ball sizes continuously (i.e. $R \to 0$ when $M \to 0$) and beyond, inside the realm of "strangelets". This is because binding comes from strong interactions and not from gravity. Of course, there *is* a Chandrasekhar mass for SQM or CFL strange matter sequences, which brings us to the question of the vacuum energy again.

In its simplest form of (1) it is well-known that stellar models at the maximum mass scale as $B^{1/2}$. This property is related to the linearity of the EOS, and holds approximately if the latter is not strict. When pairing energy is present, the free energy of the paired mixture $\Omega_{\rm CFL}$ is smaller than the unpaired version by a term quadratic both in the gap Δ and the chemical potential μ (Alford and Reddy 2003)

$$\Omega_{\rm CFL} = \Omega_{\rm free} - \frac{3}{\pi^2}\Delta^2\mu^2 + B. \qquad (2)$$

All the important thermodynamic quantities can be derived from (2), which provides an equation of state for the CFL mixture. With that ingredient it is immediate to calculate stellar sequences of cold stars composed by this self-bound version of quarks. In this approach the gap has been assumed as a constant, in fact theoretical expectations strongly suggest that a functional dependence ensures, but at this time it is not possible to state anything reasonable and quantitative about its nature. The maximum mass along the sequence is shown in Fig. 2, and increases with increasing gaps. Of course, there must be an upper limit to the pairing energy gain in nature, otherwise all matter would decay into the more bound state (as pointed out by P. Haensel during the meeting), but this limit is not obvious and must be calculated consistently for each considered model (see Fig. 1). One important point to note here is the rather high values for

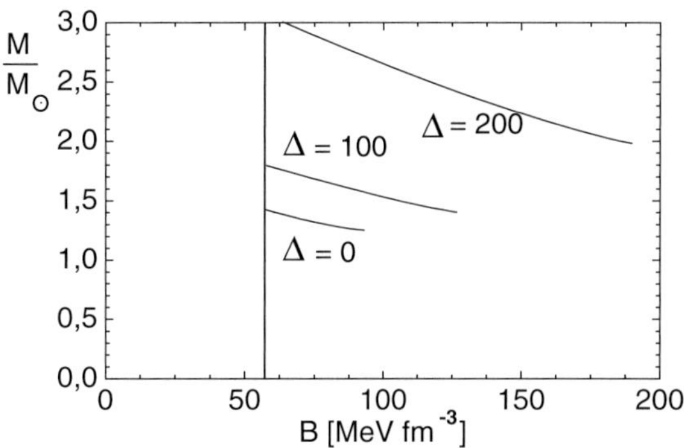

Fig. 2 The maximum mass of each stellar sequence is shown here as a function of B, for $m_s = 150$ MeV and different values of Δ, the range of the latter is the same as in Fig. 1

the maximum mass along the sequence that can be obtained for gap values deemed quite modest (i.e. $M_{\rm max} \geq 2M_\odot$ for $\Delta \sim 100$ MeV). This means that the EOS is *not* soft, but rather stiff whenever the effective vacuum energy is large enough (composed in these models by a combination of the true vacuum and the condensation energy together).

With the use of analytical general relativistic solutions (Delgaty and Lake 1998) one can go further and find the *locus* of mass maxima as a function of the radius $R_{\rm max}$. The answer is a curve indicating that, in general, larger maximum masses of self-bound sequences must have increasingly larger radii. Therefore, the observations of large compact star masses can potentially set a lower limit to their radii as well. This test is perhaps one of the simplest to perform since masses can be in some cases obtained with great precision, whereas radii are somewhat more indirect.

4 Masses and radii: recent observations

The zoo of compact star masses and radii has been growing recently, and there is now a firm expectation of finding more reliable limits to the internal structure than hitherto possible. From this point of view, important determinations are those of $M \gg 1.4M_\odot$ and $M \ll 1.4M_\odot$, because this are the limits where the self-bound and conventional models are more different. Nice et al. (2005) claim of $M = 2.1 \pm 0.2M_\odot$ ($M = 2.1^{0.4}_{0.5}M_\odot$ at 2σ level) for the compact star designated as PSR J0751+1807, and the Val Baker et al. (2005) determination of $M = 0.91 \pm 0.08M_\odot$ for SMC X-1 are just two examples of this "spreading" around the older canonical value of $1.4M_\odot$.

In fact this is one of the main reasons of why should we care about self-bound models: while it is generally believed that high masses disfavor a quark composition, it could be that *hadronic* models have a serious problem with them. This is because the appearance of hyperons (known to exist for decades) generally soften the equation of state below

$1.4M_{\odot}$ or so, a result found consistently over the years from microphysical approaches. The existence of quark cores does not help either because the maximum mass decreases with respect to the purely hadronic model. Therefore, either hyperons couple to neutrons and protons with strengths capable of giving a large extra repulsion (thus rising the maximum mass of the sequence), *or* it is acknowledged that truly exotic models, like the self-bound ones, are more compatible with the high masses (the problem with hyperons has been noted before by M. Baldo, F. Burgio and coworkers). But this extreme possibility would also predict that the big difference in radii between conventional and self-bound models would begin below $0.5M_{\odot}$, not around $1M_{\odot}$, and it is unclear how and if such low-mass stars are formed in nature. This finally means that $\sim 1M_{\odot}$ stars should show radii around 10 km, not 6–7 km as previously thought. While we can not prove that exotic matter is present in compact stars, it is also not guaranteed that microscopic EOS with all the degrees of freedom known from laboratories can fit the observations either.

5 Conclusions

We end this brief exposition about some features of self-bound models by saying that understanding of the *vacuum* is the real clue for advances in dense matter physics. This is not unlike other fields of physics, like the well-known crisis in cosmology about what the quantum vacuum should be and what actually is (Freedman and Turner 2003). The same vacuum issue, but related to matter well above the saturation density, looks even more formidable, and its understanding should solve in the wash the issues of the existence of self-bound states and the features of the resulting EOS. It is also important to remark again that there is no hadronic model devoid of problems with the high mass end: either they ignore hyperons or are solved in a mean field approach or some other scheme with its own problems and questions. There is, however, the possibility of an extreme stiffness of the equation of state, such as the hyperons do not appear at all, because the relatively low density at the center. The good news is that we can now foresee actual tests of the stiffness of the EOS in the near future.

As stated, this stiffness is very important to establish, because the very introduction of Λ particles and other hyperons would then call for very exotic interactions among them at least. Otherwise the resulting EOS become so soft that measured masses around $2M_{\odot}$ happen to lie well above the maximum masses of the respective theoretical sequences. This is why strongly repulsive interactions would be required if hyperons appear inside compact stars. High masses may be *pointing* towards the exotica rather than excluding them (Özel 2006; Alford et al. 2006b).

To conclude, we would like to quote an inspiring sentence from the English literature that may (or may not) be related to these topics, in which a bit of fantasy is always hidden

There are more things in Heaven and Earth, Horatio
Than are dreamt of in your philosophy
Hamlet, Act I, Scene V

Acknowledgements We would like to acknowledge the Organizing Committee for a great Workshop full of excellent presentations. I. Bombaci (U. of Pisa) is also acknowledged for extensive discussions and insightful ideas on dense matter that lead to improve this talk substantially. I also thank the warm hospitality and scientific advise of Dr. G. Lugones on these topics. This work has been partially supported by CNPq Agency (Brazil).

References

Alford, M., Reddy, S.: Phys. Rev. D **67**, 070424 (2003)
Alford, M., Cowan, G.: J. Phys. G **32**, 511 (2006)
Alford, M.G., Rajagopal, K., Reddy, S., et al.: Phys. Rev. D **73**, 114016 (2006a)
Alford, M., Blaschke, D., Drago, A., et al.: astro-ph/0606524 (2006b)
Bailin, D., Love, A.: Phys. Rep. **107**, 325 (1984)
Benvenuto, O.G., Horvath, J.E.: Phys. Rev. Lett. **63**, 716 (1989)
Benvenuto, O.G., Horvath, J.E., Vucetich, H.: Phys. Rev. Lett. **64**, 713 (1990)
Björken, J.D., McLerran, L.D.: Phys. Rev. D **20**, 2353 (1979)
Bodmer, A.R.: Phys. Rev. D **4**, 1601 (1971)
Chen, A., Xu, R.X.: astro-ph/0605285 (2006)
Choutko, V. (for the AMS-01 Collaboration). In: Kajita, T., Asaoka, Y., Kawachi, A., Matsubara Y., Sasaki, M. (eds.) Proceedings of the 28th International Cosmic Ray Conference, Tsukuba, Japan, 31 July–7 August, p. 1765. Universal Academic Press, Tokyo (2003)
Dai, Z.G., Peng, Q.H., Lu, T.: Astrophys. J. **440**, 815 (1995)
Degrand, T., Jaffe, R.L., Johnson, K., et al.: Phys. Rev. D **12**, 2060 (1975)
Delgaty, M.S.R., Lake, K.: Comput. Phys. Commun. **115**, 395 (1998)
Freedman, W.L., Turner, M.S.: Rev. Mod. Phys. **75**, 1433 (2003)
Gómez Dumm, D., Blaschke, D.B., Grunfeld, A.G., et al.: Phys. Rev. D **73**, 114019 (2006)
Heiselberg, H., Pethick, C.J., Staubo, E.F.: Phys. Rev. Lett. **70**, 1355 (1993)
Ichimura, M., Kamioka, E., Kitazawa, M., et al.: Nuovo Cimento A **106**, 843 (1993)
Lugones, G., Bombaci, I.: Phys. Rev. D **72**, 065021 (2005)
Lugones, G., Benvenuto, O.G., Vucetich, H.: Phys. Rev. D **50**, 6100 (1994)
Lugones, G., Horvath, J.E.: Astron. Astrophys. **403**, 173 (2003)
Lugones, G., Horvath, J.E.: Phys. Rev. D **66**, 074017 (2002)
Madsen, J.: Phys. Rev. D **71**, 014026 (2005)
Nice, D. , Splaver, E.M., Stairs, I.H., et al.: Astrophys. J. **634**, 1242 (2005)
Özel, F.: Nature **441**, 1115 (2006)
Ruster, S.B., Shovkovy, I.A., Rischke, D.H.: Nucl. Phys. A **743**, 127 (2004)
Terazawa H.: INS-Report 336 (1979)
Val Baker, A.K., Norton, A.J., Quaintrell, H.: Astron. Astrophys. **441**, 685 (2005)
van Adelsberg, M., Lai, D., Potekhin, A.Y., et al.: Astrophys. J. **628**, 902 (2005)
Weber, F.: Prog. Part. Nucl. Phys. **54**, 193 (2005)
Witten, E.: Phys. Rev. D **30**, 272 (1984)

Astrophys Space Sci (2007) 308: 435–441
DOI 10.1007/s10509-007-9315-0

ORIGINAL ARTICLE

Precession as a probe of the neutron star interior

Bennett Link

Received: 30 June 2006 / Accepted: 18 October 2006 / Published online: 21 March 2007

Abstract Strong evidence that some neutron stars precess (nutate) with long periods ($\sim$1 yr) challenges our current understanding of the neutron star interior. I describe how neutron star precession can be used to constrain the state of the interior in a new way. I argue that the standard picture of the outer core, in which superfluid neutrons coexist with type II, superconducting protons, requires revision. One possible resolution is that the protons are not type II, but type I. Another possibility is that the neutrons are normal in the outer core. I conclude with a brief discussion of the implications for detectable gravitational wave emission from millisecond pulsars.

Keywords Stars: neutron · Pulsars: general · Dense matter · Stars: rotation

PACS 97.60.Jd · 26.60.+c · 97.60.Gb

1 Introduction

There is mounting evidence that some isolated neutron stars undergo long-period precession (nutation). The strongest evidence is found in PSR B1828-11, which shows highly-periodic variations in pulse phase over a period of $\sim$500 d, accompanied by correlated changes in beam width (Stairs et al. 2000). PSR B1642-03 also shows periodic changes in pulse phase, though correlated changes in the beam have not been detected (Shabanov et al. 2001). RX J0720.9-3125 is the first X-ray pulsar to show evidence for precession, with a period of $\sim$7 yr, and correlated changes in line depth (Haberl et al. 2006). Low-level timing "noise", seen in all pulsars, is quasi-periodic in many cases and could represent precession at low amplitude (for examples see, e.g., Downs and Reichley 1983 and D'Alessandro et al. 1993). The way in which a neutron star precesses depends on the dynamics of its interior, and so observations of precession can be used, at least in principle, to constrain those dynamics. We can then ask the question: what properties must the interior have to be consistent with long-period precession? Here I describe how observations constrain the ground states of the quantum liquids of the core and the dynamics of vortices in the inner crust. I will argue that the standard picture of the outer core of a neutron star, comprising flux tubes of a type II superconductor coexisting with a neutron superfluid, is incompatible with observations of long-period precession and should be reconsidered. Finally, I will discuss the prospects for detection of gravitational waves from millisecond pulsars (MSPs) that are deformed to the extent indicated by precession measurements.

2 Evidence for precession

Precession is a rotational mode of a rigid body in which the body's angular velocity is not aligned with any principal axis. Recall that for a rigid, biaxial body of oblateness ϵ, the precession frequency ϖ, in units of the spin frequency ω, is $\varpi = \epsilon$. The motion is a superposition of a fast wobble about the angular momentum axis at frequency $\simeq\omega$, and a slow (retrograde) roll about the symmetry axis

Much of the work described here was supported by the National Science Foundation under Grant AST-00098728.

B. Link (✉)
Department of Physics, Montana State University, Bozeman, MT 59717, USA
e-mail: link@physics.montana.edu

at frequency $\omega_p = \epsilon\omega$ ($\ll \omega$). In the precessing state, the instantaneous angular velocity of the body takes some angle θ, the "wobble angle", with respect to the (conserved) angular momentum vector. For a biaxial object, θ is a constant of the motion. The symmetry axis, angular velocity and angular momentum all lie in a plane. In the body frame, the angular velocity and angular momentum vectors move in a circle about the symmetry axis, completing a cycle in one precession period $p_p = 2\pi/\omega_p$. If the object emits a beam, which sweeps past an observer as in a radio pulsar, the observer will see periodic modulation of the pulse arrival times, correlated with variations in the pulse duration, of period p_p. Figure 1 shows 13 years of timing data from PSR B1828-11. The variations in pulsar period are highly periodic, though non-sinusoidal; the data are well-fit by a sum of sinusoids with a fundamental period of 511 d and a harmonic at 256 d. The observed changes in pulse duration are highly-correlated with the timing data, and with exactly the same Fourier content, strongly suggesting that we are seeing precession. If the precession period of PSR B1828-11 is 511 d, the implied deformation is $\epsilon \simeq 10^{-8}$. The implied deformations of PSR B1642-03 and RX J0720.9-3125 are $\simeq 4 \times 10^{-9}$ and $\simeq 4 \times 10^{-8}$, respectively.

We should be cautious in the interpretation of these data. Are we really seeing precession, or something else? To be more quantitative, consider a simple model in which the star behaves approximately as a deformed rigid body. (Even though most of the interior is a liquid, let us assume for now that it co-rotates with the rest of the solid over the long time scales of precession, $\sim$1 yr.) In general, crustal stresses, magnetic stresses, or some combination of the two will render the body triaxial. The way in which the body precesses will depend on the degree of triaxiality and the form of the star's spin-down torque. If the spin-down torque depends on the angle between the star's magnetic moment and its angular velocity vector, as in the vacuum dipole model, the spin behavior can be quite complicated; triaxiality and non-linear dependence of the dipole torque on the star's angular velocity produce non-linear timing solutions so that in general the observed pulse phase, though periodic, is non-sinusoidal.

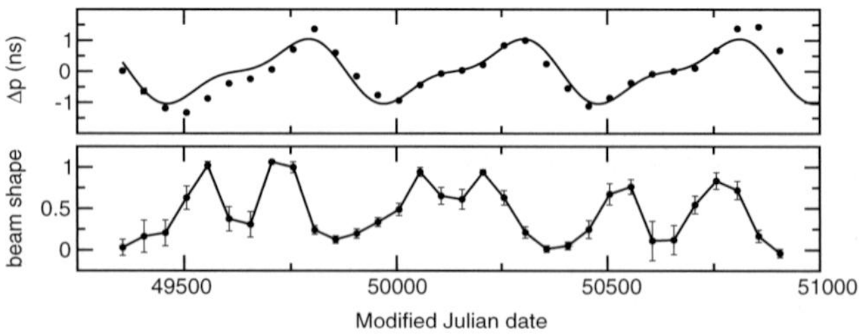

Fig. 1 Evidence for precession in PSR B1828-11 (Stairs et al. 2000). Highly-periodic period residuals (*top panel*) are seen with correlated changes in beam shape (*bottom panel*); the shape changes shown here correspond to changes in the beam width of several degrees. *The curve in the top panel* is a fit from the theoretical model described in the text

Taking the triaxiality and the strength of the dependence of the spin-down torque on the magnetic inclination as free parameters, this simple model can provide good fits to the timing data of PSR B1828-11 (Link and Epstein 2001; Akgün et al. 2006). An example fit which gives quantitative agreement with the timing data is shown as the curve in the top panel of Fig. 1. This example solution has a wobble angle of $\sim 3°$ (the characteristic angle between the star's symmetry axis and the angular momentum), consistent with the observed beam width variations of similar magnitude (Stairs et al. 2000). The triaxiality and the spin-down torque are not strongly constrained by the data.

That a simple, physically-motivated model can reproduce the main features of the timing data of PSR B1828-11 supports the precession interpretation of the spin behavior of this and other pulsars. Assuming we are, in fact, observing precession, then the inferred deformation is $\epsilon \simeq 10^{-8}$, corresponding to a "mountain" on the star of height $\sim$0.1 mm supported by crustal rigidity or some other source of stress. It is an extraordinary accomplishment in observational astronomy to measure the deformation of a neutron star to this degree of precision. And though this mountain may seem small in an absolute sense, demolition of the mountain would release $\sim 10^{39}$ erg (for PSR B1828-11).

2.1 Why do we not see more precessing pulsars?

As far as we can tell from surveys of radio pulsars, precession in isolated neutron stars is a rare phenomenon. This could be due to competition between the effects that excite precession and those that damp it.

The rotational energy of a body is minimized for a given angular momentum when the body is rotating about its major principal axis of inertia. In this state, the body does not precess. Hence, if dissipative processes act in the body they will work to damp the precession. For a neutron star, any dissipation occurring between the crust and the liquid interior, or within the crust itself, will damp precession eventually. We might ask, then, why some neutron stars precess at all. An excitation mechanism is necessary in any precessing star for which the damping time is less than the star's age. One possible excitation mechanism is structural relaxation of a brittle crust as the star spins down. Because the crust is under high pressure, cracking of the crust can occur only through shearing motions. These shearing motions necessarily break any axisymmetry that the star may have had before, and the star will then precess (Link et al. 1998; Franco et al. 2000).

The apparent rareness of precession could mean that it is only rarely excited in isolated neutron stars, and that in most stars any previous precession has already been damped. On the other hand, it could be that precession is excited sufficiently often that it never completely damps and that most neutron stars precess at low wobble angle (quasi-periodicity in timing noise could be evidence of this). In this case, stars such as PSR B1828-11, PSR B1642-03 and RX J0720.9-3125 might represent rare examples of relatively large-amplitude precession.

3 Probing the stellar interior with precession

A neutron star, of course, is not a rigid body; about 95% of its mass is in the form of various quantum liquids. In the core, superfluid neutrons are thought to coexist with superfluid protons which, by virtue of their charge, are also *superconducting*.[1] In the inner crust, the dripped neutrons are almost certainly superfluid. What are the implications for the neutron star interior? I now discuss how precession can be used to constrain the properties of the ground state of nuclear matter in the outer core, in particular, the possible states of hadronic superfluidity. I will then discuss constraints we can obtain on superfluid dynamics in the inner crust.

3.1 Precession with a quantum liquid core

Nucleon pairing calculations predict that the outer core of a neutron star consists of a neutron superfluid in coexistence with superconducting protons (see Fig. 2; for a review see Dean and Hjorth-Jensen 2003); the strong interaction forms neutron-neutron and proton-proton Cooper pairs. The superconductor is expected to be in a type II state (Baym et al. 1969a), so that the magnetic field that penetrates the core is organized in *flux tubes* (Baym et al. 1969b), long structures of microscopic cross section. The flux tubes have characteristic radii of $\sim$50 fm (1 fm $= 10^{-13}$ cm) and enormous internal magnetic fields ($B_f \sim 10^{15}$ G) which are screened to zero a short distance from each flux tube by the current distribution that minimizes the energy.

As a proto-neutron star is being born, dynamo process are likely to give the magnetic field a very complicated structure (Thompson and Duncan 1993). Shortly after, when the protons condense into a type II superconductor, the field will reorganize into flux tubes that freeze into the superconducting medium, while preserving the complicated large-scale structure that existed before (Ruderman et al. 1998; Jones 2006). The superconducting protons, other charges (e.g., electrons and muons) and the crust are all coupled together through magnetic stresses over time scales of several seconds, nearly corotating as a rigid body (Easson 1979).

[1] I assume that putative neutron stars are not strange stars in disguise.

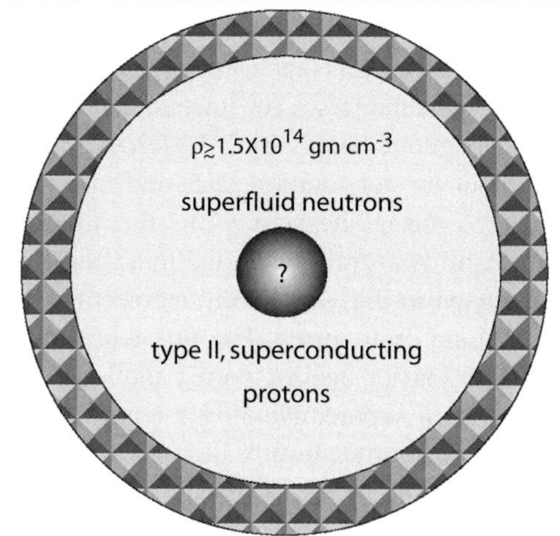

Fig. 2 The expected nucleon pairing situation in the outer core

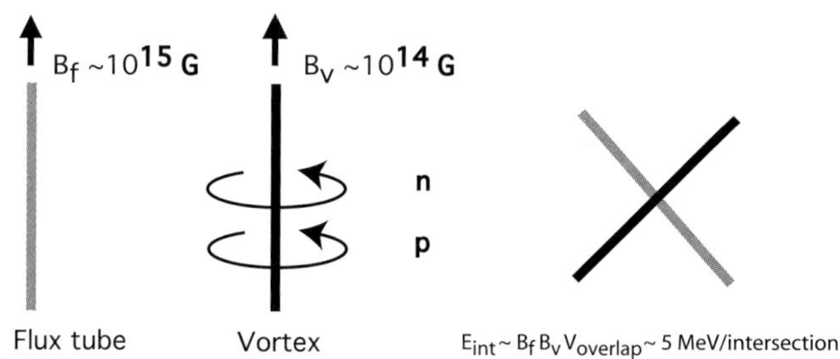

Fig. 3 *A flux tube* (*grey*) has a field in its core of order $B_f \sim 10^{15}$ G. *A vortex* (*black*) supports a flow of neutrons about it. In the outer core, where the protons are expected to be superconducting, a proton flow is entrained by the neutron flow, giving the neutron vortex a high magnetization. If a vortex approaches a flux tube, the magnetic energy is raised or lowered (depending on orientation) by $E_{\rm int} \sim 5$ MeV/intersection, pinning the vortices to the flux tubes

The rotating neutron superfluid establishes a rectilinear array of *vortex lines*, whose arrangement determines the angular momentum of the neutron fluid. Vortices are stable hydrodynamic structures; within the vortex core, of dimension $\sim$10 fm, superfluidity is destroyed; outside of the vortex, the neutron Cooper pairs circulate around the vortex core, each pair carrying exactly $\hbar$ of angular momentum. Any change in the angular momentum of the neutron fluid must involve the movement of vortices. By contrast, the superconducting proton fluid rotates almost exactly as a rigid body without the formation of vortices (Alpar et al. 1984a). This difference in rotational properties of the two fluids is due to the fact that the protons are charged and the neutrons are not.

A remarkable property of neutron vortices immersed in superconducting protons is that the neutron flow about vortices also entrains a proton flow as a consequence of Fermi liquid interactions (Alpar et al. 1984a). This entrainment effect only occurs when both hadronic species are superfluid. The proton flow about vortices gives them a very strong magnetization, of order $B_v \sim 10^{14}$ G within the vortex core (see Fig. 3). As a result, it is energetically favorable for the

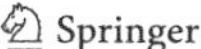

flux tubes and vortices to pin to one another (Sauls 1989). The origin of the pinning is that bringing a (magnetized) vortex close to a flux tube raises (or lowers, depending on orientation) the magnetic energy by $\sim$5 MeV per intersection. For typical neutron star rotation rates and magnetic fields, there are $\gtrsim 10^{17}$ vortex lines and $\sim 10^{31}$ flux tubes. The rectilinear vortex array is thus entangled in a spaghetti of flux tubes, as depicted in Fig. 4, with an intersection spacing of $\sim 10^{-10}$ cm along each vortex. The flux tubes represent an extremely rigid barrier against vortex motion because the flux is frozen to a superconducting medium that has stable compositional stratification. If, for example, the vortices were to push on the flux tubes as strongly as possible without breaking through, they could not move the flux tubes faster than $\sim 10^{-8}$ cm s^{-1} (Jones 2006), corresponding to a total displacement of $\sim 10^{-2}$ cm over a precession period of one year at a wobble angle of 1°.

From the standpoint of how the outer core affects precession dynamics, the star can be regarded as consisting of two components: (1) the core protons, other charges and crust, which rotate as a single "body", and (2) a neutron superfluid which rotates according to the distribution and movement of its vortices. If the system is set into precession, the angular momentum vector of the superfluid (most of the stellar mass) remains fixed to the body to the extent that the vortices remain pinned against flux tubes. As the system precesses, the angular velocity vector of the system changes its orientation with respect to the body's symmetry axes, returning to its original orientation after one precession period as for a rigid body. But unlike a rigid body, the precession frequency is no longer determined by the star's oblateness alone. Pinning of vortices anywhere dramatically increases the precession frequency (Shaham 1977). To take a mechanical analogy, the pinned vortex array acts as a gyroscope affixed to a rigid body. Such an object precesses at a rate determined by how much angular momentum resides in the gyroscope. For a neutron star, the precession frequency would be

$$\varpi = \epsilon + \frac{L_p}{I_b \omega}, \qquad (1)$$

where L_p is the angular momentum contained in the pinned superfluid (the "gyroscope") and I_b is the moment of inertia of the "body" (crust plus charges). The pinned fluid effectively acts as a huge additional contribution to the star's oblateness, since $L_p \simeq I_p \omega$ and $L_p / I_b \omega \simeq I_p / I_b \gg \epsilon$. The precession frequency is thus determined by how much of the neutron vorticity is pinned. If most of the neutron superfluid of the outer core is pinned to the flux tubes, the precession frequency will be $\varpi \simeq I_p/I_b \simeq 10$, compared to 10^{-8} observed for PSR 1828-11. Hence, perfect pinning in even a small fraction of the core gives a precession frequency that is far too fast to explain the observations. For precession to occur with a long period, the vortices must be able to closely follow the instantaneous spin vector of the rest of the star, and so cannot be perfectly pinned in any sizeable region of the star, at least, not by more than one part in 10^9 by volume; otherwise the second term in (1) would overwhelm the first.

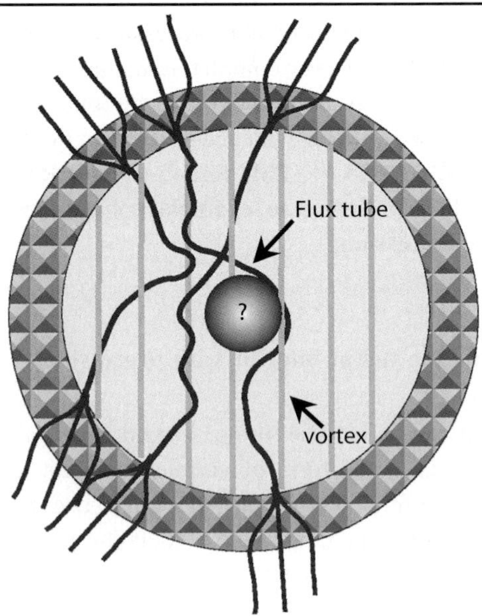

Fig. 4 In the standard picture of the outer core, the neutron vortices are entangled in the far more numerous proton flux tubes. As the star precesses, the vortices cannot push the flux tubes through the highly-conductive medium. The enormous vorticity of the vortex array acts as a gyroscope affixed to the rest of the star, giving a precession frequency that is far too fast to reconcile with observations

If there are no forces on vortex lines, they co-rotate with the net superfluid flow they all produce. On the other hand, if some force is exerted on a vortex, through a pinning interaction or drag, a hydrodynamic (Magnus) force acts on the vortex so that, in general, the vortex velocity will be different than that of the net superfluid flow. The Magnus force is a lift force arising from modification of the flow near a vortex by the global flow from all of the other vortices. If vortices are perfectly pinned to the flux tubes, their velocity will be equal to that of the charged fluid. In a precessing star, a time-dependent Magnus force is exerted on the vortex array, because the vortices are forced to move with the wobbling body to which they are pinned, and so move at a different velocity than the superfluid flow. If the velocity difference between pinned vortices and the superfluid is sufficiently large, the Magnus force can be strong enough to unpin the vortices. For example, if a star with vortices pinned to the flux tubes were suddenly excited to precess with a wobble angle of $\sim$1°, the Magnus force would be sufficient to force the vortex array through the flux tubes. This process, however, is highly dissipative (Link 2003). As a vortex cuts through a flux tube, waves are excited on the vortex which propagate away from the interaction site and eventually share their energy with the rest of the star. (Waves are also excited on the flux tubes in this process.) This process is so dissipative that the precession would damp to a very small wobble angle ($\ll 1°$) in less than an hour. At this small wobble angle, the Magnus force would be low enough that the vortices would be again pinned, the precession very fast, and of much lower amplitude than observed. This is the only possible precession state for a star in which vortices and flux tubes coexist in the outer core and are perfectly pinned together.

3.2 Precession with vortex creep

The assumption of perfect pinning is, however, an idealization. Some vortex motion with respect to pinning sites could occur through a process of *creep* by thermal activation (Alpar et al. 1984b; Link et al. 1993; Chau and Cheng 1993) or quantum tunneling (Link et al. 1993). Alpar (2005) has suggested that vortex creep could resolve the problem that precession is too fast for perfectly-pinned vortices. Even if vortex creep is accounted for, however, the precession frequency is still too fast, by a factor of $\sim 10^9$, to explain the observations (Link 2006).

The problem is that vortex creep is almost like perfect pinning. The vortices do move when a Magnus force is exerted on them (the Magnus force that arises during precession), but even under a large Magnus force they move slowly. As a consequence, *vortex creep is highly dissipative*. An analysis of precession with imperfectly pinned vortices using the concept of vortex drag (Sedrakian et al. 1999) shows that if the response of a vortex under a relatively large Magnus force is to move at low velocity, as in vortex creep, the original fast "Shaham mode" remains, but there is also a new mode, a slow mode (associated with the very slow vortex motion). The slow mode is highly over-damped, however, and so cannot complete even one cycle. *Hence, the fast (Shaham) mode is the only dynamically relevant mode* (Link 2006). As the pinning is made more effective, the slow mode disappears; its frequency becomes zero and its damping rate infinite, leaving only the fast mode.

Alpar (2005) found a slow mode for creeping vortices, and treated the damping with an assumed dissipative torque between the body and the neutron superfluid of the following linear form:

$$\mathbf{N}_{\mathrm{drag}} = -(\boldsymbol{\omega}_s - \boldsymbol{\omega}_b)/\tau \tag{2}$$

where τ is the coupling time, $\boldsymbol{\omega}_s$ is the angular velocity of the neutron superfluid and $\boldsymbol{\omega}_b$ is the angular velocity of the body. This drag torque would be appropriate for describing coupling between a solid and a normal liquid through a viscous boundary layer, but it greatly underestimates the torque on a dragged vortex array; the correct form is more complicated (Sedrakian et al. 1999). Using (2), Alpar (2005) arrived at the incorrect conclusion that the slow mode is under-damped when, in fact, it is over-damped and dynamically irrelevant. See Link (2006) for much more detail on this issue.

3.3 Implications for the ground state of the outer core

Long-period precession requires that the vortices move with little drag, that is, they do not creep, but *flow* essentially everywhere in the star, to the extent that the angular momentum of the vortex array can always nearly follow the instantaneous velocity vector of the rest of the star. In order to do this, the vortices must be able to move through the body (the charged fluid plus crust) at a speed of $\simeq R\omega\theta$, where θ is the wobble angle and R is the stellar radius (Link 2004); for PSR 1928-11, this requires that vortices move at a speed of $\sim 10^{-2}$ cm s^{-1}, which seems impossible in the core if neutron vortices interact with flux tubes; the vortices can neither creep nor push the flux tubes nearly so fast. The general class of high-drag theories that give slow (or negligible) vortex motion with respect to the background, *including vortex creep theory*, are inconsistent with observations of long-period precession.

These considerations impose severe constraints on the ground states of neutrons and protons in the outer core, in particular, the hadronic pairing situation must be such that *vortices and flux tubes do not coexist* (Link 2003). This conclusion is in direct conflict with the standard picture of the inner core in which superfluid neutrons coexist with type II superconducting protons (Fig. 4). The possibilities that might be consistent with long-period precession are: (1) normal neutrons, superconducting protons, (2) superfluid neutrons, type I protons, (3) superfluid neutrons, normal protons, (4) normal neutrons and protons, and (5) the core magnetic field has somehow been completely forced into the crust. The existence of proton superconductivity is on rather firm footing, though the *type* of superconductivity, is less clear. In a type I core, the magnetic flux could be organized in slabs or other geometries of mesoscopic dimensions, rather than the flux tubes found in a type II superconductor. (Buckley et al. 2004 claim that interaction of neutron and proton Cooper pairs will make the liquid a type I superconductor, while Alford et al. 2005 argue that this interaction is insignificant.) It might be possible for vortices to move through a type I superconductor with sufficiently low drag to allow long-period, under-damped precession; Sedrakian (2005) showed that this scenario works for two specific geometries for the magnetic field. The pairing state of neutrons is far less certain. Most pairing calculations show pairing, but at least one calculation does not (Schwenk and Friman 2004). Lacking a mechanism for complete expulsion of the core field, possibilities (1) and (2) seem the most likely within existing uncertainties regarding nucleon pairing.

4 Vortex pinning in the inner crust

Similar considerations apply in the inner crust of a neutron star. Here the pairing situation is much better understood than in the core; the neutrons are predicted to be superfluid, and so the liquid is permeated by vortices (there are no flux tubes, since the protons are bound in nuclei). Precession observations allow constraints on the dynamics of vortices in this part of the star.

Vortices of the inner crust are predicted to pin to the nuclei (Alpar 1977; Epstein and Baym 1988; Pizzochero et al. 1997; Avogadro et al. 2007; Donati and Pizzochero 2006), an effect on which many models of pulsars glitches are based (see below). Assuming, for example, that almost all of the inner-crust superfluid is pinned to nuclei and that the rest of the star rotates as a rigid body over a precession period, the precession frequency becomes $\varpi \simeq 0.01$ (Shaham 1977), again far too fast. There is a slow mode if the vortices creep, but as before it is highly over-damped. This result does not necessarily rule out vortex pinning in the inner crust; it just cannot happen in stars that are slowly precessing. An interesting open question is if coexistence of inner crust vortices with nuclei is consistent with long-period precession. Vortices must be able to move with respect to the solid with little drag. The Magnus force on vortices in the inner crust of a precessing star is, in fact, sufficient to keep the vortices unpinned (Link and Cutler 2002), but it remains to be shown if vortices can also move past nuclei with sufficiently low drag. Jones (1998) found that an infinitely long vortex moving slowly past a single nucleus tends to be temporarily trapped by the nucleus, but an adequate microscopic description of slow vortex motion past many nuclei, and the associated dissipation, is not yet available.

To summarize the conclusions so far, in a star that is slowly precessing:

1. Vortices cannot be pinned *anywhere* in a star, which requires that
2. vortices and flux tubes do not coexist in the core, and
3. vortices of the inner crust cannot be pinned and must be able to move past nuclei with little dissipation.

5 A prediction

Glitches, the sudden spin jumps observed in many neutron stars, have been attributed to large-scale unpinning of vortices somewhere in the star (e.g., Anderson and Itoh 1975; Ruderman 1976; Pines and Alpar 1985; Link and Epstein 1996; Sedrakian and Cordes 1999; Larson and Link 2002). In all such glitch theories, the vortices are pinned somewhere in the star, and the superfluid accumulates excess angular momentum as the rest of the star is spun down by the external torque; the pinned superfluid becomes a reservoir of angular momentum. Then, as a consequence of some instability, many vortices unpin, move dissipatively through the star, and deliver a torque to the crust that spins it up. All glitch models of this sort face the challenge of explaining why $\gtrsim 10^{13}$ vortices all move at once, though one possibility is that sudden heating of the crust from structural relaxation as the star spins down unpins the required number of inner crust vortices (Link and Epstein 1996; Larson and Link 2002). If we believe the basic picture that vortex pinning and unpinning is the mechanism responsible for glitches, we can make the following prediction:

A slowly-precessing neutron star cannot glitch.

This prediction follows since slow precession forbids pinning. (This observation does not rule out glitches through catastrophic vortex motion in stars that are not precessing.) If this prediction turns out to be wrong, and a precessing neutron star *does* eventually produce a glitch, then glitches have nothing to do with vortex pinning and are due to different physics, such as possible transitions between laminar and turbulent states of superfluid flow (Peralta et al. 2005).

6 Gravitational waves from neutron stars

The indication of precession observations that neutron stars are able to support deformation is interesting from the standpoint of neutron stars as potential sources of gravitational waves; a rotating neutron star with a non-zero mass quadrupole moment would be a gravitational wave source. The strain induced in a detector by a passing gravitational wave is proportional to $\epsilon\omega^2/d$, where ϵ is the fractional (quadrupolar) deformation of the star, ω is the spin rate and d is the distance. Hence, the best candidate neutron stars for detectable gravitational waves are rapidly rotating, close and highly deformed. MSPs in low-mass X-ray binary systems (LMXBs) are promising in this respect. In the future, LIGO will be tunable to have high sensitivity at the frequencies at which typical MSPs would emit gravitational waves (twice the spin frequency).

If the precession interpretation PSR B1828-11 and RX J0720.9-3125 is correct, then we know of at least two neutron stars that can support deformations of $\epsilon \gtrsim 10^{-8}$. There is no reason to expect $\epsilon \sim 10^{-8}$ to represent an upper limit. If MSPs can support quadrupolar deformations of $\epsilon \sim 10^{-7}$ or larger, they could be sources of gravitational waves detectable by LIGO-II (Cutler and Thorne 2002; see Fig. 5). Such MSPs need not precess to produce gravitational waves.

7 Conclusions

Precession of neutron stars provides a new probe of the properties of matter at supra-nuclear densities. In particular, the pairing states of the outer core can be constrained. I have argued that long-period precession is inconsistent with the standard picture (Fig. 4) of an outer core consisting of coexisting superfluid neutrons and type II, superconducting protons. Precession data might be telling us that the outer core

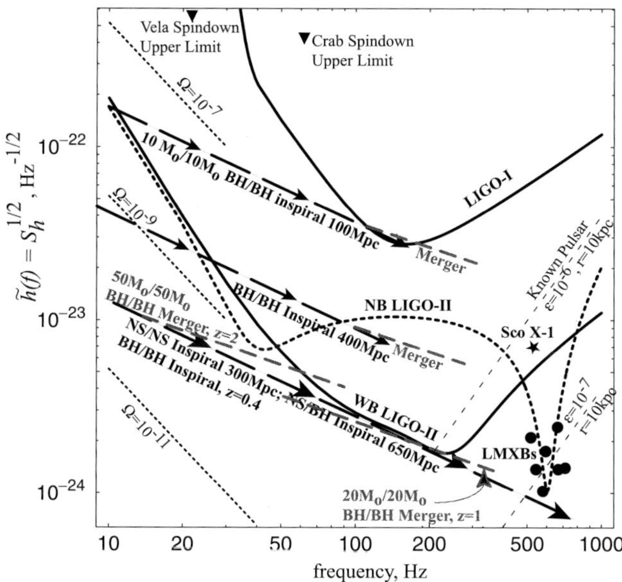

Fig. 5 The noise $\tilde{h}_s(f)$ as a function of candidate gravitational wave source frequency f for several planned LIGO interferometers. MSPs in LMXBs could be detectable sources by LIGO-II in the NB mode if the pulsars in these systems can support deformations somewhat larger than those implied by precession observations. See Cutler and Thorne (2002) for further explanation of this diagram. Figure courtesy of C. Cutler

protons are paired in a type I state, or that at least one of the hadronic fluids is normal there (the neutrons being the more likely possibility given current uncertainties). These constraints are robust provided that precession in neutron stars is real; they do not depend on the state of matter in the *inner* core. Until the issue of neutron pairing in the outer core is settled by first-principles calculations, astrophysical arguments of the type presented here are useful for constraining the various possibilities for the hadronic ground states. By similar reasoning, long-period precession requires that the neutron vortices of the inner crust are unpinned and moving through the background of nuclei with little dissipation. If a precessing star is ever seen to produce a glitch, then glitches are not related to large-scale vortex pinning and unpinning.

The precession interpretation of PSR B1828-11 and RX J0720.9-3125 requires these stars to be deformed by $\epsilon \simeq 10^{-8}$. If MSPs can support similar or larger quadrupolar deformations, they would be interesting gravitational wave sources for LIGO-II.

Acknowledgements I thank A. Sedrakian and I. Wasserman for valuable discussions.

References

Akgün, T., Link, B., Wasserman, I.: Mon. Not. Roy. Astron. Soc. **365**, 653 (2006)
Alford, M., Good, G., Reddy, S.: Phys. Rev. C **72**, 055801 (2005)
Alpar, M.A.: Astrophys. J. **213**, 527 (1977)
Alpar, M.A.: Neutron star superfluidity, dynamics and precession. In: Baykal, A., Yerli, S.K., İnam, S.Ç., Grebenev, S. (eds.) The Electromagnetic Spectrum of Neutron Stars, Part II: Mathematics, Physics and Chemistry, NATO Science Series, vol. 210, pp. 33–46. Springer, Berlin (2005)
Alpar, M.A., Langer, S.A., Sauls, J.A.: Astrophys. J. **282**, 533 (1984a)
Alpar, M.A., Anderson, P.W., Pines, D., Shaham, J.: Astrophys. J. **278**, 791 (1984b)
Anderson, P.W., Itoh, N.: Nature **256**, 25 (1975)
Avogadro, P., Barranco, F., Broglia, R.A., Vigezzi, E.: Phys. Rev. C **75**, 012805 (2007)
Baym, G., Pethick, C., Pines, D.: Nature **224**, 673 (1969a)
Baym, G., Pethick, C., Pines, D., Ruderman, M.: Nature **224**, 872 (1969b)
Buckley, K.B., Metlitski, M.A., Zhitnitsky, A.R.: Phys. Rev. Lett. **92**, 151102 (2004)
Chau, H.F., Cheng, K.S.: Phys. Rev. B **47**, 2707 (1993)
Cutler, C., Thorne, K.S.: An overview of gravitational wave sources. In: Bishop, N., Maharaj, S.D. (eds.) General Relativity and Gravitation, Proceedings of the 16th International Conference, pp. 72–111. World Scientific, Singapore (2002)
D'Alessandro, F., McCulloch, P.M., King, E.A., Hamilton, P.A., McConnell, D.: Mon. Not. Roy. Astron. Soc. **261**, 883 (1993)
Dean, D.J., Hjorth-Jensen, M.: Rev. Mod. Phys. **75**, 607 (2003)
Donati, P., Pizzochero, P.M.: Phys. Lett. B **640**, 74 (2006)
Downs, G.S., Reichley, P.: Astrophys. J. Suppl. **53**, 169 (1983)
Easson, I.: Astrophys. J. **228**, 257 (1979)
Epstein, R.I., Baym, G.: Astrophys. J. **328**, 680 (1988)
Franco, L.M., Link, B., Epstein, R.I.: Astrophys. J. **543**, 987 (2000)
Haberl, F., Turolla, R., Vries, C.P.D., Zane, S., Vink, J., Mendez, M., Verbunt, F.: Astron. Astrophys. **415**, L17 (2006)
Jones, P.B.: Phys. Rev. Lett. **81**, 4560 (1998)
Jones, P.B.: Mon. Not. Roy. Astron. Soc. **365**, 339 (2006)
Larson, M.B., Link, B.: Mon. Not. Roy. Astron. Soc. **333**, 613 (2002)
Link, B.: Phys. Rev. Lett. **91**, 101101 (2003)
Link, B.: Phys. Rev. Lett. **92**, 149002 (2004)
Link, B.: Astron. Astrophys. **458**, 881 (2006)
Link, B., Cutler, C.: Mon. Not. Roy. Astron. Soc. **336**, 211 (2002)
Link, B., Epstein, R.I.: Astrophys. J. **457**, 844 (1996)
Link, B., Epstein, R.I.: Astrophys. J. **556**, 392 (2001)
Link, B., Epstein, R.I., Baym, G.: Astrophys. J. **403**, 285 (1993)
Link, B., Franco, L.M., Epstein, R.I.: Astrophys. J. **508**, L838 (1998)
Peralta, C., Melatos, A., Giacobello, M., Ooi, A.: Astrophys. J. **635**, 1224 (2005)
Pines, D., Alpar, M.A.: Nature **316**, 27 (1985)
Pizzochero, P.M., Viverit, L., Broglia, R.A.: Phys. Rev. Lett. **79**, 3347 (1997)
Ruderman, M.: Astrophys. J. **203**, 213 (1976)
Ruderman, M., Zhu, T., Chen, K.: Astrophys. J. **492**, 267 (1998)
Sauls, J.A., Ögelman, H.: In: van den Heuvel, E.P.J. (ed.) Timing Neutron Stars. Proc. NATO ASI, p. 457. Dordrecht (1989)
Schwenk, A., Friman, B.: Phys. Rev. Lett. **92**, 082501 (2004)
Sedrakian, A.: Phys. Rev. D **71**, 083003 (2005)
Sedrakian, A., Cordes, J.M.: Mon. Not. Roy. Astron. Soc. **307**, 365 (1999)
Sedrakian, A., Wasserman, I., Cordes, J.M.: Astrophys. J. **524**, 341 (1999)
Shabanov, T.V., Lyne, A.G., Urama, J.O.: Astrophys. J. **552**, 321 (2001)
Shaham, J.: Astrophys. J. **214**, 251 (1977)
Stairs, I.H., Lyne, A.G., Shemar, S.L.: Nature **406**, 484 (2000)
Thompson, C., Duncan, R.C.: Astrophys. J. **408**, 194 (1993)

Astrophys Space Sci (2007) 308: 443–450
DOI 10.1007/s10509-007-9359-1

ORIGINAL ARTICLE

Spin-one color superconductivity in compact stars?—an analysis within NJL-type models

D.N. Aguilera

Received: 8 August 2006 / Accepted: 7 November 2006 / Published online: 15 March 2007

Abstract We present results of a microscopic calculation using NJL-type model of possible spin-one pairings in two flavor quark matter for applications in compact star phenomenology. We focus on the color-spin locking phase (CSL) in which all quarks pair in a symmetric way, in which color and spin rotations are locked. The CSL condensate is particularly interesting for compact star applications since it is flavor symmetric and could easily satisfy charge neutrality. Moreover, the fact that in this phase all quarks are gapped might help to suppress the direct Urca process, consistent with cooling models. The order of magnitude of these small gaps ($\simeq$1 MeV) will not influence the EoS, but their also small critical temperatures ($T_c \simeq 800$ keV) could be relevant in the late stages neutron star evolution, when the temperature falls below this value and a CSL quark core could form.

Keywords Spin-one color superconductors · Compact star interiors · Neutron star cooling

PACS 12.38.Mh · 24.85.+p · 26.60.+c · 97.60.Jd

1 Introduction

The most favorable places in nature where color superconducting states of matter are expected to occur are the interiors of compact stars, with temperatures well below 1 MeV and central densities exceeding the nuclear saturation density ρ_0 by several times.

Color superconductivity has been widely studied from non-perturbative low-energy QCD models where gaps of the order of magnitude of $\simeq$100 MeV has been calculated (Rapp et al. 1998; Alford et al. 1998). One of the effective models most used is the Nambu Jona-Lasinio (NJL) model that considers that the quarks interact locally by a 4-point vertex effective force and disregards the gluon degrees of freedom. The model uses an attractive interaction in the scalar meson channel that causes spontaneous chiral symmetry breaking if the coupling is strong enough. At the mean field level, the model shows how quarks acquire a dynamical constituent mass, which is proportional to the vacuum expectation value of the scalar field. The NJL model has then been extended and widely used to described successfully chiral restoration at finite temperatures and color superconductivity at finite density (for a review see Buballa 2005).

For compact star applications, color superconducting quark matter phases enforcing color and charge neutrality has been widely studied (Alford et al. 2001; Steiner et al. 2002). It has been shown that local charge neutrality may disfavor the occurrence of phases with large gaps where quarks with different flavor pair in a spin-0 condensate, like the 2SC phase (Alford and Rajagopal 2002). On the other hand, NJL-type model calculations show that the intermediate density region of the neutral QCD phase diagram, where the quark chemical potential is not sufficiently large to have the strange quark deconfined ($\mu \geq 430$–500 MeV, see Buballa and Oertel 2002; Neumann et al. 2002), might be dominated by u, d quarks (Ruster et al. 2005; Blaschke et al. 2005). If this is the case, 2-flavor quark matter phases may occupy a large volume in the core of compact stars (Grigorian et al. 2004; Shovkovy et al. 2003).

D.N.A. work and attendance to the meeting was supported by VESF-Fellowships EGO-DIR-112/2005.

D.N. Aguilera (✉)
Department of Applied Physics, Faculty of Sciences, University of Alicante, Apartado de Correos 99, 03080 Allicante, Spain
e-mail: deborah.aguilera@ua.es

While the stability of neutral 2SC pure phase is rather model dependent and might be unlikely for moderate coupling constants (Aguilera et al. 2005; Gomez Dumm et al. 2006), phases with other pairing patterns (or none at all) become important for the phenomenology of neutron stars. Then, besides other possibilities,[1] quarks could pair in spin-one condensates (Schafer 2000; Alford et al. 2000).

These condensates with small pairing gaps ($\Delta \simeq 1$ MeV), are not expected to have influence on the equation of state but could strongly affect the transport and thermal properties of quark matter and therefore leads to consequences for the phenomenology of compact stars.

An energy gap in the quasiparticle excitation spectrum introduces a suppression of the neutrino emissivity and the specific heat of the paired fermions by a Boltzmann factor $\exp(-\Delta/kT)$ for T smaller than the critical temperature T_c for pair formation. Thus, since the neutrino luminosity is strongly dominated by the neutrino emission from the core, the cooling of a neutron star during its early life will be affected by the dense matter pairing pattern. On the other hand, most of the specific heat of the star is provided by the core and it shows a discontinuity when the star temperature crosses T_c. A comparison of cooling curves with observations could provide a hint to discriminate dense matter phases. Several studies have shown that the occurrence of unpaired quarks in the core leads to rapid cooling via the direct Urca process , incompatible with the observations. Thus, phases that present no gapless modes with small pairing gaps ($\Delta \simeq 1$ MeV) prevent the direct Urca to work uncontrolled and might help to give a consistent picture of the observed data (see Page et al. 2000, 2004; Yakovlev and Pethick 2004; Grigorian et al. 2005).

On the other hand, it has been demonstrated that the CSL phase exhibit a Meissner effect and the magnetic field of a neutron star would be expelled from a CSL quark core if it does not exceed the critical magnetic field B_c that destroy the superconducting phase (Schmitt et al. 2003). This hypothesis might be consistent with recent investigations that indicate the crust confination of the magnetic field (Geppert et al. 2004; Pérez-Azorín et al. 2006).

To study this qualitatively, we present in this work a summary of the results obtained in previous works of NJL model calculations of two spin-one pairing patterns, focusing in the CSL phase, and a new analysis of features that are relevant to apply the CSL in phenomenological applications, like compact star cooling or protoneutron star evolution.

[1] No pairing (normal quark matter), pairing with displacement of the Fermi surfaces (LOFF or crystalline structure) or deformation of the Fermi surfaces, interior gap structure, gapless 2SC or gCFL, etc.

2 Brief on the model

We consider a two-flavor system of quarks, $q = (u, d)^T$, with an NJL-Lagrangian $\mathcal{L}_{\text{eff}} = \mathcal{L}_0 + \mathcal{L}_{q\bar{q}} + \mathcal{L}_{qq}$. which contains a free part $\mathcal{L}_0 = \bar{q}(i\,\partial\!\!\!/ - m)q$, a quark-antiquark interaction channel $\mathcal{L}_{q\bar{q}}$ that corresponds to the condensate for each flavor f

$$\sigma_f = \langle \bar{q}_f q_f \rangle, \tag{1}$$

and a diquark channel $\mathcal{L}_{qq}$ with color superconducting condensate matrix $\hat{\Delta}$ (components Δ^i) which we will made explicitly in Sect. 3. The condensates (1) are responsible for dynamical chiral symmetry breaking in vacuum and define the constituent quark masses as

$$M_f = m_f - 4G\sigma_f \tag{2}$$

being G the coupling constant in the meson scalar channel. In the density regime investigated, these condensates are relatively small but their finite size have consequences in the dispersion relations (see discussion in *Absence of ungapped modes* in Sect. 3.2 and Fig. 3).

After performing a linearization of $\mathcal{L}_{\text{eff}}$ in the presence of the condensates and providing the inverse of the fermion propagator in Nambu-Gorkov space

$$S^{-1}(p) = \begin{pmatrix} p\!\!\!/ + \mu_f \gamma^0 - \hat{M} & \hat{\Delta} \\ -\hat{\Delta}^\dagger & p\!\!\!/ - \mu_f \gamma^0 - \hat{M} \end{pmatrix}, \tag{3}$$

performing usual techniques of thermal field theory, the thermodynamical potential $\Omega(T, \mu_f)$ can be derived. At the mean-field level, i.e. the stationary points

$$\frac{\delta\Omega}{\delta\Delta^i} = 0, \qquad \frac{\delta\Omega}{\delta M_f} = 0, \tag{4}$$

define a set of gap equations for Δ^i and M_f. Among the solutions, the stable one is the solution which corresponds to the absolute minimum of Ω.

The parameters (current quark mass $m = m_u = m_d$, the coupling G, and three dimensional cut-off Λ) have been determined by fitting the pion mass and decay constant to their empirical values and to vacuum constituent quark mass at zero momentum, M. In this work we consider 2-flavor quark matter (u, d) assuming that the m_s is large enough to appear only at higher densities.

In this work we will also reference results obtained in nonlocal extensions of the NJL model. The idea basically is to modify the quark interactions in order to act over a certain range in the momentum space introducing momentum dependent form factors $g(p)$ in the current-current interaction terms. We will skip the details on the model but the main consequence is that the inclusion of high momenta states beyond the usual NJL-cutoff causes a reduction of the diquark condensates and a lowering of the density for the chiral phase transition (for a complete discussion see Aguilera et al. 2006).

3 Spin-one pairing for compact stars applications

3.1 Unlikely: 2SCb phase

Besides the spin-0 isospin singlet condensate (2SC) of two colors (e.g. *red* and *green*), the remaining unpaired color (say *blue*) could pair in an anisotropic spin-one channel with an expectation value of $\zeta = \langle q^T \, C\sigma^{03}\tau_2 \, \hat{P}^{(b)} q\rangle$ (see Buballa et al. 2003 for NJL and Aguilera and Blaschke 2007 for a nonlocal extension). Then, in the 2SCb phase we assume

$$\hat{\Delta}^{2\text{SCb}} = \Delta(\gamma_5\tau_2\lambda_2)(\delta_{c,r} + \delta_{c,g}) + \Delta'\big(\sigma^{03} \, \tau_2 \, \hat{P}_3^{(c)}\big)\delta_{c,b}, \tag{5}$$

and the thermodynamical potential is

$$\Omega^{2\text{SCb}}(T,\mu) = \frac{(M_f - m)^2}{4G} + \frac{|\Delta|^2}{4G_1} + \frac{|\Delta'|^2}{16G_2} - 4\sum_{\pm,i=1}^{3}\int \frac{d^3p}{(2\pi)^3}\left[\frac{E_i^{\pm}}{2} + T\ln\big(1+e^{-E_i^{\pm}/T}\big)\right]. \tag{6}$$

The dispersion relations for the 2SC-paired quarks (r, g)

$$\left(E_{1,2}^{-}\right)^2 = (E^-)^2 = (\epsilon - \mu)^2 + |\Delta|^2, \tag{7}$$

with the free particle energy $\epsilon^2 = \vec{p}^{\,2} + M^2$ and for the third color (b) quarks

$$\left(E_3^{-}\right)^2 = (\epsilon_{\text{eff}} - \mu_{\text{eff}})^2 + \left|\Delta'_{\text{eff}}\right|^2 \tag{8}$$

where the effective variables depend on the angle θ, with $\cos\theta = p_3/|\vec{p}|$, and are defined as

$$\epsilon_{\text{eff}}^2 = \vec{p}^{\,2} + M_{\text{eff}}^2, \tag{9}$$

$$M_{\text{eff}} = M\frac{\mu}{\mu_{\text{eff}}}, \tag{10}$$

$$\mu_{\text{eff}}^2 = \mu^2 + |\Delta'|^2\sin^2\theta, \tag{11}$$

$$\left|\Delta'_{\text{eff}}\right|^2 = |\Delta'|^2\left(\cos^2\theta + \frac{M^2}{\mu_{\text{eff}}^2}\sin^2\theta\right). \tag{12}$$

The antiparticles states E^+ are obtained replacing μ by $-\mu$ and the coupling constants are taken as $G_1 = 3/4G_1$ and $G_2 = 3/16G_2$, from instanton induced interactions.

In Fig. 1 we show the results of solving the gap equations (4) for M_f, Δ and Δ'. The gaps Δ' are strongly μ-dependent rising functions and typically of the order of magnitude of keV, at least two orders of magnitude smaller than the corresponding 2SC gaps. For the nonlocal extension, the gaps are even smaller, not larger than $\Delta' \approx 0.05$ MeV in the μ-range shown in Fig. 1. Such small gaps will have no influence on the equation of state and in this pattern the are no quarks that remain unpaired. Nevertheless, it is unlikely that the blue quark pairing could survive the constraint imposed by charge neutrality: the Fermi seas of the up and down quarks should differ by about 50–100 MeV and this is much larger than the magnitude of the gap in the symmetric case.

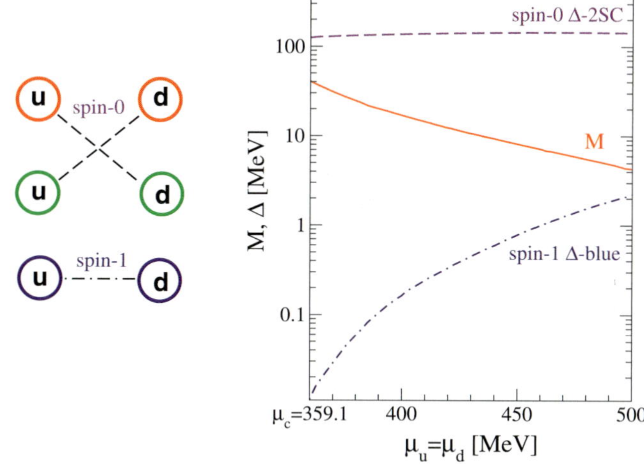

Fig. 1 Spin-0 2SC + spin-1 of the *blue* quarks pairing pattern (*left panel*) and the corresponding dynamical mass and energy gaps (*right panel*) for symmetric two-flavor quark matter, $\mu_u = \mu_d$. Parameters: $\Lambda = 595.5$ MeV, $G\Lambda^2 = 3.37$, $m = 5.56$ MeV. Fixed $M = 380$ MeV

3.2 Likely: color spin locking phase in compact stars

Single flavor spin-1 pairs are good candidates since they are inert against large splittings in the quark Fermi levels for different flavors caused by charge neutrality. They have been introduced first in Schafer (2000) and Alford et al. (2000) and their properties have been investigated later more in detail, see Schmitt (2004) and more recently Alford and Cowan (2006). Among many possible pairing patterns (polar, planar, A, CSL), the transverse Color-Spin-Locking phase (CSL) has been demonstrated to be the ground state of a spin-1 color superconductor at $T = 0$ having the largest pressure (Schmitt 2004). Their non-relativistic limit reproduces the B-phase in ^{3}He, which locks angular momentum and spin and is also the most stable phase at $T = 0$. We will show the features that make the CSL likely to occur in the interior of compact stars.

Single flavor and color neutral condensates In the CSL condensates the color and spin are locked

$$\langle q_f^T \, C\gamma^3\lambda_2 \, q_f\rangle = \langle q_f^T \, C\gamma^1\lambda_7 \, q_f\rangle = \langle q_f^T \, C\gamma^2\lambda_5 \, q_f\rangle \equiv \eta_f \tag{13}$$

in an antisymmetric antitriplet in the color-space and an axial vector in the spin-space. A scheme of the pairing is shown in Fig. 2, on the left. One can see immediately, that since Cooper pairs in the CSL phase are single flavor the

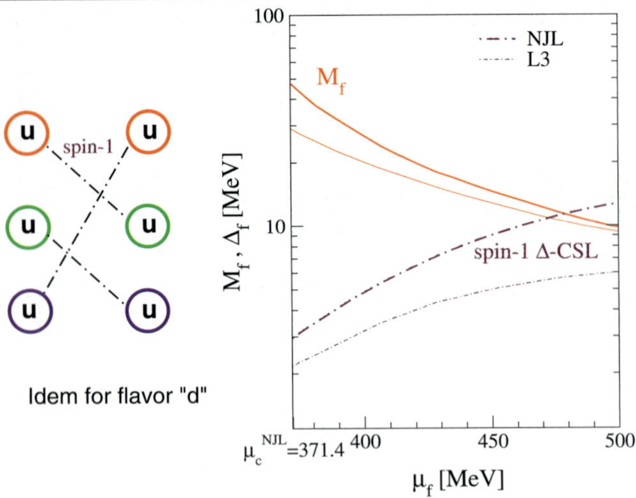

Fig. 2 Spin-1 CSL pairing (*left panel*) and the corresponding dynamical mass and energy gaps (*right panel* as a function of a single flavor chemical potential μ_f, with $f = u, d$). For the NJL model (*thick lines*) the same parameterization as Fig. 1 is used. For the nonlocal extension (label L3), parameters are fixed to $M = 380$ MeV, see Aguilera et al. (2006)

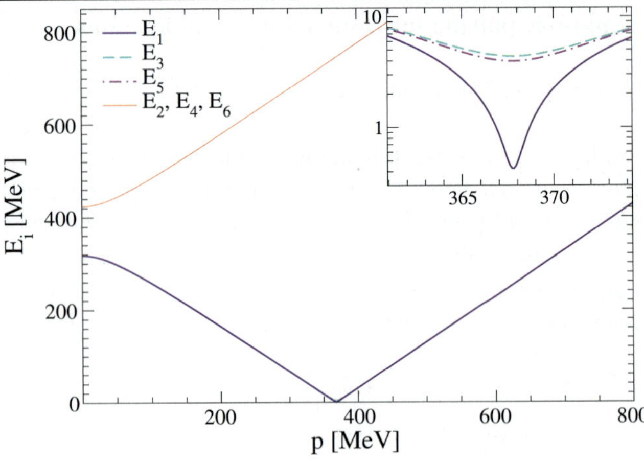

Fig. 3 Excitation energies in the CSL phase as a function of the momentum at $\mu_c = 371.4$ MeV, corresponding to the Fig. 2. Particle modes E_1, E_3, E_5 are zoomed in the *upper corner on the right* showing that there are no gapless modes. Antiparticle modes E_2, E_4, E_6 are superimposed and shown as thin lines. NJL model

results will not be affected by charge neutrality and thus we have overcome one of the most restrictively constraints of quark matter in compact stars. Moreover, there are equal number of color antitriplets ($\bar{r}$, $\bar{g}$, $\bar{b}$) and color neutrality is automatically fulfilled. The CSL diquark gaps are

$$\Delta_f = 4H_v\eta_f, \tag{14}$$

where $H_v = 3/8G$ from Fierz-transformation of single gluon exchange.

One finds that the different flavors (u, d) decouple (Aguilera et al. 2005) and the thermodynamic potential is given by the sum $\Omega_q(T, \mu) = \Omega_u(T, \mu_u) + \Omega_d(T, \mu_d)$ where

$$\begin{aligned}\Omega_f(T, \mu_f) &= \frac{1}{8G}(M_f - m)^2 + \frac{3}{8H_v}|\Delta_f|^2 \\ &\quad - \sum_{k=1}^{6}\int \frac{d^3p}{(2\pi)^3}\left(E_{f;k} + 2T\ln\left(1 + e^{-E_{f;k}/T}\right)\right).\end{aligned} \tag{15}$$

Absence of ungapped modes Defining the effective variables as modified by the factor M_f/μ_f that counts for finite mass effects

$$\varepsilon_{f,\mathrm{eff}}^2 = \vec{p}^{\,2} + M_{f,\mathrm{eff}}^2, \tag{16}$$

$$M_{f,\mathrm{eff}} = \frac{\mu_f}{\mu_{f,\mathrm{eff}}}M_f, \tag{17}$$

$$\mu_{f,\mathrm{eff}}^2 = \mu_f^2 + |\Delta_f|^2, \tag{18}$$

$$\Delta_{f,\mathrm{eff}} = \frac{M_f}{\mu_{f,\mathrm{eff}}}|\Delta_f| \tag{19}$$

we can express the dispersion relation E_1 for the first particle mode in the standard form

$$E_{f;1}^2 = (\varepsilon_{f,\mathrm{eff}} - \mu_{f,\mathrm{eff}})^2 + \Delta_{f,\mathrm{eff}}^2, \tag{20}$$

while for the other particle modes the following approximation can be used

$$E_{f;3,5}^2 \simeq (\varepsilon_f - \mu_f)^2 + c_{f;3,5}^{(1)}|\Delta_f|^2. \tag{21}$$

The dispersion relations for the antiparticles ($k = 2, 4, 6$) can be obtained replacing μ by $-\mu$ and the corresponding coefficients $c_{f;3,5}^{(1)}$ by $c_{f;4,6}^{(1)}$ (see Aguilera et al. 2005 for a complete treatment).

The results of solving the gap equations (4) for the CSL phase are shown in Fig. 2 (right panel). The dynamical quark mass and the diquark gap are plotted as functions of μ_f. Their corresponding excitation energies E_i, $i = 1$–6 for at the critical chemical potential μ_c, the onset for the CSL phase, are shown in Fig. 3.

From Fig. 2 we see that the CSL gaps are strongly μ_f-dependent functions in the considered domain. Since the constituent mass in vacuum M determines the μ_c at which the chiral phase transition takes place (and thus the onset for the superconducting phase), the low density region is qualitatively determined by the parameterization. This is crucial for the later matching of the quark sector with a high density nuclear matter EoS and for the construction of stable configurations of hybrid stars. Models with an onset of the quark matter phase at high densities might not give stable quark cores (Buballa et al. 2004).

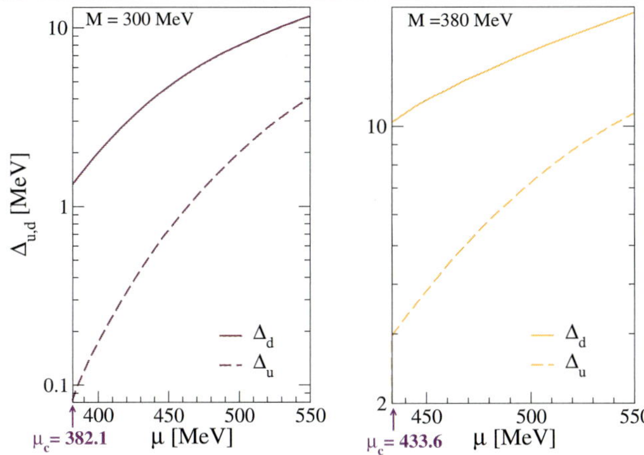

Fig. 4 Neutral matter: Δ_d, Δ_u as a function of μ. Two different parameterizations are considered: $M = 300$ MeV on the *left* and 380 MeV on the *right*. Note the different scales for Δ in both cases showing that Δ is very sensitivity to the variation of the constituent mass M and the corresponding parameterization

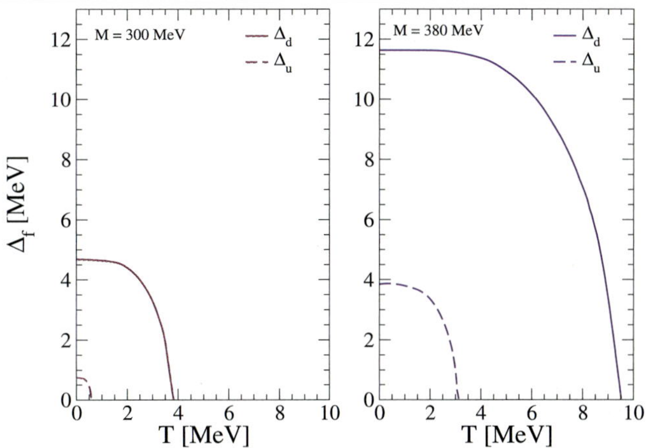

Fig. 5 Neutral matter: Temperature dependence of the CSL gaps

From Fig. 3 we have learned that there are no gapless modes as a direct consequence of keeping the finite size of the mass of the u, d quarks in (19). This may play a crucial role suppressing the neutrino emissivities and preventing the direct Urca process to work. With unpaired quark species, the direct Urca is so efficient that cool down the star too fast, in disagreement with observational data.[2]

3.3 Neutral CSL and critical temperatures

It is straightforward now to consider charge neutral matter introducing an electron chemical potential μ_e, with $\mu_u = \mu - \frac{2}{3}\mu_e$ and $\mu_d = \mu + \frac{1}{3}\mu_e$, such that the total charge vanishes. Therefore, the effect of introducing charge neutrality is that the energy gap for the two flavors splits in two branches: Δ_d, Δ_u, where $\Delta_u < \Delta_d$ for a given μ as it shown in Fig. 4. The difference between them could be as large as a factor ≈ 10 for low densities. The onset for neutral CSL matter is displayed in both parameterizations considered.

We analyze the temperature dependence of the gaps in Fig. 5. They show a decreasing behavior that ends at the critical temperature T_c where a second order phase transition to normal (unpaired) quark matter occurs. It is worth to notice that the critical temperature T_c for the CSL phase deviates from the BCS relation ($T_c \approx 0.57\Delta$ $(T = 0)$), as it has been demonstrated for weak coupling in Schmitt (2004), and follows

$$T_{cf} \approx 0.82\Delta_f \ (T = 0). \tag{22}$$

[2]Although data might be also consistent with models that allow direct Urca in quark matter, lowering the core temperature but retain the surface temperature high enough due to e.g. Joule heating in the crust (Pons et al. 2006).

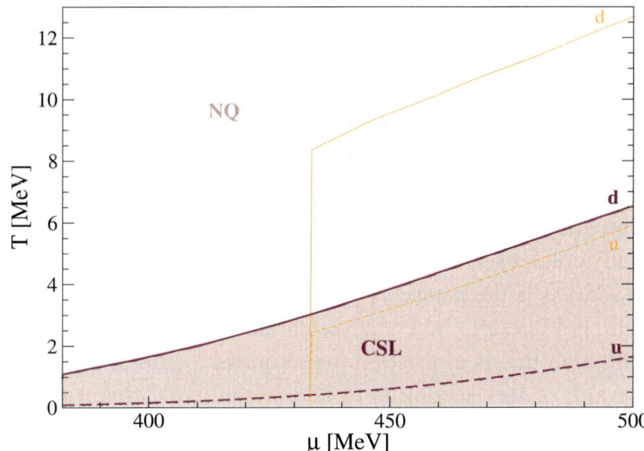

Fig. 6 Two flavor neutral matter at intermediate densities: critical temperatures are shown for the d (*solid line*) and for u (*dashed line*) quarks as a function of μ. The *thick lines* correspond to the parameterization of NJL with $M = 300$ MeV and the *thin lines* to $M = 380$ MeV

Our calculations confirm this result and are shown in the phase diagram of Fig. 6. After the critical potential μ_c for the occurrence of CSL phase, ($\mu_c = 382.1$ MeV for the $M =$ 380 MeV parameterization), two lines labeling T_c for the two flavor species appear: as quark matter cool down, first the d quarks (solid line) will condense and then, at lower temperatures, the phase transition for the u quarks (dashed line) will take place. Thus, when the temperature falls below $\simeq$1 MeV is likely that the two flavor quark matter will be in the CSL phase.

Although a comparison of the CSL free energy with the energy for other possible pairings remains to be done,[3] and will decide whether the CSL phase is preferable, there are many indications that point in the same direction. First, NJL model calculations show that when a moderate coupling

[3]Cristallyne phases, phases with consistent treatment of the strange quark mass, etc.

constant for the usual 2SC phase is considered, two flavor normal quark matter dominates the intermediate region of the phase diagram (Ruster et al. 2005; Blaschke et al. 2005). Then, recently in Alford and Cowan (2006), single pairing NJL model calculations performed with massless $u-d$ quarks and the mass of the s quark taken through an effective chemical potential, show that for large strange quark mass, CSL dominates the phase diagram at low temperature and a second order phase transition to unpaired quark matter occurs as the temperature increases. So, we expect that when the densities in the interior of a compact star are high enough to allow for two-flavor quark matter but not so high to have the strange quark deconfined, and the temperature has fallen below T_c, a CSL quark core might develop. Moreover, for such small gaps the pressure is expected to be approximately the one of the normal $u-d$ quark matter and it has been obtained that hybrid stars with a relatively large normal quark matter core can be stable (Grigorian et al. 2004).

3.4 Neutral CSL and magnetic field

Let's suppose that in the core of a neutron star the density is high enough that a quark core could be formed and has already condensed to the CSL phase. The final question we try to address is the nature of the interaction between the magnetic field present in a neutron star and the CSL quark core. Although this is a question that requires a careful analysis, the aim of this section is to review the work done on this matter and present estimations derivate from the CSL gaps calculated before.

For the interaction of the CSL phase with an external magnetic field B it has been stated that the CSL phase exhibits an electromagnetic Meissner effect (Schmitt et al. 2003) since all the gluons and the photon acquires a non-vanishing Meissner mass proportional to the quark chemical potential (multiplied by the appropriate gauge coupling) (Schmitt et al. 2004). The formation of the Cooper pairs breaks the symmetry (color-spin) $SU(3)_c \times SO(3)_J$ to $SO(3)_{c+J}$, that is a global symmetry. Concerning electromagnetism, it was shown that there is no symmetry subgroup left of $SU(3)_c \times U(1)_{em}$ different from the trivial and thus there is no combination of color and electric charge for which the Cooper pairs are invariant.

This is in contrast to the spin-0 2SC and CFL phases that although the condensate has non-zero electric charge, there is a residual local symmetry $\tilde{U}(1)$ with an associated lineal combination of the photon and a gluon ("rotated photon") that remains massless. The "rotated" $\tilde{B}$ field can penetrate and propagate in the 2SC and in the CFL phase (Alford et al. 2000). Therefore, 2SC and CFL phases are not superconductors from the electromagnetic point of view and the magnetic field $\tilde{B}$ will penetrate without the restriction to be quantized to flux tubes and is stable over very long time scales. On the other hand, B cannot penetrate in the CSL phase unless it exceeds the critical magnetic field B_c that destroy the superconducting phase.

Similarly to the ordinary superconductors, we can estimate the penetration length as the inverse of the Meissner mass of the photon (γ) and/or the gluon (a) that would penetrate the matter following (Schmitt et al. 2004)

$$\lambda_{\gamma,a} \simeq \frac{1}{m_{\gamma,a}} \simeq \frac{1}{(e,g)\mu} \tag{23}$$

where e, g are their corresponding coupling constants in dense matter. At densities in neutron stars, photons are weakly coupled $e^2/4\pi \simeq 1/137$ while gluons are strongly coupled $g^2/4\pi \approx 1$ and taking a typical value of $\mu \simeq$ 400 MeV we obtain[4] that the penetration length is $\lambda \simeq$ 1–10 fm.

The coherence length is proportional to the inverse of the energy gap $\xi \simeq 1/\Delta$ and from our previous results we obtained that being $\Delta \simeq 1$ MeV $\approx 1/200$ fm. Then, the ratio

$$\frac{\lambda}{\xi} \simeq \frac{1\text{–}10}{200} \ll 1/\sqrt{2}, \tag{24}$$

confirms that CSL is a Type-I superconductor and the magnetic field would be expelled from macroscopic regions if it does not exceed the critical field to destroy the condensate. This may be consistent with recent investigations that state that surface temperature anisotropies inferred from the observations are due to the crust confination of the magnetic field (Geppert et al. 2004; Pérez-Azorín et al. 2006). Moreover, studies of the light curves of neutron stars have shown evidence of precession and if this is confirmed, this might be inconsistent with Type II superfluity. Because of the strong interaction between the rotational vortices and the flux tubes in the latter, Type I superconductivity may be required (Link 2003).

We can estimate the critical field B_c using the Ginzburg-Landau approach following Bailin and Love (1983),

$$B_c^2 = 8\mu p_F \frac{(k_B T t)^2}{7\zeta(3)}. \tag{25}$$

Assuming that due to the low mass of u, d, the Fermi momentum $p_F \simeq \mu$ and if the $T \ll T_c$ then the parameter $t = (T - T_c)/T_c \simeq 1$. We can also express $T_c \approx 0.82\,\Delta$ $(T=0)$ and consider the μ-dependence of the gaps to obtain $B_c(\mu) \simeq 0.8\,\mu\,\Delta(\mu)$ and in relevant units

$$B_c(\mu) = 1.6\left(\frac{\mu}{400\ \text{MeV}}\right)\left(\frac{\Delta(\mu)}{\text{MeV}}\right) \times 10^{16}\ \text{G}. \tag{26}$$

[4] Heaviside-Lorentz units are used: $\mu_0 = \epsilon_0 = 1$, $\hbar = c = k = 1$, $e = \sqrt{4\pi/137} \approx 0.3$, $g \approx 3.5$ and thus $\text{GeV}^2 = 5 \times 10^{19}$ G.

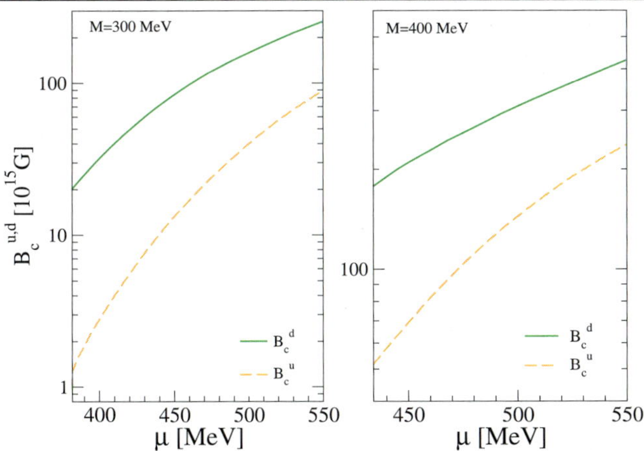

Fig. 7 Estimated critical B as a function of μ for CSL neutral matter. Note the different scales for the two parameterizations used

Therefore for typical values of $\mu = 400$ MeV and CSL gaps not exceeding $\Delta = 1$ MeV, we have that $B_c \simeq 10^{16}$ G. In Fig. 7 we display $B_c(\mu)$ taking into account the density dependence of our previous NJL model results. As the figures show, B_c is highly density dependent, and due to the different μ_u, μ_d there is a splitting in the critical field for the two flavors: $B_c^u \leq B_c^d$.

What are the consequences for compact stars? First, due to the strong density dependence, we expect that if a CSL quark core forms, it begins to grow from the center of the star. Since it is a Type I superconductor, the magnetic field will be expelled from it. If B is not so large that the CSL phase persist against the magnetic field ($B \leq B_c^u$), then, the characteristic times for the expulsion of the field over macroscopic regions and for the quark core grow start to compete. According to Ouyed et al. (2004), the magnetic field expulsion time over a sizable region ($\simeq$10 m) can be evaluated as proportional to the electrical conductivity in the normal phase $\sigma_{\rm el} \simeq 10^{23-24}$ s^{-1} (Shovkovy and Ellis 2003),

$$\tau_{\rm exp} \simeq \left(\frac{\sigma_{\rm el}}{10^{23}\ {\rm s}^{-1}}\right)\left(\frac{\delta}{10\ {\rm m}}\right)^2\left(\frac{B}{10^{15}\ {\rm G}}\right)\ {\rm s} \tag{27}$$

getting typical values of seconds. An estimation of how fast the quark core develops is unknown, but if it is much smaller than $\tau_{\rm exp}$ it might not have time to expel B. Since CSL and the magnetic field do not coexist, B would be frozen in an mixed state composed of alternating regions of normal material with flux density H_c and superconducting material exhibiting Meissner screening. This would be also the case for intermediate fields, comparable with B_c. For the case $B_u^c \leq B \leq B_d^c$, although B will try to destroy the u-condensates, d-currents will cancel the field inside, restoring the superconducting phase. As a results, we speculate that the field structure will be rearranged to give the mixed state described above.

The spin down estimates from highly magnetized neutron stars (*magnetars*) show that the magnetic field could be as large as $B \simeq 10^{15}$ G (Woods and Thompson 2004). If one consider that they could reach even higher values (one order of magnitude more?) as a results of flux conservation into confined regions, the scenario of a CSL quark core under the stress of magnetic fields that are of the order of B_c seems not to be unrealistic.

4 Summary and outlook

We summarize here the interesting features that the CSL phase presents for compact star applications:

- since Cooper pairs are single flavor they are not affected by charge neutrality;
- color neutrality is automatically fulfilled;
- there are no gapless modes. This has important consequences e.g. in neutrino emissivities, suppressing the direct Urca process that leads to a rapid cooling;
- their small gaps ($\Delta \simeq 1$ MeV) will not influence on the EoS, and the pressure is expected to be approximately the one of the normal ud-quark matter without pairing. Similar to the latter, stable hybrid star configurations could be obtained with a relatively large quark matter core;
- it exhibits a Meissner effect and the magnetic field of a neutron star would be expelled from a CSL quark core if it does not exceed the critical magnetic field B_c that destroy the superconducting phase. This hypothesis might be consistent with recent investigations that indicate the crust confination of the magnetic field. Moreover, the hypothesis of a Type I superconductor is supported by the inferred precession of some compact objects.

This study presents a qualitative analysis of the spin-1 pairing and cannot be taken as conclusive to decide whether this phase could be realized in neutron star cores. However, the features listed show that CSL could be consistent with the phenomenology of compact stars in many aspects.

Acknowledgements D.N.A. thanks J. Pons and S. Reddy for their interest in this work and for fruitful discussions on previous version of this manuscript. CSL model calculations have been done in collaboration with D. Blaschke, M. Buballa, N. Scoccola and H. Grigorian. D.N.A. thanks also D. Page for interesting discussions during the conference.

References

Aguilera, D.N., Blaschke, D.: PEPAN Lett. **4**(3), 351 (2007)
Aguilera, D.N., Blaschke, D., Buballa, M., et al.: Phys. Rev. D **72**, 034008 (2005a)
Aguilera, D.N., Blaschke, D., Grigorian, H.: Nucl. Phys. A **757**, 527 (2005b)

Aguilera, D.N., Blaschke, D., Grigorian, H., Scoccola, N.N.: Phys. Rev. D **74**, 114005 (2006)
Alford, M.G., Cowan, G.A.: J. Phys. G **32**, 511 (2006)
Alford, M.G., Rajagopal, K.: J. High Energy Phys. **0206**, 031 (2002)
Alford, M.G., Rajagopal, K., Wilczek, F.: Phys. Lett. B **422**, 247 (1998)
Alford, M.G., Berges, J., Rajagopal, K.: Nucl. Phys. B **571**, 269 (2000)
Alford, M.G., Bowers, J.A., Rajagopal, K.: J. Phys. G **27**, 541 (2001) [Lect. Notes Phys. **578**, 235 (2001)]
Alford, M.G., Bowers, J.A., Cheyne, J.M., et al.: Phys. Rev. D **67**, 054018 (2003)
Bailin, D., Love, A.: Phys. Rep. **107**, 325 (1984)
Blaschke, D., Fredriksson, S., Grigorian, H., et al.: Phys. Rev. D **72**, 065020 (2005)
Buballa, M.: Phys. Rep. **407**, 205 (2005)
Buballa, M., Oertel, M.: Nucl. Phys. A **703**, 770 (2002)
Buballa, M., Hosek, J., Oertel, M.: Phys. Rev. Lett. **90**, 182002 (2003)
Buballa, M., Neumann, F., Oertel, M., et al.: Phys. Lett. B **595**, 36 (2004)
Geppert, U., Küker, M., Page, D.: Astron. Astrophys. **426**, 267 (2004)
Gomez Dumm, D., Blaschke, D., Grunfeld, A.G., et al.: Phys. Rev. D **73**, 114019 (2006)
Grigorian, H., Blaschke, D., Aguilera, D.N.: Phys. Rev. C **69**, 065802 (2004)
Grigorian, H., Blaschke, D., Voskresensky, D.: Phys. Rev. C **71**, 045801 (2005)
Link, B.: Phys. Rev. Lett. **91**, 101101 (2003)
Neumann, F., Buballa, M., Oertel, M.: Nucl. Phys. A **714**, 481 (2003)
Ouyed, R., Elgaroy, O., Dahle, H., Keranen, P.: Astron. Astrophys. **420**, 1025 (2004)
Page, D., Prakash, M., Lattimer, J.M., Steiner, A.: Phys. Rev. Lett. **85**, 2048 (2000)
Page, D., Lattimer, J.M., Prakash, M., Steiner, A.: Astrophys. J. Suppl. Ser. **155**, 623 (2004)
Pérez-Azorín, J.F., Miralles, J.A., Pons, J.A.: Astron. Astrophys. **451**, 1009 (2006)
Pons, J., Link, B., Miralles, J.A., Geppert, U.: Phys. Rev. Lett. (2007, in press), arXiv:astro-ph/0607583 (2006)
Rapp, R., Schafer, T., Shuryak, E.V., et al.: Phys. Rev. Lett. **81**, 53 (1998)
Ruster, S.B., Werth, V., Buballa, M., et al.: Phys. Rev. D **72**, 034004 (2005)
Schafer, T.: Phys. Rev. D **62**, 094007 (2000)
Schmitt, A.: arXiv:nucl-th/0405076 (2004)
Schmitt, A.: Phys. Rev. D **71**, 054016 (2005)
Schmitt, A., Wang, Q., Rischke, D.: Phys. Rev. Lett. **91**, 242301 (2003)
Schmitt, A., Wang, Q., Rischke, D.H.: Phys. Rev. D **69**, 094017 (2004)
Shovkovy, I.A., Ellis, P.J.: Phys. Rev. C **67**, 048801 (2003)
Shovkovy, I., Hanauske, M., Huang, M.: Phys. Rev. D **67**, 103004 (2003)
Steiner, A.W., Reddy, S., Prakash, M.: Phys. Rev. D **66**, 094007 (2002)
Woods, P.M., Thompson, C.: arXiv:astro-ph/0406133 (2004)
Yakovlev, D., Pethick, C.J.: Annu. Rev. Astron. Astrophys. **42**, 169 (2004)

Astrophys Space Sci (2007) 308: 451–455
DOI 10.1007/s10509-007-9356-4

ORIGINAL ARTICLE

Exotic bulk viscosity and its influence on neutron star r-modes

Debarati Chatterjee · Debades Bandyopadhyay

Received: 30 June 2006 / Accepted: 23 October 2006 / Published online: 16 March 2007

Abstract We investigate the effect of exotic matter in particular, hyperon matter on neutron star properties such as equation of state (EoS), mass-radius relationship and bulk viscosity. Here we construct equations of state within the framework of a relativistic field theoretical model. As hyperons are produced abundantly in dense matter, hyperon–hyperon interaction becomes important and is included in this model. Hyperon–hyperon interaction gives rise to a softer EoS which results in a smaller maximum mass neutron star compared with the case without the interaction. Next we compute the coefficient of bulk viscosity and the corresponding damping time scale due to the non-leptonic weak process including Λ hyperons. Further, we investigate the role of the bulk viscosity on gravitational radiation driven r-mode instability in a neutron star of given mass and temperature and find that the instability is effectively suppressed.

Keywords Neutron stars · Dense matter · r-mode oscillation

PACS 97.60.Jd · 26.60.+c · 04.40.Dg

1 Introduction

The investigation of spin frequencies from burst oscillations of eleven nuclear-powered millisecond pulsars showed that the spin distribution had an upper limit at 730 Hz (Chakrabarty et al. 2003, 2005). The fastest rotating neutron star discovered recently has a spin frequency 716 Hz (Hessels et al. 2006). In this respect, the study of r-modes in rotating neutron stars has generated great interest in understanding the absence of very fast rotating neutron stars in nature. The r-modes are subject to Chandrasekhar–Friedman–Schutz gravitational radiation instability in rapidly rotating neutron stars (Andersson 1998, 2003; Friedman and Morsink 1998; Lindblom et al. 1998; Andersson et al. 1999; Stergioulas 2003) which may play an important role in regulating spins of young neutron stars as well as old, accreting neutron stars in low mass X-ray binaries (LMXBs). Another aspect of the r-mode instability is that this could be a possible source for gravitational radiation (Andersson 2003; Gondek-Rosinska et al. 2003; Nayyar and Owen 2006) which may shed light on the interior of a neutron star.

It was realised that the r-mode instability could be effectively suppressed by bulk viscosity due to exotic matter in neutron star interior when the compact star cools down to a temperature $\sim 10^9$ K. The coefficient of bulk viscosity due to non-leptonic weak processes involving exotic matter such as hyperon and quark matter was calculated by several authors (Jones 2001; Lindblom and Owen 2002; Haensel et al. 2002; van Dalen and Dieperink 2004; Drago et al. 2005; Madsen 1992, 2000). It was noted that relaxation times of the non-leptonic processes were within a few orders magnitude of the period of perturbations and gave rise to large bulk viscosity coefficient. This led to the complete suppression of the r-mode instability.

In this paper, we investigate the effect of hyperon matter including hyperon–hyperon interaction on bulk viscosity coefficient and the r-mode stability. In Sect. 2 of this paper, we discuss the model used to calculate equations of state, bulk viscosity coefficient and the corresponding time scale.

D. Chatterjee · D. Bandyopadhyay (✉)
Saha Institute of Nuclear Physics, 1/AF Bidhannagar, Kolkata 700064, India
e-mail: debades.bandyopadhyay@saha.ac.in

Results of our calculation are analysed in Sect. 3. Section 4 gives the outlook.

2 Model

2.1 Equation of state

We describe the β-equilibrated and charge neutral hyperon matter within the framework of a relativistic field theoretical model where baryon–baryon interaction is mediated by the exchange of scalar and vector mesons and hyperon–hyperon interaction is taken into account by two strange mesons—scalar f_0 (hereafter denoted as σ^*) and vector ϕ (Schaffner et al. 1993; Schaffner and Mishustin 1996). The Lagrangian density for hyperon matter including non-interacting electrons and muons is written as

$$\begin{aligned}\mathcal{L}_B = &\sum_B \bar{\Psi}_B \big(i\gamma_\mu \partial^\mu - m_B + g_{\sigma B}\sigma + g_{\sigma^* B}\sigma^* \\ &- g_{\omega B}\gamma_\mu\omega^\mu - g_{\phi B}\gamma_\mu\phi^\mu - g_{\rho B}\gamma_\mu \mathbf{t}_B \cdot \boldsymbol{\rho}^\mu\big)\Psi_B \\ &+ \frac{1}{2}\big(\partial_\mu\sigma\partial^\mu\sigma - m_\sigma^2\sigma^2\big) - U(\sigma) \\ &+ \frac{1}{2}\big(\partial_\mu\sigma^*\partial^\mu\sigma^* - m_{\sigma^*}^2\sigma^{*2}\big) \\ &- \frac{1}{4}\omega_{\mu\nu}\omega^{\mu\nu} + \frac{1}{2}m_\omega^2\omega_\mu\omega^\mu \\ &- \frac{1}{4}\boldsymbol{\rho}_{\mu\nu}\cdot\boldsymbol{\rho}^{\mu\nu} + \frac{1}{2}m_\rho^2\boldsymbol{\rho}_\mu\cdot\boldsymbol{\rho}^\mu \\ &- \frac{1}{4}\phi_{\mu\nu}\phi^{\mu\nu} + \frac{1}{2}m_\phi^2\phi_\mu\phi^\mu. \end{aligned} \tag{1}$$

The Dirac bispinor Ψ_B represents isospin multiplets for baryons $B = N, \Lambda, \Sigma$ and Ξ. Here vacuum baryon mass is m_B, and isospin operator t_B while $\omega_{\mu\nu}$, $\phi_{\mu\nu}$, and $\rho_{\mu\nu}$ are field strength tensors. Neutrons and protons do not couple with σ^* and ϕ mesons. The scalar self-interaction term (Boguta and Bodmer 1977) is given by

$$U(\sigma) = \frac{1}{3}g_2\sigma^3 + \frac{1}{4}g_3\sigma^4. \tag{2}$$

We perform the calculation in the mean field approximation. The total energy density and pressure are respectively given by

$$\begin{aligned}\varepsilon = &\frac{1}{2}m_\sigma^2\sigma^2 + \frac{1}{3}g_2\sigma^3 + \frac{1}{4}g_3\sigma^4 + \frac{1}{2}m_{\sigma^*}^2\sigma^{*2} \\ &+ \frac{1}{2}m_\omega^2\omega_0^2 + \frac{1}{2}m_\phi^2\phi_0^2 + \frac{1}{2}m_\rho^2\rho_{03}^2 \\ &+ \sum_B \frac{2J_B+1}{2\pi^2}\int_0^{k_{F_B}} \big(k^2 + m_B^{*2}\big)^{1/2} k^2\, dk \end{aligned}$$

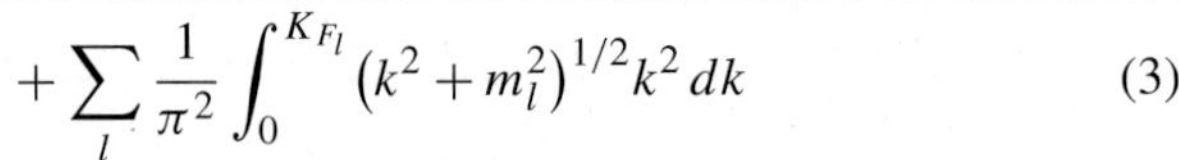

$$+ \sum_l \frac{1}{\pi^2}\int_0^{K_{F_l}} \big(k^2 + m_l^2\big)^{1/2} k^2\, dk \tag{3}$$

and

$$\begin{aligned}P = &-\frac{1}{2}m_\sigma^2\sigma^2 - \frac{1}{3}g_2\sigma^3 - \frac{1}{4}g_3\sigma^4 \\ &- \frac{1}{2}m_{\sigma^*}^2\sigma^{*2} + \frac{1}{2}m_\omega^2\omega_0^2 + \frac{1}{2}m_\phi^2\phi_0^2 + \frac{1}{2}m_\rho^2\rho_{03}^2 \\ &+ \frac{1}{3}\sum_B \frac{2J_B+1}{2\pi^2}\int_0^{k_{F_B}} \frac{k^4\, dk}{(k^2 + m_B^{*2})^{1/2}} \\ &+ \frac{1}{3}\sum_l \frac{1}{\pi^2}\int_0^{K_{F_l}} \frac{k^4\, dk}{(k^2 + m_l^2)^{1/2}}, \end{aligned} \tag{4}$$

where k_{F_B}, J_B, I_{3B} and n_B are Fermi momentum, spin and isospin projection and number density of baryon B respectively. The last terms in energy density and pressure are due to leptons. The effective baryon mass is defined as $m_B^* = m_B - g_{\sigma B}\sigma - g_{\sigma^* B}\sigma^*$. The charge neutrality and chemical equilibrium conditions are given by $Q = \sum_B q_B n_B - n_e - n_\mu = 0$ and $\mu_i = b_i\mu_n - q_i\mu_e$ where μ_n, μ_e and μ_i are respectively the chemical potentials of neutrons, electrons and baryon i and b_i and q_i are baryon and electric charge of baryon i. The chemical potential of baryons B is given by (Chatterjee and Bandyopadhyay 2006a) $\mu_B = (k_{F_B}^2 + m_B^{*2})^{1/2} + g_{\omega B}\omega_0 + g_{\phi B}\phi_0 + I_{3B}g_{\rho B}\rho_{03}$.

In this calculation we adopt nucleon-meson coupling constants of GM set (Glendenning and Moszkowski 1991). The coupling constants are obtained by reproducing properties of saturated nuclear matter such as binding energy -16.3 MeV, saturation density $n_0 = 0.153$ fm^{-3}, asymmetry energy 32.5 MeV, effective nucleon mass 0.78 and incompressibility 240 MeV. The hadronic masses used in this calculation are $m_\sigma = 550$ MeV, $m_\omega = 783$ MeV, $m_\rho = 770$ MeV and $m_N = 938$ MeV. On the other hand, hyperon-vector meson coupling constants are determined using SU(6) symmetry of the quark model (Schaffner and Mishustin 1996; Dover and Gal 1984; Schaffner et al. 1994) and the scalar σ meson coupling to hyperons is calculated from hyperon potential depths in normal nuclear matter such as $U_\Lambda^N(n_0) = -30$ MeV (Dover and Gal 1984; Chrien and Dover 1989), $U_\Xi^N(n_0) = -18$ MeV (Fukuda et al. 1998; Khaustov et al. 2000) and a repulsive potential depth for Σ hyperons $U_\Sigma^N(n_0) = +30$ MeV (Friedman et al. 1994; Batty et al. 1997; Bart et al. 1999) as obtained from hypernuclei data. The hyperon-σ^* coupling constants are determined from double Λ hypernuclei data (Schaffner et al. 1993; Schaffner and Mishustin 1996).

We obtain the EoS solving equations of motion along with the expression for effective baryon mass and charge neutrality and beta equilibrium constraints (Schaffner and Mishustin 1996; Chatterjee and Bandyopadhyay 2006a).

2.2 Bulk viscosity coefficient, damping time scales and critical angular velocity

Now we discuss the bulk viscosity coefficient in young neutron stars which cool down to temperatures $\sim 10^9$–10^{10} K after their births. As the system goes out of chemical equilibrium due to pressure and density variations associated with the r-mode oscillations, microscopic reaction processes in particular non-leptonic weak interaction processes including exotic particles might restore the equilibrium (Nayyar and Owen 2006; Jones 2001; Lindblom and Owen 2002; Chatterjee and Bandyopadhyay 2006a). Here we calculate the real part of bulk viscosity coefficient (ζ) in terms of relaxation times of microscopic processes (Landau and Lifshitz 1999; Lindblom and Owen 2002)

$$\zeta = \frac{P(\gamma_\infty - \gamma_0)\tau}{1+(\omega\tau)^2}, \tag{5}$$

where the difference of infinite (γ_∞) and zero (γ_0) frequency adiabatic indices is given by (Lindblom and Owen 2002; Nayyar and Owen 2006)

$$\gamma_\infty - \gamma_0 = -\frac{n_b^2}{P}\frac{\partial P}{\partial n_n}\frac{d\bar{x}_n}{dn_b}. \tag{6}$$

In the co-rotating frame, the angular velocity (ω) of (l,m) r-mode is related to angular velocity (Ω) of a rotating neutron star as $\omega = \frac{2m}{l(l+1)}\Omega$ (Andersson 2003). In this calculation, for $l=m=2$ r-mode, it is $\omega = \frac{2}{3}\Omega$. The relaxation time ($\tau$) for the non-leptonic process

$$n + p \rightleftharpoons p + \Lambda, \tag{7}$$

is given by Lindblom and Owen (2002)

$$\frac{1}{\tau} = \frac{(kT)^2}{192\pi^3} k_{F_\Lambda} \langle |M_\Lambda|^2 \rangle \frac{\delta\mu}{n_b \delta x_n}, \tag{8}$$

where k_{F_Λ} is the Fermi momentum for Λ hyperons, $\langle |M_\Lambda|^2 \rangle$ is the angle averaged squared matrix element and

$$\frac{\delta\mu}{n_b\delta x_n} = \alpha_{nn} - \alpha_{\Lambda n} - \alpha_{n\Lambda} + \alpha_{\Lambda\Lambda}, \tag{9}$$

with $\alpha_{ij} = \frac{\partial \mu_i}{\partial n_j}_{n_k, k\neq j}$. We obtain expressions for α_{ij} from the baryon chemical potential and equations of motion for meson fields (Chatterjee and Bandyopadhyay 2006a). For example,

$$\alpha_{\Lambda n} = \frac{\partial \mu_\Lambda}{\partial n_n} = \frac{g_{\omega\Lambda} g_{\omega N}}{m_\omega^2} - \frac{m_\Lambda^* g_{\sigma\Lambda}}{\sqrt{k_{F_\Lambda}^2 + m_\Lambda^{*2}}}\frac{\partial\sigma}{\partial n_n} - \frac{m_\Lambda^* g_{\sigma^*\Lambda}}{\sqrt{k_{F_\Lambda}^2 + m_\Lambda^{*2}}}\frac{\partial\sigma^*}{\partial n_n}. \tag{10}$$

Similarly, we compute other components of α_{ij}. It is to be noted here that nucleons do not couple with strange mesons, i.e. $g_{\sigma^* N} = 0$.

The hyperon bulk viscosity damping timescale (τ_h) contributes to the imaginary part of the r-mode frequency and is calculated in the following way (Nayyar and Owen 2006; Lindblom et al. 1998, 1999)

$$\frac{1}{\tau_h} = -\frac{1}{2E}\frac{dE}{dt}, \tag{11}$$

where E is the energy of the perturbation in the co-rotating frame of the fluid

$$E = \frac{1}{2}\alpha^2\Omega^2 R^{-2}\int_0^R \epsilon(r) r^6\, dr. \tag{12}$$

The derivative of E with respect to time is given by

$$\frac{dE}{dt} = -4\pi \int_0^R \zeta(r) \langle |\nabla \cdot \delta \boldsymbol{v}|^2 \rangle r^2\, dr, \tag{13}$$

where we need to know the energy density $\epsilon(r)$ and bulk viscosity $\zeta(r)$ profiles of a neutron star. Similarly, we estimate the damping time scale (τ_U) corresponding to the bulk viscosity due to the modified Urca process including only nucleons using the bulk viscosity coefficient as given by Sawyer (1989). It was noted that gravitational radiation drives the r-modes unstable. So the gravitational radiation time scale (τ_{GR}) has a negative contribution to the imaginary part of the r-mode frequency. Further, the overall r-mode time scale (τ_r) is defined as

$$\frac{1}{\tau_r} = -\frac{1}{\tau_{GR}} + \frac{1}{\tau_B} + \frac{1}{\tau_U}. \tag{14}$$

The r-mode is stable below the critical angular velocity which is obtained from the solution of $\frac{1}{\tau_r} = 0$. The critical angular velocity depends both on the mass of the neutron star and its temperature.

3 Results

We perform this calculation using the GM set (Glendenning and Moszkowski 1991) as it has been described in Sect. 2. It is found that Λ hyperons appear first at 2.6n$_0$ followed by Ξ^- hyperons at 2.9n$_0$. However no Σ hyperons appear in the system because of strongly repulsive Σ-nuclear matter interaction (Chatterjee and Bandyopadhyay 2006a). Further, we find that after the appearance of negatively charged Ξ hyperons, number densities of electrons and muons decrease in the system. This is attributed to the charge neutrality condition. Due to the appearance of additional degrees of freedom in the form of hyperons, the EoS of hyperon matter becomes softer compared with that of nucleons-only matter.

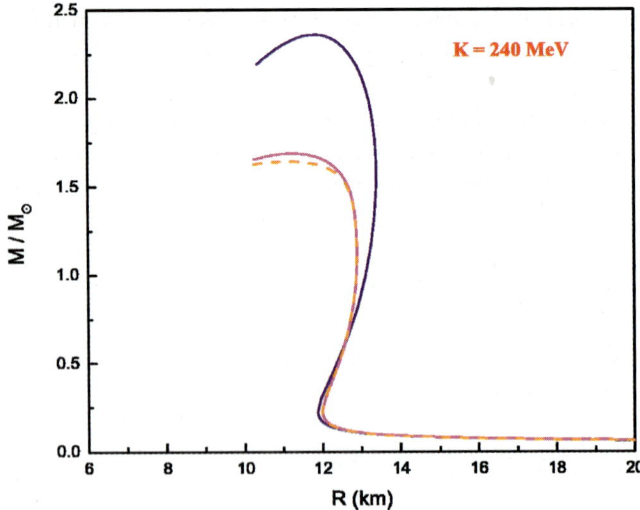

Fig. 1 Mass-radius relationship of non-rotating neutron stars including nucleons-only matter (*bold solid line*) and hyperon matter with (*dashed line*) and without (*solid line*) hyperon–hyperon interaction

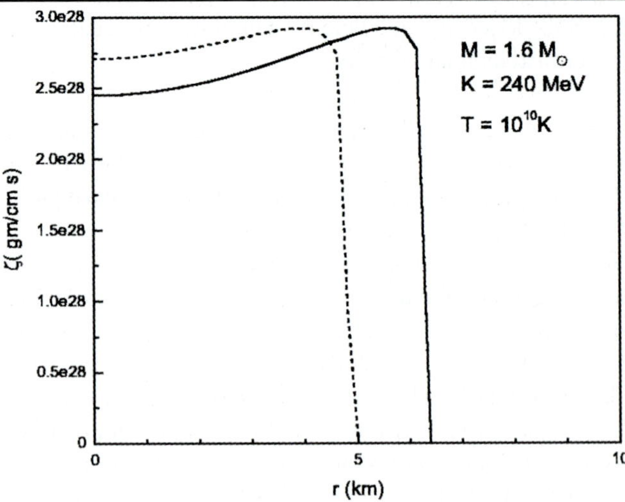

Fig. 2 Bulk viscosity coefficient as a function of radial distance for a non-rotating star (*solid line*) and as a function of equatorial distance for a rotating star (*dashed line*) of mass 1.6 $M_\odot$ including hyperon matter with hyperon–hyperon interaction

On the other hand, the interplay between the attractive σ^* field and repulsive ϕ field which becomes dominant with increasing density, makes the EoS softer initially and stiffer at higher densities than the EoS without hyperon–hyperon interaction. This behaviour of the EoS is reflected in the calculation of maximum mass and the corresponding radius of the neutron star. In Fig. 1, we plot the mass-radius relationship of non-rotating neutron stars calculated by solving Tolman–Oppenheimer–Volkoff equation for equations of state including nucleons-only matter (bold solid line) and hyperon matter with (dashed line) and without (solid line) hyperon–hyperon interaction. Here the highest maximum mass corresponds to the stiffest EoS of nucleons-only matter. The maximum masses corresponding to the EoS with and without hyperon–hyperon interaction are 1.64 $M_\odot$ and 1.69 $M_\odot$ respectively. Using the rotating neutron star (RNS) model of Stergioulas (Stergioulas and Friedman 1995), we find that the maximum masses corresponding to the EoS with and without hyperon–hyperon interaction are respectively 1.95 $M_\odot$ and 2.00 $M_\odot$.

Next we discuss the hyperon bulk viscosity due to the non-leptonic process given by (7). The bulk viscosity coefficient is dependent on the EoS as it is evident from the expression in (5). The temperature dependence of the bulk viscosity enters through the relaxation time as given by (8). The bulk viscosity coefficient reaches a maximum and then drops with increasing baryon density as found by Chatterjee and Bandyopadhyay (2006a). It has been also noted that the bulk viscosity increases as temperature decreases. Therefore, the large value of ζ might be effective in suppressing r-mode instability as neutron stars cool down to a few times 10^9 K.

The neutron star which was rotating rapidly at its birth, slows down due to gravitational wave emission. A rotating neutron star has less central energy density than its non-rotating counterpart because of the centrifugal force. Consequently, hyperon thresholds are sensitive to rotation periods of compact stars. In the calculation of critical angular velocity, we consider a rotating neutron star of mass 1.6 $M_\odot$ and the corresponding central baryon density is $3.9n_0$ which is well above the threshold of Λ hyperons. This star is rotating at an angular velocity 2952 s^{-1}. It slowed down from its Keplerian angular velocity 5600 s^{-1} when the central baryon density was below the Λ hyperon threshold. The calculation of critical angular velocity requires the knowledge of energy density and bulk viscosity profiles in compact stars as it is evident from (11–14). In Fig. 2, the hyperon bulk viscosity profile (dashed line) of the rotating neutron star of mass 1.6 $M_\odot$ is shown as a function of equatorial distance along with that of a non-rotating (solid line) star of same mass. We note that the peak of the profile shifts towards the center in the case of the rotating neutron star. Earlier calculation of critical angular velocity was performed using energy density and bulk viscosity profiles of non-rotating neutron stars. Recently, Nayyar and Owen (2006) studied the effect of rotation on critical angular velocity. We calculate the critical angular velocity using energy density and bulk viscosity profiles of the rotating star. In Fig. 3 the critical angular velocity is plotted versus temperature. We find that the r-mode instability is suppressed by hyperon bulk viscosity. Here we do not consider the impact of superfluidity of baryons on bulk viscosity. However, several calculations (Haensel et al. 2000, 2001, 2002; Nayyar and Owen 2006) showed that superfluidity of baryons might suppress the bulk viscosity and the damping of the r-mode instability.

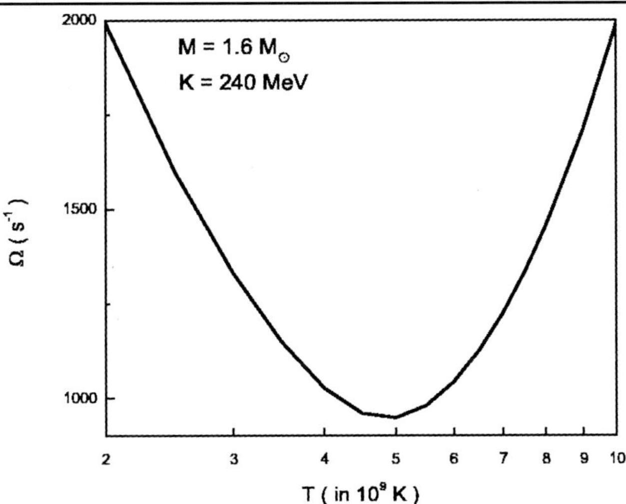

Fig. 3 Critical angular velocity as a function of temperature for a rotating star of mass 1.6 $M_{\odot}$ including hyperon matter with hyperon–hyperon interaction

4 Outlook

Other forms of matter such as K^- condensed matter and quark matter may appear and compete with the Λ hyperon threshold in neutron star interior. In an earlier calculation we showed that the condensate of K^- mesons appeared at $\sim$twice normal nuclear matter density followed by the appearance of Λ hyperons (Banik and Bandyopadhyay 2001). It would be worth investigating how the bulk viscosity coefficient due to the process (7) is modified in this situation. Also, we are studying whether the non-leptonic process involving K^- condensate $n \rightleftharpoons p + K^-$ could give rise to bulk viscosity coefficient as large as that of the non-leptonic process (7) and damp the r-mode instability effectively (Chatterjee and Bandyopadhyay 2006b).

Acknowledgements D.B. thanks the Alexander von Humboldt Foundation for support and acknowledges the warm hospitality at Frankfurt Institute for Advanced Studies (FIAS) and Institute for Theoretical Physics, J.W. Goethe University, Frankfurt am Main where a part of this work was completed.

References

Andersson, N.: Astrophys. J. **502**, 708 (1998)
Andersson, N.: Class. Quantum Gravity **20**, R105 (2003)
Andersson, N., Kokkotas, K.D., Schutz, B.F.: Astrophys. J. **510**, 846 (1999)
Banik, S., Bandyopadhyay, D.: Phys. Rev. C **64**, 055805 (2001)
Bart, S., et al.: Phys. Rev. Lett. **83**, 5238 (1999)
Batty, C.J., Friedman, E., Gal, A.: Phys. Rep. **287**, 385 (1997)
Boguta, J., Bodmer, A.R.: Nucl. Phys. A **292**, 413 (1977)
Chakrabarty, D., et al.: Nature **424**, 42 (2003)
Chakrabarty, D.: In: Rasio F., Stairs I. (eds.) The Proceedings of Binary Radio Pulsars. ASP Conference Series, vol. 328, p. 279 (2005)
Chatterjee, D., Bandyopadhyay, D.: Phys. Rev. D **74**, 023003 (2006a)
Chatterjee, D., Bandyopadhyay, D.: (2006b), astro-ph/07022593
Chrien, R.E., Dover, C.B.: Annu. Rev. Nucl. Part. Sci. **39**, 113 (1989)
Drago, A., Lavagno, A., Pagliara, G.: Phys. Rev. D **71**, 103004 (2005)
Dover, C.B., Gal, A.: Prog. Part. Nucl. Phys. **12**, 171 (1984)
Friedman, J.L., Morsink, S.M.: Astrophys. J. **502**, 714 (1998)
Friedman, E., Gal, A., Batty, C.J.: Nucl. Phys. A **579**, 518 (1994)
Fukuda, T., et al.: Phys. Rev. C **58**, 1306 (1998)
Glendenning, N.K., Moszkowski, S.A.: Phys. Rev. Lett. **67**, 2414 (1991)
Gondek-Rosinska, D., Gourgoulhon, E., Haensel, P.: Astron. Astrophys. **412**, 777 (2003)
Haensel, P., Levenfish, K.P., Yakovlev, D.G.: Astron. Astrophys. **357**, 1157 (2000)
Haensel, P., Levenfish, K.P., Yakovlev, D.G.: Astron. Astrophys. **372**, 130 (2001)
Haensel, P., Levenfish, K.P., Yakovlev, D.G.: Astron. Astrophys. **381**, 1080 (2002)
Hessels, J.W.T., et al.: Science **311**, 1901 (2006)
Jones, P.B.: Phys. Rev. Lett. **86**, 1384 (2001); Phys. Rev. D **64**, 084003 (2001)
Khaustov, P., et al.: Phys. Rev. C **61**, 054603 (2000)
Landau, L.D., Lifshitz, E.M.: Fluid Mechanics. Butterworth–Heinemann, Oxford (1999)
Lindblom, L., Owen, B.J.: Phys. Rev. D **65**, 063006 (2002)
Lindblom, L., Owen, B.J., Morsink, S.M.: Phys. Rev. Lett. **80**, 4843 (1998)
Lindblom, L., Mendell, G., Owen, B.J.: Phys. Rev. D **60**, 064006 (1999)
Madsen, J.: Phys. Rev. D **46**, 3290 (1992)
Madsen, J.: Phys. Rev. Lett. **85**, 10 (2000)
Nayyar, M., Owen, B.J.: Phys. Rev. D **73**, 084001 (2006)
Sawyer, R.F.: Phys. Rev. D **39**, 3804 (1989)
Schaffner, J., Mishustin I.N.: Phys. Rev. C **53**, 1416 (1996)
Schaffner, J., et al.: Phys. Rev. Lett. **71**, 1328 (1993)
Schaffner, J., et al.: Ann. Phys. (N.Y.) **235**, 35 (1994)
Stergioulas, N.: Living Rev. Relativ. **6**, 3 (2003)
Stergioulas, N., Friedman, J.L.: Astrophys. J. **444**, 306 (1995)
van Dalen, E.N.E., Dieperink, A.E.L.: Phys. Rev. C **69**, 025802 (2004)

Astrophys Space Sci (2007) 308: 457–465
DOI 10.1007/s10509-007-9382-2

ORIGINAL ARTICLE

Nucleon superfluidity versus thermal states of isolated and transiently accreting neutron stars

K.P. Levenfish · P. Haensel

Received: 18 July 2006 / Accepted: 6 November 2006 / Published online: 15 March 2007

Abstract The properties of superdense matter in neutron star (NS) cores control NS thermal states by affecting the efficiency of neutrino emission from NS interiors. To probe these properties we confront the theory of thermal evolution of NSs with observations of their thermal radiation. Our observational basis includes cooling isolated NSs (INSs) and NSs in quiescent states of soft X-ray transients (SXTs). We find that the data on SXTs support the conclusions obtained from the analysis of INSs: strong proton superfluidity with $T_{cp}^{\max} \gtrsim 10^9$ K should be present, while mild neutron superfluidity with $T_{cn}^{\max} \approx 2 \times (10^8–10^9)$ K is ruled out in the outer NS core. Here $T_{cn}^{\max}$ and $T_{cp}^{\max}$ are the maximum values of the density dependent critical temperatures of neutrons and protons. The data on SXTs suggest also that: (i) cooling of massive NSs is enhanced by neutrino emission more powerful than the emission due to Cooper pairing of neutrons; (ii) mild neutron superfluidity, if available, might be present only in inner cores of massive NSs. In the latter case SXTs would exhibit dichotomy, i.e. very similar SXTs may evolve to very different thermal states.

Keywords Neutron stars · Nucleon superfluidity

PACS 97.60.Jd · 26.60.+c

K.P. Levenfish (✉)
Ioffe Physical Technical Institute, St. Petersburg, Russia
e-mail: ksen@astro.ioffe.ru

P. Haensel
N. Copernicus Astronomical Center, Warsaw, Poland
e-mail: haensel@camk.edu.pl

1 Introduction

Neutron stars are very compact; their cores contain matter with density ρ a few times larger than the standard nuclear matter density $\rho_0 = 2.8 \times 10^{14}$ g cm^{-3}. Many properties of this matter cannot be calculated precisely or studied in laboratory experiments. However, these properties can be constrained by comparing neutron star theory with observations; see e.g., Yakovlev and Pethick (2004) and Page et al. (2006), for recent reviews.

In this paper, we describe current constraints on composition and superfluidity of neutron star cores, which can be obtained by comparing calculated thermal states of neutron stars with observations of thermal radiation from middle-aged INSs and NSs in SXTs in quiescent states. INSs are thought to cool gradually from initial hot states via neutrino emission from NS interiors (at age $\lesssim 10^5$ years) and via surface photon emission (at the elder age; the so called neutrino and photon cooling stages, respectively). NSs in SXTs will be assumed to support their warm states owing to deep crustal heating (pycnonuclear reactions) in accreted matter (see Brown et al. 1998). Their thermal energy is partly emitted by neutrinos and partly by the surface radiation. In both cases, thermal states of NSs are very sensitive to composition and superfluidity of matter in their cores. Although INSs and SXTs are different objects, their observations allow one to test the same physics of superdense matter (Yakovlev et al. 2003, 2004).

Composition and *superfluidity of baryons* in NS cores are main regulators of NS thermal states. These properties are thought to be the same for a given density in all NS cores. The composition determines dominant processes of neutrino cooling in NSs. Superfluidity, if it appears, reduces rates of neutrino processes at work, but opens an additional neutrino mechanism associated with Cooper pairing of baryons. Superfluidity affects also the NS heat capacity. Other regulators of NSs thermal states (composition of NS heat-blanketing envelopes, strength and geometry of NS magnetic field, etc.) are also important but, taken alone, do not allow one to reconcile theory of NS thermal state with observations (see, e.g., Yakovlev and Pethick 2004). Moreover, they may vary from one star to another. We will neglect them, for clarity.

2 Composition of NS cores

An NS core can be divided into the outer core ($\rho \lesssim$ a few ρ_0) and the inner one (higher ρ). The outer core consists mostly of neutrons, with a small (a few %) admixture of protons and leptons. Low-mass NSs have only the outer core. In the absence of nucleon superfluidity, they cool down via the modified Urca processes and weaker neutrino processes of nucleon-nucleon bremsstrahlung.

More massive NSs possess also the inner core whose composition is largely unknown. According to different hypotheses, the inner core may contain nucleons, hyperons, pion or kaon condensates, or quarks. Even if composed of nucleons and leptons, superdense matter can have a large fraction of protons. In each of these cases, neutrino emission processes of direct Urca type become allowed, much more powerful than the basic modified Urca processes. The most powerful is the nucleon (or hyperon) direct Urca process. Neutrino reactions with pions (kaons) are about two (four) orders of magnitude weaker.

In the presented figures we compare various theoretical predictions of thermal states of INSs and NSs in SXTs with observations. For INSs, we plot the redshifted effective surface temperature T_S^∞ versus NS age t. For SXTs, we show the redshifted surface luminosity L_S^∞ of NSs in quiescence versus time-averaged mass accretion rate $\dot{M}$. We assume that NSs in SXTs are in thermal steady-states, with the deep crustal heating (for a corresponding $\dot{M}$) balanced by neutrino and photon emission; see Yakovlev et al. (2003) for details. The deep crustal heating is calculated using the model of Haensel and Zdunik (1990). In the majority of cases NS interiors are nearly isothermal owing to high thermal conductivity, with the main temperature gradients located in the heat-blanketing envelope near the NS surface. We omit technical details because of space restrictions. Observations of INSs are the same as in Kaminker et al. (2002). The SXTs Aql X-1, 4U 1608-52, SAX 1808.4-3658 and Cen X-4 are described in Yakovlev et al. (2003). The data on XTE J2123-058, KS 1731-260, RX J1709-2639, SAX 1810.8-2609 and 1H 1905+000 are taken from Tomsick et al. (2004), Cackett et al. (2006), Jonker et al. (2003, 2004, 2006), respectively.

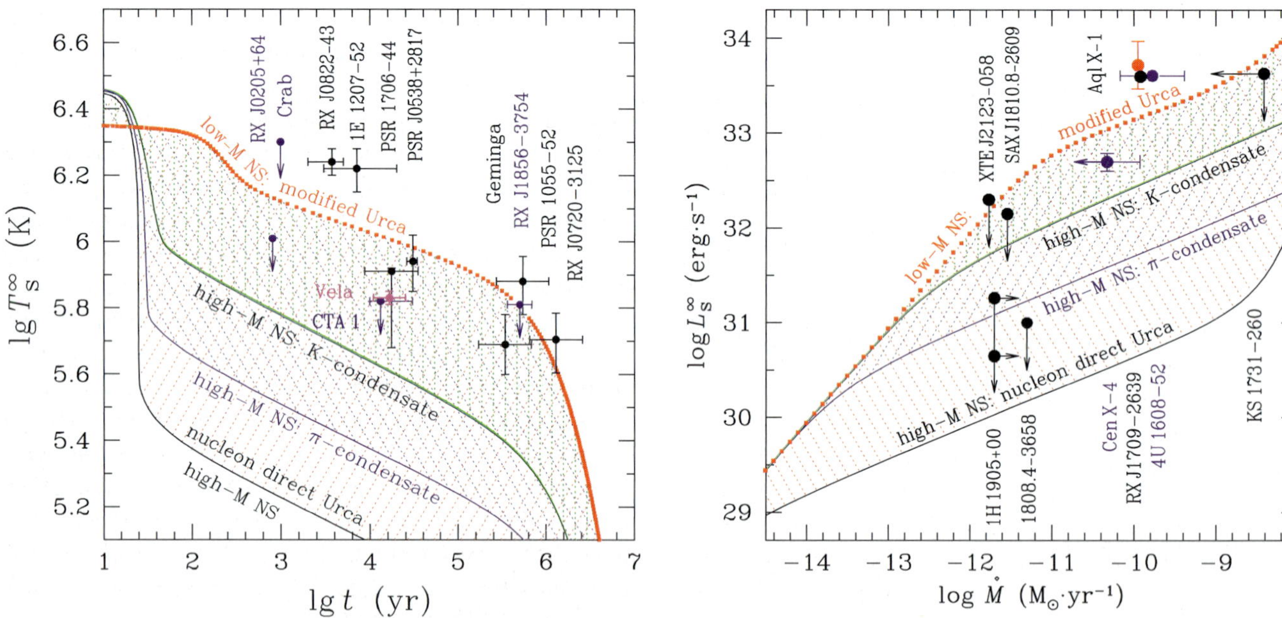

Fig. 1 Thermal states of nonsuperfluid NSs compared to observations. *Left*: effective surface temperatures of INSs, redshifted for a distant observer, versus their age. *Right*: redshifted photon luminosities of NSs in SXTs in quiescence versus time-averaged mass accretion rate. The *dotted curves* refer to the basic NS model (a non-superfluid low-mass NS which cools slowly through the modified Urca process). *Three solid curves on each panel* display scenarios with the enhanced neutrino cooling (maximum-mass NSs with inner cores containing—from *top* to *bottom*—kaon condensates, pion condensates, and nucleons with large proton fraction, sufficient to open direct Urca process). Hatched regions between the *basic curve* and any *solid curve* can be filled by curves of NSs with different masses, from $\sim 1\, M_\odot$ to the maximum one, for a corresponding NS composition

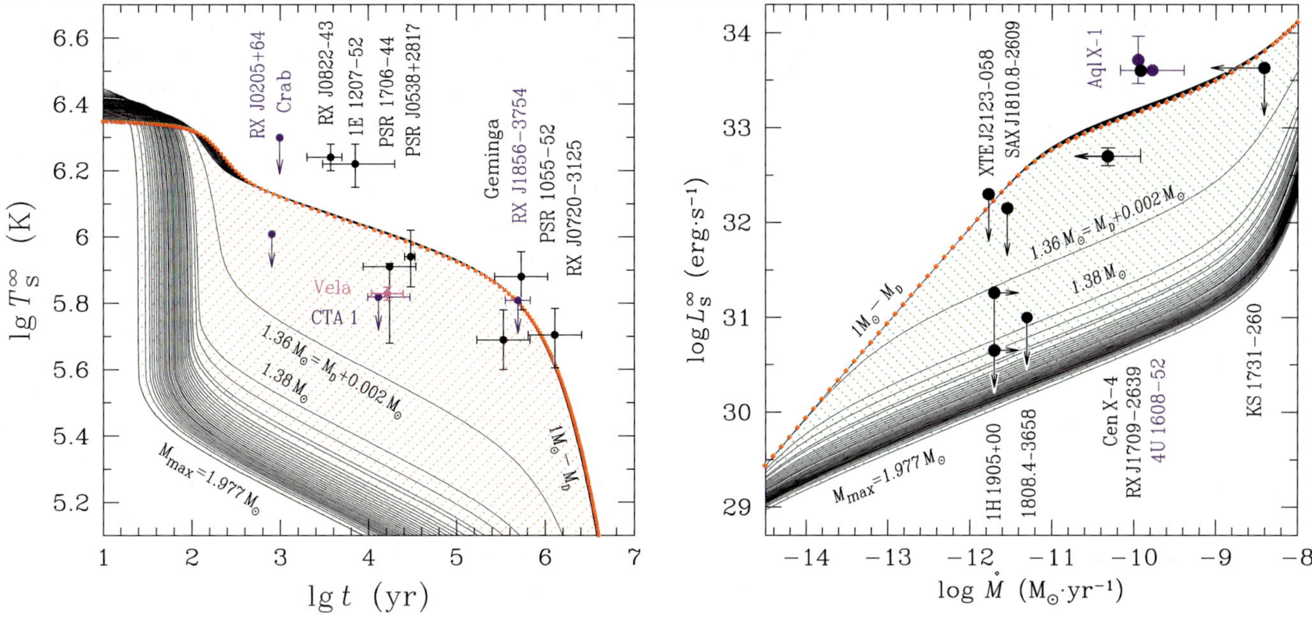

Fig. 2 Thermal states of nonsuperfluid INSs and SXTs with nucleon cores based on an EOS of Prakash et al. (1988). This EOS opens the nucleon direct Urca process at $\rho \ge \rho_D = 2.8\rho_0$, i.e., in inner cores of NSs with $M \ge M_D = 1.358\,M_\odot$. The *thick dotted basic curves* are the same as in Fig. 1. From *top* to *bottom*, *thin solid curves* correspond to NSs with masses growing from $1\,M_\odot$ to $M_{\max} = 1.977\,M_\odot$, with a step of $0.02\,M_\odot$. An *upper dense bundle of curves in each panel* contains low-mass stars with $M < M_D$, while a *less dense bottom bundle* contains NSs with $M > 1.06\,M_D$

Figure 1 shows thermal states of nonsuperfluid NSs with three different compositions in the inner core. In both panels, the upper (basic) dotted curve corresponds to low-mass NSs which possess no inner core and cool slowly via the modified Urca process from the outer core. The basic curve is almost independent of NS mass M as long as the inner core is absent. For higher M, we obtain noticeably colder NSs, cooling via enhanced neutrino emission from the inner core. Their surface temperature strongly depends on the composition in the inner core. The coldest is the maximum-mass NS (solid lines).

The coldest INSs observed to date are consistent with all three neutrino emission scenarios (see Yakovlev et al. 2003, for details of the models). At present, the data on SXTs contain colder sources and seem to be more restrictive. From the upper limits on the thermal luminosity of SAX 1808.4-3658 (Campana et al. 2002) and 1H 1905+000 (Jonker et al. 2006), one can infer that a dominant process in superdense matter should be more powerful than direct-Urca-type processes with kaons or pions. However, these results should be taken with caution—some issues of theory and observations of SXTs still have to be clarified (Sect. 5).

In what follows we will limit ourselves to the simplest nucleon models of NS cores descibed, e.g., in Yakovlev and Pethick (2004). Results presented in Figs. 2, 3, 4, 5 and 6 are obtained using our generally relativistic code of NS thermal evolution. We assume (for clarity of our study) the absence of light-element accreted envelopes on NS surfaces and neglect the effects of NS magnetic fields. NS models in Figs. 2–4 are based on a moderately stiff EOS proposed by Prakash et al. (1988). This EOS opens the nucleon direct Urca process at $\rho \ge 2.8\rho_0$, i.e., in the inner cores of massive NSs with $M \ge M_D = 1.358\,M_\odot$.

According to Fig. 2, models of non-superfluid NSs with enhanced emission in the inner cores cannot explain the data on INSs and SXTs. First, they are unable to interpret hottest sources. Second, a transition between widely spaced hot and cold NSs thermal states occurs within an unrealistically narrow NS mass range $\sim 0.01\,M_\odot$. As we show below, including superfluidity relieves these shortcomings.

3 Superfluidity of nucleon matter

Microscopic theories of dense nucleon matter predict that below some critical temperature, neutrons and protons in NS cores are superfluid. However, the critical temperatures T_{cn} and T_{cp} as a function of density are very uncertain; see, e.g., Lombardo and Schulze (2001). Therefore, at present, it seems reasonable to rely on a few general points of recent theories of nucleon superfluidity:

- Proton pairing occurs in the $^1S_0$ state and persists from $\sim 0.5\rho_0$ up to a few ρ_0.
- Critical temperature for protons, $T_{cp}(\rho)$, may be rather high, with the maximum $\gtrsim 10^9$ K somewhere between ρ_0 and $2\rho_0$.

- Neutrons form pairs in a $^3P_2$ state; this pairing is, typically, weaker and persists to higher densities than the proton one.
- Critical temperature for neutrons, $T_{cn}(\rho)$, has maximum somewhere between ρ_0 and few ρ_0; as a rule, this maximum is shifted toward higher ρ relatively to the maximum of $T_{cp}(\rho)$.

These features of $T_{cp}(\rho)$ and $T_{cn}(\rho)$ can be simulated with phenomenological models. We adopt the models shown in Fig. 7.

Strong proton superfluidity. According to several authors (see Kaminker et al. 2002 for references) a nucleon NS model with the open direct Urca in the inner core and strong proton pairing in the outer core can explain available observations of INSs. Here we show that this model can explain also the data on SXTs. Our results are displayed in Fig. 3.

The effect of strong proton superfluidity is twofold. First, this superfluidity suppresses neutrino emission from NSs, making them hotter at a given age or mass accretion rate. This brings thermal states of slowly cooling low-mass NSs into the agreement with observations of hotter sources. In massive NSs with enhanced cooling, strong proton superfluidity may spread out the opening of the direct Urca process over some density range. As a result, the enhanced cooling sets in gradually with the growth of NS mass (not as sharp at $M = M_D$, as in Fig. 2). This allows one to interpret the colder sources as massive NSs of different masses.

Strong proton superfluidity, with $T_{cp} \gtrsim 7 \times 10^9$ K, appears in hot NSs with large neutrino luminosity, so that the neutrino emission due to Cooper pairing of protons is unimportant (Yakovlev et al. 2001). However, this superfluidity suppresses the powerful Urca processes. Now, low-mass NSs mainly cool via a much weaker neutron-neutron bremsstrahlung (unaffected by proton pairing). It slows down the cooling and let the low-mass stars be hotter at a given t or $\dot{M}$. In this way one can interpret observations of the hotter sources without invoking any reheating mechanism. As noted by Yakovlev et al. (2004), the presence of light elements (H, He) in the NS heat-blanketing envelope facilitates interpretation of the hotter INSs. Such an envelope is more heat transparent and let the NS look hotter for a given inner temperature. Moreover, in that case even a weaker superfluidity (with $T_{cp} \gtrsim 10^9$ K) allows us to interpret the hotter sources; we have checked that this is also true for SXTs.

Thermal evolution of slowly cooling low-mass NSs is almost independent of the assumed EOS (Page and Applegate 1992), as well of the NSs mass. At a given age or mass accretion rate, all low-mass stars have nearly the same inner temperature. Consequently, the appearance of superfluidity affects all these stars in the same way; cf. Figs. 2 and 3 (and Fig. 4 below). This property holds as long as proton pairing is strong in the entire NS core.

The situation is different in more massive NSs. The impact of proton superfluidity on these stars depends also on how far proton pairing extends into the inner core and how steep is the slope of the $T_{cp}(\rho)$ profile in this superfluid re-

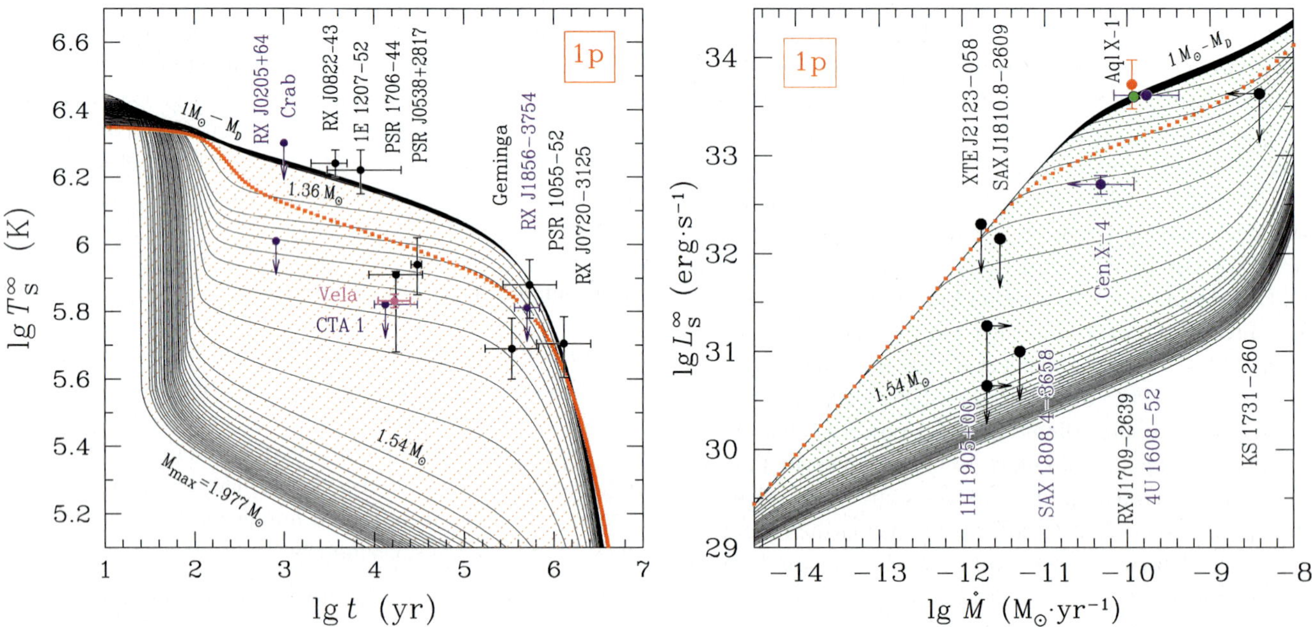

Fig. 3 Same as in Fig. 2 but for strong proton superfluidity in the NS core (model "1p" in Fig. 7). In this case, $T_{cp}(\rho)$ has maximum $\approx 6.9 \times 10^9$ K at $\rho \sim 2\rho_0$, remains $\gtrsim 5 \times 10^8$ K at $\rho \lesssim 3.2\rho_0$, and dies out at $\rho \sim 3.3\rho_0$. Accordingly, the entire cores of NSs with $M \lesssim 1.49\,M_\odot$ are strongly superfluid, while the inner cores of NSs with $M \gtrsim 1.52\,M_\odot$ have nonsuperfluid central kernels

gion (Yakovlev and Haensel 2003). In Figs. 3 and 4 the inner core of NS models with $M_D \le M \le M_{max}$ may occupy densities from $2.8\rho_0$ to $9.2\rho_0$ being superfluid at $\sim(2.8–3.3)\rho_0$ and normal at higher ρ. Accordingly, in medium-mass NSs with $M \sim (1.36–1.52)\,M_\odot$ the inner core is entirely superfluid. Superfluidity is mild near the outer boundary of the inner core, and weakens rapidly toward the core center. The more massive the star, the weaker proton superfluidity in its central part and the less it suppresses the direct Urca process, thus letting the star cool down faster. In this way thermal states of medium-mass INSs become dependent of M (Kaminker et al. 2001).

In the innermost central kernels of most massive NSs ($M > 1.52\,M_\odot$) proton superfluidity dies out, and the direct Urca process reaches its full power. This powerful fast cooling, even in a small part of the core, renders the superfluid NS as cold as the nonsuperfluid one; cf. Figs. 2 and 3.

Mild neutron superfluidity in the outer NS core. Let us now add mild neutron superfluidity and assume first that this superfluidity is located in the outer core. We have adopted model "1nt" shown in Fig. 7. Cooling of INSs with superfluidities "1nt" and "1p" was studied by Kaminker et al. (2001) and was shown to be inconsistent with hotter sources. We have tested this statement against the data on SXTs with the same conclusion. Results are displayed in Fig. 4.

In cooling INSs the effect of mild neutron superfluidity is very spectacular. First, neutron pairing reduces the heat capacity of NSs, because the heat is stored mostly in neutrons. The reduction amounts to a factor of several in low-mass NSs. This makes low-mass NSs very cold after appearing of such pairing at $t \gtrsim 3 \times 10^4$ yr. The cooling is additionally accelerated by powerful neutrino emission, triggered by Cooper pairing of neutrons. In result, the low-mass NSs models become unable to interpret four old sources: Geminga, RX J1856-3754, PSR 1055-52 and RX J0720-3125.

In the most massive NSs with $M > 1.54\,M_\odot$ the effect of neutron superfluidity is opposite. In these stars the mild pairing appears earlier, at $t \sim 10^2$ yr, when neutrino emission due to the direct Urca process is much stronger than the Cooper pairing neutrino emission (which is, therefore, insignificant). On the other hand, additional suppression of the direct Urca process by neutron superfluidity makes the most massive NSs a bit hotter in the advanced stage of neutrino cooling era.

As for transiently accreting NSs in SXTs, their thermal states become independent of the NS heat capacity as soon as they reach the steady-state regime (Yakovlev et al. 2003, 2004). Hence, a strong reduction of the heat capacity does not affect NS thermal states at the photon-dominated stage. That is why the slope of the dotted curve at $\dot{M} \lesssim 10^{-12}\,M_\odot\,\mathrm{yr}^{-1}$, corresponding to this stage, remains unchanged. However, at the neutrino stage, neutrino emission due to Cooper pairing of neutrons makes low-mass SXTs much colder (which unables one to interpret such hot sources as Aql X-1 and RX 1709-2639). The same effect in

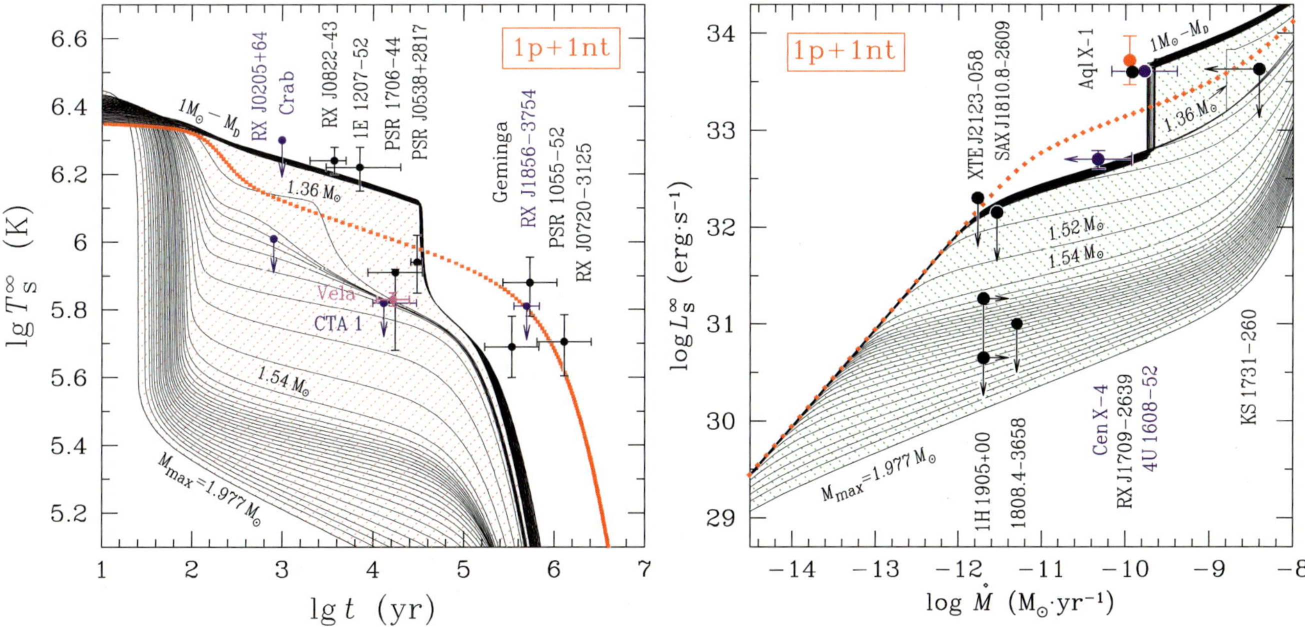

Fig. 4 Same as in Fig. 3 but with an addition of mild neutron superfluidity in the outer NS core (model "1nt" in Fig. 7). In this case, $T_{cn}(\rho)$ has maximum at $\rho \sim 2\,\rho_0$ and dies out at $\sim 5.7\,\rho_0$ (i.e., neutrons are nonsuperfluid in the central kernels of NSs with $M > 1.89\,M_\odot$). The maximum of the $T_{cn}(\rho)$ curve is wide—$T_{cn}(\rho)$ remains within $(2.0–3.3) \times 10^8$ K at $\rho \sim (0.72–3.2)\rho_0$. This density range corresponds to central densities of low-mass NSs ($M < 1.358\,M_\odot$) and moderately massive NSs (with $M \approx (1.36–1.48)\,M_\odot$)

lightest among the massive SXTs hampers the explanation of KS 1721-260. Similarly to high-mass INSs, high-mass SXTs become a bit hotter; neutron superfluidity, in its turn, slightly broadens the span of their allowed thermal states, stretched before by the proton pairing.

Mild neutron superfluidity in the inner NS core. The presence of mild neutron superfluidity in the outer NS core contradicts observations of INSs and SXTs, as it renders low-mass NSs too cold. However, mild pairing might occur in the inner core which is present only in massive NSs. Let us explore this case.

From Fig. 4 we see that neutrino emission due to Cooper pairing of neutrons may initiate a very fast cooling and, thus, may serve as a fast cooling agent (at least, for INSs).

In INSs, such a possibility was studied by Page et al. (2004) and Gusakov et al. (2004). In particular, the latter authors consider nucleon NS models based on the EOS of Douchin and Haensel (2001) which prohibits the direct Urca process. Their cooling scenario is based on strong proton pairing in the outer NS core and mild neutron pairing in the inner core (phenomenological models "p1" and "nt1", respectively; see Fig. 7). The model "nt1" has a specific dependence $T_{cn}(\rho)$ which keeps neutron pairing weak in the outer core. Let us remind that weak superfluidity, with $T_{cn} \lesssim 2 \times 10^8$ K, does not appear in low-mass NSs in the neutrino cooling era: these NSs are too hot. Accordingly, in this era neutron pairing "nt1" affects only massive NSs.

Gusakov et al. (2004) compared such cooling scenario (called by Page et al. 2004 a "minimal cooling model", for its simplicity, and the minimal number of its ingredients) with observations of INSs, and found it marginally consistent with them. The scenario can explain those INSs whose thermal emission is detected with confidence, as illustrated in the left panel of Fig. 5. However, these authors noted that the model will fail if the two sources, RX J0205+64 and CTA 1 (expected to be thermally emitting INSs), turn out to be much colder than their presently established upper limits, or if very cold INSs are detected in future.

Since NSs in SXTs are expected to be rather massive, owing to accretion of matter from their companions, it is worth to test the "minimal cooling scenario" against the data on SXTs. That is done in the right panel of Fig. 5. One can see that the model successfully reproduces hotter SXTs, including the frequently bursting sources Aql X-1, RX 1709-2639 and 4U 1608-52. However, the upper limits for at least two sources, 1H 1905+000 and SAX 1808.4-3658, definitely fall far below the predictions of the model.

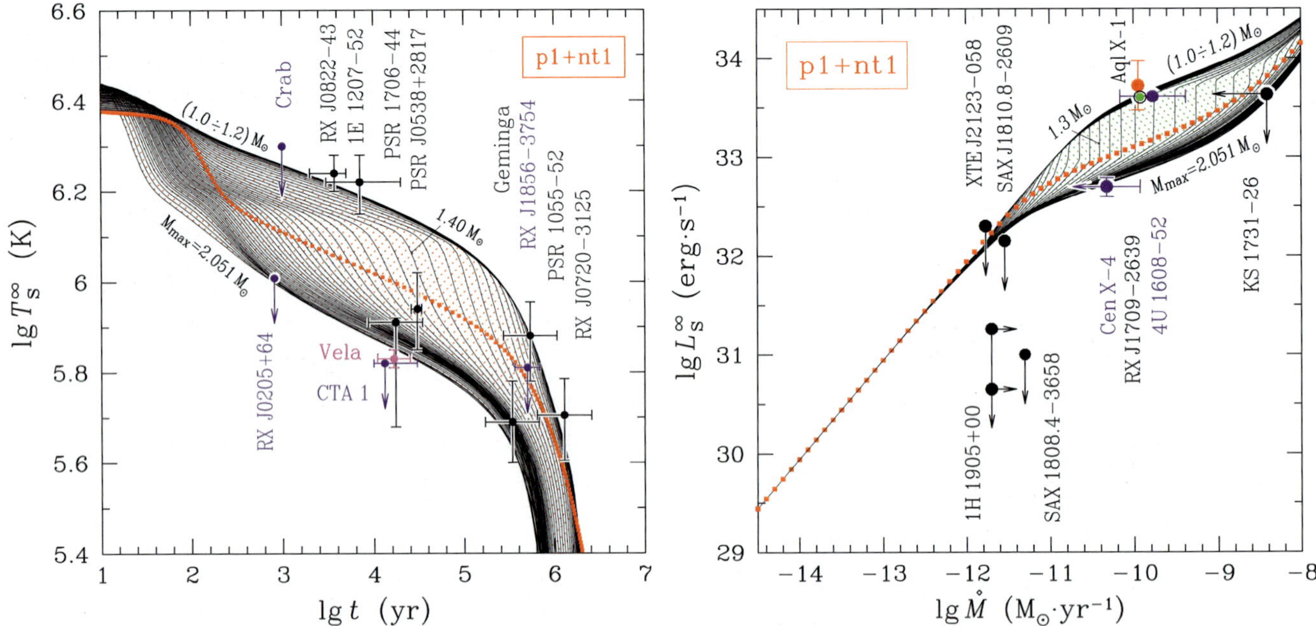

Fig. 5 Thermal states of superfluid NSs based on the EOS of Douchin and Haensel (2001) which prohibits the direct Urca process in NS cores. The figure illustrates the so called "minimal cooling scenario" in which Cooper pairing of neutrons operates as enhanced cooling agent. The scenario assumes strong proton pairing in the outer NS core and mild neutron pairing in the inner core (models "p1" and "nt1" in Fig. 7, respectively). In *both panels* the *dotted basic curve* refers to the same cooling scenario as in Fig. 1. From *top* to *bottom*, *thin solid curves* show NSs models with masses growing from 1 $M_\odot$ to $M_{\max} = 2.051\, M_\odot$, with a step of 0.02 $M_\odot$. Pairing "p1" has a wide maximum at $\rho \sim 1.6\rho_0$ and dies out at $\sim 3.8\,\rho_0$, i.e. in the central kernels of NSs with $M \gtrsim 1.49\, M_\odot$. Strong pairing with $T_{cp} \approx (2.0\text{–}6.9) \times 10^9$ K persists up to $\rho \sim 3.2\,\rho_0$ and extends over entire cores of NSs with $M \lesssim 1.27\, M_\odot$. Neutron pairing of "nt1" is weak in the outer core, at $\rho \lesssim 3.1\,\rho_0$ (i.e., in NSs with $M \lesssim 1.24\, M_\odot$), has a sharp maximum of mild strength in the inner core, at $\rho \sim 4.7\,\rho_0$, and dies out rapidly as ρ approaches $\sim 6.7\,\rho_0$. Mild pairing with $T_{cn} \approx (2.0\text{–}6.0) \times 10^8$ K extends over the density interval $\sim (3.1\text{–}6.4)\rho_0$, which corresponds to the inner cores of NSs with $M \approx (1.22\text{–}1.95)\, M_\odot$ or to spherical layers around the nonsuperfluid central kernels of NSs with $M \approx (1.97\text{–}2.05)\, M_\odot$

The transient source 1H 1905+000 was recently observed with Chandra (Jonker et al. 2006); the upper limit for its quiescent thermal luminosity seems to be firm. On the contrary, the value of $\dot{M}$ for this object is quite uncertain; some physical arguments allow one to believe it to be higher than $10^{-12}\,M_\odot\,\mathrm{yr}^{-1}$. For the frequently bursting transient SAX 1808.4-3658 the mass accretion rate is known more accurately, while the upper limit of its quiescent thermal luminosity is less certain. It varies from $\sim 6\times 10^{29}$ erg s^{-1} in Campana et al. (2002) to $\sim 4\times 10^{31}$ erg s^{-1} in estimations of P. Shtykovski (private communication; see details in Yakovlev et al. 2004 and Yakovlev and Pethick 2004). Nevertheless, both these upper limits are clearly much below the predictions of the minimal cooling model. In present paper we have adopted the same upper limit on bolometric luminosity of SAX 1808.4-3658 as in Yakovlev et al. (2003).

Thus, despite many theoretical and observational uncertainties, it seems that the "minimal cooling model" is ruled out by the data on SXTs. Therefore, the mechanism of the enhanced NS cooling should be more powerful than the process of Cooper pairing of neutrons.

Nevertheless, mild neutron superfluidity localized in the inner NS core is not prohibited. If it is available, together with proton superfluidity, the enhanced cooling should be of the direct Urca type. If, however, proton pairing does not extend to the inner core, neutron superfluidity remains the only regulator of cooling of massive NSs. This case was studied by Gusakov et al. (2005) in NS models with open direct Urca process. Because of the lack of space, we do not illustrate this case. We just note that this model easily fits hot and cold sources but scarcely explains quite a large span of intermediate sources, for example, Vela and Cen X-4.

4 Dichotomy of thermal evolution of SXTs?

As mild neutron superfluidity might exist in the inner NS core, let us outline its effect on thermal evolution of SXTs. In contrast to INSs, thermal states of NSs in SXTs depend not only on NS properties but also on the accretion rate.

As seen from Fig. 5 (right), heating curves for medium-mass SXTs drop abruptly when the mass accretion rate becomes lower than some threshold value specific for an NS of a given mass. As we already know, this drop (the vertical segment of a curve) signals the onset of neutron superfluidity in the NS core when the deep crustal heating becomes insufficient to keep the core temperature above the maximal critical temperature of neutrons.

Let us remind that NSs are thermally inertial objects. A thermal state of the star as a whole may change noticeably on time scales of $\sim 10^4$ yr (Colpi et al. 2001). We also recall that heating curves refer to steady states which NSs reach in a few millions years after the onset of the accretion stage in SXTs. The abrupt drops of heating curves indicate that thermal evolution of SXTs exhibits *dichotomy*. This means that two NSs with the same mass may evolve to very different steady states if their mass accretion rates are slightly different. The right panel of Fig. 6 illustrates how a change of $\dot{M}$ by one percent from some "threshold" value entails a

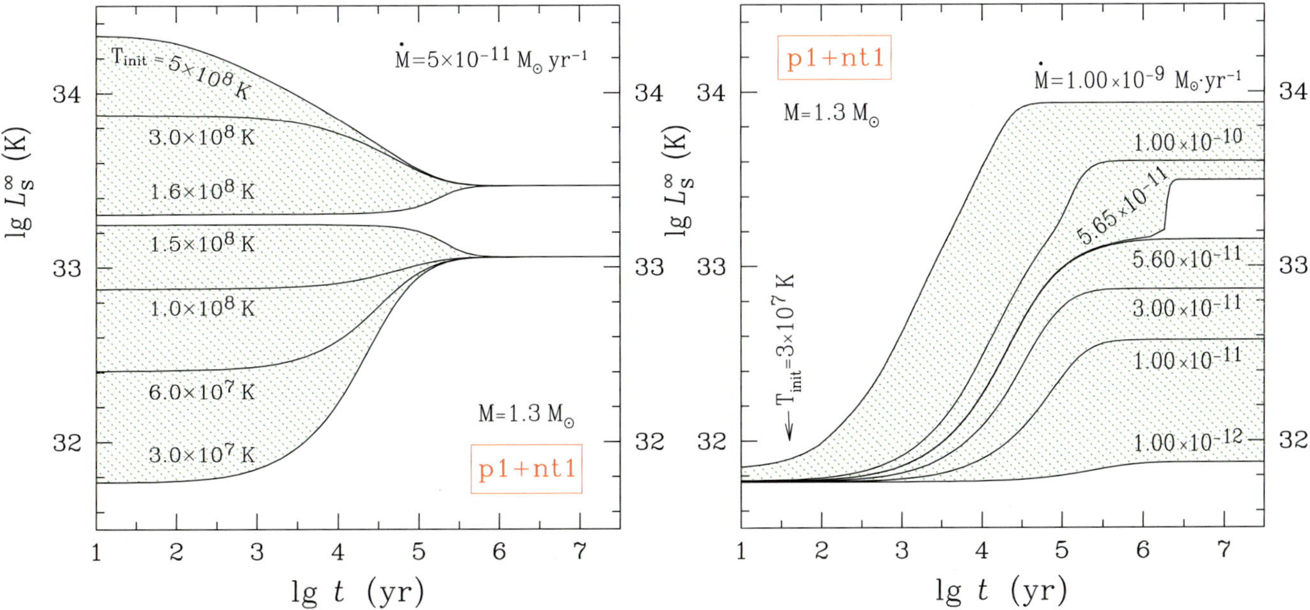

Fig. 6 Thermal evolution of NSs ($M = 1.3\,M_\odot$) in SXTs from an initial state to their final steady states. *Left*: evolution, at a given mass accretion rate, for several initial values of the inner stellar temperature. *Right*: evolution from a given initial state for several values of the mean mass accretion rate

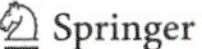

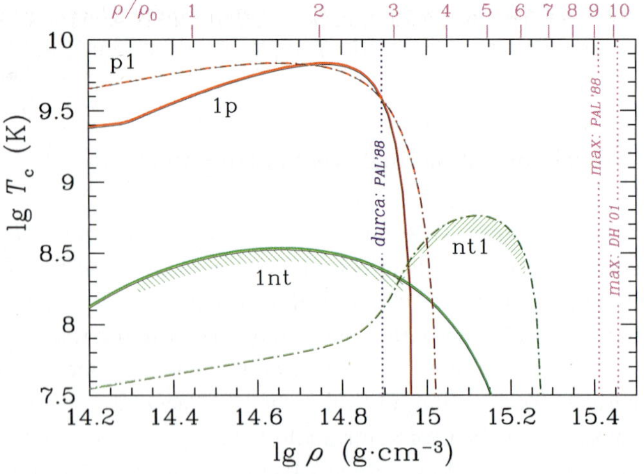

Fig. 7 Phenomenological models of neutron and proton superfluidities. *Solid curves*: models "1nt" and "1p" from Kaminker et al. (2001), used in the cooling scenarios with the direct Urca process in Figs. 3 and 4. *Dot-dashed curves*: models "nt1" and "p1" from Gusakov et al. (2004), adopted in the "minimal cooling" scenario in Fig. 5. *Vertical dotted lines*: the maximum-mass NS central densities for EOSs by Prakash et al. (1988) and by Douchin and Haensel (2001), used in the former and the latter scenarios, respectively, as well as the direct Urca threshold (for *PAL'88*)

change of the steady state thermal luminosity by a factor of $\sim$1.5.

Moreover, a tiny mismatch of masses of two NSs, or a small difference of their inner temperatures before the onset of the accretion stage, may also induce similar dichotomy. The latter effect is illustrated in the left panel of Fig. 6. The former one can be seen in Fig. 5 (and Fig. 4): at some "threshold" $\dot{M}$, NSs with masses different by one percent ($\sim$0.02 $M_\odot$) may show thermal luminosities which differ by factor of $\sim$3. One can also speculate that the dichotomy can be caused by a variable crust composition, which can change the efficiency of deep crustal heating at a given $\dot{M}$ (see Haensel and Zdunik 1990, 2003). Concluding, whatever causes the dichotomy, the presence of mild neutron superfluidity in the NS core may considerably complicate an interpretation of thermal emission from SXTs.

5 Discussion

We have shown that the data on quiescent thermal emission from SXTs require the presence of strong proton superfluidity and the absence of mild neutron superfluidity in the outer NS core (i.e., in low-mass NSs). The mechanism which controls fast neutrino cooling of massive NSs should be more powerful than that due to Cooper pairing of neutrons and, probably, than the direct-Urca-type processes with kaons and pions. We have noticed that the presence of mild neutron superfluidity in the inner NS core may cause the dichotomy of thermal evolution of SXTs.

Further studies are required to confirm or reject our inferences. The data on SXTs are very uncertain. Much work is needed to constrain these data and clarify the nature of variability of some SXTs in quiescence. There are still many challenges to respond to in order to build a realistic theory of deep crustal heating of SXTs. Note that current models of accreted crust and deep crustal heating are based on one particular cold liquid drop model of atomic nuclei and one model of pycnonuclear burning. The actual properties of highly neutron-rich nuclei surrounded by free neutrons in the inner NS crust, and the rate of pycnonuclear fusion of such nuclei (Coulomb barrier penetrability, astrophysical S-factors) are known with considerable uncertainty (see, e.g., Yakovlev et al. 2006). In this study we have neglected neutrino emission due to Cooper pairing of neutrons in the NS crust in order to show directly the effect of proton and neutron superfluidity in the NSs core on thermal states of SXTs.

Acknowledgements We express our gratitude to D.G. Yakovlev for sharing with us his expertise, and for his help in the preparation of this paper. We thank A.D. Kaminker and M.E. Gusakov for providing us with the table of EOS without dUrca. K.L. is grateful the organizers of the conference "Isolated Neutron Stars: from the Interior to the Surface" for financial support. This work is supported by Polish MEiN (grant No. 1P03D-008-27), by Russian Foundation for Basic Research (grants 05-02-16245, 05-02-22003), by the Russian Federal Agency for Science and Innovations (grant NSh 9879.2006.2), and by Russian Science Support Foundation.

References

Brown, E.F., Bildstein, L., Rutledge, R.E.: Astrophys. J. **504**, L95 (1998)

Cackett, E.M., Wijnands, R., Linares, M., et al.: Mon. Not. Roy. Astron. Soc. **369**, 407 (2006)

Campana, S., Stella, L., Gastadello, F., et al.: Astrophys. J. **575**, L15 (2002)

Douchin, F., Haensel, P.: Astron. Astrophys. **380**, 151 (2001)

Colpi, M., Geppert, U., Page, D., et al.: Astrophys. J. **548**, L175 (2001)

Gusakov, M.E., Kaminker, A.D., Yakovlev, D.G., et al.: Astron. Astrophys. **423**, 1063 (2004)

Gusakov, M.E., Kaminker, A.D., Yakovlev, D.G., Gnedin, O.Y.: Mon. Not. Roy. Astron. Soc. **363**, 555 (2005)

Haensel, P., Zdunik, J.L.: Astron. Astrophys. **227**, 117 (1990)

Haensel, P., Zdunik, J.L.: Astron. Astrophys. **404**, L33 (2003)

Jonker, P.G., Mendez, M., Nelemans, G., et al.: Mon. Not. Roy. Astron. Soc. **341**, 823 (2003)

Jonker, P.G., Wijnands, R., van der Klis, M.: Mon. Not. Roy. Astron. Soc. **349**, 94 (2004)

Jonker, P.G., Bassa, C.G., Nelemans, G., et al.: Mon. Not. Roy. Astron. Soc. **368**, 1803 (2006)

Kaminker, A.D., Haensel, P., Yakovlev, D.G.: Astron. Astrophys. **373**, L17 (2001)

Kaminker, A.D., Yakovlev, D.G., Gnedin, O.Y.: Astron. Astrophys. **383**, 1076 (2002)

Lombardo, U., Schulze, H.-J.: In: Blaschke, D., Glendenning, N.K., Sedrakian, A. (eds.) Physics of Neutron Stars Interiors, pp. 30–53. Springer, Berlin (2001)

Page, D., Applegate, J.H.: Astrophys. J. **394**, L17 (1992)

Page, D., Lattimer, J.M., Prakash, M., Steiner, A.W.: Astrophys. J. Suppl. **155**, 623 (2004)

Page, D., Geppert, U., Weber, F.: Nucl. Phys. A **777**, 497 (2006)
Prakash, M., Ainsworth, T.L., Lattimer, J.M.: Phys. Rev. Lett. **61**, 2518 (1998)
Tomsick, J.A., Gelino, D.M., Halpern, J.P., et al.: Astrophys. J. **610**, 933 (2004)
Yakovlev, D.G., Haensel, P.: Astron. Astrophys. **407**, 259 (2003)
Yakovlev, D.G., Pethick, C.J.: Annu. Rev. Astron. Astrophys. **42**, 169 (2004)
Yakovlev, D.G., Kaminker, A.D., Gnedin, O.Y., Haensel, P.: Phys. Rep. **354**, 1 (2001)
Yakovlev, D.G., Levenfish, K.P., Haensel, P.: Astron. Astrophys. **407**, 265 (2003)
Yakovlev, D.G., Levenfish, K.P., Potekhin, A.Y., et al.: Astron. Astrophys. **417**, 169 (2004)
Yakovlev, D.G., Weischer, M., Gasques, L.: Mon. Not. Roy. Astron. Soc. **371**, 1322 (2006)

Astrophys Space Sci (2007) 308: 467–469
DOI 10.1007/s10509-007-9298-x

ORIGINAL ARTICLE

The drift model of "magnetars"

Nature of AXPs, SGRs and radio pulsars with very long periods

I.F. Malov · G.Z. Machabeli

Received: 20 June 2006 / Accepted: 29 August 2006 / Published online: 20 March 2007

Abstract It is shown that the drift waves near the light cylinder can cause the modulation of emission with periods of order several seconds. These periods explain the intervals between successive pulses observed in AXPs, SGRs and radio pulsars with long periods. The model under consideration gives the possibility to calculate real rotation periods P of host neutron stars. It is shown that $P \leq 1$ s for the investigated objects. The magnetic fields at the surface of the neutron star are of order 10^{11}–10^{13} G and equal to the fields usual for the known radio pulsars.

Keywords Anomalous X-ray pulsars · Soft gamma repeaters · Radio pulsars · Magnetic fields · Drift waves

PACS 97.60Gb

1 Models of AXPs and SGRs

It follows from the model of the magneto-dipole slowing down, that magnetic fields B at the surface of a neutron star in AXPs and SGRs must be 10^{14}–10^{15} G, two orders of magnitude higher than fields in "normal" pulsars. It was suggested that X-ray radiation took its energy from a magnetic reservoir (the magnetar model). However this reservoir is exhausted during several hundreds years, if we take into account the necessity to inject relativistic particles into ambient SNRs (Kouveliotou et al. 1998). To avoid this difficulty it is necessary to postulate the existence of magnetic fields $B \sim 10^{16}$ G inside a neutron star (Thompson and Duncan 1996).

It is well known that the necessary stage to generate pulsar radio emission is creation of electron-positron pairs. But a gamma-quantum will convert in very strong magnetic fields ($B \gg 10^{12}$ G) into two other gamma-quanta (Baring and Harding 1998). Therefore AXPs and SGRs must be radio quiet objects. However Shitov et al. (2000) detected radio emission from SGR 1900+14 and Malofeev et al. (2005) registered pulsed radio signals from AXP 1E2259+586.

So there is the alternative: either we do not understand how radio pulsars radiate or magnetic fields of AXPs and SGRs are much less than 10^{14}–10^{15} G.

The braking index n is equal to 3 for the magneto-dipole slowing down. However the data of Shitov et al. (2000) have given $n = 0.20 \pm 0.47$ for SGR 1900+14. Hence, the basic suggestion on the magneto-dipole braking is not correct.

Other models were put forward to explain the peculiarities of AXPs and SGRs, namely accretion (see, for example, Marsden et al. 2001), magnetic white dwarfs (Paczynski 1990; Usov 1993), strange stars (Dar and De Rujula 2000; Usov 2001), free precession of a neutron star (Shaham 1977), but all these models collide with many difficulties as well.

In this report we discuss a drift model for describing the "magnetar" phenomenon using usual values of magnetic fields at the surface of a neutron star $B_s \sim 10^{12}$ G.

I.F. Malov (✉)
P.N. Lebedev Physical Institute, Russian Academy of Sciences, Pushchino, Moscow region, PRAO FIAN, 142290, Russia
e-mail: malov@prao.ru

G.Z. Machabeli
Abastumani Astrophysical Observatory, Al. Kazbegi Avenue 2a, 380060 Tbilisi, Georgia
e-mail: g_machabeli@hotmail.com

2 The drift model

As was shown by Kazbegi et al. (1991), transverse electromagnetic drift waves could be generated in the magnetosphere. These waves propagate almost perpendicular to the magnetic field and are characterized by the frequency

$$\omega_0 = \mathrm{Re}\,\omega = k_x u_x^b \tag{1}$$

and the increment

$$\Gamma = \mathrm{Im}\,\omega \approx \left(\frac{n_b}{n_p}\right)^{1/2} \gamma_p^{1/2} k_x \frac{u_x^b}{\gamma_b^{1/2}}. \tag{2}$$

Here k_x, k_φ, k_r are the components of the wave vector in the cylindrical coordinate system, u_x is the drift velocity and γ_b is the Lorentz-factor of the primary beam.

These waves cause variations of curvature of field lines:

$$K \approx \frac{1 - k_\varphi r B_r / B_\varphi}{r}. \tag{3}$$

If $k_\varphi r \gg 1$ the change of K may be significant. As far as radiation is emitted along a tangent to the local direction of magnetic field the change of its curvature leads to the change of the radiation direction.

This model gives the possibility to explain the observed periods $P_{\mathrm{obs}} = 5$–12 s of AXPs and SGRs and some other their peculiarities (Malov et al. 2003), to calculate the real values of the rotation periods ($P = 11$–500 ms) and the magnetic fields at the surface of AXPs and SGRs ($\log B = 11.2$–12.8) (Malov 2006) and explain their high quiescent X-ray emission and gamma-ray bursts (Malov and Machabeli 2005). The real positions of AXPs and SGRs are shown in Fig. 1.

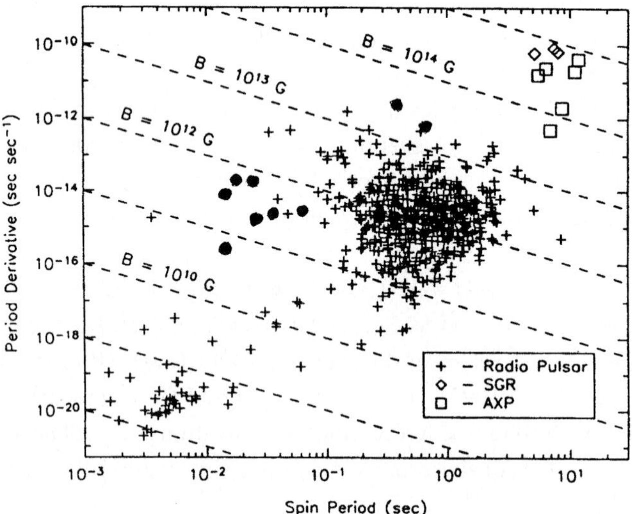

Fig. 1 Location of AXPs and SGRs on the diagram $(dP/dt) - P$ in the frame of our model (*black circles*) and the "magnetar" model (Woods and Thompson 2006)

In our model we can expect a modulation of observed emission with the rotation period. The detection of such modulation will be the good evidence of the vitality of this model.

3 Radio pulsars with very long periods (Lomiashvili et al. 2006)

Using our model and the scheme presented in Fig. 2 we can calculate real values of pulsar parameters (Table 1).

4 Discussion

One of the main characteristics of observed emission is the stability of pulse periods. The drift waves are stabilized due to the neutron star rotation and the permanent injection of relativistic particles in the region of their generation. Moreover as is shown by Gogoberidze et al. (2005), the nonlinear induced scattering leads to a transfer of waves from higher

Table 1 Radio pulsars with observed long periods

Pulsar	PSR J2144-3933	PSR J1847-0130	PSR J1814-1744
P_{obs} (s)	8.5	6.7	4.0
P(s)	0.85	1.12	0.5
$(dP/dt)_{-15}$	0.048	210	190
$B_s \times 10^{12}$ G	0.2	16	6.9
dE/dt (10^{32} erg/s)	0.032	61	300
$\beta_0 \approx \delta$ (deg)	7	5	5
θ (deg)	1.5	3	2
W_{10}/P	0.1	0.3	0.2

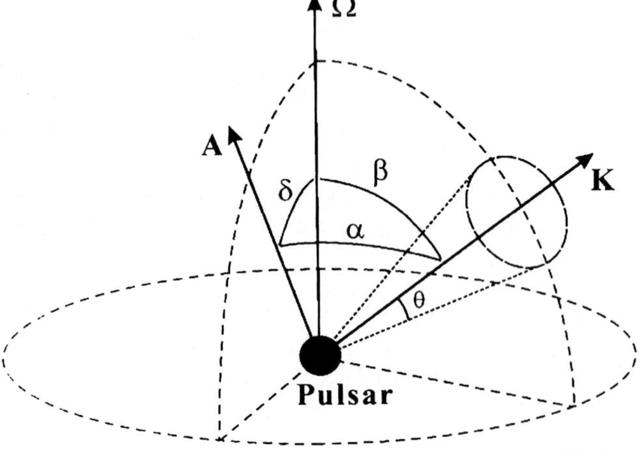

Fig. 2 Geometry under consideration. **K** is emission axis, **A** is observer's one. Angles δ and θ are constant, while β and α are oscillated with time

to lower frequencies. As the result one eigenmode becomes dominant. So the wave energy accumulates in waves with the certain azimuthal number m, characterizing the lowest frequency. This means that the period of the modulation and the interval between observed pulses must be rather stable.

In fact, the spectral energy of the drift waves with smaller periods is much less than that of the mode with the period $P = P_{dr}^{\max}$.

We have used the suggestion on the small angles between rotation axes and magnetic moments of neutron stars in AXPs and SGRs. In fact observed X-ray pulses in these objects are quite wide, and this indicates that they are nearly aligned rotators.

Recently discovered transient radio pulsars (McLaughlin et al. 2006) may belong to the population of objects described by our model. Indeed, 5 of them have rather long visible periods ($P > 4$ s) and one of them has the surface magnetic field obtained in the magneto-dipole model $B_s = 5 \times 10^{13}$ G $> B_{cr}$. Precession, star-quakes or other reasons can lead to the fulfillment of the condition $\alpha_{\min} < \theta$ for a short time and to an appearance of a number of visible pulses.

Thus, our model can be used for the explanations of all objects with long visible periods between observed pulses and with high values of magnetic fields estimated by the magneto-dipole formula.

References

Baring, M.G., Harding, A.K.: Astrophys. J. **507**, L55 (1998)

Dar, A., De Rujula, A.: In: Greco, M. (ed.) Results and Perspectives in Particle Physics, vol. XVII, p. 13 (2000)

Gogoberidze, G., Machabeli, G.Z., Melrose, D.B., Luo, Q.: Mon. Not. Roy. Astron. Soc. **360**, 669 (2005)

Kazbegi, A.Z., Machabeli, G.Z., Melikidze, G.I.: Aust. J. Phys. **44**, 573 (1991)

Kouveliotou, C., Dieters, S., Strohmayer, T.: Nature **393**, 235 (1998)

Lomiashvili, D., Machabeli, G., Malov, I.: Astrophys. J. **637**, 1010 (2006)

Malofeev, V.M., Malov, O.I., Teplykh, D.A., et al.: Astron. Rep. **49**, 242 (2005)

Malov, I.F.: Astron. Rep. **50**, 398 (2006)

Malov, I.F., Machabeli, G.Z.: Astron. Rep. **49**, 459 (2005)

Malov, I.F., Machabeli, G.Z., Malofeev, V.M.: Astron. Rep. **47**, 232 (2003)

Marsden, D., Lingefelter, R.E., Rothschild, R.E., Higdon, J.C.: Astrophys. J. **550**, 397 (2001)

McLaughlin, M.A., Lyne, A.G., Lorimer, D.R., et al.: Nature **431**, 817 (2006)

Paczynski, B.: Astrophys. J. **365**, L9 (1990)

Shaham, J.: Astrophys. J. **214**, 251 (1977)

Shitov, Yu.P., Pugachev, V.D., Kutuzov, S.M.: In: Kramer, M., Wex, N., Wielebinski, N. (eds.) IAU Colloq. 177: Pulsar Astronomy - 2000 and Beyond, Bonn. Astron. Soc. Pac. Conf. Ser., vol. 202, p. 685 (2000)

Thompson, C., Duncan, R.C.: Astrophys. J. **473**, 322 (1996)

Usov, V.: Astrophys. J. **410**, 761 (1993)

Usov, V.V.: Phys. Rev. Lett. **871**, 1001 (2001)

Woods, P.M., Thompson, C.: In: Lewin, W.H.G., van der Klis, M. (eds.) Compact Stellar X-ray Sources. Cambridge Astrophys. Series, No. 39, p. 547. Cambridge Univ. Press, Cambridge (2006)

Astrophys Space Sci (2007) 308: 471–475
DOI 10.1007/s10509-007-9294-1

ORIGINAL ARTICLE

Importance of Compton scattering for radiation spectra of isolated neutron stars

Valery Suleimanov · Klaus Werner

Received: 26 June 2006 / Accepted: 1 September 2006 / Published online: 22 March 2007

Abstract Model atmospheres of isolated neutron stars with low magnetic field are calculated with Compton scattering taking into account. Radiation spectra computed with Compton scattering are softer than computed with Thomson scattering at high energies ($E > 5$ keV) for hot ($T_{\rm eff} > 10^6$ K) atmospheres with hydrogen-helium composition. Compton scattering is more significant to models with low surface gravity. Compton scattering is less important to models with solar abundance of heavy elements.

Keywords Radiative transfer · Scattering · Atmospheres · Neutron stars · X-rays

PACS 95.30.Jx · 95.85.Nv · 97.10.Ex · 97.60.Jd

1 Introduction

At present time the model atmospheres of neutron stars (NS) with various chemical composition are widely used for interpretation of spectra of isolated NSs (see review by Pavlov et al. 2002). Such model atmospheres have been calculated by many authors for magnetized NSs as well as for non-magnetic ones (see review by Zavlin and Pavlov 2002 for details).

Spectra of light elements (H and He) model atmospheres without magnetic field show strong deviations from corresponding blackbody spectra due to the strong dependency of the true opacity on the photon energy and significant contribution of the electron scattering in the opacity. Therefore, the Compton effect can change the emergent spectra most hot ($T_{\rm eff} > 10^6$ K) H and He NS atmospheres.

It is well known (e.g. London et al. 1986) that Compton scattering changes significantly the emergent spectra of X-ray bursting NSs. But the effect of the Compton scattering is not carefully investigated for isolated NSs that have smaller $T_{\rm eff}$ ($< 10^7$ K). Here we present a set of NS model atmospheres with various chemical compositions which were calculated taking the Compton effect into consideration.

2 The method of modeling

We computed model atmospheres of hot isolated NSs subject to the constraints of hydrostatic and radiative equilibrium assuming planar geometry using standard methods (e.g. Mihalas 1978).

The model atmosphere structure for a hot isolated NS with effective temperature $T_{\rm eff}$, surface gravity g and given chemical composition is described by the hydrostatic equilibrium equation,

$$\frac{dP_{\rm g}}{dm} = g - 4\pi \int_0^\infty H_\nu \frac{k_\nu + \sigma_e}{c} d\nu, \qquad (1)$$

V. Suleimanov (✉)
Astronomy Department, Kazan State University,
Kremlevskaya 18, 420008 Kazan, Russia
e-mail: suleimanov@astro.uni-tuebingen.de

Present Address:
V. Suleimanov
Institut für Astronomie und Astrophysik, Sand 1,
72076 Tübingen, Germany

K. Werner
Institut für Astronomie und Astrophysik, Sand 1,
72076 Tübingen, Germany

where

$$g = \frac{GM_{\rm NS}}{R_{\rm NS}^2\sqrt{1 - R_{\rm S}/R_{\rm NS}}}, \tag{2}$$

$R_{\rm S} = 2GM_{\rm NS}/c^2$ is the Schwarzschild radius of the INS, k_ν is opacity per unit mass due to free-free, bound-free and bound-bound transitions, $\sigma_{\rm e}$ is the electron (Thomson) opacity, H_ν is Eddington flux, $P_{\rm g}$ is the gas pressure, and the column density m is determined as $dm = -\rho\, dz$. Variable ρ denotes the gas density and z is the vertical distance.

Compton scattering is taken into account in the radiation transfer equation using the Kompaneets operator (Kompaneets 1957; Zavlin and Shibanov 1991; Grebenev and Sunyaev 2002):

$$\frac{\partial^2 f_\nu J_\nu}{\partial \tau_\nu^2} = \frac{k_\nu}{k_\nu + \sigma_{\rm e}}(J_\nu - B_\nu) - \frac{\sigma_{\rm e}}{k_\nu + \sigma_{\rm e}}\frac{kT}{m_{\rm e}c^2} \times x\frac{\partial}{\partial x}\left(x\frac{\partial J_\nu}{\partial x} - 3J_\nu + \frac{T_{\rm eff}}{T}xJ_\nu\left(1 + \frac{CJ_\nu}{x^3}\right)\right), \tag{3}$$

where $x = h\nu/kT_{\rm eff}$ is the dimensionless frequency, $f_\nu(\tau_\nu) \approx 1/3$ is the variable Eddington factor, J_ν is the mean intensity of radiation, B_ν is the black body (Planck) intensity, T is the local electron temperature, and $C = c^2h^2\ /\ 2(kT_{\rm eff})^3$. The optical depth τ_ν is defined as $d\tau_\nu = (k_\nu + \sigma_{\rm e})\, dm$.

These equations have to be completed by the energy balance equation

$$\int_0^\infty k_\nu(J_\nu - B_\nu)d\nu - \sigma_{\rm e}\frac{kT}{m_{\rm e}c^2} \times \left(4\int_0^\infty J_\nu\, d\nu - \frac{T_{\rm eff}}{T}\int_0^\infty xJ_\nu\left(1 + \frac{CJ_\nu}{x^3}\right)d\nu\right) = 0, \tag{4}$$

the ideal gas law $P_{\rm g} = N_{\rm tot}kT$, where $N_{\rm tot}$ is the number density of all particles, and also by the particle and charge conservation equations. We assume local thermodynamical equilibrium (LTE) in our calculations, so the number densities of all ionization and excitation states of all elements have been calculated using Boltzmann and Saha equations. We take into account the pressure effects on the atomic populations using the occupation probability formalism (Hummer and Mihalas 1988) as it is described by Hubeny et al. (1994).

For solving the above equations and computing the model atmosphere we used a version of the computer code ATLAS Kurucz (1970, 1993), modified to deal with high temperatures; see Ibragimov et al. (2003) for further details. This code was also modified to account for Compton scattering (Suleimanov and Poutanen 2006; Suleimanov et al. 2006)

The scheme of calculations is as follows. First of all, a starting model using a grey temperature distribution is calculated. The calculations are performed with a set of 98 depth points $m_{\rm i}$ distributed logarithmically in equal steps from $m \approx 10^{-7}$ g cm^{-2} to $m_{\rm max}$. The appropriate value of $m_{\rm max}$ is found from the condition $\sqrt{\tau_{\nu,\text{b-f,f-f}}(m_{\rm max})\tau_\nu(m_{\rm max})} > 1$ at all frequencies. Here $\tau_{\nu,\text{b-f,f-f}}$ is the optical depth computed with only true opacity (bound-free and free-free transitions, without scattering) taken into consideration. Satisfying this equation is necessary for the inner boundary condition of the radiation transfer.

For the starting model, all number densities and opacities at all depth points and all frequencies (we use 300 logarithmically equidistant frequency points) are calculated. The radiation transfer equation (3) is non-linear and is solved iteratively by the Feautrier method (Mihalas 1978, see also Zavlin and Shibanov 1991; Pavlov et al. 1991; Grebenev and Sunyaev 2002). Between iterations we calculate the variable Eddington factors f_ν and h_ν, using the formal solution of the radiation transfer equation in three angles at each frequency.

We used the usual conditions at the outer boundary and at the inner boundary of the radiation transfer equation. The outer boundary condition is found from the lack of incoming radiation at the NS surface, and the inner boundary condition is obtained from the diffusion approximation $J_\nu \approx B_\nu$ and $H_\nu \approx 1/3 \times \partial B_\nu/\partial\tau_\nu$.

The boundary conditions along the frequency axis are usual too (Zavlin and Shibanov 1991; Grebenev and Sunyaev 2002). The condition at the lower frequency boundary ($\nu_{\rm min} = 10^{14}$ Hz, $h\nu_{\rm min} \ll kT_{\rm eff}$) $J_\nu = B_\nu$ means that at the lowest energies the true opacity dominates the scattering $k_\nu \gg \sigma_{\rm e}$, and therefore $J_\nu \approx B_\nu$. The condition at the higher frequency boundary ($\nu_{\rm max} \approx 3 \times 10^{19}$ Hz, $h\nu_{\rm max} \gg kT_{\rm eff}$) was found from the requirement that there is no photon flux along frequency axis at the highest energy.

The solution of the radiative transfer equation (3) was checked for the energy balance equation (4) together with the surface flux condition

$$4\pi\int_0^\infty H_\nu(m = 0)\, d\nu = \sigma T_{\rm eff}^4 = 4\pi H_0. \tag{5}$$

The temperature distribution was obtained iteratively by using temperature correction procedures (Kurucz 1970), modified for Compton scattering.

Our method of calculation was tested on a model of X-ray bursting neutron star atmospheres (Pavlov et al. 1991; Madej et al. 2004). We found that our models are in good agreement with these calculations.

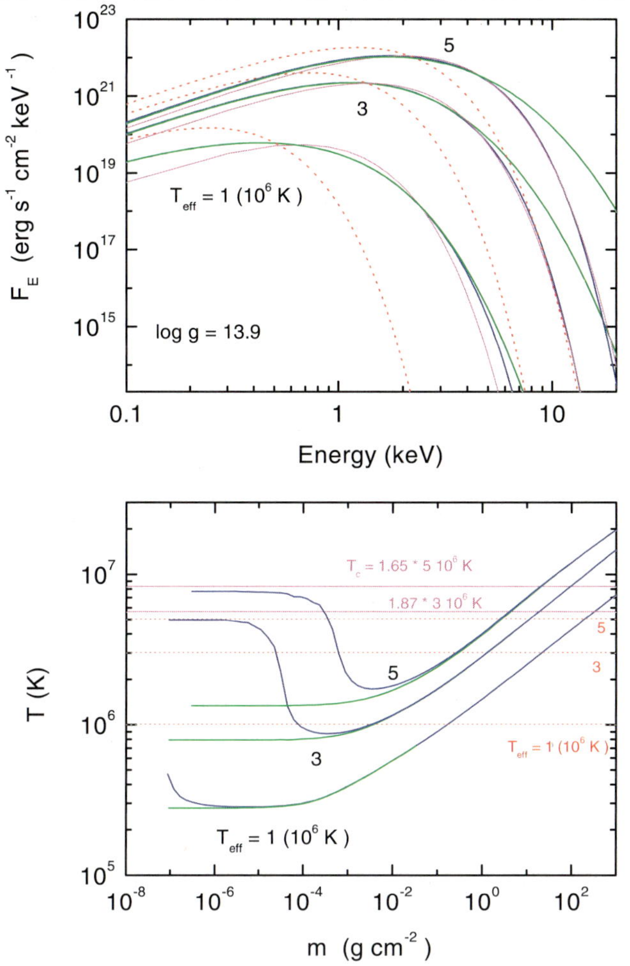

Fig. 1 *Top panel*: emergent (unredshifted) spectra of pure H low gravity NS model atmospheres. *Blue curves*—with Compton effect, *green curves*—without Compton effect, *red dotted curves*—blackbody spectra, *thin magenta curves*—diluted blackbody spectra with hardness factors 2.9, 1.87 and 1.65 for models with $T_{\rm eff} = 1$, 3 and 5×10^6 K. *Bottom panel*: temperature structures of the corresponding model atmospheres. Effective and color temperatures are shown by *red dotted* and *magenta lines* respectively

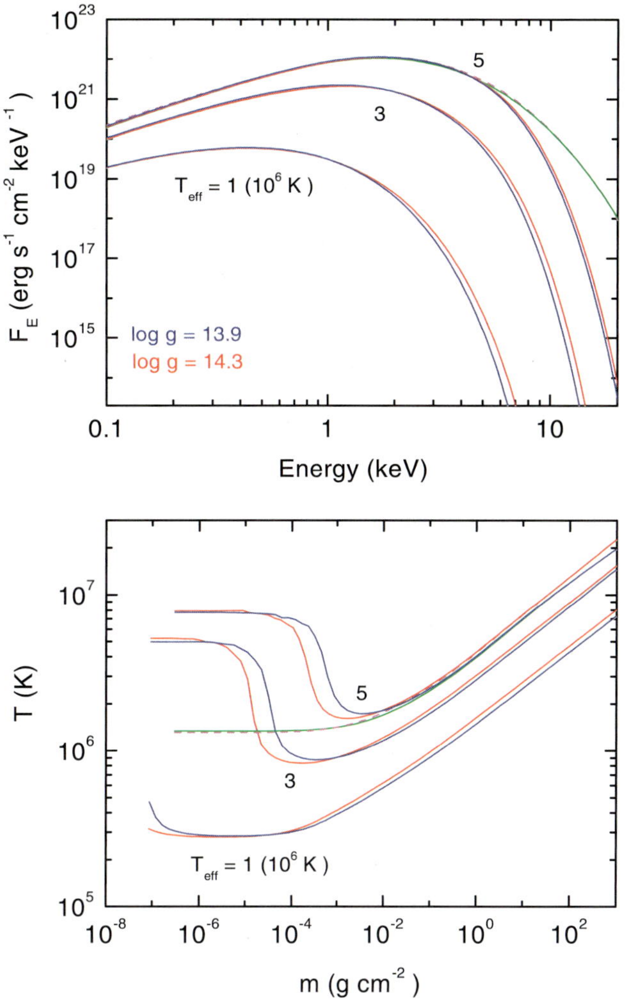

Fig. 2 *Top panel*: emergent (unredshifted) spectra of pure H NS model atmospheres with different surface gravities (*red curves*—high gravity models, *blue curves*—low gravity models). For comparison the model spectra without Compton effect are shown for hottest model (*green curve*—low gravity model, *magenta dashed curve*—high gravity model). *Bottom panel*: temperature structures of the corresponding model atmospheres

3 Results

Using this method the set of hydrogen and helium NS model atmospheres with effective temperatures 1, 3, and 5×10^6 K, and surface gravities $\log g = 13.9$ and 14.3 were calculated. Models with Compton scattering and Thomson scattering (to comparison) were computed. Part of the obtained results in Figs. 1, 2 and 3 are discussed.

The Compton effect is significant to emergent model spectra of hot ($T_{\rm eff} \geq 3 \times 10^6$ K) hydrogen model atmospheres at high energies (Fig. 1). The hard emergent photons lost energy and heat upper layers of the atmospheres due to interactions with electrons. As a result the high energy tails of the emergent spectra become close to Wien spectra and chromosphere-like structures with temperatures close to color temperatures of the Wien spectra in the upper layers of the model atmospheres appear. Moreover, all emergent model spectra of high temperature atmospheres can be presented in first approximation as diluted blackbody spectra with color temperatures close to Wien tail color temperatures:

$$F_{\rm E} = \frac{\pi}{f_{\rm c}^4} B_{\rm E}(T_{\rm c}), \qquad T_{\rm c} = f_{\rm c} T_{\rm eff}, \tag{6}$$

where $f_{\rm c}$ is hardness factor. These results are similar to those obtained for model atmospheres and emergent spectra of X-ray bursting NS in LMXBs (London et al. 1986; Lapidus et al. 1986; Madej 1991; Pavlov et al. 1991).

The Compton scattering effect on the emergent model spectra of high gravity atmospheres is less significant (Fig. 2). The reason is a relatively small contribution of

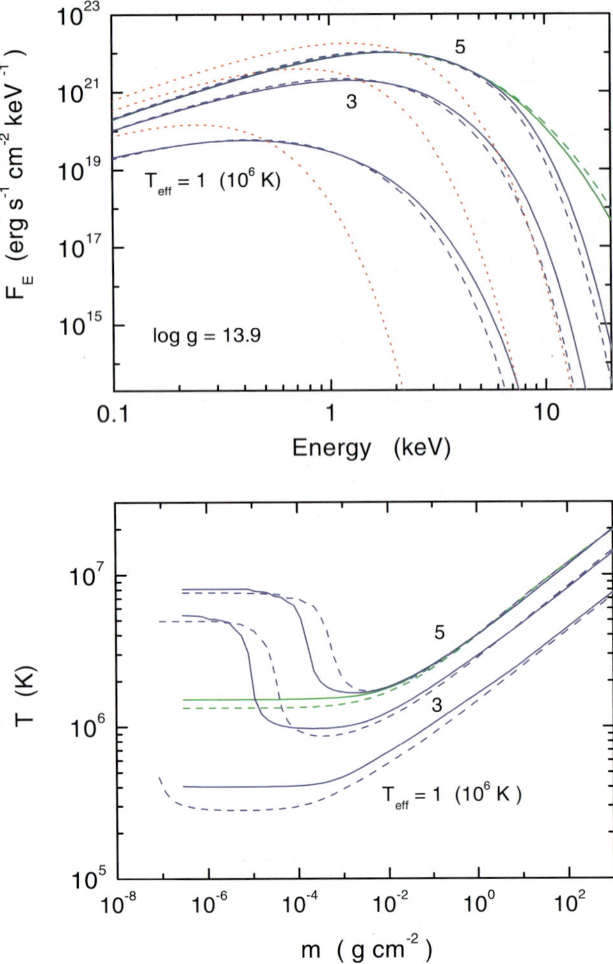

Fig. 3 *Top panel*: emergent (unredshifted) model spectra of pure He low gravity NS model atmospheres. For comparison the model spectra of pure H atmospheres are shown by *dashed curves*. The model spectra of hottest atmospheres without Compton effect are shown by *green curves*. *Bottom panel*: temperature structures of the corresponding model atmospheres

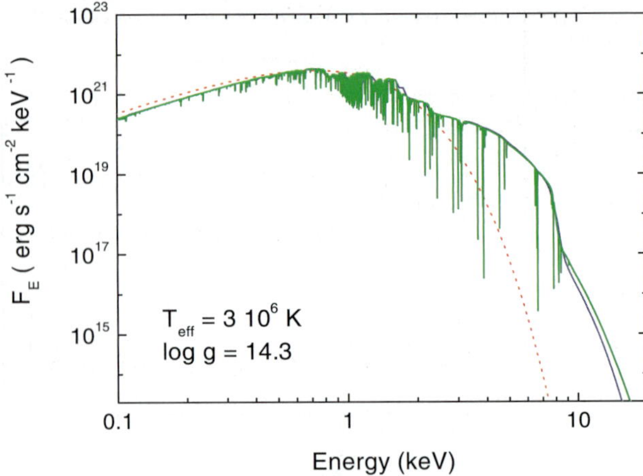

Fig. 4 Emergent (unredshifted) model spectra of high gravity NS atmospheres with solar abundance of 15 most abundant heavy elements with (*blue curve*) and without (*green curve*) Compton scattering. The *red dotted curve* is the corresponding blackbody spectrum

electron scattering to the total opacity in high gravity atmospheres than in low gravity ones. The matter density in the high gravity models is larger, and the opacity coefficient (in cm^2/g) is independent of the density for electron scattering and proportional to the density for free-free transitions. The color temperatures of the Wien tails in the emergent model spectra of the high gravity atmospheres are higher than for emergent spectra of the low gravity models. But in the observed isolated NS spectra these differences can be changed due to different gravitational redshifts.

The Compton scattering effect on helium model atmospheres is also less significant than on hydrogen model atmospheres with the same T_{eff} and $\log g$ (Fig. 3). The reason is the same as in the case of high gravity models. The ratio of electron scattering to true opacity is smaller in the helium model atmospheres.

We also computed one NS model atmosphere including heavy elements with solar chemical abundance and $T_{\text{eff}} = 3 \times 10^6$ K and $\log g = 14.3$ (see Fig. 4). The model was calculated with Thomson and Compton scattering and we found the Compton effect on the emergent spectrum is very small.

4 Conclusions

Emergent spectra of light elements NS model atmospheres with $T_{\text{eff}} > 10^6$ K are changed by the Compton effect at high energies ($E > 5$ keV), and spectra of the hottest ($T_{\text{eff}} \geq 3 \times 10^6$ K) model atmospheres can be described by diluted blackbody spectra with hardness factors $\sim 1.6 - 1.9$. But at the same time the spectral energy distribution of these models are not significantly changed at the maximum of the SED (at energies 1–3 keV), and effects on the color temperatures are not large.

The Compton effect is less significant to He model atmospheres and to high gravity model atmospheres. Differences in the observed spectra for models with different gravities are changed by different gravitational redshifts. Emergent model spectra of NS atmospheres with solar metal abundances are changed by Compton effects very slightly.

Acknowledgements The work by Valery Suleimanov was supported by DFG (grant We 1312/35-1) and partially supported by Russian Foundation of Basic Research (grant 05-02-17744).

References

Grebenev, S.A., Sunyaev, R.A.: Astron. Lett. **28**, 150 (2002)
Hubeny, I., Hummer, D.G., Lanz, T.: Astron. Astrophys. **282**, 151 (1994)
Hummer, D.G., Mihalas, D.: Astrophys. J. **331**, 794 (1988)

Ibragimov, A., Suleimanov, V., Vikhlinin, A., et al.: Astron. Rep. **47**, 186 (2003)
Kompaneets, A.S.: Sov. Phys. JETP **4**, 730 (1957)
Kurucz, R.: SAO Spec. Rep. **309**, 1 (1970)
Kurucz, R.: CD-ROMs. Smithsonian Astrophysical Observatory, Cambridge (1993)
Lapidus, I.I., Sunyaev, R.A., Titarchuk, L.G.: Sov. Astron. Lett. **12**, 383 (1986)
London, R.A., Taam, R.E., Howard, W.M.: Astrophys. J. **306**, 170 (1986)
Madej, J.: Astrophys. J. **376**, 161 (1991)
Madej, J., Joss, P.C., Rozanska, A.: Astrophys. J. **602**, 904 (2004)
Mihalas, D.: Stellar Atmospheres. Freeman, San Francisco (1978)
Pavlov, G.G., Shibanov, Y.A., Zavlin, V.E.: Mon. Not. Roy. Astron. Soc. **253**, 193 (1991)
Pavlov, G.G., Zavlin, V.E., Sanwal, D.: In: Becker, W., Lesch, H., Trümper, J. (eds.) The Proceedings of the 270 WE-Heraeus Seminar on Neutron Stars, Pulsars, and Supernova Remnants, MPE Report **278**, 273. Garching bei München (2002)
Suleimanov, V., Poutanen, J.: Mon. Not. Roy. Astron. Soc. **369**, 2036 (2006)
Suleimanov, V., Madej, J., Drake, J.J., et al: Astrophys. Astron. **455**, 679 (2006)
Zavlin, V.E., Shibanov, Y.A.: Sov. Astron. **35**, 499 (1991)
Zavlin, V.E., Pavlov, G.G.: In: Becker, W., Lesch, H., Trümper, J. (eds.) The Proceedings of the 270 WE-Heraeus Seminar on Neutron Stars, Pulsars, and Supernova Remnants, MPE Report **278**, 263. Garching bei München (2002)

Astrophys Space Sci (2007) 308: 477–479
DOI 10.1007/s10509-007-9312-3

ORIGINAL ARTICLE

Search for fast optical activity of SGR 1806-20 at the SAO RAS 6-m telescope

G. Beskin · V. Debur · V. Plokhotnichenko · S. Karpov · A. Biryukov · L. Chmyreva · A. Pozanenko · K. Hurley

Received: 18 July 2006 / Accepted: 29 October 2006 / Published online: 20 March 2007

Abstract The region of SGR 1806-20 localization was observed during its gamma-ray activity in 2001. The observations have been performed on the 6-meter telescope of the Special Astrophysical Observatory, using the Panoramic Photometer-Polarimeter (PPP). The search for variability was performed on the 10^{-6}–10 s time scale, and its results were compared to the properties of corresponding X-ray flares.

Keywords Methods: data analysis · Objects: SGR 1806-20

PACS 95.75.-z · 95.75.Wx · 95.85.Kr · 97.60.Jd

1 Introduction

There is now increasing evidence that the soft repeater SGR 1806-20 hosts a magnetar, i.e. a neutron star with an anomalously high magnetic field $B > 10^{14}$ Gs (Duncan and Thompson 1992; Woods and Thompson 2004). The SGR 1806-20 is the most active among soft gamma repeaters and is characterized by the emission of short sporadic flashes of soft gamma rays with characteristic durations of 10 ms–1 s and luminosities of 10^{39}–10^{42} erg/s. (Gogus et al. 2001; Hurley 2000). They are detected during sporadic periods of the object's activity lasting from days to months. Pulsations of the persistent X-ray flux of SGR 1806-20 with a period of 7.47 s were discovered (Kouveliotou 1998). The culmination of the long period of its activity that started in the end of 2003 was a giant flare of 27th December 2004 (Borkowski et al. 2004; Mazets et al. 2004; Golenetskii et al. 2004). During the main spike that lasted 0.2–0.5 s, about 10^{47} ergs above 50 keV was emitted for a distance of 15 kpc (Palmer et al. 2004; Mereghetti et al. 2005; Schwartz et al. 2005; Terasawa et al. 2005). The long part of the flare showed a pulsation with the period of 7.57 s during about 300 s (Borkowski et al. 2004).

Unfortunately, even the minimal estimation of the distance to SGR 1806-20 is 6 kpc (while the most reasonable one is 15 kpc), and the absorption reaches $A_V \sim 30$ m (Eikenberry et al. 2004; McClure-Griffiths and Gaensler 2005), so there is not much hope of detecting its optical emission, even though its infrared counterpart seems to have been found, with a magnitude of $K = 21.6$ m (Israel et al. 2005).

However, we have carried out a set of observations of the location of SGR 1806-20 in the optical band with 1 μs temporal resolution to try to detect very short and strong optical spikes. The epoch of the observations has been chosen in relation to the increase of γ-ray activity of the source, according to HETE data (Ricker et al. 2001a, 2001b).

We have reported the preliminary results of our monitoring in earlier papers (Beskin et al. 2001, 2003). In the present work we describe the process of observation, the equipment, and we present a more detailed analysis of the acquired data.

This work has been supported by the Russian Foundation for Basic Research (grant No 04-02-17555), Russian Academy of Sciences (program "Evolution of Stars and Galaxies"), and by the Russian Science Support Foundation. The authors would also like to thank the anonymous referee for his/her valuable comments.

G. Beskin (✉) · V. Debur · V. Plokhotnichenko · S. Karpov · A. Biryukov · L. Chmyreva · A. Pozanenko · K. Hurley
Special Astrophysical Observatory of RAS, Nizhniy Arkhyz 369167, Karachaevo-Cherkessia, Russia
e-mail: beskin@sao.ru

2 Observations, hardware and software

The field of SRG 1806-20 has been observed using the 6-meter telescope of the Special Astrophysical Observatory. Observations were carried out with Panoramic Photometer-Polarimeter (PPP) with high time resolution (Plokhotnichenko et al. 2003) in the telescope prime focus on June 20 2001 (2 days after HETE trigger), and on August 22 2001 (15 hours after HETE trigger) (Ricker et al. 2001a, 2001b).

The main part of PPP is the Positional Sensitive Detector (PSD), which consists of a vacuum tube with a standard S20 photocathode, a set of microchannel plates and a four-electrode anode. The pixel size of the detector is $0''.21$, the FOV is a circle of about $1'.5$, and the time resolution (dead time) is 1 μs (Debur et al. 2003).

For the determination of the photon arrival times and coordinates, and for the storage of the whole data set, a special "time-code" converter "Quantochron 4-48" connected to the PC in realtime has been used. The information on the observations is summarized in Table 1.

The search for variability has been performed in the region centered on the IPN localization of the source (Hurley et al. 1999). The statistical properties of the photon lists have been studied separately for 9 square boxes with the size of $6''.5$ covering the localization region (see Fig. 1).

As an indicator of the variability we used the function y_2, defined as the ratio between the distributions of the photon arrival time intervals for the source and background (Schvartsman 1977). It is:

$$y_2(\tau) = \frac{P_s(\tau) - P_b(\tau)}{P_b(\tau)}, \tag{1}$$

where τ is the interval between the times of arrival of successive photons, while $P_s(\tau)$ and $P_b(\tau)$ are the distribution functions of τ for the source and background boxes, respectively.

We used the following definitions and relations: $I(t)$—intensity variation during the flare and its mean value:

$$\langle I \rangle = \frac{1}{\langle \tau \rangle} = \frac{1}{T}\left[I_0 + \int_0^T I(t)\,dt \right], \tag{2}$$

where τ is the mean interval between the arrival times of successive photons, T—mean time between the flares and I_0—persistent emission intensity; $I_{\max}$—maximum intensity of the flare; characteristic time scale of the variability:

$$\tau_{\text{var}} = \frac{1}{I_{\max}} \int_0^T I(t)\,dt; \tag{3}$$

relative amplitude of the intensity variations $A \equiv \frac{I_{\max}}{\langle I \rangle}$, and relative power of the variable emission component

$$S = \frac{1}{\langle I \rangle T} \int_0^T I(t)\,dt = \frac{I_{\max}\tau_{\text{var}}}{\langle I \rangle T} = FA, \tag{4}$$

where $F = \tau_{\text{var}}/T$—the flares duty cycle.

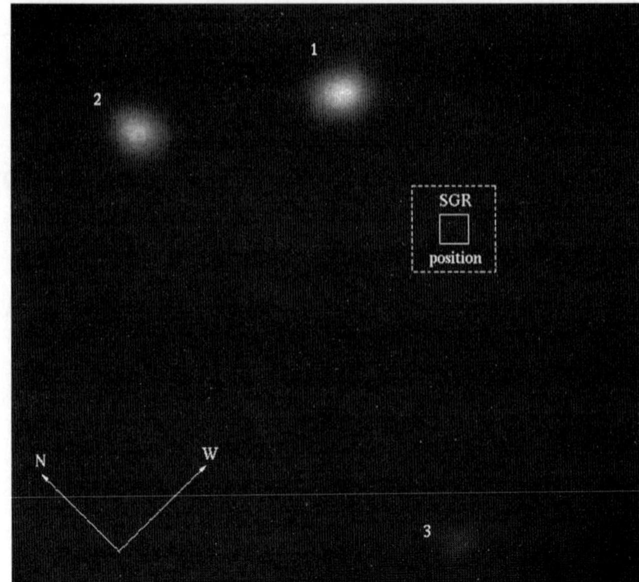

Fig. 1 Observed field of SGR 1806-20. Numbered stars are listed in Table 2

The function y_2 is related to the variability parameters defined above (Schvartsman 1977) as

$$y_2(\tau \ll \langle \tau \rangle, \tau_{\text{var}}) \approx \frac{kS^2}{F}. \tag{5}$$

where $k = \frac{1}{\langle I \rangle T}\int_0^T I^2(t)\,dt$—flare shape coefficient. For example, sinusoidal flares have $k = 0.25$, triangular ones have $k = \frac{2}{3} - 2\beta + 2\beta^2 \approx 0.2$–$0.7$.

From the registered photon lists it is easy to compute $y_2(\tau)$, from which S follows:

$$S(\tau) \approx \sqrt{\frac{y_2(\tau)F}{k}}. \tag{6}$$

The absence of variability is rejected if $S(\tau) > 3\sigma_S(\tau)$, where

$$\sigma_S(\tau) = \left(\frac{1}{P_b(\tau)N} \right)^{\frac{1}{4}}, \tag{7}$$

and N is the number of photons detected (Plokhotnichenko 1992, 1983).

3 Results of observations

Using the arrival times of 10^6 photons, detected in each subbox of the SGR 1806-20 localization, the distribution function of intervals between the following photons has been built. The mean background intensity has been 152 counts/s with nearly Poissonian statistics.

Table 1 Observations details

Date	Start time (UT)	Exposure time (s)	Zenith distance	Filter
20.06.2001	19^h58^m	4544.6	67.°1	B
22.08.2001	20^h31^m	800.33	77.°7	B

Table 2 USNO-A2.0 stars in the field of SGR 1806-20

N	R.A. J2000*	DEC. J2000	Bmag	Rmag
1	$18^h08^m39.^s42$	−20°24′11.″74	14.6	13.7
2	$18^h08^m41.^s13$	−20°24′02.″29	15.5	14.6
3	$18^h08^m41.^s47$	−20°25′12.″61	16.5	15.9

*Monet D.G., 1998, BAAS 30, Vol. 4, 120.03

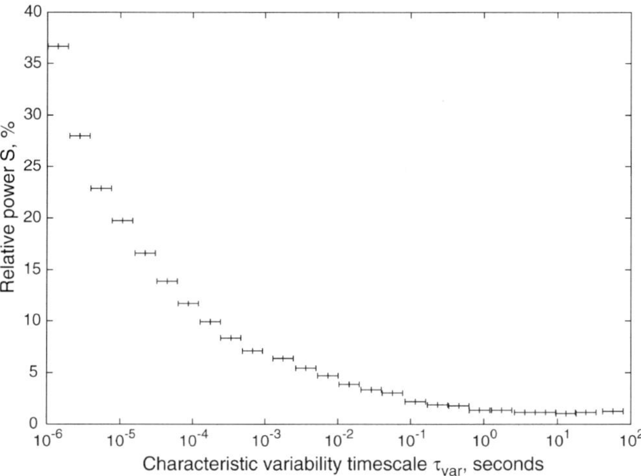

Fig. 2 Upper limits for the relative power of the variable emission

Figure 2 shows the upper limits for the relative power S of the variable emission component derived using (1–7). The model of triangular flares with the duty cycle of 0.1 was used.

We used the information on flares detected in gamma and X-ray bands (see, for example, Woods and Thompson 2004 and Götz et al. 2006).

For the photometric calibration, the stars listed in Table 2 have been used.

So, the upper limits for flashes with the duration of $\sim 10^{-2}$ s (similar to the parameters of γ-ray flashes) will be about 20 m. For very rare flashes with the duty cycle of 10^{-4}, the upper limits will be thirty times better—we should be able to detect events of 23–24 m.

Unfortunately, even having good upper limits for the variable optical component, due to huge absorption it is impossible to derive a reasonable estimate of the luminosity. Assuming $A_V \sim 20$ m and a distance of 6 to 15 kpc (Eikenberry et al. 2004; McClure-Griffiths and Gaensler 2005), we have for the upper limits of flaring optical luminosity of SGR 1806-20 values ranging from 7×10^{46} erg/s to 4×10^{47} erg/s. These values are similar to the peak luminosity of the object in the γ-ray band, i.e. to the values registered during the giant flare. This means that very short and very rare "giant optical flares" could be detected.

References

Beskin, G., Debur, V., Panferov, A., et al.: GCN 1129 (2001)
Beskin, G., Debur, V., Panferov, A., et al.: AIPS Conf. Proc. **662**, 583 (2003)
Borkowski, J., Götz, D., Mereghetti, S., et al.: GCN 2920 (2004)
Debur, V., Arkhipova, T., Beskin, G., et al.: Nucl. Instrum. Methods Phys. Res. A **513**, 127 (2003)
Duncan, R.C., Thompson, C.: Astrophys. J. **392**, L9 (1992)
Eikenberry, S.S., Matthews, K., LaVine, J.L., et al.: Astrophys. J. **616**, 506 (2004)
Gogus, E., Kouveliotou, C., Woods, P.M., et al.: Astrophys. J. **558**, 228 (2001)
Golenetskii, S., Aptekar, R., Mazets, E., et al.: GCN 2923 (2004)
Götz, D., Mereghetti, S., Molkov, S., et al.: Astron. Astrophys. **445**, 313 (2006)
Hurley, K., Kouveliotou, C., Cline, T., et al.: Astrophys. J. **523**, L37 (1999)
Hurley, K.: In: AIP Proc., 5th Huntsville Symp. on Gamma-Ray Bursts (2000)
Israel, G.L., Covino, S., Mereghetti, S., et al.: Astron. Telegr. **378** (2005)
Kouveliotou, C., Dieters, S., Strohmayer, T., et al.: Nature **393**, 295 (1998)
Mazets, E., Golenetskii, S., Aptekar, R., et al.: GCN 2922 (2004)
McClure-Griffiths, N.M., Gaensler, B.M.: Astrophys. J. **630**, L161 (2005)
Mereghetti, S., Götz, D., von Kienlin, A., et al.: Astrophys. J. **624**, L105 (2005)
Palmer, D.M., Barthelmy, S., Gehrels, N., et al.: Nature **434**, 1107 (2005)
Plokhotnichenko, V.: Commun. Spec. Astrophys. Obs. **38**, 29 (1983)
Plokhotnichenko, V.: PhD thesis (1992)
Plokhotnichenko, V., Beskin, G., Debur, V., et al.: Nucl. Instrum. Methods Phys. Res. A **513** 167 (2003)
Ricker, D., Lamb, D., Woosley, S., et al.: GCN 1068 (2001a)
Ricker, G., Lamb, D., Woosley, S., et al.: GCN 1089 (2001b)
Shvartsman, V.: Commun. Spec. Astrophys. Obs. **19**, 5 (1977)
Schwartz, S.J., Zane, S., Wilson, R.J., et al.: Astrophys. J. **627**, L129 (2005)
Terasawa, T., Tanaka, Y.T., Takei, Y., et al.: Nature **434**, 1110 (2005)
Woods, P.M., Thompson, C.: astro-ph/0406133 (2004)

Astrophys Space Sci (2007) 308: 481–485
DOI 10.1007/s10509-007-9357-3

ORIGINAL ARTICLE

Instabilities in rotating relativistic stars driven by viscosity

Motoyuki Saijo · Eric Gourgoulhon

Received: 30 June 2006 / Accepted: 27 October 2006 / Published online: 20 March 2007

Abstract We investigate the instability driven by viscosity in rotating relativistic stars by means of an iterative approach. We focus on polytropic rotating equilibrium stars and impose an $m = 2$ perturbation in the lapse. We vary both the stiffness of the equation of state and the compactness of the star to study these factors on the critical value T/W for the instability. For a rigidly rotating star, the criterion T/W, where T is the rotational kinetic energy and W the gravitational binding energy, mainly depends on the compactness of the star and takes values around 0.13–0.16, which slightly differ from that of Newtonian incompressible stars ($\sim$0.14). For differentially rotating stars, the critical value of T/W is found to span the range 0.17–0.25. The value is significantly larger than in the rigidly rotating case with the same compactness of the star. Finally we discuss the possibility of detecting gravitational waves from viscosity-driven instabilities using ground-based interferometers.

Keywords Instabilities · Stars: rotation · Relativity · Gravitational waves

PACS 04.40.Dg · 04.25.Dm · 04.30.Db · 97.10.Kc

M. Saijo (✉)
School of Mathematics, University of Southampton,
Southampton, SO17 1BJ, UK
e-mail: ms1@maths.soton.ac.uk

E. Gourgoulhon
Laboratoire de l'Univers et de ses Théories, UMR 8102 du CNRS, Observatoire de Paris, 92195 Meudon Cedex, France
e-mail: eric.gourgoulhon@obspm.fr

1 Introduction

Stars in nature are usually rotating and subject to nonaxisymmetric rotational instabilities. An analytical treatment of these instabilities exists only for incompressible equilibrium fluids in Newtonian gravity (e.g., Chandrasekhar 1969; Tassoul 1978). For these configurations, global rotational instabilities arise from non-radial toroidal modes $e^{im\varphi}$ ($m = \pm1, \pm2, \dots$) when $\beta \equiv T/W$ exceeds a certain critical value. Here φ is the azimuthal coordinate, while T and W are the rotational kinetic and gravitational binding energies. In the following we will focus on the $m = \pm2$ bar-mode, since it is the fastest growing mode when the rotation is sufficiently rapid.

There exist two different mechanisms and corresponding timescales for bar-mode instabilities. Rigidly rotating, incompressible stars in Newtonian gravity are *secularly* unstable to bar formation when $\beta \gtrsim \beta_{\rm sec} \simeq 0.14$. This instability can grow in the presence of some dissipative mechanisms, such as viscosity or gravitational radiation, and the growth time is determined by the dissipative timescale, which is usually much longer than the dynamical timescale of the system. By contrast, a *dynamical* instability to bar formation sets in when $\beta \gtrsim \beta_{\rm dyn} \simeq 0.27$. This instability is independent of any dissipative mechanism, and the growth time is the hydrodynamic timescale of the system.

In the absence of thermal dissipation there are two dissipative mechanisms that can drive the secular bar instability; they are viscosity and gravitational radiation. The instability driven by viscosity sets in when a mode has a zero-frequency in the frame rotating with the star, and the first unstable mode in terms of m is the $m = 2$ bar mode. The quasi-static evolution of the star due to the viscosity-driven instability deforms the Maclaurin spheroid to the Jacobi ellipsoid in Newtonian incompressible stars. Throughout the

deformation process the circulation of a given closed curve at the beginning of deformation varies but the angular momentum is conserved. On the other hand, the instability induced by gravitational radiation sets in when the backward going mode is dragged forward in the inertial frame, and the modes are all unstable when m exceeds a certain value. In incompressible Newtonian stars the quasi-static evolution of the star due to the gravitational wave-driven instability deforms a Maclaurin spheroid into a Dedekind ellipsoid. Throughout the deformation process the star's angular momentum varies but the circulation of a given closed curve at the beginning of deformation is conserved.

The purpose of the paper is twofold. Firstly, we investigate the critical value T/W of the viscosity-driven instability for a rigidly rotating compressible star. The argument that the viscosity-driven instability can deform a star from a Maclaurin spheroid to a Jacobi ellipsoid is valid if the star has no internal energy. Otherwise the total energy can be converted into the internal energy without any emission from the star. Here we assume that the cooling timescale of the star is shorter than the thermal heating timescale so that the thermal energy generated by viscosity is immediately radiated away. Therefore the picture of the deformation process due to viscosity is quite similar to the case of incompressible stars. The stars are considered as compressible bodies, and therefore it is worthwhile taking the compressibility into account to see whether there is a significant effect on the critical value of the instability. In this respect the present work extends that of Gondek-Rosińska and Gourgoulhon (2002) to the compressible case.

Our other main purpose in this paper is to investigate the effect of differential rotation on the secular bar instabilities driven by viscosity. For high viscosity or a strong magnetic field, the star maintains rigid rotation. However, in nature, the star may rotate differentially, as is the case for the Sun. Stellar collapses and mergers may also lead to differentially rotating stars. For the coalescence of irrotational binary neutron stars, the presence of differential rotation may temporarily stabilize the "hypermassive" remnant. Therefore it is worthwhile taking differential rotation into account to study instabilities driven by viscosity in rotating relativistic stars.

This paper is organized as follows. In Sect. 2 we present the iterative evolution approach to determine the stability due to viscosity in relativistic rotating stars. We discuss our numerical results in Sects. 3 and 4, focusing on the instability driven by viscosity in rigidly and differentially rotating stars. In Sect. 5 we briefly summarize our findings. Throughout this paper, we use the geometrized units with $G = c = 1$ and adopt polar coordinates (r, θ, φ) with the coordinate time t. A more detailed discussion is presented in Saijo and Gourgoulhon (2006).

2 Iterative evolution approach

We follow an iterative evolution approach (Bonazzola et al. 1996, 1998) to investigate the viscosity-driven instability in rotating relativistic stars. The physical viewpoint of this approach can be understood in Newtonian gravity by considering the transition between a rigidly rotating incompressible axisymmetric star (Maclaurin spheroid) to a nonaxisymmetric star (Jacobi ellipsoid). The above deformation process is driven by viscosity, since viscosity causes the circulation to vary but keeps the angular momentum constant in the Newtonian incompressible star. From a computational viewpoint, the key theme of this approach is that, instead of performing a time evolution of the star to investigate its stability, we treat the iteration number as an evolutional time and determine the stability of the star with respect to this iteration. The advantage of this approach is that there is no restriction on the evolutional time-step even in a star with high compactness. The main uncertainty of this approach is whether one can treat the iterative number as time. However there is a correspondence. In the Newtonian incompressible star, Gondek-Rosińska and Gourgoulhon (2002) investigate the difference between the exact critical value of the bifurcation point (e.g., Chandrasekhar 1969) and the critical value computed by their numerical code. They find that the results agree within the round-off error.

To determine the stability of a rotating relativistic star driven by viscosity, we follow a computational procedure to construct an equilibrium configuration until we reach a relative error in the enthalpy norm of 1.5×10^{-7}. At this iteration step, we put the following $m = 2$ perturbation in the logarithmic lapse ν (the lapse defines the proper time between consecutive layers of spatial hypersurfaces) to enhance the growth of the bar mode instability as

$$\nu = \nu_{\mathrm{eq}}(1 + \varepsilon_{\mathrm{amp}} \sin^2\theta \cos 2\varphi), \tag{1}$$

where ν_{eq} is the logarithmic lapse in the equilibrium, and $\varepsilon_{\mathrm{amp}}$ is the amplitude of the perturbation. We diagnose the maximum logarithmic lapse of the $m = 2$ coefficients $\hat{\nu}_2$ in terms of mode decomposition as

$$q = \max |\hat{\nu}_2|. \tag{2}$$

We also define the logarithmic derivative of q in the iteration step $\mathcal{N}_i$ as

$$\frac{\dot{q}}{q} = \frac{q_i - q_{i-1}}{q_{i-1}}, \tag{3}$$

where q_i denotes q at the iteration step $\mathcal{N}_i$. We then determine the stability of the star as follows. When the diagnostic q grows exponentially after we impose a bar mode perturbation in the logarithmic lapse, we conclude that the star is

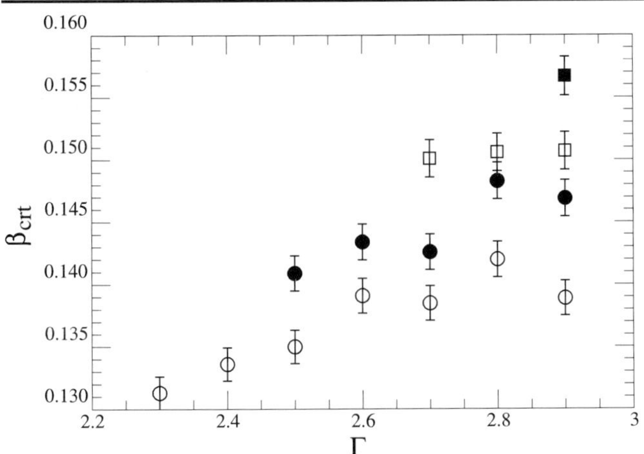

Fig. 1 Critical value of β as a function of an adiabatic index Γ for four different compactness of rigidly rotating stars (see Table II of Saijo and Gourgoulhon 2006). Open circles, filled circles, open squares, and filled squares refer to the compactness M/R of 0.01, 0.05, 0.1, and 0.15, respectively. The star whose adiabatic index is $\Gamma = \Gamma_{\rm low} - 0.1$, where $\Gamma_{\rm low}$ is the lowest Γ of an unstable star for any compactness in this figure, is stable. Error bars at 99% significance level are estimated from our convergence tests in terms of perturbation amplitudes and collocation points (see Saijo and Gourgoulhon 2006)

unstable. On the other hand when the diagnostic decays after we introduce the perturbation, the star is stable. Finally we determine the critical value of β as the minimum one in the unstable branch. We also confirm our argument that in all equilibrium stars there is a continuous transition between stable and unstable stars as a function of β (Saijo and Gourgoulhon 2006).

3 Rigidly rotating stars

We study the critical value of β for the viscosity-driven instability in rigidly rotating stars (see Fig. 1). We find that relativistic gravitation tends to stabilizes the star, and that the critical value of β for each compactness is almost insensitive to the stiffness of the equation of state. Our computational results (Fig. 1) show that the critical β is ~0.137 for $M/R = 0.01$ (where M is the gravitational mass, R the circumferential radius), ~0.145 for $M/R = 0.05$, ~0.150 for $M/R = 0.1$, and ~0.157 for $M/R = 0.15$, respectively. The critical value of β monotonically increases when increasing the compactness of the star. For the case of Newtonian compressible stars it was shown in Fig. 3 of Bonazzola et al. (1996) that the critical value of β is ~0.134, and is not very sensitive to the stiffness of the star.

4 Differentially rotating stars

We will now investigate the threshold for the viscosity-driven instability in differentially rotating stars. First we show the result of a fixed rotation profile throughout the evolution (Table 1). We find that both relativistic gravitation and differential rotation tend to stabilize the star. The critical value is $\beta_{\rm crt} \sim 0.13$–$0.16$ for a rigidly rotating star depending on the compactness of the star, while $\beta_{\rm crt} \sim 0.18$–$0.25$ for a differentially rotating star with a moderate degree of differential rotation.

Table 1 Critical value of the viscosity-driven instability in differentially rotating relativistic stars. We choose $\Gamma = 2$ for the polytropic equation of state and $\hat{A}_{\rm rot} = 1$ as the degree of differential rotation

R_p/R_e [a]	$H_{\rm max}$ [b]	$\beta_{\rm crt}$	M/R
0.4458	7.594×10^{-3}	0.1828	0.01000
0.3985	3.830×10^{-2}	0.1999	0.05000
0.3820	7.492×10^{-2}	0.2186	0.1000
0.3457	1.116×10^{-1}	0.2354	0.1500
0.2982	1.581×10^{-1}	0.2496	0.2000

[a] R_p/R_e: Ratio of the polar proper radius to the equatorial proper radius

[b] $H_{\rm max}$: Maximum logarithmic enthalpy

Next we study the variation of the rotation profile as viscosity also plays a significant role in changing the angular momentum distribution of the star. In order to mimic this process, after we impose a perturbation we vary slightly the parameter which represents the degree of differential rotation:

$$\hat{A}_{\rm rot}^{-1} = \hat{A}_{\rm rot}^{-1\,(\rm eq)}[1 - \epsilon_{\rm rot}(\mathcal{N} - \mathcal{N}_{\rm ptb})], \tag{4}$$

where $\hat{A}_{\rm rot}$ is the degree of differential rotation, $\hat{A}_{\rm rot}^{(\rm eq)}$ the degree of differential rotation in the equilibrium state, $\epsilon_{\rm rot}$ the degree of the variation of the rotation profile which we set to be 1.0×10^{-4}, $\mathcal{N}$ the iteration number, and $\mathcal{N}_{\rm ptb}$ the iteration number when we impose the perturbation. Since the viscosity only affects the local interaction between fluid components, the total angular momentum is conserved even when the viscosity plays a role. Therefore we also vary the central angular velocity $\Omega_{\rm c}$ in the following manner

$$\Omega_{\rm c} = \Omega_{\rm c}^{(\rm eq)}[1 - \epsilon_{\rm omg}(\mathcal{N} - \mathcal{N}_{\rm ptb})], \tag{5}$$

where $\Omega_{\rm c}^{(\rm eq)}$ is the central angular momentum in the equilibrium state, and $\epsilon_{\rm omg}$ the degree of the variation of the central angular velocity required in order to (approximately) conserve the total angular momentum.

Taking into account the change of rotational profile, we show our numerical results for $\beta_{\rm crt}$ in Fig. 2. We find that all stars with $\beta \approx \beta_{\rm crt}$ and a fixed rotation profile become unstable. In fact, the plateau stage of $\dot{q}/q$ for a fixed rotation profile in Fig. 7 of Saijo and Gourgoulhon (2006) has a continuous increase in our Fig. 2. Therefore, we estimate two relevant timescales, the growth time of the bar mode due to viscosity and the variation time of the rotation profile due to viscosity, and compare them.

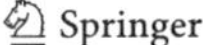

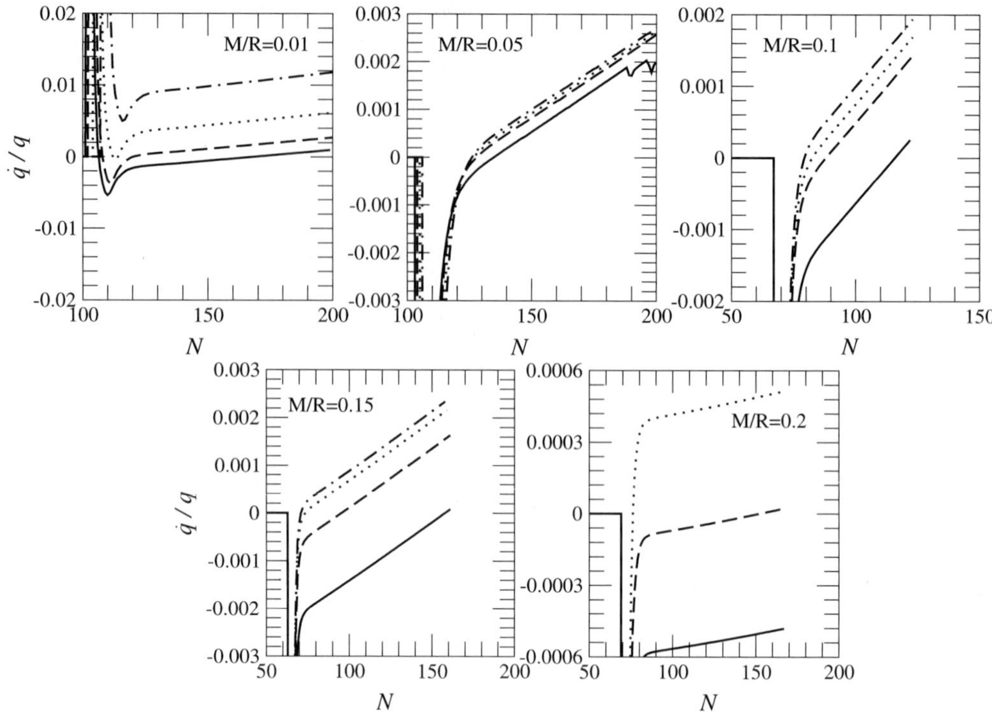

Fig. 2 Diagnostic $\dot{q}/q$ as a function of iteration step $\mathcal{N}$ for five different differentially rotating stars. *Solid*, *dashed*, *dotted*, and *dashed lines* denote $\beta = (0.1825, 0.1828, 0.1834, 0.1844)$ and $\varepsilon_{\rm omg} = 0.8 \times 10^{-4}$ for $M/R = 0.01$; $\beta = (0.1993, 0.1996, 0.1998, 0.1999)$ and $\varepsilon_{\rm omg} = 0.9 \times 10^{-4}$ for $M/R = 0.05$; $\beta = (0.2180, 0.2184, 0.2185, 0.2186)$ and $\varepsilon_{\rm omg} = 1.3 \times 10^{-4}$ for $M/R = 0.1$; $\beta = (0.2348, 0.2352, 0.2353, 0.2354)$ and $\varepsilon_{\rm omg} = 1.1 \times 10^{-4}$ for $M/R = 0.15$; and $\beta = (0.2494, 0.2495, 0.2496)$ and $\varepsilon_{\rm omg} = 1.0 \times 10^{-4}$ for $M/R = 0.2$, respectively. Note that $\dot{q}/q$ is always increasing around the critical value of β in differentially rotating stars. We see that variations of rotation profile due to viscosity unstabilize the star

Based on the analytical estimation of the two time-scales (Saijo and Gourgoulhon 2006), the timescales used in our numerical results should be described as

$$\tau_{\rm ang} \approx \varepsilon_{\rm omg}^{-1}, \tag{6}$$

$$\tau_{\rm bar} \approx \varepsilon_{\rm omg}^{-1} \left(\frac{\Omega_{\rm c} - \Omega_{\rm s}}{\Omega_{\rm c}} \right) \left(\frac{\beta_{\rm sec}}{\beta - \beta_{\rm sec}} \right), \tag{7}$$

where $\Omega_{\rm s}$ is the equatorial surface angular velocity, and $\beta_{\rm sec}$ is the critical value $\beta_{\rm crt}$ of the secular bar mode instability driven by viscosity. We confirm that differential rotation tends to stabilize the star. Also, (7) shows that the timescale of the bar growth becomes short as the viscosity reduces the degree of differential rotation. The critical value $\beta_{\rm crt}$ changes from the value computed for a fixed rotation profile, but the change is roughly of the same order of magnitude as for differential rotation profile (4), which means $\approx\varepsilon_{\rm omg}$ $(\approx\varepsilon_{\rm rot})$.

5 Conclusion

We have studied the viscosity-driven instability in both rigidly and differentially rotating polytropic stars by means of iterative evolution approach in General Relativity. We have focused on the threshold of the instability driven by viscosity.

We find that relativistic (rather than Newtonian) gravitation stabilizes the star, preventing the onset of a viscosity-driven instability. Also, the critical value T/W is not sensitive to the stiffness of the polytropic equation of state for a given compactness of the star. In a previous study devoted to compressible stars, Bonazzola et al. (1998) investigated a sequence of mass-shedding stars and showed that relativistic gravitation does stabilize rigidly rotating polytropic stars. In our study we have calculated the value of $\beta_{\rm crt}$ for rigidly rotating stars.

We have also found that differential rotation stabilizes a star against the instability driven by viscosity. If we fix the compactness of the star, we find a significant increase of the critical value $\beta_{\rm crt}$, which supports the above statement. We also confirmed the above statement by changing the rotation profile due to viscosity and found that differential rotation still significantly stabilizes the instability driven by viscosity.

Finally, let us mention the characteristic amplitude and the frequency of gravitational waves emitted throughout the secular bar instability driven by viscosity. They produce quasi-periodic gravitational waves detectable in the ground based interferometers. The characteristic amplitude h of gravitational waves estimated from the evolution of a Jacobi-ellipsoid to a Maclaurin spheroid is [(4.2) of Lai and Shapiro 1995]

$$h \approx 9.1 \times 10^{-21} \times \left(\frac{30\ \text{Mpc}}{d}\right)\left(\frac{M}{1.4M_\odot}\right)^{3/4}\left(\frac{R}{10\ \text{km}}\right)^{1/4} f^{-1/5}, \quad (8)$$

where d is the distance to the source and the characteristic frequency $f = \Omega/\pi$ is $f \gtrsim 1000$ Hz, depending on the magnitude of stellar rotation. Note that the frequency increases throughout this process. Although the likely frequency of the source is slightly higher than the highest sensitive frequencies of the ground based detectors, we may have a chance to detect viscosity-driven instabilities if they occur in the Virgo cluster.

Acknowledgements It is our pleasure to thank Ian Jones and the anonymous referee for their careful and critical reading of our manuscript.

References

Bonazzola, S., Frieben, J., Gourgoulhon, E.: Astrophys. J. **460**, 379 (1996)
Bonazzola, S., Frieben, J., Gourgoulhon, E.: Astron. Astrophys. **331**, 280 (1998)
Chandrasekhar, S.: Ellipsoidal Figures of Equilibrium. Yale University Press, New York (1969), Chap. 5
Gondek-Rosińska, D., Gourgoulhon, E.: Phys. Rev. D **66**, 044021 (2002)
Lai, D., Shapiro, S.L.: Astrophys. J. **442**, 259 (1995)
Saijo, M., Gourgoulhon, E.: Phys. Rev. D **74**, 084006 (2006)
Tassoul, J.: Theory of Rotating Stars. Princeton University Press, New Jersey (1978), Chap. 10

Astrophys Space Sci (2007) 308: 487–491
DOI 10.1007/s10509-007-9313-2

ORIGINAL ARTICLE

10 years of *RXTE* monitoring of anomalous X-ray pulsar 4U 0142+61: long-term variability

Rim Dib · Victoria M. Kaspi · Fotis P. Gavriil

Received: 5 July 2006 / Accepted: 12 July 2006 / Published online: 21 March 2007

Abstract We report on 10 yr of monitoring of the 8.7-s Anomalous X-ray Pulsar 4U 0142+61 using the *Rossi X-Ray Timing Explorer (RXTE).* This pulsar exhibited stable rotation from 2000 until February 2006: the RMS phase residual for a spin-down model which includes ν, $\dot{\nu}$, and $\ddot{\nu}$ is 2.3%. We report a possible phase-coherent timing solution valid over a 10-yr span extending back to March 1996. A glitch may have occurred between 1998 and 2000, but it is not required by the existing data. We also report that the source's pulse profile has been evolving in the past 6 years, such that the dip of emission between its two peaks has been getting shallower since 2000, almost as if the profile is recovering to its pre-2000 morphology, in which there was no clear distinction between the peaks. These profile variations are seen in the 2–4 keV band but not in 6–8 keV. Finally, we present the pulsed flux time series of the source in 2–10 keV. There is evidence of a slow but steady increase in the source's pulsed flux since 2000. The pulsed flux variability and the narrow-band pulse profile changes present interesting challenges to aspects of the magnetar model.

Keywords Pulsars: individual (4U 0142+61) · Stars: neutron · X-rays: stars

PACS 97.60

This work was supported by the Natural Sciences and Engineering Research Council (NSERC) PGSD scholarship to R.D. F.P.G. holds a National Research Council Research Associateship Award at NASA Goddard Space Flight Center. Additional support was provided by NSERC Discovery Grant Pgpin 228738-03 NSERC Steacie Supplement Smfsu 268264-03, FQRNT, CIAR, and CFI. V.M.K. is a Canada Research Chair.

R. Dib (✉) · V.M. Kaspi
Department of Physics, McGill University, Montreal, QC H3A 2T8, Canada
e-mail: rim@physics.mcgill.ca

V.M. Kaspi
e-mail: vkaspi@physics.mcgill.ca

F.P. Gavriil
X-Ray Astrophysics Laboratory, NASA Goddard Space Flight Center, Greenbelt, MD 20771, USA
e-mail: gavriil@milkyway.gsfc.nasa.gov

1 Introduction

1.1 Anomalous X-ray pulsars (AXPs)

The existence of magnetars—young, isolated neutron stars powered by the decay of an ultrahigh magnetic field—is now well supported by several lines of evidence (Woods and Thompson 2006). There are at least two flavors of magnetars: SGRs and AXPs. They both exhibit: X-ray pulsations with a luminosity of 10^{34-36} erg s^{-1}, periods ranging from 5–12 s, period derivatives of 10^{-13}–10^{-11}, and surface dipolar magnetic fields of 0.6–7 $\times$ 10^{14} G. In the magnetar model, the pulsed X-rays are the result of a combination of surface thermal emission and resonant scattering in the magnetosphere (Thompson et al. 2002). For more on AXPs and SGRs, see reviews by V. Kaspi and S. Mereghetti (this volume).

1.2 AXP 4U 0142+61

4U 0142+61 is an 8.7-s AXP. It has $\dot{P} \cong 0.2 \times 10^{-11}$, implying a surface dipole magnetic field of 1.3×10^{14} G.[1]

[1] Magnetic fields here are calculated via $B \equiv 3.2 \times 10^{19} \sqrt{P\dot{P}}$ G, where P is the spin period in seconds and $\dot{P}$ is the period derivative.

It is known to pulsate in the optical band (Kern and Martin 2002; Dhillon et al. 2005), and has been detected in the near-IR (Hulleman et al. 2004), in the far-IR (Wang et al. 2006), and in hard X-rays (Den Hartog et al. 2006; Kuiper et al. 2006). It has a soft X-ray spectrum well fitted by a combination of a blackbody and a power law (see, for example, White et al. 1996). AXP 4U 0142+61 rotates with high stability (Gavriil and Kaspi 2002). However, Morii et al. (2005) reported a timing glitch in 1999 on the basis of an *ASCA* observation in which the value of the frequency is marginally discrepant with that predicted by the ephemeris reported by Gavriil and Kaspi (2002).

Here we report on continued *RXTE* monitoring observations of this source. Our observations are described in Sect. 2. Our timing, pulsed morphology, and pulsed flux analysis are presented, respectively, in Sects. 3, 4, and 5.

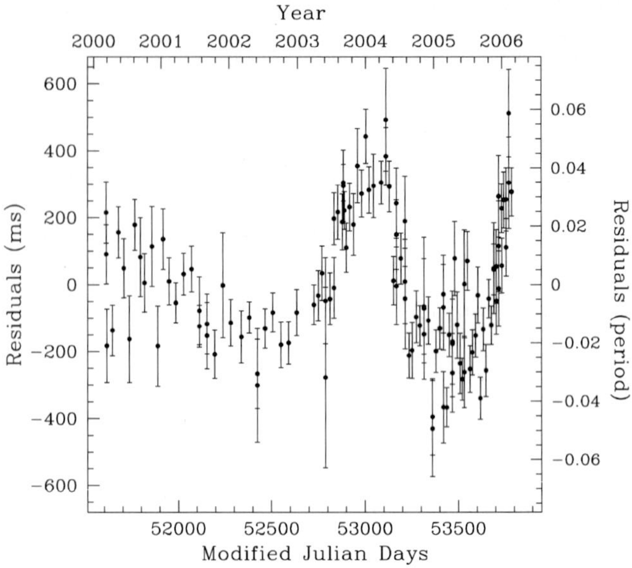

Fig. 1 Arrival time residuals for 4U 0142+61 for the post-gap period, using the post-gap ephemeris given in Table 1. The residuals have RMS 2.3% of the pulse period

2 Observations

We used 136 *RXTE/PCA* observations of various lengths in our analysis: (a) 4 very closely spaced *RXTE* Cycle 1 observations, (b) 14 short Cycle 2 observations spanning a period of a year, (c) 1 Cycle 3 observation (followed by a 2-yr gap with no observations), (d) 118 observations taken regularly from 2000 to 2006 as part of a long-term monitoring program spanning *RXTE* Cycles 5 to 10.

For each observation, photon arrival times were barycentered and binned with 31.25-ms time resolution.

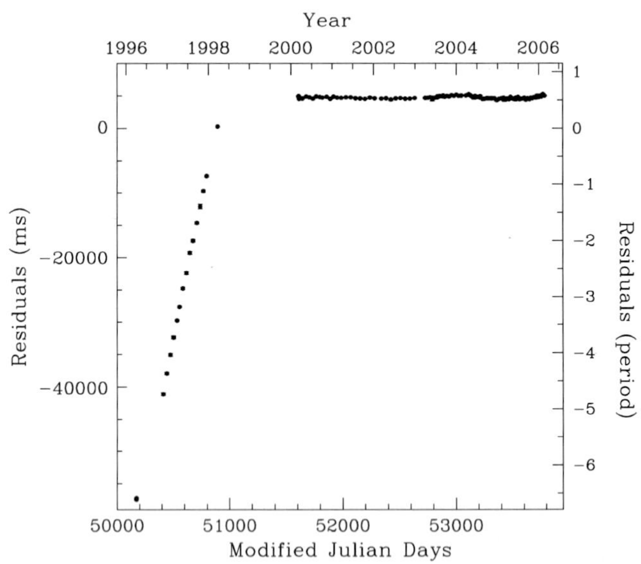

Fig. 2 Arrival time residuals for 4U 0142+61 for all Cycles using the post-gap ephemeris

3 Phase-coherent timing

Each binned time series was folded at the pulse period. Resulting pulse profiles were cross-correlated with a high S/N template. This returned an average pulse time of arrival (TOA) for each observation corresponding to a fixed pulse phase. The pulse phase at any time can be expressed as a Taylor expansion polynomial. The TOAs were fitted to the polynomial using the pulsar timing software package TEMPO.[2]

We report a phase-coherent timing solution that spans the post-gap (i.e. after 2000) 6-yr period up until February 2006 (MJD 53787) including all data in *RXTE* Cycles 5–10. The parameters of our best-fit spin-down model are shown in Table 1. The phase residuals are shown in Fig. 1. The best-fit post-gap ephemeris does not, however, fit the pre-gap TOAs well. Figure 2 shows a clear systematic deviation in the pre-gap residuals obtained after subtracting the post-gap ephemeris. This could indicate that a glitch occurred at some time during the gap. However, by using 6 frequency derivatives, we found a possible ephemeris that fits the entire Cycle 1 to Cycle 10 range (see Table 1, Fig. 3).

The existence of our overall ephemeris cannot rule out the possibility of the glitch having occurred in 1999 (Morii et al. 2005): if a fully recovered glitch with a short relaxation time occurred in the *RXTE/PCA* observing gap between Cycles 3 and 5, only a random phase jump would be observed. To investigate this, we added an arbitrary but constant time

[2]http://pulsar.princeton.edu/tempo.

Table 1 Spin Parameters for 4U 0142+61

Parameter	Pre-gap ephemeris spanning Cycles 1 to 3	Post-gap ephemeris spanning Cycles 5 to 10	Possible overall ephemeris spanning all Cycles
MJD range	50170.693–50893.288	51610.636–53787.372	50170.693–53787.372
TOAs	19	118	137
ν (Hz)	0.115096869(3)	0.1150969337(3)	0.1150969304(2)
$\dot{\nu}$ (10^{-14} Hz s^{-1})	−2.659(3)	−2.6935(9)	−2.6514(7)
$\ddot{\nu}$ (10^{-23} Hz s^{-2})	–	4.17(10)	−1.7(2)
$d^3\nu/dt^3$ (10^{-31} Hz s^{-3})	–	–	3.62(12)
$d^4\nu/dt^4$ (10^{-39} Hz s^{-4})	–	–	8.7(3)
$d^5\nu/dt^5$ (10^{-46} Hz s^{-5})	–	–	−5.01(13)
$d^6\nu/dt^6$ (10^{-54} Hz s^{-6})	–	–	6.6(4)
Epoch (MJD)	51704.000025	51704.000025	51704.000000
RMS residual	0.019	0.023	0.019

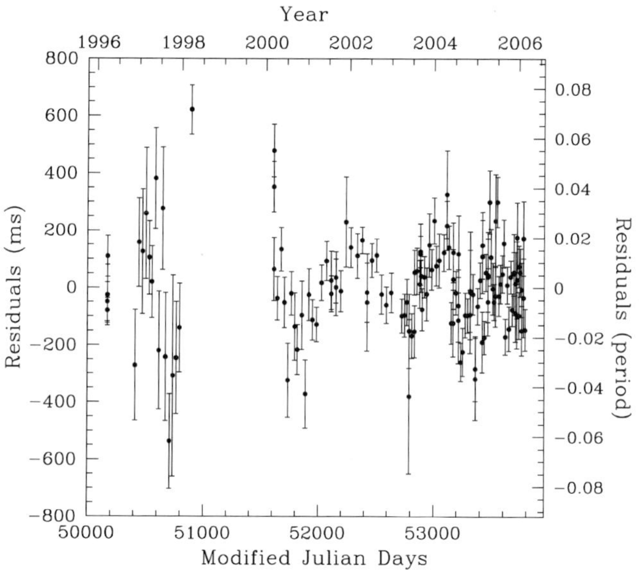

Fig. 3 Arrival time residuals for all Cycles using the overall ephemeris (see Table 1). The residuals have RMS 1.9% of the pulse period

jump to all the post-gap TOAs. We were still able to find a new ephemeris that connected the TOAs through the two-year gap. This indicates that our overall ephemeris is not unique. Hence, we cannot rule out the possibility of a phase jump between Cycles 3 and 5, and therefore a glitch in 1999 cannot be ruled out. However, it is also not required by the data. Indeed, the frequency of the discrepant *ASCA* observation reported by Morii et al. (2005) is consistent with the frequency predicted by the overall ephemeris shown in Table 1 to within 2σ.

4 Pulse profile changes

To search for pulse profile changes, the phase-aligned profiles were averaged for each *RXTE* Cycle. This was done in three energy bands. The average profiles in all bands are presented in Fig. 4. The ratios of the Fourier amplitudes of the pulse profiles in all bands as a function of *RXTE* Cycle are presented in Fig. 5.

In Fig. 4, the pulse profile changes are clear: for 2–10 and 2–4 keV, in Cycles 1 and 2, the smaller peak is not well defined. After the two-year gap, in Cycle 5, the dip between the peaks is much more pronounced. The peaks start to merge back in subsequent Cycles almost as if the profile is recovering to its original morphology. In 6–8 keV, the smaller peak is not as obvious, indicating that it has a softer spectrum relative to the larger peak.

Given the spectrum of the source (see White et al. 1996), which is fitted to a two-component model consisting of power law and thermal emission, the 2–4 keV band includes both thermal and power-law photons while the 6–8 keV band contains negligibly few thermal photons. In Fig. 5, the fact that the ratio of the harmonics is dropping only in 2–4 keV suggests that only the thermal component of the spectrum is evolving.

5 Pulsed flux

For each observation, the pulsed flux, $F_{\rm rms}$ (in counts/s/PCU), was calculated by taking the square root of the average of the squares of the deviations from the mean number of counts in the pulse profile. We omitted PCU 0 from this

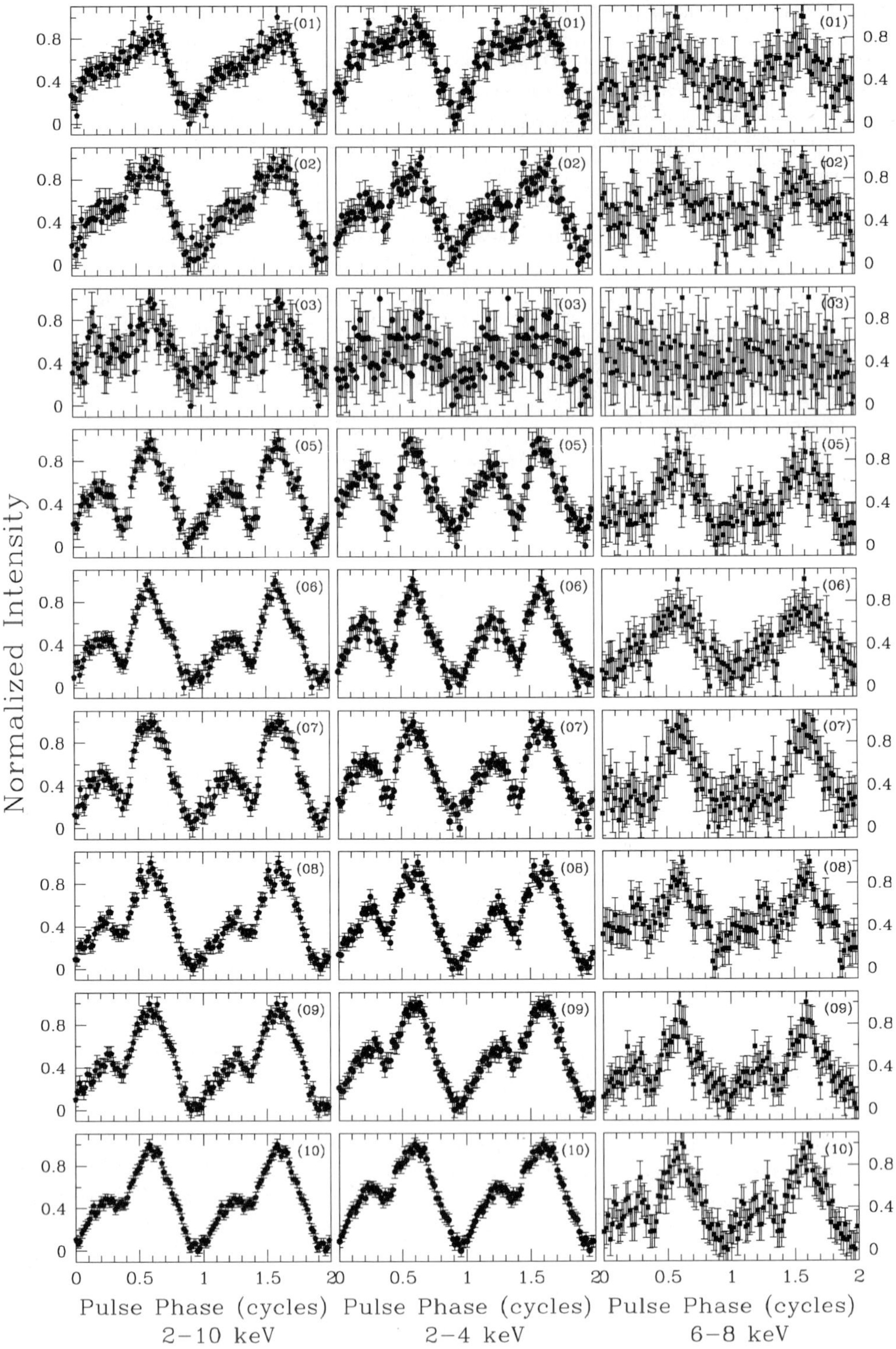

Fig. 4 Average pulse profiles in all RXTE Cycles in 2–10, 2–4, and 6–8 keV. In a given band, the different profile qualities are due to different net exposure times. The Cycle number is shown in the *top right corner* of each plot

analysis because of the uncertainties in its response due to the loss of the propane layer.

The pulsed flux series for 4U 0142+61 shows a slow but steady increase since 2000 in 2–10 keV (see Fig. 6). There are hints that the change is also present in 2–4 keV and not in 6–8 keV but our statistics do not let us confirm this. We verified there are no comparable trends in the long-term light curves of the other AXPs observed as part of this monitoring program. For further discussion of the pulsed flux evolution of this source, see Dib et al. (2007, submitted).

From Thompson et al. (2002), increases in the twist angle of the field lines in the magnetosphere can cause luminosity increases as well as pulse profile changes. It is tempting to also attribute the increase in the pulsed

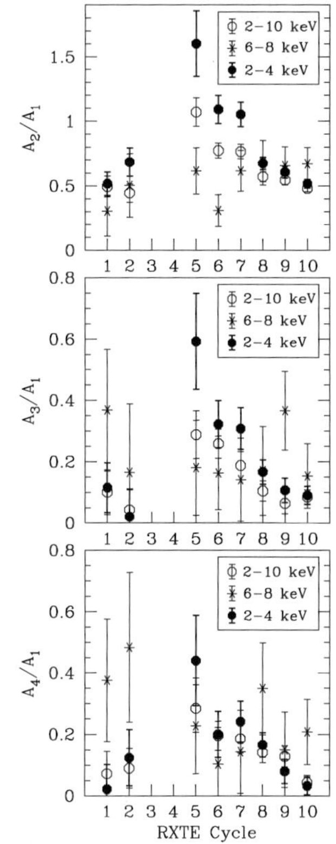

Fig. 5 Ratios of the Fourier amplitudes of the pulse profiles in all three energy bands as a function of *RXTE* Cycle. *Top*: ratio of the 2nd amplitude (A2) to that of the fundamental (A1). Middle: (A3/A1). *Bottom*: (A4/A1)

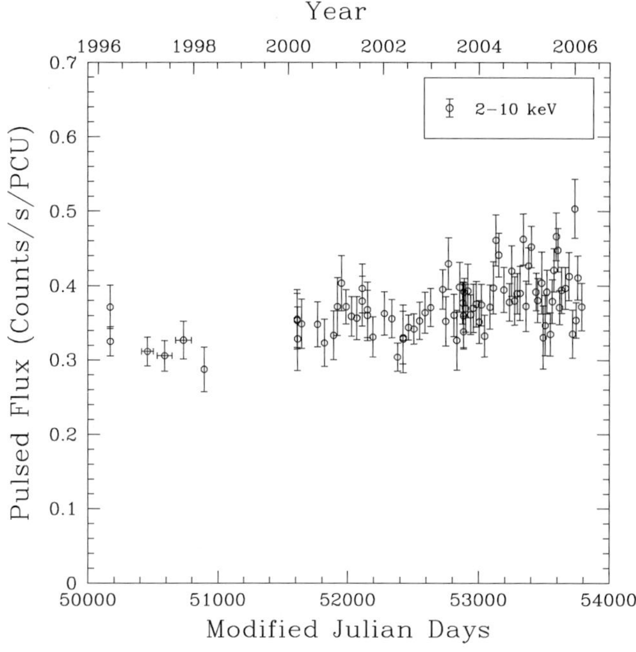

Fig. 6 Pulsed flux time series for 4U 0142+61 in the 2–10 keV band

flux of this source to an increase in the twist angle. However, Thompson et al. (2002) also predict that an increase in the twist angle is accompanied by an increase in the spin down rate, which is not what we are currently observing: in the post-gap ephemeris presented in Table 1, $\ddot{\nu}$ is positive, i.e. the magnitude of the spin down rate is decreasing. Thus, another explanation is needed.

Finally, it is interesting to note that in April 2006, after the last *RXTE* Cycle included in this analysis, the pulsar appears to have entered an extended active phase: a single burst accompanied by a pulse profile change was detected from the pulsar on April 06 (Kaspi et al. 2006). A series of four bursts was later detected on June 25 (Dib et al. 2006). This interesting turn of events is presently under careful study.

References

Den Hartog, P.R., Hermsen, W., Kuiper, L., et al.: Astron. Astrophys. **451**, 587 (2006)
Dib, R., Kaspi, V., Gavriil, F., et al.: Astron. Telegr. **845**, 1 (2006)
Dib, R., et al.: Astrophys. J. (2007, submitted)
Dhillon, V.S., Marsh, T.R., Hulleman, F., et al.: Mon. Not. Roy. Astron. Soc. **363**, 609 (2005)
Gavriil, F.P., Kaspi, V.M.: Astrophys. J. **567**, 1067 (2002)
Hulleman, F., van Kerkwijk, M.H., Kulkarni, S.R.: Astron. Astrophys. **416**, 1037 (2004)
Kaspi, V., Dib, R., Gavriil, F.: Astron. Telegr. **794**, 1 (2006)
Kern, B., Martin, M.: Nature **417**, 527 (2002)
Kuiper, L., Hermsen, W., Den Hartog, P.R., et al.: Astrophys. J. **645** 556 (2006)
Morii, M., Kawai, N., Shibazaki, N.: Astrophys. J. **622**, 544 (2005)
Thompson, C., Lyutikov, M., Kulkarni, S.R.: Astrophys. J. **574**, 332 (2002)
Wang, Z., Chakrabarty, D., Kaplan, D.L.: Nature **440**, 722 (2006)
White, N.E., Angelini, L., Ebisawa, K., et al.: Astrophys. J. Lett. **463**, L83 (1996)
Woods, P.M., Thompson, C.: In: Lewin, W.H.G., van der Klis, M. (eds.) Compact Stellar X-Ray Sources (2006)

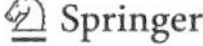

Astrophys Space Sci (2007) 308: 493–497
DOI 10.1007/s10509-007-9354-6

ORIGINAL ARTICLE

Exotic ion H_3^{++} in strong magnetic fields

Ground state and low-lying states

Juan C. López Vieyra · Alexander V. Turbiner · Nicolais L. Guevara

Received: 3 July 2006 / Accepted: 16 November 2006 / Published online: 24 March 2007

Abstract

1. The exotic system H_3^{++} (which does not exist without magnetic field) exists in strong magnetic fields:
   (a) In triangular configuration for $B \approx 10^8$–10^{11} G (under specific external conditions)
   (b) In linear configuration for $B > 10^{10}$ G
2. In the linear configuration the positive z-parity states $1\sigma_g$, $1\pi_u$, $1\delta_g$ are bound states
3. In the linear configuration the negative z-parity states $1\sigma_u$, $1\pi_g$, $1\delta_u$ are repulsive states
4. The H_3^{++} molecular ion is the most bound one-electron system made from protons at $B > 3 \times 10^{13}$ G

Possible application: The H_3^{++} molecular ion may appear as a component of a neutron star atmosphere under a strong surface magnetic field $B = 10^{12}$–10^{13} G.

Keywords Molecular ion H_3^{++} · Strong magnetic field

PACS 31.15.Pf · 31.15.-p · 31.10.+z · 32.60.+i · 97.10.Ld

1 Introduction

In the 70's the first theoretical arguments were given indicating that the physics of atoms and molecules in a strong magnetic field can exhibit a wealth of new and unexpected phenomena (Kadomtsev-Kudryatsev-Ruderman). In particular, the possibility of formation of unusual chemical compounds not existing without a magnetic field was emphasized. In practice, the atmosphere of neutron stars, which is characterized by enormous surface magnetic fields $\sim 10^{12}$–10^{13} G (up to 10^{15}–10^{16} G for magnetars), provides a valuable paradigm where this physics can be realized.

Recently, the observational data (collected by the Chandra X-ray spatial observatory) of the soft X-ray spectrum of the isolated neutron star 1E1027.4-5209 revealed for the first time certain irregularities (Sanwal et al. 2002). These irregularities might be interpreted as absorption features at $\sim$0.7 KeV and $\sim$1.4 KeV of a possible atomic-molecular nature assigned to the content of the neutron star atmosphere. In this context the study of the atomic and molecular, traditional and exotic, systems in strong magnetic fields deserves a special attention.

Among observed features of one-electron atomic-molecular systems when the magnetic field increases are (see Turbiner and López Vieyra 2006):

- The *total* and *binding* energies increase. (The systems become more and more bound.)
- Drastic decrease in the electron localization length in both directions, transverse and longitudinal, leading to a decrease in the equilibrium distance in molecular systems. (The systems become more and more compact.)
- The electronic cloud takes a needle-like shape forming in most cases the quasi-one-dimensional systems with very unusual physical properties.

J.C. López Vieyra (✉) · A.V. Turbiner · N.L. Guevara
Instituto de Ciencias Nucleares, UNAM,
Mexico, DF 04510, Mexico
e-mail: vieyra@nucleares.unam.mx

A.V. Turbiner
e-mail: turbiner@nucleares.unam.mx

N.L. Guevara
e-mail: nicolais@nucleares.unam.mx

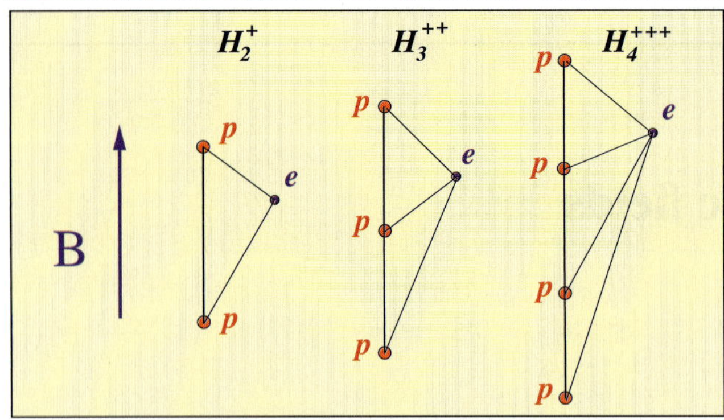

Fig. 1 One-electron molecular systems in a uniform magnetic field $\mathbf{B} = (0, 0, B)$: the traditional system H_2^+, and the exotic ions H_3^{++} and H_4^{+++}

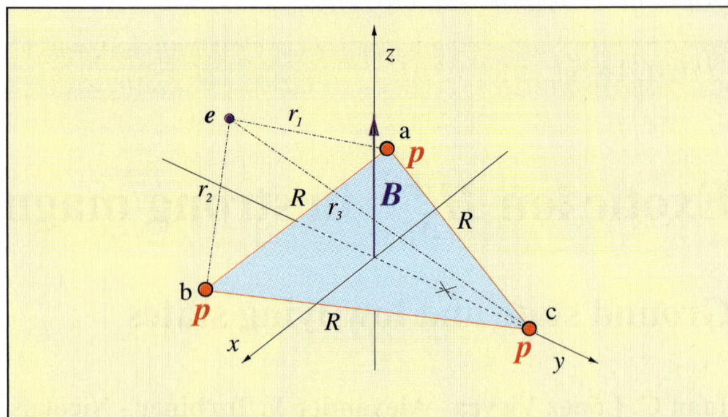

Fig. 2 The H_3^{++} molecular ion in triangular (equilateral) configuration perpendicular to the magnetic field **B**

2 Goal

We study molecular systems, traditional and exotic, in strong magnetic fields in the context of the variational method with physically adequate trial functions (see e.g. Turbiner and López Vieyra 2006). Eventually, these functions are modifications of the celebrated Heitler-London, Hund-Mulliken and Guillemin-Zener functions. Our goal is to find the domain of existence of the systems and then to obtain their description with sufficiently high accuracy in the whole range of magnetic fields where the non-relativistic approximation is valid: $B = 0$–4.414×10^{13} G.

In particular, we have carried out a detailed study in the Born-Oppenheimer approximation of the ground state for the one-electron systems: (ppe)-system—the traditional molecular ion H_2^+ (the simplest molecular system) and for the systems $(pppe)$ and $(ppppe)$ giving rise to the exotic molecular ions H_3^{++} and H_4^{+++}, respectively, in linear configuration parallel to the magnetic field (Fig. 1), and for different spatial configurations of the charged centers. The goal of the present article is to give a brief review of the properties of the H_3^{++} molecular ion in a strong magnetic field.

3 Formulation of the problem

In the Born-Oppenheimer approximation of zero order (where protons are infinitely massive), the Schrödinger equation for a system of one-electron and a certain number of protons in a magnetic field **B** is

$$[(\mathbf{p} - \mathbf{A})^2 + V_c(r_1, r_2, \ldots)]\Psi(\mathbf{r}) = E\Psi(\mathbf{r}), \tag{1}$$

where $\mathbf{p} = -i\nabla$ is the *momentum of the electron*, **A** is a *vector potential* associated with the constant uniform magnetic field **B**, and $V_c(r_1, r_2, \ldots)$ describes the *Coulomb interaction* between the charged particles. For the $(pppe)$-system (see Figs. 1, 2):

$$V_c = \frac{2}{R_{ab}} + \frac{2}{R_{ac}} + \frac{2}{R_{bc}} - \frac{2}{r_1} - \frac{2}{r_2} - \frac{2}{r_3}. \tag{2}$$

In particular, for a uniform magnetic field in the z-direction, $\mathbf{B} = (0, 0, B)$ we can choose the vector potential in a *gauge*:

$$\mathbf{A} = B\big((\xi - 1)y,\ \xi x,\ 0\big), \tag{3}$$

where ξ is a gauge parameter. Observables are, of course, gauge invariant and do not depend on the value of ξ. In variational calculations for a fixed class of trial functions the variational energy (which is approximate) can depend on the gauge, there can exist an optimal value of ξ giving a minimal variational energy. Therefore one can consider ξ as a variational parameter.

4 H_3^{++} (triangular)

1. The system $(pppe)$ in triangular equilateral configuration in a magnetic field which is perpendicular to the plane of the triangle (see Fig. 2) develops a well pronounced minimum for 10^8 G $< B < 10^{11}$ G. Therefore the system H_3^{++} can exist for those magnetic fields. It is unstable towards deviations of the protons out of the plane perpendicular to the magnetic field. Thus, the H_3^{++} ion exists in such a configuration if this configuration is supported externally.
2. As the magnetic field increases the binding energy increases and the size of the triangle diminishes.

5 H_3^{++} (linear)

1. The system H_3^{++} in linear configuration exists for $B \gtrsim 10^{11}$ G.

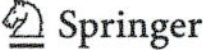

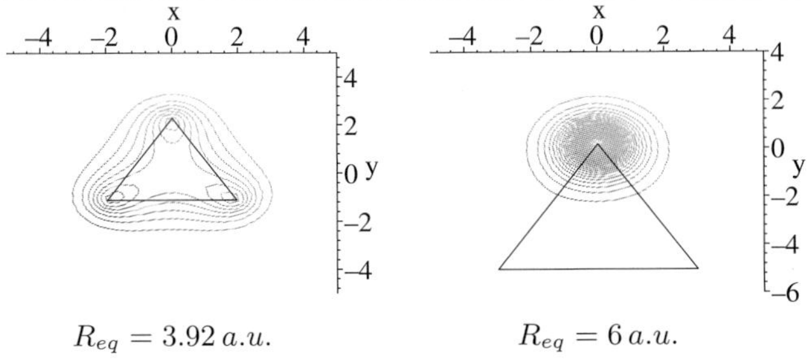

Fig. 3 Evolution of the electronic distribution for the H_3^{++} ion in triangular configuration as function of the size of the triangle (of side R) in a magnetic field $B = 10^9$ G. *The plot on the left* describes the bound state and *the plot in the right* corresponds to the situation when the dissociation $H_3^{++} \to H + p + p$ occurs. Figure from López Vieyra and Turbiner (2002)

2. As the magnetic field grows the binding energy increases while the internuclear distance decreases.
3. For $B > 10^{13}$ G, H_3^{++} is the most stable one-electron system made from protons. In particular, the binding energy of H_3^{++} for $B = 4.414 \times 10^{13}$ G is $E_b^{H_3^{++}} = 55.23$ Ry, compared with $E_b^{H_2^+} = 54.50$ Ry, $E_b^{H\text{-atom}} = 32.77$ Ry for the H_2^+ molecular ion and the H-atom respectively.

6 Ground state properties

Based on the assumption that the dynamics of the one-electron Coulomb system in a strong magnetic field is governed by the ratio of transverse to longitudinal sizes of the electronic cloud $\mathcal{X} = r_t/r_l$, we construct approximations of some characteristics of the ground state of the H_3^{++} ion in parallel configuration.

For strong magnetic fields the transverse size of the electronic cloud r_t shrinks drastically as $\sim B^{-1/2}$ being almost coincident with the Larmor radius. The longitudinal size r_l also contracts but at a much slower rate, $\sim(\log B)^{-1}$, at large magnetic fields. Thus, as a first step we write approximation formulas for the transverse (r_t) and longitudinal (r_l) sizes:

$$r_t = \frac{r_t^0}{(1+\alpha_t^2 B^2)^{1/4}} \left(\frac{1+a_t B^2}{1+b_t B^2}\right),$$

$$r_l = \frac{r_l^0}{1+\alpha_l \log(1+\beta_l^2 B^2 + \gamma_l^2 B^4)} \left(\frac{1+a_l B^2}{1+b_l B^2}\right).$$

Their parameters are found by fitting the results of calculations. The data for r_t, r_l are obtained variationally using the strategy described in Turbiner et al. (1999, 2005a). The parameters of the fit are given in Table 1. The form of the electronic cloud for $B = 10^{11}$ G and $B = 10^{12}$ G is shown in Fig. 4.

Table 1 The parameters of the fit of the transverse size r_t and of the longitudinal size r_l of the electron cloud of the H_3^{++} ion (in a.u.) (see text)

r_t parameters		r_l parameters	
r_t^0	0.645875	r_l^0	1.94408
α_t	0.048196	α_l	0.279956
a_t	0.0097061	β_l	0.0191558
b_t	0.0230488	γ_l	0.000008
		a_l	0.0066712
		b_l	0.0136894

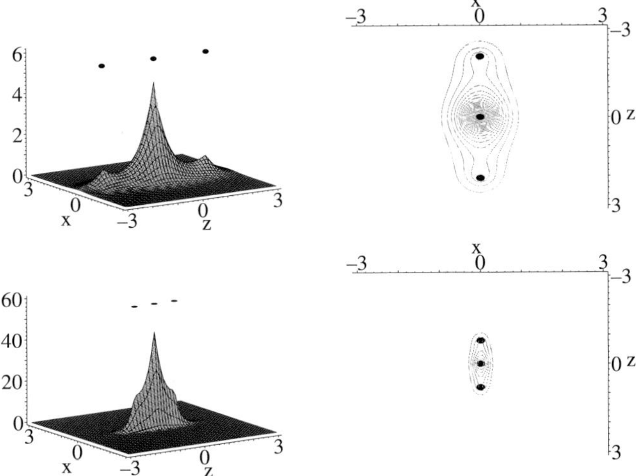

Fig. 4 Illustration of the evolution of the normalized electronic distribution $\psi^2(x, y=0, z)/\int \psi^2(x,y,z)\,d\mathbf{r}$ for the ion H_3^{++} in linear configuration aligned with a magnetic field at (*top*) $B = 10^{11}$ G and (*bottom*) $B = 10^{12}$ G. The position of the protons is illustrated by *bullets*. Figure from Turbiner et al. (2005a)

According to the main assumption, the binding energy E_b is approximated by the following formula

$$E_b = A\mathcal{X}_l^2 + B\mathcal{X}_l + C, \quad \mathcal{X}_l = \log \mathcal{X},$$

where A, B, C are parameters, which are found by making fit of the results of calculations done in Turbiner et al. (1999, 2005a). The parameters of the fit are given in Table 2. It is worth emphasizing that the parameters of $\mathcal{X}$ ($\mathcal{X}_l$) are already fixed by the fits of r_t, r_l, respectively. The formula for E_b agrees with the perturbative expansion in powers B^2 (at small B) and gives a correct asymptotic expansion at large B. Figure 5 shows the fit of the binding energy.

In the same way, we assume that the interproton equilibrium distance is mostly defined by the longitudinal size of the electron cloud, which is slightly modified by including terms depending on $\mathcal{X}_l = \log \mathcal{X}$. The equilibrium distance is approximated by the formula

$$R_{eq} = r_l(c_0 + c_1\mathcal{X}_l + c_2\mathcal{X}_l^2).$$

Table 2 The parameters of the fit of the binding energy of the H_3^{++} molecular ions (all in Ry)

System	A	B	C
H_3^{++}	12.8455	20.4849	3.95821

Table 3 The dimensionless parameters of the fit for the equilibrium distance $R_{\rm eq}$ (in a.u.) of the H_3^{++} ion

System	c_0	c_1	c_2
H_3^{++}	4.48200	2.25814	0.380948

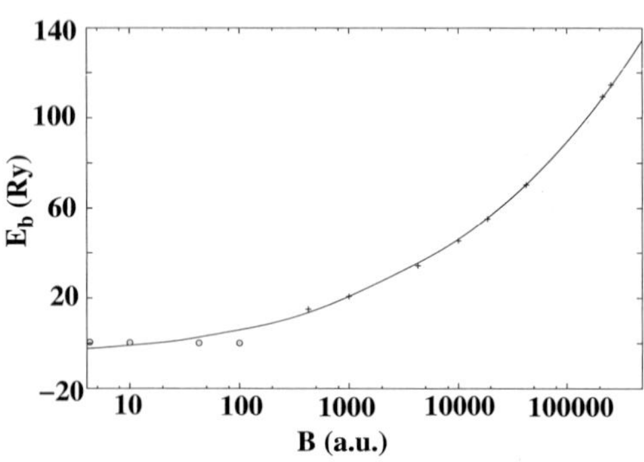

Fig. 5 H_3^{++} ion: the fit of the binding energy. Only calculated values which are indicated by *crosses* are used for fitting, while calculated values shown by *circles* are not taken into account. ($B = 1$ a.u. $\equiv 2.35 \times 10^9$ G)

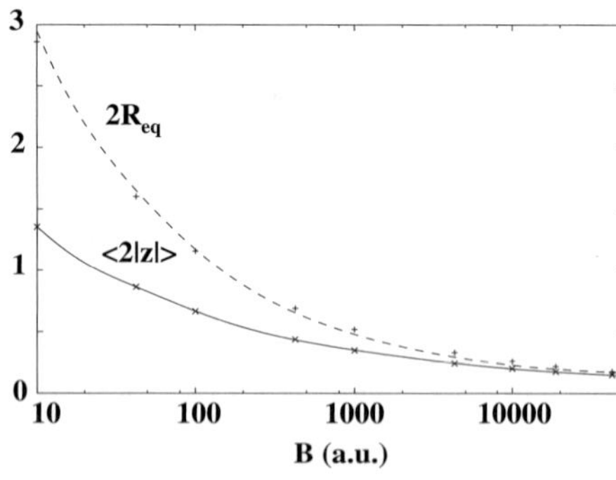

Fig. 6 H_3^{++} ion: the fit of longitudinal size of the electron cloud $2\langle|z|\rangle$ (*solid line*) and of the equilibrium distance $L_{\rm eq} = 2R_{\rm eq}$ (*dashed line*). Calculated values are indicated by *crosses*. ($B = 1$ a.u. $\equiv 2.35 \times 10^9$ G)

The parameters of the fit are given in Table 3. It is worth mentioning that the parameters c_i, $i = 0, 1, 2$ decrease very fast with i. This might be considered as an indication of adequateness of the approximation formula. Figure 6 shows the fit of the equilibrium distance $R_{\rm eq}$ as well as a similar fit for the expectation value $\langle 2|z|\rangle$ (see Turbiner et al. 2005b).

In the fits, some irregularities can be seen in the region $(5\text{–}50)\times 10^{10}$ G, near the threshold of appearance of the H_3^{++} ion (see Figs. 5, 6). One of the reasons for these irregularities can be related to highly increased technical difficulties we encountered when exploring this region. This could lead to a loss of accuracy. The overall quality of the fit for the range 10^{11}–4.414×10^{13} G is very high, 1–5%.

It is worth mentioning that at large magnetic fields the behavior of the binding energy and the equilibrium distance is similar for all possible existing one-electron molecular systems at strong magnetic fields (see Turbiner et al. 2005b for more details).

7 Excited states of H_3^{++}

For all magnetic fields $B \geq 10^{11}$ G:

1. Positive z-parity states $1\sigma_g$, $1\pi_u$, $1\delta_g$ are bound
2. Negative z-parity states $1\sigma_u$, $1\pi_g$, $1\delta_u$ are repulsive
3. The following hierarchy holds:

$$E_T^{1\sigma_g} < E_T^{1\pi_u} < E_T^{1\delta_g}, \qquad R_{\rm eq}^{1\sigma_g} < R_{\rm eq}^{1\pi_u} < R_{\rm eq}^{1\delta_g}$$

For example, at $B = 10^{12}$ G:

$$E_b^{1\sigma_g} = 15.16 \text{ Ry}, \qquad E_b^{1\pi_u} = 10.17 \text{ Ry},$$

$$E_b^{1\delta_g} = 8.28 \text{ Ry},$$

$$R_{\rm eq}^{1\sigma_g} = 0.345 \text{ a.u.}, \qquad R_{\rm eq}^{1\pi_u} = 0.497 \text{ a.u.},$$

$$R_{\rm eq}^{1\delta_g} = 0.601 \text{ a.u.}$$

8 Conclusion

The absorption features at $\sim$700 eV and $\sim$1400 eV discovered in radiation of the neutron star 1E1207.4-5209 by Chandra (Sanwal et al. 2002) and confirmed by XMM-Newton can be reproduced if one assumes that the main abundance in the atmosphere is H_3^{++} and the magnetic field strength is $(4 \pm 2) \times 10^{14}$ G. Photodissociation $H_3^{++} \to H + 2p$ corresponds to absorption feature at $\sim$700 eV, and photoionization $H_3^{++} \to e + 3p$ contributes to absorption feature at $\sim$1400 eV. Photodissociation $H_3^{++} \to H_2^+ + p$ corresponds to absorption feature at 80–150 eV, and it is not seen being beyond of Chandra detector acceptance (for details see Turbiner and López Vieyra 2004).

References

López Vieyra, J.C., Turbiner, A.: Phys. Rev. A **66**, 023409 (2002)
Sanwal, D., Pavlov, G.G., Zavlin, V.E., Teter, M.A.: Astrophys. J. **574**, L61 (2002)

Turbiner, A.V., López Vieyra, J.C.: Mod. Phys. Lett. A **19**, 1919 (2004), astro-ph/0404290

Turbiner, A.V., López Vieyra, J.C.: Phys. Rep. **424**, 309 (2006)

Turbiner, A., López, J.C., Solis, H.U.: J. Exp. Theor. Phys. Lett. **69**, 844 (1999)

Turbiner, A.V., López Vieyra, J.C., Guevara, N.L.: Phys. Rev. A **72**, 023403 (2005a)

Turbiner, A.V., Kaidalov, A.B., López Vieyra, J.C.: Collect. Czechoslov. Chem. Commun. **70**, 1133 (2005b)

Astrophys Space Sci (2007) 308: 499–503
DOI 10.1007/s10509-007-9355-5

ORIGINAL ARTICLE

The ion H_3^+ in a strong magnetic field

Linear configuration

Juan C. López Vieyra · Alexander V. Turbiner · Nicolais L. Guevara

Received: 3 July 2006 / Accepted: 16 November 2006 / Published online: 15 March 2007

Abstract A first detailed study of the low-lying electronic states of the H_3^+ molecular ion in linear configuration, parallel to a magnetic field, is carried out for $B = 0$–4.414×10^{13} G in the Born-Oppenheimer approximation. The variational method is employed with a single, physically adequate trial function which includes, in particular, explicitly the electronic correlation term in the form $\exp(\gamma r_{12})$, where γ is a variational parameter. The state of the lowest total energy (ground state) depends on the magnetic field strength. It evolves from spin-singlet ${}^1\Sigma_g$ for small magnetic fields $B \lesssim 5 \times 10^8$ G to weakly-bound spin-triplet ${}^3\Sigma_u$ for intermediate fields and eventually to spin-triplet ${}^3\Pi_u$ state for $B \gtrsim 5 \times 10^{10}$ G.

Keywords Molecular ion H_3^+ · Strong magnetic field

PACS 31.15.Pf · 31.15.-p · 31.10.+z · 32.60.+i · 97.10.Ld

1 Formulation of the problem

In the Born-Oppenheimer approximation of zero order where protons are infinitely massive, the Hamiltonian which describes the two-electron-three-proton system (*pppee*) in a constant uniform magnetic field $\mathbf{B} = (0, 0, B)$ is

$$\mathcal{H} = \sum_{\ell=1}^{2} (\hat{\mathbf{p}}_\ell + \mathcal{A}_\ell)^2 - \sum_{\substack{\ell=1,2\\ \kappa=a,b,c}} \frac{2}{r_{\ell,\kappa}} + \frac{2}{r_{12}} + \frac{2}{R_+} + \frac{2}{R_-} + \frac{2}{R_+ + R_-} + 2\mathbf{B}\cdot\mathbf{S}, \qquad (1)$$

with protons situated along magnetic line (so called *parallel configuration*, see Fig. 1). Here $\hat{\mathbf{p}}_\ell = -i\nabla_\ell$ is the 3-vector of the momentum of the ℓth electron; r_{12} is the interelectronic distance and $\mathbf{S} = \hat{s}_1 + \hat{s}_2$ is the operator of the total spin. $\mathcal{A}_\ell$ is a vector potential which corresponds to the constant uniform magnetic field $\mathbf{B}$. It is chosen in the symmetric gauge,

$$\mathcal{A}_\ell = \frac{1}{2}(\mathbf{B} \times \mathbf{r}_\ell) = \frac{B}{2}(-y_\ell, x_\ell, 0). \qquad (2)$$

Finally, the Hamiltonian can be written as

$$\mathcal{H} = \sum_{\ell=1}^{2} \left(-\nabla_\ell^2 + \frac{B^2}{4}\rho_\ell^2\right) - \sum_{\ell,\kappa} \frac{2}{r_{\ell\kappa}} + \frac{2}{r_{12}} + \frac{2}{R_+} + \frac{2}{R_-} + \frac{2}{R_+ + R_-} + B(\hat{L}_z + 2\hat{S}_z), \qquad (3)$$

where $\hat{L}_z = \hat{l}_{z_1} + \hat{l}_{z_2}$ and $\hat{S}_z = \hat{s}_{z_1} + \hat{s}_{z_2}$ are the z-components of the total angular momentum and total spin, respectively, and $\rho_\ell^2 = x_\ell^2 + y_\ell^2$.

2 Goal

Recently, extended studies of possible one-electron molecular systems in a strong magnetic field were performed (see

This work was supported in part by PAPIIT grant **IN121106** (Mexico).

J.C. López Vieyra (✉) · A.V. Turbiner · N.L. Guevara
Instituto de Ciencias Nucleares, UNAM,
Mexico, DF 04510, Mexico
e-mail: vieyra@nucleares.unam.mx

A.V. Turbiner
e-mail: turbiner@nucleares.unam.mx

N.L. Guevara
e-mail: nicolais@nucleares.unam.mx

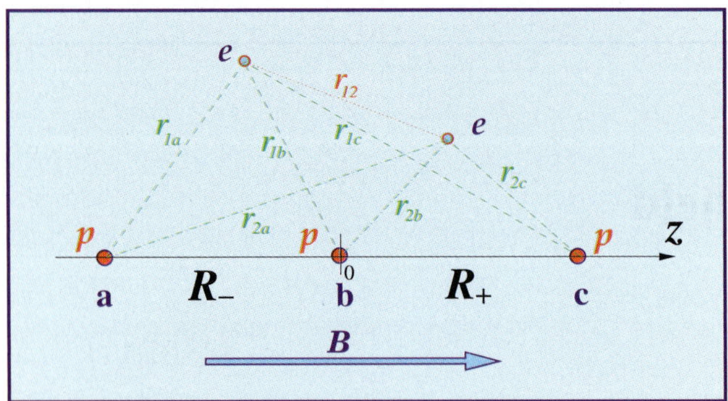

Fig. 1 The H_3^+ molecular ion in parallel configuration in a uniform constant magnetic field $\mathbf{B} = (0, 0, B)$

Turbiner and López-Vieyra 2006). Our present goal is to carry out a detailed study of the ground state and the low-lying excited states of the $2e$ molecular ion H_3^+ in parallel configuration in a magnetic field where the non-relativistic approximation is valid: $B = 0 - 4.414 \times 10^{13}$ G. In the past a single, semi-quantitative attempt to study the H_3^+ ion was carried out (Warke and Dutta 1977). This work contains many numerical errors (see a discussion in Turbiner et al. 2006).

3 Variational calculus

Take $\psi_{\text{trial}}(x, \{\alpha\})$ and find a potential for which it is an exact solution

$$V_{\text{trial}}(x, \{\alpha\}) = \frac{\Delta \psi_{\text{trial}}}{\psi_{\text{trial}}}, \qquad E_{\text{trial}} = 0, \tag{4}$$

where $\{\alpha\}$ are variational parameters. So, we know the Hamiltonian for which ψ_{trial} is the exact eigenfunction

$$H_{\text{trial}} \psi_{\text{trial}} = [p^2 + V_{\text{trial}}] \psi_{\text{trial}} = 0,$$

then

$$\begin{aligned} E_{\text{var}} &= \min_{\{\alpha\}} \int \psi^*_{\text{trial}} H \psi_{\text{trial}} \\ &= \int \psi^*_{\text{trial}} \underbrace{H_{\text{trial}} \psi_{\text{trial}}}_{=0} + \int \psi^*_{\text{trial}} (H - H_{\text{trial}}) \psi_{\text{trial}} \\ &= 0 + \int \psi^*_{\text{trial}} (V - V_{\text{trial}}) \psi_{\text{trial}} \text{“} + \cdots \text{”} \end{aligned} \tag{5}$$

- Hence, the variational energy can be interpreted as the sum of the first two terms of a perturbation theory with perturbation potential

$$(V - V_{\text{trial}}). \tag{6}$$

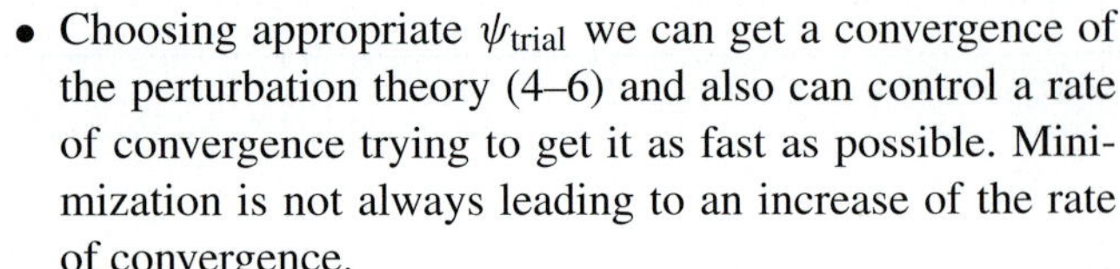

- Choosing appropriate ψ_{trial} we can get a convergence of the perturbation theory (4–6) and also can control a rate of convergence trying to get it as fast as possible. Minimization is not always leading to an increase of the rate of convergence.
- In practice, ψ_{trial} is chosen in such a way to contain as much as possible physical properties of the problem we study as well.

4 Trial function

Following the guidelines mentioned above, we choose a trial function of the form

$$\begin{aligned} \psi^{(\text{trial})} &= (1 + \sigma_e P_{12})(1 + \sigma_N P_{ac})(1 + \sigma_{N_a} P_{ab} + \sigma_{N_a} P_{bc}) \\ &\quad \rho_1^{|m|} e^{im\phi_1} e^{\gamma r_{12} - B\beta_1 \rho_1^2/4 - B\beta_2 \rho_2^2/4} \\ &\quad e^{-\alpha_1 r_{1a} - \alpha_2 r_{1b} - \alpha_3 r_{1c} - \alpha_4 r_{2a} - \alpha_5 r_{2b} - \alpha_6 r_{2c}} \end{aligned} \tag{7}$$

where $\sigma_e = \pm 1$ stands for spin singlet ($S = 0$) and triplet states ($S = 1$), while $\sigma_N = \pm 1$ stands for nuclear gerade and ungerade states, respectively.

For S_3-permutationally symmetric case $\sigma_N = \sigma_{N_a} = \pm 1$. P_{ij}, $i, j = a, b, c$ are the operators which interchange the two protons i and j, and α_{1-6}, β_{1-2} and γ are variational parameters. The operator P_{12} interchanges electrons $(1 \leftrightarrow 2)$,

5 Classification of states

$$^{2S+1}M_p$$

$2S + 1$ is the electronic total spin multiplicity, it is 1 for spin-singlet ($S = 0$) and 3 for spin-triplet ($S = 1$); M represents the total magnetic quantum number, $m = 0, -1, -2$ it is denoted by Σ, Π, Δ, respectively; p (the spatial parity) denotes gerade ($p = +1$), ungerade ($p = -1$) states.

6 Results (for details see Turbiner et al. 2006)

1. It is found that in the Born-Oppenheimer approximation, for the system ($pppee$) in parallel configuration in a magnetic field ranging in $B = 0$–4.414×10^{13} G, the total energy curves display a well pronounced minimum at finite internuclear distances at $R_+ = R_- = R_{\text{eq}}$ (see Fig. 1) for the lowest states with magnetic quantum numbers $m = 0, -1, -2$, total spin $S = 0, 1$ and parity $p = \pm 1$.
2. For all studied states, as the magnetic field increases the internuclear distance R_{eq} decreases and the system becomes more compact, while the total energies of spin-singlet states increase and of spin-triplet states decrease.

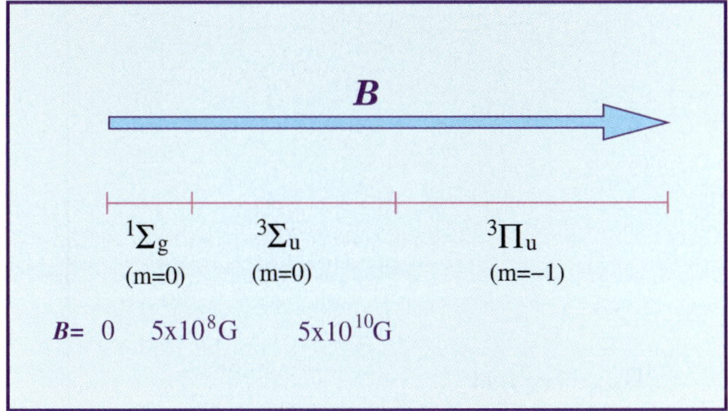

Fig. 2 Ground state evolution for the H_3^+-ion in parallel configuration as a function of the magnetic field strength

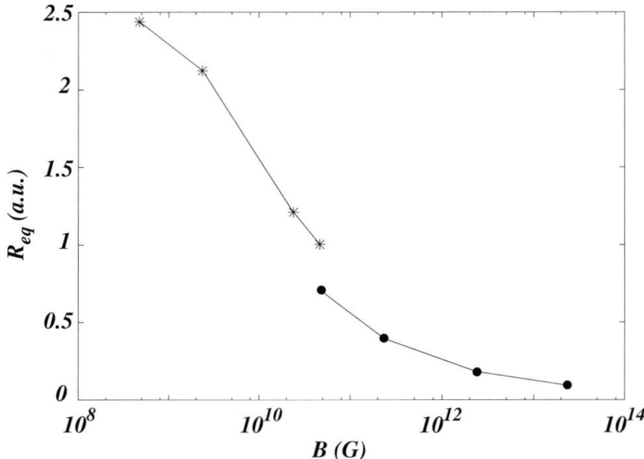

Fig. 3 Equilibrium distance for the ground state: $^3\Sigma_u$ (*stars*) and $^3\Pi_u$ (*bullets*)

3. The state of the lowest total energy (ground state) depends on the magnetic field strength. It evolves from spin-singlet $^1\Sigma_g$ for small magnetic fields $B \lesssim 5 \times 10^8$ G to weakly-bound spin-triplet $^3\Sigma_u$ state for intermediate fields and eventually to spin-triplet $^3\Pi_u$ state for $B \gtrsim 5 \times 10^{10}$ G (see Figs. 2 and 6).
4. For $B \lesssim 5 \times 10^8$ G the ground state $^1\Sigma_g$ is unstable towards small deviations from linearity. This indicates to a "limited" existence of the molecular ion H_3^+ in the state $^1\Sigma_g$ for these magnetic fields. One can state that it exists if in some way the linear configuration of the protons is supported externally.
5. For $B \gtrsim 5 \times 10^8$ G the energy well corresponding to the ground state supports at least one longitudinal vibrational state.
6. The equilibrium distance of the ground state decreases monotonously as the magnetic field increases showing a discontinuous behavior at $B \approx 5 \times 10^{10}$ G. This discontinuity corresponds to the transition from $^3\Sigma_u$ as the ground state to the $^3\Pi_u$ ground state (see Fig. 3).

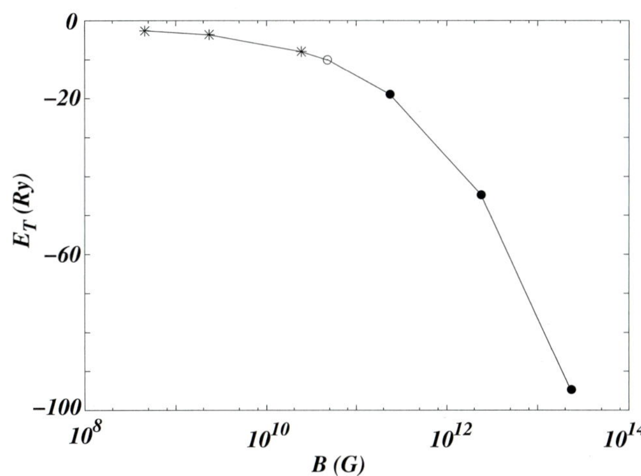

Fig. 4 Ground state energy of H_3^+ viz. magnetic field: $^3\Sigma_u$ (*stars*) and $^3\Pi_u$ (*bullets*), a point of crossing of these states is marked by *circle*

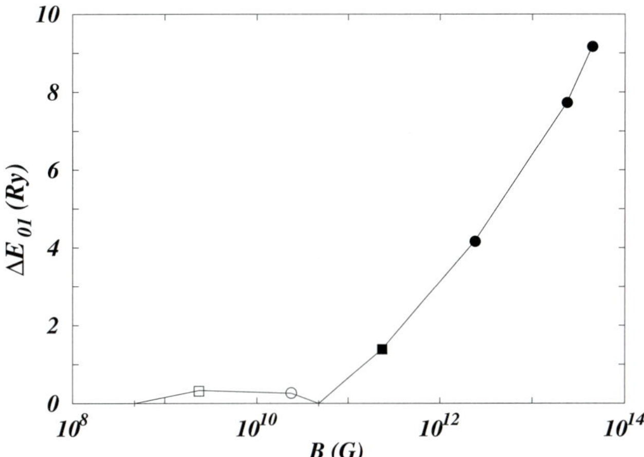

Fig. 5 Transition energy $\Delta E_{01} = E_T^{(1)} - E_T^{(0)}$ for the two lowest lying states of the H_3^+ molecular ion as a function of the magnetic field strength ($E_T^{(0)}$ and $E_T^{(1)}$ are the total energies of the ground and the first excited states, respectively). Transition energies are indicated by □ for $^3\Sigma_u \to {}^3\Sigma_g$, ○ for $^3\Sigma_u \to {}^3\Pi_u$, ■ for $^3\Pi_u \to {}^3\Sigma_u$, and • for $^3\Pi_u \to {}^3\Delta_g$

7. The total energy of the ground state decreases monotonously and smoothly as the magnetic field grows (see Fig. 4).
8. The transition energy from the ground state to the lowest excited state is relatively small for $B \lesssim 5 \times 10^{10}$ G ($\Delta E_{01} < 1$ Ry) while for $B \gtrsim 5 \times 10^{10}$ G it increases monotonously as the magnetic field grows reaching $\Delta E_{01} \simeq 9$ Ry for the Schwinger limit 4.414×10^{13} G (see Fig. 5).
9. The dissociation energy corresponding to the channel $H_3^+(^3\Sigma_u) \to H_2^+(1\sigma_g) + H(1s)$ (with electrons in spin-triplet state) varies from 0.03 Ry to 0.06 Ry for the magnetic fields in the range 5×10^8 G $\lesssim B \lesssim 5 \times 10^{10}$ G.

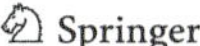

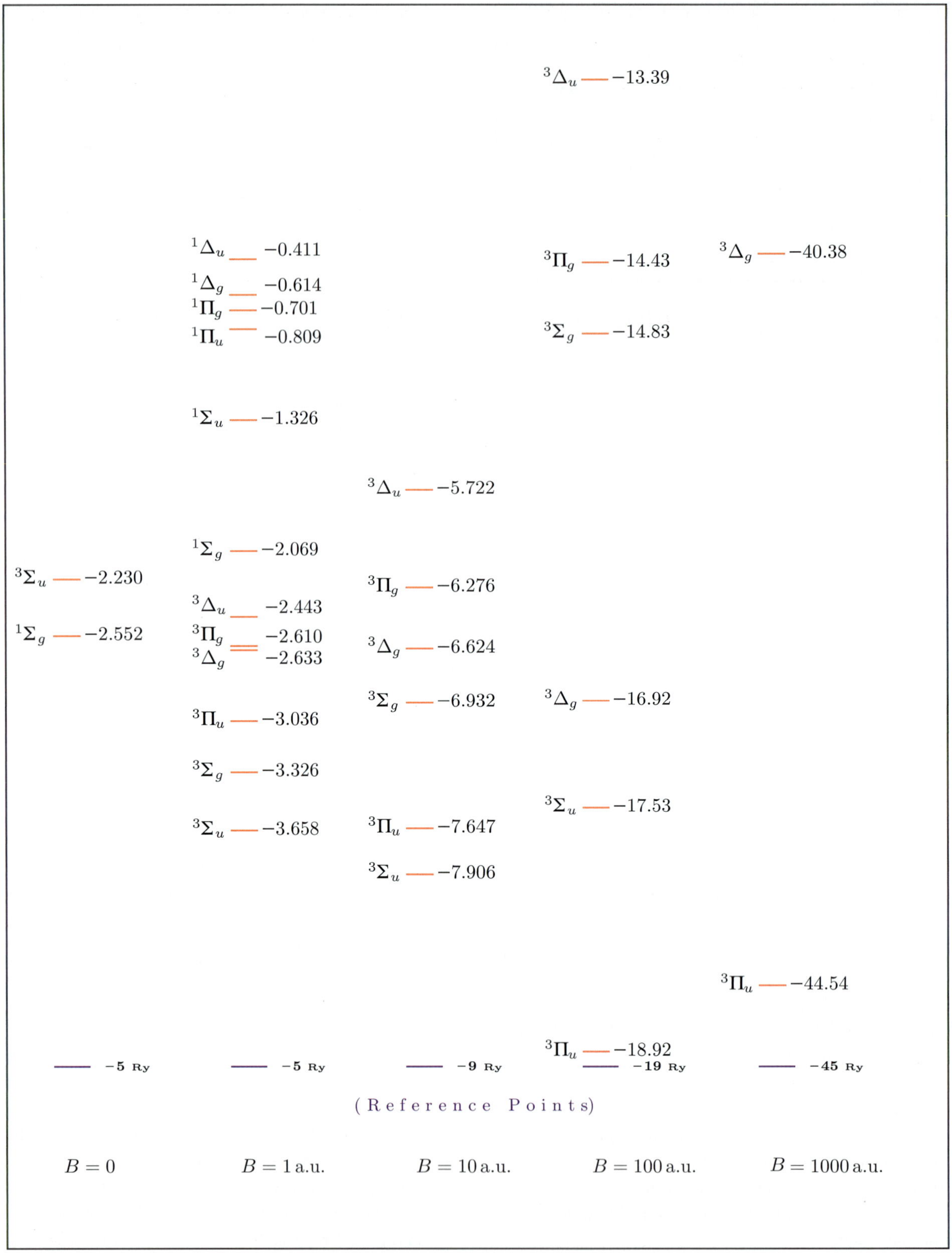

$B = 0$:
$^3\Sigma_u$ — -2.230
$^1\Sigma_g$ — -2.552
— -5 Ry

$B = 1$ a.u.:
$^1\Delta_u$ — -0.411
$^1\Delta_g$ — -0.614
$^1\Pi_g$ — -0.701
$^1\Pi_u$ — -0.809
$^1\Sigma_u$ — -1.326
$^1\Sigma_g$ — -2.069
$^3\Delta_u$ — -2.443
$^3\Pi_g$ — -2.610
$^3\Delta_g$ — -2.633
$^3\Pi_u$ — -3.036
$^3\Sigma_g$ — -3.326
$^3\Sigma_u$ — -3.658
— -5 Ry

$B = 10$ a.u.:
$^3\Delta_u$ — -5.722
$^3\Pi_g$ — -6.276
$^3\Delta_g$ — -6.624
$^3\Sigma_g$ — -6.932
$^3\Pi_u$ — -7.647
$^3\Sigma_u$ — -7.906
— -9 Ry

$B = 100$ a.u.:
$^3\Delta_u$ — -13.39
$^3\Pi_g$ — -14.43
$^3\Sigma_g$ — -14.83
$^3\Delta_g$ — -16.92
$^3\Sigma_u$ — -17.53
$^3\Pi_u$ — -18.92
— -19 Ry

$B = 1000$ a.u.:
$^3\Delta_g$ — -40.38
$^3\Pi_u$ — -44.54
— -45 Ry

(Reference Points)

Fig. 6 Total energy (in Ry) of the low-lying electronic states of the H_3^+ ion in a magnetic field (in a.u.) in linear, parallel configuration. In field-free case the $^1\Sigma_g$ state is unstable towards deviations from linearity, and the true ground state is realized by an equilateral triangular configuration. For $B = 0$ the only stable linear configuration corresponds to the $^3\Sigma_u$ state, and its total energy is -2.2322 Ry (see e.g. Preiskorn et al. 1991). The energy scale is the same for all magnetic fields but reference points depend on them. The atomic unit for the magnetic field is defined as $B_0 = 2.35 \times 10^9$ G $\equiv 1$ a.u.

10. The ion $H_3^+(^3\Pi_u)$ is stable towards the decays $H_3^+ \to H_2^+(^1\pi_u) + H(1s)$, $H_3^+ \to H_2^+(1\sigma_g) + H(2p_{-1})$ and $H_3^+ \to H_2(^3\Pi_u) + p$ for magnetic fields $B \gtrsim 5 \times 10^{10}$ G. For all three channels the dissociation energy grows monotonously as the magnetic field increases. For the dominant channel $H_3^+ \to H_2(^3\Pi_u) + p$ it reaches 30.3 Ry at the Schwinger limit 4.414×10^{13} G, while for the channel $H_3^+ \to H_2^+(^1\pi_u) + H(1s)$, at this field strength, about 35 Ry are required for dissociation.

7 Conclusion

The existence of the H_3^+ ion is demonstrated for all studied magnetic fields up to the Schwinger limit. Its ground state evolves, as the magnetic field increases, from $^1\Sigma_g$ to $^3\Sigma_u$ and, eventually, to $^3\Pi_u$. For large magnetic fields $B \gtrsim 5 \times 10^{10}$ G the H_3^+ ion is stable and in particular it does not decay to $H_2 + p$.

References

Preiskorn, A., Frye, D., Clementi, E.: J. Chem. Phys. **94**, 7204 (1991)
Turbiner, A., López Vieyra, J.C.: Phys. Rep. **424**, 309 (2006)
Turbiner, A.V., Guevara, N.L., López Vieyra, J.C.: The H_3^+ molecular ion in a magnetic field: linear parallel configuration, physics/0606083
Warke, C.S., Dutta, A.K.: Phys. Rev. A **16**, 1747 (1977)

Astrophys Space Sci (2007) 308: 505–511
DOI 10.1007/s10509-007-9310-5

ORIGINAL ARTICLE

X-ray intensity-hardness correlation and deep IR observations of the anomalous X-ray pulsar 1RXS J170849-400910

N. Rea · G.L. Israel · T. Oosterbroek · S. Campana · S. Zane · R. Turolla · V. Testa · M. Méndez · L. Stella

Received: 30 July 2006 / Accepted: 18 October 2006 / Published online: 20 March 2007

Abstract We report here on X-ray and IR observations of the Anomalous X-ray Pulsar (AXP) 1RXS J170849-400910. First, we report on new XMM-Newton, Swift-XRT and Chandra observations of this AXP, which confirm the intensity–hardness correlation observed in the long term X-ray monitoring of this source. These new X-ray observations show that the AXP flux is rising again, and the spectrum hardening. If the increase of the source intensity is indeed connected with the glitches and a possible bursting activity, we expect this source to enter in a bursting active phase around 2006–2007. Second, we report on deep IR observations of 1RXS J170849-400910, taken with the VLT-NACO adaptive optics, showing that there are many weak sources consistent with the AXP position. Neither star A or B, as previously proposed by different authors, might yet be conclusively recognised as the IR counterpart of 1RXS J170849-400910. Third, using Monte Carlo simulations, we re-address the calculation of the significance of the absorption line found in a phase-resolved spectrum of this source, and interpreted as a resonant scattering cyclotron feature.

Keywords Neutron stars · Pulsars · Magnetars · X-ray · 1RXS J170849-400910

PACS 97.10.Sj · 97.60.Jd · 97.60.Gb · 98.38.Jw · 98.70.Qy

1 Introduction

AXPs are a small group of neutron stars (NSs) which stand apart from other known classes of X-ray sources. At the moment there are 7 confirmed AXPs plus 2 candidates. These X-ray pulsars share many peculiarities: they are all (but one) radio-quiet (Camilo et al. 2006; Burgay et al. 2006), exhibit X-ray pulsations with spin periods in a small range of values (∼5–12 s), they have a large spin-down rate ($\dot{P} \approx 10^{-13}$–10^{-10} s^{-1}), a rather high X-ray luminosity ($L_X \approx 10^{34}$–10^{36} erg s^{-1}), and faint IR counterparts with $K_s \sim 20$–22 magnitudes (for a recent review see Woods and Thompson 2004 and Kaspi 2007). The nature of their X-ray emission was intriguing all along. In fact, it is too high to be produced by the loss of rotational energy alone, but on the other hand, no hints for a companion star were found, neither through deep observations at other wavelength, nor

N.R. is supported by an NWO Post-Doctoral Fellowship. S.Z. thanks the Particle Physics and Astronomy Research Coucil, PPARC, for support through an Advanced Fellowship.

N. Rea (✉) · M. Méndez
SRON Netherlands Institute for Space Research, Sorbonnelaan, 2, 3584CA, Utrecht, The Netherlands
e-mail: N.Rea@sron.nl

G.L. Israel · V. Testa · L. Stella
INAF—Astronomical Observatory of Rome, via Frascati 33, 00040 Monteporzio Catone (Rome), Italy

T. Oosterbroek
Science Payload and Advanced Concepts Office, ESA ESTEC, Postbus 299, 2200 AG, Noordwijk, The Netherlands

S. Campana
INAF—Astronomical Observatory of Brera, via Bianchi 46, 23807 Merate (Lc), Italy

S. Zane
Mullard Space Science Laboratory, University College London Holmbury St. Mary, Dorking Surrey RH5 6NT, UK

R. Turolla
Physics Department, University of Padua, via Marzolo 8, 35131 Padova, Italy

timing the X-ray pulsations with the hope of finding Doppler shifts (Israel et al. 2003a; Mereghetti et al. 1998).

At present, the model which is most successful in explaining the peculiar observational properties of AXPs is the "magnetar" model. In this scenario AXPs are thought to be isolated NSs endowed with ultra-high magnetic fields ($B \sim 10^{14}$–10^{15} Gauss) and their X-ray emission powered by magnetic field decay (Duncan and Thompson 1992; Thompson and Duncan 1993, 1996). This idea is strongly supported by the estimate of AXPs' magnetic field through the classical dipole braking formula, $B \sim 3.2 \times 10^{19}\sqrt{P\dot{P}}$ Gauss, which gives in all cases values above the electron critical magnetic field ($B_{\rm QED} \sim 4.4 \times 10^{13}$ Gauss).

Alternative scenarios, invoking accretion from a fossil disk remnant of the supernova explosion (van Paradijs et al. 1995; Chatterjee et al. 2000; Alpar 2001), are still open possibilities although encounter increasing difficulties in explaining the data.

1RXS J170849-400910 was discovered with ROSAT (Voges et al. 1996) and later on a ~11 s modulation was found in its X-ray flux with ASCA (Sugizaki et al. 1997). Early measurements suggested that it was a fairly stable rotator with a spin period derivative of $\sim 1.9 \times 10^{-11}$ s^{-1} (Israel et al. 1999). However, in the last four years the source experienced two glitches, with different post-glitch recoveries (Kaspi et al. 2000; Dall'Osso et al. 2003; Kaspi and Gavriil 2003). Searches for optical/IR counterparts ruled out the presence of a massive companion (Israel et al. 1999). Very recently, two different objects were proposed by different groups, as being 1RXS J170849-400910 IR counterpart, and there still is an open debate on which one is the AXP counterpart (Israel et al. 2003b; Safi-Harb and West 2005; Durant and van Kerkwijk 2006). A diffuse (~8′) radio emission at 1.4 GHz was recently reported, possibly associated with the supernova remnant G346.5-0.1 (Gaensler et al. 2000).

Pulse phase spectroscopy analysis of two *Beppo*SAX observations of 1RXS J170849-400910 (Israel et al. 2001; Rea et al. 2003) revealed (i) a large spectral variability with the spin-phase, (ii) a strong energy dependence of the pulse profile shape, and (iii) shifts in the pulse phase between the low and the high energy profiles. High variability of the pulse shape with energy is now detected at higher energies, up to ~220 keV (Kuiper et al. 2006).

By analysing a *Beppo*SAX observation taken in 2001 (the longest pointing ever performed on this source), Rea et al. (2003) reported the presence of an absorption line at ~8 keV in a phase-resolved spectrum. Interpreting the feature as a cyclotron line due to resonant scattering yields a neutron star magnetic field of either 9.2×10^{11} G or 1.6×10^{15} G, in the case of electron or proton scattering, respectively.

In Sect. 2 we report on new 1RXS J170849-400910 XMM-Newton, Chandra and Swift observations which were used together with the previous ROSAT, ASCA and *Beppo*SAX observations to monitor the X-ray spectrum and flux of the AXP. Then we report on deep infrared VLT-NACO observations of this AXP. In Sect. 3 we carefully re-address the calculation of the significance of the absorption line found around 8 keV during a long *Beppo*SAX observation (Rea et al. 2003), we then summarise and discuss all the results in Sect. 4.

2 Observations and results

In this section we report on the observations and the data analysis of the four new X-ray and the IR observations. The results on the X-ray timing analysis are reported below in the text while the X-ray spectral parameters may be found in Table 1. All the X-ray spectra were fit by an absorbed blackbody plus a power-law component (but see also Rea et al. in this volume for a different spectral modelling). Giving the very high statistics we have in the XMM–Newton observation, and assuming the absorption does not vary along the line of sight, we fixed for all the spectra the absorption at the XMM–Newton value (`phabs` XSPEC model: $N_H = (1.36 \pm 0.04) \times 10^{22}$ cm^{-2}; abundances from Anders and Grevesse 1989)

2.1 XMM–Newton

1RXS J170849-400910 was observed with XMM–Newton between 2003 August 28th and 29th, for ~50 ks. The MOS

Table 1 Best fit values of the spectral parameters obtained for about ten years X-ray monitoring of 1RXS J170849-400910. Fluxes (and percentages of fluxes) are unabsorbed, in units of 10^{-10} erg s^{-1} cm^{-2} and in the 0.5–10 keV energy range. The N_H was fixed at the XMM-Newton value of 1.36×10^{22} cm^{-2} for all the observations. See text for details. Errors are at 90% confidence level

	ASCA 1996	SAX 1999	SAX 2001	Chandra 2002	XMM 2003	Chandra 2004	Swift 2005
kT (keV)	$0.41^{+0.01}_{-0.01}$	$0.465^{+0.002}_{-0.017}$	$0.424^{+0.003}_{-0.006}$	$0.475^{+0.0}_{-0.02}$	$0.456^{+0.007}_{-0.004}$	$0.43^{+0.01}_{-0.01}$	$0.430^{+0.015}_{-0.017}$
Gamma	$2.51^{+0.11}_{-0.11}$	$2.65^{+0.08}_{-0.03}$	$2.45^{+0.04}_{-0.03}$	$2.47^{+0.11}_{-0.1}$	$2.792^{+0.008}_{-0.012}$	$2.77^{+0.03}_{-0.08}$	$2.62^{+0.04}_{-0.02}$
Flux	$1.5^{+0.1}_{-0.08}$	$1.23^{+0.04}_{-0.05}$	$1.30^{+0.013}_{-0.015}$	$1.06^{+0.02}_{-0.02}$	$0.87^{+0.004}_{-0.002}$	$1.30^{+0.04}_{-0.06}$	$1.43^{+0.01}_{-0.04}$
PL Flux (%)	82 ± 9	73 ± 4	69 ± 3	74 ± 3	84 ± 1	83 ± 3	71 ± 3
χ^2_ν (d.o.f.)	1.05 (71)	1.07 (148)	1.19 (215)	0.93 (430)	1.14 (221)	0.95 (147)	1.11 (182)

cameras were operated in Prime Partial Window Mode, while the PN camera was in Prime Small Window Mode, all with the medium optical photons blocking filter. Since a higher background affected the last ∼10 ks of the observation, we used only the data during intervals in which the count rate above 10 keV was less than 0.35 counts s^{-1}. The source events and spectra were extracted within a circular region of 27″ centred on the peak of the point spread function of the source. This non standard radius was used because the source was located near the edge of the chip. The background was obtained from a source-free region of 27″. In order to determine the spin period of 1RXS J170849-400910 we barycentered the events arrival times and obtained, through a phasefitting technique, a best spin period of $P_s = 11.00170 \pm 0.00004$ s (all errors are at the 90% confidence level). We found that the pulsed fraction of the X-ray signal (defined as the amplitude of the best-fitting sine wave divided by the, background corrected, constant level of the emission) is energy-dependent, and it varies from $39.0 \pm 0.5\%$ in the 0.5–2.0 keV range to $29 \pm 1.5\%$ in the 6.0–10.0 keV range. These values are consistent with those reported for the pre-glitches *Beppo*SAX observation (Israel et al. 2001) while both are larger than those reported for the post-glitches *Beppo*SAX observation (Rea et al. 2003). Detailed results for this observation are reported in Rea et al. (2005a).

2.2 Chandra

1RXS J170849-400910 was observed by the *Chandra* Advanced CCD Imaging Spectrometer (ACIS), first for ∼30 ks with the High Energy Transmission Grating Spectrometer (HETGS) on 2002 September 9th, then for ∼30 ks in Continuous Clocking (CC) mode on 2004 July 3. For a more detailed description of the instruments and on the data processing we defer to the *Chandra* X-ray Center (CXC) documents.[1] Detailed results for these two observations are reported in Rea et al. (2005a) and Campana et al. (2007).

2.2.1 *High energy transmission grating spectrometer*

The High Energy Transmission Grating Spectrometer (HETGS) employs two sets of transmission gratings: the Medium Energy Gratings (MEGs) with a range of 2.5–31 Å (0.4–5.0 keV) and the High Energy Gratings (HEGs) with a range of 1.2–15 Å (0.8–10.0 keV). The HETGS spectra were imaged by ACIS-S, an array of 6 CCD detectors normally read-out every 3.2 s. The HETGS/ACIS-S combination provides an undispersed (zeroth order) image and dispersed spectra from the gratings. The various orders overlap and are sorted using the intrinsic energy resolution of the ACIS CCDs: $\Delta\lambda = 0.012$ Å for the HEG and 0.023 Å for the MEG.

The MEG and HEG first order count rate were only 0.5 and 0.2 cts s^{-1}, we therefore did not expect the dispersed spectrum to be affected by pileup, while the zeroth-order image was not used in our spectral analysis because highly affected by photon pileup. We used the standard CIAO tools to create detector response files for the MEG and HEG +1 and −1 order spectra. These were combined when the +/− order spectra were added for the HEG and MEG separately. We binned the data at 0.08 Å with a minimum of 30 counts per bin. To look for high-resolution spectral features, the data were binned at 0.015 Å for the HEG and 0.03 Å for the MEG. We also created background files for the HEG and MEG spectra using the standard CIAO tools.

2.2.2 *Continuous clocking*

In order to avoid pile-up effects, in this second observation the source was observed in the Continuous Clocking (CC) mode (CC33_FAINT; time resolution 2.85 ms). The source was positioned in the back-illuminated ACIS-S3 CCD on the nominal target position. A detailed description on the analysis procedures, such as extraction regions, corrections and filtering applied to the source events and spectra can be found in Rea et al. (2005b).

In order to perform a timing analysis we corrected the events arrival times for the barycenter of the solar system (with the CIAO `axbary` tool) using the provided ephemeris. We used for the timing analysis only the events in the 0.3–8 keV energy range and the standard *Xronos* tools (version 5.19). One fundamental peak plus one harmonic were present in the power-spectrum. A period of 11.00223 ± 0.00005 s was detected referred to MJD 53189. The pulse profile did not change with respect to the previous detection and the 0.3–8 keV PF is $35.4 \pm 0.5\%$. Being the CC mode not yet spectrally calibrated, the Timed Exposure (TE) mode response matrices (rmf) and ancillary files (arf) are generally used for the spectral analysis. We defer to Rea et al. (2005b) for a detailed description on the extraction procedures of the spectral matrices.

2.3 Swift

1RXS J170849-400910 was observed with the *Swift* satellite a few times between 2005 January 29th and March 29th, being a calibration source for the timing accuracy and for the wings of the Point Spread Function of the X-Ray Telescope (XRT). Here we focus on data taken in Window Timing (WT) and Photon Counting (PC) mode longer than 1 ks. We extracted data from two WT observations. The extraction region is computed automatically by the analysis software and

[1] http://asc.harvard.edu/udocs/docs/docs.html; http://asc.harvard.edu/ciao/.

is a box 40 pixels along the WT strip, centred on source, encompassing $\sim$98% of the Point Spread Function in this observing mode. We extracted photons from PC data from an annular region (3 pixels inner radius, 30 pixel outer radius) in order to avoid pile-up contamination. We consider standard grades 0–2 in WT and 0–12 in PC modes. Background spectra were taken from close-by regions free of sources. The photon arrival times were corrected to the Solar system barycenter. A period search led to a clear detection of the neutron star spin period at $P = 11.0027 \pm 0.0003$ s, derived with a phase fitting techniques. This period is consistent with the extrapolation from known ephemerides at a constant period derivative (Kaspi and Gavriil 2003; Dall'Osso et al. 2003). We found a PF of $31 \pm 2\%$, $39 \pm 3\%$, $29 \pm 4\%$ and $35 \pm 7\%$ in the 0.2–10 keV, 0.2–2 keV, 2–4 keV and 4–10 keV energy bands, respectively. Detailed results are reported in Campana et al. (2007).

2.4 IR observation: VLT–NACO

A deep observation of the 1RXS J170849-400910 field, was taken on 2003 June 20th from the Very Large Telescope using the NAOS–CONICA adaptive optics. We defer to Israel et al. 2004 for details in the data reduction. In Fig. 1 we present the K_s band field around the Chandra 0.8″ position of 1RXS J170849-400910 (Israel et al. 2004). Besides sources A and B proposed by Israel et al. (2003b), Safi-Harb and West (2005) and Durant and van Kerkwijk (2006), as the possible IR counterparts to this AXP, many further faint sources (e.g. Star C and Star D) were detected strongly consistent with the Chandra uncertainty region. Unfortunately, H band images were obtained under poor sky conditions, and thus the $H - Ks$ colour could not be determined for these other faint objects. Nonetheless, we note that the Ks magnitudes of e.g. stars C and D (20.3 ± 0.2 and 21.7 ± 0.3, respectively) are in better agreement with the IR magnitudes typical of AXPs, than star A and B previously proposed (17.61 ± 0.07 and 18.78 ± 0.05, respectively). Deep images in the L' band were also obtained, but no object was detected within the Chandra uncertainty region at a limiting magnitude of $L' \sim 17.8$, the deepest limit ever obtained for an AXPs in this band. Detailed results and discussion will be reported in Israel et al. (2006).

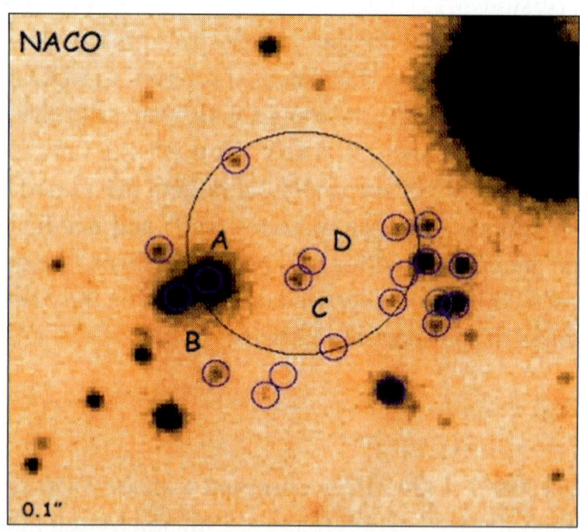

Fig. 1 VLT-NACO image in the K_s band of the field of 1RXS J170849-400910. The Chandra 0.8″ 90% error circle is over-plotted (Israel et al. 2003b). We marked all the faint sources being consistent at 3σ with the AXP position. Note that our astrometry (following Israel et al. 2003b) is slightly different from the one reported by Durant and van Kerkwijk (2006), this is due to a different catalogue used for the astrometry

3 On the absorption line at 8.1 keV

During the post glitches *Beppo*SAX observation in 2001, evidence for an absorption feature was found (see Fig. 2). This feature was not detected during the XMM observation 3 years later, the only observation by now with a comparable statistics. The XMM–Newton upper limit for the line depth is 0.15 at 95% confidence level, which compared with the value found by Rea et al. (2003); 0.8 ± 0.4 at 90% confidence level, leave only a very small chance that the two measurements are consistent.

We then undertook a careful re-analysis of the *Beppo*SAX data. This re-analysis resulted in the finding that the phases at which the absorption line was strongest were given incorrectly in the published version of Rea et al. (2003), as we noticed earlier: in particular, the line is strongest close to the pulse minimum in the 0.1–2 keV band (or the pulse maximum in the 6–10 keV band). We nevertheless found that the reported estimate of the significance is sound and not much influenced by different choices in the background subtraction (annular regions, circular regions far from the source or using blank field files) or by different extraction

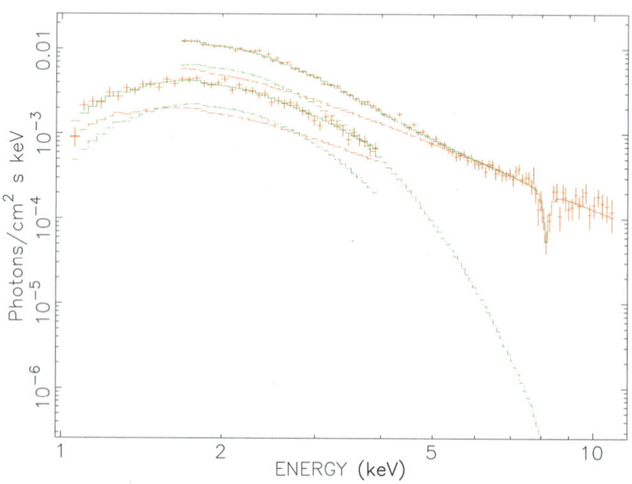

Fig. 2 Unfolded phase-resolved spectrum presenting the absorption line discovered in 1RXS J170849-400910 (Rea et al. 2003, 2004)

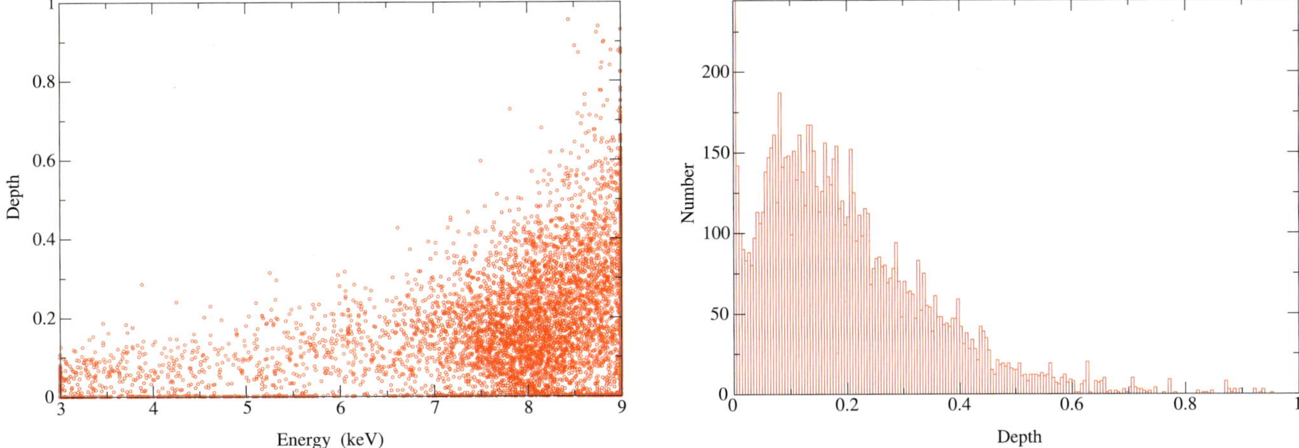

Fig. 3 Results of the Monte Carlo simulation of 10^4 spectra. *Left panel*: Depth versus energy of the detected lines. *Right panel*: Number of spectra for which a line at 8.1 keV was detected, as a function of the line depth

regions for the source. The re-analysis of the *Beppo*SAX data made varying the extraction radius, the criterion for the background subtraction and the spectral binning factor, results in basically the same line properties, which strengthen our confidence in the robustness of the result.

Using an F-test method and taking into account the six trials we made in the phase resolved spectra, we derive a confidence level for the absorption line of $\sim 4\sigma$. Note that even if we take into account all the possible energies at which the feature could lie in the LECS plus MECS energy range, the confidence level is still $\gtrsim 3.5\sigma$. However, despite being the most common method in astrophysics to derive the significance of the emission or absorption spectral features, Protassov et al. (2002) pointed out that the F-test may be inappropriate in these circumstances, leading sometimes to incorrect significance estimates.

Following the recipe of Protassov et al. (2002), in order to further investigate on this significance issue, we ran a Monte Carlo simulation of 10^4 spectra fixing only the continuum model (parameters reported in Rea et al. 2003) and the same number of photons of the phase resolved spectrum which showed the line in the 2001 *Beppo*SAX observation. The results of the simulation is shown in Fig. 3: in the left panel each red circle represents one of the 10^4 simulated spectra for which a line was detected, here we plot the depth of the lines as a function of the line energy. On the other hand, in the right panel we report the number of spectra, among the 10^4 simulated spectra, for which the statistical fluctuation reported the presence of a line at 8.1 keV, as the function of the line depth. From this simulation we found 32 spectra with depth >0.8 in 10^4 points. We can then reliably say that the probability of the line being a fluctuation is $<0.32\%$. In summary, we confirm the detection of the 8.1 keV absorption line in the *Beppo*SAX data made by Rea et al. (2003) at 99.68% confidence level (see Fig. 3). Note that the non homogeneous coverage of red circles over the entire spectral range mirrors the energy dependency of the *Beppo*SAX spectral matrices.

The interpretation of the absorption feature as a cyclotron scattering line proposed by Rea et al. (2003) was based on the following criteria: (1) a Gaussian line gives a bad fit and does not reproduce the asymmetrical shape of the observed feature; (2) the best fitting model is the *XSPEC cyclabs* model; (3) the line strength is highly phase dependent; (4) no atomic edges or absorption lines are known to lie around 8.1 keV (at least without assuming ad hoc shifts possibly due to the high gravitational redshift or to Zeeman effects in such strong magnetic field); (5) the relation between the line energy and width agrees with that of cyclotron scattering features discovered in other classes of sources (see Fig. 5 in Rea et al. 2003); (6) the magnetic field inferred from the line energy, either in the case of an electron or proton cyclotron resonance, is reasonably consistent with what is expected for a normal neutron star ($\sim 10^{12}$ G) or for a magnetar ($\sim 10^{15}$ G), both being still open possibilities. Then, if this feature is real, all the above points hint toward the cyclotron nature of the absorption feature at 8.1 keV.

Keeping always in mind the possibility that the absorption line in the *Beppo*SAX spectrum might be due to statistical fluctuations, in Rea et al. (2005a) we discuss physical mechanisms which could be responsible for the appearance of a transient cyclotron line in this source in the context of the magnetar scenario.

4 Discussion

4.1 Intensity–hardness correlation

The long-term evolution of the source flux and the spectral hardness are shown in Fig. 4, where different observations,

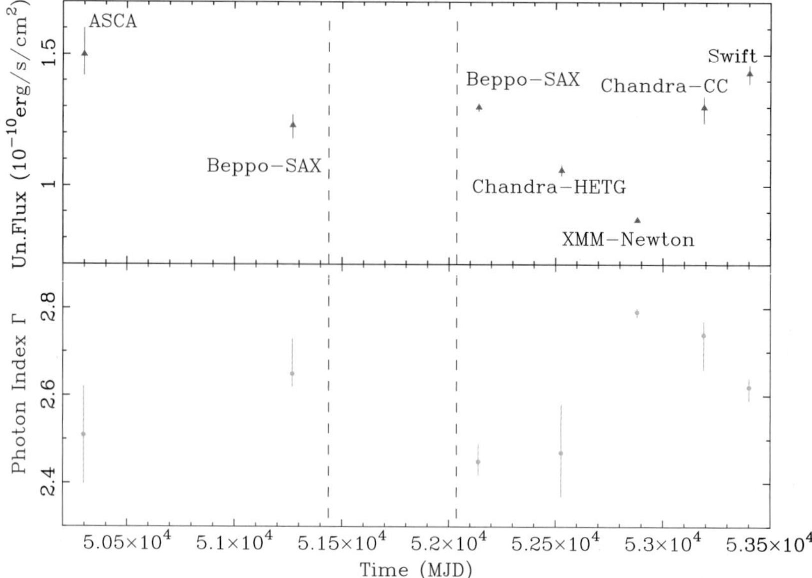

Fig. 4 Correlated photon index and X-ray flux changes with time. *Dashed line* represents the two glitch epochs. All fluxes are unabsorbed and calculated in the 0.5–10 keV energy band keeping fixed the $N_H = 1.36 \times 10^{22}$ cm^{-2}

spanning nearly ten years, were collected. Comparing the two panels, there seems to be a correlation between the photon index the source X-ray flux. The spectrum became progressively harder as the flux rose in correspondence of the two glitches and then softened as the luminosity dropped, following the glitch recovery. This is suggestive of a scenario in which the mechanism responsible for the glitches is also at the basis of the enhanced emission and of the spectral hardening.

The similarity of the second glitch of 1RXS J170849-400910 with the one discovered during the bursting activity of 1E 2259+586 (Kaspi et al. 2003), after which a similar exponential recovery was seen, suggests that bursts likely occurred in 1RXS J170849-400910 as well, but the sparse observations did miss them, as already suggested by Kaspi and Gavriil (2003). Moreover, the spectral parameters and the flux changes after the recovery of the glitch strengthens this idea in comparison with what was reported for the post-bursts fading of 1E 2259+586 (Woods et al. 2004).

This may be interpreted as the onset of a twist, which grew, culminated in the glitches, and then decayed. A twisted external field, in fact, is in an unstable magnetostatic equilibrium and evolves towards a pure dipole field which represents the configuration of minimal energy (see Rea et al. 2005a for a detailed interpretation).

The last Chandra and Swift observations show that the source is slowly increasing again its flux and hardening its emission. Whether the suggested correlation with the glitching and possibly bursting activity holds, we expect the source to re-enter an active state around 2006–2007. If confirmed for all the magnetars, the X-ray monitoring might be an excellent tool to foresee the activity of magnetars.

4.2 On 1RXS J170849-400910 IR counterpart

Based on the deep VLT–NACO observation, we believe that the identification of the IR counterpart of 1RXS J170849-400910 is still an open issue, mainly due to the very crowded region in which this source is located. Note that the 2.5σ variability recently reported for source B by Durant and van Kerkwijk (2006) does not seem to be a conclusive word on the IR counterpart of this source, especially considering that source B has IR magnitudes much brighter than all other AXPs.

4.3 On the possible cyclotron line

Following Protassov et al. (2002) we re-addressed the issue of the significance of the absorption line discovered by *Beppo*SAX. We end up with a 99.68% confidence level for its existence. The fact that the source was not completely recovered by the second glitch at the time the line appeared, make the correlation between line appearance, glitching activity and flux enhancement possibly intriguing and might suggests that the conditions for line formation were met at the epoch of the *Beppo*SAX pointing. Future long X-ray observations are needed in order to follow the source in its slow flux increase and possibly re-detect the source in such state.

Acknowledgements We thank Gordon Garmire for having observed 1RXS J170849-400910 with *Chandra* within his Guarantee Time, and Cees Bassa for having recognised in the different catalogues used for the astrometry, the reason for the shift between our IR NACO field and the one reported by Durant and van Kerkwijk (2006). We also acknowledge F. Haberl, L. Kuiper and the anonymous referee for useful comments.

References

Alpar, A.: Astrophys. J. **554**, 1245 (2001)
Anders, E., Grevesse, N.: Geochimica Cosmochimica Acta **53**, 197 (1989)
Burgay, M., Rea, N., Israel, G.L., et al.: Mon. Not. Roy. Astron. Soc. **372**, 410 (2006)
Camilo, F., Ransom, S., Halpern, J., et al.: Nature **442**, 892 (2006)
Campana, S., Rea, N., Israel, G.L., Zane, S., Turolla, R.: Astron. Astrophys. **463**, 1047 (2007)
Chatterjee, P., Hernquist, L., Narayan, R.: Astrophys. J. **534**, 373 (2000)
Dall'Osso, S., Israel, G.L., Stella, L., et al.: Astrophys. J. **499**, 485 (2003)
Durant, M., van Kerkwijk, M.H.: Astrophys. J. **463**, 1082 (2006)
Duncan, R.C., Thompson, C.: Astrophys. J. **392**, L9 (1992)
Gaensler, B.M., Stappers, B.W., Frail, D.A., et al.: Mon. Not. Roy. Astron. Soc. **318**, 58 (2000)
Israel, G.L., Covino, S., Stella, L., et al.: Astrophys. J. **518**, L107 (1999)
Israel, G.L., Oosterbroek, T., Stella, L., et al.: Astrophys. J. **560**, L65 (2001)
Israel, G.L., Stella, L., Covino, S., et al.: IAU Symposium 218, preprint astro-ph/0310482 (2003a)
Israel, G.L., Covino, S., Perna, R., et al.: Astrophys. J. **589**, L93 (2003b)
Israel, G.L., et al.: in preparation (2006)
Kaspi, V.M., Lackey, J.R., Chakrabarty, D.: Astrophys. J. **537**, L31 (2000)
Kaspi, V.M., Gavriil, F.P.: Astrophys. J. **596**, L71 (2003)
Kaspi, V.M., Gavriil, F.P., Woods, P.M., et al.: Astrophys. J. **588**, L93 (2003)
Kaspi, V.M.: Astrophys. Space Sci., DOI 10.1007/s10509-007-9309-y (2007)
Kuiper, L., Hermsen, W., den Hartog, P.R., Collmar, W.: Astrophys. J. **645**, 556 (2006)
Mereghetti, S., Israel, G.L., Stella, L.: Mon. Not. Roy. Astron. Soc. **296**, 689 (1998)
Protassov, R., van Dyk, D.A., Connors, A., et al.: Astrophys. J. **571**, 545 (2002)
Rea, N., Israel, G.L., Stella, L., et al.: Astrophys. J. **586**, L65 (2003)
Rea, N., Israel, G.L., Stella, L.: Nucl. Phys. B **132**, 554 (2004)
Rea, N., Oosterbroek, T., Zane, S., et al.: Mon. Nat. Roy. Astron. Soc. **361**, 710 (2005a)
Rea, N., Tiengo, A., Mereghetti, et al.: Astrophys. J. **627**, L133 (2005b)
Rea, N., Zane, S., Lyutikov, M., Turolla, R.: Astrophys. Space Sci., DOI 10.1007/s10509-007-9306-1 (2007)
Safi-Harb, S., West, J.: Adv. Space Res. **35**(6), 1172–1176 (2005)
Sugizaki, M., et al.: Publ. Astron. Soc. Jpn. **49**, L25–L30 (1997)
Thompson, C., Duncan, R.C.: Astrophys. J. **408**, 194 (1993)
Thompson, C., Duncan, R.C.: Astrophys. J. **473**, 322 (1996)
van Paradijs, J., Taam, R.E., van den Heuvel, E.: Astron. Astrophys. **299**, L41 (1995)
Voges, W., et al.: IAU Circ. **6420**, 2 (1996)
Woods, P.M., Thompson, C.: preprint astro-ph/0406133 (2004)
Woods, P.M., et al.: Astrophys. J. **605**, 378 (2004)

Astrophys Space Sci (2007) 308: 513–517
DOI 10.1007/s10509-007-9329-7

ORIGINAL ARTICLE

RX J1856.5-3754 as a possible strange star candidate

Jillian Anne Henderson · Dany Page

Received: 16 November 2006 / Accepted: 21 November 2006 / Published online: 24 March 2007

Abstract RX J1856.5-3754 has been proposed as a strange star candidate due to its very small apparent radius measured from its X-ray thermal spectrum. However, its optical emission requires a much larger radius and thus most of the stellar surface must be cold and undetectable in X-rays. In the case the star is a neutron star such a surface temperature distribution can be explained by the presence of a strong toroidal field in the crust (Pérez-Azorín et al.: Astron. Astrophys. 451, 1009 (2006); Geppert et al.: Astron. Astrophys. 457, 937 (2006)) We consider a similar scenario for a strange star with a thin baryonic crust to determine if such a magnetic field induced effect is still possible.

Keywords RX J1856.5-3754 · Strange star · Neutron star

PACS 26.60.+c · 97.60.Jd · 95.30.Tg

1 Introduction

Quark stars have long ago been proposed as an alternative to neutron stars (Itoh 1970). The "strange matter hypothesis" (Witten 1984) gave a more precise theoretical formulation for their existence: that at zero pressure three flavor quark matter, i.e. with u, d and s quarks, has a lower density per baryon than nuclear matter and would hence be the true ground state of baryonic matter. These stars are now called "strange stars" (Alcock et al. 1986; Haensel et al. 1986) and share many similarities with neutron stars: they can have similar masses, have similar radii in the observed range of masses, similar cooling histories and, to date, it has been practically impossible to conclusively prove or disprove their existence (for recent reviews, see Weber 2005; Page et al. 2006; Page and Reddy 2006).

One possible distinctive property of a strange star could be a small radius. Given the impossibility to treat quark-quark interactions from first principles, i.e. starting from Q.C.D., at densities relevant for compact stars, only simplified models are possible and results are, naturally, model dependent. However, several classes of such models do predict small radii, $5 < R < 10$ km at masses $\sim 1.4 M_{\odot}$ (see, e.g., Dey et al. 1998; Hanauske et al. 2001), and *all* models predict very small radii, $\lesssim 5$ km, at masses $\ll 1 M_{\odot}$. Hence, measurement of a compact star radius giving a radius $\ll 10$ km directly allows a claim for a strange star candidate.

The "Magnificent Seven" (Haberl 2007) arouse great expectations to measure compact star radii with high enough accuracy to put strong constraints on the dense matter equation of state. In particular, fits of the observed soft X-ray thermal spectrum of RX J1856.5-3754 (Pons et al. 2002) pointed to a very small radius and lead to the claim that this object may be a strange star (Drake et al. 2002). However, observations in the optical band allowed the identification of the Rayleigh–Jeans tail of a second component of the surface thermal emission, corresponding to a lower temperature and much larger radius than the component detected in the X-ray band. An interpretation of these results is that the surface temperature of the star is highly non-uniform (Pons et al. 2002; Trümper et al. 2004), possibly

This work was partially supported by PAPIIT, UNAM, grant IN119306. J.A.H. studies at UNAM and travel to London are covered by fellowships from UNAM's Dirección General de Estudios de Posgrado.

J.A. Henderson (✉) · D. Page
Instituto de Astronomía, Universidad Nacional Autónoma de México, Ciudad Universitaria, Mexico, D.F., CP 04510, Mexico
e-mail: hendersj@astroscu.unam.mx

D. Page
e-mail: page@astroscu.unam.mx

due to the presence of a strong magnetic field. Models of surface temperature distribution with purely poloidal magnetic fields (Page 1995; Geppert et al. 2004) do predict non-uniform surface temperature distributions, but such inhomogeneities are not strong enough to produce such small X-ray emitting regions surrounded by large cold regions detectable in the optical band as observed. However, inclusion of a toroidal component of the magnetic field, confined to the neutron star crust, has a dramatic effect (Pérez-Azorín et al. 2006; Geppert et al. 2006; see also Pons 2007; Page 2007): this field component inhibits heat from the stellar core to flow to the surface through most of the crust, except for small domains surrounding the magnetic axis, and results in highly non-uniform surface temperature distributions producing good fits to the observed thermal spectra, from the optical up to the X-ray band.

These models of small hot regions, detected in the X-ray band, surrounded by large cold regions, detected in the optical band, which allow to reproduce the entire observed thermal spectrum and results in large radii for the star are in contradiction with the proposed strange star interpretation of RX J1856.5-3754, which was based on the small radius detected in the X-ray band. Here we want to push the discussion one step further: are these highly non-uniform surface temperature distributions, assuming they are real, incompatible with a strange star model? We consider strange stars having a thin crust, composed of normal baryonic matter, with a strong magnetic field. Since a strange star crust can, at most, reach the neutron drip density, it is much thiner than the crust of a normal neutron star and the specific question is: can such a thin layer produce the surface temperature distributions deduced from observation?

2 The strange star models

We will consider strange star models built on the MIT-bag inspired equation of state of (Farhi and Jaffe 1984) which has three parameters: the QCD coupling constant α_s, the bag constant B and the strange quark mass m_s (u and d quarks are treated as massless). Figure 1 illustrates four families of such strange matter models: by varying the parameters, these equations of state allow the production of a wide range of models, from very compact stars up to very large ones. It is important to notice from this figure that, depending on the assumed parameters of the model, strange stars can have large radii and thus, although a small radius measurement is a strong argument in favor of a strange star, a large radius is *not* an argument against a strange star.

On top of the quark matter, a thin crust can exist as long as the electron density within it is smaller than that in the quark matter (Alcock et al. 1986). Such a baryonic crust is, however, much thinner than a neutron star crust as illustrated in Fig. 2.

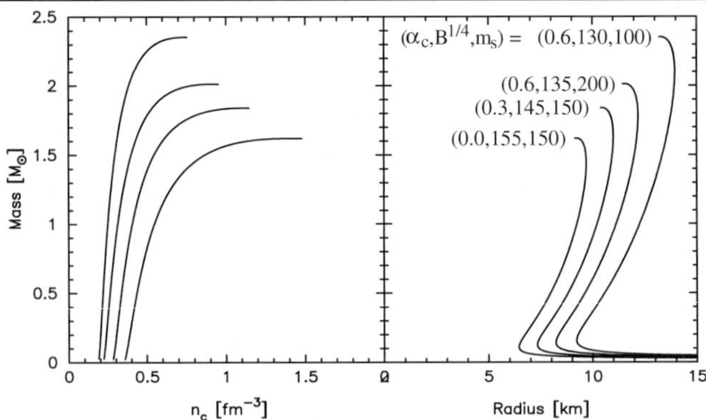

Fig. 1 Mass vs. central density (*left*) and radius (*right*) for strange stars built with four MIT-bag inspired models of strange matter (model parameters are indicated in the *right panel*) and covered by a thin baryonic crust (see Fig. 2)

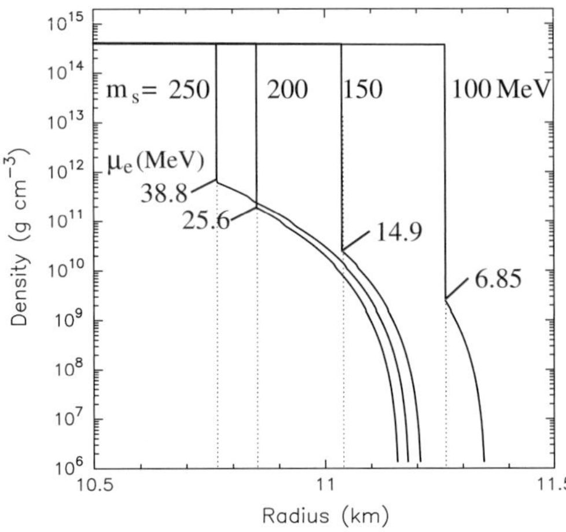

Fig. 2 Density profile of the upper layers of four strange stars built with four MIT-bag inspired models of strange matter, using $\alpha_s = 0.3$, $B^{1/4} = 140$ MeV and four different strange quark masses as indicated. Indicated are also the electron chemical potentials μ_e at the quark surface, which is the parameter determining the maximum baryonic crust density, i.e., the baryonic crust thickness. The crust equation of state is from (Haensel et al. 1989)

Following the neutron star models presented by Geppert et al. (2006) we consider dipolar magnetic fields with three components (Fig. 3): a poloidal one maintained by currents in the quark core, $\mathbf{B}^{\text{core}}$, a poloidal one maintained by currents in the baryonic crust, $\mathbf{B}^{\text{crust}}$, and a toroidal one maintained by currents in the baryonic crust, $\mathbf{B}^{\text{tor}}$. The separation between currents in the crust and in the core is motivated by the likely fact that quark matter forms a Maxwell superconductor (Alford 2001; Page and Reddy 2006): at the moment of the phase transition, occurring very early in the life of the star, superconductivity will prevent any current in the crust from penetrating the core while currents in the core will become supercurrents and be frozen there. Moreover, flux expulsion due to the star's spin-down can also signif-

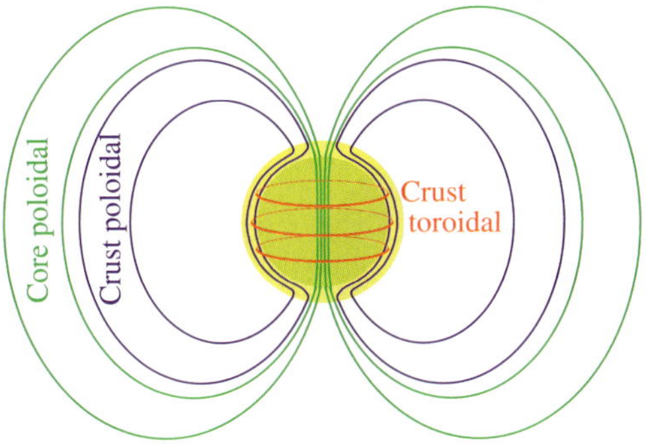

Fig. 3 The three components of the magnetic field considered in this work

icantly increase the crustal field at the expense of the core field.

The importance of the crustal field is that its field lines are forced to be closed within the crust and hence it has a very large meridional component B_θ, compared to the core component. Due to the classical Larmor rotation of electrons, a magnetic field causes anisotropy of the heat flux and the heat conductivity becomes a tensor whose components perpendicular, $\kappa_\perp$, and parallel, $\kappa_\parallel$, to the field lines become

$$\kappa_\perp = \frac{\kappa_0}{1+(\Omega_B \tau)^2} \quad \text{and} \quad \kappa_\parallel = \kappa_0 \qquad (1)$$

where κ_0 is the conductivity in the absence of a magnetic field, Ω_B the electrons cyclotron frequency and τ their collisional relaxation time. The large values of B_θ in the crust have the effect of inhibiting radial heat flow except in regions close to the magnetic axis where B_r dominates over B_θ (Geppert et al. 2004).

3 Strange stars - results

We performed heat transport calculations using the 2D code described in Geppert et al. (2004) which incorporates the thermal conductivity anisotropy described by (1). Details of the crust microphysics are as described in Geppert et al. (2004, 2006). We consider a strange star model with a radius of ~11 km and a baryonic crust of thickness ~250 m for a $1.4 M_\odot$ mass. The two poloidal components of the magnetic field are parametrized by $B_0^{\rm core}$ and $B_0^{\rm crust}$ which are the strengths of the corresponding field components at the surface of the star along the magnetic axis so that, ideally, $B_0^{\rm core} + B_0^{\rm crust}$ would be the dipolar field estimated from the star's spin-down. Notice that the maximum value of $B^{\rm core}$ in the crust is only slightly larger than $B_0^{\rm core}$ while *maximum values of $B^{\rm crust}$ are up to almost 100 times larger than $B_0^{\rm crust}$* due to its large tangential component, $B_\theta^{\rm crust}$, resulting from the compression of the field within the narrow crust. The strength of the toroidal field is parametrized by $B_0^{\rm tor}$, defined as the maximum value reached by $B^{\rm tor}$ within the crust. We keep $B_0^{\rm core}$ at 10^{13} G and vary $B_0^{\rm crust}$ and $B_0^{\rm tor}$.

We display in Fig. 4 the resulting crustal temperature profiles for several typical values of $B_0^{\rm crust}$ and $B_0^{\rm tor}$. One sees that, independently of the strength of the poloidal component, $B_0^{\rm tor}$ needs to reach 10^{15} G to have a significant effect, a result similar to what was obtained by Geppert et al. (2006) for the neutron star case. However, independently of the value of $B_0^{\rm tor}$, once $B_0^{\rm crust}$ reaches 10^{13} G highly non-uniform temperature profiles develop in the thin strange star crust: such profiles are sufficiently non-uniform to produce the wanted surface temperature distribution, i.e. small hot regions surrounded by extended cold ones.

4 Discussion and conclusions

The optical + X-ray spectrum of RX J1856.5-3754, when fitted with blackbodies, requires two components with very different temperatures and emitting areas. The implied highly non-uniform surface temperature distribution can be physically justified by the introduction of a very strong magnetic field, whose supporting currents are mostly located within the star's crust. We have shown here that, similarly to the neutron star case, such field configurations can be found in the case of a strange star, despite the shallowness of its baryonic crust. However, the strength of these fields, either the toroidal component $\mathbf{B}^{\rm tor}$ or the crust anchored poloidal one $\mathbf{B}^{\rm crust}$ must reach strengths close to, or above, 10^{15} G to produce the desired temperature anisotropy. Similarly to the neutron star case, such surface temperature distributions impose severe, but not unrealistic, restrictions on the orientation of either the observer or the magnetic field symmetry axis with respect to the rotation axis to explain the absence of pulsations (Braje and Romani 2002; Geppert et al. 2006).

That the thin strange star crust can support such huge field strengths is an open question, but a positive answer seems doubtful as some simple estimates illustrate. The magnetic shear stress, $B_r B_\theta / 4\pi$, reaches 10^{26} dyne cm^{-2} in the crust when $B_0^{\rm crust}$ reaches 10^{13} G, which is comparable or higher to the maximum value sustainable by a crust of thickness $\Delta \sim 200$–300 m (Ruderman 2004): violent readjustments of the crust are expected but have yet to be observed in any of the "Magnificent Seven". Moreover, the ohmic decay time in the low density crust is relatively short, less than 10^5 yr (Page et al. 2000), and the presence of such a strong field in a $\sim 10^6$ yr old star would require an initial crustal field about two orders of magnitude higher when the star was young. (However, the highly non-linear evolution of coupled strong poloidal and toroidal magnetic fields remains to be studied under such conditions.)

We have also adopted the very ingenuous assumption that the surface emits as a perfect blackbody. A condensed

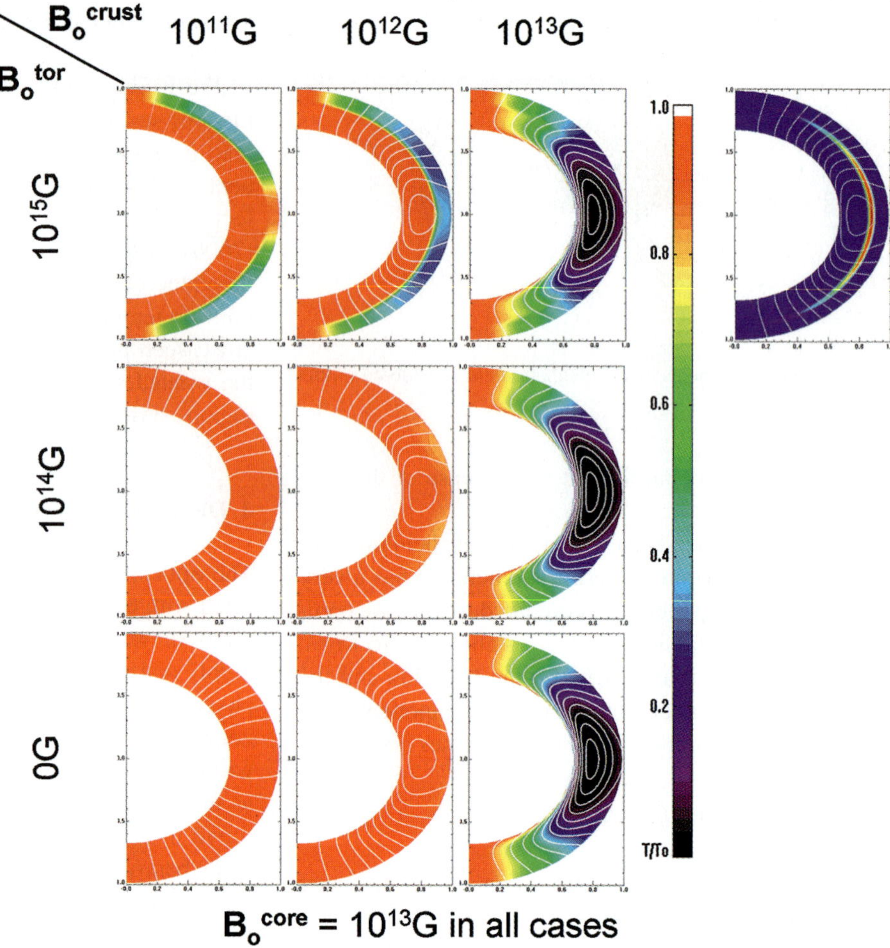

Fig. 4 Crustal temperature profiles for various strange star magnetic field structures according to the model parameters (see text for details). The radial aspect of the crust has been stretched by a factor of 15 in order to clearly show the thermal structure. The *upper right profile* shows the field lines of the poloidal component $\mathbf{B}^{\mathrm{core}} + \mathbf{B}^{\mathrm{crust}}$ (*white lines*) and the intensity distribution of the toroidal component $\mathbf{B}^{\mathrm{tor}}$ (*in colours*)

matter surface may simulate a blackbody spectrum (Turolla et al. 2004; Pérez-Azorín et al. 2005; van Adelsberg et al. 2005) but such models still require strong fields to produce a non-uniform temperature distribution (Pérez-Azorín et al. 2006). However, other interpretations are possible, such as a thin atmosphere atop a solid surface (Motch et al. 2003; Ho et al. 2007) which may be able to reproduce both the optical and X-ray spectra without invoking strongly non-uniform temperatures and can be applied as well to strange stars with a crust as to neutron stars since they only consider the very surface of the star.

In conclusion, the crustal field scenario, which is successful when applied to neutron star models in order to explain the observed thermal spectrum properties of RX J1856.5-3754, can be "successfully" applied to a strange star model but requires such a huge magnetic field confined within such a thin crust that its applicability is doubtful. It is hence difficult to conciliate the observed, from optical to X-ray, properties of RX J1856.5-3754 with a strange star interpretation unless these are due exclusively to the emitting properties of its surface.

Acknowledgements We thank M. Küker and U. Geppert for allowing us to use their 2D transport code.

References

Alcock, C., Farhi, E., Olinto, A.: Astrophys. J. **310**, 261 (1986)
Alford, M.: Annu. Rev. Nucl. Part. Sci. **51**, 131 (2001)
Braje, T.M., Romani, R.W.: Astrophys. J. **580**, 1043 (2002)
Dey, M., Bombaci, I., Dey, J., et al.: Phys. Lett. B **438**, 123 (1998)
Drake, J.J., Marshall, H.L., Dreizler, S., et al.: Astrophys. J. **572**, 996 (2002)
Farhi, E., Jaffe, R.L.: Phys. Rev. D **30**, 2379 (1984)
Geppert, U., Küker, M., Page, D.: Astron. Astrophys. **426**, 267 (2004)
Geppert, U., Küker, M., Page, D.: Astron. Astrophys. **457**, 937 (2006)
Haberl, F.: Astrophys. Space Sci. **308**. DOI 10.1007/s10509-007-9342-x (2007)
Haensel, P., Zdunik, J.L., Schaeffer, R.: Astron. Astrophys. **160**, 121 (1986)
Haensel, P., Zdunik, J.L., Dobaczewski, J.: Astron. Astrophys. **222**, 353 (1989)
Hanauske, M., Satarov, L.M., Mishustin, I.N., et al.: Phys. Rev. D **64**, 3005 (2001)
Ho, W.C.G., Kaplan, D.L., Chang, P., et al.: Astrophys. Space Sci. **308**. DOI 10.1007/s10509-007-9366-2 (2007)
Itoh, N.: Prog. Theor. Phys. **44**, 291 (1970)
Motch, C., Zavlin, V.E., Haberl, F.: Astron. Astrophys. **408**, 323 (2003)
Page, D.: Astrophys. J. **442**, 273 (1995)
Page, D.: Astrophys. Space Sci. **308**. DOI 10.1007/s10509-007-9316-z (2007)
Page, D., Geppert, U., Zannias, T.: Astron. Astrophys. **360**, 1053 (2000)

Page, D., Geppert, U., Weber, F.: Nucl. Phys. A **777**, 497 (2006)
Page, D., Reddy, S.: Annu. Rev. Nucl. Part. Sci. **56**, 327 (2006)
Pérez-Azorín, J.F., Miralles, J.A., Pons, J.A.: Astron. Astrophys. **433**, 275 (2005)
Pérez-Azorín, J.F., Miralles, J.A., Pons, J.A.: Astron. Astrophys. **451**, 1009 (2006)
Pons, J.A., Walter, F.M., Lattimer, J.M., et al.: Astrophys. J. **564**, 981 (2002)
Pons, J.: Astrophys. Space Sci. **308**. DOI 10.1007/s10509-007-9336-8 (2007)
Ruderman, M.: In: Baykal, A., Yerli, S.K., Gilfanov, M., Grebenev, M. (eds.) The Electromagnetic Spectrum of Neutron Stars (e-print: astro-ph/0410607)
Trümper, J.E., Burwitz, V., Haberl, F., et al.: Nucl. Phys. B Proc. Suppl. **132**, 560 (2004)
Turolla, R., Zane, Z., Drake, J.J.: Astrophys. J. **603**, 265 (2004)
van Adelsberg, M., Lai, D., Potekhin, A.Y., et al.: Astrophys. J. **628**, 902 (2005)
Weber, F.: Prog. Nucl. Part. Phys. **54**, 193 (2005)
Witten, E.: Phys. Rev. D **30**, 272 (1984)

Astrophys Space Sci (2007) 308: 519–523
DOI 10.1007/s10509-007-9295-0

ORIGINAL ARTICLE

Chandra monitoring of the candidate anomalous X-ray pulsar AX J1845.0-0258

Cindy R. Tam · Victoria M. Kaspi · Bryan M. Gaensler · Eric V. Gotthelf

Received: 14 July 2006 / Accepted: 18 August 2006 / Published online: 20 March 2007

Abstract The population of clearly identified anomalous X-ray pulsars has recently grown to seven, however, one candidate anomalous X-ray pulsar (AXP) still eludes reconfirmation. Here, we present a set of seven Chandra ACIS-S observations of the transient pulsar AX J1845.0-0258, obtained during 2003. Our observations reveal a faint X-ray point source within the ASCA error circle of AX J1845.0-0258's discovery, which we designate CXOU J184454.6-025653 and tentatively identify as the quiescent AXP. Its spectrum is well described by an absorbed single-component blackbody ($kT \sim 2.0$ keV) or power law ($\Gamma \sim 1.0$) that is steady in flux on timescales of at least months, but fainter than AX J1845.0-0258 was during its 1993 period of X-ray enhancement by at least a factor of 13. Compared to the outburst spectrum of AX J1845.0-0258, CXOU J184454.6-025653 is considerably harder: if truly the counterpart, then its spectral behavior is contrary to that seen in the established transient AXP XTE J1810-197, which softened from $kT \sim 0.67$ keV to ~ 0.18 keV in quiescence. This unexpected result prompts us to examine the possibility that we have observed an unrelated source, and we discuss the implications for AXPs, and magnetars in general.

Keywords Pulsar · AXP · Neutron star · Magnetar · AX J1845.0-0258

PACS 97.60.Gb · 97.60.Jd · 95.85.Nv

C.R. Tam (✉) · V.M. Kaspi
Department of Physics, Rutherford Physics Building, McGill University, 3600 University Street, Montreal, QC H3A 2T8, Canada
e-mail: tamc@physics.mcgill.ca

B.M. Gaensler
Harvard-Smithsonian Center for Astrophysics, 60 Garden Street, Cambridge, MA 02138, USA

Present address:
B.M. Gaensler
School of Physics, University of Sydney, Sydney, NSW 2006, Australia

E.V. Gotthelf
Columbia Astrophysics Laboratory, Columbia University, 550 West 120th Street, New York, NY 10027-6601, USA

1 Introduction to AX J1845.0-0258

The 6.97-s X-ray pulsar AX J1845.0-0258 was discovered serendipitously in archival ASCA observations from 1993 (Gotthelf and Vasisht 1998; Torii et al. 1998) during a period of apparent outburst. Its slow spin period, soft spectrum ($kT \sim 0.64$ keV) and positional coincidence with the newly discovered supernova remnant G29.6+0.1 (Gaensler et al. 1999) were suggestive of the small but growing class of anomalous X-ray pulsars (AXPs, see review by V. Kaspi, this volume[1]). However, without an estimate of $\dot{P}$, and thus B, the AXP identification could not be confirmed, so further attempts were made to re-detect the pulsar and pulsations. Unfortunately, it was not seen in a 1997 observation from the ASCA Galactic Plane Survey (Torii et al. 1998), and a 1999 pointed follow-up observation with ASCA found a possible counterpart, AX J184453-025640,[2] that was almost 10 times fainter, too faint for

[1]For a summary of AXP properties, see the online catalog http://www.physics.mcgill.ca/~pulsar/magnetar/main.html.

[2]The position of AX J184453-025640 has been corrected since the publication of Vasisht et al. (2000); see Tam et al. (2006) for the best position.

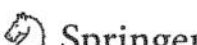

a measurement of pulsations or a spectrum (Vasisht et al. 2000). Chandra, XMM-Newton and BeppoSAX observations during 2001–2003 revealed a point source coincident with AX J184453-025640 and similar in brightness, but with a slightly harder absorbed spectrum ($kT \sim 1.0$ keV) than that seen for AX J1845.0-0258 in 1993 (Israel et al. 2004).

Presented here are the results of a Chandra X-ray Observatory monitoring campaign, conducted in 2003, with the goal of characterizing the spectral and timing properties of AX J1845.0-0258 in a post-outburst state.

2 Chandra observations

Between June and September 2003, we obtained seven observations with Chandra ACIS-S in timed exposure mode. The first six were taken in 1/8 subarray mode in order to achieve high time resolution (0.4 s); the seventh observation was full field. Since the 1993 ASCA position of AX J1845.0-0258 had a large (3′ radius) uncertainty, we centered our observations at the Chandra HRC position of a possible counterpart (G. Israel, private communication). All data processing was performed using the CIAO 3.2.2 and CALDB 3.0.3 software packages.

One bright point source in the 3′ ASCA error circle was found and designated CXOU J184454.6-025653 (see Fig. 1). This is likely the counterpart to AX J184453-025640, and possibly AX J1845.0-0258. There was no evidence of extended emission.

Two additional fainter point sources were detected inside the 1993 error circle of AX J1845.0-0258 but outside the 1999 error circle of AX J184453-025640 (see Fig. 1). One of them, CXOU J184507.2-025657, was found coincident with a bright near-infrared object from the 2MASS All Sky Survey; this challenges an AXP interpretation, since all confirmed near-IR counterparts to AXPs are very faint ($K \sim 20$ mag) and AX J1845.0-0258 is known to be highly absorbed ($N_H \geq 6 \times 10^{22}$ cm^{-2}, Gotthelf and Vasisht 1998).

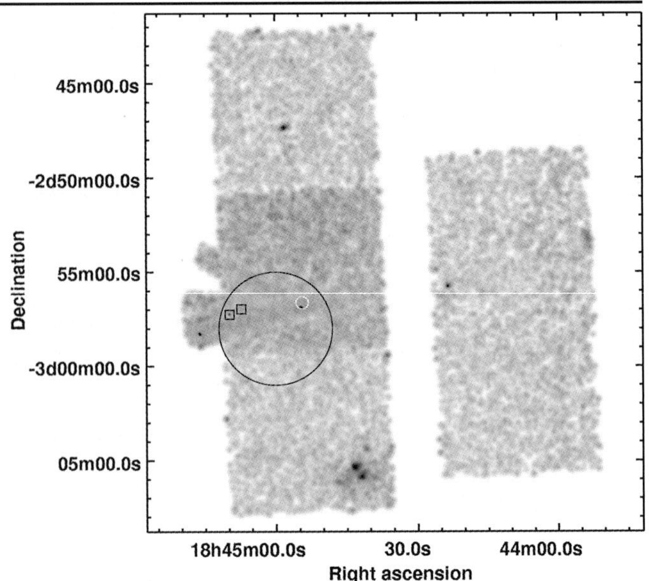

Fig. 1 Combined 2–10 keV Chandra ACIS-S image of the field surrounding AX J1845.0-0258, originally published in Tam et al. (2006). The potential counterpart CXOU J184454.6-025653 falls within the 1993 and 1999 ASCA error circles of AX J1845.0-0258 (*black circle*, Gotthelf and Vasisht 1998) and AX J184453-025640 (*white circle*, Vasisht et al. 2000), respectively. Also indicated are two fainter point sources, CXOU J184507.2-025657 (*right box*) and CXOU J184509.7-025715 (*left box*)

Table 1 CXOU J184454.6-025653 spectral parameters. Errors reflect 90% confidence region. The absorbed flux is given for the 2–10 keV energy range, and we determine its uncertainty by fixing N_H and kT or Γ at the best-fit value and adopting the fractional uncertainty on the normalization

Model	N_H (10^{22} cm^{-2})	kT (keV) or Γ	F (10^{-13} erg s^{-1} cm^{-2})
BB	$5.6^{+1.6}_{-1.2}$	$2.0^{+0.4}_{-0.3}$	2.6 ± 0.2
PL	$7.8^{+2.3}_{-1.8}$	$1.0^{+0.5}_{-0.3}$	2.8 ± 0.2

3 Timing and spectral analysis

Consider two cases: (1) CXOU J184454.6-025653 is the counterpart to AX J1845.0-0258, and (2) CXOU J184454.6-025653 is unrelated to AX J1845.0-0258. The results of the following analysis were originally published in Tam et al. (2006).

Case 1: *CXOU J184454.6-025653 is the counterpart.* We extracted light curves from CXOU J184454.6-025653 at the highest possible time resolution (0.4 s for six observations, 3.2 s for the seventh) from each data set, in three energy ranges: 1–10, 1–3, and 3–10 keV. A fast Fourier transform (FFT) was performed on barycentered event data, however, no evidence for pulsations was seen in any of the resulting power density spectra. For the longest observation and the frequency range 0.0880–0.1436 Hz, we find a 95% confidence upper limit on the pulsed amplitude of 80% in the 1–10 keV range.

The individual observations contained insufficient counts to adequately fit a spectrum, so we summed the extracted spectra into one combined spectrum. Using XSPEC 11.3.1, we found that the background-subtracted combined spectrum was equally well fit to a single-component absorbed thermal blackbody or power law: Table 1 lists the best-fit spectral parameters, and Fig. 2 shows the data fit to a blackbody. Assuming the blackbody model for now, we measured

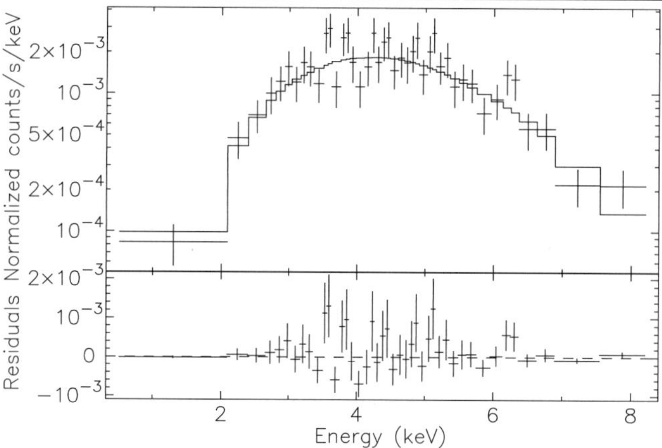

Fig. 2 The spectrum of CXOU J184454.6-025653 shown with its best-fit blackbody model

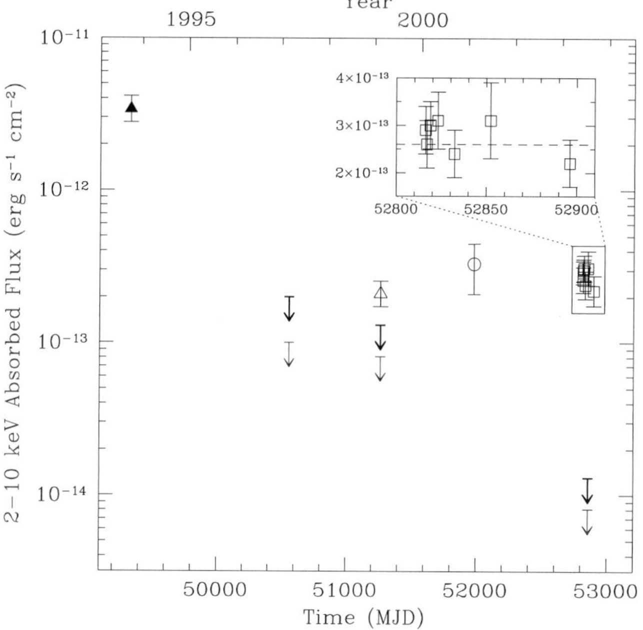

Fig. 3 The 10-year flux history of AX J1845.0-0258, originally presented in Tam et al. (2006). *The filled triangle* is the original 1993 ASCA detection of AX J1845.0-0258; *the open triangle* is the 1999 ASCA detection of AX J184453-025640, and assumes the 1993 outburst spectrum. *The circle* is the BeppoSAX detection of a possible counterpart, and assumes the spectrum given by Israel et al. (2004). *Squares* indicate the Chandra detections of CXOU J184454.6-025653 reported here, and assume the best-fit blackbody spectrum. We also represent the observed background levels as upper limits—in case the detections made were of unrelated objects—that assume the spectrum of AX J1845.0-0258 in outburst (*thick arrows*) and XTE J1810-197 in quiescence (*thin arrows*). The Chandra points are magnified in *the inset plot*, where the flux measured from the combined data set is indicated by *the dashed line*

the combined absorbed 2–10 keV flux to be $2.6 \pm 0.2 \times 10^{-13}$ erg s^{-1} cm^{-2}; we also estimated the unabsorbed flux to be 2.5–4.0 $\times\, 10^{-13}$ erg s^{-1} cm^{-2}, taking into account the large uncertainties on N_H. Since the flux is roughly consistent with that of AX J184453-025640, we speculate that we have detected the same object.

Fixing N_H and kT at their best-fit blackbody values but allowing the normalization to vary, we fitted the data from the seven individual observations and found that the observed 2–10 keV flux at each epoch was consistent with CXOU J184454.6-025653 being constant over the 4-month Chandra observing period, at the combined flux value. The inset plot of Fig. 3 shows this.

Case 2: CXOU J184454.6-025653 is unrelated to AX J1845.0-0258. We extracted light curves and spectra from the two additional faint sources detected in the 3′ error region, but in all instances there were not enough counts to detect pulsations or fit a spectral model. However, we noticed that most of the photons from CXOU J184507.2-025657 were below 2 keV, which contradicts what is known of AX J1845.0-0258, namely that it is highly absorbed. Because of this and the aforementioned evidence in Sect. 2, we consider CXOU J184507.2-025657 an unlikely counterpart to AX J1845.0-0258. For CXOU J184509.7-025715, the data were insufficient for us to draw meaningful conclusions about this source as candidate.

AX J1845.0-0258 may not have been re-detected at all, falling below the 3σ background flux level. We estimated an upper limit on a hypothetical point source to be $\sim$ 8–13 $\times\, 10^{-15}$ erg s^{-1} cm^{-2} (2–10 keV), based on a variety of likely spectral models (see Fig. 3).

4 A transient AXP?

Whether CXOU J184454.6-025653 is truly the counterpart or not, its flux in 2003 is a factor of $\sim$ 13 smaller than AX J1845.0-0258's in 1993. However if AX J1845.0-0258 has not been re-detected at all, then this factor grows significantly larger to $\sim$260–430, respresenting an unprecedented range in variability for AXPs. Figure 3 outlines the 10-year flux evolution of AX J1845.0-0258 and its potential counterparts.

Comparable flux variability on large time scales has been seen in at least one other AXP. The 5.5-s transient AXP (TAXP) XTE J1810-197 was also discovered when it was in a high state, in 2003 (Ibrahim et al. 2004), and has since faded back towards its "quiescent" flux level (Gotthelf and Halpern 2005), as measured in archival ROSAT observations from 1993 (Gotthelf et al. 2004). The pre-outburst source flux, which is nearly 2 orders of magnitude lower than its peak outburst flux, is much fainter than that of any non-transient AXPs, bringing to mind the question of how many more TAXPs have gone undetected in the Galaxy.

TAXPs are not accounted for in the framework of the magnetar model (Thompson and Duncan 1995, 1996), which attributes their persistent high-energy emission to continual heating of and stresses on the magnetar's crust. The source of this crustal stress and heating is the gradual decay of its ultra-high ($\sim 10^{15}$ G) magnetic field, and can be used to predict an X-ray luminosity that is well matched by that seen in quiescent non-transient AXPs (Thompson et al. 2002). X-ray bursts, like those now observed in four AXPs including XTE J1810-197 (Woods et al. 2005), are thought to result from the sudden fracturing of the magnetar's surface and reconfiguring of its field lines. So the question remains: if these common elements link transient and non-transient AXPs as magnetars, what is the cause of their differences?

Spectrally, we previously saw that AX J1845.0-0258 was not unlike other AXPs, which typically have soft spectra (recall $kT \sim 0.64$ keV during outburst). For this reason, the observed hardness of the Chandra source ($kT \sim 2.0$ keV) brings into question the proposed association with AX J1845.0-0258, and an overall AXP interpretation. Moreover, XTE J1810-197, the *bona fide* TAXP, was observed to be harder in outburst than quiescence ($kT \sim 0.67$ keV compared to $kT \sim 0.18$ keV, respectively, from Gotthelf et al. 2004), which is the opposite to what we have witnessed if CXOU J184454.6-025653 is indeed a TAXP.

5 Alternate endings

Given the uncertainty in the identity of CXOU J184454.6-025653, it seems prudent to consider other plausible alternatives. We argue on the basis of key observable properties, such as its relatively hard spectrum, intrinsic luminosity $L_X \approx 10^{33}(d/5\ \mathrm{kpc})^2$, and apparent stability on time scales of days to weeks.

Active galactic nuclei. The measured photon index $\Gamma \sim 1.0$ from the power law model is not unlike that seen for an active galactic nucleus (AGN, Watanabe et al. 2004; Nandra et al. 2005). Using Chandra ACIS-I, Ebisawa et al. (2005) studied the faint X-ray emission from an "empty" Galactic plane region that was conveniently centered only 1° away from our target, meaning that they might have local properties in common such as N_H. From their models of Galactic source populations, we estimate a $\sim 2\%$ likelihood that a circular region $3'$ in radius would contain a coincident AGN, $3 \times 10^{-13}\ \mathrm{erg\,s^{-1}\,cm^{-2}}$ (2–10 keV) or brighter. Predicted optical/IR magnitudes fall at the limits of what current observatories are capable of, which will make it difficult to conclusively confirm or rule out an AGN interpretation through such means.

Galactic sources. Winds from massive stars have similar spectral and flux properties, as do some high-mass X-ray binaries (Muno et al. 2004). These systems, however, would tend to be bright in optical/IR, which disagrees with the faint upper limit set by Israel et al. (2004) of $H > 21$ mag.

Another group of Galactic objects with similar properties are cataclysmic variables (CVs). According to Muno et al. (2004), the IR emission of CVs at comparable distances and extinctions to our Chandra source ought to be relatively faint, roughly $K \approx 22$–25 mag. Therefore, it seems clear that optical/IR observations alone will be insufficient to identify this source.

6 Conclusions

We have observed and analysed the Chandra point source CXOU J184454.6-025653, which may be the transient X-ray pulsar and candidate AXP AX J1845.0-0258. If it is the counterpart, then either AX J1845.0-0258 is not actually an AXP, or AXPs are much more diverse in their spectral and flux characteristics during quiescence than previously thought. If it is not the counterpart and AX J1845.0-0258 is an AXP, then the exhibited flux variability presents a challenge to our current understanding of AXPs as magnetars, and hints at a much larger population of faint AXPs that remain undetected.

Acknowledgements Thanks to G. Israel for providing details on the position and flux of the possible counterpart to AX J1845.0-0258. V.M.K. is a Canada Research Chair and acknowledges funding from NSERC via a Discovery Grant and Steacie Supplement, the FQRNT, and CIAR. B.M.G. is an Alfred P. Sloan Fellow and acknowledges support from Chandra GO grant GO3-4089X, awarded by the SAO.

References

Ebisawa, K., Tsujimoto, M., Paizis, A., et al.: Astrophys. J. **635**, 214 (2005)
Gaensler, B.M., Gotthelf, E.V., Vasisht, G.: Astrophys. J. **526**, L37 (1999)
Gotthelf, E.V., Halpern, J.P.: Astrophys. J. **632**, 1075 (2005)
Gotthelf, E.V., Halpern, J.P., Buxton, M., et al.: Astrophys. J. **605**, 368 (2004)
Gotthelf, E.V., Vasisht, G.: New Astron. **3**, 293 (1998)
Ibrahim, A.I., Markwardt, C.B., Swank, J.H., et al.: Astrophys. J. **609**, L21 (2004)
Israel, G., Stella, L., Covino, S., et al.: In: Camilo, F., Gaensler, B. (eds.) Proceedings of IAU Symposium, vol. 218, p. 247. PASP, San Francisco (2004)
Muno, M.P., Arabadjis, J.S., Baganoff, F.K., et al.: Astrophys. J. **613**, 1179 (2004)
Nandra, K., Laird, E.S., Adelberger, K., et al.: Mon. Not. Roy. Astron. Soc. **356**, 568 (2005)
Tam, C.R., Kaspi, V.K., Gaensler, B.M., et al.: Astrophys. J. **652**, 548 (2006)
Thompson, C., Duncan, R.C.: Mon. Not. Roy. Astron. Soc. **275**, 255 (1995)
Thompson, C., Duncan, R.C.: Astrophys. J. **473**, 322 (1996)

Thompson, C., Lyutikov, M., Kulkarni, S.R.: Astrophys. J. **574**, 332 (2002)
Torii, K., Kinugasa, K., Katayama, K., et al.: Astrophys. J. **503**, 843 (1998)
Vasisht, G., Gotthelf, E.V., Torii, K., et al.: Astrophys. J. **542**, L49 (2000)
Watanabe, C., Ohta, K., Akiyama, M., et al.: Astrophys. J. **610**, 128 (2004)
Woods, P.M., Kouveliotou, C., Gavriil, F.P., et al.: Astrophys. J. **629**, 985 (2005)

Astrophys Space Sci (2007) 308: 525–529
DOI 10.1007/s10509-007-9330-1

ORIGINAL ARTICLE

Spin-down of young pulsars with a fallback disk

X.-D. Li · Z.-B. Jiang

Received: 13 June 2006 / Accepted: 22 August 2006 / Published online: 21 March 2007

Abstract By performing Monte-Carlo calculations of the spin-down of young pulsars surrounded by a supernova fallback disk, we present possible constraints on the propeller torques exerted by the disks on neutron stars.

Keywords Neutron stars · Magnetic fields · Fallback disks

PACS 97.60.Bw · 97.60.Gb · 97.60.Jd

1 Introduction

The rotational power of pulsars is generally assumed to decrease with time in a power-law,

$$I\dot{\Omega} = -K\Omega^n \tag{1}$$

where I is the moment of inertia, Ω the angular velocity, and n the braking index of the neutron star respectively. In the standard magnetic dipole radiation (MDR) model $n = 3$. But the values of n measured so far are all less than 3 (Lyne et al. 1988; Kaspi et al. 1994; Lyne et al. 1996; Livingstone et al. 2006). Various models have been put forward to reconcile this discrepancy, one of which assumes that some young pulsars could be surrounded by a supernova fallback disk (Michel and Dessler 1981, 1983; Chattterjee et al. 2000; Alpar 2001; Menou et al. 2001; Marsden et al. 2001; Blackman and Perna 2004; see also Wang et al. 2006). In this paper we construct a model for fallback disk assisted spin-down in young radio pulsars. We describe the model in Sect. 2. In Sect. 3 we present the statistical characteristics of young pulsars with surrounding fallback disks by Monte Carlo simulation. Our discussion and conclusions are presented in Sect. 4.

This work was supported by the National Natural Science Foundation of China.

X.-D. Li (✉) · Z.-B. Jiang
Department of Astronomy, Nanjing University,
Nanjing 210093, China
e-mail: lixd@nju.edu.cn

2 The model

We consider a young neutron star surrounded by a disk formed through fullback of part of the supernova ejecta. We take the inner disk radius to be at the magnetospheric radius R_{m}, where the magnetic pressure is balanced by the ram pressure in the disk. According to Sturrock and Smith (1968) and Roberts and Sturrock (1973), at the inner radius of the disk, the field lines changes from closed to open, so that $B \propto r^{-3}$ for $R_* \le r \le R_{\mathrm{m}}$, and $B \propto r^{-2}$ for $R_{\mathrm{m}} \le r \le R_{\mathrm{LC}}$. So the magnetic spin down torque is then

$$T_{\mathrm{mag}} = \frac{1}{2} B_*^2 R_*^6 R_{\mathrm{m}}^{-2} R_{\mathrm{LC}}^{-1} \tag{2}$$

where $R_{\mathrm{LC}} = c/\Omega$ is the light cylindrical radius, B_* and R_* are the magnetic field strength and radius of the neutron star respectively.

Since no mass accretion occurs in radio pulsars, the inner disk radius must be larger than the so called corotation radius, i.e. $R_{\mathrm{m}} > R_{\mathrm{co}} \equiv (GM/\Omega^2)^{1/3}$. The infalling material is stopped at the magnetosphere by the centrifugal barrier, which prevents material from accreting onto the neutron star. The ejected material may carry away the angular momentum of neutron stars due to the propeller effect (Illarionov and Sunyave 1975), and decelerate its spin. There are many proposed propeller spin-down models, with large differences in the spin-down efficiency (Davies and Pringle 1980). The spin-down torque by the propeller action can be expressed in a general form as,

$$T_p = -\dot{M} R_m^2 \Omega_K(R_m) \left[\frac{\Omega}{\Omega_K(R_m)} \right]^\gamma \tag{3}$$

where $\dot{M}$ the mass inflow rate within the disk, $\Omega_K(R_m)$ is the Keplerian angular velocity at R_m, γ is a parameter reflecting the various mechanisms and efficiencies for the propeller effect with its value ranging from -1 to 2 (Alpar et al. 2001; Mori and Ruderman 2003; Eksi and Alpar 2003; Eksi et al. 2005; Ertan et al. 2006).

Note that in (3) $\dot{M}$ changes with time. Cannizzo et al. (1990) showed that after the supernova fallback, there is a transient spreading phase of timescale t_0 during which the accretion is nearly constant, $\dot{M}(t) \simeq \dot{M}(t_0)$, after which $\dot{M}$ declines in a power law with time,

$$\dot{M}(t) = \dot{M}(t_0)(t/t_0)^{-\alpha}, \tag{4}$$

with $t_0 \simeq 300$ s and $\alpha = 1.25$ (Francischelli et al. 2002).

The evolution of the star's rotation is determined by the sum of the energy outflow and the propeller spin-down torque,

$$\begin{aligned} I\dot{\Omega} &= T_{mag} + T_p \\ &= -\frac{B_*^2 R_*^6}{2R_m^2 R_{LC}} - \dot{M}(t) R_m^2 \Omega_K(R_m) \left[\frac{\Omega}{\Omega_K(R_m)} \right]^\gamma . \end{aligned} \tag{5}$$

The braking index n, the second braking index m, and the characteristic age T_{Ch} are defined correspondingly as

$$n(t) \equiv \frac{\ddot{\Omega}\Omega}{\dot{\Omega}^2}, \qquad m(t) \equiv \frac{\dddot{\Omega}\Omega^2}{\dot{\Omega}^3}, \qquad T_{Ch} \equiv -\frac{\Omega}{2\dot{\Omega}}. \tag{6}$$

In the following we use the measured values of these parameters for specific pulsars to constrain the possible range of the parameter γ in (5), as well as the magnetic field strength B_*, initial accretion $\dot{M}(t_0)$ and the initial spin period P_i of the pulsars.

3 Monte Carlo simulations

To examine in what extent the fallback disks affect the evolution of young radio pulsars, we did Monte Carlo simulations of the evolution of 2×10^6 neutron stars based on the spin-down model presented in Sect. 2. For the initial parameters of newborn neutron stars, we chose the initial magnetic fields B_* so that $\log B_*$ is distributed normally with a mean of 12.5 and a standard deviation of 0.3. We assumed that all pulsars were born with a surrounding supernova fallback disk, the initial masses of which $\log(\dot{M}_0 t_0/M_\odot)$ were distributed uniformly between -6 and -2. We set the initial spin periods P_i to be distributed uniformly between 10 ms and 100 ms. We stopped the calculations at a fiducial time of 10^4 yr, and counted the number of pulsars with $R_{co} < R_m < R_{LC}$ within 10^3 and 10^4 yr for various propeller models. The emerging rates of the disk-fed pulsars are 21.1%, 24.0%, 38.7%, 51.6% for $\gamma = -1, 0, 1, 2$ respectively at the age of 10^3 years, and are 9.3%, 9.7%, 15.4%, 25.3% for $\gamma = -1, 0, 1, 2$ at the age of 10^4 years. Obviously a disk is easier to survive for the propeller mechanisms with relatively large γ. The reason is that more efficient spin-down can cause the neutron star's spin period to increase more rapidly, so that the condition $R_{LC} > R_m$ can be satisfied for longer time.

In Fig. 1 we plot the histogram of the spin periods for disk-fed pulsars. The solid lines correspond to the age of 10^3 yr and the hatched regions for 10^4 yr. We note that the average value of $\log P$ increases with γ. In the case of $\gamma = 2$, most neutron stars have $P > 1$ s, in contrast with the periods of the pulsars with measured ages $< 10^4$ yr. Moreover the periods predicted are so large that the neutron stars may have extinguished or very low radio emission because they may have already passed the death line for radio pulsars. This is not unexpected, since the $\gamma = 2$ subsonic propeller torque actually works only when $R_m \leq R_{co}$ (Davies and Pringle 1980), which may not be applicable for radio pulsars. The $\gamma = 1$ model has the same problem, though not as severe as the $\gamma = 2$ one. For the other two models P are most likely to be distributed around 0.1 s, longer than the initial values by roughly one order of magnitude. Since 10^3–10^4 yr are much shorter than the current ages of most radio pulsars, these periods can be practically taken as the initial periods for pulsars driven by MDR. Recent pulsar analysis by Vranesevic et al. (2004) have shown that about 40% of the pulsars may be injected into the population with initial periods of 0.1–0.5 s (see also Vivekanand and Narayan 1981). The fallback disk involved evolution may be one of the reasons to account for this fact.

The distributions of the calculated braking indices n are shown in Fig. 2 at the ages of 10^3 and 10^4 yr. The solid, dashed, and dotted lines correspond to $\gamma = -1$, 0, and 1 respectively. In the $\gamma = 1$ case the values of n are strongly clustered around $\sim 1 - 2$. But for $\gamma = -1$ and 0, the distributions are much wider, though in the latter small values of n seem to be more populated. Current measured values of n for five pulsars are all less than 3. If this trend is confirmed by future measurements for other young pulsars, models with $\gamma \leq -1$ could be excluded.

Similar as in Fig. 2, we show the histogram of the ratio of T_{Ch} and T at ages of $T = 10^3$ and 10^4 years in Fig. 3. It is well known that some SNR ages deviate from the characteristic ages of the pulsars inside, which can be (partly) attributed to disk-involved neutron star evolution (e.g., Marsden et al. 2001). From the figure we see that when $\gamma = -1$,

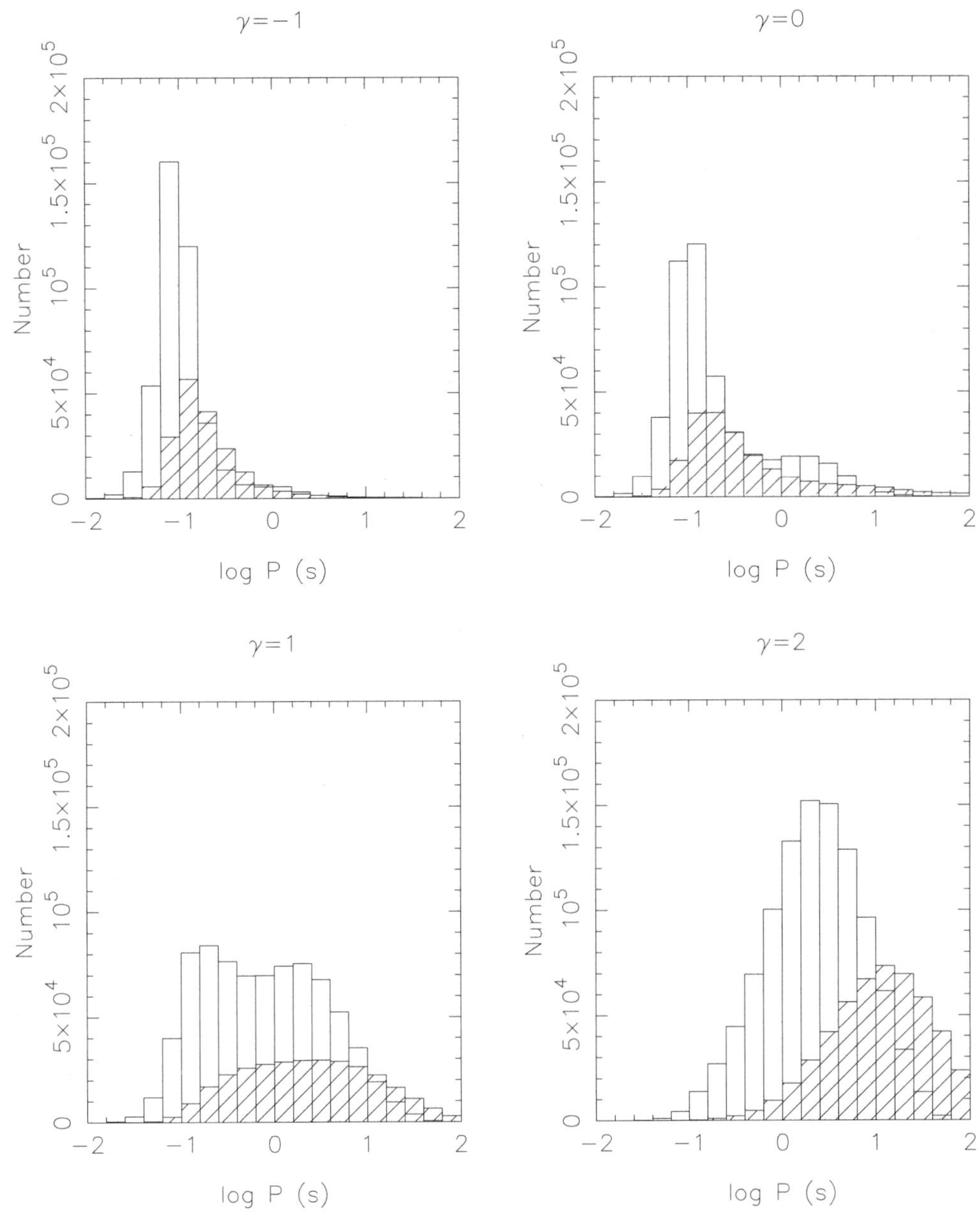

Fig. 1 Histograms of the periods for disk-fed pulsars in the propeller phase for various γ models. The *solid lines* correspond to the age of 10^3 years and *hatched regions* for 10^4 years

T_{Ch} is most likely to be larger than T, while T_{Ch} is generally less than T when $\gamma = 1$, with $\gamma = 0$ lying between them.

4 Discussion and conclusions

A disk-involved spin-down model for young pulsars is presented in this paper. The influence of the fallback disks on the spin evolution of pulsars is demonstrated in two aspects. Firstly it causes the magnetic fields of neutron stars to redistribute and deviate from the pure magnetic dipolar form. For example, for the Crab pulsar, to reproduce its age and brake index, we have $B_* = 1.5 \times 10^{11}$–7.1×10^{12} G and $\dot{M}(t_0) = 3.5 \times 10^{29}$–$1.5 \times 10^{25}$ gs^{-1} in the $\gamma = 1$ model; for PSR B0540–69 $B_* = 1.4 \times 10^{11}$–1.6×10^{11} G, and $\dot{M}(t_0) = 6.9 \times 10^{20}$–$1.1 \times 10^{29}$ gs^{-1} when $\gamma = 2$. Secondly it carries away the rotational energy of neutron stars with the operation of the propeller effect. The motivation of our work is two fold. Firstly, we try to examine whether the disk-involved pulsar evolution can reproduce the observed timing characteristics of young radio pulsars. Secondly, we may derive the overall features of pulsar populations if the fallback disks indeed exist. Although there have been many investigations on the propeller effects, the spin-down torque by a disk on rapidly rotating neutron stars are still unknown and there are many controversies in the literature. We wish that the observed quantities of radio pulsars may present valuable information about the propeller efficiency. The available data of young pulsars and our Monte Carlo simulations

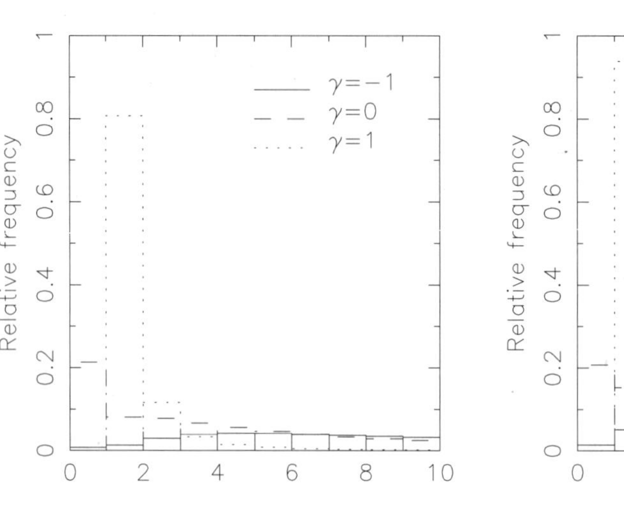

Fig. 2 Histograms of the braking indices n

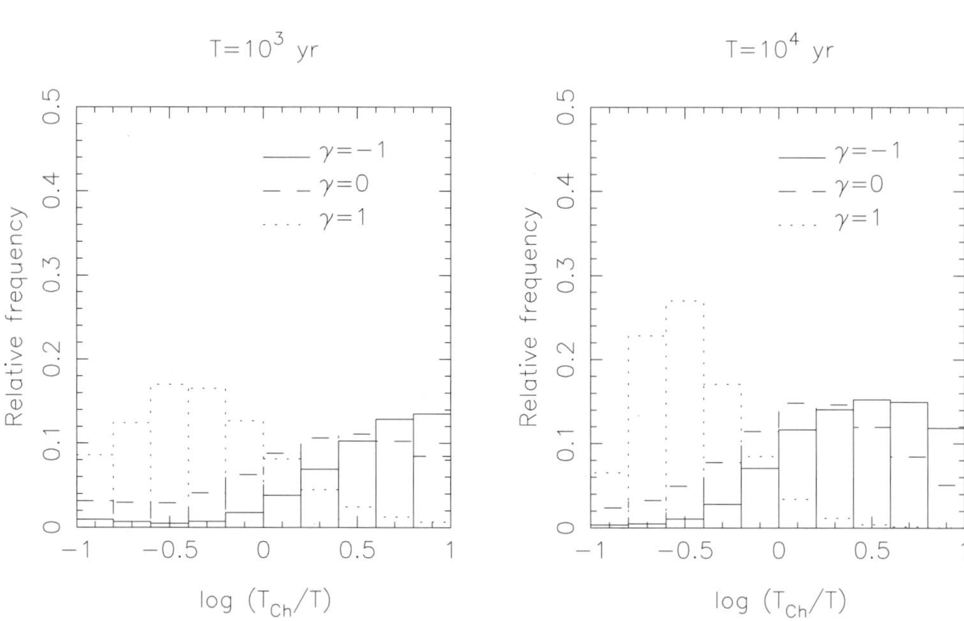

Fig. 3 Histograms of the ratio of T_{Ch} and the true ages T

show that the models with $\gamma \leq -1$ and $\gamma \geq 2$ are probably not viable, implying that the propeller effect may operate through exchange of angular momentum rather through exchange of energy between magnetized neutron stars and the accreting gas (see, however, Wang and Robertson 1985). More timing measurements of young pulsars are required before a firm conclusion is reached.

References

Alpar, M.A.: Astrophys. J. **554**, 1245 (2001)
Alpar, M.A., Ankay, A., Yazgan, E.: Astrophys. J. **557**, L61 (2001)
Blackman, E.G., Perna, R.: Astrophys. J. **601**, L71 (2004)
Cannizzo, J.K., Lee, H.M., Goodman, J.: Astrophys. J. **351**, 38 (1990)
Chattterjee, P., Hernquist, L., Narayan, R.: Astrophys. J. **534**, 373 (2000)
Davies, R.E., Pringle, J.E.: Mon. Not. Roy. Astron. Soc. **191**, 599 (1980)
Eksi, K.Y., Alpar, M.A.: Astrophys. J. **599**, 450 (2003)
Eksi, K.Y., Hernquist, L., Narayan, R.: Astrophys. J. **623**, L41 (2005)
Ertan, Ü., Erkut, M.H., Eksi, K.Y., et al.: Preprint, astro-ph 0606259 (2006)
Francischelli, G.J., Wijers, R.A.M.J., Brown, G.E.: Astrophys. J. **565**, 471 (2002)
Illarionov, A.F., Sunyave, R.A.: Astron. Astrophys. **39**, 185 (1975)
Kaspi, V.M., Manchester, R.N., Siegman, B., et al.: Astrophys. J. **422**, L83 (1994)
Livingstone, M.A., Kaspi, V.M., Gotthelf, E.V., et al.: Astrophys. J. **647**, 1286 (2006)
Lyne, A.G., Pritchard, R.S., Smith, F.G.: Mon. Not. Roy. Astron. Soc. **233**, 667 (1988)
Lyne, A.G., Pritchard, R.S., Smith, F.G., et al.: Nature **381**, 497 (1996)

Marsden, D., Lingenfelter, R.E., Rothschild, R.E.: Astrophys. J. **547**, L45 (2001)
Menou, K., Perna, R., Hernquist, L.: Astrophys. J. **554**, L63 (2001)
Michel, F.C., Dessler, A.J.: Astrophys. J. **251**, 54 (1981)
Michel, F.C., Dessler, A.J.: Nature **303**, 48 (1983)
Mori, K., Ruderman, M.A.: Astrophys. J. **592**, L75 (2003)
Roberts, D.H., Sturrock, P.A.: Astrophys. J. **181**, 161 (1973)
Sturrock, P.A., Smith, S.M.: Solar Phys. **5**, 87 (1968)
Vivekanand, M., Narayan, R.: J. Astrophys. Astron. **2**, 315 (1981)
Vranesevic, N., Manchester, R.N., Lorimer, D.R., et al.: Astrophys. J. **617**, L139 (2004)
Wang, Y.-M., Robertson, J.A.: Astron. Astrophys. **151**, 361 (1985)
Wang, Z., Chakrabarty, D., Kaplan, D.L.: Nature **440**, 772 (2006)

Astrophys Space Sci (2007) 308: 531–534
DOI 10.1007/s10509-007-9353-7

ORIGINAL ARTICLE

Search for radio pulsations in four anomalous X-ray pulsars and discovery of two new pulsars

Marta Burgay · Nanda Rea · GianLuca Israel · Andrea Possenti · Luciano Burderi · Tiziana Di Salvo · Nichi D'Amico · Luigi Stella · Elisa Nichelli

Received: 12 July 2006 / Accepted: 29 August 2006 / Published online: 21 March 2007

Abstract We have performed deep searches for radio pulsations from four southern anomalous X-ray pulsars (AXPs) to investigate their physical nature in comparison with the rotation powered pulsars. The data were acquired using the Parkes radio telescope with the 1.4 GHz multibeam receiver. No pulsed emission with periodicity matching the X-ray ephemeris have been found in the observed targets down to a limit of ∼0.1 mJy. A blind search has also been performed on all the 13 beams of the multibeam receiver (the central beam being pointed on the target AXP), leading to the serendipitous discovery of two new radio pulsars and to the further detection of 18 pulsars. Also a search for single dispersed pulses has been performed in the aim to detect signals similar to those of the recently discovered rotating radio transients.

M. Burgay (✉) · A. Possenti
INAF, Osservatorio Astronomico di Cagliari, Loc. Poggio dei Pini, Strada 54, 09012 Capoterra, CA, Italy
e-mail: burgay@ca.astro.it

N. Rea
SRON, Netherlands Institute for Space Research, Sorbonnelaan, 2, 3584 CA, Utrecht, The Netherlands

G.L. Israel · L. Stella · E. Nichelli
INAF, Osservatorio Astronomico di Roma, via di Frascati 33, Monte Porzio Catone, Roma, Italy

L. Burderi · N. D'Amico
Dipartimento di Fisica dell'Università di Cagliari, S.P. Monserrato-Sestu Km 0.700, 09042 Monserrato, CA, Italy

T. Di Salvo
Dipartimento di Scienze Fisiche e Astronomiche dell'Università di Palermo, via Archirafi 36, 90134 Palermo, Italy

Keywords AXPs: individual (1E 1048.1-5937, 1RXS J170849.0-400910, 1E 1841-045, AX J1845-0258) · Pulsar

PACS 97.60.Jd · 97.60.Gb

1 Introduction

In the last two decades, a significant amount of observational and theoretical effort has been dedicated towards the understanding of an unusual class of X-ray pulsars, namely the Anomalous X-Ray pulsars (AXPs) and the Soft Gamma-Ray Repeaters (SGRs). These relatively bright X-ray sources were soon recognized as a distinct class of neutron stars with respect to the well known radio pulsars or X-ray binary populations. In particular, they are not pulsating in the radio band, their rotational energy loss alone is not sufficient to power the observed X-ray luminosity, and there is no evidence for a companion star.

Instead, the X-ray emission from these sources is thought to be powered by the decay of the ultra-strong magnetic field according to the magnetar theory (Damour and Taylor 1992; Thorsett and Dewey 1993; Thompson and Duncan 1995). This is suggested by the fact that, for an isolated neutron stars characterised as a magnetic dipole, their inferred magnetic fields are extremely high ($B \sim 10^{14}$–10^{15} Gauss). These field strengths lies above the so-called "quantum critical field" $B_c = 4.4 \times 10^{13}$ G, above which the radio emission of pulsars is expected to be suppressed by processes such as photon splitting, which may inhibit pair-production cascades (Baring and Harding 1998). However, the efficacy of this process is put in doubt by the discovery of high magnetic field radio pulsars (Camilo et al. 2000; McLaughlin et al. 2003) and transient radio pulsed emission from at least one AXP (Camilo et al. 2006).

In the past years searches for radio pulsations from AXPs and SGRs have always given negative results (e.g. Crawford et al. 2002). However, in two peculiar circumstances radio emission have been detected (a) following the rare and highly energetic giant flare events from SGRs (Frail et al. 1999; Cameron et al. 2005; Gaensler et al. 2005), and (b) during the outburst of the only confirmed example of a transient AXP, XTE J1810-197. In the former case, the radio emission slowly fades on a timescale of weeks and in the later, the radio emission, initially discovered as a VLA point source (with angular resolution of 6″) of flux density 4.5±0.5 mJy at 1.4 GHz (Halpern et al. 2005), has also been recently found to be pulsed at the X-ray period (Camilo et al. 2006).

With this picture in mind we have undertaken a systematic deep search for radio pulsations in three confirmed (1RXS J170849.0-400910, 1E 1048.1-5937, 1E 1841-045) and one candidate AXP (AX J1845-0258), visible from the southern hemisphere using the Parkes radio telescope. Observations and data analysis are described in Sect. 2 while in Sect. 3 we report on the obtained results, discussed in Sect. 4.

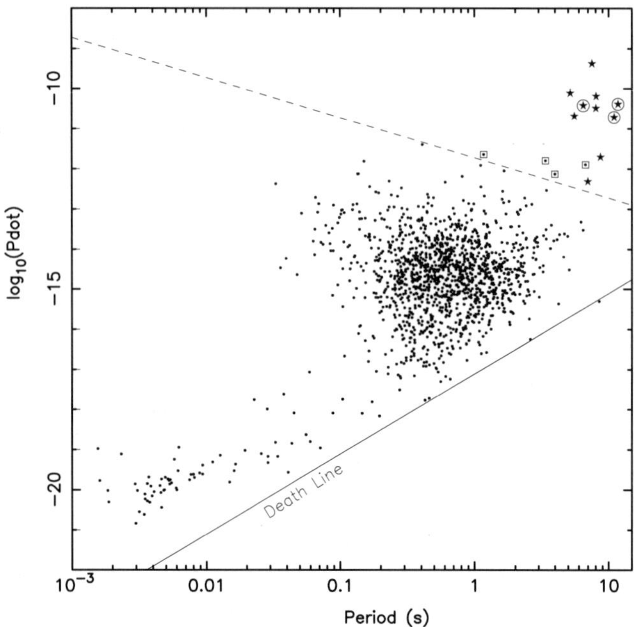

Fig. 1 P–$\dot{P}$ diagram for radio pulsars (*dots*) and magnetars (*stars*). The *dashed line* represents the limit for radio emission quenching due to photon splitting (Baring and Harding 1998). The four sources surrounded by a square emit in radio besides being above this line. The three confirmed AXPs of our sample are surrounded by a circle

2 Observations and data analysis

In Table 1 the sample of observed AXPs is presented. For each object the name, period and period derivative of X-ray pulsations, the derived surface magnetic field, the estimated distance D and the association with a SNR or with a nebula are listed.

The radio observations have been performed between October 12 and 15 1999 using the 13 beams of the multibeam receiver (Staveley-Smith et al. 1996) of the Parkes radio telescope (NSW Australia) at a frequency of 1374 MHz, with the central beam pointed at the target AXP. In order to mitigate the effects of the dispersion of the signals in the interstellar medium (ISM), the total 288 MHz bandwidth is split into 96 frequency channels each 3-MHz wide. The outputs from each channel are summed in polarisation pairs, high-pass filtered and 1-bit sampled every 1.0 or 1.2 ms. Each source has been pointed for 2.8 hours and for all but 1E1841-0450 the observation has been performed twice.

Data from the central beam of the multibeam receiver (pointed on the AXPs) have been analysed using the programme pdm: the code takes as input a period P and a dispersion measure DM and folds the time series according to a number of trial values around the input ones, searching for the combination of P and DM for which the signal-to-noise ratio S/N is maximised. The period range searched for each source has been obtained from the X-ray ephemeris and their

Table 1 For each of the four AXPs of our sample columns 2 and 3 report the most recent period and period derivative from the X-ray timing (1E1048: Tiengo et al. 2005, 1RXS J1708: Rea et al. 2005, 1E 1841: Gotthelf et al. 2002, AXJ 1845: Gotthelf and Vasisht 1998), column 4 is the surface magnetic field derived from the dipole formula $B = 3.2 \times 10^{19}\sqrt{P\dot{P}}$, columns 5 and 6 report the distance estimated and the relative dispersion measure derived adopting the Cordes and Lazio (2002) model for the distribution of free electrons in the ISM (the values obtained using Taylor and Cordes (1993) model are similar) and last column shows the association with supernova remnant or with hydrogen bubble (Gaensler et al. 2001; 2005). Numbers in parentheses are the errors on the last quoted digit

Name	P (s)	$\dot{P}$ (10^{-11} s s^{-1})	B (10^{14} G)	D (kpc)	DM$_{nom}$ (pc cm^{-3})	Association
1E 1048.1-5937	6.456109(5)	3.3	4.7	5	279.7	HI bubble
1RXS J170849.0-400910	11.00170(4)	1.9	4.6	8	742.5	–
1E 1841-045	11.77505(5)	4.1	7.0	7	529.7	G27.4+0.0
AX J1845-0258	6.9712(1)	–	–	8	646.1	G29.6+0.1

errors (see Table 1). A number of trial DM values, ranging from 0 up to the value giving a maximum broadening of the pulse of 10% of the period, has been explored.

The observations relative to all the 13 beams have been subsequently analysed with the blind search code `pmsearch` (an FFT based code; see (Manchester et al. 2001) to search for possible radio counterparts of the AXPs pulsating at a different period with respect to the one measured in X-rays and to search for new radio pulsars in the vicinity of our main targets.

Finally a code sensitive to strong single dispersed pulses (Cordes and McLaughlin 2003) has been applied to the 13 beams of each pointing with the purpose to discover signals similar to those seen in the class of the rotating radio transients (RRATs, McLaughlin et al. 2006). The aim was to test if AXPs have RRAT-like emission and to search single pulses from nearby sources.

3 Results

No radio pulsation has been found in the four AXPs observed down to a limit of ∼0.06–0.1 mJy (depending on the source; see Table 2).

Table 2 For each observed AXP we report here the period range and maximum DM explored in the `pdm` search (columns 2 and 3) and the upper limits on the flux densities (column 4)

Name	P range (s)	$\mathrm{DM_{max}}$ ($\mathrm{pc\,cm^{-3}}$)	S_{min} mJy
1E 1048.1-5937	6.0438–7.2524	62754	0.11
1RXS J170849.0-400910	9.6247–12.3734	107069	0.06
1E 1841-045	10.3026–13.2451	114611	0.06
AX J1845-0258	6.0999–7.8426	67861	0.06

Folding the time series for 1E 1048.1-5937 with a trial period of 6.16 s and a DM of 334 $\mathrm{pc\,cm^{-3}}$, the pulsar J1058-5957 (Kramer et al. 2003) has been detected through its tenth sub-harmonic, confirming the reliability of the detection algorithm used.

Analysis of the data collected in all the 13 beams of the Parkes multibeam receiver used during this work led to the serendipitous discovery of two new radio pulsars, both detected in the same beam and hence having the provisional names J1712-3943-1 and J1712-3943-2. The first pulsar has a spin period of 0.78 s and a dispersion measure of 525 $\mathrm{pc\,cm^{-3}}$, while the second source has a period of 92.5 ms and a DM of 713 $\mathrm{pc\,cm^{-3}}$. The average pulse profiles of the new pulsars, obtained summing the data relative to the detection and confirmation observations, are shown in Fig. 2.

The `pmsearch` code also detected 18 previously known pulsars present in the pointed areas.

Finally no new RRAT-like signal has been found in the single pulse search over the 52 observed regions, while two previously known pulsars (J1705-3950 and J1844-0433) have been detected also through their single pulses.

4 Discussion and conclusions

No radio pulsation has been found from the four southern AXPs observed, down to a limit of $S_{\mathrm{min}} \sim 0.1$ mJy (Table 2).

Comparing the upper limits on the luminosity at 1400 MHz (defined as $L_{1400} = S_{\mathrm{min}} \times D^2$) for our targets with the luminosity of the observed radio pulsar population in the Galactic field we note that the limits reached by our search are a factor of six lower than the median of the population. Note however that, if we compare our results with the *intrinsic* luminosity distribution of the radio pulsars we

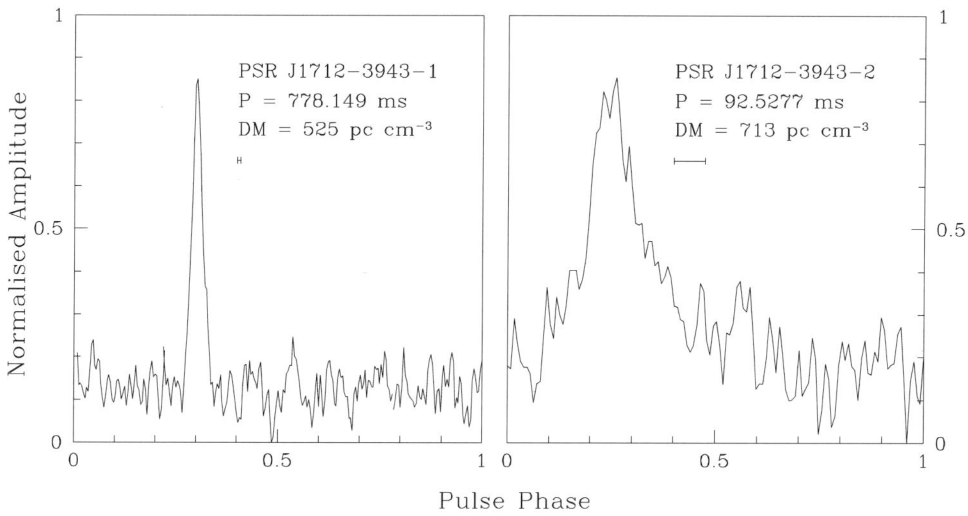

Fig. 2 Mean 1374-MHz pulse profiles of the newly discovered pulsars J1712-3943-1 (*left*) and J1712-3943-2 (*right*) obtained by adding the data of the discovery and confirmation observations. The maximum of each profile is placed at phase 0.3 (the range on the x-axis of each panel being 0 to 1). For each profile, the pulsar name, period and DM are given. The small horizontal bar drawn under the DM indicates the effective resolution of the profile, calculated by adding the bin size, the sampling time and the effects of interstellar dispersion in quadrature

obtain a probability of ∼76% that our observations are not deep enough to detect a radio pulsed signal from our targets.

In considering the causes of the non detection of a radio pulsed signal from our four targets, besides the luminosity bias, we must take into account the possibility that, although the X-ray beam is pointing toward us, the radio beam, usually narrower, is not. Assuming a pulse duty cycle of ∼5% typical of long period pulsars and similar to that of the radio pulsed signal detected by Camilo et al. (2006) in the transient AXP XTE J1810-197 (∼4% at 1.4 GHz), we can calculate, that the probability that such a narrow radio beam misses the earth is ≤77%. The composite probability that the beams of all four AXPs are not pointing toward us is hence ≤34%.

The non detection of RRAT-like bursts from any of these AXPs, despite our long exposures, seems to weaken the hypothesis that RRAT bursts might be related to the short bursts observed from the magnetars leaving us with other plausible conjectures of a relation with other classes of neutron stars such as middle aged radio pulsars displaying giant pulses (Reynolds et al. 2006; Weltevrede et al. 2006) or with X-ray Dim Isolated Neutron Stars (Popov et al. 2006).

The only case of a detection of radio pulsations from an AXP concerns the only confirmed transient magnetar XTE J1810-197 (Camilo et al. 2006). Radio emission from this source is strongly related with the occurrence of an outburst of its X-ray emission (Halpern et al. 2005), as well as an IR enhancement (Rea et al. 2004). Furthermore, whereas the X-ray flux is decaying exponentially with timescale of a few hundreds days (Gotthelf and Halpern 2005), XTE J1810-197 radio emission is still on more than 3 years after the X-ray outburst. Interestingly the sole other possible transient AXP is the candidate AX J1845-0258, one of our targets. Our radio observations of this source were performed more than six years after its possible X-ray outburst occurred in 1993, hence unfortunately nothing can be safely concluded from our upper limits, in favor or against the possible radio and X-ray correlation during the outbursts of this source. However, assuming that AX J1845-0258 experienced, after the X-ray outburst, a phase of radio emission similar to that of XTE J1810-197, our null detection implies that the fading of the radio emission has a time scale of the order of few years: in particular, if AX J1845-0258 at the onset of its putative radio emission phase had a similar luminosity as XTE J1810-197, this would imply a decrease in L_{1400} of a factor of ∼20 over six years.

Acknowledgements M.B., A.P., L.B., T.D.S. and N.D.A. received support from the Italian Ministry of University and Research (MIUR) under the national program *PRIN-MIUR 2005*. NR is supported by an NWO post-doctoral fellow. The Parkes radio telescope is part of the Australia Telescope which is funded by the Commonwealth of Australia for operation as a National Facility managed by CSIRO.

References

Baring, M.G., Harding, A.K.: Astrophys. J. **507**, L55 (1998)

Cameron, P.B., Chandra, P., Ray, A., et al.: Nature **434**, 1112 (2005)

Camilo, F., Kaspi, V.M., Lyne, A.G., et al.: Astrophys. J. **541**, 367 (2000)

Camilo, F., Ransom, S.M., Halpern, J.P., Reynolds, J., Helfand, D.J., Zimmerman, N., Sarkissian, J.: Nature **442**, 892 (2006)

Cordes, J.M., Lazio, T.J.W.: preprint, astro-ph/0207156 (2002)

Cordes, J.M., McLaughlin, M.A.: Astrophys. J. **596**, 1142 (2003)

Crawford, F., Pivovaroff, M.J., Kaspi, V.M., Manchester, R.N.: In: Slane, P., Gaensler, B. (eds.) Neutron Stars in Supernova Remnants A Sensitive Targeted Search Campaign at Parkes to Find Young Radio Pulsars at 20 cm. ASP Conf. Ser., vol. 271, pp. 37 (2002)

Damour, T., Taylor, J.H.: Phys. Rev. D **45**, 1840 (1992)

Frail, D.A., Kulkarni, S.R., Bloom, J.S.: Nature **398**, 127 (1999)

Gaensler, B.M., Slane, P.O., Gotthelf, E.V., et al.: Astrophys. J. **559**, 963 (2001)

Gaensler, B.M., Kouveliotou, C., Gelfand, J.D., et al.: Nature **434**, 1104 (2005)

Gaensler, B.M., McClure-Griffiths, N.M., Oey, M.S., et al.: Astrophys. J. **620**, L95 (2005)

Gotthelf, E.V., Halpern, J.P.: Astrophys. J. **632**, 1075 (2005)

Gotthelf, E.V., Vasisht, G.: New Astron. **3**, 293 (1998)

Gotthelf, E.V., Gavriil, F.P., Kaspi, V.M., et al.: Astrophys. J. **564**, L31 (2002)

Halpern, J.P., Gotthelf, E.V., Becker, R.H., et al.: Astrophys. J. **632**, L29 (2005)

Kramer, M., Bell, J.F., Manchester, R.N., et al.: Mon. Not. Roy. Astron. Soc. **342**, 1299 (2003)

Manchester, R.N., Lyne, A.G., Camilo, F., Bell, J.F., Kaspi, V.M., D'Amico, N., McKay, N.P.F., Crawford, F., Stairs, I.H., Possenti, A., Kramer, M., Sheppard, D.C.: Mon. Not. Roy. Astron. Soc. **328**, 17 (2001)

McLaughlin, M.A., Stairs, I.H., Kaspi, V.M., et al.: Astrophys. J. **591**, L135 (2003)

McLaughlin, M.A., Lyne, A.G., Lorimer, D.R., et al.: Nature **439**, 817 (2006)

Popov, S.B., Turolla, R., Possenti, A.: Mon. Not. Roy. Astron. Soc. **369**, L23 (2006)

Reynolds, S.P., Borkowski, K.J., Gaensler, et al.: Astrophys. J. **639**, L71 (2006)

Rea, N., Testa, V., Israel, G.L., et al.: Astron. Astrophys. **425**, L5 (2004)

Rea, N., Oosterbroek, T., Zane, S., et al.: Mon. Not. Roy. Astron. Soc. **361**, 710 (2005)

Staveley-Smith, L., Wilson, W.E., Bird, T.S., et al.: Proc. Astron. Soc. Aust. **13**, 243 (1996)

Taylor, J.H., Cordes, J.M.: Astrophys. J. **411**, 674 (1993)

Thompson, C., Duncan, R.C.: Mon. Not. Roy. Astron. Soc. **275**, 255 (1995)

Thorsett, S.E., Dewey, R.J.: Astrophys. J. **419**, L65 (1993)

Tiengo, A., Mereghetti, S., Turolla, R., et al.: Astron. Asrophys. **437**, 997 (2005)

Weltevrede, P., Stappers, B.W., Rankin, J.M., et al.: Astrophys. J. **645**, L149 (2006)

Astrophys Space Sci (2007) 308: 535–539
DOI 10.1007/s10509-007-9345-7

ORIGINAL ARTICLE

On the iron interpretation of the 6.4 keV emission line from SGR 1900+14

Alaa I. Ibrahim · Hisham Anwer · Mohamed H. Soliman · Nicholas Mackie-Jones · Kalvir S. Dhuga · William C. Parke · Jean H. Swank · Tilan Ukwatta · M.T. Hussein · T. El-Sherbini

Received: 21 July 2006 / Accepted: 25 September 2006 / Published online: 20 March 2007

Abstract A 6.4 keV emission line was discovered in an unusual burst from the soft gamma repeater SGR 1900+14 with the Rossi X-ray Timing Explorer (RXTE). The line was detected in part of a complex multipeak precursor that preceded the unusual burst of 1998 August 29, i.e. two days after the giant flare of August 27 from the source. The origin of the line was not firmly identified and two possible interpretations were equally plausible including (a) Kα fluorescence from a small iron rich material that was ejected to the magnetosphere during the August 27 flare, and (b) proton or α-particle cyclotron resonance. If the iron scenario was correct, we expect to find evidence for the line during the intervening interval between the flare and the August 29 burst, i.e. on August 28. Here we present the results of the August 28 burst observation, taken with RXTE. We detect a total of seven bursts whose individual and joint spectra do not show evidence for spectral lines. We also investigated a sample of nine bursts before and after the August 29 burst (from 1998 June to December) that do not reveal evidence for a spectral line near 6.4 keV or elsewhere. These results disfavor the iron scenario and make the proton/α-particle cyclotron resonance interpretation more plausible. The appearance of the emission line in part of a complex burst and its absence from the studied sample indicate that the line is likely due to a transient phenomenon that may depend on the burst morphology, energetics and the properties of the emission region.

Keywords Gamma-rays: observations · X-rays: observations · Pulsars: individual SGR 1900+14 · Pulsars: general

PACS 95.85.Pw · 95.85.Nv · 96.12.Hg · 97.60.Gb

A.I. Ibrahim (✉) · H. Anwer · M.H. Soliman · M.T. Hussein · T. El-Sherbini
Department of Physics, Faculty of Science, Cairo University, Cairo, Egypt
e-mail: alaa@gwu.edu

A.I. Ibrahim · H. Anwer · M.H. Soliman · M.T. Hussein · T. El-Sherbini
Center for Advanced Interdisciplinary Sciences, Astrophysics Division, Faculty of Science, Cairo University, Cairo, Egypt

A.I. Ibrahim · N. Mackie-Jones · K.S. Dhuga · W.C. Parke · T. Ukwatta
Department of Physics, George Washington University, Washington, DC 20052, USA

J.H. Swank
Astrophysics Science Division, NASA Goddard Space Flight Center, Greenbelt, MD 20771, USA

1 Introduction

SGR 1900+14 entered an active outburst episode in the summer of 1998 that started in May and culminated in August 27 when it emitted a giant flare (Hurley et al. 1999). Two days later, RXTE detected another unusual but less energetic burst from the source that showed a number of unique features including a complex, long (0.9 s) precursor, a strong 3.5 s peak, and a 1000 s long pulsating tail modulated at the neutron star 5.16 s rotational period (Ibrahim et al. 2001). A 6.4 keV emission line was also discovered in the first two peaks of the precursor where the spectrum was harder than the rest of the precursor (Strohmayer and Ibrahim 2000). The line was narrow (<1 keV), had a considerable equivalent width ($\sim$400 eV), and was significant at the 4σ level. The proposed interpretations were iron fluorescence or proton/α-particle cyclotron resonance. The iron

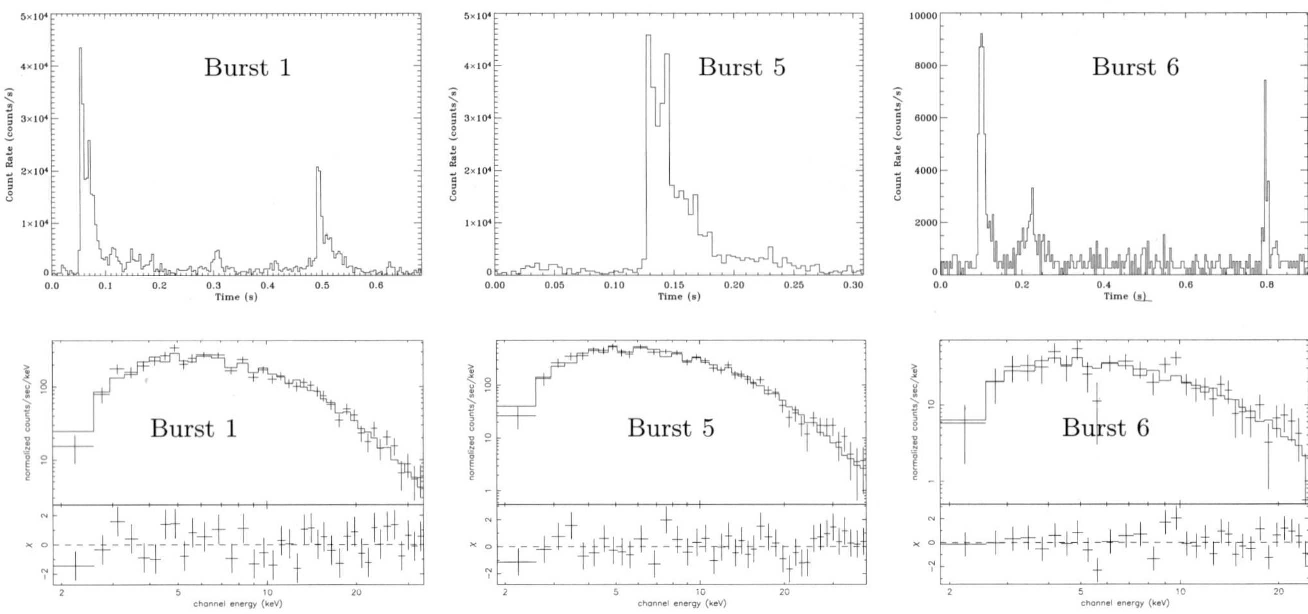

Fig. 1 Three bursts from SGR 1900+14 emitted on 1998 August 28. The *top panels* show the light curves of the bursts and the *lower panels* show the individual spectral continua and the fit residuals. No evidence of a spectral feature near 6.4 keV is present

scenario due to a fossil/fall back disk near the SGR (e.g. Marsden et al. 2001) is now implausible due to the optical and IR observations of the source that rule out such disks (Kaplan et al. 2002). Here we focus on the iron scenario that could result from material ejected to the magnetosphere by the August 27 giant flare. If the line is indeed due to such a mechanism, there should be evidence for the emission line during the August 28 activity. We study the RXTE observation of the bursts emitted during that day. We also search for the line in a sample of bursts during that active period of the source that spans from 1998 May to December.

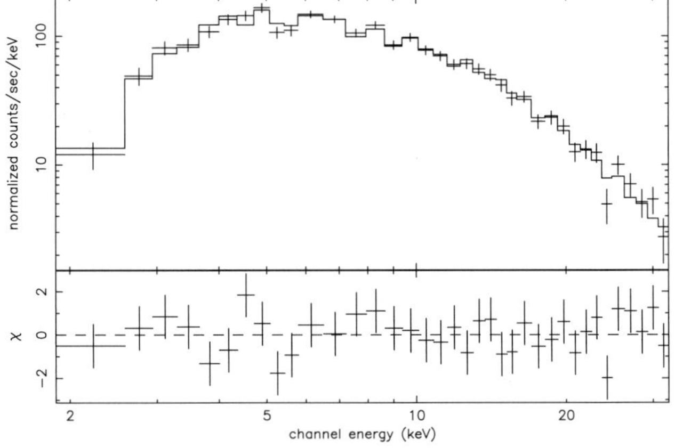

Fig. 2 The joint spectrum of all seven bursts of August 28 fitted to the absorbed Bremss model. The fit residuals show no evidence of a spectral feature near 6.4 keV

2 Observation and data analysis

2.1 The 1998 August 28 observation

We obtained the RXTE Proportional Counter Array (PCA) data of the August 28 observation and used HEASoft/Ftools 6.06 to reduce the data and obtain the burst light curves and spectra. The 5 individual detectors of the PCA were operating normally and their signals were jointly co-added.

The data were recorded in the event-mode configuration, with 125 μs time resolution and 64 energy channels, covering 2–60 keV. We found a total of 8 bursts one of which had count rate exceeding 1.5×10^5 counts s^{-1} and was excluded due to significant pileup and dead-time effects. Four other bursts were too dim (containing less than 400 counts) and were excluded from the individual spectral fits. The light curves of the remaining three bursts are shown in Fig. 1. We accumulated the individual spectra of such three bursts and performed the spectral fitting with XSPEC 11 where we found good fits with an optically thin thermal bremsstrahlung model (Bremss), with photoelectric absorption (this is the same continuum model that was used to fit the August 29 burst). The residuals of the spectral fits showed no evidence for a spectral line near 6.4 keV (see Fig. 1). To amplify a possibly weak feature we added all seven bursts together and fit the joint spectrum to the Bremss model. We obtained an acceptable fit with no evidence for a line spectrum at 6.4 keV or elsewhere as shown in Fig. 2. The spectral parameters of the fits are given in Table 1.

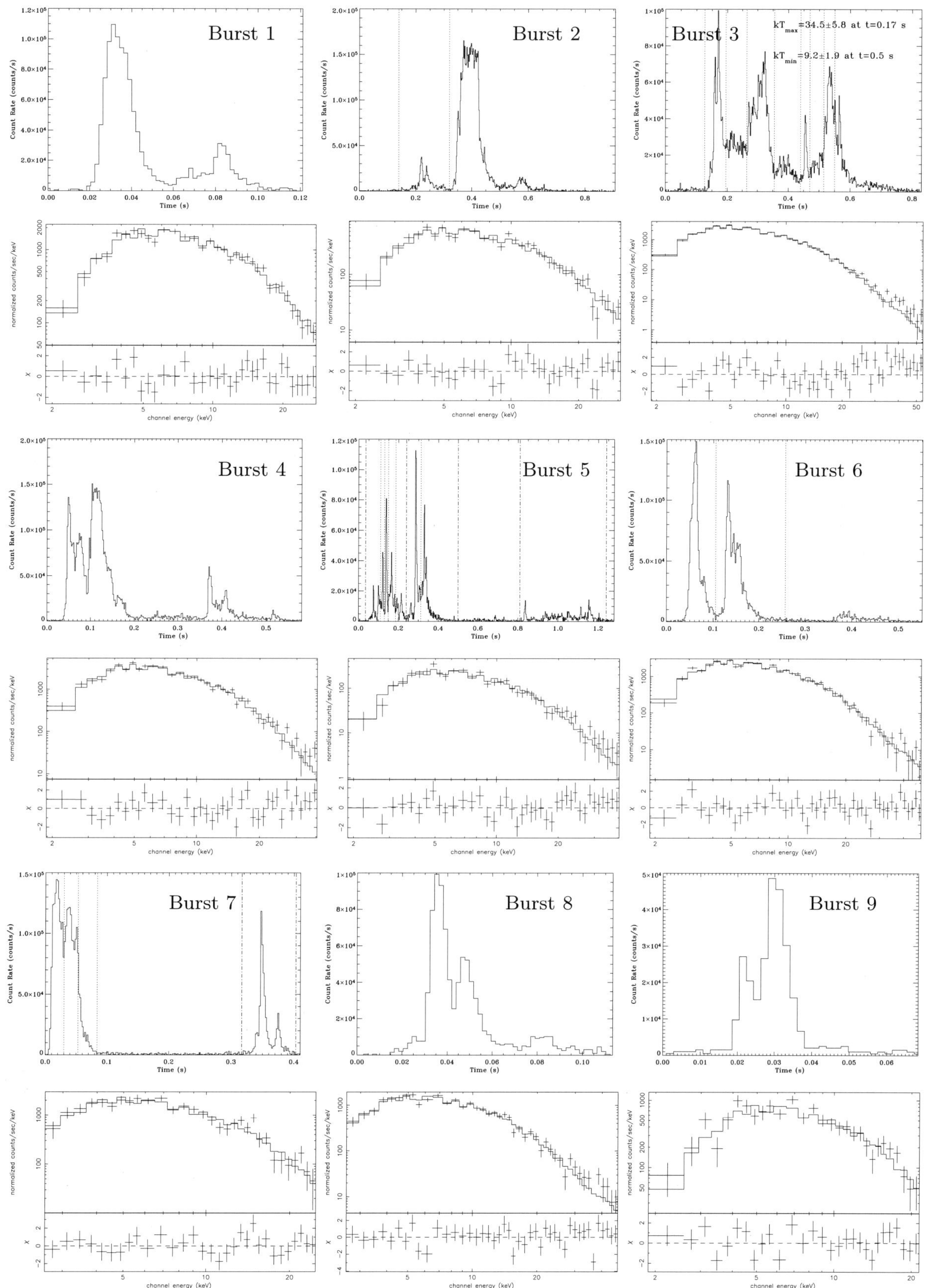

Fig. 3 The light curves and spectral continua of 1998 June to December burst sample. The spectrum under the light curve represents the whole burst for bursts 1, 3, 8, and 9; the first interval for burst 2; and the last interval for burst 4, 5, 6, and 7. Intervals where the count rate exceeded 90 000 counts s^{-1} were omitted

Table 1 Spectral parameters of SGR 1900+14 bursts

Burst	Interval	Date	Time[a] $T - T_0$ (s)	kT (keV)	N_H (10^{22} cm^{-2})	χ^2_ν	dof	Flux[b] ($\times 10^{-8}$) erg cm^{-2} s^{-1}
1	whole	28 Aug 1998	7493479.457	68.8 ± 22.8	7.32 ± 1.33	1.01	35	1.37
5	whole	28 Aug 1998	7494961.878	29.1 ± 5.55	9.63 ± 1.52	0.80	38	2.11
6	whole	28 Aug 1998	7495836.739	14.5 ± 8.95	26.8 ± 11.4	1.37	26	0.39
1 to 7	whole	28 Aug 1998	–	69.0 ± 15.9	6.83 ± 0.88	0.83	34	73.0
1	whole	02 Jun 1998	0.0	68.8 ± 21.5	10.2 ± 1.37	0.91	31	10.2
2	first	30 Aug 1998	7642977.103	62.5 ± 23.8	7.15 ± 1.59	0.93	33	3.24
3	whole	31 Aug 1998	7748171.200	16.2 ± 0.67	6.55 ± 0.39	1.79	45	7.62
4	last	01 Sep 1998	7816236.764	29.2 ± 5.24	5.58 ± 1.19	0.94	41	7.03
5	last	01 Sep 1998	7833155.470	47.4 ± 16.4	8.69 ± 1.82	0.85	38	1.18
6	last	01 Sep 1998	7834521.308	23.4 ± 2.25	7.66 ± 0.85	1.02	43	8.99
7	last	01 Sep 1998	7835905.930	47.3 ± 12.9	10.1 ± 1.71	1.28	38	8.53
8	whole	27 Nov 1998	15351931.767	35.2 ± 6.58	10.2 ± 1.48	1.12	41	7.16
9	whole	21 Dec 1998	17401208.206	17.1 ± 5.07	14.3 ± 3.35	1.14	27	3.10

[a] $T_0 = 139434282.8126764297$ s, RXTE Mission Elapsed Time

[b] Flux is 2–50 keV, unabsorbed

2.2 The 1998 June–December observations

We then selected a sample of bursts that covers the full range of the source's activity as observed with RXTE. We found that the great majority of the bursts were emitted during August and September while a few bursts were scattered in June, October, November and December.

We did not find bursts that match the light curve morphology of the August 29 precursor and selected the sample such that it contains the various temporal structures (see Fig. 3). We divided the complex bursts into small intervals and fit each interval separately as well as the whole burst. We did not find emission or absorption lines with this time-resolved spectroscopy but noticed significant changes in the spectral hardness from burst to burst and within the same burst. As shown in Table 1, the Bremss kT varied among all individual bursts from 68.8 ± 22.8 to 14.5 ± 8.95 keV while N_H varied from $7.32 \pm 1.33 \times 10^{22}$ to $26.8 \pm 11.4 \times 10^{22}$ cm^{-2}, showing anti-correlation with kT. Within the same burst (e.g. burst 3 of Fig. 3), we also find a significant spectral evolution as indicated on the light curve panel.

3 Conclusion

We have examined a constraining observation for the 6.4 keV emission line from SGR 1900+14. In light of the proposed interpretations by Strohmayer and Ibrahim (2000), the lack of evidence for the line in the August 28 observation argues against the iron scenario and favors the proton/α-particle interpretation. The 9 additional bursts we studied had different light curve morphology but none of them matched the duration or the structure of the August 29 precursor in which the line was detected. However, burst 3 of Fig. 3 exhibited a significant spectral evolution where the Bremss kT varied between 34.5 ± 5.8 keV near the onset of the burst to 9.2 ± 1.99 keV towards its end. This pattern is similar to that seen in the August 29 precursor where kT varied from 42.6 ± 9.2 keV at the beginning to 11.8 ± 0.9 keV near the end. Spectral variability in SGR 1806-20 bursts have also been seen with RXTE (Strohmayer and Ibrahim 1999) and Integral observations (Götz et al. 2004). Such strong variabilities together with the transient appearance of the 6.4 keV emission line from SGR 1900+14 and the 5 keV and 20 keV absorption lines from SGR 1806-20 (see Ibrahim et al. 2002, 2003, 2007) point to a likely dependence of the spectral features on the burst emission properties and temporal profile. Further work on time-resolved spectroscopy of the entire data set is needed to understand such a relationship.

References

Götz, D., Mereghetti, S., Mirabel, I.F., et al.: Astron. Astrophys. **417**, L45 (2004)

Hurley, K., Cline, T., Mazets, E., et al.: Nature **397**, 41 (1999)

Ibrahim, A.I., Safi-Harb, S., Swank, J.H., et al.: Astrophys. J. **574**, L51 (2002)
Ibrahim, A.I., Strohmayer, T.E., Woods, P.M., et al.: Astrophys. J. **558**, 237 (2001)
Ibrahim, A.I., Swank, J.H., Parke, W.C.: Astrophys. J. **584**, L17 (2003)
Ibrahim, A.I., et al.: Astrophys. Space Sci. **308**. DOI 10.1007/s10509-007-9311-4 (2007)
Kaplan, D.L., Kulkarni, S.R., Frail, D.A., et al.: Astrophys. J. **566**, 378 (2002)
Marsden, D., Lingenfelter, R.E., Rothschild, R.E., et al.: Astrophys. J. **550**, 397 (2001)
Strohmayer, T.E., Ibrahim, A.I.: In: Meegan, C.A., Preece, R.D., Koshut, T.M. (eds.) Fourth Huntsville Symp. on Gamma-Ray Bursts, Huntsville, AL. AIP Conf. Proc., vol. 428, p. 947. AIP, New York (1999)
Strohmayer, T.E., Ibrahim, A.I.: Astrophys. J. **537**, L111 (2000)

Astrophys Space Sci (2007) 308: 541–543
DOI 10.1007/s10509-007-9378-y

ORIGINAL ARTICLE

Mountains on neutron stars

B. Haskell

Received: 23 August 2006 / Accepted: 11 October 2006 / Published online: 20 March 2007

Abstract The aim of this work is to compare a neutron star with an accreted crust and one with a non-accreted crust, and estimate which one is potentially a better source of gravitational waves (i.e. can sustain a larger "mountain"). To do this we present a new formalism, and find that a non-accreted crust can sustain a slightly larger "mountain". We also discuss the importance of relativistic effects.

Keywords Dense matter · Stars neutron

PACS 97.60.Jd · 26.60.+c

1 Introduction

The possibility that the crust of a rotating neutron star may sustain a "mountain", thus creating a source of gravitational waves, has recently sparked much interest. This is mainly due to the suggestion by Bildsten (1998), that the gravitational wave torque due to such a mechanism may be dictating the spin equilibrium period for the LMXBs. A recent estimate of how large a quadrupole one may expect was given by Ushomirsky et al. (2000). From their work it appears that the quadrupole may be big enough to balance the accretion torque, if the breaking strain of the crust (a very poorly constrained parameter) is taken to be near the maximum of the possible values.

Many more questions can still be asked, the main one being if there is a difference between the "mountain" that one can build on the non-accreted crust of a rotating neutron star and one that can be built on the accreted crust of a neutron star in a binary system (such as the LMXBs). The case of isolated neutron stars is particularly interesting as current gravitational wave detectors are beginning to put astrophysically relevant limits on the ellipticities of known pulsars. We will discuss this issue, and present a new scheme for calculating the maximum quadrupole which allows us to have better control of the boundary conditions at the base of the crust than in the scheme of Ushomirsky et al. (2000). In addition we shall drop the Cowling approximation made in Ushomirsky et al. (2000).

B. Haskell (✉)
School of Mathematics, Southampton University, Southampton, SO17 1BJ, UK
e-mail: haskellb@soton.ac.uk

2 Accreted vs. non-accreted crusts

The equation of state and composition of the crust can be quite different in a non-accreted crust and in an accreted crust, where the original material has been replaced by accretion. The equation of state can determine the thickness of the crust, while the composition determines the shear modulus (Ogata and Ichimaru 1990):

$$\mu = 0.1194\left(\frac{3}{4\pi}\right)^{1/3}\left(\frac{1-X_n}{A}n_b\right)^{4/3}(Ze)^2 \qquad (1)$$

where X_n is the fraction of neutrons outside nuclei, n_b is the baryon density, Z is the proton number and A is the atomic number. As we are interested in understanding which kind of crust can sustain a "larger mountain" we shall evaluate the expression in (1) for the case of a non-accreted crust, using the EOS and composition from Douchin and Haensel (2001), and for an accreted crust, using the EOS from Haensel and Zdunik (1990). The result is in Fig. 1.

 Astrophys Space Sci (2007) 308: 541–543

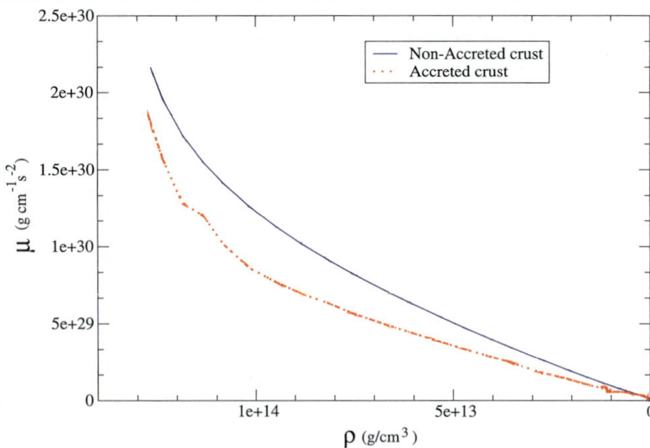

Fig. 1 The shear modulus μ, for an accreted and a non-accreted crust. A slightly larger μ for the non-accreted crust suggests that it may be able to sustain a larger "mountain" than an accreted crust

3 The formalism

We shall assume that the reference shape of the star (i.e. the unstrained configuration) is spherical. We then perturb the spherical background

$$x^a \longrightarrow x^a + \xi^a \tag{2}$$

where $\xi^a = \xi_r(r) Y_{lm} r^a + \xi_\perp \nabla^a Y_{lm}$. One can then write the stress tensor in the crust as

$$\tau_{ab} = -p g_{ab} + \mu \sigma_{ab} \tag{3}$$

where p is the pressure, μ is the shear modulus and σ is the stress tensor due to the deformation. We now solve the Newtonian equations of hydrostatic equilibrium in the crust and impose the magnitude of the perturbation, i.e. the "size" of the mountain, at the surface, which can be expressed in terms of a variable that labels the deformed isobaric surfaces

$$\varepsilon_s = -\frac{\delta p}{r \partial_r p}. \tag{4}$$

In order to evaluate the maximum deformation that the star can sustain, we can apply the following procedure:

- Solve equations of hydrostatic equilibrium in the crust $\nabla^a \tau_{ab} = \rho \nabla_b \phi$.
- Assume that the crust will crack if $\bar{\sigma} \geq \sigma_{\max}$ where $2\bar{\sigma} = \sigma_{ab}\sigma^{ab}$ (Von Mises criterion). $\sigma_{\max} \approx 10^{-5}$–$10^{-2}$ is the breaking strain, a very poorly constrained parameter.
- Increase the "size" of the mountain at the surface until the crust breaks.

We now need to discuss the boundary conditions at the base of the crust. In a fully Newtonian framework one should perturb the core of the star and then match the tractions at the base of the crust. However we shall also consider the case of an unperturbed, spherical core with a perturbed crust. As we shall see this is not a bad approximation for a thick crust and, as the core is unperturbed, allows us to consider both a

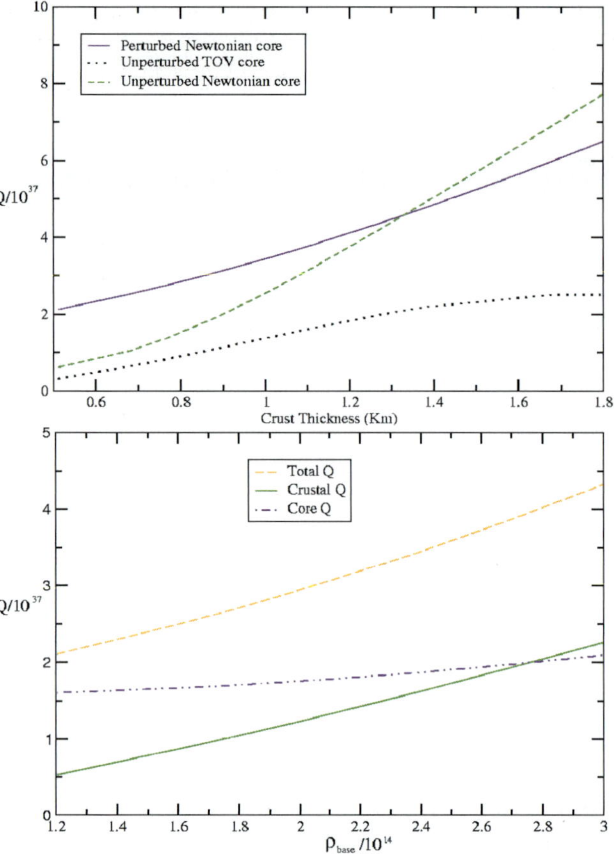

Fig. 2 As the crust becomes thicker (or alternatively the density at the base of the crust is taken to be higher), the main contribution to the quadrupole comes from the crust, thus allowing us to take the core to be unperturbed

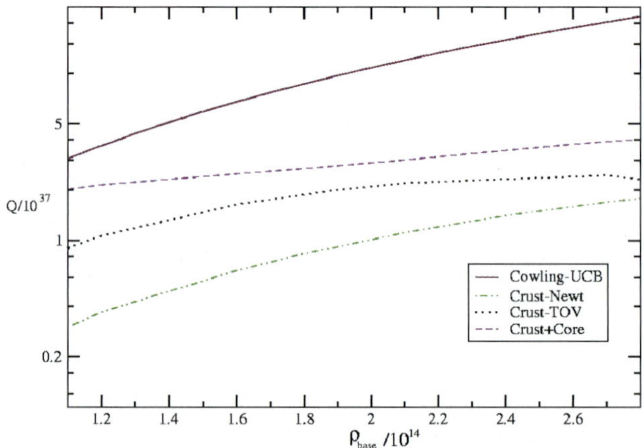

Fig. 3 The maximum quadrupole plotted with respect to the density at the base of the crust. We present the three models discussed in the text and the results of Ushomirsky et al. (2000), marked as UCB. As we can see relativistic effects can play a part, as they determine general properties such as the crust thickness or the stellar radius and mass

Table 1 The stellar models we consider to compare stars with an accreted and with a non-accreted crust

	Accreted crust	Non-accreted crust	Accreted crust
Mass	$1.4M_{\odot}$	$1.4M_{\odot}$	$1.6M_{\odot}$
Radius	12.56 km	12.3 km	12.3 km
Crust thickness	1.76 km	1.5 km	1.5 km

Table 2 Maximum quadrupole for two stars of equal mass, one with an accreted crust and one with a non-accreted crust. The last column shows a star with an accreted crust of the same thickness as that of the non-accreted crust, but a different mass. The base of the crust is taken to be at 1.3×10^{14} g/cm^3 as in Douchin and Haensel (2001). ϵ is the ellipticity defined as $\frac{I_x - I_y}{I_0}$

	Accreted crust	Non-accreted crust	Accreted crust
Mass	$1.4M_{\odot}$	$1.4M_{\odot}$	$1.6M_{\odot}$
Radius	12.56 km	12.3 km	12.3 km
Crust thickness	1.76 km	1.5 km	1.5 km
$Q_{\max}$	$1.8 \times 10^{39}(\frac{\sigma}{10^{-2}})$ g cm^2	$3.1 \times 10^{39}(\frac{\sigma}{10^{-2}})$ g cm^2	$1.6 \times 10^{39}(\frac{\sigma}{10^{-2}})$ g cm^2
ϵ	1.3×10^{-6}	2.4×10^{-6}	1.1×10^{-6}

Newtonian and relativistic core. We shall then be considering three models:

- A fully Newtonian star, perturbed crust and *perturbed* core.
- A fully Newtonian star, perturbed crust but *unperturbed* core.
- A star with a Newtonian perturbed crust but a *relativistic unperturbed* core.

4 Results

Let us now apply our formalism to a specific stellar model, for which we take the EOS to be an $n = 1$ polytrope and for simplicity we take a constant ratio $\mu/\rho = 10^{16}$ cm^2/s^2. The mass of the star is $M = 1.4M_{\odot}$. Rather than the deformation, we will evaluate the maximum quadrupole that the star can sustain. The results are presented in Figs. 2 and 3.

5 Results for accreted vs. non-accreted crusts

From our polytropic model we have determined that if the crust is thick enough we can make the approximation that the core is unperturbed. This allows us to consider the case of the stellar models in Table 1 with an unperturbed relativistic core (the EOS is taken from Douchin and Haensel (2001)). It is important that we can consider relativistic equations of hydrostatic equilibrium, as it would not seem appropriate to use a realistic equation of state if we are neglecting relativistic effects in the core. We can thus calculate the maximum quadrupole in the different cases and determine which kind of crust, accreted or non-accreted, can sustain a larger "mountain". The results are in Table 2.

The upper limits set by LIGO are still larger than the values of ϵ we have calculated, but are getting closer, and will soon be better than those set by electromagnetic observations (e.g. spindown). Gravitational wave observations (from LIGO and certainly from Advanced LIGO) may thus allow us to gain more insight into this problem and possibly place constraints on the crustal breaking strain.

6 Conclusions

The upper limits set by LIGO are still larger than the values of ϵ we have calculated, but are getting closer, and will soon be better than those set by electromagnetic observations (e.g. spindown). Gravitational wave observations (from LIGO and certainly from Advanced LIGO) may thus allow us to gain more insight into this problem and possibly place constraints on the crustal breaking strain.

References

Bildsten, L.: Astrophys. J. **501**, L89 (1998)
Douchin, F., Haensel, P.: Astron. Astrophys. **380**, 151 (2001)
Haensel, P., Zdunik, J.L.: Astron. Astrophys. **229**, 117 (1990)
Ogata, S., Ichimaru, S.: Phys. Rev. A **42**, 4867 (1990)
Ushomirsky, G., Cutler, C., Bildsten, L.: Mon. Not. Roy. Astron. Soc. **319**, 902 (2000)

Astrophys Space Sci (2007) 308: 545–549
DOI 10.1007/s10509-007-9308-z

ORIGINAL ARTICLE

Optical spectroscopy of the radio pulsar PSR B0656+14

S. Zharikov · R.E. Mennickent · Yu. Shibanov · V. Komarova

Received: 27 June 2006 / Accepted: 23 August 2006 / Published online: 21 March 2007

Abstract We have obtained the spectrum of a middle-aged PSR B0656+14 in the 4300–9000 Å range with the ESO/VLT/FORS2. Preliminary results show that at 4600–7000 Å the spectrum is almost featureless and flat with a spectral index $\alpha_\nu \simeq -0.2$ that undergoes a change to a positive value at longer wavelengths. Combining with available multiwavelength data suggests two wide, red and blue, flux depressions whose frequency ratio is about 2 and which could be the 1st and 2nd harmonics of electron/positron cyclotron absorption formed at magnetic fields $\sim 10^8$ G in upper magnetosphere of the pulsar.

Keywords Pulsars · Spectroscopy · PSR B0656+14

PACS 97.60.Jd

Based on observations collected at the European Southern Observatory, Paranal, Chile (ESO Programme 074.D-0512A). The work was partially supported by DGAPA/PAPIIT project IN101506, CONACYT 48493, RFBR (grants 05-02-16245, 05-02-22003) and Nsh 9879.2006.2. R.E.M. was supported by Fondecyt 1030707.

S. Zharikov (✉)
OAN IA UNAM, Ensenada, Mexico
e-mail: zhar@astrosen.unam.mx

R.E. Mennickent
Universidad de Concepcion, Concepcion, Chile

Yu. Shibanov
Ioffe Physical Technical Inst., RAS, St. Petersburg, Russia

V. Komarova
Special Astrophysical Observatory, RAS, Nizhnii Arkhyz, Russia

1 Introduction

Multiwavelength observations of radio pulsars are an important tool for the study of not yet clearly understood radiative mechanisms and spectral evolution of rotation-powered isolated neutron stars (NSs). Optical observations are an essential part of these studies. Among a dozen optically identified NSs, only seven pulsars have parallax based distances and, therefore, minimal uncertainties ($\lesssim$10%) in luminosities. Their optical spectra are shown in Fig. 1. A reliable optical spectrum has been obtained only for the young and bright Crab pulsar (Sollerman et al. 2000). Other pulsars are fainter and are represented mainly by broadband photometric points. Published optical spectra of PSR B0540-69 (Hill et al. 1997; Serafimovich et al. 2004) are strongly contaminated by a bright pulsar nebula (Serafimovich et al. 2004), while a tentative spectrum of Geminga (Martin et al. 1998) is much noisier than available photometric fluxes. Mignani et al. (2004) reported on the Vela pulsar spectral observations but these data are not published yet.

In Fig. 2 we show the evolution of luminosity L and radiation efficiency $\eta = \mathrm{L}/\mathrm{L}_{\mathrm{sd}}$ (L_{sd} is spin-down luminosity) demonstrated by these seven pulsars using the data of Table 1 from Zharikov et al. (2006). We note significantly non-monotonic dependencies of η_{Opt} and η_{X} versus pulsar age with a pronounced minimum at the beginning of the middle-age epoch ($\simeq 10^4$ yr) and comparably higher efficiencies of younger and older pulsars.

Owing to its relative brightness and proximity, the middle-aged PSR B0656+14 is one of isolated NSs most intensively studied in different wavelengths. It was discovered in radio by Manchester et al. (1978), then identified in X-rays with Einstein, and observed in details with ROSAT, ASCA, Chandra and XMM (for references see Shibanov et al. 2005, 2006). The X-ray emission can be described

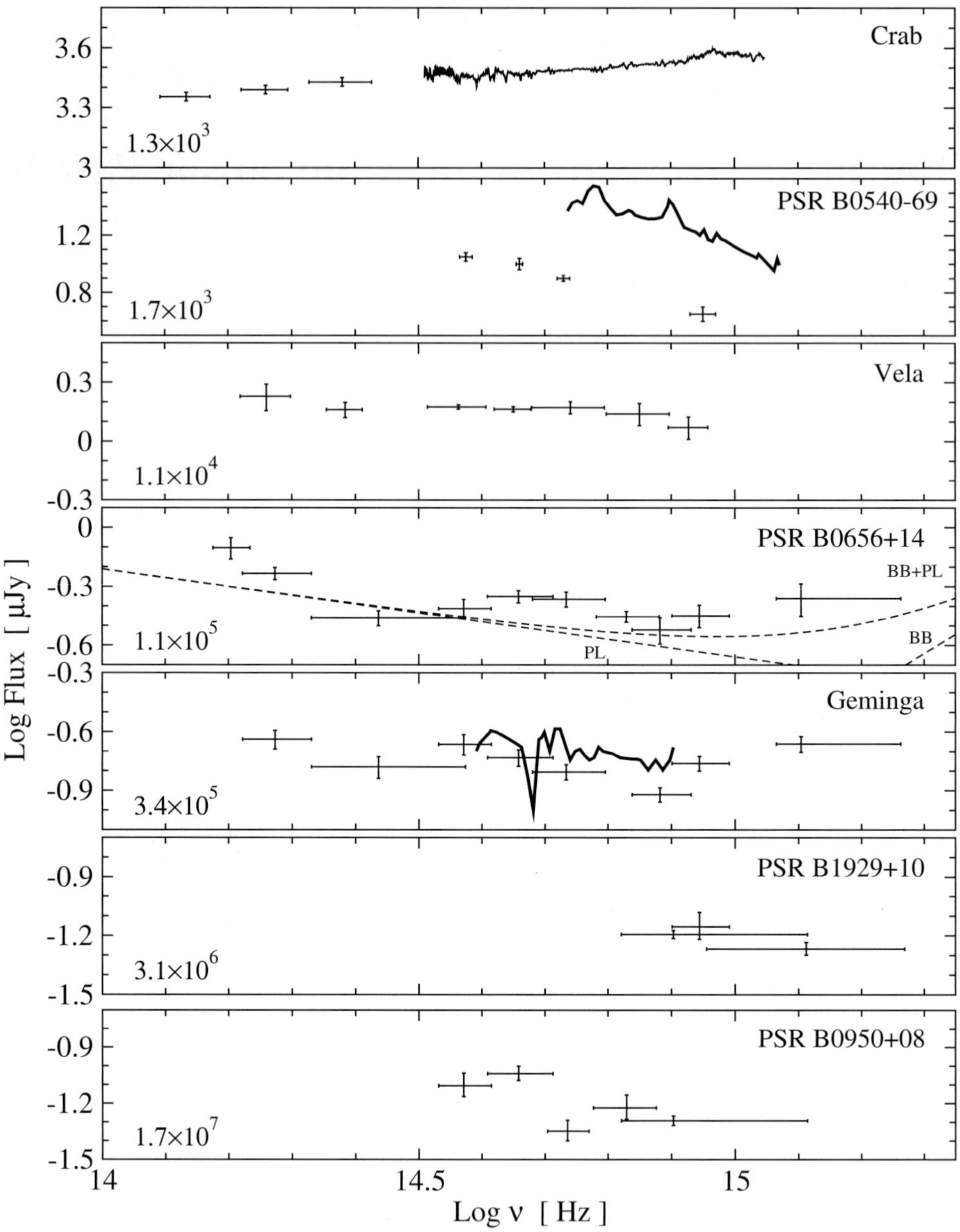

Fig. 1 Optical spectra of seven pulsars of different characteristic age indicated in the *left-bottom corner* of each panel in years. The youngest Crab pulsar is at the *top* and the oldest PSR B0950-08 is at the *bottom*. For PSR B0656+14 *dashed lines* show the low energy extensions of the blackbody (BB, $T = 0.84$ MK) and power law (PL, $\alpha = 0.45$) X-ray spectral components and their sum (Koptsevich et al. 2001)

as a combination of thermal radiation from the entire surface of a cooling NS and from hotter polar caps heated by relativistic particles of magnetospheric origin. An excess over the hot thermal component at energies ≥ 2 keV was interpreted as nonthermal radiation from the pulsar magnetosphere (Greiveldinger et al. 1996). The pulsar has been also marginally detected in γ-rays (≥ 50 MeV) by Ramanamurthy et al. (1996).

In the optical PSR B0656+14 was identified by Caraveo et al. (1994) with the ESO/NTT telescopes in the V band. It was then studied in UV with the HST/FOC in the F130LP, F430W, F342W and F195W bands (Pavlov et al. 1997) and with the HST/WFPC in the F555W band (Mignani et al. 1997). Detailed photometric studies in the optical-NIR were performed by Kurt et al. (1998), Koptsevich et al. (2001), Komarova et al. (2003), and Shibanov et al. (2006). The studies showed that the bulk of the optical radiation is of nonthermal origin. This was confirmed by the detection of coherent optical pulsations with the radio pulsar period in the B band (Shearer et al. 1997), in a wide 400–600 nm passband (Kern et al. 2003) and in NUV (Shibanov et al. 2005). The pulse profiles are rather sharp with a high pulse fraction as expected for nonthermal emission mechanisms. The phase integrated multiwavelength spectrum (Fig. 4) shows (Koptsevich et al. 2001; Shibanov et al. 2006) that the NIR-optical-UV spectral energy distribution is, in a first approach, compatible with the low energy extension of the sum of the X-ray thermal blackbody (BB) spectral compo-

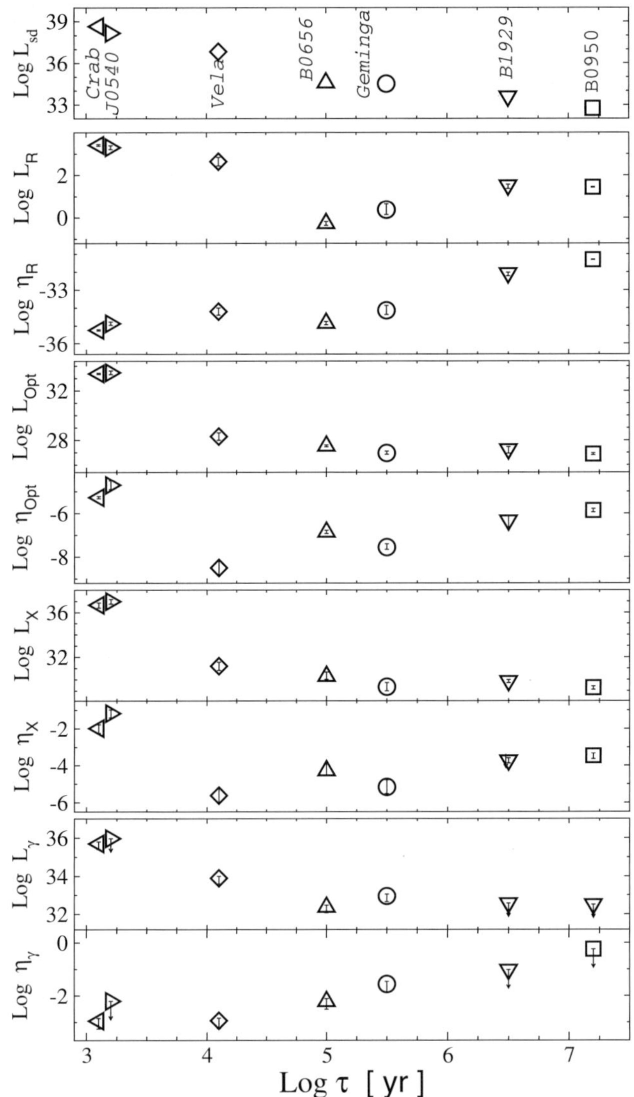

Fig. 2 *From top to bottom:* Evolution of the spin-down, radio, optical, X-ray, and γ-ray luminosities and respective efficiencies demonstrated by the 7 optical pulsars from Fig. 1

nent from the whole NS surface and the power law (PL) component dominating in the high energy tail. The BB extension does not contribute at longer wavelengths where the optical-NIR fluxes are in a good agreement with the PL alone (Fig. 1). This indicates a common origin of the nonthermal optical and X-ray emission, which is strongly supported by a good coincidence in phase and shape of the pulse profiles in the optical and the X-ray tail (Shibanov et al. 2005). The same origin of the nonthermal optical and X-ray photons is likely to be a general property for other pulsars detected in both ranges, as follows from a strong correlation between respective efficiencies (Fig. 3) found by Zharikov et al. (2004, 2006).

In the optical emission of PSR B0656+14 there is an apparent, $\simeq (3\text{–}5)\sigma$, flux excess over the PL "continuum" at $\text{Log}(\nu) \simeq 14.7$ (Fig. 1). This could reveal an additional, 3rd, spectral component to the BB+PL discussed above. Here we present first results of the optical spectroscopy of the pulsar partially motivated by more detailed studies of the excess. In a broader sense, these results also allow us to consider, for the first time, optical properties of a middle-aged ordinary pulsar at the spectroscopic level, as it has been achieved so far only for the Crab pulsar.

2 Observations and data analysis

The spectrum of PSR B0656+14 was obtained on November–December 2004 and February 2005 during several observational runs of the ESO program 074.D-0512A using the VLT/UT1 telescope in a service mode. The FORS2[1] instrument was used in a long slit spectroscopic setup with the grating GRIS_300V and the filter GG435, which cover the wavelength interval of about 4300–9600 Å and provide a

[1] www.eso.org/instruments/fors

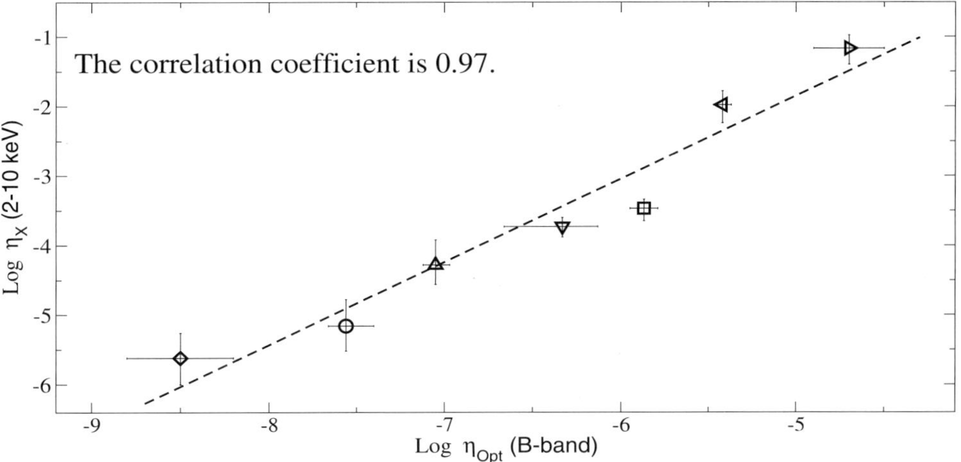

Fig. 3 Relationship between the B-band optical and 2–10 keV X-ray efficiencies for the same 7 pulsars as in Figs. 1, 2

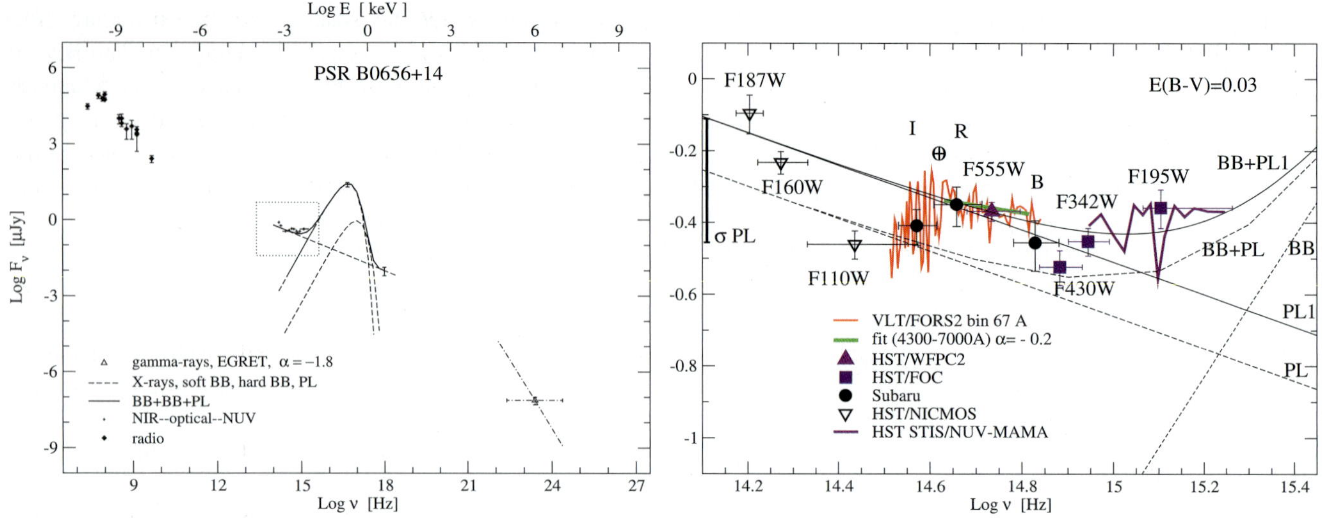

Fig. 4 *Left:* Unabsorbed multiwavelength spectrum of PSR B0656+14 from the radio through γ-rays. The *box* marks the range zoomed in the right panel. *Right:* The spectral and photometric fluxes of PSR B0656+14 in the NIR-optical-NUV range obtained with different telescopes and instruments, as notified in the plot. *Black dashed lines* show low energy extensions of the soft blackbody (BB) and power law (PL) spectral fits and their sum (BB+PL) obtained in X-rays. The *black solid lines* show the same but with the PL normalization shifted up by a factor of 1.4 (PL1) to fit the upper edge of its 1σ error bar shown at the *left side* of the plot. These are possibly a better match for the optical and NUV spectra than the *dashed lines*. The *symbol* ⊕ marks the Earth atmospheric absorption band near 7600 Å

medium spectral resolution of 3.35 Å/pixel. The slit width was 1″ and its position angle was selected in a such way as to obtain also spectra of several nearby stars for a sure pulsar astrometric referencing and wavelength/flux calibration. Eighteen 1400 s science spectroscopic exposures were taken with a total exposure time of 25200 s at a mean seeing of 0.6″. Standard reference frames (biases, darks, in each observational run, while the slit and slitless observations of spectrophotometric standards (Feige110, LTT3218 and LTT1788) for the flux calibration were carried out in separate runs on the same nights.

A combination of the MIDAS and IRAF packages was used for standard CCD data reduction, cosmic-ray track removing, spectra extraction, and subsequent data analysis. A faint, R = 24.65, pulsar is at a limit of spectroscopic capability of the VLT. Nevertheless, excellent seeing conditions allowed us to resolve its spectrum even at each individual exposure, albeit with a low signal to noise ratio S/N. These exposures were co-added. The spectrum was then extracted with a 3 pixel wide extraction slit (0.2″/pix) centered on the pulsar. The backgrounds were extracted with a 6 pixel wide slit centered above and below the center of the pulsar spectrum. The correction factor for the PSF and sensitivity function were obtained from the Feige110 standard observations. The S/N of the resulting spectrum was about 4 (per pixel) in the 4450–5500 Å range and declined to ~ 1 near/above 8000 Å, due to higher sky backgrounds and a drop in sensitivity towards longer wavelengths. We binned the spectral flux in 20 pixel bins (67 Å) to get S/N near/above 15 and 4, respectively, making the flux accuracy to be comparable with that of available photometric data.

3 Results and discussion

The binned and dereddened with $A_V = 0.093$ spectrum of the pulsar is shown in Fig. 4 (red curve in the right panel). We show also available multiwavelength data (see Shibanov et al. (2006) for the data and A_V description). The spectrum is in a good agreement with the broadband VRI fluxes, while it is somewhat higher than the B band flux. It does not show any strong nebular emission lines that could be responsible for the apparent VR excess, mentioned above as a 3rd component, while the presence of weak features can not be completely ruled out. The continuum of the dereddened spectrum in 4600–7000 Å range has a power law shape $\propto \nu^{-0.2\pm0.2}$ (green line) in agreement with the nonthermal nature of the bulk of the optical emission. Its slope is close to the value expected in this range from the BB+PL extended from X-rays. The slope likely undergoes a change to a positive value between the R and I bands, as follows also from the photometric data.

Within uncertainties the bluest end of the optical and the reddest end of the NUV (Shibanov et al. 2005) spectra are compatible with each other, suggesting a smooth connection of both. However, the F430W photometric flux in the gap between them drops below this connection with a significance of about 2σ of the flux confidence level. Unless this is a of some unknown systematics in calibration, it suggests a spectral dip in the pulsar emission centered near $\simeq 4300$ Å ($\simeq 14.83$ in Log(ν)).

The optical and NUV spectra and two NIR photometric points, F160W and F187W, almost perfectly match the BB+PL extension from X-rays, if the PL normalization is taken to be a factor of 1.4 higher (solid lines) than its best X-ray fit value (dashed lines). The change is within 1σ uncertainty of the fit. Considering the solid line version as a new optical "continuum level" we find an additional and more significant flux depression in the red part of the spectrum overlapping the I and F110W bands and centered near 9000–10000 Å (Log(ν) $\simeq 14.5$).

Additional spectral studies are necessary to confirm the suggested "red" and "blue" features and to measure their shapes and wavelengths more accurately. Nevertheless, if they are real, an approximate blue-to-red frequency ratio is $\simeq$2. This indicates that they can be the 1st and 2nd harmonics of an electron/positron cyclotron absorption formed in the upper magnetosphere of the pulsar at an effective altitude where the magnetic field B $\simeq 10^8$ G. This is $\simeq$360 km, assuming a dipole NS field with a surface value of 4.66×10^{12} G, as derived from spin-down measurements. The absorbing $e^{\pm}$ have to be cooled enough and provide a sufficient optical depth above a source of the nonthermal continuum. The source altitude is, therefore, $<$360 km and much below the light cylinder radius of $\simeq 18 \times 10^3$ km, likely suggesting its polar cap origin. The features are broad, as expected from the magnetospheric field inhomogeneity. Tentative ($\leq 2\sigma$) absorption features in the NUV spectrum at Log(ν) $\simeq 15$ and 15.1 (Shibanov et al. 2005) may be the 3rd and 4th harmonics, respectively, which are fainter as the cyclotron harmonic intensity decreases with its number n.

Similar, albeit less significant, spectral features are likely seen in the photometric and spectral data of another middle-aged pulsar Geminga (Fig. 1). The absence of strong nebular lines suggests that the features are also of the NS magnetospheric origin. To explain the Geminga spectrum Martin et al. (1998) applied a toy model of an ion cyclotron absorption at B $\simeq 10^{11}$ G in the inner magnetosphere of the NS combined with the BB and PL components. At this field a low n ion cyclotron frequencies indeed fall in the optical range. However, it is likely that the nonthermal optical emission is generated in the upper magnetosphere where the magnetic field is by orders of magnitude weaker and any ion cyclotron absorption of the respective optical continuum is negligible. In this case the electron/positron cyclotron absorption or scattering appears to be more plausible interpretation.

The optical spectrum of the young Crab-pulsar is featureless and has a different (positive) slope (Fig. 1). The ten times older Vela-pulsar has a flat and also featureless spectrum (Mignani, private communication). The spectroscopy of PSR B0656+14 demonstrates that optical spectra of middle-aged pulsars can be distinct from those of younger ones by the presence of unusual spectral features or slope changes. To study this new spectral observations of PSR B0656+14 in the NIR and 3000–5000 Å ranges are needed. A question, whether the observed difference in pulsar spectra is simply caused by different pulsar geometry or by a change of physical conditions in the emission region with age, demands quantitative modeling physical processes in pulsar magnetospheres.

References

Caraveo, P., Bignami, G.F., Mereghetti, S.: Astrophys. J. Lett. **422**, 82 (1994)

Greiveldinger, C., Camerini, U., Fry, W., et al.: Astrophys. J. Lett. **465**, 35 (1996)

Hill, R.J., Dolan, J.F., Bless, R.C., et al.: Astrophys. J. Lett. **486**, 99 (1997)

Kern, B., Martin, C., Mazin, B., Halpern, J.P.: Astrophys. J. **597**, 1049 (2003)

Komarova, V., Shibanov, Yu., Zharikov, S., et al.: In: Cusumano, G., Massaro, E., Mineo, T. (eds.) Proc. of the Workshop "Pulsars, AXPs and SGRs observed with BeppoSAX and Other Observatories", Marsala, Italy, 23–25 September 2002. 77 Aracne Editrice (2003)

Koptsevich, A.B., Pavlov, G.G., Zharikov, S.V., et al.: Astron. Astrophys. **370**, 1004 (2001)

Kurt, V.G., Sokolov, V.V., Zharikov, S.V., et al.: Astron. Astrophys. **333**, 547 (1998)

Manchester, R.N., Lyne, A.G., Taylor, J.H., et al.: Mon. Not. Roy. Astron. Soc. **185**, 409 (1978)

Martin, C., Halpern, J., Schiminovich, D.: Astrophys. J. Lett. **494**, 211 (1998)

Mignani, R., Caraveo, P.A., Bignami, G.F.: Messenger **87**, 43 (1997)

Mignani, R.P., de Luca, A., Caraveo, P.A.: In: Camilo, F., Gaensler, B.M. (eds.) Proc. of the IAU Symposium no. 218 "Young Neutron Stars and Their Environments", part of the IAU General Assembly, p. 391, 14–17 July 2003. Astronomical Soc. of the Pacific, Sydney (2004)

Pavlov, G.G., Welty, A.D., Cordova, F.A.: Astrophys. J. **489**, L75 (1997)

Ramanamurthy, P., Fichtel, C., Harding, A., et al.: Astron. Astrophys. Suppl. Ser. **120**, 115 (1996)

Serafimovich, N.I., Shibanov, Yu.A., Lundqvist, P., Sollerman, J.: Astron. Astrophys. **425**, 1041 (2004)

Shearer, A., Redfern, R.M., Gorman, G., et al.: Astrophys. J. Lett. **487**, 181 (1997)

Shibanov, Yu.A., Sollerman, J., Lundqvist, P., et al.: Astron. Astrophys. **440**, 693 (2005)

Shibanov, Y.A., Zharikov, S.V., Komarova, V.N., et al.: Astron. Astrophys. **448**, 313 (2006)

Sollerman, J., Lundqvist, P., Lindler, D., et al.: Astrophys. J. **537**, 861 (2000)

Zharikov, S.V., Shibanov, Yu.A., Mennickent, R.E.: Astron. Astrophys. **417**, 1017 (2004)

Zharikov, S., Shibanov, Yu., Komarova, V.: Adv. Space Res. **37**, 1979 (2006)

Astrophys Space Sci (2007) 308: 551–555
DOI 10.1007/s10509-007-9370-6

ORIGINAL ARTICLE

On the peculiarities in the rotational frequency evolution of isolated neutron stars

Anton Biryukov · Gregory Beskin · Sergey Karpov

Received: 29 June 2006 / Accepted: 9 November 2006 / Published online: 29 March 2007

Abstract The measurements of pulsar frequency second derivatives have shown that they are 10^2-10^6 times larger than expected for standard pulsar spin-down law, and are even negative for about half of pulsars. We explain these paradoxical results on the basis of the statistical analysis of the rotational parameters ν, $\dot{\nu}$ and $\ddot{\nu}$ of the subset of 295 pulsars taken mostly from the ATNF database. We have found a strong correlation between $\ddot{\nu}$ and $\dot{\nu}$ for both $\ddot{\nu} > 0$ and $\ddot{\nu} < 0$, as well as between ν and $\dot{\nu}$. We interpret these dependencies as evolutionary ones due to $\dot{\nu}$ being nearly proportional to the pulsars' age. The derived statistical relations as well as "anomalous" values of $\ddot{\nu}$ are well described by assuming the long-time variations of the spin-down rate. The pulsar frequency evolution, therefore, consists of secular change of $\nu_{\rm ev}(t)$, $\dot{\nu}_{\rm ev}(t)$ and $\ddot{\nu}_{\rm ev}(t)$ according to the power law with $n \approx 5$, the irregularities, observed within a timespan as a timing noise, and the variations on the timescale larger than that—several decades.

Keywords Methods: data analysis · Methods: statistical · Pulsars: general

PACS 97.60.Jd · 97.60.Gb · 97.10.Kc · 98.62.Ve

This work has been supported by the Russian Foundation for Basic Research (grant No 04-02-17555), Russian Academy of Sciences (program "Evolution of Stars and Galaxies"), and by the Russian Science Support Foundation. The authors would also like to thank the anonymous referee for valuable comments.

A. Biryukov (✉) · G. Beskin · S. Karpov
SAI MSU, 13, Universitetsky pr., Moscow 119992, Russia
e-mail: eman@sai.msu.ru

1 Introduction

The spin-down of radio pulsars is due to the conversion of their rotation energy into emission. According to the "classical" approach, their rotational frequencies ν evolve obeying the spin-down law $\dot{\nu} = -K\nu^n$, where K is a positive constant that depends on the magnetic dipole moment and the moment of inertia of the neutron star, and n is the braking index. The latter can be determined observationally from measurements of ν, $\dot{\nu}$ and $\ddot{\nu}$ as $n = \nu\ddot{\nu}/\dot{\nu}^2$. For a simple vacuum dipole model of pulsar magnetosphere $n = 3$; the pulsar wind decreases this value to $n = 1$; for multipole magnetic field $n \geq 5$ (Manchester and Taylor 1977). At the same time the measurements of pulsar frequency second derivatives $\ddot{\nu}$ have shown that their values are much larger than expected for standard spin-down law and are even negative for about half of all pulsars. The corresponding braking indices range from -10^6 to 10^6 (D'Alessandro et al. 1993; Chukwude 1993; Hobbs et al. 2004).

It was found that the significant correlations between $|\ddot{\nu}|$ ($|\ddot{P}|$) and $\dot{\nu}$ ($\dot{P}$) demonstrate the increase of the absolute values of the ν and P second derivatives for younger (with faster slow-down) pulsars (Cordes and Downs 1985; Arzoumanian et al. 1994; Lyne 1999). The anomalously high and negative values of $\ddot{\nu}$ and n may be interpreted as a result of low-frequency terms of the "timing noise"—a complex variations of pulsars rotational phase within a timespan (D'Alessandro et al. 1993).

It is clear that the timespan of observations is by no means intrinsic to the pulsar physics. Indeed, the variations of rotational parameters may take place on larger timescales as well. However, the timescale of observations naturally divides it into two separate classes of manifestations—the residuals in respect to the best fit for the timing solution (the

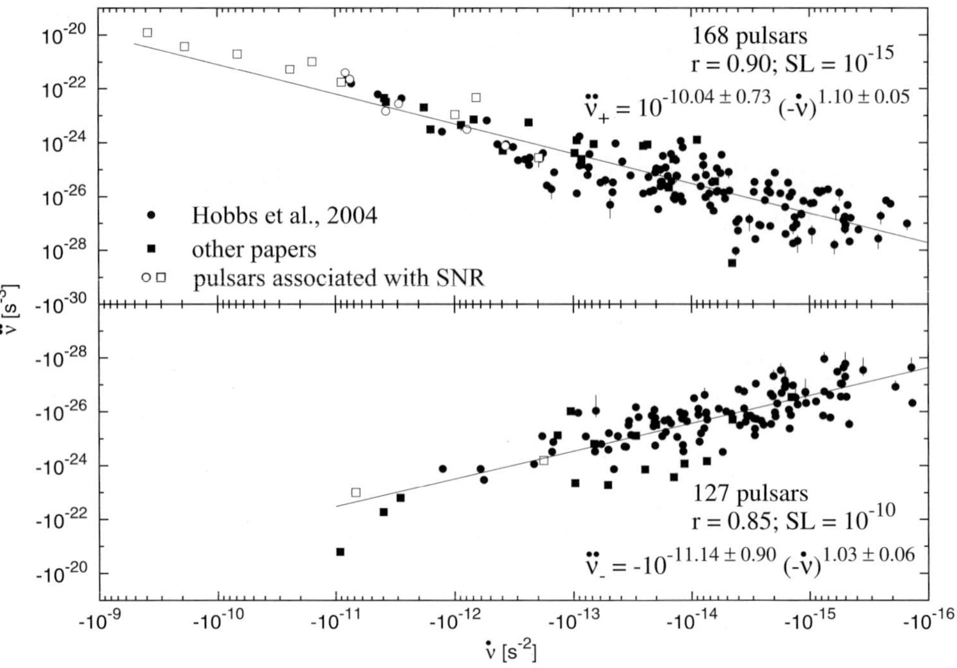

Fig. 1 The $\ddot{\nu}-\dot{\nu}$ diagram for 295 pulsars. The figure shows the pulsars from the work (Hobbs et al. 2004) as *circles*, and the objects measured by other groups as *squares*. *Open symbols* represent the pulsars associated with supernova remnants, and therefore—relatively young ones. Analytical fits for both positive and negative branches are shown as *solid lines*. Measurement errors shown as an *error bars*

"timing noise") and the systematic shift of the best fit coefficients (i.e. in the measured values of ν, $\dot{\nu}$, $\ddot{\nu}$) relative to some mean or expected value from the model. The latter effect may be called the "large timescale timing noise".

Up to date we know nearly 200 pulsars for which the timespan of observations is greater than 20 years, and the values of their $\ddot{\nu}$ still turn out to be anomalously large (Hobbs et al. 2004).

For example, for the *PSR B1706-16* pulsar, variations of $\ddot{\nu}$ with an amplitude of 10^{-24} s^{-3} have been detected on a several years timescale (see Fig. 7 in Hobbs et al. 2004), with the value of $\ddot{\nu}$ depending on the time interval selected. However, the fit over the entire 25 year timespan gives a value of $\ddot{\nu} = 3.8 \times 10^{-25}$ s^{-3} with a few percent accuracy (which leads to a braking index $\approx 2.7 \times 10^3$).

In the current work we provide observational evidence of the nonmonotonic evolution of pulsars on timescales larger than the typical contemporary timespan of observations (tens of years), using the statistical analysis of measured ν, $\dot{\nu}$ and $\ddot{\nu}$. We estimate the main parameters of such long timescale variations and discuss their possible relation to the low-frequency terms of the timing noise. We have also derived the parameters of pulsar secular spin-down.

2 Statistical analysis of the ensemble of pulsars

Our statistical analysis is based on the assumption that numerous measurements of the pulsar frequency second derivatives reflect their evolution on the timescale larger than the duration of observations, and uses the parameters of 295 pulsars.

From 389 objects of ATNF catalogue (Manchester et al. 2005) with known $\ddot{\nu}$ we have compiled a list of "ordinary" radio pulsars with $P > 20$ ms and $\dot{P} > 10^{-17}$ s s^{-1}, excluding recycled, anomalous and binary pulsars, and with relative accuracy of second derivative measurements better than 75%. It has been appended with 26 pulsars from other sources (D'Alessandro et al. 1993; Chukwude 1993). The parameters of all pulsars have been plotted on the $\ddot{\nu}-\dot{\nu}$ diagram (Fig. 1).

The basic result of the statistical analysis of this data is a significant correlation of $\ddot{\nu}$ and $\dot{\nu}$, both for 168 objects with $\ddot{\nu} > 0$ (correlation coefficient $r \approx 0.90$) and for 127 objects with $\ddot{\nu} < 0$ ($r \approx 0.85$). Both groups follow nearly linear laws, however they seem to be not exactly symmetric relative to $\ddot{\nu} = 0$. We divided both branches into 6 intervals of $\dot{\nu}$, computed the mean values and the standard deviations of $\ddot{\nu}_\pm$ in each, and rejected the hypothesis of their symmetry with a 0.04 significance level. The absolute values of $\ddot{\nu}_+$ are systematically larger than the corresponding $\ddot{\nu}_-$ (the difference is positive in 5 intervals of 6). Also, the difference between the analytical fits $|\ddot{\nu}_+|$ and $|\ddot{\nu}_-|$ is positive over the (-10^{-11})–(-10^{-15}) s^{-2} interval of $\dot{\nu}$. These are the arguments in favour of a small positive asymmetry of the branches.

We found an obvious correlation of $\dot{\nu}$ with the characteristic age $\tau_{\mathrm{ch}} = -\frac{1}{2}\frac{\nu}{\dot{\nu}}$ ($r = 0.96$, $\dot{\nu} \sim \tau_{\mathrm{ch}}^{-1.16\pm0.02}$). These parameters are nearly proportional, which leads to a significant correlation of τ_{ch} both with $\ddot{\nu}$ ($r = 0.85$ for the positive branch and $r = 0.75$ for the negative one) and with n ($r = 0.75$ and $r = 0.76$ correspondingly).

The correlations found are fully consistent with the results published in (Cordes and Downs 1985; Arzoumanian et al. 1994; Lyne 1999), as well as in (Urama et al. 2006). However, the branches with $\ddot{\nu} > 0$ and $\ddot{\nu} < 0$ in those works were not analyzed separately from each other (not as $|\ddot{\nu}|$).

Young pulsars confidently associated with supernova remnants are systematically shifted to the left in Fig. 1 (open symbols). The order of their physical ages roughly corresponds to that of their characteristic ages. This means that any dependence on $\dot{\nu}$ or $\tau_{\rm ch}$ reflects the dependence on pulsar age.

The $\ddot{\nu}-\dot{\nu}$ diagram (Fig. 1) may be interpreted as an evolutionary one. In other words, each pulsar during its evolution moves along the branches of this diagram while increasing the value of its $\dot{\nu}$ (which corresponds to the increase of its characteristic age). However, there is an obvious contradiction: for the negative branch, $\dot{\nu}$, being negative, may only decrease with time (since $\ddot{\nu}$ is formally the derivative of $\dot{\nu}$), and the motion along the negative branch may only be backward! This contradiction is easily solved by assuming non-monotonic behaviour of $\ddot{\nu}(t)$, which has an irregular component ($\delta\ddot{\nu}$) along with the monotonic one ($\ddot{\nu}_{\rm ev}$), where the subscript "ev" marks the evolutionary value. In such interpretation the value of $\ddot{\nu}_{\rm ev}$ must be positive, which will lead to a positive asymmetry of the branches. Moreover, the evolutionary increasing of $\dot{\nu}$ implies a positive evolutionary value of $\ddot{\nu}$.

The characteristic timescale T of such variations must be much shorter than the pulsar life time and at the same time much larger than the timescale of the observations. As it evolves, a pulsar repeatedly changes sign of $\ddot{\nu}$, in a spiral-like motion from branch to branch, and spends roughly half its lifetime on each one. The asymmetry of the branches reflects the positive sign of $\ddot{\nu}_{\rm ev}(t)$, and therefore, secular increase of $\dot{\nu}_{\rm ev}(t)$ (i.e. all pulsars in their secular evolution move to the right on the $\ddot{\nu}-\dot{\nu}$ diagram). Systematic decrease of branches separation reflects the decrease of the variations amplitude and/or the increase of its characteristic timescale.

Any well known non-monotonic variations of $\dot{\nu}(t)$, like glitches, microglitches, timing noise or precession, will manifest themselves in a similar way on the $\ddot{\nu}-\dot{\nu}$ diagram and lead to extremely high values of $\ddot{\nu}$ (Shemar and Lyne 1996; Stairs et al. 2000). However, their characteristic timescales vary from weeks to years, and they are detected immediately. But here the variations on much larger timescales are discussed, and their study is possible only statistically, assuming the ergodic behaviour of the ensemble of pulsars.

3 Non-monotonic variations of pulsar spin-down rate on large timescales

Variations of the pulsar rotational frequency may be complicated—periodic, quasi-periodic, or completely stochastic. Generally, it may be described as a superposition

$$\nu(t) = \nu_{\rm ev}(t) + \delta\nu(t), \quad (1)$$

where $\nu_{\rm ev}(t)$ describes the secular evolution of pulsar parameters and $\delta\nu(t)$ corresponds to irregular variations. Similar expressions describe the evolution of $\dot{\nu}$ and $\ddot{\nu}$ after a differentiation. The $\delta\ddot{\nu}(t)$ satisfies the obvious condition of zero mean value $\langle\delta\ddot{\nu}(t)\rangle_t \sim 0$ over the timespans larger than characteristic timescale of the variations. The amplitude of the observed variations of $\ddot{\nu}$ is related to the dispersion of this process as $\sigma_{\delta\ddot{\nu}} = A_{\ddot{\nu}} = \sqrt{\langle(\delta\ddot{\nu})^2\rangle}$.

The second derivative values on the upper $\ddot{\nu}_+$ and lower $\ddot{\nu}_-$ branches in Fig. 1 may be described as $\ddot{\nu}_\pm(t) = \ddot{\nu}_{\rm ev}(t) \pm A_{\ddot{\nu}}(t)$ for each pulsar. This equation describes some "average" pulsar, while the spread of points inside the branches reflects the variations of individual parameters over the pulsar ensemble and reaches 4 orders of magnitude.

The second derivative $\ddot{\nu}$ is the only parameter significantly influenced by the timing variations (see Sect. 4). Thus one can assume that the measured values of ν and $\dot{\nu}$ may be considered to be evolutionary ones, $\nu_{\rm ev}$ and $\dot{\nu}_{\rm ev}$ (since $\delta\nu$ and $\delta\dot{\nu}$ are small).

Using relations described above, the secular behavior $\nu(t)$ (or, $\nu(\dot{\nu})$) may be found by plotting the studied pulsar group onto the $\dot{\nu}-\nu$ diagram (Fig. 2). The objects with $\ddot{\nu} > 0$ and $\ddot{\nu} < 0$ are marked as filled and open circles, correspondingly. It is easily seen that the behavior of these two subgroups is the same, which is in agreement with the smallness of the pulsar frequency variations in respect to the intrinsic scatter of $\nu(\dot{\nu})$. However, a strong correlation between ν and $\dot{\nu}$ ($r \approx 0.7$) is seen, and

$$\dot{\nu} = -C\nu^n, \quad (2)$$

where $C = 10^{-15.26\pm1.38}$ and $n = 5.13 \pm 0.34$. So, the secular evolution of the "average" pulsar is according to the "standard" spin-down law with $n \approx 5$! This result is very interesting on its own, especially since the ν and $\dot{\nu}$ are always measured as independent values. The braking index $\approx$5 may suggest the importance of multipole components of pulsar magnetic field, or the deviation of the angle between pulsar rotational and dipole axes from $\pi/2$ (Manchester and Taylor 1977).

Note here, that the width of the fit on Fig. 2 is quite small—only about 1.5 orders of magnitude. If the spin-down is even approximately close to being described by the vacuum dipole model, then C should scale as $(B_0 \sin\alpha)^2$, where B is the polar field and α is the magnetic inclination angle. But the range of $(B_0 \sin\alpha)^2$ over the pulsar population

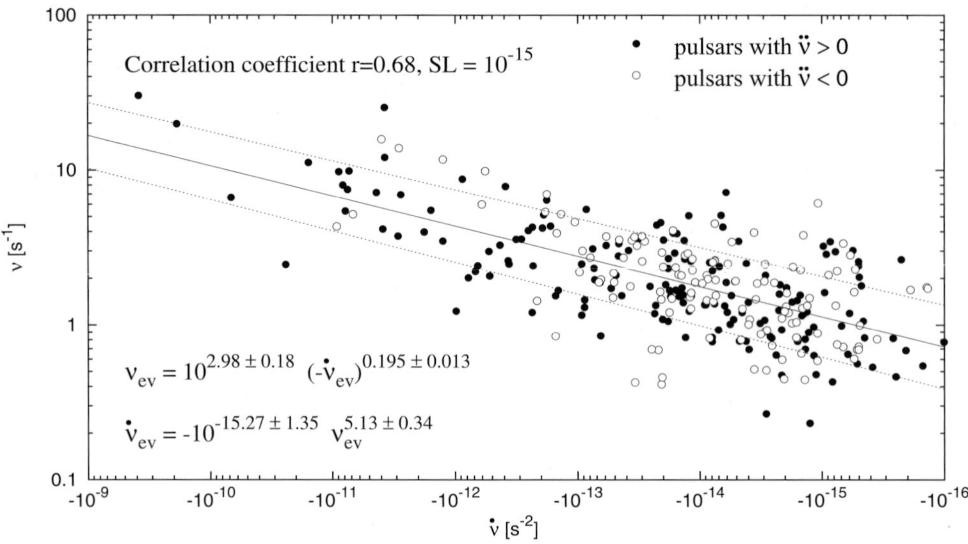

Fig. 2 The $\dot{\nu}-\nu$ diagram for the pulsars with the measured second derivative. The *filled symbols* are objects with positive $\ddot{\nu}$, the *open* ones—with negative. The behavior of both subsets is the same. The *solid line* represents the best fit, corresponding to the $n \approx 5$ braking index, the *dotted ones*—1σ range

is expected to be of many orders of magnitude. Therefore, Fig. 2 once again shows that simple vacuum dipole model is not adequate to observations. The braking index $n \approx 5$ allows the interpretation of C in terms of the quadrupole spin-down. It could be explained also in the frame of electric current mechanism of pulsar spin-down (Gurevich et al. 1993).

As has been stated above, the dependency between $\dot{\nu}$ and τ_{ch} is consistent with a power law with a slope of -1.16 ± 0.02. This value is significantly different from -1.0 which would be the case when the $\nu-\dot{\nu}$ correlation is absent and roughly consists with a slope of 5.13 in the $\dot{\nu}(\nu)$ dependency (the values are different on a 2.5σ level). But n is strongly dependent on the $\dot{\nu}-\tau$ slope value, so -1.16 seems to be more or less consistent with the measured value of n.

From (2) we may easily determine the relation between $\ddot{\nu}_{ev}$ and $\dot{\nu}$ as $\ddot{\nu}_{ev} = nC^{\frac{1}{n}}(-\dot{\nu})^{2-\frac{1}{n}}$, which is shown in Fig. 3 as a thick dashed line. The same relation may be also estimated directly by using the asymmetry of the branches seen on Fig. 1 as $\ddot{\nu}(\dot{\nu}) = \frac{1}{2}(\ddot{\nu}_+ + \ddot{\nu}_-)$, where $\ddot{\nu}_\pm$ are defined in Fig. 1. Such estimation, while being very noisy, is positive in the $(-10^{-11})-(-10^{-15})$ s^{-2} range and agrees quantitatively with the previous one.

The amplitude of the $\ddot{\nu}$ oscillations, $A_{\ddot{\nu}}$, may be easily computed in a similar way, by using $\ddot{\nu}_+$ and $\ddot{\nu}_-$, as $A_{\ddot{\nu}} = \frac{1}{2}(\ddot{\nu}_+ - \ddot{\nu}_-)$.

The behaviour of pulsars according to the derived relations is shown in Fig. 3. This simple variations model describes the observed branches, both positive and negative, rather well. The absence of negative branch objects with $\dot{\nu} > -10^{-11}$ s^{-2}, i.e. with $\tau_{ch} > 10^4$, we interpret as a prevalence of the second derivative's secular component over the varying one ($A_{\ddot{\nu}} < \ddot{\nu}_{ev}$) in this region. Older pulsars begin to change the sign of $\ddot{\nu}$ due to spin rate variations.

4 Discussion and conclusions

In general, it is impossible to estimate the amplitudes of the frequency and its first derivative variations A_ν and $A_{\dot{\nu}}$ from the amplitude of the second derivative only (the knowledge of its complete power density spectrum is needed). However, if the spectral density is relatively localized and some characteristic timescale T of the variations exists, it is possible to set some limits on it. A rough estimation is $A_\nu \sim A_{\dot{\nu}}T$, $A_{\dot{\nu}} \sim A_{\ddot{\nu}}T$ and $A_\nu \sim A_{\ddot{\nu}}T^2$. On a large timescale the variations can not lead to pulsar spin-up, so the variations of frequency first derivatives are much smaller than the secular ones, and $A_{\dot{\nu}} \sim A_{\ddot{\nu}}T \ll \dot{\nu}$, so $T \ll \dot{\nu}/A_{\ddot{\nu}}$. So, for the $-10^{-12} < \dot{\nu} < -10^{-15}$ s^{-2} range and corresponding values of $A_{\ddot{\nu}}$ from 10^{-23} s^{-3} to 10^{-26} s^{-3}, the characteristic timescale $T_{up} \sim 10^{11}$ s. Also, this characteristic timescale is obviously larger than the timespan of observations, so $50 < T < 3 \times 10^3$ years. Assuming the constancy of T during the pulsar evolution and therefore the change of A_ν with time, we get $A_\nu \sim 10^{-3}-10^{-7}$ Hz. For such a model the pulsar frequency varies with the characteristic time of several hundred years and the amplitude from 10^{-3} Hz for young objects to 10^{-7} Hz for older ones.

The physical reasons of the discussed non-monotonic variations of the pulsar spin-down rate may be similar to the ones of the timing noise on a short timescale. Several processes had been proposed for their explanation (Cordes and Greenstein 1981)—from the collective effects in the neutron star superfluid core to the electric current fluctuations in the pulsar magnetosphere. Whether these processes are able to produce long timescale variations is yet to be analyzed. On a short timescale, the pulsars show different timing behaviour. But on the long timescale their behavior seems to be alike.

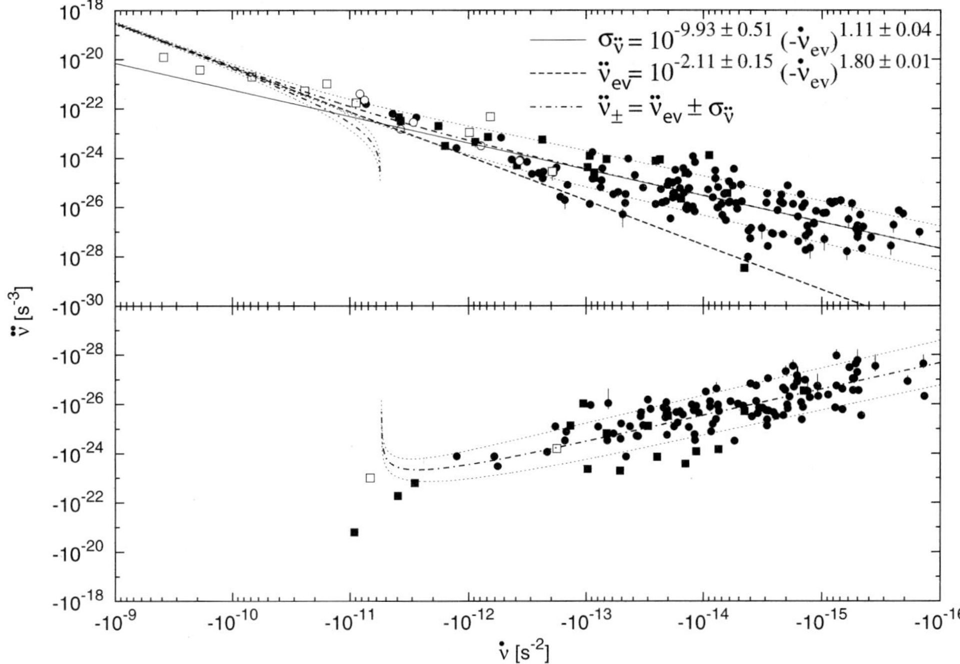

Fig. 3 The $\ddot{\nu}-\dot{\nu}$ diagram with the simple variations model. The *solid line* is the amplitude $A_{\ddot{\nu}}$ of the frequency second derivative variations, the *dashed*—the secular term $\ddot{\nu}_{ev}$, and the *dot-dashed lines* are the envelopes of the oscillations $\ddot{\nu}_{ev} \pm A_{\ddot{\nu}}$ with 1σ ranges (*dotted lines*). Each pulsar spends the majority of its life time at or very near the envelopes

The argument in favor of the similarity between the discussed variations and the timing noise is the coincidence of the timing noise $\ddot{\nu}$ amplitude extrapolated according to its power spectrum slope (Baykal et al. 1999) to the time scale of hundreds of years, with the $A_{\ddot{\nu}}$ derived from our analysis for the same $\dot{\nu}$, i.e. the same ages (see Fig. 3).

At the same time, there are several low noise pulsars with large or negative $\ddot{\nu}$. For 19 of 45 pulsars studied in (D'Alessandro et al. 1995) the timing noise is nearly absent (RMS $< 1 \times 10^{-3} P$). Six of them have anomalous $\ddot{\nu}$ measured in (Hobbs et al. 2004), which are well consistent with the $|\ddot{\nu}|-\dot{\nu}$ correlation (Cordes and Downs 1985; Arzoumanian et al. 1994) and have a wide range of $\dot{\nu}$. This shows a possible difference between timing noise and the long timescale variations described above.

In any case, the principal point is that all the pulsars evolve with long-term variations, and the timescale of such variations significantly exceeds several tens of years. That explains the anomalous values of the observed $\ddot{\nu}$ and braking indices and gives reasonable values of the underlying secular spin-down parameters.

References

Arzoumanian, Z., et al.: Astrophys. J. **422**, 671 (1994)

Baykal, A., et al.: Mon. Not. Roy. Astron. Soc. **306**, 207 (1999)

Chukwude, A.E.: Astron. Astrophys. **406**, 667 (2003)

Cordes, J.M., Greenstein, G.: Astrophys. J. **245**, 1060 (1981)

Cordes, J.M., Downs, G.S.: Astrophys. J. Suppl. Ser. **59**, 343 (1985)

D'Alessandro, F., et al.: Mon. Not. Roy. Astron. Soc. **261**, 883 (1993)

D'Alessandro, F., et al.: Mon. Not. Roy. Astron. Soc. **277**, 1033 (1995)

Gurevich, A., Beskin, V., Istomin, Ya.: Physics of the Pulsar Magnetosphere. Cambridge University Press, Cambridge (1993)

Hobbs, G., et al.: Mon. Not. Roy. Astron. Soc. **353**, 1311 (2004)

Lyne, A.: In: Arzoumanian, Z., van der Hooft, F., van den Heuvel, E.P.J. (eds.) Pulsar Timing, General Relativity and the Internal Structure of Neutron Stars. Amsterdam, p. 141 (1999)

Manchester, R.N., et al.: Astron. J. **129**, 1993 (2005)

Manchester, R.N., Taylor, J.H.: Pulsars. Freeman, San Francisco (1977)

Shemar, A.L., Lyne, A.G.: Mon. Not. Roy. Astron. Soc. **282**, 677 (1996)

Stairs, I.H., Lyne, A.G., Shemar, A.L.: Nature **406**, 484 (2000)

Urama, J.O., Link, B., Weisberg, J.M.: Mon. Not. Roy. Astron. Soc. **370**, L76 (2006)

Astrophys Space Sci (2007) 308: 557–561
DOI 10.1007/s10509-007-9371-5

ORIGINAL ARTICLE

Gravitational waves from r-modes

Paulo M. Sá · Brigitte Tomé

Received: 27 June 2006 / Accepted: 11 October 2006 / Published online: 20 March 2007

Abstract Detectability of gravitational waves emitted by newly born, hot, rapidly rotating neutron stars as they spin down due to the r-mode instability is discussed. It is shown that differential rotation induced by r-modes plays a fundamental role in the evolution of the mode's instability, making it more difficult to detect these gravitational waves.

Keywords Neutron stars · Stellar pulsation · Gravitational waves

PACS 97.60.Jd · 97.10.Sj · 04.30.Db

1 Introduction

Neutron stars are promising sources of gravitational waves (for a recent review see Kokkotas and Stergioulas 2006). In particular, neutron stars emit gravitational radiation due to r-modes, which are non-radial pulsation modes of rotating stars that have the Coriolis force as their restoring force and a characteristic frequency comparable to the rotation speed of the star (Papaloizou and Pringle 1978). These modes, which are driven unstable by gravitational radiation (Andersson 1998; Friedman and Morsink 1998), induce, at second order in the mode's amplitude, differential rotation (Rezzolla et al. 2000; Sá 2004), which plays an important role in the nonlinear evolution of the r-mode instability (Sá and Tomé 2005).

Detectability of gravitational waves emitted by newly born, hot, rapidly rotating neutron stars as they spin down due to the r-mode instability was first investigated by Owen et al. (1998), assuming a saturation amplitude of the mode of order unity. Afterwards, Arras et al. (2003) and Brink et al. (2004, 2005) showed that nonlinear mode-mode interactions limit the r-mode saturation amplitude to values much lower than unity, implying that the detection of such gravitational waves is more difficult than initially supposed. Recently, the issue of detectability of these gravitational waves was again addressed, but taking into account another nonlinear effect, namely, differential rotation induced by r-modes (Sá and Tomé 2006). In this article we discuss these recent results, showing that differential rotation may also limit the saturation amplitude of r-modes to values much lower than unity, making it more difficult to detect gravitational waves from newly born neutron stars.

2 R-modes in the linearized theory

The linearized fluid equations for a uniformly rotating, Newtonian, barotropic, perfect-fluid star in an inertial frame,

$$\partial_t \delta^{(1)}\mathbf{v} + (\delta^{(1)}\mathbf{v}\cdot\nabla)\mathbf{v} + (\mathbf{v}\cdot\nabla)\delta^{(1)}\mathbf{v} = -\nabla\delta^{(1)}U, \quad (1)$$

$$\partial_t \delta^{(1)}\rho + \mathbf{v}\cdot\nabla\delta^{(1)}\rho + \nabla\cdot(\rho\delta^{(1)}\mathbf{v}) = 0, \quad (2)$$

$$\nabla^2\delta^{(1)}\Phi = 4\pi G\delta^{(1)}\rho, \quad (3)$$

admit, at lowest order in the star's angular velocity Ω, the r-mode solution

$$\delta^{(1)}v_r = 0, \quad (4)$$

This work was supported in part by the *Fundação para a Ciência e a Tecnologia*, Portugal.

P.M. Sá (✉) · B. Tomé
Centro Multidisciplinar de Astrofísica – CENTRA, Departamento de Física, Faculdade de Ciências e Tecnologia, Universidade do Algarve, Campus de Gambelas, 8005-139 Faro, Portugal
e-mail: pmsa@ualg.pt; btome@ualg.pt

$$\delta^{(1)} v_\theta = \alpha\Omega C_l l R (r/R)^l \sin^{l-1}\theta \sin(l\phi + \omega t), \tag{5}$$

$$\delta^{(1)} v_\phi = \alpha\Omega C_l l R (r/R)^l \sin^{l-1}\theta \cos\theta \cos(l\phi + \omega t), \tag{6}$$

$$\delta^{(1)} U = \alpha\Omega^2 \frac{2C_l l}{l+1} R^2 \left(\frac{r}{R}\right)^{l+1} \sin^l\theta \cos\theta \times \cos(l\phi + \omega t), \tag{7}$$

where $C_l = (2l-1)!!\sqrt{(2l+1)/[2\pi(2l)!l(l+1)]}$, $\delta^{(1)}U \equiv \delta^{(1)}p/\rho + \delta^{(1)}\Phi$ and the frequency of the mode in the inertial frame is $\omega = -\Omega l + 2\Omega/(l+1)$ (Papaloizou and Pringle 1978). In the above expressions, $\mathbf{v}$, ρ and R are, respectively, the fluid velocity, the mass density and the radius of the unperturbed star, and ρ is related to the pressure p by an equation of state of the form $p = p(\rho)$. The quantities $\delta^{(1)}\mathbf{v}$, $\delta^{(1)}p$, $\delta^{(1)}\rho$, $\delta^{(1)}\Phi$ are, respectively, the first-order Eulerian changes in velocity, pressure, density, and gravitational potential, and α is the dimensionless amplitude of the r-mode perturbation.

When the gravitational radiation reaction force is taken into account, the hydrodynamic equations (1–3) yield velocity perturbations which have the same sinusoidal behavior and the same frequency as the gravitational radiation force-free velocity perturbations, but with an amplitude which grows exponentially (Dias and Sá 2005).

3 Nonlinear r-modes

Let us now consider r-modes within the nonlinear theory up to second order in the mode's amplitude α. At lowest order in Ω the hydrodynamic equations

$$\partial_t \delta^{(2)}\mathbf{v} + (\delta^{(2)}\mathbf{v}\cdot\nabla)\mathbf{v} + (\mathbf{v}\cdot\nabla)\delta^{(2)}\mathbf{v} + (\delta^{(1)}\mathbf{v}\cdot\nabla)\delta^{(1)}\mathbf{v} = -\nabla\delta^{(2)}U + (\delta^{(1)}\rho/\rho)\nabla(\delta^{(1)}p/\rho), \tag{8}$$

$$\partial_t \delta^{(2)}\rho + \mathbf{v}\cdot\nabla\delta^{(2)}\rho + \nabla\cdot(\rho\delta^{(2)}\mathbf{v}) + \nabla\cdot(\delta^{(1)}\rho\delta^{(1)}\mathbf{v}) = 0, \tag{9}$$

$$\nabla^2\delta^{(2)}\Phi = 4\pi G\delta^{(2)}\rho, \tag{10}$$

admit the axisymmetric time-independent solution (Sá 2004)

$$\delta^{(2)} v_r = \delta^{(2)} v_\theta = 0, \tag{11}$$

$$\delta^{(2)} v_\phi = \frac{1}{2}\alpha^2\Omega C_l^2 l^2 (l^2-1) R \left(\frac{r}{R}\right)^{2l-1} \sin^{2l-3}\theta + \alpha^2\Omega A r^N \sin^N\theta, \tag{12}$$

$$\delta^{(2)} U = -\frac{1}{4}\alpha^2\Omega^2 C_l^2 l R^2 \left(\frac{r}{R}\right)^{2l} \sin^{2l-2}\theta(\sin^2\theta - 2l^2) + \alpha^2\Omega^2 \frac{2A}{N+1} r^{N+1} \sin^{N+1}\theta, \tag{13}$$

where $\delta^{(2)}U \equiv \delta^{(2)}p/\rho + \delta^{(2)}\Phi$, A and N are constants fixed by the choice of initial data. In what follows, for

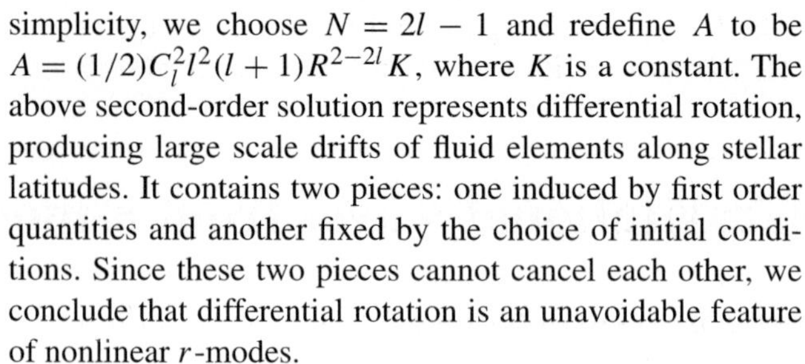
simplicity, we choose $N = 2l-1$ and redefine A to be $A = (1/2)C_l^2 l^2 (l+1) R^{2-2l} K$, where K is a constant. The above second-order solution represents differential rotation, producing large scale drifts of fluid elements along stellar latitudes. It contains two pieces: one induced by first order quantities and another fixed by the choice of initial conditions. Since these two pieces cannot cancel each other, we conclude that differential rotation is an unavoidable feature of nonlinear r-modes.

The second-order physical angular momentum of the r-mode perturbation is, at lowest order in Ω, given by Sá (2004)

$$\delta^{(2)} J = \int_V \rho\delta^{(2)} v_\phi r \sin\theta \, dV = \alpha^2\Omega \frac{2l^2 + (2K-1)l - 1}{2R^{2l-2}} \int_0^R \rho r^{2l+2} dr. \tag{14}$$

Before the discovery of the second-order solution (11–13), the physical angular momentum of r-modes was assumed to be equal to the canonical angular momentum, a quantity which is quadratic in first-order quantities and, therefore, could be evaluated with the knowledge of just the first-order r-mode solution. However, this assumption was based on the belief that, at second order, r-modes preserve the vorticity of fluid elements, which is not the case, as proven explicitly by Sá (2004) using the second-order solution (11–13).

4 Evolution of the r-mode instability

We now analyze the evolution of the r-mode instability in a newly born, hot, rapidly rotating neutron star, using the model of Sá and Tomé (2005). This model, adapted from the one proposed by Owen et al. (1998), takes into account the existence of differential rotation induced by r-modes. We focus our attention exclusively on the $l = 2$ r-mode, since it has the smallest growth timescale, and we assume that the star's mass density ρ and pressure p are related by a polytropic equation of state $p = k\rho^2$ with k such that the mass and radius of the star are, respectively, $M = 1.4M_\odot$ and $R = 12.53$ km.

Within our model of evolution of the r-mode instability, the total angular momentum of the neutron star is the sum of two components, the angular momentum of the unperturbed star and the angular momentum of the perturbation, namely,

$$J(\Omega, \alpha) = I\Omega + \delta^{(2)} J, \tag{15}$$

where $I = \tilde{I}MR^2$, with $\tilde{I} = 8\pi\int_0^R \rho r^4 dr/(3MR^2) = 0.261$, is the momentum of inertia of the unperturbed star and the angular momentum of the r-mode perturbation is given by [see (14)]

$$\delta^{(2)} J = \frac{1}{2}\alpha^2\Omega(4K+5)\tilde{J}MR^2, \tag{16}$$

with $\tilde{J} \equiv \int_0^R \rho r^6 dr/(MR^4) = 1.635 \times 10^{-2}$. Let us recall that the constant K, introduced in Sect. 3, is fixed by the choice of initial data and gives the initial amount of differential rotation associated to the r-mode. In this article we consider $-5/4 \leq K \ll 10^{13}$ (for a discussion of the full range of values of K, see Sá and Tomé 2005).

The total angular momentum of the star is a function of just two variables, Ω and α, which are determined by the system of differential equations (Sá and Tomé 2005),

$$\frac{d\Omega}{dt} = \frac{8}{3}(K+2)Q\alpha^2 \frac{\Omega}{\tau_{GR}} + \frac{8}{3}\left(K+\frac{5}{4}\right)Q\alpha^2 \frac{\Omega}{\tau_V}, \tag{17}$$

$$\frac{d\alpha}{dt} = -\left[1+\frac{4}{3}(K+2)Q\alpha^2\right]\frac{\alpha}{\tau_{GR}} - \left[1+\frac{4}{3}\left(K+\frac{5}{4}\right)Q\alpha^2\right]\frac{\alpha}{\tau_V}, \tag{18}$$

where $Q \equiv 3\tilde{J}/(2\tilde{I}) = 0.094$ and the gravitational-radiation and viscous timescales are given, respectively, by (Lindblom et al. 1998)

$$\frac{1}{\tau_{GR}} = \frac{1}{\hat{\tau}_{GR}}\left(\frac{\Omega}{\Omega_K}\right)^6, \tag{19}$$

$$\frac{1}{\tau_V} = \frac{1}{\hat{\tau}_S}\left(\frac{10^9\ \mathrm{K}}{T}\right)^2 + \frac{1}{\hat{\tau}_B}\left(\frac{T}{10^9\ \mathrm{K}}\right)^6\left(\frac{\Omega}{\Omega_K}\right)^2, \tag{20}$$

with the fiducial timescales $\hat{\tau}_{GR} = -37.13$ s, $\hat{\tau}_S = 2.52 \times 10^8$ s (Lindblom et al. 1998) and $\hat{\tau}_B = 4.52 \times 10^{11}$ s (Lindblom et al. 1999). In the above expressions, $\Omega_K = (2/3)\sqrt{\pi G\bar{\rho}} = 5612\ \mathrm{s}^{-1}$ is the Keplerian angular velocity at which the star starts shedding mass at the equator. Finally, the temperature of the star is assumed to decrease due to the emission of neutrinos via a modified URCA process,

$$\frac{T(t)}{10^9\ \mathrm{K}} = \left[\frac{t}{\tau_c} + \left(\frac{10^9\ \mathrm{K}}{T_0}\right)^6\right]^{-1/6}, \tag{21}$$

where $\tau_c = 1$ yr characterizes the cooling rate (Owen et al. 1998). The initial amplitude of the r-mode perturbation, the initial angular velocity and temperature of the star are chosen to be, respectively, $\alpha_0 = 10^{-6}$, $\Omega_0 = \Omega_K$ and $T_0 = 10^{11}$ K. The r-mode instability is active from $t \approx 0$ to $t = t_f$, where t_f lies between 3.6×10^6 s, for $K = -5/4$, and 7.1×10^6 s, for $K \gg 1$ (Sá and Tomé 2006).

Solving numerically the above system of differential equations we arrive at the conclusion that (Sá and Tomé 2005): (i) initially, the amplitude of the r-mode grows exponentially (as expected, according to the results of the linearized theory, discussed in Sect. 2) and after a few hundred seconds saturates at values which depend on K, namely, $\alpha_{sat} \propto (K+2)^{-1/2}$; (ii) the angular velocity of the star decreases to values which are quite insensitive to K, namely, $\Omega(t_f) = (0.065\text{–}0.067)\Omega_0$.

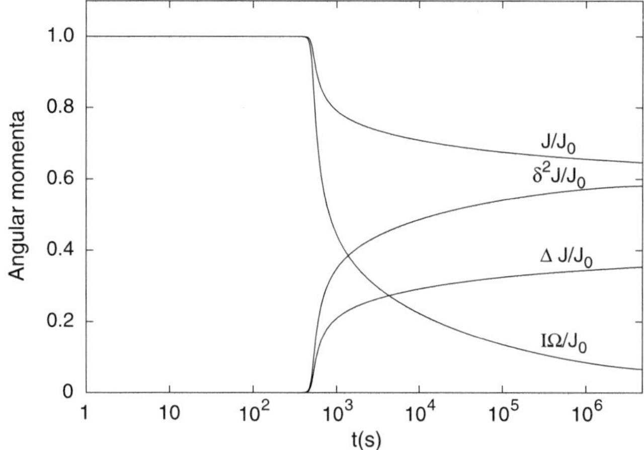

Fig. 1 Time evolution of the total angular momentum of the star, J, the angular momentum of the unperturbed star, $I\Omega$, the angular momentum of the r-mode perturbation, $\delta^{(2)}J$, and the angular momentum carried away by gravitational waves, ΔJ, for $K = 0$

Of great importance to the issue of detectability of gravitational waves from r-modes, as we will see in the next section, is the determination of the angular momentum carried away by gravitational waves, $\Delta J \equiv J_0 - J(t)$, where $J_0 \approx I\Omega_0$ is the initial angular momentum of the star. As can be seen in Fig. 1 for the case $K = 0$, despite the fact that the angular momentum of the unperturbed star decreases to about 7% of its initial value, the total angular momentum of the star decreases just to about 65% of its initial value. This is due to the fact that part of the initial angular momentum of the star (more exactly, about 58%) is transferred to the r-mode, a consequence of the rapid increase of the average differential rotation (Sá and Tomé 2005). Therefore, only about 35% of the initial angular momentum is carried away by gravitational waves.[1] For higher values of K, the amount of angular momentum carried away by gravitational waves decreases even more, namely, $\Delta J/J_0 < 1\%$ for $K > 100$.

5 Detectability of gravitational waves

As we have seen in the previous section, during the evolution of the r-mode instability, part of the initial angular momentum of a newly born neutron star is carried away by gravitational waves. The frequency of these waves is

[1] In the first investigations on the detectability of gravitational waves from r-modes (Owen et al. 1998), because differential rotation was not taken into account, almost all the angular momentum of the star was assumed to be carried away by gravitational waves (about 90% for $\alpha_{sat} = 1$, see Sá and Tomé 2005 for details).

$f = 2\Omega/(3\pi)$, while the amplitude is given by Owen et al. (1998)

$$|h(t)| = 1.3 \times 10^{-24}\alpha(t)\left(\frac{\Omega(t)}{\Omega_K}\right)^3 \frac{20\ \text{Mpc}}{D}, \tag{22}$$

where D is the distance to the source. To obtain the corresponding gravitational wave amplitude in the frequency domain we use the stationary phase approximation, $|\tilde{h}(f)| = |h(t)|/\sqrt{|df/dt|}$, where df/dt follows straightforwardly from (17),

$$\begin{aligned}\frac{df}{dt} = \alpha^2\bigg[&-8.1(K+2)\left(\frac{f}{f_{\max}}\right)^7 \\ &+ 0.02\left(K+\frac{5}{4}\right)\left(\frac{f}{f_{\max}}\right)^3 \frac{\text{s}}{t} \\ &+ 3.8 \times 10^{-9}\left(K+\frac{5}{4}\right)\frac{f}{f_{\max}}\left(\frac{t}{\text{s}}\right)^{1/3}\bigg]\ \text{Hz/s},\end{aligned} \tag{23}$$

where the maximum frequency of the gravitational wave, $f_{\max} = 1191$ Hz, corresponds to the initial value of the angular velocity of the star, $\Omega_K = 5612\ \text{s}^{-1}$. In the previous expression, the second term on the right hand side becomes smaller than the first term after just a fraction of second, while the third term only becomes comparable to the first one after $t = t_f$. Therefore, we retain only the first term on the right hand side of (23), implying that $|\tilde{h}(f)|$ is given by

$$|\tilde{h}(f)| = \frac{4.6 \times 10^{-25}}{\sqrt{K+2}}\sqrt{\frac{f_{\max}}{f}}\frac{20\ \text{Mpc}}{D}\ \text{Hz}^{-1}. \tag{24}$$

We now investigate the possibility of detecting these gravitational waves using the Advanced Laser Interferometer Gravitational Wave Observatory (in what follows referred to simply as Advanced LIGO), whose planned noise power spectral density (Abramovici et al. 1992) is well approximated, for 50 Hz $\le f \le$ 1200 Hz, by the analytical expression (Sá and Tomé 2006)

$$\begin{aligned}S_h(f) = S_1\bigg\{1 + \left(\frac{f_1}{f}\right)^7 - \frac{10}{3}\left(\frac{f}{f_2}\right)\bigg[1 - \left(\frac{f}{f_2}\right) \\ + \frac{3}{50}\left(\frac{f}{f_2}\right)^2\bigg]\bigg\},\end{aligned} \tag{25}$$

where $S_1 = 2.2 \times 10^{-47}\ \text{Hz}^{-1}$, $f_1 = 52.8$ Hz and $f_2 = 421.3$ Hz. Matched filtering is assumed in order to estimate the signal-to-noise ratio, which is given by

$$\left(\frac{S}{N}\right)^2 = 2\int_{f_{\min}}^{f_{\max}} \frac{df}{f}\left(\frac{h_c}{h_{\text{rms}}}\right)^2, \tag{26}$$

where $h_c = f|\tilde{h}(f)|$ is the characteristic amplitude of the signal and $h_{\text{rms}} = \sqrt{f S_h(f)}$ is the rms strain noise in the detector. In the above equation, $f_{\min}$ is the minimum frequency of the gravitational wave, corresponding to the final angular velocity of the star, $\Omega(t_f)$; its value is quite insensitive to K and lies in the range 77 Hz ($K = -5/4$) to 80 Hz ($K \gg 1$).

Using (24) and (25) we obtain that the signal-to-noise ratio for Advanced LIGO,

$$\frac{S}{N} = \frac{12.9}{\sqrt{K+2}}\frac{20\ \text{Mpc}}{D}, \tag{27}$$

is significant for sources located at distances up to 20 Mpc provided that the initial amount of differential rotation associated to r-modes is small ($K \approx 0$). However, the situation is quite different when the neutron star is born with substantial differential rotation. If, for instance, $K = 10^5$–10^6, then S/N is significant only if the source is located within our Galaxy.

To finish this section, let us mention that, as shown by Owen and Lindblom (2002), the signal-to-noise ratio S/N is, within a good approximation, proportional to $\sqrt{\Delta J}$, where ΔJ is the angular momentum carried away by gravitational radiation. As mentioned at the end of Sect. 4, ΔJ decreases as K increases. Therefore, our results are clear from a physical point of view: the decrease of the signal-to-noise ratio, as the initial amount of differential rotation increases (for a fixed distance to the source), is due to the fact that less angular momentum is carried away by gravitational radiation.

6 Conclusions

In this article, the influence of differential rotation on the detectability of gravitational waves from r-modes of newly born neutron stars was discussed. We have used the second-order r-mode solution of Sá (2004), which represents differential rotation, to calculate the physical angular momentum of the r-mode perturbation. This quantity, in turn, plays a crucial role in the evolution model of the r-mode instability (Sá and Tomé 2005). Within this model one can determine the time evolution of the star's angular velocity and the mode's amplitude, two quantities which are used to compute the gravitational wave amplitude (Sá and Tomé 2006). Assuming matched filtering, the characteristic amplitude of the signal (related, in a simple way, to the gravitational wave amplitude in the frequency domain) is compared to the rms strain noise in the detector, showing that the signal-to-noise ratio decreases as the initial amount of differential rotation associated to r-modes increases (Sá and Tomé 2006). Therefore, one can conclude that differential rotation induced by r-modes plays a fundamental role in the evolution of the mode's instability, making it more difficult to detect the gravitational waves emitted by newly born, hot, rapidly rotating neutron stars as they spin down due to the r-mode instability.

References

Abramovici, A., et al.: Science **256**, 325 (1992)

Andersson, N.: Astrophys. J. **502**, 708 (1998)

Arras, P., Flanagan, E.E., Morsink, S.M., Schenk, A.K., Teukolsky, S.A., Wasserman, I.: Astrophys. J. **591**, 1129 (2003)

Brink, J., Teukolsky, S.A., Wasserman, I.: Phys. Rev. D **70**, 124017 (2004)

Brink, J., Teukolsky, S.A., Wasserman, I.: Phys. Rev. D **71**, 064029 (2005)

Dias, Ó.J.C., Sá, P.M.: Phys. Rev. D **72**, 024020 (2005)

Friedman, J.L., Morsink, S.M.: Astrophys. J. **502**, 714 (1998)

Kokkotas, K.D., Stergioulas, N.: In: Mourão, A.M., Pimenta, M., Potting, R., Sá, P.M. (eds.) Proceedings of the Fifth International Workshop on New Worlds in Astroparticle Physics, Faro, Portugal, 2005, p. 25. World Scientific, Singapore (2006)

Lindblom, L., Owen, B.J., Morsink, S.M.: Phys. Rev. Lett. **80**, 4843 (1998)

Lindblom, L., Mendell, G., Owen, B.J.: Phys. Rev. D **60**, 064006 (1999)

Owen, B.J., Lindblom, L., Cutler, C., Schutz, B.F., Vecchio, A., Andersson, N.: Phys. Rev. D **58**, 084020 (1998)

Owen, B.J., Lindblom, L.: Class. Quantum Grav. **19**, 1247 (2002)

Papaloizou, J., Pringle, J.E.: Mon. Not. Roy. Astron. Soc. **182**, 423 (1978)

Rezzolla, L., Lamb, F.K., Shapiro, S.L.: Astrophys. J. **531**, L139 (2000)

Sá, P.M.: Phys. Rev. D **69**, 084001 (2004)

Sá, P.M., Tomé, B.: Phys. Rev. D **71**, 044007 (2005)

Sá, P.M., Tomé, B.: Phys. Rev. D **74**, 044011 (2006)

Astrophys Space Sci (2007) 308: 563–567
DOI 10.1007/s10509-007-9347-5

ORIGINAL ARTICLE

Giant pulses of pulsar radio emission

A.D. Kuzmin

Received: 29 June 2006 / Accepted: 22 August 2006 / Published online: 27 March 2007

Abstract We present a brief review of observational manifestations of pulsars with giant pulses radio emission, based on the survey of the main properties of known pulsars with giant pulses, including our detection of 4 new pulsars with giant pulses.

Keywords Neutron stars · Pulsars · Giant pulses

1 Introduction

Giant pulses (GPs), a short-duration outbursts, are a special form of pulsar radio emission.

This is the most striking phenomena of pulsar radio emission. Their flux densities can exceed thousands of times the mean flux density of regular pulses from the pulsar.

This rare phenomenon has been detected only in 11 pulsars out of more than 1500 known ones. The firsts of them were Crab pulsar PSR B0531+21 (Staelin and Reifenstein 1968; Gower and Argyle 1971; Argyle and Gower 1972) and the millisecond pulsar PSR B1937+21 (Wolszczan et al. 1984).

For over 20 years only these two pulsars which emit GPs were known. In the last decade GPs emission from the further 9 pulsars has been detected. They are PSR B0031-07 (Kuzmin et al. 2004), PSR J0218+42 and PSR B1957+20 (Joshi et al. 2004), PSR B0540-69 (Johnston and Romani 2003), PSR B0656+14 (Kuzmin and Ershov 2006), PSR B1112+50 (Ershov and Kuzmin 2003), PSR J1752+2359 (Ershov and Kuzmin 2005), PSR B1821-24 (Romani and Johnston 2001), PSR J1823-3021 (Knight et al. 2005).

A.D. Kuzmin (✉)
Lebedev Physical Institute, 142292 Pushchino, Moscow region, Russia
e-mail: akuzmin@prao.ru

2 General properties

The general view of a giant pulse is demonstrated in Fig. 1 for the Crab pulsar PR B0531+21. One can see one giant pulse which greatly stands out of a sequence of 160 much weaker "normal" individual pulses, emerged into the noise inside 160 pulsar periods.

Giant pulses (GPs) of pulsars are distinguished by several special properties:

The peak intensities and energies of GPs greatly exceed the peak intensity and energy of the average pulse (AP). The peak flux density of the strongest GP for Crab pulsar exceeds the peak flux density of AP by a factor of 5×10^5 (Kostyuk et al. 2003). The energy excess of GP over AP is about of 100.

Giant pulses are very short and bright. Soglasnov et al. (2004) have shown that majority of giant pulses from the millisecond pulsar B1937+21 are shorter than 15 ns. Hankins et al. (2003) found Crab pulsar pulse structure as short as 2 ns. If one interprets the pulse duration τ in terms of the maximum possible size of emitting region $r \leq c\tau$, then 2 ns corresponds to a size of emitting body of only 60 cm, the smallest entity ever detected outside our solar system.

A brightness temperature of GPs are $T_B \geq 5 \times 10^{37}$ K, for Crab pulsar B0531+21 (Hankins et al. 2003) and $T_B \geq 5 \times 10^{39}$ K, for PSR B1937+21 (Soglasnov et al. 2004)—the highest observed in the Universe.

However, these evaluations of a size of an emitting body and a brightness temperature is not unambiguous. Gil and Melikidze (2005) argued, that the apparent duration of the observed impulse $\tau_{\rm obs}$ may be shorter than the duration of

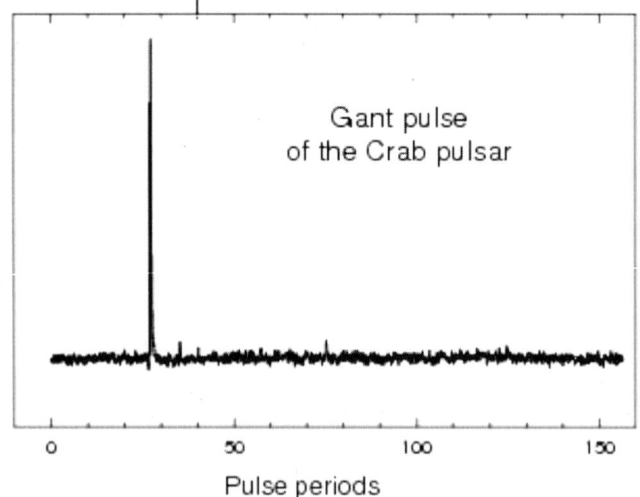

Fig. 1 Giant pulse in Crab pulsar at 111 MHz

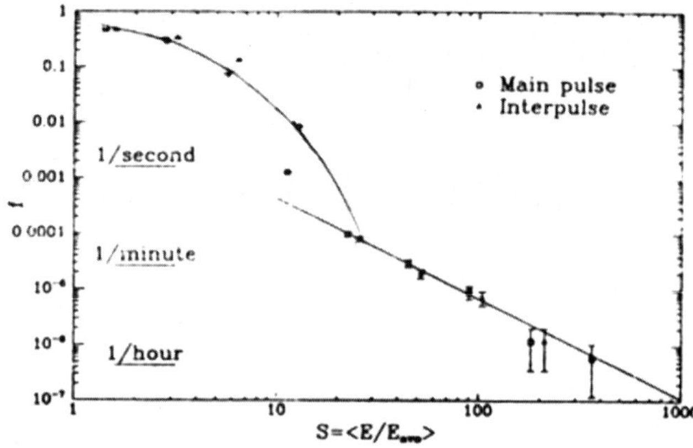

Fig. 2 Energy distribution of GPs in pulsar PSR B1937+21 (after Cognard et al. 1996)

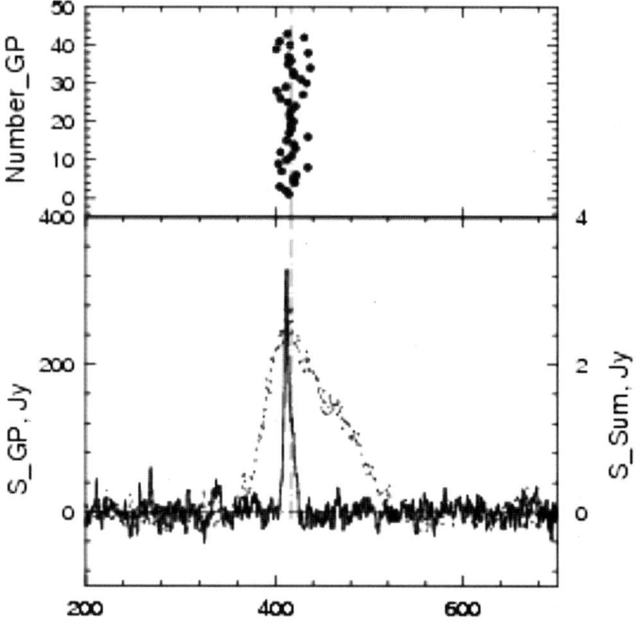

Fig. 3 (*Top*) The phase distribution of GPs; (*Bottom*) The observed GP (*bold line*) and the AP in pulsar PSR B0031-07 (after Kuzmin et al. 2004). For better seeing of an AP its plot is presented on the 100 times larger scale

the emitted one τ_{rad} as $\tau_{obs} = \tau_{rad} \times \gamma^{-2}$. For Lorentz factor $\gamma \approx 100$, the time-scale of pulse structure for Crab pulsar will transformed from observed $\tau_{obs} = 2$ ns to emitted $\tau_{rad} = 20$ µs. In its turn, the brightness temperature $T_B = S\lambda^2/2\,k\Omega$, where the solid angle of an emitting region $\Omega \propto \tau_{rad}^2$ will be γ^4 times lower, that is $T_B \approx 10^{30}$ K.

This aspect needs a further refinement.

GPs of the Crab pulsar were detected in a very wide frequency range from 23 MHz (Popov et al. 2006) up to 15 GHz (Hankins 2000).

Radio spectra of these GPs were studied by simultaneous multi-frequency observations by Sallmen et al. (1999) and Popov et al. (2006). In Sallmen et al. observations at two frequencies 1.4 and 0.6 GHz the GPs spectral indices fall between −2.2 and −4.9, which may be compared to the average pulse value for this pulsar −3.0. In Popov et al. observations at three frequencies 600, 111 and 43 MHz the GPs spectral indices fall between −1.6 and −3.1 with mean value −2.7, that also may be compared to the average pulse value for this pulsar.

Simultaneous two-frequency observations of GPs from PSR B1937+21 at 2210–2250 and 1414–1446 MHz (Popov and Stappers 2000) do not reveal any GPs to occur simultaneously in both frequency ranges. They conclude that radio spectra of detected GPs are limited in frequency at a scale of about $\Delta\nu/\nu < 0.5$.

Kinkhabwala and Thorset (2000) have estimated the average spectral properties of the GPs emission of the pulsar PSR B1937+21. At three frequencies 430, 1420 and 2380 MHz, they find a slope of −3.1 for the GPs spectrum, compared the −2.6 slope for the normal emission spectrum of this pulsar.

The distinguishing characteristic of pulsars with GPs is that above a certain threshold the pulse strength distribution is roughly a power-law. An example of the energy distribution of GPs in pulsar PSR B1937+21 is shown in Fig. 2.

GPs are clustered around a small phase window (except of Crab pulsar, for which GPs can occur anywhere within the average pulse). An example of such clustering for pulsar PSR B0031-07 is shown in the top of Fig. 3.

A common property of pulsars PSR B0531+21 and PSR B1937+21 are the extremely high magnetic field at the light cylinder $B_{LC} \approx 10^6$ G. Therefore, it was suggested that the giant pulses radio emission may depend on conditions at the light cylinder, rather than close to the stellar surface. Then, the first searches of GPs were performed in pulsars with extremely high magnetic field at the light cylinder. Five more such pulsars with GPs PSR J0218+4332 and B1957+20 (Joshi et al. 2004), B0540-69 (Johnston and Romani 2003), B1821-24 (Romani and Johnston 2001), J1823-3021A (Knight et al. 2005) were detected.

Kuzmin and co-workers have detected GPs in four pulsars PSR B0031-07, B0656+14, B1112+50 and J1752+2359 with an ordinary magnetic field at the light cylinder $B_{LC} = 10–10^3$ G. These pulsars exhibit all characteristic features of the classical GPs in PSR B0531+21 and PSR B1937+21. The peak intensities of GPs greatly exceed the peak intensity of the AP. The energy distribution of GPs has a power-law. The GPs are much narrower than the AP and their phases are clustered in a narrow window inside the AP. Moreover, the energy excess of GPs over AP is compatible or even larger than for classical pulsars with GPs.

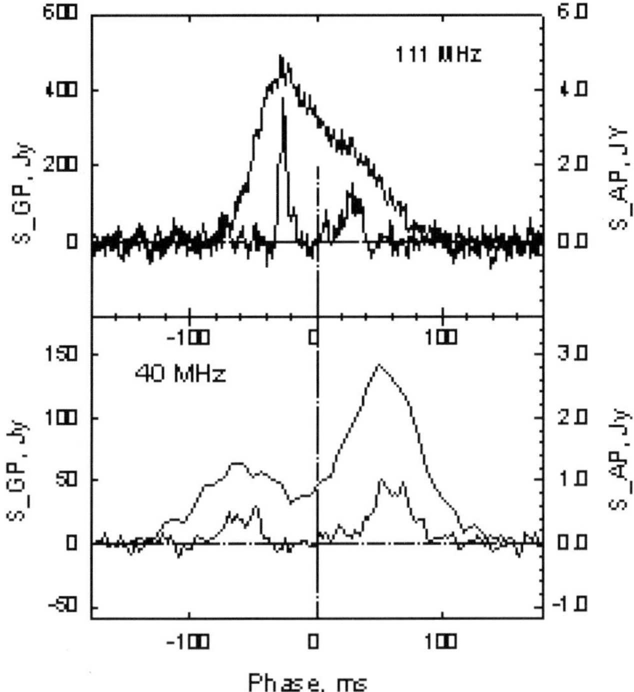

Fig. 4 The twofold GPs (*bold line*) and APs (*thin line*) at 111 and 40 MHz (after Kuzmin and Ershov 2004)

Kuzmin and Ershov (2004) have revealed that GPs in pulsar PSR B0031-07 occur in two distinct phase regions (Fig. 4.). This indicates that there are two emission regions of GPs. The separation of these regions at 40 MHz is larger than at 111 MHz. This is similar to the frequency dependence in the width of the AP, which is interpreted as a divergence of the magnetic field lines in the hollow cone model of pulsar radio emission. This suggests that the GPs from this pulsar originate in the same region as the AP, that is in a hollow cone over the polar cap instead the light cylinder region.

This approach correlate with Gil and Melikidze (2005) model that GPs are generated in the inner gap potential drop above the polar cap.

One may suggest that there are two classes of GPs, one associated with high-energy emission from the outer gaps, the other associated with polar radio emission. GPs of PSR J0218+4332, PSR B0531+21, PSR B0540-69, PSR B1821-24, PSR J1823-3021, PSR B1937+21 and PSR B1957+20 may be of the first class. GPs of PSR B0031-07, B0656+14, B1112+50 and PSR J1752+2359 may be of the second class.

In Table 1 we summarize the comparative data for all known pulsars with GPs, for which the data of energy E^{GP} or energy excess factor E^{GP}/E^{AP} has been published or may be derived. Here PSR is a pulsar name, P is pulsar period, B_{LC}—magnetic field strength at the light cylinder, Freq—observation's frequency, S_{GP}/S_{AP}—an excess of the peak flux density of a strongest GP over the peak flux density of an AP, T_B—brightness temperature of the strongest

Table 1 General properties of giant pulses

PSR	P ms	$\log B_{LC}$ G	Freq GHz	S^{GP}/S^{AP}	T_B K	E^{GP}/E^{AP}	Ref
B0031-07	943	7	0.04	400	10^{28}	15	Kuzmin and Ershov (2004)
J0218+4332	2.3	3.2×10^5	0.61	–	–	51	Joshi et al. (2004)
B0531+21	33	9.8×10^5	2.23	5×10^5	10^{34}	80	Kostyuk et al. (2003)
			5.5	–	10^{37}	–	Hankins et al. (2003)
B0540-69	50.9	3.5×10^5	1.38	5×10^3	–	–	Johnston and Romani (2003)
B0656+14	385	770	0.11	600	10^{26}	110	Kuzmin and Ershov (2006)
B1112+50	1656	4.2	0.11	80	–	10	Ershov and Kuzmin (2003)
J1752+2359	409	71	0.11	260	10^{28}	200	Ershov and Kuzmin (2005)
B1821-24	3.0	7.2×10^5	1.51	–	–	81	Romani and Johnston (2001)
J1823-3021A	5.4	2.5×10^5	6.85	–	–	64	Knight et al. (2005)
B1937+21	1.5	10×10^5	1.65	–	10^{39}	60	Soglasnov et al. (2004)
B1957+20	1.6	3.8×10^5	0.61	–	–	129	Joshi et al. (2004)

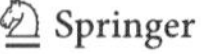

GP, E_{GP}/E_{GP}—an excess of the energy of the strongest GP over the energy of an AP.

3 Are giant pulses inherent to a special family of pulsars?

The first detected pulsars with GPs PSR B0531+21 and PSR B1937+21 are among a small group of pulsars with extremely strong magnetic field at the light cylinder $B_{LC} = 10^4–10^5$ G. This gave a rise to suggestion that GPs occur in pulsars with extremely strong magnetic field at the light cylinder. This opinion was supported by the detection of five other such pulsars. However, the successive detections of GPs in pulsars PSR B0031-07, PSR B0656+14, PSR 1112+50 and PSR J1752+2359 with ordinary magnetic field at the light cylinder have revealed that GPs can occur in ordinary pulsars too.

Johnston and Romani (2004) believe that a good working definition of GPs is that they appear associated with non-thermal high energy emission. But this association is observed only for six objects among eleven known pulsars with GPs and is not a proper indication for GPs.

Knight et al. (2005) argued that GPs may be indicated by large rotation loss luminosity $\dot{E} \propto P^{-3}\dot{P}$. But in fact a rotation loss luminosity of the known eleven pulsars with GPs differ by six order of magnitude and is not a proper indication for GPs also.

GPs occur in various types of pulsars in a wide range of periods $P = 1.5–1600$ ms and magnetic field at the light cylinder $\log B_{LC} = 4–10^6$ G over a wide range of radio frequencies.

4 Mechanisms of giant pulses radio emission

Several mechanisms of giant pulses radio emission were proposed.

Hankins et al. (2003) claimed that giant pulses radio emission from the Crab pulsar results from the conversion of electrostatic turbulence in the pulsar magnetosphere by the mechanism of spatial collapse of nonlinear wave packets.

Istomin (2004) suggests that giant pulses radio emission is generated in the electric discharge taking place due to the magnetic reconnection of field lines connecting the opposite magnetic poles.

Gil and Melikidze (2005) proposed that giant pulses are generated by means of coherent curvature radiation of charged relativistic solitons associated with sparking discharge of the inner gap potential drop above the polar cap.

Petrova (2004) argued that giant pulses and their substructure can be explaining in the terms of induced Compton scattering of pulsar radio emission of the plasma particles.

5 Summary

Giant pulses are a special form of pulsar radio emission, that is characterized by a very large excess of flux density and energy of radio emission relative to an average pulse ones, the power-low statistic of an energy distribution, giant pulses occur in a narrow-phase window of an average pulse and a short pulse time-scale as compare to an average pulse.

The flux density of giant pulses increase over a flux density of an average pulse S_{GP}/S_{AP} up to 5×10^5.

Giant pulses energy excess over an energy of an average pulse is $E_{GP}/E_{AP} \approx 50–200$ and are nearly the same for different magnetic field at the light cylinder, pulsar periods and frequencies.

A light-travel size of an emitting body indicate the smallest object ever detected outside our solar system.

Giant pulses are the brightest sources of radio emission (the brightness temperature up to 10^{39} K) among the known astronomical objects.

Giant pulses exist in various types of pulsars in a wide range of periods, magnetic field at the light cylinder and broad frequency range.

One may suggests that there are two classes of pulsars with giant pulses:—one associated with emission from the outer gaps, the other associated with polar radio emission.

Acknowledgements This work was supported in part by the Russian Foundation of Basic Research Project 05-02-16415.

References

Argyle, E., Gower, J.F.R.: Astrophys. J. **175**, L89 (1972)
Cognard, I., Shrauner, J.A., Taylor, J.H., et al.: Astrophys. J. **457**, L81 (1996)
Ershov, A.A., Kuzmin, A.D.: Astron. Lett. **29**, 91 (2003)
Ershov, A.A., Kuzmin, A.D.: Astron. Astrophys. **443**, 593 (2005)
Gil, J., Melikidze, G.I.: Astron. Astrophys. **432**, L61 (2005)
Gower, J.F.R., Argyle, E.: Astrophys. J. **171**, L23 (1971)
Hankins T.H.: In: Kramer, M., Wex, N., Wielebinsky, R. (eds.) The Proceedings "Pulsar Astronomy—2000 and Beyoind", Proc. of the IAU Coll N 177, Bonn, Germany, 1999. ASP Conference Series, vol. 165 (2000)
Hankins, T.H., Kern, J.S., Weatherall, J.C., et al.: Nature **422**, 141 (2003)
Istomin, Y.N.: In: Camilo, F., Gaensler, B.M. (eds.) The Proceedings "Young Neutron Stars and Their Environments", Sydney, Australia. ASP, IAU Symposium, vol. 218, p. 369 (2004)
Johnston, S., Romani, R.W.: Astrophys. J. **590**, L95 (2003)
Johnston, S., Romani, R.W.: In: Camilo, F., Gaensler, B.M. (eds.) The Proceedings "Young Neutron Stars and Their Environments", Sydney, Australia. ASP, IAU Symposium, vol. 218, p. 315 (2004)
Joshi, B.C., Kramer, M., Lyne, A.G., et al.: In: Camilo, F., Gaensler, B.M. (eds.) The Proceedings "Young Neutron Stars and Their Environments", Sydney, Australia. ASP, IAU Symposium, vol. 218, p. 319 (2004)
Kinkhabwala, A., Thorset, S.E.: Astrophys. J. **535**, 365 (2000)
Knight, H.S., Bailes, M., Manchester, R.N., et al.: Astrophys. J. **625**, 951 (2005)

Kostyuk, S.V., Kondratiev, V.I., Kuzmin, A.D., et al.: Astron. Lett. **29**, 387 (2003)
Kuzmin, A.D., Ershov, A.A., Losovsky, B.Ya.: Astron. Lett. **30**, 285 (2004)
Kuzmin, A.D., Ershov, A.A.: Astron. Astrophys. **427**, 575 (2004)
Kuzmin, A.D., Ershov, A.A.: Astron. Lett. **32** (2006)
Petrova, S.A.: Astron. Astrophys. **424**, 227 (2004)
Popov, M.V., Stappers, B.: Astron. Rep. **47**, 660 (2000)
Popov, M.V., Kuzmin, A.D., Ulianov, O.V., et al.: Astron. Rep. **50**, 562 (2006)
Romani, R.W., Johnston, S.: Astrophys. J. **557**, L93 (2001)
Sallmen, S., Backer, D.C., Taylor, J.H., et al.: Astrophys. J. **517**, 460 (1999)
Soglasnov, V.A., Popov, M.V., Bartel, N., et al.: Astrophys. J. **616**, 439 (2004)
Staelin, D.H., Reifenstein, E.C.: Science **162**, 1481 (1968)
Wolszczan, A., Cordes, J.M., Stinebring, D.R.: In: Reynolds, S., Stinebring, D. (eds.) The Proceedings "Millisecond Pulsars", Green Bank, USA, p. 63 (1984)

Astrophys Space Sci (2007) 308: 569–573
DOI 10.1007/s10509-007-9307-0

ORIGINAL ARTICLE

On the role of the current loss in radio pulsar evolution

V.S. Beskin · E.E. Nokhrina

Received: 26 June 2006 / Accepted: 11 August 2006 / Published online: 20 March 2007

Abstract The aim of this article is to draw attention to the importance of the electric current loss in the energy output of radio pulsars. We remind that even the losses attributed to the magneto-dipole radiation of a pulsar in vacuum can be written as a result of an Ampere force action of the electric currents flowing over the neutron star surface (see the books of Michel (*Theory of Neutron Star Magnetosphere*. University of Chicago Press (1991)) and of Beskin, Gurevich and Istomin (*Physics of the Pulsar Magnetosphere*. Cambridge Univ. Press (1993)). It is this force that is responsible for the transfer of angular momentum of a neutron star to an outgoing magneto-dipole wave. If a pulsar is surrounded by plasma, and there is no longitudinal current in its magnetosphere, there is no energy loss. It is the longitudinal current closing within the pulsar polar cap that exerts the retardation torque acting on the neutron star. This torque can be determined if the structure of longitudinal current is known. Here we remind of the solution by Beskin, Gurevich and Istomin (ed. cit.) and discuss the validity of such an assumption. The behavior of the recently observed "part-time job" pulsar B1931+24 can be naturally explained within the model of current loss while the magneto-dipole model faces difficulties.

This work was partially supported by the Russian Foundation for Basic Research (Grant no. 05-02-17700) and Dynasty fund. Elena Nokhrina thanks the Conference Organizing Committee for the PPARC/Padova University Grant, and the RFFR for the travel grant (no. 06-02-26645).

V.S. Beskin (✉)
P.N. Lebedev Physical Institute, RAS, Leninsky pr. 53, Moscow, 119991, Russia
e-mail: beskin@lpi.ru

E.E. Nokhrina
Moscow Institute of Physics and Technology, Institutsky per. 9, Dolgoprudny, 141700, Russia
e-mail: nokhrinaelena@gmail.com

Keywords Neutron stars · Magnetosphere · Pulsars

PACS 94.30.-d · 97.60.Gb

1 Magneto-dipole loss

The first idea to explain the energy loss of radio pulsars was to consider the magneto-dipole radiation (Pacini 1967). Indeed, the magneto-dipole formula gives for the radiation power

$$W_{\mathrm{m}d} = \frac{1}{6}\frac{B_0^2 \Omega^4 R^6}{c^3}\sin^2\chi \tag{1}$$

where χ is the angle between rotational and magnetic axis, R is a neutron star radius $\sim$10 km, and Ω is a pulsar angular velocity. This formula explains pulsar activity and observed energy loss for expected large magnetic field near the surface $B_0 \sim 10^{12}$ Gs.

Let us recall that the physical reason of such energy loss is the action of the torque exerted on the pulsar by the Ampere force of the electric currents flowing over the neutron star surface (Istomin 2005). The electric and magnetic fields in the outgoing magneto-dipole wave in vacuum can be found by solving the wave equations $\nabla^2\mathbf{B} + \Omega^2/c^2\mathbf{B} = 0$ and $\nabla^2\mathbf{E} + \Omega^2/c^2\mathbf{E} = 0$ with the boundary conditions stated as the fields corresponding components $\mathbf{E}_{\mathrm{t}}$ and $\mathbf{B}_{\mathrm{n}}$ being continuous through the neutron star surface. Inside the star one can consider magnetic field as homogeneous, and find the corresponding electric field using the frozen-in condition. As a result, the full solution will give us the discontinuity of $\{\mathbf{B}_{\mathrm{t}}\}$ and $\{\mathbf{E}_{\mathrm{n}}\}$ attributed to the surface charge σ_{s} and the surface current $\mathbf{J}_s = (c/4\pi)[\mathbf{n}, \{\mathbf{B}\}]$. The Ampere force exerts the torque

$$\mathbf{K} = \frac{1}{c}\int [\mathbf{r}, [\mathbf{J}_s, \mathbf{B}]]\,\mathrm{d}S \tag{2}$$

on the neutron star. The energy loss of a pulsar due to this torque is equal to (1). Thus, it is the surface current that is responsible for the angular momentum transform from a neutron star to an outgoing magneto-dipole wave (Michel 1991; Beskin et al. 1993).

A pulsar in vacuum loses its rotational energy due to angular momentum transform to the electromagnetic wave at the rate given by (1). However, this is not so if the pulsar magnetosphere is filled with plasma and there is no longitudinal current in the magnetosphere. As was shown by Beskin et al. (1993) and Mestel et al. (1999), in this case the Poynting flux through the light cylinder is equal to zero. Indeed, as the ideal conductivity condition is applicable not only inside the neutron star but outside as well there is no magnetic field discontinuity at the star surface. Consequently, there is no Ampere force acting on a pulsar and, hence, there is no energy loss. For zero longitudinal current the light cylinder is a natural boundary of the pulsar magnetosphere.

2 Current loss

In this section we remind the exact solution for the surface current within the polar cap presented in the monograph by Beskin et al. (1993). As it was shown in the previous section, the neutron star retardation is due to Ampere force $\mathbf{F}_\mathrm{A} = \mathbf{J}_s \times \mathbf{B}/c$. If the magnetosphere is filled with plasma, the surface current $\mathbf{J}_s$ is flowing within magnetic polar cap only. This surface current closes the volume longitudinal current in the magnetosphere and the return current flowing along the separatrix between open and closed field lines region.

In order to write the equation for the surface current, the several assumptions must be made. We assume that the conductivity of the pulsar surface is uniform, and the electric field $\mathbf{E}_s$ has a potential, so that the surface current can be written as $\mathbf{J}_s = \nabla\xi'$. Using the stationary continuity equation $\operatorname{div}\mathbf{J} = 0$, where $\partial J_z/\partial z$ is equal to the volume current $i_\parallel B_0$ flowing along the open field lines, one can obtain

$$\nabla^2\xi' = -i_\parallel B_0. \tag{3}$$

Making the substitution $x = \sin\theta_m$ and introducing the non-dimensional potential $\xi = 4\pi\xi'/B_0R^2\Omega$ and current $i_0 = -4\pi i_\parallel/\Omega R^2$ we get

$$(1-x^2)\frac{\partial^2\xi}{\partial x^2} + \frac{1-2x^2}{x}\frac{\partial\xi}{\partial x} + \frac{1}{x^2}\frac{\partial^2\xi}{\partial\varphi_m^2} = i_0(x,\varphi_m). \tag{4}$$

Here θ_m and φ_m are polar and azimuth angles with respect to magnetic axis.

Equation (4) needs a boundary condition. This boundary condition results from the proposition that there is no surface current outside the magnetic polar cap. This means that

$$\xi[x_0(\varphi_m), \varphi_m] = \mathrm{const} \tag{5}$$

where $x_0(\varphi_m)$ is the polar cap boundary. Indeed, let us suppose that there is no such boundary condition for the potential ξ. In this case we get $\nabla^2\xi = \text{r.h.s}$ with r.h.s. $= i_0$ inside the polar cap and r.h.s $= 0$ outside it. The solution of homogeneous equation is

$$\xi(x,\varphi_m)\big|_{x\ge 0} = \sum_{n=0}^{\infty}\left(\frac{1-x}{1+x}\right)^{n/2} f_n(\varphi_m), \tag{6}$$

$$\xi(x,\varphi_m)\big|_{x<0} = \sum_{n=0}^{\infty}\left(\frac{1+x}{1-x}\right)^{n/2} f_n(\varphi_m) \tag{7}$$

where $f_n(\varphi_m) = (a_n\cos n\varphi_m + b_n\sin n\varphi_m)$ with a_n and b_n being arbitrary constants. In this case the surface current $\mathbf{J}_s \propto \nabla\xi$ is circulating over the whole neutron star surface resulting in arbitrary energy loss. However, this means that there is a potential drop between different points of a neutron star surface, which inevitably leads to the volume current in the region of closed field lines in the magnetosphere. This contradicts to the assumption that there are no longitudinal currents flowing in the region of closed field lines. Thus, the boundary condition (5) is to be postulated. In this case the jump in the potential derivative at $x = x_0(\varphi_m)$ gives us the current flowing along the separatrix. As we see, it is defined uniquely by the longitudinal current in the region of open field lines and by condition that no longitudinal current can flow in the region of the closed field lines.

For arbitrary inclination angle χ the current i_0 can be written as a sum of its symmetric i_S and anti-symmetric i_A components. The anti-symmetric current begins playing the main role when the pulsar polar cap crosses the surface where the Goldreich–Julian charge density $\rho_{GJ} = -\boldsymbol{\Omega}\cdot\mathbf{B}/2\pi c$ changes sign. This condition can be written as

$$\chi = \frac{\pi}{2} - \sqrt{\frac{\Omega R}{c}}. \tag{8}$$

For example, taking the Goldreich–Julian current density

$$i_\mathrm{GJ}(x,\varphi_m) \approx \cos\chi + \frac{3}{2}x\cos\varphi_m\sin\chi = i_\mathrm{S} + i_\mathrm{A}x\cos\varphi_m, \tag{9}$$

we obtain the following solutions of the Dirichlet problem (4)–(5) for the symmetric and anti-symmetric volume currents respectively:

$$\xi_\mathrm{S} = \frac{i_\mathrm{S}}{4}x^2, \tag{10}$$

$$\xi_A = \frac{i_A x}{8}\left(x^2 - x_0^2\right)\cos\varphi_m. \quad (11)$$

The torque exerted by the surface current over the neutron star can be written as

$$\mathbf{K} = \frac{1}{c}\int \left[\mathbf{r}, [\mathbf{J}_s, (\mathbf{B}_0)]\right] dS \quad (12)$$

where $\mathbf{B}_0$ is the dipole field near the neutron star surface. Let us decompose the braking torque $\mathbf{K}$ over the orthogonal system of unit vectors $\mathbf{e}_m$, $\mathbf{n}_1$, and $\mathbf{n}_2$. Here $\mathbf{e}_m$ is a unit vector along the magnetic moment; vector $\mathbf{n}_1$ is perpendicular to the magnetic moment and lies in the plane of the magnetic moment and the rotational axis; vector $\mathbf{n}_2$ complements these to the right-hand triple:

$$\mathbf{K} = K_{\parallel}\mathbf{e}_m + K_{\perp}\mathbf{n}_1 + K_{\dagger}\mathbf{n}_2. \quad (13)$$

$K_{\dagger}$ plays no role in Euler equations that describe the rotational dynamics of the decelerating neutron star. As a result we have (Beskin et al. 1993):

$$K_{\parallel} = -\frac{B_0^2 \Omega R^4}{2\pi c}\int_0^{2\pi} d\varphi_m \int_0^{x_0(\varphi_m)} dx\, x^2\sqrt{1-x^2}\frac{\partial\xi}{\partial\varphi_m} \quad (14)$$

$$K_{\perp} = K_1 + K_2, \quad (15)$$

$$K_1 = \frac{B_0^2 \Omega R^4}{2\pi c}\int_0^{2\pi} d\varphi_m \int_0^{x_0(\varphi_m)} dx\, A, \quad (16)$$

$$K_2 = \frac{B_0^2 \Omega R^4}{2\pi c}\int_0^{2\pi} d\varphi_m \int_0^{x_0(\varphi_m)} dx\, x^3\cos\varphi_m\frac{\partial\xi}{\partial x} \quad (17)$$

where $A = x\cos\varphi_m \partial\xi/\partial x - \sin\varphi_m \partial\xi/\partial\varphi_m$.

The leading perpendicular torque component K_1 is equal to zero equivalently for arbitrary shape of the polar cap due to the boundary condition (5). The values $K_{\parallel}$ and $K_{\perp}$ can be written as

$$K_{\parallel} = \frac{B_0^2 \Omega^3 R^6}{c^3}\left[-c_{\parallel} i_S + \mu_{\parallel}\left(\frac{\Omega R}{c}\right)^{1/2} i_A\right], \quad (18)$$

$$K_{\perp} = \frac{B_0^2 \Omega^3 R^6}{c^3}\left[\mu_{\perp}\left(\frac{\Omega R}{c}\right)^{1/2} i_S + c_{\perp}\left(\frac{\Omega R}{c}\right) i_A\right]. \quad (19)$$

Here the coefficients $\mu_{\parallel}$ and $\mu_{\perp}$ depending on the shape of the polar cap are much less than unity, and the coefficients $c_{\parallel}$ and $c_{\perp} \sim 1$.

We can now find the derivatives of the angular velocity $\dot{\Omega}$ and of the inclination angle $\dot{\chi}$ of a neutron star through the Euler dynamics equations:

$$J_r\frac{d\Omega}{dt} = K_{\parallel}\cos\chi + K_{\perp}\sin\chi, \quad (20)$$

$$J_r\Omega\frac{d\chi}{dt} = K_{\perp}\cos\chi - K_{\parallel}\sin\chi. \quad (21)$$

For the inclination angles χ not too close to 90° (i.e., for $\cos\chi > (\Omega R/c)^{1/2}$), when the anti-symmetric current plays no role in the neutron star dynamics, we obtain

$$\frac{d\Omega}{dt} = -c_{\parallel}\frac{B_0^2 \Omega^3 R^6}{J_r c^3} i_S \cos\chi, \quad (22)$$

$$\frac{d\chi}{dt} = c_{\parallel}\frac{B_0^2 \Omega^2 R^6}{J_r c^3} i_S \sin\chi. \quad (23)$$

As a result, for homogeneous current density within open magnetic field lines region $i_S = j_{\parallel}/j_{GJ} = \text{const}$ where $j_{GJ} = c\rho_{GJ}$ we have

$$W_c = \frac{f_*^2}{4}\frac{B_0^2 \Omega^4 R^6}{c^3} i_S \cos\chi. \quad (24)$$

Here f_* is the non-dimensional area of a pulsar polar cap: $S_{cap} = f_*\pi(\Omega R/c)$. It depends on the structure of the magnetic field near the light cylinder. For a pure dipole magnetic field (and aligned rotator) $f_* = 1$, and for a magnetosphere containing no longitudinal currents f_* changes from 1.592 for the aligned rotator, $\chi = 0°$ (Michel 1991), to 1.96 for an orthogonal rotator, $\chi = 90°$ (Beskin et al. 1993). Recent numerical calculations for an axisymmetric magnetosphere with non-zero longitudinal electric current give $f_* \approx 1.23$–1.27 (Gruzinov 2005; Komissarov 2006; Timokhin 2006). If the singular point separating open and close field lines can be located inside the light cylinder, the value f_* can be even $\gg 1$. As the Goldreich–Julian charge density near the polar cap is proportional to $\cos\chi$, one can write

$$W_c \approx \frac{f_*^2}{4}\frac{B_0^2 \Omega^4 R^6}{c^3}\cos^2\chi. \quad (25)$$

On the other hand for $\chi \approx 90°$ when the anti-symmetric current plays the leading role we obtain

$$W_c \approx \frac{B_0^2 \Omega^5 R^7}{c^4}. \quad (26)$$

As we see, the energy loss of the orthogonal rotator are $\Omega R/c$ times smaller than of the aligned rotator.

3 PSR B1931+24

The recent discovery of the "part-time job" pulsars PSR B1931+24 (Kramer et al. 2006) with $\dot{\Omega}_{on}/\dot{\Omega}_{off} \approx 1.5$ shows that the current loss is indeed playing an important role in the pulsar energy loss. If we assume that in the on-state the energy loss is connected with the current loss only and in the off-state with the magneto-dipole radiation (in which case the magnetosphere must be not filled with plasma) we get

$$\frac{\dot{\Omega}_{on}}{\dot{\Omega}_{off}} = \frac{3f_*^2}{2}\cot^2\chi. \quad (27)$$

It gives $\chi \approx 60°$. On the other hand, if we assume the Spitkovsky relation for thc on-state energy loss (Spitkovsky 2006)

$$W_{\text{tot}} = \frac{1}{4}\frac{B_0^2 \Omega^4 R^6}{c^3}(1 + \sin^2\chi), \tag{28}$$

we obtain

$$\frac{\dot{\Omega}_{\text{on}}}{\dot{\Omega}_{\text{off}}} = \frac{3}{2}\frac{(1 + \sin^2\chi)}{\sin^2\chi}. \tag{29}$$

Clear, this ratio cannot be equal to 1.5 for any inclination angle. This discrepancy can be connected with that fact that all the numerical calculations produced recently contain no restriction on the longitudinal electric current magnitude. As a result, current density can be much larger than Goldreich–Julian one.

4 On the magnitude of a surface current

As we have shown, the current loss plays the major role in the pulsar dynamics. In particular, the behaviour of the pulsar B1931+24 can be naturally explained within this model. The current loss model have some important consequences:

1. The energy loss of an orthogonal rotator is $\Omega R/c$ times smaller than for an aligned rotator. This is connected with the boundary condition (5) that leads to almost full screening of the toroidal magnetic field in the open field lines region (see Beskin and Nokhrina 2004).
2. Consequently, during its evolution a pulsar inclination angle tends to $\pi/2$ where energy loss is minimal.

On the other hand, it is known for the Michel's monopole solution that in order to have the MHD flow up to infinity, the toroidal magnetic field must be of the same order as the poloidal electric field on the light cylinder. If the longitudinal current $j_{\|}$ does not exceed by $(\Omega R/c)^{-1/2}$ times the respective Goldreich–Julian current density (for the typical pulsars this factor approach the value of 10^2), the light surface $|\mathbf{E}| = |\mathbf{B}|$ for the orthogonal rotator must be located in the vicinity of the light cylinder. In this case the effective energy conversion and the current closure is to take place in the boundary layer near the light surface (Beskin et al. 1993; Chiueh et al. 1998; Beskin and Rafikov 2000).

In order for these results being not true (for example, in order for the light surface being removed to infinity) there must be a sufficient change in the current density value in the inner gap. We should emphasize that for the model with free particle escape it is hard to support the current different than the Goldreich–Julian current. Indeed, since ρ_{GJ} is the particle density needed to screen the longitudinal electric field, the value for the current must be close $j_{GJ} = c\rho_{GJ}$. In order to change this value significantly we must support the plasma inflow in the inner gap region (Lyubarskii 1992). For example, these particles can be produced in the outer gap. But for different poloidal field configuration it is obvious that the major number of field lines intersect the outer gap region outside the Alfvenic surface: as it was shown for several field configurations by Beskin et al. (1998) and Beskin and Nokhrina (2006), inside the fast magnetosonic surface the flow remains still highly magnetized. Thus, the deviation of current lines from the field lines is negligible. On the other hand, magnetized plasma can intersect the Alfvenic surface outwards only (see Beskin 2006 for more detail). Thus, the outer gap can not significantly affect the current in the vicinity of the polar cap.

Finally, it is necessary to stress that the recent numerical calculations by Gruzinov (2005), Komissarov (2006), Timokhin (2006), Spitkovsky (2006) do not include into consideration the condition that the longitudinal current density must be close to j_{GJ}. In all these works the authors assume that there may be any current flowing through the cascade region. However, if this is not so, and the current is indeed close to the Goldreich–Julian current, the structure of a magnetosphere may be different from the one obtained in the numerical simulations.

5 Conclusions

As we have seen, the current loss connected with the longitudinal current flowing in the magnetosphere plays the main role in pulsar dynamics, and recent observations of "part-time job" pulsar supports this point. This evolution includes not only a neutron star retardation but also the sufficient change in the angle between magnetic and spin axis. We have seen as well that the model of current loss depends crucially on the distribution of the electric current and its value in the inner gap. For current loss model with j_{GJ} the inclination angle grows with time so a pulsar tends to be an orthogonal rotator. In this case the energy loss is to be $\Omega R/c$ times smaller than for the aligned rotator. As a consequence, the light surface must be located in the very vicinity of the light cylinder. On the other hand, to realize the homogeneous MHD outflow up to infinity for the orthogonal rotator the current density in the open field line region is to be much larger than j_{GJ}.

Acknowledgements The authors thank A.V. Gurevich and Ya.N. Istomin for the useful discussions.

References

Beskin, V.S.: Axisymmetric Stationary Flows in Astrophysics. Fizmatlit, Moscow (2006), in Russian

Beskin, V.S., Nokhrina, E.E.: Astron. Lett. **30**, 685–693 (2004)

Beskin, V.S., Nokhrina, E.E.: Mon. Not. Roy. Astron. Soc. **367**, 375–386 (2006)
Beskin, V.S., Rafikov, R.R.: Mon. Not. Roy. Astron. Soc. **313**, 433–444 (2000)
Beskin, V.S., Gurevich, A.V., Istomin, Ya.N.: Physics of the Pulsar Magnetosphere. Cambridge University Press, Cambridge (1993)
Beskin, V.S., Kuznetsova, I.V., Rafikov, R.R.: Mon. Not. Roy. Astron. Soc. **299**, 341–347 (1998)
Chiueh, T., Li, Zh.-Yu., Begelman, M.C.: Astrophys. J. **505**, 835–843 (1998)
Gruzinov, A.: arXiv:astro-ph/0604364
Istomin, Ya.: Magnetodipole oven. In: Wass, A.P. (ed.) Progress in Neutron Star Research, pp. 27–44. Nova Science, New York (2005)
Komissarov, S.S.: Mon. Not. Roy. Astron. Soc. **367**, 19–31 (2006)
Kramer, M., Lyne, A.G., O'Brian, J.T., Jordan, C.A., Lorimer, D.R.: Science **312**, 549–551 (2006)
Lyubarskii, Yu.E.: Astron. Astrophys. **261**, 544–550 (1992)
Mestel, L., Panagi, P., Shibata, S.: Mon. Not. Roy. Astron. Soc. **309**, 388–394 (1999)
Michel, F.: Theory of Neutron Star Magnetosphere. University of Chicago Press, Chicago (1991)
Pacini, F.: Nature **221**, 567–569 (1967)
Spitkovsky, A.: arXiv:astro-ph/0603147v1
Timokhin, A.N.: Mon. Not. Roy. Astron. Soc. **368**, 1055–1072 (2006)

Astrophys Space Sci (2007) 308: 575–579
DOI 10.1007/s10509-007-9348-4

ORIGINAL ARTICLE

Force-free magnetosphere of an aligned rotator with differential rotation of open magnetic field lines

A.N. Timokhin

Received: 2 July 2006 / Accepted: 6 September 2006 / Published online: 15 March 2007

Abstract Here we briefly report on results of self-consistent numerical modeling of a differentially rotating force-free magnetosphere of an aligned rotator. We show that differential rotation of the open field line zone is significant for adjusting of the global structure of the magnetosphere to the current density flowing through the polar cap cascades. We argue that for most pulsars stationary cascades in the polar cap can not support stationary force-free configurations of the magnetosphere.

Keywords Stars: magnetic fields · Pulsars: general · MHD

PACS 97.60.Jd · 97.60.Gb

1 Introduction

Aligned rotator with the force-free magnetosphere is being considered as a first order approximation for the real pulsar magnetosphere since introduction of the model by Goldreich and Julian (1969). Any pulsar model should be tested for this simplest case. Recently substantial progress in modeling of pulsar magnetospheres was achieved. The magnetosphere of an aligned rotator was modeled using stationary (Contopoulos et al. 1999; Goodwin et al. 2004; Gruzinov 2005; Timokhin 2006) and time-dependent (Komissarov 2006; Bucciantini et al. 2006; McKinney 2006; Spitkovsky 2006) codes. The structure of the magnetosphere has been obtained even for an inclined rotator (Spitkovsky 2006). In these works the angular velocity of plasma rotation was assumed to be constant, although the case when the open field lines rotate with a constant angular velocity different from the angular velocity of the Neutron Star (NS) was addressed in some works (see e.g. Contopoulos 2005; Beskin and Malyshkin 1998). It was also implicitly assumed that the current density in the magnetosphere could adjust to any distribution required by the global structure of the magnetosphere. However, in the pulsar magnetosphere current carriers are produced mainly by the electron-positron cascades and therefore not every current density distribution can be realized (see Timokhin 2006, hereafter Paper I). Moreover, in Paper I it was shown that the current density distribution supporting the magnetospheric configuration frequently used in theoretical pulsar models, that assume the closed field line zone extending up to the Light Cylinder (LC) and the open field lines rotating with the same angular velocity as the NS, could not be realized regardless of a particular model of the polar cap cascades. Such models require presence of an electric current flowing against the accelerating electric field in some parts of the pulsar polar cap, which cannot be naturally explained.

For the natural stationary configuration of the magnetosphere, when the last closed field line lies in the equatorial plane and magnetic field lines become radial at a large distance from the LC (the so-called Y-configuration), the system has two physical "degrees of freedom", namely (i) the size of the closed field line zone and (ii) the angular velocity of rotation of the open field lines. In Paper I the angular velocity of the magnetic field lines was fixed, but the size of the closed field line zone was varied. In any of the obtained configurations the current density is much less than the Goldreich-Julian (GJ) current density along field lines passing close to the boundaries of the polar cap (see Fig. 5 in

A.N. Timokhin (✉)
Sternberg Astronomical Institute, Universitetskij pr. 13, 119992
Moscow, Russia
e-mail: atim@sai.msu.ru

Paper I). However, the current density, which could be produced by stationary polar cap cascades operating in space charge limited flow (SCLF) regime, can not be significantly less than the corresponding GJ current density (e.g. Harding and Muslimov 1998). On the other hand, the current density flowing in the polar cap of pulsar having partially screened polar gap (PSG) cannot be close to zero (Gil et al. 2003). Hence, aligned pulsar with stationary polar cap cascades operating in SCLF or PSG regime cannot have stationary force-free magnetosphere rotating with a constant velocity. As stationary SCLF and PSG are being considered as the most probable regimes for the polar cap of pulsar, it is worth to find configuration of the force-free magnetosphere compatible with them.

Here for the first time we consider self-consistently the case of a differentially rotating force-free magnetosphere of an aligned rotator in Y-configuration, i.e. we are able to explore *all* possible configurations of such magnetosphere. Our aim is to study the impact of differential rotation on the current density distribution in the force-free magnetosphere of an aligned rotator. We also try to find combinations of the angular velocity distribution and the size of the closed field line zone, which could be compatible with stationary polar cap cascade models.

2 Equations for differentially rotating force-free magnetosphere

Magnetic field lines in the closed zone are equipotential and plasma there corotates with the NS. In the open field line domain the angular velocity of plasma $\Omega_{\rm F}$ is different from the angular velocity of the NS rotation Ω. The angular velocity of the open magnetic field lines $\Omega_{\rm F}$ is determined by the potential difference in the polar cap of pulsar. In stationary cascade models $\Omega_{\rm F}$ is close to Ω for young pulsars (see Sect. 2.2 in Paper I).

The force-free magnetosphere of an aligned rotator with differential rotation can be described by a solution of the so-called pulsar equation, derived by Okamoto (1974). In notations of Paper I (see (20) there) it has the form

$$(\beta^2 x^2 - 1)(\partial_{xx}\psi + \partial_{zz}\psi) + \frac{\beta^2 x^2 + 1}{x}\partial_x\psi - S\frac{dS}{d\psi} + x^2\beta\frac{d\beta}{d\psi}(\nabla\psi)^2 = 0. \qquad (1)$$

Here $\beta \equiv \Omega_{\rm F}/\Omega$, S and ψ are normalized poloidal current and magnetic flux functions correspondingly. z axis is co-aligned with $\vec{\Omega}$. All coordinates are normalized to the LC radius of a corotating magnetosphere $R_{\rm LC} \equiv c/\Omega$. The last term in the equation takes into account contribution of differential rotation into the force balance across magnetic field lines. The current density in the open field line zone of the magnetosphere depends on differential rotation through condition at the true LC, which is at $x(\psi) = c/\Omega_{\rm F}(\psi)$ (see Sect. 2.3 in Paper I)

$$S\frac{dS}{d\psi} = 2\beta\partial_x\psi + \frac{1}{\beta}\frac{d\beta}{d\psi}(\nabla\psi)^2. \qquad (2)$$

Changes of β result in changes of the poloidal current density j being proportional to $dS/d\psi$.

So, the differential rotation contributes to the force balance perpendicular to the magnetic field lines, changes the position of the LC and affects the current density distribution, required for smooth transition of the solution trough the LC. Although, the differential rotation can modify the current density distribution, it is a quantitative question whether or not a given differential rotation $\Omega_F(\psi)$ and the corresponding current density distribution $j(\psi)$ could be supported by stationary cascades in the polar cap of pulsar. We note that $\beta(\psi)$ is a free parameter in this problem, while $j(\psi)$ is obtained from the solution.

For solution of (1) we developed an advanced version of the multigrid code described in Paper I. Now the position of the Light Cylinder is found self-consistently on each iteration step and poloidal current is calculated from (2) at the LC. The return current flowing in the current sheet between the closed and open field line zones is smeared over the interval $[\psi_{\rm last}, \psi_{\rm last} + d\psi]$,[1] and the angular velocity is continuously changing in the same interval from some value at the last open filed line $\Omega_{\rm F}(\psi_{\rm last})$ to Ω at $\psi = \psi_{\rm last} + d\psi$. Equation (1) is solved in the whole domain including the current sheet. Smearing of the return current and continuous changing of $\Omega_{\rm F}$ in the current sheet allows us to take into account the contribution of the current sheet into the force balance across magnetic field lines at the boundary between closed and open field line zones.

3 Main results

Stationary polar cap cascades without particle inflow from the magnetosphere would provide more or less constant current density distribution over the polar cap (see e.g. Harding and Muslimov 1998; Gil et al. 2003). However, in configurations of the force-free magnetosphere with constant $\Omega_{\rm F}$ the current density goes to zero near the polar cap boundaries. Here we try to find configurations of a differentially rotating force-free magnetosphere with current density, which is nearly constant over the polar cap at the NS surface. Current density in the magnetosphere rotating with a constant angular velocity decreases toward the polar cap boundary because of the boundary condition $\psi(x > x_0) = \psi_{\rm last}$ (x_0 is

[1] $\psi_{\rm last}$ is the value of the normalized magnetic flux function corresponding to the field line separating the closed and open field line domains.

the position of the point, where the last closed field line intersects the equatorial plane). In the equation for the poloidal current (see (2)) the first term on the left hand side goes to zero for ψ approaching $\psi_{\rm last}$. Hence, in order to increase the current density near the polar cap boundaries $d\beta/d\psi$ must be positive, i.e. the maximum value of β should be achieved at the polar cap boundary, for $\psi = \psi_{\rm last}$. Here we consider the case when the maximum possible value of β is 1. Such behavior of β is expected in models of the polar cap cascades.

We have performed computations for different sizes of the closed field line zone x_0. In each of these simulations the distribution of β over ψ was adjusted ad hoc in order to obtain nearly constant current density over the polar cap, which approaches zero only at field lines very close to the polar cap boundary.[2] For any size of the closed field line zone x_0 it is possible to construct a set of force-free magnetospheric configurations with the current density distribution being almost constant over the polar cap, $j(\psi) \simeq \hat{j} \equiv$ const by choosing different $\beta(\psi)$.

One of the important properties of the obtained set of solutions is that for each x_0 the constant current density $\hat{j}$ cannot be made greater than some maximum value $\bar{j}$. Distribution $j(\psi) \simeq \bar{j}$ corresponds to a distribution $\bar{\beta}(\psi)$, which achieves 1 at the polar cap boundary, $\bar{\beta}(\psi_{\rm last}) = 1$. As it was stressed above, the maximum value of β is achieved at $\psi = \psi_{\rm last}$. So, $\bar{\beta}$ differs from other distributions $\beta(\psi)$ which provide constant current density, in that it achieves the maximum possible value. Both $\bar{j}$ and $\bar{\beta}(\psi)$ depend on x_0.

The differential rotation in the magnetosphere of a young pulsar with stationary polar cap cascades is rather small (see Sect. 2.2 in Paper I). So, solutions with the minimal deviation of $\beta(\psi)$ from 1 would be of most interest for us. These turned out to be solutions with $\beta = \bar{\beta}(\psi)$, i.e. with $j \simeq \bar{j}$. One of such magnetospheric configurations with $x_0 = 0.7$ is shown in Fig. 1. Important properties of solutions with $j \simeq \bar{j}$ are:

- The value $\bar{j}$ increases with decreasing of x_0, however it does not exceeds $j_{\rm GJ}/2$, the half of the GJ current density near the NS surface (see Fig. 2).
- The angular velocity of rotation of the open magnetic field lines has smaller deviation from Ω with decreasing of x_0, see Fig. 3. As in the case of the current density distribution there exist an asymptotic distribution of $\Omega_{\rm F}(\psi)$. This asymptotic distribution deviates strongly from $\beta(\psi) \equiv 1$.
- The asymptotic form of $\bar{\beta}(\psi)$ and $\bar{j}$ for $x_0 \to 0$ matches the corresponding distributions in the split monopole configuration.

[2]We assume that the current sheet carries only the return current, so j changes the sign at $\psi_{\rm last}$.

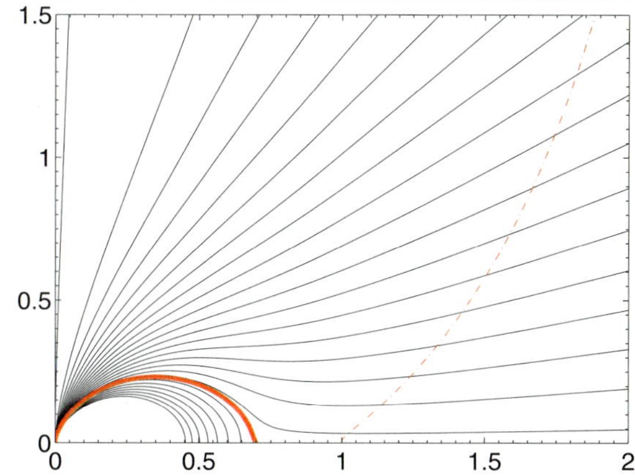

Fig. 1 Structure of the magnetosphere for $x_0 = 0.7$ and the angular velocity of the open magnetic field lines $\Omega_{\rm F}(\psi) = \Omega\bar{\beta}(\psi)$ (see text). Magnetic field lines are shown by the *thin solid lines*, the last closed field line—by the *thick solid line*, the position of the Light Cylinder is shown by the *dot-dashed line*

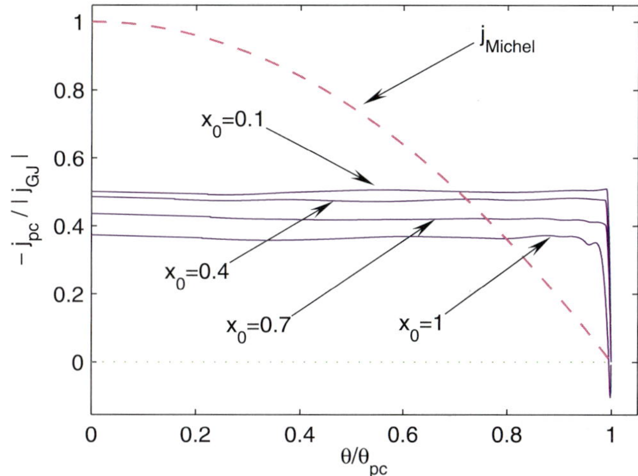

Fig. 2 Current density distributions in the polar cap of pulsar $j_{\rm pc}(\theta)$ corresponding to $j(\psi) \approx \bar{j}$ (see text) are plotted for different x_0 as functions of the colatitude θ. $j_{\rm pc}$ is normalized to the Goldreich-Julian current density $|j_{\rm GJ}|$. θ is normalized to the colatitude of the polar cap boundary $\theta_{\rm pc}$. The current density corresponding to the Michel's solution (Michel 1973) is shown by the *dashed line*. The distribution for $x_0 = 0.1$ practically coincides with the distribution for the split-monopole case

- Energy losses of the aligned rotator increases with decreasing of x_0. This increase goes a bit faster than in the case of constant $\Omega_{\rm F}$, cf. Fig. 4 here and Fig. 8 in Paper I.
- The total energy of the magnetosphere increases with decreasing of x_0.
- In none of the considered configurations the force-free condition is violated in the calculation domain, i.e. the electric field is everywhere smaller than the magnetic field.

Let us now discuss implications of these solutions for the physics of pulsars.

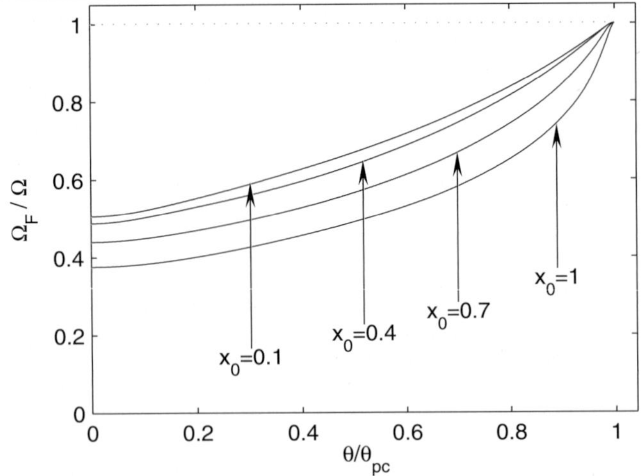

Fig. 3 Angular velocities of rotation of the open field lines $\Omega_F(\theta) = \Omega\bar{\beta}(\theta)$ (see text) for different x_0 as functions of the colatitude θ in the polar cap of pulsar. θ is normalized to the colatitude of the polar cap boundary θ_{pc}. $\Omega_F(\theta)$ is normalized to the angular velocity of the NS rotation Ω. The distribution for $x_0 = 0.1$ practically coincides with the distribution for the split-monopole case

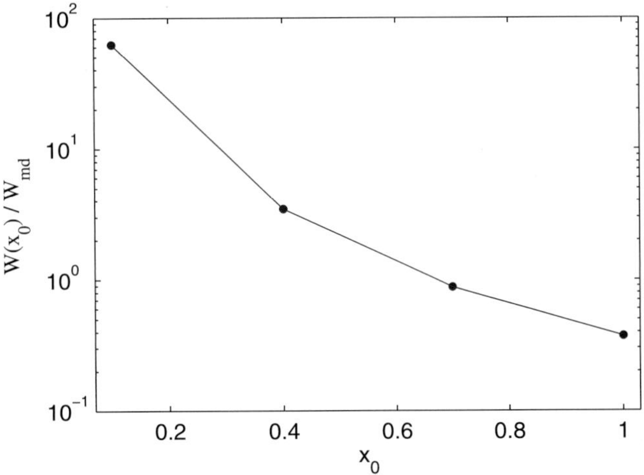

Fig. 4 Energy losses of aligned rotator normalized to the magnetodipolar energy losses as a function of x_0 for differentially rotating magnetosphere with $\Omega_F(\psi) = \Omega\bar{\beta}(\psi)$ (see text)

4 Discussion

If the polar cap cascades operate in stationary SCLF regime the current density in the polar cap of pulsar near the NS surface is very close to the local GJ current density. However, due to inertial frame dragging (i.e. due to changing of the effective Ω) the local GJ charge density decreases slower with the distance than the charge density carried by the charge-separated flow from the NS surface (see Muslimov and Tsygan 1992). This discrepancy achieves $\sim$15% at a distance of several radii of the NS. So, the current density at that distance will be by $\sim$15% less then the corresponding local GJ current density. Here we solve the problem in flat space-time. The boundary conditions on the NS surface and, hence, the physical quantities we used for normalization should be taken at some distance from the NS, where GR effects are negligible. The GR corrections will result in $\sim$15% higher ratio of j to the *local* value of j_{GJ} at the NS surface than the one obtained in our solution. However as it follows from our results, there is no configuration of the magnetosphere with almost constant current density larger than $j_{GJ}/2$. Hence, stationary force-free magnetosphere of an aligned rotator can not be supported by stationary polar cap cascades operating in SCLF regime.

Polar cap cascades operating in Ruderman and Sutherland (1975) regime with almost vacuum electric field in the gap allow larger deviations of the current density from j_{GJ} as well as larger deviation of β from 1. For older pulsar, when β could be rather small, even configuration with x_0 approaching 1 could be supported by the polar cap cascades. In this case the characteristic current density flowing through the gap must be also much less than the GJ current density.

For the partially screened polar gap (PSG) (Gil et al. 2003) the presence of a strong multipolar component of the magnetic field in the polar cap of pulsar is essential. In this model the current density at the NS surface for young pulsars should be close to the local value of j_{GJ}. However, the local GJ charge density is determined by the non-dipolar magnetic field. For this field the angle between $\vec{B}$ and $\vec{\Omega}$ can be rather large and the GJ charge density will be accordingly less. At some distance, where magnetic field becomes dipolar and the angle between $\vec{B}$ and $\vec{\Omega}$ is small, the current density of the flow would become smaller than the local GJ current density. This effect can be much stronger than the difference between j and j_{GJ} due to inertial frame dragging. Hence, for some specific configurations of the local magnetic field at the NS surface, stationary polar cap cascades could support a stationary force-free configuration of the magnetosphere.

So, aligned pulsars with polar cap cascades operating in SCLF regime, young pulsars with "Ruderman-Sutherland" cascades as well as pulsars with PSG and arbitrary surface magnetic field configuration can not have both polar cap cascades operating in stationary regime and a stationary force-free magnetosphere. From energetic point of view, configurations with non-stationary polar cap cascades would seem to be preferable. The energy of the magnetosphere increases with decreasing of x_0. The magnetosphere would try to achieve the configuration with the smallest possible energy, i.e. with the largest possible x_0, compatible with the physical conditions set by the polar cap cascades. If there are several possible self-consistent configuration of the magnetosphere and polar cap acceleration zone, the configuration with the largest $x_0$will be realized. With aging of the pulsar the conditions in the cascades change. This would result in changing of the magnetosphere configuration and, hence, in changes of energy losses relative to the corresponding magnetodipolar losses. This yields the breaking index different from 3. However, more detailed study of polar cap cascade properties is necessary in order to construct self-consistent magnetospheric configurations and to obtain the value of the breaking indexes. The current adjustment is necessary also for inclined rotator and in this case similar effects would be present too.

More detailed description of the results is given in Timokhin (2006).

Acknowledgements This work was partially supported by RFBR grant 04-02-16720, and by the grants N.Sh.-5218.2006.2 and RNP.2.1.1.5940.

References

Beskin, V.S., Malyshkin, L.M.: Mon. Not. Roy. Astron. Soc. **298**, 847 (1998)

Bucciantini, N., Thompson, T.A., Arons, J., et al.: Mon. Not. Roy. Astron. Soc. **368**, 1717 (2006)

Contopoulos, I. Astron. Astrophys. **442**, 579 (2005)

Contopoulos, I., Kazanas, D., Fendt, C.: Astrophys. J. **511**, 351 (1999)

Gil, J., Melikidze, G.I., Geppert, U.: Astron. Astrophys. **407**, 315 (2003)

Goldreich, P., Julian, W.H.: Astrophys. J. **157**, 869 (1969)

Goodwin, S.P., Mestel, J., Mestel, L., Wright, G.A.E.: Mon. Not. Roy. Astron. Soc. **349**, 213 (2004)

Gruzinov, A.: Phys. Rev. Lett. **94**, 021,101 (2005)

Harding, A.K., Muslimov, A.G.: Astrophys. J. **508**, 328 (1998)

Komissarov, S.S.: Mon. Not. Roy. Astron. Soc. **367**, 19 (2006)

McKinney, J.C.: Mon. Not. Roy. Astron. Soc. **368**, L30 (2006)

Michel, F.C.: Astrophys. J. **180**, 207 (1973)

Muslimov, A.G., Tsygan, A.I.: Mon. Not. Roy. Astron. Soc. **255**, 61 (1992)

Okamoto, I.: Mon. Not. Roy. Astron. Soc. **167**, 457 (1974)

Ruderman, M.A., Sutherland, P.G.: Astrophys. J. **196**, 51 (1975)

Spitkovsky, A.: Astrophys. J. Lett. **648**, L51 (2006)

Timokhin, A.N.: in preparation (2006)

Timokhin, A.N.: Mon. Not. Roy. Astron. Soc. **368**, 1055–1072 (2006)

Astrophys Space Sci (2007) 308: 581–583
DOI 10.1007/s10509-007-9301-6

ORIGINAL ARTICLE

Oscillations in the neutron star crust

Neutron star seismology from QPOs after flares

Lars Samuelsson · Nils Andersson

Received: 30 June 2006 / Accepted: 20 August 2006 / Published online: 15 March 2007

Abstract We investigate the spectrum of torsional modes in the neutron star crust and discuss what conclusions may be drawn about the global properties of the star from observations of such modes.

Keywords Neutron stars · Oscillations · Quasi periodic oscillations

PACS 04.40.Dg · 97.60.Jd

1 Introduction

The recent discovery of quasi periodic oscillations (QPOs) in the aftermath of giant flares in SGR 1806-20 (Israel et al. 2005) and SGR 1900+14 (Strohmayer and Watts 2005) has triggered a lot of interest in (quasi) normal modes in neutron stars. The reason is, of course, that if one can identify the excited modes invaluable information of the global properties of the neutron star, and hence of the equation of state of matter at supernuclear densities, may be obtained.

One of the proposed mechanisms responsible for producing the QPOs is torsional elastic modes in the crust. We present here an outline of a detailed description of such modes in general relativity (Karlovini and Samuelsson 2006). This work should, at least in the context of QPOs in magnetars, be viewed as a first step since the huge magnetic fields of these objects are bound to play a major role. However, recent results indicate that at least some of the observed frequencies may indeed be locked to the purely elastic mode frequencies (Glampedakis et al. 2006).

L. Samuelsson (✉) · N. Andersson
School of Mathematics, University of Southampton, Southampton SO17 1BJ, UK
e-mail: lars@soton.ac.uk

2 General axial perturbations

The formalism that we use is based on the work of Karlovini (2002). It gives a geometrical, gauge invariant description of axial perturbations of spherically symmetric (not necessarily static) backgrounds with an arbitrary matter source. The perturbation equations have the simple Maxwell-like form

$$\nabla_b\left(r^2 \sin^2\theta\, Q^{ab}\right) = \frac{8\pi G}{c^4} J^a, \tag{1}$$

$$\nabla_a J^a = 0 \tag{2}$$

where Q^{ab} is related to the metric perturbations and J^a encodes the matter perturbations.

Since we are interested in torsional motion of the relatively tenuous crust, it will suffice to consider the equations in the relativistic Cowling approximation. For the case at hand this amounts to neglecting the coupling between the gravitational waves and matter motion by setting $G = 0$ in the perturbation equations. The problem is then reduced to solving the divergence free condition of the matter current (2).

3 Elastic matter

We will not go into the intricacies of relativistic elasticity theory here. The interested reader is referred to Karlovini and Samuelsson (2003) for details and references. In this context it suffices to introduce the gauge invariant vector field K^a which essentially measures the deformation of the body with respect to its spherically symmetric relaxed state. Unlike previous treatments (e.g. the pioneering paper by Schumaker and Thorne (1983)) we do not assume that the relaxed state is locally isotropic nor that the equilibrium state (around which we perturb) is unstrained. This allows us to study effects of the so called "pasta phases" believed to exist in the deep crust of neutron stars. It would certainly be interesting to perform such a study as the effects can be quite large (Pethick and Potekhin 1998). However, our main emphasis in this communication is to show the method. Hence, from now on, we assume that the background is isotropic.

In terms of our "displacement measure" K^a, the matter current takes the simple form

$$J^a = 2r^2 \sin^2\theta(\rho + p)S^{ab}K_a \tag{3}$$

where ρ is the total energy density, p is the pressure and S^{ab} is a "metric" tensor field related to the propagation of shear waves with speed $v_\perp$,

$$S^a{}_b = \mathrm{diag}\big(-1, v_\perp^2, v_\perp^2, 0\big). \tag{4}$$

For a static, spherically symmetric background the perturbation equations reduce to a single wave equation. This equation must be accompanied by suitable boundary conditions. For the simple case treated here where we have no magnetic field, gravitational perturbations or viscosity, these are simply that the tractions vanish at the top and bottom of the crust. Hence, the crust is decoupled from the core, and for that matter, the rest of spacetime. It is important to note that almost any additional physics incorporated into this model will introduce coupling.

The simplicity induced by our assumptions allows us to treat the axial oscillations in the crust as an isolated problem depending only on the mass M_c and radius R_c of the core and the equation of state (EOS) in the crust. As the EOS in the crust is fairly well-known (compared to the EOS in the core at least) this is potentially very powerful when it comes to identifying global neutron star properties from an observed mode spectrum.

4 Results

We now leave aside the question of what the observed QPOs really are associated with and simply assume that at least a subset of them are indeed tied to elastic oscillations in the crust. At the very least this will demonstrate our method and show how one could proceed if such oscillations were observed.

The decoupled nature of our problem allows us, given an EOS of the crust, to quickly compute frequencies for a large set of M_c, R_c and l. We ignore the case $l = 1$ since the fundamental mode in this case correspond to the crust uniformly rotating around the static core. We have also only computed the fundamental mode and the first few overtones, the latter only for $l = 2$ since the overtones are quite insensitive to the angular dependence. The equation of state that

Table 1 SGR 1900+14

f [Hz]	Ref	Mode
28 ± 0.5	Strohmayer and Watts (2005)	${}_0t_2$
53.5 ± 0.5	Strohmayer and Watts (2005)	${}_0t_4$
84	Strohmayer and Watts (2005)	${}_0t_6$
155.1 ± 0.2	Strohmayer and Watts (2005)	${}_0t_{11}$

Table 2 SGR 1806-20

f [Hz]	Ref	Mode
30	Israel et al. (2005)	${}_0t_2$
92.7 ± 0.1	Israel et al. (2005)	${}_0t_6$
150	Watts (2006)	${}_0t_{9/10}$
626.46 ± 0.02	Watts and Strohmayer (2006)	${}_1t_l$

we use is presented in detail in Samuelsson (2003) and its essential parts are based on the work of Haensel and Pichon (1994) and Douchin and Haensel (2001). The shear modulus is taken from Fuchs (1936) and we assume that the temperature is zero.

The main results are presented below. For each of the flares we associate a subset of the observed lower QPO frequencies with torsional oscillation modes and search our computed data for stellar models which display all of these frequencies to some accuracy, which was chosen rather arbitrarily to be 6%, for some sequence of l's. For SGR 1900+14 we use all four observed QPOs (displayed with references in Table 1) and obtain that the excited modes have $l = 2, 4, 6, 11$. For SGR 1806-20 we assume that the frequencies lower than about 30 Hz are not directly associated with torsional oscillations and we are left with a sequence of three fundamental modes (Table 2). In addition we have a higher frequency QPO which we associate with an $n = 1$ overtone with arbitrary (but low) l.

In Figs. 1 and 2 we plot the mass-radius relation (as lines) for a few proposed EOSs. The equations of state are chosen for illustration only and include the A18-δv+UIX* model Akmal et al. (1998) with and without a deconfined quark core (APR & APRQ), two examples from Glendenning (2000) (G240 & G300) and the FPS EOS of Pandharipande and Ravenhall (1989) (FPS). On top of this we display, as blue diamonds, the allowed masses and radii if the given set of observed low frequency QPOs are associated with elastic oscillations. In the plot of SGR 1806-20 the models which have an $n = 1$ overtone is shown as magenta circles.

A few comments are appropriate here. First of all, if more realistic stellar models are considered it is likely that the allowed regions displayed in the figures will change considerably. For instance, including magnetic fields in the simplistic

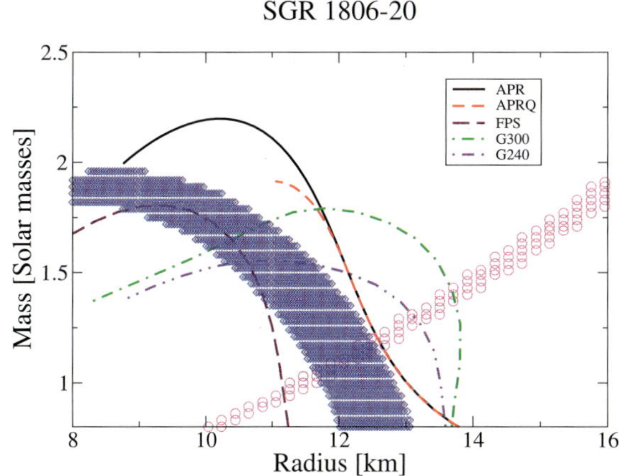

Fig. 1 For the flare in SGR 1806-20 we associate the lower frequency QPOs (approximately 30, 93 & 150 Hz) with fundamental ($n = 0$) modes with $l = 2, 6, 9$ or 10 respectively. The inclusion of the 150 Hz mode does not change the picture substantially. The higher frequency QPO (626 Hz) is associated with a $n = 1$ mode of arbitrary moderately large l. We note that the models allowed by the lower QPOs form a rather broad line that is orthogonal to the line corresponding to the models having an overtone of the right magnitude. We could in principle fit the observed high QPO by a higher order overtone which would push the allowed region towards lower mass. In any case the orthogonality is preserved and leaves us with only a small allowed region in parameter space, demonstrating the usefulness of our method

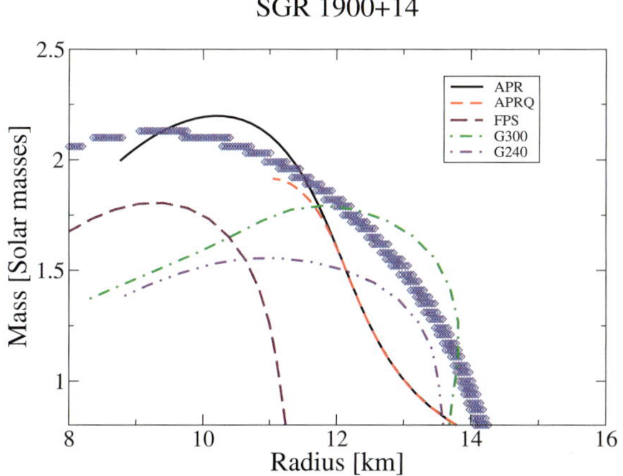

Fig. 2 For the flare in SGR 1900+14 we use all the reported QPOs (approximately 28, 53, 84 & 155 Hz) and find that the sequence of l's is (2,4,6,11). The larger number of fundamental modes compared to the situation for SGR 1806-20 gives a narrower allowed region for a given tolerance. The lack of overtones in the data means however that the parameters are much less constrained. For cases such as this it would be very useful to have additional constraints on the allowed mass and/or radius

fashion of boosting the shear modulus (Duncan 1998) will cause large systematic shifts of the allowed parameter space. A more realistic treatment will give less systematic, and hence less predictable, errors (Messios et al. 2001). However, we believe that the overall picture will not change when including more realistic physics. If the QPOs are indeed excited by crust quakes their frequencies should preferably be close to those of the crustal modes (Glampedakis et al. 2006). The abundance of observed modes could therefore provide means by which not only the mass and radius of the star is constrained, but also the magnetic field strength and configuration could be estimated independently from timing measurements.

Secondly it is worth commenting on the difference on the inferred sequences of l's compared to those of Strohmayer and Watts (2005) and Watts and Strohmayer (2006). Their sequences are deduced by using an analytic approximation due to Duncan (1998) whereas ours come from the actual computed modes. In a forthcoming paper we will show that the formula of Duncan is in fact not very accurate. The most important difference as far as the l sequence is concerned is that the scaling is much more accurately described by $(l-1)(l+2)$ rather than Duncan's $l(l+1)$.

References

Akmal, A., Pandharipande, V.R., Ravenhall, D.G.: Phys. Rev. C **58**, 1804 (1998)

Douchin, F., Haensel, P.: Astron. Astrophys. **380**, 151 (2001)

Duncan, R.C.: Astrophys. J. **498**, L45 (1998)

Fuchs, K.: Proc. Roy. Soc. A **153**, 622 (1936)

Glampedakis, K., Samuelsson, L., Andersson, N.: Mon. Not. Roy. Astron. Soc. **371**, L74 (2006)

Glendenning, N.K.: Compact Stars, 2nd edn. Springer-Verlag, New York (2000)

Haensel, P., Pichon, B.: Astron. Astrophys. **283**, 313 (1994)

Israel, G.L., Belloni, T., Stella, L., et al.: Astrophys. J. **628**, L53 (2005)

Karlovini, M.: Class. Quantum Grav. **19**, 2125 (2002)

Karlovini, M., Samuelsson, L.: Class. Quantum Grav. **20**, 3613 (2003)

Karlovini, M., Samuelsson, L.: preprint (2006)

Messios, N., Papadopoulos, D.B., Stergioulas, N.: Mon. Not. Roy. Astron. Soc. **328**, 1161 (2000)

Pandharipande, V.R., Ravenhall, D.G. In: Soyeur, M., Flocard, H., Tamain, B., Porneuf, M. (eds.) Nuclear Matter and Heavy Ion Collisions Hot Nuclear Matter. NATO ASIB Proc., vol. 205, p. 103 (1989)

Pethick, C.J., Potekhin, A.Y.: Phys. Lett. B **427**, 7 (1998)

Samuelsson, L.: PhD thesis, Stockholm University (2003)

Schumaker, B.L., Thorne, K.S.: Mon. Not. Roy. Astron. Soc. **203**, 457 (1983)

Strohmayer, T.E., Watts, A.L.: Astrophys. J. **632**, L111 (2005)

Watts, A.: private communication (2006)

Watts, A.L., Strohmayer, T.E.: Astrophys. J. **637**, L117 (2006)

Astrophys Space Sci (2007) 308: 585–589
DOI 10.1007/s10509-007-9372-4

ORIGINAL ARTICLE

Two decades of pulsar timing of Vela

Richard Dodson · Dion Lewis · Peter McCulloch

Received: 30 June 2006 / Accepted: 19 September 2006 / Published online: 20 March 2007

Abstract Pulsar timing at the Mt Pleasant observatory has focused on Vela, which can be tracked for 18 hours of the day. These nearly continuous timing records extend over 24 years allowing a greater insight into details of timing noise, micro glitches and other more exotic effects. In particular we report the glitch parameters of the 2004 event, along with the reconfirmation that the spin up for the Vela pulsar occurs instantaneously to the accuracy of the data. This places a lower limit of about 30 seconds for the acceleration of the pulsar to the new rotational frequency. We also confirm of the low braking index for Vela, and the continued fall in the DM for this pulsar.

Keywords Stars: neutron · Dense matter · Pulsars: individual (PSR B0833-45)

PACS 97.60.Jd · 97.60.Gb

1 Introduction

Mount Pleasant observatory, just outside Hobart in Tasmania, Australia, has a 14-metre dish that has been dedicated to tracking the Vela pulsar for two decades. This telescope is able to monitor the pulsar for eighteen hours every day, and therefore has caught many glitches 'in the act'. As a cross-check the older, but glitching, PSR 1644-4559 is observed for the six hours when Vela is set. There is no comparative dataset, and the conclusions we draw puts extremely tight conditions on the pulsar Equation of State (EOS) by placing a number of constraints on the models. An example of these would be that if the spin-up is very fast the crust has to have a low moment of inertia, therefore be very thin, and the coupling between the crust and the interior super-fluid has to be strong (see, for example, discussion of these issues in Epstein and Baym 1992 and Bildsten and Epstein 1989).

Three uncooled receivers are mounted at the prime focus of the 14-metre to allow continuous dispersion measure determination. Two are stacked disk, dual polarisation with central frequencies of 635 MHz and 990 MHz additional to a right hand circular helix at 1391 MHz. Bandwidths are 250 kHz, 800 kHz and 2 MHz respectively, limiting pulse broadening from interstellar dispersion to less than 1%. The output is folded for two minutes giving an integrated pulse profile of 1344 pulses. The backend to the 990 MHz receiver also has incoherent dedispersion across 8 adjacent channels allowing a study of individual pulses. Results from these systems have been reported, respectively, in McCulloch et al. (1990) and Dodson et al. (2002).

A new system, based on the PC-EVN VLBI interfaces Dodson et al. (2004). can produce TOA's with accuracy of the order of 0.1 ms every second (as opposed to every 10 seconds with the single pulse or 120 seconds with the multi-frequency systems). This interface is adapted from the Metsähovi Radio Observatories linux-based DMA, data collector, card designed for VLBI digital inputs. The two 40 MHz IFs from the 635 MHz feed provide the two polarisations, and the data are recorded in a continuous loop two hours long. This is halted by the incoherent dedispersion glitch

R. Dodson (✉)
Marie Curie Fellow Observatorio Astronómico Nacional, Madrid, Spain
e-mail: r.dodson@oan.es

D. Lewis
CSIRO, Sydney, Australia
e-mail: dion.lewis@csiro.au

P. McCulloch
University of Tasmania, Hobart, Australia

Table 1 Parameters for the glitch of MJD 53193.092. The data fit is from MJD 53171 through to 53264 (mid-June to mid-September 2004). All significant figures are given. The model fitted consists of a permanent change in the rotation frequency and deceleration (denoted with a p subscript) and a number of temporary changes in rotation frequency (denoted with a n subscript) which decay on a timescale of τ

Parameters with reference to Epoch MJD 53193		
F/Hz	$\dot{F}$/Hz s^{-1}	$\ddot{F}$/Hz s^{-2}
11.1924472071183043	−1.555028E−11	5.27E−23
ΔF_p/Hz		$\Delta \dot{F}_p$/Hz s^{-1}
2.2865E−05		−1.0326E−13
τ_n		$\Delta F_n/10^{-6}$ Hz
1 ± 0.2 mins		54
00.23 days		0.21
02.10 days		0.13
26.14 days		0.16
DM		67.74 pc cm^{-3}

monitoring program. Coherent dedispersion is performed off-line, for the data segment covering the glitch.

2 The glitch in 2004

The glitch of 2004 occurred while the telescope was recording data. Unfortunately the coherent de-dispersion system was not running at that moment and the results obtained are more or less a repeat of those in 2000. The instantaneous fractional glitch size was 2.08×10^{-6}. This is the sum of the permanent and the decaying terms. The full details are reported in Table 1, after fitting with TEMPO (Taylor et al. 1970). A similar fast decaying term as reported in the 2000 glitch can be seen in the data after the usual model is subtracted, see Fig. 1, but it is only a few sigma above the noise. This usual model consists of permanent glitch components in the frequency and frequency derivate, and other components which are co-temporal jumps in frequency which decay away. A number of these are required to fit the data and the decay timescales are denoted with τ_n. Three decaying terms have been known for sometime and these make up the usual model. In Dodson et al. (2002) a forth short term component was identified. For a fuller discussion see that paper. In Fig. 1 the longer timescale terms are subtracted, and the residuals are plotted scaled against the RMS. Time zero is the intercept of the post-glitch model with the pre-glitch model, i.e. assuming an instant spin-up. The indication of a spin-up would be negative residuals, of which there is no sign. The positive residuals seen are modelled as a later

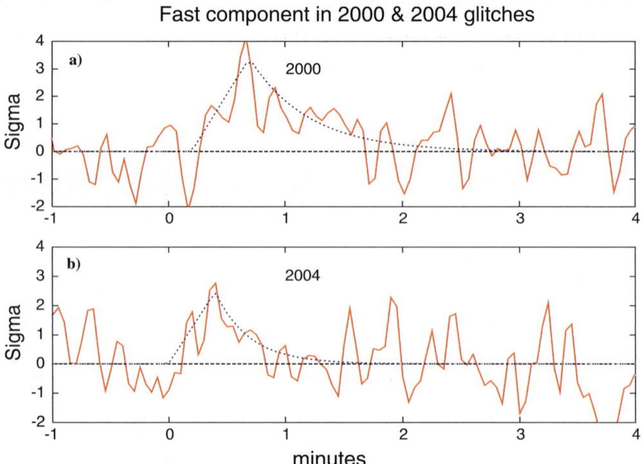

Fig. 1 The timing residuals after removing the long scale timing models, plotted as standard deviations, against the minutes relative to the assumed glitch epoch. The model of a later glitch epoch, and a very fast spin down, are overlaid. **a** is for 2000, **b** is for 2004

glitch epoch and a very fast decaying term. Figure 1a shows the glitch of 2000, where the signal was clearly above the noise, Fig. 1b shows the glitch of 2004, where the signal is barely above the noise. We present it as supporting evidence for the similar signal seen in the 2000 data, but we are unable to draw more detailed conclusions from such weak data.

3 Observations over the last twenty years

The 14-metre telescope has recorded single channel two minute folded data from July 10 1981 to October 1 2005, spanning 8857 days. It has recorded incoherently averaged ten second folded data from 1998. It observed on the day of a glitch for all ten events in that twenty year period. After upgrading to a full Az-Alt telescope in 1987 it was able to track the pulsar for 18 hours a day, and was therefore able to catch the very moment of the glitch in 1988, 1991, 2000 and 2004, as well as the first of the two in 1994. The two in 2000 and 2004 were with the incoherent single pulse system which is folded over ten seconds to give a good signal to noise. In all of these cases there is *no detectable spin up*. Figure 2 shows a montage of the recorded glitch events. Compare this to the stately, half day, spin up of the crab pulsar (Wong et al. 2001). The upper limit from the 2000 and 2004 de-dispersed data is that the spin-up occurs in less than 30 seconds. It is most unfortunate that, as yet, we have no observations with the coherent de-dispersion system, as these very fast spin up times measure directly the moment of inertia, and therefore the thickness, of the Vela pulsar crust (Epstein and Baym 1992). Figure 3 shows the fitted $\dot{F}$ over the twenty years. As an independent check of the braking index calculation of Lyne et al. (1996) we applied their method (used on the JPL data from 1969 to 1994, nine glitches) to our data (1981 to

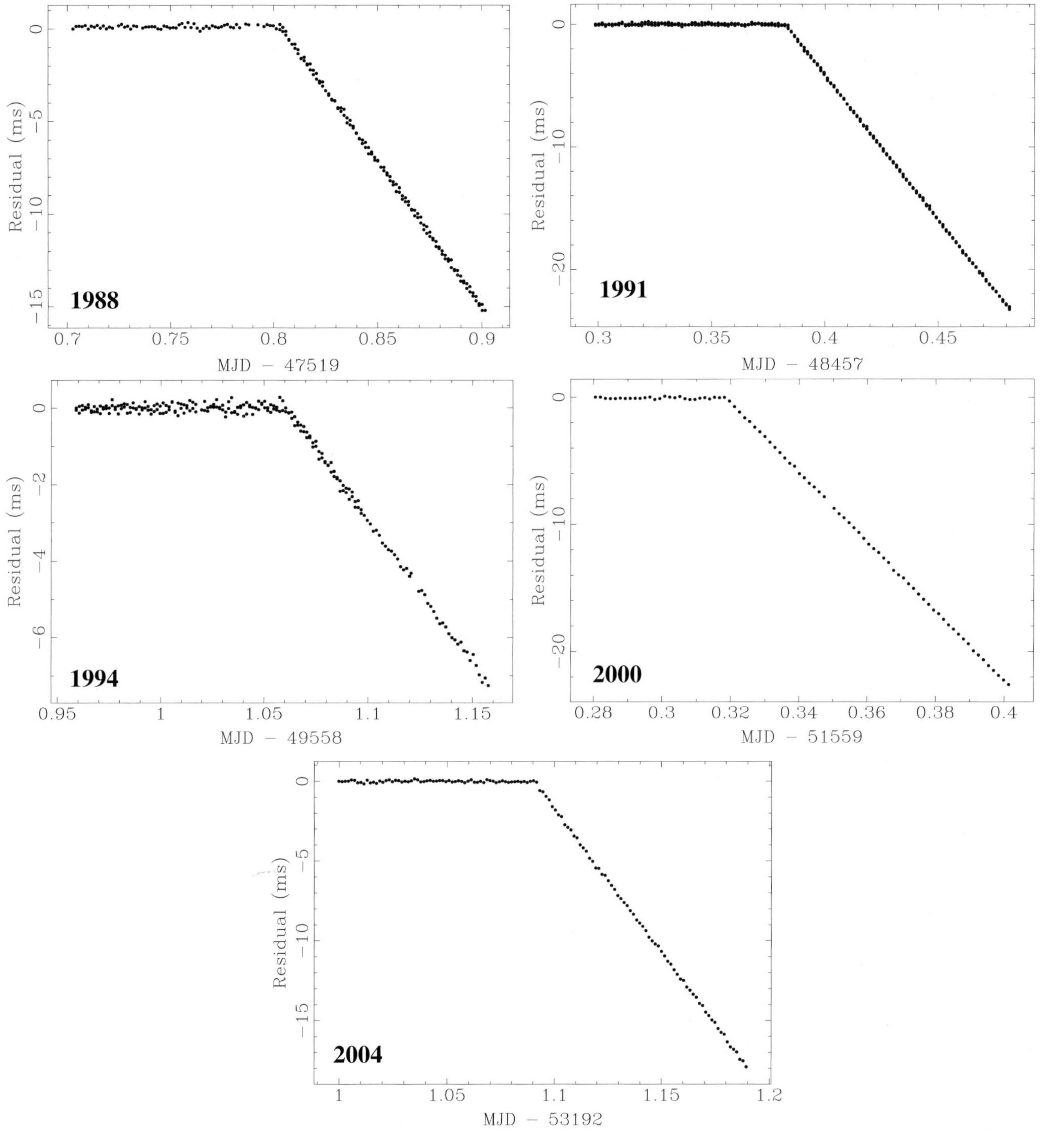

Fig. 2 Plots of all the glitches of Vela directly observed. The residuals are in milliseconds are plotted against time in days

2005, ten glitches). Here they assume that some months after the glitch the short term relaxations have decayed away, and the build-up of timing noise has not yet contaminated the rotation rate. Therefore by fitting $\ddot{F}$ to the slope $\dot{F}_{150}$, the deceleration 150 days after the glitch, the braking index, n, can be found from $F\ddot{F}\dot{F}^{-2}$. We find a braking index of 1.6 ± 0.1 for the data 150 days after the glitch. This is to be compared with the value Lyne et al. obtained by the same method of 1.5 ± 0.4. Lyne et al. improved on this value by extrapolating $\dot{F}$ back to the epoch of the glitch. We will report the full analysis, including this extrapolation, in a future paper.

The DM, which we can deduce from the multiple frequencies observed, continues to fall as reported in Hamilton

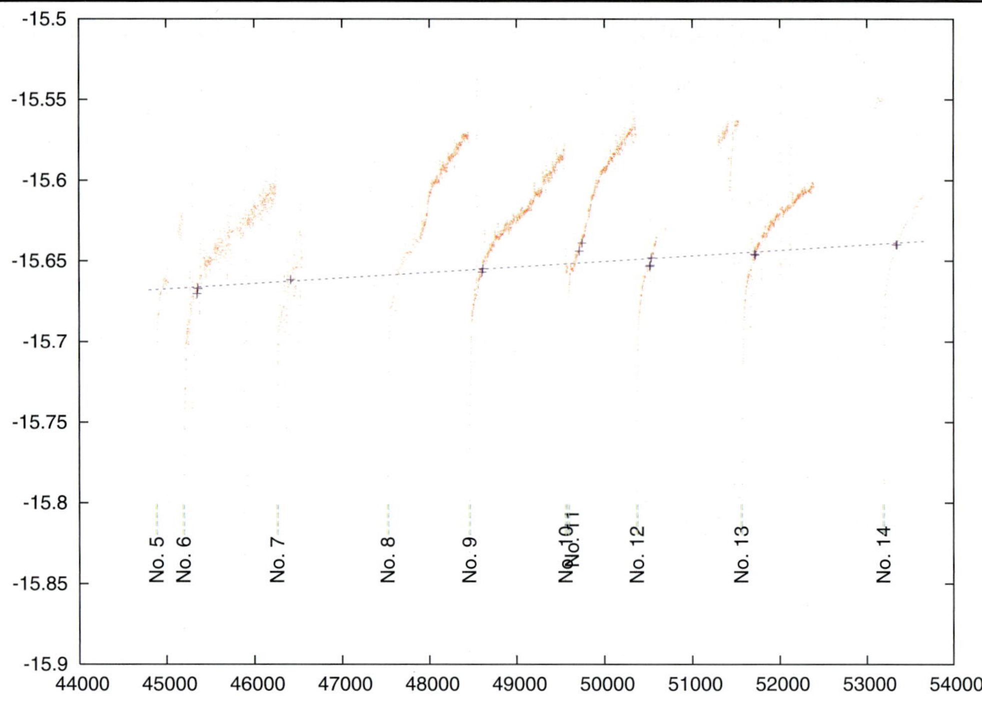

Fig. 3 Solutions for $\dot{F}$ over twenty years, with the $\dot{F}_{150}$ marked and the fitted slope for $\ddot{F}$ overlaid

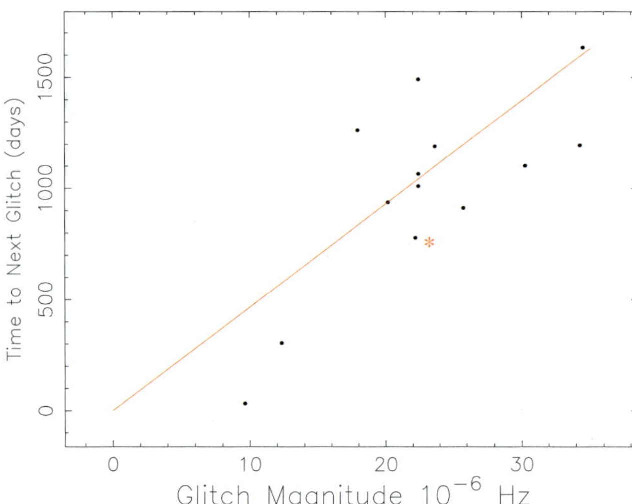

Fig. 4 Plot of the time to the next glitch vs glitch amplitude c.f. (Middleditch et al. 2006, Fig. 7), showing the weak correlation for the Vela pulsar compared to that of PSR 0537-6910. Marked with a *red star*, but not included in the fit, is the glitch of August 2006 (Flanagan and Buchner 2006)

et al. (1985). At the epoch of 53 193 it was 67.74 pc cm^{-3} and is falling by 4.3 pc cm^{-3} per century.

We don't see the strong correlation between glitch size and time between glitches in our data, as reported for PSR0537-6910 (Middleditch et al. 2006) recently. The best linear fit through the origin gives 46.5 days μHz^{-1}. The data are shown in Fig. 4.

4 Conclusions

The Vela pulsar has been timed for more than twenty years, and continues to provide new insights into the pulsar EOS by, for example, providing very low limits for the spin-up time and therefore the crust thickness. There is clearly a need to observe a glitch with higher sensitivity and time resolution to investigate both the fast decay term and to detect the spin-up. Both of these values relate directly to the pulsar EOS and will provide rich fodder for theoretical analysis by allowing the measurement of the moment of inertia of the crust. These observations could be made with the 14-metre telescope and the enhanced back-end if dedicated observing is continued. If the observation program is terminated then this effect is a ripe target for the next generation of pulsar telescopes that could monitor a large number of targets simultaneously with new beam forming techniques.

References

Bildsten, L., Epstein, R.I.: Astrophys. J. **342**, 951 (1989)

Dodson, R., Tingay, S., West, C., Phillips, C., Tzioumis, A.K., Ritakari, J., Briggs, F.: In: Bachiller, R., Colomer, F., Desmurs, J.F., de Vicente, P. (eds.) EVN on New Developments in VLBI, p. 253 (2004)

Dodson, R.G., McCulloch, P.M., Lewis, D.R.: Astrophys. J. **564**, L85 (2002)

Epstein, R.I., Baym, G.: Astrophys. J. **387**, 276 (1992)

Flanagan, C.S., Buchner, S.J.: IAU CBAT 595 (2006)

Hamilton, P.A., Hall, P.J., Costa, M.E.: Mon. Not. Roy. Astron. Soc. **214**, 5P (1985)

Lyne, A.G., Pritchard, R.S., Graham-Smith, F., Camilo, F.: Nature **381**, 497 (1996)

McCulloch, P.M., Hamilton, P.A., McConnell, D., King, E.A.: Nature **346**, 822 (1990)

Middleditch, J., Marshall, F.E., Wang, Q.D., Gotthelf, E.V., Zhang, W.: ArXiv Astrophysics e-prints (2006), see also http://pulsar.princeton.edu/tempo/index.html

Taylor, J., Manchester, R., Nice, D., Weisberg, J., Irwin, A., Wex, N., Standish, E.: Tempo, http://pulsar.princeton.edu/tempo (1970)

Wong, T., Backer, D.C., Lyne, A.G.: Astrophys. J. **548**, 447–459 (2001)

Astrophys Space Sci (2007) 308: 591–593
DOI 10.1007/s10509-007-9302-5

ORIGINAL ARTICLE

Slow glitches in the pulsar B1822-09

Tatiana V. Shabanova

Received: 21 June 2006 / Accepted: 30 June 2006 / Published online: 21 March 2007

Abstract The pulsar B1822-09 (J1825-0935) experienced a series of five unusual, slow glitches over the 1995–2004 interval. The results of further study of this unusual glitch phenomenon are presented. It is also reported the detection a new glitch of typical signature that occurred in the pulsar period in 2006 January.

Keywords Stars: neutron · Pulsars: general · Pulsars: individual: PSR B1822-09 · Stars: rotation

PACS 97.60.Jd · 97.60.Gb · 97.10.Kc · 97.10.Sj

1 Introduction

The timing observations of PSR B1822-09 obtained with the Pushchino radio telescope have revealed a new type of glitches, which has not been observed in any pulsar before. These rotation variations occurred in the form of slow glitches (Shabanova 1998; Shabanova and Urama 2000; Shabanova 2005, hereafter SH05). The present paper also reports the detection a new glitch of small size that occurred in 2006 January.

Characteristic feature of the slow glitches observed is a gradual exponential increase in the rotation frequency ν with a time-scale of 200–300 d. The curvature of the $\Delta\nu$ curve is determined by the corresponding change in the frequency derivative $\dot{\nu}$, the magnitude of which decreases by $\sim$1–2% of the initial value across the glitch. No obvious relaxation in frequency after a slow glitch is observed. The size of the slow glitches after a span of a few years is rather moderate, with magnitude of $\Delta\nu/\nu \sim 2 \times 10^{-8}$. Three slow glitches of similar amplitude have occurred in 1995 June, 1998 August and 2000 December. The third slow glitch was independently observed by Zou et al. (2004). The authors also reported a fourth smaller slow glitch that occurred in 2003.

T.V. Shabanova (✉)
Pushchino Radio Astronomy Observatory, Astro Space Center, P.N. Lebedev Physical Institute, Russian Academy of Sciences, 142290 Pushchino, Russia
e-mail: tvsh@prao.ru

2 Observations

Timing observations of the pulsar were performed with the BSA transit radio telescope at the Pushchino Observatory at frequencies around 103 and 112 MHz, using a 32×20 kHz filter bank receiver, as described in detail in SH05. The topocentric arrival times for each observation were corrected to the barycenter of the Solar System using the TEMPO software package and the JPL DE200 ephemeris. A simple spin-down model involving ν and $\dot{\nu}$ was used for fitting the barycentric arrival times. In order to study variations in the spin-down parameters of the pulsar in more detail, ν and $\dot{\nu}$ were calculated by performing local fits to the arrival time data over the intervals of $\sim$200 d.

3 Results

With respect to SH05, we extend the observational interval up to 2006 June, including two years of new observations, and present a description of the timing behavior of the pulsar over the 21-yr data span from 1985 to 2006. The timing data set includes the Pushchino data collected for the period 1991–2006 and the Hartebeesthoek data collected over the 1985–1998 interval and taken from the previously published

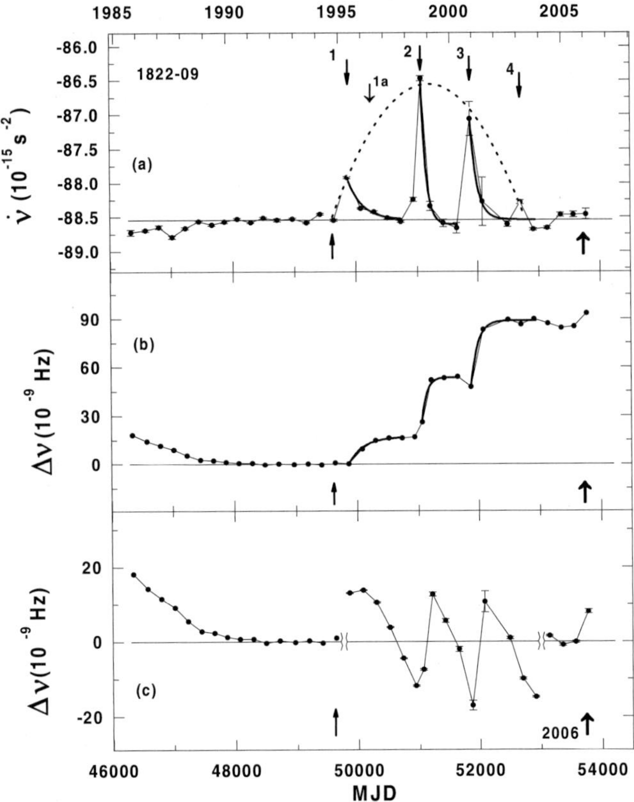

Fig. 1 $\dot{\nu}$ and $\Delta\nu$ as a function of time. **a** The rapid decreases in $\dot{\nu}$ over the interval 1995–2004 are the effect of the slow glitches. **b** $\Delta\nu$ relative to a fit to the data for the interval 1991–1994. **c** As for (**b**) but with $\Delta\nu$ relative to a new fit to the data for the 1995–2004 interval where the slow glitches occurred (*middle part of the plot*) and for the 2004–2005 interval (*right hand side of the plot*). *Arrows pointing downwards* indicate the epochs at which the slow glitches occurred while *arrows pointing upwards* indicate the normal glitches

paper (Shabanova and Urama 2000). Figure 1 gives the dependencies of $\dot{\nu}$ and the frequencies residuals $\Delta\nu$ on time, plotted as in Fig. 2 of the paper SH05, but supplemented with the four last new points.

Since 1994 the pulsar underwent a series of glitches. The first glitch that occurred in 1994 September (MJD 49615) had a typical signature and an extremely small size with the fractional increase of the rotational frequency $\sim 8 \times 10^{-10}$ (marked by the bottom arrow in Fig. 1). The next glitch occurred about a year later, in 1995 June, and initiated a series of five glitches of unusual signature, showing a slow growth in the frequency rotation during hundreds days. Here we mention five glitches because the first glitch, shown in Fig. 1, in fact represents the sum of two partially overlapped glitches of the smaller size (as was pointed out in Fig. 1 of SH05). The parameters of the slow glitches are listed in Table 1. The epochs of glitches correspond to the time at which $\dot{\nu}$ reaches its minimum value. The interval between all the glitches is approximately equal to 800 d, if the small glitch 1a is taken as the starting point. With all probability the small glitch 1a marks the starting point of a new phase in the glitching behaviour of PSR B1822-09. It is seen that rather small sizes of the glitches $\Delta\nu$ are related to large changes of $\dot{\nu}$ across the glitch, which reach ~2%. These $\Delta\dot{\nu}$ are responsible for the steepness of the front in $\Delta\nu$.

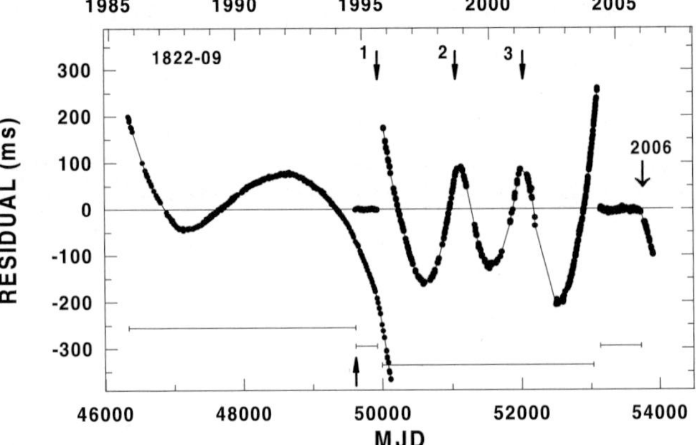

Fig. 2 Timing residuals of the pulsar over the 21-yr data span between 1985 and 2006, obtained from four independent fits for ν and $\dot{\nu}$ over four different intervals indicated in the plot by the *horizontal lines*. The three *upper arrows* indicate the epochs at which the three largest slow glitches occurred. The *bottom arrow* indicates the epoch at which the 1994 glitch of typical signature occurred. The new glitch occurred in January 2006

Table 1 Slow glitches in PSR B1822-09

No.	MJD	$\Delta\nu/\nu(10^{-9})$	$\Delta\dot{\nu}/\dot{\nu}(10^{-3})$
1	49857	12.8(2)	7.0(2)
1a	50253	4.3(2)	4.8(3)
2	51060	28.7(6)	24.2(4)
3	51879	32.0(9)	16.7(8)
4	52700	2.5(3)	2.9(3)

Fig. 1(a) shows that all the peaks of $\Delta\dot{\nu}$ lie on a curve which is the envelope of these peaks and is well described by a parabolic curve. The existence of the envelope indicates that all the slow glitches are the components of one process, the action of which ceased in the middle of 2004. The beginning of the envelope coincides with the epoch of the 1994 glitch of typical signature. It is likely that this small glitch has behaved as a trigger for the following unusual glitch-like events. Fig. 1(c) also shows that the process, responsible for the oscillatory changes in the rotation frequency, was stopped in the middle of 2004.

Figure 2 presents the timing residuals of the pulsar, plotted as in Fig. 3 of SH05, but supplemented with the last data segment from 2004 June to 2006 June. Analysis of this data segment showed that the pulsar suffered a new glitch that occurred in 2006 January 10 (MJD 53745(2)). This glitch is small, with the fractional increase $\Delta\nu/\nu = (6.6 \pm 0.5) \times 10^{-9}$. The frequency and timing residuals for this glitch are shown in the rightmost sides of Fig. 1 and Fig. 2.

Thus, a total of seven glitches have been detected in PSR B1822-09 during the 12 years since 1994. All glitches are small, with fractional increases of the rotation frequency $\Delta\nu/\nu \sim (0.8\text{–}32) \times 10^{-9}$. Five of these glitches belong to a new type of glitches that occurred in the form of slow glitches. The newly reported glitch of 2006, on the contrary, has a typical signature.

References

Shabanova, T.V.: Astron. Astrophys. **337**, 723 (1998)

Shabanova, T.V., Urama, J.O.: Astron. Astrophys. **354**, 960 (2000)

Shabanova, T.V.: Mon. Not. Roy. Astron. Soc. **356**, 1435 (2005) (SH05)

Zou, W.Z., Wang, N., Wang, H.X., et al.: Mon. Not. Roy. Astron. Soc. **354**, 811 (2004)

Astrophys Space Sci (2007) 308: 595–599
DOI 10.1007/s10509-007-9373-3

ORIGINAL ARTICLE

Short time scale pulse stability of the Crab pulsar in the optical band

S. Karpov · G. Beskin · A. Biryukov · V. Debur · V. Plokhotnichenko · M. Redfern · A. Shearer

Received: 1 July 2006 / Accepted: 11 October 2006 / Published online: 15 March 2007

Abstract The fine structure and the variations of the optical pulse shape and phase of the Crab pulsar are studied on various time scales. The observations have been carried out on 4-m William Hershel and 6-m BTA telescopes with APD photon counter, photomultiplier based 4-channel photometer and PSD based panoramic spectrophotopolarimeter with 1 μs time resolution in 1994, 1999, 2003 and 2005–2006 years. The upper limit on the pulsar precession on Dec 2, 1999 is placed in the 10 s–2 hr time range. The evidence of a varying from set to set fine structure of the main pulse is found in the 1999 and 2003 years data. No such fine structure is detected in the integral pulse shape of 1994, 1999 and 2003 years.

The drastic change of the pulse shape in the 2005–2006 years set is detected along with the pulse shape variability and quasi-periodic phase shifts.

This work has been supported by the Russian Foundation for Basic Research (grant No 04-02-17555), Russian Academy of Sciences (program "Evolution of Stars and Galaxies"), by the Russian Science Support Foundation, and by INTAS (grant No 04-78-7366).

S. Karpov (✉) · G. Beskin · V. Debur · V. Plokhotnichenko
Special Astrophysical Observatory of RAS, Nizhniy Arkhyz 369167, Karachaevo-Cherkessia, Russia
e-mail: karpov@sao.ru

A. Biryukov
Sternberg Astronomical Institute of MSU, 13 Universitetsky pr., Moscow 119992, Russia

M. Redfern · A. Shearer
National University of Ireland, Galway, University Road, Galway, Ireland

Keywords Methods: data analysis · Pulsars: general · Objects: PSR B0531+21

PACS 97.60.Jd · 97.60.Gb · 95.75.Wx

1 Introduction

Over the last 30 years the Crab pulsar has been extensively studied. The reasons for it are clear – it is the brightest pulsar seen in optics, it is nearby and young. However, the post popular groups contemporary theories of the Crab high-energy emission, the "polar cap" (Daugherty and Harding 1996) and "outer gap" (Cheng et al. 2000) ones, can't explain the whole set of observational data.

One of the main properties of the Crab emission is the very high stability of its optical pulse shape despite the secular decrease of the luminosity, related to the spin rate decrease (Pacini 1971; Nasuti et al. 1996).

At the same time the pulsars in general and the Crab itself are unstable. The instabilities manifest itself as a glitches, likely related to the changes of the neutron star crust, timing noise, powered by the collective processes in the superfluid internal parts of it, magnetospheric instabilities, results of the wisps around the pulsar, precession, et al. All these factors may influence the optical pulse structure and change it on various time scales, both in periodic and stochastic way.

However, it has been found early that the variations of the Crab optical light curve, in contrast with the radio ones, are governed by the Poissonian statistics (Kristian et al. 1970). A number of observations show the absence of non-stationary effects in the structure, intensity and the duration of the Crab optical pulses, and the restrictions on the regular and stochastic fine structure of its pulse on the time scales from 3 μs to 500 μs (Beskin et al. 1983;

Percival et al. 1993), the fluctuations of the pulse intensity (Kristian et al. 1970).

Along with the increase of the observational time spans and the accuracy of measurements the small changes of the optical pulse intensity, synchronous with the giant radio pulses, have been detected (Shearer et al. 2003). Also, the evidence for the short time scale precession of the pulsar has been detected by studying its optical light curve (Cadez et al. 2001).

All this raises the importance of the monitoring of the Crab optical emission with high time resolution.

2 Observations

We analyzed the sample of observational data obtained by our group over the time span of 12 years on different telescopes. The details of observations are summarized in Table 1. The equipment used were four-color standard photometer with diaphragms based on photomultipliers, fast photometer with avalanche photo-diodes (Shearer et al. 2003) and panoramic spectro-polarimeter based on position-sensitive detector (Debur et al. 2003; Plokhotnichenko et al. 2003). All devices provide the 1 μs time resolution.

For each data set the list of photon arrival times has been formed. They have been processed in the same way by using the same software to exclude the systematic differences due to data analysis inconsistencies. Photon arrival times have been corrected to the barycenter of the Solar System using the adapted version of *axBary* code by Arnold Rots. The accuracy of this code has been tested with detailed examples provided by Lyne et al. (2005) and is found to be better than 2 μs.

The barycentered photon lists then have been folded using both Jodrell-Bank radio ephemerides Jordan (2006) and our own fast-folding based method of timing model fitting.

The accuracy of timing model is proved to be better than at least several microseconds (see Fig. 1), which permits to fold the light curve with 5000 bin (6.6 μs) resolution.

3 Phase stability

We performed the search for timing model residuals using two longest continuous data sets of 1999 and 2005–2006 years. The data has been divided into the number of subsets of fixed length and they have been folded separately using the same base epoch. Then the sample light curves have been cross-correlated with the standard one (which has been derived for each set separately by folding the whole data) and its phase shift have been derived by fitting the maximum of the cross-correlation function with the Gaussian. The results for 1999 year set are shown in Fig. 1. No evidence for significant deviations from zero is seen, the phase is consistent with the Gaussian noise with 4.1 μs rms in the 10 s–2 hr time range.

The data of the last set of 2005–2006 years, however, show the significant quasi-periodic variations with $\sim 2.5 \times 10^{-3} P$ rms amplitude. The characteristic time scale of the variations is estimated to be roughly 0.7 d.

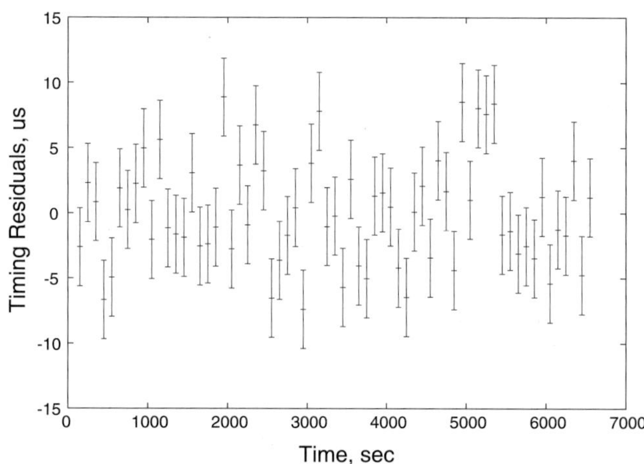

Fig. 1 Timing residuals of the Crab pulsar after applying second-order timing model (up to second frequency derivative). It corresponds to the Gaussian noise with 4.1 μs rms

Table 1 Log of observations

Date	Telescope	Instrument	Duration, s	Spectral range
Dec 7, 1994	BTA, Russia	Four-color photometer with photomultiplier	2400	U+B+V+R
Dec 2, 1999	WHT, Canary Islands	Avalanche photo-diode	6600	R
Nov 15, 2003	BTA, Russia	Avalanche photo-diode	1800	R
Dec 29, 2005–Jan 3, 2006	BTA, Russia	Panoramic spectro-polarimeter with position-sensitive detector	48 000	4000–7000 A

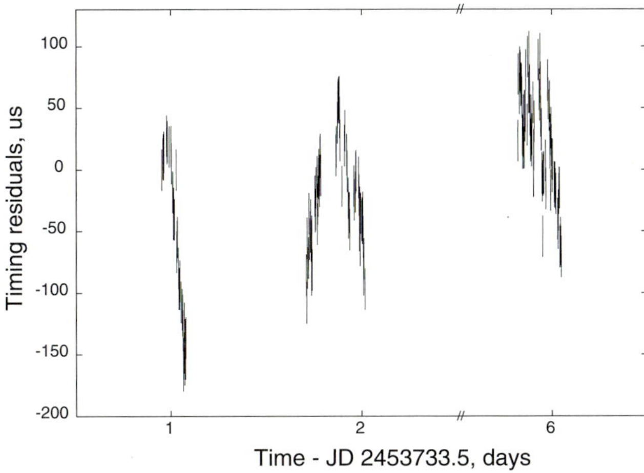

Fig. 2 Timing residuals of the Crab pulsar after applying second-order timing model. The quasi-periodic behaviour with characteristic time scale of 0.7 days is seen

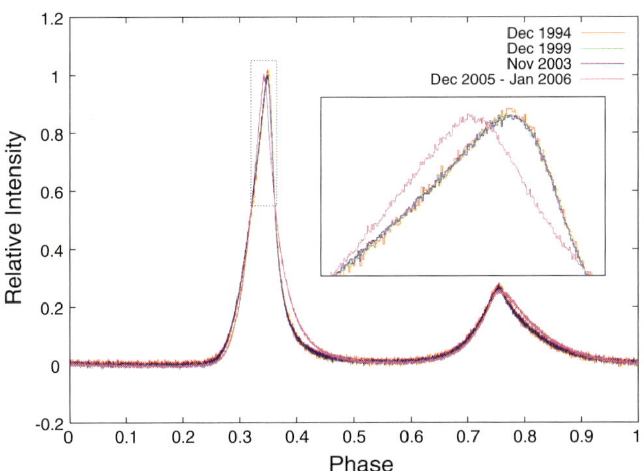

Fig. 3 Phased light curves of the Crab pulsar for all observations sets, scaled to the same pulse height

4 Pulse shape

Due to the presence of significant residuals relative to the timing model the pulse profile during the observations of 2005–2006 years can't be derived by folding the whole data set directly. Instead, we divided the data set into the one-hour segments and folded them separately applying the time shift corrections to compensate the phase residuals. The intrinsic phase shift inside each block is less than 2×10^{-4}, so the folding with 5000 bins is possible. The folded light curves have been co-added.

All the other data has been folded directly and shifted in phase to the same pulse position for the ease of comparison. All pulse profiles are shown in Fig. 3, with the off-pulse emission subtracted and pulse height scaled to the same value.

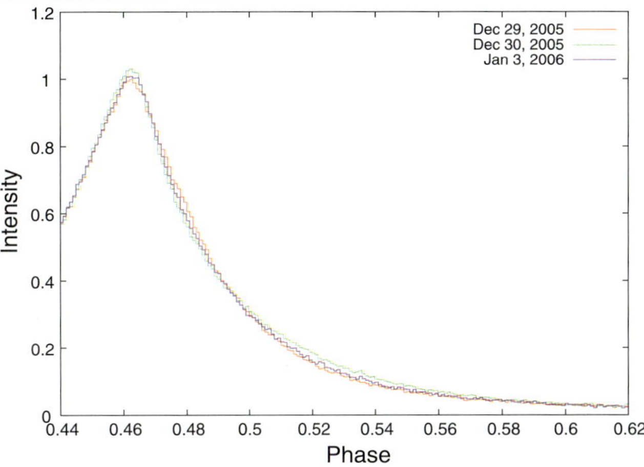

Fig. 4 Phased light curves of the Crab pulsar for the three nights of the Dec 2005–Jan 2006 set

The profiles of 1994, 1999 and 2003 years are in a perfect agreement with each other. The profile of 2005–2006 years, however, deviates from them significantly—the pulse remains of the same FWHM while its skewness is much smaller, and its shape is nearly symmetric.

We folded the data of this set for each of three observational nights separately using the same method. These profiles are shown in Fig. 4. There is the significant variation of its shape from night to night. Unfortunately the low amount of data available do not permit to track the profile shape change inside each night and check whether it is smooth or whether the shape is correlated with the timing residuals.

5 Pulse fine structure

For the first three observational set where the pulse profile is stable we performed the search for the fine structure of the main pulse. The data sets has been reduced to the same phase base point with precision better than half of the phase bin (less than 3.3 μs) and the cumulative light curve has been computed. The peak region of it is shown on Fig. 5. No statistically significant deviations from the smooth peak shape is seen. However, light curves of 1999 and 2003 years data sets alone (plotted on Fig. 6) each show the evidence of fine structure on the level of 3–5 sigma (roughly 1% of the intensity) with typical duration of 10–30 μs. Such details may give an evidence of coherent generation of optical emission, if the emission generation region is deep enough (deeper than 0.1 of light cylinder radius), due to brightness temperature exceeds 10^{12} K.

6 Discussion

The results of the last data set differs significantly from all previous results. We studied carefully the possibility of its

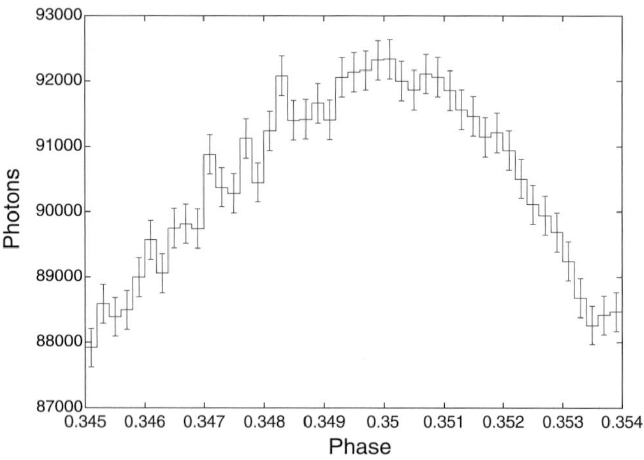

Fig. 5 The main pulse peak of the sum of light curves of 1994, 1999 and 2003 years data

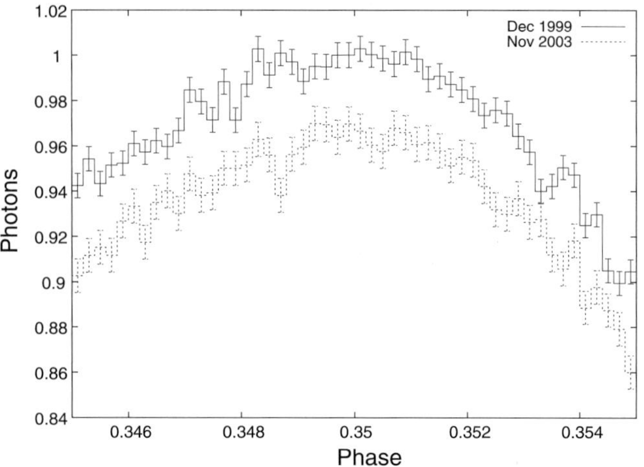

Fig. 6 The comparison of the peaks of 1999 and 2003 years. The peaks are shifted vertically for 0.03 for clearance

being the result of some hardware or data processing problem.

The data of last set has been acquired using the panoramic spectro-polarimeter based on the position-sensitive photon counter (Debur et al. 2003; Plokhotnichenko et al. 2003) in low-resolution spectral mode. There is no difference of the pulse profile in different spectral bands (derived using different parts of the spectrum detected). The detector behaviour is proved to be linear in the flux range used. The data acquisition system, "Quantochron 4-48", which records the time of arrival of the detected photons, has been checked for short time scale stability by recording the signal from the stationary 100-Hz generator. It has been processed and folded in the same way as a pulsar one (passing the unnecessary barycentering step). It shows no distortion of the signal shape larger than 1 μs. Large scale timing stability of acquisition system is ensured by means of 1 Hz and 10 kHz frequency signals from GPS receiver. The barycentering correction code passed the tests provided in Lyne et al. (2005) with the accuracy of 1 μs. The correctness of the radio ephemerides has been checked by performing the timing model fitting using our own software, the results of phase shift and folding analysis agree with ones based on the radio data. There is no small time scale (of order of 100 seconds and larger) changes of the pulse profile inside the set with amplitude comparable to the difference between the last set and previous ones.

Taking into account all these arguments we may conclude that the pulse profile change and quasi-periodic phase shifts detected in this observational set is most likely not related to the hardware or software problems of the equipment used.

The detection of the variations of both the pulse arrival times and its shape strongly supports the geometrical interpretation of the effect. It may be described as a quasi-periodic change of the pulsar beam orientation due to the strong precession commenced suddenly before the observations, but after the previous set. It may be related to the very strong glitch of the Feb-Mar 2004, or other recent change in the neutron star state (Lyne et al. 2005)

7 Conclusions

We analyzed the data of several sets of optical observations with high temporal resolution of the Crab pulsar performed by our group over the 12 last years.

No evidence for short time scale precession (like 60-sec free precession discovered in Cadez et al. 2001) is detected on the level of 10^{-5}–10^{-7} s^{-1} pulsar frequency variation on 10 s–2 hr time scale on Dec 2, 1999 (see Fig. 1), which corresponds to the precession wobble angle to be less than approximately 2×10^{-3}. Also, no signatures of short time scale timing noise is seen in this data set.

No significant fine structure is detected in the integral pulse profile of 1994, 1999 and 2003 years data set (see Fig. 5), however, each data set alone show the evidence of fine structure on the level of 3–5 sigma, which may be related to its instability on the time scale of years along with the stability of the pulse shape on the same scale.

We discovered the significant change of the time-averaged Crab pulse profile in the Dec 2005–Jan 2006 set of observations. The pulse profile also shows the variations between the nights. Also, the quasi-periodic phase shifts in respect to the second-order timing solution (up to second frequency derivative) has been detected in the data with amplitude of ~100 μs and characteristic time scale of 0.7 days. We have not found any hardware or software issue able to mimic such pulsar behaviour. These results may be interpreted as a geometric effects due to the Crab precession suddenly started between our observations of 2003 and 2005–2006 years.

References

Beskin, G.M., Neizvestnyi, S.I., Pimonov, A.A., et al.: Sov. Astron. Lett. **9**, 148 (1983)

Cadez, A., Vidrih, S., Galieie, M., et al.: Astron. Astrophys. **366**, 930 (2001)

Cheng, K.S., Ruderman, M., Zhang, L.: Astrophys. J. **537**, 964 (2000)

Daugherty, J.K., Harding, A.K.: Astrophys. J. **458**, 278 (1996)

Debur, V., Arkhipova, T., Beskin, G., et al.: Nucl. Instrum. Methods Phys. Res. A **513**, 127 (2003)

Jordan, C.A.: private communication (2006)

Kristian, J., Visvanathan, N., Vestphal, J.A., et al.: Astrophys. J. **162**, 475 (1970)

Lyne, A.G., Jordan, C.A., Roberts, M.E.: Crab Monthly Ephemeris, available on http://www.jb.man.ac.uk/~pulsar/crab.html (2005)

Nasuti, F.P., Migani, R., Caraveo, P.A., et al.: Astron. Astrophys. **314**, 849 (1996)

Pacini, F.: Astrophys. J. **163**, 17 (1971)

Percival, J.W., Biggs, J.D., Dolan, J.F., et al.: Astrophys. J. **407**, 276 (1993)

Plokhotnichenko, V., Beskin, G., Debur, V., et al.: Nucl. Instrum. Methods Phys. Res. A **513**, 167 (2003)

Shearer, A., Stappers, B., O'Connor, P., et al.: Science **301**, 493 (2003)

Astrophys Space Sci (2007) 308: 601–605
DOI 10.1007/s10509-007-9303-4

ORIGINAL ARTICLE

Using XMM-Newton to measure the spectrum of the Vela pulsar and its phase variation

Armando Manzali · Andrea De Luca · Patrizia A. Caraveo

Received: 30 June 2006 / Accepted: 4 October 2006 / Published online: 20 March 2007

Abstract We report on our analysis of two XMM-Newton observations of the Vela pulsar performed in December 2000 (total exposure time: 96.5 ks). We succeeded in resolving the pulsar spectrum from the surrounding bright nebular emission taking advantage both of the accurate calibration of the EPIC point spread function and of the Chandra/HRC surface brightness map of the nebula. This made it possible to assess to pulsar spectral shape disentangling its thermal and non-thermal components. Exploiting the photon harvest, we have also been able to perform a phase-resolved study of the pulsar emission.

Keywords Pulsars: general · Pulsars: individual (Vela) · Stars: neutron · X-rays: stars

PACS 97.60.Gb · 97.60.Jd · 98.70.Qy

1 Introduction

The Vela pulsar ($P = 89$ ms; $\tau \approx 11\,400$ y; $\dot{E} \approx 7 \times 10^{36}$ erg s^{-1}) is one of the most famous and best scrutinized neutron stars. Soft X-ray pulsed radiation from the pulsar was first detected by ROSAT (Oegelman et al. 1993). It turned out to be a difficult observation since X-rays from the neutron star itself are embedded in the bright Pulsar Wind Nebula (hereafter PWN) located near the centre of the Vela Supernova Remnant. A blackbody with temperature $\sim$1.5–1.6 $\times$ 10^6 K and a radius of 3–4 km could describe ROSAT spectrum while the nebular emission, clearly non-thermal, can be ascribed to synchrotron emission originated from the interaction of the high-energy particle pulsar wind with the interstellar medium. Chandra observations of the Vela pulsar provided high resolution images of the X-ray nebula surrounding the neutron star. The nebula turns out to be quite complex, with spectacular arc and jet-like features (Helfand et al. 2001; Pavlov et al. 2003), reminiscent of ones observed around the Crab. The thermal nature of the pulsar spectrum inferred from ROSAT data was confirmed, although the Chandra spectrum is better described by a neutron star atmosphere model (nsa), then by a blackbody one (Pavlov et al. 2001b). While high resolution spectra fail to show absorption features, a non-thermal harder tail was found in ACIS-S spectra. Analysing XMM-Newton observations Mori et al. (2002) found a spectral behaviour in agreement with the Chandra data. However in their preliminary analysis, the strong nebular contamination did not allow to detect pulsar emission at energies higher than $\sim$1 keV.

2 The data

We will use data collected by XMM-Newton and Chandra, exploiting the characteristics of both X-ray observatories. While XMM-Newton observations provide an unprecedented harvest of photons from the Vela pulsar and its PWN, the instrument point spread function (hereafter PSF, 6.6″ FWHM) does not allow to disentangle the pulsar emission from the extended one. Accounting for the bright PWN contribution is mandatory to unveil the PSR emission properties. Thus, the high resolution image obtained by the Chan-

A. Manzali (✉) · A. De Luca · P.A. Caraveo
INAF–IASF, Via Bassini 15, I-20133 Milano, Italy
e-mail: armando@iasf-milano.inaf.it

A. Manzali
Dipartimento di Fisica Nucleare e Teorica, Università degli Studi di Pavia, Via Bassi 6, I-27100 Pavia, Italy

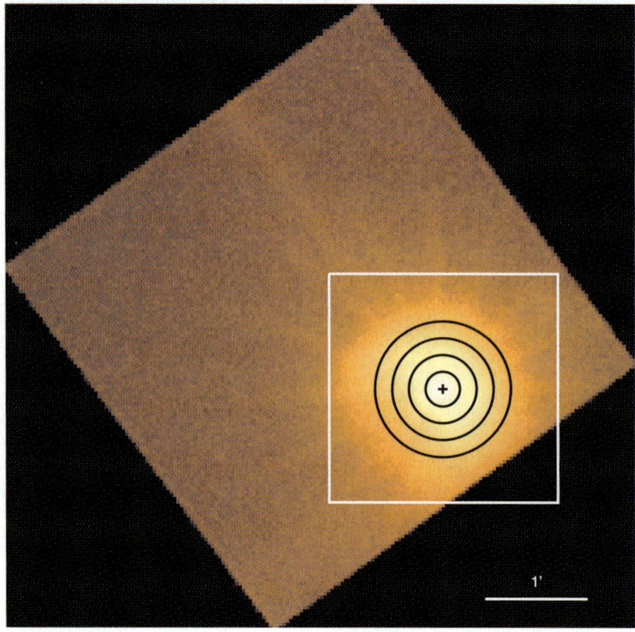

Fig. 1 EPIC-PN image of the Vela pulsar. The *cross* marks the radio position of the pulsar and the *box* marks the region covered by Chandra image (Fig. 2). The spectra extraction regions (par. 3) are also marked

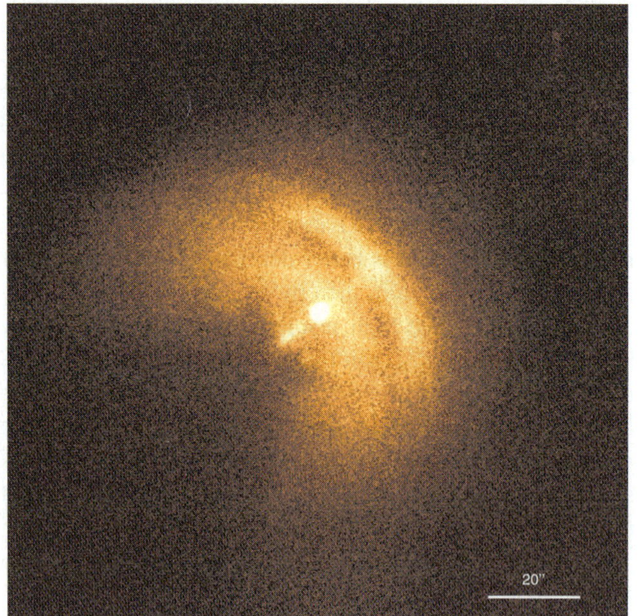

Fig. 2 Chandra/HRC image of the Vela pulsar and the surrounding nebula

dra observatory is crucial to have a clear view of the Vela pulsar and its surrounding nebular emission.

Figures 1 and 2 show an image extracted from XMM/PN and Chandra/HRC data respectively: the XMM-Newton telescopes angular resolution does not allow to resolve the point-like source from the diffuse emission neither the PWN features observed in Chandra image (see also Pavlov et al.

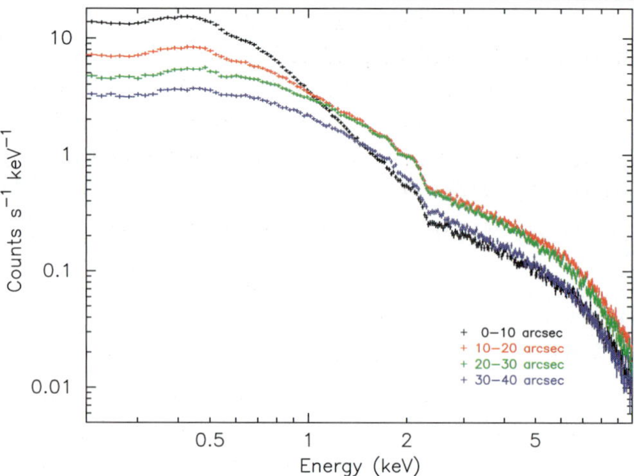

Fig. 3 Observed EPIC-PN spectra from the 4 different extraction regions

2001a; Helfand et al. 2001; Pavlov et al. 2003). In order to have a precise determination of the spatial distribution of the bright nebular emission we analysed three Chandra/HRC-Imaging public observations of the Vela pulsar.

3 Spectral analysis

To perform the spectral analysis of the XMM-Newton data on the Vela pulsar, we exploit both the spectral and temporal resolution of the EPIC-PN camera. In order to disentangle the PSR photons from the PWN ones, we take advantage of their different space distribution: while the PSR photons follow the instrument PSF, the nebular ones do not.

We estimate the nebular contribution in the EPIC-PN event file in the following way:

- we extract spectra from concentric annular regions of increasing radii;
- we determine the PSR counts contributions in each spectrum as a function of the PSF and of the Encircled Energy Fraction (EEF);
- we determine the PWN counts contribution in each spectrum using a surface brightness map of the nebular emission derived from Chandra data;
- we fit all the spectra at once with a two component (PSR + PWN) model whose normalisation ratios are fixed following the two previous steps.

We extract spectra from merged EPIC-PN event file, with flag 0 and PATTERN 0–4, from a 10″ radius circle centered on the pulsar position as well as from three annular (10″–20″, 20″–30″, 30″–40″) regions. These are drawn in Fig. 3 where the composite and varying nature of the spectra is clearly visible: while at lower energies the thermal spectrum prevails, with the flux values in the 4 different extraction regions proportional to EEF, at higher energies almost all the photons detected are produced by the PWN and the flux values are proportional to the spatial distribution of the nebula surface brightness. We do not perform background subtraction since background determination will follow directly from the fitted spectra.

The XMM-Newton telescopes radially averaged PSF profile at distance r from the aimpoint is well described by the King's profile (Gondoin et al. 1998). The Encircled Energy Fraction, which specifies the fraction of energy collected within a certain radius R from the source distribution centroid is simply the integral of the King's profile multiplied by radius over the radius, normalised at $5'$.

We used Chandra data to produce a surface brightness map of the nebular emission. Since we needed an estimate of the PWN flux in the proximity of the position of the pulsar, a two dimensional model of a point-like source flux distribution was obtained with a simulation performed with ChaRT[1] and MARX.[2] A 2D fitting was then performed on the HRC images with the addition of a spatially constant contribution from the PWN. Following the criterion by Helfand et al. (2001), the region inside $3.46''$, 26 pixel, was cutted from all images and replaced with a poissonian distribution with a mean value equal to the best fit count rate.[3]

In order to take into account the angular resolution of the XMM telescopes, the maps obtained were convolved with a kernel reproducing the King's profile. We tried Gaussian and Lorentzian kernel of different width. A Lorentzian kernel of $\Gamma = 7.125''$, corresponding to a FWHM of the distribution of $6.6''$, the nominal FWHM of the XMM mirrors, was finally chosen on the base of the minimizations of the spectral fitting residuals.

Finally, the convolved image was used to compute the encircled PWN fraction in each extraction region of our XMM image.

The XMM spectra extracted from the four annuli (Fig. 3) were then fitted simultaneously with an absorbed two component model,

$$\text{wabs}\big[(\text{PSR\_coefficient} \times \text{PSR\_model}) + (\text{PWN\_coefficient} \times \text{PWN\_model})\big].$$

where PSR_coefficient and PWN_coefficient represent the different contribution to the total flux (i.e. encircled fractions) of the point-like and diffuse emission, within each extraction region. The values of this coefficients are given in column 3 and 4 of Table 1.

[1] http://cxc.harvard.edu/chart/.

[2] http://space.mit.edu/CXC/MARX/.

[3] A more detailed description of the method and results will be published elsewhere (Manzali et al. 2007).

Table 1 Number of photons collected by the EPIC-PN in the different spectra extraction regions. Also shown are XMM-Newton telescopes EEF and PWN counts derived from Chandra surface brightness map. These values have been used to compute the PSR and PWN coefficients for the spectral fitting

Extraction region	Total number of photons	PSR EEF (%)	PWN flux (counts)
10″	671 021	0.582342	17729.6985
10″–20″	541 089	0.215818	40962.9075
20″–30″	431 813	0.0770648	42880.3005
30″–40″	293 991	0.0377001	29622.9375

The spectral fitting was performed with XSPEC v.11.3, in the energy range 0.2–10 keV with a 5% systematic error added in order to account for the uncertainties arising in the previously described processes.

The best fit parameters allow us to determine the nebular background flux and photon index for all spectra. We found the non-thermal nebular emission become softer as the distance from the pulsar increases, with photon index varying from $1.367^{+0.006}_{-0.007}$ to $1.651^{+0.007}_{-0.006}$. The result is consistent with Chandra/ACIS spectra obtained from 30 November 2000 observation (see also Kargaltsev and Pavlov 2004). The power-law normalisation yields the PWN flux in the EPIC-PN data, its value turns out to be $F_X = 5.394 \times 10^{-11} \pm 1.5 \times 10^{-13}$ erg cm^{-2} s^{-1} keV^{-1}, in the range 1–8 keV, corresponding to a luminosity of

$$L_X = 5.594 \times 10^{32} \pm 1.6 \times 10^{30} \text{ erg s}^{-1} \quad (1)$$

at the parallactic distance. Both values are in good agreement with Kargaltsev et al. (2002).

The spectral distribution of the ~500 000 Vela PSR photons detected in the inner $10''$ radius circle, can be described as the superposition of different components. A simple blackbody thermal model does not fit well the observed spectra, that appears harder than a simple planckian emission. The structure of the residuals suggests the addition of a second "pulsar" component: both a second blackbody or a power-law gave acceptable fits. Similar results were obtained with a *magnetised hydrogen atmosphere* (nsa, Pavlov et al. 1991; Zavlin et al. 1996) + *power-law* model. The interstellar column density of $\sim 2.6 \times 10^{20}$ cm^{-2}, inferred from the different models, is in good agreement with the results of Pavlov et al. (2001b) and Mori et al. (2002). For the double blackbody model we obtained $\chi^2_\nu = 1.1$ (1010 degrees of freedom).

The soft spectrum is described by a cooler ($T_{bb} = 1.06 \pm 0.03 \times 10^6$ K) component, with a radius $R_{bb} = 5.2^{+0.4}_{-0.3}$ km and a smaller and hotter spot of $T_{BB} = 2.16^{+0.06}_{-0.07} \times 10^6$ K and $R_{BB} = 0.75^{+0.08}_{-0.07}$ km. An equally good fit was obtained with a $\gamma = 3.48^{+0.08}_{-0.06}$ power-law; in this case the blackbody

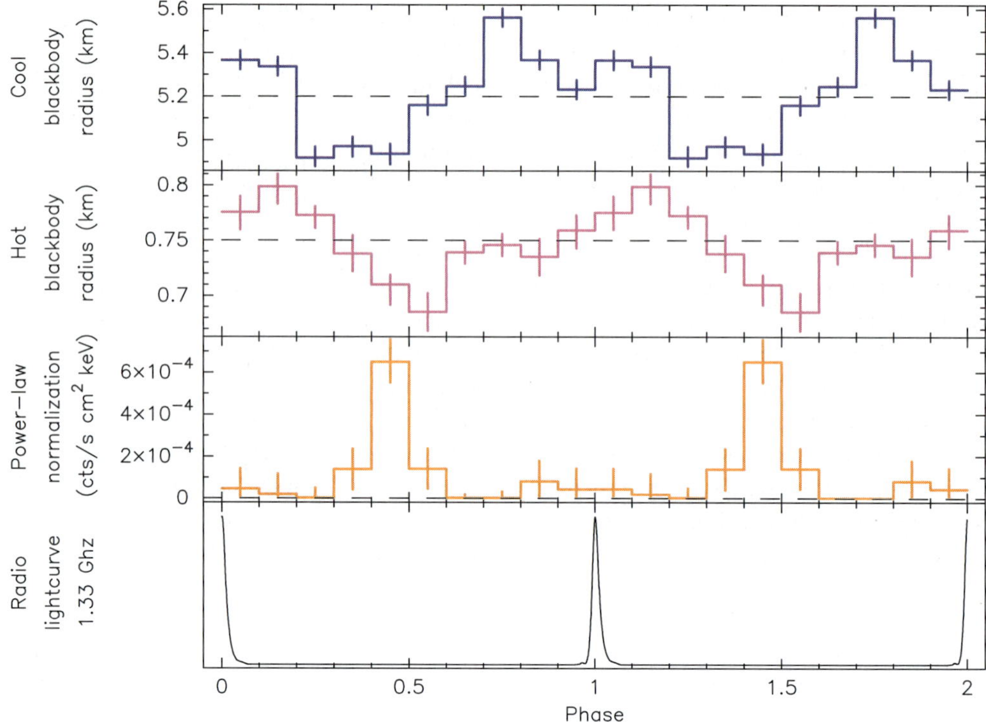

Fig. 4 Variation of the three spectral component (cool and hot blackbody radius and power-law normalisation) as a function of the rotational phase. Two rotational period and radio light curve are reported for clarity

is found to have a temperature $T = 1.49^{+0.02}_{-0.02} \times 10^6$ K and a radius $R = 2.0^{+0.6}_{-0.4}$ km. The power-law is much steeper than that observed in all other X-ray emitting pulsars.

A hydrogen atmosphere plus power-law model yields a slightly better fit ($\chi^2_\nu = 1.0$ for 1010 degrees of freedom). The surface temperature, for a radius $R = 10$ km and a magnetic field $B = 10^{12}$ G, is $T_{\rm ha} = 0.681 \pm 0.004 \times 10^6$ K; the best fit distance $D = 269^{+12}_{-14}$ pc agrees with the parallattic measurements. The power-law photon index $\gamma = 2.8 \pm 0.2$ is however steeper that the one found by Pavlov et al. (2001b).

Nevertheless we preferred to perform phase-resolved spectral fitting with the phenomenological model also adopted by Caraveo et al. (2004) and De Luca et al. (2005) in their analysis of Geminga, PSR B1055-52 and PSR B0656+14, the "Three Musketeers".[4]

4 Phase-resolved spectral analysis

Exploiting the photon harvest, we have been able to extract a spectrum from each of 10 different phase intervals and perform individual spectral fitting. Following the approach of Caraveo et al. (2004) and De Luca et al. (2005) we have compared the phase-resolved spectra with the two blackbodies best fit model using the two normalisation coefficients as free parameters.

Such an approach works well for all the phase resolved spectra but for that encompassing phase interval $0.4 < \varphi < 0.5$ which is characterized by high energy residuals impossible to account for with the two blackbodies model. Leaving the temperature of the two blackbodies and the hydrogen column density fixed to the values found above, a power-law with photon index $\gamma = 2.4 \pm 0.4$ is required to fit the spectrum. The normalisation of the power-law is ~10 times lower than the nebular emission in this phase interval, thus, its contribution to the phase-integrated spectrum is ~100 times lower than the nebular one and it cannot be observed. Fig. 4 shows the variation of the cool and hot blackbody radii, as well as the power-law normalization, during pulsar rotation.

5 Conclusions

Similarly to the "Three Musketeers", the phase-resolved spectra of the Vela pulsar could be described by a 3 component model (two blackbodies plus a power-law), although in this case the overall modulation is ~15%. The hot blackbody emission reaches its maximum in phase with the radio pulse, long known to mark the pulsar polar region. The geometry of the Vela pulsar, an high inclined rotator seen at an angle of $\sim 60°$ (see e.g. Helfand et al. 2001 and references therein), as well as the radius (Goldreich and Julian 1969) and the luminosity of the hotter blackbody component, suggests that it could be associated with a polar cap

[4]Phase-resolved spectroscopy with the nsa+power law model showed a clear anticorrelation between model parameters.

heated by downward acceleration of pairs produced by curvature photons, in agreement with Harding and Muslimov (2002). The cooler blackbody would represent the radiation from almost the totality of the star surface. The radius inferred from the phase-averaged spectrum is too small to fit in any proposed equation of state for a star composed mainly of neutron. We also observed a $\sim 10\%$ modulation of this spectral component as a function of the rotational phase. Magnetospheric reprocessing of the thermal photons emitted from the surface could provide a phase-dependent "obscuration" of a fraction of the neutron star surface, depending on magnetic field configuration and viewing geometry. The phenomenon of the magnetospheric "blanket", e.g. cyclotron resonance scattering by plasma at a few stellar radii (Ruderman 2004), originally proposed by Halpern and Ruderman (1993) as an explanation of the soft thermal emission of Geminga, could provide the physical basis for the observation of a phase-dependent emitting area.

Acknowledgements We would thank Richard Dodson and the ATNF for kindly providing ad hoc radio ephemeris for the observation date.

References

Caraveo, P.A., De Luca, A., Mereghetti, S., et al.: Science **305**, 376 (2004)
De Luca, A., Caraveo, P.A., Mereghetti, S., et al.: Astrophys. J. **623**, 1051 (2005)
Goldreich, P., Julian, W.H.: Astrophys. J. **157**, 869 (1969)
Gondoin, P., Aschenbach, B.R., Beijersbergen, M.W., et al. In: Hoover, R.B., Walker, A.B. (eds.) X-Ray Optics, Instruments, and Missions. Proc. SPIE, vol. 3444, pp. 278–289 (1998)
Halpern, J.P., Ruderman, M.: Astrophys. J. **415**, 286 (1993)
Harding, A.K., Muslimov, A.G.: Astrophys. J. **568**, 862 (2002)
Helfand, D.J., Gotthelf, E.V., Halpern, J.P.: Astrophys. J. **556**, 380 (2001)
Kargaltsev, O., Pavlov, G. In: Camilo, F., Gaensler, B.M. (eds.) IAU Symposium, p. 195 (2004)
Kargaltsev, O., Pavlov, G.G., Sanwal, D., et al. In: Slane, P.O., Gaensler, B.M. (eds.) Neutron Stars in Supernova Remnants. ASP Conf. Ser., vol. 271, p. 181 (2002)
Manzali, A., De Luca, A., Caraveo, P.A.: Astron. Astrophys. (2007, submitted)
Mori, K., Hailey, C., Paerels, F. et al. In: COSPAR, Plenary Meeting (2002)
Oegelman, H., Finley, J.P., Zimmerman, H.U.: Nature **361**, 136 (1993)
Pavlov, G.G., Shibanov, I.A., Zavlin, V.E.: Mon. Not. Roy. Astron. Soc. **253**, 193 (1991)
Pavlov, G.G., Kargaltsev, O.Y., Sanwal, D., et al.: Astrophys. J. **554**, L189 (2001a)
Pavlov, G.G., Zavlin, V.E., Sanwal, D., et al.: Astrophys. J. **552**, L129 (2001b)
Pavlov, G.G., Teter, M.A., Kargaltsev, O., et al.: Astrophys. J. **591**, 1157 (2003)
Ruderman, M.: In: Tavani, M., Pellizoni, A., Vercellone, S. (eds.) X-Ray and Gamma-Ray Astrophysics and Galactic Sources, pp. 35. ARACNE, Rome (2004)
Zavlin, V.E., Pavlov, G.G., Shibanov, Y.A.: Astron. Astrophys. **315**, 141 (1996)

Astrophys Space Sci (2007) 308: 607–611
DOI 10.1007/s10509-007-9350-x

ORIGINAL ARTICLE

A toy model for global magnetar oscillation

Implications for QPOs during magnetar flares

Kostas Glampedakis · Lars Samuelsson · Nils Andersson

Received: 13 July 2005 / Accepted: 11 October 2006 / Published online: 22 March 2007

Abstract The presence of a magnetic field in a neutron star interior results in a dynamical coupling between the fluid core and the elastic crust. We consider a simple toy-model where this coupling is taken into account and compute the system's mode oscillations. Our results suggest that the notion of pure torsional crust modes is not useful for the coupled system, instead all modes excite Alfvén waves in the core. However, we also show that among a rich spectrum of global MHD modes the ones most likely to be excited by a fractured crust are those for which the crust and the core oscillate in concert. For our simple model, the frequencies of these modes are similar to the "pure crustal" frequencies. We advocate the significant implications of these results for the attempted theoretical interpretation of QPOs during magnetar flares in terms of neutron star oscillations.

Keywords Neutron stars · Oscillations · Magnetars

PACS 04.40.Dg · 97.60.Jd

1 Introduction

The recent observational evidence (Israel et al. 2005; Strohmayer and Watts 2005; Watts and Strohmayer 2006) of quasi-periodic oscillations (QPOs) during giant flares in the soft gamma-ray repeaters (SGRs) 1806-20 and 1900+14 may constitute the first direct detection of neutron star oscillations, initiating an exciting new era for neutron star physics.

K. Glampedakis (✉) · L. Samuelsson · N. Andersson
School of Mathematics, University of Southampton,
Southampton SO17 1BJ, UK
e-mail: glampeda@sissa.it

The overall physics of SGRs and the particular features of extremely energetic events like the giant flares can be understood within the *magnetar* model, introduced by Duncan and Tompson (1992) (and Tompson and Duncan 1995) more than a decade ago. Magnetars are neutron stars endowed with ultra-strong magnetic fields ($\sim 10^{14-15}$ G), believed to manifest themselves as both SGRs and anomalous X-ray pulsars. The giant flares are thought to be the outcome of magnetic field activity which carries enormous amounts of energy from the interior to the magnetosphere. In the process, the magnetar's crust is likely to suffer straining and fracturing.

The observed QPOs are then associated with shear crustal oscillations. This is a reasonable interpretation since the fundamental toroidal crustal modes have frequencies (~30–100 Hz) that could match the observations (van Horn 1980; Duncan 1998). The relevant mode periods are well approximated by the formula (McDermott et al. 1988; Hansen and Cioffi 1980; Strohmayer 1991),

$$P_\ell^0 \approx \frac{2\pi R}{\tilde{\mu}^{1/2}}[\ell(\ell+1)]^{-1/2} \approx 60 R_6 [\ell(\ell+1)]^{-1/2}\ \text{ms} \qquad (1)$$

where $\tilde{\mu} = \mu/\rho$ is the specific shear modulus, ℓ is the usual spherical harmonic index, R is the radius and $R_6 = R/10^6$ cm. This formula is Newtonian. In order to put theory and observations face to face one needs to account for general relativistic corrections, in which case the above result is amended with a multiplicative factor which is a function of the neutron star's mass and radius (Duncan 1998) (see contribution by Samuelsson and Andersson in these proceedings for a rigorous treatment of the problem). Such comparison is presented in Table 1, where we have fixed the $\ell = 2$ mode frequency at 30 Hz and 28 Hz for 1806-20 and 1900+14, respectively. We can see a remarkable agreement, which extends to higher frequency QPOs

Table 1 Comparing the observed QPO frequencies (Israel et al. 2005; Strohmayer and Watts 2005; Watts and Strohmayer 2006) against the theoretical prediction for the frequency of the fundamental toroidal crustal mode, for various ℓ

SGR	$f_{\rm QPO}$ (Hz)	f mode (Hz)	ℓ
1806-20	18	–	–
	26	–	–
	30	30	2
	92	92	7
	150	153	12
1900+14	28	28	2
	54	51	4
	84	85	7
	155	154	13

not shown here, see Strohmayer and Watts (2005) and Watts and Strohmayer (2006) and contribution by A. Watts in these proceedings. However, it appears impossible to interpret the 18 and 26 Hz QPOs of SGR 1806-20 as crustal toroidal modes, unless extreme values for the magnetar's mass and radius are assumed (Strohmayer and Watts 2005; Watts and Strohmayer 2006).

2 Magnetic crust-core coupling

Previous work on crustal oscillations (Carroll et al. 1986; Messios et al. 2001; Piro 2005) ignored the crucial role of the magnetic field in coupling the crust to the fluid core. This key point was recently emphasised by Levin (2006).

A rough estimate for the crust-core coupling timescale is provided by the Alfvén crossing time,

$$t_A = 2R/v_A \sim 70 B_{15}^{-1} \rho_{14}^{1/2} R_6 \text{ ms} \tag{2}$$

where $B = 10^{15} B_{15}$ G, $\rho = 10^{14} \rho_{14}$ g/cm^3 and

$$v_A^2 = \frac{B^2}{4\pi\rho}. \tag{3}$$

Clearly, the above coupling timescale is comparable to the period of the fundamental crustal mode which simply means that, for parameters relevant to the crust-core interface of a magnetar, an efficient coupling to the fluid core is already established within a single oscillation of a "crustal" mode. This has crucial implications for any attempt to calculate the properties of the modes. In essence, it is no longer meaningful to think of modes confined to the crust, instead one is forced to consider *global* oscillations of the coupled crust-core system.

Accepting this generic crust-core coupling and the notion of global modes one faces an obvious puzzle: if the QPO

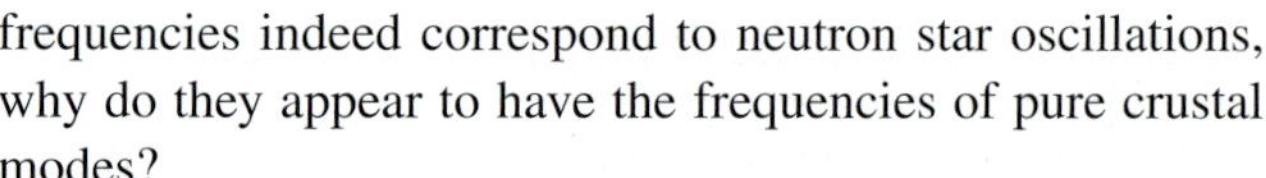
frequencies indeed correspond to neutron star oscillations, why do they appear to have the frequencies of pure crustal modes?

We offer a solution to this puzzle by first demonstrating that a magnetically coupled crust-core system admits global mode oscillations with frequencies similar to pure "crustal" frequencies, and then explaining why these modes are favoured to be excited by an initial disturbance in the crust region (say, following a starquake induced by a magnetic field eruption). A more detailed version of the present work can be found in Glampedakis et al. (2006).

3 A simple toy-model

We consider a plane-parallel "star" where the fluid core is sandwiched between two slabs of "crust", see Fig. 1 (this model is similar to the one employed by Piro 2005). The only non-trivial coordinate is z, which runs from $z = +R$ (the surface) to $z = 0$ (the core's centre) and ends up back at the surface, at $z = -R$. The crust-core interface is located at $z = \pm R_c$.

The crust is elastic, with a uniform shear modulus μ. Furthermore, uniform density, incompressibility and ideal MHD conditions are assumed everywhere. In the unperturbed configuration the magnetic field is $B_\circ \hat{\mathbf{z}}$. In terms of the fluid displacement ξ the Euler equation in the crustal region is (assuming a harmonic dependence $\xi \sim \exp(i\sigma t)$),

$$[\tilde{\mu} + v_A^2]\partial_z^2 \xi^i + \left[\sigma^2 - \tilde{\mu}\frac{\ell(\ell+1)}{R^2}\right]\xi^i = 0, \quad i = x, y. \tag{4}$$

The boundary conditions are formulated in terms of the traction components. For our combined fluid/magnetic field system the traction takes the following form in the crust region:

$$T^i = \rho[\tilde{\mu} + v_A^2]\partial_z \xi^i, \quad \text{and} \quad T^z = 0. \tag{5}$$

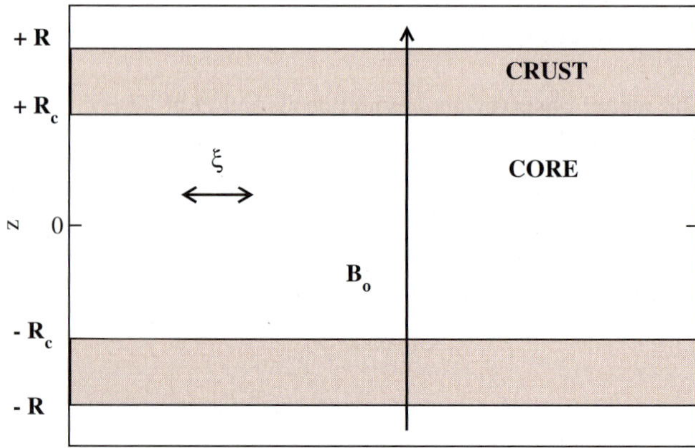

Fig. 1 A schematic illustration of the plane-parallel toy model used in this work

The corresponding expressions for the core region are obtained by setting $\tilde{\mu} = 0$.

The appropriate MHD boundary conditions are: (i) continuity of the tractions at the crust/core interface, (ii) vanishing tractions at the surface and (iii) continuity of the (normal) transverse components of the (magnetic) electric field at the interface. These translate into,

$$\xi^x(R_c^+) = \xi^x(R_c^-), \qquad \partial_z \xi^x(R) = 0. \tag{6}$$

$$\partial_z \xi^x(R_c^-) = [1 + \tilde{\mu}/v_A^2]\partial_z \xi^x(R_c^+). \tag{7}$$

4 Pure crustal modes

It is instructive to consider first the special case where the crust is decoupled from the core. As we discussed earlier, this case is artificial in the presence of a magnetic field. Nevertheless, the magnetic coupling between the crust and the fluid core can be deactivated by altering the boundary conditions at the bottom of the crust, requiring vanishing tractions (Carroll et al. 1986; Messios et al. 2001; Piro 2005).

Solving the Euler equation in the crust and imposing the above (incorrect) conditions provides the mode frequencies,

$$\sigma_n = \frac{\tilde{\mu}^{1/2}}{R}\left[\ell(\ell+1) + \left(\frac{n\pi R}{\Delta}\right)^2 \left\{1 + \frac{v_A^2}{\tilde{\mu}}\right\}\right]^{1/2} \tag{8}$$

where $n = 0, 1, 2, 3, \ldots$ and $\Delta = R - R_c$ is the crust thickness.

This prediction of our toy model for the "toroidal" crust mode frequencies is in good agreement with more rigorous results (McDermott et al. 1988; Hansen and Cioffi 1980; Strohmayer 1991; Piro 2005). For example, it predicts that the fundamental modes ($n = 0$) are oblivious to the thickness of the crust and magnetic field while having frequencies much lower than the higher overtones.

Hereafter we normalise the frequencies as $\tilde{\sigma} = \sigma R/\tilde{\mu}^{1/2}$. It is useful to note that the fundamental ($n = 0$) and first overtone ($n = 1$) quadrupole modes then take the respective values $\tilde{\sigma}_0 = 2.45$ and $\tilde{\sigma}_1 = 31.5$ (for the latter we have assumed $\Delta = 0.1R$ and $v_A = 0$).

5 Global MHD modes

Returning to the full crust-core system, we solve the Euler equations in the two regions and impose the correct boundary conditions (6–7). We find that the complete mode spectrum is determined by

$$e^{-4iR_c\beta}\left[\alpha\left(1 + \frac{\tilde{\mu}}{v_A^2}\right)\sinh(\Delta\alpha) - i\beta\cosh(\Delta\alpha)\right]^2 - \left[\alpha\left(1 + \frac{\tilde{\mu}}{v_A^2}\right)\sinh(\Delta\alpha) + i\beta\cosh(\Delta\alpha)\right]^2 = 0 \tag{9}$$

where

$$\alpha = \frac{1}{R}\left[\frac{\ell(\ell+1) - \tilde{\sigma}^2}{1 + v_A^2/\tilde{\mu}}\right]^{1/2} \quad \text{and} \quad \beta = \sigma/v_A. \tag{10}$$

Despite being very simple, the toy-model provides a set of interesting results. First note that we can obtain analytic solutions to (9) for a weak magnetic field. When $v_A^2 \ll \tilde{\mu}$ acceptable mode solutions coincide with the Alfvén frequencies:

$$\sigma_A = (k\pi/2R_c)v_A, \quad k = 0, 1, 2, 3, \ldots. \tag{11}$$

Remarkably, these frequencies also satisfy (9) when the frequency coincides with a crustal frequency, i.e. $\sigma = \sigma_n$, see (8). This triple intersection is naturally interpreted as a resonance between the crust and the core. Of course, despite having frequencies close (or identical) to the crustal frequencies, these modes are global with eigenfunctions extending to both crust and core regions.

We have determined the exact spectrum by solving (9) numerically. A part of the spectrum for $\ell = 2$ and $\Delta = 0.1R$ is shown in the left panel of Fig. 2, together with the Alfvén frequencies (11), as a function of the ratio

$$\frac{v_A^2}{\tilde{\mu}} \approx 0.04 B_{15}^2 \rho_{14}^{-1}. \tag{12}$$

We have assumed a standard value $\tilde{\mu} \approx 2 \times 10^{16}\ \mathrm{cm}^2/\mathrm{s}^2$.

Figure 2 shows that, for a given value of the ratio (12), the spectrum consists of a semi-infinite family of modes (we only show the first few). The separation between consecutive modes increases as the ratio attains higher values, i.e. if we increase the magnetic field while keeping the other parameters fixed. The most important feature to note in the spectrum is that there are always mode-frequencies comparable to the crustal frequencies σ_n. Figure 2 shows the spectrum in the vicinity of σ_0 but a similar picture arises for *all* frequencies σ_n of a given ℓ. As is apparent in the figure, for certain discrete values of the ratio $v_A^2/\tilde{\mu}$ the mode frequency *exactly* coincides with a crustal frequency σ_n as well as one of the Alfvén frequencies (11). For any other realistic value of $v_A^2/\tilde{\mu}$ there is always a mode with frequency very close (with at most a few percent deviation) to each frequency σ_n.

The coupled crust-core system obviously has a much richer mode spectrum than the decoupled elastic crust. Yet, as we discussed earlier, one can interpret most of the observed QPOs as crustal frequencies for various values of ℓ. Therefore, we need to explain why only modes with $\sigma \approx \sigma_n$ are observed. To do this we need to recall that in the standard model for the giant flares the magnetic field erupts and induces a starquake in the crust. Based on physical intuition we would expect that the initial perturbation in the crust will predominantly excite those global modes which communicate the least amount of energy to the core. To test this

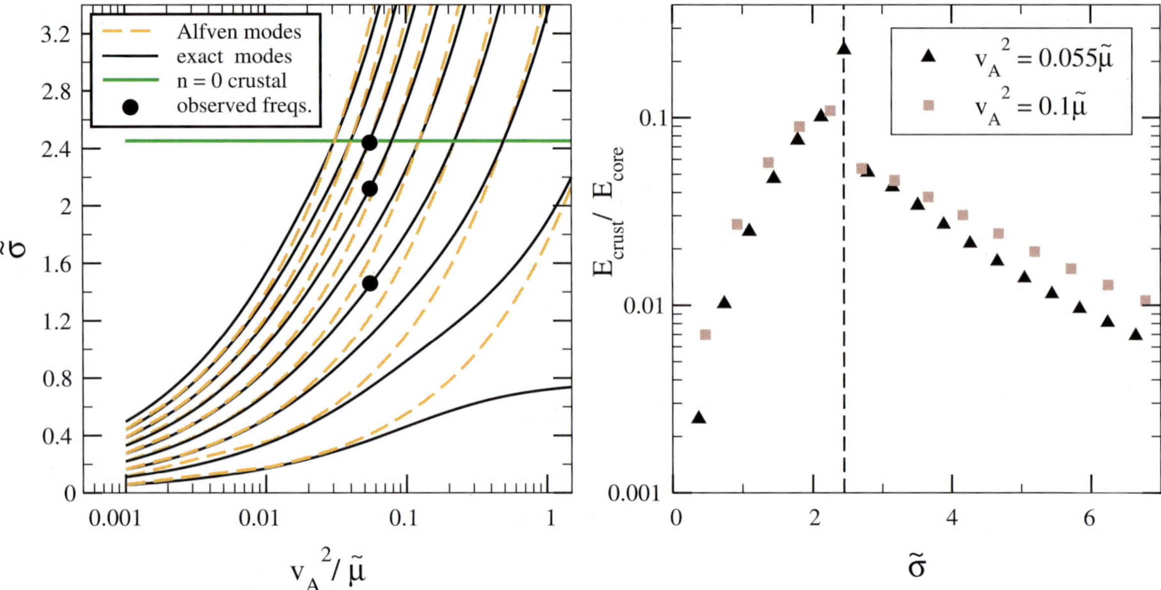

Fig. 2 *Left panel*: A part of the global mode spectrum for the toy model of a magnetar with crust-core coupling. The data correspond to $\ell = 2$ and $\Delta = 0.1R$. *Right panel*: Assessing the excitation of various modes by means of the ratio between the mode energy in the crust E_{crust} and that in the core, E_{core}. The *dashed line* labels the crustal frequency $\tilde{\sigma}_0$

idea, we calculate the total energy (kinetic + magnetic) associated with each mode and consider the ratio $E_{\mathrm{crust}}/E_{\mathrm{core}}$. As can be seen from the right panel of Fig. 2, the results corroborate our intuition. The mode which *maximises* the energy ratio, i.e. should be easier to excite from an initial crust motion, is the one nearest to the resonant frequency $\sigma_A = \sigma_0$ (analogous behaviour is found for modes in the vicinity of all higher σ_n frequencies, for any ℓ). This would explain why the other modes of the system are more difficult to excite, they are predominantly core-Alfvén modes which would be energetically more expensive to excite via the proposed mechanism. Similar conclusions can be drawn from the displacement ξ, see Glampedakis et al. (2006) for details. The oscillation amplitude is generally far greater in the core, but the modes located in the vicinity of the crust frequencies σ_n are exceptions. In those cases the crust oscillates with a comparable amplitude.

6 Low frequency QPOs

The conclusion of the discussion above is that once the crust is shaken by a starquake the core-crust system will naturally choose to vibrate in those global modes which have frequencies similar to the frequencies of the toroidal modes of the uncoupled crust, with identical values of ℓ. Hence, the model provides an explanation for all observed QPO frequencies, both intermediate (~30–155 Hz, different values of ℓ) and high frequency (~625 Hz, overtone with one node in the crust). Our model differs from previous ones only in that the modes are global, not localised to the crust, and hence have a significant amplitude in the core.

A key feature of Fig. 2 is the existence of modes with frequencies *below* the fundamental crustal frequency σ_0. This is interesting since the presence of low frequency QPOs has been confirmed in the data of both the RXTE and RHESSI satellites (Israel et al. 2005; Strohmayer and Watts 2005; Watts and Strohmayer 2006): 18, 26 and 30 Hz QPOs for the December 2004 giant flare in SGR 1806-20. It is natural to identify the 30 Hz QPO with the frequency σ_0 (Israel et al. 2005), since the higher frequency QPOs then fit the predictions from (1) quite well (see Table 1).

However, we then find ourselves left with a puzzle: what is the origin of the remaining two frequencies? There are certainly no toroidal crustal modes with frequency below σ_0. Our model offers a natural explanation for these low-frequency QPOs.

Although we cannot claim that our model is quantitatively accurate, we can still attempt to match the observed data for the low frequency QPOs of SGR 1806-20. To do this we first artificially identify $\tilde{\sigma}_0 = 2.45$ with the 30 Hz frequency (presumably, this identification can be made rigorous in a calculation for a realistic neutron star model). Then the frequencies 18 and 26 Hz correspond to modes with $\tilde{\sigma} = 1.46$ and 2.12, respectively. Some navigation in Fig. 2 leads us to the value $v_A^2 = 0.055\tilde{\mu}$ which has all three desired modes (indicated by filled circles in the figure). This value is reasonably consistent with $B_{15} \approx 1$ (assuming $\rho_{14} \approx 1$), which is the anticipated magnetic field strength. Moreover, the energy argument indicates that it is reasonable to expect these QPOs to be excited, cf. Fig. 2. Finally, we have an additional mode at roughly 22 Hz. In a sense, this could be considered a testable prediction.

7 Future work

Our model provides an answer to the question posed in Sect. 2, while being consistent with the actual QPO observations. Our results are highly suggestive of how the magnetically coupled crust-core system can support global modes with frequencies similar to the crustal frequencies of a decoupled crust. We have then shown why these modes should be the most favourable for excitation during a giant flare event.

Clearly, our model is based on drastic simplifications, however, the basic physics predicted here should survive for more realistic neutron star models. Obviously, one must consider a spherical neutron star model allowing for non-uniform density/crust elasticity and more complex magnetic field configuration, taking into account proton superconductivity in the core. In order to make firm comparisons with observations it is also essential to work in a fully general relativistic framework.

These are some of the tasks that lie ahead. These are challenging problems but the prospect of initiating neutron star seismology in the near future provides ample motivation.

References

Carrol, B.W., et al.: Astrophys. J. **305**, 767 (1986)
Duncan, R.C., Thompson, C.: Astrophys. J. **392**, L9 (1992)
Duncan, R.C.: Astrophys. J. **498**, L45 (1998)
Glampedakis, K., Samuelsson, L., Andersson, N.: Mon. Not. Roy. Astron. Soc. **371**, L74 (2006)
Hansen, C.J., Cioffi, D.F.: Astrophys. J. **238**, 740 (1980)
Israel, G.L., et al.: Astrophys. J. **628**, L53 (2005)
Levin, Y.: Mon. Not. Roy. Astron. Soc. **368**, L35 (2006)
McDermott, P.N., van Horn, H.M., Hansen, C.J.: Astrophys. J. **325**, 725 (1988)
Messios, N., Papadopoulos, D.M., Stergioulas, N.: Mon. Not. Roy. Astron. Soc. **328**, 1161 (2001)
Piro, A.L.: Astrophys. J. **634**, L153 (2005)
Strohmayer, T.E.: Astrophys. J. **372**, 573 (1991)
Strohmayer, T.E., Watts, A.L.: Astrophys. J. **632**, L111 (2005)
Thompson, C., Duncan, R.C.: Mon. Not. Roy. Astron. Soc. **275**, 255 (1995)
van Horn, H.M.: Astrophys. J. **236**, 899 (1980)
Watts, A.L., Strohmayer, T.E.: Astrophys. J. **637**, L117 (2006)

Astrophys Space Sci (2007) 308: 613–617
DOI 10.1007/s10509-007-9374-2

ORIGINAL ARTICLE

Structure of pair winds from compact objects with application to emission from bare strange stars

A.G. Aksenov · M. Milgrom · V.V. Usov

Received: 21 June 2006 / Accepted: 6 November 2006 / Published online: 20 March 2007

Abstract We present the results of numerical simulations of stationary, spherically outflowing, $e^{\pm}$ pair winds, with total luminosities in the range 10^{34}–10^{42} ergs s^{-1}. In the concrete example described here, the wind injection source is a hot, bare, strange star, predicted to be a powerful source of $e^{\pm}$ pairs created by the Coulomb barrier at the quark surface. We find that photons dominate in the emerging emission, and the emerging photon spectrum is rather hard and differs substantially from the thermal spectrum expected from a neutron star with the same luminosity. This might help distinguish the putative bare strange stars from neutron stars.

Keywords Plasmas · Radiation mechanisms: thermal · Radiative transfer

PACS 52.27.Ep · 95.30.Jx · 97.10.Me

1 Introduction

There is now compelling evidence that electron-positron ($e^{\pm}$) pairs form and flow away in the vicinity of many compact astronomical objects (radio pulsars, accretion disk coronae of Galactic X-ray binaries, soft γ-ray repeaters, active galactic nuclei, cosmological γ-ray bursters, etc.).

A.G. Aksenov (✉)
Institute of Theoretical and Experimental Physics,
B. Cheremushkinskaya, 25, Moscow 117218, Russia
e-mail: alexei.aksenov@itep.ru

M. Milgrom · V.V. Usov
Center of Astrophysics, Weizmann Institute, Rehovot 76100,
Israel

The estimated luminosity in $e^{\pm}$ pairs varies greatly depending on the object and the specific conditions: from $\sim 10^{31}$–10^{36} ergs s^{-1} for radio pulsars up to $\sim 10^{50}$–10^{52} ergs s^{-1} in cosmological γ-ray bursters.

For a wind out-flowing spherically from a surface of radius R there is a maximum (isotropic, unbeamed) pair luminosity beyond which the pairs annihilate significantly before they escape. This is given by

$$L_{\pm}^{\max} = \frac{4\pi m_e c^3 R \Gamma^2}{\sigma_{\mathrm{T}}} \simeq 10^{36}(R/10^6\ \mathrm{cm})\Gamma^2\ \mathrm{ergs\ s^{-1}}, \quad (1)$$

where Γ is the pair bulk Lorentz factor, and σ_{T} the Thomson cross section. When the injected pair luminosity, $\tilde{L}_{\pm}$, greatly exceeds this value the emerging pair luminosity, $L_{\pm}$, cannot significantly exceed $L_{\pm}^{\max}$; in this case photons strongly dominate in the emerging emission: $L_{\pm} < L_{\pm}^{\max} \ll \tilde{L}_{\pm} \simeq L_{\gamma}$. Injected pair luminosities typical of cosmological γ-ray bursts (e.g., Piran 2000), $\tilde{L}_{\pm} \sim 10^{50}$–$10^{52}$ ergs s^{-1}, greatly exceed $L_{\pm}^{\max}$. For such a powerful wind the pair density near the source is very high, and the out-flowing pairs and photons are nearly in thermal equilibrium almost up to the wind photosphere (Paczyński 1990). The outflow process of such a wind may be described fairly well by relativistic hydrodynamics (e.g., Grimsrud and Wasserman 1998; Iwamoto and Takahara 2002).

In contrast if $\tilde{L}_{\pm} \ll L_{\pm}^{\max}$ annihilation of the outflowing pairs is negligible. It is now commonly accepted that the magnetospheres of radio pulsars contain such a very rarefied ultra-relativistic ($\Gamma_{\pm} \sim 10$–10^2) pair plasma that is practically collisionless (e.g., Melrose 1995).

Recently we developed a numerical code for solving the relativistic kinetic Boltzmann equations for pairs and photons. Using this we considered a spherically out-flowing, non-relativistic ($\Gamma \sim 1$) pair winds with the total luminosity in the range 10^{34}–10^{42} ergs s^{-1}, that is $\sim(10^{-2}$–$10^{6})L_{\pm}^{\max}$ (Aksenov et al. 2004, 2005). (A brief account of the emerging emission from such a pair wind has been given by Aksenov et al. 2003.) While our numerical code can be more generally employed, the results presented in this paper are for a hot, bare, strange star as the wind injection source. Such stars are thought to be powerful sources of pairs created by the Coulomb barrier at the quark surface (Usov 1998, 2001a).

2 Formulation of the problem

We consider an $e^{\pm}$ pair wind that flows away from a hot, bare, unmagnetized, non-rotating, strange star. Space-time outside the star is described by Schwarzschild's metric with the line element

$$ds^2 = -e^{2\phi}c^2\,dt^2 + e^{-2\phi}\,dr^2 + r^2(d\vartheta^2 + \sin^2\vartheta\,d\varphi^2), \tag{2}$$

where

$$e^{\phi} = \left(1 - \frac{r_g}{r}\right)^{1/2}, \qquad r_g = \frac{2GM}{c^2} \simeq 2.95 \times 10^5 \frac{M}{M_{\odot}}\ \text{cm}. \tag{3}$$

Following Page and Usov (2002) we consider, as a representative case, a strange star with a mass of $M = 1.4M_{\odot}$ and the circumferential radius $R = 1.1 \times 10^6$ cm.

The state of the plasma in the wind may be described by the distribution functions $f_{\pm}(\mathbf{p}, r, t)$ and $f_{\gamma}(\mathbf{p}, r, t)$ for positrons (+), electrons (−), and photons, respectively, where $\mathbf{p}$ is the momentum of particles. There is no emission of nuclei from the stellar surface, so the distribution functions of positrons and electrons are identical.

We use the general relativistic Boltzmann equations for the $e^{\pm}$ pairs and photons, whereby the distribution function for the particles of type i, $f_i(|\mathbf{p}|, \mu, r, t)$, ($i = e$ for $e^{\pm}$ pairs and $i = \gamma$ for photons), satisfies

$$\frac{e^{-\phi}}{c}\frac{\partial f_i}{\partial t} + \frac{1}{r^2}\frac{\partial}{\partial r}(r^2\mu e^{\phi}\beta_i f_i) - \frac{e^{\phi}}{p^2}\frac{\partial}{\partial p}\left(p^3\mu\frac{\phi'}{\beta_i}f_i\right) - \frac{\partial}{\partial\mu}\left[(1-\mu^2)e^{\phi}\left(\frac{\phi'}{\beta_i} - \frac{\beta_i}{r}\right)f_i\right] = \sum_q(\bar{\eta}_i^q - \chi_i^q f_i). \tag{4}$$

Here, μ is the cosine of the angle between the radius-vector from the stellar center and the particle momentum $\mathbf{p}$, $p = |\mathbf{p}|$, $\beta_e = v_e/c$, $\beta_{\gamma} = 1$, and v_e is the velocity of electrons and positrons. Also, $\bar{\eta}_i^q$ is the emission coefficient for the production of a particle of type i via the physical process labelled by q, and χ_i^q is the corresponding absorption coefficient. The summation runs over physical processes that involve a particle of type i. The processes we include are listed in Table 1.

The thermal emission of pairs from the surface of strange quark matter depends on the surface temperature, T_S, alone. In our simulation we use the flux of $e^{\pm}$ pairs from the bare surface of a strange star calculated by Usov (1998, 2001a) as a boundary condition at the internal computational boundary ($r = R$). Thermal emission of photons from the surface of a bare strange star is strongly suppressed for $T_S \ll 10^{11}$ K (Usov 2001a; Cheng and Harko 2003; Jaikumar et al. 2004 and see Fig. 1), and we neglect this in our simulations.

Table 1 Physical processes included in simulations

Basic two-body interaction	Radiative variant
Møller and Bhaba scattering	Bremsstrahlung
$ee \to ee$	$ee \leftrightarrow ee\gamma$
Compton scattering	Double Compton scattering
$\gamma e \to \gamma e$	$\gamma e \leftrightarrow \gamma e\gamma$
Pair annihilation	Three photon annihilation
$e^+e^- \to \gamma\gamma$	$e^+e^- \leftrightarrow \gamma\gamma\gamma$
Photon-photon pair production	
$\gamma\gamma \to e^+e^-$	

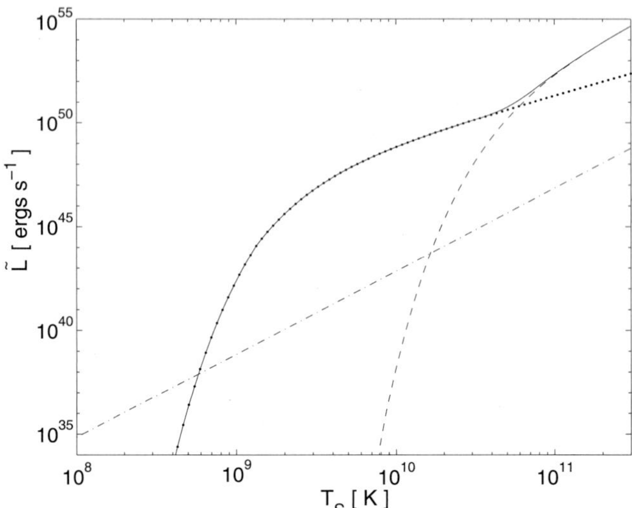

Fig. 1 Luminosities of a hot, bare, strange star in e^+e^- pairs (*dotted line*), in thermal equilibrium photons (*dashed line*), and the total (*solid line*) as functions of the surface temperature T_S. The theoretical upper limit on the luminosity in non-equilibrium photons, $10^{-6}L_{BB}$ (Cheng and Harko 2003), is shown by the dot-dashed line, L_{BB} being the blackbody luminosity

The stellar surface is assumed to be a perfect mirror for both $e^{\pm}$ pairs and photons. At the external boundary ($r = r_{\rm ext} = 1.7 \times 10^8$ cm), the pairs and photons escape freely from the studied region.

3 Computational details

Our grid in the $\{r, \mu, \epsilon\}$ phase-space is defined as follows. The r domain ($R < r < r_{\rm ext}$) is divided into $j_{\rm max}$ spherical shells. The μ-grid is uniform and made of $k_{\rm max}$ intervals $\Delta\mu_k = 2/k_{\rm max}$. The energy grids for photons and electrons are both made of $\omega_{\rm max}$ energy intervals, but the lowest energy for photons is 0, while that for pairs is $m_e c^2$.

Here we use an (r, μ, ϵ)-grid with $j_{\rm max} = 100$, $k_{\rm max} = 8$, and $\omega_{\rm max} = 13$. The shell thicknesses are geometrically spaced: $\Delta r_1 = 2 \times 10^{-4}$ cm, and $\Delta r_j = 1.3\Delta r_{j-1}$ ($1 \le j \le j_{\rm max}$). The discrete energies (in keV) of the ϵ-grid minus the rest mass of the particles are 0, 2, 27, 111, 255, 353, 436, 491, 511, 530, 585, 669, 766, and ∞. A finite-difference scheme developed for solving this problem is presented in (Aksenov et al. 2005).

4 Numerical results

In this section, we present the results for the structure of the stationary $e^{\pm}$ winds and their emergent emission. Although the pair plasma ejected from the strange-star surface contains no radiation, as the plasma moves outwards photons are produced by pair annihilation and bremsstrahlung emission. Figure 2 shows the mean optical depth for photons, from r to $r_{\rm ext}$. The contribution from $r_{\rm ext}$ to infinity is negligible for $r < 10^8$ cm, so $\tau_\gamma(r)$ is practically the mean optical depth from r to infinity for these values of r. The pair wind is optically thick [$\tau_\gamma(0) > 1$] for $\tilde{L}_{\pm} > 10^{37}$ ergs s^{-1}. The radius of the wind photosphere $r_{\rm ph}$, determined by condition $\tau(r_{\rm ph}) = 1$, varies from $\sim R$ for $\tilde{L}_{\pm} = 10^{37}$ ergs s^{-1} to $\sim 10R \simeq 10^7$ cm for $\tilde{L}_{\pm} = 10^{42}$ ergs s^{-1}. The wind photosphere is always deep inside our chosen external boundary ($r_{\rm ph} < 0.1 r_{\rm ext}$), justifying our neglect of the inward ($\mu < 0$) fluxes at $r = r_{\rm ext}$.

Figure 3 shows the emerging luminosities in $e^{\pm}$ pairs (L_e) and photons (L_γ) as fractions of the total luminosity

$$L = L_{\pm} + L_\gamma = \left(1 - \frac{r_g}{R}\right)\tilde{L}_{\pm} \simeq 0.63\tilde{L}_{\pm}. \tag{5}$$

For $L > L_{\rm eq} \simeq 10^{34}$ ergs s^{-1}, the emerging emission consists mostly of photons ($L_\gamma > L_e$). This simply reflects the fact that in this case the pair annihilation time $t_{\rm ann} \sim (n_e \sigma_{\rm T} c)^{-1}$ is less than the escape time $t_{\rm esc} \sim R/c$, so most injected pairs annihilate before they escape. The value of $L_{\rm eq}$ is about two orders of magnitude smaller than $L_{\pm}^{\rm max}$ [see (1)] that is estimated from consideration of the same processes, but without taking into account gravity of the star. However, with gravity at low luminosities ($\tilde{L}_{\pm} < 10^{36}$ ergs s^{-1}) pairs emitted by the stellar surface are mainly captured by the gravitational field, and a pair atmosphere forms. The probability of pair annihilation increases because of the increase of the pair number density in the atmosphere, and this results in decrease of the fraction of pairs in the emerging emission in comparison with the case when gravity of the star is neglected.

The number rate of emerging pairs ($\dot{N}_{\pm}$) as functions of $\tilde{L}_{\pm}$ is shown in Fig. 4. For $\tilde{L}_{\pm} > 10^{37}$ ergs s^{-1} the value of $\dot{N}_{\pm}$ is $\sim$1.5–2 times smaller than the same calculated by Aksenov et al. (2004) where gravity is neglected. This

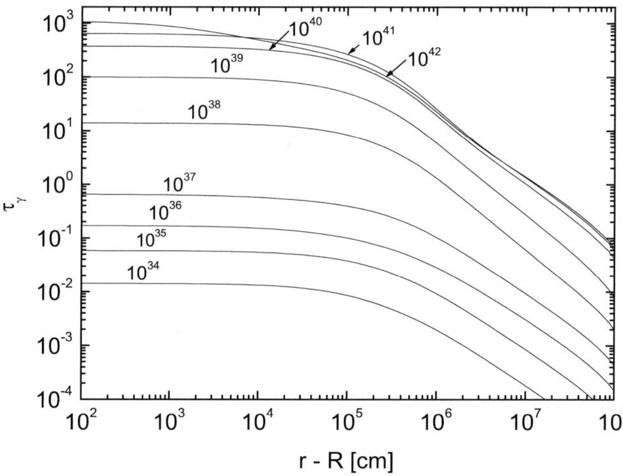

Fig. 2 The mean optical depth for photons, from r to $r_{\rm ext}$, as a function of the distance from the stellar surface, for different values of $\tilde{L}_{\pm}$, as marked on the *curves*

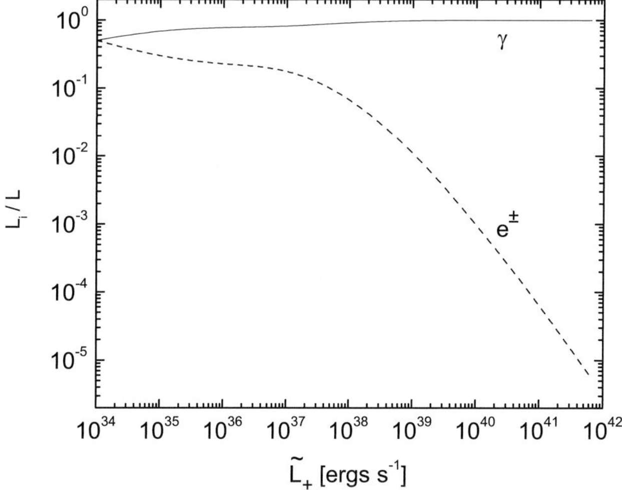

Fig. 3 The fractional emerging luminosities in pairs (*dashed line*) and photons (*solid line*) as functions of the injected pair luminosity, $\tilde{L}_{\pm}$

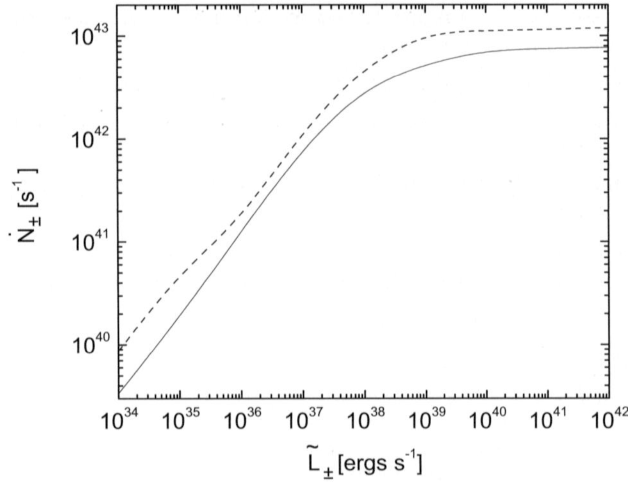

Fig. 4 Number rate of emerging pairs as functions of the injected pair luminosity (*solid line*). The result by Aksenov et al. (2004) where gravity has been neglected is shown by the *dashed line*

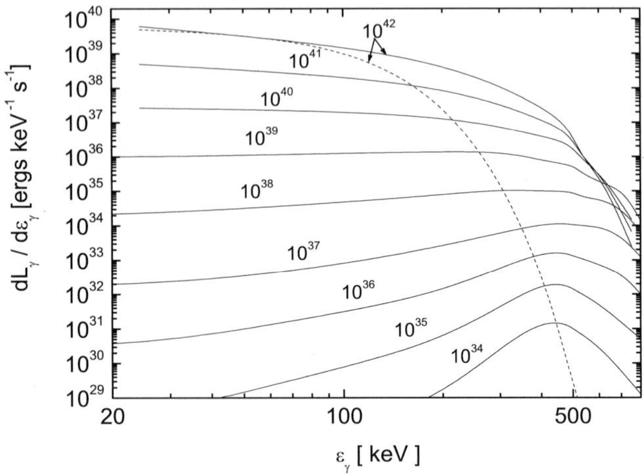

Fig. 5 The energy spectrum of emerging photons for different values of $\tilde{L}_\pm$, as marked on the *curves*. The *dashed line* is the spectrum of blackbody emission

is due to partial suppression of pair creation as the photon energies are reduced by gravitational redshift. We can see that there is an upper limit to the rate of emerging pairs $\dot{N}_e^{\max} \simeq 10^{43}$ s^{-1}.

Figure 5 presents the energy spectra of the emerging photons for different values of $\tilde{L}_\pm$. At low luminosities, $\tilde{L}_\pm \sim 10^{35}$–$10^{37}$ ergs s^{-1}, photons that form in annihilation of $e^\pm$ pairs escape from the vicinity of the strange star more or less freely, and the photon spectra resembles a very wide annihilation line with the mean energy of $\sim$400 keV (see Fig. 6). The small decrease in mean photon energy $\langle \epsilon_\gamma \rangle$ from $\sim$430 keV at $\tilde{L}_\pm \simeq 10^{34}$–$10^{35}$ ergs s^{-1} to $\sim$370 keV at $\tilde{L}_\pm \simeq 10^{37}$ ergs s^{-1} occurs because of the energy transfer from annihilation photons to $e^\pm$ pairs via Compton scattering (Aksenov et al. 2004, 2005). As a result of this transfer, the emerging $e^\pm$ pairs are heated up to the mean energy $\langle \epsilon_e \rangle \simeq 400$ keV at $\tilde{L}_\pm \simeq 10^{37}$ ergs s^{-1}. For $\tilde{L}_\pm > 10^{37}$ ergs s^{-1}, changes in the particle number due to three body processes are essential, and their role in thermalization of the outflowing plasma increases with the increase of $\tilde{L}_\pm$. We see in Fig. 5 that, for $\tilde{L}_\pm = 10^{42}$ergs s^{-1}, the photon spectrum is near blackbody, except for the presence of a high-energy tail at $\epsilon_\gamma > 100$ keV. At this luminosity, the mean energy of the emerging photons is $\sim$40 keV, while the mean energy of the blackbody photons is $\sim$30 keV.

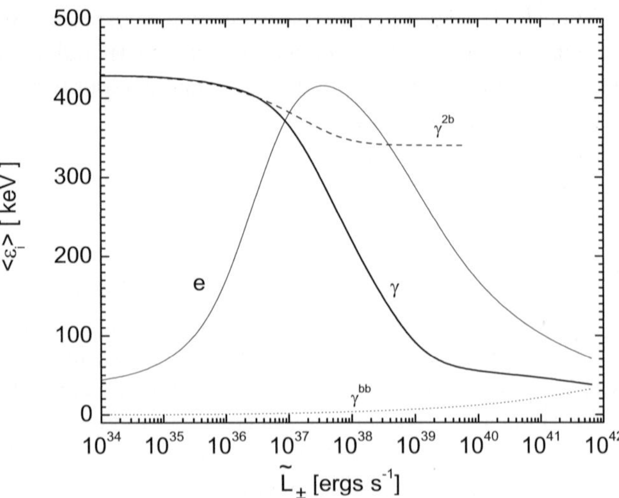

Fig. 6 The mean energy of the emerging photons (*thick solid line*) and electrons (*thin solid line*) as a function of the total injection luminosity $\tilde{L}_\pm$. For comparison, we show as the dotted line the mean energy of blackbody photons for the same energy density as that of the photons at the photosphere. Also shown as the *dashed line* is the mean energies of the emerging photons in the case when only two particle processes are taken into account

Since strange quark matter at the surface of a bare strange star is bound via strong interaction rather than gravity, such a star is not subject to the Eddington limit and can radiate in photons and pairs at the luminosity of 10^{51}–10^{52} ergs s^{-1} or even higher (see Fig. 1). Another important idiosyncrasy that we find is hard spectra and a strong anti-correlation between spectral hardness and luminosity. While at very high luminosities ($L > 10^{43}$ ergs s^{-1}) the spectral temperature increases with luminosity as in blackbody radiation, in the range of luminosities we studied, where thermal equilibrium is not achieved, the expected correlation is opposite (see Fig. 6). These differ qualitatively from the photon emission from neutron stars and provides a definite observational signature for bare strange stars.

The compact objects that underlie soft gamma-ray repeaters and gamma-ray bursters might be strange stars (Usov 2001b; Paczyński and Haensel 2005). But, it is not possible to apply our existing results to these objects as we have not included the effects of magnetic fields and rotation which can significantly change the interaction of individual particles in pair winds (Harding and Lai 2006) and lead to additional phenomenon such as generation of large-amplitude electromagnetic waves and acceleration of particles by these waves (Usov 1994; Melatos and Melrose 1996).

References

Aksenov, A.G., Milgrom, M., Usov, V.V.: Mon. Not. Roy. Astron. Soc. **343**, L69 (2003)

Aksenov, A.G., Milgrom, M., Usov, V.V.: Astrophys. J. **609**, 363 (2004)

Aksenov, A.G., Milgrom, M., Usov, V.V.: Astrophys. J. **632**, 567 (2005)

Cheng, K.S., Harko, T.: Astrophys. J. **596**, 451 (2003)

Grimsrud, O.M., Wasserman, I.: Mon. Not. Roy. Astron. Soc. **300**, 1158 (1998)

Harding, A.K., Lai, D.: Rep. Prog. Phys. **69**, 2631 (2006)

Iwamoto, S., Takahara, F.: Astrophys. J. **565**, 163 (2002)

Jaikumar, P., Gale, C., Page, D., Prakash, M.: Phys. Rev. D **70**, 023004 (2004)

Melrose, D.B.: Astron. Astrophys. **16**, 137 (1995)

Melatos, A., Melrose, D.B.: Mon. Not. Roy. Astron. Soc. **279**, 1168 (1996)

Paczyński, B.: Astrophys. J. **363**, 218 (1990)

Paczyński, B., Haensel, P.: Mon. Not. Roy. Astron. Soc. **362**, L4 (2005)

Page, D., Usov, V.V.: Phys. Rev. Lett. **89**, 131101 (2002)

Piran, T.: Phys. Rep. **333**, 529 (2000)

Usov, V.V.: Mon. Not. Roy. Astron. Soc. **267**, 1035 (1994)

Usov, V.V.: Phys. Rev. Lett. **80**, 230 (1998)

Usov, V.V.: Astrophys. J. **550**, L179 (2001a)

Usov, V.V.: Phys. Rev. Lett. **87**, 021101 (2001b)

Astrophys Space Sci (2007) 308: 619–623
DOI 10.1007/s10509-007-9375-1

ORIGINAL ARTICLE

The complex X-ray spectrum of the isolated neutron star RBS1223

Axel D. Schwope · Valeri Hambaryan · Frank Haberl · Christian Motch

Received: 4 July 2006 / Accepted: 26 September 2006 / Published online: 15 March 2007

Abstract We present a first analysis of a deep X-ray spectrum of the isolated neutron star RBS1223 obtained with XMM-Newton. Spectral data from four new monitoring observations in 2005/2006 were combined with archival observations obtained in 2003 and 2004 to form a spin-phase averaged spectrum containing 290 000 EPIC-pn photons. This spectrum shows higher complexity than its predecessors, and can be parameterised with two Gaussian absorption lines superimposed on a blackbody. The line centers, $E_2 \simeq 2E_1$, could be regarded as supporting the cyclotron interpretation of the absorption features in a field $B \sim 4 \times 10^{13}$ G. The flux ratio of those lines does not support this interpretation. Hence, either feature might be of truly atomic origin.

Keywords Stars: individual (RBS1223) · Stars: neutron · Stars: magnetic fields · X-rays: individuals (RBS1223)

PACS 97.60.Jd · 97.10.Ld

1 Introduction

RBS1223 is one out of three XDINs (X-ray dim isolated neutron stars) found in the ROSAT Bright Survey (RBS), an optical identification program of the more than 2000 X-ray sources detected in the RASS with a count rate >0.2 s^{-1} at high galactic latitude Schwope et al. (2000). The initial discovery Schwope et al. (1999) was based on the soft X-ray spectrum and the steep SED, $f_X/f_{opt} > 10^4$. Follow-up Chandra observations revealed a periodically modulated signal Hambaryan et al. (2002), and HST-observations uncovered a candidate optical counterpart at $m_{50CCD} = 28.56 \pm 0.13$ mag (Kaplan et al. 2002). Initial observations with XMM-Newton showed deviations from a Planckian energy distribution at energies below 500 eV which could be described with a Gaussian absorption line and interpreted as a cyclotron absorption line in a field of $(2-6) \times 10^{13}$ G Haberl et al. (2003). The large number of photons collected with XMM-Newton also uncovered the true spin period of 10.31 s. In Schwope et al. (2005) the light curves in different energy bands were modeled in terms of a photospheric cap model. The temperature structure was found to be roughly compatible with the crustal field models of (Geppert et al. 2004). The implied field structure deviates from a simple centered dipole model. Spin-period changes appeared likely at that time thus stimulating monitoring observations with Chandra and XMM-Newton. The spin history could be unequivocally fixed by a series of Chandra pointings (Kaplan and van Kerkwijk 2005) and revealed a spin-down of the star at a rate $\dot{P} = 1.120 \times 10^{-13}$ s s^{-1}. Under the assumption that this is due to magnetic dipole torques, a characteristic age of 1.5 Myr and a magnetic field strength of 3.4×10^{13} G were inferred. Here we describe initial results of recent monitoring observations with XMM-Newton which, due to the large number of accumulated photons, besides accurate timing also allows spectroscopy with unprecedented accuracy.

A.D. Schwope (✉) · V. Hambaryan
Astrophysikalisches Institut Potsdam, An der Sternwarte 16, 14482 Potsdam, Germany
e-mail: ASchwope@aip.de

F. Haberl · V. Hambaryan
Max-Plank Institute für Extraterrestrische Physik, Giessenbachstr., 85748 Garching, Germany

C. Motch
Observatoire Astronomique, CNRS UMR 7550, 11 rue de l'Université, 67000 Strasbourg, France

2 XMM-Newton observations of RBS1223

XMM-Newton observed RBS1223 several times between 2001 and 2006. The relevant pointings for this contribution are listed in Table 1. All but the first, which was put in small window mode (SW), were obtained in full frame mode through the thin filter. All observations were reduced in a homogeneous manner with a recent version of the SAS (V6.5). The monitoring observations set constraints on the evolution of different parameters of the star. The overall brightness did not change between 2001 and 2006 and at all occasions the star displayed its double-humped light curves discussed in detail by Schwope et al. (2005). Also the amplitude of the light curve remained unchanged. Hence, the only property which was shown to evolve so far is the spin period (Kaplan and van Kerkwijk 2005).

3 Spectral analysis of RBS1223

The constancy of the star allows a joint spectral analysis of all available observations. In order to minimise possible remaining calibration uncertainties for different camera modes we used only those observations with identical camera settings (six observations between revolutions 561–1115). This selection left us with an effective exposure time of 113 ks with almost 290 000 photons for spectral analysis. Here we concentrate on the mean spectrum of all observations. Phase-resolved spectroscopy is underway and will be presented elsewhere.

Table 1 XMM-Newton observations of RBS1223 analysed in this contribution

Rev.	AO	Day	Exp. [ks]	Mode	Countrate [s^{-1}]
377	1	2001 Dec 31	18.6	SW	2.49[a]
561	2	2003 Jan 1	27.0	FF	2.53
743	3	2003 Dec 30	30.2	FF	2.57
1015	4	2005 Jun 25	14.9	FF	2.56
1016	4	2005 Jun 26	12.9	FF	2.57
1025	4	2005 Jul 15	12.9	FF	2.56
1115	4	2006 Jan 10	14.9	FF	2.56

Exposure time is given for EPIC-pn

Mean countrates are given for the whole XMM-Newton energy range (0.15–10 keV)

[a]Countrate from Haberl et al. (2003)

An initial attempt to analyse the spectrum was made with a spectrum extracted from a merged photon event table of the six observations and with common response and effective area files. Such an approach does not take into account possible time-dependencies of the detector files. We therefore extracted a spectrum and computed the response matrix and effective area file for each observation separately. Spectral modeling was performed with XSPEC V12 following two different approaches. Firstly, the spectral parameters were determined for each spectrum individually but forcing the column density of cold interstellar absorbing matter, N_H, to be the same for all the observations (six data groups). This experiment revealed the spectral parameters of all observations to be the same within the errors. We then followed the second approach fitting all six spectra jointly with one common model which reveals one set of spectral parameters instead of six. The results of those latter experiments are listed in Table 2.

Spectral models were chosen with increasing complexity until a satisfactory fit was achieved. The fit assuming only a blackbody emission spectrum absorbed by neutral interstellar matter which gives $\chi^2_\nu \simeq 4$ (for 1247 d.o.f.) can be ruled out completely (model 1 in Table 2, Fig. 1a). This mismatch was found already by Haberl et al. (2003) when fitting data from revolutions 377 & 561. They successfully applied a model with a Gaussian absorption line superposed on the absorbed blackbody. With such a model the current fit is improved but it does no longer adequately represent the shape of the combined spectrum (model 2 in Table 2, Fig. 1b). Compared to Haberl et al. (2003) the fit with the single Gaussian gives a higher line energy, $E_1 = 390$ eV vs. 300 eV, and smaller line width, $\sigma_1 = 60$ eV vs. 100 eV.

Table 2 Spectral modeling of RBS1223. E and σ denote the central energy and width of the two Gaussians (subscript 1 and 2), while F denotes the line flux

#	N_H 10^{20} cm^{-2}	kT_{bb} eV	E_1 keV	σ_1 keV	F_1 keV cm^{-2} s^{-1}	E_2 keV	σ_2 keV	F_2 keV cm^{-2} s^{-1}	red. χ^2	#d.o.f
1	3.3	100	–	–	–	–	–	–	3.96	1250
2	3.7	93	0.39	0.06	-6×10^{-4}	–	–	–	2.05	1247
3	1.8	102	0.23	0.15	-35×10^{-4}	0.46	0.26	-7×10^{-4}	1.11	1244
4	2.0	104	0.28	0.13	-21×10^{-4}	$=2 \times E_1$	0.24	-7×10^{-4}	1.12	1245
5	1.2	100	0.20	0.17	-27×10^{-4}	0.73	$=\sigma_1$	-3×10^{-4}	1.11	1245
6	1.8	102	0.28	0.15	-35×10^{-4}	$=1.5 \times E_1$	0.26	-7×10^{-4}	1.11	1245

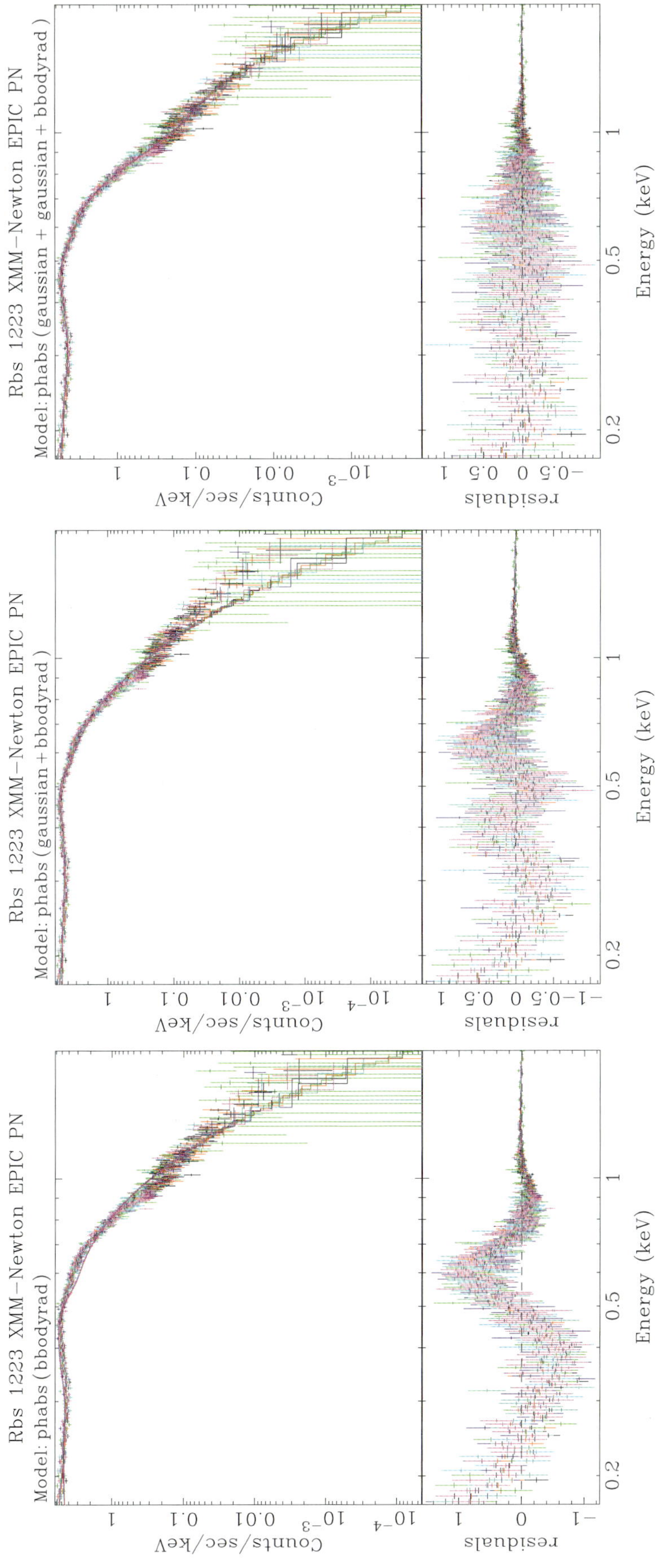

Fig. 1 Results of spectral fitting ordered according to increased complexity of the model. From *left* to *right* the models shown are an absorbed blackbody, a blackbody plus Gaussian absorption line, and a blackbody with two Gaussian absorption lines. The spectral parameters are listed in Table 2, models 1–3

Part of this change is due to the improved signal/noise of the spectrum which allows a free fit of σ_1 instead of fixing it at 100 eV as in Haberl et al. (2003). The updated calibration of the detector might also helpful. After inclusion of a second Gaussian a successful fit showing no systematic residuals was achieved ($\chi^2_\nu = 1.11$ for 1244 degrees of freedom, model 3, Fig. 1c).

Since the second Gaussian could be the higher harmonic of a fundamental at lower energies (Haberl et al. 2003 proposed an interpretation as proton cyclotron line in a field of $(2-6) \times 10^{13}$ G) we made some attempts to fit the two Gaussians with related parameters. In our model 3, both Gaussians were fitted independently, they were centered at line energies $E_2 \simeq 2 \times E_1$. Hence, if the parameters are forced to obey this relation, the achieved fit becomes equally good and the other parameters are found almost unchanged (model 4). The picture changes slightly if the width of the two lines is pre-set to the same, adjustable value (model 5). This broadens the width of the principal first line slightly from 0.15 keV to 0.17 keV and shifts the line energy of the second Gaussian from 0.46 keV to 0.73 keV at the same quality of the fit ($\chi^2_\nu = 1.11$). Nevertheless the flux in the second Gaussian is reduced by 50% in this fit. The last model listed in Table 2 (number 6) uses a different relation between the line energies, $E_2 = 1.5 \times E_1$, and was similarly successful with almost unchanged parameters compared to the free fit.

For models 1 and 2 no formal parameter errors were determined because those models did not adequately describe the spectrum. The formal errors of the parameters of fit 3 are: $\Delta N_H = 0.2 \times 10^{20}$ cm^{-2}, $\Delta kT = 2$ eV, $\Delta E_1 = 0.02$ keV, $\Delta\sigma_1 = 0.02$ keV, $\Delta F_1 = 4 \times 10^{-4}$ keV cm^{-2} s^{-1}, $\Delta E_2 = 0.03$ keV, $\Delta\sigma_2 = 0.03$ keV, and $\Delta F_2 = 4 \times 10^{-4}$ keV cm^{-2} s^{-1}. The above experiments have shown cross-talk between the parameters, hence the true uncertainties are larger than the mentioned statistical ones. For example, models 3–6 are indistinguishable on statistical grounds but some of their parameters are rather different. The main problem consists in a proper location of the continuum. The primary Gaussian is so broad, that the low-energy end of the spectrum is still affected by this feature. This allows the flux in this features to be traded against the amount of interstellar absorption and the temperature of the blackbody. At the high-energy end of the spectrum photons are detected up to 2 keV which in principle gives a better leverage for constraining the blackbody temperature. Our fitted model assumes that these photons belong to the Wien tail of the blackbody. Consequently, the newly determined blackbody temperature is higher than found before from a spectrum which missed those high-energy photons and was affected by the structure which is now described by a second Gaussian in its presumed Wien-tail.

4 Results and discussion

The joint analysis of six observations of RBS1223 with XMM-Newton has led to a slight revision of the spectral parameters. The spectrum can be described with a blackbody and two Gaussian absorption lines superimposed. The blackbody temperature is significantly higher than previously found thanks to the detection of X-ray photons up to 2 keV. Also the parameters of the primary Gaussian absorption line are changed: it appears significantly narrower and more blueshifted than before. Whether we have seen the true continuum at any energy remains an open question.

Similarly unclear is the nature of the features which were parameterised with two Gaussian absorption lines. They have a flux ratio of about 5:1, their equivalent widths of $\mathrm{EW}_1 \sim 200$ eV (for $E_1 = 0.23$ keV) and $\mathrm{EW}_2 \sim 180$ eV (for $E_2 = 0.46$ keV) are rather large. Their relative line energy and the rather large width of the lines are supportive of the proton cyclotron interpretation in a field of few times 10^{13} G (the line center of the primary and hence the derived magnetic field strength is not well constrained, in model 5 the best fit is achieved for $E_1 = 0.2$ keV, i.e. it converges to the lower bound of the pre-set parameter range), consistent with the observed spin down rate of the star. The rather high flux ratio of the two lines, nevertheless, makes an interpretation of the two lines as being harmonics of each other unlikely, since the oscillator strength of the first harmonic is smaller than that of the fundamental by a factor $\sim E/(mc^2)$, which becomes very small for proton masses (Pavlov et al. 1980).

An alternative could be to associate the two lines with different fields from e.g. the two polar caps. Since the field was shown to be somehow non-dipolar or off-centered at least (Schwope et al. 2005), different field strength could be encountered at the two caps. It is difficult to test this hypothesis on the basis of the current data and it needs to be seen if the results of the phase-resolved spectroscopy reveal any helpful indication in this respect. However, if the two lines would be the fundamentals of two different cyclotron systems from two different regions, the inferred field strengths would be $B_1 = 4.2 \times 10^{13}$ G and $B_2 = 8.5 \times 10^{13}$ G according to $B(10^{13}\mathrm{G}) = \frac{1}{1.16}(1+z)\frac{m_p}{m_e}E(\mathrm{keV})$ and assumed $z = 0.18$ (see Schwope et al. 2005), $E_1 = 0.23$ keV and $E_2 = 0.46$ keV. Hence, one field would be slightly below and one above the critical quantum field, where vacuum polarization becomes important in shaping the lines (see Lai and Ho 2003, 2004, and the discussion in van Kerkwijk et al. 2004). Since the second Gaussian line (higher field) is rather broader than the primary Gaussian, this scenario seems to be unlikely.

A possibility which needs to be explored in detail is a blend of magnetically shifted atomic transitions with proton cyclotron resonances. As the model computations of Lai and

Ho (2004) indicate, the amount of ionisation in the assumed hydrogen atmosphere will play a crucial role. An as accurate as possible temperature determination is thus very important for further progress. Due to its photometric stability on the one hand and the pronounced spin variability on the other hand, RBS1223 is a very promising target to gain further insight from even deeper X-ray spectroscopy.

Acknowledgement This work was supported in part by the German DLR under grant 50 OR 0404. Based on observations obtained with XMM-Newton, an ESA science mission, with instruments and contributions directly funded by ESA Member States and NASA.

References

Geppert, U.R.M.E., Küker, M., Page, D.: Astron. Astrophys. **426**, 267 (2004)

Haberl, F., Schwope, A.D., Hambaryan, V., et al.: Astron. Astrophys. **403**, L19 (2003)

Hambaryan, V., Hasinger, G., Schwope, A.D., et al.: Astron. Astrophys. **381**, 98 (2002)

Kaplan, D.L., van Kerkwijk, M.H.: Astrophys. J. **635**, L65 (2005)

Kaplan, D.L., Kulkarni, S.R., van Kerkwijk, M.H.: Astrophys. J. **579**, L29 (2002)

Lai, D., Ho, W.C.G.: Astrophys. J. **588**, 962 (2003)

Lai, D., Ho, W.C.G.: Astrophys. J. **607**, 420 (2004)

Pavlov, G.G., Shibanov, Iu.A., Iakovlev, D.G.: Astrophys. Space Sci. **73**, 33 (1980)

Schwope, A.D., Hasinger, G., Schwarz, R., et al.: Astron. Astrophys. **341**, L51 (1999)

Schwope, A.D., Hasinger, G., Lehmann, I., et al.: Astron. Nachr. **321**, 1–52 (2000)

Schwope, A.D., Hambaryan, V., Haberl, F., et al.: Astron. Astrophys. **441**, 597 (2005)

van Kerkwijk, M.H., Kaplan, D.L., Durant, M., et al.: Astrophys. J. **608**, 432 (2004)

Astrophys Space Sci (2007) 308: 625–629
DOI 10.1007/s10509-007-9296-z

ORIGINAL ARTICLE

High frequency oscillations during magnetar flares

Evidence for neutron star vibrations

Anna L. Watts · Tod E. Strohmayer

Received: 23 June 2006 / Accepted: 19 July 2006 / Published online: 16 March 2007

Abstract The recent discovery of high frequency oscillations during giant flares from the Soft Gamma Repeaters SGR 1806-20 and SGR 1900+14 may be the first direct detection of vibrations in a neutron star crust. If this interpretation is correct it offers a novel means of testing the neutron star equation of state, crustal breaking strain, and magnetic field configuration. We review the observational data on the magnetar oscillations, including new timing analysis of the SGR 1806-20 giant flare using data from the *Ramaty High Energy Solar Spectroscopic Imager* and the *Rossi X-ray Timing Explorer*. We discuss the implications for the study of neutron star structure and crust thickness, and outline areas for future investigation.

Keywords Magnetars · Neutron stars · Seismology

PACS 26.60.+c · 97.10.Sj · 97.60.Jd

1 Introduction

The Soft Gamma Repeaters (SGRs) are thought to be magnetars, neutron stars with magnetic fields in excess of 10^{14} G (Duncan and Thompson 1992, 1995). Decay of the strong field powers regular gamma-ray flaring activity that culminates, on rare occasions, in a giant flare with a peak luminosity in the range 10^{44}–10^{46} erg s^{-1}. The three giant flares detected to date consist of a short, spectrally hard initial peak, followed by a softer decaying tail that lasts for several hundred seconds. Pulsations with periods of a few seconds are visible in the tail and reveal the neutron star spin. Their presence is thought to be due to a fireball of ejected plasma, trapped near the stellar surface by the strong magnetic field (Thompson and Duncan 1995).

Powering the giant flares requires a catastrophic global reconfiguration of the magnetic field. The coupling between the field and the charged particles in the neutron star crust means that this is likely to be associated with large-scale crust fracturing (Flowers and Ruderman 1977; Thompson and Duncan 1995, 2001; Schwartz et al. 2005). This in turn should excite global seismic vibrations: on Earth seismologists regularly observe global modes after large earthquakes (see for example Park et al. 2005). In the SGR case, the coupling of field and crust should cause the modes to modulate the X-ray lightcurve.

Various different types of oscillation are possible, but theory suggests that the easiest to excite and observe should be the toroidal shear modes of the crust (Blaes et al. 1989). The particular harmonics excited will depend on the size, shape and speed of the fracture. The observed mode frequencies depend on the neutron star mass and radius (via gravitational redshift and the influence of the equation of state on crust structure), crustal composition, and magnetic field strength and configuration (Hansen and Cioffi 1980; Schumaker and Thorne 1983; McDermott et al. 1988; Strohmayer 1991; Duncan 1998; Messios et al. 2001; Piro 2005). Detection and identification of crustal modes would therefore probe all of these areas of neutron star physics.

In 2004, SGR 1806-20 emitted the most powerful flare ever recorded. In a landmark paper, Israel et al. (2005) reported the detection of Quasi-Periodic Oscillations (QPOs)

A.L. Watts (✉)
Max Planck Institut für Astrophysik, Karl-Schwarzschild-Str. 1, 85741 Garching, Germany
e-mail: anna@mpa-garching.mpg.de

T.E. Strohmayer
NASA Goddard Space Flight Center, Exploration of the Universe Division, Mail Code 662, Greenbelt, MD 20771, USA

in the 18–93 Hz range in the decaying tail of the flare. Prompted by this result, Strohmayer and Watts (2005) reanalysed the 1998 giant flare from SGR 1900+14 and found QPOs in the range 28–155 Hz. In both cases tentative identifications can be made with a sequence of toroidal modes. More in-depth analysis of the SGR 1806-20 flare has since revealed additional QPOs, including the first evidence for a higher radial overtone (Watts and Strohmayer 2006; Strohmayer and Watts 2006). We review all of the observational results in Sect. 2. In Sect. 3 we discuss the results in the context of the toroidal mode models, and show how they can be used to constrain neutron star parameters including the crust thickness. We conclude in Sects. 4 and 5 with a discussion of outstanding issues.

2 Observational results

2.1 SGR 1806-20

The December 2004 flare was the brightest ever recorded, with a peak luminosity of $\sim 10^{46}$ erg s^{-1}. Analysis of data from the *Rossi X-ray Timing Explorer* (RXTE) by Israel et al. (2005) revealed a transient 92.5 Hz QPO in the tail of the flare, associated with a particular rotational phase. The QPO appeared around 170 s after the main peak, at the same time as an apparent boost in unpulsed emission. The presence of weaker 18 and 30 Hz features at late times was also suggested.

The toroidal modes are labeled by the standard quantum numbers l, m and n, the first two being angular quantum numbers, the final one denoting the number of radial nodes. The 30 and 92.5 Hz features found by Israel et al. (2005) can be identified as the fundamental $l = 2$, $n = 0$ mode and the $l = 7$, $n = 0$ harmonic (see Sect. 3 for a discussion of current mode models). The 18 Hz feature, by contrast, is at too low a frequency to be identified as a toroidal mode of the crust.

The other spacecraft with high time resolution data of the SGR 1806-20 giant flare was the *Ramaty High Energy Solar Spectroscopic Imager* (RHESSI), a solar-pointing satellite that covers a wider energy band than RXTE. RHESSI's detectors are split into front and rear segments. Strong albedo flux from the Earth affected the time resolution of the rear segments (making them unsuitable for studying frequencies > 50 Hz), but when the rear segments are included the countrate recorded by RHESSI exceeded that of RXTE.

Watts and Strohmayer (2006) analysed the RHESSI data of the flare and confirmed the presence of the 92.5 Hz QPO, at the same time and rotational phase found by Israel et al. (2005). At low frequencies, where the larger RHESSI countrate gave added sensitivity, broad QPOs at 18 and 26 Hz were found 50–200 s after the main flare, at the same rotational phase as the 92.5 Hz QPO. Although there was a weaker feature at 30 Hz, it was not at the 3σ level after accounting for the number of trials so we were not able to make a robust confirmation of the Israel et al. (2005) result. Subsequent closer comparison of the two datasets, however, indicates that RHESSI's countrate is not the only factor affecting its sensitivity. As such we now believe that the 30 Hz detection is robust (see below).

Also detected in the RHESSI data was a 625 Hz QPO[1] at an energy band nominally higher than that recorded by RXTE.[2] A full discussion of the QPO properties and the significance of the detection can be found in Sect. 2 of Watts and Strohmayer (2006). Compared with the 92.5 Hz QPO the 625 Hz QPO is in a higher energy band, emerges earlier in the tail of the flare, and is at a different rotational phase. Excitingly, however, it is also at the approximate frequency expected for the radial overtone $n = 1$ toroidal modes (Piro 2005).

We have recently completed a more in-depth analysis of the now public RXTE dataset (Strohmayer and Watts 2006). We start by discussing the low frequency QPOs. In this regime the RXTE countrates are lower than RHESSI's. However, at times when the 18 and 26 Hz QPOs are particularly strong in the RHESSI dataset they are also detected in the RXTE data. We are thus able to confirm both the frequencies and the rotational phase dependence. Further analysis of the 30 Hz QPO confirms the detection made by Israel et al. (2005), and reveals that this QPO is also strongly rotational phase dependent. Rather surprisingly, given the difference in countrates, this feature is far stronger in the RXTE data than in the RHESSI dataset. There are several factors, however, that could account for this discrepancy (see the discussion in Strohmayer and Watts 2006).

Looking to higher frequencies, a rotational phase dependent search reveals additional QPOs at 150 Hz, 1840 Hz and, most excitingly, at 625 Hz (the latter appearing at the same rotational phase as the 92.5 Hz QPO). Full details

[1]Grid shadowing in RHESSI can give rise to spurious high frequency signals, and there is a known artifact at 718 Hz in the giant flare data (Gordon Hurford, private communication). The phenomenon affects certain detectors far more strongly than others since it is due to spacecraft geometry. It is also (perhaps unsurprisingly) more prominent at lower energies. The fact that the 625 Hz QPO is dependent on the rotational phase of the magnetar, despite there being comparable numbers of photons at other rotational phases, is a good indication that it is not a detector artifact. In addition it is only detected at higher energies, where shadowing should be less of an issue. However, to be sure we also tested the variation of QPO power when photons from individual detectors were removed from the analysis. The drops in power were consistent with the drop in countrate, confirming that the signal is not anomalously strong in one or two detectors. As such we are confident that the signal is indeed associated with the source.

[2]Direct comparison of recorded energies is difficult because for both spacecraft the flare was off-axis, resulting in substantial scattering within the body of the satellite.

of the QPO properties and the significance of the detection can be found in Sect. 2.1 of Strohmayer and Watts (2006). There are, it turns out, intriguing differences between the RHESSI and RXTE 625 Hz QPOs. The RXTE QPO has lower fractional amplitude, lower coherence, a different rotational phase association, and appears at later times in a lower energy band. One possibility is that we are seeing time evolution of one QPO whose photon energy and amplitude are falling over the course of the tail. The second possibility is that we are seeing two different modes. As pointed out by Piro (2005) the $n = 1$ modes are nearly degenerate in l, so there are several different modes with very similar frequencies. Despite the differences, it seems extremely unlikely that two independent instruments would detect signals at a consistent frequency unless that frequency were intrinsic to the source.

2.2 SGR 1900+14

In August 1998 SGR 1900+14 emitted a giant flare with a peak luminosity $\sim 10^{44}$ erg s^{-1} (Hurley et al. 1999). The flare was detected by RXTE, albeit with data gaps due to the configuration of the spacecraft at that time. Strohmayer and Watts (2005) started by analysing each good interval separately, and discovered a strong transient 84 Hz QPO during a ~ 1 s interval about 60 s after the main flare. Folding up data from more intervals revealed a pair of persistent QPOs at 53.5 and 155.5 Hz, with a weaker feature at 28 Hz, all at the same rotational phase as the 84 Hz signal. No QPOs were detected at other rotational phases. The scaling of the QPO frequencies is what the existing models suggest for the $n = 0$, $l = 2$, 4, 7 and 13 toroidal modes of the crust.

3 Constraining neutron star properties

Detection of signals at similar frequencies in the giant flares from two different SGRs implies that the same process is operating in both objects. In addition, the strong rotational phase dependence of *all* of the detected QPOs provides strong evidence that the modulations are associated with the stellar surface. In this section we discuss the implications of the QPOs in the light of the toroidal mode model. Existing models do have shortcomings, which we discuss in more detail in Sect. 4. However, crust mode models remain at present the most promising mechanism and we will proceed accordingly.

For a non-rotating, nonmagnetic star, Duncan (1998) estimates the frequency ν of the fundamental $l = 2$, $n = 0$ toroidal mode (denoted ${}_2t_0$) to be

$$\nu({}_2t_0) = 29.8\,\frac{(1.71 - 0.71 M_{1.4} R_{10}^{-2})^{1/2}}{0.87 R_{10} + 0.13 M_{1.4} R_{10}^{-1}}\ \text{Hz}, \tag{1}$$

where $R_{10} = R/10$ km and $M_{1.4} = M/1.4 M_\odot$. The frequencies of the higher order $n = 0$ modes are given by

$$\nu({}_lt_0) = \nu({}_2t_0)\left[\frac{l(l+1)}{6}\right]^{1/2}\left[1 + \left(\frac{B}{B_\mu}\right)^2\right]^{1/2}, \tag{2}$$

where the final factor is a magnetic correction, $B_\mu \approx 4 \times 10^{15} \rho_{14}^{0.4}$ G and $\rho_{14} \sim 1$ is the density in the deep crust in units of 10^{14} g cm^{-3}. In deriving this equation it is assumed that magnetic tension boosts the field isotropically. Field configuration and corresponding non-isotropic effects could however alter this correction dramatically (Messios et al. 2001). More recent calculations by Piro (2005) using better models of the neutron star crust confirm these frequency estimates.

The detection of a set of modes from both SGR 1806-20 and SGR 1900+14 with the expected $[l(l+1)]^{1/2}$ scaling in frequency is indicative of the presence of toroidal modes. We can then ask what can be learnt from the fact that the fundamental ${}_2t_0$ mode frequency seems to differ for the two SGRs. For SGR 1806-20 it is inferred to be 30 Hz, whereas for SGR 1900+14 it is lower, at 28 Hz. For this to be the case, the properties of the two stars must differ. Given an equation of state (EOS), (1), (2) specify the relationship between mass and magnetic field necessary to give modes at the inferred frequencies. Figure 1, taken from Strohmayer and Watts (2005), shows for several EOS the stellar parameters that give ${}_2t_0$ oscillations at the frequencies inferred for the two stars.

Several conclusions follow from Fig. 1. Firstly, unless the masses differ substantially, SGR 1806-20 must have a stronger magnetic field than SGR 1900+14. This accords

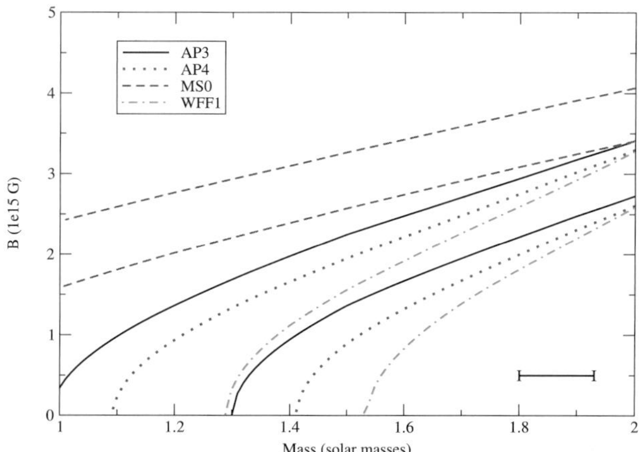

Fig. 1 Mass and magnetic field required to give the toroidal mode frequencies inferred for SGR 1806-20 and SGR 1900+14. We show results for four of the EOS from Lattimer and Prakash (2001); in order of increasing stiffness they are WFF1, AP4, AP3 and MS0. *The upper line* for each EOS indicates the parameters necessary to give the SGR 1806-20 frequencies; *the lower line* those for the lower SGR 1900+14 frequency. *The horizontal line* indicates the uncertainty in the position of the footpoints of the lines due to the width of the QPOs

with estimates of field strength from timing studies (Woods et al. 2002). Secondly, the highest masses and the hardest EOS require fields far higher than those inferred from timing, whereas the softest EOS require high masses. Clearly these inferences are all model-dependent, and could change as the models get more sophisticated. However, it illustrates the potential of neutron star seismology to constrain the equation of state.

If the 625 Hz QPOs detected in the SGR 1806-20 flare are indeed $n = 1$ toroidal crust modes then the implications are profound, since from the ratio of frequencies of the $n = 0$ and $n = 1$ modes one can deduce the thickness of the crust, ΔR (gravitational redshift factors scale out). In the limit of a thin crust ($\Delta R \ll R$) and constant shear wave speed (Hansen and Cioffi 1980; McDermott et al. 1988; Piro 2005) one can show that

$$\frac{\Delta R}{R} = \left(\frac{3}{2}\right)^{1/2} \frac{\nu({}_2t_0)}{\nu({}_lt_1)}. \quad (3)$$

So given $\nu({}_2t_0) = 30$ Hz and $\nu({}_lt_1) = 625$ Hz we find $\Delta R/R = 0.06$. A more sophisticated estimate that does not assume a thin crust or a constant shear speed gives $\Delta R/R$ in the range 0.1–0.12 (Strohmayer and Watts 2006).

Measuring the crust thickness also gives an additional constraint on the EOS (independent of redshift effects), since stellar models of a given mass computed with different EOS will in general have crusts with different depths. The article by Lattimer in this volume outlines how this could be used to constrain the nuclear symmetry energy and the nuclear force model.

4 Theoretical issues

Neutron star seismology, revealed during giant flares from magnetars, has great potential as a probe of stellar structure, composition and magnetic field geometry. There are many theoretical issues, however, that remain to be resolved. One of the main issues is that of the coupling between the crust and the fluid core due to the presence of the strong magnetic field. Global magneto-elastic modes of the whole star may be necessary to explain, in particular, the QPOs at 18 and 26 Hz, which do not sit comfortably within existing toroidal mode models. The coupling between the crust and the field will also determine how the modes modulate the lightcurve.

Modes of the neutron star crust (most likely coupled to the core via the magnetic field) remain the most promising mechanism identified to date for the QPOs. Although several other suggestions have been made, all have serious difficulties.Modes of the magnetosphere, mentioned by Levin (2006), are likely to have too high a frequency due to the high Alfvén speed in this region. Modes of the trapped fireball are unlikely since one would expect the frequency to change as the fireball shrinks: no such correlation is observed. The third possibility, raised by Alpar at this meeting, is interaction with a debris disk and a mechanism similar to that thought to generate QPOs in the accreting systems. However even if a small disk were present in these systems, it would be difficult to explain the rotational phase dependence.

The complex temporal variation of the QPOs also requires explanation. In both giant flares some of the QPOs seem to be highly transient, whereas others persist for most of the flare. We mentioned in Sect. 3 the possible evolution of the 625 Hz QPO seen during the SGR 1806-20 flare. However the 92.5 Hz QPO also shows variation in frequency and amplitude (Israel et al. 2005; Strohmayer and Watts 2006). Evolution of the magnetic field in the aftermath of the flare, relaxation of the deep crust, or evolution of the surrounding plasma are all candidates for causing such variations. Damping and excitation mechanisms are a critical area for future study.

5 Conclusions

The discovery of high frequency oscillations during giant flares from the Soft Gamma Repeaters SGR 1806-20 and SGR 1900+14 may be the first direct detection of vibrations in a neutron star crust. If this interpretation is correct it offers a new means of testing the neutron star equation of state, crustal breaking strain, and magnetic field configuration. In particular, if the mode interpretation is secure, it allows us to make the first direct estimate of the thickness of a neutron star crust. This is particularly impressive when one considers the fact that all of the data acquired so far have been obtained by chance using satellites that were observing other objects. If a rapid-slew instrument such as SWIFT could be configured to record high time resolution data throughout the tails of giant flares from known SGRs, the potential for additional discoveries is immense.

Acknowledgements A.L.W. acknowledges support from the European Union FP5 Research Training Network 'Gamma-Ray Bursts: An Enigma and a Tool'. T.E.S. thanks NASA for its support of high energy astrophysics research.

References

Blaes, O., Blandford, R., Goldreich, P., et al.: Astrophys. J. **343**, 839 (1989)
Duncan, R.C.: Astrophys. J. **498**, L45 (1998)
Duncan, R.C., Thompson, C.: Astrophys. J. **392**, L9 (1992)
Flowers, E., Ruderman, M.A.: Astrophys. J. **215**, 302 (1977)
Hansen, C.J., Cioffi, D.F.: Astrophys. J. **238**, 740 (1980)
Hurley, K., Cline, T., Mazets, E., et al.: Nature **397**, 41 (1999)
Israel, G., Belloni, T., Stella, L., et al.: Astrophys. J. **628**, L53 (2005)
Lattimer, J.M., Prakash, M.: Astrophys. J. **550**, 426 (2001)

Levin, Y.: Mon. Not. Roy. Astron. Soc. **368**, L35 (2006)
McDermott, P.N., van Horn, H.M., Hansen, C.J.: Astrophys. J. **325**, 725 (1988)
Messios, N., Papadopoulous, D.B., Stergioulas, N.: Mon. Not. Roy. Astron. Soc. **328**, 1161 (2001)
Palmer, D.M., Barthelmy, S., Gehrels, N., et al.: Nature **434**, 1107 (2005)
Park, J., Song, T.-R.A., Tromp, J., et al.: Science **308**, 1139 (2005)
Piro, A.L.: Astrophys. J. **634**, L153 (2005)
Schwartz, S.J., Zane, S., Wilson, R.J., et al.: Astrophys. J. **627**, L129 (2005)
Schumaker, B.L., Thorne, K.S., Mon. Not. Roy. Astron. Soc. **203**, 457
Strohmayer, T.E.: Astrophys. J. **372**, 573 (1991)
Strohmayer, T.E., Watts, A.L.: Astrophys. J. **632**, L111 (2005)
Strohmayer, T.E., Watts, A.L.: Astrophys. J. **653**, 593 (2006)
Thompson, C., Duncan, R.C.: Mon. Not. Roy. Astron. Soc. **275**, 255 (1995)
Thompson, C., Duncan, R.C.: Astrophys. J. **561**, 980 (2001)
Watts, A.L., Strohmayer, T.E.: Astrophys. J. **637**, L117 (2006)
Woods, P.M., Kouveliotou, C., Göğüs, E., et al.: Astrophys. J. **576**, 381 (2002)

Astrophys Space Sci (2007) 308: 631–639
DOI 10.1007/s10509-007-9318-x

ORIGINAL ARTICLE

Magnetar corona

A.M. Beloborodov · C. Thompson

Received: 28 July 2006 / Accepted: 10 November 2006 / Published online: 20 March 2007

Abstract Persistent high-energy emission of magnetars is produced by a plasma corona around the neutron star, with total energy output of $\sim 10^{36}$ erg/s. The corona forms as a result of sporadic starquakes that twist the external magnetic field of the star and induce electric currents in the closed magnetosphere. Once twisted, the magnetosphere cannot untwist immediately because of its self-induction. The self-induction electric field lifts particles from the stellar surface, accelerates them, and initiates avalanches of pair creation in the magnetosphere. The created plasma corona maintains the electric current demanded by curl **B** and regulates the self-induction e.m.f. by screening. This corona persists in dynamic equilibrium: it is continually lost to the stellar surface on the light-crossing time $\sim 10^{-4}$ s and replenished with new particles. In essence, the twisted magnetosphere acts as an accelerator that converts the toroidal field energy to particle kinetic energy. The voltage along the magnetic field lines is maintained near threshold for ignition of pair production, in the regime of self-organized criticality. The voltage is found to be about ~ 1 GeV which is in agreement with the observed dissipation rate $\sim 10^{36}$ erg/s. The coronal particles impact the solid crust, knock out protons, and regulate the column density of the hydrostatic atmosphere of the star. The transition layer between the atmosphere and the corona is the likely source of the observed 100 keV emission. The corona also emits curvature radiation up to 10^{14} Hz and can supply the observed IR-optical luminosity.

Keywords Plasmas · Stars: coronae, magnetic fields, neutron · X-rays: stars

PACS 97.60.Jd

This work was supported by NASA grant NNG-06-G107G.

A.M. Beloborodov (✉)
Physics Department and Columbia Astrophysics Laboratory, Columbia University, 538 W 120th Street, New York, USA
e-mail: amb@phys.columbia.edu

C. Thompson
Canadian Institute for Theoretical Astrophysics, University of Toronto, 60 St. George Street, Toronto, Canada
e-mail: thompson@cita.utoronto.ca

1 Introduction

At least 10% of all neutron stars are born as magnetars, with magnetic fields $B > 10^{14}$ G. Their activity is powered by the decay of the ultrastrong field and lasts about 10^4 years. They are observed at this active stage as either Soft Gamma Repeaters (SGRs) or Anomalous X-ray Pulsars (AXPs) (Woods and Thompson 2006).

Besides the sporadic X-ray outbursts, a second, persistent, form of activity has been discovered by studying the emission spectra of magnetars. Until recently, the spectrum was known to have a thermal component with temperature $k_{\mathrm{B}}T \sim 0.5$ keV, interpreted as blackbody emission from the star's surface. The soft X-ray spectrum also showed a tail at 2–10 keV with photon index $\Gamma = 2$–4. This deviation from pure surface emission already suggested that energy is partially released above the star's surface. Recently, observations by RXTE and INTEGRAL have revealed even more intriguing feature: magnetars are bright, persistent sources of 100 keV X-rays (Kuiper et al. 2006). This high-energy emission forms a separate component of the magnetar spectrum. It becomes dominant above 10 keV, has a very hard spectrum, $\Gamma \simeq 1$, and peaks above 100 keV where the spectrum is unknown. Its luminosity, $L \sim 10^{36}$ erg s^{-1}, even exceeds the thermal luminosity from the star's surface. The observed hard X-rays can be emitted only in the exterior of the star and demonstrate the presence of an active plasma corona.

This corona must form as a result of starquakes that shear the magnetar crust and its external magnetic field. Magnetospheric twists can explain several observed properties of magnetars. Thompson et al. (2002) investigated the observational consequences using a static, force-free model, idealizing the magnetosphere as a globally twisted dipole. They showed that a twist affects the spindown rate of the neutron star: it causes the magnetosphere to flare out slightly from a pure dipole configuration, thereby increasing the braking torque acting on the star. Although the calculations of force-free configurations are independent of the plasma behavior in the magnetosphere, they rely on the presence of plasma that conducts the required current $\mathbf{j}_B = (c/4\pi)\nabla \times \mathbf{B}$. This requires a minimum particle density $n_c = j_B/ec$. Even this minimum density can modify the stellar radiation by resonant scattering (Thompson et al. 2002).

The problem of plasma dynamics in the closed twisted magnetosphere has been formulated in Beloborodov and Thompson (2007; hereafter BT), and a simple solution has been found to this problem. It allows one to understand the observed energy output of the magnetar corona and the evolution of magnetic twists. It is worth mentioning that plasma behavior around neutron stars has been studied for decades in the context of radio pulsars, and that problem remains unsolved. The principle difference with radio pulsars is that their activity is caused by *rotation*, and dissipation occurs on *open magnetic lines* that connect the star to its light cylinder. Electron-positron pairs are then created on open field lines (Ruderman and Sutherland 1975; Arons and Scharlemann 1979). By contrast, the formation of a corona around a magnetar does not depend on its rotation. All observed magnetars are slow rotators ($\Omega = 2\pi/P \sim 1$ Hz), and their rotational energy is unable to feed the observed coronal emission. Practically the entire plasma corona is immersed in the *closed* magnetosphere and its heating must be caused by some form of dissipation on the closed field lines. It is this closure that facilitates the corona problem by providing both boundary conditions at the two footpoints of a field line.

The basic questions that one would like to answer are as follows. How is the magnetosphere populated with plasma? The neutron star surface has a temperature $k_B T \lesssim 1$ keV and the scale-height of its atmospheric layer (if it exists) is only a few cm. How is the plasma supplied above this layer and what type of particles populate the corona? How is the corona heated, and what are the typical energies of the particles? If the corona conducts a current associated with a magnetic twist $\nabla \times \mathbf{B} \neq 0$, how rapid is the dissipation of this current, i.e. what is the lifetime of the twist? Does its decay imply the disappearance of the corona?

An outline of the proposed model is as follows. The key agent in corona formation is an electric field $E_\parallel$ parallel to the magnetic field. $E_\parallel$ is generated by the self-induction of the gradually decaying current and in essence measures the rate of the decay. It determines the heating rate of the corona via Joule dissipation. If $E_\parallel = 0$ then the corona is not heated and, being in contact with the cool stellar surface, it will have to condense to a thin surface layer with $k_B T \lesssim 1$ keV. The current-carrying particles cannot flow upward against gravity unless a force $eE_\parallel$ drives it. On the other hand, when $E_\parallel$ exceeds a critical value, $e^\pm$ avalanches are triggered in the magnetosphere, and the created pairs screen the electric field. This leads to a "bottleneck" for the decay of a magnetic twist, which implies a slow decay.

Maintenance of the corona and the slow decay of the magnetic twist are intimately related because both are governed by $E_\parallel$. In order to find $E_\parallel$, one can use Gauss' law $\nabla \cdot \mathbf{E} = 4\pi\rho$ where ρ is the net charge density of the coronal plasma. This constraint implies that $\mathbf{E}$ and ρ must be found self-consistently. The problem turns out to be similar to the classical double-layer problem of plasma physics with a new essential ingredient: $e^\pm$ creation.

A direct numerical experiment can be designed that simulates the creation and behavior of the plasma in the magnetosphere. The experiment shows how the plasma and electric field self-organize to maintain the time-average magnetospheric current $\bar{\mathbf{j}} = \mathbf{j}_B$ demanded by the magnetic field, $(4\pi/c)\bar{\mathbf{j}} = \nabla \times \mathbf{B}$. The electric current admits no steady state on short timescales and keeps fluctuating, producing $e^\pm$ avalanches. Pair creation is found to provide a robust mechanism for limiting the voltage along the magnetic lines to $e\Phi_e \lesssim 1$ GeV and regulate the observed luminosity of the corona.

2 Mechanism of corona formation

A tightly wound-up magnetic field is assumed to exist inside magnetars (e.g. Thompson and Duncan 2001). The internal toroidal field can be much stronger than the external large-scale dipole component. The essence of magnetar activity is the transfer of magnetic helicity from the interior of the star to its exterior, where it dissipates. This involves rotational motions of the crust, which inevitably twist the external magnetosphere anchored to the stellar surface. The magnetosphere is probably twisted in a catastrophic way, when the internal magnetic field breaks the crust and rotates it. Such starquakes are associated with observed X-ray bursts (Thompson and Duncan 1995). The most interesting effect of a starquake for us here is the partial release of the winding of the internal magnetic field into the exterior, i.e., the injection of magnetic helicity into the magnetosphere.

Since the magnetic field is energetically dominant in the magnetosphere, it must relax there to a force-free configuration with $\mathbf{j} \times \mathbf{B} = 0$. Electric current $\mathbf{j}$ initiated by a starquake flows along the magnetic field lines to the exterior of the star, reaches the top of the field line, and comes back to the star at the other footpoint, then enters the atmosphere and the crust, following the magnetic field lines ($\mathbf{j} \parallel \mathbf{B}$).

The emerging currents are easily maintained during the X-ray outburst accompanying a starquake. A dense, thermalized plasma is then present in the magnetosphere, which easily conducts the current. Plasma remains suspended for some time after the starquake because of the transient thermal afterglow (Ibrahim et al. 2001). Eventually the afterglow extinguishes and the radiative flux becomes unable to support the plasma outside the star. The decreasing density then threats the capability of the magnetosphere to conduct the current of the magnetic twist. A minimal "corotation" charge density $\rho_{\text{co}} = -\mathbf{\Omega} \cdot \mathbf{B}/2\pi c$ is always maintained (Goldreich and Julian 1969), but it is far smaller than needed to supply the current $\mathbf{j} \sim (c/4\pi)(B/R_{\text{NS}})$. Can the lack of charges stop the flow of electric current? A simple estimate shows that the current cannot stop quickly because of its self-induction. A slow decay of the current generates a sufficient self-induction voltage that helps the magnetosphere to re-generate the plasma carrying the current, and the twisted force-free configuration persists with a qausi-steady $\mathbf{j} \neq 0$.

The stored energy of non-potential (toroidal) magnetic field associated with the ejected current is subject to gradual dissipation. In our model, this dissipation feeds the observed activity of the corona. The stored energy is concentrated near the star and carried by closed magnetic lines with a maximum extension radius $R_{\max} \sim 2R_{\text{NS}}$. Thus, most of the twist energy will be released if these lines untwist, and we focus here on the near magnetosphere $R_{\max} \sim 2R_{\text{NS}}$.

Consider a magnetic flux tube with cross section $S \lesssim R_{\text{NS}}^2$ and length $L \sim R_{\text{NS}}$ which carries a current $I = Sj$. The stored magnetic energy of the current per unit length of the tube is

$$\frac{\mathcal{E}_{\text{twist}}}{L} \sim \frac{I^2}{c^2}. \tag{1}$$

The decay of this energy is associated with an electric field parallel to the magnetic lines $\mathbf{E}_\parallel$: this field can accelerate particles and convert the magnetic energy into plasma energy. The decay of the twist is related to the voltage between the footpoints of a magnetic line,

$$\Phi_e = \int_A^C \mathbf{E}\, d\mathbf{l}. \tag{2}$$

Here A and C are the anode and cathode footpoints and $d\mathbf{l}$ is the line element; the integral is taken along the magnetic line outside the star (Fig. 1). The product $\Phi_e I$ represents the dissipation rate $\dot{\mathcal{E}}$ in the tube. The voltage Φ_e is entirely maintained by the self-induction that accompanies the gradual untwisting of the magnetic field.

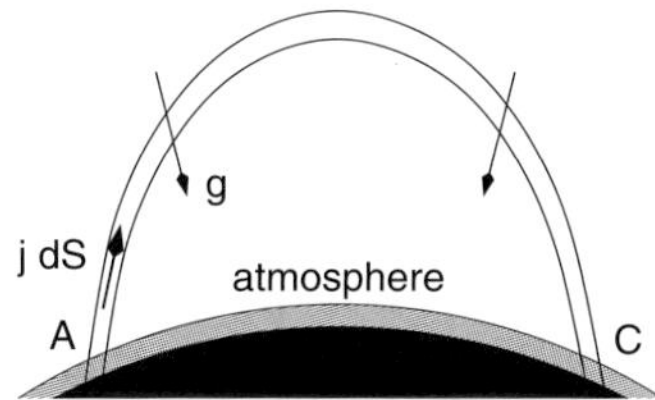

Fig. 1 Schematic picture of a current-carrying magnetic tube anchored to the star's surface. The current is initiated by a starquake that twists one (or both) footpoints of the tube. The current flows along the tube up to the magnetosphere and comes back to the star at the other footpoint. A self-induction voltage is created along the tube between its footpoints, which accompanies the gradual decay of the current. The voltage is generated because the current has a tendency to become charge-starved above the atmospheric layer whose scale-height $h = k_{\text{B}}T/gm_p$ is a few cm

A higher rate of untwisting implies a higher energy $e\Phi_e$ gained per particle in the tube. A huge magnetic energy is stored in the twisted field, and a quick untwisting would lead to extremely high Lorentz factors of the accelerated particles. There is, however, a bottleneck that prevents a fast decay of the twist: the tube responds to high voltages through the copious production of $e^\pm$ pairs. Runaway pair creation is ignited when electrons are accelerated to a certain Lorentz factor $\gamma_\pm \sim 10^3$ (see below). The created $e^\pm$ plasma does not tolerate large $E_\parallel$—the plasma is immediately polarized, screening the electric field. This provides a negative feedback that limits the magnitude of Φ_e and buffers the decay of the twist.

A minimum Φ_e is needed for the formation of a corona. Two mechanisms can supply plasma: (1) Ions and electrons can be lifted into the magnetosphere from the stellar surface. This requires a minimum voltage,

$$e\Phi_e \sim gm_i R_{\text{NS}} \sim 200 \text{ MeV}, \tag{3}$$

corresponding to $E_\parallel$ that is strong enough to lift ions (of mass m_i) from the anode footpoint and electrons from the cathode footpoint. (2) Pairs can be created in the magnetosphere if

$$e\Phi_e \sim \gamma_\pm m_e c^2 = 0.5\gamma_\pm \text{ MeV}, \tag{4}$$

which can accelerate particles to the Lorentz factor $\gamma_\pm$ sufficient to ignite $e^\pm$ creation.

If Φ_e is too low and the plasma is not supplied, a flux tube is guaranteed to generate a stronger electric field. The current becomes slightly charge-starved: that is, the net density of free charges becomes smaller than $|\nabla \times \mathbf{B}|/4\pi e$. The ultrastrong magnetic field, whose twist carries an enormous energy compared with the energy of the plasma, does not change and responds to the decrease of $\mathbf{j}$ by generating an electric field: a small reduction of the conduction

current $\mathbf{j}$ induces a displacement current $(1/4\pi)\partial\mathbf{E}/\partial t = (c/4\pi)\nabla \times \mathbf{B} - \mathbf{j}$. The longitudinal electric field $E_\parallel$ then quickly grows until it can pull particles away from the stellar surface and ignite pair creation.

The limiting cases $E_\parallel \to 0$ (no decay of the twist) and $E_\parallel \to \infty$ (fast decay) both imply a contradiction. The electric field and the plasma content of the corona must regulate each other toward a self-consistent state, and the gradual decay of the twist proceeds through a delicate balance: $E_\parallel$ must be strong enough to supply plasma and maintain the current in the corona; however, if plasma is oversupplied $E_\parallel$ will be reduced by screening. A cyclic behavior is possible, in which plasma is periodically oversupplied and $E_\parallel$ is screened. Our numerical experiment shows that such a cyclic behavior indeed takes place.

3 Electric circuit: numerical experiment

Three facts facilitate the simulation of the coronal electric circuit:

1. The ultrastrong magnetic field makes the particle dynamics one-dimensional (1D). Magnetic field lines are not perturbed by the plasma inertia, and they can be thought of as fixed "rails" along which the particles move. The particle motion is confined to the lowest Landau level and is strictly parallel to the field (the lifetime of a particle in an excited Landau state is tiny).
2. The particle motion is collisionless in the magnetosphere. It is governed by two forces only: the component of gravity projected onto the magnetic field and a collective electric field $E_\parallel$ which is determined by the charge density distribution and must be found self-consistently.
3. The star possesses a dense and thin atmospheric layer.[1] Near the base of the atmosphere, the required current $\mathbf{j}_B = (c/4\pi)\nabla \times \mathbf{B}$ is easily conducted, with almost no electric field. Therefore the circuit has simple boundary conditions: $E_\parallel = 0$ and fixed current. The atmosphere is much thicker than the skin depth of the plasma and screens the magnetospheric electric field from the star.

Our goal is to understand the plasma behavior above the screening layer, where the atmospheric density is exponentially reduced and an electric field $E_\parallel$ must develop. The induced electric field $\mathbf{E}$ and conduction current $\mathbf{j}$ satisfy the Maxwell equation,

$$\nabla \times \mathbf{B} = \frac{4\pi}{c}\mathbf{j} + \frac{1}{c}\frac{\partial \mathbf{E}}{\partial t}. \tag{5}$$

Here $\mathbf{j}$ is parallel to the direction of the magnetic field, and the force-free condition requires $\nabla \times \mathbf{B} \parallel \mathbf{B}$. Therefore, in

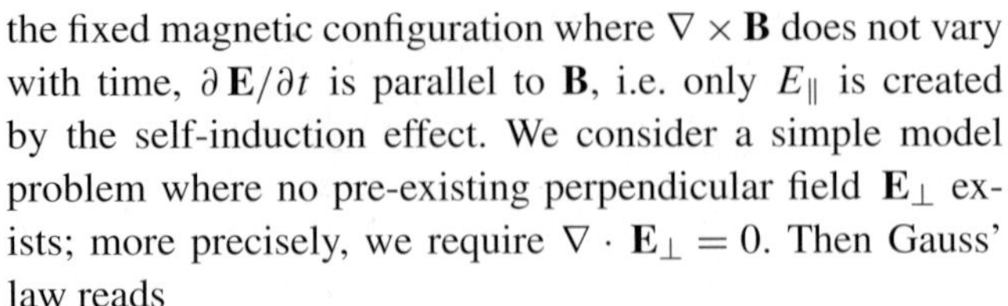

the fixed magnetic configuration where $\nabla \times \mathbf{B}$ does not vary with time, $\partial\mathbf{E}/\partial t$ is parallel to $\mathbf{B}$, i.e. only $E_\parallel$ is created by the self-induction effect. We consider a simple model problem where no pre-existing perpendicular field $\mathbf{E}_\perp$ exists; more precisely, we require $\nabla \cdot \mathbf{E}_\perp = 0$. Then Gauss' law reads

$$4\pi\rho = \nabla \cdot \mathbf{E}_\parallel = \frac{dE_\parallel}{dl}, \tag{6}$$

where l is length measured along the magnetic tube. Then the problem becomes strictly 1D since $\mathbf{E}_\perp$ has no relation to charge density and falls out from the problem.

If the conduction current j is smaller than $j_B \equiv (c/4\pi)|\nabla \times \mathbf{B}|$, then $\partial E_\parallel/\partial t > 0$ and an electric field appears that tends to increase the current. Alternatively, if $j > j_B$ then an electric field of the opposite sign develops which tends to reduce the current. Thus, j is always regulated toward j_B. This is the standard self-induction effect. The timescale of the regulation, τ, is very short, of the order of the Langmuir oscillation timescale $\omega_P^{-1} \sim 10^{-13}$ s. Local deviations from charge neutrality tend to be erased on the same plasma timescale. The net charge imbalance of the corona must be extremely small—a tiny imbalance would create an electric field that easily pulls out the missing charges from the surface. When the corotation charge density is neglected, the net charge of the corona vanishes.

Adopting a characteristic velocity $v \sim c$ of the coronal particles, the Debye length of the plasma $\lambda_D = v/\omega_P$ can be taken equal to the plasma skin depth c/ω_P. An important physical parameter of the corona is the ratio of the Debye length to the circuit size $L \sim R_{NS}$,

$$\zeta = \frac{\lambda_D}{L} = \frac{c}{L\omega_P} \ll 1. \tag{7}$$

The force-free condition $\mathbf{j}_B \times \mathbf{B} = 0$ together with $\nabla \cdot \mathbf{j}_B = 0$ requires that $j_B(l) \propto B(l)$ along the magnetic line. We can scale out this variation simply dividing all local quantities (charge density, current density, and electric field) by B. This reduces the problem to an equivalent problem where $j_B(l) = \text{const}$. Furthermore, only forces along magnetic lines control the plasma dynamics, and the curvature of magnetic lines falls out from the problem. Therefore, we can set up the experiment so that plasma particles move along a straight line connecting anode and cathode (Fig. 2). We designated this line as the z-axis, so that $l = z$.

This one-dimensional system may be simulated numerically. Consider N particles moving along the z-axis ($N \sim 10^6$ in our simulations). The electric field acting on a particle at a given z is given by

$$E(z) = 4\pi \int_0^z \rho\, dz, \tag{8}$$

[1] Such a layer initially exists on the surface of a young neutron star. Its maintenance is discussed in BT.

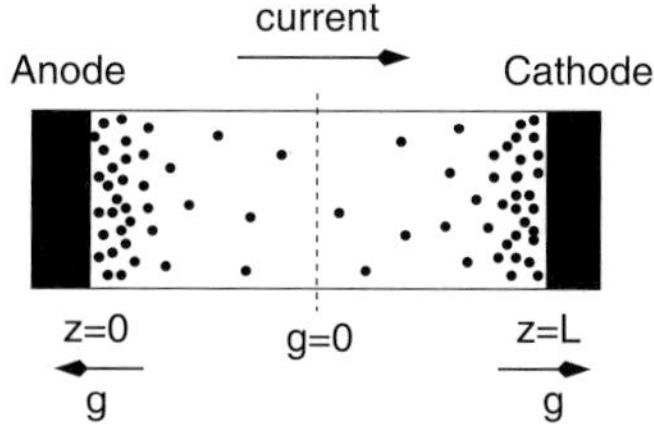

Fig. 2 Set up of numerical experiment. Thin and dense plasma layers are maintained near the cathode and anode by injecting cold particles through the boundaries of the tube (footpoints of a magnetic flux tube). The electric current is kept constant at the boundaries and the system is allowed to evolve in time until a quasi-steady state is reached. Without voltage between anode and cathode, the current cannot flow because gravity g traps the particles near the boundaries. The constant current at the boundaries, however, implies that voltage is immediately generated should the flow of charge stop in the tube. The experiment aims to find the self-induction voltage that keeps the current flowing

where $E(0) = E(L) = 0$ are the boundary conditions. Given the instantaneous positions z_i of all charges e_i we immediately find $E(z)$. To find a quasi-steady state of such a system we let it relax by following the particle motion in the self-consistent electric field.

Consider first a circuit where $e^{\pm}$ production is (artificially) forbidden, so that only the cold hydrostatic atmosphere can supply particles to the corona. We find that such a circuit quickly relaxes to a state where it acts as an ultra-relativistic linear accelerator (Fig. 3). In this state, oscillations of electric field (on the plasma timescale ω_P^{-1}) are confined to the thin atmospheric layers. A static accelerating electric field is created above the layers where the atmosphere density is exponentially reduced. The asymmetry of the solution is caused by the difference between the electron and ion masses.

This configuration is a relativistic double layer (Carlqvist 1982). It is well described by Carlqvist's solution, which has no gravity in the circuit and assumes zero temperature at the boundaries, so that particles are injected with zero velocity. According to this solution, the potential drop between anode and cathode, in the limit $e\Phi_e \gg m_i c^2$, is

$$e\Phi_e = \left[\left(\frac{m_i}{Zm_e}\right)^{1/2} + 1\right] \frac{L\omega_P}{2} m_e c, \qquad \omega_P^2 \equiv \frac{4\pi je}{m_e c}, \quad (9)$$

where Z is the ion charge number (in our experiment $Z = 1$ was assumed). The established voltage is much larger than is needed to overcome the gravitational barrier Φ_g. It does not even depend on Φ_g as long as Φ_g is large enough to prevent the expansion of the cold atmosphere. Gravity causes the transition to the linear-accelerator state, but the state itself does not depend on Φ_g.

Thus, the first result of numerical experiment is that an ion-electron circuit (without $e^{\pm}$ creation) in a gravitational field relaxes to the double layer of macroscopic size L and huge voltage Φ_e. The system does not find any state with

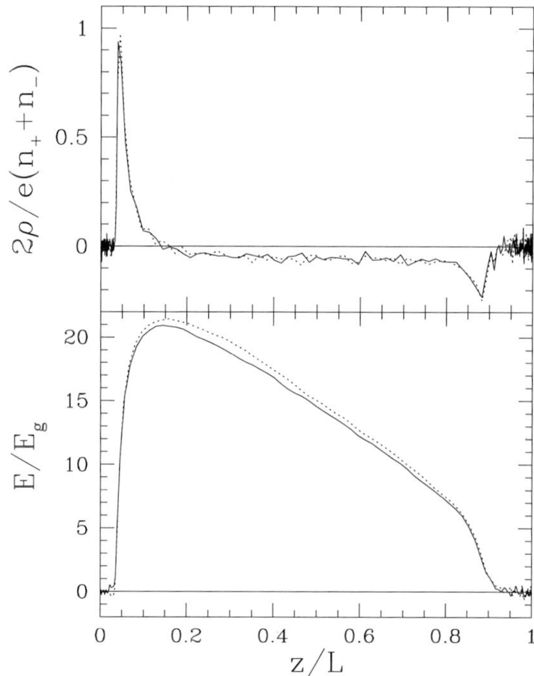

Fig. 3 Circuit without $e^{\pm}$ production. *Upper panel* shows the normalized charge density and *lower panel* shows the electric field in units of E_g defined by $eE_g = m_i g_0$. *Solid and dotted curves* correspond to two different moments of time, demonstrating that a quasi-steady state has been reached. In this simulation, $m_i = 10m_e$ and $\zeta = c/\omega_P L \simeq 0.01$. The final state is the relativistic double layer described analytically by Carlqvist (1982)

a lower Φ_e, even though it is allowed to be time-dependent and strong fluctuations persist in the atmospheric layers (reversing the sign of E).

If this double layer were maintained in the twisted magnetosphere, the twist would be immediately killed off: the huge voltage implies a large untwisting rate (Sect. 2). The electron Lorentz factor developed in the linear accelerator is (taking the real $m_i/m_e = 1836$ and $\zeta \sim 3 \times 10^{-9}$),

$$\gamma_e = \frac{e\Phi_e}{m_e c^2} \simeq \frac{20}{\zeta} \sim 6 \times 10^9. \quad (10)$$

However, new processes will become important before the particles could acquire such high energies: production of $e^{\pm}$ pairs will take place. Therefore, the linear-accelerator solution cannot describe a real magnetosphere. We conclude that pair creation is a key ingredient of the circuit that will regulate the voltage to a smaller value.

4 Pair discharge and self-organized criticality

In the magnetospheres of canonical radio pulsars with $B \sim 10^{12}$ G, $e^{\pm}$ pairs are created when seed electrons are accelerated to large Lorentz factors $\gamma_e \sim 10^7$. Such electrons emit curvature gamma rays that can convert to $e^{\pm}$

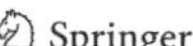

off the magnetic field. In stronger fields, another channel of $e^\pm$ creation appears. It is also two-step: an accelerated particle resonantly upscatters a thermal X-ray photon, which subsequently converts to a pair (Hibschman and Arons 2001; BT). The resonant scattering requires the electron to have a Lorentz factor,

$$\gamma_{\rm res} = \frac{B/B_{\rm QED}}{1-\cos\theta_{kB}}\frac{m_e c^2}{\hbar\omega_X} \approx 10^3 B_{15}\frac{10\ {\rm keV}}{\hbar\omega_X}, \tag{11}$$

where $\hbar\omega_X$ and θ_{kB} are the energy and pitch angle of the target photon. The distance over which an accelerated electron creates a pair is approximately equal to the free path to scattering $l_{\rm res}$. It depends only on B and the spectrum of target photons; $l_{\rm res} \ll R_{\rm NS}$ for the relevant parameter $B_{15}\hbar\omega_X < 100$ keV (BT).

When the voltage in the circuit becomes high enough to accelerate electrons to $\gamma_{\rm res}$, an $e^\pm$ breakdown develops in the magnetic flux tube. An illustrative toy model of breakdown is shown in a spacetime diagram in Fig. 4. Each pair-creation event gives two new particles of opposite charge, which initially move in the same direction. One of them is accelerated by the electric field and lost after a time $< L/c$, while the other is decelerated and can reverse direction before reaching the boundary. This reversal of particles in the tube and repeated pair creation allows the $e^\pm$ plasma to be continually replenished. In the super-critical regime (left panel in Fig. 4) more than one reversing particle is created per passage time L/c and an avalanche develops exponentially on a timescale $\sim a_{\rm res}/c$. In the near-critical regime shown in the right panel, just one reversing particle is created per passage time L/c.

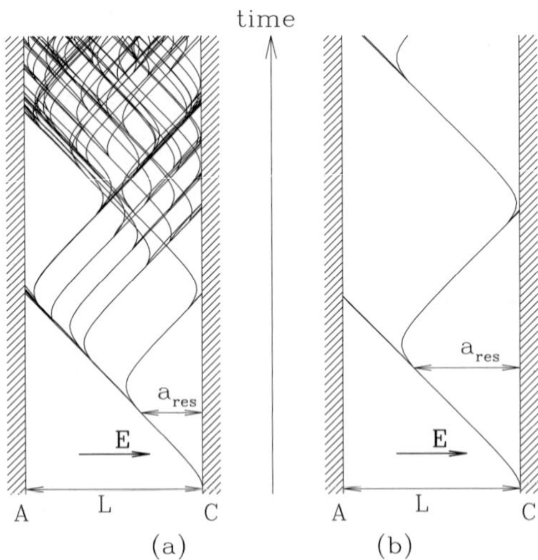

Fig. 4 Spacetime diagrams illustrating the critical character of the $e^\pm$ breakdown and formation of avalanches. The seed electron is placed at cathode and accelerated to $\gamma_{\rm res}$ after passing the distance $a_{\rm res} = (\gamma_{\rm res}-1)m_e c^2/eE$. The figure shows the worldlines of created particles. **a** $a_{\rm res}/L = 0.35$ (supercritical case). **b** $a_{\rm res}/L = 0.6$ (critical case)

This critical state is unstable: sooner or later the avalanche will be extinguished.

This toy model shows an essential property of the $e^\pm$ breakdown: it is a critical stochastic phenomenon. Above a critical voltage pair creation proceeds in a runway manner, and the current and the dissipation rate would run away if the voltage were fixed. Below the critical voltage pair creation does not ignite. The criticality parameter is $L/a_{\rm res} = e\Phi_e/(\gamma_{\rm res}-1)m_e c^2$. The tube with enforced current at the boundaries must self-organize to create pairs in the near-critical regime $e\Phi_e \sim \gamma_{\rm res} m_e c^2$ and maintain the current.

The discharging tube is similar to other phenomena that show self-organized criticality, e.g., a pile of sand on a table (Bak et al. 1987). If sand is steadily added, a quasi-steady state is established with a characteristic mean slope of the pile. The sand is lost (falls from the table) intermittently, through avalanches—a sort of "sand discharge." In our case, charges of the opposite signs are added steadily instead of sand (fixed j at the boundaries), and voltage $\Phi_e = EL$ plays the role of the mean slope of a pile. The behavior of the discharging system is expected to be time-dependent, with stochastic avalanches.

We implemented the process of pair creation in our numerical experiment. Regardless the details of this process (values of $\gamma_{\rm res}$ and $l_{\rm res}$) the circuit relaxed to the state of self-organized criticality after a few $t_{\rm dyn} = L/c$. This state is time-dependent on short timescales, but it has a well-defined steady voltage when averaged over a few $t_{\rm dyn}$ (Fig. 5),

$$e\bar{\Phi}_e \sim \gamma_{\rm res} m_e c^2. \tag{12}$$

We found that this relation applies even to circuits with $\gamma_{\rm res} m_e c^2 \gg m_i c^2$, where lifting of the ions is energetically preferable to pair creation. We conclude that *the voltage along the magnetic tube is self-regulated to a value just enough to maintain pair production and feed the current with $e^\pm$ pairs*. In all cases, the pair creation rate $2\dot{N}_+ \sim j_B/e$ is maintained.

The solution for the plasma dynamics in the corona is strongly non-linear and time-dependent. It is essentially global in the sense that the plasma behavior near one footpoint of a magnetospheric field line is coupled to the behavior near the other footpoint. Remarkably, this complicated global behavior may be described as essentially one-dimensional electric circuit that is subject to simple boundary conditions $E \approx 0$ on the surface of the star and can be studied using a direct numerical experiment.

The current is carried largely by $e^\pm$ everywhere in the tube. The ion fraction in the current depends on the ratio $e\Phi_e/m_i c^2 \simeq \gamma_{\rm res} m_e/m_i$ (Fig. 6). If this ratio is small, pairs are easily produced with a small electric field, too small to lift the ions, and the ion current is suppressed. In the opposite case, ions carry $\sim 1/2$ of the current. By coincidence, $\gamma_{\rm res} \lesssim m_i/m_e$ is typical for magnetars; it implies that ions carry $\lesssim 10\%$ of the current.

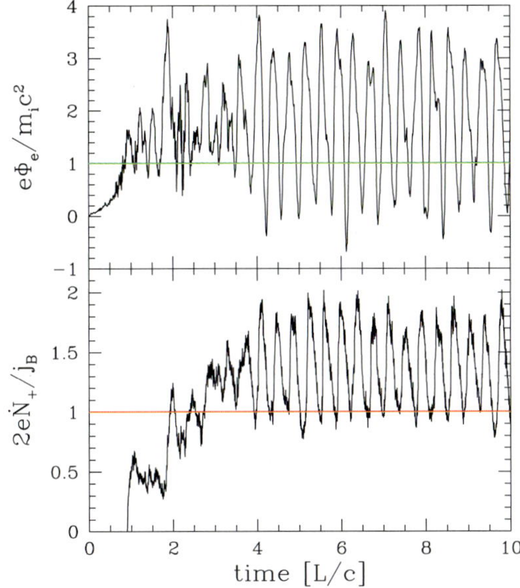

Fig. 5 Time history of one numerical experiment. The circuit parameters are: $m_i = 10m_e$, $\gamma_{\rm res} = m_i/m_e$, $\zeta = 0.01$. After a few dynamical times the voltage Φ_e stops growing and the circuit enters an oscillatory regime. During each oscillation, an increased voltage triggers $e^\pm$ discharge, then $\dot{N}_+$ drops, and Φ_e begins to grow again. The quasi-steady oscillating state (self-organized criticality) is established. *The horizontal lines* show the voltage $e\Phi_e = \gamma_{\rm res} m_e c^2$ and the minimum pair production rate $2\dot{N}_+ = j_B/e$ that can feed the required current j_B

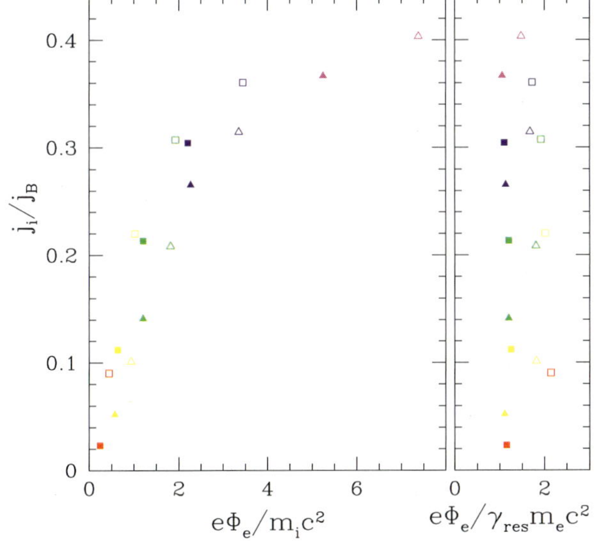

Fig. 6 Fraction of the electric current carried by ions vs. voltage. *In the left panel* voltage is taken in units of m_ic^2, and *in the right panel*—in units of $\gamma_{\rm res} m_e c^2$. Different colors and symbols show circuits with $m_i/m_e = 10, 30$ and various values of $\gamma_{\rm res}$ ranging from $m_i/5m_e$ to $5m_i/m_e$ (see BT)

5 Implications

The basic finding of this work is that an $e^\pm$ corona must be maintained around a magnetar. It exists in a state of self-organized criticality, near the threshold for $e^\pm$ breakdown. The established voltage along a twisted magnetic tube is marginally sufficient to accelerate an electron (or positron) to the energy where collisions with ambient X-ray photons spawn new pairs. This physical condition determines the rate of energy dissipation in the corona. The generation of this voltage is precisely the self-induction effect of the gradually decaying magnetic twist, and the energy release in the corona is fed by the magnetic energy of the twisted field.

The rate of energy dissipation in the twisted magnetosphere is given by $L_{\rm diss} = I\Phi_e$ where I is the net current through the corona. The current may be estimated as $I \sim j_B a^2 \sim \frac{c}{4\pi}\,\Delta\phi\,(B/R_{\rm NS})\,a^2$, where a is the size of a twisted region on the stellar surface and $\Delta\phi = |\nabla \times \mathbf{B}|(B/R_{\rm NS})^{-1} \lesssim 1$ characterizes the strength of the twist. The calculated $e\Phi_e \sim 1$ GeV implies

$$L_{\rm diss} = I\Phi_e \sim 10^{37}\,\Delta\phi\,B_{15}\left(\frac{a}{R_{\rm NS}}\right)^2\left(\frac{e\Phi_e}{\rm GeV}\right)\,{\rm erg\,s^{-1}}. \quad (13)$$

Observed luminosity $L_{\rm diss} \sim 10^{36}$ erg s^{-1} is consistent with a partially twisted magnetosphere, $a \sim 0.3R_{\rm NS}$, or a global moderate twist with $\Delta\phi\,(e\Phi_e/{\rm GeV}) \sim 0.1$.

Once created, a magnetospheric twist has a relatively long but finite lifetime. The energy stored in it is $\mathcal{E}_{\rm twist} \sim (I^2/c^2)R_{\rm NS}$. This energy dissipates in a time

$$t_{\rm decay} = \frac{\mathcal{E}_{\rm twist}}{L_{\rm diss}} \approx 3\left(\frac{L_{\rm diss}}{10^{36}\,{\rm erg\,s^{-1}}}\right)\left(\frac{e\Phi_e}{\rm GeV}\right)^{-2}\,{\rm yr}. \quad (14)$$

Note that a stronger twist (brighter corona) lives longer. If the time between large-scale starquakes is longer than $t_{\rm decay}$, the magnetar should be seen to enter a quiescent state. Such behavior has been observed in AXP J1810-197: an outburst was followed by the gradual decay on a year timescale (Gotthelf and Halpern 2005).

The voltage Φ_e implies a certain effective resistivity of the corona, $\mathcal{R} = \Phi_e/I$. This resistivity leads to spreading of the electric current *across* the magnetic lines, which is described by the induction equation $\partial\mathbf{B}/\partial t = -c\nabla \times \mathbf{E}$. The timescale of twist spreading to the light cylinder is comparable to $t_{\rm decay}$. Therefore, the impact of a starquake on spindown may appear with a delay of $\sim$ years.

Even a small ion current implies a large mass transfer through the magnetosphere over the $\sim 10^4$-yr active lifetime of a magnetar. It requires a significant excavation of the crust at the anode footpoints of twisted magnetic tubes. The bombarding relativistic electrons from the corona cause this excavation. They spall heavy ions in the uppermost crust, re-generate the light-element atmosphere on the surface, and regulate its column density to a value $\sim 100m_p/\sigma_{\rm T}$ (BT).

The radiative output from the corona likely peaks in its inner region because the coronal current is concentrated on closed field lines with a maximum extension $R_{\rm max} \lesssim 2R_{\rm NS}$

(Sect. 2). Emission from the inner corona must be suppressed above ~ 1 MeV, regardless the mechanism of emission, because photons with energy $\gtrsim 1$ MeV cannot escape the ultra-strong magnetic field. The observed high-energy radiation extends above 100 keV (Kuiper et al. 2004; Mereghetti et al. 2005; Molkov et al. 2005; den Hartog et al. 2006; Kuiper et al. 2006). There is an indication for a cut-off below 1 MeV from COMPTEL upper limits for AXP 4U 0142+61.

The possible mechanisms of emission are strongly constrained. The corona has a low density $n \sim n_c = j_B/ec \lesssim B(4\pi\, e R_{\rm NS})^{-1} \sim 10^{17} B_{15}$ cm^{-3} and particle collisions are rare, so two-body radiative processes are negligible. The up-scattering of keV photons streaming from the surface does not give the observed 100-keV emission (BT). A possible source of the 100 keV X-rays is the transition layer between the corona and the thermal photosphere (Thompson and Beloborodov 2005, BT). It is heated by the coronal beam and cooled by heat conduction toward the surface, and its temperature is regulated to $\sim 100(\ell/\ell_{\rm Coul})^{-2/5}$ keV, where $\ell/\ell_{\rm Coul} < 1$ parameterizes the suppression of thermal conductivity by plasma turbulence (BT). The emitted X-ray spectrum may be approximated as optically thin bremsstrahlung with a single temperature (Thompson and Beloborodov 2005). Its photon index below the exponential cutoff is close to -1, in agreement with spectra observed with INTEGRAL and RXTE.

The observed pulsed fraction increases toward the high-energy end of the spectrum and approaches 100% at 100 keV (Kuiper et al. 2004). If the hard X-rays are produced by one or two twisted spots on the star surface, the large pulsed fraction implies that the spots almost disappear during some phase of rotation. Then they should not be too far from each other. They cannot be, e.g., antipodal because a large fraction of the star surface is visible to observer due to the gravitational bending of light: $S_{\rm vis}/4\pi R_{\rm NS}^2 = (2\text{–}4GM/c^2 R_{\rm NS})^{-1} \approx 3/4$ (Beloborodov 2002).

The low-frequency radiation of magnetars is of a non-thermal origin and is likely to be produced in the corona. The observed infrared (K-band) and optical luminosities of magnetars are $\sim 10^{32}$ erg/s (Hulleman et al. 2004), which is far above the Rayleigh-Jeans tail of the surface blackbody radiation. The inferred brightness temperature is $\sim 10^{13}$ K if the radiation is emitted at radii $r \sim R_{\rm NS}$. One possible emission mechanism is curvature radiation by $e^{\pm}$ bunches in the corona (BT). Charges moving on field lines with a curvature radius $R_C \lesssim R_{\rm NS}$ will emit radiation with frequency $\nu_C \sim \gamma_e^3 c/2\pi R_C$, which is in the K-band ($\nu = 10^{14}$ Hz) or optical when $\gamma_e m_e c^2 \gtrsim$ GeV. The radio luminosity of a pulsar can approach $\sim 10^{-2}$ of its spindown power, and it is not implausible that the observed fraction $\sim 10^{-4}$ of the power dissipated in a magnetar corona is radiated in optical-IR photons by the same coherent mechanism.

The corona luminosity is simply proportional to the net current flowing through the magnetosphere, and is determined by the competition between sporadic twisting (due to starquakes) and gradual dissipative untwisting of the magnetosphere. During periods of high activity, when starquakes occur frequently, the magnetospheric twist may grow to the point of a global instability: when a critical $\Delta\phi \gtrsim 1$ is achieved, the magnetosphere suddenly relaxes to a smaller twist angle. A huge release of energy must then occur, producing a giant flare; a model of how this can happen is discussed by Lyutikov (2006).

A possible way to probe the twist evolution is to measure the history of spindown rate $\dot{P}$ of the magnetar. The existing data confirm the theoretical expectations: the spin-down of SGRs was observed to accelerate months to years following periods of activity (Kouveliotou et al. 1999; Woods et al. 2002). A similar effect was also observed following flux enhancement in AXPs (Gavriil and Kaspi 2004). This "hysteresis" behavior of the spindown rate may represent the delay with which the twist is spreading to the outer magnetosphere.

Since the density of the plasma corona is proportional to the current that flows through it, changes in the twist also lead to changes in the coronal opacity. The density of the plasma corona is close to its minimum $n_c = j/ec$, and it is largely made of $e^{\pm}$ pairs rather than electron-ion plasma. The opacity should increase following bursts of activity and affect the X-ray pulse profile and spectrum. Such changes, which persist for months to years, have been observed following X-ray outbursts from two SGRs (Woods et al. 2001).

Acknowledgement This research was supported by NASA grant NNG-06-G107G.

References

Arons, J., Scharlemann, E.T.: Astrophys. J. **231**, 854 (1979)
Bak, P., Tang, C., Wiesenfeld, K.: Phys. Rev. Lett. **59**, 381 (1987)
Beloborodov, A.M.: Astrophys. J. **566**, L85 (2002)
Beloborodov, A.M., Thompson, C.: Astrophys. J. **657**, 967 (2007)
Carlqvist, P.: Astrophys. Space Sci. **87**, 21 (1982)
den Hartog, P.R. et al.: Astron. Astrophys. **451**, 587 (2006)
Gavriil, F.P., Kaspi, V.M.: Astrophys. J. **609**, L67 (2004)
Goldreich, P., Julian, W.H.: Astrophys. J. **157**, 869 (1969)
Gotthelf, E.V., Halpern, J.P.: Astrophys. J. **632**, 1075 (2005)
Hibschman, J.A., Arons, J.: Astrophys. J. **554**, 624 (2001)
Hulleman, F., van Kerkwijk, M.H., Kulkarni, S.R.: Astron. Astrophys. **416**, 1037 (2004)
Ibrahim, A.I., et al.: Astrophys. J. **558**, 237 (2001)
Kouveliotou, C., et al.: Astrophys. J. **510**, L115 (1999)
Kuiper, L., Hermsen, W., Mendez, M.: Astrophys. J. **613**, 1173 (2004)
Kuiper, L., Hermsen, W., den Hartog, P.R., Collmar, W.: Astrophys. J. **645**, 556 (2006)
Lyutikov, M.: Mon. Not. Roy. Astron. Soc. **367**, 1594 (2006)
Mereghetti, S., Götz, D., Mirabel, I.F., Hurley, K.: Astron. Astrophys. **433**, L9 (2005)
Molkov, S., et al.: Astron. Astrophys. **433**, L13 (2005)

Ruderman, M.A., Sutherland, P.G.: Astrophys. J. **196**, 51 (1975)
Thompson, C., Beloborodov, A.M.: Astrophys. J. **634**, 565 (2005)
Thompson, C., Duncan, R.C.: Mon. Not. Roy. Astron. Soc. **275**, 255 (1995)
Thompson, C., Duncan, R.C.: Astrophys. J. **561**, 980 (2001)
Thompson, C., Lyutikov, M., Kulkarni, S.R.: Astrophys. J. **574**, 332 (2002)
Woods, P.M., Thompson, C.: In: Lewin, W.H.G., van der Klis, M. (eds.) Compact Stellar X-ray Sources. Cambridge University Press, Cambridge (2006), astro-ph/0406133
Woods, P.M., et al.: Astrophys. J. **552**, 748 (2001)
Woods, P.M., et al.: Astrophys. J. **576**, 381 (2002)

Astrophys Space Sci (2007) 308: 641–645
DOI 10.1007/s10509-007-9333-y

ORIGINAL ARTICLE

Intrinsic spectra of the AXPs

De-extinction from model-independent measurement of interstellar column densities

Martin Durant

Received: 22 June 2006 / Accepted: 15 July 2006 / Published online: 20 March 2007

Abstract Using direct measurements of photo-electric absorption edges, I derive the intrinsic spectra for the Anomalous X-ray Pulsars. In the past, the hydrogen column density, N_H had been found by fitting the X-ray spectra with a variety of simple continuum models. Since different models fit equally well, with different values of N_H, little is learned about the true column density or intrinsic spectra. Here, I measure the column densities in a model-independent way, and thus derive intrinsic spectra without first assuming what those spectra ought to look like. Particularly for the brightest source, 4U 0142+61, the column density can be determined accurately, and is a factor of 1.4 smaller that the typically N_H quoted. With this new value, the new intrinsic spectrum can be investigated anew, and shows a slight hint of a feature around 13 Å.

To be emphasised, this is an empirical method with the minimum of assumptions, as is appropriate for these beguiling sources, the behaviour of which has mystifies astronomers for over a decade.

This paper summarises the methods and conclusions to be published in the Astrophysical Journal by Durant and Van Kerkwijk (2006; see astro-ph/0606604).

Keywords Pulsars: anomalous X-ray pulsars · Interstellar extinction · X-ray spectroscopy

PACS 97.60

M. Durant (✉)
Astronomy & Astrophysics, University of Toronto, 60 St. George Street Room 1403, Toronto, M5S 3H8, Canada
e-mail: durant@astro.utoronto.ca

1 Introduction

Magnetars (Soft Gamma-ray Repeaters and Anomalous X-ray Pulsars) have been very exciting to research as of late, with their many and varied high-energy phenomena. They promise, with extreme magnetic, electrodynamic, energy, density and gravitational properties, to be ideal natural laboratories for unlocking the secrets of fundamental physics.

AXPs are X-ray bright, and the majority of investigations of them have thus been conducted in this band. A poorly constrained source of uncertainty in determining the true, intrinsic X-ray spectra, in particular at lower energies, is the amount of interstellar extinction. In previous studies, AXPs' X-ray spectra have been fit with simple continuum models, deriving extinction column densities with small statistical errors. Different choices of model, however, produced a wide range of columns, each with statistically acceptable fits. The column density can thus be uncertain by as much as a factor of 2 (see, for example, the values of N_H for 4U 0142+61 in Juett et al. 2002).

The most often used and quoted model for AXP spectra is the sum of a black-body and a power-law. This accounts for the peak in the spectra around 2 keV and an extended tail to higher (∼10 keV) energies. The model fails, however, at even higher energies (20–150 keV), where den Hartog et al. (2006) have found a rising power-law spectrum; and at lower energies, where an extrapolation of the model grossly over-predicts the optical emission (Hulleman et al. 2004). From a physical perspective, it is not clear why there should be a power-law component in the spectra of AXPs. Recent physical models, such as those presented by Lyutikov and Gavriil (2006) can account for a power-law like high-energy tail, but do not predict a soft part. In these models, the blackbody surface flux is modified by interactions in the outer atmosphere or magnetosphere, for example through the inverse-Compton scattering of photons off high-energy particles. This creates an extended high-energy tail but leaves the spectrum at low energies (the Rayleigh-Jeans side of the blackbody spectrum) unaffected.

Here, we attempt to measure the interstellar extinction in a model-independent way, using individual absorption edges of the elements O, Fe, Ne, Mg and Si in X-ray spectra taken with XMM-Newton.

2 Data analysis

We searched the XMM-Newton archive for observations of all the AXPs. The XMM Newton observatory (Jensen 1999) provides data from three separate instruments simultaneously, but in this work we are concerned with the Reflection Grating Spectrometer (RGS) instruments (den Herder et al. 2001), which provide high-resolution spectra in the range 6–40 Å. We found a number of long observations for the four brightest AXPs (we omitted shorter data sets with few counts). For all these, RGS is used with the same instrumental setup, thus ensuring a fair comparison of the sources

In order to improve the signal-to-noise ratio, we decided to merge the spectra from different observations of each object into averaged spectra. Here, we must raise a caveat: some AXPs—1E 1048.1-5937 in particular—are variable, and the spectral shape may be different in each observation. This, however, should not change our column estimates, since we fit in small spectral regions around each absorption edge.

For our measurements, we assume that is possible to find small regions of a spectrum around an absorption edge, over which the intrinsic spectrum is continuous and well-described by a power-law. For each region, we fit a simple analytical edge model, using the latest edge energies and structure information from high signal-to-noise spectroscopy (Takei et al. 2002; Juett et al. 2006; Ueda et al. 2005). Figure 1 shows such a fit, in this case for Oxygen in the spectrum of 4U 0142+61.

We determine our uncertainties through the 'bootstrap' Monte-Carlo method (Press et al. 1992), by resampling the data 1000 times with replacement and taking the 68% confidence region from this.

3 Results and conclusions

Table 1 shows the measured column densities to the AXPs of the metals within our sensitivity range, with appropriate uncertainties. Table 2 shows the implied hydrogen columns. Note that the iron and silicon columns were not reliably determined in any case, so these are not used to derive the hydrogen columns, and that we assume the abundances given

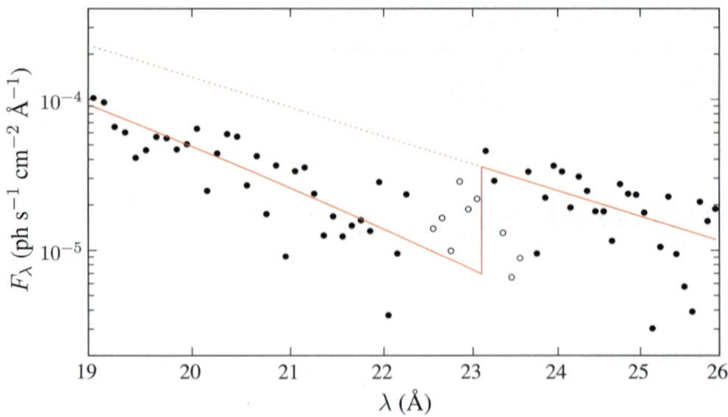

Fig. 1 Spectrum of 4U 0142+61 around the Oxygen-K edge, fit by a simple power-law and edge model. *Open symbols* have not been included in the fit, due to edge structure and narrow atomic absorption features

by Asplund et al. (2004), which fit with our relative abundances for 4U 0142+61 very nicely.

Although in some cases, the measurements do not seem significant, this in not a case where we are attempting to prove the existence of a feature which may or may not be there, but rather, it is a case where the location and shape of the feature is known a priori, and that interstellar extinction must surely exist and here our job is simply to quantify it. Thus, the equivalent hydrogen columns can be safely averaged together, and reddenings estimated through Predehl and Schmitt's (1995) relation, $A_V = 5.6(N_H/10^{22}\ \mathrm{cm}^{-2})$ mag. The main systematic uncertainty in this is that of the appropriate abundances to use, and the gas-to-dust ratio. At the very least, the relative extinctions between the sources is well-determined, with 4U 0142+61 the least extincted, followed by 1E 1048.1-5937, 1E 2259+586 and 1RXS J170849.0-400910, in that order.

With these values of column densities, it is possible to reverse the process and derive the intrinsic spectra of the AXPs. In Fig. 2, we show the de-extincted spectra derived for all four AXPs under consideration. Two things are immediately apparent: they are not consistent in shape with one-another, and there is no continuation of the >2 keV power-law at low energies (the photon indices, measured from power-law plus black-body fits, of 2.4 to 4.0 correspond to slopes of $\alpha = -0.6, -0.1, 0.4, 1.0$ [$F_\lambda \propto \lambda^\alpha$], for 1RXS J170849.0-400910, 1E 1048.1-5937, 4U 0142+61 and 1E 2259+586 respectively). The spectra as shown have indices of approximately $\alpha = -3, -2, -2$ and 0 respectively, if the low-energy tail is taken to be a power-law. The high-energy power-law clearly does not continue to low energies for any of the sources.

While there is no evidence of a rising power-law component at low energies, the spectra also do not decline as fast as would be expected if they were due to a black-body component. If the thermal emission arises from the neutron-star surface, as seems likely, this might simply reflect a range of

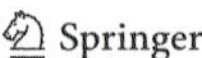

Table 1 Column densities to the AXPs. Units are 10^{17} cm^{-2} and uncertainties are 1-σ

AXP	O K	Fe L	Ne K	Mg K	Si K
4U 0142+61	28.8 ± 4.5	0.7 ± 1.4	5.3 ± 1.3	2.2 ± 0.5	2.0 ± 2.7
1E 1048.1-5937	...	...	7.1 ± 7	2.7 ± 2.4	11 ± 7
1E 2259+586	...	13 ± 6	8.5 ± 4	3.6 ± 1.4	0.6 ± 3.6
1RXS J170849.0-400910	...	...	15 ± 5	2.9 ± 1.8	2 ± 5

Table 2 Inferred hydrogen column densities and optical reddenings for the AXPs. Column $\langle N_H \rangle$ is the weighted mean of all measurements and A_V is the reddening in magnitudes of extinction in the V-band; the errors listed are statistical only. Abundances are relative to Hydrogen

AXP	N_H(O K)	N_H(Ne K)	N_H(Mg K)	$\langle N_H \rangle$	A_V
Abundance	4.6×10^{-4}	6.9×10^{-5}	3.4×10^{-5}		
4U 0142+61	6.0 ± 0.9	7.7 ± 1.9	6.5 ± 1.5	6.4 ± 0.7	3.5 ± 0.4
1E 1048.1-5937	...	10.3 ± 10	7.9 ± 7	8.7 ± 5.7	4.9 ± 3.2
1E 2259+586	...	12.3 ± 5.8	10.6 ± 4.1	11.2 ± 3.3	6.3 ± 1.8
1RXS J170849.0-400910	...	21 ± 7	9 ± 6	13.8 ± 4	7.7 ± 2.2

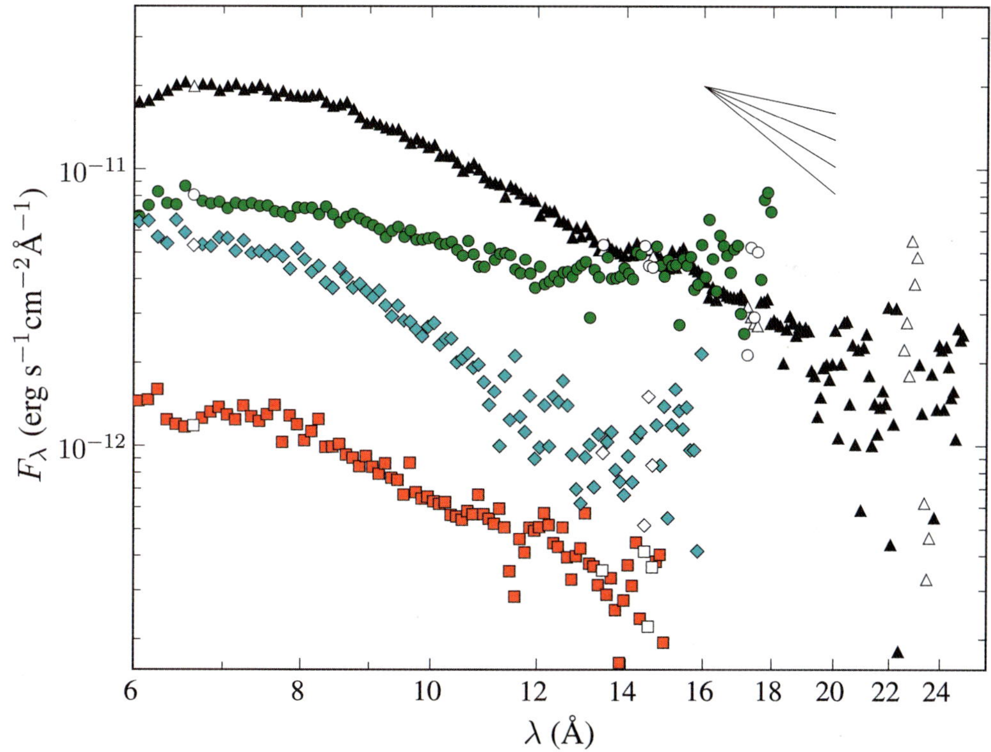

Fig. 2 Spectra for each AXP, de-extincted with the column densities given in Table 1. *Black triangles* are 4U 0142+61, *red squares* 1E 1048.15937, *green circles* 1E 2259+586, and *cyan diamonds* 1RXS J170849.0400910. *Open symbols* are affected by absorption edge structure and absorption lines. The spectra have been truncated where the signal-to-noise ratio decreases below about 1. Also shown are power laws ($F_\lambda \propto \lambda^\alpha$, $\alpha = 1, 2, 3, 4$) to guide the eye

temperatures on the surface. Alternatively, it may indicate that more realistic models are required to describe the emission, which also include the effects of magnetic field, interactions with high-energy particles in the magnetosphere, and gravitational light-bending (the latter particularly important for phase-resolved spectra).

Finally, looking in detail at the spectra, it appears that for 4U 0142+61, there is a hint of a feature in the spectrum at about 13.5 Å (this is easier to see in Fig. 3 at around 2×10^{17} Hz). One could interpret this either as a broad absorption feature at ~13.5 Å, or a broad emission feature at ~15 Å. Assuming it is cyclotron absorption (emission), i.e., $E_{\text{cyc}} = \hbar eB/mc$, this corresponds to 7.9×10^{10} G (7.1×10^{10} G) for electrons or 1.5×10^{14} G (1.3×10^{14} G) for protons. If the line is red-shifted, the inferred magnetic field strength would increase by a factor $1 + z_{\text{GR}} = (1 - 2GM/Rc^2)^{-1/2}$, equal to ~1.3 at the surface for a neutron star with $M = 1.4M_\odot$ and $R = 10$ km. In-

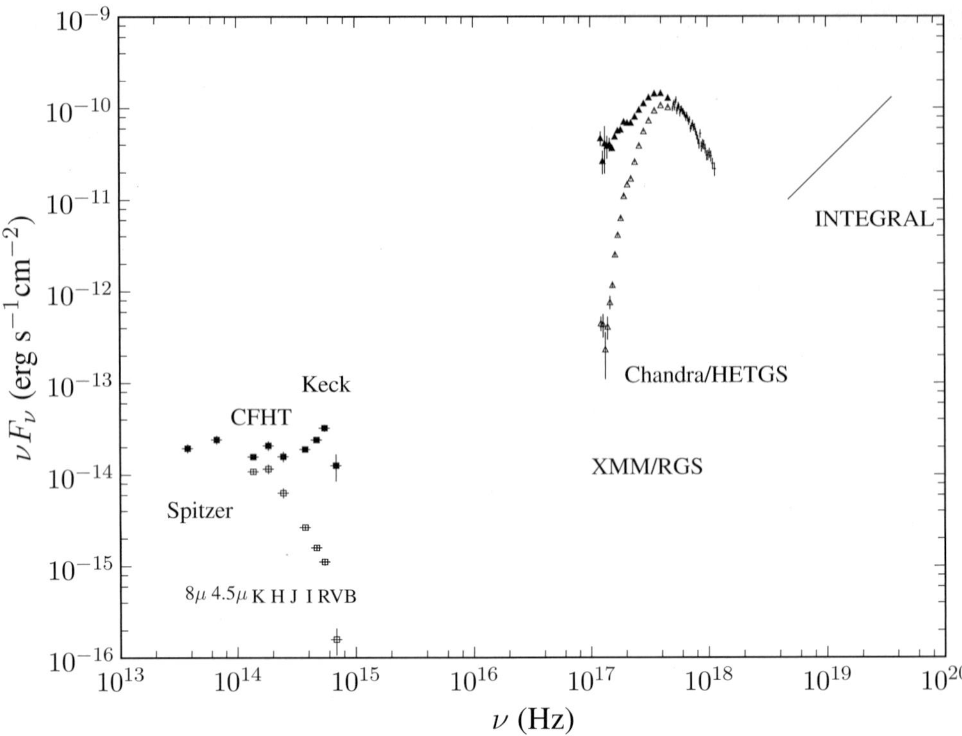

Fig. 3 Spectral energy distribution for 4U 0142+61. *Triangles* are XMM/RGS data, as observed (*open*) and de-extincted (*filled*) as described in the text. *Open squares* are observed broad-band photometry (mid-IR from Wang et al. 2006; JKH from Durant and Van Kerkwijk (2006c); BVRI from Hulleman et al. 2004). *Filled squares* are the photometric points de-reddened with $A_V = 3.5$. *Crosses* in the higher-energy X-ray part of the spectrum come from Chandra (Juett et al. 2002); extinction is not important in this region of the spectrum. XMM data is binned in frequency ($\delta\nu$) corresponding to $\delta\lambda = 1$ Å, and Chandra data to $\delta\lambda = 0.1$ Å for clarity. The different observations from which these data are derived were not taken simultaneously

triguingly, the value for proton cyclotron lines is close to the magnetic dipole field strength inferred from timing measurements, $B_{\rm dip} = 3.2 \times 10^{19}\sqrt{(P\dot{P})} = 1.3 \times 10^{14}$ G (Woods and Thompson 2004).

With our empirical estimates of the intrinsic (soft) X-ray spectra, and the revised estimate for optical/infrared reddening, we can re-examine the spectral energy distributions of the AXPs. We only consider 4U 0142+61, since this object is the only one for which we find a significantly different column density from that typically quoted. Furthermore, it has the best X-ray data, and the best broad-band coverage, from mid-infrared (Wang et al. 2006), to near-infrared and optical (Hulleman et al. 2004; Durant and Van Kerkwijk 2006c), to hard X-rays (den Hartog et al. 2006, also in these proceedings).

The soft X-ray spectrum in Fig. 3 suggestively has a slope which would meet up with the optical points if extended, in marked contrast to the often-used power-law plus blackbody model. Note, however, that the observations in different parts of the spectrum were not taken simultaneously, and the source is known to vary on some level in every spectral range (Durant and Van Kerkwijk 2006c).

In conclusion, we have attempted to measure the extinction to the AXPs without making assumptions about what their intrinsic spectral shapes might be. With our resulting best estimates, we derived intrinsic spectra, which can be compared with each other as well as with predictions, such as those from simulations and semi-analytic models that are now being produced within the magnetar framework (e.g., Lyutikov and Gavriil 2006; Fernández and Thompson 2006, pers. comm.; see also Fernández, these proceedings).

One exciting use for the new extinction measurements, is to find distances based on the infrared extinction (i.e., reddening) of field stars. Although the extinctions have not been measured very accurately, it turns out that each AXP falls in a region of dense extincting material (i.e., spiral arms), and so their distances can be well determined (see Durant and Van Kerkwijk 2006b).

References

Asplund, M., Grevesse, N., Jacques Sauval, A.: In: Bash, F.N., Barnes, T.G. (eds.) Cosmic Abundances as Records of Stellar Evolution and Nucleosynthesis. ASP Conf. Ser. (2004)

den Hartog, P., Hermsen, W., Kuiper, L., et al.: Astron. Astrophys. **451**, 587 (2006)

den Herder, J., Brinkman, A., Kahn, S., et al.: Astron. Astrophys. **365**, L7 (2001)

Durant, M., Van Kerkwijk, M.: Astrophys. J. **650**, 1082 (2006a)

Durant, M., Van Kerkwijk, M.: Astrophys. J. **650**, 1070 (2006b)

Durant, M., Van Kerkwijk, M.: Astrophys. J., **652**, 576 (2006c)

Hulleman, F., Van Kerkwijk, M., Kulkarni, S.: Astron. Astrophys. **416**, 1037 (2004)

Jensen, G.: Bull. Am. Astron. Soc. **32**, 724 (1999)

Juett, A., Marshall, H., Chakrabarty, D., et al.: Astrophys. J. **568**, L31 (2002)

Juett, A., Schulz, N., Chakrabarty, D., et al.: Astrophys. J. **648**, 1066 (2006)

Lyutikov, M., Gavriil, F.: Mon. Not. Roy. Astron. Soc. **368**, 690 (2006)

Predehl, W., Schmitt, S.A.: Astron. Astrophys. **293**, 889 (1995)
Press, W., Teukolsky, S.A., Vetterling, W., et al.: Numerical Recipes: The Art of Scientific Computing, 2nd edn. Cambridge University Press (1992)
Takei, Y., Fujimoto, R., Mitsuda, K., et al.: Astrophys. J. **581**, 307 (2002)
Ueda, Y., Mitsuda, K., Murakami, H., et al.: Astrophys. J. **620**, 274 (2005)
Wang, Z., Chakrabarty, D., Kaplan, D.: Nature **440**, 772 (2006)
Woods, P., Thompson, C.: In: Lewin, W., Van der Klis, M. (eds.) Compact Stellar X-Ray Sources (2004)

Astrophys Space Sci (2007) 308: 647–653
DOI 10.1007/s10509-007-9367-1

ORIGINAL ARTICLE

The first multi-wavelength campaign of AXP 4U 0142+61 from radio to hard X-rays

P.R. den Hartog · L. Kuiper · W. Hermsen · N. Rea · M. Durant · B. Stappers · V.M. Kaspi · R. Dib

Received: 14 July 2006 / Accepted: 3 November 2006 / Published online: 21 March 2007

Abstract For the first time a quasi-simultaneous multi-wavelength campaign has been performed on an Anomalous X-ray Pulsar from the radio to the hard X-ray band. 4U 0142+61 was an INTEGRAL target for 1 Ms in July 2005. During these observations it was also observed in the X-ray band with Swift and RXTE, in the optical and NIR with Gemini North and in the radio with the WSRT. In this paper we present the source-energy distribution. The spectral results obtained in the individual wave bands do not connect smoothly; apparently components of different origin contribute to the total spectrum. Remarkable is that the INTEGRAL hard X-ray spectrum (power-law index 0.79 ± 0.10) is now measured up to an energy of $\sim$230 keV with no indication of a spectral break. Extrapolation of the INTEGRAL power-law spectrum to lower energies passes orders of magnitude underneath the NIR and optical fluxes, as well as the low $\sim$30 μJy (2σ) upper limit in the radio band.

P.R. den Hartog (✉) · L. Kuiper · W. Hermsen · N. Rea
SRON Netherlands Institute for Space Research, Sorbonnelaan 2, 3584 CA Utrecht, The Netherlands
e-mail: hartog@sron.nl

W. Hermsen · B. Stappers
Astronomical Institute, University of Amsterdam, Kruiskade 403, 1098 SJ Amsterdam, The Netherlands

M. Durant
Department of Astronomy & Astrophysics, University of Toronto, 60 Saint George Street, MP 1404D Toronto, ON M5S 3H8, Canada

B. Stappers
ASTRON, The Netherlands Foundation for Research in Astronomy, P.O. Box 2, 7990 AA, Dwingeloo, The Netherlands

V.M. Kaspi · R. Dib
Physics Department, McGill University, 3600 University Street, Montreal, PQ H3A 2T8, Canada

Keywords Neutron stars · Magnetars · Anomalous X-ray pulsars

PACS 97.60.Jd · 97.60.Gb · 95.85.Pw · 95.85.Nv · 95.85.Kr · 95.85.Jq · 95.85.Bh

1 Introduction

Anomalous X-ray Pulsars (AXPs) are young rotating isolated neutron stars (for a review in this volume, see Kaspi 2007). Currently there are 8 AXPs known and there are a few more candidates (Woods and Thompson 2006). These objects are called anomalous, because their X-ray luminosities exceed by far the available total energy released by rotational energy loss. The energy output is believed to originate from an immense energy reservoir stored in a toroidal magnetic field within the Neutron Star. The surface magnetic fields, inferred from their periods and period derivatives, are of the order of 10^{14}–10^{15} G. Therefore, AXPs are believed to be magnetars, as originally proposed for the Soft Gamma-ray Repeaters (SGRs, see Duncan and Thompson 1992; Thompson and Duncan 1995, 1996). Both AXPs and SGRs are well studied objects in the X-ray band for energies below 10 keV. However, little was known about their persistent emission in the hard X-ray band (>10 keV). In 2004 INTEGRAL discovered hard X-ray emission from the position of 1E 1841-045 (Molkov et al. 2004). Kuiper et al. (2004) showed unambiguously that the hard X-rays originated from the AXP by extracting a pulsed hard X-ray signal from the source using archival RXTE data. After INTEGRAL discovered hard X-rays from two other AXP locations, namely from 1RXS J1708-4009 (Revnivtsev et al. 2004) and 4U 0142+61 (den Hartog et al. 2004), Kuiper et al. (2006) also showed for these and for a fourth AXP (1E 2259+586) pulsed hard X-ray emission using archival RXTE data. That means that presently already for 4 of the 7 established AXPs hard X-ray emission has been detected and this can now be considered to be a common characteristic, which is not yet understood.

In this paper we focus on the AXP 4U 0142+61. This AXP was discovered by the Uhuru X-ray observatory in the early seventies (Giacconi et al. 1972; Forman et al. 1978). The spin period of 8.7 s was found by Israel et al. (1994). They realised that the X-ray luminosity is too high to be explained by rotational energy loss. Like for the other AXPs, there is no proof for a companion, nor for an active (i.e. accreting) disk that could explain the high X-ray luminosity of 4U 0142+61. The passive (i.e. non accreting) debris disk discovered by Wang et al. (2006) does not power the X-ray emission. The X-ray luminosities recently measured with e.g. XMM-Newton and Chandra are of the order of 10^{35} erg cm^{-2} s^{-1} (2–10 keV, see Patel et al. 2003; Göhler et al. 2005), assuming a distance of 3.6 kpc (Durant and van Kerkwijk 2006a). The X-ray spectra (0.5–10 keV) of AXPs are soft and are commonly fitted with a black-body and a power-law model. The inclusion of the power-law component is required to fit excess photons with energies above ~3 keV. For 4U 0142+61, the best fit parameters are a black-body temperature of $kT \sim 0.4$ keV and a power-law photon index of $\Gamma \sim 3.4$.

4U 0142+61 was detected by den Hartog et al. (2006) in hard X-rays up to 150 keV in 1.6 Ms of INTEGRAL observations (see also Kuiper et al. 2006). The 20–150 keV flux was measured to be $(9.7 \pm 0.9) \times 10^{-11}$ erg cm^{-2} s^{-1}. The total spectrum could be fitted with a power-law model with photon index $\Gamma = 0.73 \pm 0.17$. They also revisited the Compton Gamma-Ray Observatory (COMPTEL) archives (0.75–30 MeV, see Schönfelder et al. 2002) and determined flux upper limits at the location of the AXP. These limits put constraints on the extrapolation of the hard X-ray power law. Assuming that the hard X-ray flux is stable, a spectral break has to occur in the hard X-ray regime below ~750 keV in order for the spectrum not to be in conflict with the COMPTEL upper limits. den Hartog et al. (2006) and Kuiper et al. (2006) were not able to measure such a break, but a hint was found with a 2.3σ fit improvement when a high-energy cutoff was added to the power-law model by den Hartog et al. (2006). The indication for the cutoff was found at an energy of 73 ± 15 keV.

A close connection may exist between the production of non-thermal hard X-rays and radio emission. However, until recently all AXPs were radio quiet. For AXP 1E 2259+58 the detection of radio emission has now been claimed by Malofeev et al. (2005), but the observations and analysis were difficult and confirmation is required. Halpern et al. (2005) discovered transient radio emission from the transient AXP XTE J1810-197 which appeared sharply modulated at the rotation period with peak flux densities >1 Jy (Camilo et al. 2006), orders of magnitude brighter than the reported upper limits for this or any other AXP. Gaensler et al. (2001) have observed 4U 0142+61 with the VLA (1.4 GHz), but only a 5σ upper limit of 0.27 mJy could be extracted from the data.

4U 0142+61 was the first AXP for which an optical counterpart was discovered (Hulleman et al. 2000). Kern and Martin (2002) found that the optical emission is pulsed for a considerable fraction. Hulleman et al. (2000) showed for the first time that it is not possible to understand the optical and NIR measured fluxes with respect to the X-ray fluxes. There is an optical and NIR excess that can not be explained by the Rayleigh–Jeans tail from the X-ray black body. Moreover the optical and NIR emissions seem to be non thermal and exhibit more variability than seen in the X-rays (Hulleman et al. 2004; Israel et al. 2004; Morii et al. 2005; Durant and van Kerkwijk 2006c).

We present the first quasi-simultaneous multi-wavelength observation campaign to study 4U 0142+61 from the radio up to hard X-rays.

2 Multi-wavelength campaign

For June–July 2005, 1 Ms INTEGRAL (20–300 keV) dedicated 4U 0142+61 observations were scheduled. With these observations, we tried to get, nearly simultaneously, as much wavelength-band coverage of 4U 0142+61 as possible. An approved 12 hour radio observation with the WSRT (21 cm) was rescheduled to overlap with the INTEGRAL observations. A regular RXTE (2–250 keV) monitoring observation also fell in the INTEGRAL time line. For an X-ray imaging observation a Target of Opportunity (ToO) was granted with Swift (0.2–10 keV). Finally, optical and NIR observations were requested and approved in the Directors' Discretionary Time (DDT) on Gemini North. During two nights 4U 0142+61 was imaged in the K_s and r' bands. Unfortunately it was not possible to schedule the K_s and the INTEGRAL observations contemporaneous (see Table 1).

2.1 Hard X-rays: INTEGRAL

The INTErnational Gamma-Ray Astrophysics Laboratory (INTEGRAL; Winkler et al. 2003) is ESA's currently operational hard X-ray/soft gamma-ray space telescope. For the study of AXPs the low-energy detector of IBIS, called ISGRI (20–300 keV, Lebrun et al. 2003), has proven itself to be of great importance. The serendipitous discovery of AXPs in the INTEGRAL energy band was a result of the combination of long exposure times and the 29° wide FOV of IBIS.

Table 1 Multi-wavelength campaign measurements

Obs	Time span	Exposure
WSRT	July 2, 2005	12 h
Gemini K_s	July 26, 2005	1125 s
Gemini r'	July 13, 2005	2400 s
Swift	July 11–12, 2005	7400 s
INTEGRAL	June 29–July 17, 2005	868 ks

For this work we have analysed 1 Ms of observations of 4U 0142+61 performed between June 29 and July 17, 2005. The data were screened for solar flares and erratic count rates resulting from passes through the Earth's radiation belts. After screening the net exposure was 868 ks (Table 1). The observations consist of 265 pointings (Science Windows, ScWs) which can last up to one hour. The ScWs have been analysed separately with the official INTEGRAL software OSA 5.1 (see Goldwurm et al. 2003, for IBIS-ISGRI scientific data analysis) in 20 energy bands between 20 keV and 300 keV with exponential binning. These analyses result in sky images for every ScW in 20 energy bands. The spectrum was built up by averaging the count rates from each ScW, weighted by the variance. For the conversion into flux values, the spectrum was normalized to the known total Crab spectrum (nebula and pulsar) using a curved power law as determined by Kuiper et al. (2006).

4U 0142+61 is detected up to 230 keV with a 3.0σ significance in the 150–230 keV energy band. The total spectrum shown in Fig. 1 was fitted with a power-law model resulting in a photon index $\Gamma = 0.79 \pm 0.10$. The quality of the fit is good with a reduced chi square $\chi_r^2 = 1.12$ for 16 degrees of freedom. The 20–230 keV flux is $(17.0 \pm 1.4) \times 10^{-11}$ erg cm^{-2} s^{-1}.

Our new total spectrum shows that this AXP is now detected at even higher energies than reported earlier, without an indication for a spectral break. The hint for a break found by den Hartog et al. (2006) is not confirmed in this data set.

2.2 Soft X-rays

2.2.1 *Swift*

The Swift-XRT (0.2–10 keV; Burrows et al. 2005) observation was performed on July 11–12, lasting 8500 s. Of this observation 7400 s of data were taken in the Photon-Counting mode and were analysed with the FTOOLS `xrtpipeline`, version build-14 under HEADAS 6.0 (Hill et al. 2005). Photons were extracted from an annular region (3 pixels inner radius, 30 pixels outer radius) in order to avoid pile-up contamination. Background spectra were taken from close-by source-free regions.

As mentioned in Sect. 1, AXP spectra in the X-ray domain (<10 keV) can usually be fitted satisfactorily with a black-body and a power-law model. When we use this canonical model the fit results are: $N_H = (1.01 \pm 0.10) \times 10^{22}$ cm^{-2}; $kT = (0.400 \pm 0.012)$ keV; $\Gamma = 2.7 \pm 0.3$; $\chi_r^2 = 0.97$ (dof = 573). These parameters yield an unabsorbed flux of $(3.9 \pm 0.2) \times 10^{-10}$ erg cm^{-2} s^{-1} in the 0.7–6.0 keV band, or $(7.8 \pm 0.5) \times 10^{-11}$ erg cm^{-2} s^{-1} in the more commonly used 2–10 keV band. Such a model fits the measured X-ray spectrum well, but systematically overestimates the flux at the lower X-ray energies. Furthermore it seems meaningless for extrapolation to the NIR window. In particular, the soft power-law component dominates the black-body component for energies less than ~1.5 keV, meaning that also the N_H estimate is affected and estimated too high.

Alternatively, we used a double black-body model to fit the measured spectrum, yielding an acceptable fit ($\chi_r^2 = 0.98$; dof = 573) with an excellent fit at the lower X-ray energies, but underestimating the higher X-ray energies (>4.5 keV). The two temperatures are 0.38 ± 0.02 keV and 0.78 ± 0.10 keV and the N_H is $(0.61 \pm 0.03) \times 10^{22}$ cm^{-2}. The unabsorbed flux is $(2.11 \pm 0.06) \times 10^{-10}$ erg cm^{-2} s^{-1} in the 0.7–6.0 keV band and $(6.94 \pm 0.29) \times 10^{-11}$ erg cm^{-2} s^{-1} in the 2–10 keV band. Arguably, this model is again canonical, but it is more appropriate for extrapolation to the NIR, which is shown in Fig. 1.

2.2.2 *Rossi-XTE*

4U 0142+61 monitoring data of the PCA (2–60 keV; Jahoda et al. 1996) on board the Rossi X-ray Timing Explorer (RXTE, Bradt et al. 1993) were used to create a phase-coherent timing solution valid during the multi-wavelength campaign (see Gavriil and Kaspi 2002, for a detailed description and use of the AXP-monitoring campaign). This ephemeris is essential for INTEGRAL timing analysis (cf., den Hartog et al. 2007). In this work it is used for the WSRT observation in order to reduce the number of trials in the search for a (possible) weak pulsed radio signal. Data from RXTE-observation IDs 90076 and 91070 were analysed with the pulsar-timing software package TEMPO.[1] The resulting ephemeris is valid between MJD 53251 and MJD 53619, with the following characteristics: Epoch MJD 53420.0, $\nu = 0.1150929855(7)$ Hz, $\dot{\nu} = -2.639(8) \times 10^{-14}$ Hz s^{-1} and $\ddot{\nu} = 3(2) \times 10^{-23}$ Hz s^{-2}.

2.3 Optical and NIR: Gemini

The field of 4U 0142+61 was imaged in the K_s band with the Near Infrared Imager (NIRI; Hodapp et al. 2003) on Gemini North, Hawaii, in the night of July 26th, 2005. NIRI has standard broad-band and narrow-band filters covering 1–5 μm.

[1] http://pulsar.princeton.edu/tempo.

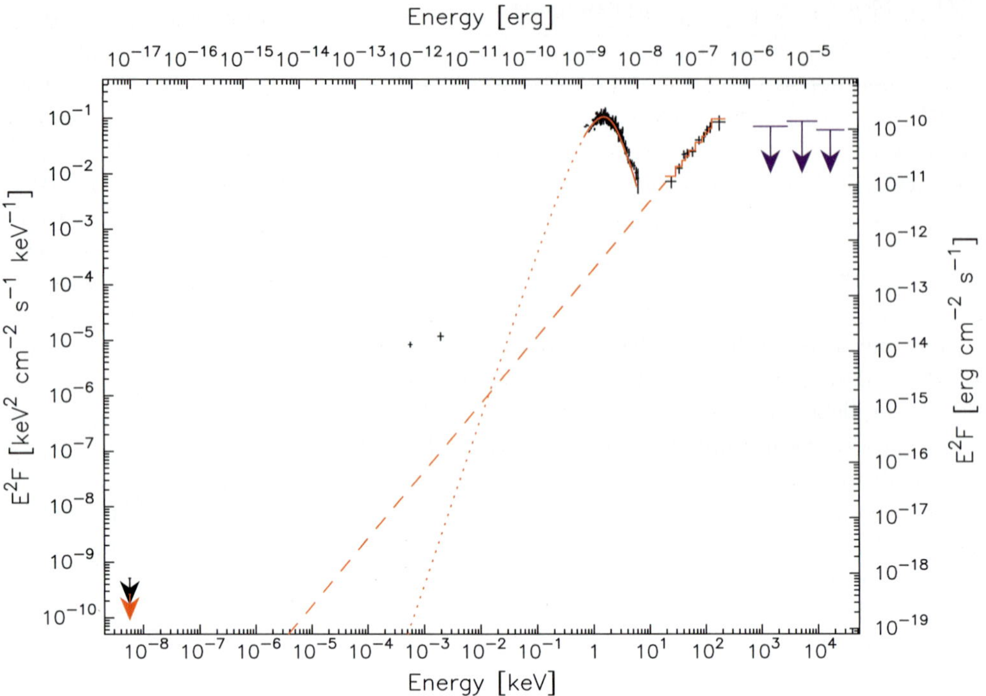

Fig. 1 Source-energy distribution of 4U 0142+61 from radio up to gamma-ray energies. For observation details, see Sect. 3. The WSRT radio limits are shown for the continuum emission (*top arrow*) and the pulsed emission (*lower arrow*). Extrapolations to lower energies are shown for the Swift-XRT double black-body fit (*dotted line*) and for the INTEGRAL power-law fit (*dashed line*). It can be seen that neither the soft X-ray nor the hard X-ray spectral fits extrapolate to the Optical and NIR fluxes. An optical-NIR excess orders of magnitude above these extrapolations is evident. The COMPTEL upper limits at MeV energies do not belong to the multi-wavelength campaign

The final image was created by subtracting dark frames from each science frame, dividing by a flat field derived from the images themselves, and then aligning and stacking all the images. The photometry was measured using the PSF-fitting package DAOPhot (Stetson 1987), and calibrated against the photometry provided for several field stars in Hulleman et al. (2004). The K_s magnitude was found to be 19.96(10).

Optical r'-band images were obtained on the night of July 13th, 2005 using the Gemini-North Multi-Object Spectrograph (GMOS-N; Hook et al. 2004). We subtracted the bias and divided by screen flats, before stacking and photometering the images. For the calibration, the photometry listed in Hulleman et al. (2004) was used, interpolating between the R and V-bands using the relationship in (Smith et al. 2002). We find for 4U 0142+61 $r' = 25.42(6)$, where the statistical uncertainty in the measurement and the uncertainty in the calibration of the photometry zero-point are similar.

2.4 Radio: WSRT

Using the Westerbork Synthesis Radio Telescope (WSRT) we have searched for both pulsed and unpulsed radio emission. An observation of 12 hour duration was carried out at a frequency of 1380 MHz (∼21 cm) with a bandwidth of 80 MHz. Using the synthesis data a map was made using standard routines in the MIRIAD[2] package. In the resulting map (rms ∼ 30 µJy) no source was detected at the location of 4U 0142+61 leading to a 3σ upper limit on its flux of ∼90 µJy. Simultaneously we also summed the signals from all 14 dishes of the WSRT coherently to form a so-called tied array. The data were then sent to the Pulsar Machine PuMa (Voûte et al. 2002) which formed a digital filter bank with 512 channels and a sampling time of 409.6 µs. As the dispersion measure in the direction of the source was unknown, we tried many trial dispersion measures and then folded each one with the RXTE ephemeris (see Sect. 2.2.2). Each resultant profile was then inspected to determine if there was a significant detection. We also performed a standard pulsar search analysis on the full data set. Neither the folding nor the search revealed any significant detection of radio pulsations. A 5σ upper limit was determined at $77 \times \sqrt{(d/1-d)}$ µJy where d is the duty cycle of the pulsar. Using $d = 0.5$, a typical value in the X-ray regime, the 3σ upper limit is ∼46 µJy. It has to be noted that for this analysis the whole observation was used. A finer analysis like performed by Halpern et al. (2005) and Camilo et al. (2006) who discovered the transient AXP XTE J1810-197 as a bright transient radio source in smaller intervals is still ongoing. However 4U 0142+61 has not exhibited the same sort of transient behaviour in X-rays like XTE J1810-197, specially it has not shown any large outburst, and therefore the radio characteristics of these sources might be very different.

3 Multi-wavelength source-energy distribution

In Fig. 1 the quasi-simultaneous multi-wavelength Source-Energy Distribution (SED) is presented covering roughly 10

[2]http://www.atnf.csiro.au/computing/software/miriad/.

orders of magnitude in photon energy. In the lower left corner the WSRT (1380 MHz) continuum (upper arrow) and timing (lower arrow) 2σ upper limits are shown. The Gemini K_s (2.15 μm) and r' (0.6 μm) data points are dereddened assuming $A_V = 3.6$ as determined by Durant and van Kerkwijk (2006b). The Swift-XRT spectrum between 0.7 keV and 6 keV is corrected for absorption with $N_{\rm H} = 0.61 \times 10^{22}$ cm^{-2}. Also shown is the corresponding double black-body fit (solid line), extrapolated towards lower energies (dotted line). The INTEGRAL flux values between 20 keV and 230 keV show the hard X-ray spectrum for this AXP. The power-law fit (solid line) is also extrapolated to lower energies (dashed line).

Significant flux variability has been reported in the optical, NIR and soft X-ray bands (Durant and van Kerkwijk 2006c), therefore in this quasi-simultaneous spectrum we can investigate better how the fluxes in the different bands relate to each other.

4 Discussion

The results for the different wave-band measurements render separately no surprises. The Swift-XRT spectrum shows a typical soft AXP spectrum and NIR and optical magnitudes are around the earlier reported values. The INTEGRAL spectrum is in agreement with previously found results, however, thanks to the high effective exposure in this dedicated observation the maximum energy up to which 4U 0142+61 could be detected is now higher, namely 230 keV. The spectrum, described with a power-law model with photon index $\Gamma = 0.79 \pm 0.10$ and luminosity 8.7×10^{34} erg s^{-1} (20–100 keV, $d = 3.6$ kpc) can be compared with the spectral results reported earlier by den Hartog et al. (2006) using an independent data set: $\Gamma = 0.73 \pm 0.17$ and luminosity 8.5×10^{34} erg s^{-1} (20–100 keV, $d = 3.6$ kpc). Kuiper et al. (2006) derived $\Gamma = 1.05 \pm 0.11$ and luminosity 8.1×10^{34} erg s^{-1} (20–100 keV, $d = 3.6$ kpc). All these findings are within errors in agreement. Therefore there is no evidence for long-term time variability at hard X-ray energies yet.

den Hartog et al. (2006) published 2σ flux upper limits for 4U 0142+61 in the 0.75–30 MeV window analysing COMPTEL data collected over the years 1991–2000. These observations are obviously not contemporaneous to the multi-wavelengths campaign reported here, but the apparent stability at hard X-ray energies seems to justify a direct comparison of the COMPTEL upper limits with the hard X-ray spectrum measured with INTEGRAL. Thus assuming that the hard X-ray emission is stable, Fig. 1 clearly shows that the hard X-ray/soft gamma-ray spectrum of 4U 0142+61 has to break between ~200 keV and 750 keV in order not to be in conflict with the COMPTEL upper limits. If we were to assume that by chance 4U 0142+61 was in a low state at MeV energies during the long COMPTEL observations, it would be remarkable that also for the other AXPs detected similarly by INTEGRAL (1E 1841-045 and 1RXS J1708-4009) no signal was found at MeV energies during the many observations that they were in the COMPTEL field-of-view spread over 10 years (Kuiper et al. 2006). We conclude that a drastic break is required.

The Swift-XRT and INTEGRAL spectra can not be fitted simultaneously with a two-component model consisting of i.e. a black-body plus only one power-law component. The excess of high-energy photons (w.r.t a single black-body spectrum) in the Swift spectrum can not be accounted for by the upcoming hard X-ray spectrum. This is evident in Fig. 1. The extrapolation of the hard X-ray spectral fit towards lower energies is already five times lower at 6 keV than measured with Swift.

The relation of the optical and NIR flux values to the soft and hard X-ray spectra is still an enigma. It was noted earlier (e.g., Hulleman et al. 2004) that extrapolations of fits to the soft X-ray spectra to the optical and NIR bands are not consistent with the measured variable flux values. This is more than evident in Fig. 1, where we show the extrapolation of the double black-body fit, which reaches optical and NIR fluxes ~4–5 orders of magnitude lower than the measured values. The very hard X-ray spectra above 10 keV, originally interpreted as being of non-thermal origin, led to suggestions of a common origin as the likely non-thermal optical and NIR emissions. However, Fig. 1 unambiguously shows that also extrapolation of the power-law spectral shape measured at hard X-ray energies passes 2–3 orders of magnitude underneath the optical and NIR data points. The hard X-ray, soft X-ray, optical and NIR components seem to have different origins.

The discovery of the hard X-ray spectra of AXPs, stimulated new attempts to search for non-thermal radio emission possibly originating from the same sites and/or production mechanisms (e.g. synchrotron radiation) in the magnetar's magnetosphere. The WSRT upper limits at 1380 MHz are however not constraining in this respect. The INTEGRAL power-law fit extrapolates also orders of magnitudes below the radio upper limits.

The physical interpretation of the hard X-ray spectrum measured by INTEGRAL is not evident. Assuming a distance of 3.6 kpc as recently determined by Durant and van Kerkwijk (2006a), the 20–230 keV flux (Sect. 2.1) translates to a luminosity of 2.6×10^{35} erg s^{-1}, which exceeds the maximum luminosity available from rotational energy loss[3] with a factor of ~2000. Moreover, the hard X-ray luminosity is comparable with the high soft X-ray luminosity

[3] $L_{\rm RE} = 1.3 \times 10^{32}$ erg s^{-1} assuming a neutron star with $R = 10$ km, $M = 1.4 M_\odot$

(1.1×10^{35} erg s^{-1}, 2–10 keV) as measured with Swift (see Sect. 2.2.1).

Over the last two years theoretical attempts have been made trying to explain the new hard X-ray emission above 10 keV from AXPs. A particularly interesting interpretation of the high-energy emission is the application of a magnetar-corona model by Thompson and Beloborodov (2005) and Beloborodov and Thompson (2007). In the last papers, a theoretical model is developed explaining the formation of a hot corona above the surface of a magnetar, so far the only model attempting to explain the multi-wavelength emission. In essence, the twisted magnetosphere acts as an accelerator that converts the toroidal-field energy to particle kinetic energy. They show numerically that the corona self organises quickly into a quasi-steady state, with a voltage of $\sim$1 GeV along the magnetic field lines. Pair production occurs at a rate just enough to feed the electric current. Interestingly, the heating rate of the corona is $\sim 10^{36}$–10^{37} erg s^{-1}, in agreement with the observed persistent, high-energy output of magnetars. Furthermore, they deduce that the static twist will decay on a timescale of 1–10 yr, setting the scale for time variability to look for in the high-energy emission. The transition layer between the atmosphere and the corona is the likely source of the observed $\sim$100 keV (e.g. bremsstrahlung) emission from magnetars. Finally, it is worth mentioning that the corona emits curvature radiation which can also supply the observed IR-optical luminosity. Of particular concern in the hot coronae model is the reported timescale of 1–10 yr. For 4U 0142+61 we do not see evidence for strong variability in the hard X-ray emission over the first years of INTEGRAL observations, nor over the longer period of RXTE monitoring observations. Future work will concentrate on this aspect, as well on detailed timing studies, including phase resolved spectroscopy. Obviously, as much INTEGRAL data as possible will be collected on 4U 0142+61 and the other hard X-ray emitting AXPs to search for the break energy in the total spectrum, a key parameter in the discrimination between the different proposed models. More detailed analysis of this campaign and additional INTEGRAL observations will be presented in forthcoming papers by Rea et al. (2007) and den Hartog et al. (2007), respectively.

Acknowledgements We acknowledge the Director of Gemini for his DDT and also the Swift team for the ToO. The results in this paper are based on observations with INTEGRAL.

We acknowledge J. Vink for the use of private INTEGRAL Cas A data.

References

Beloborodov, A.M., Thompson, C.: Astrophys. J. (2007, accepted)
Beloborodov, A.M., Thompson, C.: Astrophys. Space Sci. **308**. DOI 10.1007/s10509-007-9318-x (2007)
Bradt, H.V., Rothschild, R.E., Swank, J.H.: Astron. Astrophys. Suppl. Ser. **97**, 355 (1993)
Burrows, D.N., Hill, J.E., Nousek, J.A., et al.: Space Sci. Rev. **120**, 165 (2005)
Camilo, F., Ransom, S., Halpern, J.P., et al.: Nature **442**, 892 (2006)
den Hartog, P.R., Kuiper, L., Hermsen W., et al.: Astron. Telegr. **293**, 1 (2004)
den Hartog, P.R., Hermsen, W., Kuiper, L., et al.: Astron. Astrophys. **451**, 587 (2006)
den Hartog, P.R., Hermsen, W., Kuiper, L., et al.: Astron. Astrophys. (2007, in preparation)
Duncan, R.C., Thompson, C.: Astrophys. J. Lett. **392**, L9 (1992)
Durant, M., van Kerkwijk, M.H.: Astrophys. J. **650**, 1070 (2006a) (astro-ph/0606027)
Durant, M., van Kerkwijk, M.H.: Astrophys. J. **650**, 1082 (2006b) (astro-ph/0606604)
Durant, M., van Kerkwijk, M.H.: Astrophys. J. **652**, 576 (2006c)
Forman, W., Jones, C., Cominsky, L., et al.: Astrophys. J. Suppl. Ser. **38**, 357 (1978)
Gaensler, B.M., Slane, P.O., Gotthelf, E.V., et al.: Astrophys. J. **559**, 963 (2001)
Gavriil, F., Kaspi, V.M.: Astrophys. J. **567**, 1067 (2002)
Giacconi, R., Murray, S., Gursky, H., et al.: Astrophys. J. **178**, 281 (1972)
Göhler, E., Wilms, J., Staubert, R.: Astron. Astrophys. **433**, 1079 (2005)
Goldwurm, A., David, P., Foshini, L., et al.: Astron. Astrophys. **411**, L223 (2003)
Halpern, J.P., Gotthelf, E.V., Becker, R.H., et al.: Astrophys. J. Lett. **632**, L29 (2005)
Hill, J.E., Angelini, L., Morris, D.C., et al.: Int. Soc. Opt. Eng. **5898**, 325 (2005)
Hodapp, K.W., Jensen, J.B., Irwin, E.M., et al.: Publ. Astron. Soc. Pac. **115**, 1388 (2003)
Hook, I.M., Jørgenson, I., Allington-Smith, J.R., et al.: Publ. Astron. Soc. Pac. **116**, 425 (2004)
Hulleman, F., van Kerkwijk, M.H., Kulkarni, S.R.: Nature **408**, 689 (2000)
Hulleman, F., van Kerkwijk, M.H., Kulkarni, S.R.: Astron. Astrophys. **416**, 1037 (2004)
Israel, G.L., Mereghetti, S., Stella, L.: Astrophys. J. Lett. **433**, L25 (1994)
Israel, G.L., Stella, L., Covino, S., et al.: IAU Symp. **218**, 247 (2004)
Jahoda, K., Swank, J.H., Giles, A.B., et al.: Int. Soc. Opt. Eng. **2808**, 59 (1996)
Kaspi, V.M.: Astrophys. Space Sci. (2007) this volume
Kern, B., Martin, C.: Nature **417**, 527 (2002)
Kuiper, L., Hermsen, W., Mendez, M.: Astrophys. J. **613**, 1173 (2004)
Kuiper, L., Hermsen, W., den Hartog, P.R., et al.: Astrophys. J. **645**, 556 (2006)
Lebrun, F., Leray, J.P., Lavocat, P., et al.: Astron. Astrophys. **411**, L141 (2003)
Malofeev, V.M., Malov, O.I., Teplykh, D.A., et al.: Astron. Rep. **49**, 242 (2005)
Molkov, S., Cherepashchuck, A.M., Lutovinov, A.A., et al.: Astron. Lett. **30**, 534 (2004)
Morii, M., Kawai, N., Kataoka, J., et al.: Adv. Space Res. **35**, 1177 (2005)
Patel, S.K., Kouveliotou, C., Woods, P.M., et al.: Astrophys. J. **587**, 367 (2003)
Rea, N., Turolla, R., Zane, S., et al.: (2007, submitted)
Revnivtsev, M.G., Sunyaev, R.A., Varshalovich, D.A., et al.: Astron. Lett. **30**, 382 (2004)
Schönfelder, V., Aarts, H., Bennett, K., et al.: Astrophys. J. Suppl. Ser. **86**, 657 (1993)
Smith, J.A., Tucker, D.L., Kent, S., et al.: Astron. J. **123**, 2121 (2002)

Stetson, P.B.: Publ. Astron. Soc. Pac. **99**, 191 (1987)
Thompson, C., Duncan, R.C.: Mon. Not. Roy. Astron. Soc. **275**, 255 (1995)
Thompson, C., Duncan, R.C.: Astrophys. J. **473**, 322 (1996)
Thompson, C., Beloborodov, A.M.: Astrophys. J. **634**, 565 (2005)
Voûte, J.L.L., Kouwenhoven, M.L.A., van Haren, P.C., et al.: Astron. Astrophys. **385**, 733 (2002)
Wang, Z., Chakrabarty, D., Kaplan, D.L.: Nature **440**, 772 (2006)
Winkler, C., Courvoisier, T.J.-L., Di Cocco, G., et al.: Astron. Astrophys. **411**, L1 (2003)
Woods, P.M., Thompson, C.: In: Lewin, W., van der Klis, M. (eds.) Compact Stellar X-ray Sources. Cambridge Astrophysics Series, vol. 39, p. 547. Cambridge University Press, Cambridge (2006) (astro-ph/0406133)